FRACTURE MECHANICS: EIGHTEENTH SYMPOSIUM

Eighteenth National Symposium
on Fracture Mechanics
sponsored by
ASTM Committee E-24
on Fracture Testing
Boulder, Colorado, 25–27 June 1985

ASTM SPECIAL TECHNICAL PUBLICATION 945

D. T. Read and R. P. Reed
National Bureau of Standards
editors

ASTM Publication Code Number (PCN)
04-945000-30

ASTM 1916 Race Street, Philadelphia, PA 19103

Library of Congress Cataloging-in-Publication Data

National Symposium on Fracture Mechanics (18th: 1985:
Boulder, Colo.)
Fracture mechanics.

(ASTM special technical publication; 945)
"ASTM publication code number (PCN) 04-945000-30"
Includes bibliographies and index.
1. Fracture mechanics—Congresses. I. Read, D. T.
II. Reed, R. P. (Richard Palmer), 1934- . III. ASTM
Committee E-24 on Fracture Testing. IV. Title.
V. Series.
TA409.N38 1985 620.1′126 87-30666
ISBN 0-8031-0949-0

Library of Congress Catalog Card Number: 87-30666

NOTE

The Society is not responsible, as a body,
for the statements and opinions
advanced in this publication.

Printed in Ann Arbor, MI
February 1988

Foreword

The Eighteenth National Symposium on Fracture Mechanics was held on 25–27 June 1985 in Boulder, Colorado. ASTM Committee E-24 on Fracture Testing was the sponsor of this symposium. D. T. Read and R. P. Reed, National Bureau of Standards, served as coeditors of this publication. R. P. Reed also served as chairman of the symposium.

Related ASTM Publications

Fracture Mechanics: Seventeenth Conference, STP 905 (1986), 04-905000-30

Fracture Mechanics: Sixteenth Symposium, STP 868 (1985), 04-868000-30

Fracture Mechanics: Fifteenth Symposium, STP 833 (1984), 04-833000-30

Fracture Mechanics: Fourteenth Symposium—Volume I: Theory and Analysis, STP 791 (1983), 04-791001-30

Fracture Mechanics: Fourteenth Symposium—Volume II: Testing and Applications, STP 791 (1983), 04-791002-30

Fracture Mechanics (Thirteenth Conference), STP 743 (1981), 04-743000-30

A Note of Appreciation to Reviewers

The quality of the papers that appear in this publication reflects not only the obvious efforts of the authors but also the unheralded, though essential, work of the reviewers. On behalf of ASTM we acknowledge with appreciation their dedication to high professional standards and their sacrifice of time and effort.

ASTM Committee on Publications

ASTM Editorial Staff

Contents

POSTER SESSION—ELASTIC PLASTIC FRACTURE MECHANICS I

APPLICATIONS

Ductile/Brittle Transition

J-Integral Test Methods

Analysis

Poster Session—Elastic Plastic Fracture Mechanics II

Poster Session—Fatigue

Subcritical Crack Growth

FATIGUE I

ELASTIC-PLASTIC FRACTURE MECHANICS II

FATIGUE II

Summary

Introduction

This volume is based on the Eighteenth National Symposium on Fracture Mechanics, held in Boulder, Colorado, 25–27 June 1985, sponsored by ASTM Committee E-24 on Fracture Testing. The conference was held at the University of Colorado; the support of the CU Office of Conference Services for the conference arrangements was excellent.

The National Symposium on Fracture Mechanics has served as an annual state of the art review of current fracture research since its beginnings in 1965. The Eighteenth Symposium carried on this tradition creating on open forum for fracture researchers from the whole world. There were over eighty papers presented by formal talks and posters. Six speakers were specifically invited, including one from England, one from Japan, two from U.S. universities, one from U.S. industry, and one from a U.S. national laboratory. The submitted papers were grouped into 16 sessions. The papers on elastic plastic fracture mechanics made up four sessions, with one session on each of the allied areas of ductile-to-brittle transition and J-integral test methods. Four sessions on fatigue emphasized elevated temperature studies, including frequency and hold-time dependence, and effects of short cracks. Two sessions were held on analysis, including linear elastic and elastic-plastic analyses, and the remaining four sessions covered applications, crack arrest, micromechanisms, and subcritical crack growth.

Continuing the standard practice for ASTM Symposium publications, each paper in this volume has been peer-reviewed by knowledgable researchers in relevant subject areas. The papers accepted for this volume have been revised and carefully edited to promote significance, technical accuracy, and relevance. It therefore truly represents a broad view of the current state of fracture mechanics research. It is recommended to stimulate and aid future research, to give design and failure analysis practitioners needed insight and new approaches, and to contribute to new and improved test standards through its record of advances in basic understanding and the latest test procedures and results.

R. P. Reed

National Bureau of Standards, Boulder, CO 80303; symposium chairman and coeditor.

Micromechanisms

William W. Gerberich[1]

Metallurgical Aspects of Crack-Tip Failure Processes

REFERENCE: Gerberich, W. W., "**Metallurgical Aspects of Crack-Tip Failure Processes,**" *Fracture Mechanics: Eighteenth Symposium, ASTM STP 945,* D. T. Read and R. P. Reed, Eds., American Society for Testing and Materials, Philadelphia, 1988, pp. 5–18.

ABSTRACT: Process zone models for fracture of fiber-reinforced concrete, metal-particle reinforced glass, and polymer crazes have good experimental verification. This is because the microfracture process zone is often macroscopic in size or easily identifiable or both. Similar models have been proposed for metals and ceramic microstructures. This paper addresses how the microfracture process zone develops in such microstructures. Specifically, the microcrack evolution process may be controlled by chemistry and microstructure as well as localized stress distributions. The importance of this is that the microcrack distribution and the energy dissipation process in the remaining ligaments behind the advancing crack front control the R-curve and final fracture instability. Examples of R-curve behavior in ductile fracture and semibrittle composites are shown, and a model for brittle fracture of steel is proposed. The latter shows that a semi-cohesive process zone of variable size and strength may represent an approach to brittle fracture where weakest-link models are not applicable. Predictions of fracture toughness for ferrite/pearlite steels as a function of test temperature and grain size are obtained with such an approach.

KEY WORDS: fracture toughness, process zones, R-curves, microstructure, ductile ligaments, composites, brittle fracture, cleavage

The process zone concept was introduced two decades ago by Krafft [*1*] as a fundamental microstructural length parameter. Since then, there have been numerous applications of characteristic length parameters to ductile fracture [*2*], metal-metal composites [*3*], cleavage in the lower shelf region [*4*], and subcritical growth phenomena [*5,6*]. Although such process zones recently have been reviewed [*7*], the microscopic and macroscopic character of such zones will be addressed. This will then be applied to the ductile-brittle transition region of steels and, specifically, to classes of steel which range from the lower shelf to toughness levels where mixed modes may be becoming dominant. This is believed to be important because weakest-link hypotheses, for example, cleavage initiating at the largest carbide, cannot be applied to ductile fracture. In all probability, they are also not applicable to the upper shelf region near the transition temperature where a mixture of microvoid coalescence and cleavage or quasi-cleavage is found. In fact, the weakest-link hypothesis often might not apply to the lower shelf regime when a fine crack is present, but may be restricted to fractures emanating from blunt cracks or notches. The evidence for and treatment of multiple fracture origins rather than a single weakest link follows.

First, consider the process zone as a microscopic region where microcracks develop. This may be a single fractured carbide near the nil-ductility transition temperature (NDTT) which, as a weakest link, triggers cleavage in the surrounding grains. On the other hand, it may be a group of second-phase particles such as sulfides or oxides which nucleate multiple regions of void

[1]Professor, Chemical Engineering and Materials Science, University of Minnesota, Minneapolis, MN 55455.

growth. This zone, containing a distribution of microcracks, has been designated a semicohesive zone [6] and is depicted in Fig. 1. Here, the crack-tip opening angle is approximately equal to the crack-tip "final stretch," displacement divided by the process zone, or $\hat{\delta}/\Delta$ [8–10]. Such a zone would have to develop, at least in a two-dimensional sense, in ductile fracture where microvoids nucleate and grow at multiple origins along a growing crack front. Other types of zones are depicted in Fig. 2, as taken from Ref 7. Brittle or semibrittle zones are depicted as being filled with a series of microcracks of varying sizes or spacings within Δ. Such microstructural zones could include properly oriented cleavage microcracks or populations of the weakest intergranular facets or brittle regions between ductile ligaments in two-phase materials. Next, the theoretical concept will be illustrated and then it will be applied to microscopic and macroscopic composite data.

Theoretical Model

The theoretical concept was first presented [7] and described as a series of traction forces in two zones, one a microcracked, semicohesive zone, and a second, monolithic, cohesive zone. A depiction of equilibrium is schematically shown in Fig. 3 where the "fishhooks" are trying to pull the imaginary crack open while the "springs" in the two zones are holding it together. Note that because of the voids in the semicohesive zone, the density of springs, that is, the traction force, is less. This was first applied to organic solvent-induced slow crack growth in glassy polymers where the larger semi-cohesive region was used to predict "equilibrium" craze lengths [11]. The microfibrils of the craze represented internal traction forces and in order to produce additional crazing at the crack tip, after equilibrium, greater applied stress intensities would be required.

With such a model as depicted in Fig. 3, integrating the stress field within each zone gives

$$K_{\mathrm{I}}\left(\frac{\pi}{c}\right)^{1/2} - 2(\sigma_c - \sigma_{sc})\cos^{-1}\left(\frac{b}{a}\right) - 2\sigma_{sc}\cos^{-1}\left(\frac{c}{a}\right) = 0 \tag{1}$$

Here, σ_{sc} is the strength of the semicohesive zone and σ_c is the strength of the cohesive zone. For a ductile material, σ_c is the flow stress, whereas for a truly brittle material it may approach the theoretical strength [7]. In a yielding situation for crack initiation in an elastic-perfectly plastic material, the parameters are

$$\begin{aligned} \sigma_c &= \sigma_{ys}; \qquad & b &= c + \Delta \\ \sigma_{sc} &= \sigma_{ys}(1 - f_v); \qquad & a &= c + \Delta + R_{P\mathrm{I}o} \end{aligned} \tag{2}$$

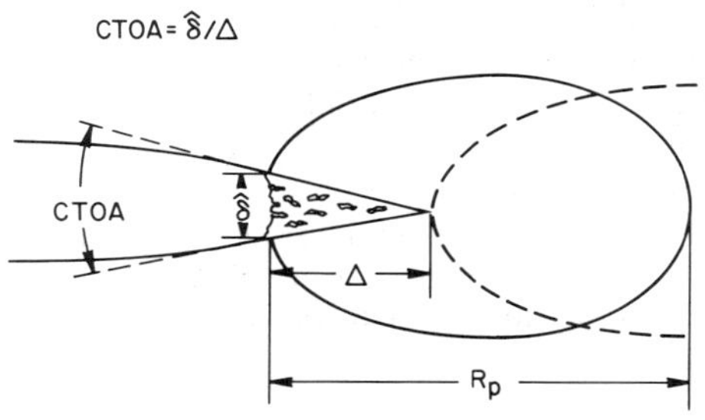

FIG. 1—*Process zone model (Refs* 8–10*) for a growing crack. The crack-tip opening angle (CTOA) may be represented by the "final stretch" displacement, δ, and the process zone, Δ.*

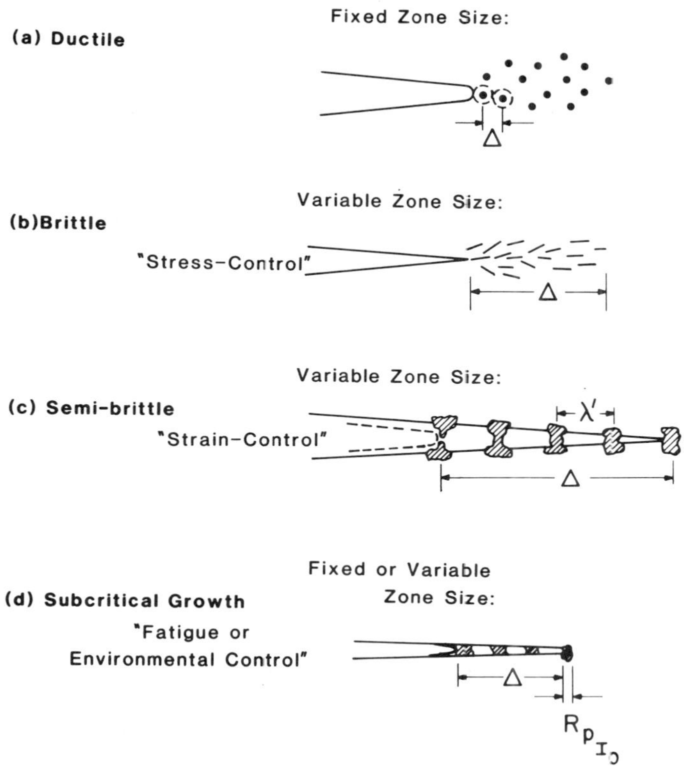

FIG. 2—*Process zone size variations, Δ, as a function of different controlling microstructural features. (Courtesy of Ref 7.)*

where

f_v = microvoid volume fraction,
Δ = process zone,
c = half-crack length, and
$R_{P_{Io}}$ = plastic zone size for nucleating the process zone.

The assumption here is that a minimum strain or stress for failure, as related to the size of $R_{P_{Io}}$, is required to initiate microvoids or microcracks in the absence of the process zone. Since in plane stress, $R_{P_{Io}}$ is given by

$$R_{P_{Io}} = \frac{\pi K_{Io}^2}{8\sigma_{ys}^2} \tag{3}$$

there are three unspecified parameters K_{Io}, Δ, and f_v. In principle, K_{Io} is the initiation value for the onset of crack growth and may be estimated by sensitive crack-tip opening angle (CTOA) nonlinearities or acoustic emission. Both Δ and f_v would be dictated by the microstructure and could be measured by either serial sectioning techniques or fractographic analysis or both. It is seen that once K_{Io}, Δ, and f_v are either measured or estimated, all parameters are known from Eqs 1, 2, and 3 for a material with a given yield strength, containing a crack of length c. Although it appears as though K_I for equilibrium with the first crack jump through Δ is strongly

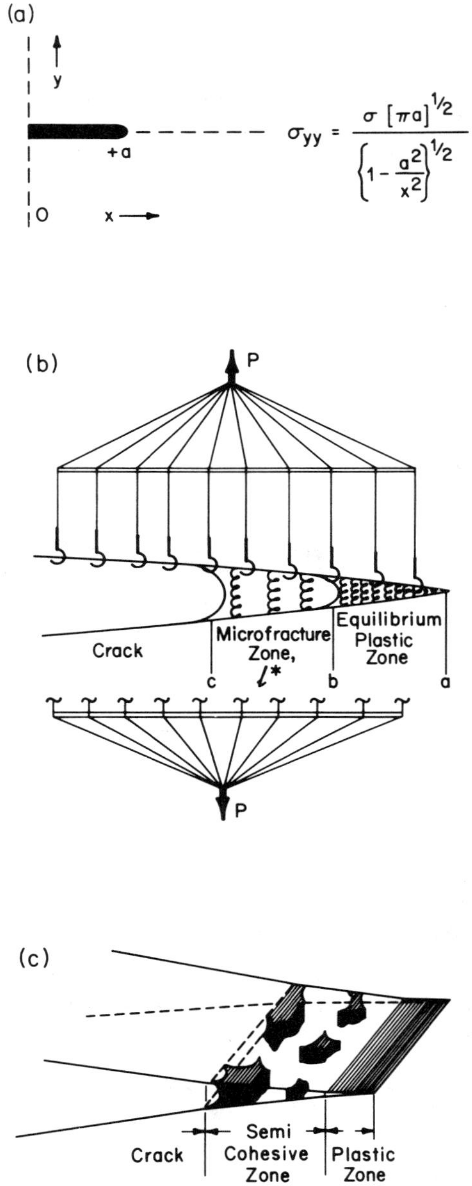

FIG. 3—*Semicohesive zone model showing* (a) *stress distribution function,* (b) *equilibrium for the three zones, and* (c) *a schematic of the ligaments in the semicohesive zone.*

crack length dependent, further inspection shows that both $\cos^{-1}(b/a)$ and $\cos^{-1}(c/a)$ are offsetting functions. In fact, one can show for $R_{P_I} \sim \Delta \leq 0.01C$ that K_I varies by less than 1%. In addition, for small nonlinear zones, this approaches the CTOA stretch model previously reported [7] which gave

$$K_{Ic} \simeq 2.9\sigma_{ys}\Delta^{1/2} \tag{4}$$

Taking Eqs 1 to 3 with $R_{P_{Io}} \sim \Delta$ and $f_v \sim 1/2$, that is, a process zone containing a 50% void fraction of microcracks, gives

$$K_{Ic} \simeq \left(\frac{c}{\pi}\right)^{1/2}\left[\sigma_{ys}\cos^{-1}\left(\frac{c}{c+2\Delta}\right) + \sigma_{ys}\cos^{-1}\left(\frac{c+\Delta}{c+2\Delta}\right)\right] \tag{5}$$

Again, within about 1%, the following two approximations

$$\cos^{-1}\left(\frac{c}{c+2\Delta}\right) \simeq \left(\frac{4\Delta}{c}\right)^{1/2}; \qquad \cos^{-1}\left(\frac{c+\Delta}{c+2\Delta}\right) \simeq \left(\frac{2\Delta}{c}\right)^{1/2}; \qquad \Delta \ll c \tag{6}$$

in conjunction with Eq 5 give

$$K_{Ic} \simeq 1.93\sigma_{ys}\Delta^{1/2} \tag{7}$$

Note that this is of the same form as Eq 4 and that it in fact fits the tool steel data of Shelton and Wronski [12] slightly better than the CTOA stretch model in Fig. 4.

An approach such as Eq 5 is reasonable for relatively low toughness materials undergoing minimal slow crack growth prior to instability. For more stable situations, one must consider the increase in the plastic zone under plane stress conditions, that is, R-curve behavior. This may be treated by allowing the plastic zone to increase with applied K_I according to Eq 3. Since K_I is increasing, the local stress or strain in the process zone is increasing and therefore it becomes easier to separate interfaces or promote cleavage. This produces a larger volume fraction of microcracks which decreases the strength of the process zone. Given Eq 2, it is possible to

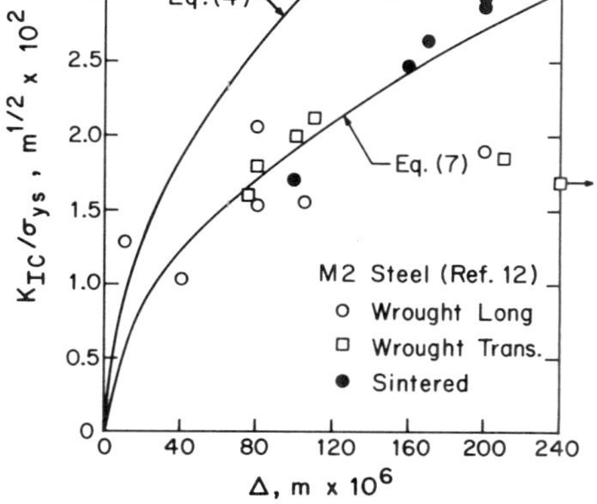

FIG. 4—*Prediction of fracture toughness,* K_{Ic}, *from CTOA stretch (Eq 4) and semi-cohesive process zone (Eq 7) models.*

treat this variable strength process zone, as has been done before [13], by allowing f_v to increase with K_I according to

$$f_v = \frac{3}{\pi} \sec^{-1}\left(\frac{K_I}{K_{Io}}\right)^{1/2} \tag{8}$$

This saturates the zone with microcracks at $K_I/K_{Io} = 4$. With Eqs 1 to 3 and 8, equilibrium is now represented by

$$K_I = \left(\frac{c}{\pi}\right)^{1/2}\left[\frac{6}{\pi}\sigma_{ys}\cos^{-1}\left(\frac{c}{c+\Delta+R_{P_I}}\right)\sec^{-1}\left(\frac{K_I}{K_{Io}}\right)^{1/2} + 2\sigma_{ys}\cos^{-1}\left(\frac{c}{c+\Delta+R_{P_I}}\right)\left\{1-\frac{3}{\pi}\sec^{-1}\left(\frac{K_I}{K_{Io}}\right)\right\}^{1/2}\right] \tag{9}$$

This may be applied by choosing parameters for K_{Io}, σ_{ys}, and Δ; as an example these were picked to be 20 MPa · $m^{1/2}$, 800 MPa, and 1 mm, respectively. The crack was allowed to grow one process zone length at a time, and K_I was recalculated after each step knowing the new crack length and plastic zone from Eq 3. This iterative process produced R-curve behavior and was repeated for initial crack lengths of 3, 10, and 20 mm. Figure 5*b* shows residual strength or the gross applied stress versus half-crack length.

For comparison, some 400-mm-wide plate data on 2024-T3 aluminum [14] sheets are shown in Fig. 5*a*. Except for one outlier, the upper and lower bounds of the experimental data are similar in shape to the theoretical model. It should be pointed out that there are two sets of aluminum data here representing transverse and longitudinal orientations, the later generally being the higher of the two. There are differences here in that the residual strengths for the

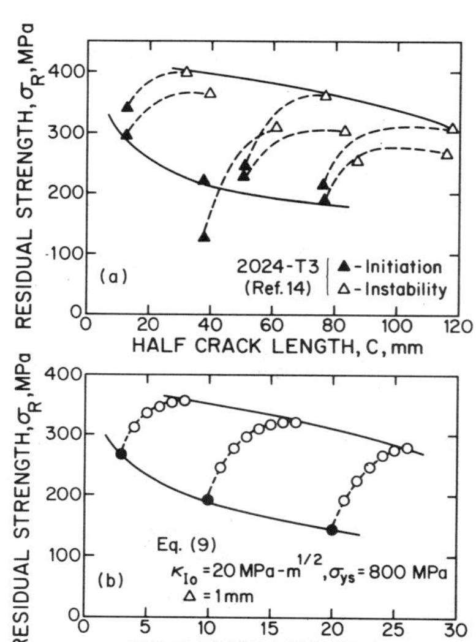

FIG. 5—*R-curve behavior* (a) *as observed in ductile fracture of wide-plate, 2024-T3 aluminum tests;* (b) *as predicted by Eq 9.*

experimental data were finite width corrected while the infinite plate solution represents gross stress. Also, there was no attempt made to match yield strength or crack sizes as it was only desired to illustrate general behavior. Using a $\sec(\pi c/w)$ finite-width correction for the aluminum data gives K_c at maximum load to increase from 80 to 106 MPa · $m^{1/2}$ for initial half-crack lengths increasing from 12.5 to 75 mm. For the model in Fig. 5*b*, an R-curve analysis using $\partial G_c/\partial c = \partial R/\partial c$ gives K_c increasing from 56 to 80 MPa · $m^{1/2}$ for initial-half crack lengths ranging from 3 to 20 mm. To summarize this section, even though the relationship between microcrack density and applied stress intensity is very artificial for ductile tearing, fairly realistic R-curve behavior may be predicted. More realistic hole-growth and strain hardening models should be attempted to evaluate how useful such an approach is for ductile fracture.

Application to a Macroscopic Process Zone (Concrete Composites)

To gain some practical insight as to how realistic such models are, it is first useful to illustrate them on a macroscopic scale where all parameters may be easily measured. The example chosen is a concrete aggregate reinforced with steel wires [*15,16*]. An accurate fit for the double-cantilever beam (DCB) specimens used has already been accomplished [*15*] using curve-fitting parameters, but it is preferred here to use the simpler infinite plate solution, although not as applicable, with no adjustable parameters. The depiction of this model is shown in Fig. 6. The value of ℓ_p, which is equivalent to Eq 3, was found to be constant for DCB specimens and equal to 0.063 m. Given that the fracture toughness of pure concrete is on the order of 1.0 MPa · $m^{1/2}$, this is equivalent to a cohesive stress of 2.5 MPa for the aggregate as observed. The only assumed value was the strength of the semicohesive zone, which was taken to be 5 MPa and would be equivalent to a process zone having a 1% volume fraction of 500-MPa strength steel fibers. Alternatively, this is the same order as the interfacial shear strength estimate as used by Wecharatana and Shah [*15*]. With both the process zone and the strain-energy release rate being measured, all parameters are known as given below (the present nomenclature and that of Fig. 6 are used)

$$c = 0.2\ \text{m} \qquad \sigma_{coh} = 2.5\ \text{MPa}$$
$$R_{P_{Io}} = \ell_p = 0.063\ \text{m} \qquad \sigma_{sc} = 5\ \text{MPa}$$
$$\Delta = \ell_f = \text{measured process zone}$$

With these values, Eqs 1 to 3 could be used to calculate K_I for increasing process zone sizes. For comparison, the strain-energy release rates were converted to K_R values. The results in Fig. 7 indicated a similar trend for measured and theoretical values. Both predict about a factor of three increase in toughness with increasing process zone size. The difference in magnitude is to be expected when comparing DCB data to infinite plate theory.

Application to a Microscopic Process Zone (Cleavage of Steel)

The reasons for applying such an approach to cleavage near the ductile-brittle transition are twofold. First, some acoustic emission observations in plain carbon steels [*17*] previously indicated that multiple discontinuous microcleavage events preceded instability, even at 77 K. If true, this precludes a weakest-link hypothesis for sharp cracks. Second, there were indications that the simplest weakest-link hypothesis, while predicting the temperature dependence of K_{Ic} in the lower shelf, did not predict the grain size dependence. The latter is illustrated below.

In Ritchie and Thompson's [*18*] recent analysis of microstructure and fracture toughness, the Ritchie-Knott-Rice (RKR) model of cleavage fracture is reviewed. Simply stated, when the stress intensity produces a local stress over a characteristic distance equal to the cleavage fracture stress, instability ensues. Its success is its simplicity and its connection to both microscopi-

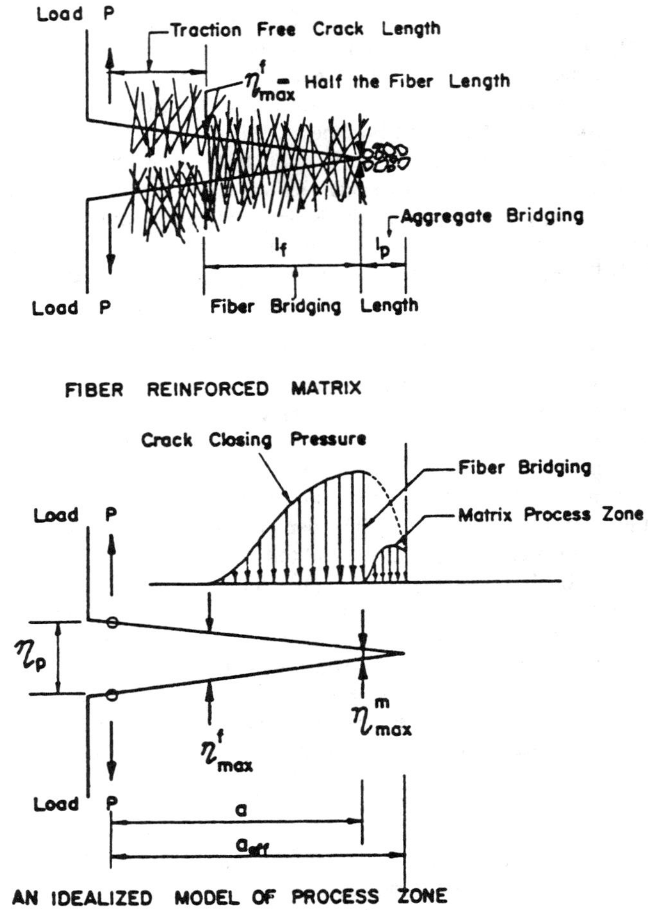

FIG. 6—*Depiction of process zone model for metal-fiber-reinforced concrete composites. (Courtesy of Ref* 15.*)*

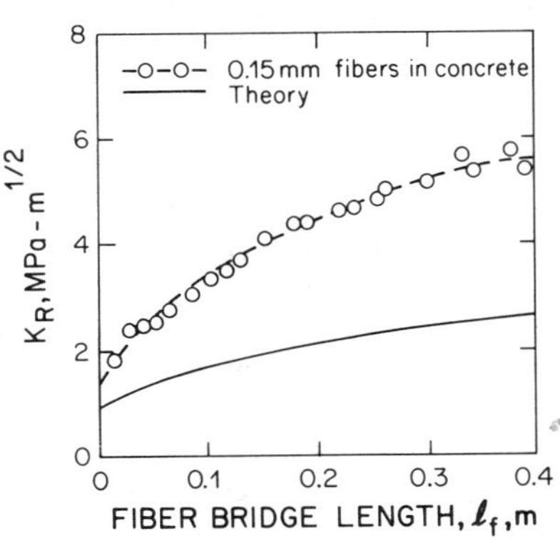

FIG. 7—*Comparison of theoretical R-curve behavior for an infinite plate to double-cantilever beam results. Resistance increases with process zone size,* $\Delta = \ell_f$.

cally measurable parameters, that is, carbide size distributions and continuum descriptions of the stress distribution. In this model [18], K_{Ic} is a function of cleavage fracture stress, σ_f^*, and yield strength, σ_{ys}, as given by[2]

$$K_{Ic} \propto \frac{\sigma_f^{*[(1+n)/2n]}}{\sigma_{ys}^{[(1-n)/2n]}} \Delta^{1/2} \tag{10}$$

Here, n is the strain-hardening exponent and the process zone, Δ, is equivalent to the RKR characteristic distance parameter, typically taken as two grain diameters. It has been shown elsewhere [19,20] that $\sigma_f^* \simeq k_f d^{-1/4}$ and that the Hall-Petch relationship typically gives $\sigma_{ys} \simeq k_y d^{-1/2}$ for test temperatures not far below room temperature. In terms of grain size, Eq 10 would thus give

$$K_{Ic} \propto \frac{k_f^{[(1+n)/2n]}}{k_y^{[(1-n)/2n]}} d^{-[(1+n)/8n]} d^{[(1-n)/4n]} (2d)^{1/2} \propto \text{const. } d^{[(1+n)/8n]} \tag{11}$$

For strain hardening exponents ranging from 1 to 0.05, this gives K_{Ic} proportional to $d^{1/4}$ up to $d^{21/8}$. Thus, in all cases K_{Ic} should increase with increasing grain size. This is at variance with experimental observations showing K_{Ic} decreasing with increasing grain size. Recent statistical models of Evans [21] and Lin, Evans, and Ritchie [22] appear to clear up this deficiency. Nevertheless, this requires a three-parameter Weibull distribution and still relies on a weakest-link hypothesis.

With regards to weakest-link models, some earlier data [17] had strongly indicated that cleavage initiated from fatigue cracks as a series of events. These started well below K_{Ic}, and it was hypothesized that the final fracture event represented a connecting up of a large series of nonpropagating cleavage microcracks. The evidence for this was acoustic emission of sufficiently large magnitude so as to not represent unfiltered noise or grip/specimen interface problems. In addition fractography of both 1000 (decarburized) and 1020 plain carbon steels indicated that the fatigue/cleavage interface was free of ductile tearing preceding instability at $-196°C$. An example of acoustic emission occurring in a compact-tension specimen of 1000 decarburized steel is given in Fig. 8. On several specimens, there was an indication of twinning, which may also be a source of acoustic emission. Thus, it cannot be stated unambiguously that all acoustic emission was associated with microcleavage events. There is further evidence, however, on the discontinuous nature of cleavage microcracking via acoustic emission (AE). Atherton [23] evaluated AE and fracture toughness above and below the DBTT in a thermomechanically processed spheroidized steel with a grain size of 1.1 μm. From -20 to $-75°C$, the yield strength increased only slightly from 690 to 745 MPa, and yet the acoustic emission activity increased by an order of magnitude. In addition, there was considerable slow crack growth in each case as indicated by the load-displacement record. The fracture surface indicated clear evidence of cleavage in the 75°C test, whereas only microvoid coalescence was observed in the 20°C test. There was no evidence of twinning. Since the test setup was identical, the increased AE had to be associated with either ferrite cleavage or carbide fracture. Furthermore, since the activity increased over a range of applied stress intensity from 40 to 90 MPa $\cdot$ m$^{1/2}$, there appears to be little question that in the transition region the weakest-link hypothesis is not viable in terms of there being a single controlling event for final instability.

Given this evidence, it is proposed that serious consideration be given to multiple origins of microcleavage existing prior to the final instability. This would apply particularly to the transition region and perhaps, in some instances, to the lower shelf. If multiple origins of microcleavage connect up at instability as the critical event, it would be the details of the size and microcleavage fraction of this process zone that are of importance. This may be modeled in the

[2]There is a transcription error in Eq 10 of Ref *18*.

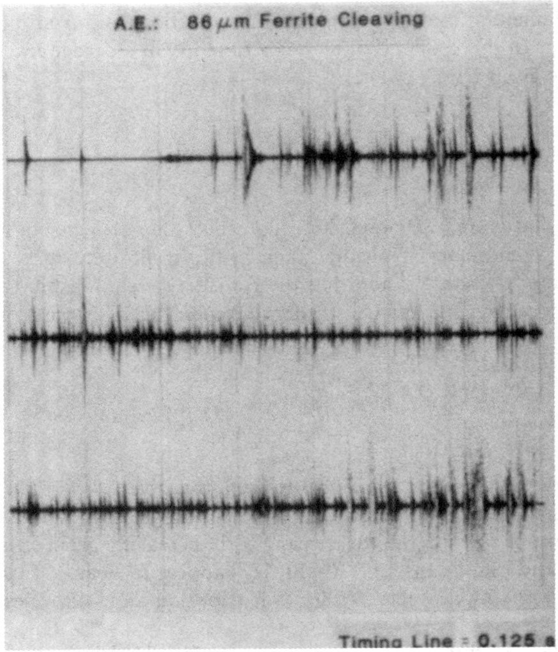

FIG. 8—*Acoustic emission during compact tension testing of coarse-grain ferrite at* −196°C. *(Courtesy of Ref* 30.*)*

same way that led to Eq 5 except that the number of cleaved grains would tend to increase with applied stress intensity. Specifically, it has been shown [*7, 13*] that the volume fraction of grains fracturing contiguous to the crack front could be given as a function of the applied K_I to the stress intensity for nucleating a coplanar microcrack, K_n. This is given by

$$f_v = \frac{n}{\pi} \sec^{-1}(K_I/K_n)^{1/2} \tag{12}$$

where n would be 4 for orthogonal cleavage planes or 3 for hexagonally oriented intergranular fracture. Since this considers only twist and not tilt misorientations, cleavage might actually be best represented by some value of n between 3 and 4. One could specify either a nucleating stress intensity for elastic behavior or a plastic constraint approach for elastic-plastic behavior. Since for low- to medium-strength steels in the transition regime, the plastic zone still extends several grain diameters for even 100 μm size grains, it makes most sense to consider the plastic constraint approach. This gives

$$f_v = \frac{n}{\pi} \sec^{-1}(\sigma_{yy}^{R_p}/\sigma^*_{f\theta=0°})^{1/2} \tag{13}$$

where $\sigma_{yy}^{R_p}$ is the stress in the plastic zone and $\sigma^*_{f\theta=0°}$ is the coplanar cleavage stress. With the maximum stress in the plastic zone taken as the plastic constraint, pcf, times the yield stress, and using the approximations in Eq 6, Eqs 1 to 3, 6, and 13 give

$$K_{\rm I} = 2\sqrt{2}\,\sigma_{\rm ys}\left[1 - \frac{n}{2\pi}\sec^{-1}\left(\frac{\text{pcf}\,\sigma_{\rm ys}}{\sigma^*_{{\rm f}\theta=0^\circ}}\right)^{1/2}\right]\left[\frac{2\Delta}{\pi}\right]^{1/2} \tag{14}$$

Since the pcf increases from zero load, what Eq 14 indicates is that there are multiple equilibria with the decrease in the strength of the process zone, the first term in brackets, being offset by the increase in the process zone size, the second term in brackets. This variable size process zone may be represented by

$$\Delta = \beta K_{\rm I}^2/\pi\sigma_{\rm ys}^2 \tag{15}$$

since the region of stress elevation due to plastic constraint must lie within the plastic zone. Using a previous formulation which has been successfully utilized to predict brittle fracture processes [*24–27*], the plastic constraint is given by

$$\text{pcf} = 1 + \alpha(K_{\rm I}/\sigma_{\rm ys}) \tag{16}$$

where $\alpha = 12.55\ \text{m}^{1/2}$. With Eqs 14 to 16, the fracture toughness becomes

$$K_{\rm Ic} = \frac{\sigma^*_{{\rm f}\theta=0^\circ}\sec^2\left\{\dfrac{2\pi}{n}\left[1 - \dfrac{\pi}{4\beta^{1/2}}\right]\right\} - \sigma_{\rm ys}}{\alpha} \tag{17}$$

This may be applied to the fracture toughness data of Ritchie et al. [*28*], which have become a standard against which many models have been compared. By using the previously given value of $\alpha = 12.55\ \text{m}^{1/2}$, $n = 3$ or 4, a value of $\sigma^*_{{\rm f}\theta=0^\circ} = 1050$ MPa and from values of $\sigma_{\rm ys}$ as a function of temperature for SA 533 B-1 steel, there is only β as an unknown.[3] From Eq 15, β must be on the order of unity and therefore this first pick was used in Eq 17. A good representation of the data in Fig. 9 is achieved up to the NDTT. It should be pointed out that changing β by a factor of two greatly affects fracture toughness levels. As such, it is a scaling parameter and further definition needs to be done.

Given this relative success, it is useful to consider in slightly more detail what is happening at the crack tip as the stress intensity increases. According to Eqs 12 and 13, when the elastic stress intensity exceeds the value of K_n for a coplanar crack to cleave, more unfavorably oriented grains along the crack front now see a resolved $K_{\rm I}^R$ such that they cleave. A similar argument applies for plastic constraint, since Eqs 13 and 16 would give

$$f_v = \frac{n}{\pi}\sec^{-1}\left(\frac{\sigma_{\rm ys} + \alpha K_{\rm I}}{\sigma^*_{{\rm f}\theta=0^\circ}}\right)^{1/2} \tag{18}$$

With $n = 3$, Eqs 12 and 18 may be compared with the number of grains cleaving along the crack front. If it is assumed that all of the acoustic emission in Fig. 8 are associated with discon-

[3]Note that in the RKR model, $\sigma_{\rm f}^*$ was taken as 1830 MPa. The value used above is substantially below this and could be due to three sources. First, the value of $\sigma_{\rm f}^*$ measured by notch bend testing using finite-element formulations says nothing about the orientation of the grain, the "weakest link," that cleaves. If it is a large grain by a large carbide, it need not be coplanar. Thus, $\sigma_{\rm f}^*$ could be measured for a tilted or twisted grain or both. The value of $\sigma_{\rm f}^*$ used here is for a coplanar cleavage plane. Second, there is no guarantee that the largest carbide that triggers failure is in the area of maximum constraint. If the pcf is not maximum where cleavage is triggered, then $\sigma_{\rm f}^*$ is overestimated. Finally, there is no guarantee that the empirical formulation which led to $\alpha = 12.55\ \text{m}^{1/2}$ in the proposed analysis is correct. Thus, the required value of $\sigma_{\rm f}^*$ to obtain a reasonable fit to the data in Fig. 9 may be in error.

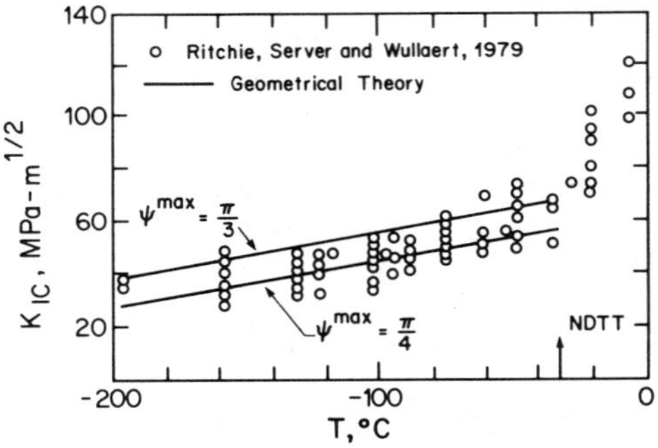

FIG. 9—*Fracture toughness prediction from Eq 17 for 533B steel as a function of test temperature in the lower shelf region.*

tinuous cleavage events, then a previously determined calibration of AE amplitude versus crack area could be used to determine the number of grains cleaving [*29*]. In fact, it may be shown that the small amplitude emission near the start of the record (at the second timing line) in Fig. 8 would represent an area of about 9000 μm^2, or slightly more than the area of an average size grain. Summing the acoustic emission over several time, and thus K_I intervals, gives the result in Fig. 10. Equation 12 was applied by allowing two rows of grains to participate in microcleavage.

It should be noted that this 25.7-mm-thick specimen had, on the average, 300 grains through the thickness. Equation 18 was applied by allowing the first row to participate only until the plastic zone reached about midway to the second row. This occurred near 18 MPa $\cdot$ $m^{1/2}$. It is seen that these two equations reasonably bracket the data if the acoustic emission represents only cleavage and only cracking along grains which are defined by a narrow wedge extending from the crack tip. Such observations clearly need more precise quantification with modern acoustic emission techniques and post-sectioning fractography studies to verify the source of the emission. Nevertheless, this discontinuous microfracture zone approach appears promising for those situations where weakest-link hypotheses might not be viable.

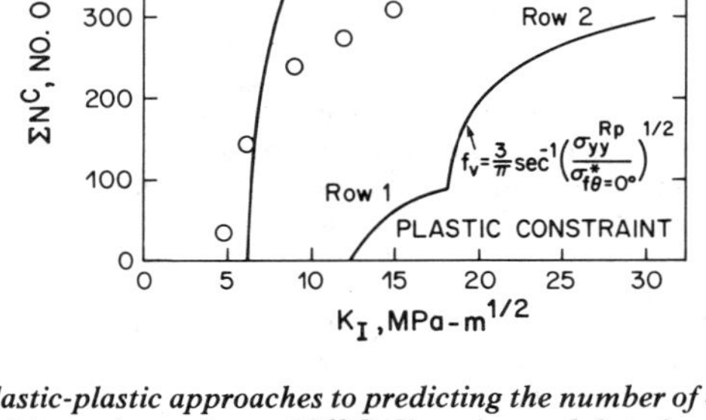

FIG. 10—*Elastic and elastic-plastic approaches to predicting the number of cleaved grains in the process zone of a coarse-grain ferritic steel tested at −196°C. Experimental data from Fig. 8.*

One final comparison is made for the 1000 data represented in Fig. 8 and other plain carbon steels heat treated to vary the carbide distribution and ferrite grain size [30]. For these ferrite/pearlite and spheroidized steels, the ferrite grain size ranged from 1.4 to 104 μm, and yield strengths from -73 to -196°C were measured to vary from 400 to 1200 MPa. The values of $\sigma^*_{f\theta=0°}$ were estimated to be only a function of grain size as given by

$$\sigma^*_{f\theta=0°} \simeq k_f d^{-1/4} \tag{19}$$

with $k_f = 65$ MPa · m$^{1/4}$. Although k_f is somewhat smaller than used elsewhere [20], the carbides in these ferrite/pearlite and spheroidized steels were very coarse. The predicted cleavage fracture stresses ranged from about 650 to 1900 MPa. With this single adjustable parameter, k_f, known yield strengths, and n, β, and α taken as 3, 1, and 12.55 m$^{1/2}$ as above, Eq 17 was used to predict K_{Ic}. The result in Fig. 11 is good at -196°C with the factor of 3 decrease in K_{Ic} with increasing grain size being predicted. There appears to be little effect of grain size on K_{Ic} at the higher temperatures, but the data are too limited to be definitive.

It should be pointed out that only specimens undergoing complete cleavage fracture, as perceived by visual and representative scanning electron microscope scans, were selected. This eliminated some of the very fine grain structures at higher temperatures due to ductile tearing. Second, the coarsest grain structures were not evaluated at the two higher test temperatures. Even with these limited data the general trends as to how grain size and test temperature affect K_{Ic} are represented. In addition, this is in agreement with the effect of grain size on K_{Ic} as observed by Curry and Knott [31] in ferritic steels.

Conclusions

1. Process zone models are useful in modeling microstructural details of discontinuous fracture processes.
2. One such model of a microfractured, semicohesive zone is shown to represent ductile R-curve behavior and stability of metal-fiber composites.
3. Acoustic emission and prior fractographic evidence suggest that weakest-link hypotheses may have limited applicability to brittle fracture of steel. They may not be applicable to cleavage fracture near the nil-ductility transition temperature and do not appear to apply in the transition region.
4. A viable alternative to weakest-link models is proposed to be a semi-cohesive process zone of variable size and strength. This is dependent upon the grain size and carbide distributions as

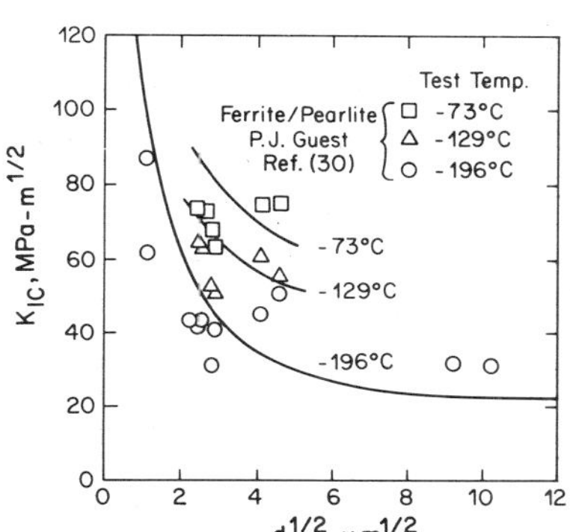

FIG. 11—*Fracture toughness predictions from Eq 17 for ferrite/pearlite steels as a function of test temperature and grain size.*

well as the flow and fracture strengths of those grains which depend upon temperature, local orientation, and plastic constraint.

Acknowledgment

This work was supported by the Materials Science Division, Basic Energy Sciences of the Department of Energy, under Contract DOE/DE-AC02-79ER 10433.

References

[*1*] Krafft, J. M., *Applied Materials Research,* Vol. 3, 1964, p. 88.
[*2*] McClintock, F. A. and Irwin, G. R. in *Fracture Toughness Testing and Its Application, ASTM STP 381,* American Society for Testing and Materials, Philadelphia, 1965, p. 84.
[*3*] Gerberich, W. W., *Journal of Materials Science,* Vol. 5, 1970, pp. 283-294.
[*4*] Ritchie, R. O., Knott, J. F., and Rice, J. R., *Journal of the Mechanics and Physics of Solids,* Vol. 21, 1973, pp. 395-410.
[*5*] Moody, N. R. and Gerberich, W. W., *Metal Science,* March 1980, pp. 95-100.
[*6*] Gerberich, W. W. and Moody, N. R. in *Fatigue Mechanisms, ASTM STP 675,* American Society for Testing and Materials, Philadelphia, 1979, pp. 292-341.
[*7*] Gerberich, W. W. in *Fracture: Interactions of Microstructure, Mechanisms and Mechanics,* J. M. Wells and J. D. Landes, Eds., The Metallurgical Society of the American Institute of Mining, Metallurgical, and Petroleum Engineers, Warrendale, PA, 1985, pp. 49-74.
[*8*] Wnuk, M. P. in *Proceedings,* International Conference on Dynamic Crack Propagation, Lehigh University, Bethlehem, PA, 1972, pp. 273-280.
[*9*] Rice, J. R. and Sorenson, E. P., *Journal of the Mechanics and Physics of Solids,* Vol. 26, 1973, p. 571.
[*10*] Smith, E., *International Journal of Fracture,* Vol. 17, 1981, pp. 443-448.
[*11*] Israel, S. J., Kantamneni, C. S., and Gerberich, W. W. in *Mechanical Behavior of Materials,* K. J. Miller and R. F. Smith, Eds., Pergamon Press, New York, 1979.
[*12*] Shelton, P. W. and Wronski, A. S., *Metal Science,* Vol. 17, No. 11, 1983, pp. 533-539.
[*13*] Gerberich, W. W. and Wright, A. G. in *Proceedings,* 2nd International Conference on Materials Degradation, Virginia Tech Printing, Blacksburg, VA, 1981, pp. 183-206.
[*14*] Knoll, A. H., "Tests of Wide Slotted Sheet Specimens of Aluminum Alloys," Report 12-60-2, Alcoa Research Laboratories, New Kensington, PA, 1960.
[*15*] Wecharatana, M. and Shah, S. P., *Cement and Concrete Research,* Vol. 13, 1983, pp. 819-829.
[*16*] Shah, S. P. in *Advances in Fracture Research,* S. R. Valluri, D. M. R. Taplin, P. Rama Rao, J. F. Knott, and R. Dubey, Eds., Pergamon Press, New York, Vol. 1, 1985, pp. 495-514.
[*17*] Gerberich, W. W., Zackay, V. F., Parker, E. R., and Porter, D. in *Ultrafine-Grain Metals,* J. J. Burke and V. Weiss, Eds., Syracuse University Press, Syracuse, NY, 1970, pp. 259-293.
[*18*] Ritchie, R. O. and Thompson, A. W., *Metallurgical Transactions A,* Vol. 16A, 1985, pp. 233-248.
[*19*] Curry, D. A. and Knott, J. F., *Metal Science,* Vol. 13, 1978, pp. 511-514.
[*20*] Gerberich, W. W., Chen, Y. T., Atteridge, D. G., and Johnson, T., *Acta Metallurgica,* Vol. 29, 1981, pp. 1187-1201.
[*21*] Evans, A. G., *Metallurgical Transactions A,* Vol. 14A, 1983, pp. 1349-1355.
[*22*] Lin, T., Evans, A. G., and Ritchie, R. O., "A Statistical Model of Brittle Fracture by Transgranular Cleavage," Report LBL-19763, Lawrence Berkeley Laboratories, University of California at Berkeley, CA, May 1985.
[*23*] Atherton, C. S., "Relation of Toughness to Microstructure in Spheroidized Steel," Report UCRL-20540, Lawrence Berkeley Laboratories, University of California at Berkeley, CA, March 1971.
[*24*] Hahn, G. T. and Rosenfield, A., *Transactions,* American Society for Metals, Vol. 59, 1966, p. 909.
[*25*] Ritchie, R. O., *Metal Science,* Aug./Sept. 1977, pp. 368-381.
[*26*] Elices, M., Sanchez, V., Bernstein, I., Thompson, A., and Pinero, J. in *Hydrogen Effects in Metals,* I. M. Bernstein and A. W. Thompson, Eds., The Metallurgical Society of the American Institute of Mining, Metallurgical, and Petroleum Engineers, Warrendale, PA, 1981, pp. 971-978.
[*27*] Gerberich, W. W. and Chen, Y. T., *Metallurgical Transactions A,* Vol. 6A, 1975, pp. 271-278.
[*28*] Ritchie, R. O., Server, W. L., and Wullaert, R. A., *Metallurgical Transactions A,* Vol. 10A, 1979, pp. 1557-1570.
[*29*] Gerberich, W. W., Atteridge, D., and Lessar, J., *Metallurgical Transactions A,* Vol. 6A, 1975, pp. 797-801.
[*30*] Guest, P. J., "The Micromechanics of Fracture of Pearlitic Structural Steels," Ph.D. thesis, UCRL Report 19106, University of California at Berkeley, CA, Dec. 1969.
[*31*] Curry, D. A. and Knott, J. F., *Metal Science,* Vol. 10, 1976, p. 1.

William R. Tyson[1] *and Bernard Marandet*[2]

Cleavage Toughness Variability and Inclusion Size Distribution of a Weld Metal

REFERENCE: Tyson, W. R. and Marandet, B., "**Cleavage Toughness Variability and Inclusion Size Distribution of a Weld Metal,**" *Fracture Mechanics: Eighteenth Symposium, ASTM STP 945,* D. T. Read and R. P. Reed, Eds., American Society for Testing and Materials, Philadelphia, 1988, pp. 19–32.

ABSTRACT: The cleavage toughness of a carbon-manganese electroslag weld metal has been measured in the transition region and related to measurements of the inclusion size distribution. Results were interpreted using the "weakest link" theory, based on the hypothesis that fracture occurs when the local stresses acting on a microcracked inclusion become large enough to propagate the microcrack.

KEY WORDS: ductile-brittle transition, cleavage, toughness variability, inclusions, data scatter, size distribution, statistics, weld metal

Variability in toughness measurements is a continuing problem in the application of fracture mechanics. A quantitative representation of the degree of variability would be particularly useful in the determination of parameters such as lower bound toughness, confidence levels of average toughness, and degree of structural reliability.

The best-known statistical theory of strength is the "weakest link" theory of Weibull [*1*]. This theory has been applied recently to fracture in the stress field at a crack tip, in which the probabilities of fracture have been summed either numerically [*2*] or analytically [*3*]. There are four essential hypotheses:

1. Potential initiating sites for fracture are microcracks of volumetric density N_v and frequency distribution $f(r)$, where r is the half-length of the microcracks.
2. The local stress to initiate fracture from a microcrack may be found from its size by the well-known Griffith relation

$$\sigma = \sqrt{\alpha' \frac{E\gamma}{(1 - \nu^2)r}}$$

where

α' = $\pi/2$ for cracked spheres or $2/\pi$ for cracked platelets,
E = Young's modulus,
ν = Poisson's ratio, and
γ = fracture surface energy of ferrite.

[1]Head, Engineering Physics Section, Physical Metallurgy Research Laboratories, Canada Centre for Mineral and Energy Technology, Ottawa, Canada K1A OG1.

[2]Head, Fracture and Welding Group, Institut de Recherches de la Sidérurgie Française, 78105 St. Germain en Laye, France.

3. The probability that a randomly chosen microcrack will fail at a stress equal to or less than σ is simply the probability that its half-length is equal to or greater than r, that is

$$p(\sigma) = \int_r^\infty f(r')dr'$$

4. Based on the hypothesis that failure of a specimen containing N microcracks follows when the first microcrack propagates, the specimen failure probability Φ is given from the theory of extreme values by

$$\ell n(1 - \Phi) = \sum_i [-N_i p(\sigma_i)] \tag{1}$$

where N_i is the number of microcracks subjected to the local stress σ_i.

The fracture probability is therefore dependent on the stress pattern within the specimen and on the size distribution of microcracks. The stress pattern may be found from the known loads and specimen geometry by analytical or numerical methods, and is reasonably well understood. The microcrack distribution, however, is difficult to measure and has often been treated as an adjustable parameter.

The first problem in determining $f(r)$ is to identify physically the microcracks. In the case of mild steels, these have been shown to be cracks in carbide particles [*4*]. For steels where carbides are very small or of very low density, the situation is less clear. Borides have been implicated in the fracture of an alloy steel [*5*]. In steels containing high densities of inclusions, such as some weld metals, there is evidence that cracked inclusions may nucleate fracture [*6*]; fracture has been also shown to initiate from an inclusion in a low-alloy steel [*7*]. It is reasonable to suppose that particles become cracked as a result of strain inhomogeneities during deformation and, although little is known on this point, that the proportion of particles cracked is not dependent on particle size. Then $f(r)$ is identical to the size distribution of particles, and $N_v = \alpha/V_p$ where V_p is the volume per particle and α is the fraction of particles which contain cracks.

Methods to determine $f(r)$ experimentally have been discussed by Ashby and Ebeling [*8*], Widgery and Knott [*9*], and Passoja and Hill [*10*]. Sometimes $f(r)$ is assumed to be of log normal form, although this is based on mathematical convenience rather than physical theory. The form of $f(r)$ is implicitly assumed to follow a power law for large r in the original Weibull treatment used subsequently by Beremin [*2*], and in slightly modified form by Evans [*3*]. A functional form reducing to a power law for large r but capable of fitting the data for all values of r was used by Wallin et al. [*11*] to fit carbide size distributions. Unfortunately, none of these are based on physical theories of particle nucleation and growth.

The objective of the present work was to measure the toughness of a carbon-manganese (C-Mn) weld metal at several temperatures within the toughness transition range, and to relate the measurements to observations of the inclusion size distribution. The simplest possible form for $f(r)$ and for the statistical theory was sought to help in the physical interpretation and application.

Experimental Methods

Experimental material was obtained as a by-product from a larger program to study the arrest of cracks initiating in a weld and propagating into base metal [*12*]. Welds joining wide plates of approximate dimension 1 m^2 by 20 mm thick were made by electro-slag welding. The composition of the as-deposited weld metal is given in Table 1. The microstructure consists of a rather irregular mixture of elongated ferrite grains of about 6.6 μm mean diameter with islands

TABLE 1—*Chemical composition of weld metal.*

Element	Weight, %
Carbon	0.099
Manganese	0.816
Silicon	0.243
Phosphorus	0.019
Sulfur	0.013
Aluminum	0.006
Molybdenum	0.166
Niobium	0.005
Vanadium	0.022
Titanium	0.002
Nitrogen (N_2)	0.008
Oxygen (O_2)	0.044

of pearlite. The 28 J Charpy V-notch (CVN) transition temperature was measured as 40°C, and the fracture appearance transition temperature (FATT) as 60°C.

For the present work, standard three-point bend fracture toughness specimens of dimensions 17.5 by 35 by 170 mm were machined transverse to the weld with a through-thickness notch in the weld metal. Thus, the crack propagated entirely through weld metal; the propagation direction was counter to the welding direction, although this condition was not felt to be important. In all, 40 specimens were tested; seven of these contained 25% side grooves.

Specimens were fatigue precracked according to the ASTM Test Method for J_{Ic}, a Measure of Fracture Toughness (E 813-81), and tested in three-point bending in a screw-driven universal testing machine. Testing was done in air (+20°C), or in a well-stirred bath of liquid Freon (−40°C and −100°C) refrigerated by nitrogen circulating through cooling coils. Load-line displacement was recorded using a clip gage mounted between knife edges attached to the upper anvil and to a bar connecting the two lower anvils.

The a-c potential drop across the crack mouth was monitored to detect the initiation of ductile tearing [*13*], which was marked clearly by a minimum in potential when tearing occurred before cleavage.

Fracture surfaces were examined in a Japan Electron Optics Laboratory (JEOL) scanning electron microscope (SEM) [at Institut de Recherches de la Sidérurgie Française (IRSID)], and in a Cambridge instrument [at Physical Metallurgy Research Laboratories (PMRL)]. Quantitative metallography (QTM) on polished sections to determine inclusion distribution was performed (at IRSID) using standard techniques [*14*].

Results

All specimens fractured by cleavage, although at −40 and +20°C this was sometimes preceded by ductile crack growth. Figure 1 shows measurements made on one specimen at +20°C where cleavage occurred only after considerable crack growth, and after maximum load was passed. In all tests at −100°C, cleavage occurred on a rising load before ductile tearing began.

Values of the J-integral were calculated according to ASTM E 813, using 9-point average values of the initial fatigue crack length, and converted to stress intensity factors using $K_J = \sqrt{JE/(1 - \nu^2)}$. The results are shown in Fig. 2. According to the validity requirements reported in Table 2, all of the results for cleavage before tearing are valid according to Method E 813, although only the three lowest values at the lowest temperature are valid according to the ASTM

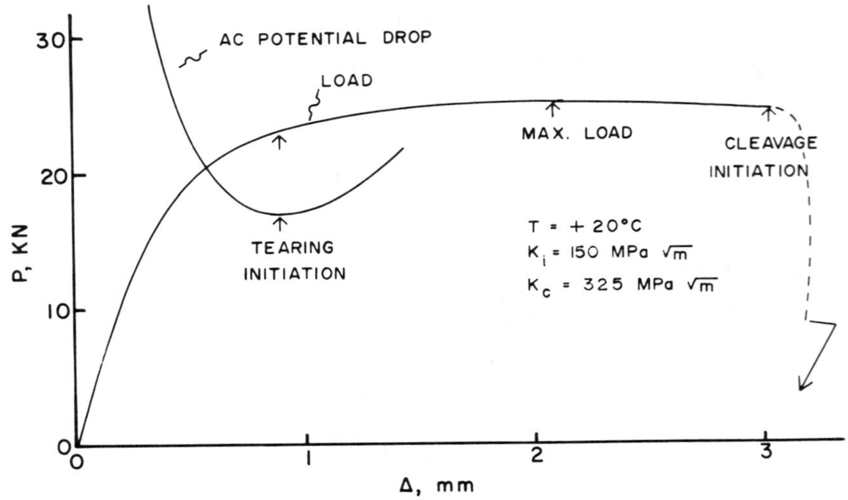

FIG. 1—*Load* P *and potential drop plotted against load-line displacement* Δ *for a test at* +20°C.

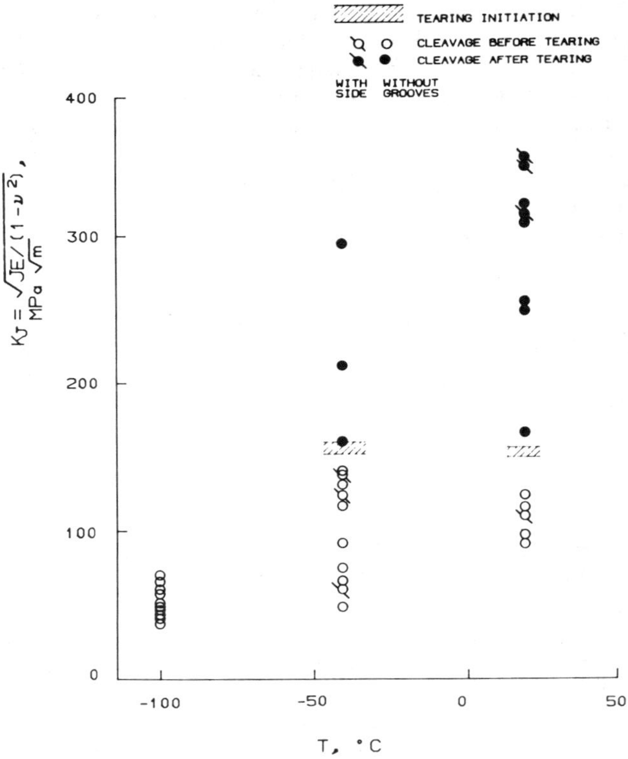

FIG. 2—*Values of fracture toughness calculated from the J-integral measured at* −100, −40, *and* +20°C.

TABLE 2—*Maximum values of valid* K *according to ASTM Method E 399.* ($B \geq 2.5(K/\sigma_y)^2$) *and E 813*[a] ($B \geq 25\ J/\sigma_y$).

T, °C	σ_y, MPa	Maximum K, MPa$\sqrt{m}$	
		E 399	E 813
−100	535	45	291
−40	460	39	270
+20	425	36	260

[a]Using $K = \sqrt{JE/(1 - \nu^2)}$.

Test Method for Plane-Strain Fracture Toughness of Metallic Materials (E 399-83). The toughness at tearing initiation was highly reproducible; all measurements fall within the scatter band shown in Fig. 2. There was no apparent influence of side grooving on toughness levels for tearing initiation or cleavage.

Fracture surfaces were examined optically and by scanning electron microscopy. Because the microscopic appearance of the fracture surface was dominated by the solidification structure of the weld metal, it was not possible to identify the macroscopic fracture origin from the pattern of cleavage markings. However, at the microscopic level it was possible to follow river markings within a cleavage facet, and in this way a number of areas similar to those shown in Refs 6 and 7 were found where it was plausible to identify inclusions as nucleating sites. Figure 3 shows two such locations. However, owing to the absence of macroscopic chevron markings (because of the dominating columnar macrostructure of the weld), it was difficult to positively identify inclusion sites such as those shown in Fig. 3 as locations where cleavage initiated rather than simply as sites where inclusions participated in cleavage propagation, possibly by microcracking ahead of the main crack front. The present observations do not exclude the possibility of multiple origins.

Inclusion size distributions were measured on polished sections by standard quantitative metallography, and by measuring the dimensions of inclusions exposed on ductile regions of fracture surfaces. A typical fractograph of the ductile region of a specimen that exhibited some ductile tearing is shown in Fig. 4. The true diameters of inclusions could be easily measured on these fractographs; a total of 331 measurements were made. From energy dispersive X-ray patterns, the inclusions were found to contain primarily silicon, manganese, aluminum, calcium, and sulfur; oxygen, of course, is not detectable by this technique. The inclusions are, presumably, complex silicates. Assuming that all of the oxygen in the weld metal is in these silicates and that they have approximately the volumetric density of silica (SiO_2), the volume fraction of inclusions was deduced to be about 0.3%. This is in reasonable agreement with the amount measured by QTM on polished sections (0.25% ± 0.07%).

To calculate the true volumetric size distribution from fractographs, it was necessary to make an assumption concerning the sampling of inclusions by the fracture path. This problem has been discussed by Widgery and Knott [9], who demonstrated that the fracture surface contains inclusions whose centers lie within a multiple c of their diameter from the mean fracture path. With this assumption, it follows that the ratio of the number of inclusions per unit area on a fracture surface (twice the number on each half) to the number on a polished section is $2c$. This ratio was found in the present work to be 10, giving $c = 5$, which compares favorably with Widgery and Knott's result of 5.8. The histogram of the volumetric size distribution measured from fracture surfaces is shown in Fig. 5 along with two fitted distribution functions to be described later. The distribution functions are shown on a log-log plot in Fig. 6.

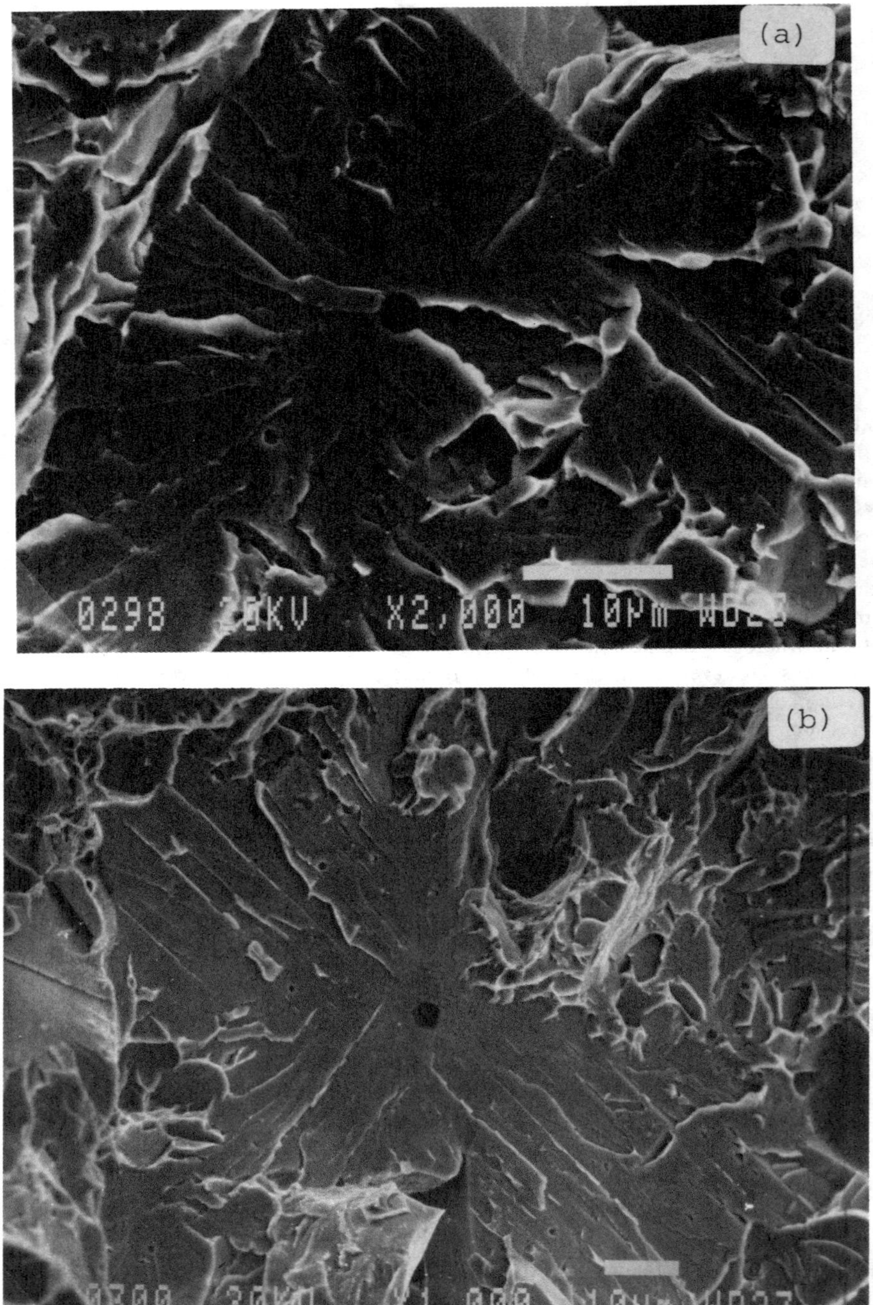

FIG. 3—*Cleavage facets containing inclusions at the focus of river markings.*

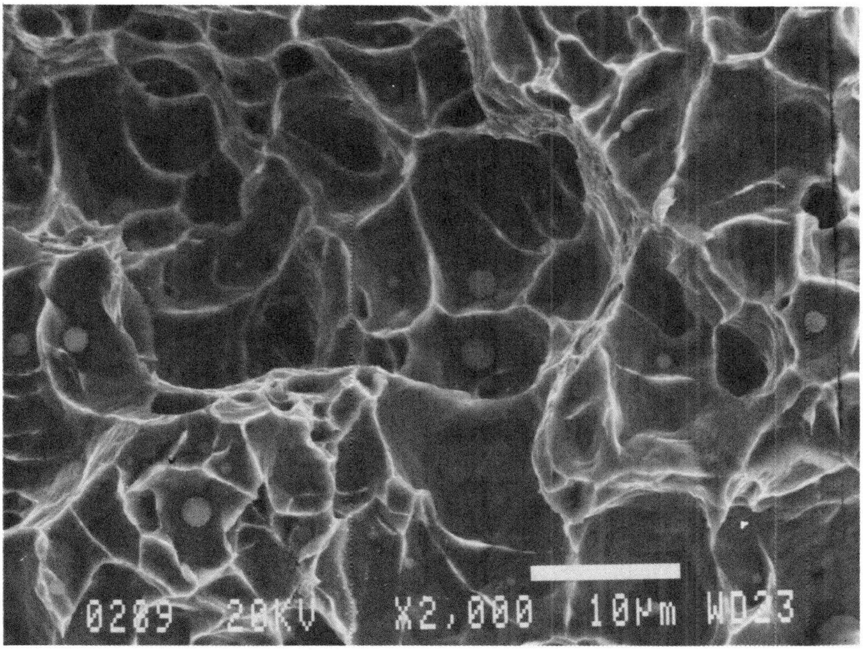

FIG. 4—*SEM fractograph of ductile tear region of specimen broken at +20°C.*

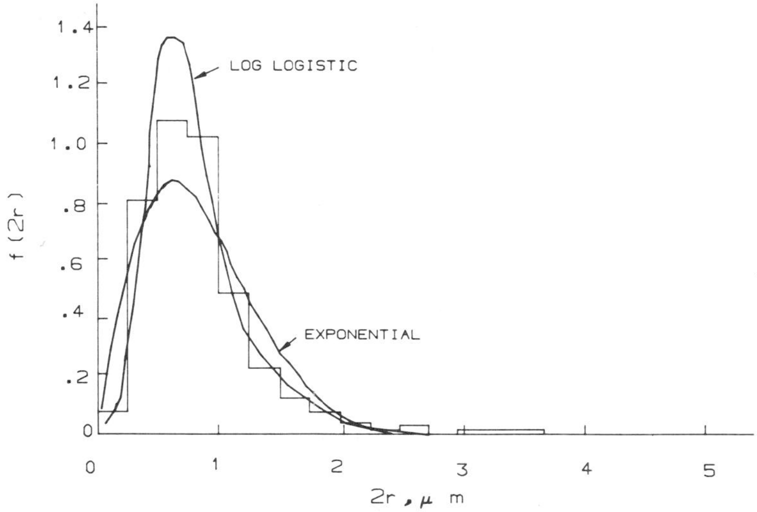

FIG. 5—*Volumetric frequency distribution* f(2r) *of true inclusion diameter* 2r *calculated from fractographic measurements (bars) and fitted by two different distribution functions.*

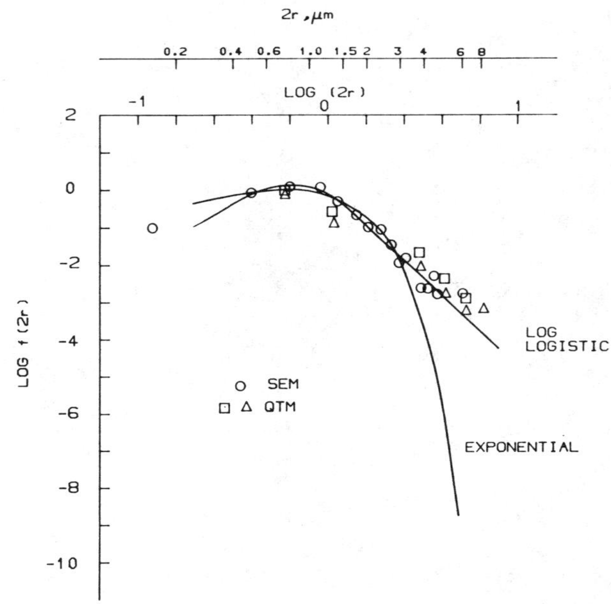

FIG. 6—*Volumetric distribution functions of inclusion diameter and fitted analytical approximations.*

Statistical Interpretation of Variability

The "weakest link" theory of cleavage fracture has been well described by Beremin [*2*] and by Evans [*3*]. The work of Evans is particularly attractive, as it is entirely analytic and offers the possibility of avoiding cumbersome numerical summations.

The essence of the theory has been described in the introduction. Evans makes the further approximation that only microcracks in the direction of macroscopic crack advance x, bounded by the curves $y = \pm\beta x^k$, participate in fracture where β and k are constants. He assumes that the elemental failure probability is given by

$$p(\sigma) = \left(\frac{\sigma - \sigma_u}{\sigma + \sigma_o}\right)^m \quad \text{for} \quad \sigma \geq \sigma_u$$
$$= 0 \quad \text{for} \quad \sigma < \sigma_u \tag{2}$$

where σ_u, σ_o, and m are parameters dependent on the microcrack distribution.

The failure probability Φ of a macroscopic specimen containing a large number of elements (microcracks) is then given by

$$\ell\text{n}\,\frac{1}{1 - \Phi} = \int p(\sigma)\,\frac{dV}{V_o} \tag{3}$$

where $dV = \beta x^k B dx$, where B is the specimen thickness, and $V_o = V_p/\alpha$ is the volume per microcrack. The variation of σ with x is given by the Hutchinson-Rice-Rosengren (HRR) crack tip stress singularity

$$\frac{\sigma}{\sigma_y} = \bar{\sigma}(\theta)\left[\frac{1 - \nu^2}{I(N)}\right]^{1/1+N}\left(\frac{K^2}{\sigma_y^2 x}\right)^{1/1+N} \quad (4)$$

where

$\sigma(\theta) \sim 2.5$,
$1 - \nu^2 \sim 0.91$, and
$I(N) \sim 4.5$;
N is the work hardening exponent.

Substituting Eq 2 in Eq 3 and integrating, using Eq 4, Evans shows that (taking $\sigma_u = 0$)

$$\ell n \frac{1}{1 - \Phi} = C_1 B \sigma_y^{(1+k)(N-1)} K^{2(1+k)} \quad (5)$$

where

$$C_1 = A \frac{\beta}{V_o}(1 + N)\frac{\Gamma[(1 + k)(1 + N)]\Gamma[m - (1 + k)(1 + N)]}{\Gamma(m)}$$

and

$$A = 2.5^{(1+N)(1+k)}\left(\frac{0.91}{4.5}\right)^{(1+k)}$$

Evans used a length ℓ in place of B in Eq 5 and proposed as a criterion for fracture that $\Phi = 0.5$, which requires that half of the strips of width ℓ contain microcracks that have propagated. However, this criterion does not enable the variability in toughness to be predicted and a single value of K results. We have used the full specimen width B, and the strict weakest-link theory in which the first microcrack to propagate causes fracture of the entire specimen. Thus, in Eq 5, we interpret Φ as the fraction of nominally identical samples of width B that have fractured at stress intensities of K or less.

Although the expression for $p(\sigma)$ used by Evans (Eq 2) is an improvement over the simple power-law expression $p(\sigma) = (\sigma/\sigma_o)^m$ that is often used, it is nevertheless limited in that it predicts an unrealistic microcrack size distribution. The distribution $f(2r)$ that is implicit in Eq 2 is of power-law form at large r, but displays no maximum at small r and in fact increases without limit as r approaches zero. This is avoided by an arbitrary cutoff at the radius corresponding to a microcrack fracture stress of σ_u, but in deriving Eq 5, Evans has used $\sigma_u = 0$.

Two alternative distributions that are better able to describe the observed inclusion size distribution and yet can be analytically integrated to yield convenient expressions for Φ were tested in this work.[3] These are

exponential

$$f(2r) = \theta\mu(2r)^{\theta-1} \exp - [\mu(2r)^\theta] \quad (6)$$

[3]A variety of analytical forms can be found in the literature to describe inclusion size distributions. The so-called "log normal" distribution is most popular, and Wallin [11] has used a distribution that reduces to an exponential form at small r and a power-law form at large r. Unfortunately, these distributions are generally not amenable to analytic integration, and have no more theoretical basis than the forms proposed here. The exponential and log logistic distributions (Eqs 6 and 7) have quite different behavior (exponential and power-law, respectively) for the large r values of particular interest here, and can reveal the effects of such different variations on the cleavage toughness within the "weakest link" model.

log logistic

$$f(2r) = \theta\mu \frac{(2r)^{\theta-1}}{[1 + \mu(2r)^{\theta}]^2} \tag{7}$$

These functions have the required form for a frequency distribution, namely

$$\int_o^\infty f(2r)d(2r) = 1$$

The parameters θ and μ were obtained from the first and second moments of the observed frequency distributions,

$$2\bar{r} = \int_o^\infty 2rf(2r)d(2r) = 0.834\ \mu m$$

$$M = \int_o^\infty (2r)^2 f(2r)d(2r) = 0.914(\mu\text{m})^2$$

The corresponding parameters are $\mu^{-1/\theta}\Gamma(1 + 1/\theta)$ and $\mu^{-2/\theta}\Gamma(1 + 2/\theta)$ for the exponential function, and $\mu^{-1/\theta}\Gamma(1 + 1/\theta)\Gamma(1 - 1/\theta)$ and $\mu^{-2/\theta}\Gamma(1 + 2/\theta)\Gamma(1 - 2/\theta)$ for the log logistic function. Values of μ and θ were thus estimated as 1.123 and 1.85 for the exponential function, and 3.10 and 3.85 for the log logistic function; these values were used in calculating the curves in Figs. 5 and 6.

The elemental failure probability $p(\sigma)$ follows directly from integration of Eqs 6 and 7. As described in the introduction

$$p(\sigma) = \int_r^\infty f(r')dr'$$

$$= \int_{2r}^\infty f(2r')d(2r')$$

where the relation between particle size and fracture stress is given by

$$r = \frac{\alpha' E\gamma}{1 - \nu^2} \frac{1}{\sigma^2}$$

with $\alpha' = \pi/2$ for spherical particles. The integrations give, for exponential and log-logistic distributions respectively

$$p(\sigma) = \exp - (\sigma_\theta/\sigma)^{2\theta} \tag{8}$$

and

$$p(\sigma) = \frac{\sigma^{2\theta}}{\sigma^{2\theta} + \sigma_\theta^{2\theta}}$$

with

$$\sigma_\theta = \mu^{1/2\theta}\left[\frac{\pi E\gamma}{(1-\nu^2)}\right]^{1/2}$$

The specimen fracture probability Φ follows, as before, from Eq 3. We find for both distributions that Φ is given by an equation identical to Eq 5 with C_1 replaced by

$$C_2 = A\,\frac{\beta}{V_o}\,\sigma_\theta^{-(1+k)(1+N)}\,\frac{(1+N)}{2\theta}\,\Gamma\left[\frac{(1+N)(1+k)}{2\theta}\right]$$

for the exponential function, and

$$C_3 = A\,\frac{\beta}{V_o}\,\sigma_\theta^{-(1+k)(1+N)}\,\frac{(1+N)}{2\theta}\,\frac{\pi}{\sin\dfrac{\pi(1+N)(1+k)}{2\theta}}$$

for the log logistic function. For the integral to remain finite in the log logistic distribution, it is further required that $2\theta > (1 + k)(1 + N)$.

It follows that, for both distributions, the failure probability is given by

$$\ell n\,\ell n\,\frac{1}{1-\Phi} = \ell n\,C + \ell n\,B + (1+k)(N-1)\ell n\,\sigma_y + 2(1+k)\ell n\,K \qquad (10)$$

where $C = C_1$ (exponential) or $C = C_2$ (log logistic).

To test Eq 10, failure probabilities at a given temperature were associated with toughness values by ranking the measured K-values in increasing order and using $\Phi_i = n_i/(M + 1)$ where n_i is the rank of the ith observation and M is the number of K values at that temperature. The results are plotted in Fig. 7. Straight lines of Slope 4 may be reasonably fit to the data for cleavage before tearing initiation, although values for cleavage after some ductile crack growth do not seem to fit the same relation. (Note, the fact that some of the K_J-values are invalid does not affect the Φ values of valid data.) Using Eq 10, we deduce that $2(1 + k) = 4$, that is, $k = 1$. The active zone is therefore wedge-shaped. For cleavage after tearing initiation, the trend of the rather limited data in Fig. 7 (solid points) indicates that rather larger K_J-values are required compared to those expected from extrapolation of the lines fitted to data for cleavage before tearing. The weakest link concept is not relevant for tearing initiation, as demonstrated by the small scatter of K_J for tearing initiation in Fig. 2, but should still apply for cleavage after tearing. However, since the model depends on the relation between local stress at a cracked inclusion and the characterizing J-value, the relation analogous to the HRR field (Eq 4) for the stress distribution ahead of the tearing crack would have to be known for predictions of Φ to be made after tearing initiation.

Values of K at $\Phi = 0.5$ were read from Fig. 7, and plotted in Fig. 8 with the values of σ_y reported in Table 2. The data fit a straight line of slope -0.234, and thus from Eq 10 we find $N = 9.5$. This is in reasonable accord with the measured value of $N = 8.25$; N was found experimentally to be constant to within 10% over the temperature range from -100 to $+25$°C. Equation 10 thus is a reasonable description of the experimental variability and yield strength dependence of the data using $k = 1$ and $N = 9.5$; $\ell n\ C$ is found from the data to be -118.4 with units of B (m), σ_y (MPa), and K (MPa $\sqrt{\text{m}}$).

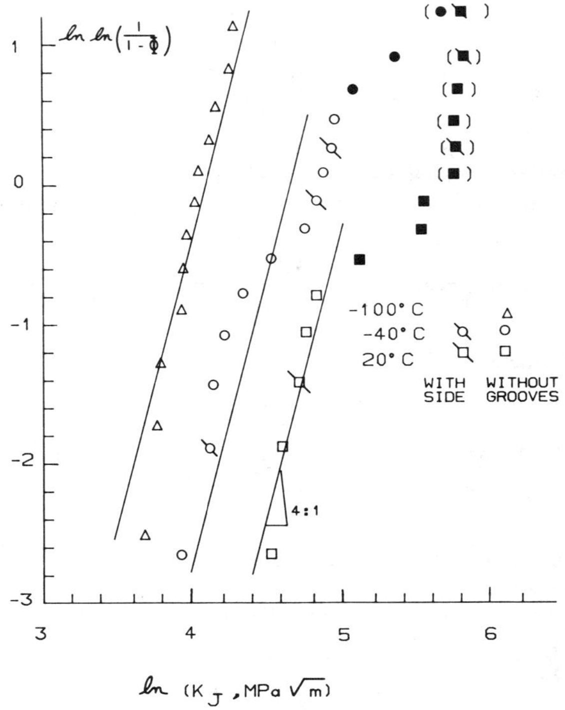

FIG. 7—*Failure probabilities Φ as a function of toughness* K_J. *Solid points refer to cleavage after tearing initiation.* K_J*-values that are invalid by ASTM E 813 are enclosed in brackets.*

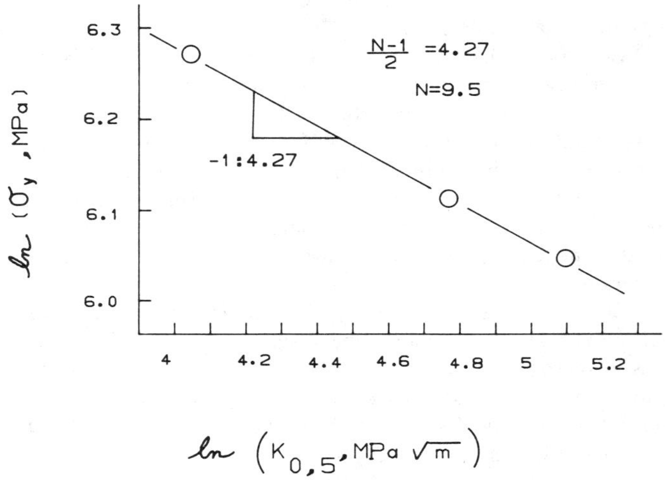

FIG. 8—*Dependence of toughness* $K_{0.5}$ *at* $\Phi = 0.5$ *on yield strength* σ_y.

Discussion

Agreement of the measured toughness variability with the observed inclusion size distribution requires that values of C calculated from $f(2r)$ be reasonably in accord with the experimental value of exp (-118.4).

For the exponential distribution, this can be achieved by choosing $\beta = 1$, $V_o = 2.86 \times 10^{-13}$ m^3, and $\sigma_\theta = 3071$ MPa. The value of V_o was found from $V_o = 1/(\alpha N_p)$ where N_p is the number of inclusions per unit volume and α (taken as 10^{-3}) is the fraction of inclusions that are cracked. The value of σ_θ is consistent with $E = 206$ GPa, $\nu = 0.3$, and $\gamma = 12.5$ J/m^2. All of these values are reasonable.

In the case of the log logistic distribution, however, agreement cannot be attained because the value of $2\theta = 7.7$ is not larger than $(1 + k)(1 + N) = 21$. This means, in effect, that the rate of decrease of $p(\sigma)$ with increasing distance x from the crack tip is not rapid enough to generate a finite failure probability. This is somewhat disturbing, as the observed size distribution is described much better, particularly at large values of $2r$, by a power-law type of distribution than by the exponential distribution (Fig. 6).

Some further insight into the problem may be gained by a calculation of the most probable fracture distance x_c given by

$$\frac{d}{dx} x^k p(\sigma) = 0$$

which, for the exponential distribution, gives

$$x_c = \left[\frac{k(1 + N)}{2\theta D^{2\theta}}\right]^{1+N/2\theta}$$

where

$$D = \frac{\sigma_\theta}{\sigma_y} \bigg/ \left[2.5\left\{\frac{0.91}{4.5}\left(\frac{K}{\sigma_y}\right)^2\right\}^{1/1+N}\right]$$

The value of x_c is found to be 6.4 μm at values of K and σ_y for $\Phi = 0.5$. The expectation value of x is several times larger than x_c, but still small: of the order of the crack tip opening displacement (CTOD). (It should be also remarked that it follows from the expression for x_c with the definition of D and Eq 10 that x_c depends only on Φ.) This indicates that the critical microcrack is to be found within a few multiples of the CTOD of the crack tip. It also indicates that selection of a suitable cut-off distance could permit the preferred log logistic distribution to be used by avoiding summation of failure probabilities at large distances from the crack tip where the HRR stress field is not applicable. This will be addressed in a separate publication.

It has been also assumed in the model that the size distribution of cracked inclusions is identical to that of the inclusions themselves; in other words, that the probability of an inclusion being cracked is independent of the inclusion size and of the stress level. This is a rather strong assumption, and its effect should be investigated further. However, experimental work to determine the size distribution of cracked inclusions would be very difficult, since the fraction of inclusions that are cracked is likely to be very small.

It is also possible that unstable cleavage is preceded by the spread of microcracks from more than one inclusion, followed by arrest at grain boundaries or other barriers. Instability would then occur after accumulation of a critical number or size of such microcracks. However, no quantitative treatment of such a model seems to be available.

Conclusions

1. Fracture toughness measurements have been made for nominally identical specimens of a carbon-manganese weld metal. Values have been found to vary by as much as a factor of six in the transition region.

2. The inclusion size distribution has been measured for this weld metal, and characterized by two different distribution functions.

3. The "weakest link" theory of cleavage fracture has been used to derive an expression for the variability and dependence on yield strength of the fracture toughness. The predicted form is in good agreement with measurements.

4. Quantitative agreement between theory and experiment has been obtained for one of the distribution functions, which unfortunately is the less good fit to the experimental size distribution.

Acknowledgments

The authors are grateful to the technical staff of IRSID for their contributions to this work. Special thanks are due to Mlle M. Thomas, M. Blanc, and M. Englebert. M.Y. Sixou provided essential input on statistical methods, and suggested the use of the log logistic function. V. Chartrand of PMRL performed the SEM X-ray analyses.

References

[1] Weibull, W., "A Statistical Theory of the Strength of Materials," *Ingeniörs vetenskapsakademiens Handlingar,* 1939, Stockholm.
[2] Beremin, F. M., *Metallurgical Transactions,* Vol. 14A, 1983, pp. 2277–2287.
[3] Evans, A. G., *Metallurgical Transactions,* Vol. 14A, 1983, pp. 1349–1355.
[4] Curry, D. A. and Knott, J. F., *Metal Science,* Vol. 10, 1976, pp. 1-6.
[5] Veistinen, M. K. and Lindroos, V. K., *Scripta Metallurgica,* Vol. 17, 1983, pp. 1147–1150.
[6] Tweed, J. H. and Knott, J. F., "Microstructure-Toughness Relationships in a C-Mn Weld Metal" in *Proceedings,* 4th European Conference on Fracture, Leoben, Austria, 1982.
[7] Rosenfield, A. R. and Shetty, D. K., *Engineering Fracture Mechanics,* Vol. 14, 1981, pp. 833–842; see also *Elastic-Plastic Fracture Test Methods: The User's Experience, ASTM STP 856,* E. T. Wessel and F. J. Loss, Eds., American Society for Testing and Materials, Philadelphia, 1985, pp. 196–209.
[8] Ashby, M. F. and Ebeling, R., *Transactions,* Metallurgical Society of the American Institute of Mining, Metallurgical, and Petroleum Engineers, Vol. 236, 1966, pp. 1396–1404.
[9] Widgery, D. J. and Knott, J. F., *Metal Science,* Vol. 12, 1978, pp. 8–11.
[10] Passoja, D. E. and Hill, D. C. in *Fractography—Microscopic Cracking Processes, ASTM STP 600,* American Society for Testing and Materials, Philadelphia, 1976, pp. 30–46.
[11] Wallin, K., Saario, T., and Torronen, K., *Metal Science,* Vol. 18, 1984, pp. 13–16.
[12] de Roo, P. and Kaplan, D., "Analyse mecanique de la rupture d'un assemblage soudé en présence d'un défaut dans le métal fondu," Institut de Recherches de la Sidérurgie Française, Report RE 1092, St. Germain en Laye, France, 1984.
[13] Marandet, B. and Sanz, G. in *Flaw Growth and Fracture, ASTM STP 631,* American Society for Testing and Materials, Philadelphia, 1977, pp. 462–476.
[14] De Hoff, R. T. and Rhines, F. N., *Quantitative Microscopy,* McGraw-Hill, New York, 1968.

Crack Arrest

Richard D. Cheverton[1] *and David G. Ball*[2]

Application of Probabilistic Fracture Mechanics to the Pressurized-Thermal-Shock Issue

REFERENCE: Cheverton, R. D. and Ball, D. G., "**Application of Probabilistic Fracture Mechanics to the Pressurized-Thermal-Shock Issue,**" *Fracture Mechanics: Eighteenth Symposium, ASTM STP 945,* D. T. Read and R. P. Reed, Eds., American Society for Testing and Materials, Philadelphia, 1988, pp. 35–50.

ABSTRACT: As a part of the Nuclear Regulatory Commission (NRC) effort to obtain a resolution to the pressurized water reactor (PWR) pressurized-thermal-shock (PTS) issue, a probabilistic approach has been applied that includes a probabilistic fracture-mechanics (PFM) analysis. The PFM analysis is performed with OCA-P, a computer code that performs thermal, stress, and fracture-mechanics analyses and estimates the conditional probability of vessel failure, $P(F|E)$, using Monte Carlo techniques. The stress intensity factor (K_I) is calculated for two- and three-dimensional surface flaws using superposition techniques and influence coefficients. Importance-sampling techniques are used, as necessary, to limit to a reasonable value the number of vessels actually calculated.

Analyses of three PWR plants indicate that (1) the critical initial flaw depth is very small (5 to 15 mm), (2) the benefit of warm prestressing and the role of crack arrest are transient dependent, (3) crack arrest does not occur for the dominant transients, and (4) the single largest uncertainty in the overall probabilistic analysis is the number of surface flaws per vessel.

KEY WORDS: probabilistic fracture mechanics, pressurized thermal shock, reactor pressure vessels

The pressurized-thermal-shock (PTS) issue is concerned with the possibility of failure of a pressurized water reactor (PWR) pressure vessel as a result of the combined effects of (1) pressure and thermal-shock loadings, (2) radiation damage to the vessel material, and (3) the existence of a sharp, crack-like defect (flaw) on the inner surface of the vessel. Thermal shock is a major contributor to the possibility of failure because it can result in relatively high tensile stresses and a reduction in fracture toughness near the inner surface, where the radiation-induced reduction in fracture toughness is the greatest; this combination of conditions introduces the possibility of initiation (incipient propagation) of very shallow inner-surface flaws.

The PTS issue has been under investigation for many years [*1–6*], but the effort was accelerated when it was recognized that (1) the presence of copper (a tramp element) and nickel (an alloying element) enhance radiation damage; (2) existing PWR vessels may have "high" concentrations of these elements; and (3) a number of PWR PTS-type transients had already occurred. A preliminary generic analysis [*7*] of the Rancho Seco transient of 1978 [*8*] indicated the possibility of vessel failure, given the necessary ingredients, long before normal end of plant life,

[1]Senior development engineer, Oak Ridge National Laboratory, Oak Ridge, TN 37831.

[2]Computing analyst, Computing and Telecommunications Division, Martin Marietta Energy Systems, Inc., Oak Ridge, TN 37831.

and the Nuclear Regulatory Commission (NRC) declared PTS an unresolved safety issue (December 1981).

Most of the early PTS analyses were of a conservative, deterministic nature; that is, a very severe PTS transient was assumed, high concentrations of copper and lower-bound fracture-toughness data were used, warm prestressing was ignored, and flaws of appropriate size were assumed to exist. Unfortunately, the results of such calculations indicated that very shallow flaws, which are difficult to detect, could lead to vessel failure before normal end of plant life [*7*].

Even though the above analyses predicted premature failure, it was generally believed that because of the "conservative" nature of the analytical approach, the probability of failure was actually very small. In an effort to obtain a better understanding of the nature and magnitude of the problem, the NRC proposed the development of a comprehensive probabilistic approach and established the Integrated Pressurized Thermal-Shock (IPTS) Program for this purpose (May 1981).

The IPTS effort included (1) the postulation of PTS transients for three specific PWR plants, (2) an estimation of their frequency of occurrence, (3) a systems analysis to determine the primary-system pressure, downcomer coolant temperature and fluid-film heat-transfer coefficient on the inner surface of the vessel, and (4) a probabilistic fracture-mechanics analysis that uses the information from Item 3 as input. Item 4 provides an estimate of the conditional probability of vessel failure, $P(F|E)$, and this can be multiplied by the frequency of the corresponding transient $\phi(E)$ and the product summed over all postulated transients to obtain the total frequency of failure $\phi(F)$ for a specific plant; that is

$$\phi(F) = \sum_{n} \phi_n(E) \cdot P_n(F|E) \qquad \text{(failures per reactor year)} \tag{1}$$

where n denotes the nth transient (event).

The individual products $\phi_n(E) \cdot P_n(F|E)$ in Eq 1 are of interest because they define the order of dominance; that is, they indicate the extent to which a transient contributes to $\phi(F)$ for a specific plant. The dominant transients might then be used in place of the "worst conceivable" for a less conservative deterministic fracture-mechanics analysis, if this "engineering" approach were preferred to the direct application of $\phi(F)$. In either case, it is necessary to calculate $P(F|E)$, and the development and application of a method for doing this in connection with the PTS issue is the primary subject of this paper.

Probabilistic Fracture-Mechanics Model

The probabilistic fracture-mechanics model developed for the IPTS program is included in the OCA-P [*9*] computer code, which accepts coolant-temperature, heat-transfer-coefficient, and pressure transients as input and performs one-dimensional thermal and stress analyses for the vessel wall, fracture-mechanics analyses for two- and three-dimensional flaws, and a probabilistic analysis.

Fracture Mechanics Model

Stress Intensity Factor—The fracture-mechanics model in OCA-P is based on linear elastic fracture mechanics (LEFM), and this allows the stress intensity factor (K_I) to be calculated using superposition techniques in conjunction with influence coefficients, which are calculated with finite-element techniques [*10*]. The application of this procedure makes it possible to perform a large number of deterministic fracture-mechanics calculations at reasonable cost, a necessary condition for performing the probabilistic analysis. (Results of thermal-shock experiments conducted with large, thick-walled steel cylinders indicate that LEFM is applicable for the problem at hand [11].)

Fracture Toughness—Material properties required for the fracture-mechanics analysis include the static crack initiation and arrest toughness values (K_{Ic} and K_{Ia}) and the nil-ductility reference temperature (RTNDT). For the probabilistic fracture-mechanics analysis, mean values of these parameters are required, and they were obtained for the vessel material as follows:

$$\bar{K}_{Ic} = 1.43\{36.5 + 3.08 \exp[0.036(T - \text{RTNDT} + 56)]\}, \qquad \text{MPa}\sqrt{\text{m}} \tag{2}$$

$$\bar{K}_{Ia} = 1.25\{29.5 + 1.34 \exp[0.026(T - \text{RTNDT} + 89)]\}, \qquad \text{MPa}\sqrt{\text{m}} \tag{3}$$

where the quantity in braces represents the American Society of Mechanical Engineers (ASME) Section XI [12] lower-bound toughness value, and T is the temperature in °C at the tip of the flaw. These expressions were obtained by letting the ASME lower-bound curves represent the mean values minus two standard deviations (2σ) and by letting

$$\sigma_{(K_{Ic})} = 0.15\bar{K}_{Ic} \quad \text{and} \quad \sigma_{(K_{Ia})} = 0.10\bar{K}_{Ia}$$

In many cases, if crack arrest takes place, it must do so at or close to upper-shelf temperatures. Crack arrest under these conditions is not well understood but has been included in an approximate manner by specifying a maximum value of K_{Ia} that corresponds to the upper portion of an upper-shelf tearing resistance curve. As illustrated in Fig. 1, which is a plot of K_I versus fractional crack depth (a/w, where a = crack depth, w = wall thickness) at a specific time in a PTS transient, if the load line (K_I versus a/w) intersects the K_{Ia} curve at $K_{Ia} < (K_{Ia})_{max}$, upper-shelf temperatures are not encountered. On the other hand, if the load line misses the rising portion of the K_{Ia} curve and then decreases, as it does for some transients, according to the model there is a possibility of crack arrest at upper-shelf temperatures.

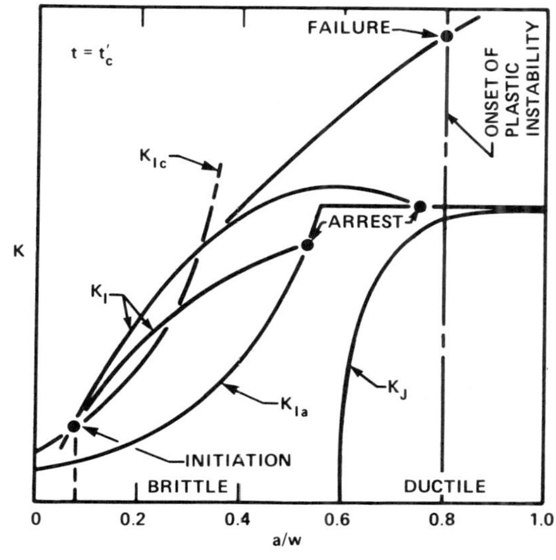

FIG. 1—*Illustration of arrest, failure, and method for selecting* $(K_{Ia})_{max}$.

The tearing-resistance curve selected for the IPTS study represents a specific high-copper, low-upper-shelf, PWR-vessel weld material that had been irradiated to a fluence of $\sim 1.2 \times 10^{19}$ neutrons (n)/cm^2 at a temperature of $\sim 290°C$ and tested at 200°C [*13*]. The upper, nearly flat portion of this curve corresponds to a K_J value of ~ 220 MPa$\sqrt{\text{m}}$, and this value was used for $(K_{Ia})_{max}$; K_J was obtained using the relation

$$K_J = \sqrt{JE} \tag{4}$$

where

J = strain energy release rate and,
E = Young's modulus.

The RTNDT is equal to the sum of an initial value (RTNDT$_0$) and an increase due to radiation damage (ΔRTNDT); that is

$$\text{RTNDT} = \text{RTNDT}_0 + \Delta\text{RTNDT} \tag{5}$$

The correlation used in the IPTS studies for the mean value of ΔRTNDT was essentially the same as that used in a preliminary NRC evaluation of PTS [*14*] and is

$$\overline{\Delta\text{RTNDT}} = 0.56[-10 + 470\ \text{Cu} + 350\ \text{Cu Ni}](F \times 10^{-19})^{0.27}, \qquad °\text{C} \tag{6}$$

or

$$\overline{\Delta\text{RTNDT}} = 0.56[283(F \times 10^{-19})^{0.194} - 48], \qquad °\text{C} \tag{7}$$

whichever is smaller
where

F = fast-neutron fluence (energy > 1 MeV), n/cm^2, and
Cu, Ni = concentrations of copper and nickel, weight %.

The attenuation of the fluence through the wall of the vessel was approximated with

$$F = F_0 e^{-0.0094\,a} \tag{8}$$

where F_0 is the fluence at the inner surface of the vessel and a is the crack depth in mm. The specific value of the coefficient in the exponent accounts to some extent for the effect of neutron space-wise spectral changes on radiation damage [*14*].

The fracture-toughness properties of the cladding material are very uncertain and are known to be dependent on the cladding-application process. The few experimental data that are available indicate that the radiation-induced reduction in fracture toughness can be similar to that for the base material [*15*]. As an expediency, which may or may not be conservative, it was assumed that for times of interest in the life of a vessel, the cladding would have essentially the same fracture-toughness properties as the base material.

Warm Prestressing—Fracture toughness is load- and temperature-history dependent, and some of the histories encountered in the IPTS studies result in fracture-toughness values greater than K_{Ic}. For instance, when $\dot{K}_I < 0$, a crack will not propagate [*11,16*], and, if following a period of $\dot{K}_I < 0$ the stress intensity factor once again increases with time, the critical value of K_I may be substantially greater than K_{Ic} [*16*]. These phenomena are generally referred to as warm prestressing (WPS) and can have a significant effect on the potential for crack propagation during a PTS transient. OCA-P includes certain WPS features. However, because of uncertainties regarding details of load histories for postulated transients, there was reluctance to take advantage of WPS in the IPTS studies except for special cases in which, following a loss of pressure, repressurization, and thus an increase in K_I, would not be possible (primary-system pipe breaks, valve failures, etc.).

Flaws—The region of the vessel of greatest concern with regard to PTS transients is that located between the ends of the core, where the fluence and thus radiation damage are substantially greater than elsewhere. Within this region, the vessels included in the IPTS studies are composed of sections of plate that are joined together with both axial and circumferential welds, as shown in Fig. 2. After the sections are joined, the inner surface of the composite cylinder is clad with ~5 mm of stainless steel.

The flaws of interest are surface cracks that extend through the cladding into the base material. Such flaws may be the result of the cladding process and thus could appear in plate sections as well as the welds that join the plates. In this regard, the plate region could contribute more to the probability of vessel failure than the welds because of the possibility of many more flaws in the plate. On the other hand, for many vessels the concentration of copper is substantially higher for the welds, whereas the concentration of nickel is about the same for the plates as for the welds. Thus, it is necessary to consider both the plate and weld regions, and these basic regions can be subdivided in accordance with possible variations in the combinations of F_0, $RTNDT_0$, copper (Cu), nickel (Ni), and flaw density (N).

As indicated in Fig. 2, both two- and three-dimensional surface flaws are included in OCA-P and were considered for the IPTS studies. Because of a tendency for short and shallow flaws to extend on the surface to effectively become long, shallow flaws [*11,17*], all initial flaws for the IPTS studies were assumed to be two dimensional. The surface length at arrest and for subsequent initiation-arrest events was specified as ~2 m (height of shell course) for axial flaws in high-copper axial welds, and was considered to be effectively infinite for all circumferential flaws and for axial flaws in plate regions (assuming low-copper plate).

The 2-m-length flaw was specified for deep axial flaws ($a/w > 0.2$) because an axial flaw in a high-copper axial weld is not likely to extend beyond the end of the weld into low-copper plate material (see Fig. 2), and because K_I values for a deep flaw of this length are substantially less than for a two-dimensional flaw [*18*]. An infinite length was selected for circumferential flaws because circumferential welds are continuous (Fig. 2), and the azimuthal variation in fluence generally is not large enough to justify otherwise. Offsetting this particular difference between circumferential and axial flaws, however, is a tendency for circumferential flaws to have lower values of K_I because of lower pressure stresses and a smaller deep-flaw bending effect [*11*]. Propagation of all flaws was judged on the basis of the K ratios (K_I/K_{Ic}, K_I/K_{Ia}) at the deepest point of the flaw.

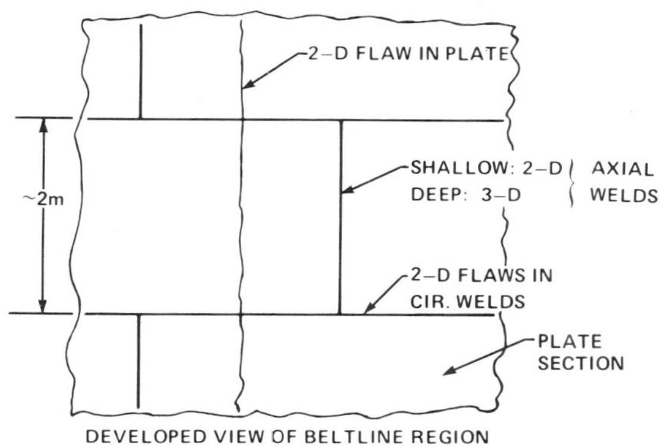

FIG. 2—*Developed view of beltline region of plate-type vessel indicating orientations and types of welds and flaws considered in IPTS studies.*

Cladding—OCA-P treats cladding on the inner surface of the vessel as a discrete region to the extent that thermal and stress effects are included. The effect of cladding on the surface extension of a flaw is not known and thus not included.

Probabilistic Model

Basic Concept—The OCA-P probabilistic model, which is similar to that developed by Gamble and Strosnider [*19*], is based on Monte Carlo techniques; that is, a large number of vessels is simulated, and each vessel is then subjected to a fracture-mechanics analysis to determine whether the vessel will fail. Each vessel is defined by randomly selected values of several parameters that are judged to have significant uncertainties associated with them. The calculated probability of vessel failure is simply the number of vessels that fail divided by the total number of vessels generated. It constitutes a conditional probability of failure, $P(F|E)$, because the assumption is made that the PTS transient (event) takes place. A logic diagram summarizing the various steps in the OCA-P probabilistic analysis is shown in Fig. 3.

Parameters Simulated—Parameters simulated for the IPTS studies were crack depth (a), fast neutron fluence (F), concentrations of copper and nickel, crack-initiation toughness (K_{Ic}), crack-arrest toughness (K_{Ia}), RTNDT_0, and ΔRTNDT. Normal distributions were assumed for each of these parameters except the crack depth; standard deviations and truncation values used in the analysis are included in Table 1.

As a convenience, but at the expense of introducing some additional uncertainty in the estimation of $\phi(F)$, RTNDT_s (the value of RTNDT at the inner surface) can be used as an independent variable in lieu of F_0, Cu, Ni, and RTNDT_0. To do this, Eq 7 is ignored, and Eq 6 is combined with Eq 8 to obtain

$$\Delta\text{RTNDT}(a) = \Delta\text{RTNDT}_s\, e^{-0.0025\, a} \tag{9}$$

where $\Delta\text{RTNDT}(a)$ is the value of ΔRTNDT at the tip of the flaw. Assuming a nominal value of RTNDT_0 and combining Eq 9 with Eq 5 allow one to use Eqs 2 and 3 directly without having to specify F_0, copper, and nickel.

An appropriate distribution function must be developed when using RTNDT_s as an independent variable. For the IPTS studies this was done by performing a Monte Carlo analysis with Eq 6, in which case F_0 and Cu were simulated, and different values of $\overline{F_0}$, $\overline{\text{Cu}}$, and $\overline{\text{Ni}}$ were included. Based on this analysis, a normal distribution with $1\sigma = 0.14\,\overline{\Delta\text{RTNDT}}$ was selected for ΔRTNDT_s. The additional uncertainty mentioned above is the result of the distribution and thus $P(F|E)$ being a function of $\overline{F_0}$, $\overline{\text{Cu}}$, and $\overline{\text{Ni}}$—particularly $\overline{\text{Cu}}$. Furthermore, for a given value of RTNDT, $P(F|E)$ is somewhat sensitive to the value of RTNDT_0. The standard deviation selected for ΔRTNDT was based on $\overline{\text{Cu}} = 0.25$ weight %, and when RTNDT_s was used as an independent variable, −18°C was selected for RTNDT_0.

Flaw Density—The number of flaws in a specific region with a depth in a specific range of crack depths Δa_i is given by

$$N_j(\Delta a_i) = N_j A_j \int_{\Delta a_i} f(a) B(a) da \tag{10}$$

where

N_j = number of flaws of all depths per unit surface area of the jth region,
A_j = surface area of the jth region,
$f(a)$ = flaw-depth density function, and
$B(a)$ = probability of nondetection.

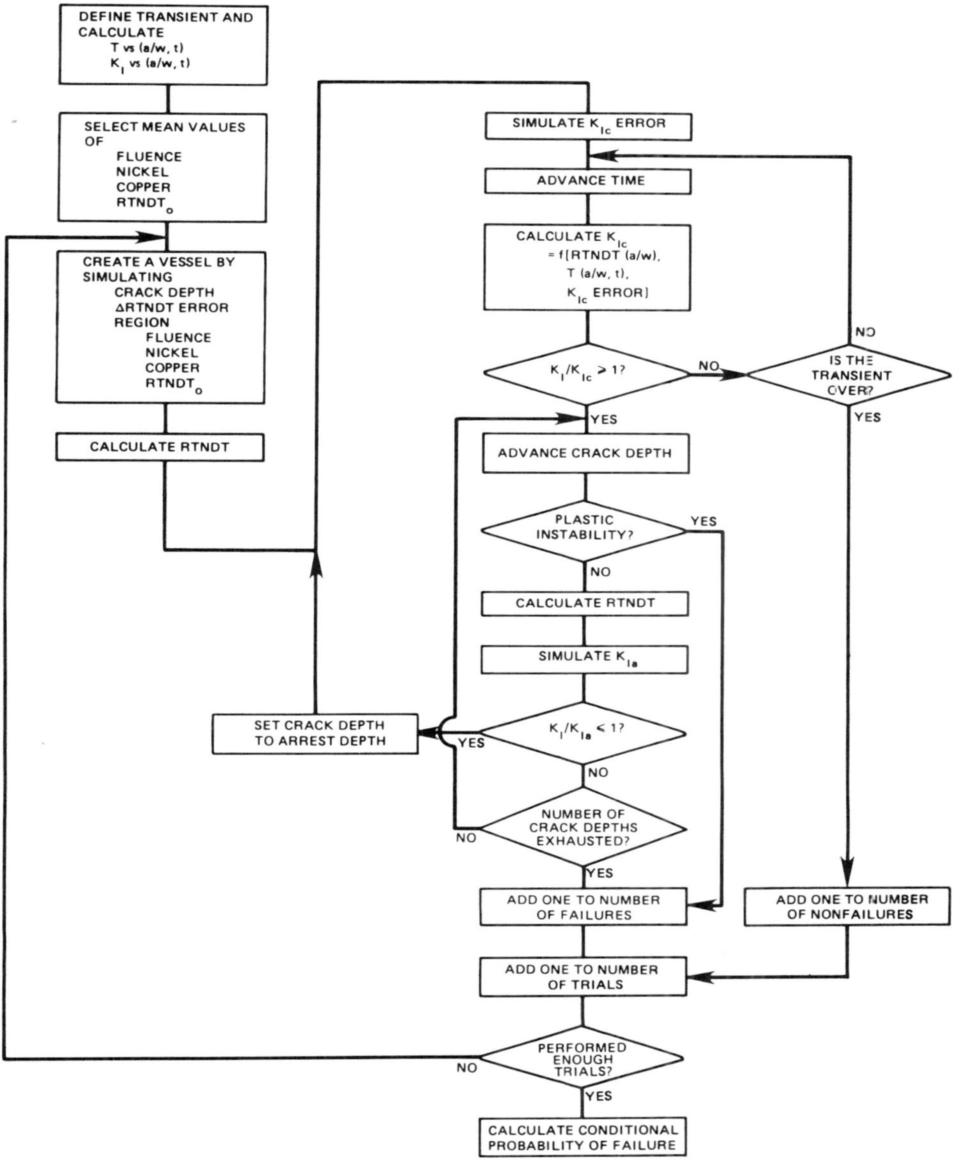

FIG. 3—*OCA-P program logic.*

The parameters N_j and $f(a)$ pertain to vessel conditions prior to preservice inspection and repair, and $B(a)$ is derived on the basis of repairing or otherwise disposing of all detected flaws.

The value of N_j and the functions $f(a)$ and $B(a)$ are not well known because most of the available inspection data do not pertain to surface flaws that extend into and through the cladding of a PWR pressure vessel. For the IPTS studies the functions $f(a)$ and $B(a)$ were those suggested in the Marshall Report [*20*] and are as follows:

$$f(a) = 0.16e^{-0.16a}, \qquad \text{mm}^{-1} \tag{11}$$

$$B(a) = 0.005 + 0.995e^{-0.113a} \tag{12}$$

where

a = crack depth, mm, and

$\int_0^w f(a)da = 1.$

Conditional Probability of Vessel Failure—The calculated probability of vessel failure must include the contributions from all regions of the vessel. For $P_j(F|E) < 0.1$

$$P(F|E) \simeq \sum_j P_j(F|E) \tag{13}$$

The total number of flaws per region is

$$n_j = N_j A_j \int_0^w f(a)B(a)da \tag{14}$$

and when $n_j \leq 1$

$$P_j(F|E) = \hat{P}_j \cdot n_j \tag{15}$$

where

$\hat{P}_j = N'_{fj}/N'_{vj}$,
N'_{fj} = number of vessels with a flaw in the jth region that fail, and
N'_{vj} = number of vessels simulated with a flaw in the jth region.

TABLE 1—*Parameters simulated in OCA-P.*

Parameter	Standard[a] Deviation, σ	Truncation
Fluence (F)	0.3 $\mu(F)$	$F = 0$
Copper	0.025%	+0.40
Nickel	0.0	...
$RTNDT_0$	9°C[b]	[b]
$\Delta RTNDT$[c]	13°C[b,c]	[b]
$\Delta RTNDT$[d]	0.14 $\mu(\Delta RTNDT)$[d]	$\pm 3\sigma$
K_{Ic}	0.15 $\mu(K_{Ic})$	$\pm 3\sigma$
K_{Ia}	0.10 $\mu(K_{Ia})$	$\pm 3\sigma$

[a]Normal distribution used for each parameter.
[b]$\sigma_{(RTNDT)} = [\sigma^2_{(RTNDT_0)} + \sigma^2_{(\Delta RTNDT)}]^{1/2}$, truncated at $\pm 3\sigma$.
[c]Accounts for uncertainty in correlation.
[d]Accounts for uncertainty in Cu, Ni, and F_0 when $RTNDT_s$ used as independent variable.

When $n_j > 1$,

$$P_j(F|E) = n_j \hat{P}_j - \frac{n_j(n_j - 1)}{2!} \hat{P}_j^2 + \frac{n_j(n_j - 1)(n_j - 2)}{3!} \hat{P}_j^3 - \cdots + (-1)^{(n_j - 1)} \hat{P}_j^{n_j} \quad (16)$$

It is apparent that Eq 15 is also appropriate with $n_j > 1$, provided that $\hat{P}_j \ll 1$.

Importance Sampling—For very small values of $P(F|E)$, the values of N'_{vj} required to achieve reasonable accuracy becomes quite large. Under some circumstances the value of N'_{vj} can be reduced by using importance sampling techniques [21]. For the IPTS studies this was done by eliminating flaw depths that did not contribute significantly to initiation and by sampling only the tails of the distribution functions for other simulated parameters. The portion of the distribution function not sampled is accounted for by multiplying the number of simulated vessels, N'_{vj}, by a correction factor. Equation 13 then becomes

$$P(F|E) = \sum_j \frac{\hat{P}_j}{\Pi_k F_{kj}} N_j A_j \int_0^w f(a)B(a)da \quad (17)$$

where F_{kj} is the correction factor for the kth simulated parameter.

Values of F_k are a function of the points on the distribution curve at which sampling is started and stopped (truncation point). For the IPTS studies, when importance sampling was used for the flaw-depth density function, only the first flaw-depth increment was omitted. Thus

$$F_k(\text{flaw depth density}) = \frac{1}{\int_{\Delta a_1}^{w} f(a)B(a)da} = 3.24$$

Table 2 includes values of F_k for several different starting points on a normal distribution curve that is truncated at 3σ.

For the IPTS studies, importance sampling was applied to three simulated parameters, including flaw depth, and for many cases the starting point on the normal distribution curves was 1.25σ. Thus, $\Pi_k F_k \simeq 300$, which represents a significant savings in computer costs for the same accuracy in $P(F|E)$.

TABLE 2—*Values of* F_k *for a normal distribution truncated at* 3σ.

Start of Sampling (No. of Standard Deviations Above Mean)	Fraction of Distribution Not Simulated	F_k
1.0	0.8422	6.3
1.25	0.8954	9.6
1.50	0.9343	15.2
1.75	0.9611	25.7
2.00	0.9784	46.4
2.25	0.9891	91.7
2.60	0.9951	204.1
2.75	0.9983	588.2

Statistical Error—The minimum number of simulated vessels required to satisfy a specified accuracy is estimated by applying the central limit theorem [*21*]. Using this approach, the 95% confidence interval about $P_j(F|E)$ is

$$\frac{\hat{P}_j N_j A_j}{\Pi_k F_{kj}} \int_0^w f(a)B(a)da \pm 2\sigma_j \tag{18}$$

where

$$\sigma_j \simeq \frac{1}{\Pi_k F_{kj}} \left[\frac{\hat{P}_j(1 - \hat{P}_j)}{N'_{vj}}\right]^{1/2} \cdot N_j A_j \int_0^w f(a)B(a)da \tag{19}$$

The value of σ corresponding to all of the vessels simulated is

$$\sigma_{P(F|E)} = \sqrt{\Sigma_j \sigma_j^2} \tag{20}$$

and the error, ϵ_j, associated with the jth region is

$$\epsilon_j = \frac{2\sigma_j \Pi_k F_{kj}}{\hat{P}_j N_j A_j \int_0^w f(a)B(a)da} \tag{21}$$

For $\hat{P}_j \ll 1$

$$\epsilon_j \simeq 2\left(\frac{1}{\hat{P}_j N'_{vj}}\right)^{1/2} = 2\left(\frac{1}{N'_{fj}}\right)^{1/2} \tag{22}$$

Considering all regions of interest, the total error ϵ is

$$\epsilon = \frac{2\sigma_{P(F|E)}}{\sum_j \frac{\hat{P}_j N_j V_j}{\Pi_k F_{kj}} \int_0^w f(a)B(a)da} \tag{23}$$

It is of interest to note (see Eq 22) that the error for a single region, ϵ_j, is only a function of N'_{fj}. According to Refs *22* and *23*, for the estimate of ϵ_j to be "reasonably" accurate, N'_{fj} should be greater than 5 (Ref *22*) or perhaps 9 (Ref *23*).

Failure Criteria—After an appropriate number of vessels has been simulated, each vessel is subjected to a fracture-mechanics analysis to determine if failure will occur during a particular transient at a specific time in the life of the plant. For the IPTS studies, failure was assumed to occur if, following an initiation event ($K_I = K_{Ic}$ at deepest point of flaw), K_I remained greater than K_{Ia} up to or beyond the point at which plastic instability occurred in the remaining ligament (Fig. 1). Thus, failure was defined as through-wall cracking, which does not necessarily imply an inability of the vessel to retain sufficient coolant to cover the core. The consequences of through-wall cracking are evaluated in a separate study [*24*].

Plants Included in IPTS Studies

Three nuclear plants (Oconee-1, Calvert Cliffs-1, and H. B. Robinson-2), each one representing a different nuclear vendor and utility, were selected for the IPTS studies. In addition, one hypothetical reactor vessel (HBR-HYPO) was included for the purpose of more appropriately illustrating the probabilistic fracture-mechanics methods of analysis. Information pertaining to these plants and their reactor vessels is provided in Table 3, and a typical set of pressure and temperature transients for a rather severe transient is shown in Fig. 4.

Trends Deduced from IPTS Studies

It is not the intent of this paper to discuss in detail the results of the IPTS studies (preliminary results are included in Refs *25* to *27*). However, some of the fracture-mechanics-oriented trends should be mentioned.

P(F | E) *versus Effective Full-Power Years (EFPY)*

Because of the accumulative effect of radiation damage (Eqs 6 and 7), $P(F|E)$ increases with reactor operating time and thus with RTNDT, as indicated in Fig. 5 for the transient in Fig. 4. It is of interest to note that for this particular transient increasing RTNDT from 40 to 100°C increases $P(F|E)$ by a factor of more than 10^4.

Relative Contribution of Plate and Weld

As indicated in Table 3, for Oconee-1, Calvert Cliffs-1, and HBR-HYPO, RTNDT for the plate region is much less than for the dominant axial weld, but the area of the plate is much greater than for the welds. The net effect, assuming the same flaw density (N_j) for all regions, is that the axial weld included in Table 3 is the dominant contributor to $P(F|E)$. For HBR-2 the plate region has the largest value of RTNDT as well as area and thus is the dominant contributor to $P(F|E)$.

TABLE 3—*Information pertaining to nuclear plants included in IPTS studies.*

Plant Designer Utility[a]	Region	Chemistry, weight% Cu	Ni	$RTNDT_0$, °C	$F_0{}^b$, 10^{19} n/cm²	Area, m²	$\overline{RTNDT}^c$, °C
Oconee-1	axial weld[d]	0.29	0.55	−7	1.1	0.2	97
B&W	cir.[e] weld[d]	0.26	0.61	−7	0.9	1.0	84
Duke Power	plate	0.12	0.60	4	1.2	50.0	46
Calvert Cliffs-1	axial weld[d]	0.21	0.87	−49	6.1	0.12	90
CE	cir. weld[d]	0.24	0.18	−62	6.1	0.64	46
BG&E	plate	0.12	0.64	−12	6.1	50.0	55
H. B. Robinson-2	axial weld[d]	0.22	0.04	−49	3.2	0.14	24
Westinghouse	cir. weld[d]	0.19	0.80	−49	2.0	0.46	39
CP&L	plate	0.12	0.10	8	4.2	50.0	49
HBR-HYPO	axial weld[d]	0.22	0.80	−18	3.2	0.14	99
	cir. weld[d]	0.22	0.80	−18	2.0	0.46	85
	plate	0.12	0.80	−18	4.2	50.0	48

[a]Babcock and Wilcox (B&W), Combustion Engineering (CE), Baltimore Gas and Electric (BG&E), Carolina Power and Light (CP&L).
[b]Inner-surface fluence at 32 EFPY (maximum in region).
[c]Inner-surface value of $\overline{RTNDT}$ at 32 EFPY.
[d]Dominant weld region only.
[e]Circ. = circumferential.

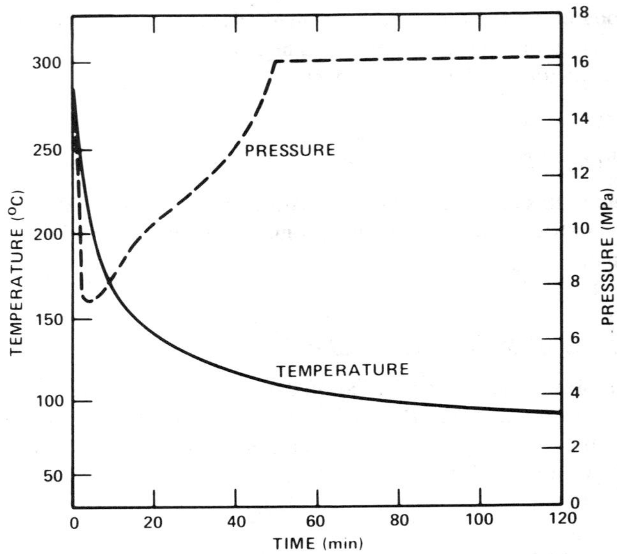

FIG. 4—*Downcomer coolant temperature and primary-system pressure versus time for a typical severe PTS transient.*

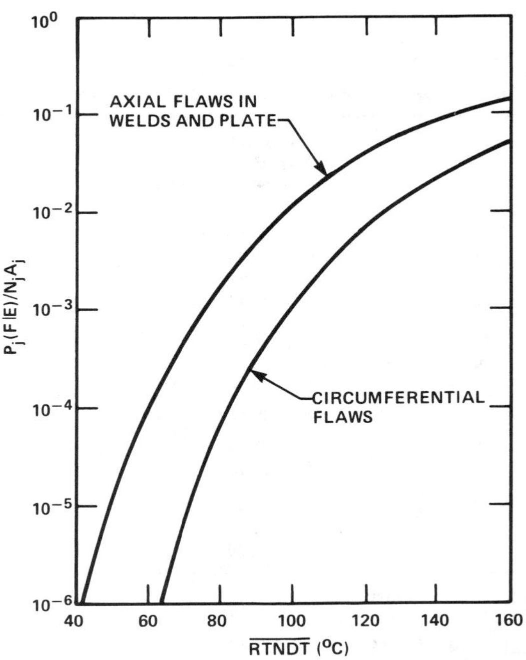

FIG. 5—$P_j(F|E)/N_jA_j$ *versus* $\overline{RTNDT}$ *for the transient in Fig. 4.*

The Role of Crack Arrest

Crack arrest can take place and prevent failure of the vessel, if the primary-system pressure is low enough, in which case the value of $(K_{Ia})_{max}$ influences $P(F|E)$ [27]. However, for high-pressure transients, such as that illustrated in Fig. 4, crack arrest does not take place, irrespective of the value of $(K_{Ia})_{max}$, because K_I is more sensitive to crack depth than K_{Ia} is. Thus, the role of crack arrest is dependent upon which transients are dominant, that is, which ones contribute the most to $\phi(F)$. For the IPTS studies, the dominant transients were high-pressure transients. Therefore, the value of $(K_{Ia})_{max}$ used in the studies was not critical.

Time of Events, Duration of Transient, and Warm Prestressing ($\dot{K}_I < 0$)

The time in the transient at which most events take place is transient dependent, and the effects of WPS and of the duration of the transient are dependent on the time of the events. This is illustrated in Fig. 6, which is a histogram of percent failures versus time of failures for two Oconee-1 dominant transients (O-A and O-B), indicating the time for each transient at which $\dot{K}_I = 0$ (incipient warm prestressing). For both transients the time of incipient WPS is early in the transient (20 to 30 min), and $\dot{K}_I$ is less than zero for the remainder of the 2-h transients. Ignoring warm prestressing, failure for one transient (O-B) occurs much later, in which case the inclusion of WPS would eliminate all indicated failures. For the other transient (O-A), about one half of the failures occurred before incipient WPS.

The factors by which $P(F|E)$, for the five most dominant transients, and $\phi(F)$ are reduced by the inclusion of WPS are indicated in Table 4. It is apparent that although the benefit may be large for some individual transients it may not be very large for $\phi(F)$, depending on which transients are dominant.

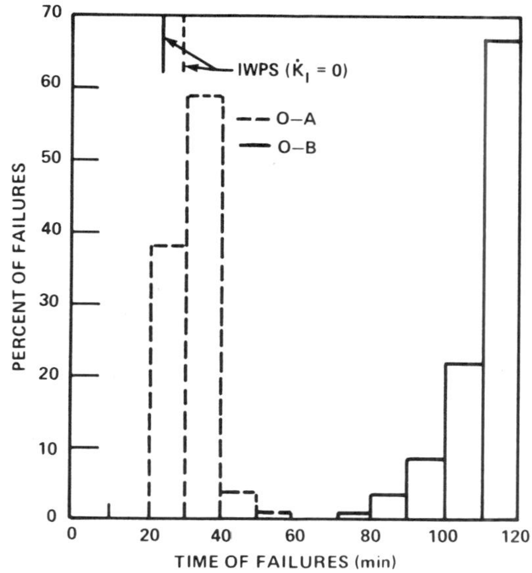

FIG. 6—*Histogram of percent failure versus time of failure for two Oconee-1 transients.*

TABLE 4—*Benefit of warm prestressing for Oconee-1, Calvert Cliffs-1, and HBR-HYPO.*

Transient (in Order of Dominance)	Plant		
	Oconee-1	Calvert Cliffs-1	HBR-HYPO
		$P(F\|E)_{WPS}/P(F\|E)$	
1	$<10^{-3}$	$<2 \times 10^{-3}$	9×10^{-3}
2	3×10^{-1}	1	$<2 \times 10^{-3}$
3	$<10^{-3}$	$<1 \times 10^{-2}$	$<2 \times 10^{-3}$
4	$<10^{-3}$	1×10^{-1}	$<2 \times 10^{-3}$
5	3×10^{-1}	$<5 \times 10^{-1}$	5×10^{-2}
		$\Phi(F)_{WPS}/\Phi(F)$	
	1×10^{-1}	2×10^{-1}	7×10^{-3}

Critical Flaw Depths

Most of the initial flaws that resulted in failure were very shallow (5 to 15 mm), and most of these resided entirely within the cladding. Such flaws are difficult to detect, so essentially no data regarding flaw surface density are available. As a result, the uncertainty in N_j and thus $P(F|E)$ and $\phi(F)$ is quite large. As a part of the IPTS studies, the uncertainty factor for $\phi(F)$ was estimated to be $\sim 10^3$, and the single largest contributor to this uncertainty was N_j.

Discussion and Conclusions

A probabilistic fracture-mechanics model was developed as a part of the NRC Integrated Pressurized-Thermal-Shock Program to help resolve the PWR pressurized-thermal-shock issue. The model is based on linear elastic fracture mechanics and considers both two- and three-dimensional flaws. Monte Carlo techniques are used to calculate the conditional probability of vessel failure, and importance sampling is used to reduce the number of vessels that must be simulated.

The flaws of concern are those that extend from the inner surface into and through the cladding, and it was found that very shallow (5 to 15 mm) flaws were responsible for most of the calculated failures. Because essentially no inspection data are available for such flaws, the number of these flaws per vessel is the single largest uncertainty in the analysis.

The cladding is included in the thermal and stress analyses, but because of a lack of radiation-damage data, it is assumed that the cladding will not prevent surface extension of the initially short and shallow flaws. If surface extension is impeded by the cladding, the calculated probability of failure will be much less. (The actual effects of cladding are under investigation [*15,29*].)

The benefit of warm prestressing ($\dot{K}_I < 0$) was investigated and was found to be dependent on the transient. Thus, the effect on the overall frequency of failure is dependent on which transients are dominant. For one of the nuclear plants considered, the inclusion of warm prestressing reduced the frequency of failure by only a factor of 5. For another the factor was ~ 100. (WPS effects with $\dot{K}_I > 0$ are under investigation [*30*]).

Crack arrest will prevent vessel failure for low-pressure but not high-pressure transients. These latter transients are the dominant contributors to the frequency of failure. Thus, to reduce the frequency of failure it is necessary to prevent propagation of the initial shallow flaws. (Crack-arrest behavior at elevated temperatures is also being investigated [*30*].)

Acknowledgment

This research was sponsored by the Office of Nuclear Regulatory Research, U.S. Nuclear Regulatory Commission under Interagency Agreements 40-551-75 and 40-552-75 with the U.S. Department of Energy under Contract DE-AC05-84OR21400 with Martin Marietta Energy Systems, Inc.

References

[*1*] Tuppeny, W. H., Jr., Siddall, W. F., Jr., and Hsu, L. C., "Thermal Shock Analysis on Reactor Vessels Due to Emergency Core Cooling System Operation," Combustion Engineering, Inc., Windsor, CT, Docket 50309, National Technical Information Service, 1968.

[*2*] Hutto, R. C., Morgan, C. D., and Van Der Sluys, W. A., "Analysis of the Structural Integrity of a Reactor Vessel Subjected to Thermal Shock," Report BAW-10018, Babcock & Wilcox Co., Lynchburg, VA, May 1969.

[*3*] Ayres, D. J. and Siddall, W. F., "Finite Element Analysis of Structural Integrity of a Reactor Pressure Vessel During Emergency Core Cooling," Report A-70-10-2, Combustion Engineering, Inc., Windsor, CT, Jan. 1970.

[*4*] Buchalet, C. A. and Bamford, W. H., "Methods of Fracture Mechanics Analysis of Nuclear Reactor Vessels Under Severe Thermal Transients," American Society of Mechanical Engineers Paper 75-WA/PVR-3, 1975.

[*5*] Cheverton, R. D., "Thermal-Shock Studies Associated with Injection of Emergency Core Coolant Following A Loss-of-Coolant Accident in PWRs," *Nuclear Safety,* Vol. 19, No. 1, Jan.-Feb. 1978.

[*6*] Cheverton, R. D., Iskander, S. K., and Whitman, G. D., "The Integrity of PWR Pressure Vessels During Overcooling Accidents" in *Proceedings,* International Meeting on Thermal Nuclear Reactor Safety, NUREG/CP-0027, Vol. 1, Feb. 1983, pp. 421–430.

[*7*] Cheverton, R. D., "Parametric Analysis of Rancho Seco Overcooling Accident," letter to M. Vagins (USNRC), March 1981, available at Nuclear Regulatory Commission Public Document Room.

[*8*] Memorandum from Sacramento Municipal Utility District to Nuclear Regulatory Commission Director of Regulatory Operations, Region IV, Subject: "Reactor Cooldown Rate Exceeds Limit Following Trip at Rancho Seco," NSIC Accession OOZO-138830, available at Nuclear Regulatory Commission Public Document Room, 31 March 1978.

[*9*] Cheverton, R. D. and Ball, D. G., "OCA-P, A Deterministic and Probabilistic Fracture-Mechanics Code for Application to Pressure Vessels," NUREG/CR-3618 (ORNL-5991), Union Carbide Corp., Nuclear Division, Oak Ridge National Laboratory, Oak Ridge, TN, May 1984.

[*10*] Ball, D. G., Bass, B. R., Bryson, J. W., Cheverton R. D., and Drake, J. B., "Stress-Intensity-Factor Influence Coefficients for Surface Flaws in Pressure Vessels," NUREG/CR-3723 (ORNL/CSD/TM-216), Martin Marietta Energy Systems, Inc., Oak Ridge National Laboratory, Oak Ridge, TN, Feb. 1985.

[*11*] Cheverton, R. D. et al., "Pressure Vessel Fracture Studies Pertaining to the PWR Thermal-Shock Issue: Experiments TSE-5, TSE-5A and TSE-6," NUREG/CR-4249 (ORNL-6163), Oak Ridge National Laboratory, Oak Ridge, TN, June 1985.

[*12*] "Flaw Evaluation Procedures, ASME Section XI, Background and Application of ASME Section XI, Appendix A, Special Report," Marston, T. U., Ed., EPRI NP-719-SR, American Society of Mechanical Engineers, Electric Power Research Institute, Palo Alto, CA, Aug. 1978.

[*13*] Letter from F. J. Loss (Naval Research Laboratory) to R. H. Bryan (Oak Ridge National Laboratory), 31 March 1981.

[*14*] NRC Staff Evaluation of Pressurized Thermal Shock, 13 Sept. 1982 draft, attachment to transmittal document Secy 82-465, 23 Nov. 1982 (available in Nuclear Regulatory Commission Public Document Room).

[*15*] Corwin, W. R., "Assessment of Radiation Effects Relating to Reactor Pressure Vessel Cladding," NUREG/CR-3671 (ORNL-6047), Oak Ridge National Laboratory, Oak Ridge, TN, July 1984.

[*16*] Loss, F. J., Gray, Jr., R. A., and Hawthorne, J. R., "Significance of Warm Prestress to Crack Initiation During Thermal Shock," Report NRL/NUREG-8165, Naval Research Laboratory, National Technical Information Service, 29 Sept. 1977.

[*17*] Cheverton, R. D. et al., "Pressure Vessel Fracture Studies Pertaining to the PWR Thermal-Shock Issue: Experiment TSE-7," NUREG/CR-4304 (ORNL-6177), Oak Ridge National Laboratory, Oak Ridge, TN, Aug. 1985.

[*18*] Cheverton, R. D. and Ball, D. G., "A Reassessment of PWR Pressure Vessel Integrity During Overcooling Accidents, Considering 3-D Flaws," *Journal of Pressure Vessel Technology,* Nov. 1984, Vol. 106, pp. 375–382.

[19] Gamble, R. M. and Strosnider, J., Jr., "An Assessment of the Failure Rate for the Beltline Region of PWR Pressure Vessels During Normal Operation and Certain Transient Conditions," NUREG/0778, June 1981.
[20] Marshall, W., "An Assessment of the Integrity of PWR Pressure Vessels," United Kingdom Atomic Energy Authority, Second Report, March 1982.
[21] Rubinstein, R. Y., *Simulation and the Monte Carlo Method,* Israel Institute of Technology, Wiley, New York, 1981.
[22] Hoel, P. G., *Introduction to Mathematical Statistics,* 3rd ed., Wiley, New York, 1962.
[23] Hald, A., *Statistical Theory with Engineering Applications,* Wiley, New York, 1952.
[24] Simonen, F. A. et al., "Status of Vessel Failure Mode Analysis: Feasibility Study and Preliminary Evaluation," Pacific Northwest Laboratory, Dec. 1983.
[25] Burns, T. J. et al., "Pressurized Thermal-Shock Evaluation of the Oconee-1 Nuclear Power Plant, NUREG/CR-3770 (ORNL/TM-9176), Oak Ridge National Laboratory, Oak Ridge, TN, 13 April 1984.
[26] Selby, D. L. et al., "Pressurized Thermal-Shock Evaluation of the Calvert Cliffs Unit 1 Nuclear Power Plant," NUREG/CR-4022 (ORNL/TM-9408), Oak Ridge National Laboratory, Oak Ridge, TN, 9 Oct. 1984.
[27] Selby, D. L. et al., "Pressurized Thermal Shock Evaluation of the H. B. Robinson Unit 2 Nuclear Power Plant," NUREG/CR-4183 (ORNL/TM-9567), Oak Ridge National Laboratory, Oak Ridge, TN, March 1985.
[28] Cheverton, R. D. and Ball, D. G., "The Role of Crack Arrest in the Evaluation of PWR Pressure Vessel Integrity During the PTS Transients," *Engineering Fracture Mechanics,* Vol. 23, No. 1, 1986, pp. 71-80.
[29] Corwin, W. R., Robinson, G. C., Nanstad, R. K., Merkle, J. G., Berggren, R. G., Goodwin, G. M., Swain, R. L., and Owings, T. D., "Effects of Stainless Steel Weld Overlay Cladding on the Structural Integrity of Flawed Steel Plates in Bending, Series 1," NUREG/CR-4015 (ORNL/TM-9390), Oak Ridge National Laboratory, Oak Ridge, TN, April 1985.
[30] Bryan, R. H., Bass, B. R., Bolt, S. E., Bryson, J. W., Edmonds, D. P., McCulloch, R. W., Merkle, J. G., Nanstad, R. K., Robinson, G. C., Thoms, K. R., and Whitman, G. D., "Pressurized-Thermal-Shock Test of 6-in.-Thick Pressure Vessels. PTSE-1: Investigation of Warm Prestressing and Upper-Shelf Arrest," NUREG/CR-4106 (ORNL-6135), Oak Ridge National Laboratory, Oak Ridge, TN, April 1985.

Susumu Machida,[1] Hiroshi Yajima,[2] and Takeshi Kanazawa[1]

Crack Arrest Testing, Results, and Applications

REFERENCE: Machida, S., Yajima, H., and Kanazawa, T., "**Crack Arrest Testing, Results, and Applications,**" *Fracture Mechanics: Eighteenth Symposium, ASTM STP 945,* D. T. Read and R. P. Reed, Eds., American Society for Testing and Materials, Philadelphia, 1988, pp. 51–72.

ABSTRACT: This paper reviews two experimental and analytical studies made in Japan on predicting the brittle crack propagation and arrest behavior using wide plate testings and the dynamic or simplified dynamic analysis technique. Also reviewed is an analysis, though statical in technique employed, made of an upper deck plating brittle failure experienced by a 10 379 gross ton bulk carrier.

KEY WORDS: brittle crack propagation and arrest, crack arrest testing, static and dynamic analysis, wide plate testing, service failure analysis

In Japan statical analyses [*1,2*] used to be employed, in characterizing brittle crack behaviors in steel plates, to relate the crack motions in a small plate of several hundred millimetres in width to those in a larger specimen simulating an actual structure.

Recently there has been a strong demand for the safety and the structural integrity of liquefied gas carriers, nuclear vessels, storage tanks, etc., in response to the gradual increase in number of such sophisticated structures. And nowadays in characterizing brittle cracks, a more rational and precise approach [*3–9*] considering the dynamic effect of the phenomenon has become available and popular. Until now, however, almost all the dynamic analyses conducted were limited to the characterization of the material resistance using the standard shape test specimen. Very few trials have been done to predict the crack behavior in the real steel structure because of the troublesomeness of the dynamic calculations.

This paper reviews two experimental and analytical studies made in Japan on predicting the brittle crack propagation and arrest behaviors using wide plate testing and the dynamic or simplified dynamic technique. Also reviewed is a statical technique analysis of an upper deck plating brittle failure experienced by a 10 379 gross ton (GT) bulk carrier [*10*].

Simplified Dynamic Calculation for Predicting Brittle Crack Propagation and Arrest in Ship Frame Structure

In this study, first the dynamic fracture toughness of the common ship hull steel was obtained on the basis of the dynamic estimation. Second, the experiment and the forecasting calculation were made of the brittle crack propagation in the uniform plate with the stiffener. The simplified static estimation as a substitute for the iterative dynamic calculations was confirmed to be promising in predicting the fluctuation of the brittle crack velocity and the crack arrest [*9*].

[1]Professor and emeritus professor, respectively, Department of Naval Architecture, Faculty of Engineering, University of Tokyo, 7-3-1, Hongo, Bunkyo-ku, Tokyo 113, Japan.

[2]Assistant chief research engineer, Nagasaki Research and Development Center, Mitsubishi Heavy Industries, Ltd., 1-1, Akunoura, Nagasaki, 850-91, Japan.

Brittle Crack Propagation Test

The test specimens were prepared from NK (Nippon Kaiji Kyokai) standard KA class ship steel plates (20 mm thickness). The specimens were two large-size double tension test specimens 1600 mm in width (named Specimen SI-1 and SI-2) and two stiffened specimens of the same size (named Specimen BS-1 and BS-2).

The shape and the size of the specimens are shown in Fig. 1. The stiffened test specimens were made by manual continuous fillet welding (one-pass layer, weld leg length 7 mm).

The crack detector (CD) gages were attached, at 150-mm spacing, to one side of the plate surfaces in order to measure the crack velocity by the time intervals at which they are cut. A strain gage was attached at the edge of the tab plate opening to pick up the transient load changes.

The load was imposed in the double tension style using the 9800-kN test rig and the 1470-kN subtension equipment. The specimen was first cooled down to the prescribed temperature distribution over the width and was then held in that condition for some time. Thereafter the plate was loaded to the predetermined stress level (118 MPa), and the brittle crack was made to initiate at the tip of the notch and run into the main plate by the subtensile shock load. With the cutting of the CD gage nearest to the crack initiation point as the trigger, the time history of each gage cutting and the strain gage output at the edge of the opening were measured and stored by the memory recorder.

The temperature distributions along the crack path are shown in Fig. 2. Within 600 or 800 mm from the crack initiation point, the temperature gradient was inevitable because the temperature of the crack initiation part was lowered to about −100°C. Beyond that area, there was also a temperature gradient in Specimen SI-1, but in the SI-2 the temperature was flat at −10°C. The stiffener type specimens BS-1 and BS-2 were also uniform in temperature.

As shown in the test results, the brittle crack was arrested at 785 mm from the crack initiation point in Specimen SI-1, and Specimen SI-2 was severed. The crack velocity in those specimens is shown in Fig. 8. In Specimen BS-1, the brittle crack stopped as a surface crack at the stiffener. In Specimen BS-2, the crack ran through the intersection, cutting the stiffener, and was arrested at the location 170 mm away from the stiffener. The measured crack velocity fluctuations in the BS-1 and BS-2 specimens are shown as the dotted lines in Fig. 9 and Fig. 10, respectively.

Brittle Crack Behavior Prediction

The transient movement of the specimen and the test rig during the crack propagation was investigated using the simple mass-spring model.

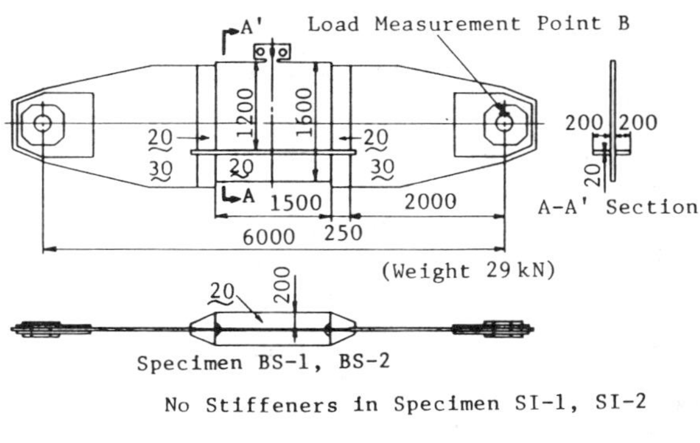

FIG. 1—*Specimen shape and dimensions.*

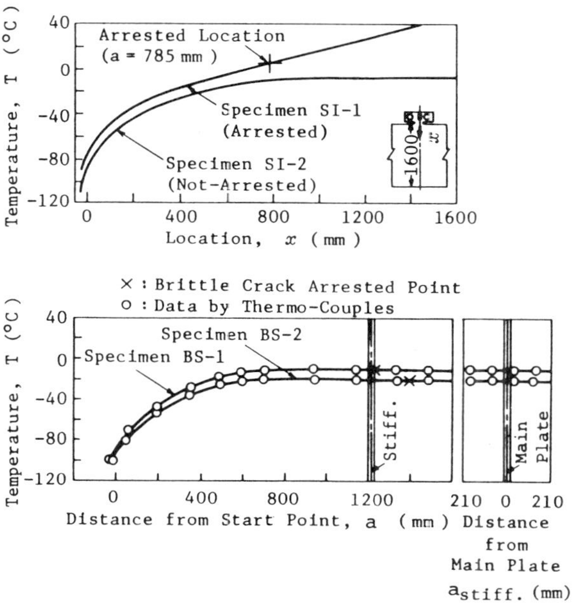

FIG. 2—*Temperature distribution.*

The test rig and the specimens were, considering the symmetry, idealized as the mass-spring system as shown in Fig. 3. In this model the crack progress was represented by the decrease in the spring constant κ_{SC}.

The equilibrium equations of the mass M_R which represents one half of the test rig and of the mass M_S representing one half of the specimen are as follows

$$M_R \ddot{\chi}_R = -\kappa_R(\chi_R - \chi_R^*) - \kappa_{SN}\{(\chi_R - \chi_R^*) - (\chi_S - \chi_S^*)\} - \gamma(\dot{\chi}_R - \dot{\chi}_S) \quad (1)$$

$$M_S \ddot{\chi}_S = -\kappa_{SC}(\chi_S - \chi_S^*) - \kappa_{SN}\{(\chi_S - \chi_S^*) - (\chi_R - \chi_R^*)\} - \gamma(\dot{\chi}_S - \dot{\chi}_R) \quad (2)$$

where χ_R^* and χ_S^* are the moment-by-moment, imaginary static equilibrium locations in the dynamic behavior, and are given by the following equations

$$\chi_R^* = P(1/\kappa_R) \quad (3)$$

$$\chi_S^* = P(1/\kappa_R + 1/\kappa_{SN}) \quad (4)$$

$$\delta = P(1/\kappa_R + 1/\kappa_{SN} + 1/\kappa_{SC}) \qquad (= \text{constant}) \quad (5)$$

The spring constants κ_{SC} and κ_{SN} were approximately given by the following equations, which were estimated by the ratio between the elongation and the load in the statically balanced specimen [11] (Fig. 4)

$$\kappa_{SC} = \text{(load)/(elongation between point 0 and } G_1)$$

$$= Ebt/(h^* + 2af) \quad (6)$$

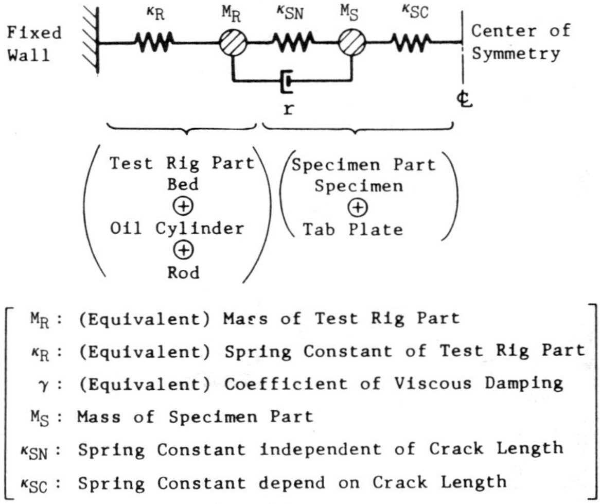

FIG. 3—*Idealized mass-spring system.*

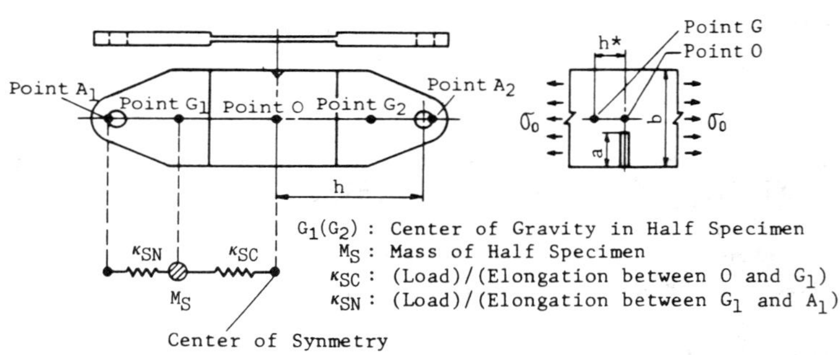

FIG. 4—*Spring constants.*

$$\kappa_{SN} = Ebt/(h - h^*) \tag{7}$$

where

$$f = \left(\frac{a}{b}\right) \Big/ \left(1 - \frac{a}{b}\right)^2 \cdot \left\{0.99 - \left(\frac{a}{b}\right)\left(1 - \frac{a}{b}\right)\left(1.3 - 1.2\frac{a}{b} + 0.7\frac{a^2}{b^2}\right)\right\} \tag{8}$$

Setting the initial values, that is, the condition at the time of the crack initiation and using the time history of the spring constant κ_{SC} which is given by the crack length record obtained by the experiment, the behavior of the mass-spring system can be revealed by numerically solving the basic Eqs 1 and 2. Using these solutions, the load P_{end} acting at the end of the specimen is expressed by

$$P_{\text{end}} = \kappa_{SN}(\chi_S - \chi_R) \tag{9}$$

Also, the energy components of this mass-spring system are given by the following equations

(In specimen)

$$\text{strain energy: } U^{(S)} = \frac{1}{2}\kappa_{SC}(\delta - \chi_S)^2 + \frac{1}{2}\kappa_{SN}(\chi_S - \chi_R)^2 \tag{10}$$

$$\text{kinetic energy: } K^{(S)} = \frac{1}{2}M_S(\dot{\chi}_S)^2 \tag{11}$$

$$\text{work done by the test rig: } W^{(S)} = -\int_0^{\pi_R} P_{\text{end}}\,d\chi_R \tag{12}$$

(In test rig)

$$\text{strain energy: } U^{(R)} = \frac{1}{2}\kappa_R(\chi_R)^2 \tag{13}$$

$$\text{kinetic energy: } K^{(R)} = \frac{1}{2}M_R(\dot{\chi}_R)^2 \tag{14}$$

$$\text{work done by the specimen: } W^{(R)} = -W^{(S)} \tag{15}$$

$$\text{work done in friction: } F^{(R)} = \int_0^{\chi_R - \chi_S} \gamma(\dot{\chi}_R - \dot{\chi}_S)\,d(\chi_R - \chi_S)$$

$$= \int_0^t \gamma(\dot{\chi}_R - \dot{\chi}_S)^2 dt \tag{16}$$

Consequently, the dissipated energy ($-Đ$) except $F^{(R)}$ can be shown by the following equations

$$-Đ = \Pi - (\Pi)_{t=0} \tag{17}$$

$$\Pi = U^{(S)} + K^{(S)} + U^{(R)} + K^{(R)} + F^{(R)} \tag{18}$$

The dynamic fracture toughness $K_Đ$ is obtained from the dissipated energy $Đ$

$$K_Đ = \left\{\frac{E}{(1-\nu^2)t}\,\frac{dĐ}{da}\right\}^{1/2} \tag{19}$$

From the test results of SI-1, the individual movements of the mass point M_S and M_R are numerically calculated using the Runge-Kutta-Gill technique.

The numerical constants for the specimen are as follows, considering the configuration in Fig. 1

$$M_S = 1.5 \times 10^3 \text{ kg} \tag{20}$$

$$\kappa_{SN} = 7.1 \times 10^9 \text{ kg/s}^2 \ (= 7.1 \times 10^6 \text{ N/mm}) \tag{21}$$

$$(\kappa_{SC})_{t=0} = 1.8 \times 10^9 \text{ kg/s}^2 \ (= 1.8 \times 10^6 \text{ N/mm}) \tag{22}$$

The numerical constants M_R, κ_R, γ governed by the test rig should not be altered as long as the same testing machine is used, but cannot be determined easily in the idealized model. Therefore, conducting the iterative calculations with arbitrary values used as the above constants, an attempt was made to find out the best set of the constants that matches the experimentally obtained P_{end} (Eq 9) fluctuation characteristics during the crack propagation.

The following set of values was found to give the best agreement with the experimental results as shown in Fig. 5 and also to simulate comparatively well the free vibration mode after the crack arrest

$$M_R = 8.0 \times 10^2 \text{ kg} \quad (23)$$

$$\kappa_R = 7.0 \times 10^8 \text{ kg/s}^2 \ (= 7.0 \times 10^5 \text{ N/mm}) \quad (24)$$

$$\gamma = 5.0 \times 10^5 \text{ kg/s} \ (= 5.0 \times 10^2 \text{ N} \cdot \text{s/mm}) \quad (25)$$

Using the preceding numerical constants for the test rig, the dissipated energy given by Eq 17 was calculated for Specimens SI-1 and SI-2. The changes in the dynamic fracture toughness K_D in Eq 19 were obtained as shown in Fig. 6.

In general, the dynamic fracture toughness K_D depends on the crack velocity and tempera-

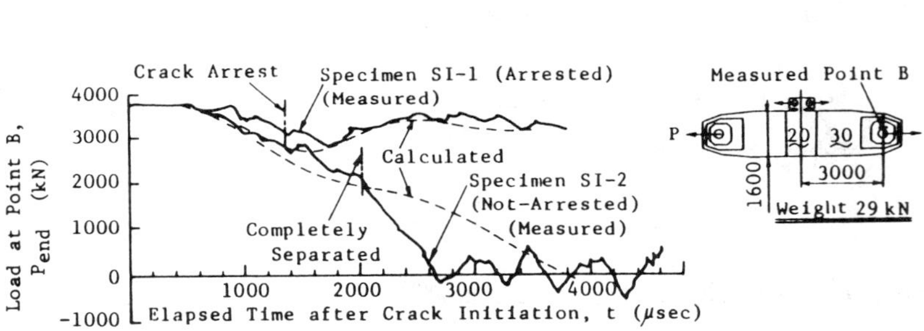

FIG. 5—*Transient load fluctuation.*

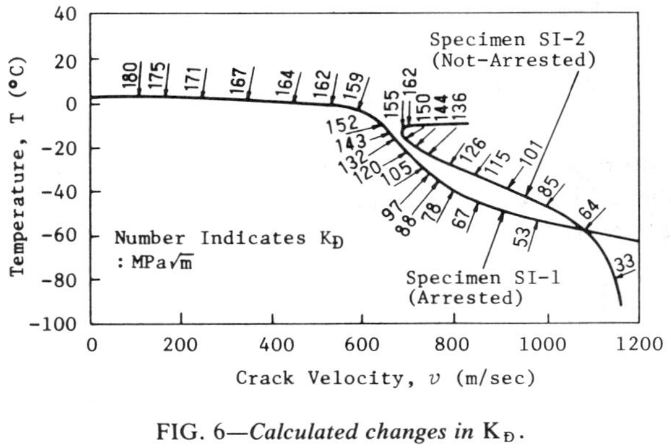

FIG. 6—*Calculated changes in* K_D.

ture. Assuming that the equations for K_D would take the forms of the Eqs 26 to 28, the unknown constants in Eqs 27 and 28 were determined by the results in Fig. 6

$$K_D = \exp\{\gamma_0 + \gamma_2(V/V_0)^2\} \tag{26}$$

$$\gamma_0 = 6.1539 + 4.6409 \times 10^{-2}T + 4.2857 \times 10^{-4}T^2 \tag{27}$$

$$\gamma_2 = 2.2005 \times 10 - 1.0048T - 2.4955 \times 10^{-2}T^2 \tag{28}$$

where

$V_0 = 5200$ m/s,
$V =$ m/s,
$K_D = 0.3101$ MPa$\sqrt{\text{m}}$, and
$T =$ °C.

The preceding equations are valid when $-55°C \lesssim T \lesssim -10°C$, and $V \lesssim 1000$ m/s. Eq 26 is given in Fig. 7.

So far, the dynamic fracture toughness of the tested steel plate was determined in relation to the crack velocity and temperature. So conducting the energy balance calculations, the brittle crack behavior can be predicted by a rather complicated procedure. However, the accurate prediction by this procedure requires the incremental dynamic crack analysis. This orthodox approach would take great amount of time and cost for calculations. Therefore instead of the iterative dynamic analysis, a substitute method that is a substantially static approach was employed to estimate the relevant parameters and the dynamic stress-intensity factor $K^{(D)}$.

This simplified approach is based on Freund's solutions [12] about the stationary crack propagation in an infinite plate and is intended to be expediently extended to nonstationary problems. Freund revealed that under the uniform remote stress acting on the semi-infinite crack,

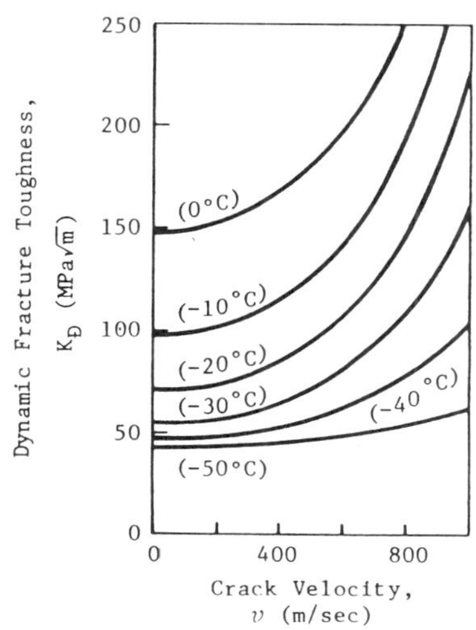

FIG. 7—K_D *versus* V *and temperature.*

the dynamic stress-intensity factor $K^{(Đ)}$ and the dynamic energy release rate $\mathcal{G}^{(Đ)}$ are given by the following equations, respectively

$$K^{(Đ)} = k_{(v)} K^{(S)} \tag{29}$$

$$\mathcal{G}^{(Đ)} = \frac{(1 - \nu^2)}{E} A_{(v)} \{K^{(Đ)}\}^2 \qquad \text{(for unit thickness)} \tag{30}$$

where $K^{(S)}$ is the static stress-intensity factor and $k_{(v)}$ and $A_{(v)}$ are functions of crack velocity.

The energy balance criterion for the propagating crack is the equality of the dynamic energy release rate $\mathcal{G}^{(Đ)}$ and the dissipated energy rate $dĐ/da$ in Eq 19. Consequently, the necessary condition during the crack propagation is given by

$$K_{Đ} = \{A_{(v)}\}^{1/2} k_{(v)} K^{(S)} \tag{31}$$

The right-side value in Eq 31 can be simplified as follows, if a bit of numerical errors are permitted

$$K_{Đ} = \sqrt{1 - \frac{V}{V_R}} \cdot K^{(S)} \tag{32}$$

V_R is the Rayleigh wave velocity ($V_R \cong 0.92\sqrt{G/\rho}$ when $\nu = 0.25$).

In Eq 33, which is modified from Eq 32, the left side of the equation depends only on temperature and crack velocity but not on the crack length, and the right side of the equation depends only on crack length

$$K_{Đ} \Big/ \sqrt{1 - \frac{V}{V_R}} = K^{(S)} \tag{33}$$

That is, neither side of this equation needs to be incrementally calculated or dynamically estimated. This provides a simple and promising tool to predict the brittle crack behaviors. For example, a graphical solution can be devised for a crack in a plain uniform plate. This procedure was used for Specimens SI-1 and SI-2, and the results are shown in Fig. 8, in which 10% scatter of $K_{Đ}$ is allowed. Except immediately before the specimen severance, it seems that the estimated velocities are in good agreement with the velocities measured.

Next, the prediction trials as to Specimens BS-1 and BS-2 were done using Eq 33. Having $K_{Đ}$ as in Eq 26, it was only necessary to estimate the static stress-intensity factor $K^{(S)}$ in Eq 33.

The K-value estimation about a crack in a stiffened plate is not easy, but could be evaluated by some manipulations. When the crack tip locates far before the plate-stiffener intersection, Ishida's solution [*13*] is available. And when the crack tip runs through the intersection, an expedient method is available in which the stress interaction along the plate-stiffener intersection is simplified as a concentrated force which is determined from the equivalent crack opening displacements in the plate and the stiffener [*14*]. When the crack tip is intermediate between the above two cases, the K-value can be estimated by the smooth interpolation. Using the above methods to estimate the static stress-intensity factor $K^{(S)}$, the crack velocity satisfying Eq 33 was obtained. The results are shown in Fig. 9 (Specimen BS-1) and in Fig. 10 (Specimen BS-2).

Since it is said that there is a lower limit of crack velocity for a brittle crack to propagate [*15*], it was assumed that the crack should immediately be arrested if the velocity decreases below 200 m/s. Considering this effect, the crack that stopped at the stiffener intersection in Specimen BS-1 and the crack that ran through the intersection in BS-2 can be well predicted by the calculation.

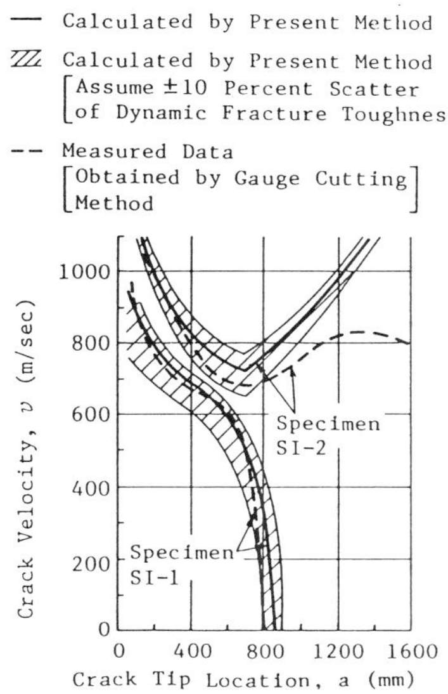

FIG. 8—*Calculated crack velocity.*

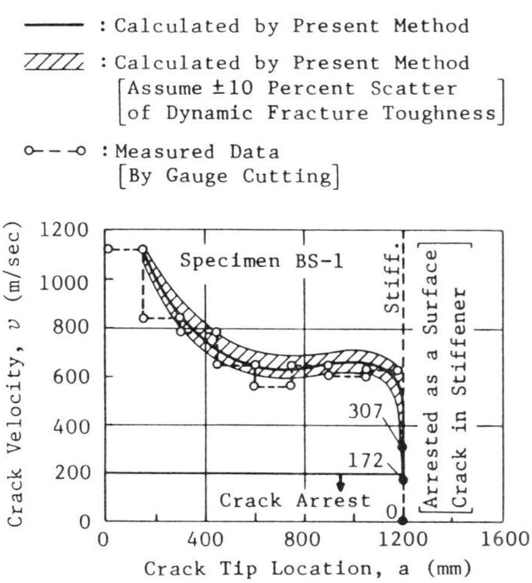

FIG. 9—*Comparison between calculated and measured crack velocity.*

Before the plate intersection, the agreement between calculated and measured crack velocity is good. However as in BS-2, immediately before the specimen severance, this simplified method does not give very good results.

Some Considerations and Conclusions

Procedures for prediction of brittle crack behaviors including a simplified one have already been presented. Now the restriction of the simplified approach needs to be made clear and confirmed.

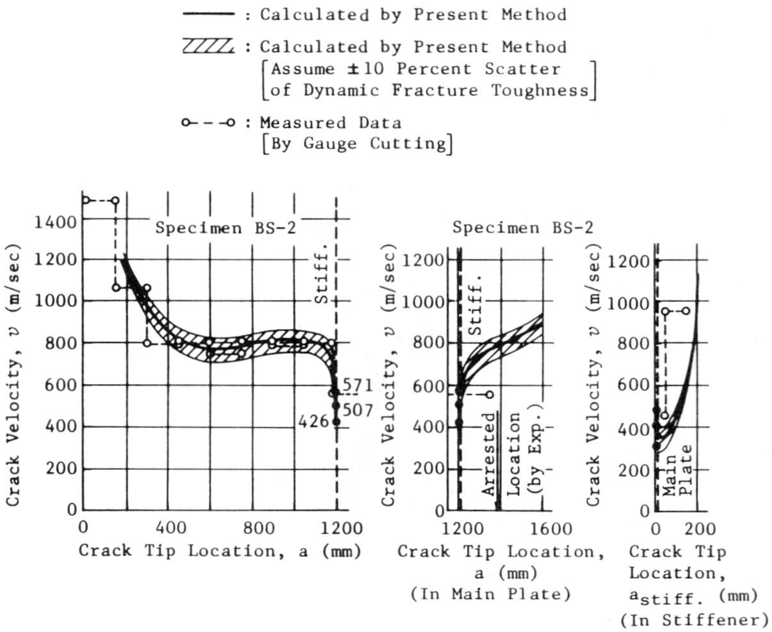

FIG. 10—*Comparison between calculated and measured crack velocity.*

Equations 29 and 30 by Freund were derived from the assumption of the stationary crack propagation in the infinite plate. And these equations were extended to the nonstationary problem in this study. The main difference between stationary and nonstationary phenomena is basically whether the crack velocity is constant or not. Another point is whether the stress waves reflected back from the specimen boundary gives a significant effect or not to the stress state of running crack tip. Therefore, when the static stress-intensity factor $K^{(S)}$ is evaluated by the simplified approach summarized in Eq 33, it is necessary to keep the condition that the boundary loads make no additional work.

As far as the experiments in this study are concerned, before the load boundaries begin to make additional work during the crack propagation, that is, before the stress waves which are transferred from the crack tip to the tub plate opening come back through the same path, the boundary characteristics could never affect the crack behaviors. So it seems that the stationary condition is almost satisfied until the return of the stress waves.

The results of this chapter are summarized as follows: Using the large-size plate specimens and the stiffened plate specimens, it was found that the brittle crack prediction about the propagation speed and the crack arrest location can be done with considerably good accuracy. The proposed simplified method without making the iterative dynamic analyses was found to be especially considerably valid except immediately before the specimen severance.

Evaluation of Duplex-Type Crack Arrester by Dynamic Technique

In the preceding sections, the authors suggested the dynamical calculation of high-velocity crack propagation and, through analyses of the results of double tension test and DCB test with the ship mild steels, decided that the crack propagation and arrest characteristics could be defined with a fair degree of accuracy using the dynamic fracture toughness [*3–6,8*]. On the basis of the conclusion obtained from these studies, the crack arrest performance of the duplex-type crack arrester was re-evaluated by dynamical technique based on the results of the duplex double tension test [7]. Accordingly, this section reviews this reevaluation of the duplex-type crack arrester performance.

Brittle Crack Propagation Test

The test specimens were prepared from NK standard KA and KE class ship steel plates (20 mm thickness). First of all, the standard double tension test was performed. In this test, commonly with the KA steel (A series) and KE steel (E series), the standard double tension test specimen consisting of a 500-mm-wide by 500-mm-long by 20-mm-thick crack propagation part, and a 200-mm-wide by 440-mm-long by 20-mm-thick crack initiation part was used. The specimen configuration and strain gage attachment position are shown in Fig. 11. The crack propagation velocity was measured from the change in voltage that took place with the severance of the CD gages attached to the crack-propagation part at 30 mm intervals. Also, by attaching the strain gages at the positions shown in Fig. 11, the change in strain that took place with the crack propagation was measured. It is to be mentioned that both series specimen were of temperature gradient type and two kinds of temperature gradients shown in Fig. 12 were employed.

Commonly with the A and E series, tests were performed with four each of the previously mentioned test specimen, varying the initial stress and temperature gradient in each instance. Measured in the tests, the strain and crack propagation velocity, respectively, gave readings as shown in Figs. 13 and 14. Figure 13 shows an example of the comparison between measured and computed (with the assumption of fixed-grip condition) variation of strain distribution during crack propagation.

The duplex double tension test (AE series) specimen was prepared by joining the KA and KE steel plates by electron-beam welding 230 mm from the entrance into the crack-propagation part. The specimen configuration is shown in Fig. 15. To reduce the residual stress prior to the

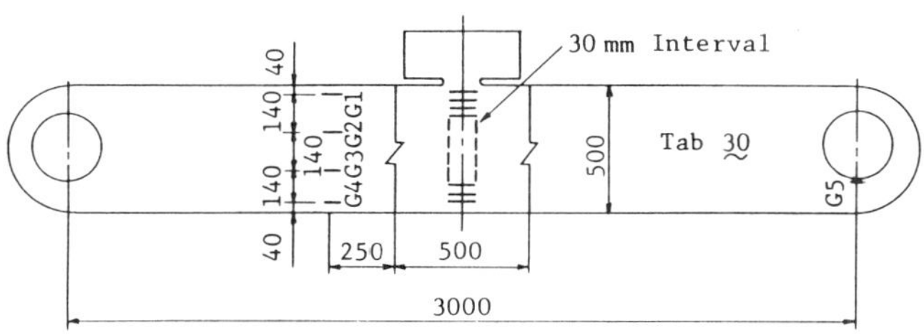

FIG. 11—*Specimen shape and gage positions A and E series.*

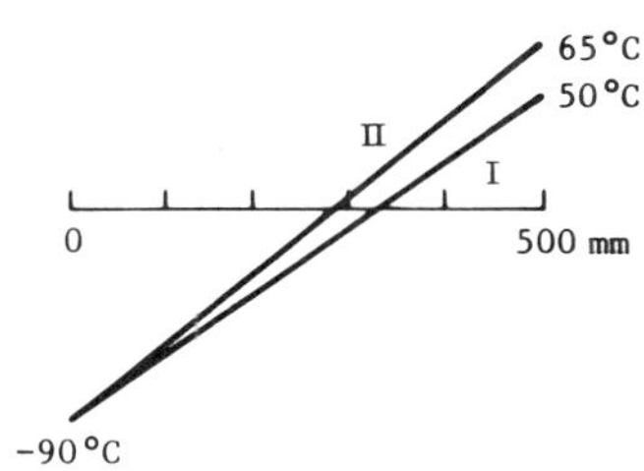

FIG. 12—*Temperature gradients.*

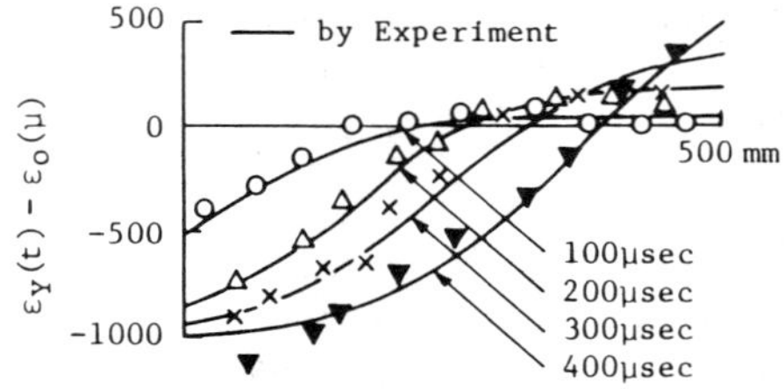

FIG. 13—*Measured and computed variations of strain distribution during crack propagation.*

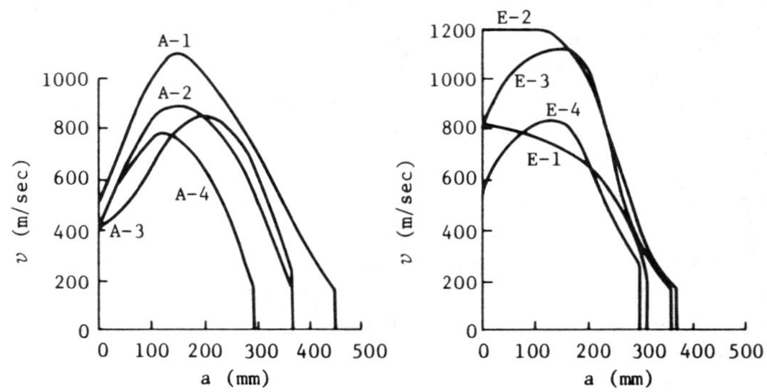

FIG. 14—*Variation of crack velocity (A and E series).*

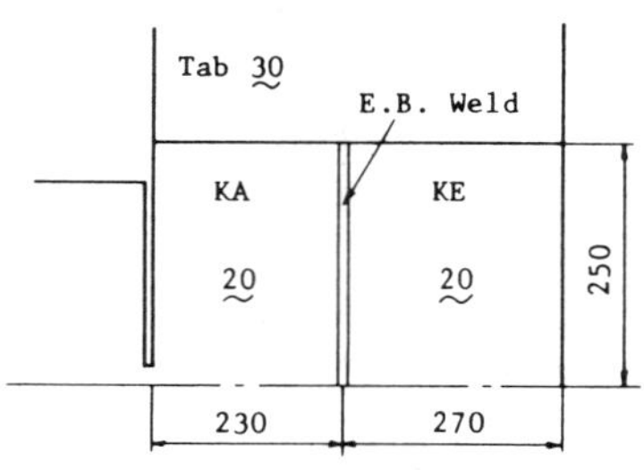

FIG. 15—*Specimen configuration (AE series) (only upper half is shown).*

test, the test specimen was pre-loaded close to the yield point. As with the A series and E series, the change in strain and the crack velocity were measured. Figure 16 shows the measured crack velocity change for the test temperature of −10°C with three levels of applied stress.

It is obvious that beyond the weld seam, the crack velocity decreased abruptly by about 200 m/s in every instance. As can be seen, the crack running into the arrester behaved in three different manners: arrested immediately after running into the arrester; arrested after running about 40 mm through the arrester; and running through the full width of the arrester. The fracture surface in the last instance was obviously that of brittle fracture, whereas the fracture surface in either of the first two instances was rough-textured and somewhat different from that of brittle fracture and also indicated fairly large shear lips.

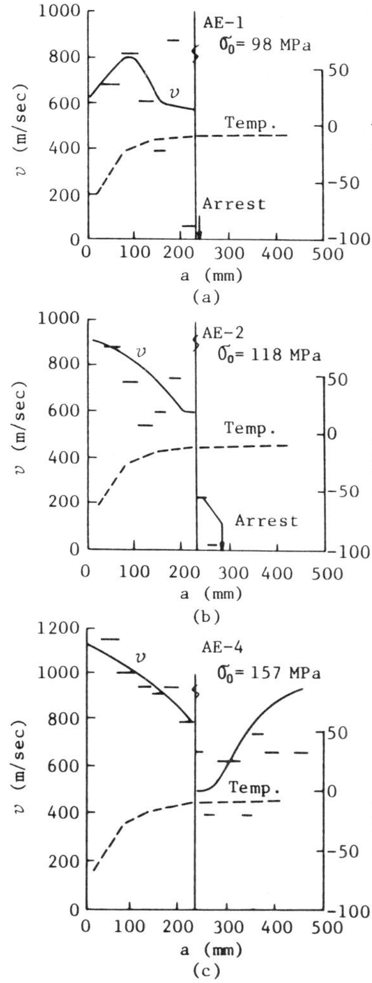

FIG. 16—*Variation of crack velocity (AE series).*

Analysis and Consideration

Assuming the test specimen to be a two-dimensional elastic body, the equations of motion governing such an elastic object will be given as follows

$$\left.\begin{aligned} \frac{\partial^2 u}{\partial t^2} &= C_1^2 \frac{\partial^2 u}{\partial x^2} + (C_1^2 - C_2^2)\frac{\partial^2 v}{\partial x \partial y} + C_2^2 \frac{\partial^2 u}{\partial y^2} \\ \frac{\partial^2 v}{\partial t^2} &= C_2^2 \frac{\partial^2 v}{\partial x^2} + (C_1^2 - C_2^2)\frac{\partial^2 u}{\partial x \partial y} + C_1^2 \frac{\partial^2 v}{\partial y^2} \end{aligned}\right\} \quad (34)$$

where u and v are displacements in x and y-directions and C_1 and C_2 are longitudinal and transverse wave velocities

Equation 34 is resolved using the finite-difference method. The mesh size was $h = 10$ mm, and the time increment was $\Delta t = 1\ \mu s$.

Now, as stated in the preceding section, the following equation results from the law of energy-balance

$$U + K + Ð = W \tag{35}$$

where

W = work done by external load,
U = strain energy,
K = kinetic energy, and
$Ð$ = dissipated energy.

Assuming that the term $Ð$ is mostly the energy dissipated in the fractured surface layers, dynamic fracture toughness $K_Ð$ is defined as follows by analogy with the static case

$$K_Ð = \left\{\frac{E}{1-\nu^2}\,\frac{dÐ}{da}\right\}^{1/2} \tag{36}$$

$K_Ð$ can be calculated by Eqs 35 and 36. In practice, however, using the crack closure energy ΔE_c given as

$$\Delta E_c = h \times \int \sigma_y(t)\{d\nu(t)\} \tag{37}$$

where $\sigma_y(t)$ is the stress in the y-direction near the crack and $\nu(t)$ is the displacement in the y-direction near the crack.

$K_Ð$ is calculated by the following equation [4,5]

$$\begin{aligned} K_Ð &= \left\{\frac{E}{1-\nu^2}\,\frac{\Delta E_c}{h}\right\}^{1/2} \\ &= \left[\frac{E}{1-\nu^2}\int \sigma_y(t)\{d\nu(t)\}\right]^{1/2} \end{aligned} \tag{38}$$

$K_Ð$ was calculated for the A series, E series, and AE series. Examples of $K_Ð$ and crack velocity plotted against the crack length a are shown in Fig. 17 for the A and E series and in Fig. 18 for the AE series. For reference, the static stress-intensity factor $\sigma_0\sqrt{2B \tan \pi a/2B}$ and temperature distribution also are indicated. With either of A and E series, $K_Ð$ tends to gradually approach $K^{(S)}$ immediately before the crack arrest. With the AE series, the crack was arrested by the arrester in AE-1 and AE-2, under the uniform temperature condition (−10°C). Although with AE-2 the crack propagation about 40 mm after running into the arrester, $K_Ð$ increased once the crack ran into the arrester. The fracture surface in AE-2 exhibited fairly large shear lips, and it appears that a considerable amount of energy was required to cause this ductile fracture.

In Figs. 19*a* and 19*b*, $K_Ð$ obtained by the basic double tension test are correlated using the temperature and the crack velocity as parameter for the KA and KE steels, respectively.

Figure 20 shows $K_Ð$ versus crack velocity relations at −10°C obtained from gradient-temperature type double-tension test for KA (A series) and KE (E series), and uniform-temperature

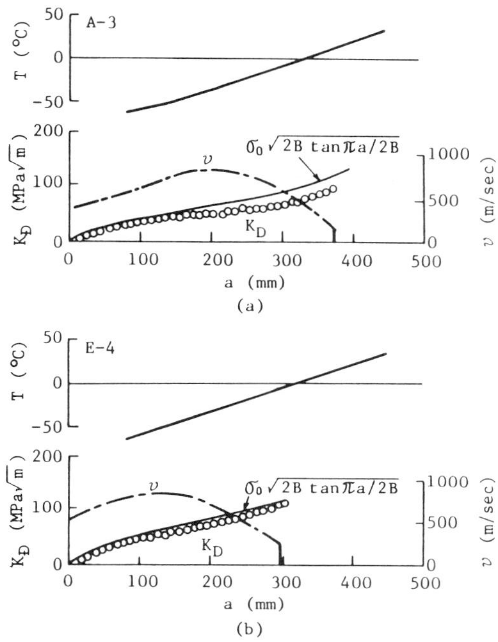

FIG. 17—*Variation of* K_D *(A and E series).*

duplex type double-tension test (AE series). Despite the difference in the above two testing methods, unique relations between K_D and crack velocity are defined for KA and KE. The horizontal arrows in the figure mean the levels of dynamic K-values corresponding to available energy (which is always balanced by K_D) at the instant that a crack in starter plate reach the arrester plate.

In AE-4 and AE-8 tests, cracks from the starter section were abruptly slowed down at their entrance into arrester section but they ran through whole arrester section with increasing speed following the K_D curve in upper right direction. In AE-2 test, a crack from the starter section was abruptly slowed down at its entrance into arrester section and it propagated with very low speed until it was arrested in the arrester plate. The crack in the arrester plate showed a fracture appearance with some degree of ductility accompanied by shear lips and quite different from usual brittle crack. The K_D value for this crack defines the low-speed region of the K_D curve for KE as shown in Fig. 20 where K_D increases as crack speed decreases, and it has a minimum value at the crack speed of about 200 m/s. In AE-1 test, the level of dynamic K-value at the entrance of arrester plate was less than minimum value of K_D of the arrester material, and the crack could not get into arrester plate and was arrested. The preceding three different crack propagation behaviors are schematically shown in Fig. 21 in terms of the characteristics of K_D curves. A crack running into arrester material with available dynamic K-value close to the minimum value of its K_D, it follows the lower-speed part of the K_D curve (Al in Fig. 21) showing fracture appearance with some extent of ductility which is different from the brittle appearance observed in higher crack velocity region.

Applying the dynamic fracture toughness curve to the double tension test performed, the brittle crack propagation and arrest behaviors were simulated by simplified approximation.

Assuming a crack to propagate in an infinite plate subjected to uniform tensile stress, the

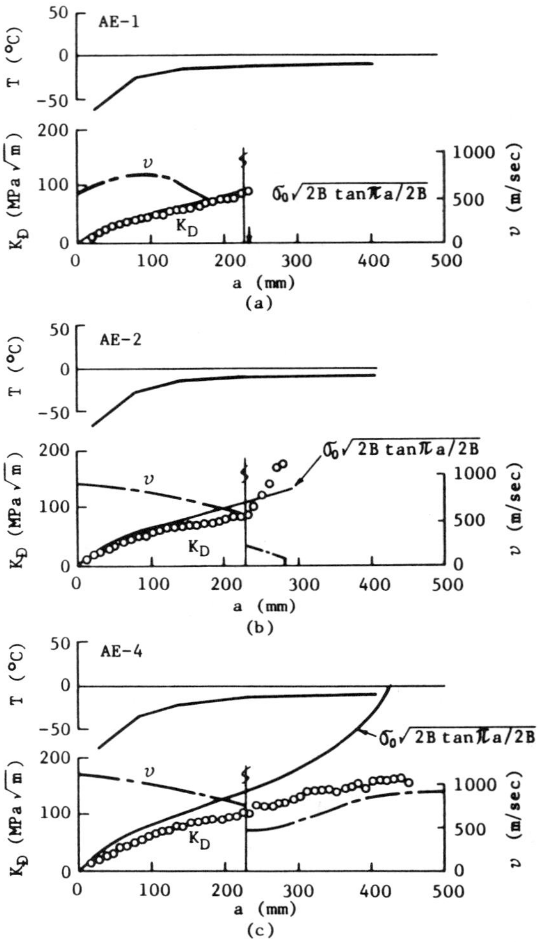

FIG. 18—*Variation of* K_D *(AE series).*

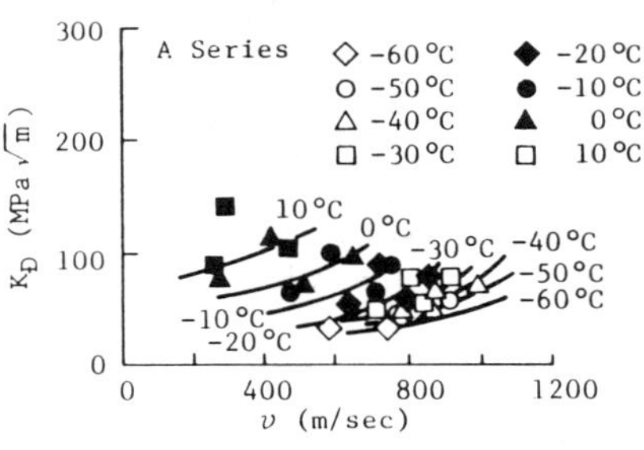

FIG. 19*a*—K_D *of KA as a function of crack velocity.*

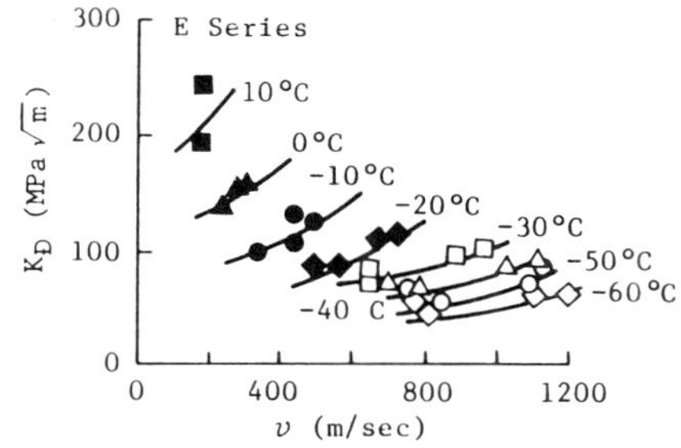

FIG. 19*b*—K_D *of KE as a function of crack velocity.*

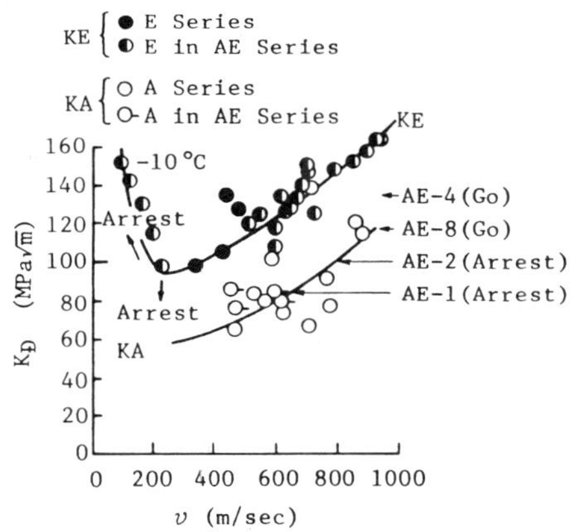

FIG. 20—K_D *versus crack velocity at −10°C for KA and KE obtained from conventional type and duplex-type double tension tests.*

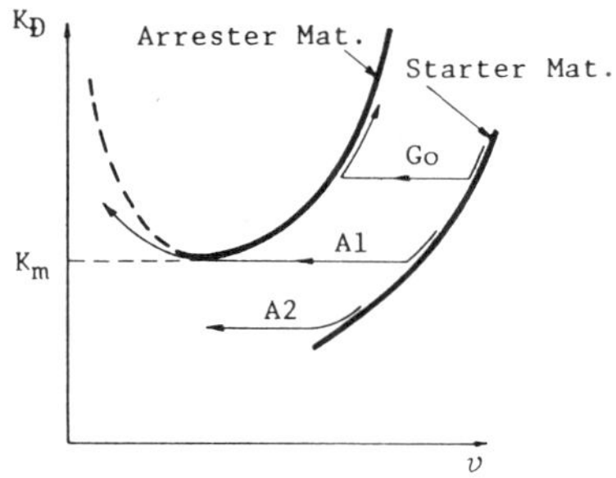

FIG. 21—*Schematic illustration for three different behaviors observed in duplex-type double tension test (AE series).*

relationship expressed by the following equation can approximately be established between the static stress-intensity factor $K^{(S)}$ and dynamic fracture toughness K_D [5,12]

$$\sqrt{1 - \frac{V}{V_R}}\, K^{(S)} = K_D(V, T) \tag{39}$$

in which V_R is the velocity of Rayleigh wave. Therefore, since T (temperature) is given as a function of a in the double tension test, the crack velocity can be obtained in terms of its relation with the crack length from the value on the left side of Eq 39 and dynamic fracture toughness value. Those applied to the results of tests performed this time are shown in Fig. 22. The measured and calculated values by comparison can be said to be in fairly good agreement. This is probably because the test specimens used were long enough and the effects of stress wave reflected from the end remained not so pronounced until the crack was arrested.

Conclusion

The duplex type crack arrester performance was evaluated using the dynamical technique, with the following results.

1. With either of KA and KE steels, the crack arrest toughness and temperature as well as the dynamic fracture toughness, crack velocity, and temperature have their own relationships.
2. As observed in the test, the crack velocity decreased once the crack ran into the crack arrester.
3. As an index of the crack arrester performance, the minimum value of the dynamic fracture toughness curve for the arrester material (K_m) may be considered. Should the crack driving force at the time of entry be K_m or less, it is highly possible that the crack will be arrested. It was found that in this instance, there were two modes of crack arrest: the immediate arrest upon entering the arrester; and the arrest after the crack slowing down by its energy being greatly absorbed by fracture surface different from the brittle fracture surface.
4. The results of simulation by simplified approximation showed a good agreement with the measured values.

Analysis of Ship Hull Damage by Statical Technique

This section reviews a static analysis performed on an upper deck plating brittle failure which was experienced by a 10 379-GT bulk carrier [10].

Damage Description

While the ship was underway, a crack was produced in the top side tank upper deck plating near midship section. The crack rapidly grew and, after propagating 1250 mm through the upper deck plating and severing the shear strake, ran 2590 mm through the side shell plating before coming to a halt. Figure 23 shows the damage. A fatigue crack, about 400 mm in length, was located at the place of crack initiation.

The atmospheric temperature and seawater temperature were 1 and 5°C, respectively. The ship was in ballast condition, and it is estimated that the upper deck plating was under the following stress

$$\sigma_{DK} = 51.3 \pm 24.5 \text{ (MPa)} \tag{40}$$

Results of Analysis

From the V-notch Charpy values of test specimens removed from the failed upper deck plating (NK standard KA class) and side shell plating (NK standard KD and KA class), K_{CA}-values

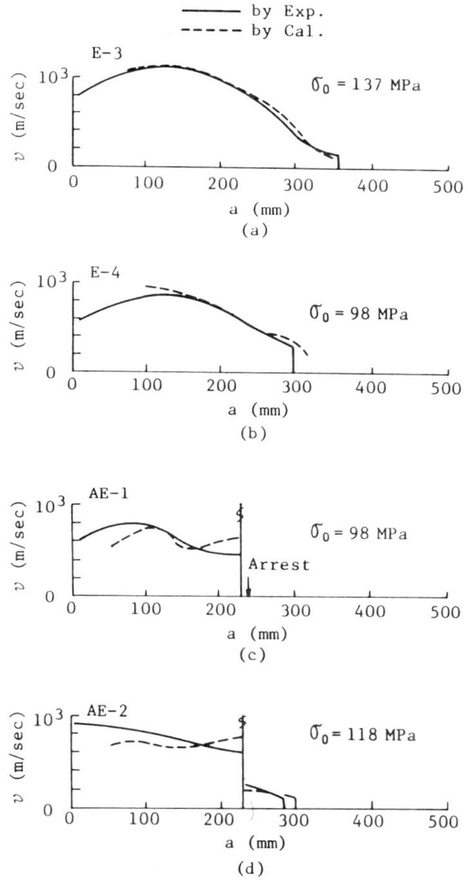

FIG. 22—*Comparison between measured and calculated crack velocity.*

at the temperature (atmospheric temperature: 1°C) prevailing at the time of failure were estimated. On the other hand, K-value at the crack tip was estimated by static analysis, applying the K_{eff} concept [*2*].

The K_{eff} concept is briefly explained as follows. Applying statical concept, we can experimentally determine crack arrest toughness K_{CA} (K-value at crack arrest) as a temperature dependent property for a given material, using the standard size (500 mm wide) gradient temperature type double tension test. But if we use very wide plate crack arrest specimen, for example, 2.5-m-wide plate which is five times as wide as the standard specimen, the result do not compare well, as shown in Fig. 24. A very wide specimen gives a very large K_{CA}-value as compared to a standard-size specimen. The reason why this difference results has not been clarified yet. Dynamic effects have been revealed to be partially responsible. For convenience we interpret that the nominal K-value computed for very wide specimen has its effective value. Plotting the nominal K_{CA}-values for very long arrested cracks obtained from very wide plate crack arrest tests on the abscissa and K_{CA} value obtained from standard-size specimen on the ordinate, we obtained experimentally the K_{eff} curve as shown in Fig. 25 [*16–18*]. The results of these analyses are indicated in Fig. 26. It appears that how the brittle crack was arrested can be explained fairly well by these calculations.

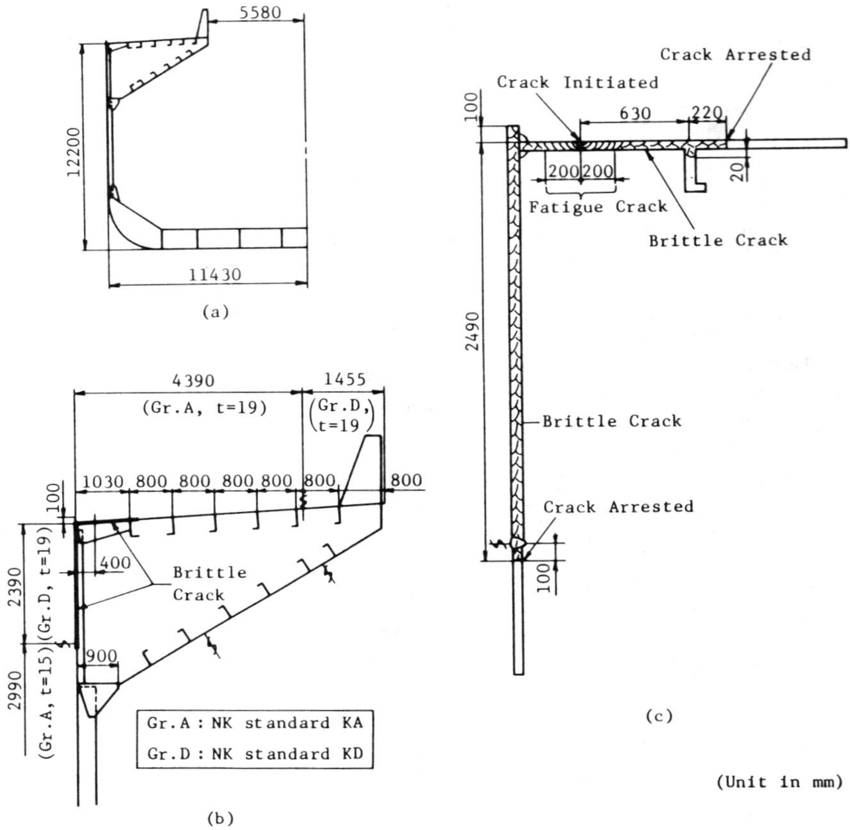

FIG. 23—*Damage plan of brittle failure in bulk carrier.*

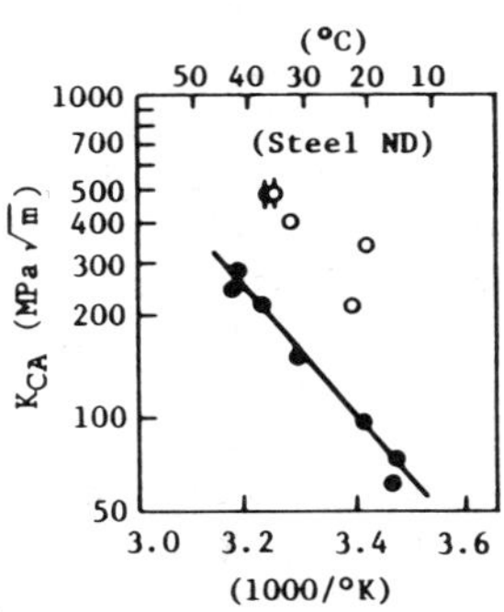

FIG. 24—*Crack arrest toughness* K_{CA} *versus temperature (log* K_{CA} — *reciprocal of absolute temperature) obtained from standard size specimen* (solid circle) *and from very wide specimen* (hollow circle).

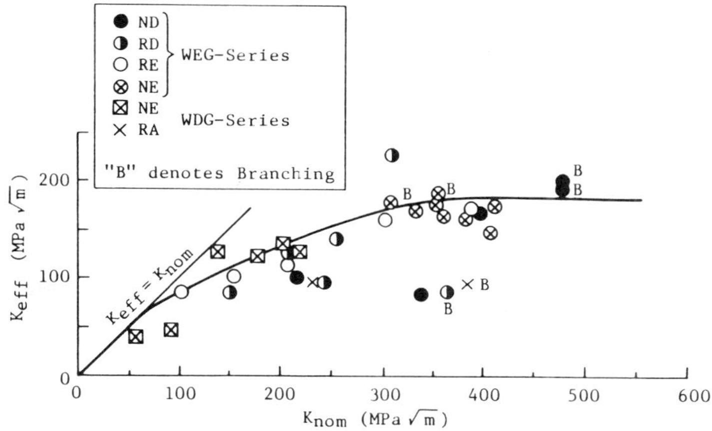

FIG. 25—*Relation between Nominal* K *and* K_{eff}.

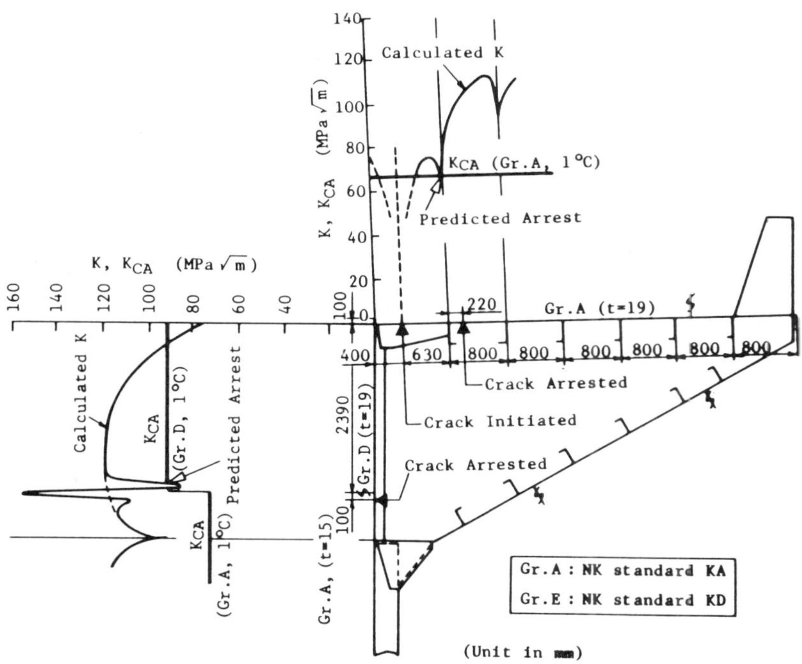

K value is obtained in statical calculation and modified by K_{eff} concept in higher K value region.

K_{CA} values are estimated as the lowest toughness from the damaged plates V-notch Charpy tests.

FIG. 26—*Calculation on brittle crack arrest in bulk carrier.*

Concluding Remarks

A dynamic analysis and a simplified dynamic analysis recently performed in Japan to predict brittle crack propagation and arrest behaviors have been reviewed herein. Also, an analysis of brittle crack propagation and arrest experienced by a bulk carrier, via the statical technique, has been reviewed.

Welded structures used in low-temperature conditions, such as liquefied natural gas (LNG) carriers, liquefied petroleum gas (LPG) carriers, pipelines, and pressure vessels, are on the increase. These structures are exposed to the danger of brittle failure even under normal service conditions, and their brittle failures generally are catastrophic. Thus, in response to the demands for assurance of safety against the brittle crack developing into a catastrophic failure, the usefulness of the crack arrester has come to attract renewed attention in a general move toward fail-safe design.

It is expected that the demands for establishing or refining the brittle crack propagation and arrest prediction technique will increase more and more henceforth.

References

[*1*] Kihara, H., Kanazawa, T., Ikeda, K., Yajima, H., et al., "Effectiveness of Crack Arrester (1st Report)," *Journal of the Society of Naval Architects of Japan,* Vol. 122, 1967.12; (2nd Report), Vol. 124, 1968.

[*2*] Machida, S. and Aoki, M., "Some Basic Considerations on Crack Arresters for Welded Steel Structures (7th Report)," *Journal of the Society of Naval Architects of Japan,* Vol. 131, 1972.6; and Kanazawa, T., Machida, S., Yajima, H., and Aoki, M., "Study on Brittle Crack Arrester," selected paper, *Journal of the Society of Naval Architects of Japan,* Vol. 11, 1973.

[*3*] Kanazawa, T., Machida, S., and Teramoto, T., "Study on Fast Fracture and Crack Arrest—The 1st Report," *Journal of the Society of Naval Architects of Japan,* Vol. 141, 1977.

[*4*] Kanazawa, T., Machida, S., and Niimura, Y., "Study on Fast Fracture and Crack Arrest—The 2nd Report," *Journal of the Society of Naval Architects of Japan,* Vol. 142, 1977.

[*5*] Kanazawa, T., Machida, S., Niimura, Y., and Yoshinari, H., "Study on Fast Fracture and Crack Arrest—The 3rd Report," *Journal of the Society of Naval Architects of Japan,* Vol. 144, 1978.

[*6*] Teramoto, T., Machida, S., and Kanazawa, T., "Study on Fast Fracture and Crack Arrest—The 4th Report," *Journal of the Society of Naval Architects of Japan,* Vol. 146, 1979.

[*7*] Kanazawa, T., Machida, S., Teramoto, T., et al., "Study on Fast Fracture and Crack Arrest—The 5th Report," *Journal of the Society of Naval Architects of Japan,* Vol. 150, 1981.

[*8*] Kanazawa, T., Machida, S., and Yoshinari, H., "Evaluation of Brittle Fracture Propagation Arrest Characteristics by DCB Testing," *Journal of the Society of Naval Architects of Japan,* Vol. 146, 1979.

[*9*] Kanazawa, T., Machida, S., Yajima, H., and Kawano, H., "Study on Brittle Crack Propagation and Arrest Behavior in Plates and Beams Structure—1st Report," *Journal of the Society of Naval Architects of Japan,* Vol. 149, 1981.

[*10*] Yajima, H. and Kawano, H., "Analysis of Ship Hull Damage," Commission No. 1, Welding Research Committee, 1-550-82, *The Society of Naval Architects of Japan,* 1982.

[*11*] Tada, H. et al., "The Stress Analysis of Crack Handbook," Del Research Corp., Hellertown, PA, 1973.

[*12*] Freund, L. B., "Crack Propagation in an Elastic Solid Subjected to General Loading—I," *Journal of the Mechanics and Physics of Solids,* Vol. 20, 1972.

[*13*] Ishida, M., "Analysis of Stress Intensity Factors for Plates Containing Random Array of Cracks," *Bulletin of the Japan Society of Mechanical Engineers,* Vol. 13, 1970.

[*14*] Kanazawa, T., Machida, S., and Doi, H., "Some Basic Consideration on Crack Arresters—The 6th Report," *Journal of the Society of Naval Architects of Japan,* Vol. 124, 1968.

[*15*] Akita, Y. and Ikeda, K., "On the Brittle Crack Propagation and Arrest with Special Reference to ESSO Test with Temperature Gradient," *Journal of the Society of Naval Architects of Japan,* Vol. 112, 1962.

[*16*] The 101 Research Committee of The Japan Shipbuilding Research Association, Report No. 78, 1968.

[*17*] Kanazawa, T., Machida, S., Yajima, H., and Aoki, M., "Effects of Specimen Size and Loading Condition on the Brittle Fracture Propagation Arrest Characteristics," *Journal of the Society of Naval Architects of Japan,* Vol. 130, 1971, pp. 343–351.

[*18*] Ishida, M., "Effect of Width and Length on Stress Intensity Factors of Internally Cracked Plates under Various Boundary Conditions," *International Journal of Fracture Mechanics,* Vol. 7, 1971, p. 301.

Alan R. Rosenfield,[1] *Paul N. Mincer,*[1] *and Charles W. Marschall*[1]

High-Temperature Crack-Arrest Toughness Measurements Using Compact Specimens

REFERENCE: Rosenfield, A. R., Mincer, P. N., and Marschall, C. W., "**High-Temperature Crack-Arrest Toughness Measurements Using Compact Specimens,**" *Fracture Mechanics: Eighteenth Symposium, ASTM STP 945,* D. T. Read and R. P. Reed, Eds., American Society for Testing and Materials, Philadelphia, 1988, pp. 73–85.

ABSTRACT: The ASTM Crack-Arrest Test, which is under development, is generally limited to providing data up to 20 to 40°C above the ductile-to-brittle transition temperature (expressed as RT_{NDT} in the ASME Boiler and Pressure Vessel Code). For this test, the geometry is a wedge-loaded compact specimen containing a brittle weld starting notch, which appears incapable of reliably generating a rapid crack above the stated temperature range.

Two alternatives have been used for obtaining data at higher temperatures: very large specimens combined with soft loading systems and modified compact specimens. This paper describes compact-specimen modifications which have extended the successful temperature range to the onset of the Charpy upper shelf, the highest relative temperature reached being RT_{NDT} + 100°C.

Specific modifications utilized are: a high-strength steel crack-starter section (duplex specimen), a low-friction loading arrangement (inverted split pins), and contoured side grooves. The high-temperature data generated using these modifications compare favorably with large-specimen results.

KEY WORDS: crack arrest, fracture toughness, fracture, temperature effects, ASTM A508 steel, ASTM A533 steel

Nomenclature

a	Crack length
B	Thickness
B_n	Thickness measured across roots of side grooves
E	Young's modulus
f	Function of crack length tabulated in Ref *4*
HAZ	Heat-affected zone
K_I	Crack-tip stress intensity
K_{IA}	Crack-arrest toughness
K_{Ia}	Stress intensity shortly after crack arrest; conservative estimate of K_{IA}
K_{Ic}	Plane-strain fracture toughness [Test Method for Plane-Strain Fracture Toughness of Metallic Materials (ASTM E 399-83)]
K_{ID}	Propagating-crack toughness
K_o	Nominal stress intensity at initiation of unstable fracture in a crack-arrest experiment
RPV	Reactor pressure vessel

[1]Research leader, master research technician, and senior research scientist, respectively, Battelle Memorial Institute, Columbus, OH 43201.

RT_{NDT} Reference nil-ductility temperature (ASME Boiler and Pressure Vessel Code)
USE Upper-shelf energy from Charpy test
w Width of compact specimen
y Displacement at $0.25w$ in front of loading-hole centerline of compact specimen

Crack arrest is a phenomenon that first received major attention in connection with World War II Liberty Ship failures [1]. More recently, the impetus for studying crack arrest has arisen from analysis of nuclear reactor pressure vessel integrity [2].

In the early days it was believed that crack-arrest resistance of steels could best be described in terms of the Charpy V-notch transition temperature. Now, however, it is generally believed that knowledge of fracture toughness is required to understand this phenomenon. As an example, the crack-arrest toughness as a lower bound to fracture resistance is embodied in the American Society of Mechanical Engineers (ASME) Boiler and Pressure Vessel Code. There is, however, some confusion as to the meaning of crack-arrest toughness. A consensus view [2,3] is that there exists a velocity-dependent propagating-crack toughness, K_{ID}, and that the minimum value of K_{ID} at any given temperature is the crack-arrest toughness, K_{IA}. However, since measurements of K_{IA} are impractical, at best, the usual procedure has been to measure the static stress intensity, K_{Ia}, very shortly after arrest [4]. It is generally believed that K_{Ia} is a conservative estimate of K_{IA}. There is also a significant amount of evidence that K_{Ia} measured using moderate-sized specimens (maximum size = 200 by 200 by 50 mm) is equal to the value of K_{Ia} obtained on very large test specimens [5,6].

As is the case with all other measures of steel toughness, K_{Ia} increases with increasing temperature in the transition region. Because K_{Ic} is also rising, it becomes increasingly difficult to initiate a rapidly moving crack and to analyze the test records using linear elasticity equations [7]. Thus it is difficult to obtain useful crack-arrest data in the upper part of the ductile-to-brittle transition region from a moderate-sized specimen.

The standard specimen being discussed by the ASTM Task Group on Crack Arrest is a blunt-notch compact design, based on the ASTM Standard Recommended Practice for R-Curve Determination (E 561–81), that uses a brittle weld at the notch root to initiate a crack [4]. The specimen is wedge-loaded in order to approximate fixed-grip conditions and to minimize dynamic effects. This design and procedure has proven useful up to a point. Typical experience with this specimen at Battelle is that unstable fracture can be initiated in this specimen only up to the temperature at which the Charpy V-notch impact energy reaches about two thirds of its maximum value. The reason is believed to be that a stable crack can be initiated easily in the notch-tip weld metal because of high local residual constraint. The crack then grows slowly into the heat-affected zone (HAZ) and becomes unstable only if the temperature is low compared to the ductile-to-brittle temperature of the HAZ [8].

The work reported in this paper was undertaken in order to extend the temperature capability of moderate-sized specimens (typically 150 mm square by 38 mm thick) to the upper-shelf dimpled-rupture region. A feasibility analysis of the literature [9] suggested that the associated K_{Ia} values for RPV steels may be in the range of 200 to 300 MPa $m^{1/2}$. In order to attain such high values, specimen and load-train modifications are required.

Test-Procedure Background

Three modifications of the procedure being considered by the ASTM Task Group on Crack Arrest (E 24.01.06) were used. We adopted a guideline of minimizing the departures from that procedure.

Duplex Specimens

The feasibility analysis suggested that measurements in the upper-shelf region may be possible using the duplex crack-arrest specimen [3]. This specimen consists of a high-strength, low-toughness crack-starter section made of SAE 4340 steel, electron-beam welded to the RPV steel being tested. Although this arrangement was developed originally using the double-cantilever-beam geometry [*10*], it also has been applied successfully to the compact geometry [*3*]. For this reason the duplex specimen is an allowable option in the proposed ASTM test method [*4*]. In addition to providing a low-toughness region for unstable-crack initiation, the high yield strength of SAE 4340 ensures that the extent of plastic displacements associated with crack initiation will be limited, thus avoiding unresolved questions as to how to take nonelastic displacements into account when calculating K_{Ia}.

Three problems arise in the use of duplex specimens, and these must be overcome if the design is to be successful in obtaining upper-shelf crack-arrest toughness levels:

1. Very high values of effective initiation stress intensity, K_o, are required. With no modification of design or starter-section material, the maximum K_o level obtainable may be less than upper-shelf K_{Ia} levels.
2. The electron-beam weld and its associated HAZ can be effective crack stoppers [*11*].
3. Specimen dimensions may have to be larger than used heretofore. Tentative requirements proposed by the ASTM Task Group on Crack-Arrest Testing [*4*] would necessitate a specimen thickness of 75 to 100 mm in order to measure a K_{Ia} value of 200 MPa $m^{1/2}$ and a thickness of about 200 mm for a K_{Ia} value of 300 MPa $m^{1/2}$ for typical RPV steels.

In the work reported in this paper it was found that the wedge-loaded duplex compact specimen can provide arrest-toughness values in excess of 200 MPa $m^{1/2}$, a value greater than obtainable using the brittle-weld design. Although our data do not satisfy the proposed ASTM size requirements, evidence is presented that they are representative of large-specimen data.

Positive Stress-Intensity Gradient

High K_{Ia} values are more likely to be obtained if the stress intensity (K_I) increases with increasing crack length. The natural tendency of K_I to decrease with increasing crack jump under wedge-loading conditions can be counterbalanced by progressively increasing the side-groove depth. According to Crosley and Ripling [*12*], the stress intensity associated with the compact specimen under fixed displacement is given by

$$K_I = Ey\left(\frac{B}{WB_n}\right)^{1/2} f\left(\frac{a}{w}\right) \tag{1}$$

Equation 1 can be used to define the side-groove profile of a constant-K wedge-loaded compact specimen

$$\frac{B_n}{B_{n0}} = \left[\frac{f\left(\frac{a_o}{w}\right)}{f\left(\frac{a}{w}\right)}\right]^2 \tag{2}$$

where the subscript 0 refers to the crack length at which deepening begins. A positive K_I gradient can be obtained by increasing the depth of side grooving faster than calculated from Eq 2, and this was done as is illustrated in Fig. 1. Note that the side-groove depth increases gradually from 25 to 60%. Limited evidence from a size-effect study suggests that 40% side grooving does not leave too small a remaining thickness [*13*] and that the relatively deeper grooving used in the present work appeared not to perturb the results.

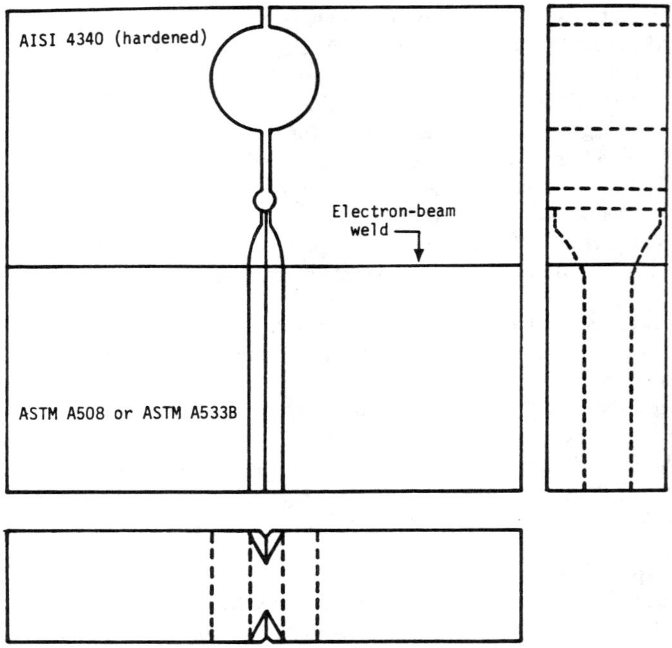

FIG. 1—*Schematic drawing of duplex-crack-arrest specimen with contoured side grooves.*

Lowered Arm Friction

Irwin designed the inverted-split-pin-loading geometry, shown in Fig. 2*b,* in order to reduce the friction between the specimen and the support plate [*14*]. The design succeeded in its purpose as was evident by the large arm motions associated with run/arrest events observed in early experiments on steel using this geometry [*15*]. Comparison with more extensive data [*6*] suggested that the inverted-split-pin K_{Ia} values, measured over the temperature range where brittle weld specimens are used, have a tendency to be significantly lower than those measured using other well-known test procedures (for example, thermal shock of large cylinders, compact specimens tested using the arrangement illustrated in Fig. 2*a,* and wide-plate tests). However, in subsequent experiments, performed as part of the ASTM Crack-Arrest Round Robin, normal and inverted split-pin data were essentially indistinguishable [*16*]. Thus inverted-split-pin loading can be utilized, particularly if it is the only option for obtaining data at very high temperatures using relatively small specimens.

Materials

Compact-specimen crack-arrest experiments were performed on the five different lots of steel listed in Table 1. These steels all were of RPV grades (either ASTM A508 or A533) and were supplied to Battelle by Oak Ridge National Laboratory as part of their Heavy-Section Steel Technology Program [*17–20*]. The objective of our experiments has been to examine the extent to which small-specimen crack-arrest data are representative of large-scale experiments. As discussed in Refs *5* and *21,* this objective has largely been met. All of the lots in Table 1 were used in large-scale experiments, except for 6R, which was specially heat-treated to provide a low ductile-to-brittle transition temperature. Steels designated 5A, 6, and 6R are from the same ingot and are identical except for heat treatment and resulting transition temperature.

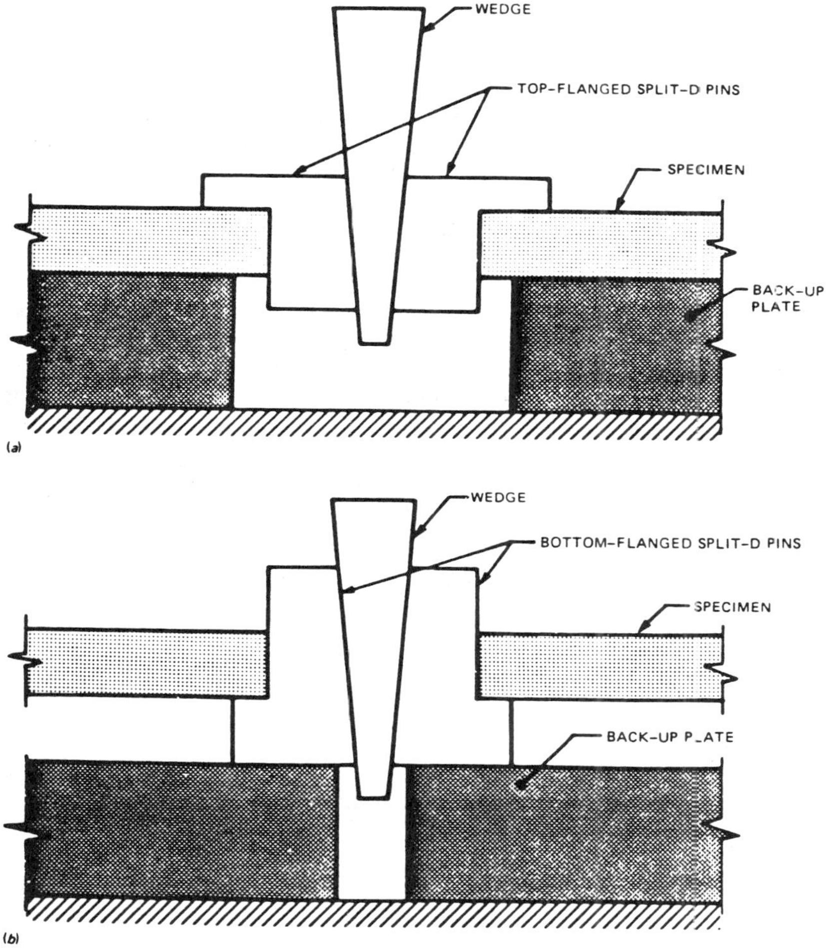

FIG. 2—*Alternative split-pin configurations developed by Irwin to vary friction in crack-arrest tests:* (a) *Conventional arrangement;* (b) *Test fixture that minimizes friction (from Ref* 14*).*

TABLE 1—*Steels used for crack-arrest experiments.*

Steel Designation				
ASTM	Project[a]	Yield Strength at Room Temperature, MPa	Ductile-to-Brittle Transition Temperature, °C	Ref
A508	TSE5A	605	10[b]	*15*
A508	TSE6	710	66[b]	*15*
A508	TSE6R	558	−29[c]	*16*
A508	PTSE1	600	91[b]	*17*
A533B1	WP1	445	−23[b]	*18*

[a]TSE refers to thermal-shock experiment; PTSE refers to pressurized-thermal-shock experiment; WP refers to wide-plate experiment.

[b]Reference nil-ductility temperature, RT_{NDT}.

[c]41 J (30 ft · lb) Charpy impact energy level.

Results

Limiting Temperature

The three methods of extending the temperature capability of the compact crack-arrest specimen were used either singly or in combination to obtain data on the five lots of project steel. Data which exemplify the current status of high-temperature crack-arrest testing are given in Fig. 3. This figure contains both Battelle wedge-loaded compact-specimen data and the result of an Oak Ridge National Laboratory (ORNL) pressurized-thermal-shock experiment [*19*]. Our data were obtained using a prolongation of the ORNL cylinder and was part of the ORNL pretest evaluation.

The low-temperature range in the figure contains compact-specimen data obtained in accordance with the typical crack-arrest test procedure [*4*], which includes a brittle weld starter-notch tip and normal pin geometry.[2] For PTSE-1 steel this procedure was successful only at and below the ductile-to-brittle transition temperature ($RT_{NDT} = 91°C$), which is about 20 to 40°C below its usual upper temperature capability. Inverted-split-pin loading of duplex specimens extended the temperature range by 60°C up to 150°C, which is the highest absolute temperature reached in Battelle crack-arrest experiments. The reason that higher temperatures have not been achieved is possibly associated with the fracture toughness transition behavior of the electron-beam weld and HAZ through which the crack must penetrate in the duplex specimen.

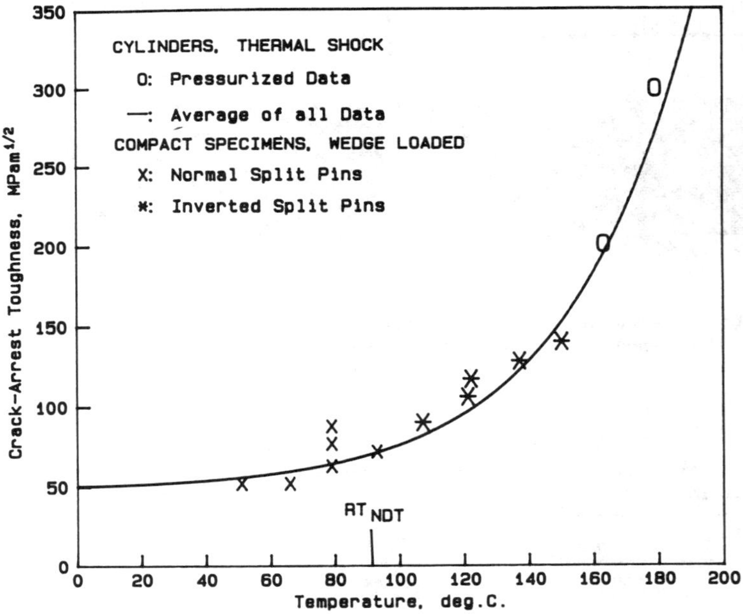

FIG. 3—*Comparison of crack-arrest toughness values obtained using different specimen designs (A508 steel from ORNL Experiment PTSE-1).*

[2]In a number of specimens from this lot, the arrested cracks were reinitiated in a subsequent test to produce another run/arrest event. Stress intensities from the subsequent arrests were obtained [*22*] and will be reported in another paper.

Because of the apparent limitation of 150°C for present-day compact specimens, temperatures corresponding to arrest in the large cylinder were not attempted. There was a 13°C difference between the highest temperature compact-specimen result and the lowest-temperature large-cylinder result.

The line drawn in Fig. 3 represents typical K_{Ia} behavior. Based on the combined results for all thermal shock experiments [*23–25*], it is also representative of the mean of data obtained using virtually all other specimen designs [*6*]. Figure 4 shows the thermal-shock and pressurized thermal-shock results used to establish the solid line drawn through them. The dashed line in the figure is the ASME K_{IR} curve and the solid line is the median cylinder-value. The median was established by an analysis of the Battelle Data Bank, which showed that the temperature dependence of K_{Ia} is most conveniently expressed as a temperature-independent K_{Ia}/K_{IR} ratio, with the median of this ratio for the cylinders being 1.63 and the standard deviation being 0.47. This result is an approximation to a more rigorous statistical analysis [*26*], which showed that the ratio is nearly temperature-independent and that the standard deviation of K_{Ia} is proportional to its median. In contrast, the usual assumption [*27*] is that the standard deviation is temperature-independent, an assumption that can cause serious errors in the calculation of lower-bound toughness values.

Table 2 gives the highest temperature results for compact specimens and compares them with Charpy upper-shelf values. Note that in all cases the lowest temperature on the Charpy upper shelf is approximately equal to, or slightly below, the highest temperature at which the compact specimen successfully provided crack-arrest data. Exact correspondence between these two temperatures would not be anticipated since the failure mechanisms in the two test specimens is quite different. Indeed, the cylinder results in Fig. 3 show that cleavage can persist in very large specimens at temperatures significantly above the Charpy upper shelf.

Figure 5 contains a comparison between the upper-shelf energy and crack-arrest toughness at the comparable temperatures listed in Table 2. While the data do not suggest a simple relation

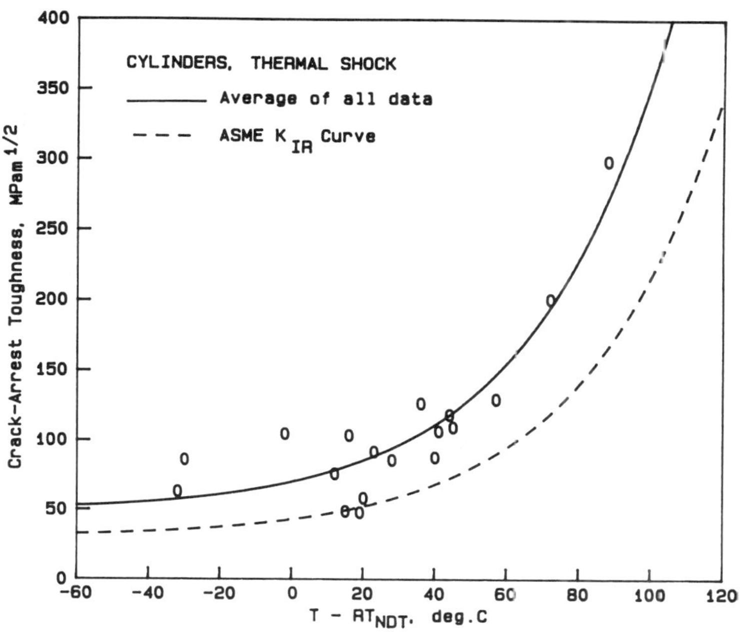

FIG. 4—*Compilation of crack-arrest toughness values for cylinders subjected to thermal shock.*

TABLE 2—*Comparison of compact-specimen crack-arrest toughness with Charpy upper-shelf impact energy.*

	Charpy Shelf		Crack-Arrest Experiments	
Lot	Energy, J	Approximate Temperature, °C	Highest Temperature Reached, °C	Highest K_{Ia} Value, MPa $m^{1/2}$
TSE5A	160	70	70	152
TSE6	106	100	121	157
TSE6R	150	70	71	260
PTSE1	110	140	150	140
WP1	170	50	56	219

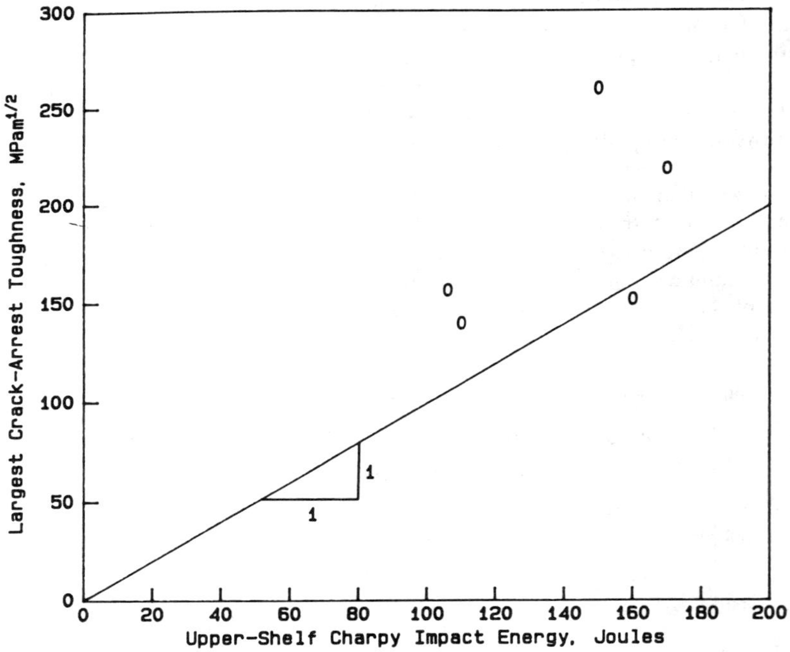

FIG. 5—*Crack-arrest toughness at onset of Charpy upper shelf.*

between the two quantities, an average value of the K_{Ia}/USE ratio is 4/3, provided both quantities are expressed in SI (metric) units. The 1:1 line in the figure represents a reasonably conservative estimate of K_{Ia} which might be made when only USE is known.

Temperature Dependence of Crack-Arrest Toughness

Figure 6 shows that the data generated using inverted-split-pin loading on duplex compact specimens are quite consistent with the cylinder data up to approximately (RT_{NDT} + 80°C); beyond this temperature the compact-specimen data again appear to be systematically low. However, this high-temperature region is one where data of any kind are extremely sparse (for example, the highest thermal-shock arrest [*19*] is at RT_{NDT} + 88°C), and the very-high-temper-

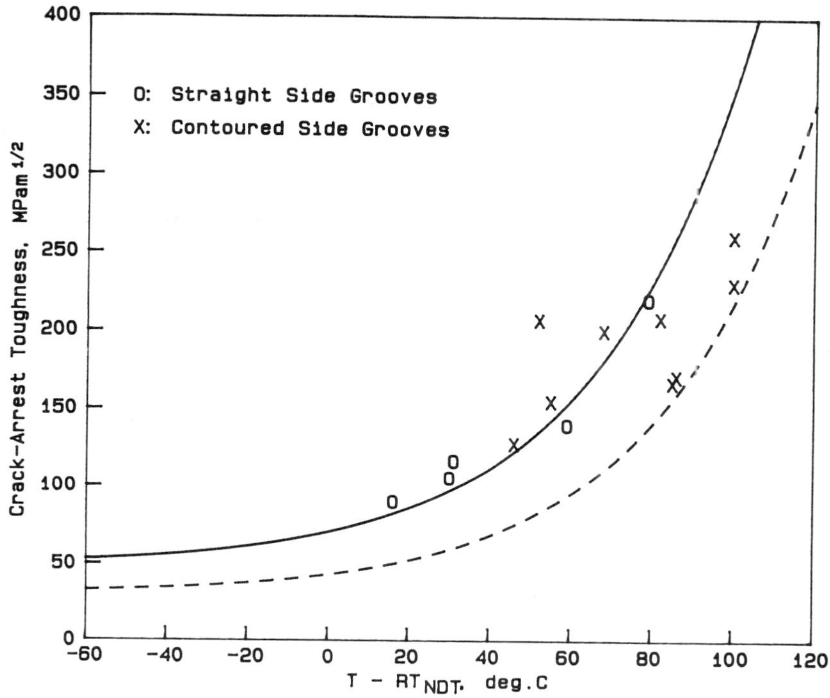

FIG. 6—*Crack-arrest data obtained using inverted-split-pin loading ASTM A508 steel duplex compact specimens (lines as on Fig. 4).*

ature compact-specimen data may be indeed representative of crack-arrest toughness values for large cylinders.

Discussion

The results in this paper suggest that the compact specimen can be modified to produce reliable high-temperature crack arrest data. Of the modifications, inverted-split-pin loading is the least understood. That geometry was originally viewed as causing problems because of the extremely low crack-arrest toughness values observed in the earliest experiments [*14*]. Those were brittle-weld specimens, which were tested in the vicinity of RT_{NDT} and exhibited both large quick openings[3] and long crack jumps. Those specimens did not satisfy the size criteria in Ref *4*, and resulted in toughness values far below K_{IR}. The reason that these low toughness values did not satisfy the size criteria was generally that the uncracked ligament after arrest was very small.

The early problems with inverted-split-pin loading reinforce the importance of understanding the significance of K_{Ia}: it is not a material property, but a conservative estimate of a material property, K_{IA}. To the extent that the early data fell far below typical K_{Ia} values, they also produced overly conservative K_{Ia} estimates.

In contrast, the duplex specimens, which were tested at high temperatures using inverted split pins, showed more typical crack-arrest behavior. Not only were the K_{Ia} values consistent with other designs, but also the crack-jump lengths and quick openings were close to those

[3]"Quick opening" describes the change in measured displacement between the time of initiation of an unstable crack and a time shortly after it has arrested.

observed with normal split-pin loading. The decrease in quick opening may result from the action of unbroken ligaments [*28*], which tend to inhibit specimen vibrations [*29*]. Ligaments are expected to play a greater role at high temperatures since they become more numerous as the test temperature is increased [*28*]. They may be modeled as pinching forces [*30*], bridging the crack, holding its opposing faces together, and restoring the restraint that is lost by changing the pin design. In effect, the ligaments tend to make the specimen behavior more static, and thus more representative of thermal-shock conditions [*17*].

With regard to the other modifications, Fig. 6 suggests that contouring may lead to increased scatter and, possibly, decreased temperature dependence of measured arrest toughness. However, more data are needed to make definitive conclusions.

Conclusions

The compact-specimen modifications described in this paper can provide crack arrest data consistent with those obtained from large cylinders up to test temperatures corresponding to the onset of the Charpy upper-shelf region. At temperatures below the upper shelf, the crack-arrest toughness can be expressed as $K_{Ia} = (1.63 + 0.47) K_{IR}$.

Acknowledgments

This research was performed under Subcontract 85X-17624C between Battelle Memorial Institute and Martin-Marietta Energy Systems. We are grateful to the technical monitor, C. E. Pugh, Oak Ridge National Laboratory, for his continuing support and encouragement.

References

[*1*] Williams, M. L., "Analysis of Brittle Behavior in Ship Plates," *ASTM STP 158,* American Society for Testing and Materials, Philadelphia, 1954, pp. 11–41.
[*2*] Smith, E., *Journal of Nuclear Materials,* Vol. 115, 1983, pp. 95–100.
[*3*] Crosley, P. B. et al., "Cooperative Test Program on Crack Arrest Toughness Measurements," Report NUREG/CR-3261, University of Maryland, College Park, MD, April 1983.
[*4*] Fourney, W. L. et al., "Crack Arrest Toughness of Ferritic Materials," *Metals Handbook,* 9th ed., Vol. 8, 1985, pp. 288–297.
[*5*] Rosenfield, A. R., "Validation of Compact-Specimen Crack-Arrest Data," *Transactions, American Society of Mechanical Engineers, Journal of Engineering Materials and Technology,* Vol. 106, 1984, pp. 207–208.
[*6*] Rosenfield, A. R., "Reactor-Pressure-Vessel Steel Crack-Arrest Data Bank," Battelle Memorial Institute Report to Oak Ridge National Laboratory, 1985.
[*7*] Marschall, C. W., Rosenfield, A. R., and Landow, M. P., "Specimen-Size Considerations in Crack-Arrest Testing of Irradiated RPV Steels," *ASTM STP 888,* American Society for Testing and Materials, Philadelphia, 1986, pp. 339–352.
[*8*] Rosenfield, A. R. et al., "Critical Experiments, Measurements and Analyses to Establish a Crack Arrest Methodology for Nuclear Pressure Vessel Steels," Battelle-Columbus Report NUREG/CR-1887, 1981.
[*9*] Rosenfield, A. R. et al., "BCL HSST Support Program" in *Heavy-Section Steel Technology Program Quarterly Progress Report for July–Sept. 1982,* NUREG/CR-2751, G. D. Whitman and R. H. Bryan, Eds., Vol. 3, 1983, pp. 7–39.
[*10*] Hahn, G. T., Hoagland, R. G., and Rosenfield, A. R., "Dynamic Crack Propagation and Arrest in Structural Steels," Ship Structure Committee Report SSC-256, 1976.
[*11*] Marschall, C. W. and Rosenfield, A. R., "Crack-Arrest Tests of Irradiated High-Copper ASTM A508 Submerged-Arc Weld Metal," *Advances in Fracture Research,* D. Francois et al., Eds., Pergamon, Oxford, U.K., 1981, pp. 2467–2475.
[*12*] Crosley, P. B. and Ripling, E. J., "A Compact Specimen for Plane Strain Crack Arrest Toughness Testing," *Journal of Testing and Evaluation,* Vol. 8, 1980, pp. 25–31.
[*13*] Rosenfield, A. R., Mincer, P. N., Marschall, C. W., and Markworth, A. J., "Recent Advances in Crack-Arrest Technology," *ASTM STP 833,* American Society for Testing and Materials, Philadelphia, 1984, pp. 149–164.

[14] Fourney, W. L., "Investigation of Damping and of Cleavage-Fibrous Transition in Reactor-Grade Steel," in *Heavy-Section Steel Technology Program Quarterly Progress Report for October–December 1981,* NUREG/CR-2141, Vol. 4, 1982, pp. 21–46.

[15] Rosenfield, A. R. et al., "BCL HSST Support Program," in *Heavy-Section Steel Technology Program Quarterly Progress Report for April–June 1982,* NUREG/CR-2751, G. D. Whitman and R. H. Bryan, Eds., Vol. 2, 1983, pp. 3–30.

[16] Barker, D. B. et al., "A Report on the Round Robin Program Conducted to Evaluate the Proposed ASTM Test Method for Crack Arrest Fracture Toughness of Ferritic Materials," University of Maryland report to Oak Ridge National Laboratory, 1986.

[17] Cheverton, R. D. et al., "Fracture Mechanics Data Deduced from Thermal-Shock and Related Experiments with LWR Pressure Vessel Material," *Transactions, American Society of Mechanical Engineers, Journal of Pressure Vessel Technology,* Vol. 105, 1983, pp. 102–110.

[18] Rosenfield, A. R. et al. in *Heavy-Section Steel Technology Program Semiannual Progress Report for January-March 1983,* NUREG/CR-3334, C. E. Pugh, Ed., Vol. 1, 1984, pp. 12–48.

[19] Bryan, R. H. et al., "The Heavy-Section Steel Technology Pressurized-Thermal Shock Experiment, PTSE-1," *Engineering Fracture Mechanics,* Vol. 23, 1986, pp. 81–97.

[20] Edmonds, D. P. and McGowan, J. J., "Preliminary Characterization of Wide-Plate Test Material at ORNL," in *Heavy-Section Steel Technology Program Semiannual Progress Report for April–September 1984,* NUREG/CR-3744, C. E. Pugh, Ed., Vol. 2, 1985, pp. 96–102.

[21] Rosenfield, A. R., *Heavy-Section Steel Technology Program Semiannual Progress Report for October 1985–March 1986,* NUREG/CR-4219, C. E. Pugh, Ed., Vol. 3, No. 1, pp. 189–198.

[22] Rosenfield, A. R. et al., "Crack Arrest Studies at BCL," in *Heavy-Section Steel Technology Program Semiannual Progress Report for October 1983–March 1984,* NUREG/CR-3744, C. E. Pugh, Ed., Vol. 1, 1985, pp. 57–72.

[23] Cheverton, R. D. and Ball, D. G., "IPTS Program Studies," *Engineering Fracture Mechanics,* Vol. 23, 1986, pp. 173–189.

[24] Pellisier-Tanon, A., Sollougoub, P., and Houssin, B., "Crack Initiation and Arrest in an SA 508 Class-3 Cylinder Under Liquid Nitrogen Thermal Shock Experiment," *Transactions,* 7th International Conference on Structural Mechanics in Reactor Technology, Vols. G and H, 1983, pp. 132–142.

[25] Cheverton, R. D., "Pressure Vessel Fracture Studies Pertaining to the PWR Thermal-Shock Issue: Experiment TSE-7," Report NUREG/CR-4304, Oak Ridge National Laboratory, Oak Ridge, TN, 1985.

[26] Rosenfield, A. R., Mincer, P. N., and Marschall, C. W., "BCL HSST Support Program" in *Heavy-Section Steel Technology Program Quarterly Progress Report for July-September, 1983,* NUREG/CR-3334, C. E. Pugh, Ed., Vol. 3, 1984, pp. 3–20.

[27] Oldfield, W. and Server, W. L., "Fracture Toughness Prediction for Pressure Vessel Steels: The Development of a Statistically Based Method," *Reference Fracture Toughness Procedures Applied to Pressure Vessel Materials,* T. R. Mager, Ed., Publication MPC-24, American Society of Mechanical Engineers, 1984, pp. 9–26.

[28] Hahn, G. T., et al., "Fast Fracture Toughness and Crack Arrest Toughness of Reactor Pressure Vessel Steel," *ASTM STP 711,* American Society for Testing and Materials, Philadelphia, 1980, pp. 289–320.

[29] Rahka, K., "Vibrations in Compact Specimens During Crack Arrest Testing," *Journal of Testing and Evaluation,* Vol. 8, 1980, pp. 318–323.

[30] Rosenfield, A. R. and Majumdar, B. S., "Micromechanisms and Toughness for Cleavage Fracture of Steel" in *International Seminar on Local Approach to Fracture,* EDF, Moret-sur-Loing, France, 1986.

DISCUSSION

C. E. Pugh[1] *(written discussion)*—This paper shows that for inverted split-pin loading, crack-arrest toughness values from monolithic weld embrittled specimens are generally lower than the values for thermal and pressurized thermal-shock cylinders (Fig. 6). On the other hand, better agreement is obtained using duplex specimens with high-yield strength starter sections. The comment is made that because the load-point displacement of the duplex specimens at initiation is predominantly elastic, the problem of taking plastic displacements into account

[1]Oak Ridge National Laboratory, Oak Ridge, TN.

for calculating K_{Ia} values is avoided. However, the reason for the undercalculation of K_{Ia} values from the monolithic specimen data is not explained. The reason could be the neglect of the plastic portion of the displacement at initiation, which the University of Maryland[2] argues, on an experimental basis, should be included in the calculation of K_{Ia}. However, neither the present paper nor the one just mentioned[2] shows any direct evidence concerning how much of the total displacement under load is elastic at the arrested crack depth, so the controversy is not resolved.

Previous load-displacement measurements with compact specimens by Landes et al.,[3] over a wide range of relative crack depths, have shown that the ratio of elastic to plastic displacements at a given total displacement appears to be approximately independent of a/w. Furthermore, Merkle[4] derived this same result mathematically, using an approximate expression for J for the compact specimen, the algebraic form of which was suggested by Andrews and Shih,[5] and the load-scaling function agreed with the data of Landes et al.[3] Therefore, the load-displacement curve for the arrested-crack depth, and thus the value of J at arrest, should be obtainable by scaling the load values for the initial crack depth down by a factor that depends only on the two crack depths. The load values for the P-Δ curve prior to initiation should be obtainable from the wedge-force analysis proposed by Milella et al.[6] In addition, the scaling factor can be checked by force measurements on the wedge after arrest.

It should be noted that in their 1981 paper "Subsize Specimens for Crack-Arrest Testing,"[7] Marschall et al. recognized the empirical scaling of CT specimen P-Δ curves according to the plotting method suggested by Andrews.[5] However, this information was applied for the purpose of avoiding the plastic displacement by cycling, rather than using it analytically. The use of a dimensionless nonlinear load-displacement curve was considered but rejected for reasons not explained. This idea deserves to be reexamined.

Alan R. Rosenfield, Paul N. Mincer, and Charles W. Marschall (authors' closure)—The discussion impinges on one of the unresolved issues in crack arrest testing: What value of displacement should be used to calculate K_{Ia}? The displacement measured shortly after crack arrest contains both elastic and plastic components. In addition, there is a quick-opening component caused by motion of the specimen arms associated with rapid crack propagation. Currently it is not clear whether plasticity and quick opening should be considered as contributing to crack arrest.

We have taken the viewpoint that only elastic energy is available to drive a rapid crack. The implications of this assumption in measuring crack initiation toughness, K_{Ic}, are discussed elsewhere.[8] For the purpose of this Closure the main result is that significantly better agreement is found between small-specimen and large-specimen crack initiation data when plasticity is

[2]Chona, R., Fourney, W. L., Link, R. E., and Sanford, R. J., "The Recoverability of Plastic Zone Energy in Crack Arrest," paper presented at the Spring Meeting of the Society for Experimental Mechanics, Las Vegas, NV, 9–14 June 1985.

[3]Landes, J. D., Walker, H., and Clarke, G. A., "Evaluation of Estimation Procedures Used in J-Integral Testing," *Elastic-Plastic Fracture, ASTM STP 668,* American Society for Testing and Materials, Philadelphia, 1979, pp. 266–287.

[4]Merkle, J. G. and Hudson, C. E., "Stable Crack Growth Estimates Based on Effective Crack Length and Crack-Opening Displacement" in *Proceedings,* USNRC-CSNI Specialists Meeting on Plastic Tearing Instability, NUREG/CP-0010 (CSNI Report No. 30), U.S. Nuclear Regulatory Commission, Washington, DC, Jan. 1980, pp. 115–142.

[5]Andrews, M. R. and Shih, C. F., "Thickness and Side-Groove Effects on J- and δ-Resistance Curves for A533B Steel at 93°C," *Elastic-Plastic Fracture, ASTM STP 668,* American Society for Testing and Materials, Philadelphia, 1979, pp. 426–450.

[6]Milella, P. P., Marschall, C. W., and Rosenfield, A. R., "A Method to Obtain the Crack-Opening Force in Transverse-Wedge Loading," *International Journal of Fracture,* Vol. 17, 1981, pp. R77–R81.

[7]Marschall, C. W., Mincer, P. A., and Rosenfield, A. R., "Subsize Specimens for Crack Arrest Testing," paper presented at ASTM Symposium on *The Use of Small-Scale Specimens for Testing Irradiated Material.*

[8]Rosenfield, A. R. and Shetty, D. K., "Cleavage Fracture of Steel in the Ductile-Brittle Transition Region," *ASTM STP 856,* American Society for Testing and Materials, Philadelphia, 1985, pp. 196–209.

discounted. By analogy it can be agreed that the same should be true for crack arrest. Our paper suggests that this is indeed the case.

In contrast, the discusser argues that plasticity should be accounted for and suggests some means for doing so. As the discusser has pointed out, we have previously considered a key-curve method for deducing load. Another possibility involves estimating the friction component from the cyclic load/displacement record.[6] Doubtless there are other possibilities. In the light of the various uncertainties, the ASTM Task Group on Crack Arrest has recommended a compromise. Specifically, half of the accumulated plasticity is to be included in the displacement value used in the linear elastic formula for calculating K_{Ia} [Proposed ASTM Method for Determining the Plane-Strain Fracture Toughness, K_{Ia}, of Ferritic Materials (E24.01.06)]. A similar compromise has been suggested for dealing with the dynamic effects associated with quick opening.

Arun Shukla[1] *and Ravinder Chona*[2]

The Stress Field Surrounding a Rapidly Propagating Curving Crack

REFERENCE: Shukla, A. and Chona, R., **"The Stress Field Surrounding a Rapidly Propagating Curving Crack,"** *Fracture Mechanics: Eighteenth Symposium, ASTM STP 945,* D. T. Read and R. P. Reed, Eds., American Society for Testing and Materials, Philadelphia, 1988, pp. 86–99.

ABSTRACT: Results are presented from experiments conducted with a cross-shaped, biaxial fracture specimen which was eccentrically loaded to obtain rapid crack propagation along a curved path. High-speed photographs of the crack-tip isochromatic patterns were analyzed to obtain the leading coefficients of the dynamic, mixed-mode, crack-tip stress field expressed in series form. Values of K_{II} associated with the propagating crack were found to be essentially zero when compared with the dominant opening mode stress intensity factor, K_I.

KEY WORDS: curving cracks, dynamic stress fields, photoelastic studies, stress intensity factors

Cracks under mixed-mode loading have been studied by several investigators [*1–6*]. All of these studies have assumed quasi-static conditions, but the need to use analysis models that incorporate the effects of inertia is well accepted since crack curving is often associated with rapid crack propagation, especially in brittle materials. To date, the major hesitation in using dynamic analysis models to address this problem has stemmed from the increased complexity of the analysis and the difficulties encountered in providing physical explanations for the crack-tip stress field parameters.

When a crack moves along a curved path, the isochromatic fringes (contours of constant maximum in-plane shear) associated with the advancing crack tip no longer retain the symmetry associated with opening mode cracking. This lack of symmetry is generally ascribed to the presence of a mixed-mode crack-tip stress field. However, it is unresolved as to whether or not the contribution from the shear mode stress field includes a Mode II singularity, K_{II}. The possibility that K_{II} is zero has been discussed qualitatively [*7*], and it has been also demonstrated, through computer-generated fringe patterns, that asymmetry in the crack-tip isochromatics can be obtained from a combination of Mode II nonsingular terms with a Mode I singularity, even in the absence of K_{II} [*8*].

This paper discusses the results from a series of experiments conducted using a cross-shaped, biaxial fracture specimen that was eccentrically loaded so as to obtain rapid crack propagation along a curved path. High-speed photographs of the crack-tip fringe patterns were recorded using a multiple spark camera of the Cranz-Schardin type [*9, 10*] and analyzed using a combination of the dynamic stress field equations presented here and the multiple-point least-squares method of Sanford and Dally [*11*]. The results showed that the apparent K_{II} was only 2 to 5% of the calculated K_I-value and that the asymmetry of the observed isochromatic fringes was due largely to the nonsingular terms in the series representation of the shear mode stress field.

[1]Associate professor, Department of Mechanical Engineering and Applied Mechanics, University of Rhode Island, Kingston, RI 02881.

[2]Assistant professor, Department of Mechanical Engineering, Texas A&M University, College Station, TX 77843.

Dynamic Stress Field Representation

Irwin [12] has shown that for a crack-tip stress field translating in the positive x-direction at a fixed speed, c, the dilatation, Δ, and rotation, ω, can be expressed without loss of generality as

$$\Delta = \frac{\partial u}{\partial x} + \frac{\partial v}{\partial y} = \alpha(1 - \lambda_1^2)\,\mathrm{Re}\,\Gamma_1(z_1)$$
$$\omega = \frac{\partial v}{\partial x} - \frac{\partial u}{\partial y} = \beta(1 - \lambda_2^2)\,\mathrm{Im}\,\Gamma_2(z_2) \qquad (1)$$

where λ_1, λ_2, z_1, and z_2 are coordinate parameters defined in Fig. 1. Using Hooke's law, we can write

$$\sigma_{xx} = \mu[\alpha(1 + 2\lambda_1^2 - \lambda_2^2)\,\mathrm{Re}\,\Gamma_1 - 2\beta\lambda_2\,\mathrm{Re}\,\Gamma_2],$$
$$\sigma_{yy} = \mu[-\alpha(1 + \lambda_2^2)\,\mathrm{Re}\,\Gamma_1 + 2\beta\lambda_2\,\mathrm{Re}\,\Gamma_2], \quad \text{and} \qquad (2)$$
$$\tau_{xy} = \mu[-2\alpha\lambda_1\,\mathrm{Im}\,\Gamma_1 + \beta(1 + \lambda_2^2)\,\mathrm{Im}\,\Gamma_2]$$

where the constants α and β have to be evaluated so as to satisfy the boundary conditions for the crack problem of interest and depend upon the choice of the stress functions, Γ_1 and Γ_2, and μ is the shear modulus.

For the opening mode case, a first pair of appropriate choices for Γ_1 and Γ_2 is

$$\Gamma_1 = Z_1 = \sum_{n=0}^{n=N} A_n z_1^{n-1/2} \quad \text{with} \quad \Gamma_2 = Z_2 = \sum_{n=0}^{n=N} A_n z_2^{n-1/2} \qquad (3)$$

The leading coefficient, A_0, is related to the opening-mode stress-intensity factor, $K_{\mathrm{I}} = A_0\sqrt{2\pi}$, and the leading term is the familiar inverse-square-root stress singularity. The symmetry condition requires that $\beta = 2\lambda_1\alpha/(1 + \lambda_2^2)$.

It has been demonstrated for the static problem [13, 14] that the stress field obtained from Eqs 2 and 3 is incomplete, and that a second stress function is required to completely describe

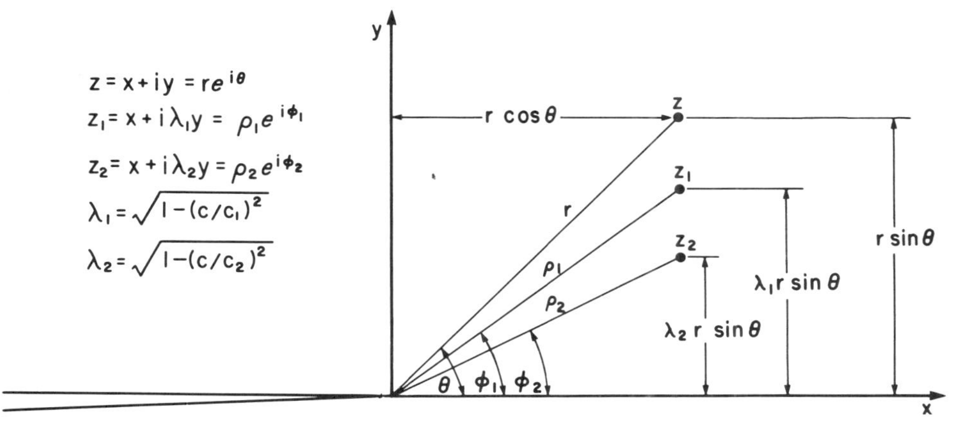

FIG. 1—*Crack-tip coordinate transformations associated with a constant crack speed analysis.*

the stress field in specimens with finite boundaries. The analogous choices for the dynamic problem are

$$\Gamma_1 = Y_1 = \sum_{m=0}^{m=M} B_m z_1^m \quad \text{with} \quad \Gamma_2 = Y_2 = \sum_{m=0}^{m=M} B_m z_2^m \tag{4}$$

In this case, the boundary condition on the crack faces requires that $\beta = \alpha(1 + \lambda_2^2)/2\lambda_2$. The leading term, B_0, gives rise to a superposed constant stress in the direction of crack propagation which is similar to the σ_{ox}-term in Irwin's static near-field equations [*15*].

The constant α can be determined from the definition of the opening mode stress intensity factor

$$K_{\text{I}} = \lim_{\substack{r \to 0 \\ \theta = 0}} \sigma_{yy} \sqrt{2\pi r} \tag{5a}$$

as

$$\alpha\mu = (1 + \lambda_2^2)/[4\lambda_1\lambda_2 - (1 + \lambda_2^2)^2] \tag{5b}$$

The general solution to the constant speed opening mode elastodynamic crack problem can be then obtained as

$$\sigma_{xx_{\text{I}}} = \frac{(1 + \lambda_2^2)}{4\lambda_1\lambda_2 - (1 + \lambda_2^2)^2} \left\{ (1 + 2\lambda_1^2 - \lambda_2^2) \operatorname{Re} Z_1 - \frac{4\lambda_1\lambda_2}{1 + \lambda_2^2} \operatorname{Re} Z_2 \right.$$
$$\left. + (1 + 2\lambda_1^2 - \lambda_2^2) \operatorname{Re} Y_1 - (1 + \lambda_2^2) \operatorname{Re} Y_2 \right\}$$

$$\sigma_{yy_{\text{I}}} = \frac{(1 + \lambda_2^2)}{4\lambda_1\lambda_2 - (1 + \lambda_2^2)^2} \left\{ -(1 + \lambda_2^2) \operatorname{Re} Z_1 + \frac{4\lambda_1\lambda_2}{1 + \lambda_2^2} \operatorname{Re} Z_2 \right. \tag{6}$$
$$\left. - (1 + \lambda_2^2) \operatorname{Re} Y_1 + (1 + \lambda_2^2) \operatorname{Re} Y_2 \right\}$$

$$\tau_{xy_{\text{I}}} = \frac{(1 + \lambda_2^2)}{4\lambda_1\lambda_2 - (1 + \lambda_2^2)^2} \left\{ -2\lambda_1 \operatorname{Im} Z_1 + 2\lambda_1 \operatorname{Im} Z_2 \right.$$
$$\left. - 2\lambda_1 \operatorname{Im} Y_1 + \frac{(1 + \lambda_2^2)^2}{2\lambda_2} \operatorname{Im} Y_2 \right\}$$

The shear mode counterpart to Eqs 6 can be obtained from the following considerations. First, Eqs 1 and 2 are general solutions that are not restricted to the opening mode crack problem. Second, the major difference between Mode I and Mode II stress fields is the antisymmetric nature of the stress field resulting from shear mode loading. A logical first step then is to select the stress functions, Γ_1 and Γ_2, as

$$\Gamma_1 = Z_1^* = \sum_{n=0}^{n=N} -iC_n z_1^{n-1/2} \quad \text{with} \quad \Gamma_2 = Z_2^* = \sum_{n=0}^{n=N} -iC_n z_2^{n-1/2} \tag{7}$$

where multiplying by $i = \sqrt{-1}$ provides the desired antisymmetry and the minus sign has been introduced for computational convenience. In this case, the leading coefficient, C_0, is related to the shear-mode stress-intensity factor, $K_{II} = C_0\sqrt{2\pi}$, and the leading term is once again an inverse-square-root stress singularity. The boundary conditions for the Mode II problem require that, for this choice of Γ_1 and Γ_2, $\beta = \alpha(1 + \lambda_2^2)/2\lambda_2$.

A second pair of choices for Γ_1 and Γ_2 that follows from finite specimen boundaries in a manner analogous to Eqs 4 is

$$\Gamma_1 = Y_1^* = \sum_{m=0}^{m=M} -iD_m z_1^m \quad \text{with} \quad \Gamma_2 = Y_2^* = \sum_{m=0}^{m=M} -iD_m z_2^m \tag{8}$$

with the boundary conditions being satisfied when $\beta = 2\lambda_1\alpha/(1 + \lambda_2^2)$. The definition of the Mode II stress-intensity factor

$$K_{II} = \lim_{\substack{r \to 0 \\ \theta = 0}} \tau_{xy}\sqrt{2\pi r} \tag{9a}$$

yields

$$\alpha\mu = 2\lambda_2/[4\lambda_1\lambda_2 - (1 + \lambda_2^2)^2] \tag{9b}$$

The shear mode counterpart to Eqs 6 is then

$$\begin{aligned}
\sigma_{xx_{II}} &= \frac{2\lambda_2}{4\lambda_1\lambda_2 - (1 + \lambda_2^2)^2}\Big\{(1 + 2\lambda_1^2 - \lambda_2^2)\,\text{Re}\, Y_1^* - \frac{4\lambda_1\lambda_2}{1 + \lambda_2^2}\,\text{Re}\, Y_2^* \\
&\quad + (1 + 2\lambda_1^2 - \lambda_2^2)\,\text{Re}\, Z_1^* - (1 + \lambda_2^2)\,\text{Re}\, Z_2^*\Big\} \\
\sigma_{yy_{II}} &= \frac{2\lambda_2}{4\lambda_1\lambda_2 - (1 + \lambda_2^2)^2}\Big\{-(1 + \lambda_2^2)\,\text{Re}\, Y_1^* + \frac{4\lambda_1\lambda_2}{1 + \lambda_2^2}\,\text{Re}\, Y_2^* \\
&\quad - (1 + \lambda_2^2)\,\text{Re}\, Z_1^* + (1 + \lambda_2^2)\,\text{Re}\, Z_2^*\Big\} \\
\tau_{xy_{II}} &= \frac{2\lambda_2}{4\lambda_1\lambda_2 - (1 + \lambda_2^2)^2}\Big\{-2\lambda_1\,\text{Im}\, Y_1^* + 2\lambda_1\,\text{Im}\, Y_2^* \\
&\quad - 2\lambda_1\,\text{Im}\, Z_1^* + \frac{(1 + \lambda_2^2)^2}{2\lambda_2}\,\text{Im}\, Z_2^*\Big\}
\end{aligned} \tag{10}$$

Finally, since the problem being considered is linear elastic, Eqs 6 and 10 can be combined to provide a general solution to the mixed-mode elastodynamic problem for a crack propagating at constant speed in a finite geometry. Similar results have been obtained by other investigators [*16,17*]. It should be noted that the leading term in Eqs 8, D_0, does not influence the stress components in Eqs 10; hence, there is no counterpart to the σ_{ox}-term for a crack propagating under pure shear mode loading.

It should be also noted that for the case of a curving crack, the x and y directions must be taken as the instantaneous tangent and normal to the crack path, respectively. Consequently, care must be exercised when selecting a region for data acquisition, if the data are to be analyzed using straight crack stress functions such as those developed here. The assumption of a locally straight crack is reasonable provided the radius of curvature of the crack path is large relative to the size of the region from which data are taken for analysis.

Experimental and Data Reduction Procedures

The experimental technique utilized to study the curving crack problem was a combination of high-speed photography and dynamic photoelasticity. A series of experiments was performed using eccentrically loaded biaxial specimens of the geometry shown in Fig. 2, in which the crack propagated along a curved path. The specimens were fabricated from a 10-mm-thick sheet of a brittle, birefringent material (Homalite 100), and the isochromatic fringe patterns at the tip of the running crack were recorded using a high-speed multiple spark gap camera of the Cranz-Schardin type. The details of the experimental procedure are not given here for the sake of brevity, and can be found in any of a large number of papers on dynamic photoelastic studies of fast fracture, for example Refs *9* and *10*.

Two of these experiments were selected for analysis and discussion in this paper. A sequence of frames showing the changes in the crack tip fringe patterns with crack propagation from each experiment is shown in Figs. 3 and 4, for Experiments 11 and 12, respectively. In both cases the fringe patterns show considerable asymmetry, indicating the presence of substantial shear mode deformations. The major difference between the two experiments is the magnitude of the load applied parallel to the original crack line. Experiment 11 had a large tension applied in this direction, while Experiment 12 had virtually no remote loading applied parallel to the initial crack direction. The fringe patterns recorded in each case confirm this observation, with the fringes of Experiment 11 (Fig. 3) showing the strong backward lean characteristic of this type of loading. The loads normal to the initial crack direction were almost the same in both experiments.

Each crack-tip fringe pattern was analyzed to determine the leading coefficients of the crack-tip stress field representation of Eqs 1 to 10 using the multiple-point, least-squares method

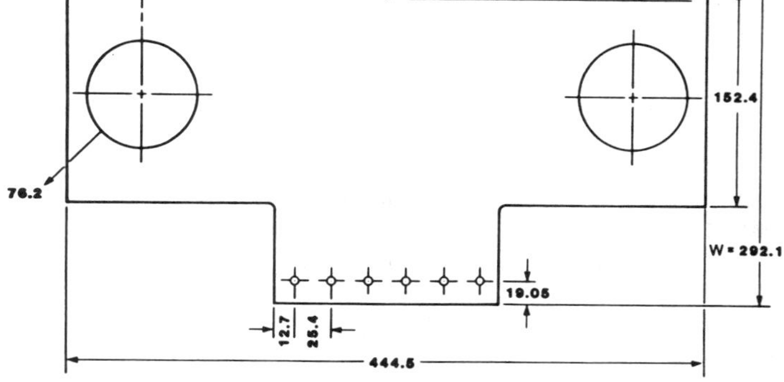

FIG. 2—*Specimen geometry and loading of the cross-shaped biaxial fracture specimen used in this study (all dimensions in mm).*

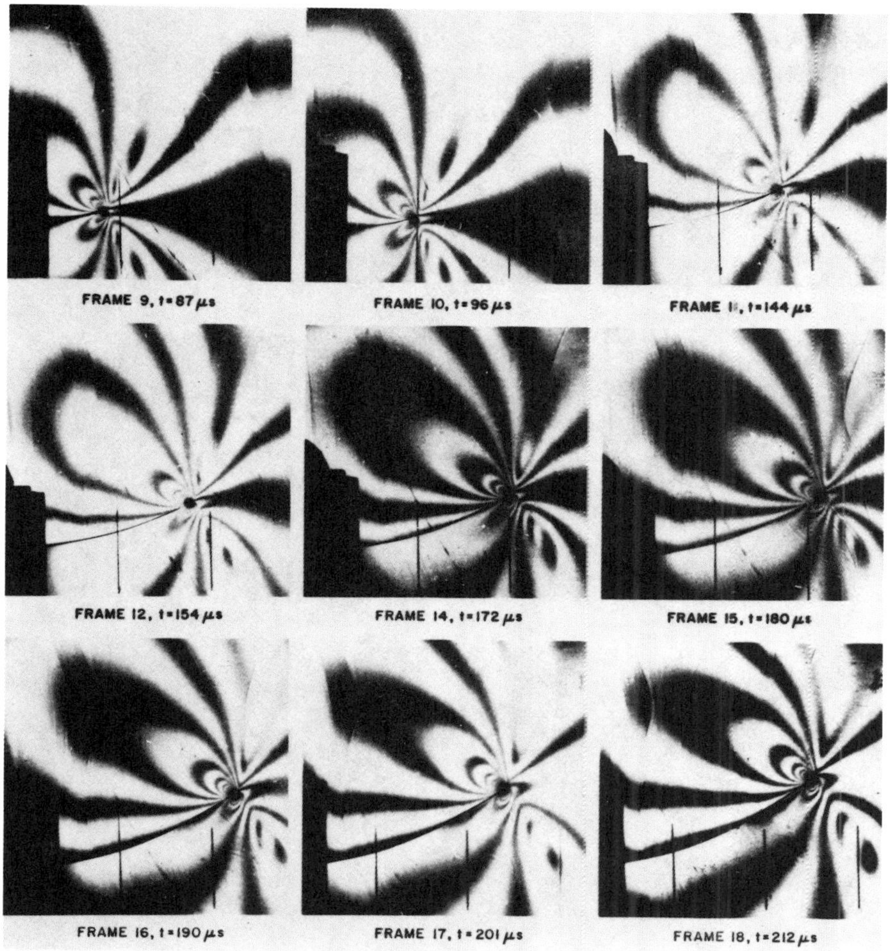

FIG. 3—*Sequence of frames showing the stress field surrounding the rapidly propagating curving crack in Experiment 11.*

developed by Sanford and Dally [*11*]. The procedure employed is best illustrated through the following example.

Consider Frame 12 from Experiment 11. The first step in the analysis procedure was to select a large number of data points from within a circular region of radius approximately equal to 13 mm, with 50 to 60 data points typically being selected. The size of the data acquisition region was kept small relative to the radius of curvature of the crack path (which ranged from 100 to 250 mm for these two experiments) and data were not taken from the fringes that were very close to the curved crack boundaries, where the assumption of a locally straight crack would not be valid. Care was also taken to exclude data points that would lie within one half the plate thickness of the crack tip, or 5 mm in this case, for the reasons discussed in Ref *18*. Otherwise, the only factor in data selection was fringe clarity. Coordinate pairs (r, θ) together with the known fringe order, n_i, at each point were then digitized and input to the analysis algorithm to obtain the coefficients of a best-fit solution in a least-squares sense.

The same data set was analyzed sequentially with models of Order 2, 3, 4, 5, and 6. (A model

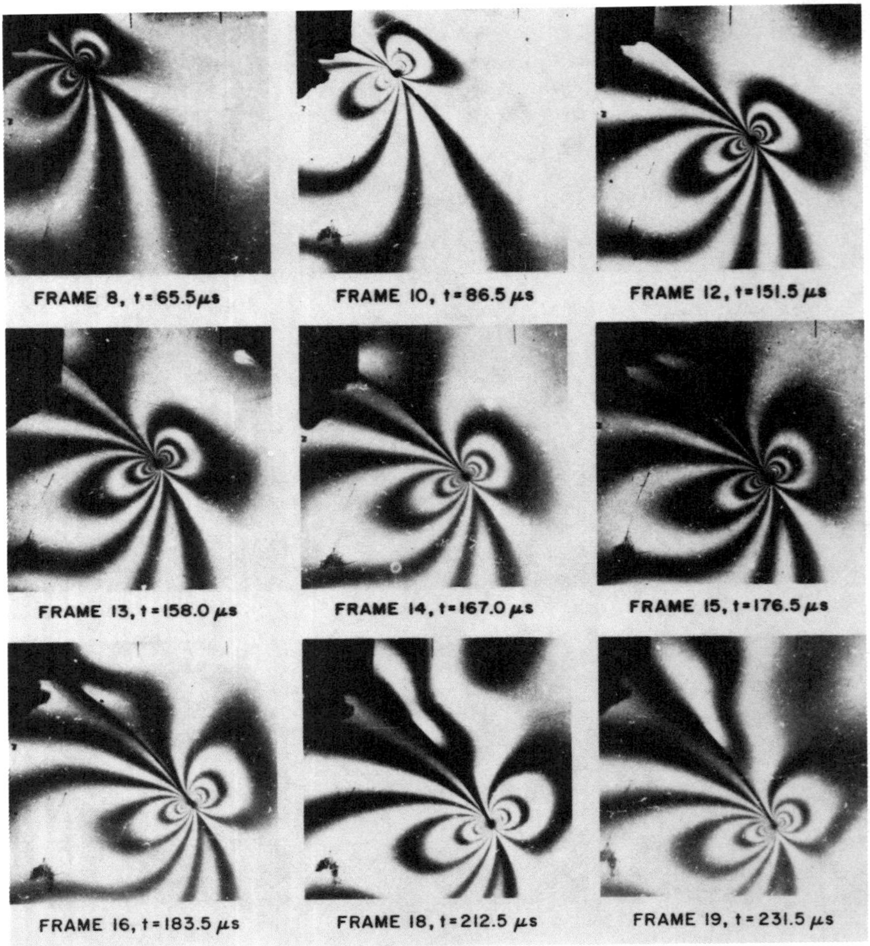

FIG. 4—*Sequence of frames showing the stress field surrounding the rapidly propagating curving crack in Experiment 12.*

of Order 2 would retain terms up to r^0, that is, A_0, C_0, and B_0; a model of Order 3 would retain terms up to $r^{1/2}$, that is, A_0, C_0, B_0, A_1, and C_1; a model of Order 4 would retain terms up to r^1, that is, A_0, C_0, B_0, A_1, C_1, B_1, and D_1; and so on.) The changes in the fringe order error [*19*] with an increase in the order of the analysis model were then examined and typically showed the trends illustrated in Fig. 5, with the error stabilizing to about 3% of the input average fringe order once the stress state had been modeled well over the region of data acquisition. The leading coefficients of each series were also found to stabilize once a good fit had been achieved, and this is also shown in Fig. 5.

Each of the coefficient sets obtained from the analysis was used to reconstruct the fringe pattern over the region of data acquisition so as to confirm visually that the stress state around the crack tip had in fact been properly modeled by the analysis model selected as being "correct." The sequence of reconstructed fringe patterns corresponding to 2- through 5-parameter models for Frame 12 of Experiment 11 is compared with the experimental pattern in Fig. 6 and the 5-parameter solution can be seen to match closely the stress state over the region of data

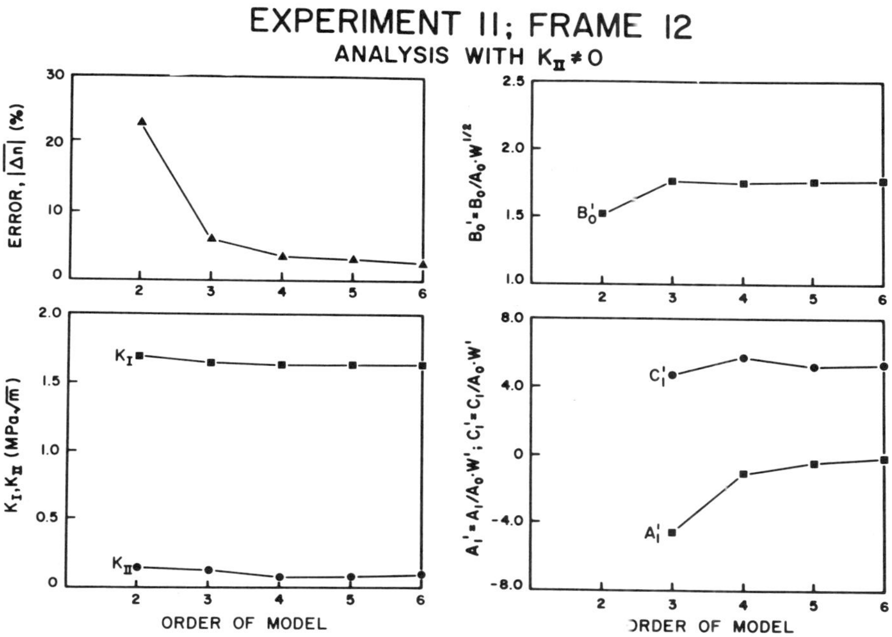

FIG. 5—*Changes in the fringe order error and leading coefficients of the series stress field representation as a function of the number of parameters retained in the analysis of Frame 12 from Experiment 11.*

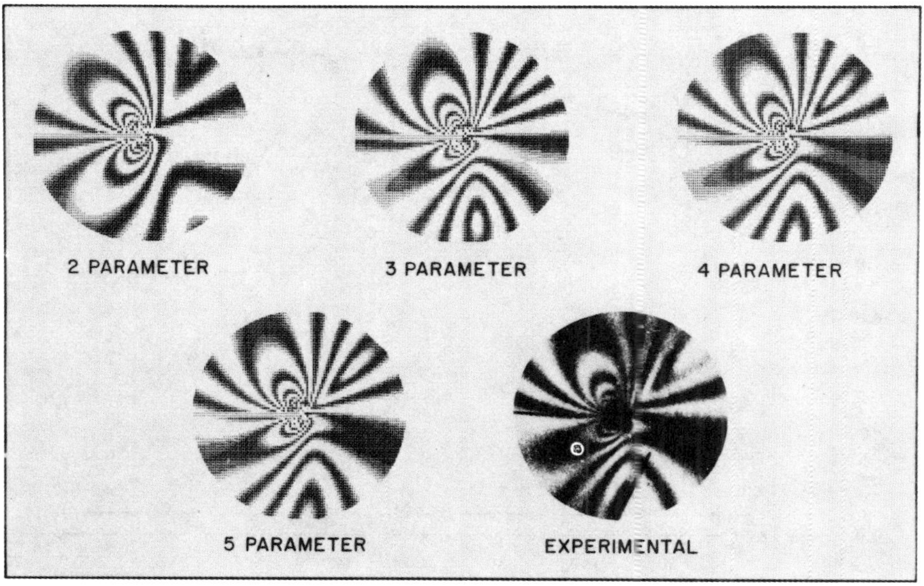

FIG. 6—*Experimental fringe pattern over the data acquisition region of radius equal to 13 mm and the reconstructed (computer generated) isochromatic patterns corresponding to increasingly higher-order analysis models for Frame 12 from Experiment 11.*

acquisition. Similar analyses were performed for each frame of the two experiments being discussed here, and the results are presented in the next section.

Results and Discussion

Figures 7 and 8 show the values of K_I and K_{II} determined from the dynamic isochromatic patterns for Experiments 11 and 12, respectively. The results have been shown as functions of the time after crack initiation, since the cracks propagated at a constant velocity of 380 m/s in both experiments. In each case, the opening mode stress-intensity factor increased monotonically as the crack propagated across the specimen, changing from about 1.5 to 1.8 MPa$\sqrt{m}$ in 225 μs. This result is consistent with the results one would expect if the crack had propagated straight across a specimen of this geometry with an initial applied load equal to that used in these two experiments [*10*]. In both cases K_{II} remained constant and very close to zero, with a value that ranged between 2 and 5% of K_I. This result is particularly interesting since it raises some questions about the influence of K_{II} as the sole factor determining crack propagation along a curved path and the major source of the strong asymmetry observed in the fringe patterns for both experiments.

The results for the σ_{ox}-term are shown in Figs. 9 and 10 in normalized form. As expected, the results for Experiment 11 correspond to a large remote tension parallel to the original crack extension direction, though B_0' does decrease as the crack propagates. In Experiment 12, on the other hand, the remote tension applied prior to crack extension was small, and B_0' remains small and essentially constant throughout the experiment. The results for the next pair of normalized nonsingular terms, A_1' and C_1', which represent the Mode I and Mode II coefficients of $r^{1/2}$, respectively, are shown in Figs. 11 and 12. The A_1'-term follows the same trend in both

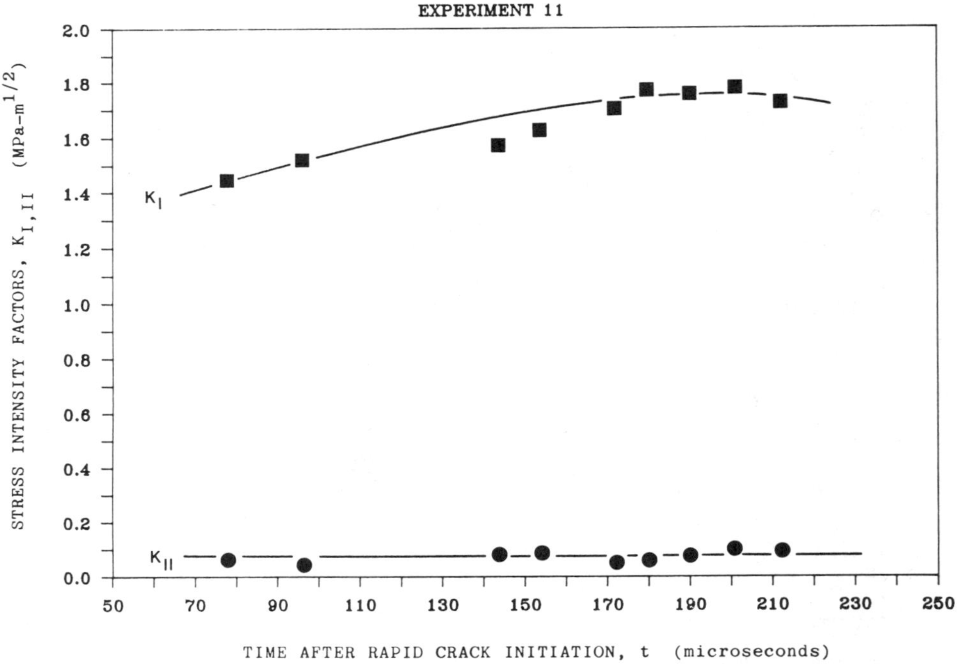

FIG. 7—*Opening and shear mode stress intensity factors for Experiment 11 as a function of time after crack initiation.*

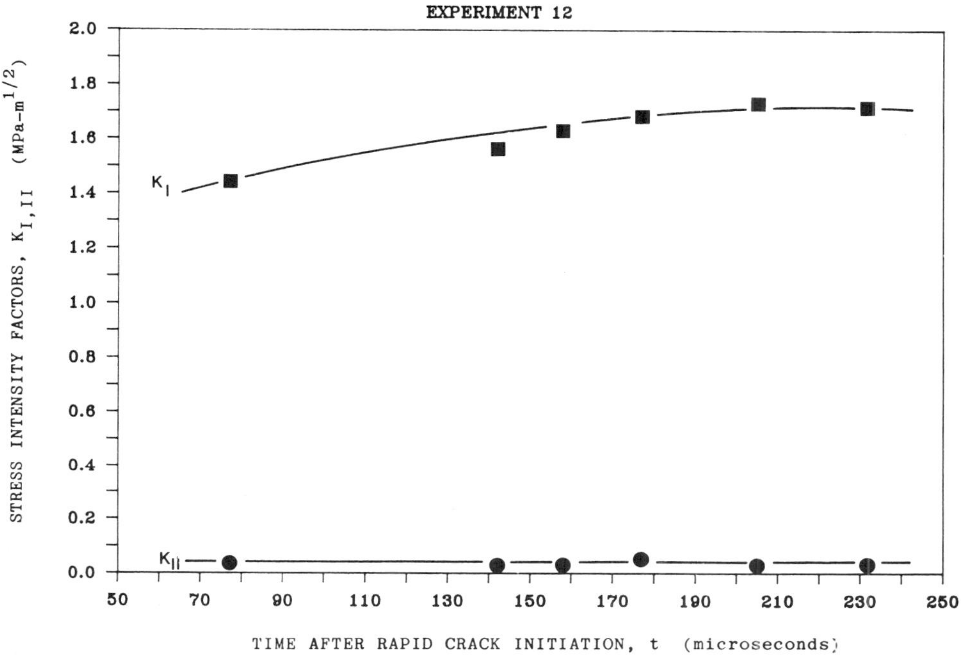

FIG. 8—*Opening and shear mode stress intensity factors for Experiment 12 as a function of time after crack initiation.*

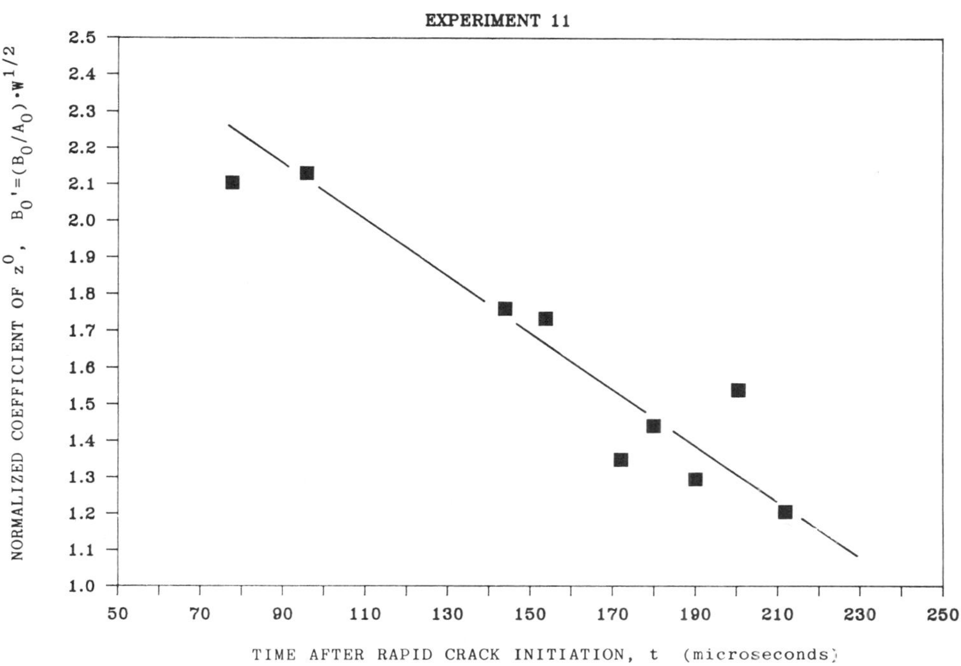

FIG. 9—*Normalized coefficient of z^0 (B_0') as a function of time for Experiment 11.*

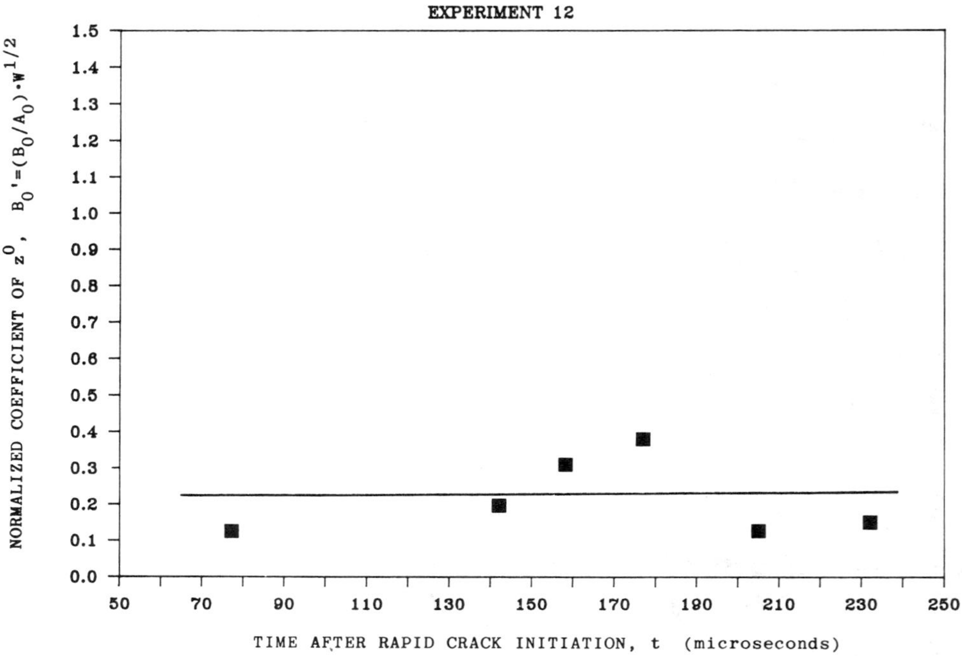

FIG. 10—*Normalized coefficient of* z^0 *(*B_0'*) as a function of time for Experiment 12.*

cases, decreasing monotonically from positive to negative values, while the Mode II counterpart, C_1', shows smaller changes with crack extension. The magnitude of the C_1'-term is large for both experiments. Its strong influence on the asymmetric portion of the crack-tip stress field relative to the contribution of the shear mode singularity can be appreciated better when one keeps in mind that C_1 has been normalized relative to A_0 in Figs. 11 and 12. Had the normalization been made with respect to the calculated shear mode singularity, C_0, which is of the order of 5% of A_0, the resulting values of C_1' would have been an order of magnitude larger.

It is also interesting to note that for Experiment 11, the values of C_1' determined from the fringe patterns (Fig. 11) show an increasing trend while the photographs in Fig. 3 show the crack moving along an increasingly curved path. The corresponding results for Experiment 12 (Figs. 4 and 12) show little change in the curvature of the crack path with time, while C_1' also shows little change. This suggests the possibility of a connection between the radius of curvature and the magnitude of the coefficient of $r^{1/2}$ in the shear mode stress field, and merits further investigation.

Several of the data sets from both Experiments 11 and 12 were reanalyzed with an algorithm in which the contribution to the stress field of K_{II} was assumed to be zero. Detailed results will not be presented here, but the general trend that was observed was for the Mode I coefficients to remain unchanged while the Mode II coefficients, particularly C_1 and D_1, took on substantially higher values in the absence of any K_{II}. The fringe order error stabilized to about the same number ($\sim$3 to 5%) as the results reported here and the reconstructed fringe patterns continued to show a good match between predicted and experimentally observed stress states once the analysis model retained terms up to $r^{3/2}$.

Finally, it should be noted that the computed K-values and the leading nonsingular terms were all found to be smoothly varying continuous functions of time (or crack extension) for both experiments, with trends that were consistent between both experiments. This is what one

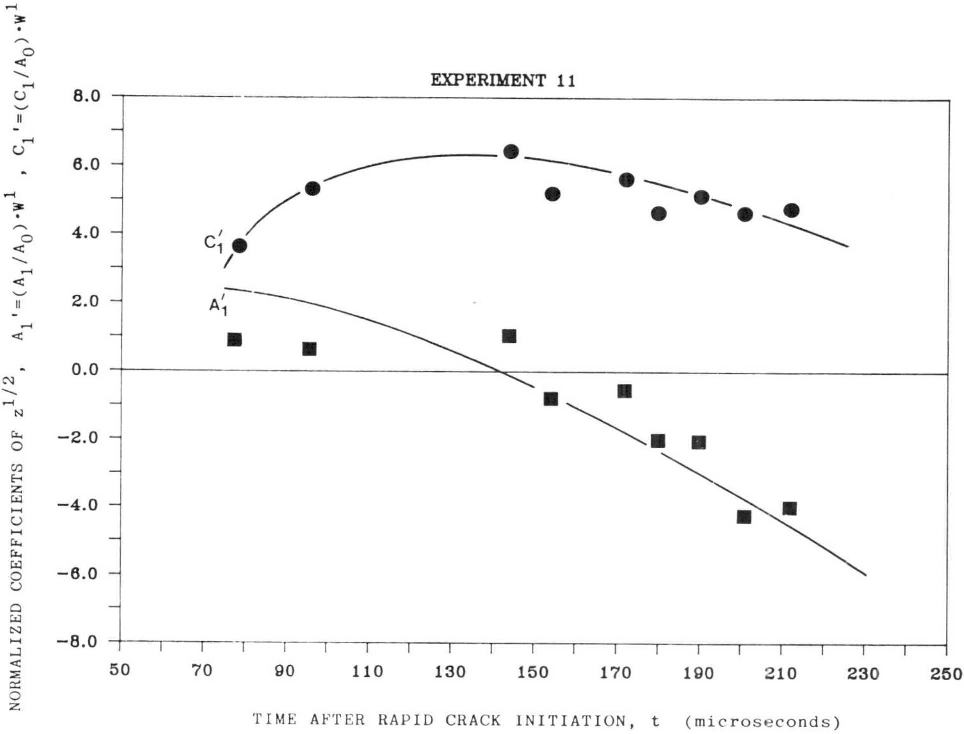

FIG. 11—*Normalized coefficients of* $z^{1/2}$ *for the opening mode* (A_1') *and shear mode* (C_1') *as functions of time for Experiment 11.*

would expect if the nonsingular coefficients being determined are associated with the crack-tip location in a specific geometry rather than just curve-fitting parameters for a particular combination of data points [*19*].

Summary

The stress field equations needed to fully describe the stress state associated with rapidly propagating curving cracks have been developed and combined with a multiple-point technique to determine with a high degree of confidence the fracture parameters of interest from photoelastic data. The results from the analysis of several frames from two dynamic fracture experiments show that the apparent K_{II}-values are small even when the fringe patterns show substantial asymmetry and the cracks are following paths whose curvature is changing rapidly. Indications of a possible connection between the radius of curvature of the crack path and the coefficient of $r^{1/2}$ in the shear mode component of the total stress field have been discussed qualitatively.

Acknowledgments

The authors would like to thank Professor George Irwin for his advice and encouragement throughout this study. This work has been supported by Grant No. MEA-8415494 from the National Science Foundation to the University of Rhode Island and Subcontract No. ORNL/

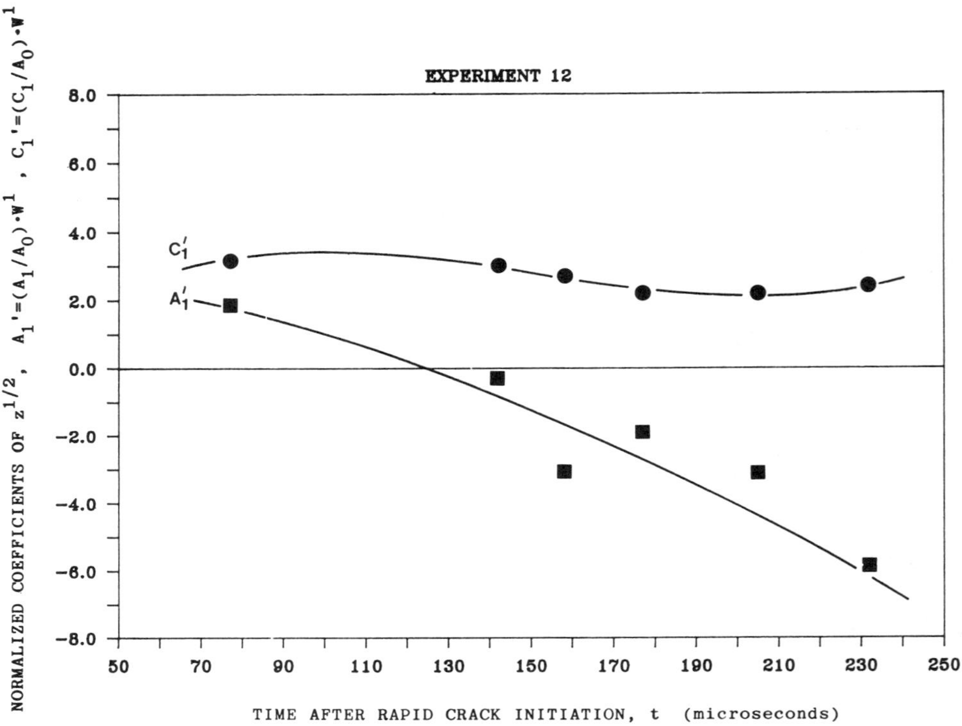

FIG. 12—*Normalized coefficients of* $z^{1/2}$ *for the opening mode* (A_1') *and shear mode* (C_1') *as functions of time for Experiment 12.*

Sub-7778 from the Oak Ridge National Laboratory Heavy-Section Steel Technology Program to the University of Maryland. The support provided by the Computer Science Center of the University of Maryland is also gratefully acknowledged.

References

[1] Smith, D. G. and Smith, C. W., "Photoelastic Determination of Mixed-Mode Stress Intensity Factors," *Engineering Fracture Mechanics,* Vol. 4, 1972, pp. 357–366.

[2] Palaniswamy, K. and Knauss, W. G., "Propagation of a Crack Under General In-plane Tension," *International Journal of Fracture,* Vol. 8, 1972, pp. 114–117.

[3] Hussain, M. A., Pu, S. L., and Underwood, J. H., "Strain Energy Release Rate for a Crack Under Combined Mode I and Mode II" in *Fracture Analysis, ASTM STP 560,* American Society for Testing and Materials, Philadelphia, 1974, pp. 2–28.

[4] Gdoutos, E. E. and Theocaris, P. S., "A Photoelastic Determination of Mixed-Mode Stress Intensity Factors," *Experimental Mechanics,* Vol. 18, 1978, pp. 87–96.

[5] Dally, J. W. and Sanford, R. J., "Classification of Stress Intensity Factors from Isochromatic Fringe Patterns," *Experimental Mechanics,* Vol. 18, 1978, pp. 441–448.

[6] Rossmanith, H. P., "Analysis of Mixed-Mode Isochromatic Crack-Tip Fringe Patterns," *Acta Mechanica,* Vol. 34, 1979, pp. 1–38.

[7] Rossmanith, H. P., "How Mixed is Dynamic Mixed-Mode Crack Propagation?—A Dynamic Photoelastic Study," *Journal of the Mechanics and Physics of Solids,* Vol. 31, 1983, pp. 251–260.

[8] Ramulu, M., Kobayashi, A. S., and Barker, D. B., "Analysis of Dynamic Mixed-Mode Isochromatics," *Experimental Mechanics,* Vol. 25, 1985, pp. 344–353.

[9] Dally, J. W., "Dynamic Photoelastic Studies of Fracture," *Experimental Mechanics,* Vol. 19, 1979, pp. 349–367.

[10] Anand, S. and Shukla, A., "Dynamic Crack Propagation and Branching Under Biaxial Loading" in *Fracture Mechanics: Seventeenth Symposium, ASTM STP 905,* J. H. Underwood et al., Eds., American Society for Testing and Materials, Philadelphia, 1986.

[11] Sanford, R. J. and Dally, J. W., "A General Method for Determining Mixed Mode Stress Intensity Factors from Isochromatic Fringe Patterns," *Engineering Fracture Mechanics,* Vol. 11, 1979, pp. 621–633.

[12] Irwin, G. R., "Constant Speed Semi-Infinite Tensile Crack Opened by a Line Force, *P*, at a Distance, *b*, from the Leading Edge," *Lehigh University Lecture Notes,* Bethlehem, PA, 1968.

[13] Liebowitz, H., Lee, J. D., and Eftis, J., "Biaxial Load Effects in Fracture Mechanics," *Engineering Fracture Mechanics,* Vol. 10, 1978, pp. 315–335.

[14] Sanford, R. J., "A Critical Re-examination of the Westergaard Method for Solving Opening-Mode Crack Problems," *Mechanics Research Communications,* Vol. 6, 1979, pp. 289–294.

[15] Irwin, G. R., "Discussion of: The Dynamic Stress Distribution Surrounding a Running Crack—A Photoelastic Analysis" in *Proceedings,* Society for Experimental Stress Analysis, Vol. 16, 1958, pp. 93–96.

[16] King, W. W., Malluck, J. F., Aberson, J. A., and Anderson, J. M., "Application of Running Crack Eigenfunction to Finite Element Simulation of Crack Propagation," *Mechanics Research Communications,* Vol. 3, 1976, pp. 197–202.

[17] Nishioka, T. and Atluri, S. N., "Path Independent Integrals, Energy Release Rates and General Solutions of Near-Tip Fields in Mixed Mode Dynamic Fracture Mechanics," *Engineering Fracture Mechanics,* Vol. 18, 1983, pp. 1–22.

[18] Rosakis, A. J. and Ravi-Chandar, K., "*On Crack Tip Stress State: An Experimental Evaluation of Three Dimensional Effects,*" Report No. SM 84-2, Graduate Aeronautical Laboratories, California Institute of Technology, Pasadena, CA, March 1984.

[19] Chona, R., Irwin, G. R., and Sanford, R. J., "Influence of Specimen Size and Shape on the Singularity Dominated Zone" in *Fracture Mechanics: Fourteenth Symposium—Vol. 1: Theory and Analysis, ASTM STP 791,* American Society for Testing and Materials, Philadelphia, 1983, pp. I-3–I-23.

Elastic-Plastic Fracture Mechanics I

J. C. Newman, Jr.,[1] *S. R. McNeill,*[2] *and M. A. Sutton*[2]

Application of the V_R Resistance Curve Method to Fracture of Various Crack Configurations

REFERENCE: Newman, J. C., Jr., McNeill, S. R., and Sutton, M. A., "**Application of the V_R Resistance Curve Method to Fracture of Various Crack Configurations,**" *Fracture Mechanics: Eighteenth Symposium, ASTM STP 945,* D. T. Read and R. P. Reed, Eds., American Society for Testing and Materials, Philadelphia, 1988, pp. 103–117.

ABSTRACT: A fracture resistance curve method based on the crack-tip-opening displacement (V_R) for a "stationary" crack has been previously developed to characterize stable crack growth and instability behavior of cracked metallic materials. In this study, the V_R-curve method has been applied to various crack configurations made of 2024-T351 and 7075-T651 aluminum alloys. Relationships between V_R, crack length, specimen width, and tensile properties have been derived from the Dugdale or strip-yield model. For arbitrarily cracked configurations, two methods of analysis, namely, the boundary-integral equation and finite-element methods, have been developed herein to analyze the strip-yield model. These analyses and the V_R-curve determined from compact specimens were used to predict stable crack growth and maximum load behavior for several crack configurations made of two aluminum alloys.

The V_R-curve method was found to correlate stable crack growth data from compact specimens up to maximum load. After the maximum load was reached in a test, the values of V_R remained nearly constant for each specimen but the constant value was dependent upon specimen type, crack length, and width. Predicted stable crack growth and maximum load behavior for compact, middle-crack and single-edge-crack tension specimens agreed within about 5% of experimental failure loads, whereas predictions made on three-hole-crack tension specimens were within about 15% of experimental failure loads.

KEY WORDS: fracture, cracks, plastic deformation, stress analysis, displacement

Nomenclature

- a Crack length, m
- a_0 Initial crack length, m
- B Specimen thickness, m
- C Constant in V_R equation (Eq 3)
- E Modulus of elasticity, N/m^2
- K Stress-intensity factor, N/m$^{-3/2}$
- K_p Stress-intensity factor due to applied load, N/m$^{-3/2}$
- K_σ Stress-intensity factor due to flow stress, N/m$^{-3/2}$
- k_s "Rigid" spring stiffness, N/m
- n Constant in V_R equation (Eq 3)
- P Applied load, N
- P_{exp} Experimental failure load, N

[1]Senior scientist, NASA Langley Research Center, Hampton, VA 23665-5225.

[2]Graduate student and assistant professor, respectively, University of South Carolina, Columbia, SC 29208.

P_{pred} Predicted failure load, N
R_0 Yield force for "rigid" spring, N
S Applied gross stress, N/m^2
u, v Displacements in x- and y-directions, respectively, m
V_c Critical one-half crack-tip-opening displacement (CTOD), m
V_i Constant in V_R equation (Eq 3)
V_p One-half CTOD due to applied load, m
V_{pw} One-half CTOD due to plastic wake, m
V_R Material crack-growth resistance in terms of CTOD, m
V_σ One-half CTOD due to flow stress, m
w One-half width or width of specimen, m
x, y Cartesian coordinates
Δa Physical crack extension, m
ρ Length of plastic zone, m
σ_0 Flow stress of material $(\sigma_{ys} + \sigma_u)/2$, N/m^2
σ_y Normal stress acting in y-direction, N/m^2
σ_{ys} Yield stress (0.2% offset), N/m^2
σ_u Ultimate tensile strength, N/m^2

The concept of linear-elastic fracture mechanics (LEFM) has been very useful in correlating fracture data for cracked plates and structural components in which the crack-tip plastic deformations are constrained to small regions [*1*]. However, for high-toughness sheet materials in which large amounts of plastic deformation and stable crack growth occur before failure, LEFM concepts are not applicable. The elastic-plastic and stable crack-growth behavior must be considered in order to characterize the fracture behavior of these materials.

The resistance curve methods [*2*] are generally used to characterize the resistance to fracture during slow-stable crack extension in metallic materials. A resistance curve provides a toughness record as a crack is driven stably into the plastic zone caused by increasing applied loads. The resistance curve is unique for the material and thickness of interest. It is insensitive to crack length, specimen width, and specimen type for through-the-thickness cracks. But is is dependent upon temperature, environmental conditions, and strain rate. The resistance curve is used to predict stable crack growth and instability behavior of cracked structural components.

In the present paper, a resistance curve method based on the crack-tip-opening displacement (V_R) for a "stationary" crack [*3,4*] has been applied to various crack configurations made of two aluminum alloys (2024-T351 and 7075-T651) tested at room temperature. The crack-tip-opening displacement was calculated at the current crack length, which may include crack extension, using a "stationary" crack solution. Two methods of analysis, namely the boundary-integral equation (BIE) [*5*] and finite-element (FE) [*6*] methods, have been developed herein to analyze the strip-yield or Dugdale model [*7*] for the crack configuration of interest. The results from these analyses are compared with previously developed collocation results [*3*]. These analyses and the V_R-curve concept were used to predict stable crack growth and maximum load behavior for several crack configurations. Comparisons are made between experimental and predicted load-against-crack extension behavior on compact, middle-crack, single-edge-crack and three-hole-crack tension specimens.

V_R Resistance Curve Method

In the V_R resistance curve method [*3,4*], the "crack drive" is written in terms of the crack-tip-opening displacement for a "stationary" crack at the current crack length. Figure 1 illustrates the definition of V_R. A cracked plate with an initial crack length a_0 is subjected to an applied load P which causes the crack to stably tear into the active plastic zone and leave behind

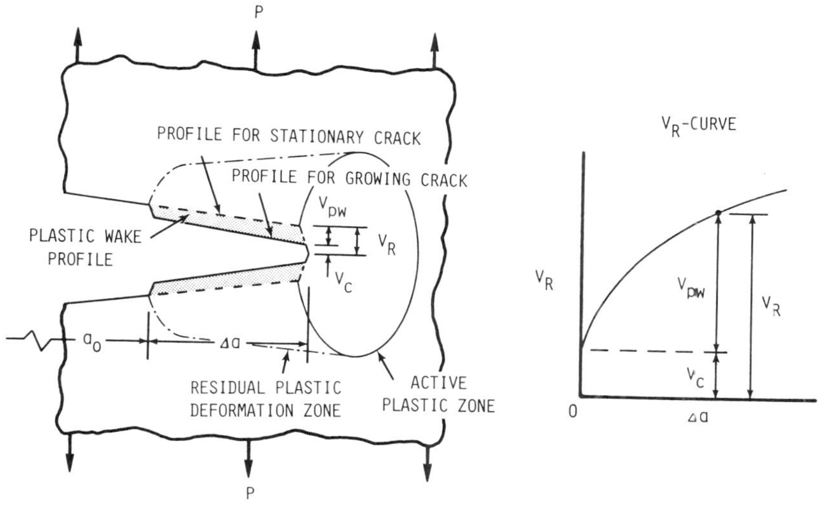

FIG. 1—*The* V_R *resistance curve concept.*

the residual plastic deformation zone [*9*]. The critical crack-tip-opening displacement (V_c) has been shown to be constant during stable crack growth [*8, 9*]. The amount of stable crack growth has also been shown to be controlled by the residual plastic deformation zone. Without this zone, any crack growth would be unstable. The "plastic wake" profile is defined as the difference in the crack surface displacements for a stationary crack and for a growing crack (shaded region). V_{pw} is the plastic-wake displacement. The displacement V_R is defined as the sum of $V_c + V_{pw}$. In the V_R hypothesis, it is assumed that V_R and V_{pw} are uniquely related to the amount of crack extension Δa. Experimental data analyzed in Refs *3* and *4* confirm this hypothesis. Thus, the V_R-curve characterizes the influence of the plastic wake on stable crack extension. Note that V_R in Fig. 1 can also be calculated from the crack-tip-opening displacement for a "stationary" crack.

For simplicity, the crack-tip-opening displacement for a stationary crack is calculated from the Dugdale model, as shown in Fig. 2. The plastic-zone size (ρ) is determined from the requirement that the stress-intensity factor (K) at the "fictitious" crack tip (Point A) must be zero. V_R is then calculated as the sum of the displacements ($V_p + V_\sigma$) at the physical crack tip due to the applied load P and due to the flow stress σ_0, respectively. Because the Dugdale model solutions are easily obtained from superposition of two elastic crack problems, the V_R-curve method can be applied to any through-the-thickness crack configuration for which these two elastic solutions have been obtained.

Many crack configurations could be used to obtain the V_R-curve. However, the compact and middle-crack tension specimens (see Fig. 3), being the most widely tested, were considered here.

Compact Specimen

From load-crack-extension (P, Δa) data calculate the plastic-zone size as

$$\rho = \frac{\pi}{8}\left(\frac{K}{\sigma_0}\right)^2 F_0 \tag{1}$$

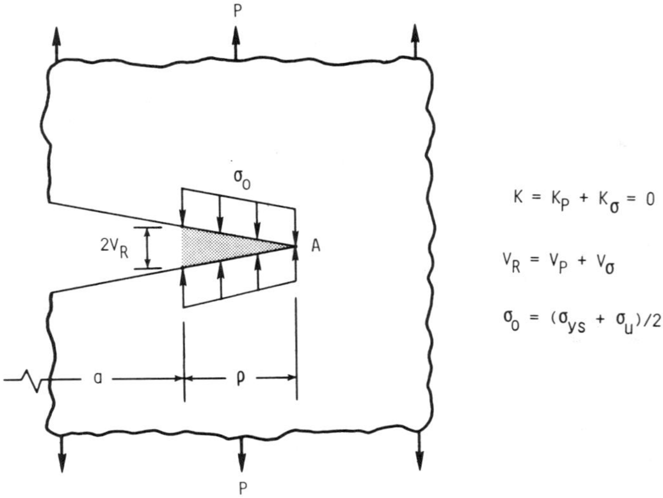

FIG. 2—*Crack-growth resistance* V_R *calculated from Dugdale model.*

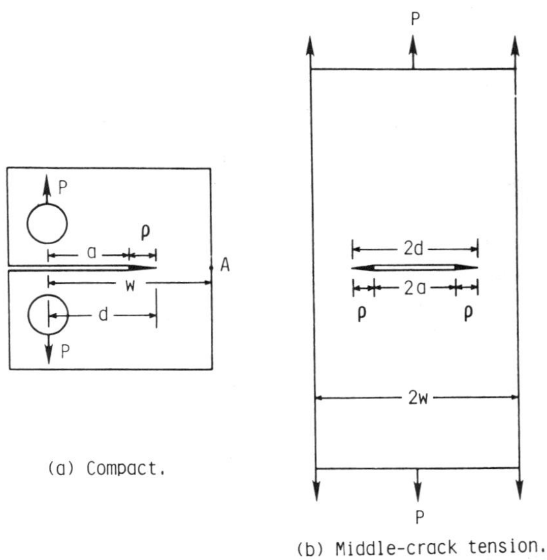

FIG. 3—*Crack configurations analyzed with Dugdale model and used to obtain* V_R *curves.*

where K is the elastic stress-intensity factor at the current load and crack length and the function F_0 is defined in Ref *10*. Next, calculate V_R as

$$V_R = P\phi(a, w, B, \rho) - \sigma_0\psi(a, w, B, \rho) \tag{2}$$

where ϕ and ψ are defined in Ref *3*. The boundary-collocation method was used to obtain these equations [*10*].

For convenience, fit the equation

$$V_R = V_i + C(\Delta a)^n \tag{3}$$

to the V_R against Δa data using a least-squares procedure described in Ref *3*. (Although V_i could be denoted as V_c, the determination of the displacement at initiation of crack extension is difficult and the fitted constant V_i might not necessarily be equal to V_c.)

Middle-Crack Tension Specimen

Using the load-crack-extension (P, Δa) data, calculate the plastic-zone size [*3*] as

$$\rho = a\left\{\frac{2w}{\pi a}\sin^{-1}\left[\sin\left(\frac{\pi a}{2w}\right)\sec\left(\frac{\pi Sf}{2\sigma_0}\right)\right] - 1\right\} \tag{4}$$

where $f = 1 + 0.22(a/w)^2$. Next, calculate V_R as

$$V_R = P\phi(a, w, B, \rho) - \sigma_0\psi(a, w, B, \rho) \tag{5}$$

where ϕ and ψ are, again, defined in Ref *3*. The collocation method was also used to obtain these equations. Next, fit the V_R equation (Eq 3) to the V_R against Δa data.

Analysis of Test Data

Stable crack growth data from the literature [*11,12*] on compact specimens made of several aluminum alloys were analyzed to determine the V_R-curve. The materials were 2024-T351 (LT), 2024-T351 (TL), and 7075-T651 (TL). All specimens were 12.7 mm thick.

McCabe and Ernst [*11*] conducted displacement-control fracture tests on a compact specimen made of 2024-T351 (LT) aluminum alloy. They determined the physical crack extension using either the unloading compliance method or visual observation. For each applied load and crack extension value (P, Δa) measured on the compact specimen, the crack-growth resistance V_R was calculated from Eq 2. The values of V_R and corresponding Δa are plotted in Fig. 4 for both compliance and visual measurements. Solid symbols denote crack extension values beyond maximum load. Up to maximum load the visual and compliance values agreed well, but beyond maximum load the results were in disagreement. The reason for this behavior is not understood. However, the compliance indicated values of crack extension may give a better "average" crack extension through the thickness of the specimen and, consequently, be a better measure of V_R.

In the past, the determination of V_R data had been restricted to crack extension values up to maximum load. Beyond maximum load, the V_R data had not agreed with the "expected" V_R-curve. However, recent evaluations of V_R data beyond maximum load indicate that V_R might be a constant, as shown by the solid circular symbols. The value of this constant would be the value of V_R at maximum load. For convenience, Eq 3 was fitted to the V_R data (open symbols), and the corresponding constants (V_i, C, and n) are given on the figure.

Figures 5 and 6, similarly, show V_R data on various width compact specimens made of 2024-T351 (TL) and 7075-T651 (TL) aluminum alloy [*12*]. The open symbols show data up to maxi-

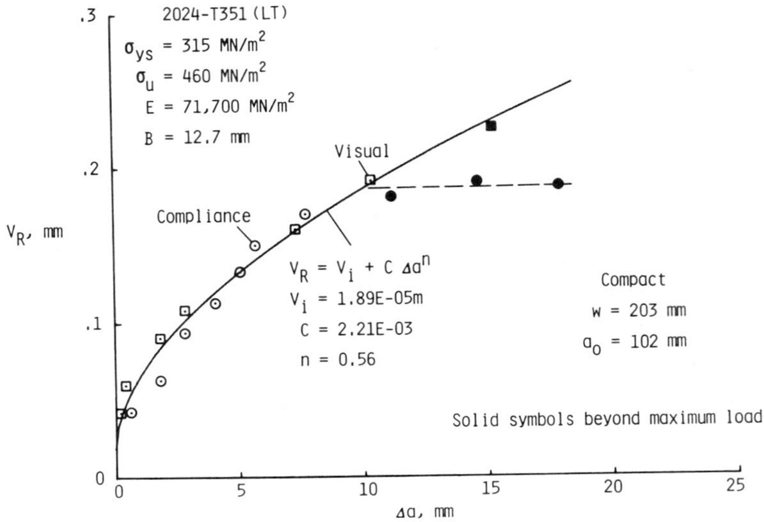

FIG. 4—*Crack-growth resistance curve for 2024-T351 (LT) aluminum alloy compact specimen.*

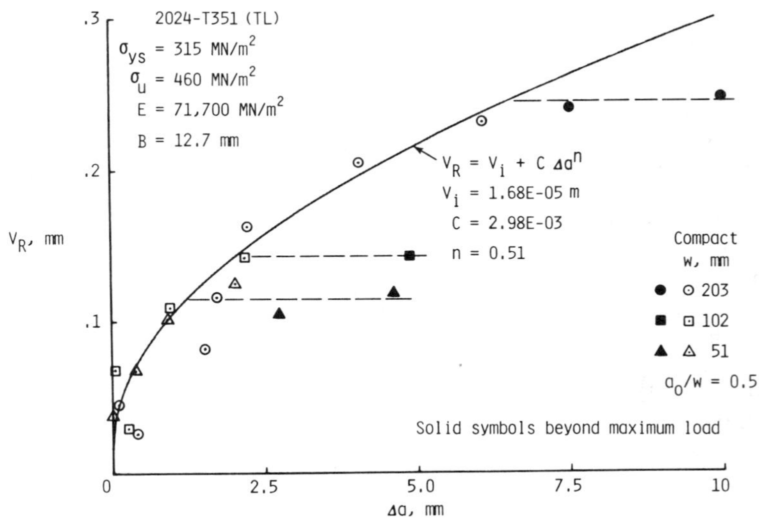

FIG. 5—*Crack-growth resistance curve for 2024-T351 (TL) aluminum alloy compact specimens.*

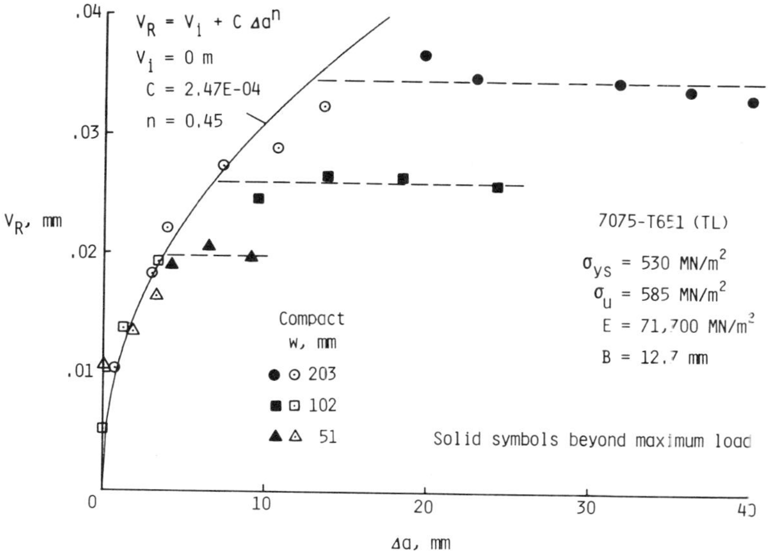

FIG. 6—*Crack-growth resistance curve for 7075-T651 (TL) aluminum alloy compact specimens.*

mum load. Solid symbols show data beyond maximum load and, again, indicate that V_R is nearly constant. But the value of the constant is a function of specimen width. The constant, however, can be approximated as the V_R value at maximum load. Again, the V_R equation (Eq 3) was fitted to the open symbol data and the corresponding constants are given on the respective figures.

Application of V_R Resistance Curve Method

The V_R-curves determined from the compact specimens in the previous section are used herein to predict stable crack growth and maximum load behavior on other crack configurations. To apply the V_R-curve method to an arbitrarily-cracked plate configuration, such as that shown in Fig. 7, a Dugdale or strip-yield model analysis needs to be conducted. Several analytical methods could be used. The boundary-collocation analysis, which was used to analyze the compact and middle-crack tension specimens, would be difficult if yielding occurred at locations other than the crack tip, such as ρ_2 in Fig. 7. To account for yielding at any location along the x-axis, the boundary-integral equation (BIE) and finite-element (FE) methods can be used to analyze the strip-yield model in an arbitrarily-cracked plate. The details of these analyses are given in Appendixes I and II.

In the following, the collocation, BIE, and FE methods are used, together with the material V_R-curve, to predict the load-against-crack extension behavior of compact, middle-crack, single-edge-crack, and three-hole-crack tension specimens. These predictions are compared with experimental data.

Compact Specimen

The V_R-curve determined from the large compact specimen made of 2024-T351 (LT) aluminum alloy material (Fig. 4) was used to calculate the P-Δa behavior on that specimen. Three methods of analysis were used: collocation, BIE, and FE. The comparison among the experimental (open symbols) and calculated results is shown in Fig. 8. Both visual and compliance-

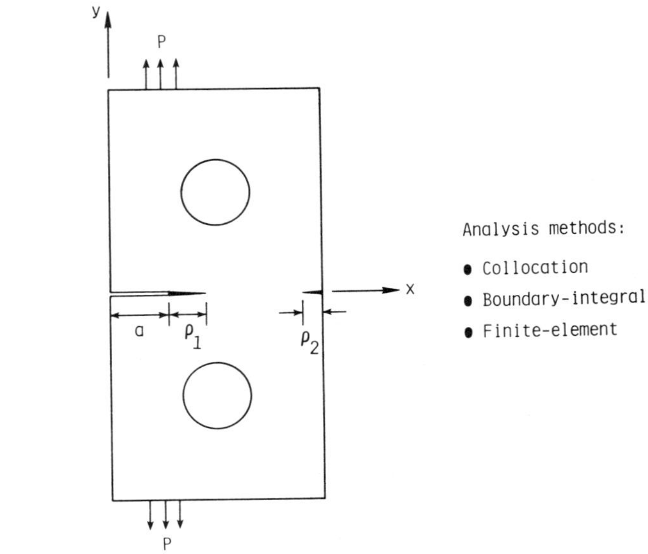

FIG. 7—*Dugdale or strip-yield model analysis of arbitrarily cracked plate.*

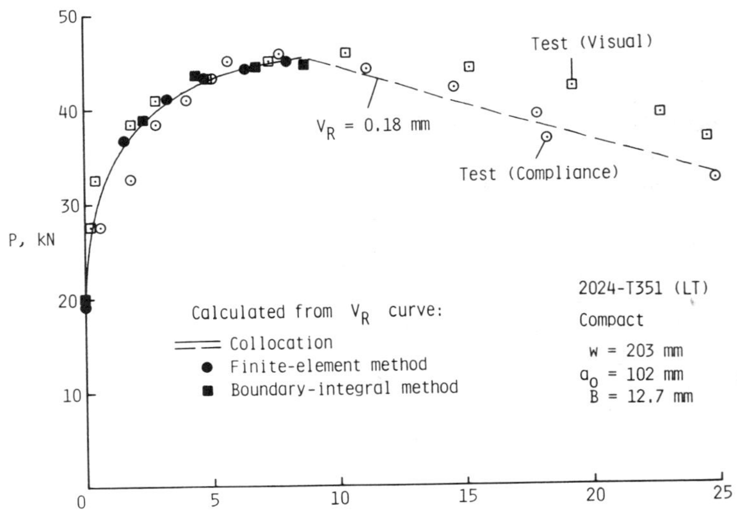

FIG. 8—*Comparison of experimental and calculated crack-growth behavior of 2024-T351 (LT) aluminum alloy compact specimen.*

indicated crack extension results are shown. Beyond maximum load the visual crack extension values were much greater than the compliance values. The BIE and FE analyses (solid symbols) were terminated at maximum load, whereas the collocation method (Eqs 1 and 2) was applied beyond maximum load (dashed curve).

The calculated and experimental results agreed well up to maximum load. Beyond maximum load, the results from the concept of a "constant" V_R agreed well with the compliance indicated values. The value of this constant (0.18 mm) was determined from the value of V_R at maximum load. Because the compliance indicated values of crack extension may give a better average crack extension through the thickness, these results lend credibility to V_R being constant after maximum load. These results also show that the relationship between load and crack extension is nearly linear after maximum load. The results from the BIE and FE analyses agreed well with the results obtained from the collocation method (solid curve).

Middle-Crack Tension Specimen

Similarly, the V_R-curve determined from the large compact specimen was also used to predict the P-Δa behavior of a middle-crack tension specimen made of 2024-T351 (LT) aluminum alloy. A comparison between experimental (open symbols) and predicted results is shown in Fig. 9. Again, the experimental and predicted results using the three analysis methods agreed well up to maximum load. Beyond maximum load, the results from the collocation method using a constant V_R agreed well with the limited data. Again, the constant V_R (0.22 mm) beyond maximum load was determined from the value of V_R at maximum load.

Single-Edge-Crack Tension Specimen

To test the general applicability of the V_R-curve concept and the analytical methods, we made single-edge-crack tension specimens from the 7075-T651 (TL) and 2024-T351 (TL) aluminum alloy materials (B = 12.7 mm) and tested them at the NASA Langley Research Center. The specimens were loaded through single pins at each end of the specimen. A schematic of the

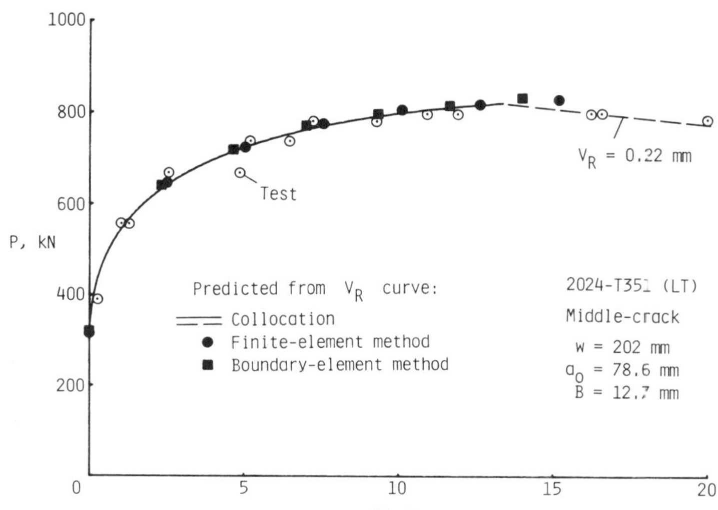

FIG. 9—*Comparison of experimental and predicted crack-growth behavior of 2024-T351 (LT) aluminum alloy middle-crack tension specimen.*

specimen is shown in Fig. 10*a*. The width (w) of the specimens was 86 mm, and the initial crack-length-to-width (a_0/w) ratio was 0.3.

For the 7075-T651 aluminum alloy specimens, the three analytical methods were used with the material V_R-curve (Fig. 6) to predict the failure load. The ratio of predicted-to-experimental failure load (P_{pred}/P_{exp}) is given in Table 1 for each analytical method. P_{exp} is the average experimental failure load on two tests. Because this material fails under small-scale yield conditions, the collocation analysis was preformed using the small-scale yield relations between plastic-zone size, CTOD, and stress-intensity factor (see Ref *3*). All analyses predicted the failure loads within 4% of the experimental failure load.

For the 2024-T351 aluminum alloy specimens, only the BIE and FE methods were used with the V_R-curve (Fig. 5) to predict failure loads. These results are also shown in Table 1. The collocation method was not used here because this material fails under large plastic deformation (plastic-zone size is of the order of the crack length). In the BIE and FE methods, yielding at Point A in Fig. 10 is accounted for in the analysis. These methods were able to predict the failure loads within 5% of the experimental failure load.

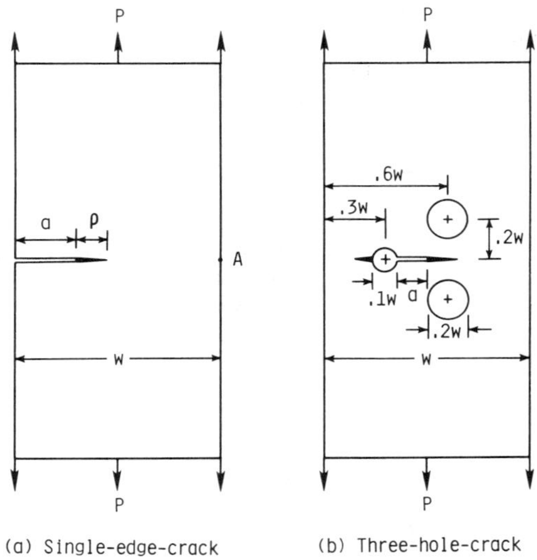

FIG. 10—*Crack configuration used to verify uniqueness of* V_R *resistance curves.*

TABLE 1—*Ratio of predicted-to-experimental failure loads on single-edge-crack tension specimens.*

Material	B, m	w, mm	a_0, mm	P_{exp},[a] kN	P_{pred}/P_{exp} Collocation	BIE[b]	FE[c]
7075-T651(TL)	12.8	86	26.5	78.5	1.01	0.96	0.98
2024-T351(TL)	12.6	86	25.8	134.4	...	1.05	1.01

[a] P_{exp} is average failure load on two tests.
[b] Boundary-integral equation method.
[c] Finite-element method.

Three-Hole-Crack-Tension Specimens

The three-hole-crack specimen was designed to simulate the failure of a cracked stiffened panel [9], see Fig. 10*b*. The stress-intensity factor solution is quite similar to that of a cracked stiffened panel. During fracture, large amounts of stable crack growth (greater than 50 mm for w = 254 mm) can be measured when the initial crack length is less than 25 mm. Several specimens were made from the 7075-T651 (TL) and 2024-T351 (TL) aluminum alloy materials (B = 12.7 mm). Because large amounts of crack extension are expected, the V_R-curves for these materials had to be extrapolated, as shown in Fig. 11. The solid curves show the extent of experimental data obtained from the large compact specimens (see Figs. 5 and 6) and the dashed curves are extrapolated to 50 mm of crack extension. These curves were used with the BIE and FE analyses to predict load-against-crack-length behavior on these specimens.

Figures 12 and 13 show a typical comparison between experimental and predicted crack-growth behavior for 7075-T651 and 2024-T351 materials, respectively. The initial crack length was 25 mm in both specimens. The symbols show the experimental data (solid symbol indicates test failure). Solid and dashed curves show the predicted results from the FE and BIE analyses, respectively. In both cases, the FE and BIE analyses agreed well with each other. For the 7075-T651 material, both methods tended to overpredict loads significantly (20%) when the crack grew beyond the centerline of the large holes. This discrepancy may be due to the extrapolated V_R-curve. For the 2024-T351 material, both methods also tended to overpredict loads (5% to 15%) for a given amount of crack extension. However, in this case, both analyses predicted plastic collapse after a small amount of crack extension (about 5 mm).

Conclusions

A fracture resistance curve method based on crack-tip-opening displacement for a "stationary" crack, the V_R-curve method, was applied to fracture of various crack configurations. The configurations were compact, middle-crack, single-edge-crack, and three-hole-crack tension

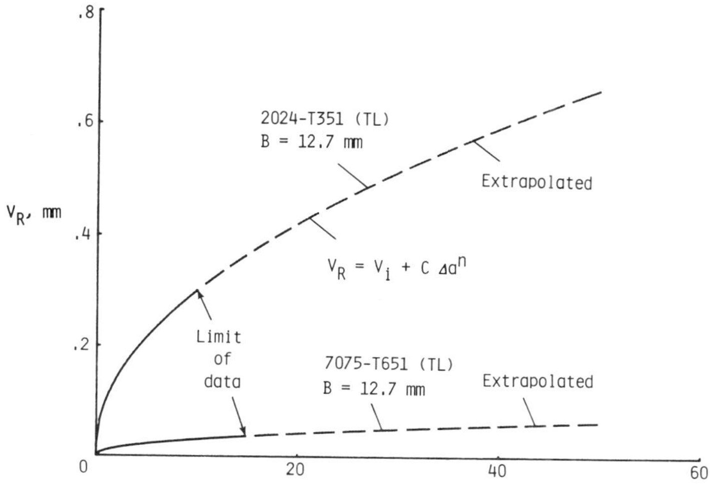

FIG. 11—*Extrapolation of* V_R *resistance curves for analysis of three-hole-crack tension specimens.*

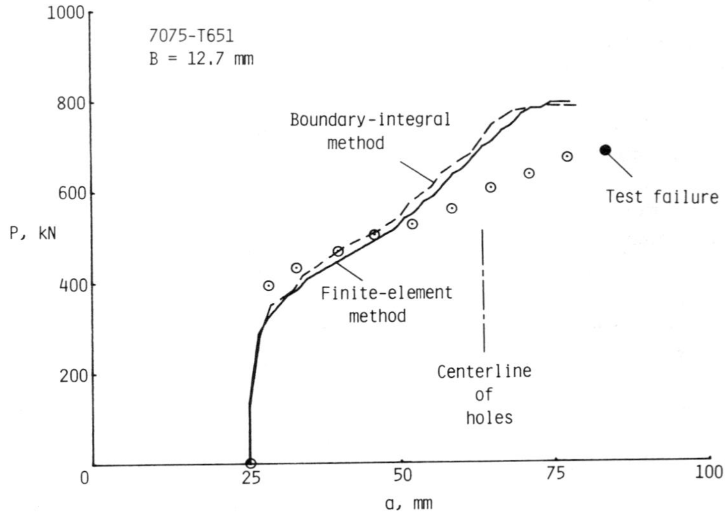

FIG. 12—*Comparison of experimental and predicted crack-growth behavior of 7075-T651 (TL) aluminum alloy three-hole-crack tension specimen.*

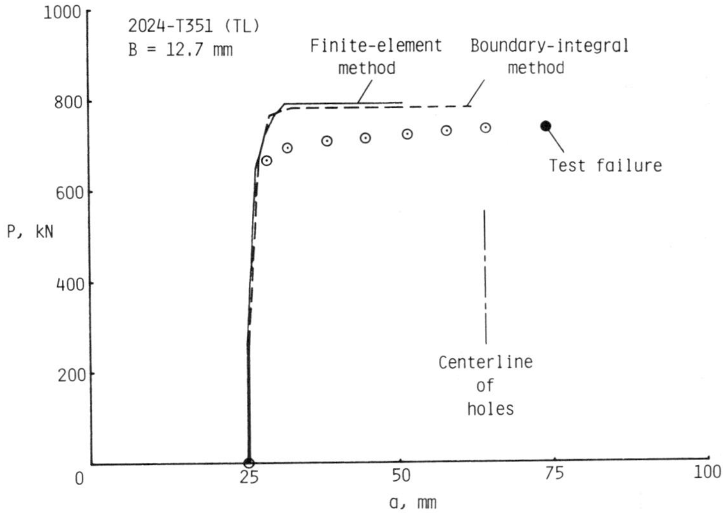

FIG. 13—*Comparison of experimental and predicted crack-growth behavior of 2024-T351 (TL) aluminum alloy three-hole-crack tension specimen.*

specimens with a thickness of 12.7 mm. The materials were 2024-T351 and 7075-T651 aluminum alloys. The results presented in this paper support the following conclusions:

1. The V_R resistance curves were insensitive to crack length, specimen width, and specimen type up to maximum load. After the maximum load was reached in a test, V_R remained nearly constant for the remainder of the test. The value of this constant was dependent upon specimen type, specimen width, and crack length.

2. The strip-yield model analyses using the boundary-integral equation and finite-element methods agreed quite well with the Dugdale model analysis of the compact and middle-crack tension specimens obtained with the collocation method.

3. The V_R resistance curve method can be used with the strip-yield analyses to accurately predict stable crack growth and instability (or residual strength) of cracked metallic materials.

APPENDIX I

Boundary-Integral Equation Analysis of the Strip-Yield Model

The procedure for using the boundary-integral equation (BIE) method to analyze an arbitrarily-cracked elastic plate with the strip-yield model and the V_R-curve concept is described herein.

In the BIE analysis [5], only the boundaries and lines-of-symmetry need to be modelled. A typical boundary model is shown in Fig. 14. On the section of the boundary containing the crack and the strip-yield plastic zone (x-axis), the number of elements are large compared to the number of elements on other boundaries. Typically, on this portion of the boundary, each element was about 1% of the specimen width. Although larger elements could have been used to achieve equivalent results, the small elements were used to model small amounts of crack extension and small plastic-zone sizes.

The possible boundary conditions, of which two had to be specified for each boundary element, were the traction in the x-direction, traction in the y-direction, u-displacement, and v-displacement. Elements along the crack surface were given traction-free boundary conditions to model the crack. At the crack tip, a zero-length element was used to achieve the step change in tractions. The traction boundary conditions for elements along the plastic zone length (ρ) were set equal to the flow stress (σ_0) in the y-direction and zero in the x-direction. The remaining boundary elements were constrained to zero displacement in the y-direction and zero traction in the x-direction.

As the loads were applied to the crack configuration, the normal stresses at each node along the x-axis (end points of each element) were monitored and compared against the flow stress (σ_0). If the stress was greater than σ_0, then the boundary condition on that element was set equal to the flow stress in the y-direction. The traction in the y-direction was made positive or negative depending upon whether the element was in tension or compression, respectively. After all

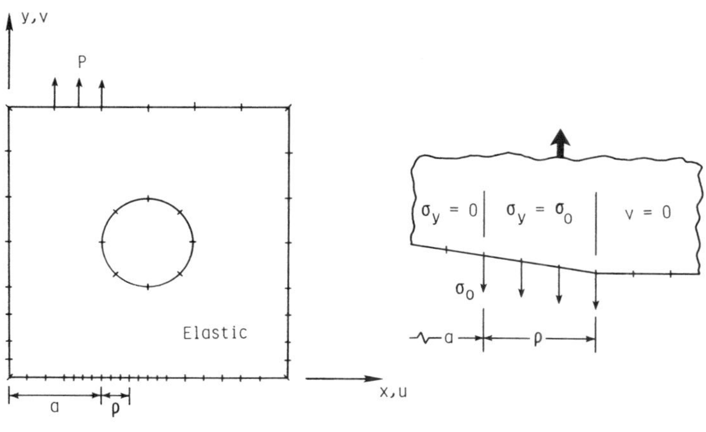

FIG. 14—*Strip-yield analysis with boundary-integral method.*

x-axis nodes along the uncracked ligament were checked, the analysis was again performed with the same applied load with the new boundary conditions. Again, the process was repeated until the plastic-zone size stabilized.

When the plastic zone stabilized, the v-displacement at the crack-tip node was compared against the value of V_R at the current value of crack extension. If the v-displacement was less than V_R, then the applied load was increased and the plastic-zone iteration process was repeated. However, if the v-displacement was greater than V_R, the crack was extended one element size. Crack extension was accomplished by moving the zero-length element at the crack tip to the next node. The analysis was again repeated with the same applied load and the new crack length.

The instability (or maximum) load was determined when the crack extended without any further increase in applied load or when the normal stresses on all nodes along the uncracked ligament (x-axis) became equal to the flow stress (plastic collapse). These calculations were performed under load-control conditions and, consequently, the analysis was terminated at maximum load. The analysis, however, could also be performed under displacement-control.

APPENDIX II

Finite-Element Analysis of Strip-Yield Model

The procedure for using the finite-element (FE) method to analyze an abitrarily cracked elastic plate with the strip-yield model and the V_R curve concept is quite similar to that described for the boundary-integral equation method in Appendix I. The procedure described herein is independent of the type of element (constant or linear strain) used to model the elastic body except that modifications to the procedure would be required if elements with mid-side nodes are used.

In the finite-element analysis [6], the cracked body of interest is modeled with elements, as illustrated in Fig. 15. The elements along the crack surface (a) and strip-yield plastic zone (ρ) were uniform, and their size was about 1% of the specimen width. The small elements were used to model small amounts of crack extension and small plastic-zone sizes.

The boundary conditions along the crack surface are traction-free. Those in the plastic zone have normal forces (R_0) acting in the y-direction to simulate uniform stress (σ_0). Note that the force at the crack tip must be 0.5 R_0 for uniform stress. All nodes are traction-free in the x-direction. The fixed boundary conditions in the y-direction along the uncracked ligament

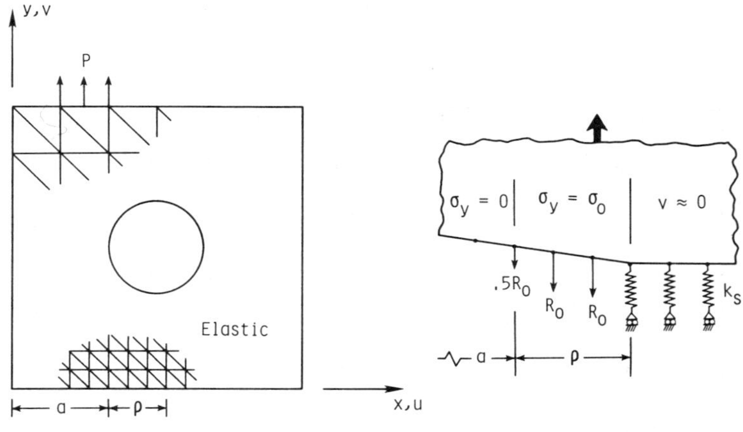

FIG. 15—*Strip-yield analysis with finite-element method.*

were approximated by using "rigid" springs. The spring stiffness (k_s) was selected as a large value (10^6 times the modulus of elasticity of the body material). The use of springs was selected because an efficient technique to modify coefficients of the elastic stiffness matrix during crack extension is available [6].

As the loads were applied to the crack configuration, the spring forces were monitored and compared against the yield force (R_0). If the force was greater than R_0, then the spring was broken (spring stiffness set to zero) and the yield force was applied to the node. The force was positive or negative depending upon whether the spring was in tension or compression, respectively. After all spring forces were checked, the analysis was repeated with the same applied load and the new boundary conditions. The process was repeated until the plastic-zone size stabilized.

When the plastic-zone size stabilized, the v-displacement at the crack-tip node was compared against the value of V_R at the current value of crack extension. If the displacement was less than V_R, then the applied load was increased and the plastic-zone iteration process was repeated. But, if the displacement was greater than V_R, the crack was extended one element size. Crack extension was simulated by removing the force ($0.5\,R_0$) at the crack-tip node and by removing one-half of the yield force at the new crack-tip node. The analysis was again repeated with the same applied load and the new crack length.

The instability (or maximum) load was determined when the crack extended without any further increase in applied load or when the yield force was applied to all nodes along the uncracked ligament (plastic collapse). Again, these calculations were performed under load-control conditions and, consequently, the analysis was terminated at maximum load. The analysis could also be performed under displacement control.

References

[1] *Current Status of Plane Strain Crack Toughness Testing of High Strength Metallic Materials, ASTM STP 410,* American Society for Testing and Materials, Philadelphia, 1966.

[2] *Elastic-Plastic Fracture Mechanics Technology, ASTM STP 896,* American Society for Testing and Materials, Philadelphia, 1985.

[3] Newman, J. C., Jr., "Prediction of Stable Crack Growth and Instability Using the V_R Curve Method," *Elastic-Plastic Fracture Mechanics Technology, ASTM STP 896,* J. C. Newman, Jr., and F. J. Loss, Eds., American Society for Testing and Materials, Philadelphia, 1985, pp. 139–166.

[4] Newman, J. C., Jr., "Evaluation of the Crack-Tip-Opening Displacement (V_R) Resistance Curve Method," *Proceedings,* Crack-Tip-Opening Displacement Workshop, Geesthacht, West Germany, 23–25 April 1985.

[5] Sutton, M. A., Liu, C. H., Chao, W. J., Ranson, W. F., Peters, W. H., and McNeill, S. R., "Two-Dimensional BIE Method in Elasticity with a Consistent Boundary Formulation," accepted for publication in *Engineering Analysis,* 1985.

[6] Newman, J. C., Jr., "Finite-Element Analysis of Fatigue Crack Propagation—Including the Effects of Crack Closure," Ph.D. thesis, Virginia Polytechnic Institute and State University, Blacksburg, VA, May 1974.

[7] Dugdale, D. S., "Yielding of Steel Sheets Containing Slits," *Journal of the Mechanics and Physics of Solids,* Vol. 8, 1960, pp. 100–104.

[8] de Koning, A. U., "A Contribution to the Analysis of Slow Stable Crack Growth," NLR MP 75035 U, National Aerospace Laboratory Report, the Netherlands, 1975.

[9] Newman, J. C., Jr., "Finite-Element Analysis of Initiation, Stable Crack Growth, and Instability Using a Crack-Tip-Opening Displacement Criterion," NASA TM-84564, National Aeronautics and Space Administration, Washington, DC, Oct. 1982.

[10] Newman, J. C., Jr. and Mall, S., "Plastic Zone Size and CTOD Equations for the Compact Specimen," *International Journal of Fracture,* Vol. 24, 1984, pp. R59–R63.

[11] McCabe, D. E. and Ernst, H. A., "A Perspective on R-Curves and Instability Theory," *Fracture Mechanics: Fourteenth Symposium, ASTM STP 791,* Vol. I: Theory and Analysis, J. C. Lewis and G. Sines, Eds., American Society for Testing and Materials, Philadelphia, 1983, pp. I561–I584.

[12] McCabe, D. E., "Data Development for ASTM E24.06.02 Round Robin Program on Instability Prediction," NASA CR-159103, National Aeronautics and Space Administration, Washington, DC, Aug. 1979.

Robert A. Riddle,[1] *Ronald D. Streit,*[1] *and Iain Finnie*[2]

The Failure of Aluminum Compact Shear Specimens Under Mixed-Mode Loading

REFERENCE: Riddle, R. A., Streit, R. D., and Finnie, I., **"The Failure of Aluminum Compact Shear Specimens Under Mixed-Mode Loading,"** *Fracture Mechanics: Eighteenth Symposium, ASTM STP 945,* D. T. Read and R. P. Reed, Eds., American Society for Testing and Materials, Philadelphia, 1988, pp. 118–133.

ABSTRACT: The failure behavior of 7075-T6 aluminum under mixed-mode loading was investigated using the compact shear specimen. This specimen is capable of providing a range of Mode I and Mode II loading conditions at the crack tip, depending upon the boundary conditions provided by the loading fixtures.

Finite-element analyses of the specimen were made which included the effects of loading fixture constraints and the material's elastic-plastic properties. The symmetric and antisymmetric components of the stress and deformation fields were used to calculate the applied J-integral values, J_I and J_{II}, corresponding to the Mode I and Mode II loading at the crack tip. Methods to obtain the critical values of the J-integral at crack initiation from the load displacement curves of the specimen are discussed.

The failure behavior of the material varied greatly depending upon the relative amounts of Mode I and Mode II loading at the crack tip. For near Mode I loading, the specimen behavior at failure loads was nearly elastic, and the failure processes appeared to be typical of those for Mode I plane-strain fracture toughness testing. A fairly significant amount of shear loading seemed to have little effect on the failure load. For near Mode II loading, the amount of plastic deformation is much more significant, with the ligament ahead of the crack tip nearly fully plastic at the failure load. The failure processes in the material, as evidenced by the fracture surface morphology, are different as compared to Mode I type failure. There is evidence of slow stable crack growth occurring at loads near the departure from linearity of the specimen load-displacement curve.

For a mixed-mode case in which Mode II dominates with a smaller amount of positive Mode I loading, the critical J-integral value is substantially larger than the value of the critical J value for Mode I loading. When it is transformed into the corresponding critical stress intensity factor values, K_{IIc} and K_{Ic}, it is found that K_{IIc} is approximately twice K_{Ic}. However, there is some question as to whether the specimen failure is controlled by a critical stress intensity value or by the applied shear stress.

The use of the J-integral as a failure criterion and as a means to predict the angle of crack growth for elastic-plastic failure is discussed.

KEY WORDS: mixed-mode loading, Mode II failure, Mode II fracture toughness of 7075 T-6 aluminum, J-integral, elastic-plastic fracture mechanics, fracture test methods

The failure behavior of 7075-T6 aluminum under combined Mode I and Mode II loading was investigated using the compact shear specimen. A limited number of additional mixed-mode fracture tests were performed on an epoxy resin. The objective of the investigation was to determine the effects of mixed Mode I and Mode II loading on the failure behavior of these materials, considering such effects as the relationship between specimen geometry and failure loads, data reduction methods, crack growth versus limit load shear failure, and the direction of crack growth for mixed mode loading.

[1]Engineer and section leader, respectively, Lawrence Livermore National Laboratory, Livermore, CA 94550.

[2]Professor, University of California at Berkeley, Berkeley, CA 94720.

Gaining a more thorough understanding of material failure behavior under mixed Mode I and Mode II loading is important in quantifying the effects of material properties on such metals manufacturing operations as hole punching, precision machining, and grinding. An understanding of mixed-mode failure behavior is also essential in predicting failure loads for structures in which shear loads predominate. The lack of standard specimens and data reduction techniques has hampered progress in obtaining mixed mode failure properties for engineering materials.

Analysis of Compact Shear Specimen

The compact shear specimen was chosen as the specimen for this investigation because of its capability of providing a range of mixed Mode I and Mode II loading conditions at the crack tip, depending upon the boundary conditions provided by the loading fixtures, and because of the fairly modest material requirements for each specimen and the required load capacity of the testing machine. Discussion of the compact shear specimen and of the effects of the loading fixtures on the conditions at the crack tip is given in an earlier paper [*1*]. Figure 1 shows the specimen with the associated coordinate system and dimensional definitions.

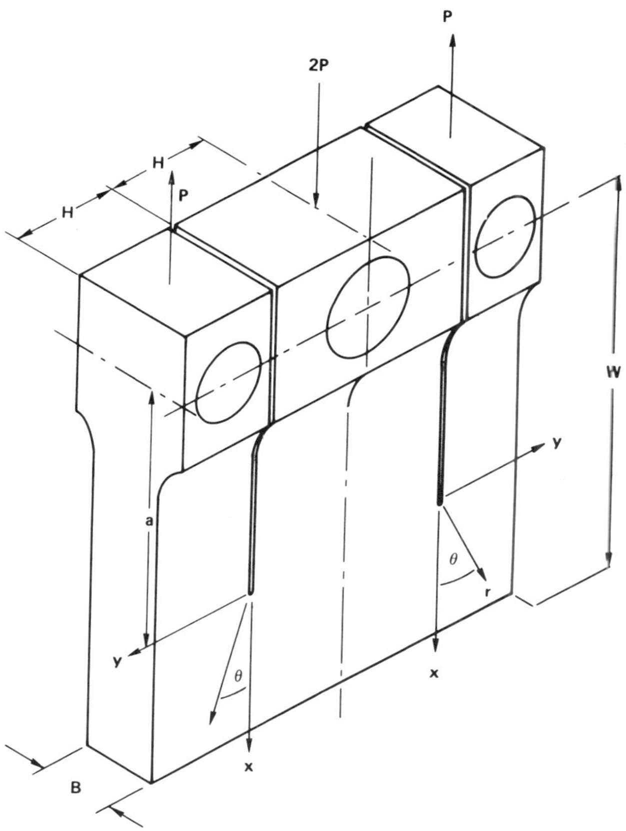

FIG. 1—*Compact shear specimen with associated coordinate system and dimensional definitions.*

Ten compact tension specimens were machined from a 2.54-cm (1-in.)-thick plate of 7075-T6 aluminum. This material has directional properties, and the Mode I fracture toughness for crack growth in the longitudinal direction (T-L) is less than that for crack growth in the transverse direction (L-T). However, the elastic modulus, as determined tensile tests, was found to be independent of rolling direction to within a few percent, as were the stress-strain curves beyond the elastic limit. Also, because the Mode I fracture toughness (K_{Ic}) values between the T-L and L-T directions were different by less than 10%, the average K_{Ic} value is used. Table 1 shows the results of the tensile and Mode I fracture toughness tests on this material. These material properties were used in the constitutive relations for the material in the finite element analysis of the specimens. The material model included the elastic modulus, Poisson's ratio, assumed to be 0.33, and values of effective plastic strain versus effective stress, as derived from the experimental data. Isotropic hardening in the material is also assumed, because with the single load cycle to failure being considered, the amount of unloading or rotation of principal stress directions should be small. For strains beyond the effective plastic strain to failure, calculated to be 0.1498, the material is assumed to harden at a rate given by the tangent modulus as defined by the last two points on the effective plastic strain versus effective stress curve.

The mesh for finite-element analysis of the specimen is seen in Fig. 2, with the boundary conditions shown schematically. Plain-strain quadrilateral elements were used. The finite-element program NIKE2D [*2*] was used in these analyses. Finite-element analysis, or some equally versatile numerical method of analysis for the specimen, was necessary because of the complex effect of the boundary conditions and the material plasticity effects on the specimen's

TABLE 1—*Summary of tension and fracture toughness test data for 7075-T6 aluminum.*

Elastic Modulus	Yield Strength	Ultimate Strength	Reduction in Area, %	Mode I Fracture Toughness
70.67 GPa (10.25×10^6 psi)	543.3 MPa (78 800 psi)	600 MPa (87 000 psi)	14.7	29.78 MPa · in.$^{1/2}$ (27.1 ksi · in.$^{1/2}$)

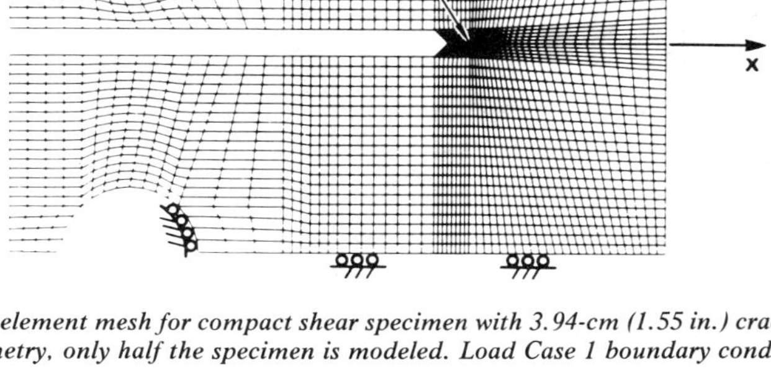

FIG. 2—*Finite-element mesh for compact shear specimen with 3.94-cm (1.55 in.) crack length. Because of specimen symmetry, only half the specimen is modeled. Load Case 1 boundary conditions.*

failure behavior. In conjunction with the finite-element analysis, a J-integral post-processor was written to characterize the crack tip loading in terms of this elastic-plastic fracture mechanics parameter. The J-integral post-processor takes the stresses, strains, and displacements along a prescribed path as calculated in the finite element analysis, and calculates a unique J-integral value which describes the energy available for crack growth in the material.

The J-integral, first applied by Rice [3] to solve problems in elastic-plastic fracture mechanics, is defined as the vector quantity

$$J_k = \int_s \left(W n_k - T_i \frac{\partial u_i}{\partial x_k} \right) ds \quad (1)$$

where the subscript k indicates the direction of potential energy release, and the subscript i takes the values 1, 2 indicating coordinate direction in the plane (using the convention of summation over repeated indices).

The J-integral path is defined as path s with increment ds, and n as the outward normal vector to curve s. W represents the quantity

$$W = \int_0^{\epsilon} \sigma_{ij} d\epsilon_{ij} \quad (2)$$

which is the strain energy density when elastic material behavior is assumed, where

σ_{ij} = stress tensor,
ϵ_{ij} = strain tensor,
T = traction vector with components $T_i = \sigma_{ij} n_j$,
u_i = displacement vector, and
x_i = cartesian coordinate vector.

The J-integral provides a unique characterization of the stress and strain singularities at the crack tip because the value of the integral is independent of the path location around the crack, provided that the path begins and ends on a traction-free surface at the crack plane. The theoretical prediction of path independence for the value of the J-integral has been substantiated in numerical calculations of J-integral values, where the J-integral values were independent of the path to within the error inherent in the approximate nature of finite-element calculations, provided that the path was taken through elements at least a few elements away from the crack tip and not right at a free surface [4]. This path independence was taken advantage of in the calculations for the J-integral values at failure loads in the compact shear specimens in that the paths were always taken through regions of elastically loaded material surrounding the plastically deformed material at the crack tip. Here W was the strain energy density and independent of the load history of the element and, therefore, much more easily calculated. Because the J-integral is path-independent and because at the required loads there was always an essentially elastic region in the specimen, taking the path through the elastic region was always possible with no sacrifice in accuracy in these instances.

The calculation of the relative amounts of Mode I and Mode II loading was accomplished using a scheme developed by Ishikawa, Kitagawa, and Okamura [5]. Using the before-described J-integral formulation, the symmetric stress, strain, and displacement quantities, which correspond to Mode I loading are used to calculate a symmetric J-integral quantity, J_{I}. The corresponding integral J_{II} corresponding to Mode II loading is calculated in a similar manner using the antisymmetric components of stress, strain, and displacements. Ishikawa et al. showed that the quantities J_{I} and J_{II} are path-independent and that

$$J_{\mathrm{I}} + J_{\mathrm{II}} = J \quad (3)$$

These theoretical predictions were confirmed by numerical results. In the finite-element analysis of the compact shear specimen, it became evident that the boundary conditions applied at the loading holes greatly affected the relative amounts of Mode I and Mode II loading at the crack tip. Four different load cases were identified with the crack tip loading varying for these from near Mode I to near Mode II.

Case 1 loading is achieved with the outer tangs in tension, the center tang in compression, and the outer tangs constrained to have zero deflection in the y-direction at the center node of the loading.

In Case 2, the loads are again applied in the same directions as those in Case 1, so that the outer tangs are in tension and the center tang is in compression, but the constraint of zero y deflection in the outer tangs is removed, allowing the tangs to rotate (that is, deflect normal to the loading direction).

In Case 3, the load directions are reversed, so that the outer tangs are in compression and the center tang is in tension. In this case, the nodes at the center in the outer tangs are constrained to have zero y deflection.

In Case 4, the load directions are the same as Case 3, but the displacement restraint at the center of the outer loading hole is removed, and the outer tangs may rotate.

Figure 3 depicts the loading cases as defined by the boundary constraints. Figure 4 the relative amount of Mode II loading for each case. Figures 5 and 6 show the loading fixtures used to get the appropriate boundary constraints for Case 3 and Case 4 loading, where the outer tangs of the compact shear specimen are in compression. The loading fixtures for Cases 1 and 2 are

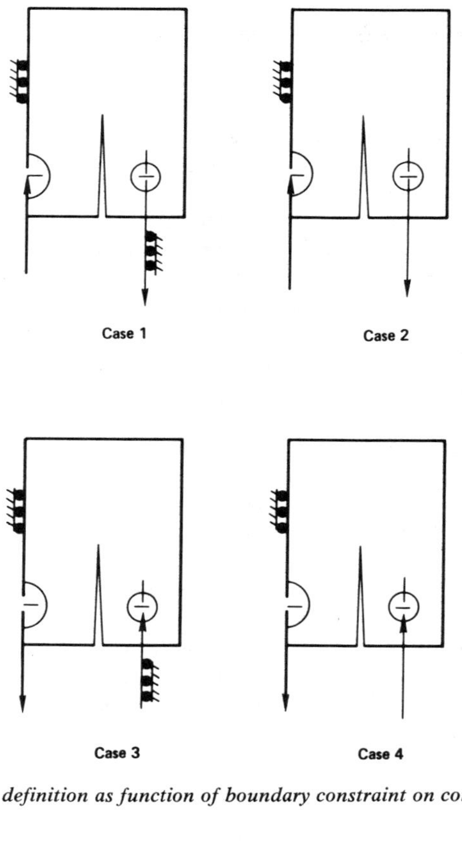

FIG. 3—*Loading case definition as function of boundary constraint on compact shear specimen.*

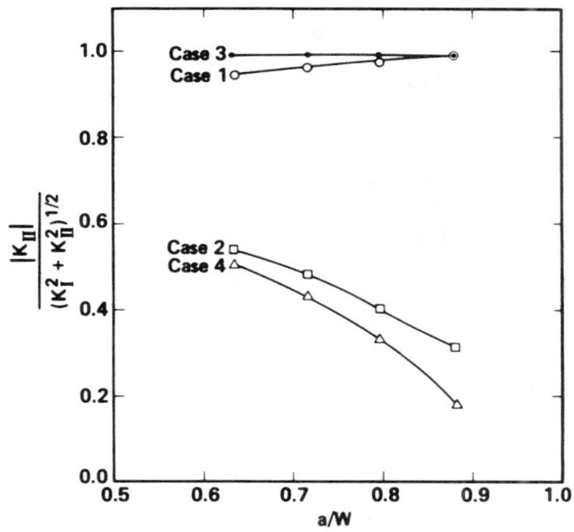

FIG. 4—*Relative amount of Mode II deformation as function of crack length and loading case,* H/W = *0.408.*

very similar, but the relative position of the specimen is changed so that the outer tangs of the specimen are now in tension. Figures 7 and 8 show computer-generated shear stress contours, demonstrating the predominantly Mode II crack tip stress field for Case 3 loading (Fig. 7) and mixed-mode loading for Case 4 (Fig. 8).

Testing the Compact Shear Specimens

In all the specimens tested, the tang width H was 2.54 cm (1.0 in.), and W, the length between the center line of the loading hole and the back edge of the specimen, was 6.22 cm (2.45 in.). Crack lengths a of 3.94 cm (1.55 in.) and 4.45 cm (1.75 in.) were tested, as shown in Table 2. The thickness in the regions of the loading holes was increased to prevent plastic deformation there and to increase the stiffness of the tang in the transverse direction. The thickness of the specimens in the crack growth region was 1.27 cm (0.5 in.).

The required thickness in the crack tip region, to ensure plane-strain deformation under Mode II loading, should be less than that required for Mode I loading. Because Mode II loading produces antisymmetric stress and displacement fields at the crack tip, the out-of-plane stress has a minimum at $\theta = 0$ and is equal to zero for both plane-strain and plane-stress deformation. The thickness requirement of mixed mode loading in the compact shear specimen was conservatively based upon the plane-strain deformation thickness criteria for Mode I loading.

In order to eliminate any effects of notch tip radius on the failure behavior of the specimen, a special fixture was made which allowed fatigue loading of the compact shear specimen in Mode I, as shown in Fig. 8 of Ref *1*. The appropriate loading angle was obtained by finite element analysis. This angle is particular to our specimen geometry and is a function of crack length. Then the specimens were cyclically loaded to produce a fatigue precrack of 0.25 cm (0.1 in.) in the plane of the notch. Slight deviations in the lengths of the machine notches, the amount of fatigue precracking, and the angle of the fatigue precrack from the desired geometry are noted. In regard to the angle of the fatigue crack, it was observed in the post-failure examination of the compact shear specimens that a very small shear lip had formed at the surface during fatigue precracking, distorting the angle of the fatigue crack at the surface. The fatigue crack angle in the interior of the material was closer to zero degrees than indicated in Table 2. It

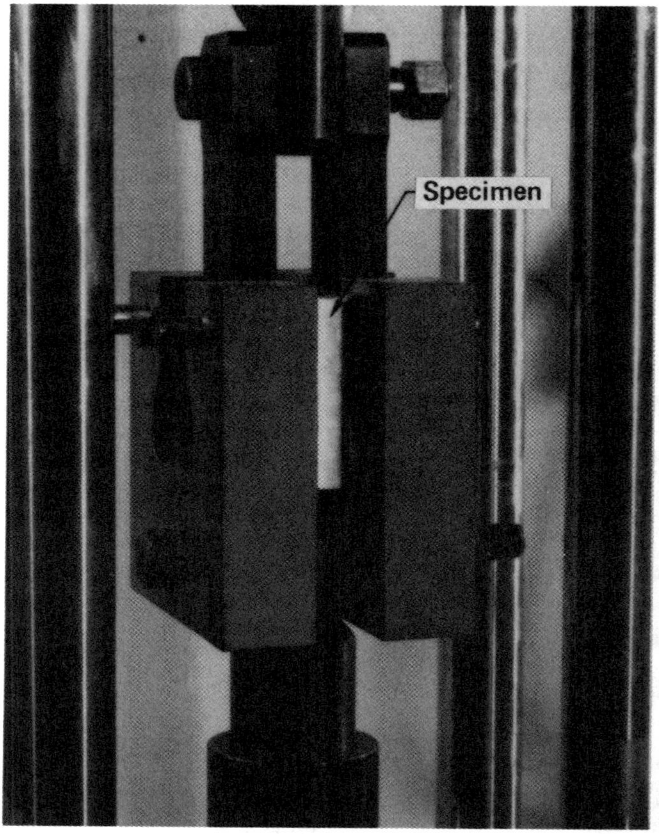

FIG. 5—*Case 3 loading conditions, side view. Rigid load fixture reduces the amount of Mode I loading at crack tip.*

was also observed that in cases in which one crack length was measurably longer than the other in a specimen, the longer crack failed first.

The load of the compact shear specimen from the center and outer tangs creates a bending moment at the crack tip that changes in sign with the direction of loading. This is in contrast to the ASTM Test Method for Plane-Strain Fracture Toughness of Metallic Materials (E 399-83) fracture specimens in which the loading is colinear. The moment in the compact shear specimen has a tendency to open the crack (Cases 3 and 4) when the outer tangs are in compression, whereas with the applied load on the outer tang in tension, the moment will cause the crack faces to close (Cases 1 and 2).

Specimen loading fixtures and a displacement measuring system were devised so that for each test a record was made of load versus displacement for displacements in the loading direction. Displacements in the transverse direction between the tangs were also measured. In considering Mode II fracture test procedures, it was determined that the relationship between the change in compliance (c) and the increase in crack length (a) was different in the compact shear specimen as compared to that for the ASTM E 399 Mode I fracture toughness specimens.

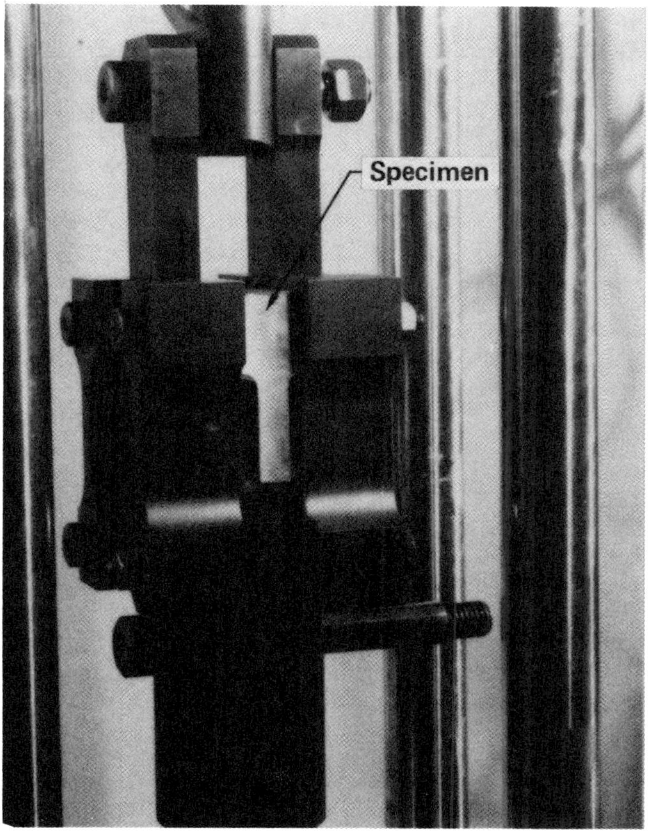

FIG. 6—*Case 4 loading conditions, side view. Pinned load fixtures allow outer tangs to rotate, increasing amount of Mode I loading at crack tip.*

For the compact shear specimens, in the range $0.6 < a/W < 0.8$, and load Cases 1 and 3, the elastic finite-element solutions and J-integral post-processor calculations show that

$$\frac{\Delta c}{c} = 0.6\left(\frac{\Delta a}{a}\right) \tag{4}$$

In contrast, computations show for the ASTM E 399 compact tension that the constant is 2.5 in the preceding equation. Therefore, the change in compliance with respect to increasing crack length in the compact shear specimens is about one fourth as sensitive as that same relationship for Mode I compact tension fracture toughness specimen. If the load at a change of compliance of 5% were used to calculate K_{IIc}, similar to that done in the ASTM E 399 Mode I fracture toughness test procedure and as has been suggested for the compact shear specimen data reduction [*6*], K_{IIc} at crack initiation would be overestimated because either a change in crack length of more than 8% or substantial plastic deformation would be allowed. However, in the testing of the 7075-T6 aluminum compact shear specimens, the stress threshold for Mode II crack growth

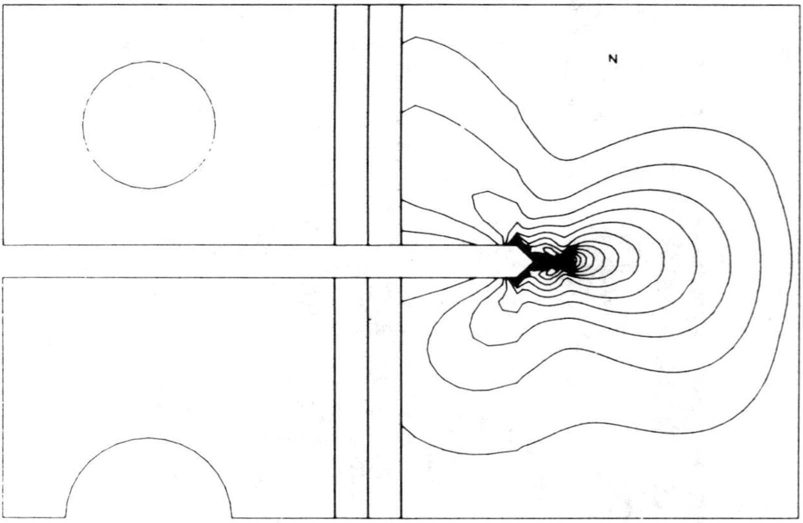

FIG. 7—*Computer-generated shear stress contours for Case 3 loading on the compact shear specimen. Crack length is 3.94 cm (1.55 in.).*

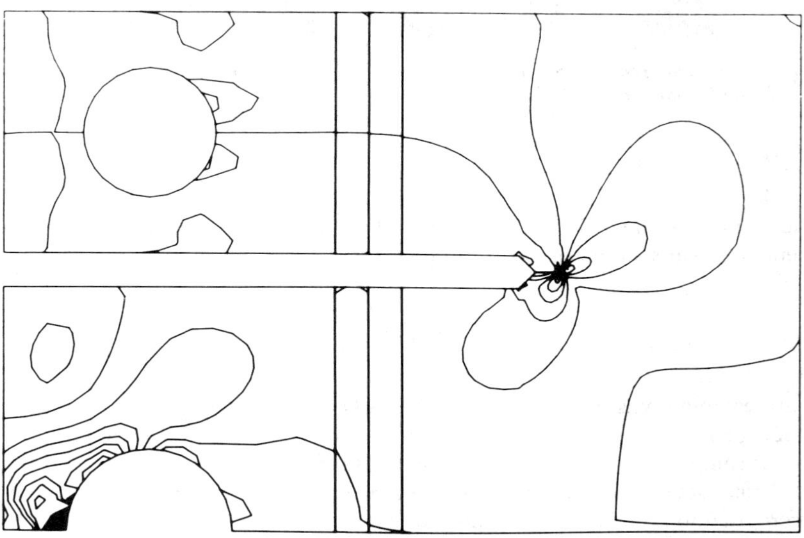

FIG. 8—*Computer-generated shear stress contours for Case 4 loading on compact shear specimen. Crack length is 3.94 cm (1.55 in.).*

TABLE 2—*Configuration of the aluminum compact shear specimens prior to final testing.*

Specimen Identification	Material Orientation	Crack Length, cm (in.) (1)		Crack Length, cm (in.) (2)		Fatigue Angle, deg (1)	Fatigue Angle, deg (2)	Load Case
1A	T-L	4.445	(1.750)	4.455	(1.754)	−3.9	−2.3	3
1B	T-L	4.493	(1.769)	4.313	(1.698)	−5.0	1.7	4
1X	L-T	4.399	(1.732)	4.392	(1.729)	1.9	−4.25	3
1Y	L-T	4.392	(1.729)	4.397	(1.731)	−3.4	−0.3	4
1D	T-L	3.871	(1.524)	3.889	(1.531)	1.65	0.65	2
1E	T-L	3.942	(1.552)	3.927	(1.546)	−1.9	−3.19	1
1G	T-L	3.937	(1.550)	3.937	(1.550)	−3.0	3.1	3
1T	L-T	3.924	(1.545)	3.937	(1.550)	2.85	1.65	2
1V	L-T	3.950	(1.555)	3.975	(1.565)	0.4	1.95	1
1W	L-T	3.912	(1.54)	3.912	(1.54)	4.05	−1.35	3

was so close to the stress for gross plastic yielding, in load Cases 1 and 3, that either the load at departure from linearity on the load-displacement record or the maximum load could be logical choices as the critical load for crack growth.

In load Cases 1 and 2, the final fracture event did not occur simultaneously in both tangs. From observation of the tests and from examination of the broken pieces of the specimens, it was evident that first the side with the longer crack failed in shear and then the second tang rotated and was broken in a more nearly Mode I type of fracture. The crack surfaces of the material which failed in predominately Mode II loading were immediately distinguishable from the crack surfaces of the material failed in predominately Mode I loading. For Mode II loading, the appearance of crack surfaces were smooth and nearly featureless, indicating a sliding Mode at fracture. In contrast, the Mode I fracture surface has a rough granular appearance in the 7075-T6 aluminum.

In Case 2 loading, there is a large negative K_I which acts to close the gap between the pre-fatigue crack surfaces. For both specimens tested in Case 2, the original shear failure began at the root of the machined notch, and not at the tip of the fatigue crack. In Case 2 loading, the large compressive stresses at the fatigue crack surfaces promote crack tip plasticity and inhibit crack growth. Table 3 shows, for Case 2 loading, that at the maximum load in the specimen, the average stress in the specimen ligament ahead of the crack tip was larger than the flow stress of the material by a small percentage where the flow stress is defined as one-half the yield plus ultimate strength of the material.

For Case 1 loading, where the applied loading is largely Mode II, with only a small negative K_I (as shown in Fig. 3), failure in the specimen originated at the tip of the fatigue precrack, and at stresses less than those for Case 2 loading. From Table 3, it is seen that Specimens 1V and 1E failed at applied shear stress levels of 97% and 99% of the flow stress of the material. The most simple criterion which would describe the failure of these specimens is that, for failure, the applied shear stress equals the flow stress of the material. However, there is evidence that the actual failure process in the specimen is more complicated: that there is either a loading range of slow stable crack growth followed by catastrophic failure which may be interpreted either as rapid crack growth or a shear instability due to gross plastic yielding in the region ahead of the crack tip. This evidence of slow stable crack growth prior to catastrophic failure comes from observations of the failure surfaces on the side of the compact shear specimen which failed in largely Mode I loading after the first side had failed in Mode II loading. Examining the fracture surfaces of the specimen which experienced this sympathetic failure showed regions of fatigue crack growth, Mode II crack growth, and Mode I crack growth, which were readily distinguish-

TABLE 3—*Aluminum compact shear specimen maximum load values and comparison of shear stress at failure with the predicted flow shear stress.*

Crack Length	Load Case	Specimen Identification	Maximum Load, kN (kips)	$\tau_{failure}$ MPa (ksi)	$\frac{\tau_{failure}}{\tau_{flow}}$	Material Orientation
a = 3.94 cm	1	IV	160.0 (36.0)	276.0 (40.0)	0.97	L-T
(1.55 in.)	1	IE	165.0 (37.0)	283.0 (41.1)	0.99	T-L
	2	IT	169.0 (38.0)	291.0 (42.2)	1.02	L-T
	2	ID	174.0 (39.0)	299.0 (43.3)	1.05	T-L
	3	IW	149.0 (33.4)	256.0 (37.1)	0.89	L-T
	3	IG	145.0 (32.5)	249.0 (36.1)	0.87	T-L
a = 4.45 cm	3	IX	116.0 (26.1)	257.0 (37.3)	0.90	L-T
(1.75 in.)	3	IA	111.0 (25.0)	246.0 (35.7)	0.86	T-L
	4	IY	33.4 (7.50)	73.8 (10.7)	0.26	L-T
	4	IB	30.8 (6.9)	68.3 (9.9)	0.24	T-L

able by the relative reflectivity of the surfaces. The amount of shear crack growth in Specimen 1V on the side of sympathetic failure was 0.05 cm (0.02 in.), while the corresponding shear crack growth in Specimen 1E was 0.150 cm (0.06 in.).

These increments of crack growth represent about 2.5% of the original crack length; based upon Eq 4 this would suggest a change in compliance of between 1 and 2%. Since larger changes in compliance were predicted in the elastic-plastic computer analysis of this load case, with no crack growth and observed in the load-displacement record of the specimen testing, it is reasonable to assume that stable shear crack growth occurs in conjunction with significant plastic deformation at the crack tip.

In Case 3 loading, only one side failed in shear in the original loading of the specimen. The failure surface of the side which failed in Mode II was smooth and looked abraded, even at higher magnification. The remaining part of the specimen was loaded in approximate Mode I loading until failure occurred between the remaining two holes. The amount of sympathetic stable crack growth was measured and is shown in Table 4 for each of the four specimens tested in load Case 3. In addition, Table 3 shows the load in the specimen at departure from linearity in the load-displacement record. An interesting feature of the sympathetic shear crack growth was that it was a minimum at the interior of the specimen and a maximum at the outside edges, in contrast to the fatigue precrack.

Analysis of the compact shear specimen based on Case 3 loading with the two crack lengths, and assuming that crack growth initiates at the load at departure from linearity and that final failure occurs when the applied shear stress is 1.05 times the flow stress in the material, predicts the amount of sympathetic shear crack growth of 0.2 cm (0.08 in.) for the compact shear specimens with an initial crack length of 3.94 cm (1.55 in.), and 0.15 cm (0.06 in.) for the specimens with an initial crack length of 4.45 cm (1.75 in.). These results support the conclusion that failure initiates at a critical value of J_{II} (or equivalently K_{II}) with stable shear crack growth and that the failure process becomes unstable when the value of the applied shear stress across the remaining ligament reaches a critical value for Case 3 type Mode II loading.

The results of loading in Case 4 were the most straightforward to interpret because the load-displacement record was linear to maximum load, where rapid crack growth occurred, and the part failed.

For purposes of comparison with other results reported in the literature, values of K_I and K_{II} were calculated from the J-integral values at maximum load and the loads at departure from linearity. These results are seen in Table 5 and Fig. 9.

The K_{IIc} value calculated from the load at departure from linearity represents a lower bound on the actual K_{IIc} value, since in the linear range of the load-displacement record no crack growth occurs. On the other hand, since there is experimental evidence to suggest that there was crack growth prior to maximum load, the K_{IIc} value based on maximum load is an upper bound on the real K_{IIc}. From the load Case 3 tests, the value of K_{IIc} is then known within certain limits.

TABLE 4—*Measured amounts of sympathetic shear crack growth and loads at departure from linearity—load Case 3.*

Crack Length	Specimen Identification	Δa, cm (in.)	Pd, kN (kips)
a = 3.94 cm (1.55 in.)	IW	0.23 (0.09)	138 (31.0)
	IG	0.13 (0.05)	120 (27.0)
a = 4.45 cm (1.75 in.)	IX	0.08 (0.03)	100 (22.5)
	IA	0.13 (0.05)	105 (23.5)

TABLE 5—*Values of K_I and K_{II} at onset of crack growth based on maximum load, and load at departure from linearity.*

Crack Length	Specimen Identification	Case Number	Material Orientation	Maximum Load, MPa $\sqrt{m}$ (ksi $\sqrt{in.}$)		Departure from Linearity, MPa $\sqrt{m}$ (ksi $\sqrt{in.}$)	
				K_I	K_{II}	K_I	K_{II}
a = 3.94 cm	1V	1	L-T	−20.4 (−18.6)	−73.5 (−66.8)	. . .	. . .
(1.55 in.)	1E	1	T-L	−21.4 (−19.5)	−76.7 (−69.8)	. . .	. . .
	1T	2	L-T	−74.0 (−67.3)[a]	−181.0 (−165.0)[a]	. . .	. . .
	1D	2	T-L	−75.3 (−68.5)[a]	−204.0 (−186.0)[a]	. . .	. . .
	1W	3	L-T	8.4 (7.6)	70.2 (63.9)	7.8 (7.1)	63.6 (57.9)
	1G	3	T-L	8.2 (7.5)	69.0 (62.8)	6.8 (6.2)	53.6 (48.8)
a = 4.45 cm	1X	3	L-T	5.4 (4.9)	62.5 (56.9)	4.8 (4.4)	55.3 (50.3)
(1.75 in.)	1A	3	T-L	5.1 (4.6)	59.0 (53.7)	4.6 (4.2)	51.7 (47.1)
	1Y	4	L-T	30.3 (27.6)	14.5 (13.2)	. . .	. . .
	1B	4	T-L	28.0 (25.5)	13.4 (12.2)	. . .	. . .

[a]These values are for comparison only, and are not critical elastic stress intensity factors because of significant plastic deformation in the specimen prior to failure.

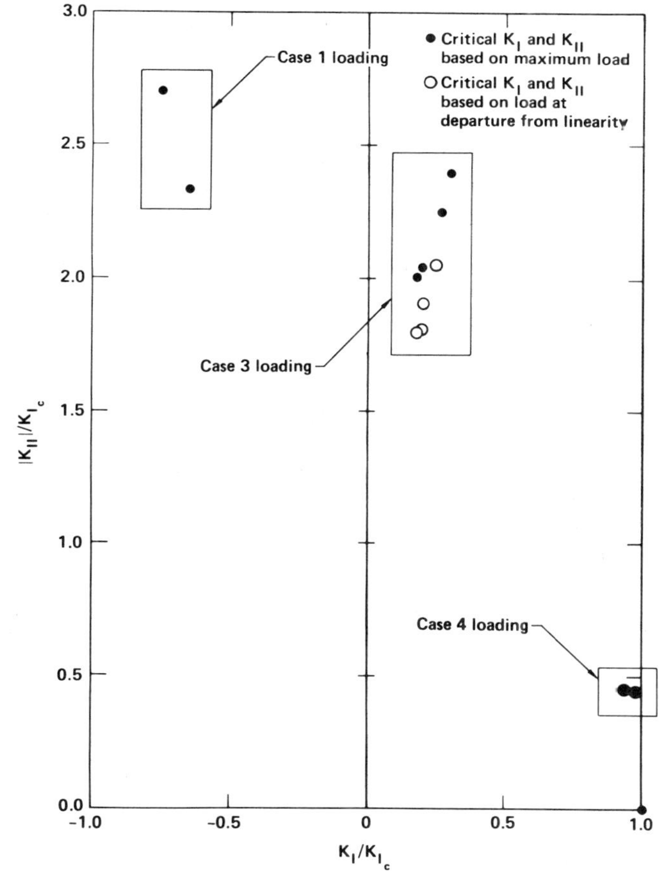

FIG. 9—*Estimates of critical stress intensity factors for mixed-mode loading in aluminum compact shear specimens.*

Discussion of Results

From Fig. 9 it is evident that the failure behavior of the material varied greatly depending upon the relative amounts of Mode I and Mode II loading in the specimen. For Case 4 loading, the material failure behavior was similar to that of Mode I loading, with the specimen behavior being nearly elastic to the onset of crack growth and the fracture surface looking dimpled. The fairly significant amount of shear loading seemed to have little effect on the failure load.

In Case 3 loading, the best estimates of K_{IIc} were obtained because the loading was predominantly Mode II, with a small amount of positive Mode I. For the 7075-T6 aluminum, K_{IIc} is nearly twice K_{Ic}. The failure process for near Mode II loading in this material begins with a combination of plastic deformation and stable crack growth.

Final failure in the specimen occurs when the applied shear stress in the remaining ligament equals the flow stress of the material. Unfortunately, the final failure load is very close in magnitude to the load for critical crack growth. This makes it difficult to a priori know the onset of crack growth from the load deflection curve.

For Case 1 loading, the loading is predominantly Mode II, but even the fairly small amount of negative Mode I loading at the crack tip greatly enhances crack tip plasticity and discourages crack growth. The results of these frictional effects are seen in the left-hand side of Fig. 9, with the values of K_{II} at failure significantly larger because of the effect of the concurrent negative K_I.

In Case 2 loading, the frictional effects from the negative Mode I loading overwhelmed any crack growth phenomena, and the results of this testing are not plotted in Fig. 9.

The results presented indicate many of the difficulties associated with the testing and data analysis of Mode II fracture specimens. Of primary concern is the similarity of the shear flow failure stress and the onset of what appears to be slow stable crack growth under Mode II loading. Although there is still some uncertainty, the crack growth observed prior to final fracture supports the assertion that Mode II crack propagation did occur, and that under conditions of Mode II which the crack propagates in a self-similar manner, that is, along $\theta = 0$. Self-similar crack propagation was not observed for the cases in a significant amount of Mode I was also present.

Many theories have been proposed which predict the angle of initial crack growth under mixed-mode loading. Both Bergkvist and Guex [7] and Streit [8] give a thorough discussion of many of these theories, and both come to the same conclusion. They conclude that the differences in the predictions of the theories are very small for near Mode I loading, and that the use of the maximum tangential stress theory is completely acceptable.

Figure 10 shows a comparison between the predicted angle of crack growth as calculated by both the maximum J-integral and the maximum tangential stress theories. The agreement between the two theories for near Mode I loading, that is, near a loading mix of 90 deg as defined in Fig. 10, is virtually exact. Not until the loading mix is at 65 deg, corresponding to a K_I/K_{II} ratio of about 2.0, do the predictions for crack angle differ by more than 1 deg. However, as the

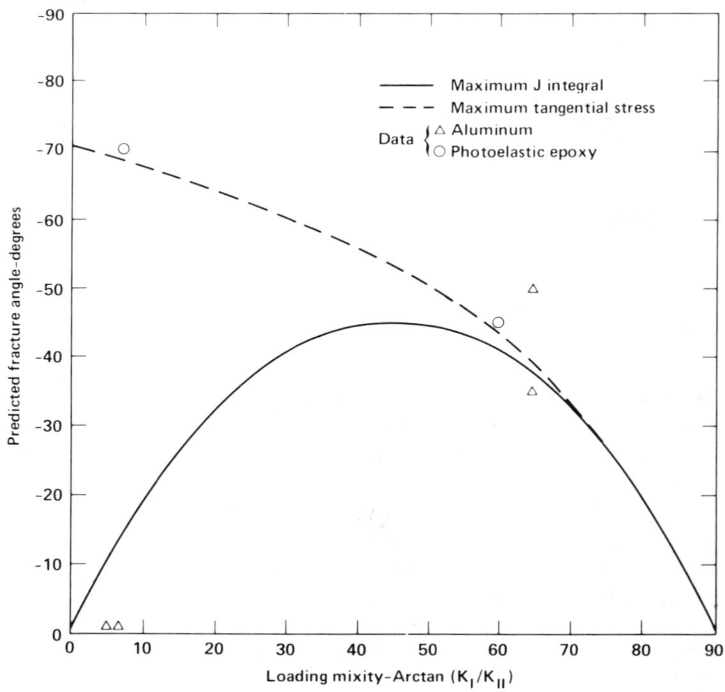

FIG. 10—*Comparison of predicted fracture angle between maximum J-integral and maximum tangential stress failure criteria. 90 deg on the abscissa corresponds to pure Mode I loading, while 0 deg corresponds to pure Mode II loading.*

amount of Mode II loading increases, the predictions of the two theories begin to differ greatly. At Mode II loading, when K_I/K_{II} equals zero, the maximum J-integral theory predicts an angle of extension of zero degrees, that is, self-similar crack growth. The maximum tangential stress theory predicts an angle of crack growth of -70 deg.

For the observed crack growth at zero degrees in the aluminum specimens loaded in Case 3, where the loading mix was about 5 deg, these data points are seen in Fig. 10 to be somewhat below the prediction of the maximum J-integral theory. The loading in Case 4 for the aluminum specimens resulted in a loading mix of about 65 deg. At this loading mix the two theories are in substantial agreement and predict crack growth at an angle of approximately 41 deg. The angle of crack growth in the aluminum specimens were 50 deg and 35 deg, close to the predicted angles from both theories.

It is possible that the tendency for the angle of crack growth in a material to follow the maximum J-integral theory depends upon the material's relative ability to plastically deform under Mode II loading. That is, the angle of crack growth in ductile materials should follow the maximum J-integral theory of predicted crack angle, whereas the crack growth angle of brittle materials, or materials in which linear elastic stress singularity dominates failure behavior, should follow the maximum tangential stress theory.

A limited number of mixed Mode fracture tests performed in other materials tend to verify this hypothesis. Compact shear specimens were fabricated from a brittle epoxy and tested in load Cases 3 and 4, and the results are also seen in Fig. 7. Under near Mode II loading, load Case 3, the angle of crack growth was -70 deg. In load Case 4, the angle of crack growth of the epoxy material was 45 deg. Further, limited fatigue tests under pure Mode II loading exhibited crack growth at -70 deg.

Further tests are needed to compare the predictive capability of these two theories for the angle of crack growth under mixed-mode loading.

Acknowledgment

This work was performed under the auspices of the U.S. Department of Energy by the Lawrence Livermore National Laboratory under Contract W-7405-ENG-48.

References

[1] Riddle, R. A., Streit, R. D., and Finnie, I., "An Evaluation of the Compact Shear Specimen for Mixed Mode Fracture Studies," *Transactions,* American Society of Mechanical Engineers, *Journal of Engineering Materials and Technology,* Vol. 105, 1983, pp. 268–270.

[2] Hallquist, J. O., "NIKE2D: An Implicit, Finite-Deformation, Finite Element Code for Analyzing the Static and Dynamic Response of Two-Dimensional Solids," Lawrence Livermore National Laboratory Report, UCRL 52678, Livermore, CA, 1979.

[3] Rice, J. R., "A Path-Independent Integral and the Approximate Analysis of Strain Concentration by Notches and Cracks," *Transactions,* American Society of Mechanical Engineers, *Journal of Applied Mechanics,* Vol. 35, 1968, pp. 379–386.

[4] Riddle, R. A., "The Application of the J-Integral to Fracture Under Mixed Mode Loading," Lawrence Livermore National Laboratory Report, UCRL 53182, Livermore, CA, June 1981.

[5] Ishikawa, H., Kitagawa, H., and Okamura, H., "J Integral of a Mixed Mode Crack and Its Application" in *Proceedings,* 3rd International Conference on Materials, Vol. 3, Cambridge, U.K., Aug. 1979, pp. 447–455.

[6] Jones, D. L. and Chisholm, D. B., "A Fractographic Study of the Edge-Sliding Mode in Fracture Toughness Testing" in *Fractography-Microscopic Cracking Processes, ASTM STP 600,* American Society for Testing and Materials, Philadelphia, 1976, pp. 235–248.

[7] Bergkvist, H. and Guex, L., "Curved Crack Propagation," *International Journal of Fracture,* Vol. 15, 1979, pp. 429–441.

[8] Streit, R. D., "The Directional Stability of Crack Propagation," UCRL 52593, Lawrence Livermore National Laboratory, Livermore, CA, 1975.

Richard A. Hays,[1] *Michael G. Vassilaros,*[1] *and John P. Gudas*[1]

Investigation of Ductile Fracture Properties of Welded Type 304 Stainless Steel Pipe and Large Plan Compact Specimens

REFERENCE: Hays, R. A., Vassilaros, M. G., and Gudas, J. P., **"Investigation of Ductile Fracture Properties of Welded Type 304 Stainless Steel Pipe and Large Plan Compact Specimens,"** *Fracture Mechanics: Eighteenth Symposium, ASTM STP 945,* D. T. Read and R. P. Reed, Eds., American Society for Testing and Materials, Philadelphia, 1988, pp. 134–150.

ABSTRACT: An experimental investigation was performed to evaluate the J-integral resistance curve behavior of 10-cm (4-in.) nominal pipe size Schedule 80 welded Type 304 stainless steel pipe. Tests were performed in four-point bending at 288°C (550°F) on 1219-mm (48-in.)-long pipe specimens with machined flaws contained in circumferential welds. Two flaw geometries were investigated. The first was a through-wall crack growing circumferentially (simple) and the second was a 360-deg part-through radial crack superimposed on a through-wall circumferential crack (complex). During the tests load, load line displacement, crack mouth opening displacement, and electrical potential drop across the crack mouth were monitored. Elastic compliance and direct current potential-drop (DCPD) techniques were used simultaneously to estimate crack extension during the tests. Pipes containing simple cracks had an average *J*-initiation of 1120 kJ/m^2 (6400 in. · lb/in.^{2}). The presence of the complex crack reduced the *J*-initiation level by approximately a factor of four and also reduced the apparent tearing modulus significantly as compared to pipes containing simple cracks.

A second part of this investigation concerned the use of laboratory-sized specimens of the same thickness as the wall of the pipe to model the J-R curve behavior of the simple through-wall cracked pipe. Large plan compact specimens were cut from a welded plate and tested at 288°C (550°F) using both elastic compliance and DCPD crack length estimation techniques. *J*-initiation values from the two types of specimens appear comparable, but the small amount of crack extension from the compact specimens limited comparison of the far-field J-R curve.

KEY WORDS: J-integral, pipe, compact specimen, elastic compliance, direct current potential drop (DCPD)

The fracture resistance of Type 304 stainless steel piping is of great concern in the nuclear power industry due to the potential presence of stress corrosion cracks associated with weldments in this material. An elastic-plastic fracture mechanics approach to assessing fracture performance is dictated due to the high toughness of the material and the large amount of plastic deformation and crack growth that occurs during loading prior to fracture instability. For this investigation, the J-integral resistance curve approach was used.

The objectives of this investigation were twofold. The first objective was to measure the J-integral resistance curve properties at 288°C (550°F) of full-scale welded Type 304 stainless steel pipe specimens subjected to loading in a four-point bend configuration. The second objective was to investigate the applicability of using compact laboratory specimens to model the pipe

[1]Materials engineer, research metallurgist, and division head, Metals and Welding Division, respectively, David W. Taylor Naval Ship Research and Development Center, Bethesda, MD 20084.

fracture behavior. This is important due to the relatively high cost of full-scale pipe fracture experiments. In order to accomplish these objectives, a series of five full-scale pipe fracture experiments on circumferentially welded, 10-cm (4-in.) nominal pipe size (NPS) Schedule 80, Type 304 stainless steel pipe, and a series of five fracture experiments on compact tension specimens were performed. The compact specimens ranged in plan size from 1T to 3T, with the thickness dimension the same as the pipe wall thickness.

Background

Fracture experiments on full-scale pipe specimens have been conducted by several investigators [*1–3*]. An apparent geometry dependence of the J-integral resistance (J-R) curve obtained from full-scale tests was reported by Wilkowski et al. [*1*]. In this work J-R curves obtained from two types of laboratory specimens, the center cracked panel and the three-point bend specimen, were compared with J-R curves obtained from 10-cm (4 in.) NPS pipes containing circumferential through-wall notches and loaded in four-point bending. The authors cited a strong dependence of the J-R curve on the degree of constraint at the crack tip and attributed the differences in results obtained from the different types of specimens to this dependence.

Vassilaros et al. [*2*] obtained relatively good agreement between full-scale pipe fracture experiments and laboratory sized compact tension specimens for a ferritic piping steel. In this case the compact specimens were cut directly from the wall of the 20-cm (8 in.) NPS pipe of the same type that was used in the fracture experiments which limited the maximum size of the compact specimens to 1T plan. Larger 2T plan specimens were also used in this work, but the specimen blanks had to be flattened prior to testing and were found to yield results inapplicable to the pipe specimens. Due to the relatively small size of the compact specimens compared to the pipe specimens, the J-R curve results were comparable over a very small range of crack extension.

Materials

The full-scale pipe fracture experiments were performed on 10-cm (4 in.) NPS (mean diameter 101.6 mm) Schedule 80, circumferentially welded Type 304 stainless steel pipe sections with an overall length of approximately 1219 mm (48 in.). The pipes were welded at Battelle Pacific Northwest Laboratories using an automatic gas tungsten arc welding procedure. The edge preparation was single "V," and the filler metal was type 308L stainless steel. The average heat input was 11.8 kJ/cm. The compact and tension specimens were machined from 914.4-mm (36-in.)-long by 8.64-mm (0.34-in.)-thick plates welded using nominally the same procedure and heat input. The edge preparation was double "V."

The tensile properties of the base and weld material at 288°C (550°F) were measured at Battelle Columbus Laboratories and are given in Table 1. The base metal tension specimens were oriented with the long axis in the direction of the weld length. The weld metal specimens were machined in the same orientation and made entirely of weld metal. It can be seen from the table that the weld metal had a yield strength approximately twice that of the base metal and a flow stress (calculated as the average of the yield and ultimate strengths) approximately 62 MPa (9 ksi) greater than the base metal. The materials exhibited similar tensile ductility properties.

Experimental Procedure

Pipe Specimen Tests

The full-scale pipe fracture specimens were loaded in four-point bending using a 1 334 000 N (300 000 lb) capacity screw-type testing machine under displacement control in a manner simi-

TABLE 1—*Tensile properties of Type 304 stainless steel base metal and Type 308L weld metal at 288°C (550°F).*

Metal Type	Yield Strength, MPa (ksi)	Ultimate Strength, MPa (ksi)	Flow Stress, MPa (ksi)	% Elongation, 25 mm	% Reduction in Area
Weld	305 (44.3)	451 (65.4)	379 (54.9)	33	74.3
	291 (42.2)	443 (64.3)	368 (53.3)	30	72.9
Average	299 (43.3)	447 (64.8)	373 (54.1)	31.5	73.6
Base	153 (22.2)	464 (67.3)	309 (44.8)	39.0	70.8
	157 (22.8)	474 (68.8)	316 (45.8)	40.5	70.8
Average	155 (22.5)	470 (68.1)	312 (45.3)	39.8	70.8

lar to that of Vassilaros et al. [*2*]. Figure 1*a* gives a schematic of the loading configuration. Five channels of data were taken during the experiments including load, cross-head deflection, load line displacement, crack mouth opening displacement (CMOD), and electrical potential drop across the crack mouth. All data were stored in digital form on magnetic media. Two crack configurations were tested as shown in Fig. 1*b*. The first was a through-wall crack growing circumferentially (simple), and the second was a through-wall crack superimposed on a 360-deg radial flaw (complex). The flaws were machined such that they were contained entirely within the weld metal and were sharpened in fatigue prior to testing. The complete pipe test matrix is given in Table 2.

J-intergral values were calculated for the pipe specimens containing both the simple and complex crack geometries using an expression published by Zahoor and Kanninen [*4*]. This formulation was used since it accounts for material hardening during loading and has a crack growth component. However, this expression cannot account for plasticity outside the crack cross sec-

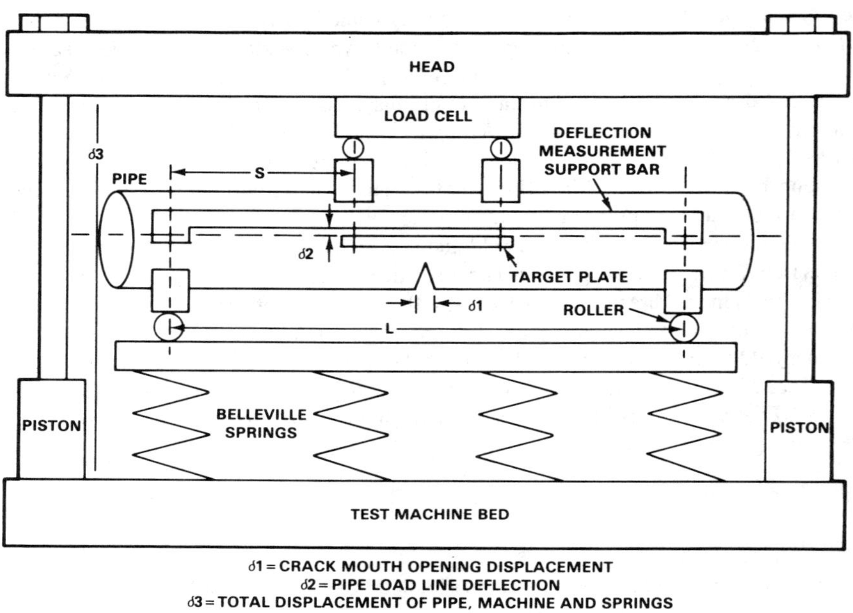

FIG. 1*a*—*Schematic of pipe loading configuration.*

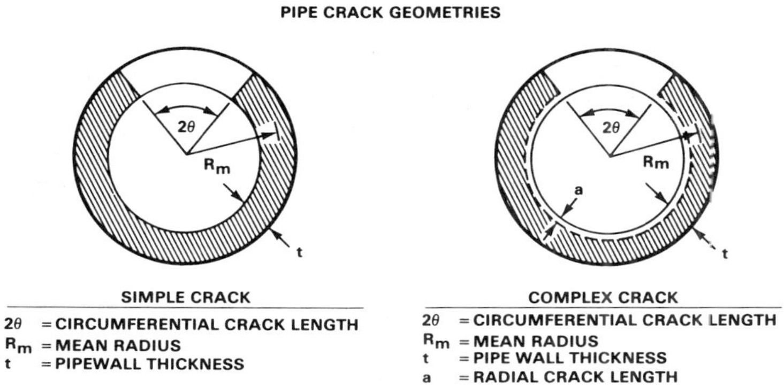

FIG. 1*b*—*Simple and complex crack geometries.*

TABLE 2—*Pipe specimen test matrix.*

Pipe Identification	Crack Geometry	Initial Crack Angle 2θ, deg	Outer Span Length L, mm (in.)	Inner Span Length S, mm (in.)	Mean Radius R, mm (in.)	Wall thickness t, mm (in.)	Radial Crack Length a, mm (in.)	a/t
GAM-100	simple	100	1067 (42)	381 (15)	52.8 (2.08)	8.64 (0.34)	...	...
GAM-200	simple	139	1067 (42)	381 (15)	52.8 (2.08)	8.64 (0.34)	...	...
GAM-900	simple	107	1067 (42)	381 (15)	52.8 (2.08)	8.26 (0.33)	...	...
GAM-400	complex	154	1067 (42)	381 (15)	54.6 (2.15)	8.64 (0.34)	3.28 (0.129)	0.380
GAM-600	complex	128	1067 (42)	381 (15)	52.8 (2.08)	8.38 (0.33)	2.21 (0.087)	0.256

tion and specimen geometry changes (ovalization) during testing. The radial component of the crack driving force was neglected for the complex crack geometry. The expression is

$$J = K^2/E + \beta \int_{\delta_0}^{\delta} (2P)d\delta + \int_{\phi_0}^{\phi} \gamma J d\phi$$

where

K = stress intensity factor,
E = elastic modulus,
$\beta = h'(\phi)/Rth(\phi)$,
$2P$ = total bending load,
δ = plastic load line deflection,
$\gamma = h''(\phi)/h'(\phi)$,
R = radius,
t = thickness,
ϕ = total crack angle, and
$h(\phi) = [\cos(\phi/4) - {}^1\!/_2 \sin(\phi/2)]$.

Crack extension during the tests was monitored using elastic compliance and direct current potential-drop (DCPD) techniques simultaneously. The elastic compliance technique used the slope of the load-versus-CMOD curve obtained during small elastic unloadings performed periodically during the tests. The specimen compliance was then compared to a compliance calibration curve to predict crack extension. The calibration curve used initially was that of Joyce [*5*] which was constructed using results from 10-cm (4 in.) NPS aluminum pipe. However, during loading and subsequent crack growth, ovalization of the crack cross section similar to that observed by Bruckner et al. [*3*] occurred with the vertical diameter of the cross section becoming longer and the horizontal diameter becoming shorter. This ovalization produced a stiffening of the specimen during loading and invalidated the calibration curve mentioned above. New calibration curves were constructed using the measured initial and final crack lengths and the measured specimen compliances for both crack geometries. Figure 2 compares the compliance calibrations used for the pipes with that of Ref *5*. The curve used for the pipe specimens containing simple flaw geometries has a lower slope than the other two curves on the figure. This indicates that when crack cross sectional ovalization takes place, a real increase in specimen compliance corresponds to a larger increase in crack length than when the crack cross section remains circular. Because of the presence of the radial flaw in the pipes containing complex crack geometry, very little cross-sectional ovalization took place during the tests. Thus the compliance calibration curve for the complex crack case appears very similar to that of Joyce [*5*].

The DCPD technique used was similar to that described by Vassilaros and Hackett [*6*] except that the relationship between crack length and potential drop after blunting was obtained by fitting an exponential equation to the data published by Wilkowski and Maxey [*7*]. After each test the specimens were heat tinted and broken open. The initial and final crack lengths were then measured at the inner and outer diameter as well as on several points along the pipe radius. Crack extension occurred fairly symmetrically at each crack tip, with the difference in final half crack lengths being an average of 2%. The average initial and final crack lengths were reported and used in the DCPD analysis.

Due to the sparseness of the crack length data near crack initiation, J-initiation was calculated as the intersection between the blunting line and a straight-line fit to all data after crack initiation. As compared with the ASTM Test Method for J_{Ic}, a Measure of Fracture Toughness

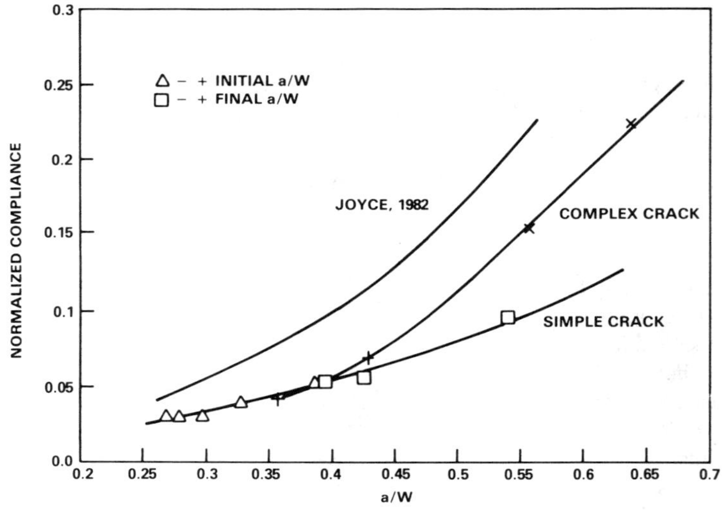

FIG. 2—*Normalized pipe compliance curve for simple and complex crack geometries.*

(E 813-81), the straight-line fit places more emphasis on points with large crack extension and tends to yield higher *J*-initiation points.

Compact Tension Specimen Tests

A series of five J-R curve tests was performed at 288°C (550°F) on modified compact tension specimens with 1T, 2T, and 3T plan geometries. The specimens blanks were cut from a flat plate and machined to a thickness of 8.64 mm (0.34 in.), which corresponds to the nominal pipe wall thickness. The notches were machined such that they were contained entirely within the weld and crack growth occurred in the direction of the weld. A schematic of the specimen design is shown in Fig. 3 and a test matrix is given in Table 3. All specimen measurement and preparation procedures detailed in ASTM E 813-81 were followed in this phase of testing. Specimens were precracked to an approximate 0.65 a/W where a is the crack length and W is the specimen width.

Methods of estimating crack length used during testing of the compact specimens were similar to those used for the pipe specimens. The computer interactive single specimen elastic compliance technique introduced by Joyce and Gudas [*8*] was used as well as the DCPD technique described by Vassilaros and Hackett [*5*]. However, due to the high curvature of the crack opening displacement versus potential-drop curve, maximum load was used as the criteria for crack initiation. Visual observations made during the tests were used to verify the onset of crack initiation. The J-integral values calculated for the compact specimens used the crack growth corrected deformation J expression published by Ernst, Paris, and Landes [*9*]. *J*-initiation values were calculated using the standard ASTM E 813 procedure.

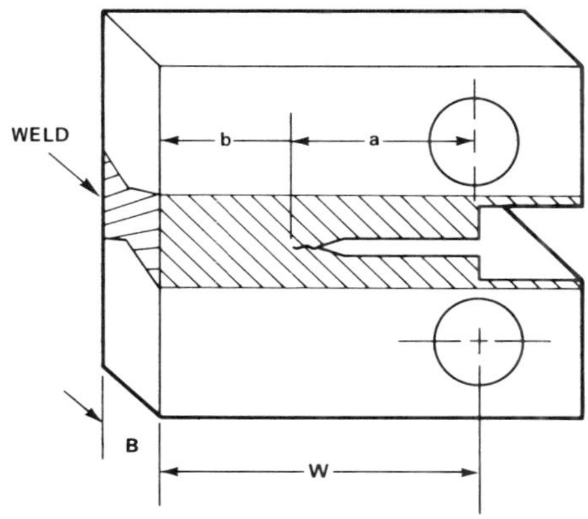

FIG. 3—*Modified compact tension specimen.*

TABLE 3—*Compact tension specimen test matrix.*

Plan Size	Number of Specimens	Specimen Width W, mm (in.)	Specimen Thickness B, mm (in.)	Initial Crack Length a, mm (in.)	Remaining Ligament b, mm (in.)	a/W	B/b
3T	1	152 (6.00)	8.60 (0.34)	99.1 (3.90)	8.64 (0.34)	0.65	0.162
2T	2	102 (4.00)	8.60 (0.34)	66.0 (2.60)	8.64 (0.34)	0.65	0.242
1T	3	50.8 (2.00)	8.60 (0.34)	33.0 (1.30)	8.64 (0.34)	0.65	0.486

J-Integral Resistance Curve Results and Discussion

Simple Crack Geometry Pipe Specimen Results

J-R curve results for pipe specimen GAM-100 which contained a simple through-wall flaw are given in Fig. 4. Although elastic compliance and DCPD techniques were used simultaneously to estimate crack lengths during the tests, elastic compliance data only are available for this specimen. This figure presents J versus average crack extension data for one crack tip. The filled symbol indicates the average optically measured crack length. As can be seen from the figure, the elastic compliance technique over-predicted the measured final crack length by approximately 7%. During the test, both crack tips remained within the weld metal. An approximate J-initiation level of 822 kJ/m^2 (4700 in. · lb/in.^{2}) was obtained for this specimen.

J-R curve results for pipe specimen GAM-200 are given in Fig. 5. Both elastic compliance and DCPD results were obtained for this test. Again, the elastic compliance technique over predicted the final crack length by a small margin. The optically measured initial and final crack lengths are used in DCPD analysis, thereby precluding overestimation or underestimation. The elastic compliance results yielded a lower estimate of J-initiation and a lower resistance curve slope than the DCPD technique for this test. The J-initiation results were 962 kJ/m^2 (5500 in. · lb/in.^{2}) and 1068 kJ/m^2 (6100 in. · lb/in.^{2}) for the elastic compliance and DCPD techniques, respectively.

Elastic compliance and DCPD results were also obtained for pipe specimen GAM-900 and are shown in Fig. 6. In this test the elastic compliance technique underpredicted the final measured crack length by approximately 15%, leading to an artificially high J-R curve. Notable in this figure is that the elastic compliance technique produced a higher J-initiation level and a lower resistance curve slope than did the DCPD technique. The lower slope is consistent with the results from pipe specimen GAM-200. The high prediction of J-initiation relative to that

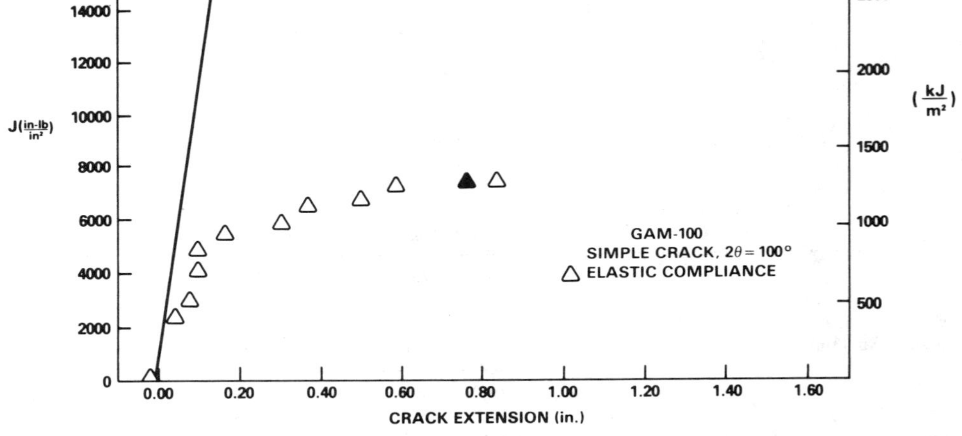

FIG. 4—*J-R curve from elastic compliance data for pipe Specimen GAM-100.*

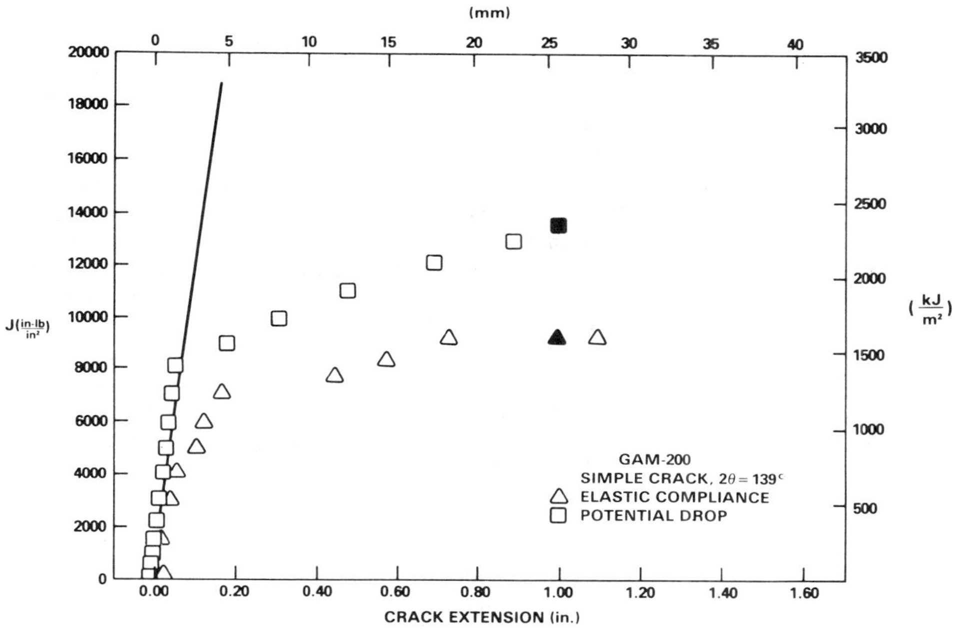

FIG. 5—*J-R curves from elastic compliance and DCPD data for pipe Specimen GAM-200.*

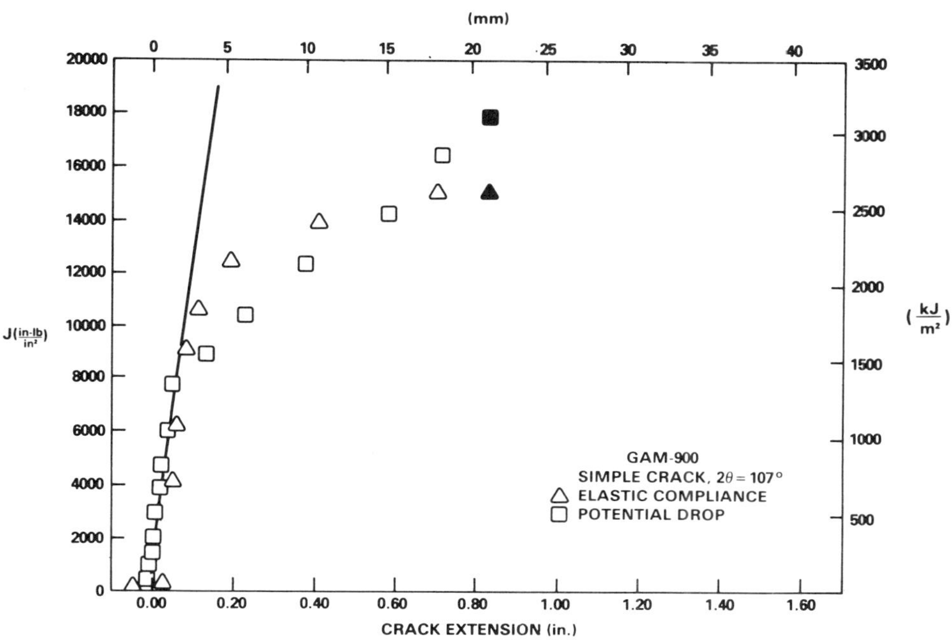

FIG. 6—*J-R curves from elastic compliance and DCPD data for pipe Specimen GAM-900.*

predicted by DCPD may be due to the insensitivity of the elastic compliance technique to crack initiation at short initial crack lengths as a result of the large distortions in the crack cross-section and overall plasticity in the region near the crack. This is reflected by the 6% change in the vertical diameter of the crack cross section for pipe specimen GAM-900 and the 2% change in the same dimension for pipe GAM-200.

Figure 7 presents a comparison of the J-R curves from elastic compliance for the pipes with the simple crack geometry. Good agreement between pipe specimens GAM-100 and GAM-200 is shown with pipe specimen GAM-900 being considerably above these two results. The good agreement is a result of the crack tips in pipe specimen GAM-100 staying entirely within the lower toughness weld metal and producing less distortion (change in vertical diameter of crack cross section of 3%). A comparison of the DCPD J-R curve results in Fig. 8 shows extremely good agreement at J-initiation for pipe specimens GAM-200 and GAM-900. The difference in resistance curve slopes may be due to the difference in initial crack lengths as suggested by Smith [*10*]. The range of initiation toughnesses measured here agrees to a large extent with those measured for 10-cm (4 in.) NPS base metal Type 304 stainless steel pipes in Ref *1*. This result indicates that the presence of the overmatching circumferential weld does not affect the initiation toughness of the pipe to a large degree.

Complex Crack Geometry Pipe Specimen Results

J-R curve results for pipe specimens GAM-400 and GAM-600 which contained the complex crack geometry are given in Fig. 9. Elastic compliance and DCPD data agree exceptionally well for GAM-400. The long initial crack angle ($2\theta = 154°$) and nearly 40% through-wall radial crack reduced specimen distortion and kept the crack plane perpendicular to the long axis of

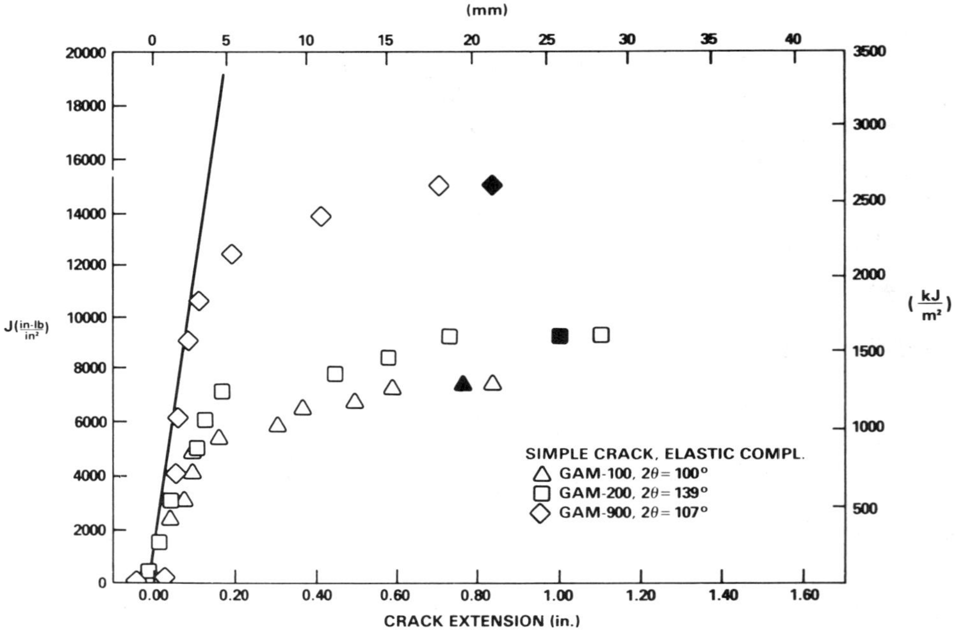

FIG. 7—*Comparison of J-R curves for pipes containing simple crack geometry from elastic compliance data.*

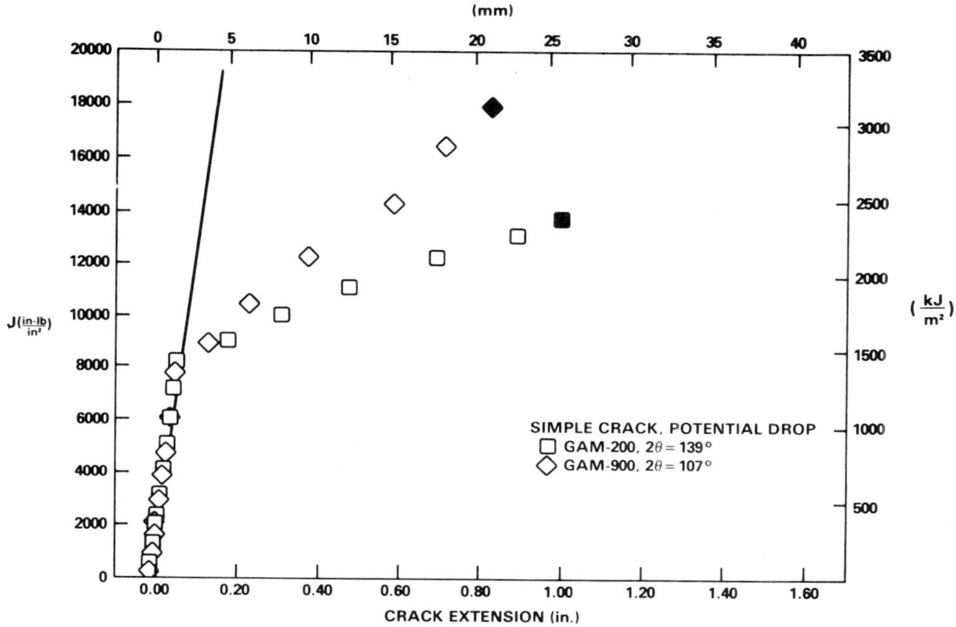

FIG. 8—*Comparison of J-R curves for pipes containing simple crack geometry from DCPD data.*

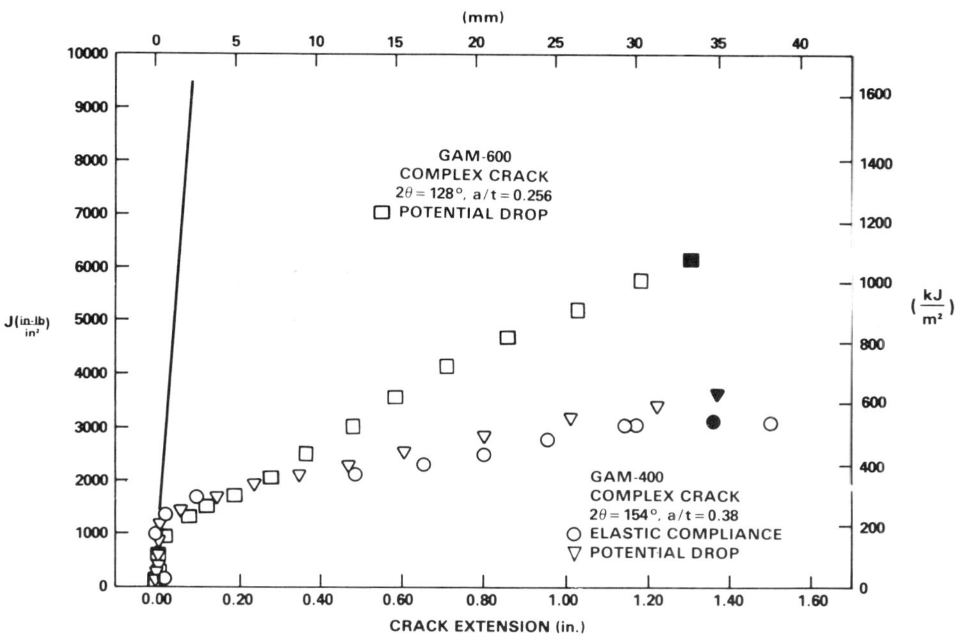

FIG. 9—*J-R curves from elastic compliance and DCPD data for pipe Specimens GAM-400 and GAM-600.*

the pipe contributing to the repeatability of the J-R curve measurement. J-initiation for this specimen was approximately 262 kJ/m^2 (1500 in. · lb/in.^{2}).

Pipe specimen GAM-600 had an initial crack angle $2\theta = 128$ deg and a radial crack depth of 25% of the wall thickness. The test results indicate a J-initiation value of approximately 228 kJ/m^2 (1300 in. · lb/in.^{2}) and a high resistance curve slope as compared with results from pipe specimen GAM-400. The difference in resistance curve slope between the two specimens is expected and primarily due to the difference in wall thickness, but the difference in initial crack angle may also play a role.

Comparison of test results from the pipes containing the two crack geometries reveals a significant lowering in fracture toughness from the pipes with the simple crack geometry to the pipes with the complex crack geometry. It must be remembered that the radial component of the crack driving force was neglected in the J expression. Representative J-R curves for pipes containing the two crack geometries from elastic compliance and DCPD data are plotted in Figs. 10 and 11, respectively. The data in Fig. 10 indicate a decrease in initiation toughness by a factor of approximately 3 and a clear reduction in resistance curve slope between the pipes containing simple and complex cracks. This is expected due to the high level of constraint imposed by the complex crack geometry and the 38% reduction in wall thickness. The data in Fig. 11 support the large reduction in initiation toughness showing a reduction of approximately 5½ between pipes containing different crack geometries. Again there is a reduction in resistance curve slope; however, it is not as large as the data shown in the previous figure. This is a result of the pipe containing the complex crack having only a 25% reduction in wall thickness.

The decrease in toughness between the pipes containing the simple and complex cracks is also extremely evident from the load-versus-displacement records. Figure 12 gives representative curves from pipes containing the two types of crack geometry. The simple crack geometry is

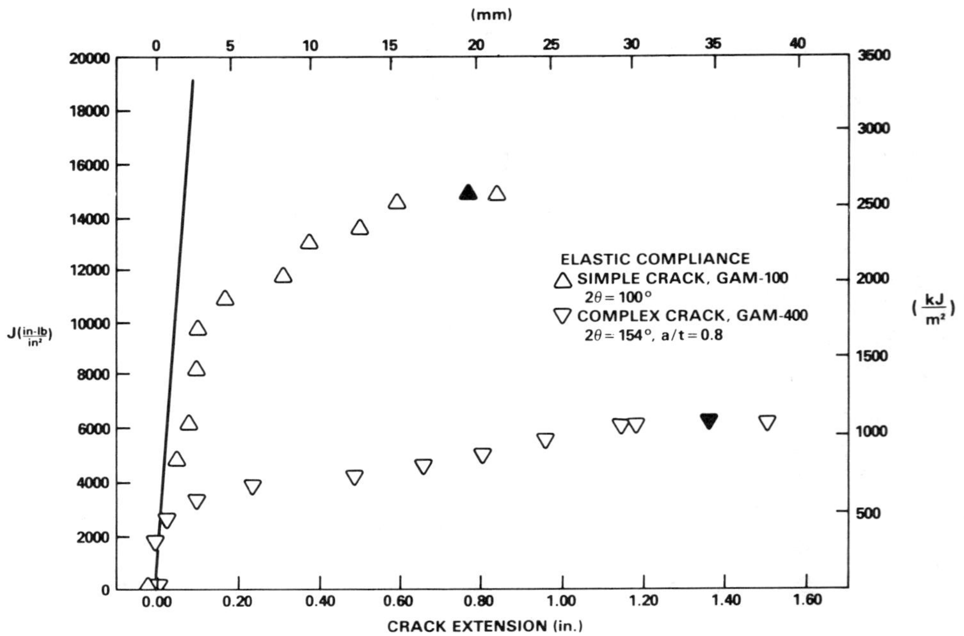

FIG. 10—*Comparison of J-R curves for pipes containing simple and complex crack geometries from elastic compliance data.*

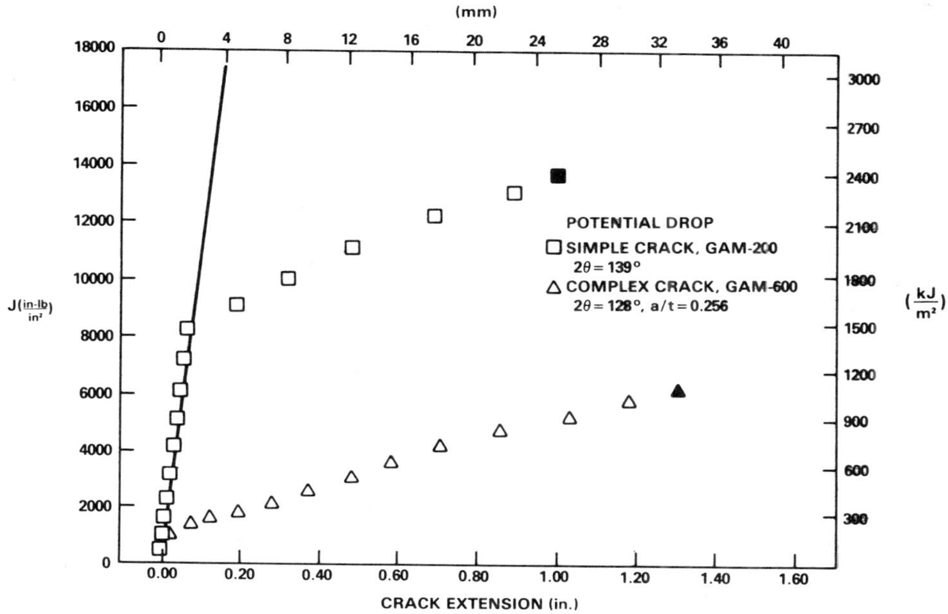

FIG. 11—*Comparison of J-R curves for pipes containing simple and complex crack geometries from DCPD data.*

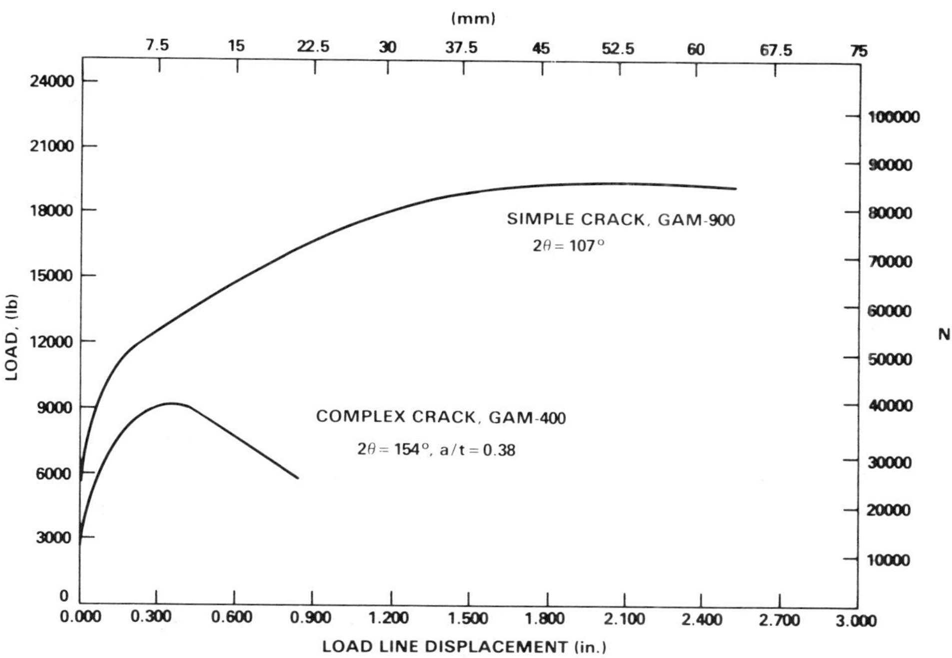

FIG. 12—*Load versus load line displacement records for pipes containing simple and complex crack geometries.*

characterized by a rather flat curve both before and after maximum load. The complex crack geometry is characterized by attainment of maximum load very early in terms of displacement and a large negative slope after maximum load.

Compact Tension Specimen Results

Both elastic compliance and DCPD techniques were used to estimate crack extension for the compact tension specimens as well as the pipe specimens. Figure 13 presents the J-R curve results from the 1T and 2T plan specimens produced using the elastic compliance technique. There is considerable variability in the curves with the *J*-initiation values ranging from 840 kJ/m^2 (4800 in. · lb/in.^{2}) to 1400 kJ/m^2 (8000 in. · lb/in.^{2}). The figure shows that the 1T plan specimen GGP-1 produced the lowest curve of the four specimens reported. Upon further inspection it can be seen that the elastic compliance technique predicted the final crack length within 5% for this specimen while under-predicting the final crack lengths of the other three specimens by an average of 30%. Correcting the J-R curves of these three specimens to meet the final crack length would produce lower curves and better agreement with the results from specimen GGP-1.

The variability in J-R curves from the compact specimens discussed previously was reduced considerably when the DCPD data was analyzed since this analysis is tied to the measured initial and final crack lengths. This is evident in Fig. 14, which is a plot of the J-R curve results from the 1T, 2T, and 3T plan specimens. *J*-initiation values range from approximately 858 kJ/m^2 (4900 in. · lb/in.^{2}) to 1120 kJ/m^2 (6400 in. · lb/in.^{2}) for these curves. It appears from this figure that there is no significant J-R curve dependence on specimen plan size.

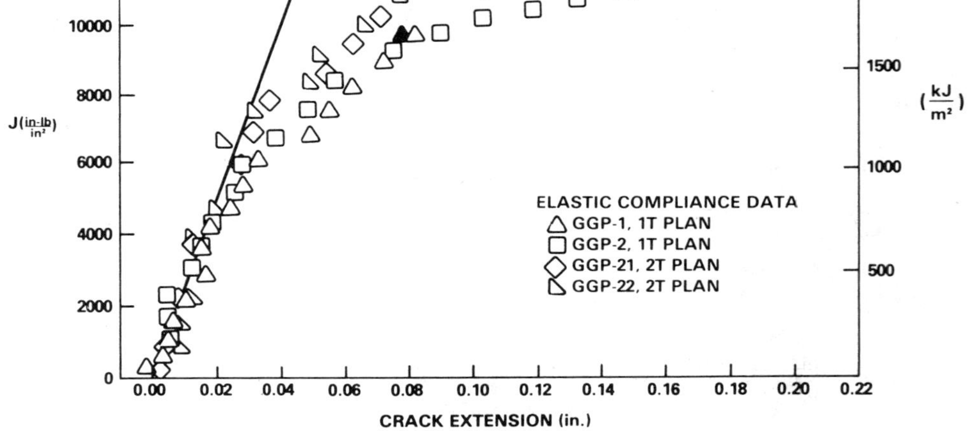

FIG. 13—*J-R curves from elastic compliance data for 1T plan and 2T plan compact Specimens.*

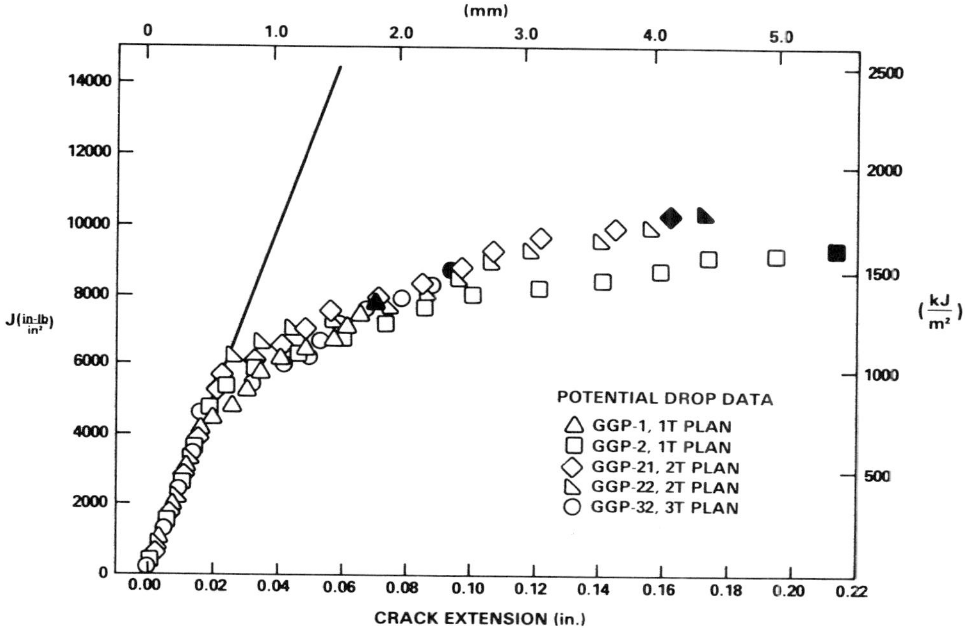

FIG. 14—*J-R curves from DCPD data for 1T plan, 2T plan, and 3T plan compact specimens.*

Comparability of Pipe and Compact Specimen J-R Curves

An important aspect of the work presented here is the use of laboratory-sized compact tension specimens to model the J-R curve behavior of the larger and considerably more expensive full-scale pipe specimens. In this investigation, pipes with simple through-wall circumferential cracks were modeled using non-side-grooved specimens of varying plan size with the thickness dimension the same as the nominal pipe wall thickness. The range of J-R curve results from elastic compliance data for both specimen types are plotted in Fig. 15. It appears from this figure that the compact specimen data validates the highest of the resistance curves from the pipe specimens. However, difference in weld geometry and the inability of the elastic compliance technique to accurately predict the final crack length for either the pipe or compact specimens leads to some uncertainty in the results. As stated previously, correction for under-prediction of the measured final crack length leads to lowering of the J-R curve after crack initiation.

Figure 16 presents the range of J-R curve results from DCPD data for the pipe and compact specimens. Very good agreement in initiation toughness between the two types of specimens can be seen from this figure. Comparison of resistance curve slopes is very difficult here due to the small amount of crack extension which occurred in the compact specimens and to the sparseness of the pipe crack extension data in the small crack extension region. It does appear, however, that the compact specimens show a higher initial slope than the pipe specimens. The slopes over the large crack extension range appear comparable between the two types of specimens.

Conclusions

The intent of the work presented here was to measure the J-integral fracture toughness of full-size welded 10-cm (4 in.) NPS Type 304 stainless steel pipe with simple and complex crack

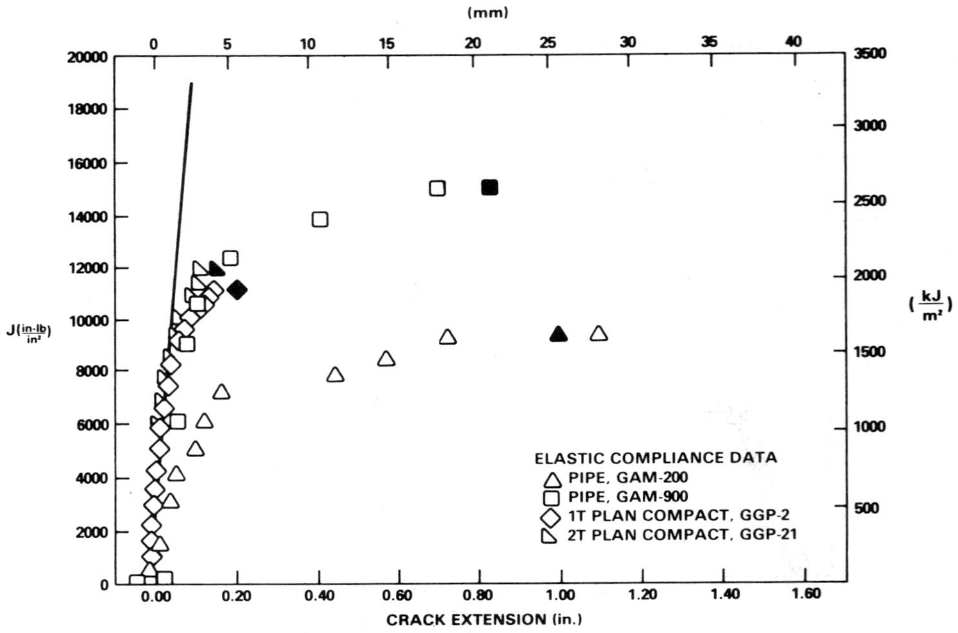

FIG. 15—*Comparison of J-R curves from elastic compliance data for pipe and compact specimens.*

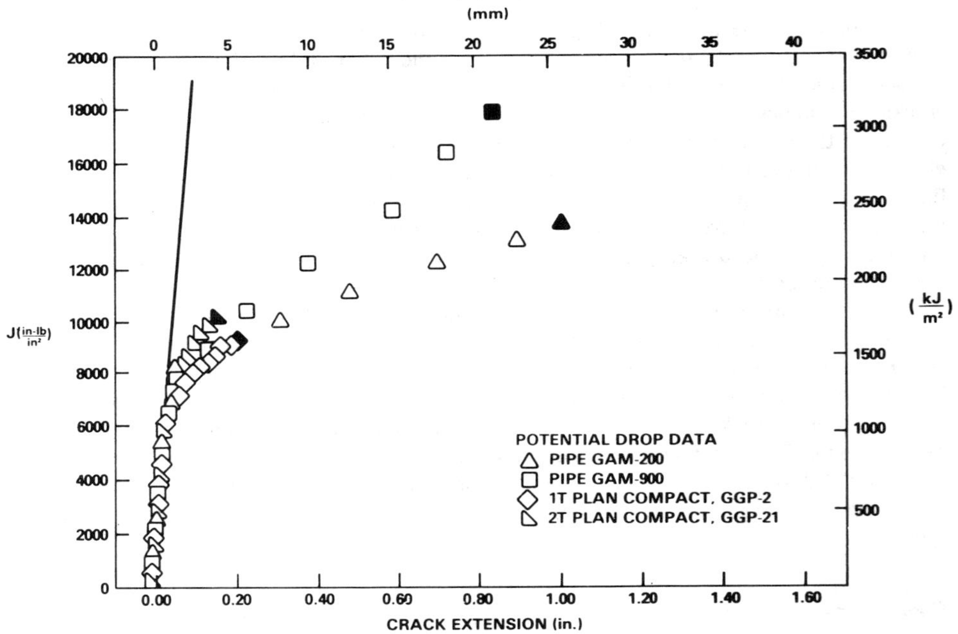

FIG. 16—*Comparison of J-R curves from DCPD data for pipe and compact specimens.*

geometries. An attempt was then made to model the fracture behavior of the pipe containing the simple crack geometry with compact laboratory specimens of the same thickness as the pipe wall. Several conclusions can be drawn from the results of this investigation:

1. Crack initiation for welded Type 304 stainless steel pipe containing simple through-wall circumferential cracks occurred at a J level of approximately 1120 kJ/m^2 (6400 in. · $lb/in.^2$).
2. Increased crack-tip constraint due to the presence of the internal notch in the complex crack geometry reduced the J level at crack initiation by approximately a factor of four as compared to the simple crack geometry.
3. The presence of the complex crack also significantly reduced the apparent tearing resistance of the pipe.
4. Ovalization of the crack cross section and large amounts of plasticity in the near-crack region lead to uncertainty in elastic compliance and DCPD crack length estimation techniques, respectively.
5. Crack initiation for compact specimens occurred at an average J level of 1050 kJ/m^2 (6000 in. · $lb/in.^2$).
6. Good agreement between the simple through-wall circumferentially cracked pipe and compact specimens using the DCPD technique indicates that crack initiation toughness measurements on laboratory-size specimens may be applicable to pipe geometries.

Acknowledgments

The authors wish to acknowledge the support of Messrs. Jack Strosnider and Milton Vagins of the U.S. Nuclear Regulatory Commission. The authors would also like to acknowledge the support of those at the David W. Taylor Naval Ship Research and Development Center who contributed to this work, especially Mr. Paul Ditta. The advice and guidance of Dr. James Joyce and Mr. Gery Wilkowski are also gratefully acknowledged.

References

[1] Wilkowski, G. M., Zahoor, A., and Kanninen, M. F., "A Plastic Fracture Mechanics Prediction of Fracture Instability in a Circumferentially Cracked Pipe in Bending—Part II: Experimental Verification on a Type 304 Stainless Steel Pipe," *Journal of Pressure Vessel Technology*, Vol. 103, Nov. 1981.

[2] Vassilaros, M. G., Hays, R. A., Gudas, J. P., and Joyce, J. A., "J-Integral Tearing Instability Analyses for 8 in. Diameter ASTM A106 Steel Pipe," NUREG/CR 3740, U.S. Nuclear Regulatory Commission Report, April 1984.

[3] Bruckner, A., Grunmach, R., Kneiffel, B., Munz, D., and Thun, G., "Fracture of Pipes with Through-Wall Circumferential Cracks in Four Point Bending," *Circumferential Cracks in Pressure Vessels and Piping*, G. M. Wilkowski, Ed., Vol. 2, PVP-Vol. 95, American Society of Mechanical Engineers, June 1982.

[4] Zahoor, A. and Kanninen, M. F., "A Plastic Fracture Mechanics Prediction of Fracture Instability in a Circumferentially Cracked Pipe in Bending—Part I: J-Integral Analysis," *Transactions*, American Society of Mechanical Engineers, *Journal of Pressure Vessel Technology*, Vol. 103, No. 4, Nov. 1981.

[5] Joyce, J. A., "Instability Tests of Compact and Pipe Specimens Utilizing a Test System Made Compliant by Computer Control," NUREG/CR-2257, U.S. Nuclear Regulatory Commission Report, March 1982.

[6] Vassilaros, M. G. and Hackett, E. M., "J-Integral R-Curve Testing of High Strength Steel Utilizing the D.C. Potential Drop Method" in *Fracture Mechanics: Fifteenth Symposium, ASTM STP 833*, American Society of Testing and Materials, Philadelphia, 1984.

[7] Wilkowski, G. M. and Maxey, W. A., "Applications of the Electrical Potential Method for Measuring Crack Growth in Specimens, Flawed Pipes and Pressure Vessels," in *Fracture Mechanics: Fourteenth Symposium, ASTM STP 791*, American Society of Testing and Materials, Philadelphia, 1983.

[8] Joyce, J. A. and Gudas, J. P. in *Elastic-Plastic Fracture, ASTM STP 668*, J. D. Landes, J. A. Begley,

and G. A. Clarke, Eds., American Society for Testing and Materials, Philadelphia, 1979, pp. 251-265.

[9] Ernst, H. A., Paris, P. C., and Landes, J. D., "Estimations on J-Integral and Tearing Modulus T from a Single Specimen Test Record" in *Fracture Mechanics (13th Conference), ASTM STP 743,* R. Roberts, Ed., American Society of Testing and Materials, Philadelphia, 1981.

[10] Smith, E., "The Geometry Dependence of the J-R Curve for Circumferential Growth of Through-Wall Cracks in Cylindrical Pipes Subject to Bending Loads," *Engineering Fracture Mechanics,* Vol. 18, No. 6, Pergamon Press, New York, 1983, pp. 1119-1123.

David T. Read[1]

J-Integral Values for Small Cracks in Steel Panels

REFERENCE: Read, D. T., **"J-Integral Values for Small Cracks in Steel Panels,"** *Fracture Mechanics: Eighteenth Symposium, ASTM STP 945,* D. T. Read and R. P. Reed, Eds., American Society for Testing and Materials, Philadelphia, 1988, pp. 151–163.

ABSTRACT: In a test program motivated by the need for a quantitative relationship between fracture toughness, flaw size, and applied loading for small flaws, to be used for fitness-for-service assessment, applied J-integral was measured as a function of applied strain in eight 14-mm-thick specimens of ASTM A710 Grade A Class 3 steel plate. All the edge cracks had lengths less than 3% of the specimen width of 82 mm. Six specimens were tested in tension; two were loaded by four-point-bending in the plane of the plate. One single-edge-cracked, transversely oriented specimen was tested at −30°C. Electrical resistance strain gage and clip-gage crack mouth opening displacement measurements were used to obtain quantities inside the J-integral. The J-integral was calculated by trapezoidal rule integration. Unloading crack mouth compliance measurements were used to obtain crack length values so that tearing effects could be observed.

Lüders strains occurring right after yield caused rapid increases in the applied J-integral values for the tension specimens. Except for the Lüders strain effect, the behavior of the applied J-integral in bending was similar to that in tension. Tearing caused a smooth exponential rise in applied J as strain increased beyond the point of initiation. The initiation toughness and tearing resistance of the panels with short cracks were equal to or greater than those of conventional three-point-bend specimens of the same thickness. No effect of temperature on applied J-integral was found from the specimen tested at −30°C. The present data support Wilson's prediction of the dependence of J on strain for a material with a bilinear stress-strain curve.

KEY WORDS: bending, fracture, J-integral, J-R curve, single-edge crack, surface crack, tension, toughness

This paper describes direct experimental evaluations of the J-integral as a function of applied strain in tensile and bending panels with small cracks. This test program was motivated by the need for a general, experimentally verified, quantitative relationship between fracture toughness, flaw size, and applied loading for small flaws, to be used for fitness-for-service assessment. The cracks used here intercept only up to 3.4% of the specimen cross-sectional area. This is small enough that considerable yielding occurs in locations remote from the crack plane. Equations that relate J-integral to area under the specimen's load-displacement curve for deeply cracked specimens are not applicable to the small cracks used here. The J-integral values for these small cracks are of interest for studies of fracture at cracks within elastic-plastic and fully plastic strain fields.

The experimental techniques used have been described in detail previously [*1*]. Including the present results, there are now enough data on the small-crack behavior of the J-integral as a function of applied strain to allow comparison to theoretical predictions, comparisons of the behavior of different alloys, comparison of tensile and bending results, and identification of the effect of ductile tearing.

[1]Physicist, Fracture and Deformation Division, National Bureau of Standards, Boulder, CO 80303.

Theoretical Background

The J-integral has been defined as a path-independent line integral on a contour surrounding a crack tip [*2*]; it has also appeared as the amplitude of the Hutchinson-Rice-Rosengren (HRR) stress-strain singularity [*3,4*]. A modified J-integral has been used in tearing studies [*5*]. The integral used in this paper is asserted to be the path-independent line integral J [*1*], taken on a contour lying far from the crack tip. No claim is intended regarding the existence or amplitudes of HRR stress-strain fields in the present specimens, or about any unique correspondence between the J-integral measured here and the initiation or extent of tearing.

Several theoretical predictions of the behavior of the J-integral in specimens with small cracks are available [*6–10*]. Figure 1 displays three of these, two for cracks in infinite plates calculated from an analysis by Wilson and one by calculated from an analysis given by Shih, German, and Kumar [*11*]. The conversions used in this figure for the normalized J-integral, j, and the normalized remote strain, e, are

$$j = JE/(\pi a \sigma_y^2) \tag{1}$$

and

$$e = \epsilon/\epsilon_y \tag{2}$$

where

J = J-integral,
E = Young's modulus,
a = crack size,
σ_{ys} = material yield strength,
σ_u = material ultimate strength,
σ_y = material flow strength $(\sigma_{ys} + \sigma_u)/2$, and
ϵ_y = nominal material flow strain, given by σ_y/E.

The two curves labeled Wilson Bilinear and Wilson Ramberg-Osgood are different predictions constructed using, bilinear and Ramberg-Osgood representations of the stress-strain curve of

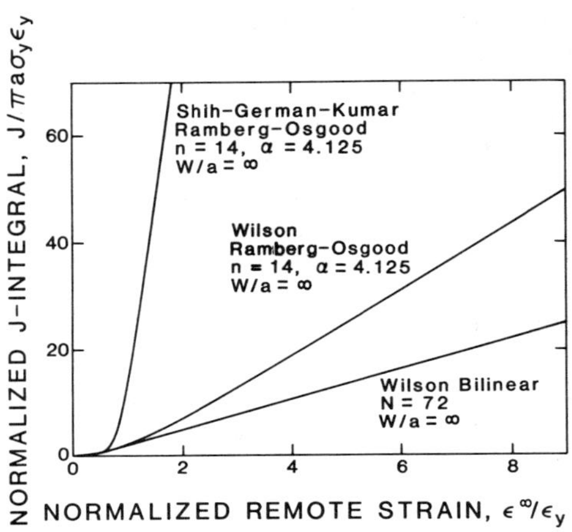

FIG. 1—*Three theoretical predictions of behavior of J-integral as a function of remote strain in specimens with small,* W/a = *8, and infinitesinal,* W/a = ∞ *cracks.*

the specimen material. The Ramberg-Osgood constants n and α were obtained by fitting the specimen material's true stress-strain curve, which was calculated from the engineering stress-strain curve. Figure 2 compares the specimen material's true stress-strain curve, obtained from a longitudinal round-bar tension specimen, to its Ramberg-Osgood and bilinear representations. The Ramberg-Osgood parameters used were $n = 14$ and $\alpha = 4.125$. A slope of $E/N = E/72$ was used for the upper part of the bilinear representation.

The Ramberg-Osgood curve (Fig. 1) for J as a function of strain makes use of upper-bound plane-stress h factors, calculated by He and Hutchinson [*12*], to relate J to strain. To apply this approach to the present specimen material, the h factors given for a series of work-hardening exponents n were plotted against $1/n$ for extrapolation to $n = 14$. The Ramberg-Osgood curve plotted in Fig. 1 is thus an extrapolated upper bound. It differs by a factor of $h/(2\pi) = 2.5$ from the curve labeled Wilson Bilinear. The bilinear curve is an analytical continuation of the linear elastic relationship for J as a function of strain. Similar curves were given previously by Merkle [*7*] and by Begley, Landes, and Wilson [*6*]. Turner [*8*] has reported curves derived from finite-element analyses that lie generally between Wilson's two curves in Fig. 1. In addition, the relationship between crack-tip opening displacement (CTOD) and strain given originally by Burdekin and described by Dawes [*9*] can be converted to give J if the constraint factor m is known. For m values in the range 1 to 2, the CTOD curve also falls generally between the two curves given by Wilson's analysis.

Unlike the other curves in Fig. 1, the curve of Shih et al. was derived for a specific finite crack size, namely, $a/W = 1/8$. The other two refer to small or infinitesimal cracks compared to the plate width. The difference between the behavior of small cracks and cracks of relative size $a/W \geqq 1/8$ has been confirmed experimentally and related to plate deformation patterns elsewhere [*13–15*]. Previous experimental results [*13–15*] indicated that the J-integral for small cracks in a high-strength quenched-and-tempered nickel steel fell between the two predictions by Wilson, closer to the upper curve.

Material

The specimen material used in this investigation was reported by the manufacturer to be ASTM A710 Grade A Class 3 alloy steel. It was received in the form of 19-mm (0.75-in.)-thick plate. The chemical composition, as supplied by a commercial source, is listed in Table 1. This

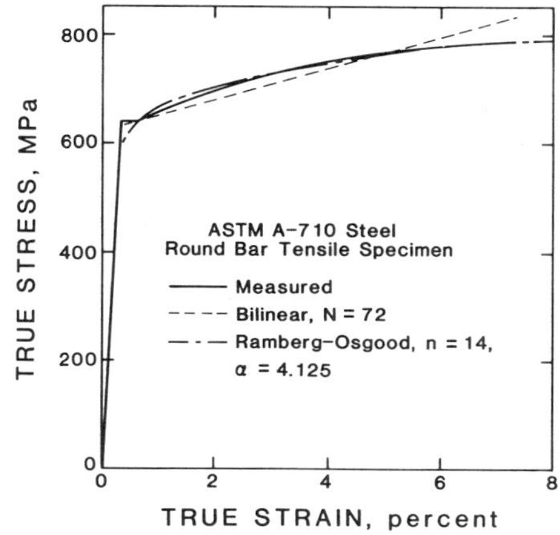

FIG. 2—*True stress-strain curve for longitudinal and transverse round-bar tension specimens of ASTM Grade A Class 3 alloy steel specimen material used in present study.*

TABLE 1—*Chemical composition, in weight percent, of ASTM A710 Grade A Class 3 alloy steel specimen material used in present study.*

Specimen Material	Weight %
Carbon	0.042
Manganese	0.54
Nickel	0.85
Chromium	0.65
Silicon	0.34
Copper	1.2
Sulfur	0.010
Phosphorous	<0.005
Niobium	<0.010
Aluminum	<0.020
Iron	balance

chemical analysis indicates less than 0.01 weight % of niobium, which is less than the specified minimum value of 0.02 weight % for A710 steel. The tensile properties of the specimen material, obtained using replicate round bar tensile specimens 6.4 mm (0.25 in.) in diameter, are listed in Table 2. The true stress-strain curve, given in Fig. 2, contains a slight yield point, some Lüders extension, and strain hardening. The tensile properties of the specimen material at room temperature were essentially the same in the transverse and longitudinal directions, except for the elongation, as can be seen from Table 2. The Lüders strain after the yield point was present in both longitudinal and transverse orientations. After its maximum, the load decreased more rapidly with extension in the longitudinal specimens than in the transverse ones, leading to the lower elongation values for the longitudinal specimens.

The fracture toughness of this specimen material at room temperature was investigated using three-point-bend specimens in the longitudinal orientation (LT) and the transverse orientation (TL). The in-plane dimensions were nominally 50 by 220 mm (2 by 9 in.). Thicknesses of 14 and 19 mm (0.55 and 0.75 in.) were used as noted. Half of the specimens (as noted) were side-grooved 12.5% on each side after fatigue precracking. The J_{Ic} and J-R results are listed in Table 3. The tests were performed according to ASTM Test Method for J_{Ic}, a Measure of Fracture Toughness (E 813-81). Items for noncompliance with the standard were the differences between compliance-measured and physical crack extension values Δa_c and Δa_p, noted next, and the

TABLE 2—*Tensile properties, at room temperature, of ASTM Grade A Class 3 alloy steel specimen material used in present study.*

Property	Value
Longitudinal	
yield strength, 0.2% offset	637 MPa (92.4 ksi)
ultimate tensile strength	732 MPa (106.2 ksi)
elongation	12.0%
reduction in area	75.5%
Transverse	
yield strength, 0.2% offset	634 MPa (93.2 ksi)
ultimate tensile strength	727 MPa (105.5 ksi)
elongation	17%
reduction in area	71%

TABLE 3—*Toughness and tearing resistance, in terms of J-integral, of ASTM Grade A Class 3 alloy steel specimen material used in present study.*

	Side-Grooved			Non-Side-Grooved		
	J_{Ic}			J_{Ic}		
Orientation	N/mm	lb/in.	T_{mat}	N/mm	lb/in.	T_{mat}
Transverse	118	674	53	183	1046	73
	120	686	49	117	669	84
	145	829	41	...	...	...
Average	128	731	48	150	857	78
Longitudinal	234	1337	103	371[a]	2120[a]	168[a]
				377[a]	2154[a]	186[a]
Average				374	2137	177

[a]19-mm-thick (0.748 in.) specimen.

specimen dimensions. The specimen size requirement was not met at higher values of J used in the tearing modulus measurements. The results of the J-integral tests indicate that the specimen material is quite tough at room temperature, and that nongrooved specimens in the LT orientation (crack grows across the rolling direction) are the toughest ($J_{Ic} = 374$ N/mm, $T_{mat} = 177$), while side-grooved specimens in the TL orientation (crack grows along the rolling direction) are the least tough ($J_{Ic} = 128$ N/mm, $T_{mat} = 48$). These values were calculated using $T = E \cdot (dJ/da)/\sigma_y^2$. In the TL orientation, the side-grooved and nongrooved specimens had similar J_{Ic} values. However in the LT orientation, the one side-grooved specimen had lower J_{Ic} and T_{mat} values than the two nongrooved specimens.

The compliance-measured crack extension values, Δa_c, differed from the respective physical crack extension values, Δa_p, by about 30%. This is attributed to friction in the conventional three-point-bend fixture used and to crack tunneling during the test. The Δa_c values were corrected by linear scaling before data reduction. No pop-ins or other signs of rapid crack advance were observed in these room-temperature tests.

Specimens, Apparatus, Instrumentation, and Techniques

Tension and four-point-bending panel specimens of the specimen material were machined with gage sections measuring 82 mm (3.22 in.) wide by 14 mm (0.55 in.) thick by 250 mm (10 in.) long. The total specimen lengths were 584 mm (23 in.) for the transverse specimens and 850 mm (33 in.) for the longitudinal specimens. The transverse specimens were made shorter because of the size of the plate of specimen material available. The grip sections at each end of the specimen were 108 mm (4.25 in.) wide to confine plastic strain to the gage section. Table 4 gives the details of the crack geometries tested. Figure 3 shows the loading methods used for the tension and bending tests. For the tension tests, mechanical wedge-grips clamped on the specimen grip sections were used, and for the bend tests, four-point loading. The outer bending points were 560 mm (22 in.) apart for the transverse specimens and 810 mm (32 in.) apart for the longitudinal specimens. The inner bending points were 250 mm (10 in.) apart for the transverse specimens and 300 mm (12 in.) apart for the longitudinal specimens. The bending points were 25-mm (1-in.)-diameter steel rollers on flat steel holders. The rollers were free to rock as the specimen bent. Loads were supplied by a displacement-controlled servohydraulic testing machine with a capacity of 1 MN (225 000 lb).

Instrumentation developed previously [1] for direct measurement of the J-integral was used,

TABLE 4—*Specimen and test types, crack sizes, and toughness and tearing resistance in terms of J-integral, for specimens used in present study.*

Abbreviation	Orientation	Crack Geometry	Loading Mode	Crack Length, mm	Relative Crack Size
DECT[a] (saw-cut crack tips)	LT[e]	double-edge	tension	1.1	0.014
SECT[b]	LT	single-edge	tension	1.9	0.023
SECT (prestrained)	LT	single-edge	tension	2.3	0.028
SECB[c]	LT	single-edge	bending	1.4	0.017
SECT	TL[f]	single-edge	tension	1.5	0.018
PTST[d]	TL	part-through surface	tension	3.7 by 13	0.034
SECB	TL	single-edge	bending	1.2	0.015
SECT (−30°C)	TL	single-edge	tension	1.2	0.014

[a]DECT = double-edge-cracked tension.
[b]SECT = single-edge-cracked tension.
[c]SECB = single-edge-cracked bend.
[d]PTST = part-through-surface tension.
[e]LT = longitudinal orientation.
[f]TL = transverse orientation.

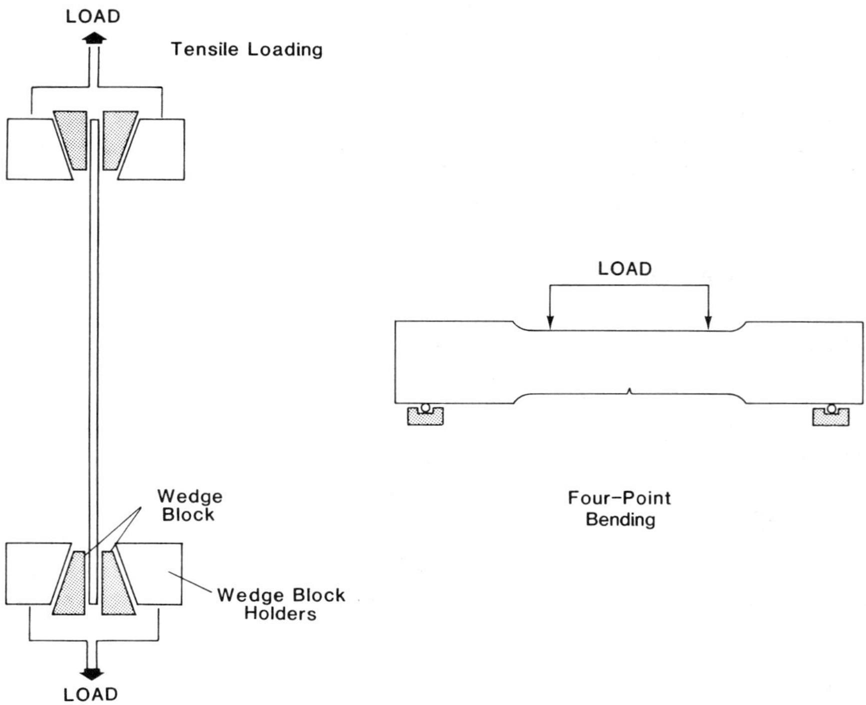

FIG. 3—*Loading methods used for tension and four-point-bend tests.*

including a load cell built into the testing machine, a crack mouth opening displacement (CMOD) gage (ring-gage), about 35 electrical resistance strain gages, and up to three linear variable differential transformer (LVDT) gages for displacement measurement. An array of five small strain gages was located near the crack mouth to give the strain profile there, as shown in Fig. 4. Some gages were always located away from the crack, for measurement of remote strain.

The strain gages, load cell, and CMOD gage were wired to conventional signal conditioning circuitry. The LVDTs contained signal conditioning circuits within their cases. The conditioned signals from all the gages were introduced to the input of a commercially available 16-bit analog-to-digital (A-to-D) conversion system. The least significant bit in the load reading corresponded to a load of 30 N, and the least significant bit in the CMOD reading corresponded to a displacement of less than 1×10^{-4} mm. The least significant bit in the strain readings corresponded to a strain of about 3×10^{-5}. The A-to-D system was linked to a commercially available microcomputer system for converting voltage readings to physical quantity values and storing these on floppy disks. The specimens were tested under displacement control. Increments of displacement were imposed and at each increment the gages were scanned and the readings were recorded, printed, and used to calculate a preliminary J-integral which was plotted along with the strain gage strains. Slight relaxations of the load, on the order of 1%, were observed after displacement increments imposed at high strains. These occurred over a period of up to 2 min after the displacement increment. The usual procedure was to impose the displacement increment, wait 2 min for the load to become constant, and then scan all instruments. The scanning itself took about 10 s. After the instrumentation readings had been recorded, the unloading compliance, CMOD per unit load, was measured during an unloading of 10 to 20% of the load. An approximate crack length was obtained. These values were corrected by linear

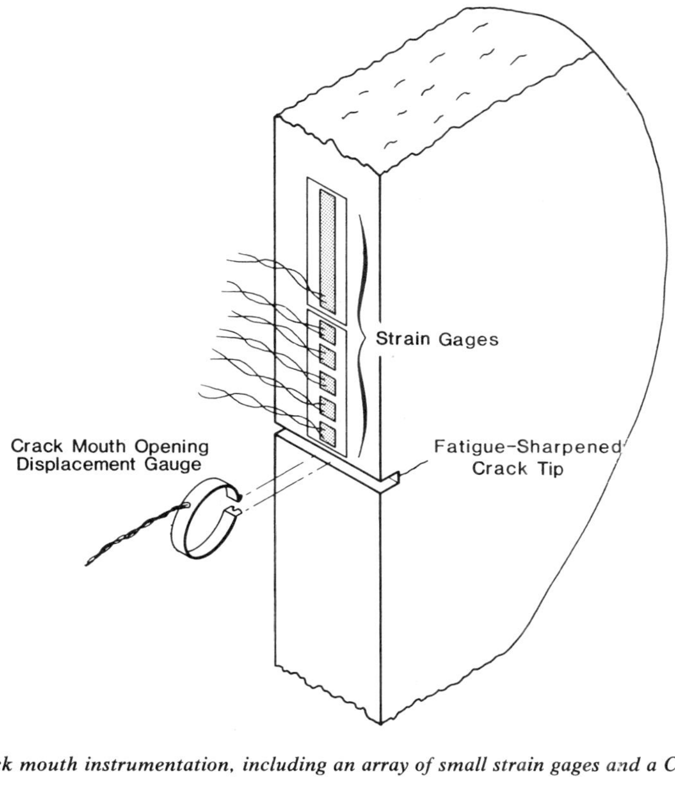

FIG. 4—*Crack mouth instrumentation, including an array of small strain gages and a CMOD gage.*

scaling after the test to give the correct initial and final crack depths. These crack depths were used to detect the initiation of tearing and in the calculation of the tearing modulus. In a typical test, on the order of 100 displacement increments were used. At some stage, the strain gages began to fail because of high imposed strains of the order of 2 to 3%. From this point, a different data analysis method had to be used.

In the post-test data analysis, all the measured strain and displacement quantities were zero-corrected by fitting the linear part of the record for each quantity to a linear dependence on the stress, then adding a constant correction term so that zero strain or displacement occurred at zero stress. The J-integral was calculated using the trapezoidal rule for displacement and stress-work integrals along both sides of the contour. Because the contours were chosen along specimen edges so that the stresses at the contour were uniaxial, an elastic-perfectly plastic approximation of the simple round-bar stress-strain curve was used for calculating the stress-work integral. Integration of strain gage strains to obtain the displacements needed for calculation of the traction-bending term of the J-integral gave some error cancellation, which would not have been obtained had the LVDT displacements been used. The LVDT displacements were used to check the integrated strain values and for gage length strain measurement after strain gage failure. Correlations within 5% were obtained experimentally between J-integral and $1 \times \sigma_y \times$ CMOD. This correlation is expected because of low geometrical constraint of the simple tension and very-short-crack bend geometries used here. After strain gage failure, this correlation was used to obtain J-integral values.

Photoelastic coatings were bonded to the surfaces of the specimens to allow photographic recording of the strain patterns in the specimens. The coatings were observed through a polarizing sheet using white light. Colored fringes corresponding to the difference between longitudinal and transverse principal strains were observed. This allowed observation of the Lüders bands.

Results and Discussion

Results on remote strain, gage length strain, stress, CMOD, J-integral, and strain as a function of distance from the crack plane on the cracked and uncracked specimen edges were obtained over the whole course of all the tests. A complete presentation of these results is too lengthy to include here and will be published elsewhere [*16*]. The key results are the plots of J-integral against remote strain. All the other results, though interesting and important in themselves, support the J-integral results.

Figure 5 shows the measured J-integral as a function of remote strain for four-point-bending with a crack length of 1.4 mm for a single-edge-cracked longitudinal specimen. This figure shows the usual parabolic dependence at low strains followed by an approximately linear dependence above yield. The deviation of the post-yield dependence from true linearity is believed to be an artifact of the experimental arrangement; perhaps it is a result of the significant curvature induced in the specimen by bending by the end of the test. The important point is the gradual rate of rise of J-integral with imposed strain above yield. This result is qualitatively similar to the Wilson results in Fig. 1.

Figure 6 shows the behavior of J-integral as a function of strain for all the specimens tested. Six of these are tensile panels, two are four-point-bending panels. This plot shows a feature of the tension tests not present in the bend results of Fig. 5, namely, the step in J-integral at a fixed value of strain near yield. From the photoelasticity results, this step in J-integral is associated with the appearance of an area of high strain that first extends at an angle from the crack tip to the opposite edge of the specimen, and then spreads along the specimen gage section. This behavior is the same as the Lüders band phenomenon in tension tests and is associated with the Lüders extension found in the tensile bar stress-strain curves for this material. This step in the J-integral versus strain behavior was not present in the bend tests; the photoelastic results for bending confirmed that no Lüders-type strains occurred in the bend tests. The measured J-inte-

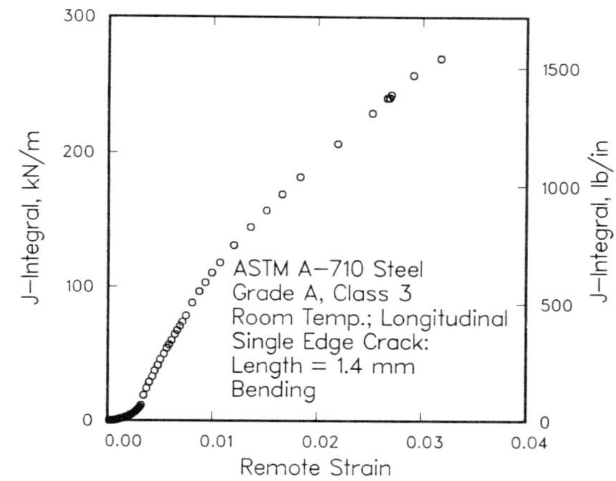

FIG. 5—*Experimentally measured J-integral as a function of applied remote strain for a four-point-bend specimen with crack length 1.4 mm (0.055 in.).*

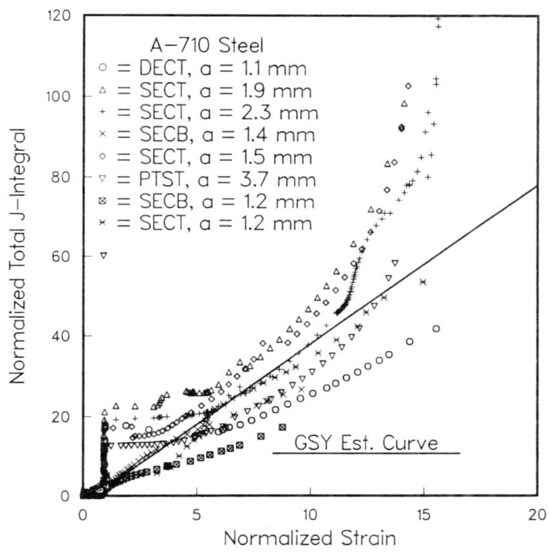

FIG. 6—*Experimentally measured total J-integral as a function of normalized remote strain for the eight specimens of this study.*

gral values agree roughly with a previously given experimental estimate for gross section yielding (GSY) for a different steel [*14*], and with the Ramberg-Osgood curve shown in Fig. 1. The agreement with the Ramberg-Osgood curve is believed to be accidental. The agreement with the previous experimental estimate confirms that the behavior of the applied J-integral is the same for different specimen materials.

The results from the tension specimens with unusual features, one with sawcut notches, one prestrained, and one tested at −30°C, are indistinguishable from the rest in Fig. 6. This is as expected for applied J-integral. The −30°C specimen had a low value of J_c (Table 5); this is consistent with expectation.

The bend test results end at strains lower than the tension test results because the specimens buckled after strains of several times the yield strain and could not be strained further. At high strains on the order of 10 times yield, tearing initiated in the tension tests, as shown by the unloading compliance data. Because of the buckling, the bend panels could not be strained enough to initiate tearing. Tearing produces the upturn in the J-integral data of Fig. 6 at strains of about 10 times yield.

Figure 7 interprets the behavior of the J-integral as a function of strain for small cracks in tension and bending specimens. Both types have a parabolic region below yield. The tension specimens show a step-like increase near yield because of Lüders strain, while the J-integral in the bending specimens increases gradually. The tensile and bending results converge after the Lüders strain region ends, and both show gradual increases of J-integral with strain. At some J-integral level, tearing begins, increasing the rate of rise of J-integral with strain.

An attempt was made to correct the data of Fig. 6 to reflect the behavior of flaws in infinitely wide plates. The correction procedure extracted only that part of the J-integral that resulted from gross section yielding. All net section yielding effects were excluded. The justification for this procedure is that through cracks in structures would be much smaller than 1 to 3% of the structural element width; therefore, net section yielding effects that may be present in these specimens would be absent in a real structure. The specific assumption used in the correction procedure was that contributions to the J-integral from strains on the specimen edge opposite the crack would not be present in a structural element with a small crack. Therefore only the strain profile in the neighborhood of the crack mouth was considered in the calculation of the corrected J-integral. The strains on the specimen edge opposite the crack were assumed to be equal to the remote strain on the cracked edge. The J-integral values resulting from this correction were termed GSY J-integral values and are plotted in Fig. 8. This plot, when compared to the total J-integral values of Fig. 6, indicates that the GSY part was only about half the total J-integral value. Although data from a surface-cracked specimen are included in Fig. 8, a more complex correction procedure is probably needed for this crack geometry because of the presence of ligament yielding.

It is the GSY J-integral values that are appropriate for comparison with the theoretical predictions discussed previously for small cracks in infinite plates. The data are compared with the Wilson bilinear result in for interior cracks in plane stress Fig. 8. This figure shows good agreement between experiment and theory. As a result of this comparison it is concluded that the

TABLE 5—*Toughness and tearing resistance, in terms of the J-integral, of tension and bend specimens with short cracks, without side grooves. Tension and bending panel tests.*

Specimen Type	Orientation,	a/W	J_c, Initiation	T_{mat}
DECT[a] (saw-cut cracks)	LT	0.014	NA[b]	NA
SECT	LT	0.023	438	400
SECT (prestrained)	LT	0.028	NA	66
SECB	LT	0.017	NA	NA
SECT	TL	0.018	338	319
PTST	TL	0.034	382	187
SECB	TL	0.015	NA	NA
SECT (−30°C)	TL	0.014	225	119

[a] Abbreviations explained in Table 4.
[b] NA = not available.

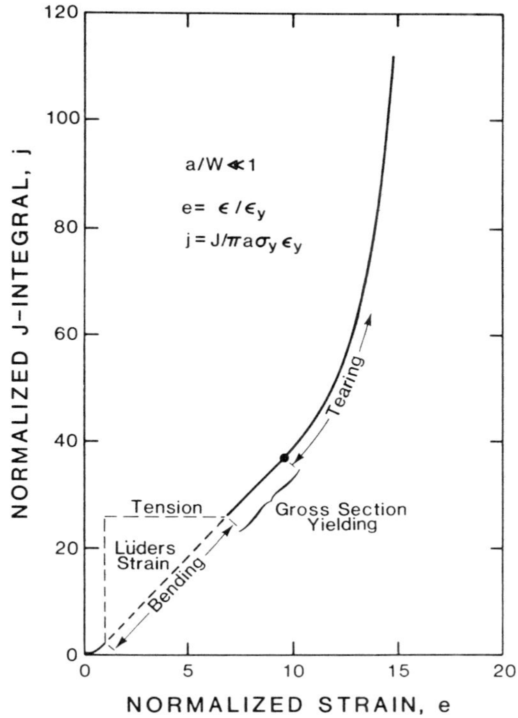

FIG. 7—*Schematic interpretation of the behavior of normalized total J-integral as a function of strain for small cracks in tension and bending specimens.*

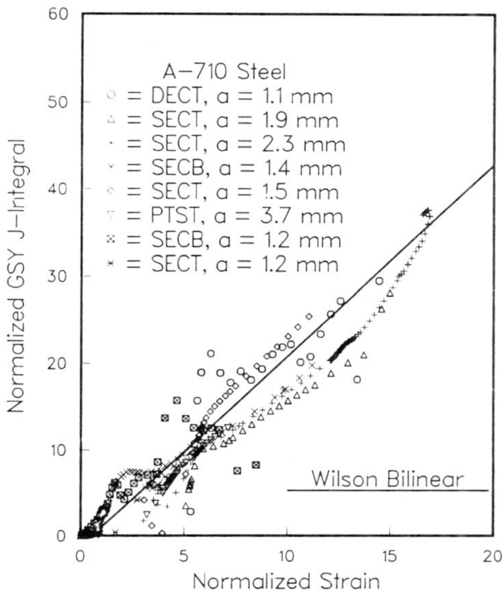

FIG. 8—*Data points: experimentally measured J-integral values corrected to give only the gross section yielding contribution. Solid curve: theoretical prediction made using Wilson's equations for a bilinear material.*

relationship given by Wilson [*10*] for a bilinear material, which is similar to expressions given earlier by Begley, Landes, and Wilson [*6*] and by Merkle [*7*], is the correct expression of the dependence of J-integral on flaw size and applied strain for small through cracks in infinitely wide plates.

The J-integral values at the initiation of tearing in the tensile panels were considerably higher than those in the side-grooved three-point-bend fracture toughness specimens, as can be seen from a comparison of Table 3 and Table 5. The initiation J-values in the tensile panels were slightly above those in the non-grooved, three-point-bend toughness specimens. The cause of this discrepancy is believed to be the low geometrical constraint in the shallow-cracked tensile and bending panels as compared to the deeply cracked, side-grooved, three-point-bend fracture toughness specimens [*17*]. The effect of specimen geometry on crack tip strain fields has been pointed out by McMeeking and Parks [*18*]. The present results indicate that the use of toughness values from three-point-bend specimens for predictions of crack initiation in elements with small cracks is conservative and may be excessively conservative for certain cases, especially if the three-point-bend specimens are side-grooved.

Conclusions

Experimental results on direct measurement of the J-integral in alloy steel panels with small relative crack sizes loaded in tension and bending allow the following conclusions to be drawn:

1. Loading in bending suppresses Lüders band effects that influence J-integral values measured on ASTM A-710 Grade A Class 3 alloy steel specimens loaded in tension.
2. When remote strain in tension is equated with remote outer fiber strain in bending, the behavior of the J-integral as a function of remote strain for panels containing small cracks is the same in bending as in tension, except for the step associated with Lüders strains in the tensile results. Comparisons of tensile results for two alloys, including a result at −30°C and results for transversely and longitudinally oriented specimens, confirm the theoretically predicted commonality of behavior of the normalized J-integral for specimens with small relative crack sizes.
3. The total J-integral measured in panels with relative crack sizes of 1 to 3% followed roughly a theoretical prediction made using formulas given by Wilson [*10*] for a Ramberg-Osgood material in plane stress; however, this agreement is believed to be accidental.
4. Correction of the J-integral results to include only the gross-section-yielding contribution reduced the J-integral values by about a factor of 2. The corrected values agreed approximately with predictions made using formulas given by Wilson [*10*] for a bilinear material in plane stress. The corrected J-integral values are believed to be more representative of the behavior of cracks in infinite panels.
5. The toughness and tearing resistance, as measured by the J-integral, of panels with small relative crack sizes were found to be significantly higher than the toughness and tearing resistance of side-grooved standard three-point-bend test specimens, and slightly higher than values from non-side-grooved specimens.

Acknowledgments

This work was supported by the Naval Sea Systems Command under Project SEA 05 R 25, and monitored by J. P. Gudas of the David W. Taylor Naval Ship Research and Development Center, Annapolis, Maryland. The experimental assistance of J. D. McColskey and helpful discussions with H. I. McHenry and J. P. Gudas are appreciated.

References

[*1*] Read, D. T., "Experimental Method for Direct Evaluation of the J-Contour Integral" in *Fracture Mechanics: Fourteenth Symposium—Volume II: Testing and Applications, ASTM STP 791,* J. C.

Lewis and G. Sines, Eds., American Society for Testing and Materials, Philadelphia, 1983, pp. II-199–II-213.

[2] Rice, J. R., "A Path Independent Integral and the Approximate Analysis of Strain Concentrations by Notches and Cracks," *Journal of Applied Mechanics*, Vol. 35, June 1968, pp. 379–386.

[3] Hutchinson, J. W., "Singular Behavior at the End of a Tensile Crack in a Hardening Material," *Journal of the Mechanics and Physics of Solids*, Vol. 16, 1968, pp. 13–31.

[4] Rice, J. R. and Rosengren, C. F., "Plane Strain Deformation Near a Crack Tip in a Power-Law Hardening Material," *Journal of the Mechanics and Physics of Solids*, Vol. 16, 1968, pp. 1–12.

[5] Ernst, H. A., "Material Resistance and Instability Beyond J-Controlled Crack Growth" in *Elastic-Plastic Fracture: Second Symposium, Volume I—Inelastic Crack Analysis, ASTM STP 803*, C. F. Shih and J. P. Gudas, Eds., American Society for Testing and Materials, Philadelphia, 1983, pp. I-191–I-213.

[6] Begely, J. A., Landes, J. D., and Wilson, W. K., "An Estimation Model for the Application of the J-Integral" in *Fracture Analysis, ASTM STP 560*, American Society for Testing and Materials, Philadelphia, 1974, pp. 155–169.

[7] Merkle, J. G., "Analytical Relations Between Elastic-Plastic Fracture Criteria," Technical Report ORNL/NUREG/TM-27, Oak Ridge National Laboratory, Oak Ridge, TN, 1976.

[8] Turner, C. E., "Elastic-Plastic Aspects of Fracture Stress Analysis: Methods for Other than Standardized Test Conditions" in *Fracture Mechanics in Design and Service, Philosophical Transactions of the Royal Society of London*, Vol. A299, 1981, pp. 73–92.

[9] Dawes, M. G., *Welding Journal Research Supplement*, Vol. 53, 1974, pp. 369s–379s.

[10] Wilson, W. K., "J-Integral Estimate for Small Edge and Interior Cracks," *Engineering Fracture Mechanics*, Vol. 20, No. 4, 1984, pp. 655–665.

[11] Shih, C. F., German, M. D., and Kumar, V., "An Engineering Approach for Examining Crack Growth and Stability in Flawed Structures," Technical Report SRD-80-94, General Electric Company, Schenectady, NY, 1980, pp. A1–A47.

[12] He, M. Y. and Hutchinson, J. W., "Bounds for Fully Plastic Crack Problems for Infinite Bodies" in *Elastic-Plastic Fracture, Volume I—Inelastic Crack Analysis, ASTM STP 803*, C. F. Shih and J. P. Gudas, Eds., American Society for Testing and Materials, Philadelphia, 1983, pp. I-277–I-290.

[13] Read, D. T. and McHenry, H. I., "Interim Progress Report: J-Integral Method for Fitness-for-Service Assessment," Technical Report NBSIR 81-1648, National Bureau of Standards, Boulder, CO, May 1981.

[14] Read, D. T., "Applied J-Integral in HY130 Tensile Panels and Implications for Fitness for Service Assessment," Technical Report NBSIR 82-1670, National Bureau of Standards, Boulder, CO, July 1982.

[15] Read, D. T., "Experimental Results for Fitness-for-Service Assessment of HY130 Weldments," Technical Report NBSIR 84-1699, National Bureau of Standards, Boulder, CO, March 1984.

[16] Read, D. T., "Applied J-Integral Values in Steel Plates," Technical Report, to be published.

[17] Li, Q.-F., "A Study About J_i and Detail in Three-Point Bend Specimens with Deep and Shallow Notches," *Engineering Fracture Mechanics*, Vol. 22, No. 1, 1985, pp. 9–15.

[18] McMeeking, R. M. and Parks, D. M., "On Criteria for J-Dominance of Crack-Tip Fields in Large-Scale Yielding," in *Elastic-Plastic Fracture, ASTM STP 668*, J. D. Landes, J. A. Begley, and G. A. Clarke, Eds., American Society for Testing and Materials, Philadelphia, 1979, pp. 175–194.

Bernard Faucher[1] *and William R. Tyson*[1]

A Study of Variability, Size, and Temperature Effects on the Fracture Toughness of an Arctic-Grade Steel Plate

REFERENCE: Faucher, B. and Tyson, W. R., **"A Study of Variability, Size, and Temperature Effects on the Fracture Toughness of an Arctic-Grade Steel Plate,"** *Fracture Mechanics: Eighteenth Symposium, ASTM STP 945,* D. T. Read and R. P. Reed, Eds., American Society for Testing and Materials, Philadelphia, 1988, pp. 164–178.

ABSTRACT: The fracture toughness of a 75-mm-thick plate of an LT60 grade steel developed by Algoma Steel has been evaluated by three-point bend tests on specimens 10, 20, and 50 mm thick. The fracture toughness measured by the value of the J-integral at onset of cleavage fracture is higher with thin specimens than with thick ones. The statistical model of Tyson and Marandet describes the temperature dependence, the experimental scatter, and, to a considerable extent, the size dependence of fracture toughness. Loss of constraint in thin specimens may be significant when the size requirements of the ASTM Test Method for J_{Ic}, a Measure of Fracture Toughness (E 813-81) are not met.

KEY WORDS: fracture toughness, J-integral, constraint, statistical failure, temperature, thickness, Lloyd's LT60, microalloyed steel, Arctic use

The exploitation of gas and oil resources in the Canadian east coast and Beaufort Sea regions requires materials with toughness properties better than those specified for North Sea use. The evaluation of fracture toughness with full-thickness specimens is inconvenient with heavy-section plates. The use of subsize specimens requires caution because of the size requirement of the various standard test methods and because the toughness may be thickness dependent in the ductile-brittle transition-temperature region. Experimental scatter in this region adds to the difficulty of measuring transition temperature reliably.

Thickness may influence fracture toughness through three factors. First, the larger constraint experienced by thicker specimens promotes plane strain fracture and a higher transition temperature [*1*]. Second, thickness may influence average toughness owing to the statistical nature of the fracture process. It has been established [*2*] that cleavage fracture of the ferrite matrix in steels initiates at cracked carbides subjected to a critical stress σ_c given (for spherical carbides) by

$$\sigma_c = [\pi E\gamma/(1 - \nu^2)r]^{1/2} \tag{1}$$

[1]Research scientist and section head, respectively, Physical Metallurgy Research Laboratories, Canada Centre for Mineral and Energy Technology, Ottawa, Canada K1A 0G1.

where

E = elastic modulus,
ν = Poisson's ratio,
γ = effective cleavage energy of the ferrite, and
r = radius of the carbide.

Cleavage requires that this critical stress exist over a critical distance [*3*] so that there is at least one cracked carbide close to the crack. The number of carbides within a given distance of the crack tip is directly proportional to the thickness; therefore, the critical distance and fracture toughness decrease with increasing thickness. Several models [*4–9*] have been proposed to quantify this statistical behavior. Third, there may be microstructural changes across the thickness of heavy-section plates, leading to different toughness properties for specimens sampling different regions of the plate.

The present work investigates the influence of these various factors on the fracture toughness of a low alloy clean steel intended for Arctic use.

Experimental Methods

Specimens of three different thicknesses have been tested in the transition temperature range. All the specimens were cut from the same 75-mm-thick plate that was partially controlled-rolled and then normalized. This microalloyed steel was produced by Algoma Steel [*10*] to meet the stringent requirements of Lloyd's LT60 specification. Its composition is given in Table 1, and some of its properties, as evaluated by Algoma [*10*], are given in Table 2.

Three-point-bend specimens were prepared according to the ASTM Standard Test Method for J_{Ic}, A Measure of Fracture Toughness (E 813-81), in three different sizes with thickness B of 10, 20, and 50 mm. About 20 specimens of each size were cut from the center of the plate, and an additional set of 10-mm-thick specimens was cut near the surface of the plate, all of them in the longitudinal-transverse (L-T) orientation. After fatigue precracking to about 60% of the width W, 60-deg V-grooves $0.1B$ deep were machined on each side of the specimens. Fracture planes for specimens of the three sizes are shown in Fig. 1. Tension specimens of the Hounsfield

TABLE 1—*Composition of Algoma LT60 steel.*

Element	Weight %	Element	Weight %
carbon	0.12	chromium	0.17
silicon	0.28	copper	0.25
manganese	1.39	molybdenum	0.001
phosphorus	0.017	nickel	0.15
sulfur	0.011	niobium	0.041
aluminum	0.027	vanadium	0.053

TABLE 2—*Properties of Algoma LT60 75-mm plate* [10].

Room temperature yield strength:	370 MPa
ultimate tensile strength:	506 MPa
elongation in 50 mm:	32%
Charpy V-notch impact energy at −60°C:	206 ± 96 J
Metallographic examination ASTM cleanness (core):	B1
ASTM grain size (core):	7½ to 8

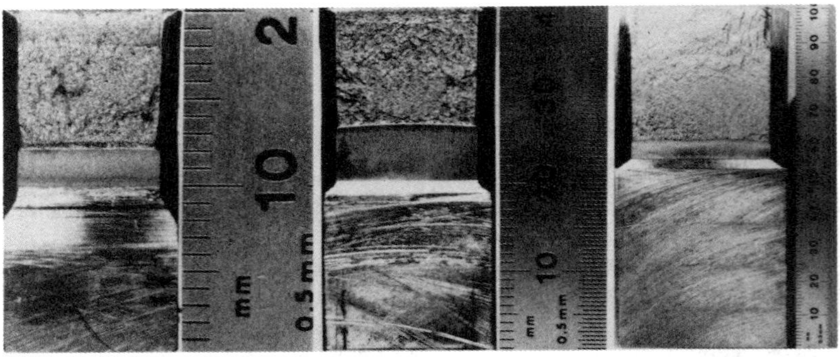

FIG. 1—*Fracture surfaces of the 10, 20, and 50-mm-thick specimens showing the machined notch, fatigue precrack, side grooves, and fractured area. Specimens broken, respectively, at −90, −70, and −60°C. (Note the different magnifications.)*

type, 16 mm long and 4.53 mm in diameter, were also made from both center and surface locations, with their axes in the rolling direction.

The three-point-bend specimens were loaded in MTS computer-controlled testing machines and a rate of stress-intensity increase of about 14, 18, and 24 MPa · $m^{1/2}$/s was applied to the 10, 20, and 50-mm-thick specimens, respectively. The time to reach 40% of the limit load was around 0.03 min for the 10- and 20-mm-thick and 0.04 min for the 50-mm-thick specimens. This rate is somewhat higher than the "static" rate defined by ASTM Standard Test Method for Plane-Strain Fracture Toughness of Metallic Materials (E 399-83) or ASTM E 813. The tests were carried out under displacement control, 10% unloadings under load control being applied to measure specimen compliance at regular displacement intervals. Before each unloading, the machine was stopped for one minute in which 10% to 15% stress relaxation occurred. Crack-length change and J-integral were subsequently evaluated from the compliance and the displacement, measured at the mouth of the crack, as described previously [*11*]. The load-line displacement was measured simultaneously to evaluate the plastic rotation factor which was found to be 0.43, 0.51, and 0.54 for the 10, 20, and 50-mm-thick specimens, respectively. Tensile tests were carried out at a constant displacement rate of 0.2 mm/s.

Specimen, loading fixture, and clip gages were enclosed in a chamber maintained at a temperature constant to ±1°C by a circulating stream of gaseous nitrogen. The clip gages were calibrated at several temperatures.

Experimental Results

Tension test results are typical of annealed low-carbon steels with upper and lower yield points and work-hardening after some Lüders strain. Figure 2 shows a stress-strain curve measured at −80°C and compared with a power-law approximation to describe workhardening, that is

$$\epsilon_p = A(\sigma/\sigma_0)^N \tag{2}$$

where

ϵ_p = true plastic strain,
σ = true stress,
A, σ_0 = empirical constants, and
N = stress exponent.

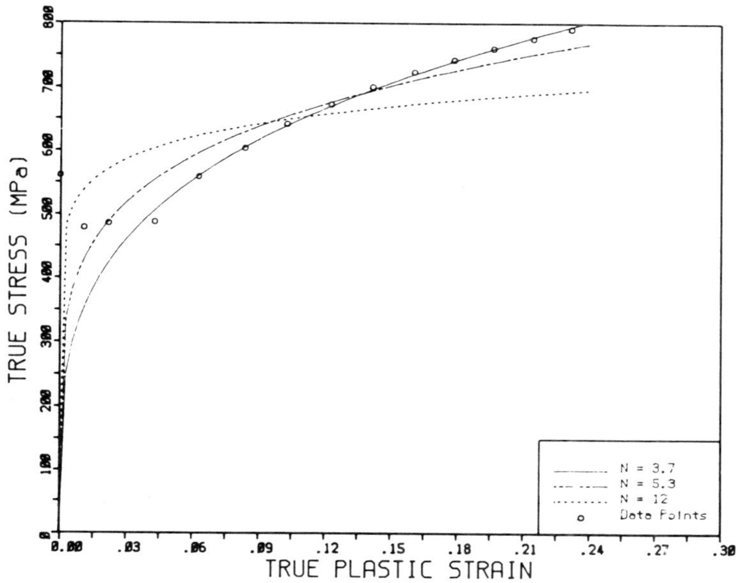

FIG. 2—*Typical tension test at −80°C. The experimental data points have been fitted with a power-law relationship using the indicated values of the stress exponent.*

Results of the tension tests (Table 3), the temperature dependence of the lower yield stress, and the ultimate strength shown in Fig. 3 indicate that there is no significant difference between the specimens taken from the surface or the center of the plate. This was also verified by metallographic examination of the longitudinal section (Fig. 4), which reveals a slightly banded microstructure with fine grains approximately 20 μm diameter at both center and surface locations.

Results of typical bend tests are shown in Fig. 5. At very low temperatures (less than about −150°C depending on the specimen size), brittle failure occurred before yielding; at higher temperatures (above about −60°C), the specimens were unloaded if they did not cleave after extensive bending, and were heat-tinted to measure crack extension. In most of the tests, only crack-tip blunting preceded cleavage fracture, even after gross plastic deformation. This was

TABLE 3—*Tension test results from Hounsfield specimens taken from center (C) and surface (S) locations.*

Temperature, °C	Upper Yield, MPa		Lower Yield, MPa		Ultimate Tensile Strength, MPa		Elongation, %		Reduction of Area, %	
	C	S	C	S	C	S	C	S	C	S
−160	800	769	702	678	763	750	49	51	66	68
−130	663	654	596	592	700	686	54	57	69	71
−100	664	613	528	510	670	643	54	57	69	71
−80	557	545	467	463	625	614	55	57	74	75
−60	525	491	435	437	614	590	55	58	74	76
−40	...	434	387	412	597	570	51	56	73	76
−10	440	409	393	384	557	541	50	58	72	77
+20	...	401	340	363	546	513	55	58	75	77

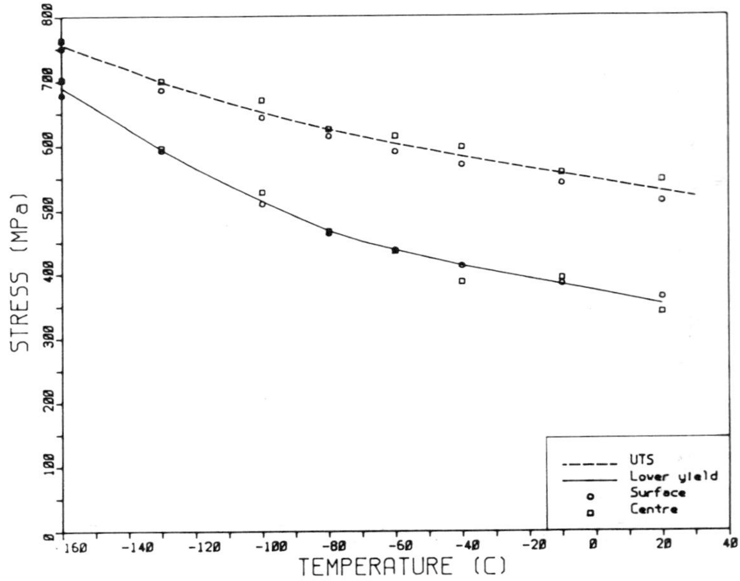

FIG. 3—*Temperature dependence of lower yield stress and of UTS measured with specimens taken from surface and center location of the plate.*

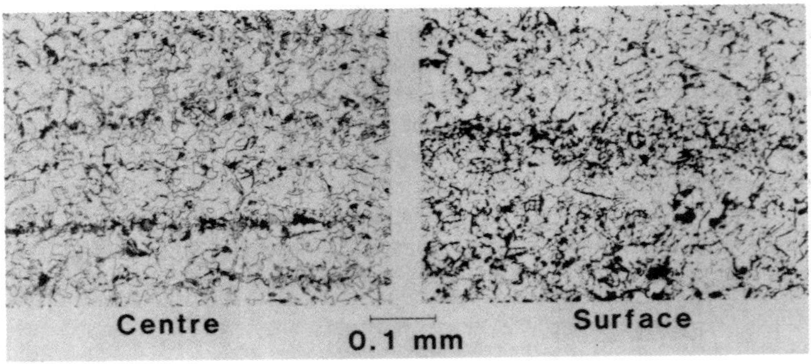

FIG. 4—*Longitudinal sections in the center and close to the surface of the plate.*

apparent from visual examination of fracture surfaces (Fig. 1) and from compliance measurements. It was confirmed also by scanning electron microscopy (Fig. 6) that ductile dimples were to be seen only at a few locations along the stretch zone (Figs. 6*c* and 6*d*). Observation of a section through a nickel-plated fracture surface of a typical specimen (Fig. 7) reveals extensive plastic deformation (Fig. 7*c*) in the process zone close to the stretch zone. Cleavage microcracks that did not propagate are observed adjacent to the fracture surface. A small amount of ductile tearing, confined to 0.1 mm at the tip of the crack, is evident upon close examination of Fig. 7*a*.

Values of the J-integral were estimated at the onset of cleavage. However, because there is negligible stable crack extension, it is not possible to evaluate J_{Ic} according to ASTM E 813. Thus, the critical values for cleavage initiation are indicated as J_c in Table 4. Table 4 also com-

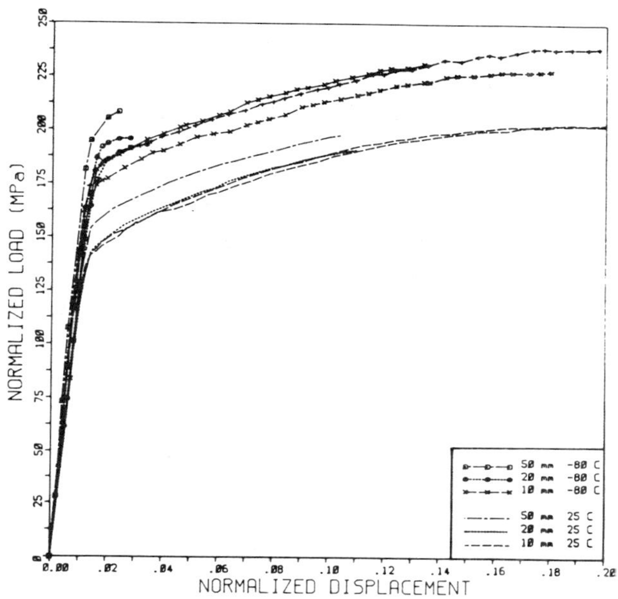

FIG. 5—*Normalized load* ($PW/(B_NEb^2)$)*-displacement* (Δ/W) *curve for three-point-bend specimens;* b *is the ligament length* (W − a) *where* a *is the crack length.*

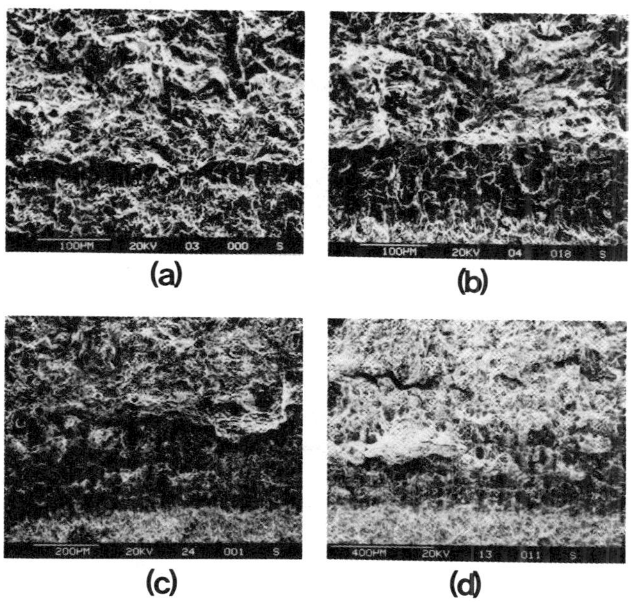

FIG. 6—*Scanning electron micrograph of fracture surfaces showing fatigue precrack, stretch zone, and cleavage area:* (a) *50-mm-thick specimen,* J_c = *45 kJ/m² at −100°C;* (b) *20 mm, 175 kJ/m² at −100°C;* (c) *10 mm, 450 kJ/m² at −100°C;* (d) *50 mm, 440 kJ/m² at −60°C.*

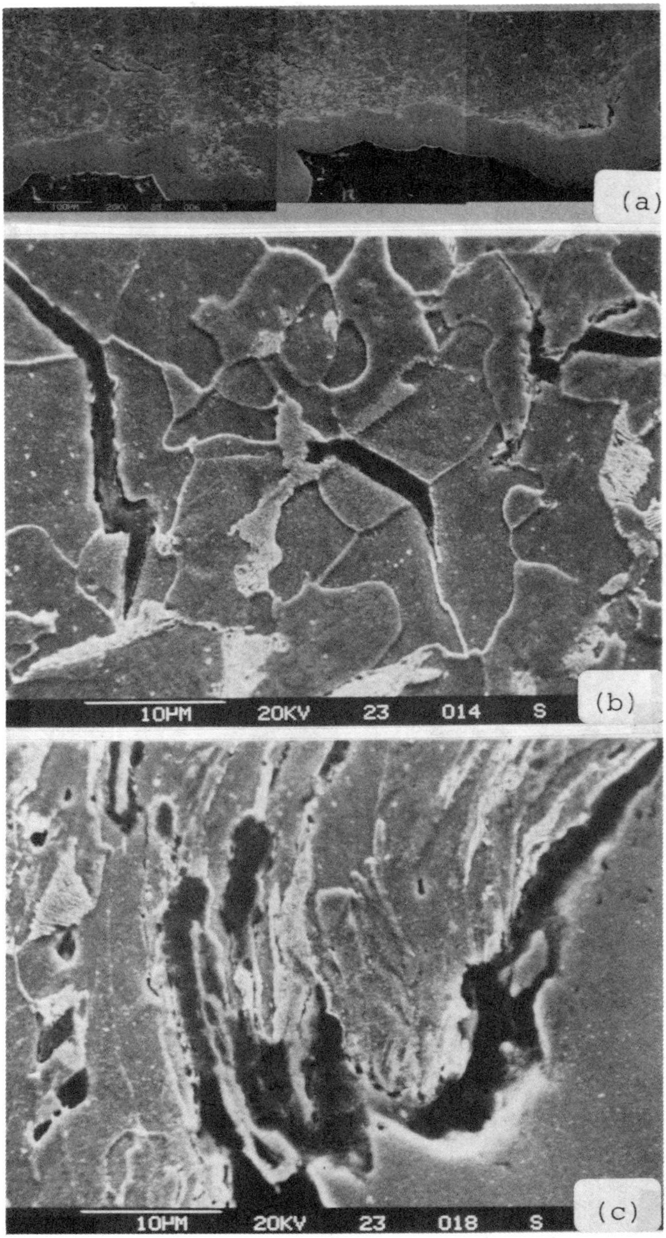

FIG. 7—*Section through the fracture surface of a 10-mm specimen which cleaved at −110°C at 293 kJ/m²*. (b) *Detail of a nonpropagating crack;* (c) *Detail of the process zone at the crack tip (extreme right of* a*).*

TABLE 4—*Fracture toughness of Algoma LT60 steel and comparison of measured stretch zone width* (Δa_m) *with ASTM estimate of crack-tip blunting* $\Delta a_b = J_c/(2\sigma_F)$. *The 10-mm-thick specimens are from surface or center location.*

	10-mm (Surface)			10-mm (Center)			20-mm			50-mm		
T, °C	J_c, kJ/m^2	Δa_m, mm	Δa_b, mm	J_c, kJ/m^2	Δa_m, mm	Δa_b, mm	J_c, kJ/m^2	Δa_m, mm	Δa_b, mm	J_c, kJ/m^2	Δa_m, mm	Δa_b, mm
−160	4	...	...	5	...	...	4	...	...	...	...	...
−150	10	...	...	3	...	...	...	...	...	...	...	...
−140	7	...	...	6	...	...	16	...	...	9	...	...
−140	45	...	...	68	...	...	...	...	...	...	...	...
−130	67	...	...	42	...	...	18	...	...	...	...	...
−120	27	...	...	22	...	...	20	...	...	31	...	...
−120	83	...	...	49	...	...	63	...	...	...	...	...
−110	177	0.08	0.15	94	...	...	25	...	...	...	...	...
−110	224	0.12	0.19	215	...	...	60	...	...	...	...	...
−110	...	...	...	293	0.22	0.24	...	...	...	...	...	...
−100	517	0.50	0.45	237	0.25	0.20	54	...	...	45	...	...
−100	715	0.72	0.62	443	0.37	0.38	175	...	...	...	...	...
−100	909	0.99	0.79	...	...	...	...	...	...	...	...	...
−90	451	0.36	0.40	...	...	...	52	...	...	50	...	...
−90	612	0.82	0.55	...	...	...	130	...	...	...	...	...
−80	530	0.49	0.53	369	0.29	0.34	143	0.10	0.13	50	...	...
−80	670	0.67	0.76	572	0.48	0.53	150	...	...	108	...	...
−80	727	1.30	0.67	...	...	...	...	...	...	170	...	...
−80	...	...	...	...	...	...	...	...	...	316	...	...
−75	...	...	...	...	...	...	135	...	...	...	...	...
−75	...	...	...	...	...	...	947	...	...	...	...	...
−70	...	...	...	...	...	...	436	...	...	288	...	...
−70	...	...	...	...	...	...	620	...	...	396	0.24	0.37
−60	555	...	...	654[a]	1.53	0.63	(880[b])	0.93	0.85	440	...	...
−60	698[a]	1.58	0.68	668[a]	0.81	0.64	...	...	...	468	...	...
−60	...	...	...	...	...	...	...	...	...	520	0.45	0.51
−60	...	...	...	...	...	...	...	...	...	625	0.76	0.61
−60	...	...	...	...	...	...	...	...	...	733	0.65	0.71
−50	...	...	...	...	...	...	...	...	...	625	...	...
−50	...	...	...	...	...	...	...	...	...	(877[b])	1.53	0.87
+25	548[a]	1.49	0.62	656[a]	1.70	0.76	(620[b])	0.72	0.70	(1400[b])	2.79	1.67

[a]No brittle failure, J_c at maximum load.
[b]No brittle failure, specimen unloaded before maximum load, J_c at unloading.

pares the crack-tip blunting Δa_m, measured on the fracture plane (Fig. 1) from the length of the stretch zone, with the value obtained from the ASTM E 813 blunting line

$$\Delta a_b = 0.5J/\sigma_F$$

where the flow stress σ_F is the average of the lower yield stress and the ultimate strength (UTS). Good agreement between these two quantities confirms that there is no significant tearing before cleavage.

The critical values of J have been converted to fracture toughness, K_J, by the relation

$$K_J = [EJ_c/(1 - \nu^2)]^{1/2}$$

where E and ν have been taken as 207 GPa and 0.3, respectively. The temperature dependence of K_J for the specimens that failed by cleavage is shown in Figs. 8 to 10. The curves that have been drawn through the data points represent a model that will be discussed later. They show the rapid increase of fracture toughness with temperature for each specimen thickness. The same curve was used for the 10-mm-thick specimens, from both the surface and the center of the plate, because there was no noticeable difference in their behavior. The figures show that there is significant variability in experimental results, even for the 50-mm-thick specimens.

Discussion

Temperature Dependence

The effect of temperature on cleavage fracture is closely associated with temperature dependence of the yield stress. As the temperature increases, the yield stress decreases and, conse-

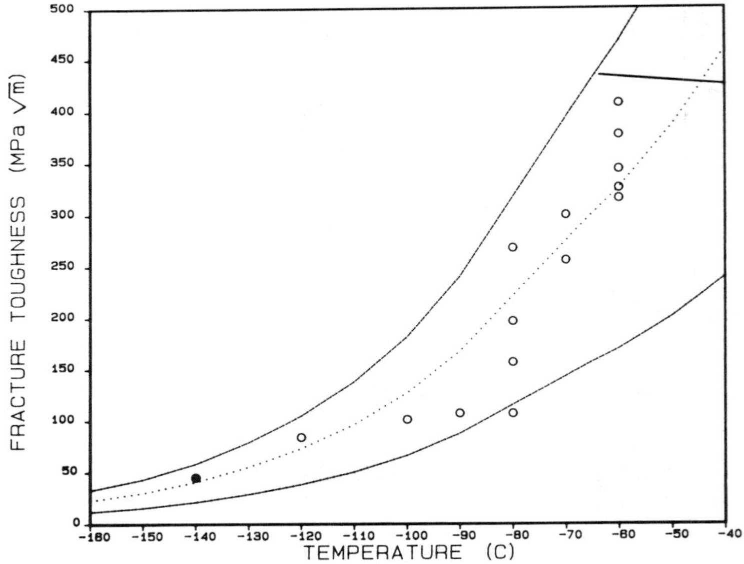

FIG. 8—*Temperature dependence of the cleavage fracture toughness of 50-mm-thick specimens. The curves evaluated from Eq 6 correspond to 50%* (dotted) *or 5 and 95%* (dashed) *failure probability. The filled-in data point meets ASTM E 399 requirements, and the short segment indicates ASTM E 813 validity limit.*

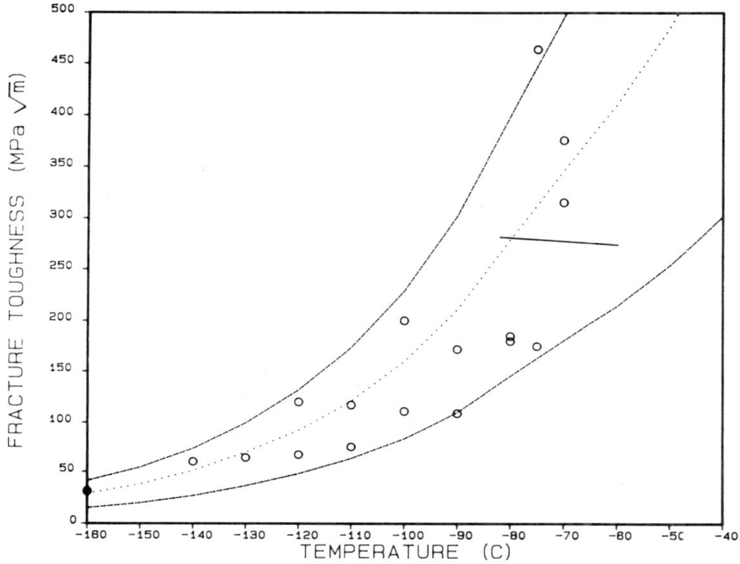

FIG. 9—*Same as Fig. 8 for 20-mm-thick specimens.*

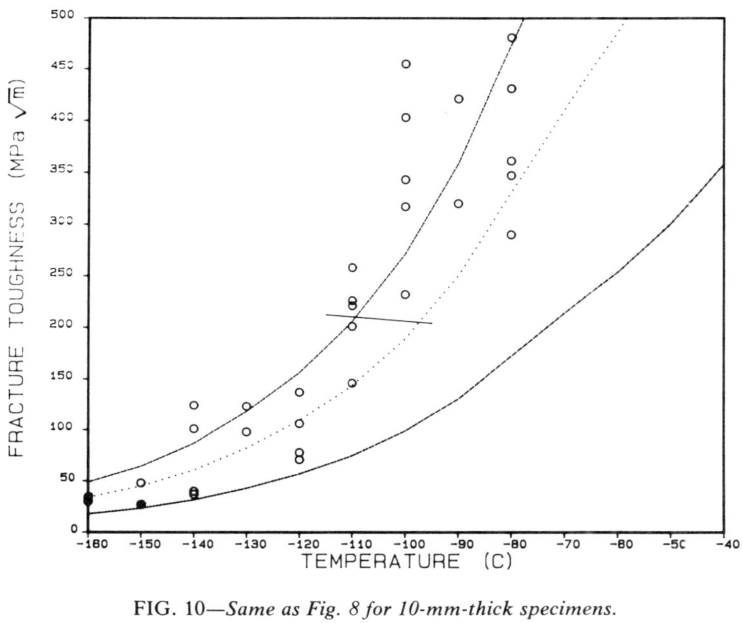

FIG. 10—*Same as Fig. 8 for 10-mm-thick specimens.*

quently, so does the stress ahead of the crack tip at a given value of J (or K). Thus, J must increase to achieve the stress given by Eq 1. This is the principal reason why most statistical models [6-9] predict between the fracture toughness K and the yield stress σ_Y a relationship of the form

$$K = K_0(\sigma_Y/\sigma_0)^{-m} \tag{3}$$

where K_0, σ_0, and m are constants. Hence, in Fig. 11 the temperature dependence of K_J has been expressed by plotting ln K_J as a function of the logarithm of the lower yield stress obtained from Fig. 3. The results indicate a good correlation with slopes m equal to 7.0, 5.1, and 5.8 for the 10, 20, and 50-mm-thick specimens, respectively.

Experimental Scatter

Variability in fracture toughness is related to the size distribution of cleavage nuclei. For example, the statistical model of Tyson and Marandet [9], based on the weakest-link concept, relates the cumulative probability of failure ϕ to the fracture toughness and yield stress by

$$\ln[1/(1 - \phi)] = CB\sigma_Y^{\alpha(N-1)/2}K^{\alpha} \tag{4}$$

where

C = material constant dependent on the characteristics of the cleavage nuclei operating in that material,
B = specimen thickness,
N = stress exponent as defined in Eq 2, and
α = parameter found experimentally to be approximately 4 [9].

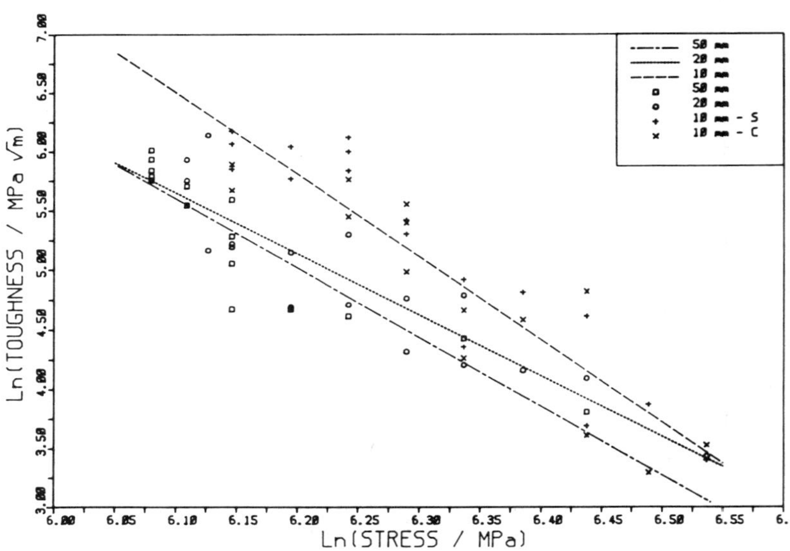

FIG. 11—*Dependence of the fracture toughness on the lower yield stress for several thicknesses.*

Although the cleavage nuclei have not been identified for the steel studied here, Eq 4 can be used to describe the experimental results with C regarded as a material parameter to be fitted using experimental data. Equation 4 may be written as

$$\ln K = -(1/\alpha)\ln(BC) - [(N-1)/2]\ln \sigma_Y + (1/\alpha)\ln \ln[1/(1-\phi)] \tag{5}$$

As interpreted by Bishop et al. [5], the fracture toughness depends on a deterministic term and a random term. The deterministic term,

$$-(1/\alpha)\ln(BC) - [(N-1)/2]\ln \sigma_Y$$

takes into account the effects of temperature (through σ_Y) and of thickness on fracture toughness, while variability is contained in the "random" term

$$(1/\alpha)\ln \ln[1/(1-\phi)]$$

A first estimation of the deterministic term, that is $(\ln BC)/\alpha$ and N, is obtained from a linear regression of $\ln K$ against $\ln \sigma_Y$. The "random" term may be taken as the difference between the measured value of $\ln K$ and the deterministic term. Then, the failure probability ϕ is obtained by ranking the random terms; for the specimen with the nth lowest random term

$$\phi = n/(1 + \Sigma\, n)$$

Linear regression of the random term against $\ln \ln[1/(1-\phi)]$ gives the value of α. With these estimates of ϕ and α, the random part is subtracted from the experimental values of $\ln K$ in Eq 5, and a better estimate of the deterministic part is obtained. After a few iterations convergence was obtained with the parameters given in Table 5. This indicates that Eq 5 adequately describes the temperature dependence of fracture toughness and the experimental scatter in cleavage fracture. This is shown in Fig. 8 where the experimental results for 50-mm specimens are compared with the predictions of Eq 5 for several failure probabilities.

The experimental values of α are very close to the value of 4 found by Tyson and Marandet [9]. The stress exponent N is about 12, which does not provide a very good fit to the stress-strain curve of Fig. 2. However, the comparison is complicated by the existence of an upper yield point and extensive Lüders strain. Curve-fitting of the bend test results (Fig. 5) yielded stress exponents in the range 10 to 15.

Thickness Effect

As explained earlier, thickness effects can be associated with three different factors: constraint, statistical variability, and microstructural inhomogeneity. The results have shown that

TABLE 5—*Experimental parameters for statistically describing fracture toughness according to Eq 5 when* K *is expressed in* $MPa\sqrt{m}$, σ_Y *in MPa, and* B *in mm.*

Specimen Size, mm	Number of Specimens	α	N	$\ln C$	Correlation Coefficient
50	15	4.03	12.6	−170	−0.996
20	17	3.49	12.5	−146	−0.995
10	30	3.26	14.1	−154	−0.995

the third factor is not important in the present case, but statistical variability and constraint remain.

The statistical model of Tyson and Marandet [9] predicts a thickness dependence of fracture toughness described by Eq 5. The results obtained for the various thicknesses in Table 5 are combined together by using the values of α and N of the 50-mm-thick specimens and an average value of the constant C calculated for 665 MPa and 50% failure probability. Thus, the fracture toughness of the LT60 steel may be obtained at any temperature (yield stress), thickness, and failure probability from the relation

$$\ln K = 41.25 + 0.248 \ln(50/B) - 5.82 \ln \sigma_Y + 0.248 \ln \ln[1/(1 - \phi)] \quad (6)$$

where the stress is expressed in MPa, thickness in mm, and fracture toughness in MPa$\sqrt{\text{m}}$. The results in Figs. 8 to 10 show excellent agreement with the experimental values as long as the specimen size meets ASTM E 813 requirements. The predicted temperature dependence of fracture toughness is compared for the three thicknesses in Fig. 12. Depending on the toughness level, a shift of about 15 deg C is obtained between the 10- and 50-mm-thick specimens. This shift calculated from Eq 6 results directly from the higher probability that larger specimens have a weak spot (large value of r in Eq 1) in the high-intensity stress field ahead of the crack tip [9].

Differences between observed and predicted behavior at toughness levels above those considered valid by ASTM E 813, shown particularly by the results for B = 10 mm (Fig. 10), may be attributed to loss of constraint. Constraint can be roughly quantified from the results shown in Fig. 5. Following Anderson et al. [12], the load at general yield P_Y may be expressed as

$$P_Y = L\sigma_Y(W - a)^2B/4W$$

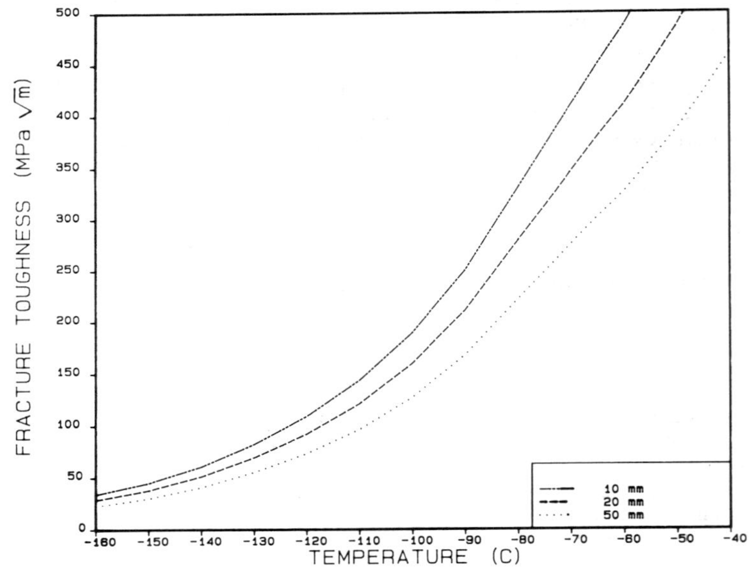

FIG. 12—*Comparison of the fracture toughness at 50% failure probability of 10, 20, and 50-mm-thick specimens.*

where L is the "constraint factor." Using the load at a plastic displacement of $0.01W$ for P_Y, L has been found to increase from 1.66 (± 0.05) for the 10-mm specimens to 1.69 and 1.81 for the 20- and 50-mm-thick specimens, respectively. Although the difference between these numbers is comparable to the experimental uncertainty, it does indicate that, even for side-grooved specimens, the constraint at yield decreases with thickness. At high toughness levels (high loads and appreciable blunting), large process zones develop which may lead to a state of plane stress. This reduces tensile stress levels in thin specimens where the process zone is a significant fraction of the section thickness, and could explain the high-toughness values measured with the 10-mm-thick specimens (and the lateral contraction of these specimens, Fig. 1) when ASTM E 813 size requirements are not met.

Conclusions

The high-toughness microalloyed steel intended for Arctic use studied in this work fractures by cleavage at temperatures lower than about −50°C for specimens 10 to 50 mm thick. However, extensive crack-tip blunting precedes cleavage in the upper part of the transition range. This leads to high values of fracture toughness even without initiation of ductile tearing. The temperature and thickness dependence and the variability of fracture toughness are well described by a statistical model. Temperature effects are related to the temperature dependence of the yield stress and thickness effects to the larger probability of finding a "weakest link" in the increased sampling volume of thick specimens. Constraint effects become significant only when ASTM E 813 size requirements are not met.

Acknowledgments

The help of the technical staff of the Physical Metallurgy Research Laboratories is gratefully acknowledged, particularly the assistance of C. Trinque in carrying out the experimental work.

References

[1] Knott, J. F., *Fundamentals of Fracture Mechanics*, Wiley, New York, 1973.

[2] Curry, D. A. and Knott, J. F., "Effects of Microstructure on Cleavage Fracture Stress in Steel," *Metal Science*, Vol. 12, 1978, pp. 511–514.

[3] Ritchie, R. O., Knott, J. F., and Rice, J. R., "On the Relationship Between Critical Tensile Stress and Fracture Toughness in Mild Steel," *Journal of the Mechanics and Physics of Solids*, Vol. 21, 1973, pp. 395–410.

[4] Landes, J. D. and Shaffer, D. H., "Statistical Characterization of Fracture in the Transition Region," *Fracture Mechanics: Twelfth Conference, ASTM STP 700*, American Society for Testing and Materials, Philadelphia, 1980, pp. 368–382.

[5] Bishop, T. A., Markworth, A. J., and Rosenfield, A. R., "Analyzing Statistical Variability of Fracture Properties," *Metallurgical Transactions A*, Vol. 14, April 1983, pp. 687–693.

[6] Beremin, F. M., "A Local Criterion for Cleavage Fracture of a Nuclear Pressure Vessel Steel," *Metallurgical Transactions A*, Vol. 14, Nov. 1983, pp. 2277–2287.

[7] Evans, A. G., "Statistical Aspects of Cleavage Fracture in Steel," *Metallurgical Transactions A*, Vol. 14, July 1983, pp. 1349–1355.

[8] Wallin, K., Saario, T., and Törrönen, K., "Theoretical Scatter in Brittle Fracture Toughness Results Described by the Weibull Distribution" in *Application of Fracture Mechanics to Materials and Structures*, G. C. Sih, E. Sommer, and W. Dahl, Eds., Martinus Nijhoff, The Hague, The Netherlands, 1984, pp. 511–518.

[9] Tyson, W. R. and Marandet, B., "Cleavage Toughness Variability and Inclusion Size Distribution for a Weld Metal," this volume, pp. 19–32.

[10] Ackert, R. J., McLean, M. K., and Ray, J. D., "LT60 Plate for Arctic Application," presented at the 23rd Annual CIM Conference of Metallurgists, Canadian Institute of Mining and Metallurgy, Quebec City, Aug. 1984.

[11] Faucher, B. and Tyson, W. R., "A Comparison of Crack-Mouth Opening and Load-Line Displacement for J-Integral Evaluation Using Bend Specimens," *Elastic-Plastic Fracture Test Methods: The User's Experience, ASTM STP 856,* E. T. Wessel and F. J. Loss, Eds., American Society for Testing and Materials, Philadelphia, 1985, pp. 278–293.

[12] Anderson, T. L., McHenry, H. I., and Dawes, M. G., "Elastic-Plastic Fracture Toughness Tests with Single-Edge Notched Bend Specimens," *Elastic-Plastic Fracture Test Methods: The User's Experience, ASTM STP 856,* E. T. Wessel and F. J. Loss, Eds., American Society for Testing and Materials, Philadelphia, 1985, pp. 210–229.

Note Added in Proof

The load readings for 50-mm-thick specimens were inadvertently overestimated by a factor of 1.112. The L-factor for 50-mm-thick specimens, therefore, should be reduced to 1.63, about the same as for thinner specimens. The J_c-values in Table 4 should also be reduced by about the same factor, which would reduce the K-values by roughly 6%. These changes would slightly modify the constant C in Table 5 and the final result of Eq 6, without affecting the validity and the trend of the main conclusions.

Elevated Temperature Fatigue

Tusit Weerasooriya[1] *and Ted Nicholas*[2]

Overload Effects in Sustained-Load Crack Growth in Inconel 718

REFERENCE: Weerasooriya, T. and Nicholas, T., "**Overload Effects in Sustained-Load Crack Growth in Inconel 718,**" *Fracture Mechanics: Eighteenth Symposium, ASTM STP 945,* D. T. Read and R. P. Reed, Eds., American Society for Testing and Materials, Philadelphia, 1988, pp. 181–191.

ABSTRACT: Crack-growth-rate experiments were conducted on compact tension (CT) specimens of Inconel 718 in air at 649°C. The loading spectrum consisted of a single 1 Hz cycle at $R = 0.1$, where R is the ratio of minimum to maximum stress-intensity factors, and a hold time which was applied at various levels of the fatigue cycle. The ratio of the hold-time amplitude to the maximum amplitude of the fatigue cycle, R_{in}, was kept at 1.0, 0.9, 0.8, or 0.5. Hold times were varied from 0 to 200 s. Tests were performed under computer-controlled constant-K conditions using values of the maximum of the fatigue cycle of 40 and 50 MPa m$^{1/2}$. Data show that for $R_{in} = 1.0$, a linear summation model works well, while at $R_{in} = 0.9$ there is a measurable retardation effect on the crack growth during the hold time. For values of R_{in} less than or equal to 0.8, the sustained-load crack growth is almost completely retarded. A simple retardation model is proposed which can fit the experimental data and is based on the concept of an overload plastic zone being produced by the fatigue cycles. It is concluded that hold times do not contribute to crack growth in this material unless their amplitude is at or near the maximum amplitude of the adjacent fatigue cycles.

KEY WORDS: crack growth, fatigue, overloads, crack growth retardation, nickel-base superalloy, elevated temperature

The recent introduction of a damage-tolerant design requirement for critical structural components in U.S. Air Force gas turbine engines has created a need for accurate crack-growth predictions [*1*]. The Engine Structural Integrity Program (ENSIP) requires that a preexisting flaw be assumed present in a component and that fracture mechanics analysis and testing demonstrate that such flaw will not grow to a catastrophic size during the lifetime of the engine. To meet this requirement, crack-growth rates must be predicted under typical mission spectra. For components such as turbine disks and spacers in advanced fighter or attack aircraft, the loading spectrum will generally consist of large amplitude, low frequency fatigue cycles with interspersed hold times. The hold times have durations of tens of seconds with amplitudes anywhere from the maximum of the fatigue cycles to less than half that amount, depending on the type of aircraft and the mission.

Concurrent with a design philosophy change from low-cycle fatigue initiation to a fracture mechanics damage-tolerant approach, engine operating temperatures have increased steadily. This has led to the development of higher temperature nickel-base superalloys and a tendency to use these alloys under conditions in which time-dependent material behavior becomes important. Thus, the effect of hold times in a design spectrum which results in crack growth during the hold times has to be considered as part of the life-prediction methodology. Although there has been substantial effort devoted to the study of elevated temperature sustained-load crack

[1]Research scientist, University of Dayton Research Institute, Dayton, OH 45469.

[2]Senior scientist, Air Force Wright Aeronautical Laboratories, (MLLN) Wright-Patterson Air Force Base, OH 45433.

growth in numerous materials [2–6], there has been little research directed at crack growth under mission spectra, particularly when such spectra involve both sustained loads and fatigue cycles with sustained loads at different levels than the maximum-load amplitudes of the fatigue cycles.

In a previous investigation, a linear cumulative damage model was applied for the prediction of crack-growth rates under simple spectrum loading consisting of fatigue cycles and hold times [7]. For Inconel 718 at 649°C, a material which exhibits a significant amount of time-dependent behavior, it was found that the linear cumulative damage model worked well when hold times were at the same level as the maximum load of the fatigue cycle. Here, the algebraic summation of the fatigue-crack-growth rate as determined from constant-load-amplitude fatigue tests and the sustained-load crack growth as determined from constant-load-crack-growth tests provided excellent correlation with experiments in which the two types of loading were combined. However, when the sustained load level was dropped to 75% of the fatigue maximum, the linear summation model broke down completely. The net growth observed experimentally appeared to be due only to the fatigue cycles, and the linear summation model grossly overpredicted the growth rate under the simple spectrum.

This investigation was undertaken to evaluate the range of validity of a linear summation model and to determine the contribution of hold-time loading in a simple spectrum consisting of combinations of fatigue cycles with hold times of various duration and amplitude. A simple analytical model is developed based on the concept that the sustained-load crack growth is retarded by the application of the overload fatigue cycles.

Experiments

A series of crack-growth-rate experiments was conducted using the simple load spectrum depicted in Fig. 1. The spectrum consists of a single fatigue cycle of frequency 1 Hz, load ratio $R = 0.1$, and maximum stress-intensity factor K_m of either 40 or 50 MPa $m^{1/2}$. Hold times, t_h, of durations varying from a few seconds to several hundred seconds were applied at amplitudes given by $K = R_{in}K_m$ where R_{in} is the ratio of the amplitude of the hold time to the maximum amplitude of the fatigue cycle. Values of R_{in} of 1.0, 0.9, 0.8, and 0.5 were used which correspond to overload ratios of 1.0, 1.11, 1.25, and 2.0, respectively, where the overload ratio is the ratio of K_m in the "overload" fatigue cycle to the baseline K of the hold-time load. All tests were conducted on CT specimens of width $W = 40$ mm and thickness $b = 10$ mm under computer-controlled constant K_m conditions. Crack lengths were determined from values of compliance calculated from the slope of the load-crack mouth opening displacement data. The displacement data were obtained during the unloading portion of the fatigue cycle with the aid of an extensometer. The test procedure is described in detail in Ref *8*.

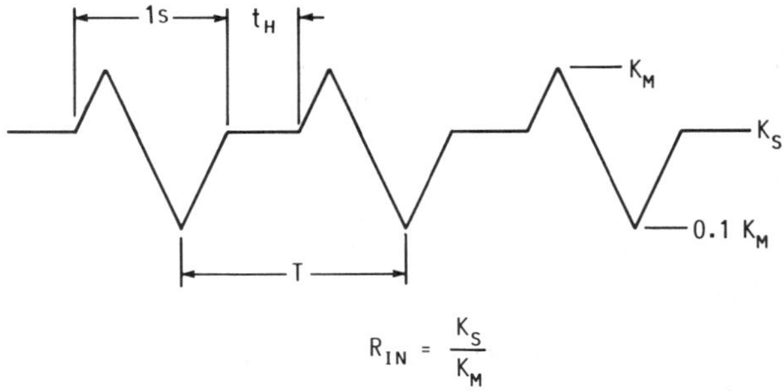

FIG. 1—*Loading spectrum consisting of 1 Hz fatigue cycle and sustained load.*

The material used in this investigation was Inconel 718. It was tested in air at 649°C and was given the standard heat treatment detailed in Table 1. The material is identical to that used in several previous investigations in which it has been shown that it exhibits significant amounts of time-dependent behavior [*7,8*]. Several test conditions were evaluated on a single test specimen under constant K control. For each condition, the crack was grown until it could be determined that the crack was growing at a constant rate within ±5% of the mean value. This was accomplished by continuously monitoring average growth rate using a least-squares linear regression fit to the crack length versus number of cycles data. When no statistically significant deviation (95% confidence band of the growth rate is within ±5% of the average growth rate) from an average growth rate was obtained from additional applied cycles, the growth rate was defined as reaching a steady-state value. This value was used in the plots of cycle block growth rate against total cycle time $T = t_H + 1$ (see Fig. 1). In most cases, the total crack extension used to establish a steady-state growth rate under a given test condition was less than 1 mm and was typically of the order of 0.5 mm. Thus, over 30 test conditions could be evaluated on a single specimen, making it possible to duplicate and verify the growth rate for any given condition. Growth rates in a single specimen were found to be reproducible to an accuracy of 10% under a given load spectrum using K as the crack-growth-rate correlating parameter. Both the reproducibility of the data and the achievement of constant growth rates under constant K conditions suggests that K is an adequate correlating parameter for this material. In addition, both experimental observations and calculation of transition times show that for this material at 650°C, K can be used as the crack-growth correlating parameter for both cyclic and sustained loading [*7,8*].

Experimental Results

Crack growth rate per cycle block, da/dN, was plotted as a function of total cycle time T for the cases of $K_m = 40$ and $K_m = 50$ MPa $m^{1/2}$ in Figs. 2 and 3, respectively. Note that a cycle time of 1 s corresponds to pure fatigue cycling at 1 Hz with no hold time. The symbols in each figure represent the experimentally determined growth rates under constant K conditions for various values of R_{in}. The three solid lines in each figure represent, from top to bottom, the analytical predictions from a linear cumulative damage model for values of R_{in} of 1.0, 0.9, and 0.8, respectively. The linear model numerically adds the growth rate for a single cycle of 1.0 Hz and the contribution from sustained-load crack growth, $\dot{a} \cdot t_H$, where $\dot{a}$ is the growth rate per unit time for the sustained-load crack growth at a stress-intensity factor value K_s. Values of $\dot{a}$ as a function of K were determined from a parallel investigation using specimens from the same material [*9*]. Several constant-load tests were conducted to obtain crack length versus time data which were reduced to $\dot{a}$ versus K and fit with a sigmoidal equation to provide numerical values of $\dot{a}$ for any K value.

The data in Figs. 2 and 3 reconfirm the previous observation that a linear cumulative damage model works well for $R_{in} = 1.0$, that is, when the hold-time load is at the maximum value of the fatigue cycle. For values of $R_{in} = 0.9$, the linear model overpredicts by more than a factor of 2; thus there appears to be some retardation of the sustained-load crack growth. For values of $R_{in} = 0.8$, there does not appear to be any contribution to the growth rate per cycle from the hold-time loading for values of hold times up to 200 s, the maximum used for both K_m values

TABLE 1—*Heat treatment for Alloy 718.*

Step	Procedure
1	anneal at 968°C (1775°F) for 1 h, then air-cool to temperature
2	age-harden at 718°C (1325°F) for 8 h, then furnace-cool at 56°C/h (100°F/h) to 621°C (1150°F)
3	age-harden at 621°C (1150°F) for a total aging time in Step 2 and Step 3 of 18 h
4	air-cool to room temperature

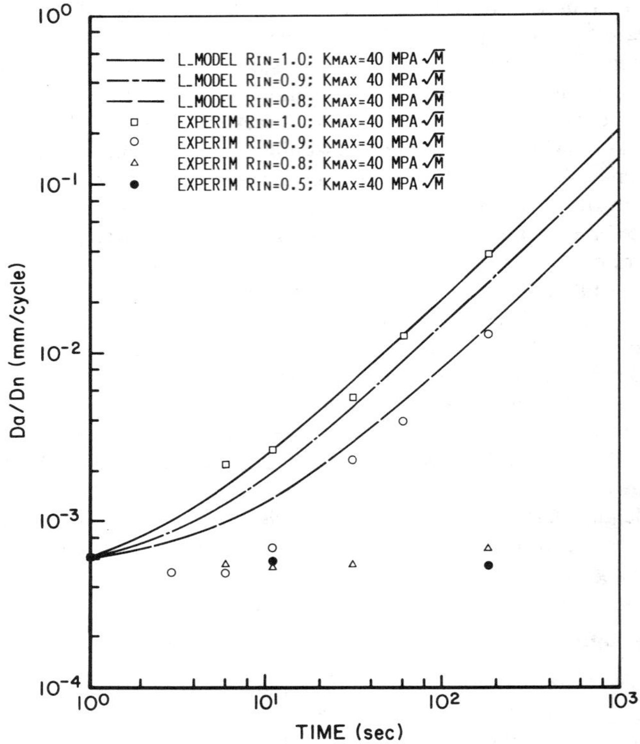

FIG. 2—*Crack-growth rate data plotted against total cycle time for* K_{max} = *40 MPa* $m^{1/2}$ *as a function of different* R_{in} *values. Linear model (L-Model) data are also given in the figure.*

considered in this investigation. The linear model grossly overpredicts the growth rate for this case. The same observation can be made for $R_{in} = 0.5$. For this latter case, results were obtained only for the first set of experiments where $K_m = 40$ MPa $m^{1/2}$. There was no reason to repeat these very time consuming experiments for $K_m = 50$ MPa $m^{1/2}$.

Analytical Model

A simple model is developed to predict the crack-growth rates under the simple spectrum of Fig. 1. The model uses a linear cumulative damage concept by summing the contributions of the single fatigue cycle and the hold-time load. The fatigue cycle is assumed to retard the sustained-load crack growth because the latter occurs within an enlarged or overload plastic zone due to the fatigue cycle. This retardation concept, similar to that used by Wheeler [*10*], is illustrated schematically in Fig. 4. The plastic zones due to the sustained load of stress-intensity factor K_s and the overload fatigue cycle of maximum stress intensity K_m have radii denoted by r_p and $\bar{r}_p$ with $\bar{r}_p \geq r_p$ since $K_s/K_m \leq 1$. If the overload is applied when the crack length is a_0, the tips of the plastic zones due to the baseline (hold-time) load and fatigue overload are at a_1 and a_2, respectively. For a crack advance of Δa from a_0, the crack will be growing in an overload plastic zone until $\Delta a = a_2 - a_1$. During this time, the growth rate will be retarded. To model the crack growth retardation, either the crack-growth rate can be modified, such as by the method proposed by Wheeler [*10*], or the applied stress intensity can be modified to an "effective" stress intensity as was done by Willenberg et al. [*11*]. The latter approach was chosen here for compu-

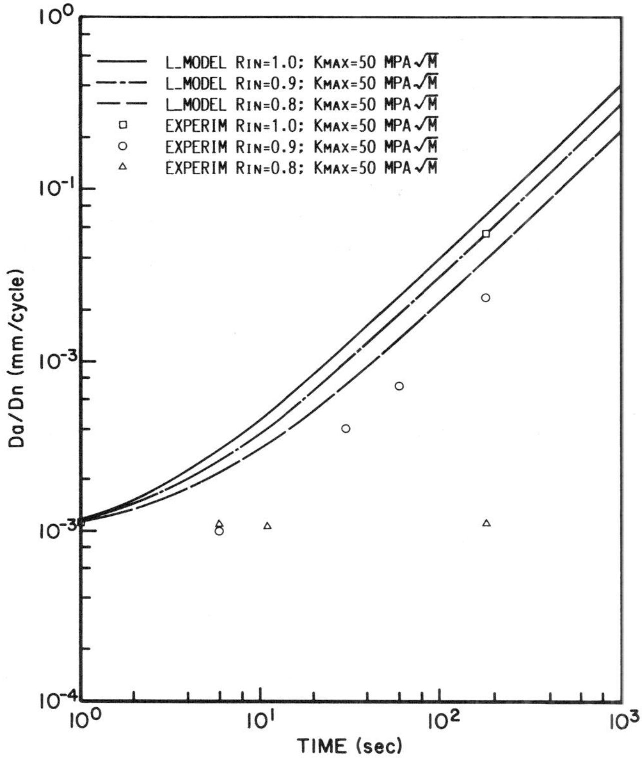

FIG. 3—*Crack-growth rate data plotted against total cycle time for* $K_{max} = 50$ *MPa* $m^{1/2}$ *as a function of different* R_{in} *values. Linear model (L-Model) data are also given in the figure.*

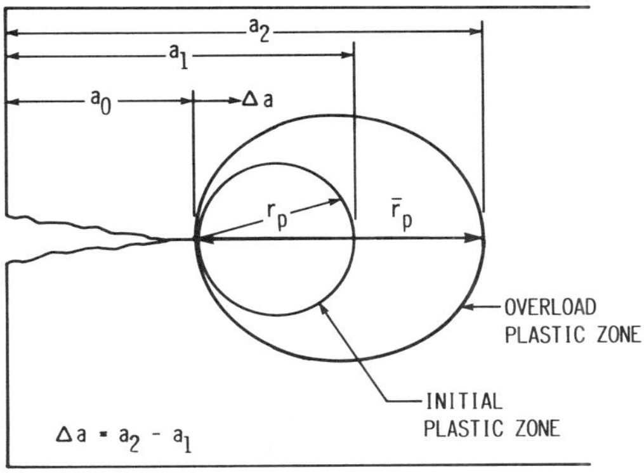

FIG. 4—*Schematic diagram of the plastic zones at the crack tip.*

tational convenience. The amount of retardation can be expressed by an effective value of stress-intensity factor, K_{eff}, which is smaller than the applied K. Using a single-valued function of $\dot{a}$ versus K to represent the sustained-load growth rate, a value of K_{eff} less than the applied K will result in a lower growth rate. For modeling purpose, K_{eff} is taken in an experimentally decaying form to represent experimental observations by Harms [9]

$$K_{eff} = K_s[1 - \alpha \exp(-\beta \Delta a)] \tag{1}$$

where α and β are parameters to be determined and Δa is the amount of crack extension after application of the overload. The parameter β is chosen such that $K_{eff} \rightarrow K_s$ when $\Delta a = \bar{r}_p - r_p$, that is, steady-state crack growth resumes when the crack has traversed the overload plastic zone. This is accomplished by setting

$$\beta \Delta a = \pi \sqrt{2} \tag{2}$$

when $\Delta a = \bar{r}_p - r_p$, resulting in a value of K_{eff}/K which is within 1% of unity as shown in Fig. 5. The plastic zone sizes are taken to be the plane-stress values given by

$$\bar{r}_p = \frac{1}{\pi}\left(\frac{K_m}{\sigma_y}\right)^2, \qquad r_p = \frac{1}{\pi}\left(\frac{K_s}{\sigma_y}\right)^2 \tag{3}$$

where σ_y is the uniaxial tensile yield stress of the material. The plane stress plastic zone sizes are used because they were found to match experimental data in a parallel investigation involving long hold times between overloads [9]. Combining Eqs 2 and 3 provides an expression for β

$$\beta = \frac{\sqrt{2}\pi^2\sigma_y^2}{K_s^2(\gamma^2 - 1)} \tag{4}$$

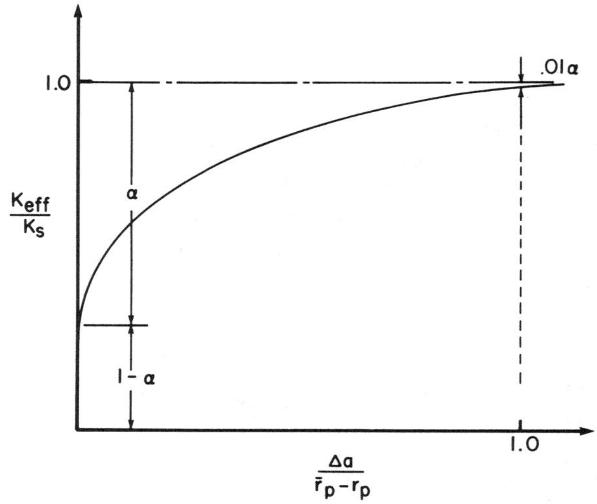

FIG. 5—*Variation of* K_{eff}/K_s *as function of normalized crack advance from the crack tip in the overload plastic zone.*

where the overload ratio γ is defined as

$$\gamma = \frac{K_m}{K_s} = \frac{1}{R_{in}} \tag{5}$$

It is important to note that β is fully defined from the test conditions and material properties and cannot be used as an adjustable parameter to fit experimental data. Thus, Eq 1 has only one parameter, α, which can be chosen to fit data. From Fig. 5, it can be seen that the value of α determines K_{eff} immediately after the overload application and therefore can be used to model the value of sustained-load crack growth rate.

Determination of Constants

Analytical predictions were made using a linear, cumulative damage model which incorporates the retarded sustained-load crack growth behavior given by Eq 1. Computations of crack extension Δa were made by numerical integration of da/dt using 1-s time increments for Δt. The crack growth rate da/dt is expressed in the form of a modified sigmoidal equation fit to experimental sustained-load crack growth data [*9*]. The stress intensity K_s in the sigmoidal equation is replaced by K_{eff} from Eq 1. The only parameter which had to be determined was α, and this was accomplished by trial and error by choosing a range of values for α. The results of these trials are shown in Figs. 6 and 7 for $K_m = 40$ and $K_m = 50$ MPa $m^{1/2}$, respectively, and

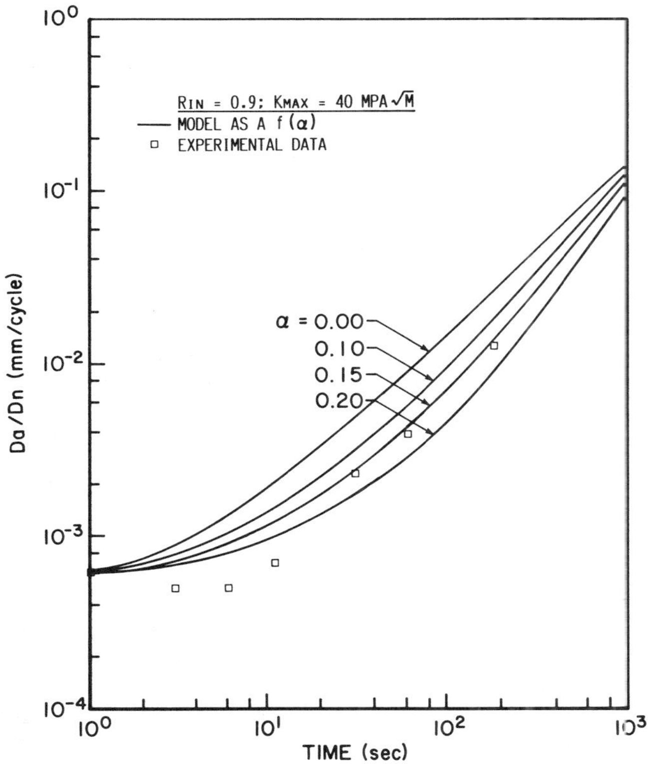

FIG. 6—*Calculated crack growth for different values of* α *for* $K_{max} = 40$ *MPa* $m^{1/2}$ *and* $R_{in} = 0.9$. *Data are also shown in the figure.*

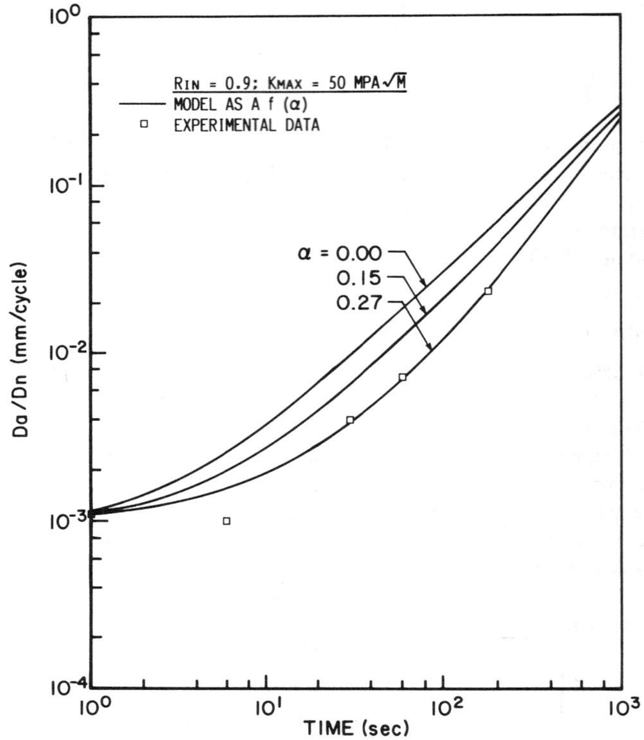

FIG. 7—*Calculated crack-growth rate for different values of* α *for* $K_{max} = 50$ *MPa* $m^{1/2}$ *and* $R_{in} = 0.9$. *Data are also shown in the figure.*

for the case of $R_{\text{in}} = 0.9$. The plots for $\alpha = 0$ represent the linear summation model for the case of no retardation. It can be seen that values of $\alpha = 0.15$ and $\alpha = 0.27$ fit the experimental data for the two cases very well, respectively. It is noted in both cases that the model overpredicts the growth rates for very short hold times below approximately 10 s.

For the case of $R_{\text{in}} = 0.8$, there appeared to be almost complete retardation for hold times up to the maximum of 200 s used in the experiments. In this situation, α could not be determined accurately, as can be seen in Fig. 8. Complete retardation is equivalent to a zero growth rate which corresponds to a value of $K_{\text{eff}} = K^*$, which is the threshold value of K in the $\dot{a}$ versus K representation of the hold-time data. In our modeling of the data using a modified sigmoidal equation, $K^* = 24$ MPa $m^{1/2}$. Values of $\alpha = 0.25$ and $\alpha = 0.4$ represent complete retardation for the cases of $K_m = 40$ and $K_m = 50$ MPa $m^{1/2}$, respectively. To represent the data in Fig. 8, values of 0.246 and 0.392, respectively, were chosen.

Calculations were made of the overload plastic zone size, defined as $\bar{r}_p - r_p$, through which distance a crack must grow under sustained load before it returns to its steady-state growth rate. For the four cases corresponding to the two values of K_m and for $R_{\text{in}} = 0.8$ and 0.9 for each case, values of this overload zone size ranged from approximately 0.1 to 0.5 mm. Note in Figs. 2 and 3 that all of the experimentally measured growth rates per cycle block were well below these values. Thus, for the simple spectrum used, the sustained-load crack growth was always within the overload plastic zone and occurred at a reduced rate due to retardation. In fact, most of the sustained-load crack growth occurs for small values of $\Delta a/(\bar{r}_p - r_p)$, or near the origin of the x-axis shown in Fig. 5. For these experimental conditions, the parameter α is the primary factor

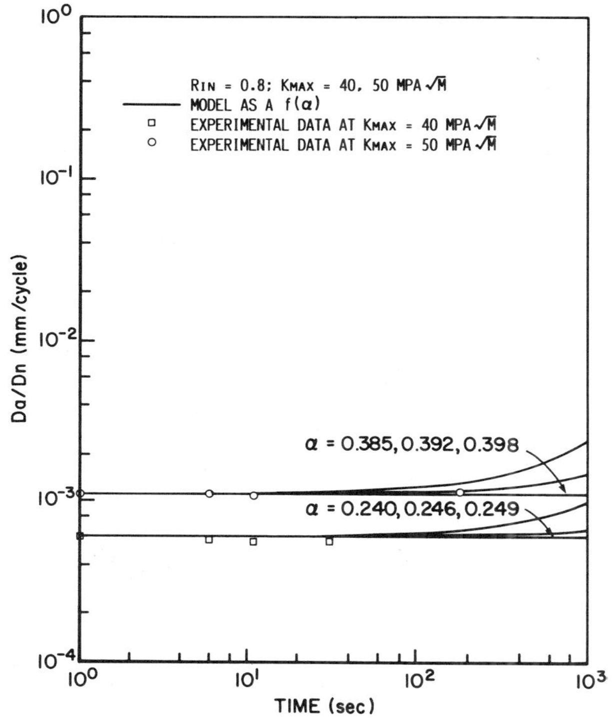

FIG. 8—*Calculated crack-growth rate for various* α *for* R_{in} = *0.8. Data are also shown in the figure.*

governing the growth rate in the analytical model. The parameter β and the shape of the function K_{eff}/K_s shown in Fig. 5 are of secondary importance for short hold times under approximately 100 s.

Delay Times

In studies of fatigue crack growth retardation due to overloads, the concept of a number of delay cycles is commonly used. Here, we introduce the equivalent concept of a delay time for sustained-load crack growth. Figure 9 presents plots of the calculated value of Δa_s, the growth during the sustained portion of the load, as a function of the hold times, t_H, for the cases where $R_{in} = 0.9$. Also shown in dashed lines is the crack extension that would occur if there were no retardation. The delay times, or the net gain in time because of the retarded growth rates, are approximately 200 s, as shown in Fig. 9. Delay times of approximately 4000 and 8000 s were calculated for the cases where $R_{in} = 0.8$ and $K_m = 40$ and 50 MPa m$^{1/2}$, respectively. In a parallel investigation on the same material, delay times of approximately 4000 and 20 000 s were experimentally measured for $R_{in} = 0.83$ and $R_{in} = 0.67$, respectively, for similar values of K_m [9].

Discussion

For the simple spectrum consisting of a fatigue cycle and a hold time, for the material used in this investigation, the hold time does not contribute to crack growth except when its amplitude is close to that of the maximum of the fatigue cycle. If the sustained-load crack growth is consid-

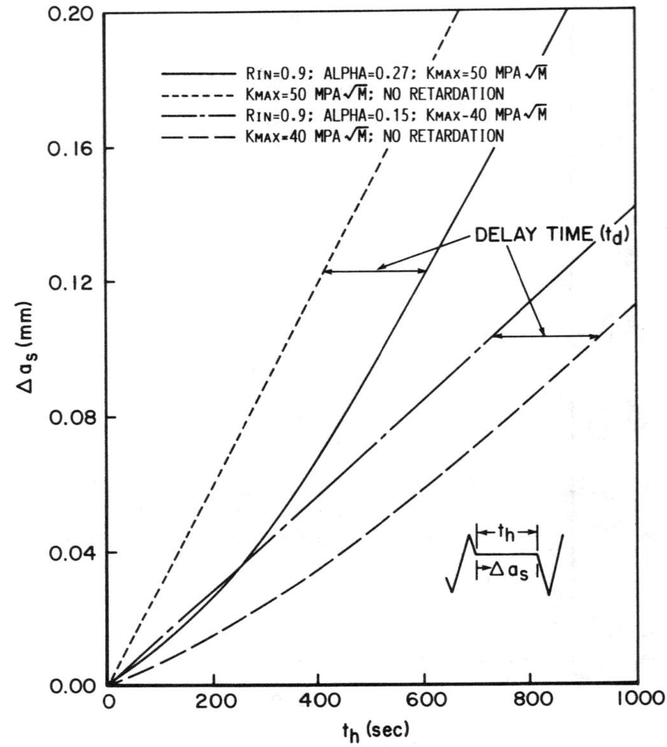

FIG. 9—*Calculated crack advance during sustained load after the application of the overload as a function of the sustained load time for* R_{in} *= 0.9 and* K_{max} *= 40 and 50 MPa* $m^{1/2}$ *as predicted from the retardation model. Predicted delay times* (t_d) *are also shown in the figure.*

ered to be retarded by the single-cycle overload, it appears that overload ratios greater than approximately 1.2 severely retard growth for hold times as long as 10^3 s.

A single-parameter retardation model for sustained-load crack growth combined with a linear summation model which adds cyclic and (retarded) sustained load contributions appears to be adequate for predicting crack-growth rates under the simple load spectrum used in this investigation. The model requires determination of the parameter α as a function of both overload ratio and stress intensity. Prior work by Harms [*9*] indicated that simpler models were inadequate in representing overload effects during sustained-load crack growth. For this complex material behavior involving both cycle-dependent and time-dependent crack growth, the model correlated data covering a range in hold times up to 200 s, although the model was slightly conservative for short hold times up to approximately 10 s. It appears that in addition to the retardation of the sustained-load crack growth due to a fatigue overload, there is also a slight retardation of the fatigue crack growth due to the sustained load. This latter phenomenon was not modeled in this investigation.

Most of the retardation modeled in these experiments covered short hold times corresponding to small crack extensions compared to the size of the overload plastic zone. Thus, the initial growth rates directly after an overload appear to be modeled well. In spite of the fact that overloads are repetitively applied before the crack has grown even a small distance through the overload plastic zone, a steady-state growth rate was achieved in all experiments under constant K conditions. This as well as transition time calculations [*7*] provide further evidence for the use of K as a correlating parameter for sustained-load crack growth in this material and specimen geometry.

Conclusions

A simple one-parameter model has been shown to adequately correlate crack growth data for a limited range of conditions involving sustained loading interrupted by periodic fatigue overloads. The models parameter is a function of both stress intensity and overload ratio. Nonetheless, a physically based approach has been demonstrated which provides direction for further modeling efforts to predict crack-growth rates under typical gas turbine engine spectrum loading.

References

[1] Military Standard: Engine Structural Integrity Program (ENSIP), MIL-STD-1783 (USAF), 30 Nov. 1984.

[2] Donath, R. C., Nicholas, T., and Fu, L. S., *Fracture Mechanics: Thirteenth Conference, ASTM STP 743,* Richard Roberts, Ed., American Society for Testing and Materials, Philadelphia, 1981, pp. 186–206.

[3] Bain, K. R. and Pelloux, R. M., *Superalloys 1984: Proceedings,* 5th International Symposium on Superalloys, Maurice Gell et al., Eds., The Metallurgical Society of the American Institute of Mining, Metallurgical, and Petroleum Engineers, 1984, pp. 389–396.

[4] Sadananda, K. and Shahinian, P., *Metallurgical Transactions,* Vol. 6A, 1975, pp. 1741–1749.

[5] Scarlin, R. B., *Materials Science and Engineering,* Vol. 30, 1977, pp. 55–62.

[6] Shahinian, P. and Sadananda, K., *Superalloys 1984: Proceedings* of the Fifth International Symposium on Superalloys, Maurice Gell et al., Eds., The Metallurgical Society of the American Institute of Mining, Metallurgical, and Petroleum Engineers, 1984, pp. 741–750.

[7] Nicholas, T. and Weerasooriya, T., *Fracture Mechanics: Seventeenth Volume, ASTM STP 905,* J. H. Underwood et al., Eds., American Society for Testing and Materials, Philadelphia, 1986, pp. 155–168.

[8] Nicholas, T., Weerasooriya, T., and Ashbaugh, N. E., *Fracture Mechanics: Sixteenth Symposium, ASTM STP 868,* M. F. Kanninen and A. T. Hopper, Eds., American Society for Testing and Materials, Philadelphia, 1985, pp. 167–180.

[9] Harms, K. E., "Overload Effects on Sustained Load Crack Growth at Elevated Temperature," M.S. Thesis, Air Force Institute of Technology, Wright-Patterson Air Force Base, OH, Dec. 1984.

[10] Wheeler, D. E., *Journal of Basic Engineering,* Vol. 94, 1972, pp. 181–186.

[11] Willenberg, J., Engle, R. M., Jr., and Wood, R. A., "A Crack Growth Retardation Model Using an Effective Stress Concept," AFFDL-TM-71-1-FBR, Wright-Patterson Air Force Base, OH, Jan. 1971.

Larry P. Zawada[1] *and Ted Nicholas*[1]

The Effect of Loading History on Closure Behavior in Rene' 95

REFERENCE: Zawada, L. P. and Nicholas, T., **"The Effect of Loading History on Closure Behavior in Rene' 95,"** *Fracture Mechanics: Eighteenth Symposium, ASTM STP 945,* D. T. Read and R. P. Reed, Eds., American Society for Testing and Materials, Philadelphia, 1988, pp. 192–205.

ABSTRACT: Experimental observations of closure in fatigue crack growth as a function of loading history are presented for a high-strength nickel-base superalloy, Rene' 95, at 649°C. Fatigue crack propagation (FCP) tests were conducted under either constant load amplitude (increasing ΔK) or under threshold load shedding (decreasing ΔK) test conditions. Closure loads were determined from load-offset displacement traces that were recorded with the aid of a crack mouth opening displacement extensometer. Plots of K_{cl} versus K_{max} revealed that K_{cl} remained constant for both the constant load and decreasing load FCG tests, indicating that K_{cl} was not a consequence of the loading history. Similar plots for side-grooved specimens and smooth specimens revealed identical behavior, thus ruling out plasticity-induced closure as the operating closure mechanism.

KEY WORDS: crack closure, crack growth mechanisms, elevated temperature, nickel-base superalloy, fatigue crack propagation, fracture mechanics, oxidation, threshold

Elber first introduced the concept of closure in 1970 [*1,2*]. According to Elber, a fatigue crack in a body experiencing positive cyclic loading remains fully open only at high load levels. At lower load levels, interference and physical contact between mating fracture surfaces behind the crack tip allow the crack tip to remain closed during part of the loading as well as the unloading phase of the cycle. Linear elastic fracture mechanics (LEFM) states that a crack cannot propagate while it remains closed, so the effect of closure is actually to reduce the nominal applied stress-intensity range ΔK ($= K_{max} - K_{min}$) to some lower effective ΔK ($K_{eff} = K_{max} - K_{cl}$) value whenever K_{min} is less than K_{cl}. Elber's proposed concept was based on the development of a plastic wake of residual deformation behind the crack tip and is referred to as plasticity-induced closure. Subsequently, other mechanisms of crack closure based on interference of mating surfaces have been proposed. They include asperity-induced closure [*3,4*] and oxide-induced closure [*5,6*]. The underlying relationship between these three different mechanisms is that they all shield the crack tip from the full stress range and hence lower the crack-tip driving force ΔK.

According to many researchers [*7–9*], the closure load depends on the type of loading history employed and the mechanism or mechanisms of closure contributing to its development. Therefore, an experimental investigation was undertaken to identify the mechanisms of closure in an elevated-temperature environment with particular emphasis placed on load-history effects. The primary goal of these experiments was to evaluate plasticity- and oxide-induced closure, primarily as a function of loading history.

[1]Materials engineer and senior scientist, respectively, Materials Laboratory, Air Force Wright Aeronautical Laboratories, AFWAL/MLLN, Wright-Patterson Air Force Base, OH 45433.

Experimental Approach

The material tested was a powder metallurgy nickel-base superalloy, Rene' 95, which was consolidated by either hot isostatic pressing (HIP) or extrusion plus isothermal forging. The chemical composition was determined by wet analysis and is listed in Table 1. Experiments were conducted on subcompact tension specimens with a nominal thickness (B) of 5 mm and a width (W) of 20 mm. Fatigue crack growth (FCG) testing was conducted at 649°C in laboratory air using a 25-kN servohydraulic test machine. A resistance heated furnace was used, and temperature was monitored using K-type thermocouples near the crack plane on the specimen. All tests were conducted at a cyclic frequency of 1 Hz and a stress ratio ($R = K_{min}/K_{max}$) of 0.1. An IBM 9000 microcomputer controlled the test machine and recorded test data. A clip gage with quartz rod extensions was used to measure crack mouth opening displacement, from which specimen compliance was obtained. Crack length was determined from the compliance measurements. In addition, test records of load versus displacement and load versus differential displacement were recorded at periodic intervals. The latter provides improved resolution for the determination of closure loads, which were always determined from the unloading portion of the fatigue cycle. Closure load was determined by manually marking the load-differential displacement plot where the curve deviated from a straight line.

Tests were conducted under conditions of either constant load (increasing ΔK) or under computer-controlled load shedding for threshold tests (decreasing ΔK). During threshold testing, the stress-intensity range, ΔK, was decreased according to the equation suggested by Saxena et al. [*10*]

$$\Delta K = \Delta K_0 e^{C(a-a_0)} \tag{1}$$

where

ΔK_0 = initial stress intensity range,
C = constant with a unit of reciprocal length,
a = crack length at any given time, and
a_0 = initial crack length.

In this investigation, the value of C was -0.2 mm^{-1}. Using this value of C allowed for two decreasing ΔK tests to be conducted on each specimen. Henry[2] has demonstrated that values of

TABLE 1—*Chemical composition of Rene' 95.*

Element	Specification Range, weight %	Analysis, weight %
Al	3.30 to 3.70	3.40
Ti	2.30 to 2.70	2.3
Cb	3.30 to 3.70	3.4
Cr	12.00 to 14.00	15.22
Co	7.00 to 9.00	10.37
Mo	3.30 to 3.70	3.30
W	3.30 to 3.70	3.3
C	0.04 to 0.09	0.068
Zr	0.03 to 0.07	0.047
B	0.006 to 0.015	0.01
Ni	balance	balance

[2]M. F. Henry, private communication, General Electric, Co., Schenectady, NY.

C from $-0.08\ mm^{-1}$ (ASTM proposed standard) down to $-1.2\ mm^{-1}$ produce no variation in near-threshold fatigue crack propagation (FCP) rates and threshold values in Rene' 95 at 649°C.

Specimens of Rene' 95 subjected to two different heat treatments were used to produce different grain sizes which might affect surface roughness induced closure. A standard sub-solvus heat treatment of 1136°C/1 h was used to produce a grain size of 5 μm, while a super-solvus heat treatment of 1163°C/1 h was used to produce a larger grain size of 30 μm. All solution heat treatments were followed by an aging heat treatment 760°C (1400°F)/8 h. This second solution treatment places into solid solution all of the gamma prime from the matrix and grain boundary triple points. Upon cooling, the gamma prime precipitates back out as a very fine distribution of matrix precipitates. These fine distributions of gamma prime have been shown to enhance coarse, planar slip [*11*]. It is believed that this coarse, planar slip increases cyclic defect tolerance by reducing the accumulated fatigue damage by slipband reversibility and consumption of energy in the crack-tip stress field by slipband cracking. This results in transgranular crack propagation and retarded FCP rates. It is thought that by varying grain size and deformation mechanisms through heat treating, there may be a difference in surface roughness resulting in different K_{cl} values. This would be true if roughness induced closure were the dominating closure mechanism. Table 2 lists the heat treatments, grain sizes, and the corresponding mechanical properties which were determined at 649°C.

To evaluate the influence of surface plasticity-induced closure, smooth specimens and specimens with side grooves were employed. The side grooves were 15% of the thickness and simulated a plane strain condition [*12*]. The 5-mm-width smooth specimens are considered to be thinner than a pure plane-strain specimen and should exhibit some surface plane-stress deformation.

Auger spectroscopy was used to estimate the thickness of oxides present on the surfaces of the fractured specimens. A Physical Electronic Industries Thin Film Analyzer Model M540 was used for this measurement. The Auger spectrometer sputtering rate was calibrated to a known thickness of nickel oxide instead of the standard tantalum oxide to improve the accuracy of the measurement. An ionized argon beam sputtered an area 3 mm in diameter while counts of intensity versus sputtering time were made at spaced time intervals.

Experimental Results and Observations

Representing Closure Data

In the literature, closure data have been reported on many different types of plots. By choosing the appropriate type of diagram, one can demonstrate that closure increases, decreases, or remains constant depending on the type of loading history employed and choice of independent variable. Therefore it is important to verify those methods of representing closure data which best represent the effects of loading history and closure mechanisms. To determine this, we constructed a total of six different plots. The independent variables were nondimensional crack

TABLE 2—*Mechanical properties of Rene' 95 at 649°C.*

Heat Treatment	$\sigma_{0.2}$, MPa	σ_{UTS}, MPa	Hardness, (R_c)	Grain Size, μm
1163°C/1 h[a]	1109	1438	47.8	30
1136°C/1 h[a]	1208	1464	48.3	5

[a]All solution heat treatments were followed by an aging heat treatment of 760°C (1400°F)/8 h.

length (a/W) and K_{max}, while the dependent variables were percent closure (P_{cl}/P_{max}), closure load (P_{cl}), and K_{cl}.

The measured closure loads were first plotted on a normalized plot of percent closure versus nondimensional crack length and are shown in Fig. 1. The percent closure for all eight constant load tests, shown in Fig. 1*a*, decreased linearly as a function of a/W, with no observable difference between the large grain material and the small grain material. The data can all be represented by a scatter band as shown in the plot. When the decreasing load fatigue crack growth tests were plotted on the same type of plot, the percent closure linearly increased with increasing values of a/W as shown in Fig. 1*b*. This is the reverse of the constant load tests. In addition, the two curves are not continuous. Superposition of the scatter band from the constant load tests clearly demonstrates that one cannot use this type of plot to correlate closure load data from tests representing different loading histories.

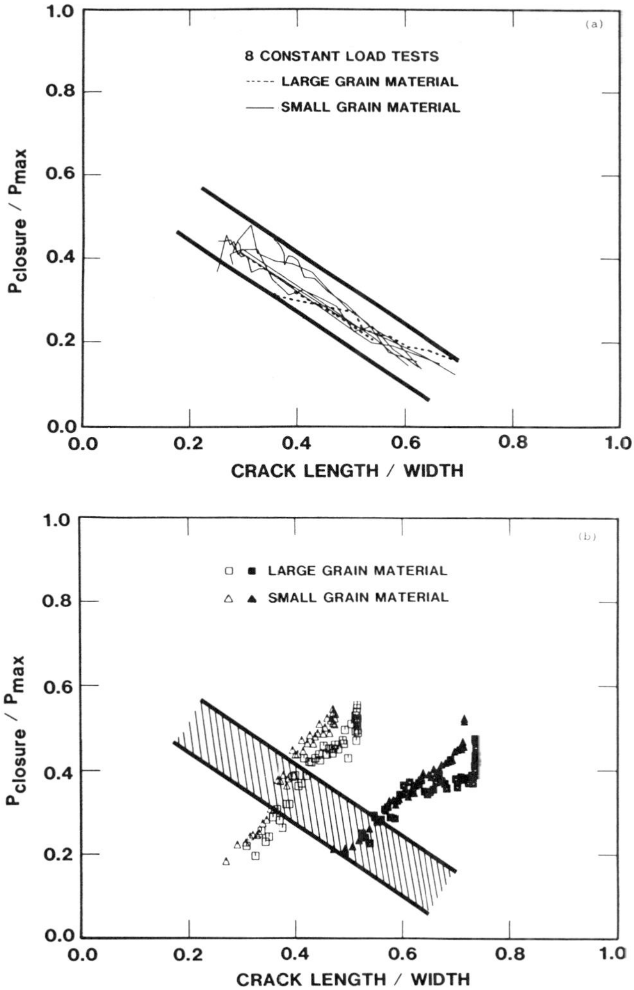

FIG. 1—P_{cl}/P_{max} *as a function of* a/W *for FCP tests:* (a) *constant load tests;* (b) *decreasing load tests. (Shaded region represents the constant load tests.)*

Since the closure load should not change abruptly if the maximum load is suddenly increased, as in going from the end of one threshold test to the beginning of another, the closure load should be a better parameter to represent closure behavior. The closure data were next plotted as closure load versus K_{max} as shown in Fig. 2. The constant load tests are represented by the shaded region which represents the scatter band of all eight tests. Here the closure loads for the constant load tests decreased as a function of increasing K_{max} until one reaches a K_{max} of 30 MPa$\sqrt{m}$, upon which the closure load levels off to a lower limit of approximately 0.5 kN. For the decreasing load tests, the closure load decreased as K_{max} was lowered, with this trend continuing up to threshold. In addition, the first threshold tests have a higher P_{cl} than the second threshold tests which were conducted at the longer initial crack lengths and lower loads. This indicates that there is a difference in closure due to different starting maximum loads. However, the closure loads decrease as the crack length increases for both loading histories, indicating some dependence on a/W. The use of K_{max} as the independent variable appears inadequate because, since P_{cl} appears to be a continuous function for different loading histories, K_{max} changes abruptly when going from the end of one threshold test to the start of the next.

It appears logical to try to consolidate all the data by plotting P_{cl} against a/W. Such a plot is presented in Fig. 3 where it can be seen that P_{cl} linearly decreases with increasing values of a/W for both the constant load tests and the decreasing load tests. An important observation is that the absolute closure loads measured for the decreasing load threshold tests did not depend on the initial maximum load but were instead primarily a function of a/W for these two different loading histories. Closure load is therefore a single-valued and continuous function of a/W and appears to be independent of loading history.

Although percent closure was found to not correlate with a/W, it was also plotted against K_{max}. Figure 4 demonstrates how percent closure varied as a function of K_{max}. The constant load tests are represented by the heavy solid lines. Notice that percent closure decreased as K_{max} increased until reaching $K_{max} = 30$ MPa$\sqrt{m}$, at which point percent closure then leveled off to a lower limit of approximately 0.2. The threshold tests followed the same trend from an initial ΔK of 30 MPa$\sqrt{m}$ down to threshold. Superimposing data from all the tests shows that there

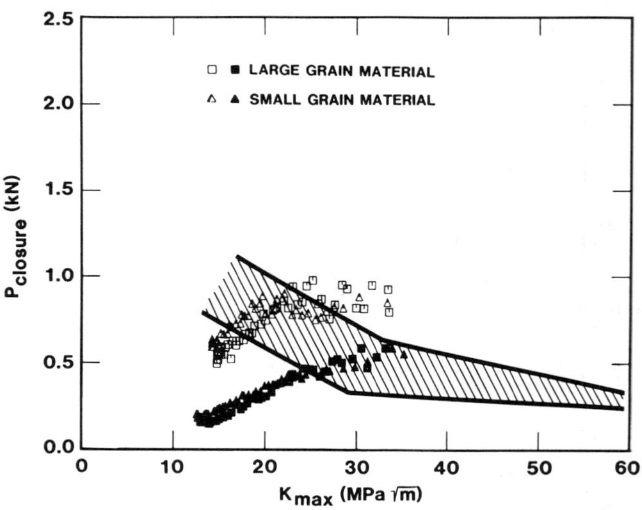

FIG. 2—*Dependence of* P_{cl} *on applied* K_{max}. *Shaded region represents constant load tests while the symbols represent decreasing load tests.*

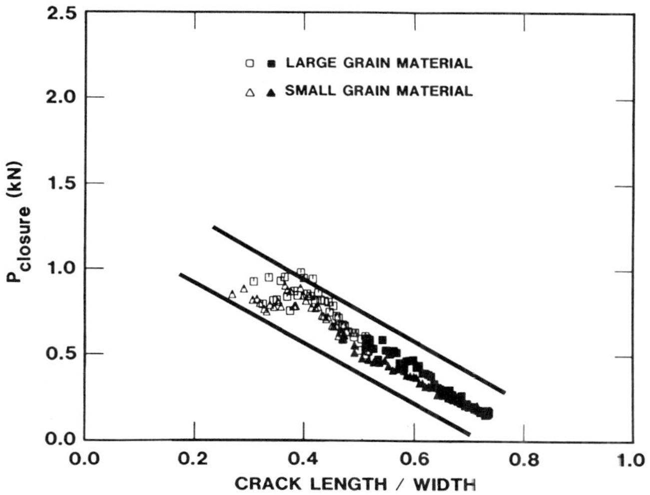

FIG. 3—*Dependence of* P_{cl} *on* a/W. *The heavy solid lines represent the constant load tests.*

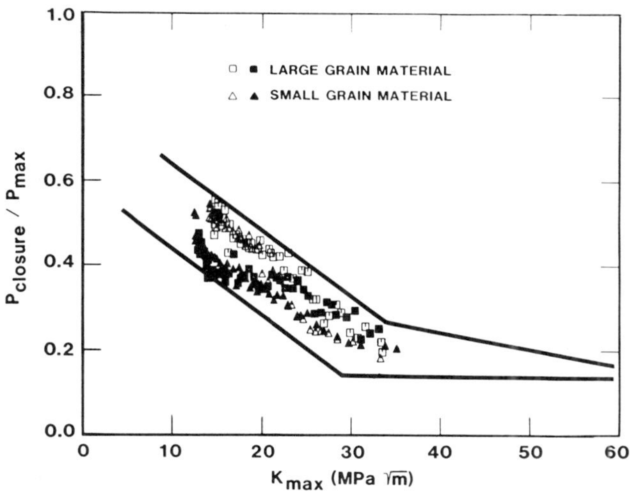

FIG. 4—*Relation between* P_{cl}/P_{max} *and* K_{max}, *for all that test data.*

was no difference due to loading histories. This implies that percent closure is solely dependent on the K_{max} that is applied.

The final method of representing closure was to plot K_{cl} versus K_{max}. The rational for using this type of diagram is that both axes utilize the LEFM stress-intensity parameter ΔK. Illustrated in Fig. 5*a* are the results for the constant load tests. The K_{cl} level remained relatively constant and independent of K_{max}, with no distinguishable difference between the two grain size materials. K_{cl} also remained constant for the decreasing load tests, as shown in Figs. 5*b* and 5*c*. The first threshold tests had slightly higher K_{cl} values than the second threshold tests,

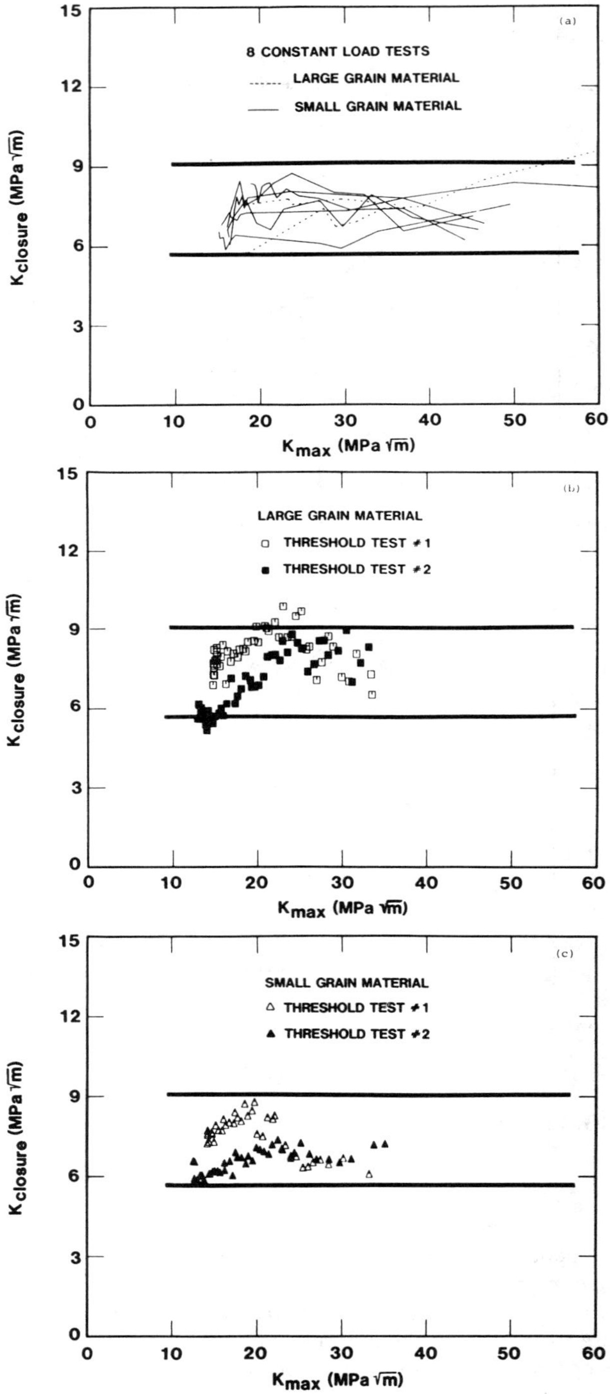

FIG. 5—K_{cl} *as a function of* K_{max}*:* (a) *constant load tests;* (b) *large grain material;* (c) *small grain material.*

with this difference increasing slightly as threshold was approached. For all of the test data, it can be observed that K_{cl} remained constant, between 7 and 8 MPa$\sqrt{m}$. This was true for both the constant load tests and the decreasing load threshold tests. Plots of K_{cl} versus a/W were not included since K_{cl} was shown to remain relatively constant and would result in plots which are nearly identical to Figs. 5*a*, 5*b*, and 5*c*.

Effects of Specimen Thickness

To evaluate whether or not the three-dimensional stress state affects the measured closure loads, fatigue crack growth rate tests were conducted on side-grooved and smooth HIP Rene' 95 subcompact tension specimens. Using the same correlating parameters as in Fig. 5, K_{cl} results in Fig. 6 demonstrate that for one side-grooved specimen and two smooth specimens, K_{cl} results were indistinguishable and remained constant as K_{max} increased. Both the side-grooved and smooth specimens exhibited constant K_{cl} values of approximately 5 MPa$\sqrt{m}$. To further evaluate the effect of specimen thickness, crack growth rate data were plotted for the smooth and side-grooved specimens. The results (Fig. 7) show that the growth rates from the two types of specimens are nearly indistinguishable and certainly exhibit less than the $\times 2$ scatter normally associated with fatigue crack propagation testing.

Optical observations were made of the fracture surfaces to determine the shape of the crack as it grew and whether or not the growth rates were similar along the surface and at the maximum depth point in the interior. It was found that both of the smooth specimens had a thumbnail-shaped crack front while the side-grooved specimen possessed a perfectly flat crack front. It is believed that the thumbnail shape of the smooth specimen results from the state of plane stress at the surface. This plane-stress condition produces larger plastic zone sizes than those points in the interior which are experiencing primarily plane strain. If plasticity-induced closure mechanisms were operative, the closure where the crack meets the surface would be stronger than the closure at the maximum depth point [*13*]. This can explain the occurrence of a curved crack front, but K_{cl} was constant for both types of specimens. This further proves that surface plasticity does not effect bulk closure load measurements and that plasticity-induced closure is not the primary operating closure mechanism.

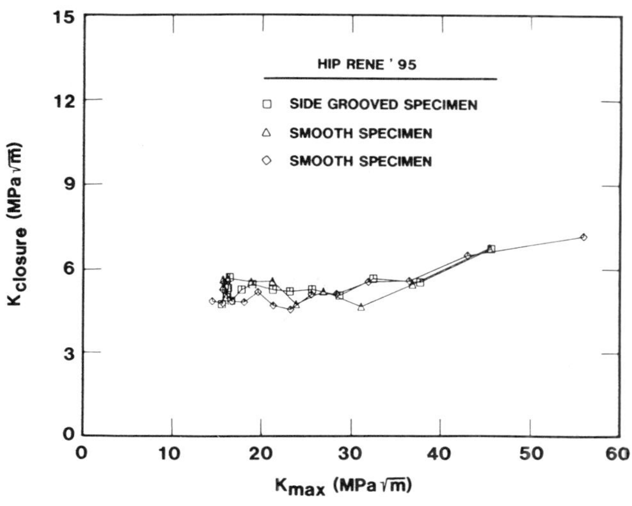

FIG. 6—K_{cl} *versus* K_{max} *for HIP Rene' 95 in side-grooved and smooth specimens.*

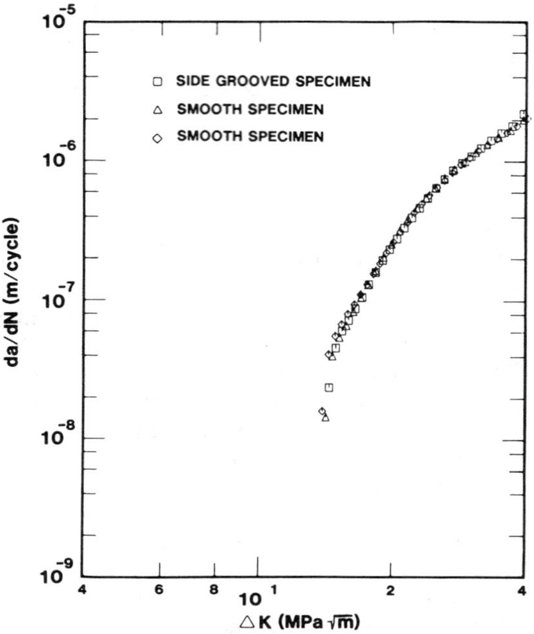

FIG. 7—*Fatigue crack growth behavior of HIP Rene' 95 for smooth and side-grooved specimens.*

Threshold Test Results

Fatigue crack growth rates as a function of applied ΔK for the two decreasing load threshold tests for the small and large grain material are shown in Figs. 8*a* and 8*b*, respectively. Also shown in the plots are da/dN versus ΔK_{eff} where the closure load data were used to calculate ΔK_{eff}. One can observe that in both materials, the first threshold tests corresponding to a shorter initial crack length have slightly higher apparent threshold values than the second threshold tests when plotted against ΔK. The crack growth rates are also lower for the first test in both materials when applied ΔK is used as the correlating parameter. However, these differences are essentially eliminated when one plots growth rates against ΔK_{eff}.

It was also observed from these data that the larger grain size material has a slower FCP rate and higher apparent threshold value than the small grain material. This was true whether ΔK or ΔK_{eff} was used as the correlating parameter. Fatigue crack growth rates for the small grain material decreasing load tests were also compared to those of the constant load tests in order to investigate the influence of loading history on growth rates near threshold stress-intensity levels. By directly comparing the FCP rates of Figs. 7 and 8*a* it can be easily shown that the growth rates are identical for both loading histories when compared against ΔK. This verifies that using a value of $C = -0.2\ \text{mm}^{-1}$ for the decreasing load threshold tests did not alter fatigue crack propagation behavior.

Another observation from the threshold testing was with respect to the nature of the crack length versus number of cycles (a versus N) data. In all of the threshold tests, the slope of the a-N curves was discontinuous. As ΔK was decreased continuously in the threshold tests, the corresponding crack growth rate also decreased. At one point, however, the crack growth suddenly arrested, and no further growth was observed within experimental uncertainty. Cycling was continued after crack arrest occurred for at least 1×10^5 cycles to verify statistically that crack propagation had ceased. This abrupt change in behavior was also observed by Ashbaugh

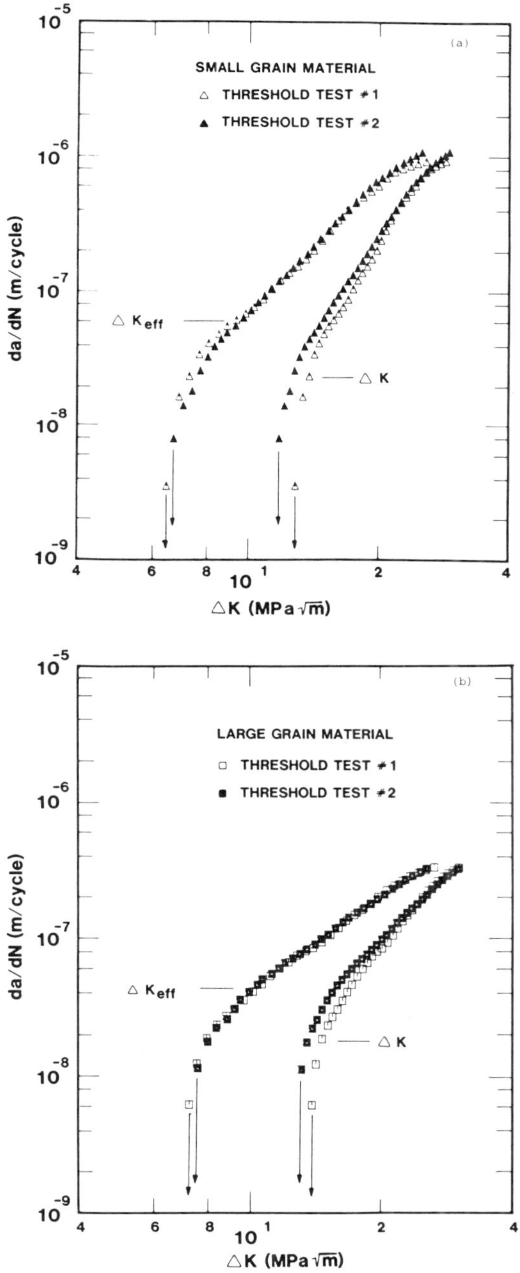

FIG. 8—*Fatigue crack growth behavior as a function of* K *and* K_{eff} *for decreasing load threshold testing:* (a) *small grain material;* (b) *large grain material.*

[14] in threshold tests in Inconel 718 at 649°C. In our case, crack arrest occurred at ΔK levels corresponding to crack growth rates of approximately 8×10^{-9} mm/cycle. Beyond this point, calculated crack growth rate data are merely an artifact of curve fitting an *a-N* "curve" of zero slope with statistical scatter. Data points corresponding to so-called crack growth beyond this point of apparent arrest have thus been eliminated and replaced with an arrow in Fig. 8 to indicate no further crack growth.

Observations on Oxide-Induced Closure

All specimens tested in this experiment developed thick oxide bands on the fracture surfaces. The heaviest oxidation occurred on the decreasing load threshold test specimens due to the long times at elevated temperature. Oxide thicknesses were determined for the second threshold tests just behind the arrested crack tip and in the center of the specimen. Auger electron spectroscopy was used to determine the oxide thickness, and these results were corrected for the volume of base metal consumed to produce a volume of oxide. The corrected value of oxide thickness for two fracture surfaces was determined to be 0.536 μm. In comparison, crack-tip opening displacements (CTOD) were calculated using the relationship

$$\text{CTOD} = 0.49(K_{\text{max}}^2/\sigma_{0.2}E)$$

Using the highest values of K_{max} at threshold CTOD was calculated to be 0.48 μm. This CTOD value is actually smaller than the oxide thickness measurements, indicating that the mechanism of oxide-induced crack closure will be operating when FCP is conducted near threshold stress intensity levels.

Oxide thickness measurements were also made along the length of the specimen in which two threshold tests were performed. In addition to the prior measurements cited at the location just behind the crack tip of the second threshold test, two other locations of progressively shorter crack lengths were made. These consisted of a region just behind the crack arrest region of the first threshold test, and a high ΔK region of the first threshold test. These two regions also correspond to successively longer exposure times of the free surface to oxidation. The oxide thicknesses were found to be only a function of temperature and exposure time. Scanning electron microscopy (SEM) observations and auger electron spectroscopy results verify that the oxides were diffusion controlled and not due to fretting.

Discussion

In all, six different methods of presenting closure data were attempted. Two of them, $P_{\text{cl}}/P_{\text{max}}$ versus a/W and P_{cl} versus K_{max}, were found to be inadequate for correlating closure data for both increasing and decreasing ΔK histories. It was observed that K_{cl} remains essentially constant from the plot of K_{cl} versus K_{max}, and it was also noted that K_{cl} versus a/W would show the same features, that is, constancy. It can be easily shown that the other two methods of representing closure data, plotting P_{cl} versus a/W and $P_{\text{cl}}/P_{\text{max}}$ versus K_{max}, provide no additional independent information because they are single-valued functions independent of loading history. Consider the manner in which K_{cl} is calculated from P_{cl} using an equation of the form

$$K_{\text{cl}} = P_{\text{cl}}f(a/W) \tag{2}$$

where $f(a/W)$ is the geometry dependence in the stress-intensity solution for the CT specimen. If K_{cl} is a constant, say $\bar{K}$, then

$$P_{\text{cl}} = \bar{K}/f(a/W) \tag{3}$$

which shows that P_{cl} will depend only on the ratio a/W. Thus, a plot of P_{cl} versus a/W would yield a single-valued function for all loading histories. Similarly, if we note that $P_{cl}/P_{max} = K_{cl}/K_{max}$, then a plot of K_{cl}/K_{max} or $\bar{K}/K_{max}$ against K_{max} will also yield a single-valued function. Thus, the fact that several methods of presenting data appear to provide a single-valued correlating function, they all turn out to be equivalent when K_{cl} is a constant. We recommend the use of a plot of K_{cl} versus K_{max} as providing the most information, since both parameters involve the use of fracture mechanics and should be able to evaluate closure data for different specimen geometries and loading histories.

The observation that K_{cl} is constant for all loading conditions implies that plasticity-induced closure is not the operative mechanism in this material at the test temperature of 649°C. One would expect K_{cl} to increase with K_{max} because higher values of K_{max} lead to higher stresses and strains at the crack tip and, hence, larger plastic wakes behind the crack tip [*2,13*]. The independence of K_{cl} on K_{max} thus rules out plasticity as the cause of closure. Further, the identical results for both closure loads and crack growth rates for the smooth as well as side-grooved specimens give additional credence to the postulate that plasticity-induced closure is not the relevant mechanism. One would expect more surface plasticity in smooth specimens than in side-grooved (plane strain) specimens because of the plane-stress state along the surface. Hence, one would expect higher closure loads in the smooth specimens than in the side-grooved specimens if plasticity induced closure were present. No such observations were made in these experiments.

Roughness-induced closure was considered a possible operative mechanism in these experiments. The use of two different solution heat treatments produced average grain sizes which differed by a factor of six. However, both the large and small grain size material possessed the same values of K_{cl}. Examination of the fracture surfaces using SEM revealed that the large grain material demonstrated purely transgranular crack propagation at higher stress intensity levels, while near threshold the failure mechanism transferred to mixed mode and finally to 100% intergranular just before crack arrest occurred in the threshold tests. It is speculated that this change in fracture mechanism may be the cause of slightly decreasing K_{cl} levels near threshold, as observed in Fig. 5*b*. In contrast to this, the fine grain material remained approximately 75% intergranular throughout the entire range of stress intensity and showed little change in K_{cl} near threshold. Precision sectioning of the fracture surfaces parallel to the crack propagation direction and perpendicular to the fracture surface revealed that the peak-to-valley dimensions of the surface asperities were different for the two grain size materials. In addition, the number of asperities per unit area were much higher for the smaller grain size material than for the larger grain size material. From all of these observations, there is insufficient evidence to establish whether or not roughness-induced closure is the governing mechanism.

Experimental evidence generated in this investigation appears to support the premise of oxide induced closure being the operative mechanism. The constancy of K_{cl} with K_{max} probably results in oxide wedging when the CTOD from fracture mechanics calculations is less than the thickness of the net residual oxide scale thickness. The very close correlation of the measured oxide thickness with the CTOD calculated from the average (constant) value of K_{cl} is prime evidence of oxide-induced closure being present. The sudden arrest of propagating cracks in threshold tests can also be interpreted as a consequence of oxide induced closure. Allison [*15*] has made similar observations of abrupt crack arrest in threshold tests on titanium alloys at 510°C in laboratory air. He attributes the arrest to the convergence of the decreasing CTOD under decreasing ΔK with the thickness of the fracture surface oxide until a value of ΔK_{eff} is reached corresponding to the intrinsic threshold for the material. Similar observations of crack arrest during threshold testing in oxidizing environments have been made by Ashbaugh [*14*] for Inconel 718 and by Henry (see footnote 2) for Rene' 95. In the experiments of Allison and Henry, the crack arrest phenomenon was not present when tests were performed in inert environment or vacuum. This evidence suggests that oxide-induced closure plays a role in arresting the crack during threshold testing. Future tests are planned in vacuum to further validate this hypothesis.

Conclusions

Crack closure data were obtained in Rene' 95 at 649°C under conditions of fatigue cycling under both increasing and decreasing ΔK. From all of the test results, it was found that K_{cl} was approximately constant and essentially independent of loading history. Of the three possible closure mechanisms (plasticity, roughness, oxides), plasticity-induced closure was ruled out because of the constancy of K_{cl} and the lack of variation in closure load when going from a smooth specimen to a side-grooved specimen. Oxide-induced closure appeared to be the primary mechanism, especially from results of oxide thickness measurements. Roughness-induced closure could not be ruled out as playing a role, but variations in grain size and fracture morphology did not significantly affect the closure behavior.

Acknowledgment

This experimental research was performed under Materials Laboratory in-house research project 2302P1, Air Force Wright Aeronautical Laboratories (AFWAL), Wright-Patterson Air Force Base, OH. The authors would like to express their appreciation to Mr. David Johnson of the University of Dayton Research Institute for his significant contributions in the experimental testing.

References

[*1*] Elber, W., "Fatigue Crack Closure Under Cyclic Tension," *Engineering Fracture Mechanics,* Vol. 2, 1970, pp. 37–45.

[*2*] Elber, W., "The Significance of Fatigue Crack Closure" in *Damage Tolerance in Aircraft Structures, ASTM STP 486,* American Society for Testing and Materials, Philadelphia, 1971, pp. 230–242.

[*3*] Walker, N. and Beevers, C. J., "A Fatigue Crack Closure Mechanism in Titanium," *Fatigue of Engineering Materials and Structures,* Vol. 1, 1979, pp. 135–148.

[*4*] Minakawa, K. and McEvily, A. J., "On Crack Closure in Near-Threshold Region," *Scripts Metallurgica,* Vol. 15, 1981, pp. 633–636.

[*5*] Ritchie, R. O., Suresh, S., and Moss, C. M., "Near-Threshold Fatigue Crack Growth in 2 1/4Cr-Mo Pressure Vessel Steel in Air and Hydrogen," *Journal of Engineering Materials and Technology, Transactions,* American Society of Mechanical Engineers, Vol. 102, 1980, pp. 293–299.

[*6*] Suresh, S., Zamiski, G. F., and Ritchie, R. O., "Oxidation and Crack Closure. An Explanation for Near Threshold Corrosion Fatigue Crack Growth Behavior," *Metallurgical Transaction,* Vol. 12A, 1981, pp. 1435–1443.

[*7*] Katchner, M. and Kaplan, M., "Effect of R-Factor and Crack Closure on Fatigue Crack Growth Rate for Aluminum and Titanium Alloy," *Fracture Toughness and Slow Stable Cracking, ASTM STP 559,* American Society for Testing and Materials, Philadelphia, 1974, pp. 264–282.

[*8*] James, M. N. and Knott, J. F., "An Assessment of Crack Closure and the Extent of the Short Crack Regime in Q1N (HY80) Steel," *Fatigue and Fracture of Engineering Materials and Structures,* Vol. 8, No. 2, 1985, pp. 177–191.

[*9*] Gray, G. T., III, Williams, J. C., and Thompson, A. W., "Roughness-Induced Crack Closure: An Explanation for Microstructurally Sensitive Fatigue Crack Growth," *Metallurgical Transactions A,* Vol. 14A, March 1983, pp. 421–433.

[*10*] Saxena, S., Hudak, S. J., Jr., Donald, J. K., and Schmidt, D. W., "Computer-Controlled Decreasing Stress Intensity Technique for Low Rate Fatigue Crack Growth Testing," *Journal of Testing and Evaluation,* Vol. 6, 1978, pp. 167–174.

[*11*] VanStone, R. H. and Henry, M. F., "Processing Effects on Microstructure and Fatigue Properties of Nickel-Base Superalloys," Naval Air Development Center Report NADC-80124-60, Naval Air Systems Command, Washington, DC, 1982.

[*12*] Shih, C. F., deLorenzi, H. G., and Andrews, W. R., "Elastic Compliances and Stress-Intensity Factors for Side-Grooved Compact Specimens," *International Journal of Fracture,* Vol. 13, 1977, pp. 544–548.

[13] Banerjee, S., "A Review of Crack Closure," Report AFWAL-TR-84-4031, Air Force Wright Aeronautical Laboratories, Wright-Patterson Air Force Base, OH, 1984.
[14] Ashbaugh, N. E., "Time Dependent and Cycle Dependent Threshold Evaluation in a Nickel-Base Superalloy," presented at the 18th National Symposium on Fracture Mechanics, Boulder, CO, American Society for Testing and Materials, Philadelphia, June 1985.
[15] Allison, J. E. and Williams, J. C., "Near-Threshold Fatigue Crack Growth Phenomena at Elevated Temperature in Titanium Alloys," *Scripta Metallurgica,* Vol. 19, 1985, pp. 773–778.

George K. Haritos,[1] Theodore Nicholas,[2] and Gerald O. Painter[3]

Evaluation of Crack Growth Models for Elevated-Temperature Fatigue

REFERENCE: Haritos, G. K., Nicholas, T., and Painter, G. O., **"Evaluation of Crack Growth Models for Elevated-Temperature Fatigue,"** *Fracture Mechanics: Eighteenth Symposium, ASTM STP 945,* D. T. Read and R. P. Reed, Eds., American Society for Testing and Materials, Philadelphia, 1988, pp. 206–220.

ABSTRACT: To date, only two models have been proposed for predicting crack growth rates in aircraft engine materials under typical operating conditions. Moreover, modeling to date has been performed on materials which exhibit limited amounts of time-dependent behavior. This investigation evaluates the predictive and interpolative capabilities of these models when applied to IN718, at 649°C. At this temperature, this material exhibits significant time-dependent behavior. Experimental data were generated and used to assess the predictive and interpolative characteristics of each model. These data covered variations in frequency, stress ratio, and hold time. Values for the coefficients in the functional relationships between the constants and the test variables were obtained for the hyperbolic sine equation (SINH) model. Similar functional relationships were incorporated into the modified sigmoidal equation (MSE) model. Several important conclusions were drawn about the capabilities of these two models to predict crack-growth rates in turbine-engine materials at elevated temperature.

KEY WORDS: elevated-temperature fatigue (materials), fracture (materials), crack-growth-rate modeling, time-dependent behavior

Over the past several years, a number of structural problems have plagued aircraft turbine engines in the U.S. Air Force inventory. These problems have led the Air Force to adopt a new design philosophy under the Engine Structural Integrity Program (ENSIP) [*1*]. This new approach assumes that flaws exist in structural components at the time they enter into service. Life-prediction analysis and testing must demonstrate that the flaw will not grow to a catastrophic size during an inspection interval or during the lifetime of the engine. This damage-tolerant approach, which is based on fracture mechanics, replaces the previous design methodology used for retiring engine structural components based on low-cycle fatigue (LCF). Under LCF, components were retired from service when a statistical lower bound (-3σ) of the distribution of LCF crack-initiation times was reached. This means that statistically, 999 out of 1000 components retired from service have some (undefined) useful life remaining to crack initiation.

Fracture mechanics has been also recently utilized under a new concept, coined retirement-for-cause (RFC), to extend the useful life in existing components by retaining the component in service until a crack has indeed been detected. Again, life prediction analysis must demonstrate

[1] Associate professor, Department of Aeronautics and Astronautics, Air Force Institute of Technology, Wright-Patterson Air Force Base, OH 45433; currently, program manager, Aerospace Sciences, Air Force Office of Scientific Research, Bolling Air Force Base, Washington, DC 20332. Major, U.S. Air Force.

[2] Senior scientist, Materials Laboratory, Air Force Wright Aeronautical Laboratories, Wright-Patterson Air Force Base, OH 45433.

[3] Aerospace engineer, Ballistic Missile Office, Norton Air Force Base, CA 92409. Captain, U.S. Air Force.

that a crack just below the detectable size will not grow to a catastrophic size during an inspection interval. Clearly, safe application of both a damage-tolerant design approach and RFC depends upon two factors: the ability to accurately predict the fatigue crack-growth rate in the material, and reliable nondestructive evaluation techniques.

The crack-growth rate in most turbine-engine alloys is characterized by the linear elastic fracture mechanics (LEFM) stress-intensity factor. It is influenced by a number of parameters, such as frequency, temperature, stress ratio, and sustained-load hold time. To date, only two models have been proposed for predicting crack-growth rates over a wide range of these parameters: one which utilizes a hyperbolic sine equation (SINH model) and one which is based on a modified sigmoidal equation (MSE model). Both of these models are described in detail in Ref *2*. The ability of these models to predict crack-growth behavior in the time-dependent region has not been fully tested since the modeling performed to date has concentrated mainly on materials exhibiting limited amounts of time-dependent behavior. In a prior investigation [*2*], the two models were compared in a limited sense for their predictive capabilities on AF115, an advanced nickel-base superalloy whose database was used in the development of the MSE model. This investigation evaluates the predictive and interpolative capabilities of the two existing models when applied to a turbine-engine disk material, IN718, at 649°C. At this temperature, this nickel-base superalloy exhibits significant time-dependent behavior over ranges of frequency and hold time typical of turbine-disk applications.

Experimental Program

A closed loop computer-controlled test system was used to collect da/dN versus ΔK data. The test procedures are described in detail by Ashbaugh [*3*]. Experiments were conducted to generate a limited amount of data spanning a range of frequency, stress ratio, and hold time. First, crack-growth-rate data were generated at the test temperature of 649°C under a baseline condition of frequency (1 Hz), stress ratio (0.1), and hold time (0). Additional data were then generated under conditions of lower frequency, higher stress ratio, and longer hold time. In each case, complete da/dN versus ΔK curves were obtained [from combinations of constant load range and decreasing ΔK tests on compact tension (CT) specimens], including data in the near-threshold regime. This limited set of data was used to determine the coefficients of the functional relationships between the model constants and the test variables. Additional sets of experimental data were generated to evaluate the interpolative and predictive characteristics of each model. Finally, comparisons were made with sets of data points covering significant ranges of frequency and hold time. These data were all obtained at one or two specific values of ΔK, under constant-K, computer-controlled test conditions in a previous investigation [*4*].

Test Matrix

A three-dimensional representation of the test matrix is shown in Fig. 1. Each of the test variables is plotted on a separate axis. The four points plotted as solid circles represent the four tests used in formulating the predictive models. The coordinates of each point are the values of the test parameters for that particular test. The two points plotted as solid squares represent the two proof tests which were performed to evaluate the interpolative capability of the models. The point at the origin of the coordinate axes represents the baseline test. Moving from the baseline outward along any axis changes only the value of the variable represented by that axis. Thus, the three points lying on the axes away from the origin represent tests in which only one test parameter is changed from the baseline value. All six tests were performed at a temperature of 649°C.

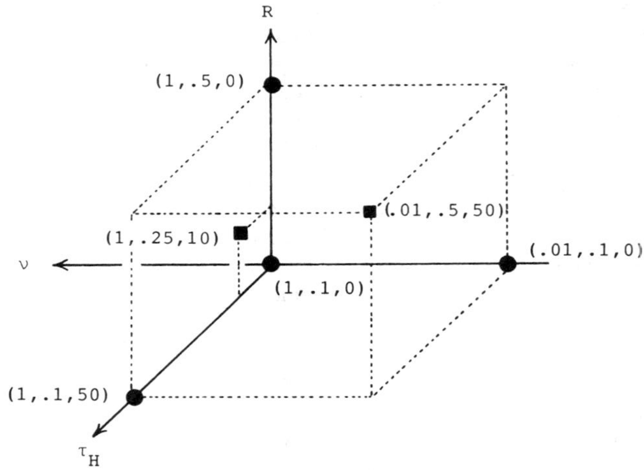

FIG. 1—*Graphic representation of test matrix defining coordinates* ν, R, τ_H.

Test Specimens and Procedures

Compact-tension type specimens used for this experimental program were machined from plates of Inconel 718. IN718 is a nickel-base superalloy with a high fracture toughness, and is commonly used in the fabrication of jet engine hot-section parts. The CT specimen geometry had a width of 40 mm and a thickness of 10 mm. The specimens were precracked at room temperature in accordance with the ASTM Test Method for Plane-Strain Fracture Toughness of Metallic Materials (E 399) under computer-controlled conditions with continuous load shedding until the final precracking K_{max}, which was smaller that the starting K_{max} for the elevated-temperature tests, was obtained. The crack length was determined from compliance measurements using an extensometer with quartz rods for the elevated temperature tests. Constants for the compliance-crack length calibration were determined from surface crack-length measurements which were corrected for crack-front curvature from visual measurements on the fracture surface. Crack-length data were reduced to crack-growth rate via the Incremental Polynomial Method suggested in the ASTM Test Method for Constant-Load-Amplitude Fatigue Crack Growth Rates Above 10^{-8} m/Cycle (E 647-81) to take advantage of its data-smoothing feature. The crack-growth rates were computed from crack length versus number of cycles (a versus N) data using a seven-point sliding polynomial curve-fitting technique.

Crack Growth Modeling

Typical da/dN versus ΔK data plotted on log-log graph paper have a sigmoidal shape. The curve is characterized by a slow crack-growth region, called the threshold region, a linear region, and a rapid growth region. Data are usually not obtained far into the rapid growth region because test specimens tend to break soon after having entered this region, and the data here are of limited use in life prediction because of the very high growth rates.

To predict the crack growth rate for any set of test parameters, a crack growth curve is required for that set of parameters. For calculation purposes, a mathematical description of the sigmoidal-shaped curve has to be developed. To describe mathematically the growth rate for all possible sets of parameters, a method has to be developed to determine the position and shape of each curve corresponding to each set of parameters. For the two models described here, this procedure has centered about describing the location of the inflection point as a function of the test parameters. Procedures have also been developed to describe the slope at the inflection point, the location of the threshold, and the general shape of the curve as a function of the test parameters. The two models have been described in the literature [*2,5*] and are described next in terms of their original formulation and their application to the data in this program.

SINH Model

The SINH crack growth rate interpolative model is based on the hyperbolic sine function [*6,7*]. Its shape closely approximates the general sigmoidal shape of da/dN versus ΔK data. The SINH equation is given by

$$\log(da/dN) = C_1 \sinh\{C_2[\log(\Delta K) + C_3]\} + C_4 \tag{1}$$

Here, da/dN is the crack growth rate per cycle and ΔK is the range of the stress intensity factor. The model parameters C_3 and C_4 give the horizontal and vertical location of the inflection point, respectively, while C_1 and C_2 "stretch" the curve vertically or horizontally, respectively. The slope of the curve at the inflection point is found to be C_1C_2.

In the applications of this equation to the prediction of crack growth rates, the following assumptions have been made:

C_1 = constant (0.5 for most materials),

$C_2 = f_2(\nu, \tau_H, R, T)$,

$C_3 = f_3(\nu, \tau_H, R, T)$, and

$C_4 = f_4(\nu, \tau_H, R, T)$.

Since all testing in this program was carried out at a single temperature, the model parameters $C_j, j = 2, 3, 4$, are not functions of the temperature T for the modeling described herein.

In the development of the SINH model, functional relationships between these model parameters and the test parameters were established [*6*]. For application to a number of materials, these functional relationships were found to be linear when plotted on log-log scale. In its simplest form, the SINH model assumes that the changes due to ν, τ_H, and R are independent of each other, and therefore their effects can be accounted for by linear superposition. Using this hypothesis, the values of the model parameters C_2, C_3, and C_4 which account for variations in each of the test parameters are given as (see Ref *6*)

$$C_j = C_{j\,\mathrm{baseline}} + b_{1j} \log \nu + b_{2j} \log(\tau_H + 1) + b_{3j} \log[(1 - R)/0.9]; \qquad j = 2, 3, 4 \tag{2}$$

where the baseline condition is taken here as $\nu = 1$, $\tau_H = 0$, and $R = 0.1$. Under these conditions, the last three terms on the right-hand side account for the changes in C_j caused by the variation of each test parameter from the baseline condition. The 0.9 factor in the last term was introduced in order for that term to vanish at the baseline value of $R = 0.1$. The addition of unity to τ_H (hold time in seconds) in the next-to-last term satisfies the same condition and simultaneously alleviates the mathematical difficulty that arises if $\log \tau_H$ is used when τ_H equals zero. This introduces a slight nonlinearity in the relationship between the constants and τ_H for short hold times. Otherwise, the functional relationships between the model parameters C_j and the test variables are as suggested in Ref *6*.

To determine the constants b_{1j}, b_{2j}, and b_{3j} (introduced in Eq 2) associated with each of the model parameters $C_j, j = 2, 3, 4$, the following procedure was used. As shown in Fig. 1, the test matrix was designed in such a way as to have on each of the axes a unique combination of two data sets, each involving the baseline set (origin) and another set in which only one of the three parameters (ν, R, τ_H) has been varied. Consider for instance the data sets on the frequency (ν) axis. In this case, the right-hand side of Eq 2 reduces to the first two terms since only ν was varied. The constant b_{1j} was then determined in three steps: (1) the best SINH curve was fitted to the baseline data set; (2) the best SINH curve was fitted to the set in which ν was varied; and (3) the constants b_{1j} of Eq 2 are found by using these two C_j values to determine the equation of the straight line relating the C_j to the test parameters. This process was repeated for the other data-set groups to obtain b_{2j} and b_{3j} in Eq 2. The SINH computer code [7] was used to automate this procedure. This code employs an iterative scheme using least-squares regression to determine the equation parameters for each data set, starting with user-supplied initial guesses for the location of the inflection point.

MSE Model

The modified sigmoidal equation (MSE) model was developed in the course of an investigation aimed at predicting crack growth rates in alloy AF115. It is based on a sigmoidal equation which, in general, may be written as (see Ref *8*)

$$da/dN = e^{B}(\Delta K/\Delta K^*)^{P}[\ell n(\Delta K/\Delta K^*)]^{Q}[\ell n(\Delta K_c/\Delta K)]^{D} \tag{3}$$

where ΔK^* and ΔK_c represent the threshold and critical values of ΔK, respectively. These two terms, as well as the four coefficients B, P, Q, and D are, in general, functions of the test parameters T, R, ν, and τ_H. However, these coefficients interact in a complex manner in controlling the shape and location of the sigmoidal curve, which makes it extremely difficult to express them as functions of the test parameters. To alleviate this difficulty, alternate coefficients were introduced which results in a MSE model given as (see Ref *8*)

$$da/dN = e^{B'}(\Delta K/\Delta K_i)^{P}[\ell n(\Delta K/\Delta K^*)]^{Q}[\ell n(\Delta K_c/\Delta K)]^{D} \tag{4}$$

A new variable, ΔK_i, the value of ΔK at the inflection point of the curve, has been introduced. Two additional parameters, da/dN_i and $(da/dN_i)'$, are introduced in subsequent equations because of their physical significance and for computational convenience. The parameter da/dN_i represents the crack growth rate per cycle at the inflection point, while $(da/dN_i)'$ represents the slope of the curve at the same point. The coefficients Q and D control the shape of the curve between the inflection point and the lower and upper asymptotes, respectively. When D is set equal to minus Q, the curve becomes symmetric. B' is the vertical distance by which the inflection point is displaced from a da/dN value of unity. P controls the rotation of the curve in the vicinity of the inflection point. In general, the coefficients D, P, and B' are given by (see Ref *8*)

$$D = -[Q^{1/2}\,\ell n(\Delta K_i/\Delta K_c)/\ell n(\Delta K_i/\Delta K^*)]^2 \tag{5}$$

$$P = (da/dN_i)' - Q/\ell n(\Delta K_i/\Delta K^*) + D/\ell n(\Delta K_c/\Delta K_i) \tag{6}$$

$$B' = \ell n(da/dN_i) - Q\,\ell n[\ell n(\Delta K_i/\Delta K^*)] - D\,\ell n[\ell n(\Delta K_c/\Delta K_i)] \tag{7}$$

It should be pointed out that these three coefficients are related to parameters which have physical significance, namely, ΔK_i, da/dN_i, and $(da/dN_i)'$, the horizontal position, vertical position, and slope of the inflection point, respectively, and a single shaping parameter Q.

Comparison of Models

By analogy, it was previously noted that for the SINH model, the coefficients C_3, C_4, and the product of C_1 and C_2 represent the horizontal and vertical position of the inflection point and the slope of the curve at that point, respectively, for that particular equation. Thus, if one were to fit experimental data with either a SINH or a MSE model, and if it were required that both models have the same location and slope for their inflection points, then the following relations would hold

$$C_3 = -\log(\Delta K_i) \tag{8}$$

$$C_4 = \log(da/dN_i) \tag{9}$$

and

$$(da/dN_i)' = C_1 C_2 \tag{10}$$

Since the SINH model is well developed in the form of a computer program with built-in regression analyses, the coefficients C_2, C_3, and C_4 (C_1 = constant) can be determined as linear functions of the test variables by fitting multiple sets of experimental data. These same functions could then be used for the MSE model by applying Eqs 8–10. Alternatively, the functional forms for relating the six MSE constants to the test parameters could be established. However, since the MSE model is not developed to the point of a production code, all required regression analyses have to be performed separately. It should be re-emphasized here that the MSE model was specifically developed for the alloy AF115 and has not been previously applied to IN718. Also, the originators of the MSE model [*8*] had access to a large data set, and they were thus able to include some higher order dependencies, that is, terms involving products of functions of the test parameters. The data set used in this investigation was limited by design.

Application to Experimental Data

To represent the experimental data for this program, it was decided that the functional relationships (linear forms) used for the SINH model parameters (Eq 2) would be also used in this investigation to represent the MSE model constants ΔK^*, ΔK_i, da/dN_i, and $(da/dN_i)'$. It was intended that this approach would establish whether these simple linear forms are valid for either model, as well as their range of applicability. These general equations, which should be applicable to both small and large data sets, have assumed functional forms as

$$\log(\Delta K^*) = \log(\Delta K^*)_{\text{baseline}} + d_1 \log[(1 - R)/0.9] + d_2 \log \nu + d_3 \log(\tau_H + 1) \tag{11}$$

$$\log(\Delta K_i) = \log(\Delta K_i)_{\text{baseline}} + d_4 \log[(1 - R)/0.9] + d_5 \log \nu + d_6 \log(\tau_H + 1) \tag{12}$$

$$\log(da/dN_i) = \log(da/dN_i)_{\text{baseline}} + d_7 \log[(1 - R)/0.9] + d_8 \log \nu + d_9 \log(\tau_H + 1) \tag{13}$$

$$(da/dN_i)' = (da/dN_i)'_{\text{baseline}} + d_{10} \log[(1 - R)/0.9] + d_{11} \log \nu + d_{12} \log(\tau_H - 1) \tag{14}$$

Note that the last three equations (Eqs 12 to 14) have the same form as in the SINH model, while Eq 11 uses the same functional form for the threshold ΔK^*. In addition, forms have to be established for the remaining two parameters, ΔK_c and Q. These parameters were assumed as

$$K_c = \begin{cases} \text{constant} & \text{(nonsymmetric curve)} \\ \Delta K_i^2/\Delta K^* & \text{(symmetric curve)} \end{cases} \tag{15}$$

$$Q = d_{13} \qquad \text{(a constant)} \tag{16}$$

In general, the constants d_1 through d_{12} can be determined by linear regression, while d_{13} is established graphically via an iterative process.

For the data sets used in this program, there were only two data curves used to resolve a change due to varying a particular test parameter in Eqs 11–16. Therefore, the linear regression approach reduced to solving a set of simultaneous equations. For example, in the case of Eq 13, there are three simultaneous equations to be solved. The value of da/dN_i is known for the baseline data set and for the three data sets where ν, R, and τ_H are varied. When these values are substituted into Eq 13 along with the appropriate values for the test parameters for each data set, three equations in the three unknowns d_7, d_8, and d_9 are obtained.

Modeling Procedure

The experimental data sets for the baseline condition and the three other conditions, each involving variation of one parameter, were fit with the SINH model. The fits were obtained using a computer program [7] which provides least-squares minimization routines to establish the constants using as a starting point a user-supplied estimate for the approximate location of the inflection point. The output provides all of the coefficients for Eq 2.

Two distinct approaches were followed in fitting the data with the MSE model. First, the data were fitted with nonsymmetric curves. For this case, the upper asymptote ΔK_c was taken as a constant. The location of the inflection point for each data set was taken from the SINH model fit (see Eqs 8 and 9). The values of ΔK^* were obtained from Eq 11 by a best (trial-and-error) fit to the experimental data. In this first approach, the slope of the curve at the inflection point $(da/dN_i)'$ was obtained manually from the data plots. Since the inflection point and lower asymptote (ΔK^*) are determined separately, and the upper asymptote (ΔK_c) is taken as a fixed value, the curves are nonsymmetric ($Q \neq -D$). The lower shaping coefficient Q was assumed to be a constant, since it was found graphically via an iterative process that adequate MSE fits could be obtained for all of the experimental data sets with a constant Q. It was originally attempted to model Q using an expression similar to those given in Eqs 11–14. However, this resulted in unreasonable values of Q for the proof tests, that is, negative Q's. It was concluded that Q was sensitive to change near the baseline, but was relatively insensitive to values of the test parameters away from the baseline. Therefore, the value of Q found to best represent the baseline condition was also used for the remaining data sets. The value of D, the other shaping coefficient, is determined from Eq 5 since Q, ΔK_i, ΔK_c, and ΔK^* have been already established. This first attempt, fitting the data with a nonsymmetric MSE curve, led to results that were considered unsatisfactory, especially when attempting to fit the data obtained from the proof tests. It was concluded that the combination of a fixed inflection point, a fixed threshold ΔK^*, and a constant value for the critical stress-intensity range ΔK_c was overconstraining the curve, resulting in a poor fit of the experimental data. An alternate procedure seemed warranted.

In this second approach, the location of the inflection point was taken from the results of the SINH fit. In addition, the slope of the inflection point from the SINH fit was used (see Eq 10). Also, the MSE model was assumed to be a symmetric curve ($Q = -D$). Using the same expression for the threshold ΔK^* as in the first approach, this requires that ΔK_c cannot be set constant or further constrained. The symmetry of the curve and the fixing of both ΔK^* and ΔK_i require that ΔK_c be the value given by Eq 16. Using this second procedure, results which were considerably better and deemed quite satisfactory were obtained.

Results and Discussion

All of the experimental data are presented in tabular form in Ref 9. The data sets used to establish the coefficients for the functional relationships between the model constants and the test parameters (solid circles in Fig. 1) are plotted along with the best-fit SINH and MSE curves in Figs. 2 and 3, respectively.

The baseline data ($\nu = 1$ Hz, $R = 0.1$, $\tau_H = 0$ s) and the data obtained when each of the test parameters were varied (in turn) from their baseline values (ν to 0.01 Hz, R to 0.5, and τ_H to 50 s) are shown in Fig. 2 along with the best-fit SINH curves. The location of the inflection points on these curves and their slopes were calculated by making use of the iterative least-squares regression scheme incorporated in the SINH computer code. This procedure requires that an initial guess for the location of the inflection point be supplied by the user. The final calculated location of the inflection point was found to be dependent upon this initial guess, and much more so in cases where the range of the data fitted did not exhibit sufficient amounts of curvature (that is, when most of the data points lie on the linear portion of the da/dN versus ΔK curve). This point has been discussed in more detail in Ref *10*.

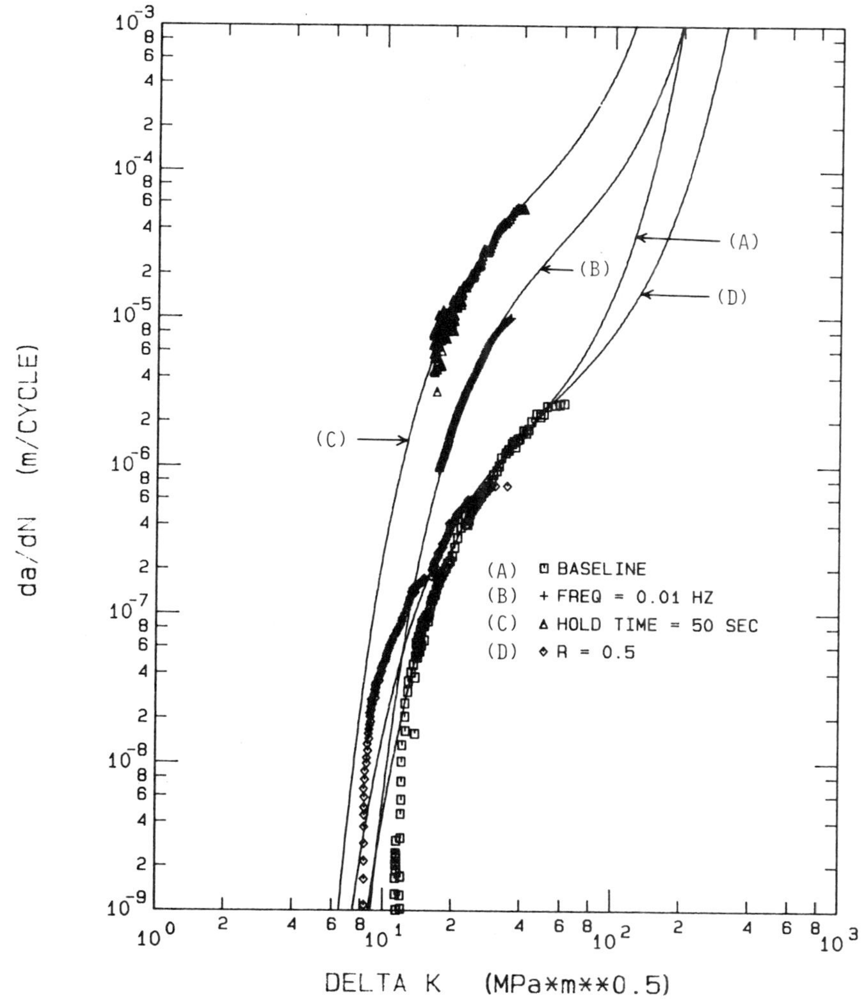

FIG. 2—da/dN *versus* ΔK *for baseline conditions with superimposed SINH model fits.*

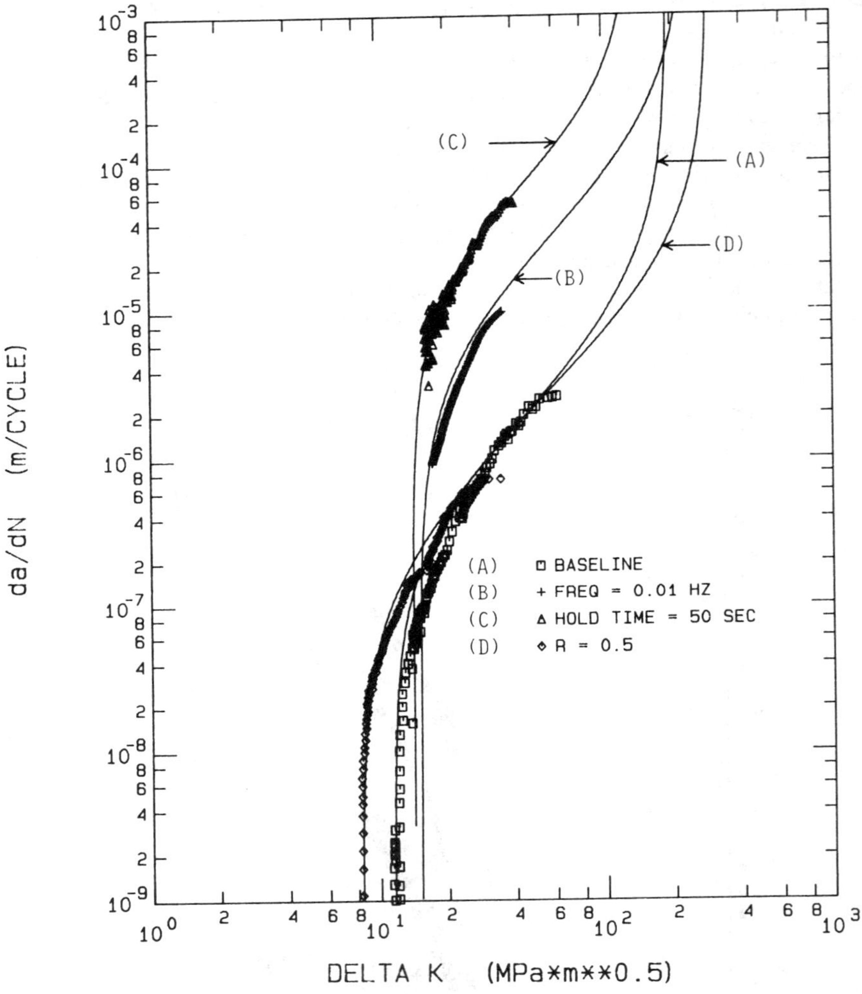

FIG. 3—da/dN *versus* ΔK *for baseline conditions with superimposed MSE model fits.*

The experimental results and resultant SINH model fits to the data show that the da/dN versus ΔK curve shifts when each of the test variables is varied from its baseline value. As seen in Fig. 2, the crack growth rate *increases* with lower frequency and longer hold time. There is no significant change in the position or shape of the curve for variations in stress ratio, although the threshold value decreases for increases in R.

The values of the coefficients needed for expressing the model parameters $C_j, j = 2, 3, 4$ of the SINH model in terms of the experimental parameters ν, R, and τ_H were established according to the procedure outlined previously. C_1 was set equal to 0.5 for all cases. The remaining coefficients are as follows

$$C_2 = 3.197 - 1.1675 \log\left(\frac{1 - R}{0.9}\right) - 0.1505 \log \nu + 0.2053 \log(\tau_H + 1) \quad (17)$$

$$C_3 = -1.762 - 0.6243 \log\left(\frac{1-R}{0.9}\right) + 0.00717 \log \nu + 0.0677 \log(\tau_H + 1) \quad (18)$$

$$C_4 = -2.530 + 1.0131 \log\left(\frac{1-R}{0.9}\right) - 0.5231 \log \nu + 0.8127 \log(\tau_H + 1) \quad (19)$$

The constant C_2 is dimensionless. In all of the equations relating the constants to the test parameters, units of MPa $\cdot$ m$^{1/2}$ have been used for ΔK while units of mm/cycle have been used for da/dN.

The data sets just described were also fitted with symmetric MSE curves as shown in Fig. 3. As explained previously, the locations of the inflection points (coordinates: ΔK_i, da/dN_i) were those established using the SINH model, while the slope at each of the inflection points, $(da/dN_i)'$, was computed from Eq 10 using the C_1 and C_2 values (SINH model) corresponding to each data set. Accordingly, from Eqs 8 through 10 and 17 through 19 with $C_1 = 0.5$, the following functional relationships were obtained for these MSE-model curves

$$\log(K_i) = 1.762 + 0.6243 \log\left(\frac{1-R}{0.9}\right) - 0.00717 \log \nu - 0.0677 \log(\tau_H + 1) \quad (20)$$

$$\log(da/dN_i) = -2.530 + 1.0131 \log\left(\frac{1-R}{0.9}\right) - 0.5231 \log \nu + 0.8127 \log(\tau_H + 1) \quad (21)$$

$$(da/dN_i)' = 1.5985 - 0.58375 \log\left(\frac{1-R}{0.9}\right) - 0.07525 \log \nu + 0.10265 \log(\tau_H + 1) \quad (22)$$

A functional relationship, Eq 11, similar to those just given was also assumed for the threshold value of ΔK, ΔK^*. The constants were established using a trial-and-error regression procedure and best estimates of the threshold from the experimental data. The resulting expression was

$$\log(\Delta K^*) = 1.0607 - 0.55659 \log\left(\frac{1-R}{0.9}\right) - 0.05995 \log \nu + 0.05259 \log(\tau_H + 1) \quad (23)$$

A constant value of $Q(= -D) = 1.0$ was selected by trial and error as the one which produced the curvature that best fitted the experimental data. All of the curves in Fig. 3 were drawn using this value.

A comparison of Figs. 2 and 3 reveals that under the conditions imposed on the predictive models in this investigation, these two models provide essentially the same fit to the experimental data, except possibly in the threshold regime of the stress intensity factor range (ΔK^*) where the MSE model, through the incorporation of the functional relationship for ΔK^* (Eq 23), exhibits additional flexibility in predicting the experimental data.

The interpolative capabilities of the SINH and MSE models were evaluated by using them, in the forms defined by Eqs 17 through 19 and 20 through 23, respectively, to predict the experimental data collected from the two proof tests. The first proof test ($\nu = 1$ Hz, $R = 0.25$, $\tau_H = 10$ s) represents a relatively short departure from the baseline data (Fig. 1), varying only R and τ_H by a small amount. On the other hand, all three test parameters are varied by significant amounts from the baseline data for the second proof test ($\nu = 0.01$ Hz, $R = 0.1$, $\tau_H = 50$ s). Therefore, the application of the models in predicting the crack growth rate in this case represents a considerable amount of extrapolation.

The experimental results from the first proof test are plotted along with the two model predictions in Fig. 4. In this case, the models predict the experimental crack growth rates quite accu-

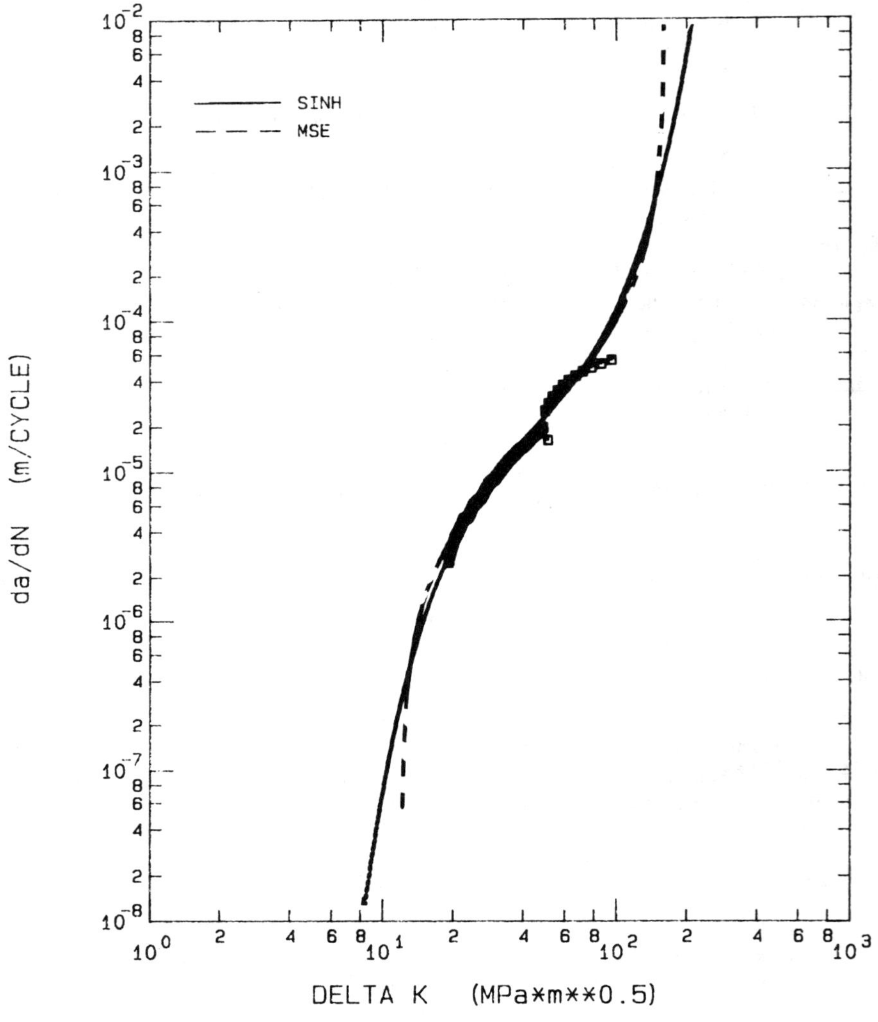

FIG. 4—*Comparison of predicted and experimental results for Proof Test 1 (*R *=* 0.25, $\tau_H = 10$ s, $\nu = 1$ Hz*).*

rately. Both predictions are within the observed experimental scatter range for this type of experiment. Also, both models provide essentially the same prediction since both use the same values for the location of the inflection point and its slope.

The data collected during the second proof test and the model predictions are shown in Fig. 5. Both models predict quite similar behavior, especially in the intermediate growth rate regime, although neither model fared very well in this case. The predicted crack growth rates differ from those observed by a factor of four at intermediate values of ΔK near the inflection point. At higher ΔK values the difference increases to over an order of magnitude. Both models conservatively predict higher crack growth rates than those observed. This seems to indicate that interaction effects are present which tend to retard the crack growth rates. Accounting for such higher-order effects is beyond the scope of the present investigation.

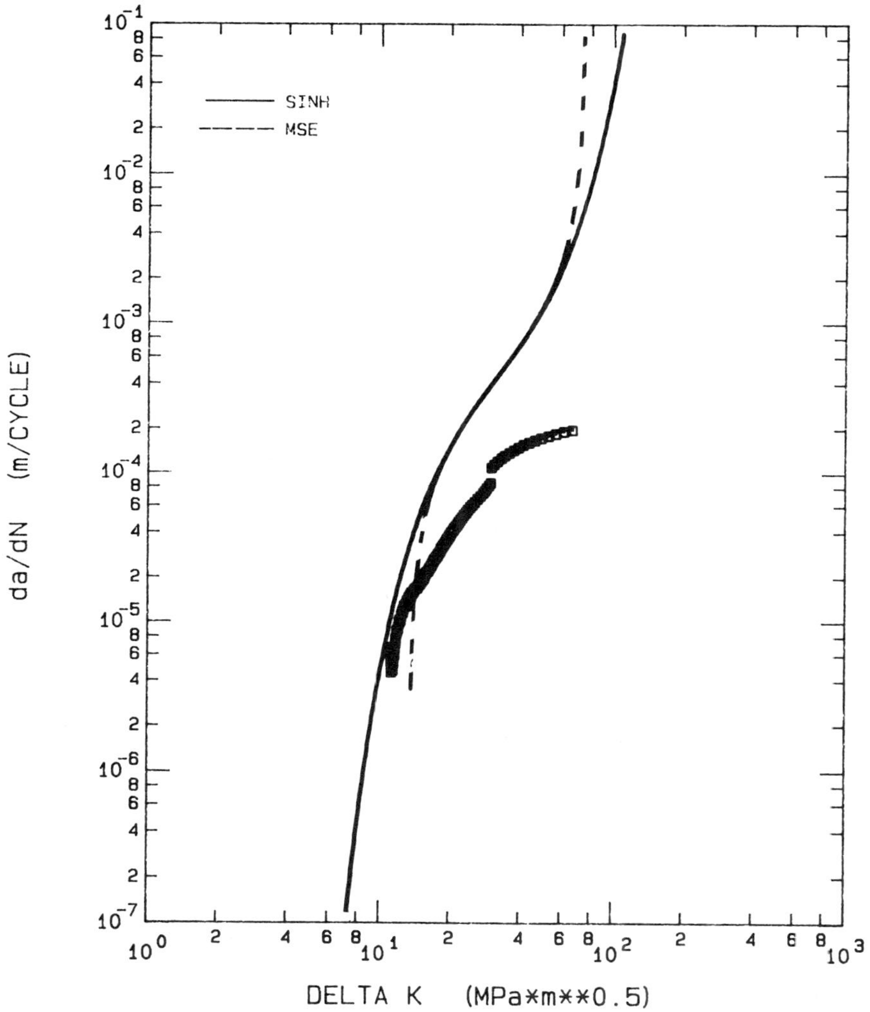

FIG. 5—*Comparison of predicted and experimental results for Proof Test 2 (*R *= 0.5,* τ_H *= 50 s,* ν *= 0.01 Hz).*

The effectiveness of the models in predicting the crack growth per cycle (da/dN) as a function of hold time duration (τ_H) and frequency (ν) was also examined. The data utilized for these comparisons were generated under a different project (Ref *4*) using a different batch of material. The experimental data are offset by approximately a factor of two in crack growth rate from the results of the present investigation, but this is due primarily to experimental scatter directly attributable to the material batch variation. What is of interest here is the ability of the models to predict *trends* in da/dN with changing hold time duration and frequency. All of these experimental data were collected under constant ΔK (36 MPa $\cdot$ $m^{1/2}$) and constant stress ratio (R = 0.1) conditions.

The crack growth per cycle variation with hold-time duration is plotted along with the SINH and MSE predictions in Fig. 6. The frequency here was held constant at 1 Hz and hold times

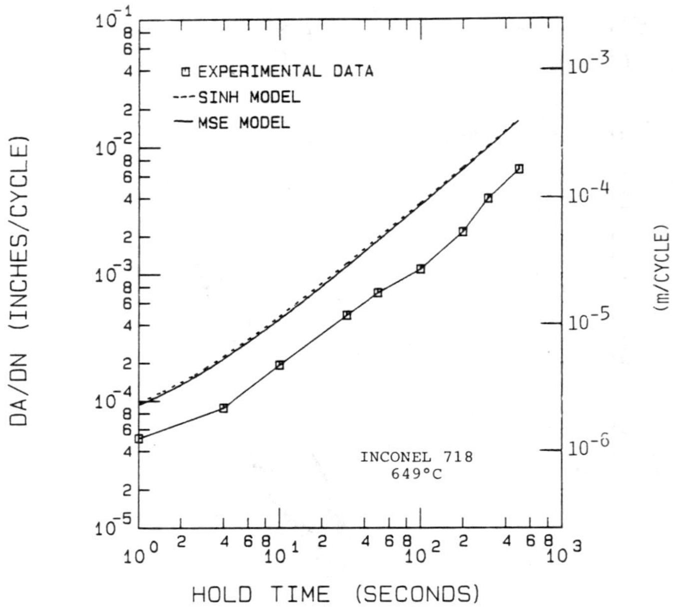

FIG. 6—*Comparison of predicted and experimental results for variations in hold time* (ΔK = *36 MPa* $\cdot$ $m^{1/2}$, $R = 0.1$, $\nu = 1$ *Hz).*

ranging from 1 to 500 s were applied. The two models yield essentially the same prediction, and both predict the trend of the experimental data (higher crack growth per cycle at higher hold-time duration) very well.

Both models, as well as the data, predict essentially purely time-dependent material behavior, that is, a slope of unity for hold times in excess of approximately 10 s. The trend of the data is very nearly linear, and the models, being based on linear functional relationships, are able to predict this trend closely.

The comparison of model predictions with experimental data for variations with frequency is presented in Fig. 7. These data were generated under constant ΔK with $\tau_H = 0$ and $R = 0.1$ and for frequencies ranging from 0.001 to 10 Hz. The predicted variation of growth rate with frequency is essentially linear and almost identical for the two models. Over this range of frequencies, however, the behavior of da/dN on a log-log scale is bilinear. At very low frequencies, the slope of the curve is very nearly -1 with the crack growth being purely time-dependent (intergranular), while between approximately 0.01 and 10 Hz this material exhibits mixed-mode growth behavior (intergranular-transgranular). Above 10 Hz (not tested here) the crack growth becomes purely cycle-dependent (transgranular) at this temperature. Clearly then, the linear models are not able to predict this bilinear behavior closely. However, the general trend of the data is adequately predicted over a significant range of frequency.

Conclusions

Two interpolative crack growth models, SINH and MSE, can be used to predict crack growth rates in materials exhibiting time-dependent behavior over ranges of frequency, stress ratio, and sustained load hold time. Fitting of experimental data and determination of model constants are accomplished most easily by working with the location and slope of the inflection point for these sigmoidal curves. The MSE model has an additional constant which allows for

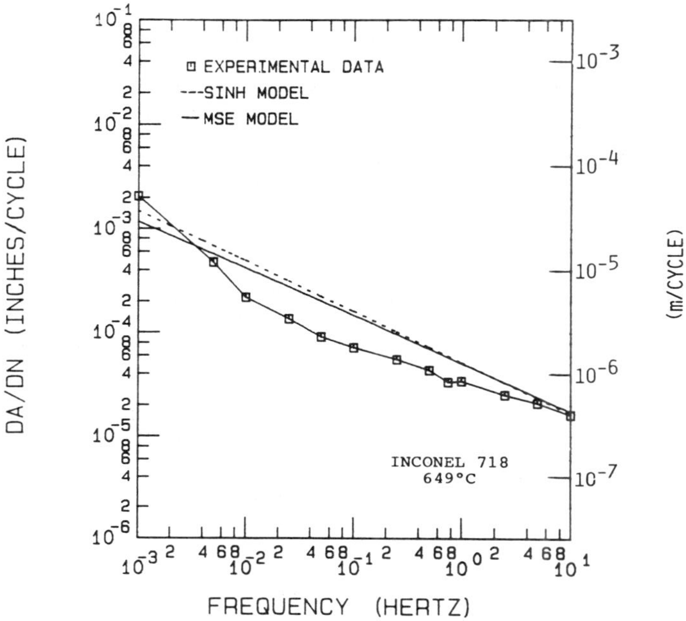

FIG. 7—*Comparison of predicted and experimental results for variations in frequency* (ΔK = *36 MPa* $\cdot$ $m^{1/2}$, R = *0.1*, τ_H = *0*).

matching values of the threshold from experimental data. If both models use the same values for the location and slope of the inflection point, their predictive capabilities are similar. In the application of these models to Inconel 718 at 649°C, predictions for values of test parameters close to those used for determining model constants were quite accurate. Extrapolation to test conditions further from those of the baseline tests were less accurate. Both models predict nearly linear variations of crack growth rate with either frequency or hold time under conditions of fixed ΔK, and all other test parameters held constant. If material behavior is bilinear under these conditions, as when undergoing a change in crack growth fracture mechanism, then neither model has the flexibility to model this behavior in its present form.

Acknowledgments

This work was part of a M.S. thesis of the third author at the Air Force Institute of Technology, Wright-Patterson AFB, Ohio. The work was supported, in part, by the AFWAL Materials Laboratory under in-house project 2302P1. The authors would like to express their appreciation to Mr. David Maxwell of the University of Dayton Research Institute for his support in the experimental program, to Major Mike Heil of AFIT for his help in the computations, and to Mr. James Larsen of the AFWAL Materials Laboratory for his assistance in the use of the SINH model computer program.

References

[1] Military Standard: Engine Structural Integrity Program (ENSIP), MIL-STD-1783, U.S. Air Force, Nov. 30, 1984.

[2] Nicholas, T., Haritos, G. K., and Christoff, J. R., *Journal of Propulsion and Power*, Vol. 1, 1985, pp. 131–136.

[3] Ashbaugh, N. E., "Time Dependent and Cycle Dependent Threshold Evaluation in a Nickel-Base Superalloy," presented at Eighteenth National Symposium on Fracture Mechanics, American Society for Testing and Materials, Boulder, CO, June 1985.
[4] Nicholas, T., Weerasooriya, T., and Ashbaugh, N. E. in *Fracture Mechanics: Sixteenth Symposium, ASTM STP 868,* M. F. Kanninen and A. T. Hopper, Eds., American Society for Testing and Materials, Philadelphia, 1985, pp. 167–180.
[5] Larsen, J. M. and Nicholas, T., "Cumulative Damage Modeling of Fatigue Crack Growth in Turbine Engine Materials," *Engineering Fracture Mechanics,* in press.
[6] Larsen, J. M., Schwartz, B. J., and Annis, C. G., "Cumulative Damage Fracture Mechanics Under Engine Spectra," Technical Report AFML-TR-79-4159, NTIS Acc. No. ADA084934, Wright-Patterson Air Force Base, OH, Jan. 1980.
[7] Schwartz, B. J., Larsen, J. M., and Annis, C. G., "Cumulative Damage Fracture Mechanics Under Engine Spectra: FORTRAN User's Manual for Super SINH," Pratt and Whitney Aircraft Group, Government Products Divisions, West Palm Beach, FL, Jan. 1980.
[8] Utah, D. A., "Crack Growth Modeling in an Advanced Power Metallurgy Alloy," Technical Report AFML-TR-80-4098, NTIS Acc. No. ADA093992, Air Force Materials Laboratory, Wright-Patterson Air Force Base, OH, Feb. 1980.
[9] Painter, G. O., "Evaluation of Interpolative Modeling Concepts for Fatigue Crack Growth at Elevated Temperature," M.S. thesis, Air Force Institute of Technology, NTIS Acc. No. ADA 153222, Wright-Patterson Air Force Base, OH, Dec. 1984.
[10] Christoff, J. R., "Evaluation of Fatigue-Creep Crack Growth in an Engine Alloy," M.S. thesis, Air Force Institute of Technology, NTIS Acc. No. ADA 136956, Wright-Patterson Air Force Base, OH, Dec. 1983.

Poster Session—Analysis

Biswajit Mukherjee[1] *and Dave McCluskey*[1]

Crack Arrest Fracture Toughness of 7075-T6 Aluminum Alloy

REFERENCE: Mukherjee B. and McCluskey, D., **"Crack Arrest Fracture Toughness of 7075-T6 Aluminum Alloy,"** *Fracture Mechanics: Eighteenth Symposium, ASTM STP 945*, D. T. Read and R. P. Reed, Eds., American Society for Testing and Materials, Philadelphia, 1988, pp. 223–236.

ABSTRACT: The plane-strain fracture toughness K_{Ic} represents the resistance of a material against crack initiation. The material toughness in the case of a run-arrest event involves a special "material property" called the crack arrest toughness K_{Ia}, which within certain limitations, is a measure of the ability of a material to arrest a fast running crack under linear elastic conditions. It has been indicated that K_{Ia} is a material property independent of specimen geometry, but this is not a universal belief. Whether or not K_{Ia} is a material property is still being debated. The objectives of this project were to measure the crack arrest fracture toughness of 7075-T6 aluminum alloy and to determine the effect of specimen size and thickness on K_{Ia}. The tests were performed in accordance with a proposed ASTM procedure developed for high-toughness and medium-strength steels. Specimens that were used in this program exceeded the proposed minimum linear elastic size requirement of ferritic steel specimens by a factor of 4 between 2 and 10. K_{Ia} showed a slight dependence on test parameters such as stress intensity at crack initiation, specimen thickness, and crack jump lengths. However, K_{Ia} was not a strong function of the stress intensity at crack initiation K_0. The mean value of K_{Ia} was an adequate representation of the crack arrest toughness value within the test matrix. The average K_{Ia} was only 10% lower than the average plane-strain fracture toughness K_{Ic}.

KEY WORDS: fracture, crack arrest, aluminum, fracture toughness, crack jump

The published literature of the 1970s on the development of crack arrest methodology is complex and appears contradictory to the uninitiated. However, two recent reviews go a long way toward explaining the issues clearly [*1,2*]. Two opposing views for the arrest of rapidly propagating cracks have been advanced [*2*]. In one view, crack arrest is considered as the reverse of crack initiation. During a run arrest event, a crack arrests at a toughness value K_{Ia}, which is considered as a material property [*1,3*]. In ferritic steel, K_{Ia} is relatively independent of K_0, the stress intensity at initiation of rapid crack propagation. K_{Ia} is a plain-strain linear elastic fracture mechanics (LEFM) parameter and can be characterized in term of the static value of the crack driving force at the arrested crack length a few milliseconds after the arrest event.

The opposing view is the dynamic approach. As indicated in Ref *2*, in linear elastic conditions, a dynamically computed value of stress intensity, K, drives a moving crack while a material and crack speed dependent critical value of K, conventionally denoted by K_{ID}, is the resistance parameter. In this approach, the K_{ID} versus crack velocity ($K_{ID} - \dot{a}$) relationship is regarded as a material property. Some investigators [*2*] (but not all [*4*]), have found K_{ID} values to be reasonably geometry independent. According to the dynamic analysis, both K_{ID} and K_{Ia} can depend on K_0, and the nature of that relationship is governed by the $K_{ID} - \dot{a}$ material property and by the size of the crack jump relative to the size of the specimen and structure.

[1]Unit head and technologist respectively, Ontario Hydro Research Division, 800 Kipling Ave., Toronto, ON Canada M8Z 5S4.

Kanninen has indicated where the static and dynamic approaches are compatible and where they are not [2]. K_{ID} at the instant of crack arrest ($\dot{a} = 0$) or at the low crack speed region ($\dot{a} \approx$ 100 m/s) is denoted by K_{IA}. It can be shown from the analysis of a crack in an infinite medium that the statically measured K_{Ia} is a good approximation of K_{IA} provided the crack jump lengths are short relative to the specimen geometry so that reflected stress waves do not return to interfere with the moving crack tip. However, Kanninen indicates that dynamic treatment may be necessary when the crack propagation path parallels a nearby free surface as in double cantilever beam or compact crack arrest specimens.

A significant amount of work has been conducted to develop test specimens [5,6], and a test method entitled Proposed ASTM Standard Test Method for Determining the Plane Strain Crack Arrest Fracture Toughness, K_{Ia}, of Ferritic Steels—Draft 86-03, to measure crack arrest fracture toughness of ferritic steel. Due to the combination of low yield and high toughness of low-carbon and low-alloy ferritic steels, specimen dimensions required to keep the load and displacements within the elastic limit are large. For SA533 B ferritic steel with a static yield stress of 480 MPa, specimens with width W = 165 mm, thickness B = 50 mm, and height $2H$ = 168 mm are required to satisfy specimen size requirements.

Marschall et al. [7] have investigated the use of subsize specimens for crack arrest testing in ferritic steel. A wedge and a split-pin assembly were used to apply load on the specimens. A wedge was forced into a split pin, which applied an opening force, causing a run-arrest segment of crack extension. These experiments indicated that subsize specimens often failed by ductile tear. Those specimens which failed by rapid fracture contained significant plastic deformation when the load on the wedge was plotted against the crack opening displacement. The inclusion of the plastic component in the linear elastic computation of crack arrest yielded nonconservative toughness values. An incremental load cycling technique was used to minimize the effect of plastic displacement prior to crack initiation. Subsize ferritic crack arrest specimens are useful for studying radiation effects in pressure vessel steel because of space limitation in reactors.

Specimen size requirements given in the proposed ASTM test method are based on experimental results on A533 B steel specimens of varying size. The purpose of these requirements is to ensure that the ferritic steel specimens remain linear elastic during the loading procedure. For high-strength, low-toughness material such as 7075-T6 aluminum alloy (σ_{yield} = 503 MPa, $\sigma_{ultimate}$ = 572 MPa), the size requirements are satisfied by a relatively smaller specimen than that required for ferritic steel. For example, it was possible to create a run-arrest event in a 7075-T6 aluminum specimen of W = 5.05 cm, $2H$ = 6 cm, and B = 0.635 cm without using an incremental load cycling technique. The linearity of the load displacement plot showed that the specimen remained elastic up to the maximum load. Therefore, this alloy can be used to investigate the effect of specimen dimensions on crack arrest toughness using specimens which are proportionately larger than the size just mentioned, without violating linear elastic conditions.

The objectives of this project were to measure plane-strain crack arrest toughness of 7075-T6 aluminum alloy according to the proposed ASTM test procedure and to determine the effect of specimen size and thickness on crack arrest toughness. 7075-T6 aluminum is used extensively in the aerospace industry, and, as such, crack arrest toughness values are of interest. Since it is possible to measure plane-strain fracture toughness (K_{Ic}) values in this alloy using a reasonable specimen size (W = 5 cm, B = 2.5 cm), a further objective was to compare plane-strain crack arrest toughness (K_{Ia}) with plane-strain crack initiation toughness (K_{Ic}).

Experimental Procedure

A proposed ASTM test method for measuring crack arrest fracture toughness of ferritic materials was utilized in this test program. The procedure involves testing of modified compact specimens that have been notched by machining. A wedge and a split-pin assembly are used to apply a load at the crack line. The wedge is forced into a split pin, which applies an opening force, causing a run-arrest segment of crack extension.

The test fixture is shown in Fig. 1. It was designed for use in a 250-kN servohydraulic testing machine. A hold-down plate was included to facilitate wedge extraction during incremental load cycling. The wedge, split pin, baseplate, and specimen hole were lubricated using thin sheets of trifluoroethylene resin. A displacement gage was used to measure the crack mouth opening displacement at 0.25 *W* from the load line.

A compact specimen geometry (Fig. 2) recommended by the ASTM Task Group on Crack Arrest was used. Table 1 shows the different specimen dimensions and thicknesses that were used in this program. Specimen dimensions ranged from approximately 16.8 by 16.8 cm to 6 by 6 cm and from 5.1 to 0.6 cm thickness.

The starter notch was spark-machined to produce an included angle of 30 deg and a root radius of approximately 0.4 mm. The specimens were side-grooved to a depth of $B/8$ per side.

All crack arrest and fracture toughness specimens were machined from a single 5-cm-thick 7075-T6 aluminum plate. All specimens were machined in the T-L orientation as per the ASTM Test Method for Plane-Strain Fracture Toughness of Metallic Materials (E 399-83). The longitudinal axes of the tension specimens were parallel to the longitudinal axis of the plate. Table 2 compares the specified mechanical properties and chemical composition of the material to measured values.

Tension tests were conducted using the ASTM Methods of Tension Testing of Metallic Materials (E 8-85). Tension specimens 1.28 cm in diameter were pulled in a tension machine. Load and extension over a 5-cm gage length were recorded on an *x-y* recorder. A typical measured stress-strain curve is shown in Fig. 3. The loading rate for the tension tests was 3000 MPa/min.

All crack arrest tests were conducted at ambient temperature. The maximum allowed displacement for the first loading cycle was determined prior to each test. Each specimen was loaded at a crosshead speed of approximately 5 mm/min until a run-arrest event was observed or the specified displacement was reached. In all of these tests, a run-arrest event occurred in

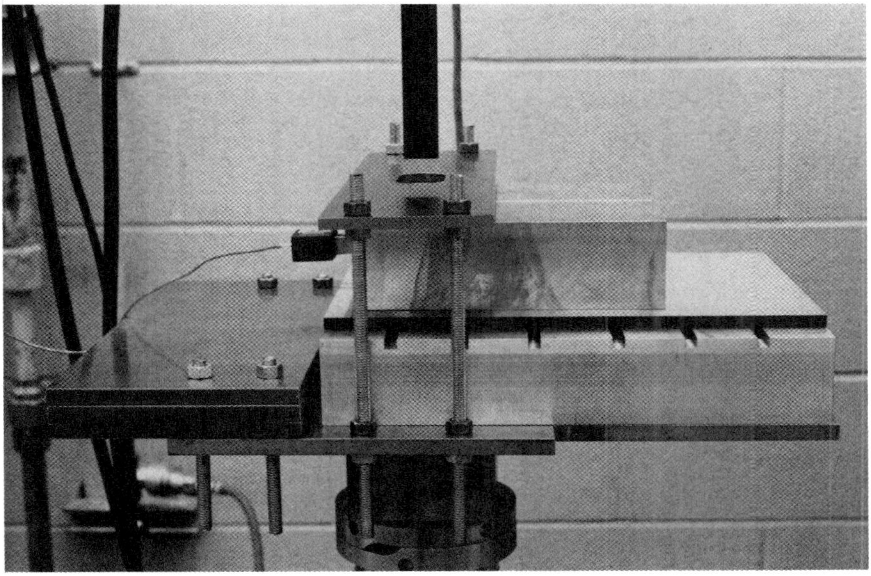

FIG. 1—*Test specimen loading arrangement.*

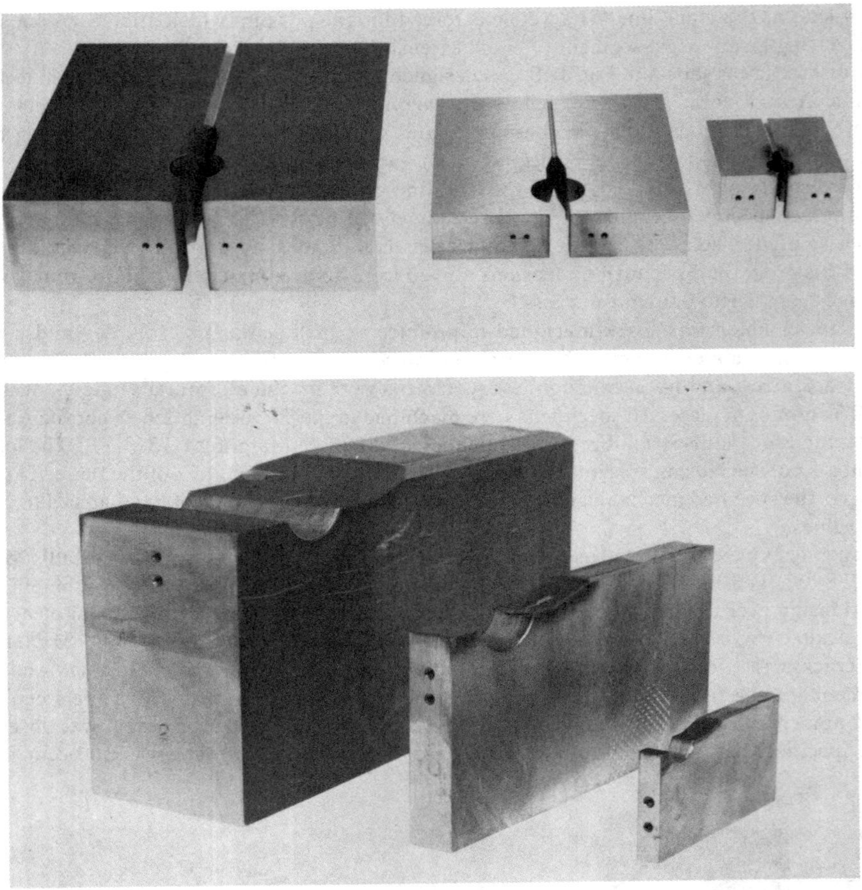

FIG. 2—*Compact crack arrest specimens used in the present study.*

the first loading cycle. The occurrence of unstable crack extension was clearly identified both audibly and as an abrupt load drop on the wedge load versus crack mouth displacement test record (Fig. 4).

The arrested crack front was marked by spraying the crack with a dye penetrant before breaking the specimen open. It is important to ensure that the fracture surface is dry before breaking the specimen; otherwise, the freshly fractured surface will also become covered with the dye and the arrested crack front will not be delineated. The arrested crack length was defined as the average of three measurements taken at the center of the specimen and midway between the center and the bottom of the side groove on each side.

Calculation of the effective stress intensity at initiation K_0, is based on measurements of the machined notch length and opening displacement at the onset of unstable crack growth. The value of crack arrest toughness K_a, is based on the measurement of crack length and opening displacements, after arrest. K_0 and K_a were calculated from

$$K = E\delta f(x)\sqrt{B/B_N}/\sqrt{W} \qquad \text{MPa}\sqrt{\text{m}}$$

TABLE 1—*Specimen dimensions and arrest toughness values.*[a]

Specimen Number	Specimen Dimensions			K_0, MPa $\sqrt{m}$	K_{Ia}, MPa $\sqrt{m}$	a_0, cm	a_a, cm	i, mm	f, mm
	W, cm	$H \times 2$, cm	B, cm						
2	13.71	16.8	5.08	33.68	25.73	5.06	7.46	0.626	0.649
1	13.67	16.8	5.08	35.17	25.94	5.08	7.98	0.655	0.703
7	10.12	12.1	2.54	38.99	26.31	3.82	6.46	0.63	0.686
10	10.13	12.1	1.27	33.21	21.37	3.82	6.78	0.539	0.599
5	5.06	6	2.54	33.46	23.2	1.91	3.11	0.383	0.407
4	5.04	6	1.27	33	23.02	1.91	3.15	0.376	0.411
6	5.05	6	1.27	36.11	24.97	1.91	3.31	0.411	0.471
11	5.07	6	0.635	34.99	21.75	1.91	3.49	0.395	0.443
12	5.05	6	0.635	47.14	19.75	1.9	4.14	0.525	0.572
13	5.06	6	0.635	40.16	25.11	1.91	3.47	0.453	0.505
14	5.05	6	0.635	35.46	23.89	1.91	3.2	0.403	0.434
ASTM E 399 Plane-Strain Fracture Toughness:									
				K_{Ic}					
15	5	6.25	2.5	26.17					
16	5	6.25	2.5	27.42					
17	5	6.25	2.5	26.98					
18	5	6.25	2.5	26.56					

[a] i = displacement at initiation.
f = displacement at crack arrest.
W = specimen width.
H = specimen height.
K_0 = stress intensity at initiation.
K_{Ia} = stress intensity after crack arrest.
a_0 = initial crack length.
a_a = crack length after arrest.

TABLE 2—*Comparison of measured and specified properties of 7075-T6 aluminum.*

	σ_y, MPa	σ_{UTS}, MPa	Chemical Composition, %
Specified	503	572	Zn 5.1-6.1, Mg 2.1-1.9, Cu 1.2-2.0, Cr 0.18-0.40
Measured	522	593	Zn 5.6, Mg 2.0, Cu 1.6, Cr 0.18

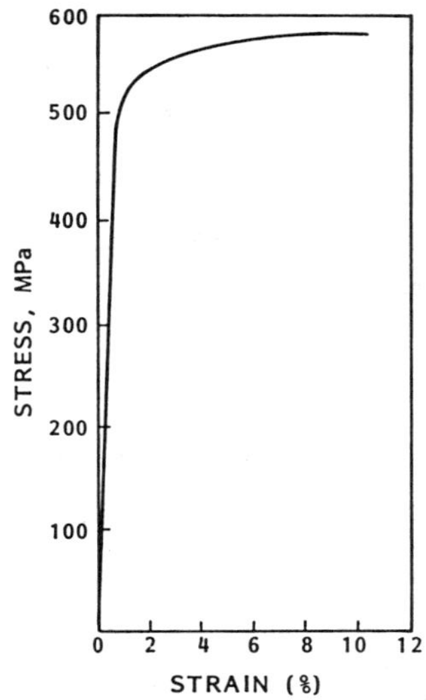

FIG. 3—*Stress-strain curve for 7075-T6 aluminum alloy.*

where

$$F(x) = \frac{2.24(1.72 - 0.9x + x^2)\sqrt{1 - x}}{9.85 - 0.17x + 11x^2}$$

and $x = a/w$.

Plane-strain fracture toughness tests were conducted according to the ASTM E 399 test standard. Compact tension specimens ($B = 2.5$ cm) were machined from the same plate and orientation as that of the crack arrest specimens. The specimens were precracked and pulled to fracture in a servohydraulic testing machine at a rate of 0.6 MPa $\sqrt{\text{m}}$/s. The load on the specimen was plotted against front face displacements and the fracture toughness values were determined as per ASTM E 399. All four specimens provided valid plane-strain fracture toughness values at room temperature.

Results and Discussion

Wedge load-displacement records from four tests are shown in Fig. 4. Linearity of the records prior to crack jump show that the specimens remained linear-elastic during loading and the test rig was free of friction effects. Figure 5 shows the fracture surface of a 5.08-cm-thick specimen.

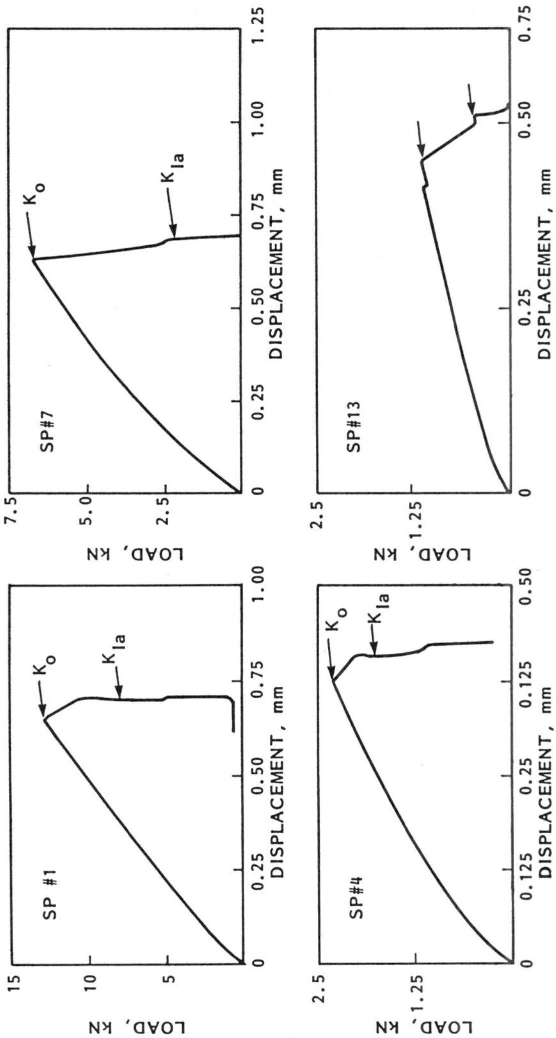

FIG. 4—*Run-arrest test record showing a plot of applied load and displacement at knife edges.*

FIG. 5—*Fracture surface of a 5-cm-thick specimen.*

Unbroken ligaments were not observed on the fracture surface of the test specimens. Results from all tests are given in Table 1.

Test results showing the dependence of K_{Ia} on various test parameters are shown in the following figures:

1. K_{Ia} versus K_0 (Fig. 6)
2. K_{Ia} versus specimen thickness (Fig. 7)
3. K_{Ia} versus crack jump/W (Fig. 8)
4. K_{Ia} versus final crack length/W (Fig. 9)
5. K_{Ia} versus remaining ligament/W (Fig. 10)

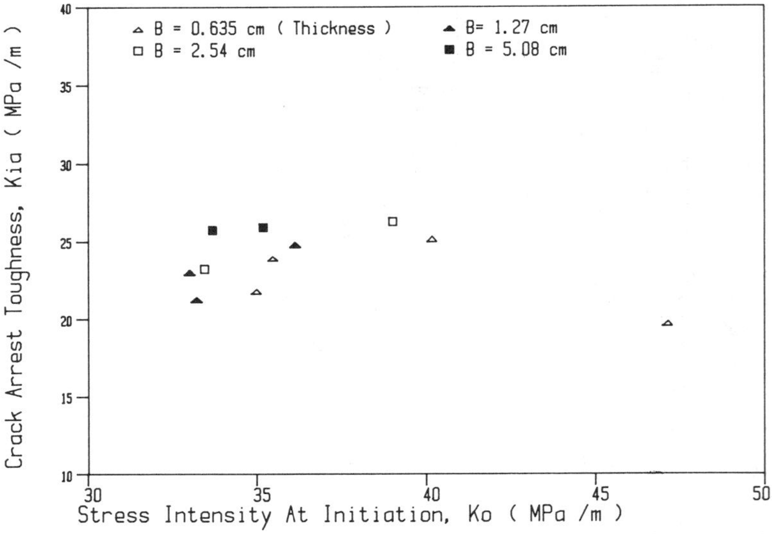

FIG. 6—*Dependence of crack arrest toughness* K_{Ia} *on initiation stress intensity* K_0.

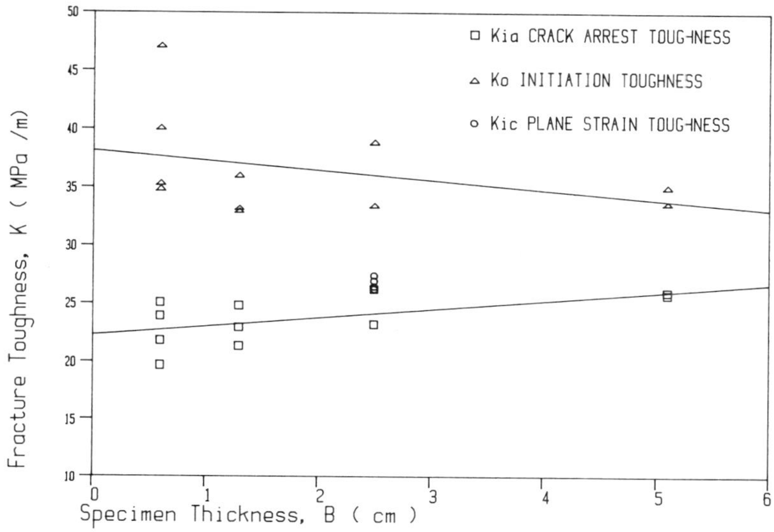

FIG. 7—*Dependence of crack arrest toughness* K_{Ia} *on specimen thickness* B.

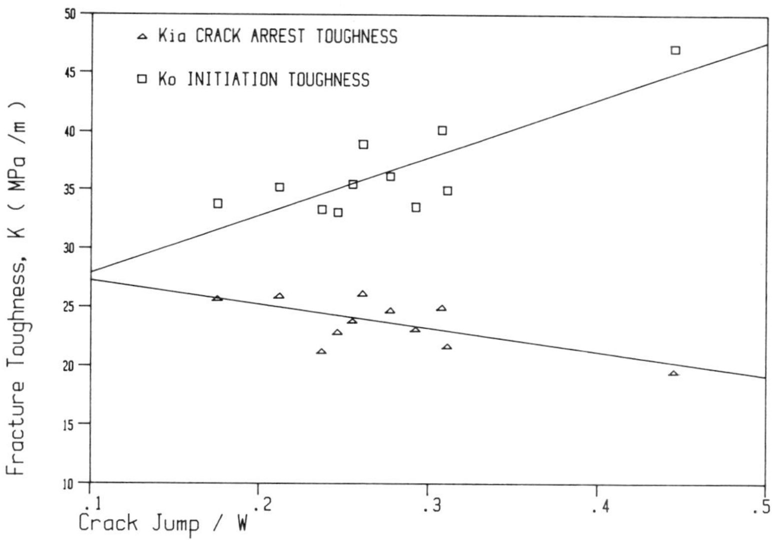

FIG. 8—*Dependence of crack arrest toughness* K_{Ia} *on normalized crack jump* ($\Delta a/W$).

K_0 values are also plotted on all of these figures for reference purposes. Statistical data on K_{Ia} and its dependence on K_0, Δa, B, $\Delta a/W$, a_a, a_a/W, the remaining ligament, and normalized remaining ligament are given in Table 3.

Data show that scatter in K_{Ia} values in 7075-T6 aluminum alloy is significantly less than that observed for SA533-B steel [*8*]. A set of 50 K_{Ia} values measured at 0°C using identical specimens gave a standard deviation of 9.9. In comparison, the standard deviation for 7075-T6 aluminum with three specimen geometries and a sample size of eleven is 2.11. This shows

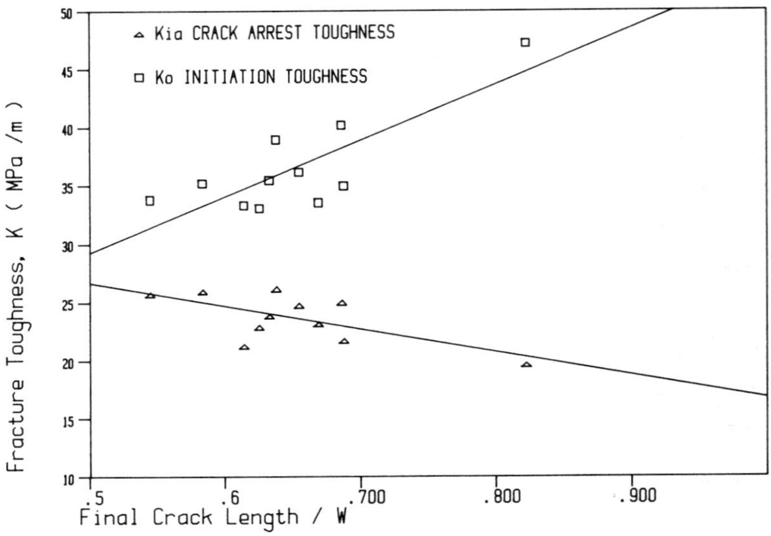

FIG. 9—*Dependence of crack arrest toughness* K_{Ia} *on normalized final crack length* (a_a/W).

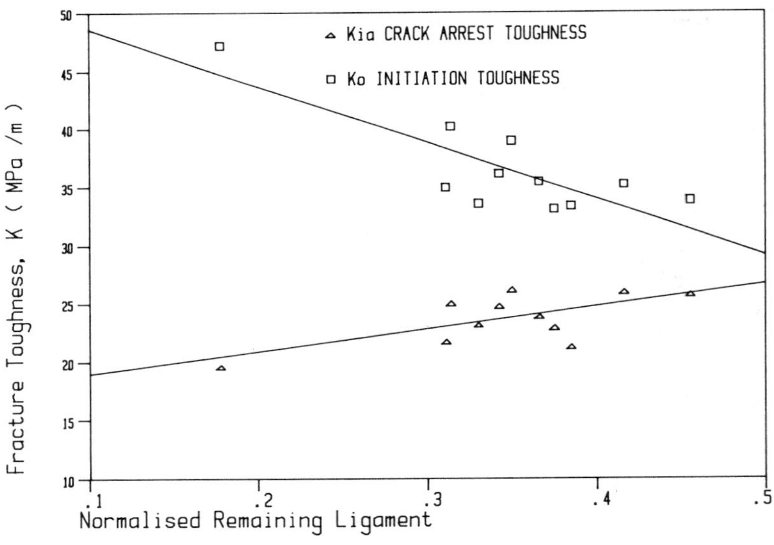

FIG. 10—*Dependence of crack arrest toughness on normalized remaining ligament.*

that reliable and consistent values of K_{Ia} can be obtained using the proposed ASTM test procedure if minimum linear elastic requirements of the proposed ASTM test method are met or exceeded.

Figure 7 shows the initiation and arrest toughness values plotted against specimen thickness. It shows that for a fixed root radius, thinner specimens produced higher crack initiation toughness values. This is due to increased notch root plasticity effects in thin specimens. It also shows that crack arrest toughness decreases with decreasing specimen thickness. Thin specimens produced higher initial stress intensity, larger crack jump, and a lower arrest toughness. Specimen

TABLE 3—*Statistical data on* K_{Ia} *and its dependence on test parameters.*

Plot File Name	Test Parameters	Mean Value	Linear Fit	Standard Deviation
...	K_{Ia} (MPa m)	23.71	...	2.11
...	K_0 (MPa m)	36.49	...	4.22
...	K_{Ic} (MPa)	26.78	...	0.54
ALCRA 1/2	K_{Ia} versus K_0 (Fig. 6)	...	$K_{Ia} = 28.87 - 0.14K_0$	1.93
ALCRA 7	K_{Ia} versus B (cm) (Fig. 7)	...	$K_{Ia} = 22.25 + 0.74B$	1.61
ALCRA 5	K_{Ia} versus Δa (cm)	...	$K_{Ia} = 23.06 + 0.33\ (\Delta a)$	2.02
ALCRA 10/11	K_{Ia} versus $\Delta a/W$ (Fig. 8)	...	$K_{Ia} = 29.25 - 20.19\ (\Delta a/W)$	1.50
ALCRA 12	K_{Ia} versus a_a (cm)	...	$K_{Ia} = 20.93 + 0.58\ a_a$	1.69
ALCRA 8/9	K_{Ia} versus a_a/W (Fig. 9)	...	$K_{Ia} = 36.64 - 19.86\ (a_a/W)$	1.50
ALCRA 6	K_{Ia} versus $(W - a_a)$ (cm)	...	$K_{Ia} = 21.89 + 0.67\ (W - a_a)$	1.65
ALCRA 13/14	K_{Ia} versus $(W - a_a)/W$ (Fig. 10)	...	$K_{Ia} = 17 + 19.33\ (W - a_a)/W$	1.53

No. 11 produced the largest K_0 and the smallest K_{Ia}. Exclusion of this data pair from the analysis does not change the trend in the data, except that the equations of the linear fit given in Table 3 change slightly.

Figure 6 shows that crack arrest toughness K_{Ia} decreases with K_0. In a recent paper, Lin and Hoagland [9] observed the same trend in the arrest toughness data of 7075-T6 aluminum alloy with K_0. Figure 11 plots normalized crack arrest toughness, K_{Ia}/K_{Ic}, against normalized initiation toughness K_0/K_{Ic} for the T-L direction from the present work and also from Ref *9* for comparison. Although Lin and Hoagland [9] produced a larger range of initial stress intensity factor K_0 (37.3 to 77.6 MPa$\sqrt{m}$ versus 33.0 to 47.1 in the present work), the trend in two independent sets of data is similar. Four points in Lin and Hoagland's data where the proposed maximum crack jump criterion was not satisfied gave the four lowest values of K_{Ia}. These four points give the K_{Ia} data in Ref *9* a stronger dependence on K_0 than that found in the present paper. (See Fig. 11.)

Figures 8 and 9 show the effect of normalized crack jump length and normalized final crack length on K_{Ia}. These show that $\Delta a/W$ and a/W depend on starting K_0. K_{Ia} decreases linearly with increasing $\Delta a/W$ and a/W. In comparison, a nonlinear and stronger reduction of K_{Ia} with ($\Delta a/W$) was observed in Ref *9*, possibly due to a large crack jump in some of their tests. Normalized K_0 from Ref *9* is plotted against normalized crack jump (T-L direction) in Fig. 12 for comparison with the present work. Again the trend in two sets of data becomes similar if four data points with normalized crack jump greater than 0.5 and actual crack jump greater than 54 mm are not considered. Note that these are the same four points which did not satisfy the proposed maximum crack jump criteria [*9*].

Finally, K_{Ia} and K_0 are plotted against normalized remaining ligament ($W - a/W$) in Fig. 10. It shows that the specimens with higher K_0 produced larger jumps, left smaller ligaments, and resulted in small K_{Ia} values.

Crosley et al. [*8*] have discussed the source and significance of scatter in K_{Ia} values in SA533 B steel. They reasoned that with K_{Ia}, the horizontal fit (mean value) is as good as a least-square linears fit of K_{Ia} versus K_0. The standard deviation of the mean value is essentially equal to the deviation about the fitted line (9.9 versus 9.4 for SA533B-1, 0°C Materials Research Laboratory Data). They also showed that the trend showing a decrease in K_{Ia} with increasing arrest crack length in SA533B is a consequence of data scatter.

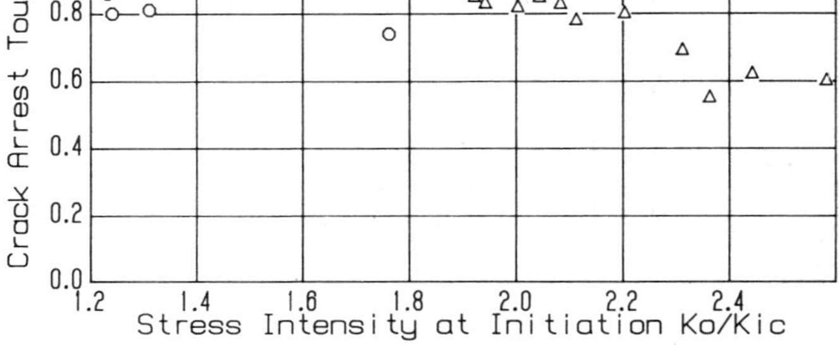

FIG. 11—*Comparison of normalized* K_{Ia} *versus* K_0 *with data published in Ref* 9.

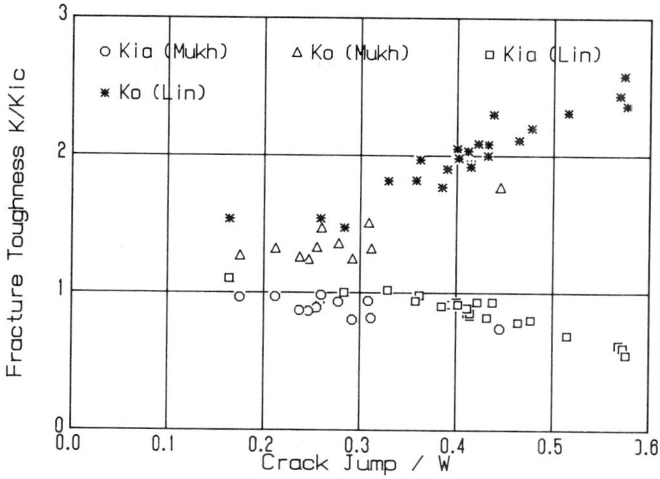

FIG. 12—*Comparison of normalized* K_{Ia} *versus crack jump with data published in Ref* 9.

Using the same reasoning, it can be argued that within the test matrix, K_{Ia} is not a strong function of K_0, since the standard deviation of the mean value is 2.11 compared to a standard deviation of 1.93 when K_{Ia} is regressed with K_0.

Conclusions

1. K_{Ia} values in 7075-T6 aluminum decreases slightly with K_0. However, the mean value of K_{Ia} is an adequate description of the crack arrest toughness behavior within the test matrix provided the proposed maximum crack jump criterion is satisfied.
2. K_{Ia} decreased with increasing crack jump.
3. The scatter in K_{Ia} values was less than that observed in SA533B steel.
4. The average K_{Ia} value was 10% lower than the average plane-strain fracture toughness K_{Ic}.

References

[1] Ravi-Chandar, K. and Knauss, W. G., "An Experimental Investigation Into Dynamic Fracture: I. Crack Initiation and Arrest," *International Journal of Fracture*, Vol. 25, 1984, pp. 247–262.
[2] Kanninen, M. F., "Application of Dynamic Fracture Mechanics for the Prediction of Crack Arrest in Engineering Structures," *International Journal of Fracture*, Vol. 27, 1985, pp. 299–312.
[3] Crosley, P. B. and Ripling, E. J., "Crack Arrest Studies," EPRI Report NP-1225, Final Report, Electric Power Research Institute, Palo Alto, CA, Nov. 1979.
[4] Dally, J. W., Fourney, W. L., and Irwin, G. R., "On the Uniqueness of the Stress Intensity Factor-Crack Velocity Relationship," *International Journal of Fracture*, Vol. 27, 1985, pp. 159–168.
[5] Hahn, G. T., Hoagland, R. G., Rosenfield, A. R., and Barnes, C. R., "A Cooperative Test Program for Evaluating Crack Arrest Testing Methods" in *Crack Arrest Methodology and Applications, ASTM STP 711*, G. T. Hahn and M. F. Kanninen, Eds., American Society for Testing and Materials, Philadelphia, 1980, pp. 248–269.
[6] Crosley, P. B. and Ripling, E. J., "Comparison of Crack Arrest Methodologies" in *Crack Arrest Methodology and Applications, ASTM STP 711*, G. T. Hahn and M. F. Kanninen, Eds., American Society for Testing and Materials, Philadelphia, 1980, pp. 211–227.
[7] Rosenfield, A. R., Mincer, P. N., Marschall, C. W., and Markworth, A. J., "Recent Advances in Crack-Arrest Technology," *Fracture Mechanics: Fifteenth Symposium, ASTM STP 833*, R. J. Sanford, Ed., American Society for Testing and Materials, Philadelphia, 1984, pp. 149–164.

[*8*] Crosley, P. B., Fourney, W. L., Hahn, G. T., Hoagland, R. G., Irwin, G. R., and Ripling, E. J., "Final Report on Cooperative Test Program on Crack Arrest Toughness Measurement," USNRC Report NUREG/CR-3261, Nuclear Regulatory Commission, Washington, DC, April 1983.

[*9*] Lin, C. and Hoagland, R. G., "Crack Arrest Behavior of a High Strength Aluminum Alloy" in *Fracture Mechanics: Sixteenth Symposium, ASTM STP 868,* M. F. Kanninen and A. T. Hopper, Eds., American Society for Testing and Materials, Philadelphia, 1985, pp. 467–489.

Eiji Sugie,[1] *Masanori Matsuoka,*[2] *Toshiya Akiyama,*[3] *Kiyoshi Tanaka,*[4] *and Masao Tsukamoto*[5]

Propagating Shear Fracture in Natural Gas Transmission Pipelines

REFERENCE: Sugie, E., Matsuoka, M., Akiyama, T., Tanaka, K., and Tsukamoto, M., "**Propagating Shear Fracture in Natural Gas Transmission Pipelines,**" *Fracture Mechanics: Eighteenth Symposium, ASTM STP 945,* D. T. Read and R. P. Reed, Eds., American Society for Testing and Materials, Philadelphia, 1988, pp. 237–246.

ABSTRACT: The propagating shear fracture in natural gas transmission pipelines can be theoretically analyzed, and the crack velocity and propagation distance can now be calculated. The propagation theory has a good consistency with the experimental results for 1219 mm (48 in.) outside diameter (OD) × 18.3 mm (0.72 in.) wall thickness (WT) line pipes, and it predicts the notch ductilities of line pipes required for arresting the fracture according to the type of the gas, the acceptable fracture length, the pipe size, and the design stress of the pipeline.

KEY WORDS: propagating shear fracture, pipeline, natural gas transmission, notch ductility, crack propagation, line pipe

Nomenclature

A_p, A_s, A_v Ligament area of specimen of pre-cracked DWTT, standard DWTT, and Charpy test, respectively, mm^2

D_p Pre-cracked DWTT energy, *kJ*

D_s Standard DWTT energy, *kJ*

C_v Full-size Charpy energy, *J*

D Diameter of pipe, mm

E Notch ductility, *kJ* or *J*

L Crack propagation distance, m

L_0 Initial distance, m

$2L$ Fracture length acceptable to pipeline designers, m

P Crack tip pressure, MPa

P_0 Initial or design pressure, MPa

T Time from crack initiation, s

T_0 Initial time, s

t Wall thickness of pipe, mm

V_a Accoustic velocity in pressurized gas, m/s

V_c Crack velocity, m/s

V_m Gas decompression velocity, m/s

[1]Senior researcher, Mizushima Technical Research Laboratories, Kawasaki Steel Corporation, Mizushima, Japan.

[2]Senior mechanical engineer, Kakogawa Works, Kobe Steel Ltd., Kakogawa, Japan.

[3]Manager, Keihin Works, Nippon Kokan K. K., Kawasaki, Japan.

[4]Senior research engineer, R and D Laboratories 2, Nippon Steel Corporation, Sagamihara, Japan.

[5]Research engineer, Technical Research Laboratories, Sumitomo Metal Industries Ltd., Ibaraki, Japan.

A, x, y, z Numerical constants
α, β Numerical constants
σ Design stress of pipeline, N/mm^2
$\bar{\sigma}$ Flow stress = (YS + TS)/2, N/mm^2

High Strength Line Pipe Research Committee (HLP), organized by the Iron and Steel Institute of Japan in 1978, has investigated the propagating shear fracture in pipelines by carrying out seven full-scale burst tests and has established a propagation theory which gives an estimation for the crack velocity and the propagation distance of the fracture [*1–4*]. The present paper summarizes the method to estimate the crack behavior of the propagating shear fracture in natural gas transmission pipelines. The theoretical estimation is compared with the experimental results for 1219 mm (48 in.) outside diameter (OD) × 18.3 mm (0.72 in.) wall thickness (WT) line pipes, and applied to establish a method to predict the notch ductilities of line pipes required for arresting the fracture according to the type of the gas, the pipe size and the design conditions which pipeline designers should take into account.

Propagation Theory

The propagating shear fracture in natural gas transmission pipelines is controlled by the conflict between the gas decompression velocity, V_m, and the crack velocity, V_c. Between V_m and V_c, there exists the following correlation [*1*]

$$V_m = \frac{1}{T} \int^{T} V_c dT \tag{1}$$

Equation 1 induces the following equation which shows that the crack velocity change can be calculated from V_m and V_c [*1*]

$$\frac{dV_c}{dT} = \frac{dV_c/dP}{dV_m/dP} \times \frac{1}{T} \times (V_c - V_m) \tag{2}$$

The propagation distance (L) is estimated from the following equation if the initial values of L and T (L_0 and T_0) are given

$$L = L_0 + \int_{T_0}^{T} V_c dT \tag{3}$$

The crack velocity, V_c, is a function of the crack tip pressure, P, as follows [5] and α and β were obtained as shown by Eqs 5 and 6 through experiments [*3*]

$$V_c = \alpha \times \bar{\sigma} \times \sqrt{\frac{1}{D_p/A_p}} \times \left(\frac{P}{P_a} - 1\right)^{\beta} \tag{4}$$

$$\alpha = 0.0212 \tag{5}$$

$$\beta = 0.393 \tag{6}$$

$$P_a = 0.380 \times \frac{t}{D} \times \bar{\sigma} \times \cos^{-1} \exp\left(\frac{-3.81 \times 10^{10}}{\sqrt{Dt}} \times \frac{D_p/A_p}{\bar{\sigma}^2}\right) \tag{7}$$

$$\frac{D_p}{A_p} = 0.121 \times \left(\frac{D_s}{A_s}\right)^{0.679} \tag{8}$$

$$\frac{D_p}{A_p} = 0.00235 \times \left(\frac{C_v}{A_v}\right)^{0.465} \tag{9}$$

The gas decompression velocity, V_m, is a function of the crack tip pressure, P, and in the case of the single-phase gas, the relationship between them is as follows [1]

$$\frac{P}{P_0} = (0.755 + 0.245\ V_m/360)^{5.08} \tag{10}$$

In the case of the dual-phase gas such as a rich natural gas, the gas decompression velocity, V_m, should be determined experimentally or theoretically according to the gas compositions [3].

The combination of Eqs 1 to 9 and the gas decompression velocity such as Eq 10 allows to estimate the crack velocity and the propagation distance through computer calculation. The method is shown in Fig. 1 schematically. The initial values, L_0 and T_0, should be determined experimentally or from the following equations [3]

$$L_0 = D \tag{11}$$

$$T_0 = D/V_a \tag{12}$$

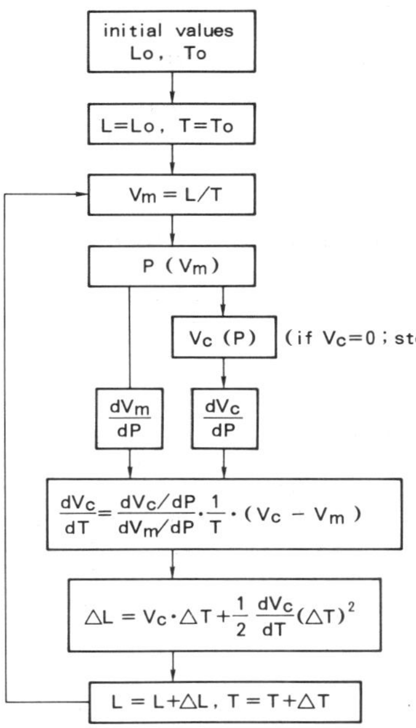

FIG. 1—*Schematic representation of calculation method.*

Following Fig. 1, the crack velocity and the propagation distance can be estimated time after time, if the pipe size (D and t), the flow stress ($\bar{\sigma}$), and the notch ductility (D_p, D_s, or C_v) of the pipe and the initial gas pressure (P_0) are given as input data.

Experiments and Discussion

HLP has carried out seven full-scale burst tests as shown in Table 1 [4]. The five tests were carried out with air as the pressurizing gas which simulates the single-phase natural gas, and the two tests with rich natural gas which represents the dual-phase natural gas. The test pipes are 1219 mm (48 in.) in OD and 18.3 mm (0.72 in.) in WT.

The gas decompression velocity curves obtained are shown in Fig. 2. In C1 and C2 tests, the pressure plateaus which feature the dual-phase gas are observed.

In Fig. 3, all the test results are illustrated. Notch ductilities (D_p, D_s, and C_v) of the test pipes were varied one by one [3,4]. Only the values of C_v are shown in Fig. 3. The crack initiated at the crack initiation pipe (CIP), propagated in both sides and was arrested.

According to the calculation method shown in Fig. 1, the crack velocity and the propagation

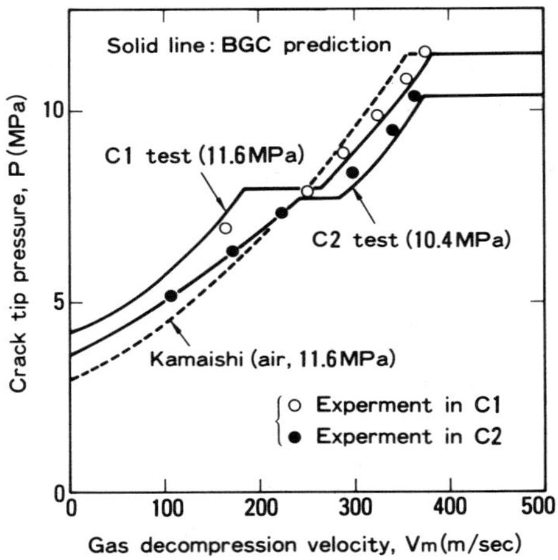

FIG. 2—*Gas decompression curves for single-phase and dual-phase gases. For dual-phase gas, predictions given by British Gas Corporation (BGC) are also shown.*

TABLE 1—*Test conditions of full-scale burst tests.*

Test site	Kamaishi		BGC[a]	
Test pipes	API5L X70 48 in. OD × 0.72 in. WT[b]			
Pressurizing gas	air		rich natural gas	
Pressure	11.6 MPa		11.6 MPa	10.4 MPa
Hoop stress	0.80 SMYS		0.80 SMYS	0.72 SMYS
Test temperature	+3 ~ +12°C		−5°C	
Series	A	B	C	
Number of tests	3	2	2	

[a]Test site of British Gas Corporation.
[b]1219 mm × 18.3 mm.

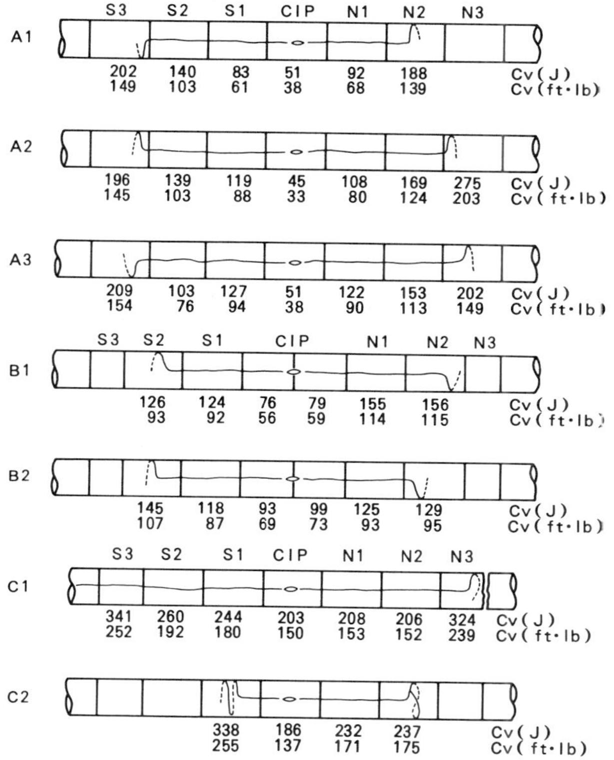

FIG. 3—*Results of full-scale burst tests.*

distance were estimated. Initial values, L_0 and T_0, were determined from experimental measurements at the outlet of CIP.

In Figs. 4 and 5, the estimations are compared with the experimental measurements. In Fig. 4, only the estimation from D_p is shown, but the estimation from D_s or C_v using Eqs 8 and 9 gave similar results. Except in A1 test, the estimation is successful because the estimated distances differ less than 12.5 m which is the ordinary length of a pipe. In A1 test, the differences of notch ductilities of successive pipes were exceptionally large as shown in Fig. 3. In such a case, the estimation will be conservative compared with actual fracture in pipelines [*3*]. In Fig. 5, the estimations for C1 and C2 tests are shown not only from D_p but also from D_s and C_v. The estimation is very successful in all the cases.

The estimation of the propagation distance enables to give the required notch ductilities for arresting propagating shear fracture in pipelines. A pipeline was considered to be constructed of pipes with uniform ductility, and the propagation distance was calculated. The results are shown in Fig. 6. Twice the propagation distance is regarded as the acceptable fracture length which the pipeline designers should take into account, and the three kinds of notch ductilities (D_p, D_s, and C_v) for calculation are regarded as the required notch ductilities. The initial values, L_0 and T_0, were estimated from Eqs 11 and 12. This figure indicates that the required notch ductility values are dependent on the type of the gas, the design stress and the acceptable fracture length in the case of X70 line pipes of 1219 mm (48 in.) OD and 18.3 mm (0.72 in.) WT [*4*].

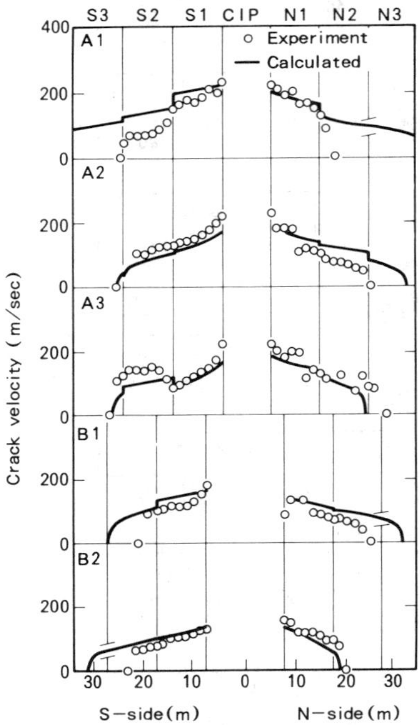

FIG. 4—*Crack propagation estimated from* D_p *in comparison with experiments in A-series and B-series tests.*

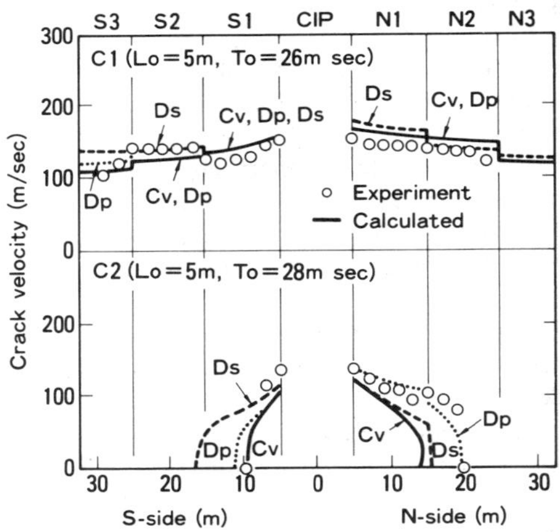

FIG. 5—*Crack propagation estimated from* D_p, D_s, *and* C_v *in comparison with experiments in C-series tests.*

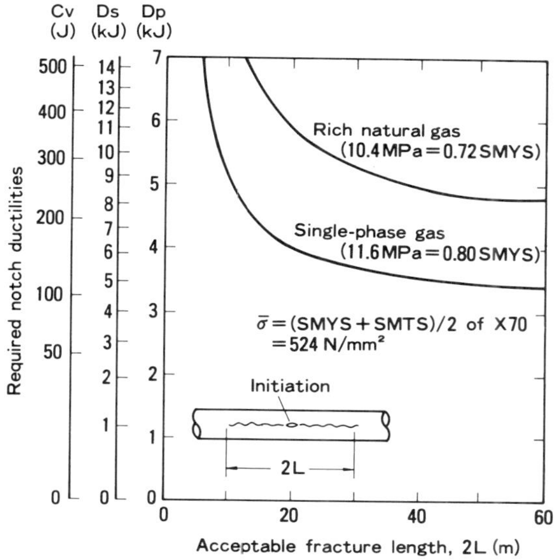

FIG. 6—*Required notch ductilities of X70 line pipes, 1219 mm (48 in.) OD and 18.3 mm (0.72 in.) WT, for arresting the propagating shear fracture in pipelines.*

Application

The propagation theory expressed by Eqs 1 to 12 is experimentally verified only on the X70, 1219 mm (48 in.) OD × 18.3 mm (0.72 in.) WT pipes. But its application to pipes of different grade and with different size is one of the pipeline engineers' concerns.

For convenience, the required notch ductilities are often expressed by the following equation [5]

$$E = AD^x t^y \sigma^z \tag{13}$$

The typical one is the Battelle's equation in which E denotes the full-size Charpy energy as follows [5]

$$C_v = 2.84 \times 10^{-5} \times D^{0.333} \times t^{0.333} \times \sigma^2 \tag{14}$$

In order to obtain the values of A, x, y, and z from the present propagation theory, the propagation distance is calculated for all the possible combinations of the following design variables.

D = 609, 914, 1219, 1422 mm (24, 36, 48, 56 in.)

t = 9.5, 12.7, 19.1, 25.4 mm (3/8, 1/2, 3/4, 1 in.)

σ = 72% of 258, 298, 345, 397 N/mm^2 (52, 60, 70, 80 ksi)

The design stress (σ) is assumed to be 72% of the specified minimum yield strength of API 5L X52, X60, X70, and X80, and the flow stress, $\bar{\sigma}$, is assumed to be the mean value of the specified minimum yield and tensile strengths of each grade.

The gas decompression velocity, V_m, is assumed to be the same as the curves in Fig. 2. For the single-phase gas (Eq 10)

$$\frac{P}{P_0} = (0.755 + 0.245\ V_m/360)^{5.08} \tag{15}$$

For the dual-phase gas (C2 test in Fig. 2)

$0 \leqq V_m < 245$

$$\frac{P}{P_0} = 0.3252 + 0.5829\left(\frac{V_m}{377}\right) + 0.1564\left(\frac{V_m}{377}\right)^3 + 0.009149\left(\frac{V_m}{377}\right)^5 \tag{16}$$

$245 \leqq V_m < 286$

$$\frac{P}{P_0} = 0.749 \tag{17}$$

$286 \leqq V_m < 377$

$$\frac{P}{P_0} = 1.048 - 1.050\left(\frac{V_m}{377}\right) + 1.326\left(\frac{V_m}{377}\right)^3 - 0.3235\left(\frac{V_m}{377}\right)^5 \tag{18}$$

$377 \leqq V_m$

$$\frac{P}{P_0} = 1 \tag{19}$$

The acoustic velocity, V_a, in Eq 12 is 360 m/s for the single-phase gas and 377 m/s for the dual-phase gas. The initial pressure, P_0, in Eqs 15 to 19 is calculated from the following equation

$$P_0 = 2t\sigma/D \tag{20}$$

The propagation distance, L, corresponding to one of the notch ductilities, D_p, is calculated for each case. By a try and error method, the values of D_p corresponding to $L = 10$ m and $L = 20$ m are determined. The length $2L$ is regarded as the acceptable fracture length, and, consequently, the required D_p is calculated for each of the combinations of varied D, t, σ, and $2L$, corresponding to the two types of gases. The answers of D_p for 256 cases are obtained. Regression analysis to these values gives the values of A, x, y, and z, and the equations obtained are as follows. For the single-phase gas and $2L = 20$ m

$$D_p = 7.36 \times 10^{-8} \times D^{0.316} \times t^{1.16} \times \sigma^{2.04} \tag{21}$$

For the single-phase gas and $2L = 40$ m

$$D_p = 11.50 \times 10^{-8} \times D^{0.261} \times t^{1.17} \times \sigma^{2.01} \tag{22}$$

For the dual-phase gas and $2L = 20$ m

$$D_p = 1.87 \times 10^{-8} \times D^{0.338} \times t^{1.08} \times \sigma^{2.40} \tag{23}$$

For the dual-phase gas and $2L = 40$ m

$$D_p = 3.75 \times 10^{-8} \times D^{0.261} \times t^{1.11} \times \sigma^{2.33} \tag{24}$$

In Eqs 21 to 24, D_p can be substituted by D_s or C_v using Eqs 8 and 9. Employing that $A_p = 67.1 \times t$, $A_s = 71.1 \times t$, and $A_v = 80$ mm^2, the equations are as follows

$$D_p = 0.449 \times t^{0.321} \times D_s^{0.679} \tag{25}$$

$$D_p = 0.0206 \times t \times C_v^{0.465} \tag{26}$$

The required notch ductilities are shown in Fig. 7. In this figure, D and t are assumed to be 1219 mm (48 in.) and 18.3 mm (0.72 in.), respectively. Only the design stress, σ, is varied, and four curves are shown corresponding to Eqs 21 to 24. Three kinds of notch ductilities (D_p, D_s, and C_v) are correlated by Eqs 25 and 26. The Battelle's equation (Eq 14), which corresponds only to C_v, is also shown by a broken line for comparison. This figure suggests that the required notch ductilities increase very steeply as the design stress increases, and that the type of the gas is very predominant. But the experimental verification of Fig. 7 is left for future work.

Conclusion

The crack velocity and the propagation distance of the propagating shear fracture in natural gas transmission pipelines can be estimated by the method proposed here according to the type of the gas, the size and the properties of the composing line pipes, and the design stress of the pipeline. The estimation showed good consistency with the results of the full-scale burst tests on the 1219 mm (48 in.) OD and 18.3 mm (0.72 in.) WT X70 pipes. The theoretical simulation by means of the method gave simplified equations to predict the notch ductilities of line pipes required for arresting the fracture according to the type of gas, the pipe size, and the design conditions of pipelines. These equations can be used when deciding the required toughness for pipelines.

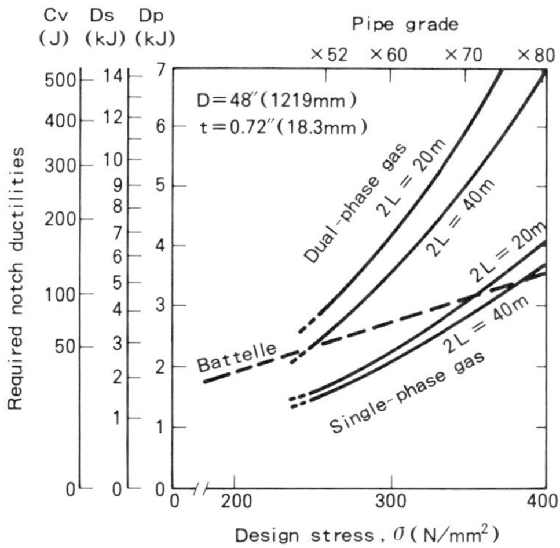

FIG. 7—*Design stress dependency of required notch ductilities for arresting the propagating shear fracture in pipelines.*

Acknowledgment

This investigation was made as one of the collaborate research activities run by High Strength Line Pipe Research Committee, Iron and Steel Institution of Japan, Tokyo, Japan.

References

[1] Sugie, E., Matsuoka, M., Mimura, H., and Kawaguchi, Y., *Journal of Pressure Vessel Technology, Transactions,* American Society of Mechanical Engineers, Vol. 104, Nov. 1982, pp. 338–343.
[2] Sugie, E., Kaji, H., Taira, T., Ohashi, M., and Sumitomo, Y., *Journal of Energy Resources Technology, Transactions,* American Society of Mechanical Engineers, Vol. 106, March 1984, pp. 55–62.
[3] Sugie, E., Matsuoka, M., Akiyama, T., Tanaka, K., and Kawaguchi, Y., "Notch ductilities of Rich Natural Gas Transmission Line Pipes for Arresting Propagating Shear Fracture," presented at the Forth International Offshore Mechanics and Arctic Engineering Symposium, American Society of Mechanical Engineers, Dallas, 17–21 Feb. 1985.
[4] Sugie, E., Matsuoka, M., Akiyama, T., Tanaka, K., and Kawaguchi, Y., "Notch ductility Requirements of Line Pipes for Arresting Propagating Shear Fracture," presented at 1985 Pressure Vessels and Piping Conference and Exhibition, American Society of Mechanical Engineers, New Orleans, 23–26 June 1985.
[5] Maxey, W. A., "Fracture Initiation, Propagation, and Arrest," 5th Symposium on Line Pipe Research, American Gas Association, 1974, J-1.
[6] Eiber, R. J. and Maxey, W. A., "Fracture Propagation Control Methods," 6th Symposium on Line Pipe Research, American Gas Association, 1979, L-1-16.

Jin C. Yu[1]

Stress-Intensity Factors for an Edge Crack in a Stiffened Sheet

REFERENCE: Yu, J. C., "**Stress-Intensity Factors for an Edge Crack in a Stiffened Sheet,**" *Fracture Mechanics: Eighteenth Symposium, ASTM STP 945,* D. T. Read and R. P. Reed, Eds., American Society for Testing and Materials, Philadelphia, 1988, pp. 247–258.

ABSTRACT: This paper presents the two-dimensional fracture mechanics solutions of a through-the-thickness edge crack in a stiffened sheet. The sheet is assumed to be under the plane-stress condition, and the stiffeners are taken as one-dimensional rods discretely attached to the sheet. The fastener forces are determined by enforcing the displacement compatibility at fastener locations and by the equilibrium condition for broken stiffeners. The problem is then reduced to the solution of a set of two simultaneous integral equations. The solution procedure employed allows direct evaluation of the unknowns, leading to efficient determination of the fastener forces as well as the stress intensity factors. Numerical examples are presented.

KEY WORDS: fracture mechanics, edge crack, stress intensity factor, stiffened sheet, integral equation

The role of stiffeners in increasing residual strength in a cracked sheet has been the subject of numerous studies (see Refs *1–3,* for example). Typically in these studies, load transfers from the sheet to stiffeners through either discrete attachment points or bonded interfaces. The sheet is assumed to be infinitely large and homogenous—fastener holes, if any, are of small size and are ignored in the formulation. The stiffeners are usually treated either as one-dimensional rods or as strips of rectangular cross section with boundary conditions partially satisfied. (Swift [*4*] accounts for stiffener bending rigidity and fastener flexibility.) If a crack is located near the sheet boundary, however, the effect of the edge on the crack-tip behavior becomes significant, and the infinite-sheet model is no longer appropriate.

This paper presents two-dimensional linear elastic solutions to the problem of an edge crack in a stiffened sheet with discrete fasteners. A system of equations is formulated in terms of fastener forces and influence coefficients, based on displacement compatibility and, in the case of broken stiffeners, on equilibrium conditions for the broken sections. Through superposition of known elastic solutions, the influence coefficients are expressed by simple integrals involving two unknown functions that satisfy a set of two simultaneous integral equations. A solution procedure is then employed to solve directly for the unknowns. This permits determination of fastener forces and stress intensity factor through simple integration.

Mechanical Model

The problem considered consists of an edge crack in a stiffened semi-infinite sheet as shown in Fig. 1. A uniform tension is applied remote from the crack, parallel to the stiffeners which are discretely fastened to the sheet. The sheet is assumed to be in a state of generalized plane

[1]Principal engineer, Douglas Aircraft Company, McDonnell Douglas Corporation, Long Beach, CA 90846.

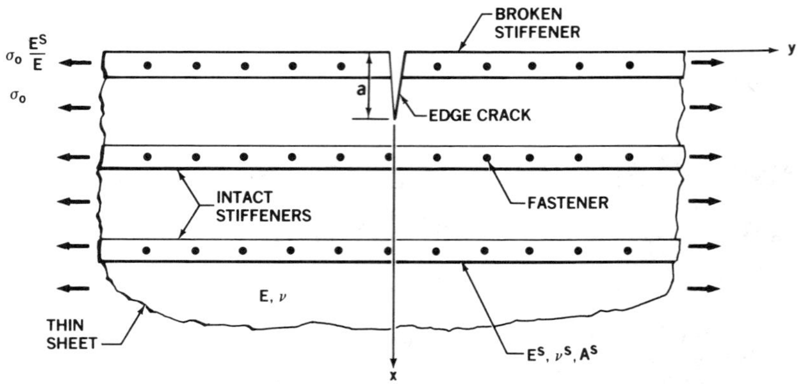

FIG. 1—*An edge crack in a stiffened panel with broken stiffener.*

stress; the stiffeners are regarded as one-dimensional rods. It is further assumed that fasteners are located symmetrically about the crack; hence, only a quarter-space is needed for analysis.

Equations for Determining Fastener Forces

Displacements at each fastener location may be conveniently expressed in terms of influence coefficients, defined as the displacement at a material point due to a unit force at any given location. Let A_{ijkl} be the y-direction displacements at the jth fastener in the ith stiffener due to unit force F_{kl} at the lth fastener in the kth stiffener; with B_{ij} the y-direction displacements at the jth fastener in the ith stiffener due to unit far-field tension stress. Then, the total displacement at this location is given by

$$V_{ij} = -\sum_{k=1}^{\infty}\sum_{l=1}^{\infty} A_{ijkl}F_{kl} + B_{ij}\sigma_0 \tag{1}$$

where the negative sign indicates that the fastener forces are in the negative y-direction. Similarly, the displacements at the same location in the stiffeners can be expressed as

$$V_{ij}^{s} = \sum_{l=1}^{\infty} A_{ijil}^{s}F_{il} + B_{ij}^{s}\sigma_0\frac{E^s}{E} \tag{2}$$

where A_{ijkl}^{s} and B_{ij}^{s} are the influence coefficients associated with fastener forces and remote stiffener stress, $\sigma_0 E^s/E$, respectively.

For the case where all stiffeners are intact, the equations can be obtained by requiring that the displacement at each fastener location in the sheet be the same as in the stiffener. Thus

$$\sum_{k=1}^{\infty}\sum_{l=1}^{\infty} A_{ijkl}F_{kl} + \sum_{l=1}^{\infty} A_{ijil}^{s}F_{il} = \left(B_{ij} - B_{ij}^{s}\frac{E^s}{E}\right)\sigma_0 \tag{3}$$

where $i, j = 1, 2, \ldots, \infty$.

Because of symmetry, the maximum load in each stiffener occurs in the interval across the x-axis. When the maximum load—say, in the ith stiffener—exceeds the stiffener strength, the stiffener fails and the equilibrium condition for the broken half-stiffener is given by

$$\frac{\sigma_0 E^s A^s}{E} + \sum_{l=1}^{\infty} F_{il} = 0 \tag{4}$$

By enforcing the condition that the elongation in the broken section between any two consecutive fastener locations be the same as in the sheet, we arrive at a system of relations

$$V_{ij} - V_{ij-1} = V_{ij}^s - V_{ij-1}^s \qquad j = 2, 3, \ldots, \infty \tag{5}$$

Equations 4 and 5 are used in place of those equations associated with the ith stiffener in Eq 3 to obtain the modified equations for determining fastener forces. Obviously, this approach can be extended to multiple stiffener failures as well.

Influence Coefficients

To evaluate A_{ijkl}, consider an edge crack of length a in a half-space subjected to two fastener loads F acting at $(h, \pm d)$ in the sketch shown in Fig. 2. We first decompose this problem into two parts: Part A is an uncracked half-space under loads F, and Part B is an edge crack in a half-space with pressure loading, $\sigma_y(x)$, on the crack faces, where $\sigma_y(x)$ is the stress on $y = 0$ in Part A.

Solution methods for Part A are summarized in the Appendix, where the y-component stress on $y = 0$ is given by

$$\sigma_y(x) = \frac{Fd}{2\pi}\left\{\frac{(1-\nu)(x-h)^2 + (3+\nu)d^2}{[(x-h)^2 + d^2]^2} - \frac{(1+3\nu)(x+h)^2 - 4\nu(x^2 - h^2) - (5-\nu)d^2}{[(x+h)^2 + d^2]^2} + \frac{4(1+\nu)xh[3(x+h)^2 - d^2]}{[(x+h)^2 + d^2]^3}\right\} \tag{6}$$

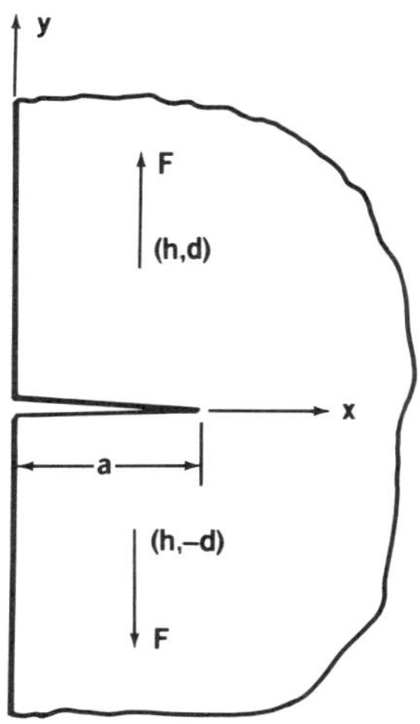

FIG. 2—*An edge crack in a half-space under a pair of point loads.*

in which fastener forces F are taken as concentrated loads. Also, the y-component displacement is given by

$$
\begin{aligned}
V_A(x, y) = {} & \frac{(1+\nu)(3-\nu)F}{16\pi E}\left\{\left(\frac{x_1}{r_0}-1\right)\log\frac{\left(\frac{x_1}{r_0}-1\right)^2+\left(\frac{y_1}{r_0}\right)^2}{\left(\frac{x_1}{r_0}-1\right)^2+\left(\frac{y_2}{r_0}\right)^2}\right. \\
& -\left(\frac{x_1}{r_0}+1\right)\log\frac{\left(\frac{x_1}{r_0}+1\right)^2+\left(\frac{y_1}{r_0}\right)^2}{\left(\frac{x_1}{r_0}+1\right)^2+\left(\frac{y_2}{r_0}\right)^2}+\left(\frac{x_2}{r_0}-1\right)\log\frac{\left(\frac{x_2}{r_0}-1\right)^2+\left(\frac{y_1}{r_0}\right)^2}{\left(\frac{x_2}{r_0}-1\right)^2+\left(\frac{y_2}{r_0}\right)^2} \\
& -\left(\frac{x_2}{r_0}+1\right)\log\frac{\left(\frac{x_2}{r_0}+1\right)^2+\left(\frac{y_1}{r_0}\right)^2}{\left(\frac{x_2}{r_0}+1\right)^2+\left(\frac{y_2}{r_0}\right)^2} \\
& -\frac{4(1-\nu)}{3-\nu}\left(\frac{y_1}{r_0}\right)\left[\tan^{-1}\frac{\frac{2y_1}{r_0}}{\left(\frac{x_1}{r_0}\right)^2+\left(\frac{y_1}{r_0}\right)^2-1}+\tan^{-1}\frac{\frac{2y_1}{r_0}}{\left(\frac{x_2}{r_0}\right)^2+\left(\frac{y_1}{r_0}\right)^2-1}\right] \\
& \left.+\frac{4(1-\nu)}{3-\nu}\left(\frac{y_2}{r_0}\right)\left[\tan^{-1}\frac{\frac{2y_2}{r_0}}{\left(\frac{x_1}{r_0}\right)^2+\left(\frac{y_2}{r_0}\right)^2-1}+\tan^{-1}\frac{\frac{2y_2}{r_0}}{\left(\frac{x_2}{r_0}\right)^2+\left(\frac{y_2}{r_0}\right)^2-1}\right]\right\} \\
& +\frac{F}{\pi E}\left\{\frac{(1+\nu)^2hx}{2}\left[\frac{x_2^2-y_1^2}{\rho_2^4}-\frac{x_2^2-y_2^2}{\rho_3^4}\right]-\frac{2(1-\nu)x_2^2yd}{\rho_2^2\rho_3^2}-\frac{1-\nu^2}{4}\log\left(\frac{\rho_2^2}{\rho_2^2}\right)\right\}
\end{aligned}
\tag{7}
$$

where

$x_1 = x - h$, $x_2 = x + h$, $y_1 = y - d$, $y_2 = y + d$,
$\rho_2^2 = x_2^2 + y_1^2$, $\rho_3^2 = x_2^2 + y_2^2$, and
r_0 = radius of fastener holes.

Here, fastener forces are represented by uniformly distributed forces along the hole diameter so as to avoid logarithmic singularity, incurred when $i = k$, $j = l$.

We approach Part B by first considering System I—a Griffith crack with pressure $q(x)$ on the crack faces (Fig. 3a); and System II—an uncracked half-space bounded by $x > 0$ with normal pressure $p(y)$ on the edge $x = 0$ (Fig. 4a). Here, functions, $p(y)$ and $q(x)$ are symmetric with respect to planes $x = 0$ and $y = 0$. Then, require that p and q simultaneously satisfy the following integral equations [5]

$$
\begin{aligned}
&p(y) + \frac{2}{\pi}\frac{y}{\sqrt{y^2 + a^2}}\int_0^a \frac{q(x)\sqrt{a^2 - y^2}}{x^2 + y^2}\left(\frac{2y^2}{x^2 + y^2} - \frac{2a^2 + y^2}{y^2 + a^2}\right)dx = 0 \\
&q(x) - \frac{4}{\pi}x\int_0^\infty \frac{p(y)y^2}{(x^2 + y^2)^2}dy = \sigma_y(x)
\end{aligned} \tag{8}
$$

Physically, this implies that the normal stress on $x = 0$, resulting from action of $q(x)$, is nullified by $p(y)$, and that the combination of $q(x)$ and y-direction stresses in the location of crack faces caused by $p(y)$ recovers $\sigma_y(x)$. Consequently, superimposing solutions for Systems I and II gives rise to solutions for Part B. Thus, we can write the y-displacement in Part B in the manner

$$
V_B(x, y) = \int_0^a q(\xi)v_1(\xi, x, y)d\xi + \int_0^\infty p(\eta)v_2(\eta, x, y)d\eta \tag{9}
$$

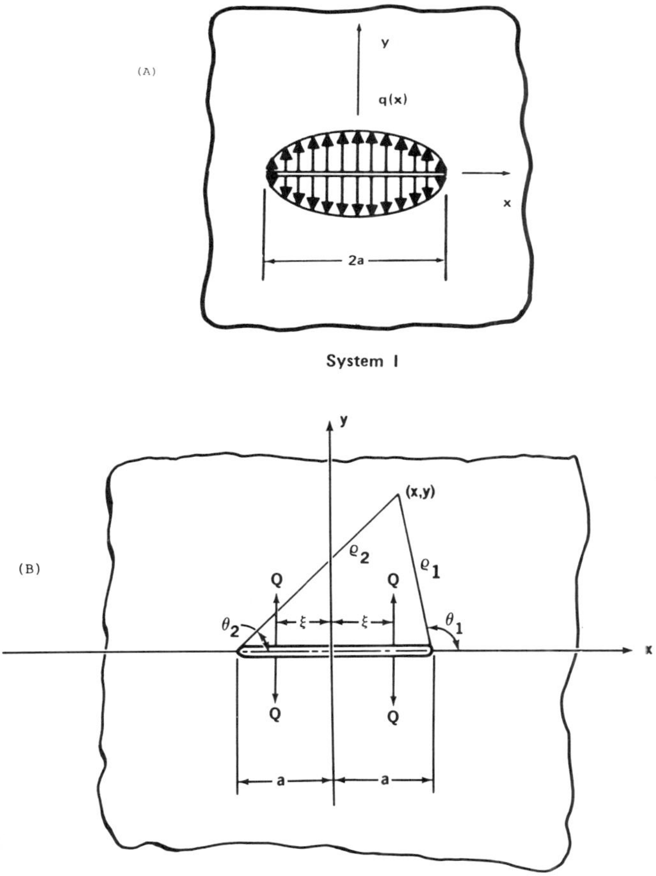

FIG. 3—*A crack in an infinite sheet.* (a) *Symmetric pressure distribution* q(x) *on the crack surfaces.* (b) *A double pair of splitting forces acting on crack faces.*

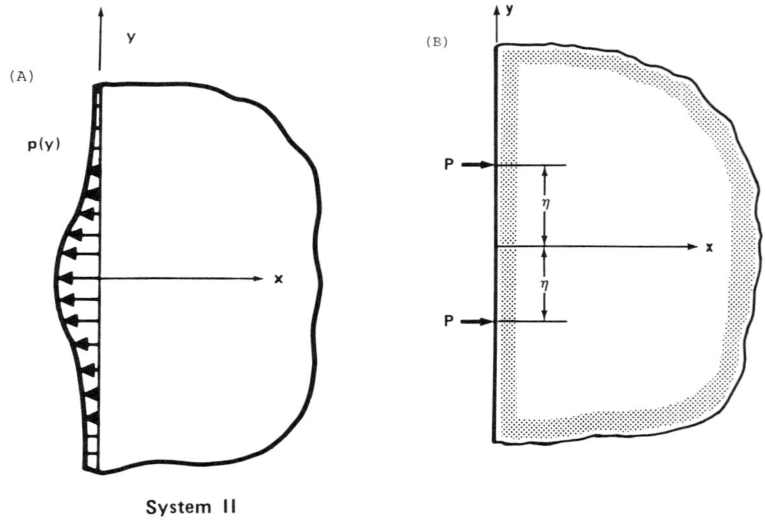

FIG. 4—*An uncracked half-plane.* (a) *Symmetric pressure distribution* p(y) *on the straight edge.* (b) *A double pair of point loads acting on the edge.*

where v_1 represents the displacements due to double pairs of splitting forces of a unit magnitude acting at $x = \xi$ on a Griffith crack of length $2a$ (Fig. 3*b*) and can be derived from Ref *6* as

$$v_1(\xi, x, y) = \frac{1}{\pi E}\left\{\log \frac{a^2 - \xi^2 + \rho_1\rho_2 + 2\sqrt{\rho_1\rho_2(a^2 - \xi^2)}\sin\phi}{a^2 - \xi^2 + \rho_1\rho_2 - 2\sqrt{\rho_1\rho_2(a^2 - \xi^2)}\sin\phi}\right.$$

$$\left. - \frac{2(1+\nu)y\sqrt{a^2 - \xi^2}}{\sqrt{\rho_1\rho_2}} \frac{x(x^2 + y^2 - \xi^2)\cos\phi - y(x^2 + y^2 + \xi^2)\sin\phi}{(x^2 - y^2 - \xi^2) + 4x^2y^2}\right\} \tag{10}$$

where

$$\phi = \frac{1}{2}(\theta_1 + \theta_2)$$

and v_2 is the displacements due to a pair of unit forces acting at $y = \eta$ on the surface $x = 0$ (Fig. 4*b*) and is given by Ref *7*

$$v_2(\eta, x, y) = \frac{1}{\pi E}\left\{(1+\nu)\left[\frac{x(y-\eta)}{x^2 + (y-\eta)^2} + \frac{x(y+\eta)}{x^2 + (y+\eta)^2}\right]\right.$$

$$\left. -(1-\nu)\left[\frac{x^2 - (y-\eta)^2}{x^2 + (y-\eta)^2}\tan^{-1}\left(\frac{y-\eta}{x}\right) + \frac{x^2 - (y+\eta)^2}{x^2 + (y+\eta)^2}\tan^{-1}\left(\frac{y+\eta}{x}\right)\right]\right\} \tag{11}$$

In view of the foregoing analysis, it is seen that influence coefficients A_{ijkl} can be determined by combining displacements in Part A and Part B for unit loads

$$A_{ijkl} = V_A(x_{ij}, y_{ij}) + V_B(x_{ij}, y_{ij}) \tag{12}$$

with coordinates $(h, \pm d)$ in Part A being the locations of the lth fastener pair in the kth stiffener.

Similarly, influence coefficients B_{ij} can be determined by letting the pressure on crack faces of problem Part B be the applied tensile stresses σ_0. If the solutions of Eq 8 with $\sigma_y(x)$ replaced by σ_0 are denoted by $p_0(y)$ and $q_0(x)$, we can write the expression for B_{ij} as

$$B_{ij} = \frac{y_{ij}}{E} + \int_0^a q_0(\xi)v_1(\xi, x_{ij}, y_{ij})d\xi + \int_0^\infty p_0(\eta)v_2(\eta, x_{ij}, y_{ij})d\eta \tag{13}$$

By taking stiffeners as one-dimensional rods, influence coefficients A^s_{ijkl} are simply given by

$$A^s_{ijil} = \frac{1}{E^s A^s}\{y_{ij}H(y_{il} - y_{ij}) + y_{il}H(y_{ij} - y_{il})\} \tag{14}$$

where $H(x)$ is the Heaviside function. The coefficient B_{ij} is simply given by

$$B^s_{ij} = \frac{y_{ij}}{E^s} \tag{15}$$

Solution Method for Integral Equations

Solution procedures for Eq 8 begin with the elimination of one of the variables. By substituting $p(y)$ from the first equation into the second, and upon interchanging the integral signs, integrating over the variable y, we arrive at

$$\bar{q}(x) + \int_0^1 \bar{q}(\xi)L(x, \xi)d\xi = \bar{\sigma}_y(x) \qquad 0 < x < 1 \tag{16}$$

where the dimensions have been normalized with crack length a and

$$L(x, \xi) = \frac{4x\sqrt{1 - \xi^2}}{\pi(\xi^2 - x^2)^2}\left\{-\frac{3x^4 + x^2\xi^2 - 4\xi^2}{(1 - x^2)^2} - \frac{2\xi^2(\xi^2 + x^2)}{(\xi^2 - x^2)\sqrt{1 - \xi^2}}\log\left(\frac{1 - \sqrt{1 - \xi^2}}{1 + \sqrt{1 + \xi^2}}\right)\right.$$
$$\left. + \left[\frac{4x^2\xi^2}{\xi^2 - x^2} - \frac{x^6 + x^4(2 - \xi^2) + 2\xi^2(x^2 - 2)}{(1 - x^2)^2}\right] \times \frac{1}{\sqrt{1 - x^2}}\log\left(\frac{1 - \sqrt{1 - x^2}}{1 - \sqrt{1 + x^2}}\right)\right\}$$

it can be shown that the kernel L is finite everywhere for $0 < x < 1$, $0 < \xi < 1$, including $x = \xi$. But $\partial L/\partial \xi$ becomes unbounded as $\xi \rightarrow 1$ due to the term $\sqrt{1 - \xi}$. To circumvent this problem, let $L^* = L\sqrt{1 - \xi}$. Thus

$$\bar{q}(x) + \int_0^1 \frac{L^*(x, 1 - y)\bar{q}(1 - y)}{\sqrt{y}}dy = \bar{\sigma}_y(x) \tag{17}$$

Applying Gaussian quadrature formula (see Ref *8*) to the integral

$$\int_0^1 \frac{f(x)}{\sqrt{x}}dx \cong \sum_{k=1}^{N} W_k f(x_k) \tag{18}$$

Eq 17 may be approximated by finite sum as

$$\bar{q}(1 - x_i) + \sum_{k=1}^{N} W_k L^*(1 - x_i, 1 - x_k)\bar{q}(1 - x_k) = \bar{\sigma}_y(1 - x_i) \tag{19}$$

where $i = 1, 2, 3, \ldots, N$ and W_k, x_k are the weight coefficients and zeroes, respectively, for the Jacobian polynomials of order N. The system of algebraic equations can then be solved for the q's by a standard procedure. Once this is done, $p(y)$ can be readily calculated from the first of Eq 8.

Tests showed that convergence of $\bar{q}(x)$ was dictated largely by the shape of function $\bar{\sigma}_y(x)$; see Eq 6. In the case of remote tension load, for which $\bar{\sigma}_y(x)$ is a constant, only eight terms ($N = 8$) were needed to reach convergence. On the other hand, for fastener loads, the speed of convergence is dependent on the ratio of the distance from the fastener to the crack, d, and the crack length, a. For $d/a > 0.5$, eight terms were sufficient to yield solutions within a few percent error. As d/a decreased, more subdivisions were required; however, no converged values of $\bar{q}(x)$ were obtained for $d/a < 0.05$.

Evaluation of Stress Intensity Factors

The crack-tip stress intensity factors can be expressed in terms of solution $\bar{q}(x)$ as (see Ref *9*)

$$K_\text{I} = \frac{2F}{\sqrt{\pi a}} \int_0^1 \frac{\bar{q}(x)dx}{\sqrt{1 - y^2}} \cong \frac{2F}{\sqrt{\pi a}} \sum_{k=1}^{N} \frac{W_k \bar{q}(1 - x_k)}{\sqrt{2 - x_k}} \tag{20}$$

Numerical values of K-factors are shown in Fig. 5 as functions of h for various d. Results for the special case of $d = 0$ from Ref *10* are also included for comparison.

Once the fastener forces have been solved from Eq 3, their contributions to the stress-intensity factors are summed, along with that due to far-field stress, to render

$$K_\text{I} = 2\sigma_0\sqrt{a} \int_0^1 \frac{\bar{q}_0(x)dx}{\sqrt{1 - x^2}} + \sum_{k=1}^{\infty} \sum_{l=1}^{\infty} \frac{2F_{kl}}{\sqrt{\pi a}} \int_0^1 \frac{\bar{q}_{kl}(x)dx}{\sqrt{1 - x^2}} \tag{21}$$

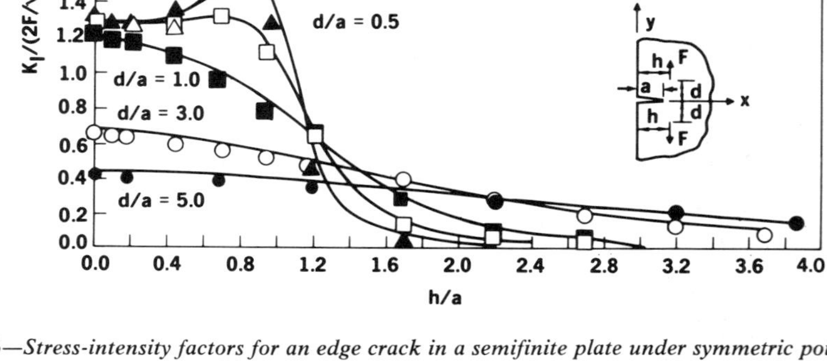

FIG. 5—*Stress-intensity factors for an edge crack in a semifinite plate under symmetric point loads.*

Stress-intensity factors for an edge crack in a typical skin/stringer structure are plotted against the crack length as shown in Fig. 6 for illustration. This sample problem contains three stiffeners, each with five fasteners. Results for the case where the first stiffener has failed are also presented for comparison. Numerical values are calculated by approximating the integrals in Eq 17 with the eight-point quadrature formula. With this approach, the first term in Eq 21 yields a value of $1.11\sigma_0\sqrt{\pi a}$.

Concluding Remarks

This paper presents an analytical approach to linear elastic fracture analysis of an edge-cracked stiffened sheet. The solution method employed allows straightforward computation of the crack-tip stress-intensity factors, and can be easily implemented in the residual strength analysis procedure for structures of this type. The degree of accuracy for the results depends on the ratio d/a; for $d/a > 1$, the errors generally fall within 5% of convergent values.

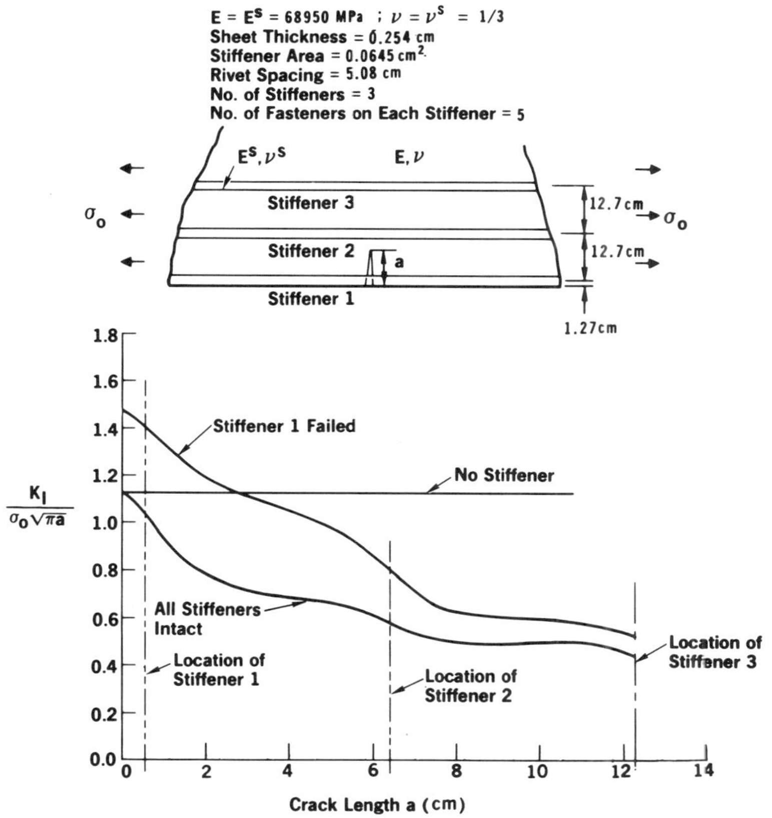

FIG. 6—*Mode I stress-intensity factors for an edge crack in a sheet with two stiffeners under far-field tension.*

APPENDIX

Half-Space Subjected to Concentrated Loads

In an infinite plate under a pair of point loads acting at $x = h$, $y = \pm d$ as shown in Fig. 7, the stress distributions can be written from known solutions for plane stress condition as

$$\begin{aligned} \sigma_x &= \sigma_x^*(h, d) + \sigma_x^*(h, -d) \\ \sigma_y &= \sigma_y^*(h, d) + \sigma_y^*(h, -d) \\ \tau_{xy} &= \tau_{xy}^*(h, d) + \tau_{xy}^*(h, -d) \end{aligned} \tag{22}$$

where

$$\sigma_x^*(h, d) = \frac{F}{4\pi}\left\{\frac{(1-\nu)(y-d)}{(x-h)^2+(y-d)^2} - \frac{2(1+\nu)(y-d)(x-h)^2}{[(x-h)^2+(y-d)^2]^2}\right\}$$

$$\sigma_y^*(h, d) = \frac{-F}{4\pi}\left\{\frac{(3+\nu)(y-d)}{(x-h)^2+(y-d)^2} + \frac{2(1+\nu)(y-d)(x-h)^2}{[(x-h)^2+(y-d)^2]^2}\right\}$$

$$\tau_{xy}^*(h, d) = \frac{F}{4\pi}\left\{-\frac{(3-\nu)(x-h)}{(x-h)^2+(y-d)^2} + \frac{2(1+\nu)(x-h)^3}{[(x-h)^2+(y-d)^2]^2}\right\}$$

Also, the displacements in the y-direction take the form

$$v = v^*(h, d) + v^*(h, -d) \tag{23}$$

where

$$v^*(h, d) = \frac{-F}{4\pi E}\left\{\frac{1}{2}(3-\nu)(1+\nu)\log[(x-h)^2+(y-d)^2] + \frac{(1+\nu)(x-h)^2}{(x-h)^2+(y-d)^2}\right\}$$

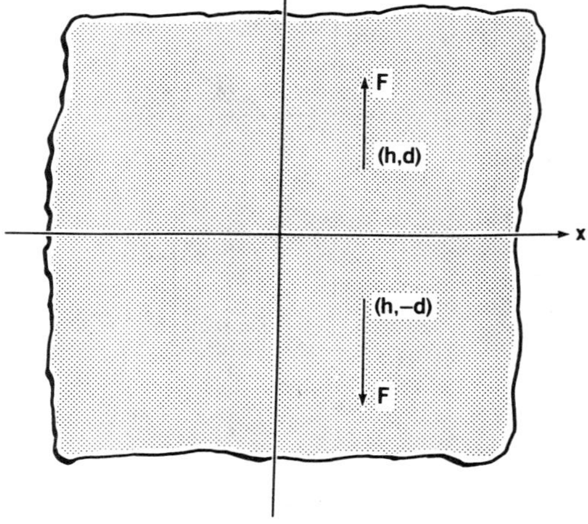

FIG. 7—*An infinite plate subjected to point loads.*

Consider now the equilibrium of the region $x > 0$. The components of stress must satisfy equations of motion and must have the same singularity at the points $(h, \pm d)$ as defined by Eq 22. Further, if we assume that the boundary $x = 0$ is free from traction, we must have

$$\sigma_x = \tau_{xy} = 0 \quad \text{on} \quad x = 0 \tag{24}$$

With this in mind, we may write for the stress components

$$\begin{aligned}
\sigma_x &= \sigma_x^*(h, d) + \sigma_x^*(h, -d) + \sigma_x^*(-h, d) + \sigma_x^*(-h, -d) + \frac{\partial^2 \phi}{\partial y^2} \\
\sigma_y &= \sigma_y^*(h, d) + \sigma_y^*(h, -d) + \sigma_x^*(-h, d) + \sigma_x^*(-h, -d) + \frac{\partial^2 \phi}{\partial x^2} \\
\tau_{xy} &= \tau_{xy}^*(h, d) + \tau_{xy}^*(h, -d) + \tau_{xy}^*(-h, d) + \tau_{xy}^*(-h, -d) + \frac{\partial^2 \phi}{\partial x \partial y}
\end{aligned} \tag{25}$$

where ϕ is the Airy stress function which satisfies the biharmonic equation $\nabla^4 \phi = 0$. Values of ϕ that ensure satisfaction of the Eq 24 condition by stress components in Eq 25 can be obtained by means of Fourier transform techniques (see Ref *11*). Using this approach, we find that

$$\begin{aligned}
\frac{\partial^2 \phi}{\partial y^2} &= -\frac{2}{\pi} \int_0^\infty \bar{p}(\xi)(1 + \xi x) e^{-\xi x} \cos(\xi y) d\xi \\
\frac{\partial^2 \phi}{\partial x^2} &= -\frac{2}{\pi} \int_0^\infty \bar{p}(\xi)(1 - \xi x) e^{-\xi x} \cos(\xi y) d\xi \\
\frac{\partial^2 \phi}{\partial x^2 \partial y} &= \frac{2x}{\pi} \int_0^\infty \xi \bar{p}(\xi) e^{-\xi x} \sin(\xi y) d\xi
\end{aligned} \tag{26}$$

where

$$\bar{p}(\xi) = \frac{F}{2} [(1 + \nu) h \xi - (1 - \nu)] e^{-\xi h} \sin(\xi d)$$

The y-component displacement in the half-space can be expressed in the form

$$v = v^*(h, d) + v^*(h, -d) + v^*(-h, d) + v^*(-h, -d) + v_\phi \tag{27}$$

where

$$v_\phi = \frac{-2}{\pi E} \int_0^\infty \bar{p}(\xi)[1 - \nu - (1 + \nu)\xi x] e^{-\xi x} \sin(\xi y) \frac{d\xi}{\xi}$$

pertains to the stress function ϕ.

References

[*1*] Romualdi, J. P., Frasier, J. T., and Irwin, G. R., "Crack-Extension-Force Near a Riveted Stiffener," Naval Research Laboratory, Washington, DC, Report 4956, 1957.

[*2*] Bloom, J. M. and Sanders, J. L., Jr., "The Effect of a Riveted Stringer on the Stress in a Cracked Sheet," *Journal of Applied Mechanics*, Vol. 33, 1966, pp. 561–570.

[*3*] Greif, R. and Sanders, J. L., Jr., "The Effect of a Stringer on the Stress in a Cracked Sheet," *Journal of Applied Mechanics*, Vol. 32, 1965, pp. 59–66.

[4] Swift, T., "Fracture Analysis of Stiffened Sheet" in *Damage Tolerance of Metallic Structures: Analysis Methods and Applications, ASTM STP 842,* American Society for Testing and Materials, Philadelphia, 1984, pp. 69–107.

[5] Irwin, G. R., "The Crack-Extension Force for a Crack at a Free Surface Boundary," Naval Research Laboratory, Washington, DC, Report 5120, 1958.

[6] Westergaard, H. M., "Bearing Pressure and Cracks," *Journal of Applied Mechanics,* Vol. 6, 1939, pp. A49–A53.

[7] Timoshenko, S. P. and Goodier, J. N., *Theory of Elasticity,* 3rd ed., McGraw-Hill, New York, 1970.

[8] Krylov, V. I., *Approximate Calculation of Integrals,* MacMillan, New York, 1962.

[9] Sih, G. C., Paris, P. C., and Erdogan, F., "Crack Tip Stress Intensity Factors for Plane Extension and Plate Bending Problems," *Journal of Applied Mechanics,* Vol. 29, 1962, p. 306.

[10] Hartranft, R. J. and Sih, G. C., "Solving Edge and Surface Crack Problems by an Alternating Method" in *Methods of Analysis and Solutions of Crack Problems,* G. C. Sih, Ed., Noordhoff, Amsterdam, The Netherlands, 1973.

[11] Sneddon, I. N., *Fourier Transform,* McGraw-Hill, New York, 1951.

Paul W. Tan,[1] Ivatury S. Raju,[1] and James C. Newman, Jr.[2]

Boundary Force Method for Analyzing Two-Dimensional Cracked Plates

REFERENCE: Tan, P. W., Raju, I. S., and Newman, J. C., Jr., **"Boundary Force Method for Analyzing Two-Dimensional Cracked Plates,"** *Fracture Mechanics: Eighteenth Symposium, ASTM STP 945,* D. T. Read and R. P. Reed, Eds., American Society for Testing and Materials, Philadelphia, 1988, pp. 259–277.

ABSTRACT: The Boundary Force Method (BFM) was formulated for the two-dimensional stress analysis of complex crack configurations. In this method, only the boundaries of the region of interest are modeled. The boundaries are divided into a finite number of straight-line segments; at the center of each segment, concentrated forces and a moment are applied. This set of unknown forces and moments is calculated to satisfy the prescribed boundary conditions of the problem. The elasticity solution for the stress distribution due to concentrated forces and a moment applied at an arbitrary point in a cracked infinite plate is used as the fundamental solution. Thus, the crack need not be modeled as part of the boundary.

The formulation of the BFM is described, and the accuracy of the method is established by analyzing several crack configurations for which accepted stress-intensity factor solutions are known. The crack configurations investigated include Mode I and mixed-mode (Modes I and II) problems. The results obtained are, in general, within $\pm 0.5\%$ of accurate numerical solutions.

The versatility of the method is demonstrated through the analysis of complex crack configurations for which limited or no solutions are known.

KEY WORDS: cracks, stress-intensity factor, stress-concentration factor, stress functions elasticity, holes

Nomenclature

a	Crack length or one-half crack length
C	Resultant couple
E	Young's modulus
$[F]$	Influence coefficient matrix
F_x, F_y	Resultant forces in x- and y-directions, respectively
H, W	Height and width of rectangular plates, respectively
K	Stress-intensity factor
L_i	Length of segment i
M	Concentrated moment
N	Number of segments
p_i, q_i, m_i	Unit loads and moment on the ith segment
P, Q	Concentrated forces in y- and x-directions, respectively
$\{P\}$	Vector of unknown forces and moments
$\{R\}$	Externally applied load vector
S	Remote applied stress
u, v	Displacements in x- and y-directions, respectively

[1]Research scientist and senior scientist, respectively, Analytical Services and Materials, Inc., Hampton, VA 23666.

[2]Senior scientist, NASA Langley Research Center, Hampton, VA 23665.

x, y	Cartesian coordinates
z	Complex variable, $z = x + iy$
z_0	Location of forces and moment, $z_0 = x_0 + iy_0$
β	Angle of inclination of crack
κ	Material constant $= 3 - 4\nu$ for plane strain $= (3 - \nu)/(1 + \nu)$ for plane stress
λ_x, λ_y	Projection of sub-arc onto x- and y-axes, respectively
μ	Shear modulus
ν	Poisson's ratio
σ_x, σ_y, σ_{xy}	Cartesian stresses
σ_n, σ_{nt}	Normal and shear stresses on boundary
ϵ_x, ϵ_y, ϵ_{xy}	Cartesian strains
ϕ^*, Ω^*, ϕ_0	Complex stress functions

Damage-tolerant design criteria for structural members require the ability to predict crack-growth rates and fracture strengths. In these predictions, stress-intensity factors are the most important parameters. Therefore, for accurate predictions, accurate stress-intensity factors are needed.

In two-dimensional analyses, many methods have been used to determine stress concentration factors for holes or notches and stress-intensity factors for cracks. Three of the most popular numerical methods are the Finite Element Method (FEM), the Boundary Collocation Method, and the Boundary Element Method (BEM).

FEM has enjoyed widespread use during the past three decades. However, in the field of fracture mechanics, FEM is cumbersome to use. In the finite-element approach, relatively large numbers of elements are needed to accurately model the crack region and, for complex configurations with cracks and notches, the number of elements needed to accurately model these problems can be extremely large. (Thus, a correspondingly large effort is required to construct such a model.) Moreover, in crack-growth-rate predictions, stress-intensity factors must be found for various crack lengths. Since a new mesh must be generated for each new crack length to accurately model the crack tip, a large amount of time is needed in the FEM for modeling.

In the collocation method, only the boundaries of the region of interest need to be modeled. Compared with the FEM, where the entire region must be modeled, the collocation method presents an attractive alternative. In this technique, however, the basic stress functions need to be changed for different classes of problems. Therefore in the collocation method, a large amount of time can be spent in developing and formulating new stress functions for each class of problems.

With the BEM, only the boundaries of the region of interest are modeled. The boundaries are discretized into a series of line segments. In this paper, a more general method, the Boundary Force Method (BFM), is formulated. The BFM is a form of an indirect BEM that models only the boundaries as in the collocation method, but does not require different stress functions for different classes of problems. Before the details of the BFM are given, previous work in indirect boundary element methods related to the BFM is briefly described.

One of the earliest indirect formulations of BEM by Nisitani [*1*] was called the "body force" method. In his method, the unknowns were constant body force densities in the x- and y-directions applied on each segment of the discretized boundaries. The boundary conditions were satisfied in terms of tractions. For cracked bodies, the crack was also modeled as a very slender elliptical notch [*2–4*].

Isida [*5*] improved on the accuracy of the body force method by satisfying the boundary conditions in terms of resultant forces. The unknowns in his technique were the body force densities in the x- and y-directions on each segment of the discretized boundaries. Again, cracks were modeled as very slender elliptical notches.

Erdogan and Arin [*6*] introduced a boundary method to analyze an elastic domain containing a crack. The stress-free conditions on the crack faces were satisfied exactly by using the analytical solution for concentrated forces in an infinite plate with a crack. Thus, the crack faces did not have to be modeled. The unknowns in this analysis were the concentrated forces in the x- and y-directions applied again on the boundaries.

In the present BFM, the stress-free conditions on the crack faces are exactly satisfied by using Erdogan's analytical solution for concentrated forces and a moment in an infinite plate with a crack [7], thus eliminating the need to model the crack. The unknowns in the BFM are the concentrated forces in the x- and y-directions and a moment on each segment of the discretized boundaries.

Briefly, the essential differences between the methods of Nisitani [*1*], Isida [*5*], Erdogan and Arin [*6*], and the present BFM can be grouped into three categories: the fundamental solutions, the treatment of the boundary conditions, and the treatment of the crack faces. These techniques and their differences are summarized in Table 1.

In this paper, first the formulation of the BFM is presented and the fundamental solution is briefly reviewed. Studies were performed on a center cracked and a single-edge cracked specimen to investigate the rate of convergence of the BFM. The accuracy of the BFM was established by analyzing crack configurations for which accurate numerical solutions are known. These were a crack between two circular holes in an infinite plate and a slant single-edge crack in a finite plate subjected to uniaxial tension. To demonstrate the versatility and "ease of use," the BFM was also used to obtain stress-intensity factor solutions for two crack configurations for which very limited or no solutions are available. These were a crack emanating from a semicircular edge notch and a four-hole cracked specimen subjected to uniaxial tension.

Analysis

In this section, first the formulation of the BFM is presented. Next, the fundamental solution for concentrated forces and a moment at an arbitrary point in an infinite plate is summarized.

TABLE 1—*Comparison of various indirect boundary element techniques.*

Investigator	Fundamental Solutions Used: Without Crack	Fundamental Solutions Used: With Crack	Treatment of Crack Faces	Treatment of Boundary Conditions (BC)
Nisitani [*1*]	Timoshenko's solution [*8*] for point load in an infinite plate	same as without crack	modeled as a very slender elliptical notch: $b/a \rightarrow 0$; b = minor axis; a = major axis	BC are satisfied in terms of stresses
Isida [*5*]	Kolosov-Muskhelishvili solution [*9*] for point loads in an infinite plate	same as without crack	modeled as a very slender elliptical notch $b/a \rightarrow 0$	BC are satisfied in terms of resultant forces
Erdogan and Arin [*6*]	Kolosov-Muskhelishvili solution [*9*] for point loads in an infinite plate	Erdogan's solution [*7*] for point loads in an infinite plate with a crack	no modeling of the crack faces needed	BC are satisfied in terms of resultant forces
Present method	Kolosov-Muskhelishvili solution [*9*] for point loads in an infinite plate	Erdogan's [*7*] solution for point loads in an infinite plate with a crack	no modeling of the crack faces needed	BC are satisfied in terms of resultant forces and moment

Then a simple and useful numerical method for evaluating some of the influence coefficients is presented.

Formulation

The Boundary Force Method is a numerical technique which uses the superposition of a finite number of sets of concentrated forces and moments in an infinite plate to obtain the solution to the boundary-value problem of interest.

The BFM uses the elasticity solution for concentrated forces and a moment in an infinite plate as the fundamental solution. For plates without a crack, Muskhelishvili's solution [*9*] for a pair of concentrated forces Q, P and a moment M in an infinite plate is used as the fundamental solution. For a plate with a crack, Erdogan's solution [*7*] for a pair of concentrated forces Q, P and a moment M in an infinite plate with a crack is used as the fundamental solution. With Erdogan's solution [*7*] the crack faces need not be modeled as part of the boundary since the stress-free conditions on the crack faces are exactly satisfied.

Because the fundamental solution satisfies all the equations of elasticity in the interior, the only remaining conditions to be satisfied are the conditions on the boundary. The given boundary conditions are satisfied by applying sets of concentrated forces and moments along an "imaginary" boundary traced on an infinite plate, corresponding to the actual configuration. The imaginary boundary is discretized into a number of segments. At the center of each segment, a pair of concentrated forces and a moment are applied at a small distance δ away from the imaginary boundary. The magnitudes of the forces and moment on each segment are determined to approximately satisfy the boundary conditions.

To illustrate the method, consider the problem of a finite plate with a crack subjected to uniaxial tension as shown in Fig. 1. The solution to this problem is obtained by the superposition of concentrated forces and moments in an infinite plate with a crack as shown in Fig. 2. The dashed lines in Fig. 2 correspond to the imaginary boundary of the plate shown in Fig. 1.

The first step in the BFM is to discretize the imaginary boundary into a finite number of segments. As an example, Fig. 2 shows an idealization of 16 segments (four in each quadrant) on the imaginary boundary. On each segment i, a concentrated force pair P_i, Q_i and a moment M_i are applied at a small distance δ_i on the outward normal from the midpoint of the segment.

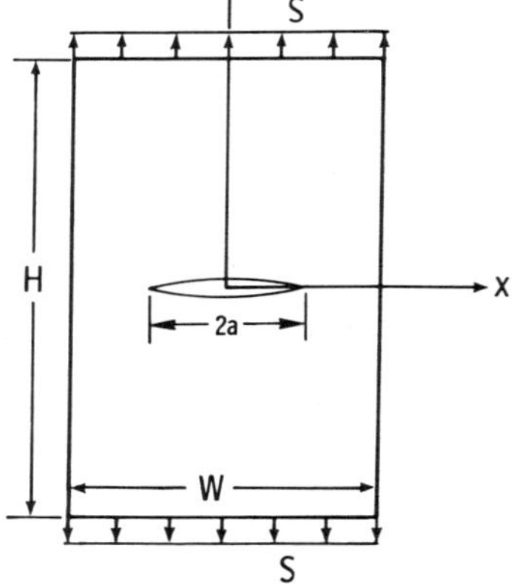

FIG. 1—*Crack in finite plate subjected to uniaxial tension.*

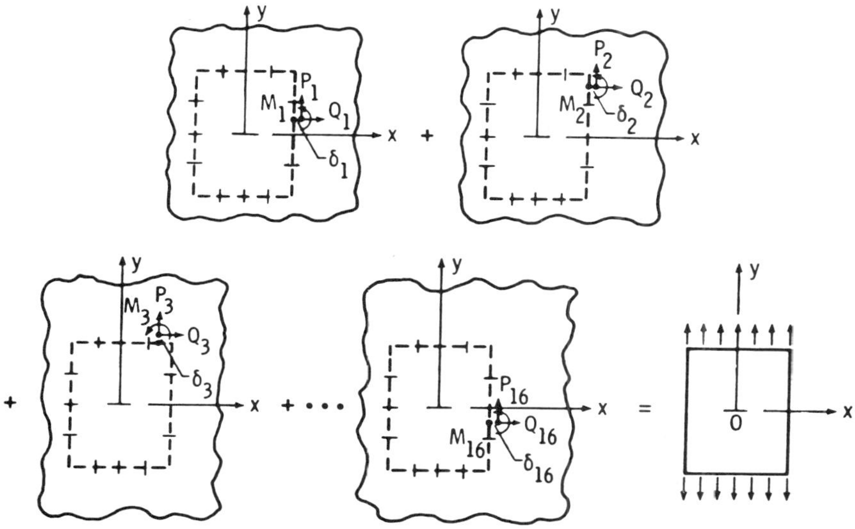

FIG. 2—*Superposition of unknown forces* P_i, Q_i *and moment* M_i *on 16 boundary segments* (i = *1 to 16).*

The distance δ_i was chosen to be one-fourth of the length of the segment i. This offset was used to eliminate the stress singularities in the computation of the stresses on the boundaries.

The concentrated forces P_i, Q_i and moment M_i on the ith segment create resultant forces R_{x_j} in the x-direction and R_{y_j} in the y-direction, and a resultant couple C_j on the jth segment. Thus, the resultant forces and couple for segment j due to the forces and moments on all segments are

$$R_{x_j} = \sum_{i=1}^{16} (F_{x_j p_i} P_i + F_{x_j q_i} Q_i + F_{x_j m_i} M_i)$$

$$R_{y_j} = \sum_{i=1}^{16} (F_{y_j p_i} P_i + F_{y_j q_i} Q_i + F_{y_j m_i} M_i)$$

$$C_j = \sum_{i=1}^{16} (C_{j p_i} P_i + C_{j q_i} Q_i + C_{j m_i} M_i)$$

Here $F_{x_j p_i}$, $F_{y_j p_i}$, $C_{j p_i}$, etc. are called influence coefficients and are defined as follows:

$F_{x_j p_i}$ = force in the x-direction created on the jth segment due to unit load p_i acting in the y-direction on the ith segment,

$F_{y_j p_i}$ = force in the y-direction created on the jth segment due to unit load p_i acting in the y-direction on the ith segment, and

$C_{j p_i}$ = resultant couple created on the jth segment due to unit load p_i acting in the y-direction on the ith segment.

with similar definitions for $F_{x_j q_i}$, $F_{y_j q_i}$, $C_{j q_i}$, $F_{x_j m_i}$, $F_{y_j m_i}$, and $C_{j m_i}$.

The resultant forces and couples are assumed to act at the center of the corresponding segments. The resultants on all segments can then be written as

$$[F]_{48\times48}\{P\}_{48\times1} = \{R\}_{48\times1} \tag{1}$$

where

$[F]$ = "influence coefficient" matrix,
$\{P\}$ = vector of unknown forces and moments, and
$\{R\}$ = vector of externally applied resultant forces and moments.

The influence coefficient matrix $[F]$ can be expressed as

$$[F] = \begin{bmatrix} F_{x1p1} & \cdots & F_{x1p16} & | & F_{x1q1} & \cdots & F_{x1q16} & | & F_{x1m1} & \cdots & F_{x1m16} \\ F_{y1p1} & \cdots & F_{y1p16} & | & F_{y1q1} & \cdots & F_{y1q16} & | & F_{y1m1} & \cdots & F_{y1m16} \\ C_{1p1} & \cdots & C_{1p16} & | & C_{1q1} & \cdots & C_{1q16} & | & C_{1m1} & \cdots & C_{1m16} \\ \cdots & & \cdots & | & \cdots & & \cdots & | & \cdots & & \cdots \\ \cdots & & \cdots & | & \cdots & & \cdots & | & \cdots & & \cdots \\ \cdots & & \cdots & | & \cdots & & \cdots & | & \cdots & & \cdots \\ F_{x16p1} & \cdots & F_{x16p16} & | & F_{x16q1} & \cdots & F_{x16q16} & | & F_{x16m1} & \cdots & F_{x16m16} \\ F_{y16p1} & \cdots & F_{y16p16} & | & F_{y16q1} & \cdots & F_{y16q16} & | & F_{y16m1} & \cdots & F_{y16m16} \\ C_{16p1} & \cdots & C_{16p16} & | & C_{16q1} & \cdots & C_{16q16} & | & C_{16m1} & \cdots & C_{16m16} \end{bmatrix} \quad (2)$$

The coefficient matrix $[F]$ is fully populated and nonsymmetric.

The external loading on the plate boundaries (uniform stress S in the y-direction at $y = \pm H/2$, see Fig. 1) can be replaced with concentrated forces equal to SL_k, where L_k is the length of the kth segment as shown in Fig. 3. The forces in the x-direction and the moment on these boundaries are zero. The force pair and moment acting on each segment of the boundaries $x = \pm W/2$ are also zero since these boundaries are stress free. This set of concentrated force pairs and moments forms the load vector $\{R\}$ on the right-hand side of Eq 1 and can be written as

$$\{R\} = \{R_{x1}, R_{y1}, C_1, \ldots\ldots R_{x16}, R_{y16}, C_{16}\}$$

$$\{R\} = \{0, 0, 0, \ldots SL_3, \ldots SL_{14}, \ldots 0, 0\}$$

where

R_{x_i} = total force in x-direction on the ith segment due to the external tractions,
R_{y_i} = total force in y-direction on the ith segment due to the external tractions, and
C_i = total moment on the ith segment due to the external tractions.

Because the influence coefficient matrix $[F]$ and the external applied load vector $\{R\}$ are known, the unknown force and moment vector $\{P\}$ can be obtained by solving the system of linear algebraic simultaneous equations. This set of forces P_i, Q_i, and M_i, acting on the imaginary boundary in the infinite plate (see Fig. 2), will approximately satisfy the specified boundary conditions. Therefore, the stress state within the region bounded by the imaginary boundary is approximately equal to the stress state in the interior of the finite cracked plate. The stresses and displacements at any point within the plate and the stress-intensity factors at the crack tips can be computed with the system of forces P_i, Q_i and moment M_i ($i = 1$ to 16). A more detailed formulation of the BFM and the advantages of considering symmetry are given in [*10*].

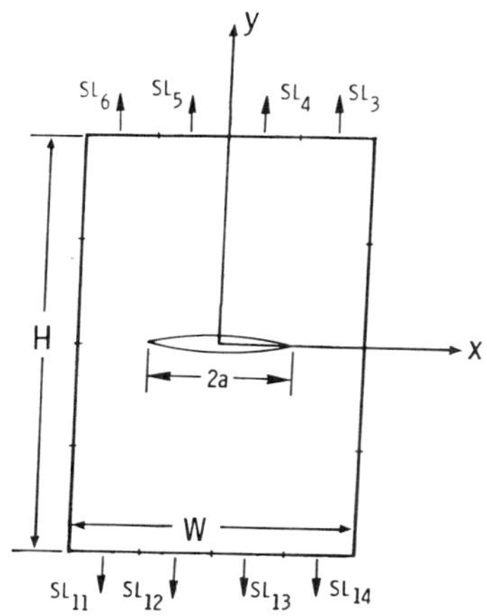

FIG. 3—*Replacements of tractions by concentrated forces.*

Fundamental Solution

The BFM formulation uses the elasticity solution for a pair of concentrated forces and a moment in an infinite plate with a crack. Such a solution was formulated by Erdogan [7] for linear, isotropic, and homogeneous materials. The solution is presented in Appendix I.

Influence Coefficient Calculations

The influence coefficients in Eq 2 are the resultant forces and moment on segment j due to unit forces and moment applied to segment i and are calculated using numerical integration of stresses given in Eq 5. While the force resultants on the jth segment due to unit loads p_i, q_i and moment m_i on the ith segment are easily obtained in closed form from Eq 7, the moment resultants C_{jp_i}, C_{jq_i}, and C_{jm_i} caused some difficulty, since the resultant moment terms C_{jp_i}, C_{jq_i}, and C_{jm_i} can be obtained in closed form only by double integrations of the stress functions. However, after the first integration, the resulting functions consist of a long series of multivalued terms. Hence, a second integration of these functions would be unwieldy and cumbersome, if not impractical. For this reason, the resultant moment C, created by concentrated forces and moment in an infinite plate, is approximated by dividing each segment into two or more parts, and computing the resultant forces on each subdivision. These resultant forces are then multiplied by the corresponding moment arm about the center of the segment. For example, consider a unit load p_1 acting at z_0 in an infinite plate as shown in Fig. 4. The segment j, between z_1 and z_2, is divided into two parts ($k = 1$ and 2). The resultant forces created by the unit load p_1 at z_0, along each sub-arc k, are $F^k_{xjp_1}$ and $F^k_{yjp_1}$ in the x- and y-directions, respectively. Thus, the resultant couple C_{jp_1} across the jth segment between z_1 and z_2, due to the unit load p_1, is

$$C_{jp_i} = \sum_{k=1}^{2} (F^k_{xjp_1} \lambda^k_{yj} + F^k_{yjp_1} \lambda^k_{xj}) \tag{3}$$

where the influence coefficients F_{xjp_1} and F_{yjp_1} were defined above and the superscript k refers to the kth subdivision. The terms

λ_{xj}^k = x-distance from the center of the kth subdivision to the center of the jth segment, and
λ_{yj}^k = y-distance from the center of the kth subdivision to the center of the jth segment.

as shown in Fig. 4.

The resultant couples C_{jq_i} and C_{jm_i}, due to unit load force q_i and unit moment m_i, can be obtained in a similar fashion.

The accuracy of the coefficients C_{jp_i} increases as the number of the subdivisions k becomes large. From numerical convergence studies on several C_{jp_i}, C_{jq_i}, and C_{jm_i} coefficients, a value of $k = 4$ was found to be sufficient in most cases. This numerical procedure simplified the computations considerably and reduced the tedious integrations and cumbersome algebraic manipulations.

Results and Discussion

First, the accuracy of the BFM is compared with some other boundary methods, where only forces are used as the unknowns, and the convergence characteristics of the BFM are studied by applying the method to two standard crack configurations. Then the BFM is used to analyze Mode I and mixed-mode problems for which accurate solutions are available in the literature. Finally, the method is used to obtain stress-intensity factor solutions for complex crack configurations for which no solutions are available.

Accuracy and Convergence Studies

In previous boundary methods [*1–6*], the boundary conditions were satisfied in terms of tractions or resultant forces alone. In the BFM, the boundary conditions are satisfied in terms of resultant forces and moments. The BFM, therefore, has an additional degree of freedom per segment compared to previous boundary methods. To compare the accuracy obtained by satis-

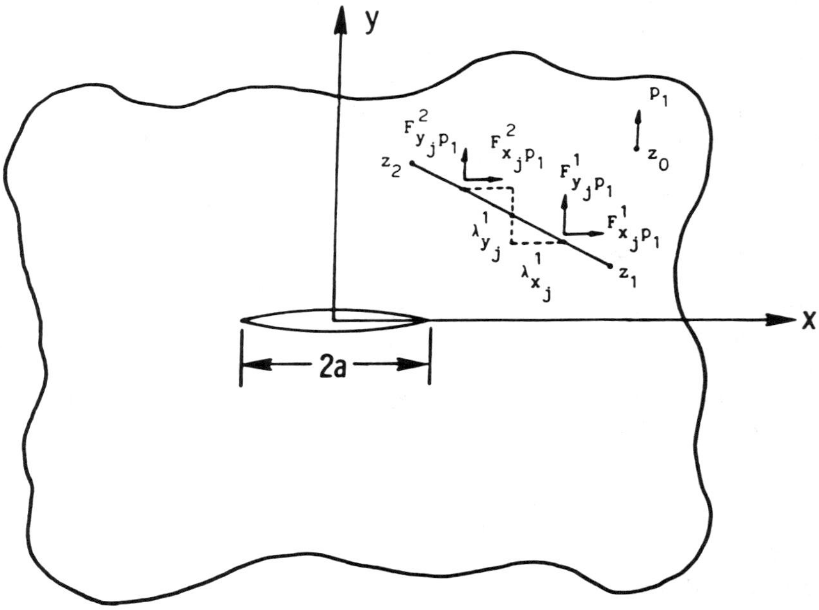

FIG. 4—*Subdivision of a segment.*

fying the boundary conditions in terms of resultant forces and moments to that of resultant forces only, a simple configuration was analyzed using the two methods. A center-cracked plate with a long crack ($2a/W = 0.8$) was analyzed. Because of symmetry in the problem, only one quarter of the plate was modeled using equal size segments. The relative error in the stress-intensity factors is used for comparison, where the relative error is defined as

$$\text{Relative error} = \left| \frac{K_{\text{computed}} - K_{\text{ref}}}{K_{\text{ref}}} \right|$$

K_{computed} is the stress-intensity factor computed by either method and K_{ref} is the reference value taken from the literature.

Figure 5 presents the relative error as a function of the number of degrees of freedom used in the modeling. For the same number of degrees of freedom, the method that includes the moments as unknowns (BFM) yields a solution which is more accurate than the method in which only forces are used as unknowns. The improved accuracy can be attributed to better satisfaction of the boundary conditions when the moment is included. Thus, the BFM is expected to yield a more accurate solution and also have a faster convergence rate than methods where only forces are used. In the above example, the present method yields a very accurate solution (within 1% of the reference solution) with as few as 24 degrees of freedom.

Next, the convergence of the BFM is studied by increasing the number of segments in the idealization. The two problems used, for which very accurate solutions are available, were a center-cracked specimen and a single-edge cracked specimen. As before, the relative error in the stress-intensity factor is used as the convergence parameter.

For the case of the center-crack tension specimen subjected to uniaxial tension as shown in Fig. 6, two extreme crack-length-to-width ratios, $2a/W = 0.2$ and 0.8, were considered. Because of symmetry, only one quarter of the plate was modeled. Again, equal size segments were used in the models. The relative error in the stress-intensity factors as a function of the number of degrees of freedom is presented in Fig. 6. For the small crack ($2a/W = 0.2$), the solution converged much faster than for the large crack length ($2a/W = 0.8$). This trend shows the effects of the external boundaries on the stress-intensity factor. In both cases, as few as 36 degrees of freedom are sufficient to yield a very accurate solution.

Figure 7 shows the convergence of the relative error for the single edge crack specimen subjected to uniaxial tension. Again for this configuration, two extreme crack-length-to-width ratios, a/W of 0.2 and 0.6, were considered. Because of symmetry, only the top half of the plate

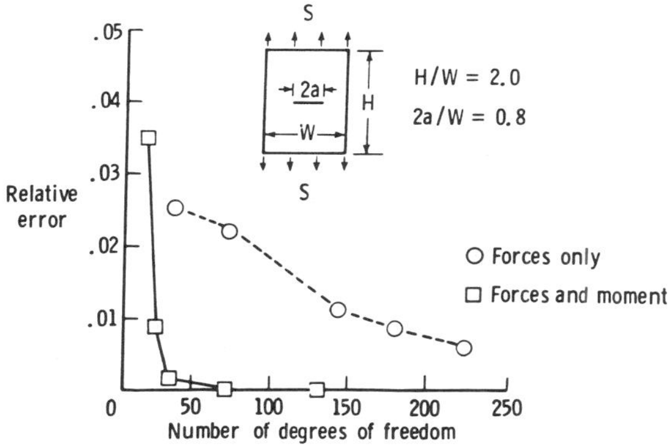

FIG. 5—*Relative error in stress-intensity factor using either "force and moment" method or "force" method.*

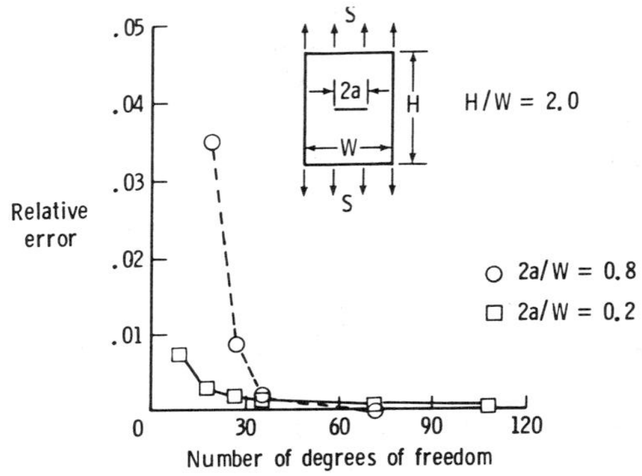

FIG. 6—*Convergence of relative error in stress-intensity factor for center-crack tension specimen.*

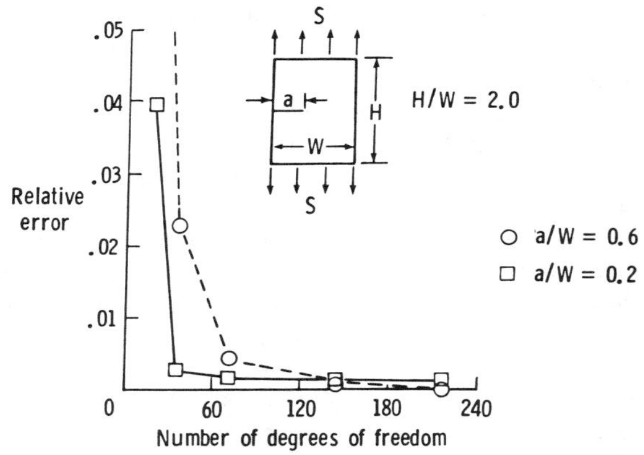

FIG. 7—*Convergence of relative error in stress-intensity factor for single edge-crack tension specimen.*

was modeled. Equal size segments were used on the boundaries. For the short crack ($a/W = 0.2$), the rate of convergence of the solution is somewhat faster than that for the long crack ($a/W = 0.6$). For both crack lengths, about 72 degrees of freedom are sufficient to yield a very accurate solution (within 1% of the reference solution of Ref *11*).

For the two crack configurations just discussed, equal size boundary segments were used. However, modeling with equal size segments is usually not optimal. Large segments could be used in those regions where the stress gradients are not severe while small segments could be used where the stress gradients are severe. Such a graded model could yield solutions which are more accurate with fewer segments than a model with equal size segments. To construct such graded models, a radial-line method for generating the mesh points on the boundaries was developed and is presented in Appendix II. Using a model generated by the radial-line method, a very accurate solution can be found for considerably fewer degrees of freedom [*10*].

Comparisons With Other Solutions

Several complex crack configurations for which accurate solutions are available in the literature were analyzed to establish the accuracy of the BFM. These configurations can be divided into two categories: opening mode (Mode I) problems and mixed-mode (Mode I and Mode II) problems. For both the Mode I and mixed-mode problems, the radial-line method (see Appendix II) was used to generate the mesh points with the origin of the radial-lines chosen to be at the crack tip. A value of 10 was used for the dividing factor *DF*.

Mode I Problems—Consider a large plate ($H/R = W/R = 64$) with a crack located between two circular holes as shown in Fig. 8. The plate is subjected to remote, uniaxial tension. Again, because of symmetry, only one quarter of the plate was modeled using about 40 segments (about 120 degrees of freedom).

The stress-intensity correction factors for a range of crack-length-to-hole-radius ratios are presented in Fig. 8. Results calculated by Newman [*12*] for an infinite plate using the collocation technique are shown for comparison. For all crack lengths considered, the agreement between the BFM results and the collocation results is excellent (within 1%). The small difference is probably due to the large but finite plate used in the present analysis, whereas an infinite plate was used in Ref *12*. In Fig. 8, the dashed line at unity represents the limiting solution as the d/R ratio approaches infinity. For larger crack lengths, $a/R > 2.5$, the influence of the holes is negligible. For small a/R values, the crack tips are shielded by the holes and thus the stress intensity factor is much lower than the solution for the plate without holes.

Mixed-Mode Problems—The common mixed-mode crack problem of a slant single-edge crack in a finite plate subjected to uniaxial tension (see Fig. 9) was analyzed to demonstrate the validity of the BFM for mixed-mode problems. Because of lack of symmetries, the complete plate is modeled with 50 segments or 150 degrees of freedom.

Three angles of inclination, $\beta = 45$, 67.5, and 90 deg were considered. For each angle β, several a/W ratios were considered. Figures 9 and 10 present the Mode I and Mode II stress-intensity correction factors, respectively, for each of the cases considered. These stress-intensity correction factors are compared with the results obtained by Wilson using collocation techniques [*11*]. The agreement between the present results and those obtained by Wilson are generally within 0.5% except for the Mode II stress-intensity corrections factors for $a/W = 0.6$. For this case, the present results are within 1.5% of Wilson's solution [*11*].

New Solutions

In this subsection, the BFM is applied to two configurations where few or no stress-intensity factor solutions are available, an edge crack emanating from a semicircular hole and a four-hole crack specimen. The radial-line method was used to generate the mesh points on the boundaries

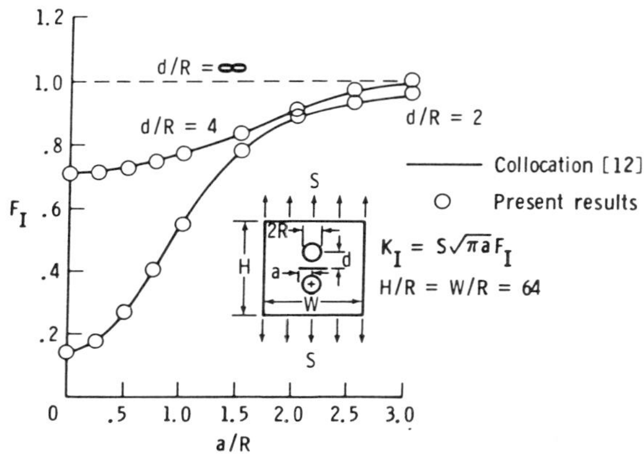

FIG. 8—*Comparison of stress-intensity correction factors for a crack located between two holes.*

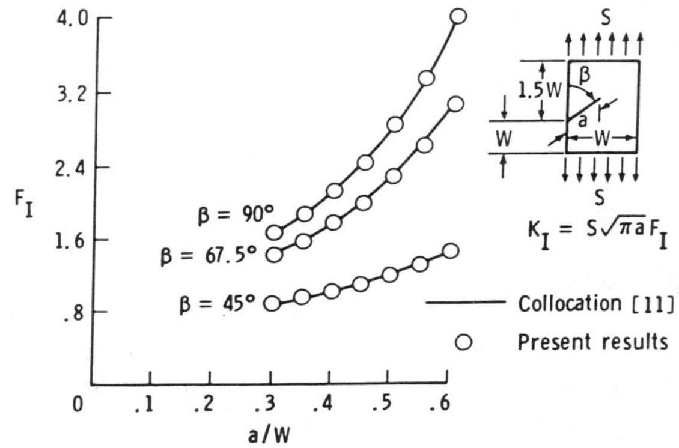

FIG. 9—*Mode I stress-intensity correction factors for a slant single-edge crack.*

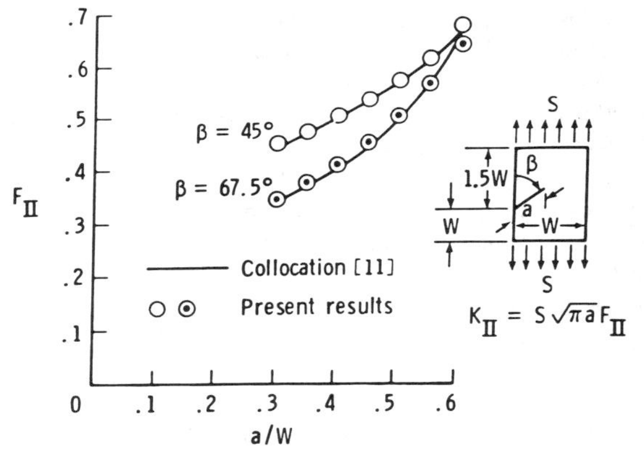

FIG. 10—*Mode II stress-intensity correction factors for a slant single-edge crack.*

and the crack tip was chosen as the origin of the radial lines. Again a value of 10 was used for the dividing factor DF.

Edge Crack Emanating from a Semicircular Hole—Figure 11 shows an edge crack emanating from a semicircular hole. Very few solutions for the range of crack lengths considered here are available in the literature. Thus, the solutions given below will add considerably to the available solutions for this crack configuration.

Due to symmetry, only the upper half of the plate was modeled. The stress-intensity correction factors obtained for three values of notch-depth-to-width ratios ($R/W = 0.25$, 0.125, and 0.0625) are presented in Table 2 and Fig. 11. In each case, the notch radius R was held constant and the value of W was increased to obtain the desired R/W ratio. For each R/W ratio, the crack length was varied from about 5 to 50% of the plate width. For short cracks, the notch boundary had a significant influence on the stress-intensity factor. However, as the crack tip approached the midsection of the plate, the stress-intensity factor approached the solution for an edge crack without the notch. Thus, the notch boundary had no influence on the stress-intensity factor when the crack is halfway through the plate, as shown in Table 2.

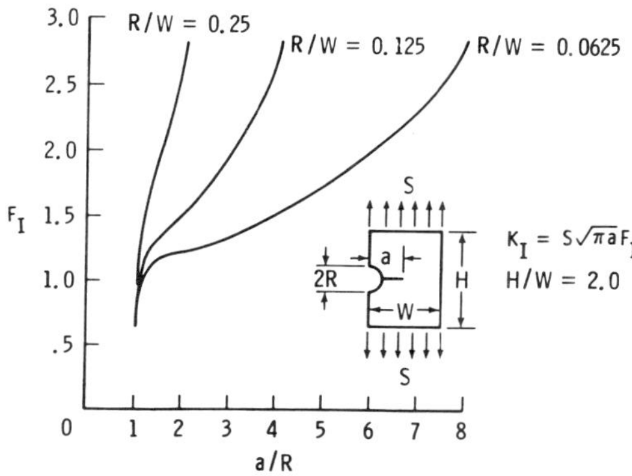

FIG. 11—*Stress-intensity correction factors for an edge crack emanating from a semicircular hole.*

TABLE 2—*Stress-intensity correction factors for an edge crack emanating from a semicircular hole.*

a/R	$F_I(R/W = 0.25)$	$F_I(R/W = 0.125)$	$F_I(R/W = 0.0625)$
1.04	0.8860	0.6937	0.6431
1.05	0.9709	0.7612	0.7018
1.06	1.0392	0.8125	0.7527
1.08	1.1531	0.8995	0.8322
1.10	1.2468	0.9665	0.8929
1.15	1.4068	1.0819	0.9951
1.20	1.5259	1.1546	1.0568
1.30	1.6946	1.2410	1.1234
1.50	1.9648	1.3319	1.1751
1.75	2.3356	1.4144	1.1993
2.00	2.8183[a]	1.5006	1.2273
3.00	...	1.9802	1.3390
4.00	...	2.8189[a]	1.4982
6.00	...	...	1.4982
8.00	...	...	2.8200[a]

[a]These values correspond to an a/W ratio of 0.5; F_I for single edge crack with $a/W = 0.5$ is 2.8153 [*11*].

Four-hole Crack Specimen—Stringers are widely used in aircraft structures as stiffening members to retard or arrest propagating cracks. The four-hole crack specimen shown in Fig. 12, which simulates the effect of a stringer on a propagating crack, was analyzed. Due to symmetry, only one quarter of the specimen was modeled.

The stress-intensity correction factors obtained for a range of crack length-to-width ratios ($0 \leq a/W \leq 0.9$) are presented in Table 3 and Fig. 12. The solution for a similar specimen with no holes (a center-crack tension specimen) is also shown in Fig. 12 for comparison. The results for the four-hole specimen show that the stress-intensity factor increases (higher than for the center-crack specimen) as the crack tip approaches the inner edge of the hole. However, as the crack tip approaches the centerline of the holes, the stress-intensity factor decreases (lower than for the center-crack specimen) until a minimum value is obtained at about $a/W = 0.55$. Thus,

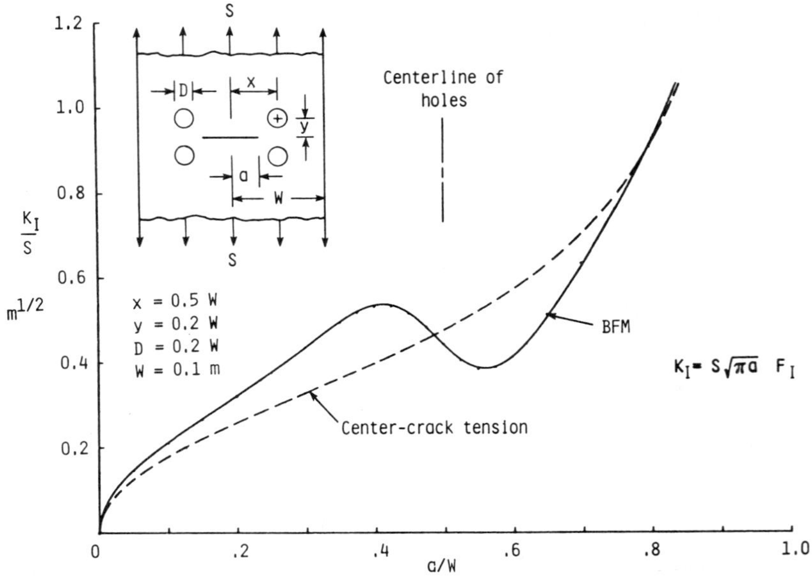

FIG. 12—*Stress-intensity correction factor for a four-hole crack specimen.*

TABLE 3—*Stress-intensity correction factors for four-hole crack specimen.*

a/W	F_I
0	1.169
0.05	1.174
0.10	1.191
0.15	1.222
0.20	1.268
0.25	1.332
0.30	1.409
0.35	1.480
0.375	1.498
0.40	1.489
0.425	1.446
0.45	1.361
0.475	1.242
0.50	1.108
0.525	0.989
0.55	0.917
0.575	0.904
0.60	0.946
0.65	1.127
0.70	1.342
0.75	1.558
0.80	1.795
0.85	2.105
0.90	2.599

a propagating crack will be retarded or arrested as it approaches the centerline of the holes. This drop in the stress-intensity factor is due to the shielding of the crack tip by the holes from the externally applied stress field.

Concluding Remarks

The Boundary Force Method was formulated for the two-dimensional analysis of complex crack configurations. In this method, the boundaries of the crack configuration were modeled by straight-line segments. At the center of each of the segments, a horizontal force, a vertical force, and a moment were applied. These sets of forces and moments are treated as the unknowns in the problem. These unknowns are determined such that the boundary conditions are satisfied approximately along the boundaries of the configuration. The BFM uses the fundamental elasticity solution of a pair of concentrated forces and a moment at an arbitrary point in an infinite plate with a crack. Therefore, the boundary conditions on the crack faces are satisfied exactly and, hence, the crack faces are not modeled as part of the boundary as in previous boundary element methods.

To verify the accuracy of the method, the BFM was used to analyze several crack configurations for which exact or accurate stress-intensity factor solutions are available in the literature. The crack configurations considered included Mode I and mixed-mode (Modes I and II) problems. The method yielded stress-intensity factors which were in excellent agreement with those in the literature for both the Mode I and mixed-mode problems. Results showed that for the same degree of accuracy, significantly fewer degrees of freedom were required in the BFM with unknown forces and moments compared to unknown forces only. In general, about 150 degrees of freedom were required to obtain very accurate solutions to complex crack configurations with the BFM with unknown forces and moments.

The versatility of the BFM was demonstrated by analyzing two complex crack configurations for which limited or no solutions are available, an edge crack emanating from a semicircular hole and the four-hole crack specimen. For each configuration, several crack length-to-width ratios were analyzed and stress-intensity factors presented. For each configuration, the stress-intensity factors for several crack lengths were obtained with minimal modeling effort since only the boundaries were modeled.

The BFM, therefore, appears to be a versatile tool to obtain very accurate stress-intensity factors for complex crack configurations. The method requires a minimal amount of modeling; thus, stress-intensity factor analyses for several crack lengths can be performed with ease.

APPENDIX I

Fundamental Solution

The BFM formulation uses the elasticity solution for a pair of concentrated forces and a moment in an infinite plate with a crack. Such a solution was formulated by Erdogan [7] for linear, isotropic, and homogeneous materials. The solution is presented below.

Stress Functions

Consider an infinite plate with a crack subjected to concentrated forces Q, P and a moment M at an arbitrary point $z_0 = x_0 + iy_0$ as shown in Fig. 13. The complex variable stress functions [7] are

$$\phi^*(z) = -\frac{T}{z - z_0} + \phi_0(z)$$

$$\Omega^*(z) = \frac{\kappa T}{z - \bar{z}_0} + \frac{\bar{T}(\bar{z}_0 - z_0) + im}{(z - \bar{z}_0)^2} + \phi_0(z)$$

$$\phi_0(z) = \frac{1}{2\pi\sqrt{z^2 - a^2}}\left\{\frac{T}{z - z_0}[I(z) - I(z_0)] - \frac{\kappa T}{z - \bar{z}_0}[I(z) - I(\bar{z}_0)] - [\bar{T}(\bar{z}_0 - z_0) + im]\left[\frac{I(z) - I(\bar{z}_0)}{(z - \bar{z}_0)^2} - \frac{J(\bar{z}_0)}{z - \bar{z}_0}\right]\right\} \quad (4)$$

where

$$m = \frac{M}{2\pi}$$

$$I(z) = \pi\left[\frac{1}{\sqrt{z^2 - a^2}} - z\right]$$

$$I(z_0) = \pi\left[\frac{1}{\sqrt{z_0^2 - a^2}} - z_0\right]$$

$$J(\bar{z}_0) = \pi\left[\frac{\bar{z}_0}{\sqrt{\bar{z}_0^2 - a^2}} - 1\right]$$

$$T = \frac{Q + iP}{2\pi(1 + \kappa)}$$

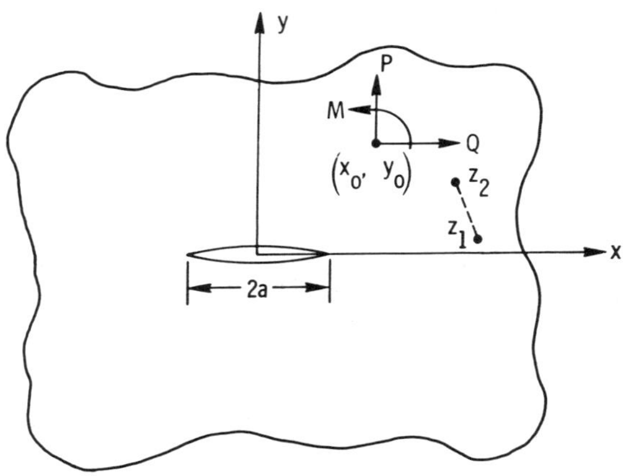

FIG. 13—*Concentrated loads and moment in an infinite plate with a crack.*

Stresses

The stresses at any point $z = x + iy$ are obtained from the stress functions as

$$\begin{aligned} \sigma_x + \sigma_y &= 2[\phi^*(z) + \overline{\phi^*(z)}] \\ \sigma_y - \sigma_x + i2\sigma_{xy} &= 2[(\bar{z} - z)\phi^{*\prime}(z) - \phi^*(z) + \bar{\Omega}^*(z)] \end{aligned} \tag{5}$$

where the barred quantities are the complex conjugates and the primed quantities represent the derivatives with respect to z. From Eqs 4 and 5, the stresses are singular at the concentrated load, that is, the stresses become infinite as z approaches z_0. In addition, the stresses also become infinite as z approaches $\pm a$, the stresses have the required square root singularity at the crack tips, and the stresses vanish as $r \rightarrow \infty$.

Displacements, Forces, and Moments

The displacements u and v at any point $z = x + iy$ are obtained from the stress functions as

$$2\mu(u + iv) = \kappa \int_0^z \phi^*(z)dz - \int_0^{\bar{z}} \Omega^*(\bar{z})d\bar{z} - (z - \bar{z})\overline{\phi^*(z)} \tag{6}$$

where μ is the shear modulus. The resultant forces F_x, F_y and the resultant moment M_0 across the arc z_1 to z_2 (see Fig. 13) due to the concentrated forces (Q, P) and moment M can be obtained either by integrating the stresses in Eq 5 from z_1 to z_2, or by using the stress functions in the following equations:

$$F_x + iF_y = -i\left[\int \phi^*(z)dz + \int \Omega^*(\bar{z})d\bar{z} + (z - \bar{z})\overline{\phi^*(z)}\right]\Bigg|_{z_1}^{z_2} \tag{7}$$

$$M_0 = \text{Re}\left[\iint \{\phi(z) + \bar{\Omega}(z)\}dzdz - z\int \{\phi(z) + \bar{\Omega}(z)\}dz + z(z - \bar{z})\phi(z)\right]\Bigg|_{z_1}^{z_2} \tag{8}$$

Stress-Intensity Factor

The stress intensity factors for concentrated forces Q and P and moment M applied at an arbitrary point, $z_0 = x_0 + iy_0$, in an infinite plate with a crack (see Fig. 13) are

$$\begin{aligned} K = K_I + iK_{II} &= 2\sqrt{2\pi}\lim_{z\rightarrow a}[\sqrt{(z - a)}\phi^*(z)] \\ &= \frac{1}{2\sqrt{\pi a}}\frac{1}{(1 + \kappa)}\left\{(Q + iP)\left[\left(\frac{a + z_0}{\sqrt{z^2 - a^2}} - 1\right) - \kappa\left(\frac{a + \bar{z}_0}{\bar{z}^2 - a^2} - 1\right)\right]\right. \\ &\quad \left. + \frac{a[(Q - iP)(\bar{z}_0 - z_0) + i(1 + \kappa)M]}{(\bar{z}_0 - a)\sqrt{\bar{z}_0^2 - a^2}}\right\} \end{aligned} \tag{9}$$

APPENDIX II

Radial Line Method

This appendix describes a systematic procedure, called the radial-line method, for modeling the boundaries of a crack configuration. In the BFM, the accuracy of the stress-intensity factor obtained depends on how well the boundary conditions are approximated. Because the boundaries near the crack tip are subjected to higher stress gradients than the boundaries far from the crack tip, smaller segments are needed to accurately model the boundary conditions near the crack tip. Therefore, a radial-line method that will generate smaller line segments on the boundaries near the crack tip and larger line segments on the boundaries away from the crack tip was developed. With the radial-line method of modeling, a significant reduction in the number of degrees of freedom is realized without sacrificing accuracy. This procedure can be best described using the following example.

Consider a center-crack tension specimen with a crack length $2a$ as shown in Fig. 14. Because of symmetry, only one quarter of the plate needs to be modeled. The boundary of this quadrant can be divided into two sections: Boundary 1 (the vertical line A to B) and Boundary 2 (the horizontal line B to C). Because the stress-intensity factor is the quantity of interest here, let the crack tip at $x = a$ be the point from which all radial lines originate. To determine the segment sizes on Boundary 1, first the distance between the origin of the radial lines (the crack tip) and the starting point z_1 (Point A) is computed. Label this distance r_1. The size of the first segment (z_1 to z_2) is chosen to be a fraction of the distance r_1. Thus, the distance from z_1 to z_2 is set equal to r_1/DF where DF is referred to as the dividing factor. Because the factor DF is assumed to be known, the distance to point z_2 can be determined. Next, r_2 the distance from the origin of the radial lines (the crack tip) to z_2 is computed. The size of the second segment, z_2 to z_3, is a fraction of the distance r_2, so that the size of the second segment is equal to r_2/DF. The dividing factor DF is assumed to be identical throughout the entire modeling. This procedure is repeated until the entire Boundary 1 is discretized. The procedure has to be modified near the end points, such as Point B in Fig. 14. If the end point of the last segment exceeds the end point of the boundary (Point B), the end point of the boundary is assigned as the end point of the last segment on that boundary. The same radial-line procedure is repeated for Boundary 2 where the starting point of Boundary 2 is the end point of Boundary 1.

In the radial-line method, the size of the segments or the distribution density of the segments is determined by the choice of the origin of the radial line and the value of the dividing factor DF. Both can be chosen arbitrarily. In all the problems investigated in this paper, the origin of the radial-line was chosen to be the crack tip, since the stress-intensity factor is the quantity of interest. A DF value of 10 was used in all the problems.

The most significant advantage of the radial-line method can be demonstrated in the case of a very small crack emanating from a semicircular notch. Because of the high stress gradient near the crack tip, small segments are needed on portions of the semicircular boundary nearest the crack tip (see Fig. 12). If equal size segments are used, a large number of segments are needed on this curved boundary. However, the portion of the curved boundary that is away from the crack tip does not have high stress gradients, and thus small-size segments on that portion of the boundary are unnecessary. The radial-line method generates small segments near the crack

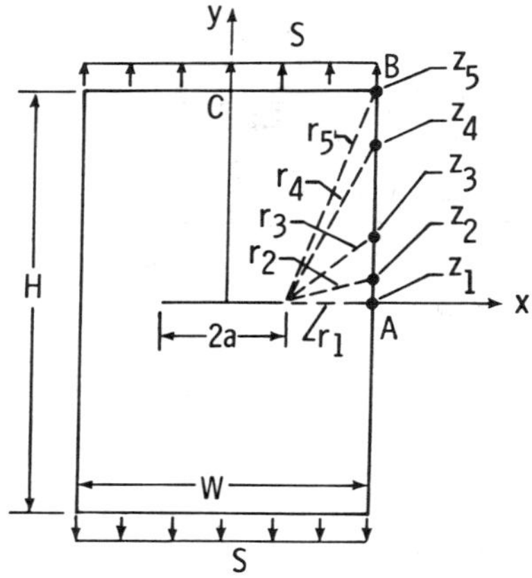

FIG. 14—*Radial-line method for generating mesh points.*

tip and large segments away from the crack tip. This type of modeling will significantly reduce the number of degrees of freedom necessary to model the curve boundary without sacrificing accuracy.

References

[1] Nisitani, H., "Two-Dimensional Problems Solved Using a Digital Computer," *Journal of Japan Society of Mechanical Engineers*, Vol. 70, No. 580, 1967, pp. 627–635.

[2] Nisitani, H. and Murakami, Y., "Stress-Intensity Factors of an Elliptical Crack or a Semi-Elliptical Crack Subject to Tension," *International Journal of Fracture*, Vol. 10, No. 3, Sept. 1974, pp. 353–368.

[3] Nisitani, H., "Interference Effects Among Cracks or Notches in Two-Dimensional Problems" in *Proceedings*, International Conference on Fracture Mechanics and Technology, Vol. 2, 1977, pp. 1127–1142.

[4] Murakami, Y. and Nisitani, H., "A Method for Calculating Stress-Intensity Factors of an Arbitrarily Shaped Finite Plate with a Crack" in *Proceedings*, International Conference on Fracture Mechanics and Technology, Vol. 2, 1977, pp. 1171–1185.

[5] Isida, M., "A New Procedure of the Body Force Method with Applications to Fracture Mechanics," *Numerical Methods in Fracture Mechanics; Proceedings*, 1st International Conference, Swansea, Wales, U.K., Jan. 1978, pp. 81–94.

[6] Erdogan, F. and Arin, K., "A Half Plane and a Strip with an Arbitrarily Located Crack," *International Journal of Fracture*, Vol. 11, No. 2, 1975, pp. 191–204.

[7] Erdogan, F., "On the Stress Distribution in a Plate with Collinear Cuts Under Arbitrary Loads" in *Proceedings*, 4th U.S. National Congress of Applied Mechanics, 1962, pp. 547–553.

[8] Timoshenko, S. and Goodier, J. N., *Theory of Elasticity*, 3rd ed., McGraw-Hill, New York, 1970, p. 127.

[9] Muskhelishvili, N. I., *Some Basic Problems of the Mathematical Theory of Elasticity*, 2nd ed., Noordhoff International Publishing, Amsterdam, The Netherlands, 1977.

[10] Tan, P. W., "The Boundary Force Method for Stress Analysis of Arbitrarily Shaped Plates with Notches and Cracks," D.Sc. dissertation, George Washington University, Washington, DC, 1985.

[11] Rooke, D. P. and Cartwright, D. J., *Compendium of Stress Intensity Factors*, Hillingdon Press, Uxbridge, Middlesex, U.K., 1976.

[12] Newman, J. C., Jr., *An Improved Method of Collocation for the Stress Analysis of Cracked Plates with Various Shaped Boundaries*, NASA TN D-6376, National Aeronautics and Space Administration, Washington DC, 1971.

Hans L. Groth[1]

Prediction of Failure Loads of Adhesive Joints Using the Singular Intensity Factor

REFERENCE: Groth, H. L., **"Prediction of Failure Loads of Adhesive Joints Using the Singular Intensity Factor,"** *Fracture Mechanics: Eighteenth Symposium, ASTM STP 945,* D. T. Read and R. P. Reed, Eds., American Society for Testing and Materials, Philadelphia, 1988, pp. 278–284.

ABSTRACT: A method to predict failure of adhesive joints is described. The fracture criterion is based on material-induced singularities that appear at the terminus of the joint. The criterion can be interpreted physically in terms of the elastic strain energy density. The criterion is formulated in terms of a singular intensity factor Q that is analogous to the stress intensity factor for the case when cracks are present. The strengths of the singularities are dependent on the elastic properties of the adhesive and the adherends and on the geometry of the bond terminus, but are postulated to be independent of loading and global geometry.

For the case of an adhesive joint, the different materials inducing the singularities may be the adherend and the adhesive. The criterion assumes that no cracks are present in the vicinity of the material-induced singularity and that fracture will occur as a cohesive fracture in the adhesive layer. It has been shown earlier that the criterion predicts with good accuracy the fracture load of test specimens loaded in combined tension and shear. The fracture load of the single-lap joint is predicted and compared with experimentally obtained values.

The presented method of predicting fracture in adhesive joints may be applicable to general joint geometries.

KEY WORDS: adhesive joints, finite element analysis, fracture criterion, material-induced singularities

Nomenclature

- ϵ Adhesive layer thickness
- E Young's modulus of elasticity
- ξ y/ϵ
- h Length of one finite-element (FE) side
- L Overlap length
- λ Singularity parameter
- P_{crit} Critical load of the joint
- P Load applied in the FE model
- ν Poisson's ratio
- Q Singular intensity factor
- Q_{crit} Critical value of Q
- σ Tensile stress
- t Adherend thickness
- τ Shear stress

[1] Research associate, Department of Aeronautical Structures and Materials, Royal Institute of Technology, S-100 44 Stockholm, Sweden.

$\sqrt{t}/L$ Joint factor
u Displacement

In Ref *1* and in the present paper a method of predicting failure of an adhesive joint is described. The criterion is based on singular stresses that appear at the interface wedge between two different materials. For the case of an adhesive joint, the different materials may be the adherend and the adhesive. The criterion assumes that no cracks are present in the vicinity of the material-induced singularity. The criterion is formulated in terms of a singular intensity factor Q that is analogous to the stress-intensity factor for the case when cracks are present.

In Ref *1* it was shown that the criterion predicted with good accuracy the fracture load of test specimens loaded in combined tension and shear. In a subsequent paper [*2*], the fracture load of a simple, commonly used joint, the single-lap joint, was predicted. The results calculated with the method presented in Ref *2* showed good agreement with expected trends given in Ref *3*.

In the present paper, a comparison between this failure criterion and a test series of overlap joints is made. It is shown that the singular stress criterion accurately predicts the failure load of the single-lap joint.

Theory

Consider the terminus of an adhesive joint in Fig. 1, with adhesive layer thickness ϵ and elastic constants E and ν, where E and ν are the Young's modulus of elasticity and the Poisson's ratio, respectively.

The joint transfers load from an adherend with elastic constants E_1, ν_1 to another adherend with the elastic constants E_2, ν_2. The joint boundary AB is assumed to be parallel to the y-axis along $x = 0$. It can now be shown, assuming plane conditions, that if Point A is approached from the $+y$-axis, the normal stress in the y-direction, σ_y, will behave according to Ref *4* for small values of $\xi + 1/2$, being the distance from Point A, as

$$\sigma_y = Q(1/2 + \xi)^{-\lambda}, \qquad (\lambda > 0) \tag{1}$$

where Q and λ are constants and $\xi = y/\epsilon$. Q is dependent on the magnitude of the loading and global geometry and λ (for a given form of the joint terminus) on the elastic constants E, ν, and E_1, ν_1. $\xi = 0$ in the center of the adhesive layer in Fig. 1.

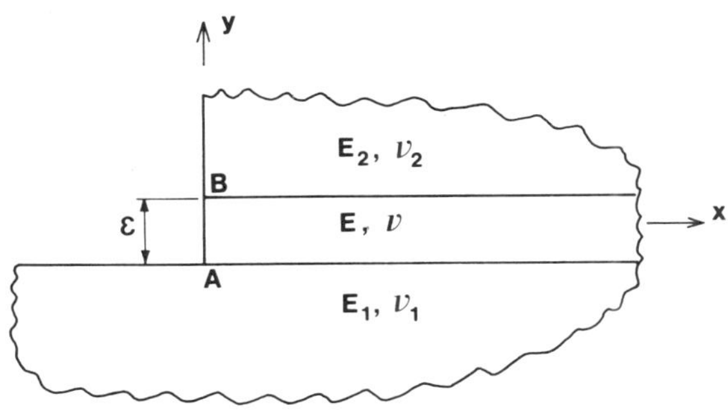

FIG. 1—*Terminus of an adhesive joint with adhesive layer thickness and adhesive material properties* E *and* ν. *Adherend material properties are* E_i, ν_i, *and* E_2, ν_2. *A and B are points where singular stresses occur. Note that* y = *0 in the center of the adhesive layer.*

To obtain σ_y, when Point B is approached along the y-axis, Q and λ in Ref *1* are replaced by Q' and λ', respectively, and $1/2 + \xi$ by $1/2 - \xi$, being the distance from Point B, giving

$$\sigma_y = Q'(1/2 - \xi)^{-\lambda'}, \qquad (\lambda' > 0) \tag{2}$$

where Q' and λ' are constants similar to Q and λ, respectively. λ' depends on E, ν, and E_1, ν_1. Since the boundary A-B is free from tractions, one can express the vertical displacement in the vicinity of A and B (u_y and u_y') in terms of Q, λ and Q', λ' simply by integrating the normal strain. In this way one obtains

$$u_y = \frac{Q\epsilon(1/2 + \xi)^{1-\lambda}}{E^*(1 - \lambda)} + u_0 \tag{3}$$

$$u_y' = \frac{-Q'\epsilon(1/2 - \xi)^{1-\lambda'}}{E^*(1 - \lambda')} + u_0' \tag{4}$$

In Eqs 3 and 4, recall that ξ is y nondimensionalized with respect to ϵ, and that E^* is equal to $E/(1 - \nu^2)$ in the case of plane strain and equal to E in the case of plane stress. Finally u_0 and u_0' are the displacements of Points A and B, respectively.

It is now assumed that cohesive fracture is governed by Q and Q' in some way. More specifically, it is assumed that fracture will occur when the following relation is satisfied

$$Q = Q_{\text{crit}} \quad \text{or} \quad Q' = Q'_{\text{crit}} \tag{5}$$

whichever happens first.

Often the adherends are made of the same material, much stiffer and considered as rigid compared to the adhesive, and if the local geometry is similar, then $\lambda = \lambda'$. In this case the fracture criterion may be written as

$$\max(Q, Q') = Q_{\text{crit}} \tag{6}$$

The parameter Q_{crit} is a material parameter analogous to the K_{Ic} (fracture toughness) in fracture mechanics. If the adherends can be considered as rigid, compared to the adhesive, the values of λ and λ' will be equal for a similar local geometry. Their values are then determined by the Poisson's ratio for the adhesive only if the wedge angles are equal.

Reference *5* gives a relation for λ as a function of the Poisson's ratio of the adhesive, if the adherends are assumed rigid. The values of λ and λ' are found to be equal to 0.32 when the Poisson's ratio is equal to 0.35; this is the case for the adhesive used in the testing in the present paper and for a local wedge angle of 90 deg at the singular points.

A state of plane-strain conditions is assumed in the adhesive layer. The assumption of plane strain may be justified because the adherends are considered as rigid compared to the adhesive and because the joint thickness is usually small compared to the joint width.

Numerical Model and Numerical Treatment

The equations for Q and Q', that is, Eqs 3 and 4, may be discretized to be suitable for numerical treatment. Consider the finite-element mesh in Fig. 2 at the terminus of an adhesive layer. The terminus has N nodes, numbered from 1 to N, at $x = 0$. If we rearrange Eqs 3 and 4 we get

$$Q = \frac{E(1 - \lambda)}{\epsilon(\xi + 1/2)^{1-\lambda}(1 - \nu^2)} (u_y - u_0) \tag{7}$$

$$Q' = \frac{E(1 - \lambda')}{\epsilon(\xi - 1/2)^{1-\lambda'}(1 - \nu^2)} (u_y' - u_0') \tag{8}$$

where $u_0 = u_N$ is the displacement at Node N and u_0' is the displacement at Node 1. If we then make the discretization, we obtain (u_0 and u_0' are denoted u_N and u_1, respectively)

$$Q_i = \frac{E(1 - \lambda)}{\epsilon(\xi_i + 1/2)^{1-\lambda}(1 - \nu^2)} (u_{yi} - u_N) \tag{9}$$

where i goes from Node $N - 1$ to Node 1, and

$$Q_i' = \frac{E(1 - \lambda')}{\epsilon(\xi_i - 1/2)^{1-\lambda'}(1 - \nu^2)} (u_{yi}' - u_1) \tag{10}$$

where i goes from Node 2 to Node N.

It is now assumed that the expression for Q_i converges as $\xi \rightarrow -1/2$ and the expression for Q_i' converges as $\xi \rightarrow 1/2$. The finite-element model used is shown in Fig. 3. The complete joints were modeled, because of no symmetry, using special preprocessors that created the input files. By this procedure it was simple to change the adherend thickness.

The FE-program BASIS was used [*6*]. The elements were triangular constant-strain elements and rectangular bilinear elements. The number of elements at the terminus was taken to be 32 and 64, respectively, and convergence was studied. Some preliminary investigations showed that it was necessary to make a very fine mesh, as shown in Fig. 3, at the terminus of the joint, in order to evaluate proper values of Q and Q'.

Assuming h^2 convergence [*7*], one obtains

$$\frac{u' - u}{u^2 - u} = \frac{O(h^2)}{O(h/2)^2} = 4 \tag{11}$$

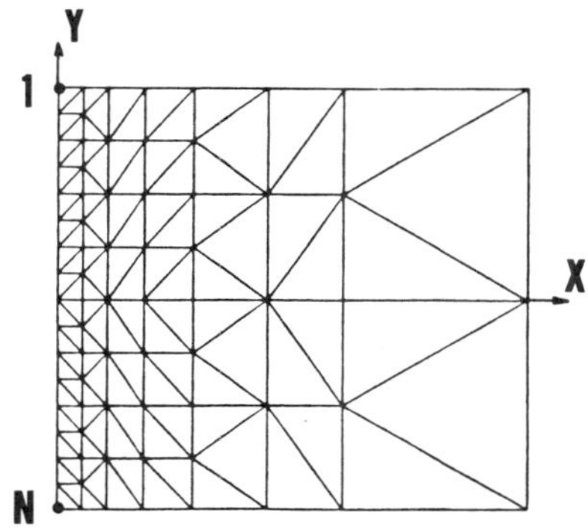

FIG. 2—*Finite-element mesh at the terminus of the adhesive layer.*

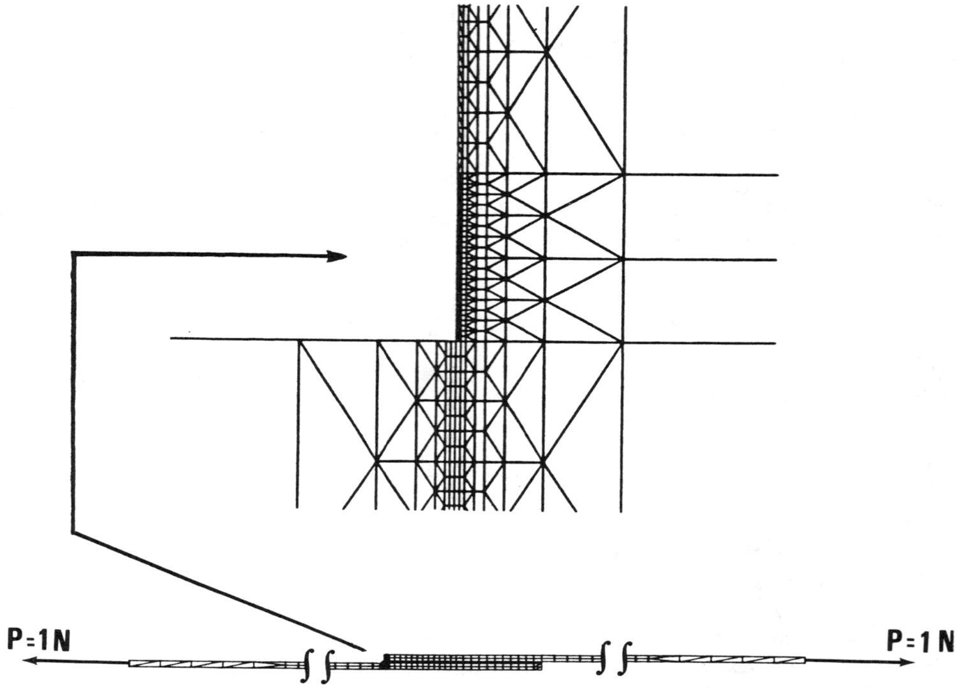

FIG. 3—*Finite-element model of the single-lap joint with applied loads. The model has 939 nodes and 998 elements.*

which gives

$$u = \frac{4u^2 - u^1}{3} \tag{12}$$

where

u = "exact" solution,
u^2 = solution from the model with 64 elements at the adhesive terminus, and
u' = solution with 32 elements at the adhesive terminus.

The length of an element side is equal to h. The single-lap-joint FE model, shown in Fig. 3, was used in present paper to predict the fracture load for some single-lap-joint test specimens.

The critical load of a joint can be calculated by elastic scaling from the relation

$$P_{\text{crit}} = \frac{Q_{\text{crit}}}{\text{Max}(Q, Q')_{\text{fem}}} P_{\text{fem}} \tag{13}$$

where

Q_{crit} = critical value of the singular intensity factor for the actual adhesive/adherend configuration,
P_{fem} = load applied in the FE model, and
$\text{Max}(Q, Q')_{\text{fem}}$ = singular intensity factor calculated by the FE model.

Testing Single-Lap Joints

A small test series of single-lap joints was carried out. In the present study both steel and aluminum adherends were used. The adhesive in all test specimens was a one-component, hot-curing epoxy, EC-2214, considered a relatively brittle adhesive with $E = 3.3$ GPa and $\nu = 0.35$.

The test results can be found in Table 1. Five specimens of each adherend thickness were tested. If the mean fracture load/mm joint width, for $t = 4.0$ mm, is chosen and the corresponding value of the $\max(Q, Q')$ from the finite-element analyses is obtained equal to 0.146, the critical value of the singular intensity factor is found to be 76.8 MPa. If we then try to predict the other fracture loads for the adherend thicknesses 2.0 and 8.0 mm using this calculated value, the corresponding critical loads are 380.2 and 698.2 N, respectively. These critical loads should be compared to 367.7 and 699.3 N, respectively, from Table 1. The prediction shows good agreement with the experiments.

Another failure criterion found in the literature may be used for the single-lap-joint geometry in the prediction of the failure load of this particular type of geometry, namely, the joint factor. The joint factor [*3*] is defined as $\sqrt{t}/L$, where t is the adherend thickness and L is the overlap length. The mean shear stress in the single-lap joint is proportional to the joint factor. The joint factor appears in the analytical solutions for the shear stress in the joint [*8,9*].

The joint factor results are also summarized in Table 1. It can be seen that the presented Q-criterion and the joint factor both give good agreement for this particular geometry. Note that the change from steel to aluminum adherends gives a very good prediction for the singular intensity factor and a poorer one for the joint factor. This change of adherend materials is not taken into account in the joint factor prediction.

The approach with the joint factor is limited to the single-lap-joint geometry, but this is not the case for the singular intensity factor. For other types of joints, as for instance scarf joints and stepped-lap joints, the singular intensity factor may be valuable to predict the structure or joint failure loads.

Conclusions

It is shown in both Ref *1* and the present paper that the method of predicting the failure loads of adhesively bonded joints gives good agreement with experiments made. It is believed that the method can be applied to general adhesive joint geometries and to other bimaterial applications.

Further investigations have to be made to obtain data for different values of the strength of the singularity, both for different material combinations and local geometry.

TABLE 1—*Fracture test load results for single-lap joints compared with analytical and numerical calculations: mean fracture load per mm specimen width (N).*[a]

Adherend Thickness, mm	Experiments	Joint Factor	Singular Intensity Factor
2.0	367.7	371.8	380.2
4.0	525.8	525.8	525.8
8.0	699.3	743.6	698.2

[a]Overlap length = 25 mm.
Adherend thickness 2.0 and 4.0 mm (steel adherends).
Adherend thickness 8.0 mm (Aluminum adherends).

Acknowledgment

The author would like to thank Dr. Per Gradin at Det norske Veritas for valuable discussions during the preparation of this paper. Dr. Gradin is currently at the Swedish Plastics and Rubber Institute, PGI, in Sundsvall, Sweden.

References

[*1*] Gradin, P. A. and Groth, H. L., "A Fracture Criterion for Adhesive Joints in Terms of Material-Induced Singularities" in *Proceedings,* 3rd International Conference on Numerical Methods in Fracture Mechanics, Pineridge Press, Swansea, Wales, U.K., 1984, pp. 711-720.

[*2*] Groth, H. L., "A Method to Predict Fracture in an Adhesive Bonded Joint" in *Proceedings,* International Adhesion Conference, University of Nottingham, U.K., 12-14 Sept. 1984, pp. 20.1-20.6.

[*3*] deBruyne, N. A., "The Strength of Glued Joints," *Aircraft Engineering,* Vol. 16, 1944, pp. 115-118.

[*4*] Bogy, D. B., "Edge-Bonded Dissimilar Orthogonal Elastic Wedges Under Normal and Shear Loading," *Journal of Applied Mechanics,* Sept. 1968, pp. 460-466.

[*5*] Gradin, P. A., "A Fracture Criterion for Edge-Bonded Bimaterial Bodies," *Journal of Composite Materials,* Vol. 16, Nov. 1982, pp. 448-456.

[*6*] Kjellberg, S., Falk, U., Merazzi, S., Stehlin, P., Ivansson, S., Andersson, B., and Palmberg, B., *BASIS User Manual,* The Aeronautical Research Institute of Sweden, Structures Department, Stockholm, Sept. 1983.

[*7*] Zienkiewicz, O. C., *The Finite Element Method,* 3rd ed., McGraw-Hill, New York, 1977, pp. 34-36.

[*8*] Volkersen, O., "Die Nietkraftverteilung in zugbeanspruchten Nietverbindungen mit konstanten Laschenquerschnitten," *Luftfahrtforschung,* Vol. 15, 1938, pp. 41-47 (in German).

[*9*] Adams, R. D. and Wake, W. C., *Structural Adhesive Joints in Engineering,* Elsevier Applied Science Publishers, London, 1984, pp. 19-23.

Joseph A. Kapp[1]

Stress-Intensity Factors and Displacements for Arc Bend Specimens Using Collocation

REFERENCE: Kapp, J. A., "**Stress-Intensity Factors and Displacements for Arc Bend Specimens Using Collocation,**" *Fracture Mechanics: Eighteenth Symposium, ASTM STP 945*, D. T. Read and R. P. Reed, Eds., American Society for Testing and Materials, Philadelphia, 1988, pp. 285–300.

ABSTRACT: Stress-intensity factors, crack mouth opening displacements, and load-line displacements were determined for the arc bend-chord support specimen using boundary collocation. Several inner radius (r_1) to outer radius (r_2) ratios were studied at each of two span (S) to width (W) ratios. The actual solution matrix is for $S/W = 3.0$, $r_1/r_2 = 0.8$, 0.6, 0.4, and for $S/W = 4.0$, $r_1/r_2 = 0.8$, 0.7, 0.6. Also, the three-point bend single-edge specimen [SE(B)] at both $S/W = 3.0$ and $S/W = 4.0$ was studied to examine the limiting behavior of the generated solutions. For all of the $S/W = 3.0$ cases, solutions were obtained for crack length-to-width ratios (a/W) from 0.2 to 0.6, and in the $S/W = 4.0$ cases, a/W was varied from 0.2 to 0.5. The solutions were obtained by the linear superposition of a pure bending stress, a pure shear stress, and a uniform normal stress on an annular segment. The magnitudes of the individual components depended upon the size of the annular segment and the S/W ratio. Wide-range expressions were fit to the numerical solutions to make them applicable over a wider range of testing conditions and for inclusion in future revisions of the ASTM Standard for Plane-Strain Fracture Toughness Testing (E 399-83).

KEY WORDS: fracture mechanics, fracture toughness testing, boundary collocation, arc-shaped fracture testing specimens

Underwood et al. [1] have suggested that the arc bend-chord support specimen [AB(C)] shown in Fig. 1 would be a good addition to the currently available fracture toughness specimens. In order to begin the standardization process, an accurate K solution is the minimum requirement. Additionally, crack mouth opening displacements and load-line displacement solutions would be useful in future utilization of this specimen for fatigue crack propagation and J_{Ic} testing, for example. The previous study [1] included these solutions generated by finite elements. This paper presents the results of the analysis of the AB(C) specimen using boundary-value collocation.

A wide variety of solutions were generated in an attempt to determine the effects of the three primary variables of the specimen: the crack length effect (a/W), the radius ratio effect (r_1/r_2), and the effect of span(S/W). It was decided to limit consideration of one of these variables (S/W) to just two values, $S/W = 3.0$ and $S/W = 4.0$. These were chosen because the present standard bending specimen has $S/W = 4.0$. However, allowing only $S/W = 4.0$ limits the AB(C) specimen to relatively thin-walled cylinders ($r_1/r_2 > 0.6$). A shorter relative span($S/W = 3.0$) was studied in order to include thicker-walled cylinders. Cylinders with ra-

[1]Materials research engineer, U.S. Army Armament, Munitions, and Chemical Command, Armament Research and Development Center, Large Caliber Weapon Systems Laboratory, Benet Weapons Laboratory, Watervliet, NY 12189-5000.

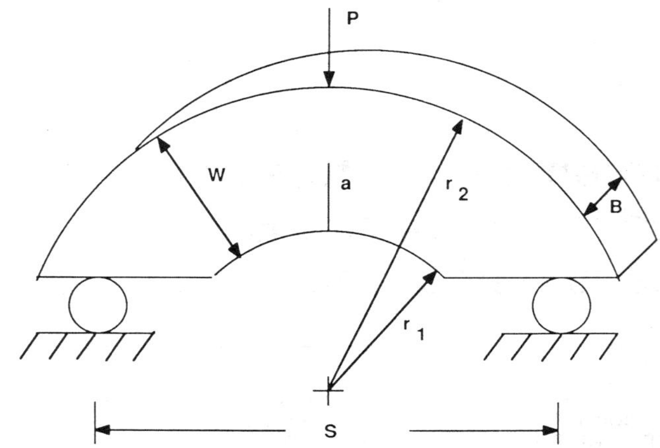

FIG. 1—*Arc bend-chord support specimen.*

dius ratios (r_1/r_2) greater than about 0.4 can be tested in this configuration. Three different radius ratios were studied for each of the two span ratio conditions, and several crack lengths were studied for each specimen geometry. Finally, wide-range expressions were fit to the numerical data to interpolate the solution between the specific conditions studied.

Method of Solution

The problem was solved using plane elasto-static boundary collocation of a homogeneous isotropic body, a method which is well documented in other sources [*2–4*]. In short, the general solution to a class of problems had to be found. Normally this solution is an infinite series solution that exactly satisfies some boundary condition of the particular problem considered. For crack problems, the general solution is the William's stress function [*5*], which satisfies the traction-free condition of the crack surfaces. For the solution of a particular problem, the numerical value of the stress function and its derivative along the boundary of the body can be determined using known solutions for uncracked bodies. The coefficients in the general solution that will make the truncated general solution satisfy the boundary conditions can then be obtained. This solution is normally obtained in the least-squares sense by knowing the boundary conditions at an overdetermined number of boundary locations. The results reported here were generated using the least-squares scheme outlined by Hussain et al. [*4*].

The particular problem shown in Fig. 1 was solved by analyzing the annular ring segment shown in Fig. 2. The value of θ_0 was chosen large enough to include the entire AB(C) specimen. This was needed to determine the load-line displacements. The magnitude of the various components were determined in order that they resolve to the reaction force at either roller support. These components are

$$\tau = \frac{P/BW}{2(\cos\theta_0 + \tan\theta_0 \sin\theta_0)} \tag{1}$$

$$\sigma = \frac{(P/BW)\tan\theta_0}{2(\cos\theta_0 + \tan\theta_0 \sin\theta_0)} \tag{2}$$

$$M = \frac{PW}{4}\left(\frac{S}{W} - \frac{(r_1 + r_2)}{(r_2 - r_1)}\sin\theta_0\right) \tag{3}$$

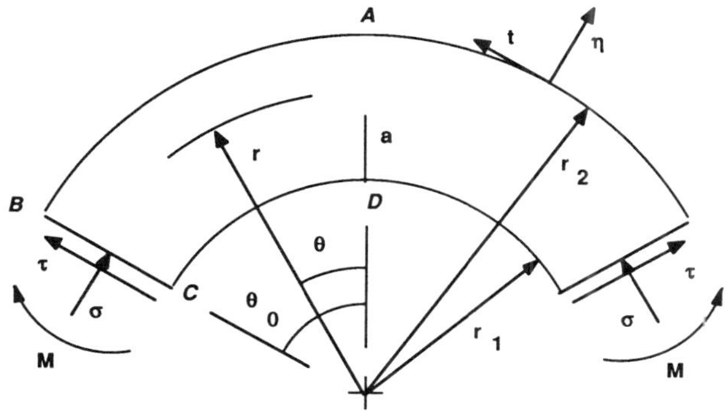

FIG. 2—*Annular segment used to analyze the arc bend-chord support specimen.*

where τ is a uniform shearing stress and σ is a uniform normal stress. The other variables in the equations are shown in Figs. 1 and 2.

Since the problem was symmetric about the plane of the crack, the stress function ϕ and its derivative with respect to the outward normal $\partial\phi/\partial\eta$ only needed to be determined along the boundary *ABCD*. For the case of pure bending, a ring segment was developed by Gross and Srawley [*6*]. The other two loading conditions have not been developed and will be reported here. For the case of a uniform normal compressive stress σ acting as shown in Fig. 2, the stress function and its derivative are

Along *DC*

$$\phi = 0 \tag{4}$$

$$\frac{\partial\phi}{\partial\eta} = 0 \tag{5}$$

along *CB*

$$\phi = \sigma(r(r_1 - r) - \frac{1}{2}(r_1^2 - r^2) \tag{6}$$

$$\frac{\partial\phi}{\partial\eta} = 0 \tag{7}$$

along *BA*

$$\phi = \sigma(r_2(r_1 - r_2)\cos(\theta_0 - \theta) - \frac{1}{2}(r_1^2 - r_2^2) \tag{8}$$

$$\frac{\partial\phi}{\partial\eta} = \sigma(r_1 - r_2)\cos(\theta_0 - \theta) \tag{9}$$

For the case of a uniform shearing stress τ that acts out from the inner radius, the stress function and its derivative are

Along DC

$$\phi = 0 \tag{10}$$

$$\frac{\partial \phi}{\partial \eta} = 0 \tag{11}$$

along CB

$$\phi = 0 \tag{12}$$

$$\frac{\partial \phi}{\partial \eta} = \tau(r_1 - r) \tag{13}$$

along BA

$$\phi = -\tau(r_1 - r_2)r_2 \sin(\theta_0 - \theta) \tag{14}$$

$$\frac{\partial \phi}{\partial \eta} = -\tau(r_1 - r_2)\sin(\theta_0 - \theta) \tag{15}$$

For the limiting case of a straight three-point bend specimen, it is more convenient to derive the stress function using the geometry shown in Fig. 3. In this case, the resultant shearing forces V should not be assumed to result in a uniform shear stress, but in the well-known parabolically varying distribution [*7*]. Using the solution in Ref *7* for a uniform rectangular section, the stress function and its derivative are

Along DC

$$\phi = 0 \tag{16}$$

$$\frac{\partial \phi}{\partial \eta} = 0 \tag{17}$$

along CB

$$\phi = 0 \tag{18}$$

$$\frac{\partial \phi}{\partial \eta} = \frac{-6V}{W^3}\left[-\frac{1}{3}(a^3 + x^3) + \frac{1}{2}(a^2 - x^2)(2a - W) - a(a - W)(a + x)\right] \tag{19}$$

along BA

$$\phi = \frac{-(y - s)6V}{W^3}\left\{\frac{1}{2}(2a - W)[a^2 - (W - a)^2] - \frac{1}{3}[a^3 + (W - a)^3] - W[a(a - W)]^3\right\} \tag{20}$$

$$\frac{\partial \phi}{\partial \eta} = 0 \tag{21}$$

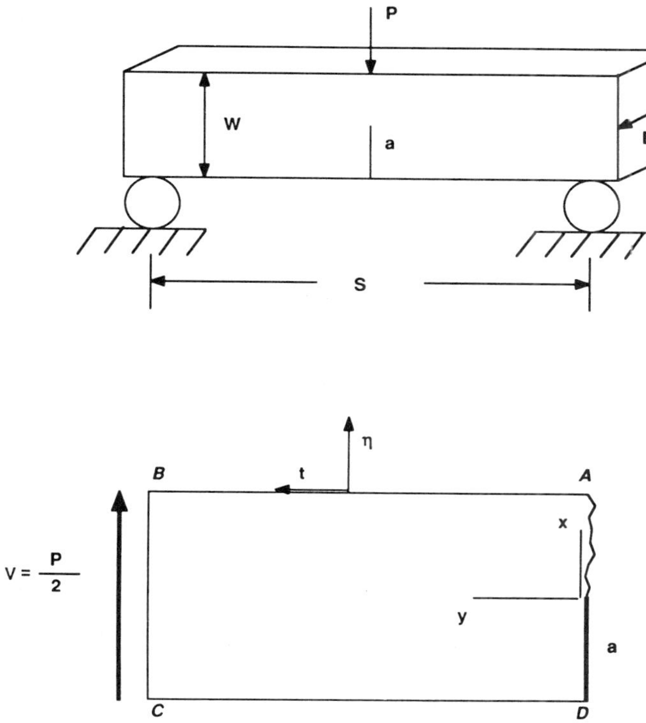

FIG. 3—*Geometry of straight rectangular bend specimen.*

If the purpose of the analysis was to establish only K and crack mouth opening displacements, the choice of the parameter θ_0 was arbitrary as long as the resulting radial cut was far enough removed to satisfy St. Vennant's principle. Since load-line displacements were also desired, it was necessary to choose θ_0 such that the position of the roller support was somewhere contained in the region *ABCD* in Fig. 2. This requirement was included since Williams's stress function [5] is based on a coordinate system with its origin at the crack tip, thus the displacement of the crack tip is zero. In this coordinate system, the displacement at Point A is the only displacement relative to the crack tip. The total load point displacement was constructed as the displacement of the roller support relative to the crack tip plus the displacement of Point A relative to the crack tip. This was accomplished by choosing θ_0 such that chord length $2r_2 \sin \theta_0$ was approximately 10% greater than the Span S. With θ_0 at this value, load-line displacements were easily calculated using the coordinate system translation outlined above. It was necessary to make the chord length larger than S to eliminate some end effects that occurred in the curved geometry. No such end effects occurred for the straight specimen (Fig. 3), and in this instance the shear force V was applied at the location of the roller supports.

When using boundary collocation, convergence of the solution must be demonstrated. Convergence problems arise at extremes in the parameters a/W, r_1/r_2, and S/W. Solutions were obtained using $r_1/r_2 = 0.6$, $S/W = 4$ and $a/W = 0.2$ or 0.8. A large number of equally spaced boundary stations (m) were established, and the number of coefficients ($2N$) in the stress function was systematically increased. Adequate convergence (less than 0.1%) of both K and crack mouth displacements was obtained when $m = 120$, $2N = 60$, and $a/W = 0.2$. At the other extreme, convergence was not obtained even when $m = 240$ and $2N = 160$. In the interest of prudent computer usage, it was decided to use $m = 120$ and $2N = 60$ and limit the reported solutions to crack lengths whose accuracy can be assured by approaching the appropriate limiting solutions.

Results

Normalized stress-intensity factors appear in Table 1, normalized crack mouth displacements in Table 2, and normalized load-line displacements in Table 3. In each table, the published solutions for the straight geometry are also given.

Discussion of these results yields a few general findings. The straight three-point bend solutions generated here for $S/W = 4$ agree well with previously published data [*8–10*] with the exception of the load-line displacements where disagreement of as much as 4% is observed at $a/W = 0.5$. This may be because the total deflection given by Haggag and Underwood [*10*] is a constant term (deflection due to shear and bending with no crack) plus a term that varies with crack length, the later being taken from Tada [*9*]. Recent work [*11*] has shown that integration of the K solution using Irwin's equation for the three-point bend specimen does not agree well with the deflections reported in Refs *9* and *10*. The results to be published by Underwood et al. [*11*] agree nominally within 1% with the results reported here.

The $S/W = 3$ solutions for the straight specimen show a consistent trend with published results. The collocation solutions are always somewhat less than the values given in Refs *8*, *9*, and *10*, again with the exception of the load-line deflection. This can be explained by the realization that the values given in the first row of each table are calculated from wide-range expressions that assume a given S/W dependence and were fit to results for the specific case of $S/W = 4$. Since this is the first time to the author's knowledge that collocation has been used for the case of $S/W = 3$, and since the $S/W = 4$ solutions agree very well with the previous solutions, it is safe to assume that the collocation solutions reported here for $S/W = 3$ are as accurate as the $S/W = 4$ solutions. This suggests that the effect of S/W is not exactly taken care of with the wide-range expressions given in Refs *8*, *9*, and *10*. It also implies that wide-

TABLE 1—*Normalized stress-intensity factors,* $KB\sqrt{W}/P$.

$S/W = 4$

r_1/r_2	a/W = 0.2	0.3	0.4	0.5
1.0 (Ref *8*)	4.70	6.08	7.93	10.65
1.0 (this study)	4.66	6.07	7.91	10.65
0.8	4.86	6.23	8.03	10.77
0.7	4.94	6.28	8.11	10.88
0.6	5.03	6.39	8.20	10.88

$S/W = 3$

r_1/r_2	a/W = 0.2	0.3	0.4	0.5	0.6
1.0 (Ref *8*)	3.52	4.56	5.95	7.99	11.31
1.0 (this study)	3.40	4.43	5.78	7.76	11.06
0.8	3.50	4.53	5.87	7.86	11.16
0.6	3.62	4.63	5.98	7.98	11.28
0.4	3.67	4.64	6.07	8.09	11.33

TABLE 2—*Normalized total crack mouth opening displacements,* $EB\delta_{cm}/P$.

$S/W = 4$				
	a/W			
r_1/r_2	0.2	0.3	0.4	0.5
1.0 (Ref *9*)	7.07	12.36	20.83	35.67
1.0 (this study)	6.83	12.22	20.76	36.75
0.8	7.47	12.64	21.28	36.09
0.7	7.54	12.89	21.69	36.78
0.6	7.72	13.26	22.41	37.75

$S/W = 3$					
	a/W				
r_1/r_2	0.2	0.3	0.4	0.5	0.6
1.0 (Ref *9*)	5.31	9.27	15.62	26.75	48.27
1.0 (this study)	5.04	9.09	15.32	26.07	47.00
0.8	5.19	9.01	15.64	26.62	47.85
0.6	5.53	9.78	16.14	27.31	49.01
0.4	5.78	10.01	16.34	27.76	50.20

TABLE 3—*Normalized load-line displacements,* $EB\delta_{LL}/P$.

$S/W = 4$				
	a/W			
r_1/r_2	0.2	0.3	0.4	0.5
1.0 (Ref *10*)	23.44	28.74	37.05	53.46
1.0 (this study)	23.83	29.38	39.09	55.46
0.8	23.61	29.93	39.74	57.28
0.7	24.16	30.43	40.75	59.77
0.6	23.59	31.37	41.83	59.81

$S/W = 3$					
	a/W				
r_1/r_2	0.2	0.3	0.4	0.5	0.6
1.0 (Ref *10*)	11.52	14.50	19.49	28.39	45.51
1.0 (this study)	11.64	15.13	20.36	29.18	46.38
0.8	11.87	15.40	20.34	29.15	47.28
0.6	12.89	16.51	21.39	30.93	49.12
0.4	10.84	16.96	17.97	33.54	47.57

range expressions must account for more than the simple S/W dependence that are derived from limiting solutions or that separate wide-range expressions should be developed for each S/W configuration.

The effect of radius ratio (r_1/r_2) on the obtained results are as expected from previous results [*6*]. As r_1/r_2 decreases, the geometry becomes more of a curved beam, and stresses at r_1 become higher. This results in an elevation in K and, correspondingly, in crack mouth opening and load-line displacement when the cracks are shallow relative to straight beam specimens. As crack length increases, the effect of curvature should diminish, and in the limit when the crack approaches through thickness, disappears. This is the case for all the solutions, with the only exception being the load-line displacement of the very thick cylinder ($r_1/r_2 = 0.4$). In this case, we were not able to have the chord length $2r_2 \sin \theta_0$ exceed S by 10%. This probably resulted in some end effects on the displacement of the roller support location. This could not be avoided using the method of solution outlined above. Therefore, the load-line displacement of the thick cylinder ($r_1/r_2 = 0.4$) should be considered suspect. Nevertheless, this end effect appears only in the load-line displacement. It is clear that the proper, consistent trend in both K and crack mouth opening is occurring, and these solutions, even for the thick cylinder, should be considered accurate.

Wide-Range Expressions

The technique of fitting wide-range expressions to numerical data has been outlined by Srawley [12] for K solutions and by Saxena and Hudak [13] for displacements. Basically, the method is to find a nondimensional form of K or displacement that has finite limits as a/W approaches both zero and one. For K, the form to use is [12]

$$\frac{KB\sqrt{W}}{P(S/W)} \quad \frac{(1 - a/W)^{3/2}}{\sqrt{a/W}} = g_K(r_1/r_2, a/W) \cdot f_K(a/W) \tag{22}$$

$$\lim_{a/W \to 1} \quad g_K \cdot f_K = 0.995$$

$$\lim_{\substack{a/W \to 0 \\ r_1/r_2 \to 1}} \quad g_K \cdot f_K = 2.987 \tag{23}$$

The function $g_K(r_1/r_2, a/W)$ is included to account for effects of r_1/r_2. The nondimensional form given in Eq 22 is plotted in Fig. 4 for $S/W = 4$ and Fig. 5 for $S/W = 3$. It is clear that this representation does not account totally for the effect of S/W; therefore it was decided to fit two different expressions, one each for $S/W = 3$ and $S/W = 4$.

The function g_K was found by using the limiting solution ($r_1/r_2 = 1$) as a reference K solution. By taking the ratio of $KB\sqrt{W}(1 - a/W)^{3/2}/[P(S/W)\sqrt{a/W}]$ of the curved specimen to the straight specimen at constant a/W values, g_K can be easily constructed over a range of r_1/r_2

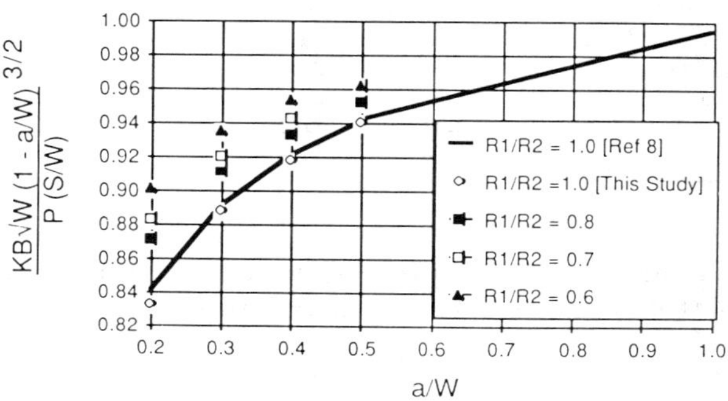

FIG. 4—*Normalized stress-intensity factor solution;* S/W = *4.0.*

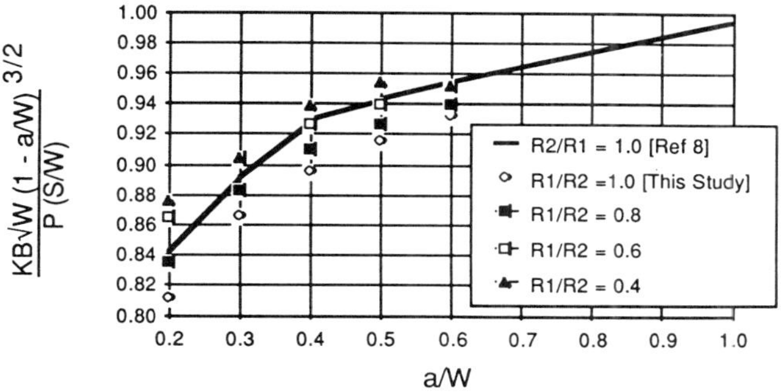

FIG. 5—*Normalized stress-intensity factor solution;* S/W = *3.0.*

and a/W. The constructed or calculated g_K can then be easily fit using multivariable least squares. Once g_K was found it was factored out, and $f_K(a/W)$ was also found by least squares. For either fit it was assumed that g_K and f_K were both polynomials.

The results of the fitting are

For $S/W = 4$

$$\frac{KB\sqrt{W}}{P(S/W)} \quad \frac{(1 - a/W)^{3/2}}{\sqrt{a/W}} = [1 + (1 - r_1/r_2)][0.34 - 0.82a/W + 0.48(a/W)^2] \cdot F(a/W)$$

$$F(a/W) = 2.65 - 5.07(a/W) + 6.34(a/W)^2 - 2.93(a/W)^3 \tag{24}$$

For $S/W = 3$

$$\frac{KB\sqrt{W}}{P(S/W)} \quad \frac{(1 - a/W)^{3/2}}{\sqrt{a/W}} = \{1 + (1 - r_1/r_2)[0.21 - 0.40a/W + 0.19(a/W)^2]\} \cdot F(a/W)$$

$$F(a/W)\; 2.517 - 4.431a/W + 5.028(a/W)^2 - 2.120(a/W)^3 \tag{25}$$

The agreement between Eqs 24 and 25 and the collocation data is shown in Tables 4 and 5. The following accuracy statements can be made from this comparison. For $S/W = 4$, Eq 24 agrees with collocation results within $\pm 0.5\%$ for $r_1/r_2 \geq 0.6$ and $a/W \geq 0.2$ and within $\pm 0.5\%$ for $r_1/r_2 > 0.6$ and $0.4 < a/W < 0.5$. For $S/W = 3$, Eq 25 agrees with collocation results within $\pm 1.0\%$ for $r_1/r_2 \geq 0.4$ and $a/W \geq 0.2$ and within $\pm 0.8\%$ for $r_1/r_2 \geq 0.4$ and $0.4 \leq a/W \leq 0.6$.

For crack mouth opening displacements, a similar procedure can be followed, but it is necessary to use different limits. Briefly, it can be shown that for crack mouth opening the following nondimensional form is appropriate [*13*]

$$\frac{EB\delta_{cm}}{P(S/W)} \quad \frac{(1 - a/W)^2}{(a/W)} = g_{cm}(r_1/r_2, a/W) \cdot f_{cm}(a/W) \tag{26}$$

$$\lim_{a/W \to 1} \quad g_{cm} \cdot f_{cm} = 3.95 \tag{27}$$

$$\lim_{\substack{a/W \to 0 \\ r_1/r_2 \to 1}} \quad g_{cm} \cdot f_{cm} = 8.736 \tag{28}$$

TABLE 4—*Normalized stress-intensity factors (S/W = 4).*

$$\frac{KB\sqrt{W}(1 - a/W)^{1.5}}{P(S/W)\sqrt{a/W}}$$

r_1/r_2		a/W 0.0	0.2	0.3	0.4	0.5	1.0
1.0	collocation	2.987	1.863	1.623	1.435	1.331	0.994
	Eq 24	2.650	1.866	1.620	1.449	1.334	0.990
	relative error	−0.113	0.002	−0.002	−0.003	0.002	−0.004
0.8	collocation	...	1.946	1.664	1.475	1.346	0.994
	Eq 24	...	1.939	1.665	1.475	1.347	0.990
	relative error	...	−0.004	0.001	...	0.001	−0.004
0.7	collocation	...	1.974	1.680	1.490	1.360	0.994
	Eq 24	...	1.975	1.687	1.487	1.354	0.990
	relative error	...	0.001	0.004	−0.002	−0.005	−0.004
0.6	collocation	...	2.013	1.708	1.507	1.360	0.994
	Eq 24	...	2.012	1.709	1.500	1.360	0.990
	relative error	...	−0.001	0.001	−0.004	...	−0.004

TABLE 5—*Normalized stress-intensity factors (S/W = 3).*

		$\frac{KB\sqrt{W}(1 - a/W)^{1.5}}{P(S/W)\sqrt{a/W}}$						
		a/W						
r_1/r_2		0.0	0.2	0.3	0.4	0.5	0.6	1.0
1.0	collocation	2.987	1.815	1.580	1.416	1.296	1.204	0.994
	Eq 25	2.517	1.815	1.583	1.413	1.294	1.211	0.994
	relative error	−0.157	...	0.002	−0.002	−0.002	0.005	...
0.8	collocation	...	1.867	1.611	1.439	1.311	1.214	0.994
	Eq 25	...	1.865	1.617	1.436	1.308	1.220	0.994
	relative error	...	−0.001	0.002	−0.002	−0.002	0.005	...
0.6	collocation	...	1.935	1.652	1.465	1.330	1.228	0.994
	Eq 25	...	1.915	1.651	1.459	1.323	1.229	0.994
	relative error	...	−0.010	−0.001	−0.004	−0.005	0.001	...
0.4	collocation	...	1.958	1.652	1.485	1.349	1.234	0.994
	Eq 25	...	1.965	1.685	1.482	1.338	1.238	0.994
	relative error	...	0.003	0.020	−0.002	−0.008	0.004	...

The collocation results using the nondimensional form of Eq 26 are plotted in Figs. 6 and 7 for $S/W = 4$ and $S/W = 3$, respectively. Using the procedure outlined above, the wide-range expressions for crack mouth opening are

For $S/W = 4$

$$\frac{EB\delta_{cm}}{P(S/W)} \quad \frac{(1 - a/W)^2}{(a/W)} = \{1 + (1 - r_1/r_2)[0.70 - 1.94a/W + 1.23(a/W)^2]\} \cdot F(a/W) \tag{29}$$

$$F(a/W) = 7.36 - 13.16a/W + 20.62(a/W)^2 - 10.87(a/W)^3$$

For $S/W = 3$

$$\frac{EB\delta_{cm}}{P(S/W)} \quad \frac{(1 - a/W)^2}{(a/W)} = \{1 + (1 - r_1/r_2)[0.21 - 0.10a/W + 0.11(a/W)^2]\} \cdot F(a/W) \tag{30}$$

$$F(a/W) = 7.09 - 10.66a/W + 12.88(a/W)^2 - 5.37a/W)^3$$

Comparisons between the collocation results and Eqs 29 and 30 are given in Tables 6 and 7. The following accuracy statements can be made: For $S/W = 4$, Eq 29 agrees with collocation

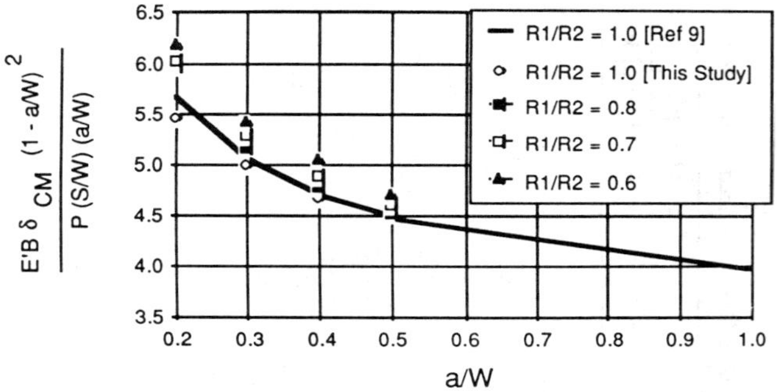

FIG. 6—*Normalized crack mouth opening displacement solution;* S/W = *4.0.*

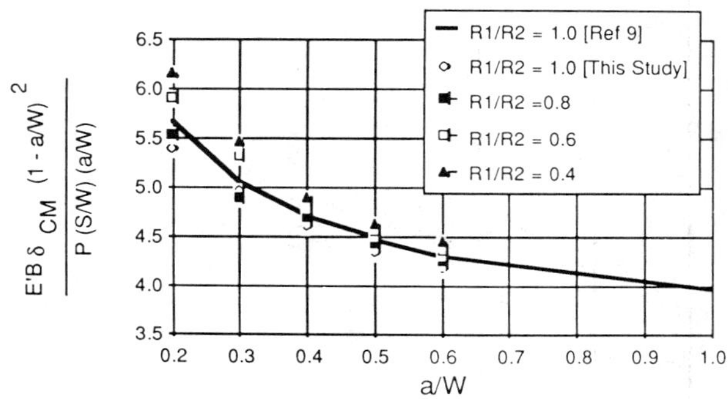

FIG. 7—*Normalized crack mouth opening displacement solution;* S/W = *3.0.*

TABLE 6—*Normalized crack mouth opening displacements (S/W = 4).*

		$\frac{E'B\delta_{cm}(1 - a/W)^2}{P(S/W)(a/W)}$					
		a/W					
r_1/r_2		0.0	0.2	0.3	0.4	0.5	1.0
1.0	collocation	8.736	5.460	4.991	4.672	4.594	3.950
	Eq 29	7.360	5.466	4.974	4.700	4.576	3.950
	relative error	−0.158	0.001	−0.003	0.006	−0.004	...
0.8	collocation	...	5.973	5.161	4.787	4.511	3.950
	Eq 29	...	5.861	5.202	4.813	4.611	3.942
	relative error	...	−0.019	0.008	0.005	0.022	−0.002
0.7	collocation	...	6.031	5.264	4.881	4.597	3.950
	Eq 29	...	6.058	5.316	4.870	4.628	3.938
	relative error	...	0.004	0.010	−0.002	0.007	−0.003
0.6	collocation	...	6.181	5.419	5.043	4.719	3.950
	Eq 29	...	6.256	5.429	4.927	4.645	3.934
	relative error	...	0.012	0.002	−0.023	−0.016	−0.004

TABLE 7—*Normalized crack mouth opening displacements (S/W = 3).*

		$\frac{E'B\delta_{cm}(1 - a/W)^2}{P(S/W)(a/W)}$						
		a/W						
r_1/r_2		0.0	0.2	0.3	0.4	0.5	0.6	1.0
1.0	collocation	8.736	5.374	4.950	4.596	4.435	4.177	3.950
	Eq 30	7.090	5.430	4.906	4.543	4.309	4.171	3.940
	relative error	−0.189	0.010	−0.009	−0.012	−0.008	−0.001	−0.003
0.8	collocation	...	5.532	4.904	4.693	4.436	4.254	3.950
	Eq 30	...	5.632	5.073	4.682	4.423	4.263	3.940
	relative error	...	0.018	0.034	−0.002	−0.003	0.002	−0.003
0.6	collocation	...	5.901	5.323	4.842	4.552	4.357	3.950
	Eq 30	...	5.833	5.230	4.820	4.537	4.355	3.940
	relative error	...	−0.011	−0.016	−0.005	−0.003	...	−0.003
0.4	collocation	...	6.163	5.451	4.903	4.627	4.463	3.950
	Eq 30	...	6.035	5.407	4.959	4.651	4.447	3.940
	relative error	...	−0.021	−0.008	0.011	0.005	−0.004	−0.003

results within ±2.5% for $r_1/r_2 \geq 0.6$ and $a/W \geq 0.2$ and within ±2.5% for $r_1/r_2 \geq 0.6$ and $0.4 \leq a/W \leq 0.5$. For $S/W = 3$, Eq 30 agrees with collocation results within ±2.5% for $r_1/r_2 \geq 0.4$ and $a/W \geq 0.2$ and within ±1.5% for $r_1/r_2 \geq 0.4$ and $0.4 \leq a/W \leq 0.6$.

For total load-line displacements, the appropriate normalized form is [11]

$$\frac{EB\delta_{LL}(1 - a/W)^2}{P(S/W)} = g_{LL}(r_1/r_2, a/W) \cdot f_{LL}(a/W) \tag{31}$$

$$\lim_{a/W \to 1} g_{LL} \cdot f_{LL} = 0.9875 \tag{32}$$

$$\lim_{\substack{a/W \to 0 \\ r_1/r_2 \to 1}} g_{LL} \cdot f_{LL} = \begin{matrix} 1.195(S/W = 4) \\ 1.010(S/W = 3) \end{matrix} \tag{33}$$

The collocation results normalized according to Eq 31 are plotted in Fig. 8 for $S/W = 4$ and Fig. 9 for $S/W = 3$. Since we have found that the load-line displacements are not as accurate as the stress-intensity factors or the crack mouth opening displacements, it was decided to forego fitting wide-range expressions to these numerical results.

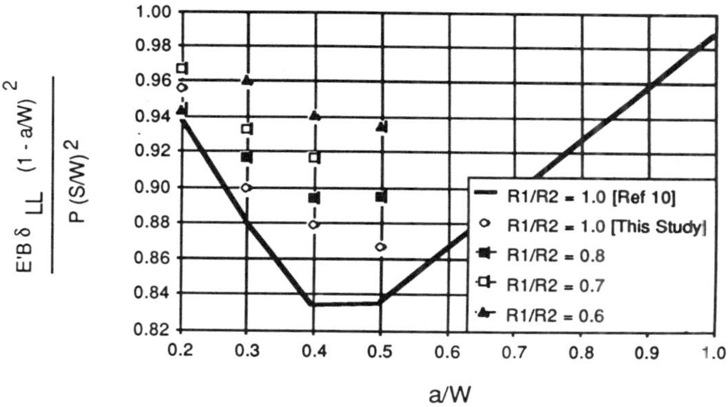

FIG. 8—*Normalized load-line displacement solution;* S/W = *4.0.*

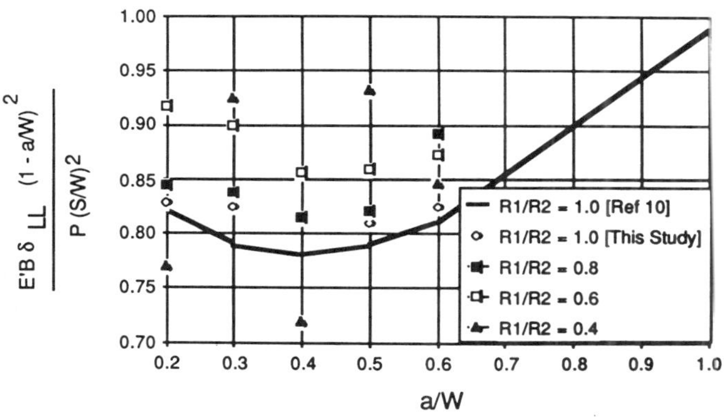

FIG. 9—*Normalized load-line displacement solution;* S/W = *3.0.*

Summary and Conclusions

The arc bend-chord support specimen [AB(C)] has been studied using collocation over a wide-range of possible testing geometries. Stress-intensity factors, crack mouth opening displacements, and total load-line deflections were obtained from the analysis. Wide-range equations were fit to the stress-intensity factor and crack mouth opening displacements solutions to make the solutions more suitable for possible inclusion in future drafts of ASTM E 399. Expressions were not fit to the load-line solutions because these solutions were not considered as accurate as the stress-intensity factors or the crack mouth opening displacements.

References

[*1*] Underwood, J. H., Kapp, J. A., and Witherell, M. D., "Fracture Testing Using Arc Bend Samples," *Fracture Mechanics: Seventeenth Volume, ASTM STP 905,* American Society for Testing and Materials, Philadelphia, 1986.

[*2*] Gross, B. and Mendelson, A., *International Journal of Fracture Mechanics,* Vol. 8, No. 3, Sept. 1972, pp. 267–276.

[*3*] Gross, B., Roberts, E. J., and Srawley, J. E., *International Journal of Fracture Mechanics,* Vol. 4, No. 3, Sept. 1968, pp. 267–276 (Errata, Vol. 6, 1970, p. 87).

[*4*] Hussain, M. A., Lorensen, W. E., Kendall, D. P., and Pu, S. L., "A Modified Collocation Method for C-Shaped Specimens," Technical Report R-WV-T-X-6-73, Benet Weapons Laboratory, Watervliet, NY, Feb. 1973.

[*5*] Williams, M. L., *Journal of Applied Mechanics,* Vol. 24, No. 1, 1957, pp. 109–114.

[*6*] Gross, B. and Srawley, J. E. in *Development in Fracture Mechanics Test Methods Standardization, ASTM STP 632,* American Society for Testing and Materials, Philadelphia, 1977, pp. 39–56.

[*7*] Seely, F. B. and Smith, J. D., *Advanced Mechanics of Materials,* 2nd ed., Wiley, New York, 1952.

[*8*] Gross, B. and Srawley, J. E., "Stress Intensity Factors for Single-Edge Notch Specimens in Bending or Combined Bending and Tension," NASA TND-2603, National Aeronautics and Space Administration, Washington, DC, Jan. 1965.

[*9*] Tada, H., Paris, P., and Irwin, G., *The Analysis of Cracks Handbook,* Del Research Corp., Hellertown, PA, 1973.

[*10*] Haggag, F. M. and Underwood, J. H., *International Journal of Fracture,* Vol. 26, 1984, pp. R63–R65.

[*11*] Underwood, J. H., Kapp, J. A., and Baratta, F. I., *International Journal of Fracture,* Vol. 28, 1985, pp. R41–R45.

[*12*] Srawley, J. E., *International Journal of Fracture,* Vol. 26, 1976, pp. 475–476.

[*13*] Saxena, A. and Hudak, S. J., Jr., *International Journal of Fracture,* Vol. 14, No. 5, Oct. 1978, pp. 453–468.

George T. Sha[1] *and Chien-Tung Yang*[2]

Determination of Mixed Mode Stress-Intensity Factors Using Explicit Weight Functions

REFERENCE: Sha, G. T. and Yang, C.-T., **"Determination of Mixed Mode Stress-Intensity Factors Using Explicit Weight Functions,"** *Fracture Mechanics: Eighteenth Symposium, ASTM STP 945*, D. T. Read and R. P. Reed, Eds., American Society for Testing and Materials, Philadelphia, 1988, pp. 301–330.

ABSTRACT: An efficient finite-element methodology for evaluating the stress-intensity factors (K_{I}, K_{II}) and the weight functions (h_{I}, h_{II}) of a self-similar propagating mixed-mode crack has been established with a single colinear virtual crack extension. This is accomplished through the use of a symmetric mesh in the crack-tip neighborhood that provides the decoupling characteristics for the field parameters of stresses, strains, displacements, tractions, and strain energy release rates into Mode I and Mode II components within the symmetric mesh zone. The application of this newly established technique to the three-point bend specimen has been studied with the mixed fracture mode introduced by deviations from normal test conditions of either loading or support eccentricity. These eccentricity effects on K_{I} and K_{II} values for three-point bend specimens are detailed in this paper.

The angular dependence of the strain energy release rate, $G(\theta)$, as a function of eccentricity, is also provided herein to account for the testing irregularities associated with three-point bend tests, which have span-to-width ratio $s/W = 4$ and different crack lengths $0.1 \leq a/W \leq 0.8$. By assuming the crack propagation taking place at the angle (θ_{cr}), with maximum strain energy release rate (G_{max}), the predictions are also provided for expressing θ_{cr} angles as a function of both loading and support eccentricities. It is found that the load eccentricity has a more severe effect than that of the support eccentricity as indicated in our numerical results for three-point bend specimen with $s/W = 4$.

Also presented explicitly are the characteristics of the Mode I and Mode II weight functions for three-point bend specimens (equivalent to edge crack geometry) with $s/W = 4$ and $0.1 \leq a/W \leq 0.8$ at the crack faces and other key traction application surfaces. The explicit weight function information at these key traction application boundaries is very useful to the fracture mechanics practitioner and designer for guiding traction application locations and directions that produce the least stress-intensity factor.

For a given crack geometry, the explicit weight functions depend on the constraint conditions. The method of obtaining the required stress-intensity factors of an asymmetric crack geometry under the constraint conditions, which are different from the available weight function for the same asymmetric crack geometry, is also made available here. This is accomplished by combining the predetermined explicit weight functions with self-equilibrium forces, which include both the applied surface tractions and the reaction forces induced by constraints.

KEY WORDS: mixed fracture mode, stress intensity factors, strain energy release rates, explicit weight functions, self-similar crack propagation, crack propagation criterion, eccentric loading effect, ASTM three-point bend specimen

[1]Staff research scientist/engineer, Allison Gas Turbine Division, General Motors Corporation, P.O. Box 420, Indianapolis, IN 46206-0420.

[2]Research engineer, Center for the Advancement of Computational Mechanics, School of Civil Engineering, Georgia Institute of Technology, Atlanta, GA 30332.

Bueckner's weight function concept [1] is, in fact, the Green's function for the stress-intensity factors of a cracked body. Repeated finite-element calculations of stress-intensity factors can be eliminated by using the load-independent characteristics of weight functions of a given geometry and constraint conditions. An efficient finite-element methodology for the explicit determination of weight functions has been established for Mode I cracks [2–4] with Rice's [5] displacement derivative representation. By using a symmetric mesh in the crack-tip neighborhood, all field parameters, such as stresses, strains, displacements, and tractions within the symmetric mesh zone, can be decoupled into Mode I and Mode II components [6, 7]. With these decoupling characteristics of field parameters, Rice's displacement derivative representation of weight function vector for symmetric cracks has been extended to the mixed fracture mode by Sha and Yang [8]. The primary purpose of this paper is to apply this newly established finite-element technique for nodal weight function evaluation to three-point bend specimens with eccentric loading, which introduces the mixed fracture mode due to the deviations from the normal testing conditions of either loading eccentricity or support eccentricity, as shown in Fig. 1. These eccentricity effects on both K_I and K_{II} values are provided in detail for the eccentric three-point bend specimen with a span-to-width ratio of $s/W = 4$. For a given eccentricity, the load eccentricity is found to be more severe than that of support eccentricity, as indicated by our numerical results for $s/W = 4$. At nodal location (x_i, y_i) within the structure of interest with crack length a and inclination angle β with respect to the loading axis, the decoupled weight function vectors $(h_{I(II)})$ can be represented as

$$h_{I(II)}(x_i, y_i, a, \beta) = \frac{H}{2K^*_{I(II)}} \cdot \frac{\partial U_{I(II)}(x_i, y_i, a, \beta)^*}{\partial a} \tag{1}$$

where

H = effective modulus with $H = E$ (Young's modulus) for plane-stress state, and $H = E/(1 - \nu^2)$ for plane-strain stress state with ν = Poisson's ratio,

$K^*_{I(II)}$ = Mode I (Mode II) stress intensity factors under * loading system, and

$\dfrac{\partial U_{I(II)}(x_i, y_i, a, \beta)^*}{\partial a}$ = decoupled partial displacement derivative for Mode I or Mode II deformation with respect to the virtual crack extension da under * loading system.

Equation 1 permits explicit determination of nodal weight functions for the entire structure, containing either an asymmetric crack or a symmetric crack with eccentric/asymmetric loading and a single colinear virtual crack extension (VCE) under the * loading system. The Mode I/Mode II weight function vectors, $h_{I(II)i}(x_i, y_i, a, \beta)$, consist of $h_{I(II)x}(x_i, y_i, a, \beta)$ and $h_{I(II)y}(x_i, y_i, a, \beta)$ components, which are defined as

$$h_{I(II)x}(x_i, y_i, a, \beta) = \frac{H}{2K_{I(II)}} \cdot \frac{\partial U_{I(II)x}(x_i, y_i, a, \beta)}{\partial a} \tag{2a}$$

$$h_{I(II)y}(x_i, y_i, a, \beta) = \frac{H}{2K_{I(II)}} \cdot \frac{\partial U_{I(II)y}(x_i, y_i, a, \beta)}{\partial a} \tag{2b}$$

where $U_{I(II)x}$ and $U_{I(II)y}$ are the displacement components along the global x and y axes for Mode I or Mode II deformation. This paper concerns the studies made for the three-point bend specimen with $\beta = 0$ for a symmetric crack subjected to eccentric loading.

With this weight function concept, the stress-intensity factors (K_I, K_{II}) can be expressed as a

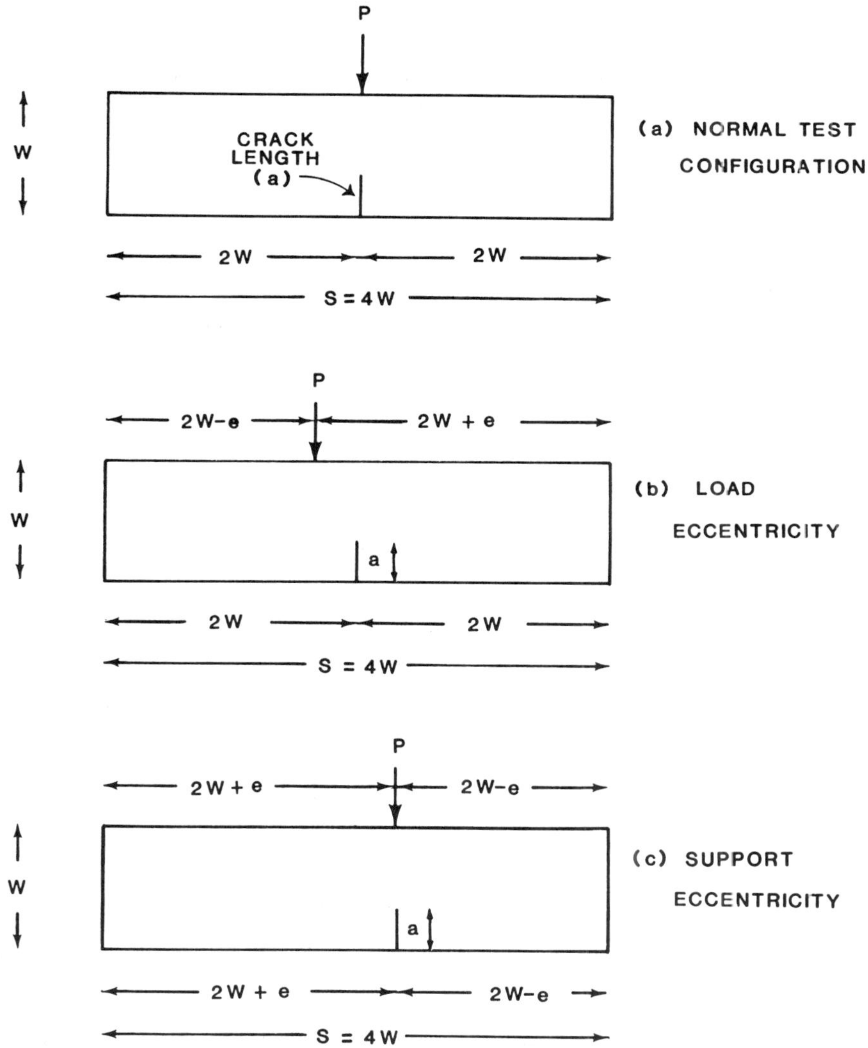

FIG. 1—*Eccentric loading conditions for three-point bend specimen associated with testing irregularities.*

sum of a worklike product between the applied load and weight functions at their point of applications as [8]

$$K_{\mathrm{I(II)}} = \int_s t \cdot h_{\mathrm{I(II)}} ds + \int_v f^b \cdot h_{\mathrm{I(II)}} dv \tag{3}$$

where

t = applied surface tractions,
f^b = body force,

s = surface with applied tractions,
v = volume,
$\boldsymbol{h}_{\mathrm{I(II)}}$ = explicit weight function vectors for Mode I ($\boldsymbol{h}_{\mathrm{I}}$) or Mode II ($\boldsymbol{h}_{\mathrm{II}}$), and
$K_{\mathrm{I(II)}}$ = stress-intensity factors for Mode I (K_{I}) or Mode II (K_{II}).

The rigorous proof of Eq 3 for Mode I symmetric cracks has been made by Rice [*5*], whereas the extension from Mode I cracks to mixed-mode cracks as indicated in Eq 3 was made by Sha and Yang [*8*] with a single colinear virtual crack extension technique [*9,10*] coupled with singular elements [*11,12*].

Without body-force loading, only the line integral is involved in the $K_{\mathrm{I(II)}}$ calculations, as shown in Eq 3 for a two-dimensional eccentric crack problem. However, by applying the linear superposition principle, the $K_{\mathrm{I(II)}}$ calculations for a given crack geometry can be evaluated from the line integral alone with the weight function concept even with body-force loading by combining the resultant "uncracked" stress components $\sigma(r_s)$ with the crack-face weight functions $\boldsymbol{h}_{\mathrm{I(II)}r_s}(r_s, a)$ at the crack-face location as

$$K_{\mathrm{I(II)}} = \int_0^a \sigma(r_s) \cdot \boldsymbol{h}_{\mathrm{I(II)}r_s}(r_s, a) dr_s = \sum_{i=1}^{N} f_i(r_s) \cdot \boldsymbol{h}_{\mathrm{I(II)}i}(r_s, a) \tag{4}$$

where

r_s = radial distance from the crack tip along the crack-face,
$\boldsymbol{h}_{\mathrm{I(II)}r_s}(r_s, a)$ = Mode I/Mode II crack-face weight function vectors with crack length a,
$f_i(r_s)$ = consistent nodal forces equivalent to applied "uncracked" stress field at the prospective crack-face location, and
N = total discretized nodes along crack-face locations with "uncracked" stress field applications.

The resultant "uncracked" stress field $\sigma(r_s)$ can include body-force loading, thermal loading, applied traction at the crack face and any other locations, residual stresses, and any of their loading combinations.

The key advantage of the weight function concept is that once the weight functions are predetermined explicitly for a given crack geometry and constraint conditions, accurate stress-intensity factors under any other loading conditions can be obtained easily and inexpensively according to Eq 3 or Eq 4. It should be noted that the explicit weight functions depend strongly on constraint conditions for a given crack geometry. An analytical method is also provided in this paper in evaluating the needed stress-intensity factors for given loading and constraint conditions that are different from the available predetermined weight functions of the same crack geometry. Basically, this is accomplished by including all self-equilibrium forces, which consist of applied surface tractions and reaction forces due to constraints, in Eq 3 or Eq 4 for K_{I} and K_{II} calculations.

The weight function characteristics at the potential traction application boundaries for the eccentric three-point bend specimen are also provided herein. The explicit weight function components along the traction boundaries in the direction normal to the traction application boundaries of a three-point bend specimen are presented with analytical expressions for a wide range of crack-to-width ratios ($0.1 \leq a/W \leq 0.8$). These weight functions for calculating the stress-intensity factors under remote tension, remote bending for edge crack geometry, and four-point bend specimens have been used for assessing the eccentric loading effects on K_{I} and K_{II} values in order to illustrate the power of weight function concepts. The use of explicit weight functions at the left-hand side (LHS) face and right-hand side (RHS) face locations as shown in Fig. 2 to evaluate K_{I} stress-intensity factors under remote tension and bending of edge crack geometry, which is equivalent to the three-point bend specimen, has been made with good

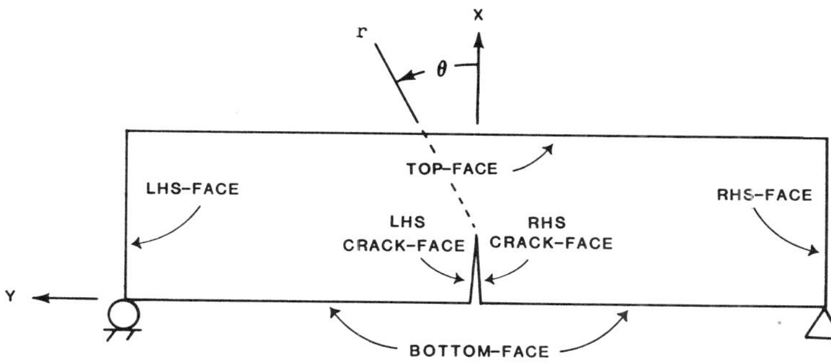

FIG. 2—*Nomenclatures, coordinates, and constraints used for a three-point bend specimen with uniform thickness* (B).

agreement with available K_I solutions in the literature. The detailed results for four-point bend specimens will be reported elsewhere shortly.

By assuming the crack propagation direction of a quasi-static crack under the given loading and constraint conditions taking place at a critical angle (θ_{cr}) with the maximum strain-energy release rate (G_{max}), the predicted θ_{cr} angles as a function of loading eccentricity are also presented in this paper. The experimental verification of these θ_{cr} predictions as a function of eccentric loading is not available yet.

Numerical Results and Discussion

An efficient finite-element evaluation for the mixed fracture mode of both stress intensity factors (K_I and K_{II}) and decoupled weight function vectors (h_I and h_{II}) has been established recently with a symmetric mesh in crack-tip neighborhood coupled with a single colinear VCE technique [8]. The application of this newly established technique to the ASTM three-point bend specimen with deviations from normal testing conditions is made; analytical results are presented herein showing the effects of testing irregularities on the stress intensity factors (K_I and K_{II}) and on the crack propagation direction of a quasi-static crack. These deviations from normal testing conditions, including the effects of loading and support eccentricities, as shown in Fig. 1, on K_I and K_{II} values, are studied for three-point bend specimens with $0.1 \leq a/W \leq 0.8$ under eccentric loading conditions. The nomenclatures, coordinates, and constraints used in the finite-element analyses of eccentric three-point bend specimens used in this paper are shown in Fig. 2.

Because the intended purpose is to demonstrate the eccentric loading effects on the stress intensity factors (K_I and K_{II}) with the weight function concept, the entire three-point bend specimen is idealized instead of only half the specimen, as required by the Mode I symmetric loading case. The typical finite-element mesh for a three-point bend specimen, with eccentric loading and eccentric supporting, is shown in Figs. 3*a* and 3*b*, respectively. The degenerated quarter-point quadratic elements, with $\sqrt{r}$ displacement variations, were assembled symmetrically with respect to the crack axis (x) in the crack-tip vicinity. These singular elements were surrounded by the conventional 8-noded quadratic elements for the remainder of the 2-D crack geometry. The virtual crack extension of an amount Δa was simulated by advancing the crack-tip node by Δa in the direction colinear with the symmetric crack under eccentric loading.

The geometric definition of eccentricity (e) of the three-point bend specimen is shown in Figs. 1*b* and 1*c* for load eccentricity and support eccentricity, respectively. The dimensionless $K_I B\sqrt{W}/P$ values, which are obtained by combining the predetermined Mode I explicit weight

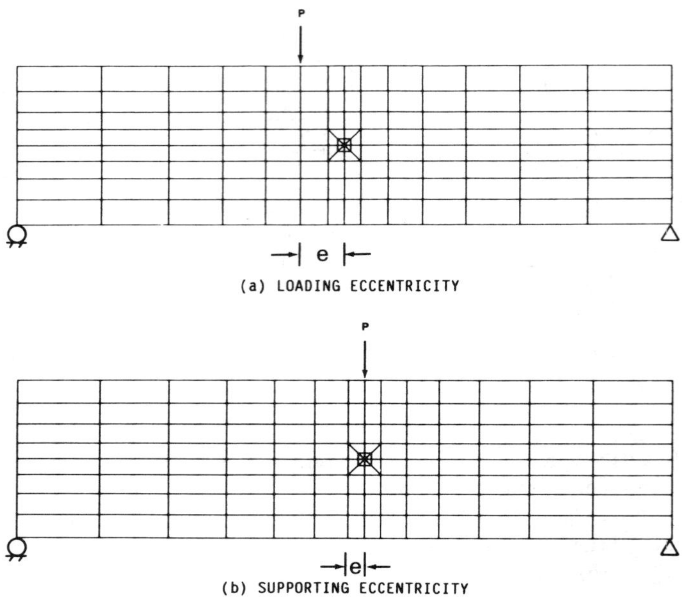

FIG. 3—*Typical finite element meshes for* (a) *loading eccentricity and* (b) *supporting eccentricity for a three-point bend specimen with* s/W = 4.

function component h_{Ix} with the three-point bend force at the $e = 0$ location, according to Eq 4, compare well with that of Srawley [13], as shown in Fig. 4 for $0.1 \leq a/W \leq 0.8$. The numerical computations concerning the effects of load eccentricity of ASTM three-point bend specimens with $s/W = 4$ on the dimensionless $K_I B\sqrt{W}/P$ and $K_{II} B\sqrt{W}/P$ values for $e/W \leq$ 20% are shown in Figs. 5*a* and 5*b*, respectively, with the weight function concept. As anticipated, the load eccentricity effects on the dimensionless $K_I B\sqrt{W}/P$ and $K_{II} B\sqrt{W}/P$ values for the ASTM three-point bend geometry increase with increasing a/W ratios. The K_I and K_{II} values as a function of loading-eccentricity, as shown in Fig. 5, are obtained from the predetermined explicit weight function components h_{Ix} and h_{IIx}, respectively, at the top-face location, according to Eq 4. In principle, the load-eccentricity effects on K_I and K_{II} values, up to 200% eccentricity to width ratio (e/W), can be obtained from the predetermined h_{Ix} and h_{IIx} components at the top-face location. It is found that the dimensionless weight function components, such as $\sqrt{W} \cdot h_{Ix}$ and $\sqrt{W} \cdot h_{Iy}$, are invariant with respect to the similar three-point bend geometry (same s/W and a/W ratios) regardless of its actual size. Assuming two three-point bend specimens of similar crack geometries with two different widths ($W^{(1)}$ and $W^{(2)}$) and with weight function vectors $h^{(1)}_{I(II)}$ and $h^{(2)}_{I(II)}$, respectively, at a similar location, the following invariant relationship holds for the dimensionless weight functions

$$\sqrt{W^{(1)}} \cdot h^{(1)}_{I(II)} = \sqrt{W^{(2)}} \cdot h^{(2)}_{I(II)} = H_{I(II)} \tag{5}$$

A similar relation holds for each of the weight function components ($h_{I(II)x}$ and $h_{I(II)y}$). In principle, the normalization factor $\sqrt{W}$, which is chosen in this paper as shown in Eq 5, can be replaced by $\sqrt{s}$ (s = span) serving equally well as the dimensionless weight function vectors. The $\sqrt{W}h_{Ix}$ and $\sqrt{W}h_{IIx}$ components of explicit weight functions along the top-face location, as a function of y/W ratios for three-point bend specimens with $0.1 \leq a/W \leq 0.8$ and $s/W = 4$, are

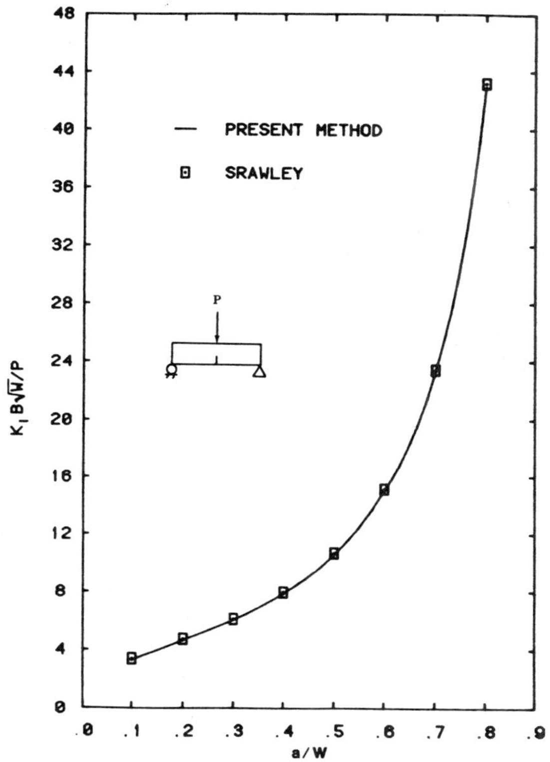

FIG. 4—*Dimensionless* $K_I B\sqrt{W}/P$ *values as a function of* a/W *for a three-point bend specimen with* s/W = *4 and 0.1* ≤ a/W ≤ *0.8.*

shown in Fig. 6. It is found that h_{Ix} and h_{IIx} explicit weight function components along the top face of the ASTM three-point bend specimen can be analytically represented by

$$\sqrt{W}h_{Ix} = \sum_{n=1}^{N} A_n \left(\frac{y}{W}\right)^{n-1} \quad \text{for} \quad -2 \leq \frac{y}{W} \leq +2 \tag{6}$$

and

$$\sqrt{W}h_{IIx} = (\operatorname{sgn} y) \sum_{n=1}^{N} B_n \left(\frac{y}{W}\right)^{n-1} \quad \text{for} \quad -2 \leq \frac{y}{W} \leq +2 \tag{7}$$

The least-squares coefficients A_n and B_n are given in Tables 1 and 2, respectively, for a three-point bend specimen with $s/W = 4$ and $0.1 \leq a/W \leq 0.8$. The associated errors in $\sqrt{W}h_{Ix}$ and $\sqrt{W}h_{IIx}$ values, as represented by Eqs 6 and 7, respectively, are less than 1%.

The crack geometry for the three-point bend specimen with support eccentricity, which varies in geometry for different e values, is geometrically distinguishable from that of load eccentricity with the same e value, as shown in Fig. 1. Therefore, the explicit weight function components for a given e value under load and supporting eccentricities differ due to the distinctness in crack geometry.

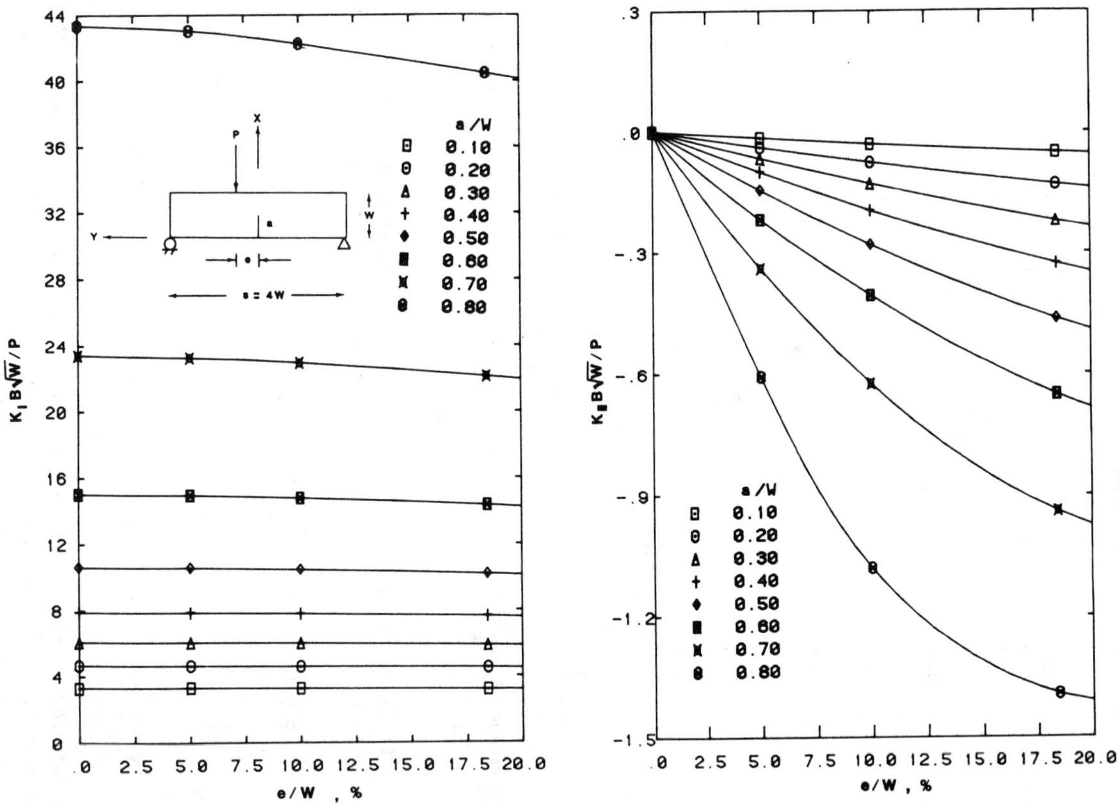

(a) $K_I B\sqrt{W}/P$ vs. e/W

(b) $K_{II} B\sqrt{W}/P$ vs. e/W

FIG. 5—*Effects of load eccentricity* ($e/W \leq 20\%$) *on* (a) *the reduction in* $K_I B\sqrt{W}/P$ *values, and* (b) *introduced* $K_{II} B\sqrt{W}/P$ *values, for a three-point bend specimen with* $s/W = 4$ *and* $0.1 \leq a/W \leq 0.8$.

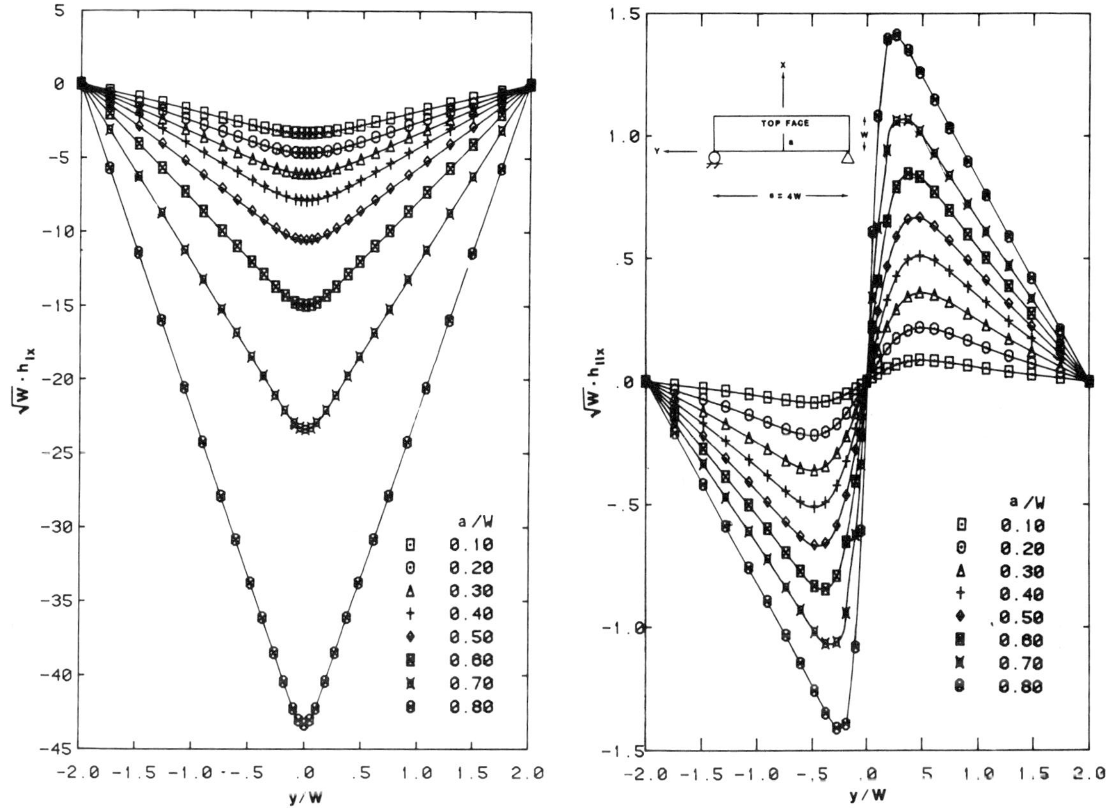

(a) $\sqrt{W} \cdot h_{Ix}$ vs. y/W

(b) $\sqrt{W} \cdot h_{IIx}$ vs. y/W

FIG. 6—*Dimensionless weight function components of* (a) $\sqrt{W}h_{Ix}$ *and* (b) $\sqrt{W}h_{IIx}$ *along the top-face locations as a function of* y/W *ratio for an ASTM three-point bend specimen with* $s/W = 4$ *and* $0.1 \leq a/W \leq 0.8$.

TABLE 1—*Least-squares fitted coefficients of Eq 6 for the Mode I dimensionless explicit weight function component* $\sqrt{W}h_{Ix}$ *as a function of* y/W *of a three-point bend specimen with* s/W = *4, 0.1* ≤ a/W ≤ *0.8.*

a/W	A_1	A_2	A_3	A_4	A_5	A_6	A_7
0.1	−0.33114D + 01	−0.12612D + 00	0.41299D + 01	−0.39706D + 01	0.18715D + 01	−0.40531D + 00	0.28718D − 01
0.2	−0.46525D + 01	−0.16284D + 00	0.57307D + 01	−0.56498D + 01	0.28321D + 01	−0.69099D + 00	0.63312D − 01
0.3	−0.60596D + 01	−0.20523D + 00	0.79290D + 01	−0.86450D + 01	0.49782D + 01	−0.14631D + 01	0.17318D + 00
0.4	−0.78916D + 01	−0.20897D + 00	0.11217D + 02	−0.13731D + 02	0.89247D + 01	−0.29623D + 01	0.39511D + 00
0.5	−0.10589D + 02	−0.11207D + 00	0.16382D + 02	−0.22417D + 02	0.16031D + 02	−0.57642D + 01	0.82154D + 00
0.6	−0.15014D + 02	0.41178D + 00	0.24044D + 02	−0.36047D + 02	0.27604D + 02	−0.10451D + 02	0.15492D + 01
0.7	−0.23428D + 02	0.23652D + 01	0.35211D + 02	−0.57053D + 02	0.46136D + 02	−0.18168D + 02	0.27732D + 01
0.8	−0.43487D + 02	0.93209D + 01	0.51276D + 02	−0.88454D + 02	0.74621D + 02	−0.30280D + 02	0.47252D + 01

TABLE 2—*Least-squares fitted coefficients of Eq 7 for the Mode II dimensionless explicit weight function component* $\sqrt{W}h_{IIx}$ *as a function of* y/W *of a three-point bend specimen with* s/W = *4 and 0.1* ≤ a/W ≤ *0.8.*

a/W	0.10	0.20	0.30	0.40	0.50	0.60	0.70	0.80
B_1	0.610276D − 04	0.190666D − 03	0.205088D − 03	0.272693D − 04	−0.587073D − 03	−0.154598D − 02	−0.759759D − 02	−0.157217D − 01
B_2	0.301725D + 00	0.763043D + 00	0.129749D + 01	0.200185D + 01	0.296894D + 01	0.466389D + 01	0.811288D + 01	0.165648D + 02
B_3	0.144071D + 00	0.314308D + 00	0.601990D + 00	0.443897D + 00	−0.535528D − 01	−0.396981D + 01	−0.182959D + 02	−0.723143D + 02
B_4	−0.178392D + 01	−0.435230D + 01	−0.829272D + 01	−0.125549D + 02	−0.194951D + 02	−0.206036D + 02	0.386033D + 01	0.158762D + 03
B_5	0.285160D + 01	0.706801D + 01	0.145029D + 02	0.242815D + 02	0.428608D + 02	0.577007D + 02	0.395343D + 02	−0.203714D + 03
B_6	−0.217510D + 01	−0.549706D + 01	−0.124244D + 02	−0.226736D + 02	−0.442376D + 02	−0.664372D + 02	−0.645721D + 02	0.157922D + 03
B_7	0.889343D + 00	0.231072D + 01	0.589135D + 01	0.116726D + 02	0.248109D + 02	0.400267D + 02	0.453669D + 02	−0.727953D + 02
B_8	−0.185360D + 00	−0.501654D + 00	−0.148441D + 01	−0.318347D + 01	−0.728615D + 01	−0.123839D + 02	−0.153857D + 02	0.183627D + 02
B_9	0.151418D − 01	0.436643D − 01	0.155586D + 00	0.359961D + 00	0.878108D + 00	0.155310D + 01	0.205363D + 01	−0.195137D + 01

In an attempt to assess the relative severity of load and support eccentricities, the effects of support eccentricity on K_I and K_{II} values are also studied for three-point bend specimens with $0.1 \leq a/W \leq 0.8$, $s/W = 4$, and $e/W = 3/16$. It is found that for a given eccentricity e, the load eccentricity affects K_I and K_{II} values more severely than the support eccentricity, as shown in Fig. 7 for $0.1 \leq a/W \leq 0.8$ and $e/W = 3/16$.

The Mode I and Mode II explicit crack-face weight functions (h_I and h_{II}) can be classified into primary and secondary components with respect to crack orientation relative to the loading direction. The crack-face weight function components h_{Iy} and h_{IIx} are the primary components, which are singular, as contrasted to the secondary components (h_{Ix} and h_{IIy}), which are nonsingular in crack-tip neighborhood with respect to the coordinate systems and crack orientation, as shown in Fig. 2 for a three-point bend specimen. The dimensionless primary crack-face weight function components $\sqrt{W}h_{Iy}$ and $\sqrt{W}h_{IIx}$ as a function of r_s/a ratios are shown in Figs. 8 and 9, respectively, with $0.1 \leq a/W \leq 0.8$. These primary crack-face weight function components are found to be represented accurately by

$$\sqrt{W}h_{Iy}^{Z}\left(\frac{r_s}{a}, \frac{a}{W}\right) = \sum_{n=1}^{N_I} C_n^Z\left(\frac{a}{W}\right)\cdot\left(\frac{r_s}{a}\right)^{\left(\frac{n}{2}-1\right)} \tag{8}$$

$$\sqrt{W}h_{IIx}^{Z}\left(\frac{r_s}{a}, \frac{a}{W}\right) = \sum_{n=1}^{N_{II}} D_n^Z\left(\frac{a}{W}\right)\cdot\left(\frac{r_s}{a}\right)^{\left(\frac{n}{2}-1\right)} \tag{9}$$

where

$C_n^Z\left(\frac{a}{W}\right), D_n^Z\left(\frac{a}{W}\right)$ = least-squares fitting coefficients that are a function of crack length (a/W) only,

N_I, N_{II} = maximum number of terms required to achieve accurate dimensionless crack-face weight function components,

$Z = R$, indicates RHS crack face, and
$= L$, indicates LHS crack face.

$N_I = 4$ and $N_{II} = 4$ are found to provide accurate $\sqrt{W}h_{Iy}(r_s/a, a/W)$ and $\sqrt{W}h_{IIx}(r_s/a, a/W)$ representations of error less than 1% for a three-point bend specimen with $s/W = 4$ and $0.1 \leq a/W \leq 0.8$. These least-squares coefficients $C_n^Z(a/W)$ and $D_n^Z(a/W)$ of Eqs 8 and 9 are given in Tables 3 and 4, respectively. As reflected in these tabulated $C_n^Z(a/W)$ and $D_n^Z(a/W)$ data, the $\sqrt{W}h_{Iy}^R$ data for a given a/W value are related to $\sqrt{W}h_{Iy}^L$ data by a translation of ($C_2^R + C_2^L$) and by reflection of other C_n^Z coefficients with sign changes. The $\sqrt{W}h_{IIx}^R$ and $\sqrt{W}h_{IIx}^L$ values are related to each other by reflection only, with sign changing of $D_n^Z(a/W)$ coefficients. The other two components, namely, h_{Ix} and h_{IIy}, of secondary crack-face weight function components, are nonsingular in the crack-tip neighborhood [*8*], as shown in Figs. 10*a* and 10*b*, respectively, with $0.1 \leq a/W \leq 0.8$. The cubic spline interpolation technique has been used successfully to obtain the primary explicit crack-face weight function components of any crack length from an adequate set of crack-face weight functions of three-point bend geometry with discrete crack lengths. Consequently, the needed K_I and K_{II} values of the induced mixed fracture mode can be evaluated by combining the "uncracked" stress field and the interpolated crack-face weight function components according to Eq 4. It is noted that if the dimensionless weight function components are used in Eq 3, the $1/\sqrt{W}$ correction factor is needed in Eqs 3 and 4 to achieve the correct K_I value with the dimensionless weight functions $H_{I(II)}$ for eliminating the size independence of $h_{I(II)}$ values.

The dimensionless Mode I weight function component $\sqrt{W}h_{Iy}$ along both the RHS-face and

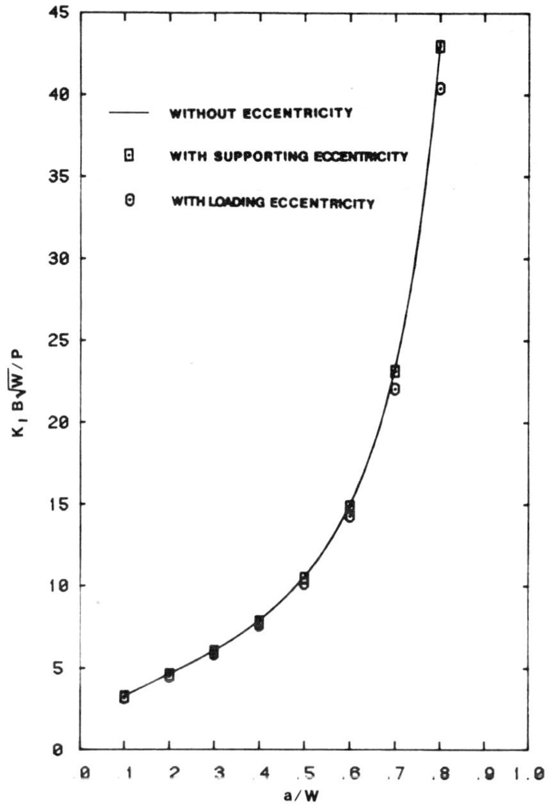

(a) $K_I B\sqrt{W}/P$ vs. a/W

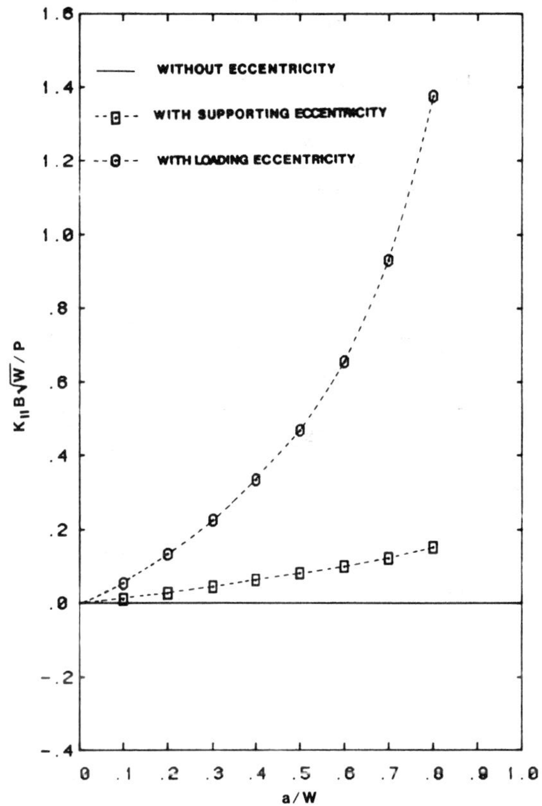

(b) $K_{II} B\sqrt{W}/P$ vs. a/W

FIG. 7—*Effects of loading and supporting eccentricities on the dimensionless stress intensity factors of* (a) $K_I B\sqrt{W}/P$ *and* (b) $K_{II} B\sqrt{W}/P$ *values as a function of* a/W *ratios for a three-point bend specimen with* s/W = *4,* e/W = *3/16, and 0.1* ≤ a/W ≤ *0.8.*

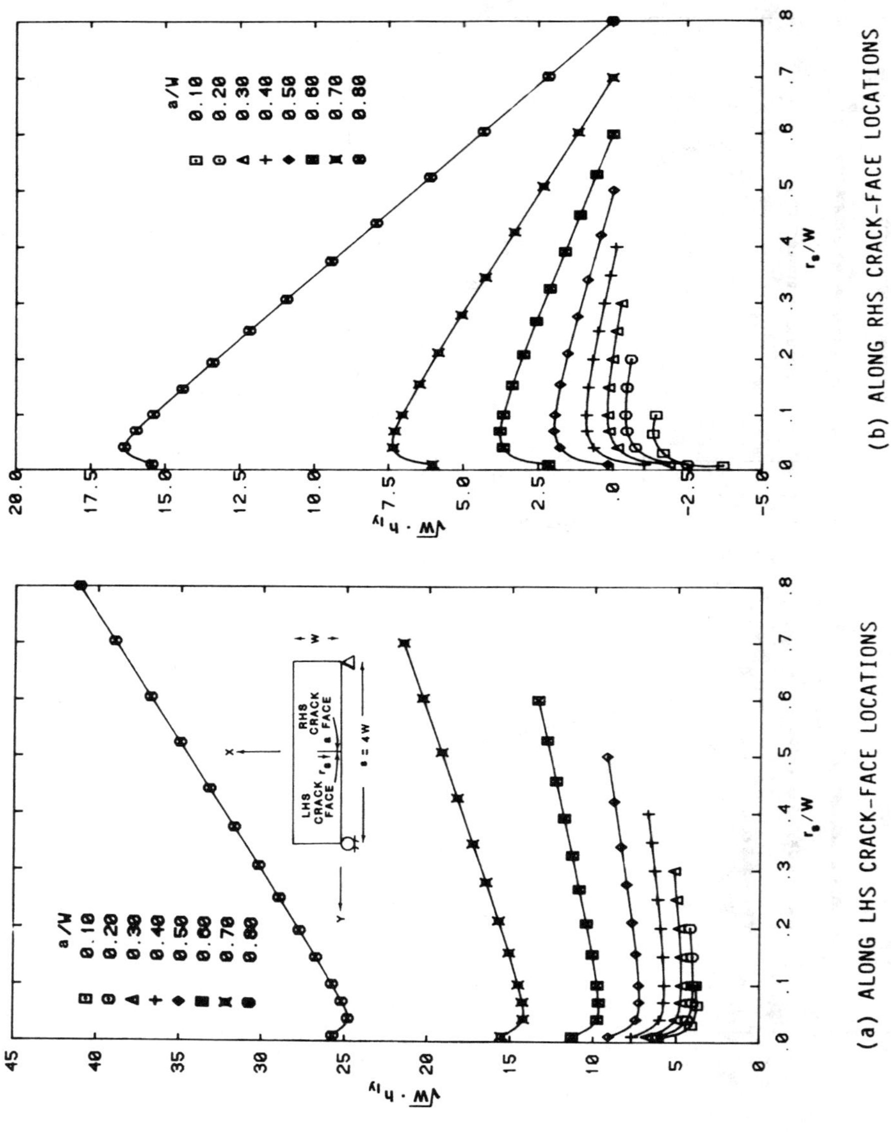

FIG. 8—*Dimensionless explicit crack-face primary weight function component,* $\sqrt{W} \cdot h_{Iy}$*, along* (a) *LHS crack-face and* (b) *RHS crack-face locations as a function of* r_s/W *ratios for a three-point bend specimen with* $s/W = 4$ *and* $0.1 \leq a/W \leq 0.8$.

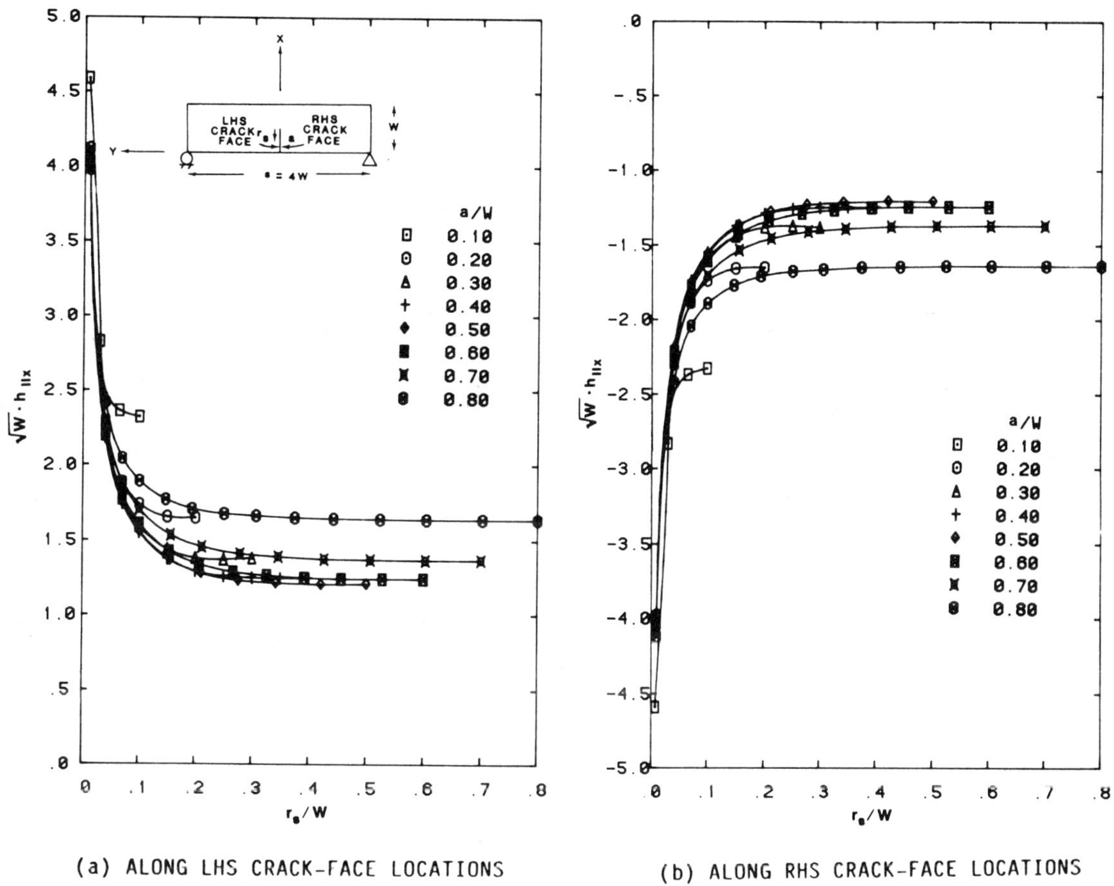

FIG. 9—*Dimensionless explicit crack-face primary weight function component,* $\sqrt{W}h_{IIx}$, *along* (a) *LHS crack-face and* (b) *RHS crack-face locations as a function of* r_s/W *ratios for a three-point bend specimen with* $s/W = 4$ *and* $0.1 \leq a/W \leq 0.8$.

TABLE 3—*Least-squares fitted coefficients of Eq 8 for the Mode I crack-face dimensionless explicit weight function component* $\sqrt{W}h_{Iy}$ *as a function of crack-face distance to crack length ratios* (r_s/a) *for an ASTM three-point bend specimen with* s/W = 4 *and* 0.1 ≤ a/W ≤ 0.8.

	LHS Crack Face				RHS Crack Face			
a/W	C_1^L	C_2^L	C_3^L	C_4^L	C_1^R	C_2^R	C_3^R	C_4^R
0.1	0.11837D + 01	0.18182D + 01	−0.41963D + 00	0.12921D + 01	−0.11837D + 01	0.62125D + 00	0.41965D + 00	−0.12921D + 01
0.2	0.90259D + 00	0.18260D + 01	0.87040D + 00	0.60535D + 00	−0.90258D + 00	0.17710D + 01	−0.87032D + 00	−0.60540D + 00
0.3	0.74491D + 00	0.24165D + 01	0.13666D + 01	0.64093D + 00	−0.74491D + 00	0.24740D + 01	−0.13666D + 01	−0.64093D + 00
0.4	0.63910D + 00	0.33561D + 01	0.16142D + 01	0.11211D + 01	−0.63910D + 00	0.32568D + 01	−0.16142D + 01	−0.11211D + 01
0.5	0.56581D + 00	0.47305D + 01	0.19023D + 01	0.20120D + 01	−0.56581D + 00	0.44492D + 01	−0.19023D + 01	−0.20120D + 01
0.6	0.50730D + 00	0.70058D + 01	0.21678D + 01	0.37341D + 01	−0.50730D + 00	0.64222D + 01	−0.21678D + 01	−0.37341D + 01
0.7	0.45631D + 00	0.11317D + 02	0.24561D + 01	0.73021D + 01	−0.45631D + 00	0.10245D + 02	−0.24561D + 01	−0.73021D + 01
0.8	0.40410D + 00	0.21571D + 02	0.27417D + 01	0.16388D + 02	−0.40410D + 00	0.19567D + 02	−0.27417D + 01	−0.16388D + 02

TABLE 4—*Least-squares fitted coefficients of Eq 9 for the Mode II crack-face dimensionless explicit weight function component* $\sqrt{W}h_{IIx}$ *as a function of crack-face distance to crack length ratios* (r_s/a) *for an ASTM three-point bend specimen with* $s/W = 4$ *and* $0.1 \leq a/W \leq 0.8$.

	LHS Crack Face				RHS Crack Face			
a/W	D_1^L	D_2^L	D_3^L	D_4^L	D_1^R	D_2^R	D_3^R	D_4^R
0.1	0.87459D + 00	0.17837D + 01	−0.18303D + 01	0.14998D + 01	−0.87461D + 00	−0.17836D + 01	0.18302D + 01	−0.14997D + 01
0.2	0.73040D + 00	0.87164D + 00	−0.93249D + 00	0.98491D + 00	−0.73040D + 00	−0.87155D + 00	0.93238D + 00	−0.98485D + 00
0.3	0.64853D + 00	0.49579D + 00	−0.45471D + 00	0.69407D + 00	−0.64853D + 00	−0.49581D + 00	0.45468D + 00	−0.69405D + 00
0.4	0.57955D + 00	0.34445D + 00	−0.15695D + 00	0.48642D + 00	−0.57955D + 00	−0.34444D + 00	0.15693D + 00	−0.48641D + 00
0.5	0.52697D + 00	0.26063D + 00	0.13345D + 00	0.29201D + 00	−0.52697D + 00	−0.26070D + 00	−0.13342D + 00	−0.29202D + 00
0.6	0.48665D + 00	0.19447D + 00	0.51114D + 00	0.55797D − 01	−0.48665D + 00	−0.19453D + 00	−0.51114D + 00	−0.55797D − 01
0.7	0.45058D + 00	0.18162D + 00	0.94962D + 00	−0.20851D + 00	−0.45058D + 00	−0.18147D + 00	−0.94961D + 00	0.20850D + 00
0.8	0.41737D + 00	0.21447D + 00	0.16246D + 01	−0.61812D + 00	−0.41737D + 00	−0.21428D + 00	−0.16246D + 01	0.61810D + 00

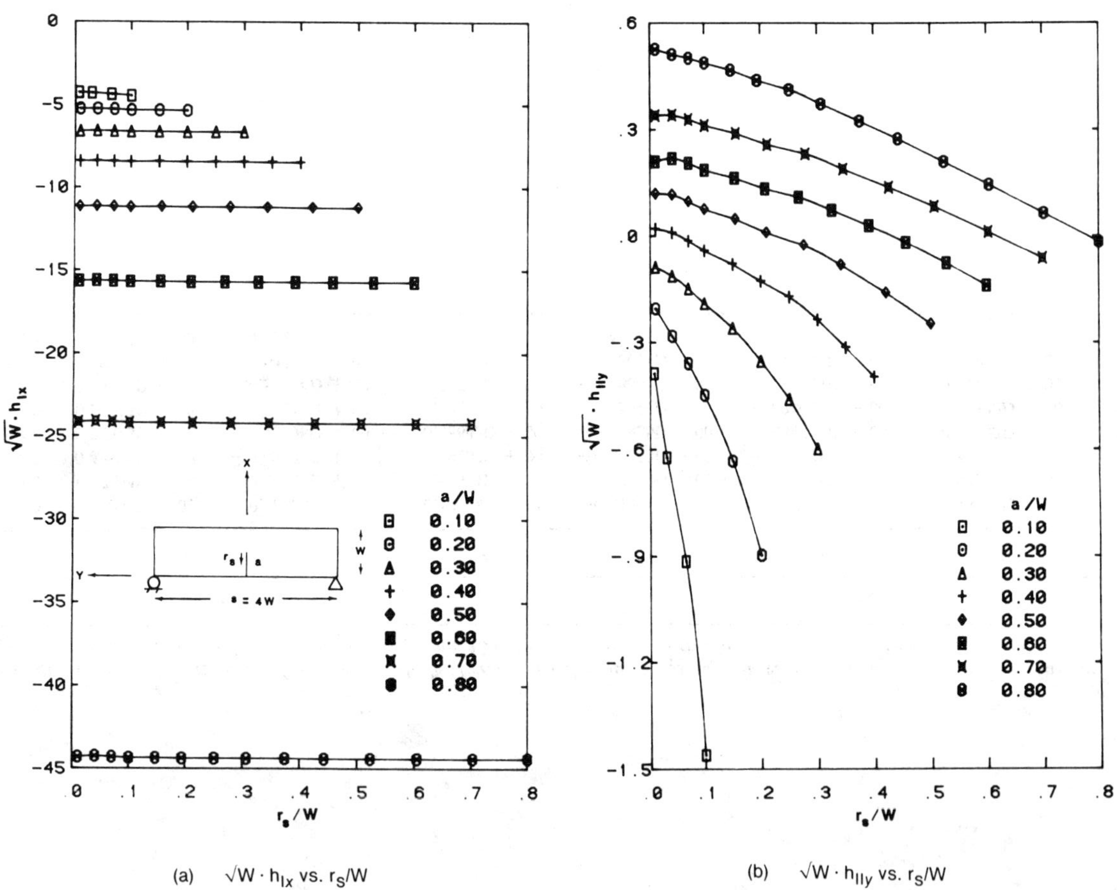

FIG. 10—Nonsingular secondary dimensionless explicit weight function components of (a) $\sqrt{W} \cdot h_{Ix}$ *and* (b) $\sqrt{W} \cdot h_{IIy}$ *in crack-tip neighborhood as a function of* r_s/W *ratios for a three-point bend specimen with* $s/W = 4$ *and* $0.1 \leq a/W \leq 0.8$.

the LHS-face locations as a function of x/W ratios is shown in Fig. 11 for different crack lengths ($0.1 \leq a/W \leq 0.8$). On the other hand, the dimensionless $\sqrt{W}h_{IIy}$ weight function component is shown in Fig. 12 as a function of x/W ratios for different crack lengths ($0.1 \leq a/W \leq 0.8$) with identical results holding for both the RHS-face and LHS-face locations. This implies that $K_{II} = 0$ is assured whenever the symmetric loading is applied simultaneously at RHS-face and LHS-face locations in the y-direction, with respect to the symmetric edge crack geometry. In other words, a relatively small K_{II} value can be induced by applying asymmetric loading along either the RHS-face or the LHS-face location. The stress-intensity factors for the edge crack geometry, which is equivalent to ASTM three-point bend specimen, have been determined from the predetermined explicit weight functions component of h_{Iy} for a wide crack range ($0.1 \leq a/W \leq 0.8$) under both remote tensile and bending conditions with the applied tractions along RHS-face and LHS-face locations, as shown in Fig. 2. These calculated stress-intensity factors with the weight function concept agree well with those of Brown and Srawley [*14*], as shown in Fig. 13.

It is noted that the explicit weight functions depend strongly on the constraint conditions for a given crack geometry. This nonuniqueness of the explicit weight functions results in concern about the ability to calculate the stress-intensity factor under different constraint conditions. It is found that by combining the predetermined explicit weight functions with the self-equilibrium forces at their application locations, accurate K_I and K_{II} values can be achieved easily under different constraint conditions. These self-equilibrium forces consist of both the applied surface tractions and the reaction forces induced from constraint conditions.

The edge crack geometry corresponding to a three-point bend specimen with $s/W = 4$ under two distinct constraint conditions, as shown in Fig. 14, is chosen to calculate the K_I and K_{II} values under different constraint conditions. Suppose one wants to calculate K_I and K_{II} values under constraint conditions of Fig. 14*b* with unit concentrated point force (P) in the $+x$-direction at location T with $x = W$ and $y = 1.073\ W$ on the top-face (BC) from the available explicit weight functions of the same crack geometry under constraints of Fig. 14*a*. The self-equilibrium forces under the constraint conditions of Fig. 14*b* are shown in Fig. 14*c*. The K_I and K_{II} calculations of Fig. 14*b*, with weight function concept, require the dimensionless explicit weight function components H^{T}_{Ix}, H^{T}_{IIx} at Point T of the top face; H^{C}_{Ix}, H^{C}_{IIx}, H^{C}_{Iy}, and H^{C}_{IIy} at Point C; and H^{D}_{Iy}, H^{D}_{IIy} at Point D as

$$K_{I(II)} = 1/\sqrt{W}[P_x \cdot H^{T}_{I(II)x} - F^{C}_{x} \cdot H^{C}_{I(II)x} + F^{C}_{y} \cdot H^{C}_{I(II)y} + F^{D}_{y} \cdot H^{D}_{I(II)y}] \tag{10}$$

The calculated K_I and K_{II} values under loading and constraint conditions, as shown in Fig. 13*b*, with weight function concept of Eq 10 in conjunction with the linear superposition principle, compare well with direct calculations with the virtual crack extension technique for the edge crack geometries with $s/W = 4$, and $0.1 \leq a/W \leq 0.8$, as seen from Fig. 15. Therefore, a methodology is established as demonstrated in the previous example for calculating the stress intensity factors (K_I, K_{II}) from the available explicit weight functions for the same crack geometry with different constraints.

The physical interpretation of explicit nodal weight functions at i's location, as shown in Eq 4, in relationship to stress-intensity factors can be interpreted as the quantified attribution to the stress-intensity factor values under unit traction loading application at the same i's location. An efficient finite-element method for explicit determination of nodal weight function vectors for the entire structure of interest has been established recently for Mode I cracks [*4*] and mixed-mode cracks [*8*]. A map of nodal weight function vectors and their components, with respect to the global coordinates for given crack geometry and constraint conditions, especially at the traction application locations, can provide the design guidance regarding where to apply a given traction vector, or vectors, that causes the least stress intensity factor, or stress intensity ranges, for a propagating crack. The application of a weight function concept to a damage tolerant design concept obviously deserves further attention.

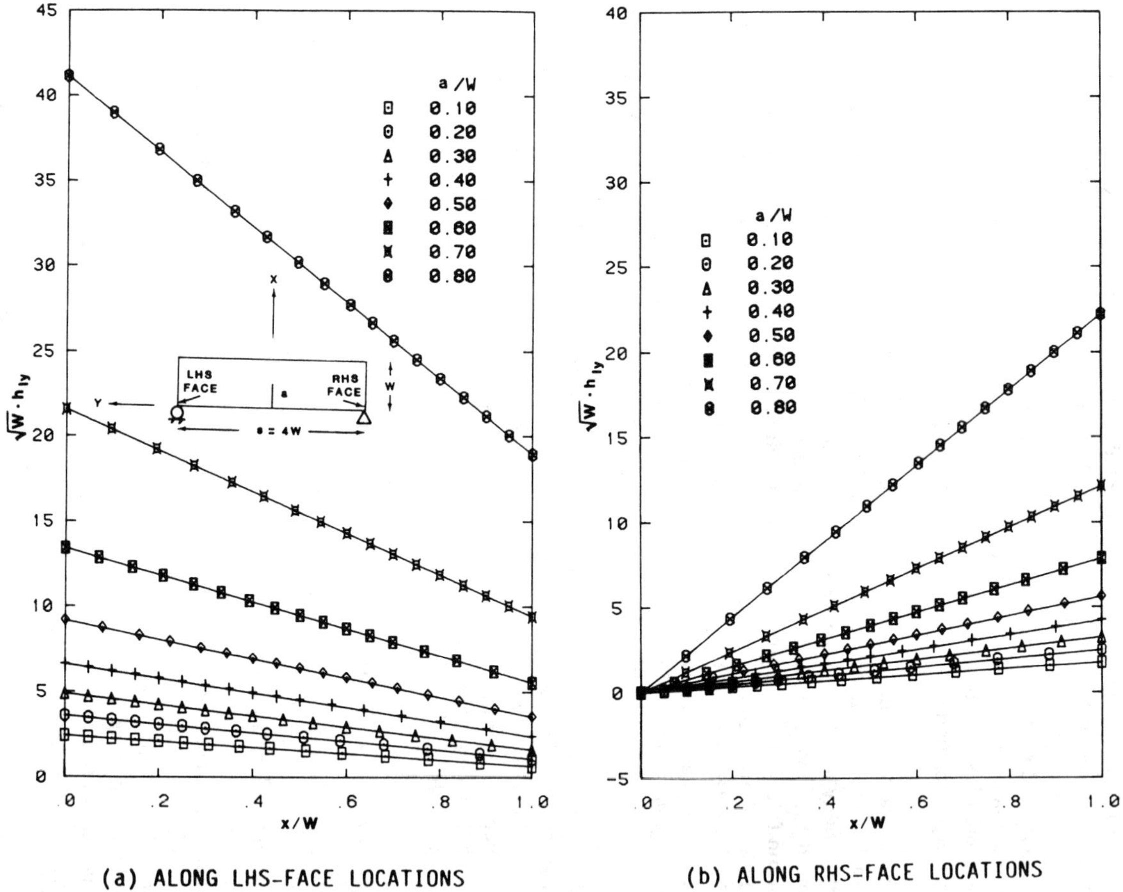

FIG. 11—*Dimensionless explicit weight function component,* $\sqrt{W}h_{Iy}$, *along* (a) *LHS-face and* (b) *RHS-face locations as a function of* x/W *ratios for a three-point bend specimen with* s/W = *4 and 0.1* ≤ a/W ≤ *0.8.*

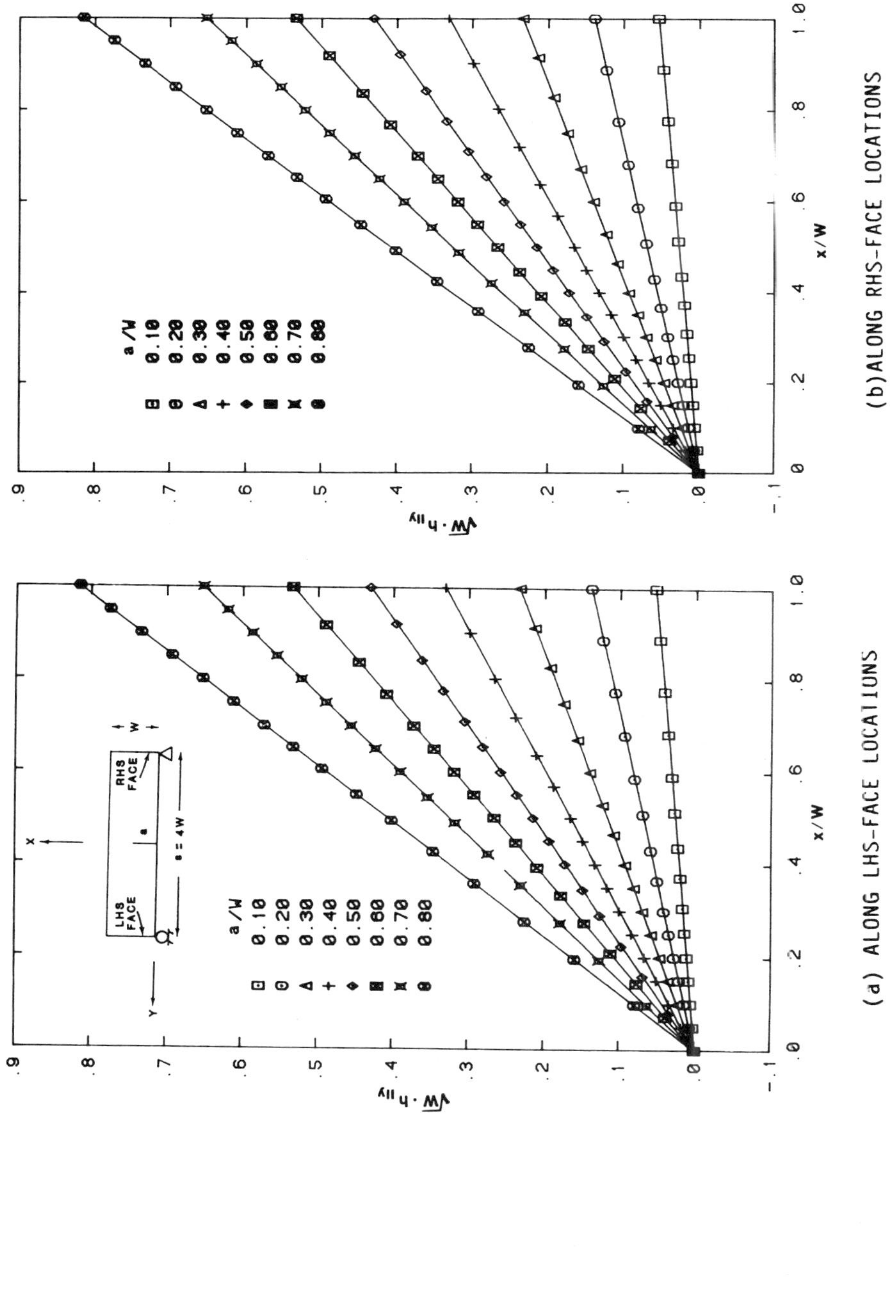

FIG. 12—*Dimensionless explicit weight function component* $\sqrt{W} \cdot h_{IIy}$ *along* (a) *LHS-face and* (b) *RHS-face locations as a function of* x/W *ratios for a three-point bend specimen with* s/W = *4 and 0.1* ≤ a/W ≤ *0.8.*

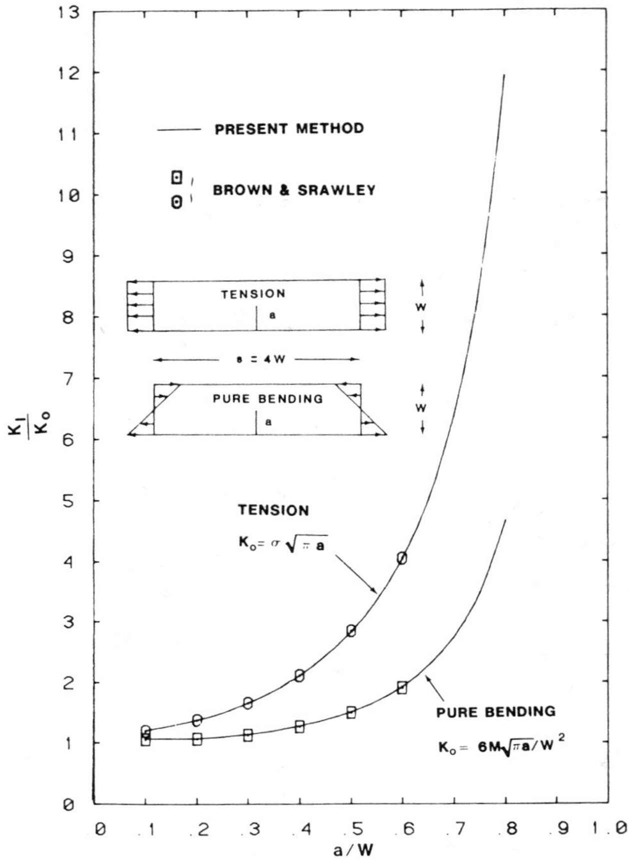

FIG. 13—*Normalized stress intensity factors obtained from the dimensionless explicit weight function component* $\sqrt{W}h_{Iy}$ *along LHS-face and RHS-face locations coupled with remote tension and bending conditions are compared with that of Brown and Srawley's results for edge crack geometry (equivalent to a three-point bend specimen).*

The angular dependence of the strain energy release rate, $G(\theta)$, can be expressed as

$$G(\theta) = G_1 \cos\theta + G_2 \sin\theta \tag{11}$$

where

G_1 = strain energy release rate at $\theta = 0$ deg, colinear with crack,
G_2 = strain energy release rate at $\theta = \pi/2$, normal to the crack, and
θ = angle defined in Fig. 2.

As indicated by Sha and Yang [*8*], G_1 can be obtained efficiently from the VCE technique in the direction colinear with the propagating crack in a self-similar manner, whereas G_2 cannot be obtained directly from the VCE in the direction normal to the crack due to the numerical error contained in any noncolinear VCE approach. However, one can indirectly and accurately evaluate G_2 from the decoupled K_I and K_{II} values using a symmetric mesh in the crack-tip neighborhood, coupled with the VCE technique for mixed fracture mode problems. Analyti-

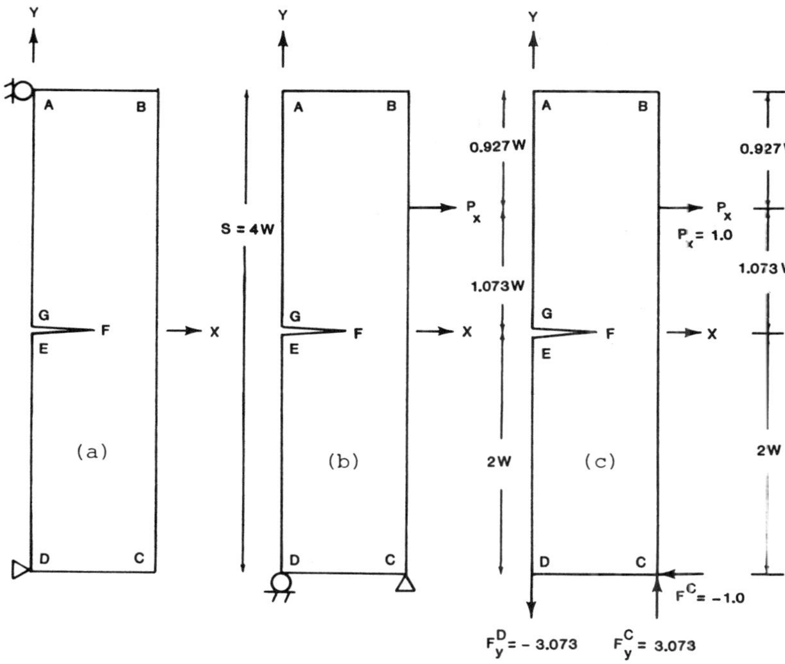

FIG. 14—*Calculating the stress-intensity factors (*K_I *and* K_{II}*) from the predetermined explicit weight functions under different constraint conditions.* (a) *Constraint conditions of available explicit weight functions.* (b) *Constraints and loading conditions for intended* K_I, K_{II} *calculations.* (c) *Self-equilibrium forces of* (b).

cally, the strain-energy release rates G_1 and G_2 are related to the decoupled $K_{\rm I}$ and $K_{\rm II}$ values from the J-integral concept, shown by Hellen et al. [*15*] and Nishioka et al. [*16*] as

$$G_1 = \frac{K_{\rm I}^2 + K_{\rm II}^2}{H} \tag{12a}$$

$$G_2 = -\frac{2K_{\rm I}K_{\rm II}}{H} \tag{12b}$$

Therefore, $G(\theta)$ values can be obtained using Eqs 11 and 12 for any 2-D mixed fracture mode problems for cracks propagating in a self-similar manner with one VCE. The application of this proposed approach to the three-point bend specimen with symmetric crack and with eccentric loading has been made in an attempt to predict the eccentricity effect on the cracking paths of a quasi-static crack.

Assuming that the crack propagation direction of a quasi-static crack takes place at θ_{cr} with the maximum strain-energy release rate G_{max} from Eq 11, one can predict θ_{cr} and G_{max}, as

$$\theta_{cr} = \tan^{-1}\left(\frac{-2K_{\rm I}K_{\rm II}}{K_{\rm I}^2 + K_{\rm II}^2}\right) = \tan^{-1}\left(\frac{G_2}{G_1}\right) \tag{13}$$

$$G_{max} = \frac{1}{H}(K_{\rm I}^4 + 6K_{\rm I}^2K_{\rm II}^2 + K_{\rm II}^4)^{1/2} \tag{14}$$

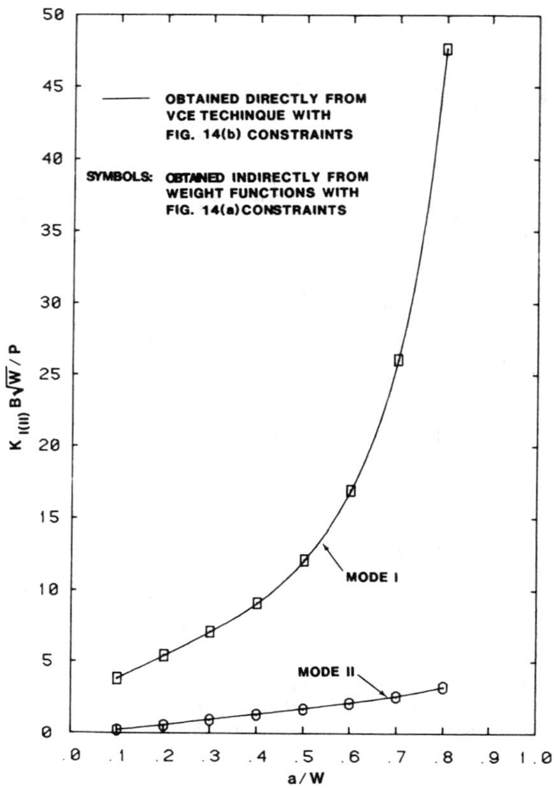

FIG. 15—*Comparison between dimensionless* $K_{I(II)}B\sqrt{W}/P$ *values obtained from directed approach (VCE technique with 14*b *constraints) and from indirect approach (explicit weight functions with 14*a *constraints) as a function of* a/W *ratios with* P_x *concentrated point force loading.*

The effect of eccentricity conditions on $G(\theta)$ is shown in Fig. 16 as a function of θ. Note the sinusoidal nature of $G(\theta)$ versus θ with both the reduction in G_{max} value and the shifting in θ_{cr} location as a result of eccentricity effects.

The plot of θ_{cr} as a function of loading eccentricity (e/W) is shown in Fig. 17 for a three-point bend specimen with $s/W = 4$ and $0.1 \leq a/W \leq 0.8$. The experimental verification of the analytical predictions on angles θ_{cr} due to eccentric loading has not been made yet. To compare the relative severity of the loading and support eccentricities, the finite-element studies with the weight function concept have been made for a three-point bend specimen with $s/W = 4$ and $e/W = 3/16$. As indicated in Fig. 18 for θ_{cr} as a function of a/W, one can conclude that the effect of load eccentricity on K_I and K_{II} values is more severe than that of support eccentricity.

Conclusions

By applying the colinear VCE technique to the symmetric mesh in a crack-tip neighborhood, an efficient finite-element method for evaluating both the decoupled stress-intensity factors (K_I and K_{II}) and the decoupled explicit weight function vectors (h_I, h_{II}) has been established recently by Sha and Yang [*8*] for the mixed-mode fracture problems. The application of this newly established technique to a three-point bend specimen with eccentric loading, which introduces

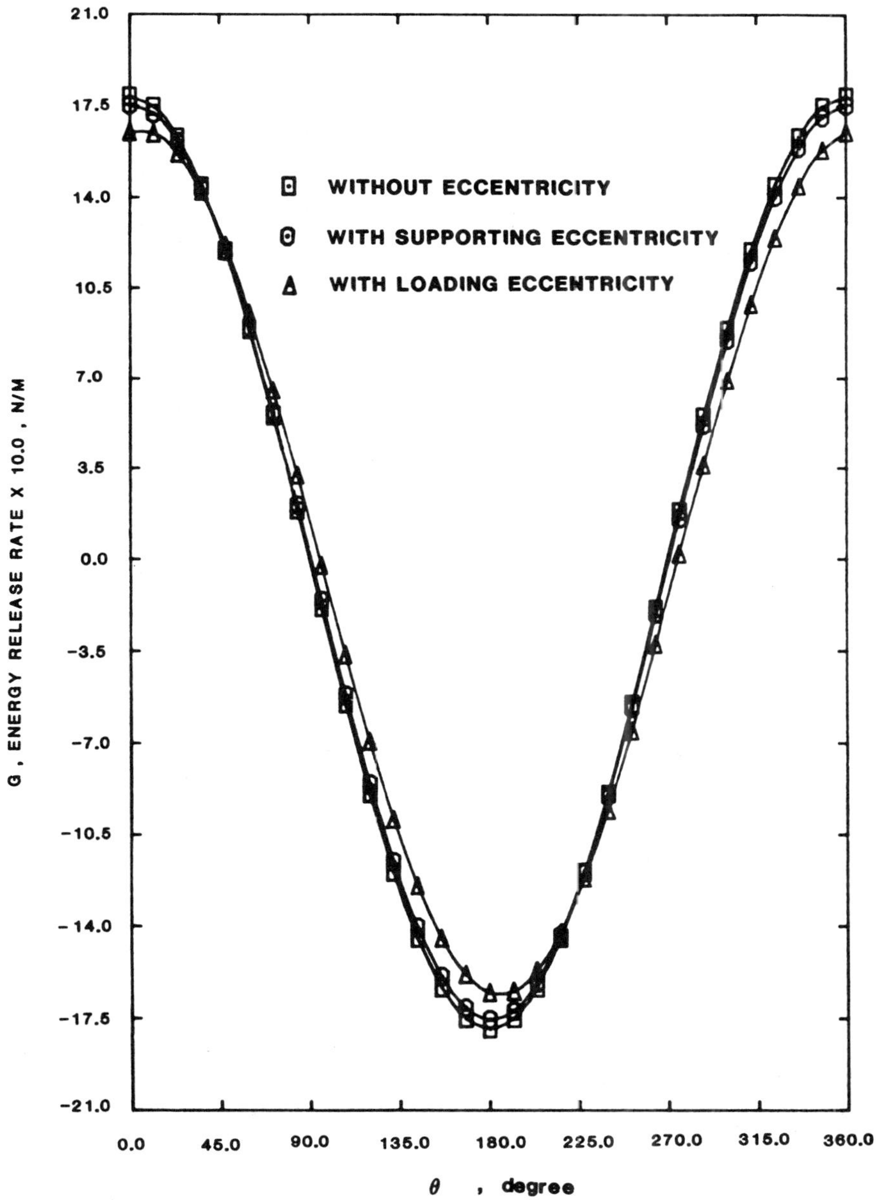

FIG. 16—*Sinusoidal characteristics of* G(θ) *for* $0 \leq \theta \leq 2\pi$ *with and without eccentricity effects for a three-point bend specimen with* s/W = *4,* a/W = *0.5, and* e/W = *0 or* 3/16.

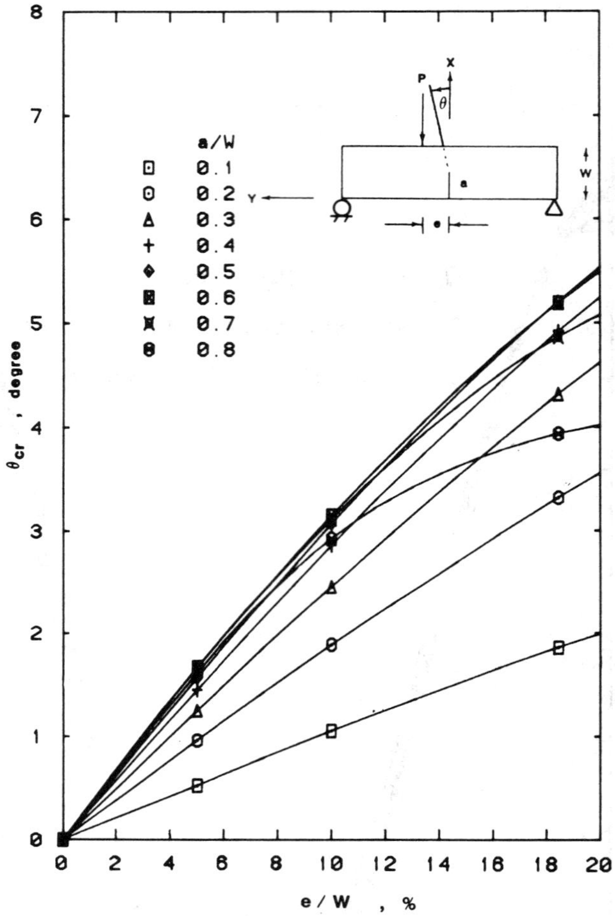

FIG. 17—θ_{cr} *as a function of* e/W *ratios for a three-point bend specimen with* s/W = *4 under loading eccentricity.*

the mixed fracture mode, is made in this paper to account for the eccentricity effects on K_{I} and K_{II} values with the weight function concept.

The additional finite-element efforts needed to obtain the explicit weight function for the entire structure of interest with an asymmetric crack are not much greater than the normal finite-element evaluations of the stress-intensity factors for given loading and constraint conditions with the colinear VCE technique coupled with symmetric mesh in crack-tip neighborhood. In general, the payoff is relatively large in providing the capability to calculate the stress intensity factors under any loading condition, including the residual stress effects. These needed stress intensity factors for life-prediction purposes can be obtained easily and inexpensively with Eq 3 or Eq 4 from the pre-determined explicit weight function vectors coupling with the applied tractions at their corresponding locations in conjunction with linear superposition principle. The load-independent characteristics of the explicit weight functions, which serve as a universal function for the stress-intensity factors, are largely responsible for the payoff of avoiding repeated calculations of $K_{\text{I(II)}}$ values under different loading conditions for a given crack geometry and constraints. The explicit weight functions for a given crack geometry are found to be non-

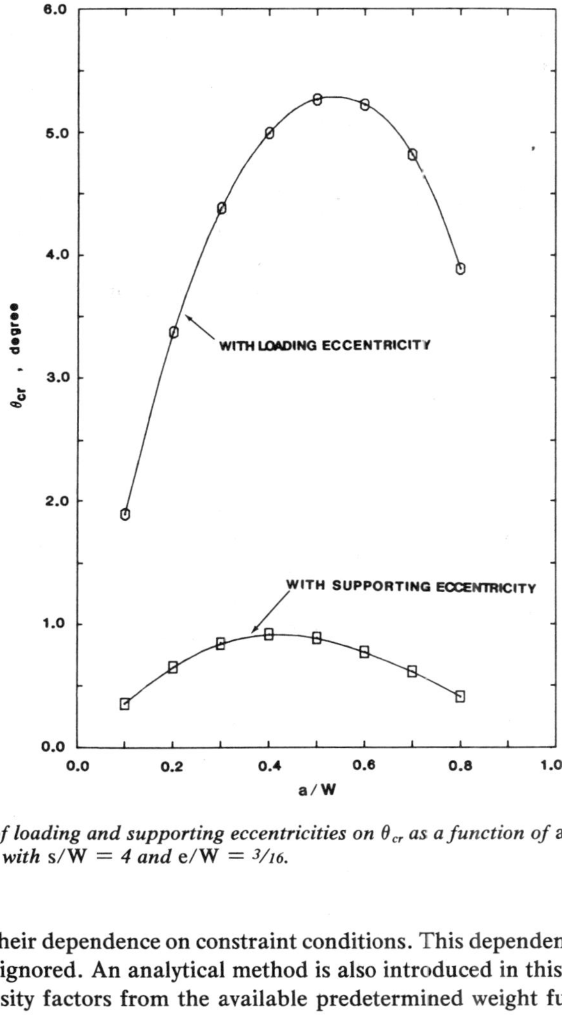

FIG. 18—*Effects of loading and supporting eccentricities on* θ_{cr} *as a function of* a/W *ratios for a three-point bend specimen with* s/W = *4 and* e/W = *3/16.*

unique because of their dependence on constraint conditions. This dependence for a given crack geometry has been ignored. An analytical method is also introduced in this paper for calculating the stress-intensity factors from the available predetermined weight functions, which are under different constraint conditions for the same crack geometry. This is done by including all self-equilibrium forces, which include the applied tractions and the reaction forces due to the imposed constraint, coupled with the predetermined explicit weight function components according to Eq 3 or Eq 4 at their locations with self-equilibrium forces.

The characteristics of the explicit weight function components along the traction application boundaries, such as the crack-face, top-face, RHS-face, and LHS-face locations, as shown in Fig. 2, are presented in this paper. With the exception of the crack-face location, only the explicit weight function component normal to the traction-application boundaries is provided with analytical expressions to facilitate K_I and K_{II} calculations with Eq 3. The calculated K_I values, which are obtained with the h_{Iy} explicit weight function component, coupled with remote tension and bending tractions at the LHS-face and RHS-face locations, compare well with those of Brown and Srawley [*14*], as shown in Fig. 13.

The eccentric loading effects, which are associated closely with the testing irregularities of the three-point bend specimen on the K_I and K_{II} values with the weight function concept, have been studied and the results are presented herein. As shown in Fig. 1, there are two kinds of eccentricity, namely, the load eccentricity and support eccentricity for a three-point bend specimen. For a given eccentricity (e), it is found that the support eccentricity is less severe than the load eccentricity in terms of the effects on K_I and K_{II} values, as shown in Fig. 7 with $e/W = 3/16$. As anticipated, the effects of load eccentricity on K_I and K_{II} values increases with increasing e/W and a/W ratios. At relatively low load eccentricity, ($e/W \leq 5\%$) for a three-point bend specimen with $a/W \leq 0.8$, the reductions in K_I value, as shown in Fig. 19, are negligibly small (less than 0.6%). However, at large crack length to width ratio, say $a/W = 0.8$, with large load eccentricity of $e/W = 20\%$, the reduction of the K_I value can be as much as 7.5%. The reduction in K_I values for a three-point bend specimen as a function of a/W ratios ($0.1 \leq a/W \leq 0.8$) with $s/W = 4$ and $e/W = 3/16$ under loading and supporting eccentricity conditions is shown in Fig. 20.

As a result of eccentric loading of the three-point bend specimen, the Mode II stress-intensity factors are produced along with the reduction in K_I values, as shown in Fig. 5, as a function of

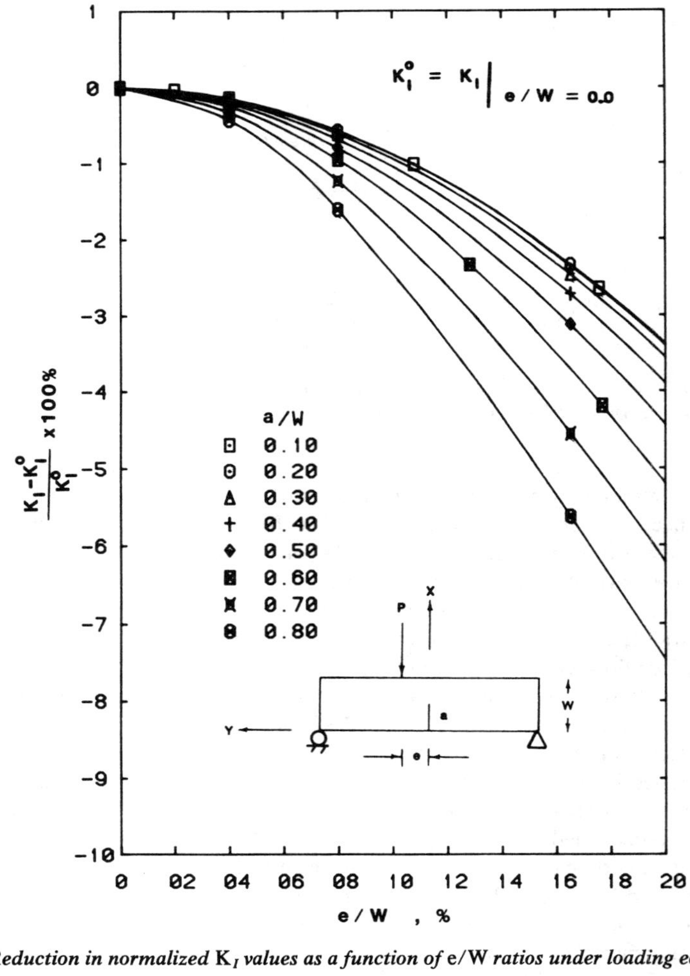

FIG. 19—*Reduction in normalized* K_I *values as a function of* e/W *ratios under loading eccentricity condition.*

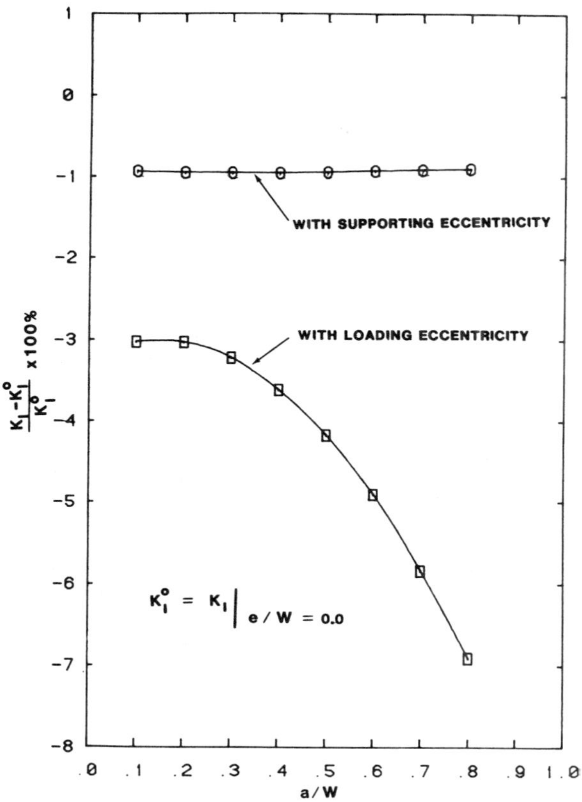

FIG. 20—*Effects of loading and supporting eccentricities on the reductions in the normalized* K_I *values for a three-point bend specimen with* s/W = 4 *and* e/W = 3/16.

e/W ratios. Assuming propagation of a quasi-static asymmetric crack taking place at the θ_{cr} direction with the maximum strain energy release rate (G_{max}), the analytical prediction of θ_{cr} are provided in this paper as a function of loading eccentricity (e/W). At $e/W = 10\%$ of load eccentricity, θ_{cr} less than 3 deg can result for $0.1 \leq a/W \leq 0.8$. However, at $e/W = 3/16$, $\theta_{cr} =$ 5.2 deg is predicted at $a/W = 0.5$, as shown in Fig. 18 for the θ_{cr} versus a/W plot. The experimental verification on θ_{cr} predictions as a function of a/W ratios is not available yet.

From the finite-element results for a three-point bend specimen with $s/W = 4$, as reported in this paper, one can conclude that at relatively large load eccentricity, with $e/W = 20\%$, the upper error boundary can be related in terms of a maximum reduction in K_I of about 7.5% and a maximum θ_{cr} of 5.5 deg for $a/W \leq 0.8$.

In summary, an efficient finite-element evaluation of the decoupled explicit weight functions for the mixed fracture mode has been achieved by using the symmetric mesh in crack-tip neighborhood in conjunction with the colinear virtual crack extension technique. This symmetric mesh approach has been applied to ASTM three-point bend specimen with eccentric loading. The effects of eccentric loading on the K_I and K_{II} values are detailed in this paper using the explicitly predetermined weight functions for the ASTM three-point bend specimen with $s/W = 4$ and $0.1 \leq a/W \leq 0.8$.

References

[1] Bueckner, H. F., "Field Singularities and Related Integral Representations" in *Mechanics of Fracture I, Method of Analysis and Solutions of Crack Problems,* G. C. Sih, Ed., Noordhoff, Leyden, The Netherlands, 1972.

[2] Parks, D. M. and Kamentzky, E. M., "Weight Functions from Virtual Crack Extension," *International Journal for Numerical Methods in Engineering,* Vol. 14, 1979, pp. 1693-1706.

[3] Vanderglas, M. L., "Weight Functions for Plane and Axisymmetric Crack Problems" in *Numerical Methods in Fracture Mechanics,* D. R. J. Owen and A. R. Luxmoore, Eds., *Proceedings,* 2nd International Conference held at University College, Swansea, Wales, U.K., July 7-11, 1980, pp. 51-65.

[4] Sha, G. T., "Stiffness Derivative Finite Element Technique to Determine Nodal Weight Functions with Singularity Elements," *Engineering Fracture Mechanics,* Vol. 19, No. 4, 1984, pp. 685-699.

[5] Rice, J. R., "Some Remarks on Elastic Crack-Tip Stress Fields," *International Journal of Solids Structures,* Vol. 8, 1972, pp. 751-758.

[6] Ishikawa, H., Kitagawa, H., and Okamura, H., "J Integral of a Mixed Mode Crack and Its Application," Third International Conference on Fracture Mechanics, Cambridge, U.K., Vol. 3, Aug. 1979, pp. 447-455.

[7] Ishikawa, H., "A Finite Element Analysis of Stress Intensity Factors for Combined Tensile and Shear Loading by Only a Virtual Crack Extension," *International Journal of Fracture,* Vol. 16, 1980, pp. 243-246.

[8] Sha, G. T. and Yang, C.-T., "Weight Function Calculations for Mixed Mode Fracture Problems with the Virtual Crack Extension Technique," *Engineering Fracture Mechanics,* Vol. 21, 1985, pp. 1119-1149.

[9] Parks, D. M., "A Stiffness Derivative Finite Element Technique for Determination of Crack Tip Stress Intensity Factors," *International Journal of Fracture,* Vol. 10, 1974, pp. 487-501.

[10] Hellen, T. K., "On the Method of Virtual Crack Extensions," *International Journal for Numerical Methods in Engineering,* Vol. 9, 1975, pp. 187-207.

[11] Henshell, R. D. and Shaw, K. G., "Crack Tip Elements are Unnecessary," *International Journal for Numerical Methods in Engineering,* Vol. 9, 1975, pp. 495-507.

[12] Barsoum, R. S., "On the Use of Isoparametric Finite Elements in Linear Fracture Mechanics," *International Journal for Numerical Methods in Engineering,* Vol. 10, 1976, pp. 25-37.

[13] Srawley, J. E., "Wide Range Stress Intensity Factor Expression for ASTM E 399 Standard Fracture Toughness Specimens," *International Journal of Fracture,* Vol. 12, 1976, pp. 475-476.

[14] Brown, W. F. and Srawley, J. E. in *Current Status of Plane Strain Crack Toughness Testing of High Strength Metallic Materials, ASTM STP 410,* American Society for Testing and Materials, Philadelphia, 1966.

[15] Blackburn, W. S. and Hellen, T. K., "Calculation of Stress Intensity Factors in Three Dimensions by Finite Element Methods," *International Journal for Numerical Methods in Engineering,* Vol. 11, 1977, pp. 211-229.

[16] Nishioka, T. and Atluri, S. N., "Path-Independent Integrals, Energy Release Rates, and General Solutions of Near-Tip Fields in Mixed-Mode Dynamic Fracture Mechanics," *Engineering Fracture Mechanics,* Vol. 18, No. 1, 1983, pp. 1-22.

Poster Session—Elastic Plastic Fracture Mechanics I

Christian Thaulow,[1] Mons Hauge,[2] Åsmund Gunleiksrud,[2] and Anthonius Johannes Paauw[2]

Single-Specimen Test Measurement of J_i and J-Δa with a Pulsed D-C Potential-Drop Technique

REFERENCE: Thaulow, C., Hauge, M., Gunleiksrud, Å., and Paauw, A. J., "**Single-Specimen Test Measurement of J_i and J-Δa with a Pulsed D-C Potential-Drop Technique,**" *Fracture Mechanics: Eighteenth Symposium, ASTM STP 945,* D. T. Read and R. P. Reed, Eds., American Society for Testing and Materials, Philadelphia, 1988, pp. 333–346.

ABSTRACT: Fracture mechanics testing [K_{Ic}, J-integral, and crack-tip opening displacement (CTOD)] of a QT-steel (AISI 4137 H) has been carried out in order to compare the testing methods. The tests were performed at room temperature, and the material toughness was varied by applying annealing temperatures from 300 to 650°C.

The J-initiation values were successfully determined with a specially developed d-c potential-drop (PD) method. The technique is based upon pulsed current and reference electrodes. Only the potential-drop difference caused by a current difference is measured; this eliminates the thermoelectric effect and offset voltage associated with PD equipment.

On the basis of the PD measurements both the J-initiation value and the R-curve (J-Δa) could be determined from a single-specimen test.

The R-curves could be fitted to a third-order polynomials, but even if the initiation values were nearly constant for replicate specimens, the slope of the R-curve would vary significantly. The mechanisms of crack initiation and growth are different, and the observed results indicate that the crack growth properties, as characterized with the R-curve, of the present steel quality vary from one specimen to the next.

The J-initiation values determined with the multispecimen technique recommended in ASTM E 813 showed considerably more scatter in comparison with the present PD technique. The results suggest that the J-initiation value should be determined on each specimen and not calculated with extrapolation techniques.

KEY WORDS: potential drop, J-integral, quenched and tempered steel

Nomenclature

- a Crack length
- A Area under load-displacement curve
- J J-integral
- K Stress intensity factor
- P Load
- u Load-point displacement
- U Dimensionless potential drop

[1]Professor, Department of Materials and Processes, The Norwegian Institute of Technology, Trondheim, Norway.

[2]Research engineers, Division of Materials Processes, The Foundation for Scientific and Industrial Research at The Norwegian Institute of Technology, Trondheim, Norway.

V Measured potential drop
σ_y Yield or 0.2% proof strength
η Geometry factor used in calculations from A to J
σ_m Ultimate tensile stress
%Ra Reduction in area
%El Elongation
J_{Ic} J-integral at initiation of ductile crack extension according to ASTM E 813 procedure
J_i J-integral at initiation of ductile crack extension according to presented potential-drop technique

J-integral testing is often introduced as an alternative to crack-tip opening displacement (CTOD) testing, partly because the mathematical expression is well suited for computer-based analysis and partly because of the potential to convert elastic-plastic J-values to K_{Ic} values.

The fundamental basis for the J-integral is related to the energy at the instant of fracture initiation. Modern structural steels have a high resistance to brittle fracture, and the critical value of J is therefore usually based upon the initiation of ductile crack growth.

Several methods have been proposed, and many experiments carried out, in order to measure the ductile crack initiation value in single-specimens tests. But no widely accepted technique has been developed so far, and both the ASTM Test Method for J_{Ic}, a Measure of Fracture Toughness (E 813-81) and the British Standards Institute Methods for Crack Opening Displacement (COD) Testing (BS 5762:1979) recommend the multiple-specimen technique in determining the initiation value.

In the practical application of the CTOD method, the problem of ductile crack initiation is avoided by tolerating ductile crack growth ($CTOD_u$, $CTOD_m$). In J-testing, detailed requirements have been specified in order to standardize the multiple-specimen technique.

In the present investigation the crack initiation and growth have been measured with both the multiple-specimen and a pulsed d-c potential-drop (PD) technique. The PD technique offers the potential for determining both the initiation value and the R-curve for each specimen tested, without introducing any partial unloading during the test.

Experimental

An AISI 4137 H quality steel, Table 1, was quenched from 845°C and subsequently tempered in the temperature range of 300 to 650°C. The yield strength decreased with increasing tempering temperature, Fig. 1, while the Charpy toughness increased, Fig. 2.

Standard fracture mechanics three-point bend specimens (28 by 56 by 275 mm) were machined and tested at room temperature in accordance with ASTM E 813-81 and BSI BS 5762.

TABLE 1—*Chemical composition of the steel.*

Element	Weight, %
Carbon	0.34
Silicon	0.36
Manganese	0.93
Sulfur	0.033
Phosphorous	0.019
Chromium	0.76
Molybdenum	0.19
Nitrogen	0.007

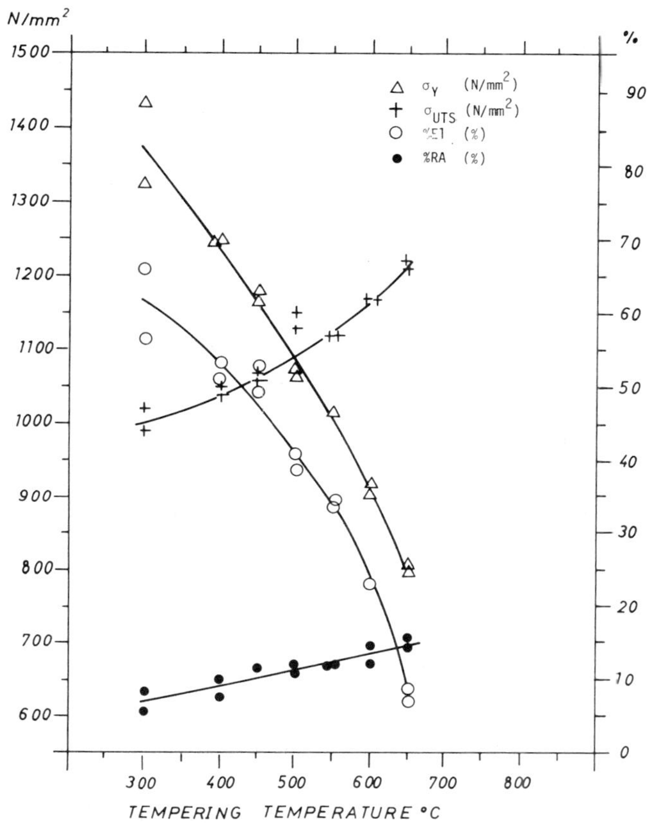

FIG. 1—*Mechanical properties dependent on the tempering temperature.*

Potential-Drop Technique

The PD technique is based on two simple concepts: pulsed current and reference electrodes. With pulsed current, only the potential-drop difference caused by a current difference is measured. This eliminates thermoelectric and offset voltage effects associated with normal PD measurements.

To reduce the influence of random electronic noise, the current difference and thus the PD difference was made as great as possible. With a given current source this is most effectively accomplished by measuring the PD difference by reversing the direction of the current.

A pair of reference electrodes is used to measure the potential drop outside the cracked area; that is, a "dummy" with the same overall geometry but without a notch is connected to the specimen. We now define a dimensionless quantity U as

$$U = V_a^1 - V_a^2 / V_0^1 - V_0^2 \tag{1}$$

where V_a^i and V_0^i refer to the potential drop measurements between the crack electrodes and reference electrodes, respectively, and $i = 1,2$ refers to the direction of the current.

Provided all four potentials are measured within sufficiently short time, temperature drift can be neglected. It is further assumed that V^1 and V^2 are measured at the same current level in

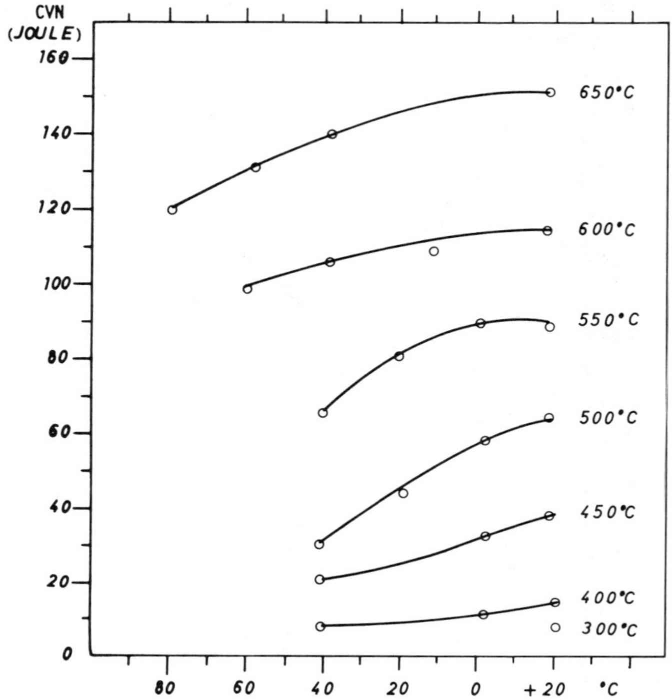

FIG. 2—*Charpy energy transition curves related to tempering temperature.*

either direction. With these assumptions satisfied, U will be independent of the absolute value of current, specimen conductivity, and temperature.

The system is based on a low-cost, single-board microcomputer (AIM 65, Rockwell International) and has been successfully applied for long-term crack depth measurements in corrosion fatigue experiments [*1*]. The system has been modified for measurements of crack initiation and crack growth in fracture mechanics testing [*2*], Fig. 3.

The four potentials (Eq 1) are measured at predetermined time intervals without interrupting the loading of the specimen. The current is switched on for one second, with a current level of approximately ±150 A.

The PD for each subsequent measurement (expressed as the dimensionless quantity U, Eq 1) is plotted on the load-displacement record and the initiation of ductile crack growth determined where a sharp shift in the potential was observed, Fig. 4. After initiation, the potential drop is often observed to be linear with respect to crack growth; hence, a linear relationship can be expected between the PD and Δa.

The PD at crack initiation is then related to the initial crack size, with addition for crack blunting, while the final crack size is measured after the test is completed and the specimen broken open in liquid nitrogen. On this basis an R-curve can be established for each single specimen and compared with "multiple specimen" results.

Results

Valid K_{Ic} results were obtained for tempering temperatures up to 550°C, Fig. 5, and the additional tests with PD measurements were performed only on specimens tempered to 600 and 650°C.

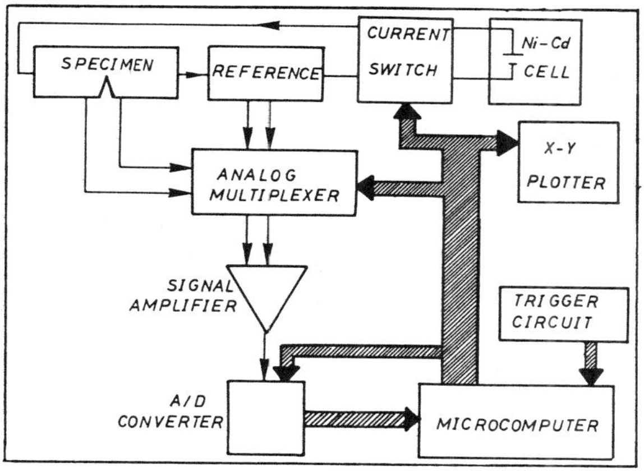

FIG. 3—*Schematic presentation of the setup for the pulsed current PD technique.*

J-Integral

The J_{Ic}-values, measured in accordance with ASTM E 813, are listed in Table 2. The R-curves are presented in Fig. 6.

The R-curves depend very much on which specimens are selected for the regression analysis, giving a maximum effect when all the specimens tempered to 650°C—including specimens with brittle fracture—are compared with the ductile, unloaded specimens.

Potential-Drop Measurements

The crack initiation value was determined on eight specimens, Table 3. The most distinct and least distinct records are shown in Fig. 7. The remaining PD measurements were unsatisfactory due to insufficient electrical isolation between the specimens and the test machine.

J-Δa values were calculated for each PD value, in accordance with the PD technique; hence, an R-curve could be drawn for each specimen, Figs. 8 and 9.

A successful fit to a third-order polynomial relationship of the form

$$J = ax^3 + bx^2 - cx + d \tag{2}$$

was obtained for two of the specimens, Fig. 10. For the other specimens, the crack extension was too limited to establish an R-curve.

CTOD

The ductile crack growth was measured on photocopies of the fracture surfaces, with ×4 magnification, giving favorable conditions for a quantification of the first evidence of crack growth, although it was difficult to differ between crack-tip blunting and crack growth. The CTOD values were calculated in accordance with BSI BS 5762, and CTOD_i determined from the CTOD_u-Δa relationships, Figs. 11 and 12.

Crack-tip blunting according to the J-integral curves is about 0.05 mm for the 600°C tempered specimens and 0.10 mm for the 650°C tempered specimens, giving CTOD_i values of 0.04 and 0.09 mm, respectively.

Limited ductile crack growth in the range of 0.2 to 0.3 mm will not be detected by normal

Calculating J_i:

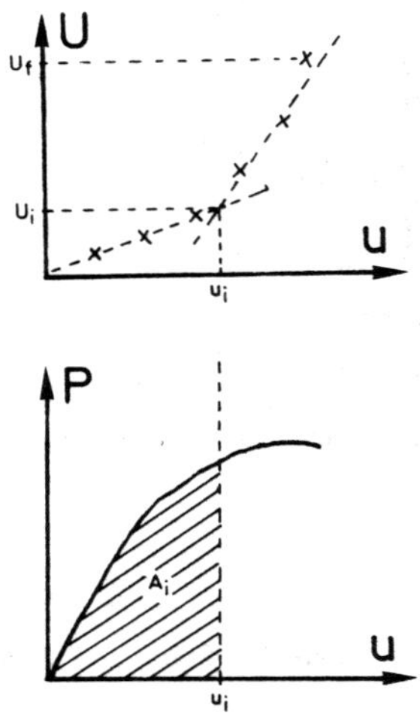

A sharp shift at P.D.-displacement record give the initiation value (u_i, U_i)

Area measurement at Load-displacement curve give Area A_i at initiation (u_i)

Hence:

$$J_i = \frac{\eta A_i}{B(W-a_o)}$$

Calculating J-R curve:

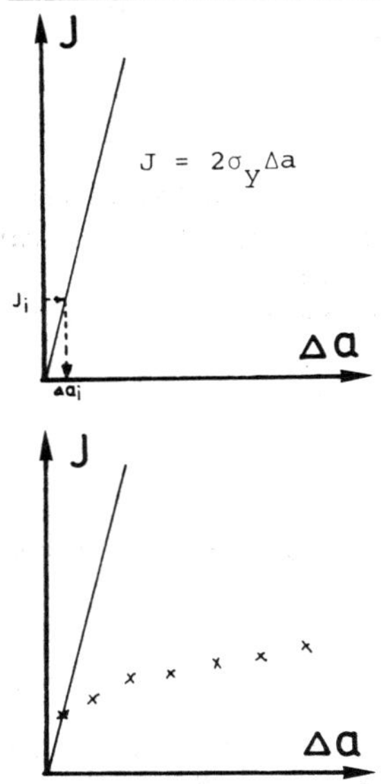

Δa_i is calculated from a known blunting line at J_i

$$a_i = a_o + \Delta a_i$$

Assume a linear relationship between P.D. and crack length after initiation

Hence:

$$a = a_i + \frac{a_f - a_i}{U_f - U_i}(U - U_i)$$

$$\Delta a = a - a_o$$

J and Δa is calculated at each potential drop measurement.

FIG. 4—*A single-specimen PD method; procedure for measurements and calculations.*

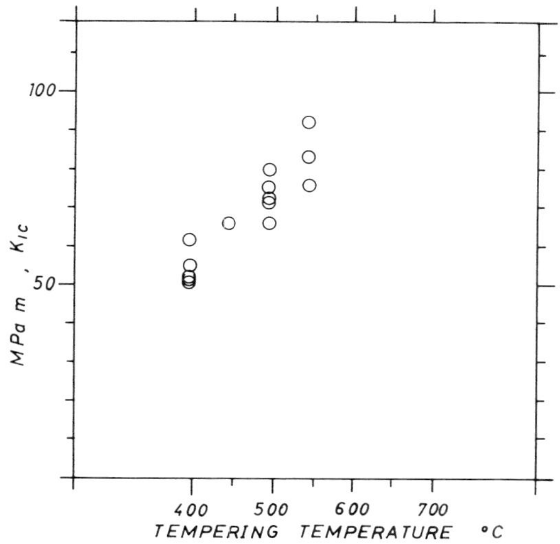

FIG. 5—*Valid K_{Ic} results, according to BSI Methods of Test for Plain Strain Fracture Toughness (K_{Ic}) of Metallic Materials (BS 5447:1977), related to tempering temperature.*

TABLE 2—*J_{Ic} values calculated in accordance with ASTM E 813.*

Tempering Temperature, °C	Total Number of Specimens	Number of Valid Results	Regression Line $J/\Delta a$	Regression Line J_0	Regression Line Correlation Coefficient	J_{Ic}, kJ/m
600	6	3	319	23	0.999	28[b]
650	12	7	366	86	0.928	123
650[a]	5	4	299	217	0.997	275

[a]Only including unloaded specimens, no brittle fracture.
[b]Only three valid data points, not valid J_{Ic}.

visual inspection; hence, typical $CTOD_c$ values recorded will be 0.07 mm at 600°C tempering and 0.14 mm at 650°C.

Discussion

In the present J-integral standard, detailed requirements have been specified in order to standardize the multiple-specimen technique. But several investigations and round-robins have documented that this is a difficult matter, and different alternatives to the standardized procedure have been proposed. The suggestions include alternative power-law relationships to the ASTM linear regression line, giving the best fit for the material investigated, in a similar way as presented in Fig. 10.

Another proposal is to introduce an engineering initiation value, related to some defined and

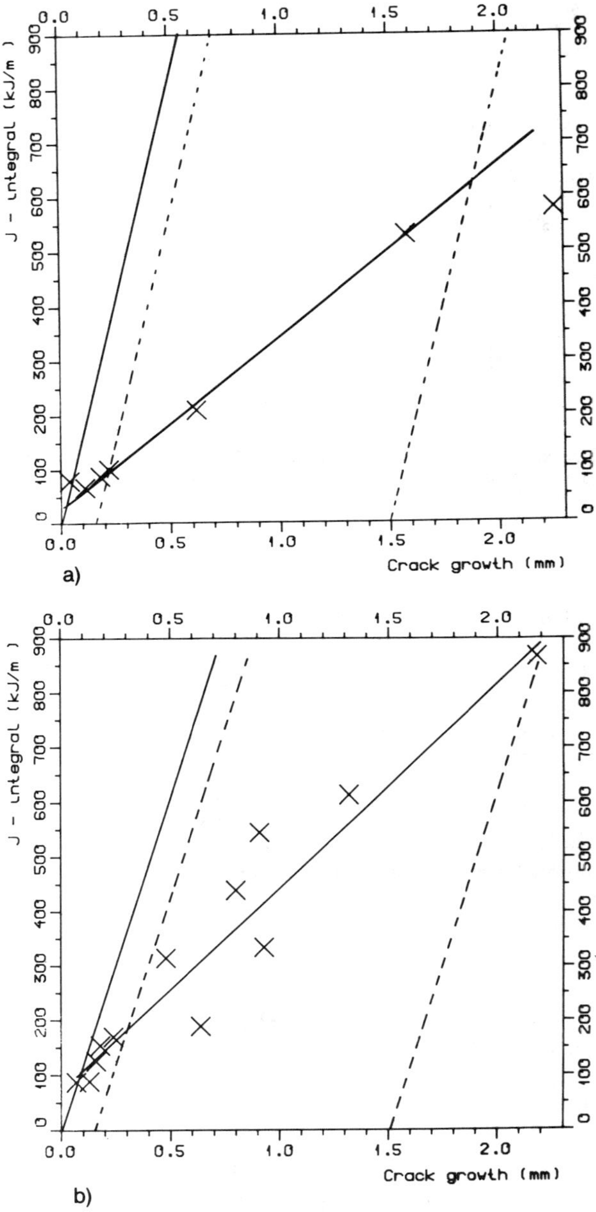

FIG. 6—*R-curves for the test series presented in Table 2.*

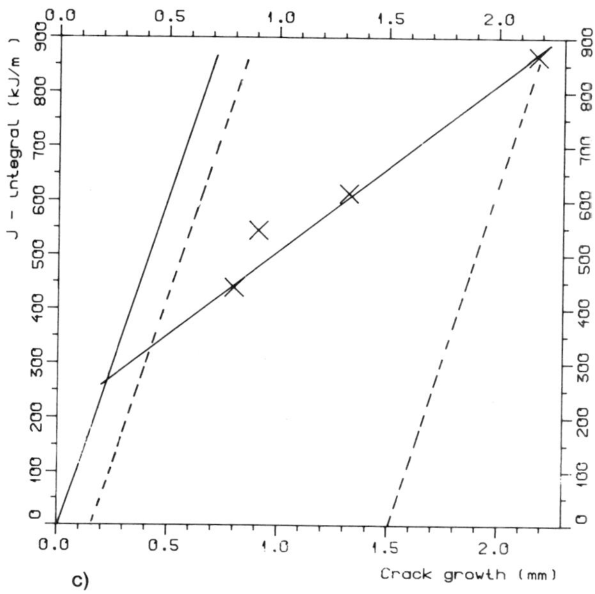

FIG. 6—*Continued.*

TABLE 3—*J_{Ic} determined from PD measurements (see Fig. 4).*

Tempering Temperature, °C	J_{Ic}, kJ/m	Average J_{Ic}, kJ/m
600	36[b], 56, 73	55
650	103[b], 109[b], 110, 131[b], 133	117

[b]Brittle fracture.

easily measurable ductile crack growth (in the same way as the 0.2% yield stress), in order to avoid the measurements close to crack initiation; see Fig. 6*b*.

One basic reason for the difficulties experienced with the multiple-specimen technique is probably due to the different micromechanisms involved in ductile crack initiation and the subsequent growth. Hence, extrapolation techniques based upon crack growth will not necessarily be representative for the initiation value.

Due to this experience, the PD method described has special advantages when the testing is performed in the transition area. In this temperature range, a "sharp shift" of the PD-load displacement curve should be fairly easy to detect. During testing of more ductile materials at upper-shelf temperatures, problems may occur with defining the initiation value. A more sophisticated calibration of the PD-crack extension relationship using more specimens will take care of such problems. A calibration procedure is presented in Fig. 13.

The present results have shown that the ASTM regression curve technique is unrealistic. The

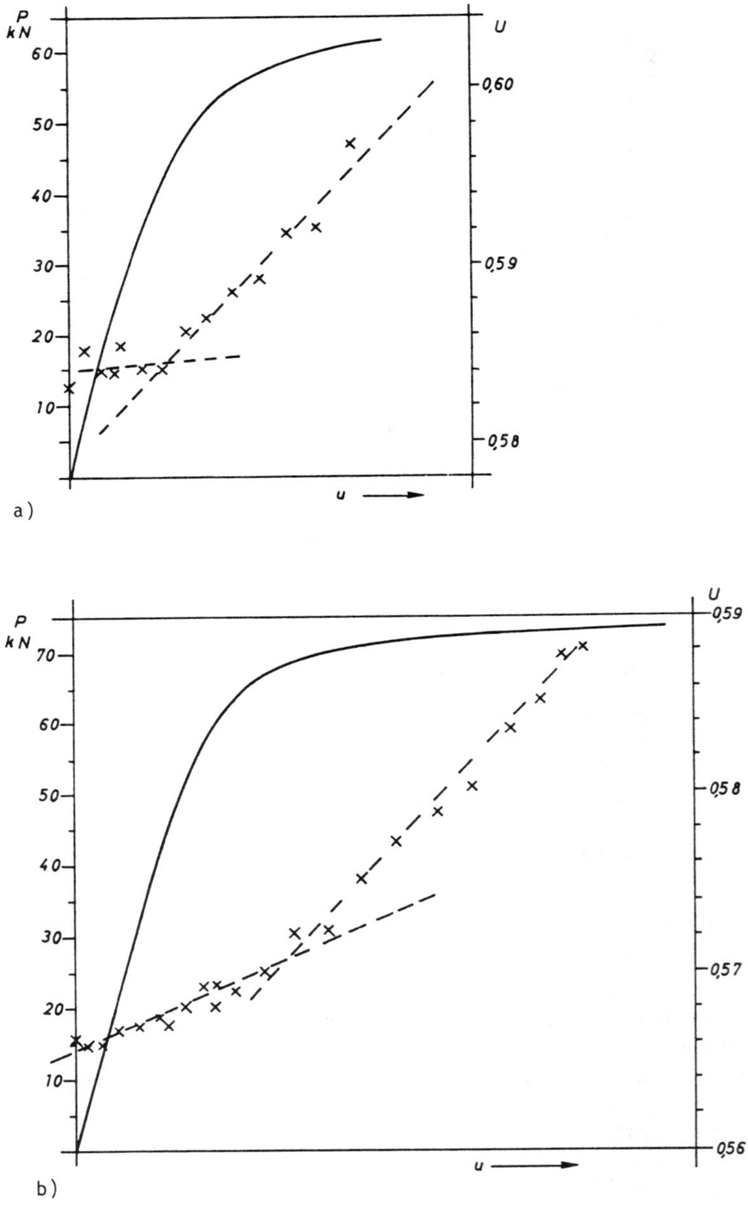

FIG. 7—*Examples of the most distinct* (a) *and least distinct* (b) *PD measurements of crack initiation. The proposed initiation point is indicated.*

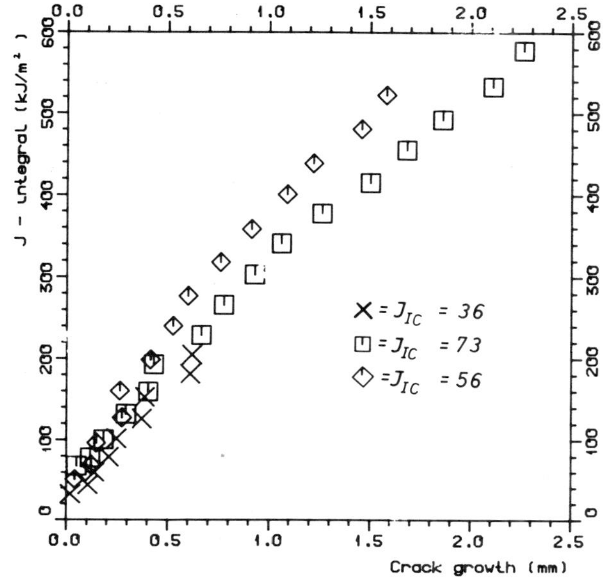

FIG. 8—*R-curves based on single-specimen PD measurements. Tempering temperatures of 600°C.* J_{Ic} *values refer to Table 3.*

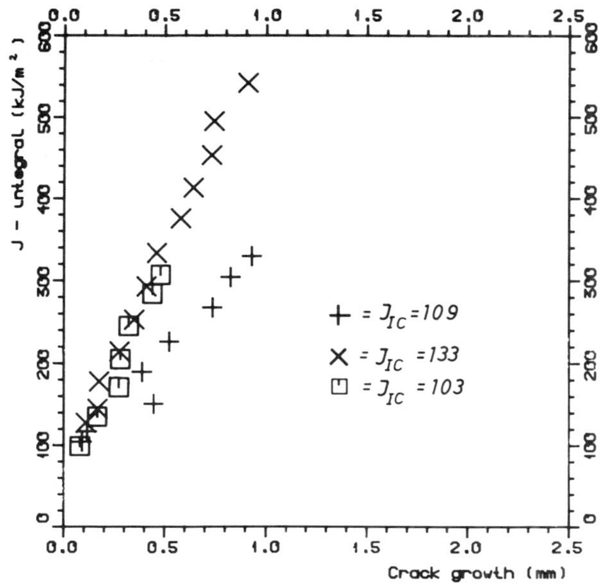

FIG. 9—*R-curves based on single-specimen PD measurements. Tempering temperature of 650°C.* J_{Ic} *values refer to Table 3.*

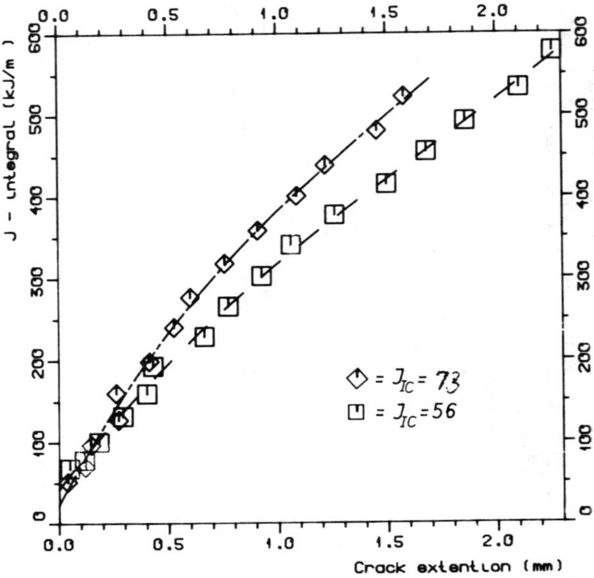

FIG. 10—*Third order polynomial curve fit for R-curves. Tempering temperature of 600°C.* J_{Ic} *values refer to Table 3.*

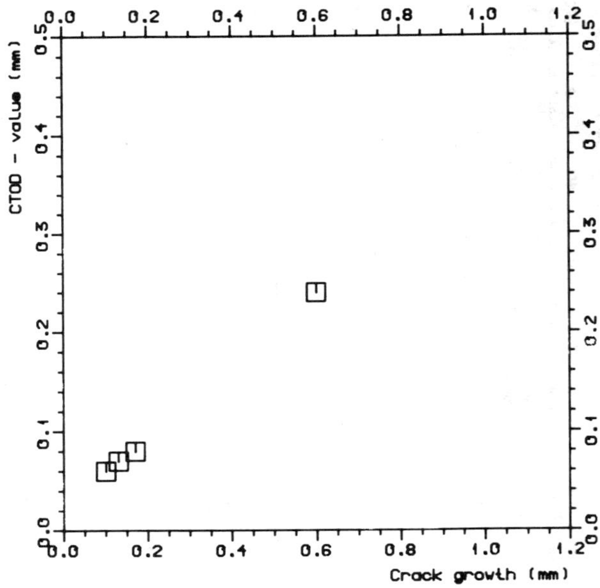

FIG. 11—*CTOD*$_u$ *values versus crack growth. Tempering temperature of 600°C.*

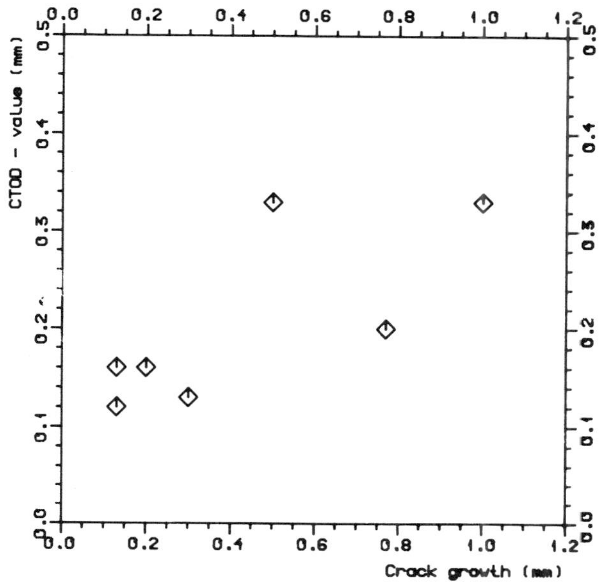

FIG. 12—*CTOD*$_u$ *values versus crack growth. Tempering temperature of 650°C.*

J_{Ic} values depend to a large degree upon the specimens selected for the regression analysis, Table 2, where a dramatic change in the initiation value is recorded when specimens exhibiting brittle fracture are excluded. The PD measurements, however, revealed a moderate scatter in the initiation values, independent of the tendency to final brittle fracture or not, Table 3. The R-curves however, could vary, indicating different resistance to crack growth (in a transition area), Fig. 9.

The ASTM procedure is therefore not applicable in the transition regime as noted in the E 813 standard. However, when an initiation of crack growth can be detected and ductile crack growth recorded prior to a fast fracture, a J_i value should be able to describe the material crack initiation resistance. The restriction on the ASTM procedure is caused by problems in obtaining a reliable regression line; there is no physical explanation for it. The problems are probably caused by the scatter in crack growth resistance as observed in the investigation.

The present PD single-specimen technique includes both the traditional "sharp" shift in potential at crack initiation and a quantification of the crack growth.

The validity of an R-curve based on PD measurements is to some degree dependent on the crack-tip blunting relation. In the present investigation, the ASTM E 813 blunting relation is used, but later investigations have shown that the blunting line slope is material-dependent.

Practical experience with the pulsed d-c potential-drop method during fracture mechanics testing is limited, but the results so far, and work in progress, indicate that the method should be further explored.

Conclusion

1. The J-initiation value should be determined on each individual specimen and not calculated from multiple-specimen techniques.

2. The pulsed current potential-drop technique presented in this paper has the possibility for determining both the crack initiation and the R-curve in single-specimen tests.

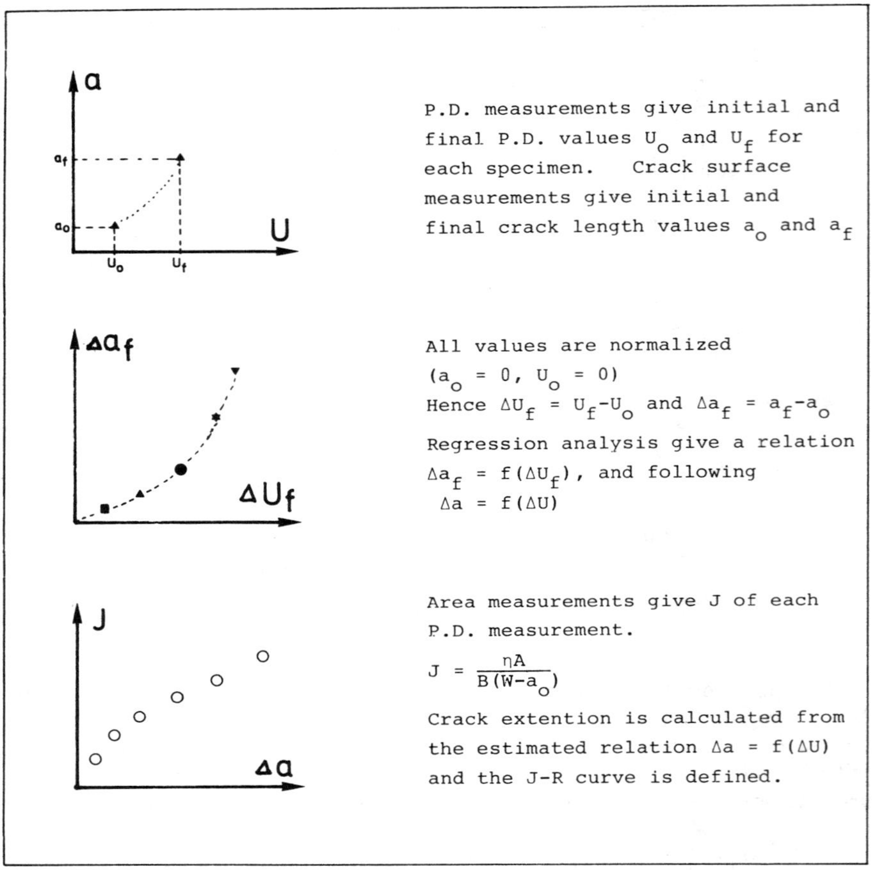

FIG. 13—*A multiple-specimen PD method. Measurements and calculations.*

References

[1] Bardal, E., Berge, T., Grøvlen, M., Haagensen, P. J., and Førre, B. M., "A Pulsed d.c. Method for the Measurement of Low Fatigue Rates of Crack Growth," *Advances in Crack Length Measurements,* Engineering Materials Advisory Services (EMAS) Ltd., United Kingdom, 1982.

[2] de Vries, M. I. (Chairman), "Working Party Crack Initiation—Crack Initiation Round Robin," European Group on Fracture, Task Group 1 (EGF, TG1), Netherlands Energy Research Foundation (ECN), The Netherlands, 1982.

[3] Hauge, M., "J-integral Testing, European Group on Fracture (EGF) Round Robin," SINTEF document, The Foundation for Scientific and Industrial Research, Norway, 1985.

Mark T. Kirk[1] and Edwin M. Hackett[1]

An Evaluation of J-R Curve Testing Using Three-Point Bend Specimens

REFERENCE: Kirk, M. T. and Hackett, E. M., "**An Evaluation of J-R Curve Testing Using Three-Point Bend Specimens,**" *Fracture Mechanics: Eighteenth Symposium, ASTM STP 945,* D. T. Read and R. P. Reed, Eds., American Society for Testing and Materials, Philadelphia, 1988, pp. 347–373.

ABSTRACT: This investigation concerns the accuracy of load-line deflection measurement techniques used in J-integral/R-curve testing of three-point bend specimens. Four methods of load-line deflection measurement were investigated. One method employed a calibrated transducer that gave a direct reading of load-line deflection. The other three methods mathematically related load-line deflection to some other displacement (that is, crack mouth opening displacement, crosshead displacement) measured during testing. Photographs taken at regular intervals during loading provided an absolute measure of load-line deflection. In comparison with this reference, the other methods had equal accuracy with varying degrees of precision. In no case, however, did the measurements from the laboratory techniques deviate from the photographic reference by more than 0.052 mm (0.002 in.) on average. The small differences between the load-line deflections measured by the various laboratory techniques cause less scatter in the J-R curves and J_{Ic} values than attributable to material variability. Further, J-R curves over the first 1.75 mm (0.068 in.) of crack extension from both bendbars and compacts were shown to be coincident for the alloys tested.

KEY WORDS: load-line deflection measurement, three-point bend, J-integral/R-curve, high-strength steel

Nomenclature

- b Uncracked ligament
- β Empirical coefficient expressing fraction of b around which the remaining ligament is assumed to hinge
- E Young's modulus
- I Moment of inertia
- d_{CMOD} Crack mouth opening displacement
- D_{LL} Load-line deflection
- D_{LL}^{d} Load-line deflection calculated using d_{CMOD}
- D_{LL}^{drc} Load-line deflection calculated using d_{CMOD} with a rotation correction
- D_{LL}^{corr} Load-line deflection calculated using an experimentally developed correlation
- h_0 Original height of the horizontal reference bar support point above the roller
- J_{Ic} Mode I critical fracture toughness
- m Slope of a plot of D_{LL} versus d_{CMOD}
- P Load
- S Span between rollers
- W Specimen width

[1]Mechanical engineer and materials engineer, respectively, David W. Taylor Naval Ship Research and Development Center, Annapolis, MD 21402-1198.

Resistance curve characterizations of materials made using three-point bend specimens (hereafter bendbars) frequently exhibit more scatter in interlaboratory data comparisons than do characterizations made using compact tension specimens [*1*]. The lack of a convenient location for attachment of a deflectometer on the load line has led to a lack of agreement regarding the proper way to measure load-line deflection. Consequently, several different techniques for measuring this quantity have arisen. These fall into two categories: direct methods that measure load-line deflection using some transducers, and indirect methods that infer load-line deflection from some other displacement quantity measured during testing. These inconsistent measurement techniques have been blamed for the lack of comparability of interlaboratory bendbar data.

Joyce and Hackett [*2*] employed a direct method, measuring load-line deflection with a thin elastic bar loaded in four-point bending by pins protruding from the bendbar. Calibration of this transducer, or "flex bar," prior to each test provided a relationship between displacement at the center of the bar and the output of the strain gage bridge affixed to the bar. Using this calibration and readings from the flex bar taken during the fracture test allowed the determination of load-line displacement.

A technique described in the proposed J-R curve testing standard [*3*] requires that a record of the testing machine crosshead displacement be made during fracture testing. Load-line deflection is determined by subtracting from these values the crosshead displacement attributable to elastic compression and brinelling of the testing machine and test fixtures. The load versus crosshead displacement relation thus derived incorporates a component of elastic compliance unique to an individual test setup. Consequently, a new relation must be determined each time any component of the setup is changed.

Because they require taking another channel of data beyond crack mouth opening displacement (d_{CMOD}) and load typically required in compact tension tests, both methods discussed heretofore somewhat complicate J-R curve testing of bendbars. Alternatively, a relationship that infers load-line displacement from d_{CMOD} could be used. This relationship could be determined using the same assumptions of rigid rotation about a plastic hinge-point used to derive crack-tip opening displacement from d_{CMOD} [*4*] or using an experimentally derived correlation. Such a relationship, if justifiable, could considerably simplify the data acquisition requirements of bendbar testing.

This investigation focuses on the accuracy of the methods of approximating load-line deflection discussed above (that is, via a flex bar, via a correction to the testing machine crosshead displacement, or via the d_{CMOD} value). Photographs taken during testing will provide an absolute reference against which all load-line deflection measurements will be compared. Also, blunt notched specimens will be tested to determine the accuracy of the measurements at large deflections in the absence of crack extension. Further, J-R curves from the bendbar tests will be compared with those obtained from compact tension specimens of the same material.

Experimental Procedure

Material and Test Specimens

J-R characterizations were developed using specimens from both a 38.1-mm (1.5-in.)-thick 3Ni steel plate and a 25.4-mm (1-in.)-thick ASTM A-710 steel plate. All specimens were notched in the T-L orientation. Table 1 presents both the mechanical properties and the chemical compositions for these materials. Bendbars, compact tension specimens, and an unnotched calibration bar were tested. Figure 1 shows drawings of all specimens with nominal dimensions for each. The Type B bendbars, having the deeper initial notch [a_0 = 33 mm (1.299 in.)], were not precracked prior to testing.

TABLE 1—*Chemical and mechanical property data.*

Element	A-710, GES	3Ni, FYB
C	0.050	0.153
Mn	0.620	0.330
P	0.011	0.012
S	0.004	0.013
Cu	1.050	0.033
Si	0.300	0.180
Ni	0.880	2.550
Cr	0.720	1.660
Mo	0.185	0.370
V	...	0.003
Ti	...	>0.001
σ_{YS}[a]	614 MPa (89 ksi)	650 MPa (93 ksi)
σ_{UTS}[a]	731 MPa (106 ksi)	731 MPa (106 ksi)
% elongation[a]	28%	23%
% reduction in area	74%	63%

[a]All mechanical properties measured using 12.88-mm (0.505-in.)-diameter tension specimens having a gage length of 50.80 mm (2 in.). Specimens were pulled transverse to the plate rolling direction.

Three-Point Bend Testing

Testing was performed in a 266 kN (60 kip) screw-driven testing machine using a modified version of the computer-interactive procedure of Joyce and Gudas [*6*] to control data acquisition. Figure 2, a schematic of the test setup, indicates that four channels of data were recorded. A clip gage attached slightly above the specimen's front face was used to determine d_{CMOD} while load and crosshead displacement came from the load cell and the linear variable differential transducer (LVDT) on the testing machine, respectively. Figure 3 shows a detail of the flex bar and how it is attached to the bend specimen. It is calibrated prior to testing by applying a known displacement at its center and observing the resulting signal from the strain gage bridge. This calibration allows determination of load-line deflection during the fracture toughness test. Photographs of the specimen were taken on Kodak Plus-X film using a Nikon FE camera with a 50-mm Nikkor macro lens. Each photo was taken as the test came out of an unloading so that the crack length would be known. The Type B bendbars were not unloaded during the test because the presence of the blunt notch prevented significant crack extension. These specimens were photographed at roughly equivalent intervals of load or load-line deflection. Subsequent to testing, the specimens were heat-tinted to mark the final extent of crack growth and broken open at liquid nitrogen temperatures.

Compact Tension Testing

Compact tension tests were performed on both materials using the computer-interactive procedure of Joyce and Gudas [*5*]. This procedure follows the guidelines set forth in the ASTM Test Method for J_{Ic}, a Measure of Fracture Toughness (E 813-81).

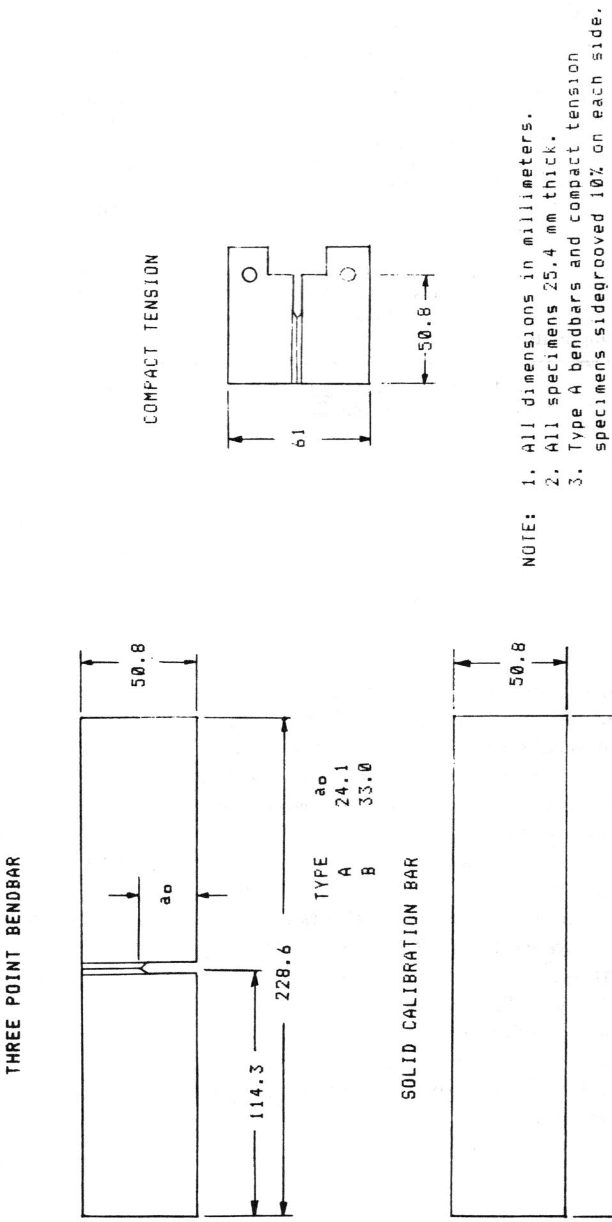

FIG. 1—*Diagrams of specimens tested.*

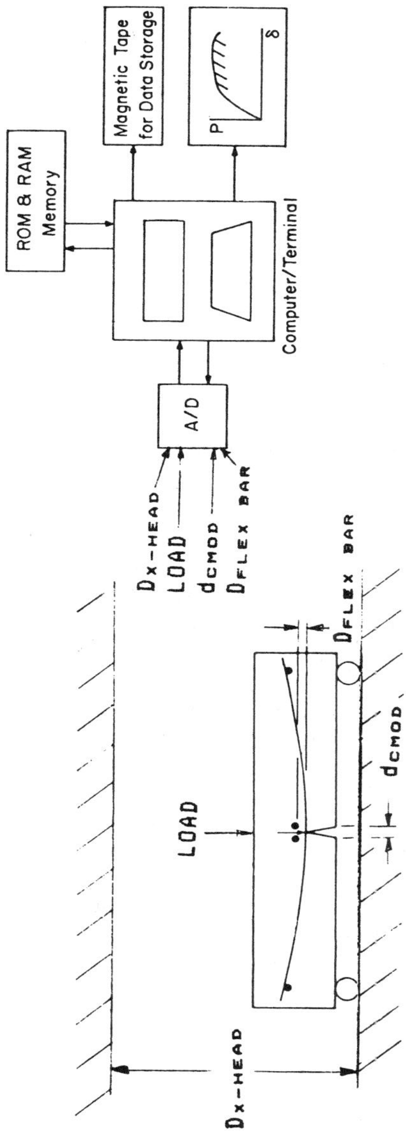

FIG. 2—*Schematic of computer-interactive test apparatus.*

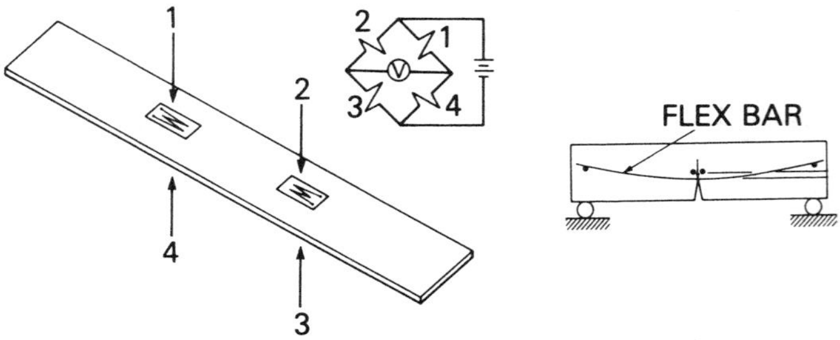

FIG. 3—*Schematic of flex bar and its attachment to three-point bend specimen.*

Load-Line Deflection Measurements

Measured from the Photographs

Bendbars were photographed during testing. Load-line deflection measured from enlargements of the photographs using a digitizing tablet served as a reference against which all other measures of load-line deflection were compared. The distance between the loading tup and the horizontal reference bar was measured to determine the load-line deflection. This distance was then reduced by the distance between the loading tup and the horizontal reference bar before loading began measured off the reference photograph. Figure 4 illustrates this procedure. A fixed reference block was placed in the lower left-hand corner of each photograph to check for photographic distortion and to provide a reference to a fixed coordinate system. This block was rigidly attached to the testing machine and did not change position during loading. The ratio of the X and Y lengths and the orthogonality of the lines was checked in each enlarged photograph to ensure that the effects of photographic distortion were negligible.

Two factors limit the accuracy of photographic load-line deflection measurements: the resolution of the digitizing tablet and motion of the horizontal reference bar. The tablet was accurate to ± 0.0635 mm (± 0.0025 in.), and the magnification of the photographs was $\times 2$, thus making the overall accuracy ± 0.0318 mm (± 0.00125 in.), assuming consistent placement of the digitizing cross hair. To minimize error resulting from inconsistent cross hair placement, each point was digitized multiple times and the average of the repeated measurements was calculated. Repeated measurement was terminated when oscillation of the average value subsided to less than 0.0508 mm (0.002 in.).

Because the horizontal reference bar is mounted on pins protruding from the specimen, specimen deformation also limits the accuracy of the photographic load line deflection measurements. Assuming that the specimen rotates as two rigid halves, as shown in Fig. 5, allows estimation of influence of this downward motion. This assumption leads to the following relation between percent error in the photographic measurement and the amount of load line deflection

$$\text{Percent error} = h_0\{1 - \cos[\sin^{-1}(2D_{LL}/S)]\} \times 100/D_{LL} \tag{1}$$

where

h_0 = original height of reference bar support point from specimen front face, 33.375 mm (1.313 in.),

S = span between rollers, 203.2 mm (8 in.), and

D_{LL} = load-line deflection.

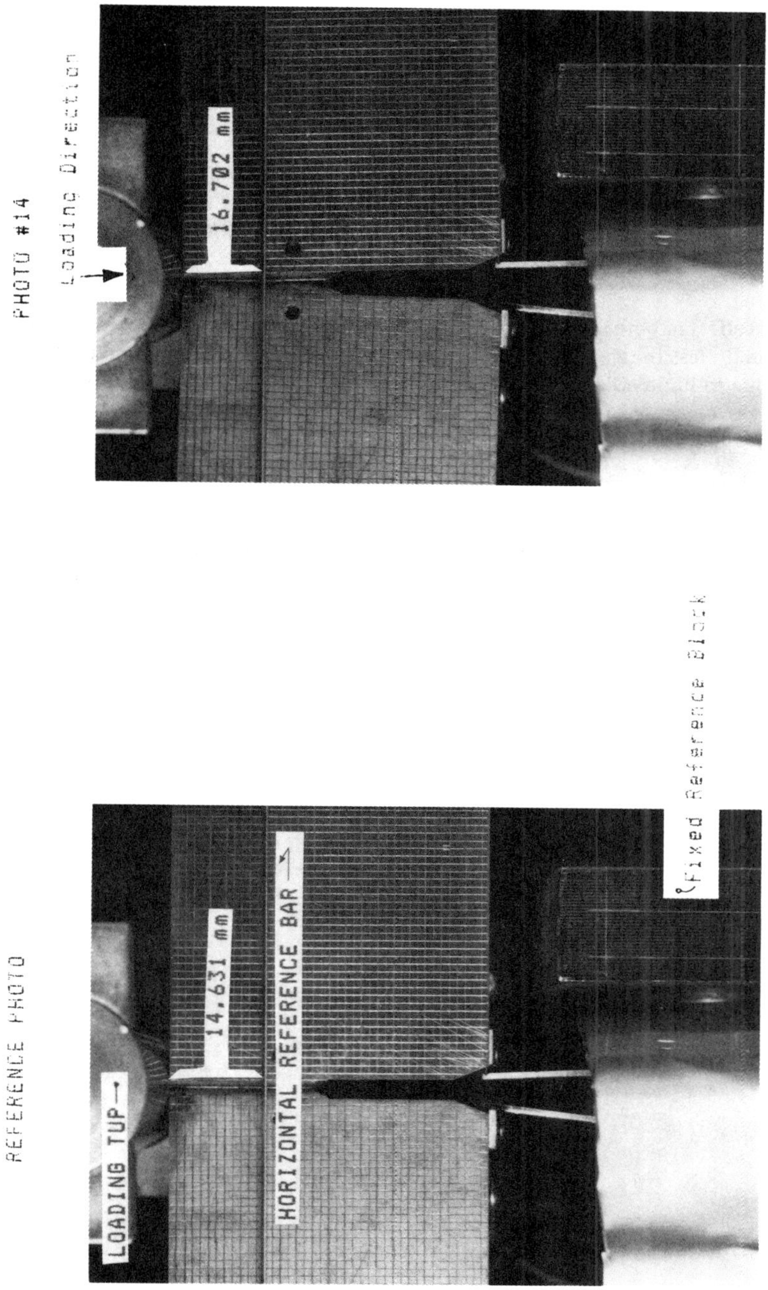

FIG. 4—*Procedure for determination of load-line displacement from photographs. Photographic displacement* = D_{LL}^{14} = *16.702 mm−14.631 mm = 2.071 mm (1 mm = 0.039 in.).*

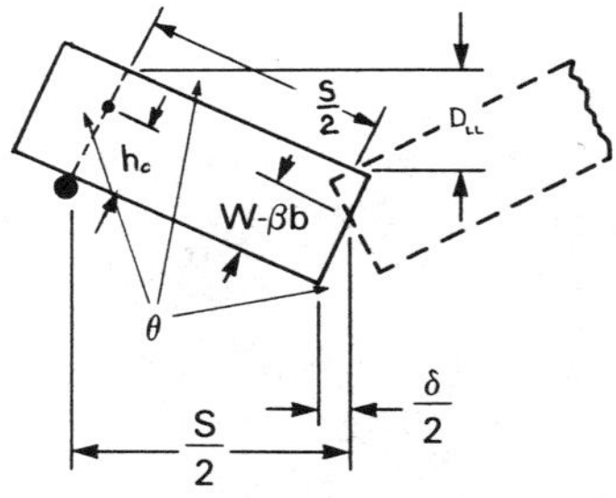

FIG. 5—*Diagram showing terms used in equations based on rigid rotation assumptions.*

Figure 6 shows the effect of both sources of error as a function of load-line displacement. Generally the error is small, less than 3%, except at load-line deflections less than 1 mm (0.039 in.), where the error due to limited digitizing pad resolution becomes quite large. For this reason, load-line deflections measured from the photographs of less than 1 mm (0.039 in.) were excluded from subsequent analysis.

Measured from the Flex Bar

Prior to each test, a series of known displacements was applied to the center of the flex bar; the value of these displacements and the output of the transducer were recorded. A line was

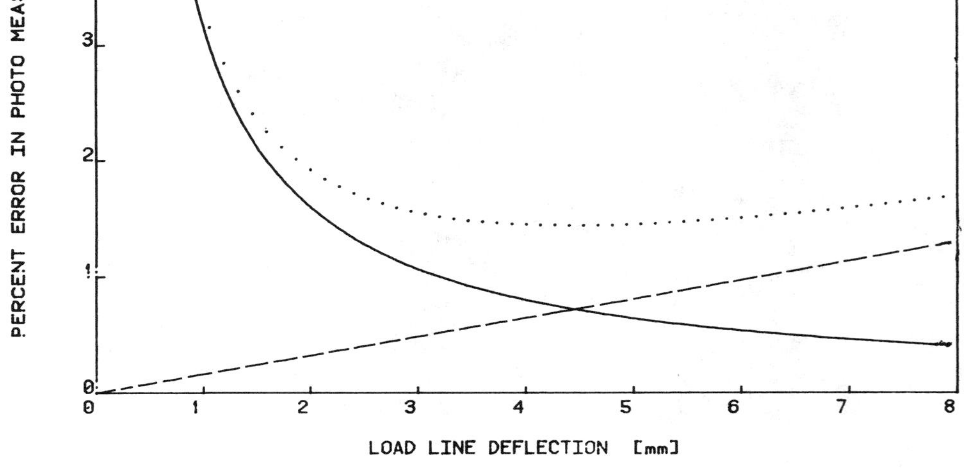

FIG. 6—*Expected error in load-line displacement measurements from the photographs due to limited digitizing pad resolution and horizontal reference bar motion.*

then fit through these data using the method of least squares. The flex bar output was converted to load-line displacement during the test by multiplying it by the slope of this line.

Calculated from the Crosshead Displacement

To determine the load-line deflection from measurements of crosshead displacement, the displacement attributable to machine compliance must be subtracted. This was done using the procedure outlined in the proposed J-R curve testing standard [3]. A solid bar of A-710 having the same overall dimensions as the test specimens was loaded to 125% of the maximum load encountered during any of the testing. Load and crosshead displacement were recorded and the crosshead displacement was reduced by that attributable to the bar, given by Euler beam theory as

$$D_{BAR} = PS^3/48EI \tag{2}$$

where

P = load applied at midspan,
S = span between roller supports,
E = Young's modulus, and
I = moment of inertia.

to determine the crosshead displacement due only to machine compliance. Figure 7 shows these results graphically. The nonlinearity in the calibration at 11 kN was reproducible from test to

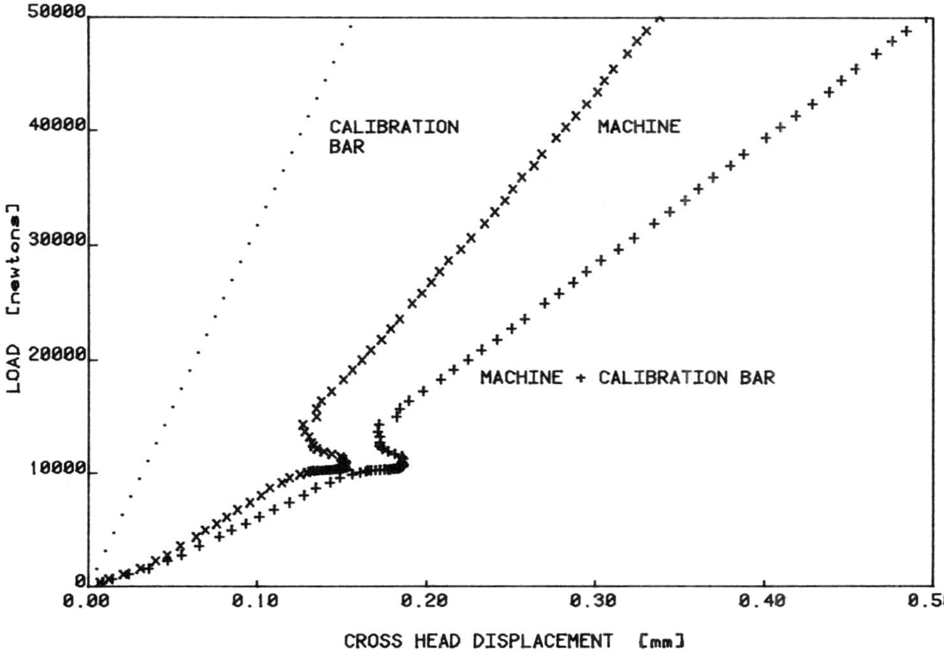

FIG. 7—*Load versus testing machine crosshead displacement characteristic.*

test and appears to be characteristic of the testing machine used, being approximately the weight of the crosshead.

Calculated from Clip Gage Measurements of d_{CMOD}

Load-line deflection may be determined from d_{CMOD} by assuming that the bendbar rotates as two rigid halves about a plastic hinge. The following relation holds if the angle of specimen rotation remains small (see Fig. 5)

$$D^{d}_{LL} = Sd_{CMOD}/2(W - \beta b) \quad (3)$$

where

S = span between roller supports,
W = specimen width,
β = an empirical coefficient expressing fraction of b around which the remaining ligament is assumed to hinge, and
b = uncracked ligament.

At larger deflections, where this assumption becomes invalid, then

$$D^{drc}_{LL} = [S/2 - Wd_{CMOD}/2(W - \beta b)] \tan\{\sin^{-1}[d_{CMOD}/2(W - \beta b)]\} \quad (4)$$

A value of $\beta = 0.6$ was used, as suggested by Garwood [*6*]. Figure 8 gives these equations in graphical form. The effects of specimen rotation are not significant until d_{CMOD} exceeds 0.254 mm (0.010 in.).

An alternative to a relationship between D_{LL} and d_{CMOD} based on rigid rotation assumptions would be to experimentally determine the relation between these two displacements. Figure 9 shows load-line deflection measured from the photographs as a function of d_{CMOD} for all specimens tested. The data are seen to be relatively insensitive to material, sidegrooving, amount of specimen rotation, and degree of plastic deformation prior to crack initiation. Consequently, a linear relationship shows promise for determining load line deflection from d_{CMOD} data. In order to use these data to predict D_{LL} from d_{CMOD} a zero intercept line of the form

$$D^{corr}_{LL} = md_{CMOD} \quad (5)$$

was fit to the data in Fig. 9 using the method of least squares. To avoid a biased prediction, the data from a given specimen were not used to determine the m-coefficient (slope) used on that specimen's d_{CMOD} data. Table 2 gives the slope values used for the specimens tested. The slope of this line is relatively insensitive to material, sidegrooving, amount of specimen rotation, degree of plastic deformation prior to crack initiation, and amount of crack growth within the range of conditions considered in this investigation. It should be noted that Eqs 3 to 5 will not provide reliable results if the bulk of the data lie in the small-displacement (linear elastic) regime.

Results and Discussion

Load versus load-line deflection curves measured using the flex bar are shown in Figs. 10 and 11 for an A-106 and a 3Ni steel specimen, respectively. The data from the photographic reference are also shown. These data are visually representative of that determined using the other techniques.

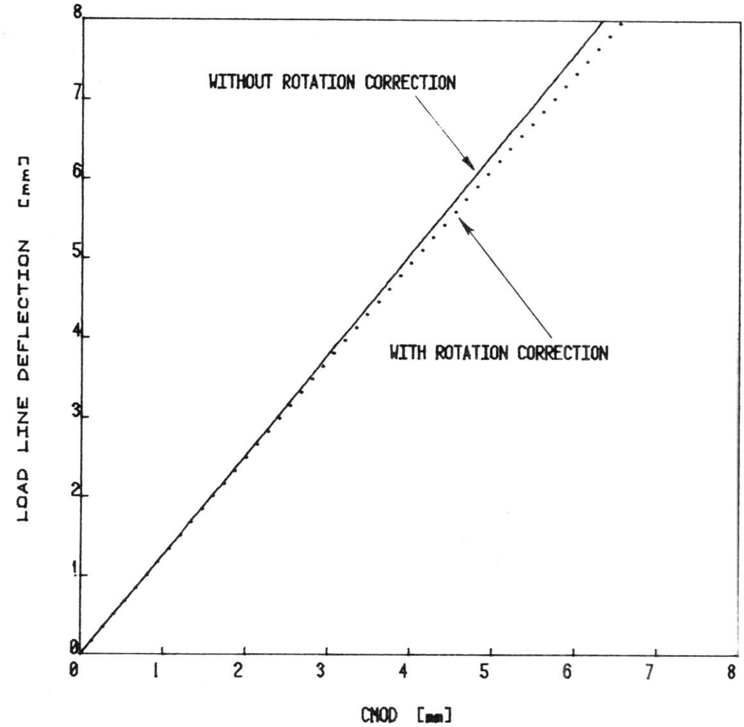

FIG. 8—*Relation between load-line deflection and* d_{CMOD} *determined using rigid rotation assumptions.*

To facilitate easier comparison of the load-line deflection measured from the photographs with displacement measurements from each technique, these data were plotted against each other. Figure 12 shows the comparison plots; an intercept of zero and a slope of one indicates perfect agreement with the photographic reference. Least-squares analysis given in Table 3 shows that the best-fit values for the various methods do not deviate very far from this ideal. In each case the fitted slope and intercept showed minor dependencies on material and crack extension. Because these were small and of relatively constant magnitude in each comparison, subsequent analysis treated all data in Fig. 12 as one group and did not distinguish between different specimens.

Because of the error in the photographic measurements shown in Fig. 6, it is not appropriate to rank the various load-line deflection measurements with respect to their ability to match the ideal slope and intercept values. Rather, if the data in Fig. 12 fall within a region about the ideal line determined by the error bounds in Fig. 6, they cannot be said to deviate significantly from the photographic reference values. Figure 13 shows this acceptable data region. To determine how well the data from each technique fall within this region, the residual, or vertical distance of each point from the acceptable data region, was calculated for each point. Points lying within the acceptable data region had no residual. The average residual for each measurement technique was then calculated as follows

$$\overline{\text{RESIDUAL}} = \sum_{i=1}^{n} [r_i^2]/n \tag{6}$$

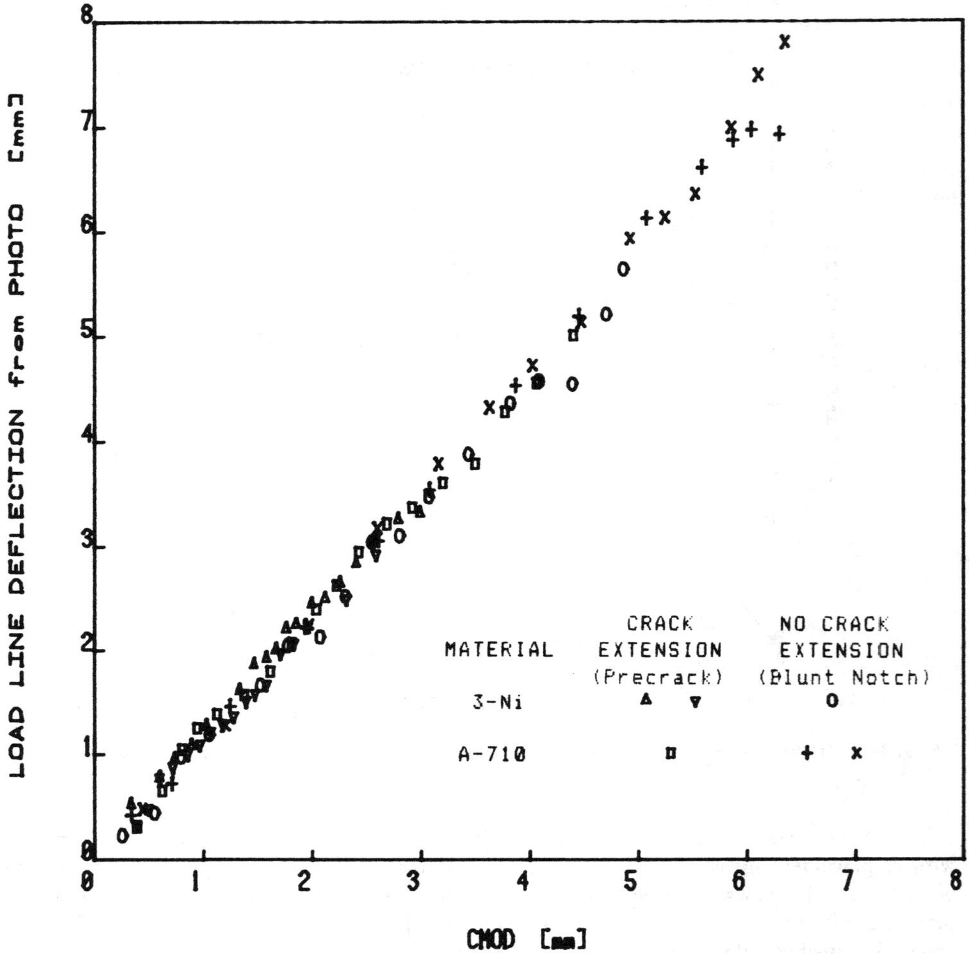

FIG. 9—*Relation between load-line deflection and* d_{CMOD} *determined using empirical correlation.*

TABLE 2—*Empirical correlation slopes relating load-line displacement to crack mouth opening displacement.*

Specimen Deleted	*m*, slope
GES-40	1.1630
FYB-511	1.1584
FYB-512	1.1624
GES-44	1.1609
GES-45	1.1459
FYB-503	1.1694

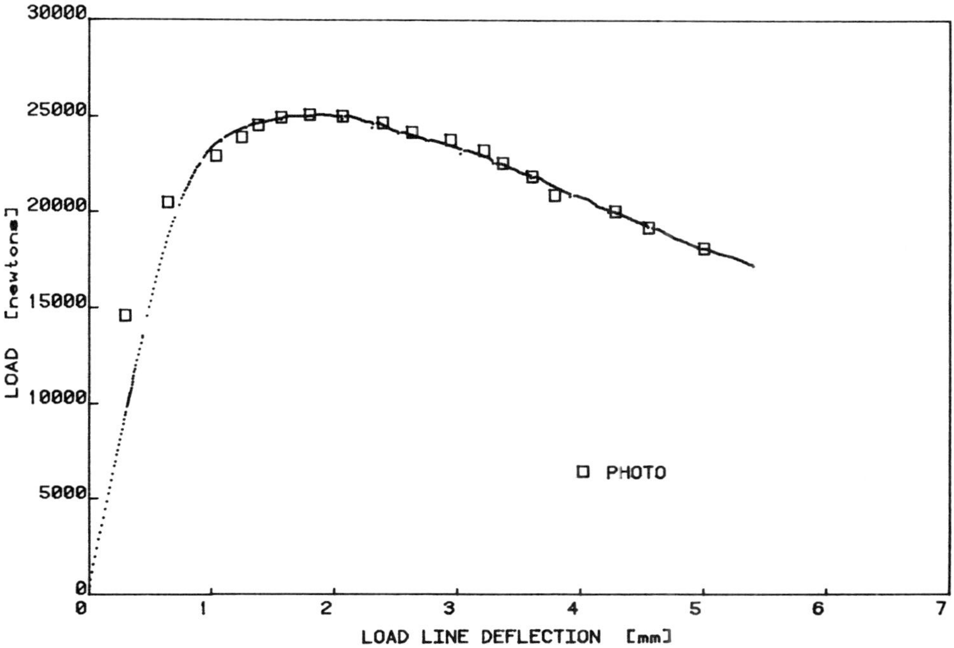

FIG. 10—*Load versus load-line deflection determined using the flex bar compared with the photographic data (A-710 Specimen GES-40).*

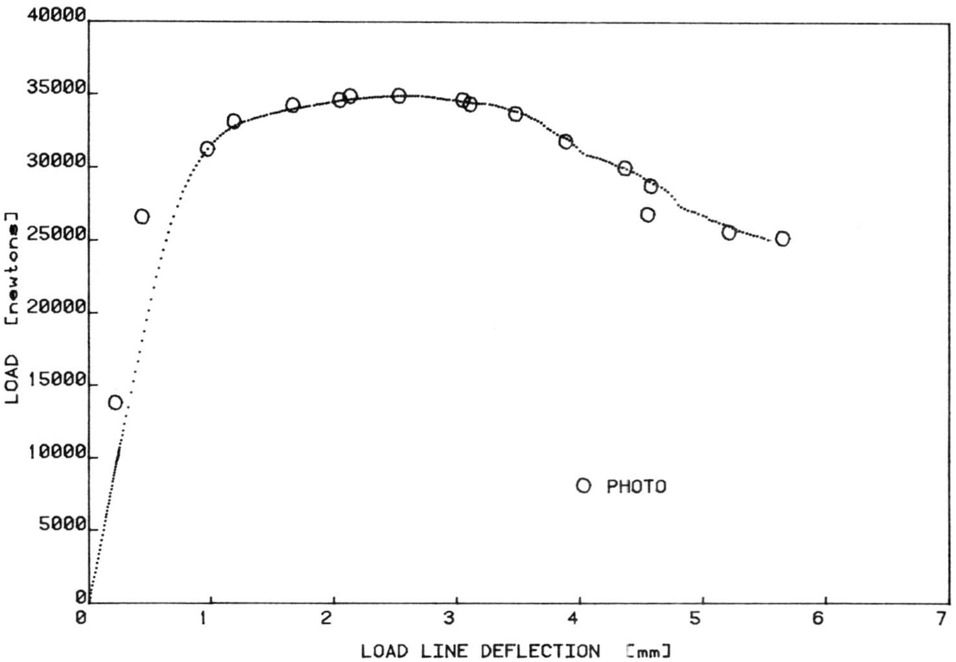

FIG. 11—*Load versus load-line deflection determined using the flex bar compared with the photographic data (3Ni Specimen FYB-503).*

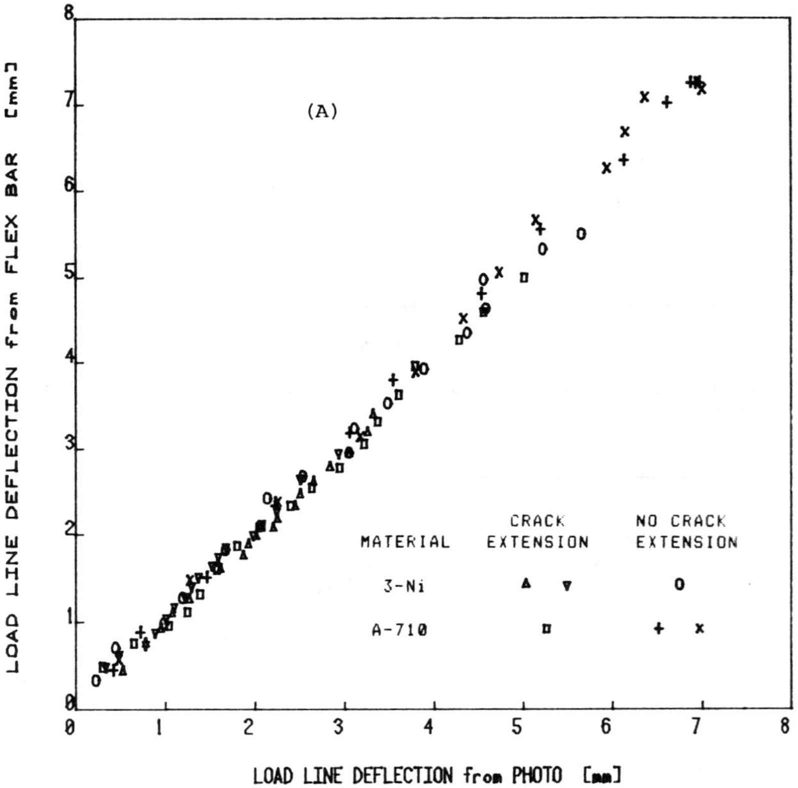

FIG. 12—*Comparison of load-line deflection measured using various experimental techniques to the photographic load-line deflection measurements for all specimens tested.* (a) *flex bar;* (b) *crosshead displacement corrected for machine compliance;* (c) d_{CMOD} *and a rigid rotation assumption;* (d) d_{CMOD} *and a rotation corrected rigid rotation assumption; and* (e) *experimentally derived correlation.*

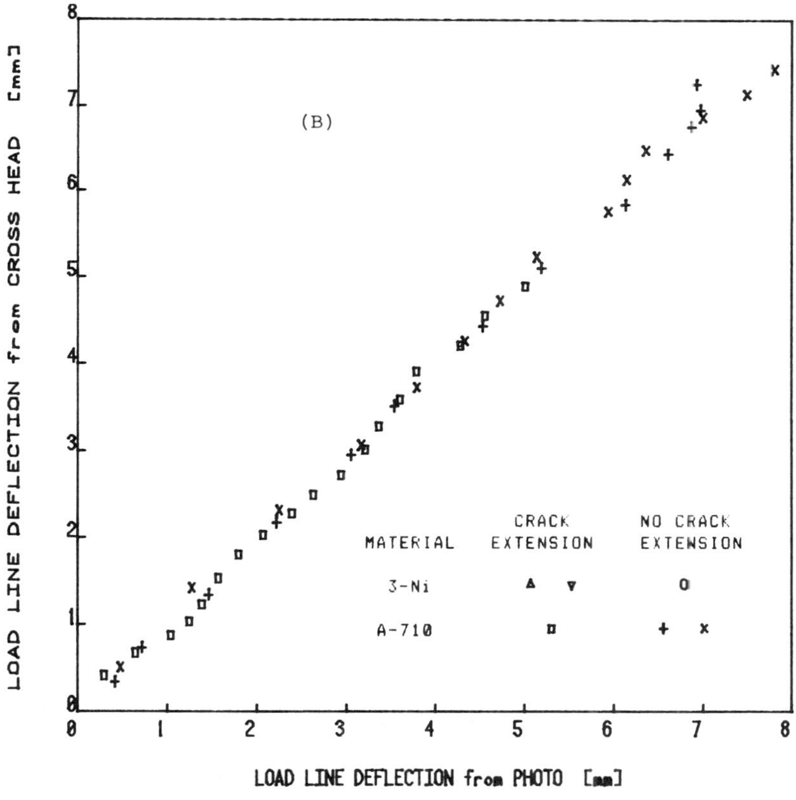

FIG. 12—*Continued.*

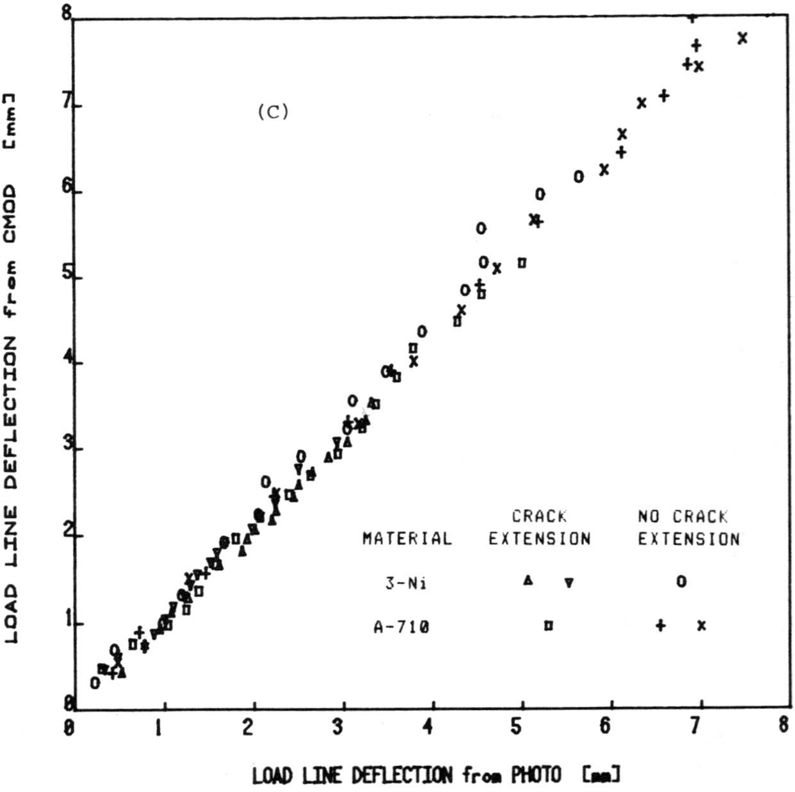

FIG. 12—*Continued.*

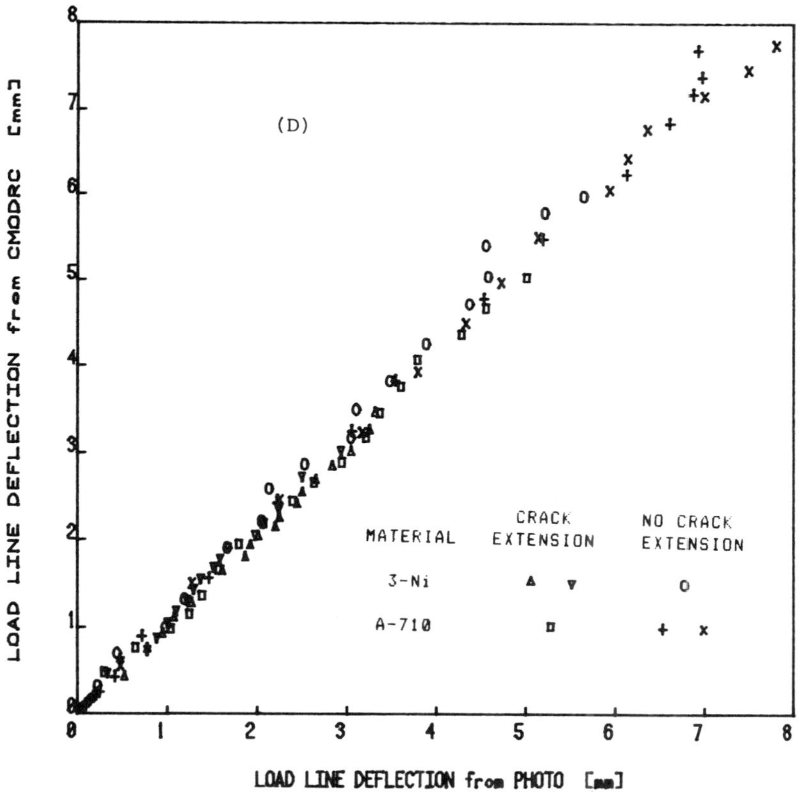

FIG. 12—*Continued.*

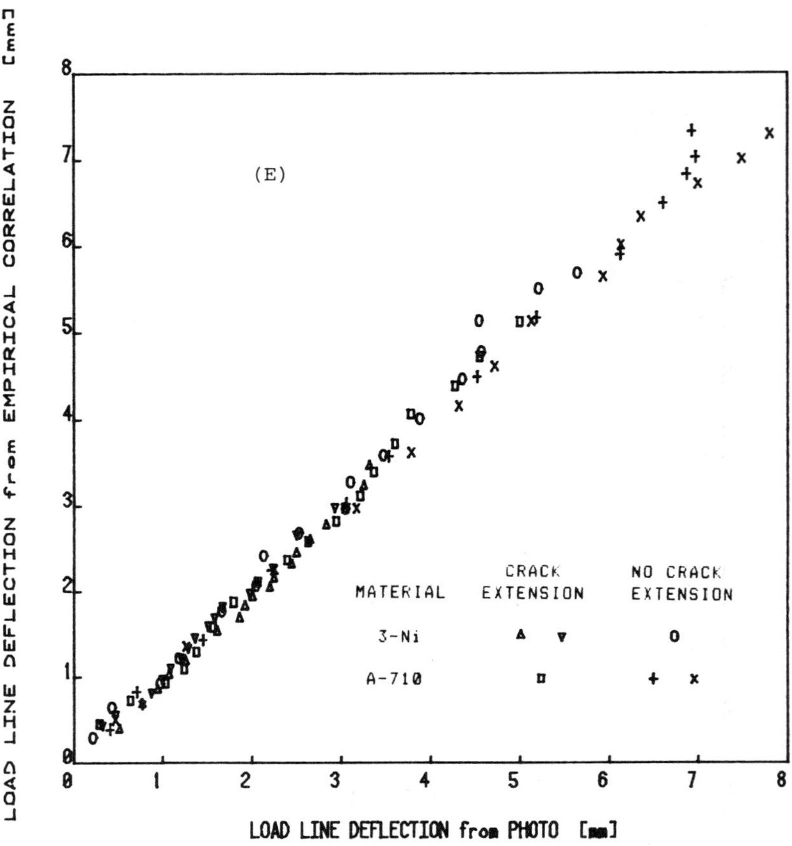

FIG. 12—*Continued.*

TABLE 3—*Least-squares coefficients for comparisons of deflection measurements to photographic reference.*

Measurement Technique	Intercept, mm[a]	Slope
Flex bar	−0.0032	1.0549
Crosshead	−0.0025	0.9956
D_{LL}^{d}	0.0004	1.0559
D_{LL}^{drc}	0.0020	1.0237
D_{LL}^{corr}	0.0023	0.9824

[a]1 mm = 0.039 in.

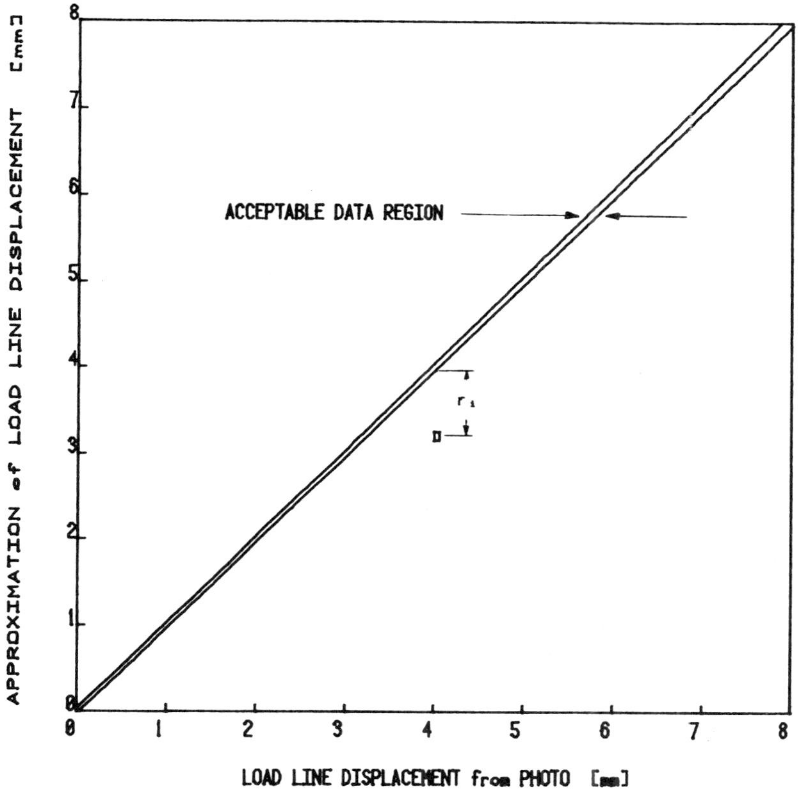

FIG. 13—*Plot showing the acceptable data region and the residual of a data point lying outside of the region.*

where

r_i = residual of the ith data point, vertical distance from the acceptable data region shown in Fig. 13, and
n = total number of data points.

Table 4 gives the average residual values for each measurement technique tried; all are small, with the largest being 0.052 mm (0.0020 in.). Because the average residuals are so close for every technique, varying only 0.009 mm (0.0004 in.) from lowest to highest, it must be concluded that all techniques match the photographic data equally well over the range of load-line deflections observed in this investigation.

Because the result of a J-integral test is either a J-R curve or a J_{Ic} value, it is important to show how these small differences in load-line deflection measurements among the various techniques translate into differences in fracture property characterizations. To this end both J-R curves and J_{Ic} values were calculated using the formulas and procedures described in ASTM E 813 and the tentative J-R curve test procedure [3] for the three bendbars that experienced crack extension. Figure 14 shows the J-R curves overlayed with 95% confidence bands from two specimens of the same material. Because the data scatter from the various load-line deflection measurement techniques is less than the material scatter, shown by the 95% bands, differences in J-R

TABLE 4—*Average residuals of each measurement technique from the photographic reference.*

Measurement Technique	$\overline{\text{RESIDUAL}}$, mm (in.)
Flex bar	0.052 (0.0020)
Crosshead	0.042 (0.0016)
D^{d}_{LL}	0.047 (0.0019)
D^{drc}_{LL}	0.046 (0.0018)
D^{corr}_{LL}	0.043 (0.0017)

curves due to the different techniques can be considered insignificant. Table 5 gives J_{Ic} values for these same tests. Comparing the ranges of J_{Ic} for the different load-line deflection measurements to the ranges given for each material leads to the same conclusion as made for the J-R curve data. That is, small differences among the various load line deflection measurements tried remain small when the data are used to calculate a J_{Ic} value.

A disparity has at times been noted between the fracture mechanics values determined using bendbar and compact tension specimens. Figure 15 shows the J-R characterizations for the three bendbars that experienced crack extension overlayed with data from two compact tension specimens of the same alloy. Good agreement between the J-R data from the two geometries exists over the acceptable data range given in the tentative J-R curve test procedure [*3*], which limits the maximum crack extension to 10% of the initial remaining ligament, or approximately 1.75 mm (0.070 in.). At longer crack extensions, larger deviations between the J-values from the two geometries occur. While this may be attributed to uncertainties in bendbar load-line deflection measurements, uncertainties in the J-integral formulation or differences in constraint between the two geometries may also play a role.

TABLE 5—J_{Ic} *values.*[a]

Specimen	Measurement Technique	J_{Ic} kPa · m (in. · lb/in.^{2})
GES-40 (A-710)	flex bar	188 (1075)
	crosshead	184 (1050)
	D^{d}_{LL}	204 (1164)
	D^{drc}_{LL}	202 (1155)
	D^{corr}_{LL}	193 (1104)
FYB-511 (3-Ni)	flex bar	148 (846)
	D^{d}_{LL}	155 (888)
	D^{drc}_{LL}	154 (881)
	D^{corr}_{LL}	146 (836)
FYB-512 (3-Ni)	flex bar	137 (780)
	D^{d}_{LL}	144 (822)
	D^{drc}_{LL}	143 (816)
	D^{corr}_{LL}	135 (768)

[a]A-710 material scatter range: 188 to 234 kPa · m (1075 to 1335 in. · lb/in.^{2}).

3-Ni material scatter range: 137 to 148 kPa · m (780 to 846 in. · lb/in.^{2}).

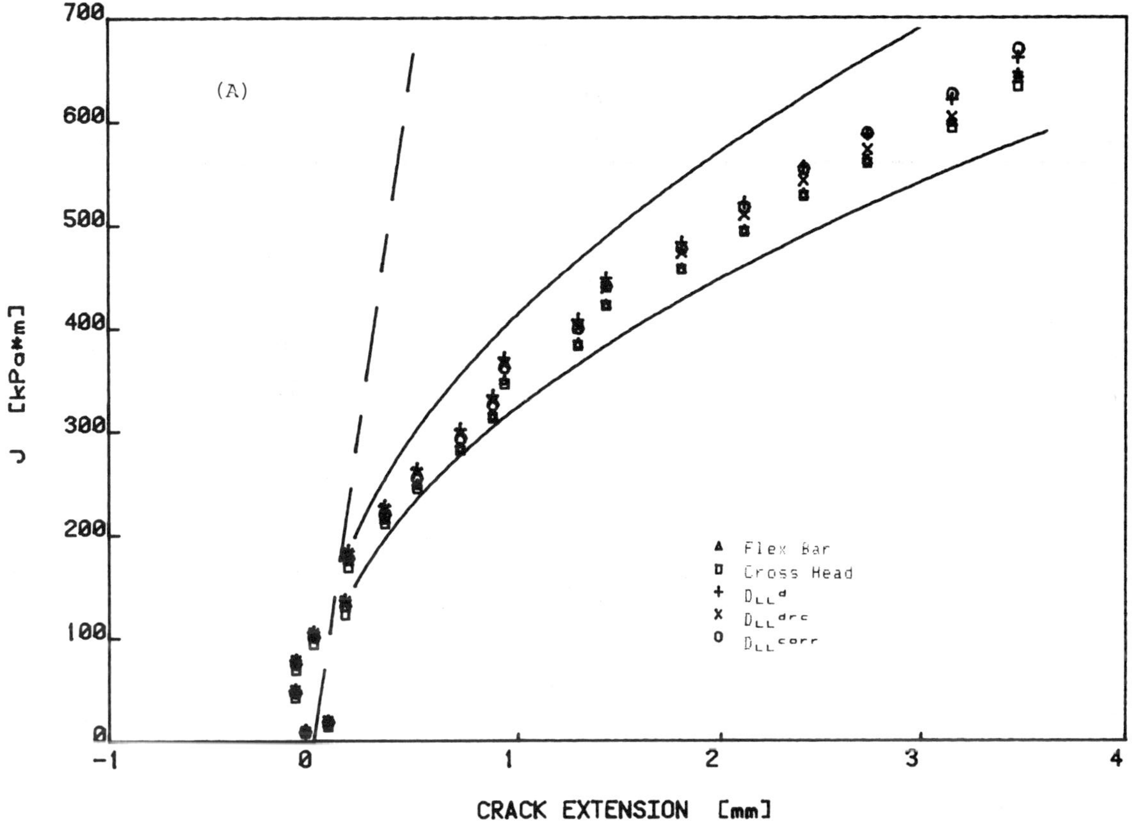

FIG. 14—*J-R curves for three bendbars showing the effect of different load-line deflection measurements (points) and material scatter (95% confidence bands):* (a) *A-710 bendbar GES-40;* (b) *3Ni bendbar FYB-511; and* (c) *3Ni bendbar FYB-512.*

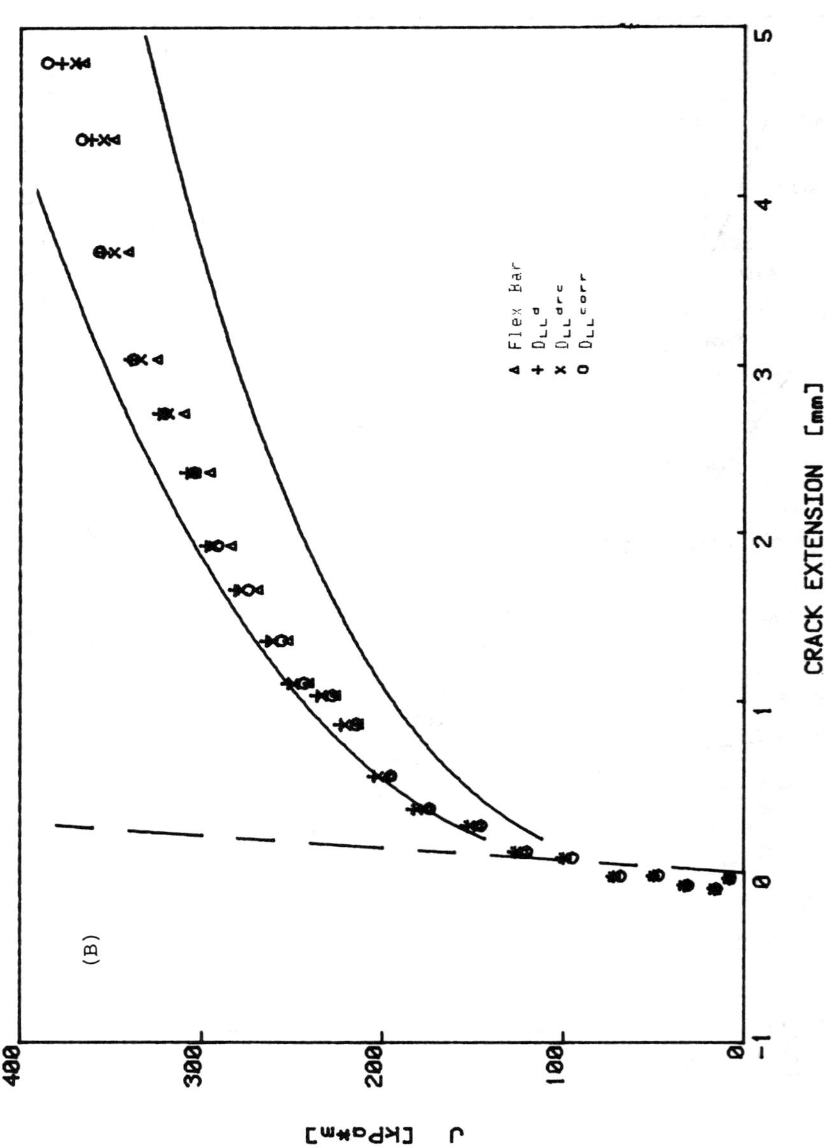

FIG. 14—*Continued.*

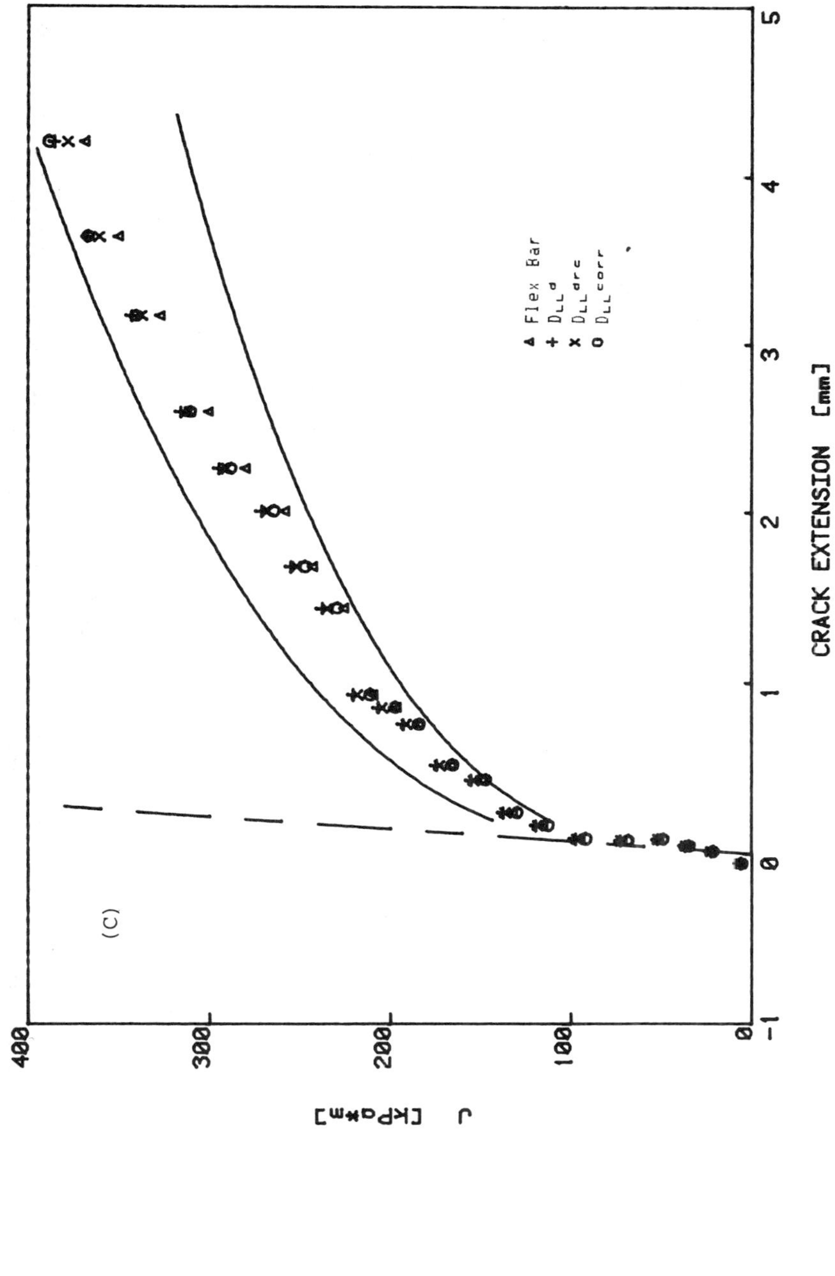

FIG. 14—*Continued.*

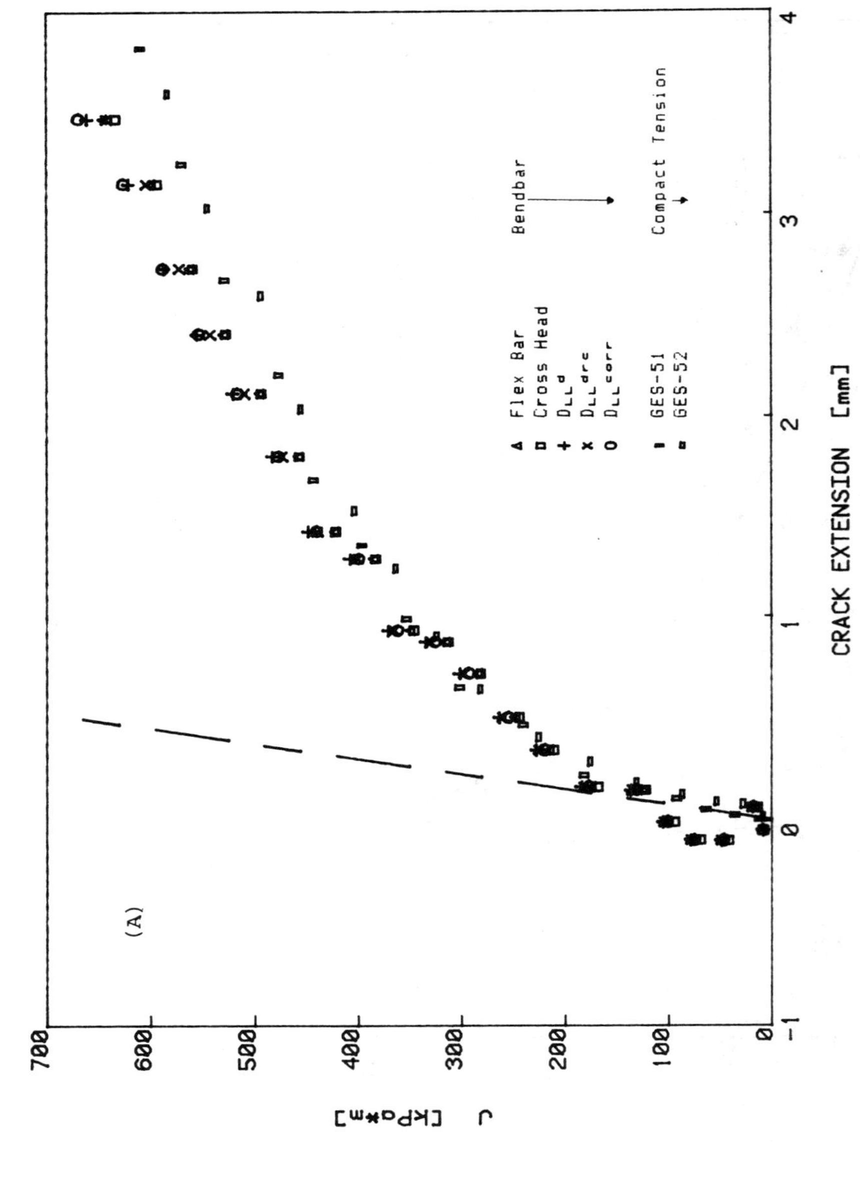

FIG. 15—*Comparison of J-R curves determined using bendbars with J-R curves determined using compact tension specimens of the same material:* (a) *A-710 bendbar GES-40;* (b) *3Ni bendbar FYB-511; and* (c) *3Ni bendbar FYB-512.*

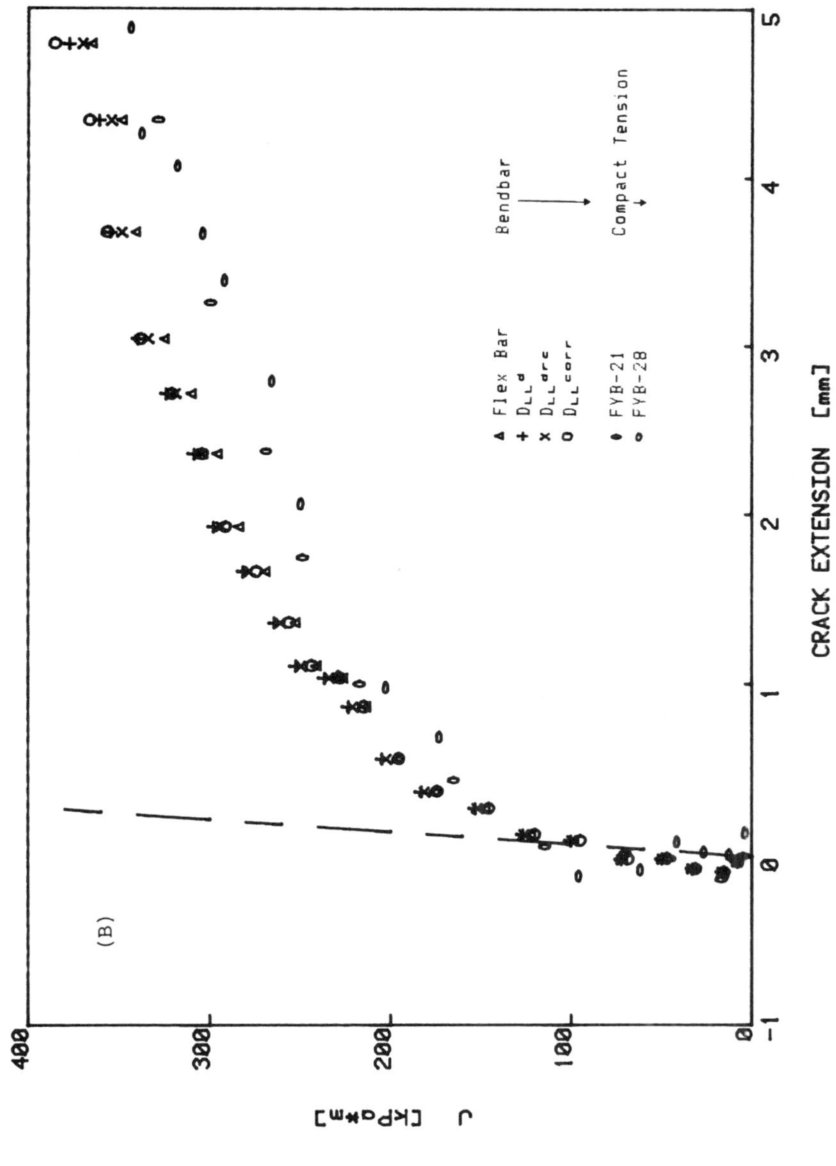

FIG. 15—*Continued.*

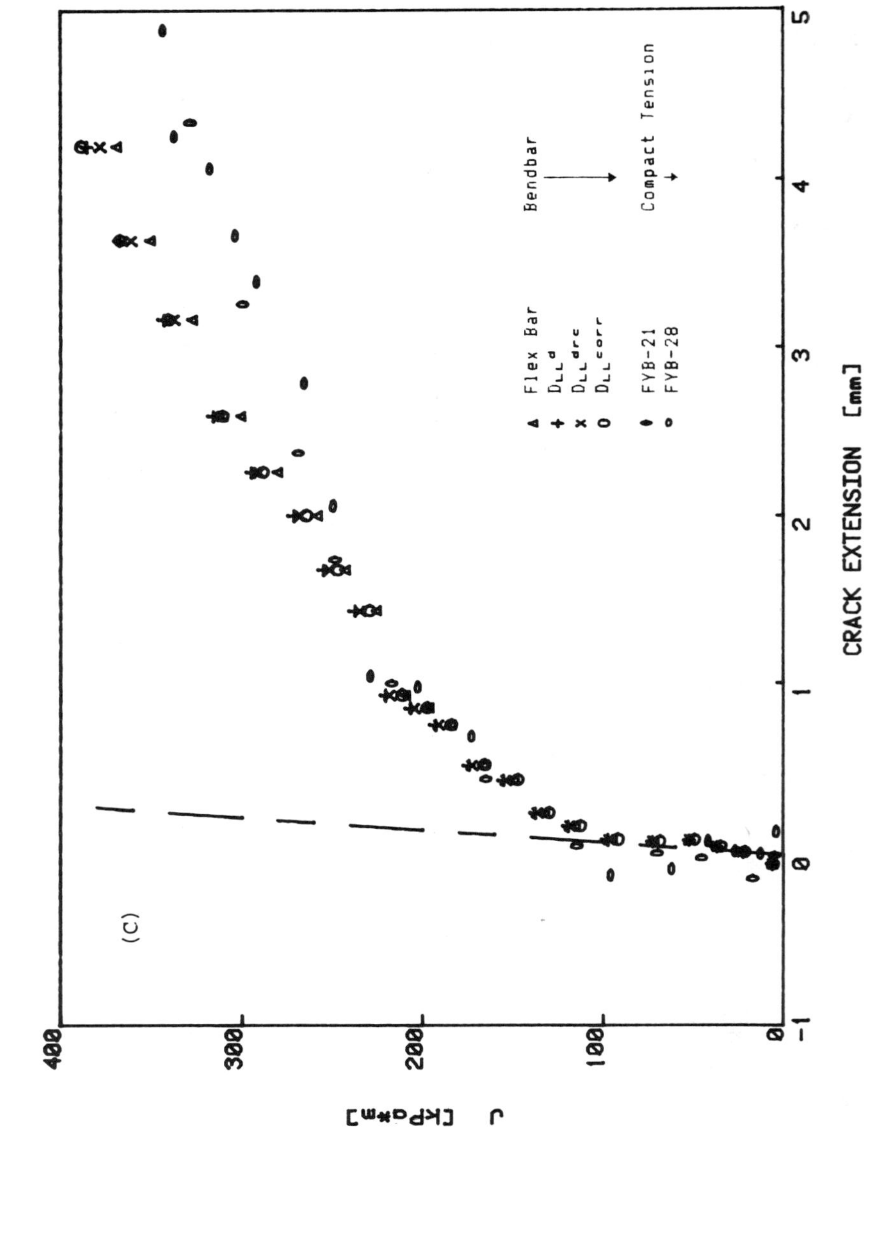

FIG. 15—*Continued.*

Conclusions

The following conclusions may be drawn from this work:

1. Statistical comparisons of load-line deflection measurements with photographic reference data show that all techniques match the photographic data equally well [within 0.052 mm (0.002 in.)] over the range of load-line deflections observed [0 to 8 mm (0 to 0.315 in.)] in this investigation. However, for a new material with which the experimentalist has no prior experience, only the flex bar or corrected crosshead displacement techniques could be employed. These are direct measurement techniques, whereas those that relate load-line deflection to crack mouth opening displacement involve some assumption or empirical correlation that must be proven or developed for a new alloy.

2. The small differences among the various load-line deflection measurement techniques investigated cause small differences in J-R curves and J_{Ic} values that are within the magnitude of scatter due to material variability. Thus, it does not appear that interlaboratory differences previously observed can be explained based on differences in load-line deflection measurement technique alone.

3. Bendbar and compact J-R curve characterizations are coincident over the first 1.75 mm (0.068 in.), or 10% of the initial remaining ligament, of crack extension for the alloys tested. At longer crack extensions, differences in J determined by the two geometries may be due to uncertainties in bendbar load-line deflection, uncertainties in the J formulation, or constraint differences between bendbars and compact tension specimens.

References

[1] Gudas, J. P. and Davis, D. A., "Evaluation of the Tentative J_I-R Curve Testing Procedure by Round Robin Tests of HY130 Steel," *Journal of Testing and Evaluation,* Vol. 10, No. 6, Nov. 1982, pp. 252–262.

[2] Joyce, J. A. and Hackett, E. M., "Dynamic J-R Curve Testing of a High Strength Steel Using the Key Curve and Multispecimen Techniques" in *Fracture Mechanics: Seventeenth Volume,* ASTM STP 905, American Society for Testing and Materials, Philadelphia, 1986, pp. 741–774.

[3] Albrecht, P., Andrews, W. R., Gudas, J. P., Joyce, J. A., Loss, F. J., McCabe, D. E., Schmidt, D. W., and VanDerSluys, W. A., "Tentative Test Procedure for Determining the Plane Strain J-R Curve," *Journal of Testing and Evaluation,* Vol. 10, No. 6, Nov. 1982, pp. 245–251.

[4] Wells, A. A. in *Proceedings,* Canadian Congress of Applied Mechanics, Calgary, AB, Canada, 1971, pp. 59–77.

[5] Joyce, J. A. and Gudas, J. P., "Computer Interactive J_{Ic} Testing of Navy Alloys" in *Elastic-Plastic Fracture, ASTM STP 668,* American Society for Testing and Materials, Philadelphia, 1979, pp. 451–468.

[6] Garwood, S. J., "The Effect of Specimen Geometry on the Crack Growth Resistance of API 5LX65 Steel," Report 61/1978/E, The Welding Institute, 1978.

Juergen Heerens,[1] Karl-Heinz Schwalbe,[1] and Alfred Cornec[1]

Modifications of ASTM E 813-81 Standard Test Method for an Improved Definition of J_{Ic} Using New Blunting-Line Equation

REFERENCE: Heerens, J., Schwalbe, K.-H., and Cornec, A., "**Modifications of ASTM E 813-81 Standard Test Method for an Improved Definition of J_{Ic} Using New Blunting-Line Equation,**" *Fracture Mechanics: Eighteenth Symposium, ASTM STP 945,* D. T. Read and R. P. Reed, Eds., American Society for Testing and Materials, Philadelphia, 1988, pp. 374–389.

ABSTRACT: A test procedure for determining the fracture resistance of ductile metallic materials under stable crack growth conditions is described. The procedure consists of the determination of a J-Δa curve with particular emphasis on the near-initiation region. In our proposal, J_{Ic} is defined as the fracture resistance at 0.2-mm stable ductile tearing. It is determined by the intersection of the J-Δa curve with an intercept line which is parallel to and 0.2 mm off the blunting line. The blunting-line formula is improved by accounting for the strain hardening behavior of the material. A simple determination of the blunting-line equation is shown, and its applicability is checked using a broad range of materials.

KEY WORDS: blunting line, resistance against crack growth, ductile tearing, crack initiation, fracture toughness, stretch zone

Nomenclature

a	Crack length
a_0	Fatigue precrack length
Δa	Amount of crack growth
Δa_B	Δa caused by crack-tip blunting
Δa_{DT}	Δa caused by ductile tearing
Δa^*_{DT}	Amount of Δa_{DT} at which J_{Ic} is defined
B	Specimen thickness
d_n	Factor relating crack-tip opening displacement to J-integral and σ_0
d^*_n	Defined as $d_n = E/\sigma_0$
E	Young's modulus
J_0	J-integral at onset of ductile tearing
n	Strain hardening exponent
SZH	Stretch zone height
SZW	Stretch zone width
W	Specimen width
δ	Crack-tip opening displacement
δ_{45}	δ defined in Fig. 4
$\epsilon_{0.2}$	Strain at $\sigma_{0.2}$

[1]Research engineer, head of department, and research engineer, respectively, GKSS Research Center Geesthacht, Max-Planck-Strasse, D-2054 Geesthacht, West Germany.

ϵ_u Strain at σ_u
σ_0 Reference stress of power law, see Fig. 13
$\sigma_{0.2}$ 0.2% proof stress
σ_F Flow stress: average of $\sigma_{0.2}$ and σ_u
σ_u Ultimate tensile strength
σ_Y Yield strength, general
ν Poisson's ratio

Practical experience with the ASTM Standard Test Method for J_{Ic}, a Measure of Fracture Toughness (E 813-81) has exposed several drawbacks to the method and has led various groups to work out improved methods; for example, the American Society for Testing and Materials (ASTM), the Central Electricity Generating Board, the European Group on Fracture, the French Fracture Group, and the German Fracture Group. Several presentations [*1-6*] at the recent CSNI[2] Workshop on Ductile Fracture Test Methods, Paris, France, 17–19 April 1985, gave an overview on these and related activities, most of which were stimulated by the ASTM Symposium on User's Experience with Elastic-Plastic Fracture Toughness Test Methods, Louisville, Kentucky, 18–22 April 1983.

A critical examination of some properties of ASTM E 813-81 and our investigation of the initiation of ductile tearing induced us to propose four new elements to be included in ASTM E 813:

(*a*) a blunting-line formulation accounting for strain hardening of the material to be tested,
(*b*) a curve fit to the J-Δa data,
(*c*) a definition of J_{Ic} as the resistance against crack growth at a specified amount of ductile tearing, and
(*d*) appropriate data point distribution.

The basis outline of these items has already been presented elsewhere [*7,8*]. The present paper describes the final version of our proposal, including additional evidence for the validity of our assumptions and a procedure for a simple determination of the theoretical blunting line.

J_{Ic} Construction According to ASTM E 813-81

For the purpose of determining J_{Ic}, the E 813-81 method treats the J-R curve as two consecutive straight lines, the first of which represents the blunting phase of the crack-tip behavior (the so-called blunting line), while the second straight line represents the ductile tearing type of crack extension, Fig. 1. In order to avoid excessive experimental effort, the blunting line is not

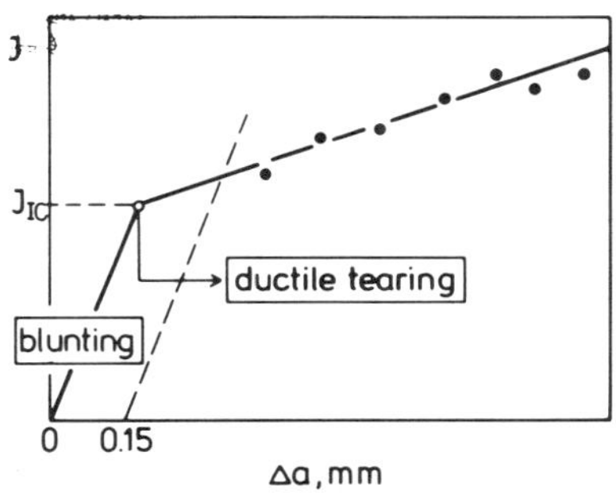

FIG. 1—*Schematic of* J_{Ic} *determination according to ASTM E 813-81.*

[2]Committee on the Safety of Nuclear Installations, OECD Nuclear Energy Agency.

supposed to be measured during a J_{Ic} test; rather, it is represented by a simple theoretical analysis: The crack tip opening displacement δ, for an elastic-ideally plastic material having the yield strength σ_Y, can be approximated by

$$\delta \approx \frac{J}{\sigma_Y} \tag{1}$$

Assuming a semicircular shape of the crack-tip profile, a crack advance, Δa_B, by blunting can be defined as one half of δ, that is

$$\Delta a_B \approx \frac{J}{2\sigma_Y} \tag{2}$$

Strain hardening is accounted for by setting

$$\sigma_Y \rightarrow \sigma_F = 0.5(\sigma_{0.2} + \sigma_u) \tag{3}$$

Thus, Eq 2 together with Eq 3 defines a theoretical blunting line. Due to the limited experimental accuracy, the data points which are going to be used for establishing the ductile tearing straight line must exhibit at least 0.15-mm ductile tearing as indicated in Fig. 1 by the exclusion line. The extrapolation of the ductile tearing straight line to the blunting line is supposed to hit the blunting line at the point of initiation of ductile tearing. Three problems emerge from this construction:

1. The actual J-Δa curve often exhibits a pronounced curvature in the initiation region, Fig. 2. Thus, J_{Ic} is not located on the actual J-Δa curve.
2. According to Fig. 2, J_{Ic} is greater than J_0, the true initiation value of J, the difference depending on material and specimen size (for the latter point see next item). The problem is not that J_{Ic} is greater than J_0; the problem is that this difference is undefined.
3. Size effects on J_{Ic} can be observed even if the actual J-Δa data are unaffected by varying the specimen size. Figure 3 shows an extreme example: although both tests are valid in the spirit of E 813-81, the J_{Ic} values can vary by a factor of 3. It can also be seen that the R-curve data points form a common scatterband. The reason for this size effect is that a straight line had to be fitted to different ranges of a curved line. In other words, as a consequence of the straight-line fit, data points well beyond initiation affect J_{Ic}, which should be representative of the initiation behavior.

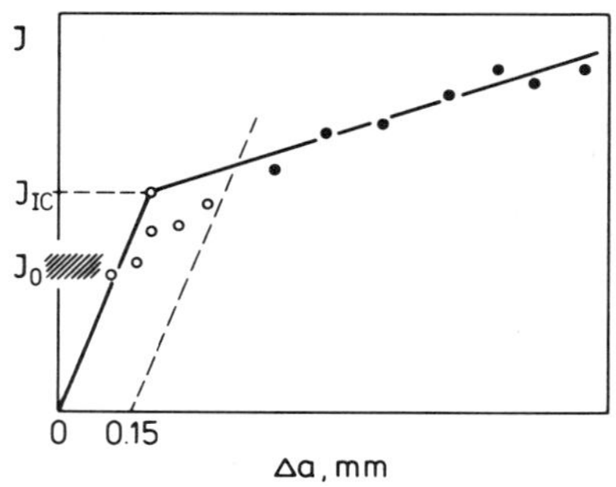

FIG. 2—*Schematic of* J_{Ic} *determination including data points close to initiation which are not valid according to the standard.*

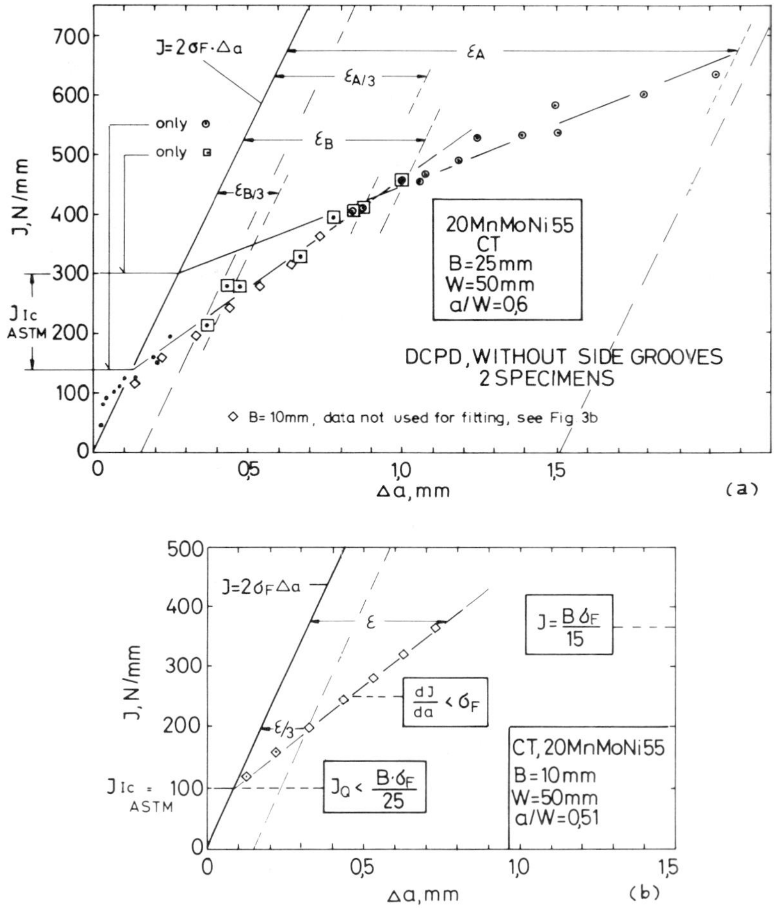

FIG. 3—J-Δa *curve for a reactor pressure vessel steel.* (a) *25-mm-thick specimen (effect of data grouping rule on* J_{IcASTM}*).* (b) *10-mm-thick specimen.*

New Proposal

A critical examination of ASTM E 813-81 on the background of our extensive experimental and theoretical investigation of crack growth, including the blunting phase [*9–15*], caused us to suggest some modifications. The simple philosophy underlying this proposal is to find a well-defined condition close to initiation of ductile tearing. Thus, such a method must be able to distinguish clearly between the two processes, blunting and ductile tearing.

The present report concentrates only on the definition of J_{Ic}; all other details of E 813 (except the data point distribution) are unaffected. The suggested modifications are as follows:

1. More appropriate formulation of the blunting line.
2. Power law fit to the J-Δa data points.
3. Definition of J_{Ic} as a point on the actual J-R curve close to initiation, avoiding extrapolation.
4. Appropriate data point distribution to avoid extrapolation.

Details are presented along with background information in the following sections.

Blunting

The small crack extension Δa_B associated with crack-tip blunting can be identified as the width of a stretched zone (SZW) on the fracture surface. In order to determine initiation of ductile tearing, the preceding event—crack-tip blunting—must be established appropriately.

The experimental determination of the J-Δa_B relationship is believed to be unsuitable, for the following reasons:

1. The multiple-specimen method would be very costly since scanning electron microscopy (SEM) would be needed. In addition, the interlaboratory reproducibility of SZW measurements is not encouraging due to the different ways of interpreting the fractographical features [*16*].
2. Indirect techniques, like the partial unloading or the potential drop techniques, don't seem to be able to reproduce SZW measurements due to the required high degree of accuracy.

Therefore, the initial part of a J-R curve, the steep blunting part, should be represented by a theoretical relationship. This is also the philosophy of E 813, but the SZW data are in general not matched by Eq 2, which tends to overestimate Δa_B [*17–19*]; consequently, J_{Ic} is overestimated. This effect increases with increasing strain hardening because of inappropriate consideration of hardening in Eq 2.

This problem can be immediately solved if advantage is taken of the properties of the Hutchinson-Rice-Rosenfield (HRR field); see Ref *20*. If the crack-tip opening displacement (CTOD) is defined as δ_{45}, Fig. 4, it is according to Shih [*20*] given by

$$\delta_{45} = d_n \frac{J}{\sigma_0} \tag{4}$$

A power law hardening material with the yield stress σ_0 and the hardening exponent n is anticipated. The factor d_n is presented graphically for plane strain and plane stress in Ref *20* as a function of n and σ_0/E. The factor d_n can also be expressed analytically so that for plane strain

$$d_n = 1.185\left[\frac{2}{\sqrt{3}}\,\sigma_0\,\frac{1+\nu}{E}\,\frac{1-n}{n^{n/(1+n)}}\right]^{1+n}\frac{E}{\pi\sigma_0(1+n)(1-\nu^2)} \tag{5}$$

can be obtained [*14*]. Except for vanishing strain hardening, Eq 5 yields values which are very close to those plotted in Ref *20*. Comparing the values calculated using Eqs 4 and 5, respectively, with the ASTM blunting line shows immediately that the flow stress, σ_F, in Eq 3 is an oversimplification of the strain hardening effects on the crack-tip behavior.

As demonstrated in detail in Appendices I and II, the crack extension Δa_B due to blunting can be set equal to 0.4 times twice the stretch zone height, 2SZH, which in turn equals δ_{45}

$$\Delta a_B = 0.4\delta_{45} \tag{6}$$

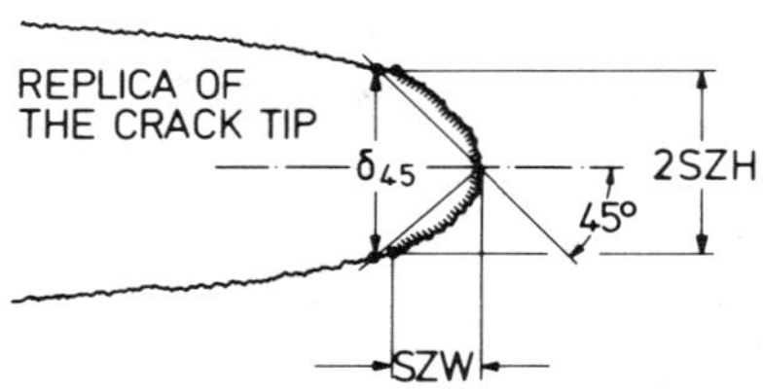

FIG. 4—*Replica of crack-tip profile showing definition of SZW, 2 SZH, and* δ_{45}.

with δ_{45} predicted by Eq 4. This blunting-line equation is in very good agreement with experimental results; see Appendix I.

Initiation and Definition of J_{Ic}

Since J_{Ic} is regarded as a measure of initiation toughness, it should be taken at the true initiation point, J_0. This would have the advantage that J_0 is fairly size and geometry independent [12,13]. With this definition, the following three problems are noted:

1. The transition from blunting to ductile tearing is gradual, Fig. 5; see also Fig. 10 in Appendix I. The first indications of ductile tearing are noticed at the J value J_0, the appropriate SZW is SZW_{min} in Fig. 5. With increasing deformation, ductile tearing is initiated at an increasing number of locations along the crack front. At the same time, the stretch zone increases in size at those locations where no ductile tearing has been initiated. When the process of initiation of ductile tearing terminates, that is, when ductile tearing occurs along the whole crack front for the first time, the stretch zone ceases to grow and it attains its maximum width, SZW_{end}. From this it follows that J_0 must be clearly defined according to the initiation events at the crack tip to avoid large scatter caused by individual interpretation of crack initiation.
2. In order to get as closely as possible to initiation, very minute amounts of crack growth have to be measured, which requires the use of a scanning electron microscope.
3. If the material is very inhomogeneous, the meaning of J_0 is difficult to define.

Although we may apply the J_0 philosophy to a number of fundamental studies, for a standard test method we propose to define J_{Ic} at a certain amount of ductile tearing, Δa^*_{DT}, which is close enough to initiation to ascertain size and geometry independent values, but large enough to ensure reliable measurement with both direct (light microscope) and indirect (partial unloading, key curve, potential drop methods) techniques. According to our experience and to recent discussions in several working groups

$$\Delta a^*_{DT} = 0.2 \text{ mm} \tag{7}$$

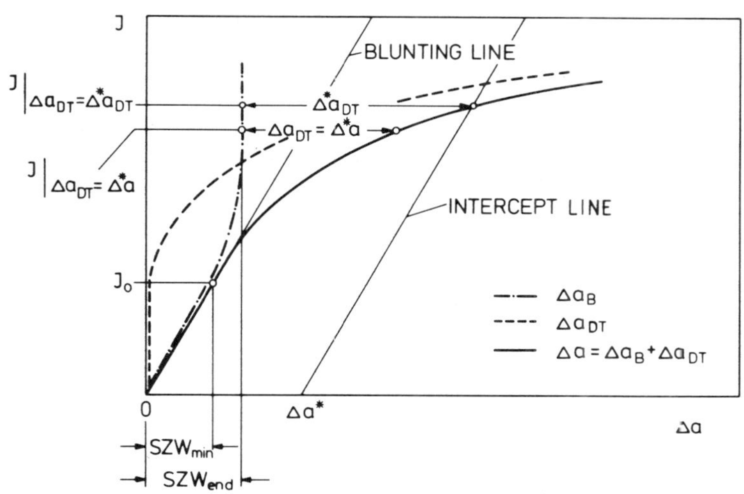

FIG. 5—*Schematic showing proposed construction for determining* J_{Ic} *from* J-Δa *data.*

seems to suit both of the above requirements. For the meaning of Δa^*_{DT}, see Fig. 5. For the determination of the J-Δa data, both the single- and multiple-specimen methods should be admitted, and in order to obtain compatible results the total crack growth, that is, blunting plus ductile tearing

$$\Delta a = \Delta a_B + \Delta a_{DT} \tag{8}$$

should be measured. The reason for this is that the single-specimen methods measure the total crack growth. Consequently, a procedure is needed for distinguishing between both contributions. Since Δa_B can be well modeled by Eq 6, Δa^*_{DT} can be approximately determined by intersecting the J-Δa data with an intercept line parallel to Eq 6 that represents the blunting line. The distance between both straight lines equals $\Delta a^*_{DT} = 0.2$ mm and the ordinate of the intersection point is supposed to represent J_{Ic} if the validity requirements are met.

Experimental Requirements for Determining J-Δa *Data*

As stated above, both the single- and multiple-specimen techniques can be used to generate the J-Δa data.

Single Specimen Technique—The techniques currently used, like the partial unloading, potential-drop and key curve techniques, measure the total crack extension, that is, ductile tearing and plasticity effects such as blunting. Careful calibration is needed if additional plasticity effects like thickness contraction at the crack tip in highly ductile materials affect the measurement. In any case, it is recommended to mark the final crack extension in each specimen tested by the usual techniques in order to have a calibration point.

The single-specimen method has the advantage that only one specimen is needed for the whole J-Δa curve and that it provides more information regarding variations of the material properties. Furthermore, continuous crack growth monitoring during a test is needed if J has to be corrected for crack growth.

Multiple-Specimen Technique—Since one specimen provides only one point on the J-Δa curve, several specimens must be tested in order to obtain a complete J-Δa curve. Crack growth shall be measured including the SZW to obtain compatibility with the single-specimen technique. The obvious advantage of this technique is that no highly sophisticated equipment is needed; a light microscope is sufficient for accurate measurements on the fracture surface.

As to the accuracy of the Δa measurement, a margin of ± 0.1 mm or $\pm 15\%$ in Δa, whichever is larger, is regarded as an appropriate requirement.

Figure 6 shows schematically how the error in Δa affects the J_{Ic} value. A minimum of five data points are required to establish a J-Δa curve with the data point distribution shown in Fig. 7. At least one J-Δa point must be located between E1 and E2 and E3 and E4, respectively.

To obtain a smooth curve, a power-law fit

$$J = A \Delta a^B \tag{9}$$

to the experimental data points is recommended.

Reasonable results have also been obtained using the equation

$$J = A \Delta a^B + C \Delta a \tag{10}$$

Examples

The modified procedure previously described has been applied to the three materials listed in Table 1. Both the single- and multiple-specimen methods were employed to generate the J-Δa data; for details of the crack length measurements see Ref *9*. Except for the proposed modifica-

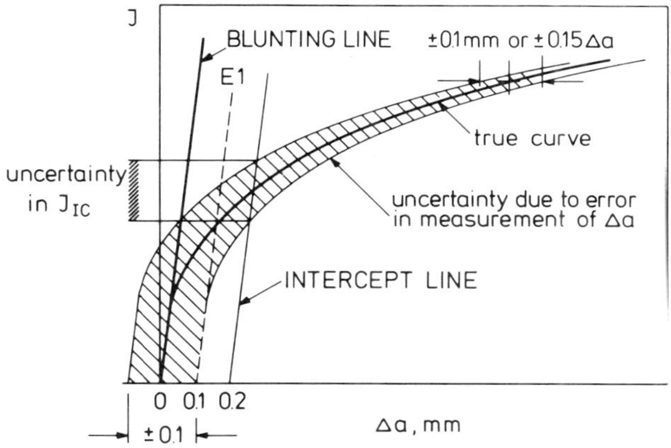

FIG. 6—*Effect of uncertainty in* Δa *measurement on* J_{Ic}.

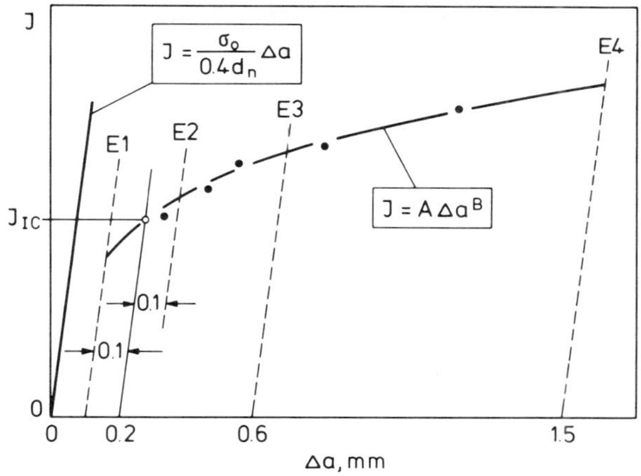

FIG. 7—*Required data point distribution. The lines E1 through E4 are parallel to the blunting line.*

TABLE 1—J_{Ic} *data of three materials from Table 2, N/mm.*

Material	J_{Ic}, ASTM E 813-81 Single-Specimen Technique[a]	J_{Ic}, Modified Single Specimen[b]	J_{Ic}, Modified Multiple Specimen	J_0
20 MnMoNi 5 5	140 to 300[c] (100 to 300)	198 ± 20	220	100 to 130
Al2024 FC	26	27 ± 3	35	8 to 12
A572	80 to 145	133 ± 15	145	60 to 80

[a]Variation in J_{Ic} caused by linear fit and data grouping rule.
[b]Variation in J_{Ic} caused by scatter of data points.
[c]Lowest value (100 N/mm) determined on 10-mm thick specimen, Fig. 3*b*. Range of 140 to 300 N/mm determined from one data set, Fig. 3*a*, variation due to exploitation of data grouping rule.

tions, the tests were performed and evaluated following ASTM E 813-81. The data points were fitted by Eq 10. Table 1 gives three different critical J values:

(*a*) J_{Ic} according to E 813-81,
(*b*) J_{Ic} according to our proposal, and
(*c*) J_0, the J value at initiation of ductile tearing.

By varying the specimen size, the J_{Ic} (E 813-81) of 20 MnMoNi 5 5 could be varied by a factor of 3 (see also Fig. 3), which is due to the straight-line fit to a curved relationship and the extrapolation to that straight line to the blunting line. Similarly, by exploiting the possibilities of the data grouping rule, the valid J_{Ic} (E 813-81) of A572 exhibits the range given in Table 1. On the other hand, in both cases (as in all other cases studied) the modified J_{Ic} method yielded unequivocal J_{Ic} values since according to our procedure J_{Ic} represents a point on the R-curve, Fig. 8.

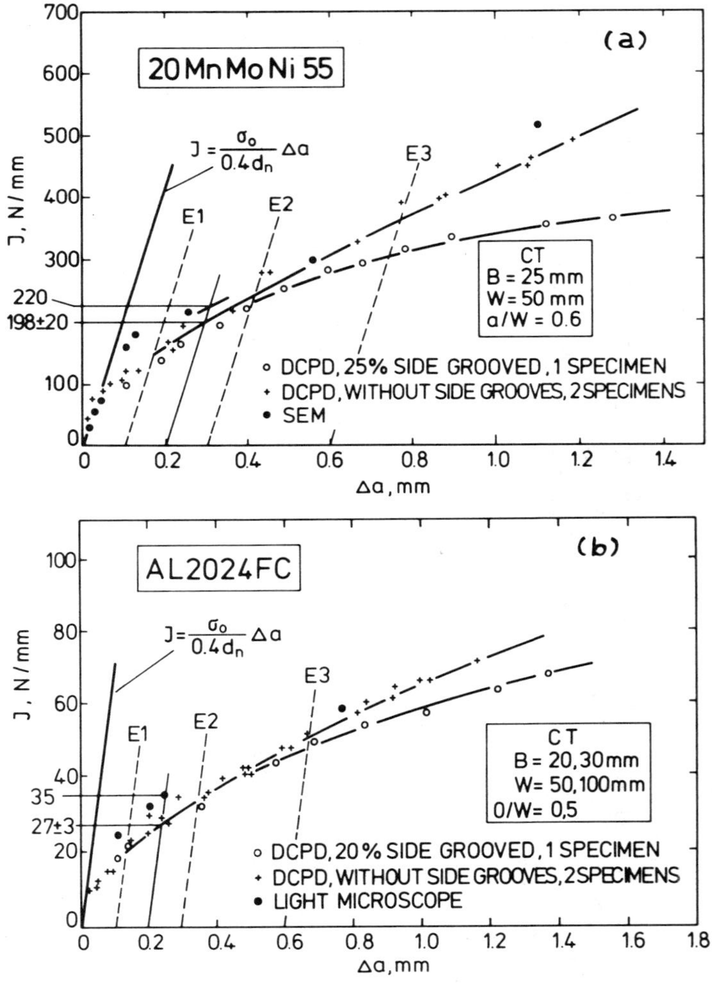

FIG. 8—J-Δa *curves with proposed* J_{Ic} *construction.* (a) *Nuclear reactor pressure vessel steel 20 MnMoNi 5 5, data points taken from Fig. 3a.* (b) *Soft aluminum alloy 2024-FC.* (c) *Ferritic steel A572; material provided by ASTM round-robin on the draft CTOD method.*

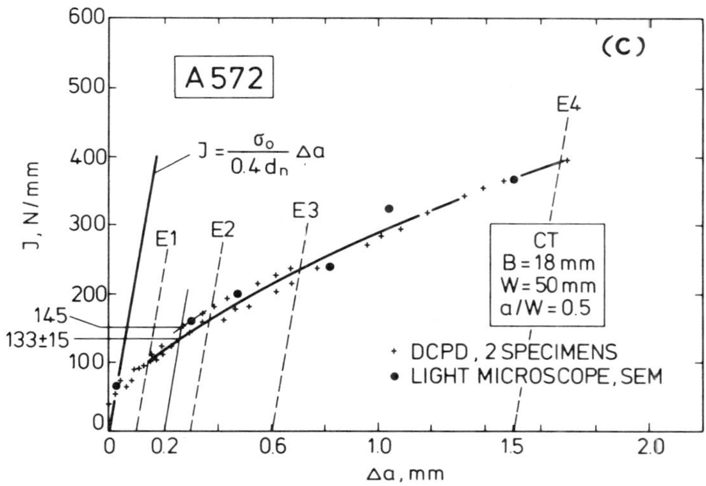

FIG. 8—*Continued.*

From Fig. 8*a* and 8*b* it can be seen that the J-Δa curves are only little affected by side grooving and hence by constraint within the crack growth range 0.4 mm. Thus, the choice Δa^*_{DT} = 0.2 mm appears reasonable in that no substantial size effects are to be expected.

A further interesting feature can be noted from Table 1. The initiation value J_0 is far below any J_{Ic}, which means that the resistance against crack growth increases rapidly just after initiation.

From an engineering viewpoint, the exploitation of the high crack growth resistance associated with some crack growth is justified if the amount of crack growth is small (J_{Ic} is supposed to be an engineering measure of initiation toughness) and if the resulting J_{Ic} is size independent.

Conclusions

A critical examination of ASTM E 813-81 led to the following modifications:

1. J_{Ic} is defined as the resistance against crack growth at a specified amount, 0.2 mm, of ductile tearing.
2. By this definition, J_{Ic} represents a point on the actual R-curve.
3. A new blunting-line equation is proposed which accounts for strain hardening in the spirit of the HRR field solutions.
4. This blunting-line equation can be easily obtained using standard tension test data without a complete stress-strain curve.
5. The resulting J_{Ic} value is unequivocal and independent of variables that may affect J_{Ic} according to E 813-81.

Acknowledgments

The authors gratefully acknowledge the contribution of the discussions in the Working Party "Kennwertermittlung und Anwendung der Bruchmechanik" of the German Fracture Group, and would also like to thank Mrs. A. Kaschemekat for typing the manuscript.

APPENDIX I

Crack-Tip Blunting Observations

Crack-tip profiles in the austenitic steel X6 CrNi 18 11 and in the nuclear reactor pressure vessel steel 20 MnMoNi 5 5 were observed and measured using the infiltration technique. In all cases the stretch zone at the crack tip was marked by refatiguing the specimens so that comparisons could be made between the replicas and the original fracture surfaces. In order to obtain the crack-tip profile, the replicas were cut perpendicular to the fracture surface and parallel to the side surface of the specimens. Subsequently, the replicas were coated with a thin gold layer and examined in an SEM. The following quantities were measured (see Fig. 4).

On the fracture surface the SZW as measured across the specimen thickness in the range of $0.3B$ to $0.7B$ where B is the specimen thickness. These values were transferred to the appropriate replica where they were used to locate the gage points for measuring the double stretch zone height, 2SZH, in the SEM. The reason for doing this is simply the circumstance that 2SZH can be measured more conveniently on the replica than on the original fracture surface. In addition, δ_{45} was determined by applying the 45-deg construction to each replica. The results of these measurements are plotted in Fig. 9. The following findings can be deduced:

1. The δ_{45} values determined on the replicas are very close to 2SZH.
2. The stretch zone relation SZW/2SZH seems to be a constant equal to 0.4. This is in agreement with finite-element calculations by McMeeking [*21*] and experimental observations by Gibson [*22*].

It can also be shown that δ_{45} measured on the replicas (or 2SZH) can be predicted by Eq 4. The main problem in using Eq 4 is given by the fact that relationships of this kind are based on power-law hardening materials, whereas actual materials don't behave like ideal power laws.

Appendix II provides guidance on how to determine the constants σ_0 and n from standard tension test data. Figure 9 shows excellent agreement between experimental and predicted δ_{45} values.

Using the findings

$$2\text{SZH} \approx \delta_{45}^{\text{exp}} \approx \delta_{45}^{\text{theor}} = d_n \frac{J}{\sigma_0}$$

$$\frac{\text{SZW}}{2\text{SZH}} \approx 0.4$$

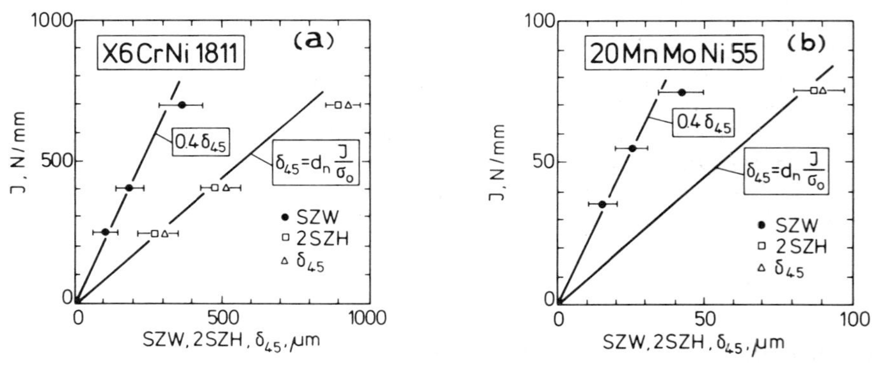

FIG. 9—*Stretch zone measurements on crack tip replicas.* (a) *Austenitic steel X6 CrNi 18 11.* (b) *Nuclear reactor pressure vessel steel 20 MnMoNi 5 5.*

a blunting-line equation can be derived

$$\text{SZW} = \Delta a_B = 0.4\, d_n \frac{J}{\sigma_0}$$

$$= 0.4\, d_n^* \frac{J}{E} \tag{11}$$

The validity of Eq 11 has been checked by SZW measurements on six different materials, the tensile properties of which are listed in Table 2. Two of these materials were taken from a Japanese round-robin exercise [*23*]. Figures 10 and 11 show excellent agreement between experimental and by Eq 11 predicted values, whereas the E 813-81 blunting line (Eq 2) overpredicts SZW and hence Δa_B in all cases.

TABLE 2—*Tensile properties of materials investigated.*

Material	$\sigma_{0.2}$, N/mm²	σ_u, N/mm²	E, N/mm²
X6 CrNi 18 11	240	622	190 000
20 MnMoNi 5 5	478	612	210 000
A 572[a]	400	592	210 000
Al 2024 FC	75	217	71 000
HT 60[b]	569	667	210 000
HT 80[b]	765	814	210 000

[a]Material provided by the ASTM CTOD round-robin organized by D. E. McCabe.
[b]SZW results taken from Ref *23*.

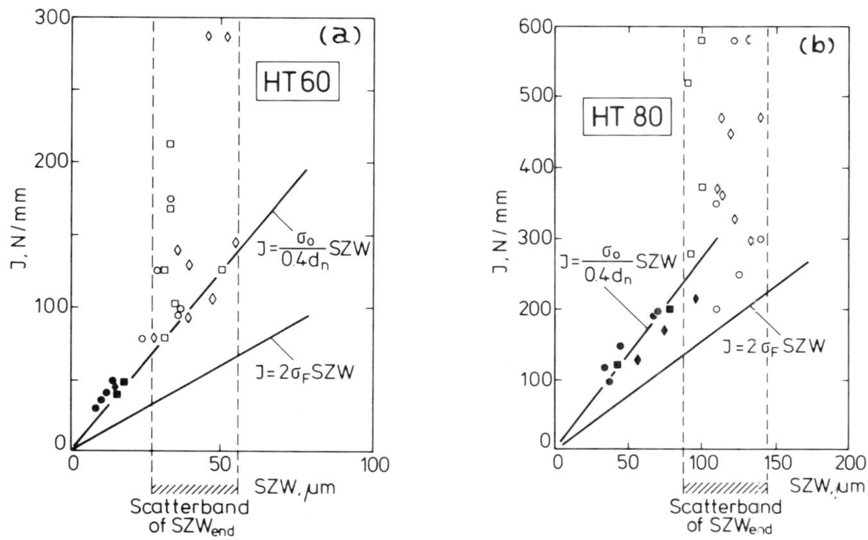

FIG. 10—*Stretch zone measurements from a Japanese round robin* [23]. *The different symbols indicate the individual participants. Filled symbols—stretch zone formation only; open symbols—stretch zone formation plus first events of ductile tearing.* (a) *Steel HT 60.* (b) *Steel HT 80.*

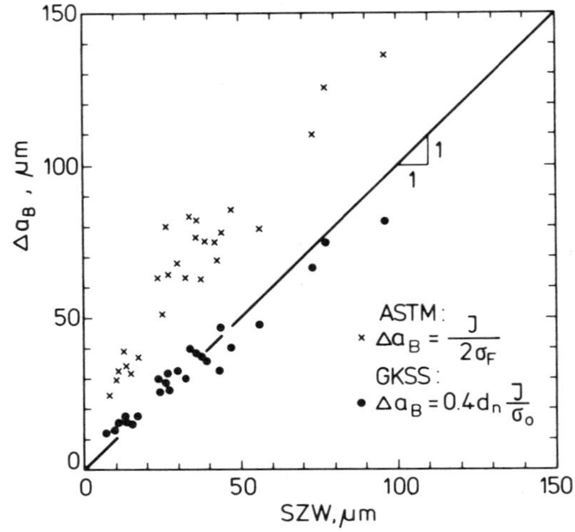

FIG. 11—*Predicted and measured stretch zone width for all materials of Table 2 except the austenitic steel X6 CrNi 18 11, which falls outside the range of the graph.*

APPENDIX II

Steps for Determination of Blunting Line

Since the HRR field is based on power-law hardening materials, the stress-strain curve of a given real material must be fitted by a power law in order to obtain the constants of that power law. The point-wise determination of a stress-strain law with a subsequent power-law fit would be the ideal way for determining the relevant constants. However, this may be regarded as an additional complication of the J_{Ic} test and could thus be used as an argument against adopting HRR field-based blunting-line formulation. On the other hand, an estimate of sufficient accuracy can be done using the ordinary tension test data: If E, $\sigma_{0.2}$, σ_u, and ϵ_u are known, the hardening exponent n can be determined by

$$n = \frac{\log[\sigma_u(1 + \epsilon_u)/\sigma_{0.2}]}{\log[\ell n(1 + \epsilon_u)/\epsilon_{0.2}]} \tag{12}$$

with

$$\epsilon_{0.2} = \frac{\sigma_{0.2}}{E} + 0.002 \tag{13}$$

If only E, $\sigma_{0.2}$, and σ_u are available, the relationship

$$\frac{\sigma_{0.2}}{\sigma_u} = \frac{1}{1 + \epsilon_{0.2}}\left[\frac{e \cdot \ell n(1 + \epsilon_{0.2})}{n}\right]^n \tag{14}$$

serves for determining n using the plot of Fig. 12 ($e = 2.7183$).

In this relationship the unknown strain ϵ_u was determined with the condition that the slope of any true stress-strain curve, n_u, at the maximum engineering stress (or maximum load of a tension specimen) has always the same value as the true strain at that point, that is

$$n_u = \ell n(1 + \epsilon_u) \tag{15}$$

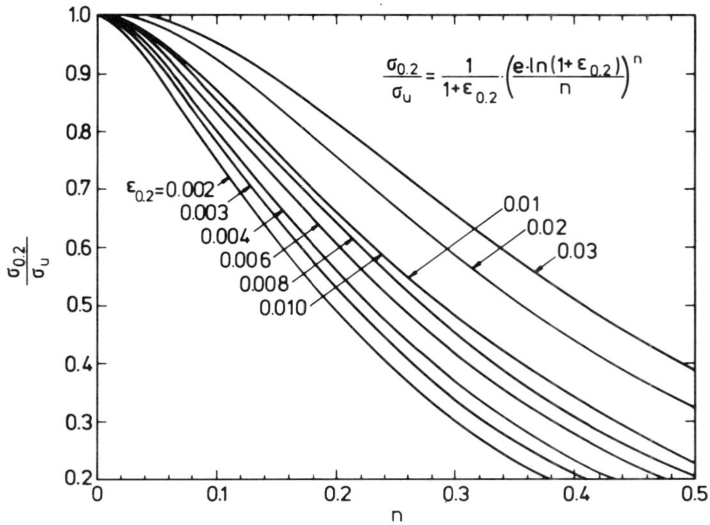

FIG. 12—*Diagram for determining* n *from* $\sigma_{0.2}$, σ_u, *and* $\epsilon_{0.2}$.

where ϵ_u is the engineering strain at the maximum engineering stress σ_u. Using this condition in analogy with the determination of the hardening exponent n in Eq 12, the relationship Eq 14 follows immediately.

It should be noted that the hardening exponent n from Eq 14 normally differs from Eq 15 for real stress-strain laws. The hardening exponent n from Eq 14 can be smaller or greater than n_u from Eq 15. Both quantities are equal only for an exact piece-wise power law. But in the space of double logarithmic axes of stress and strains, this has only very small effects on the hardening exponents determined either from Eq 12 or Eq 14.

Once the hardening exponent n is determined, the fictive yield stress σ_0 follows from Fig. 13

$$\frac{\sigma_0}{\sigma_{0.2}} = 10^{\left\{\frac{n \cdot \log[(E \cdot \epsilon_{0.2})/\sigma_{0.2}]}{n-1}\right\}} \qquad (16)$$

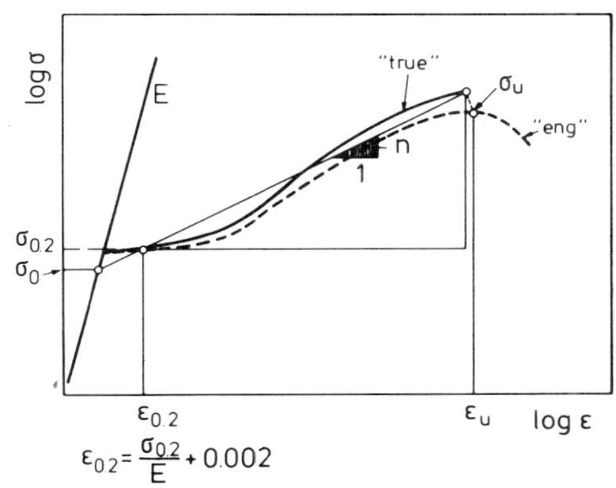

FIG. 13—*Schematical stress-strain curve showing approximation of actual data by power law.*

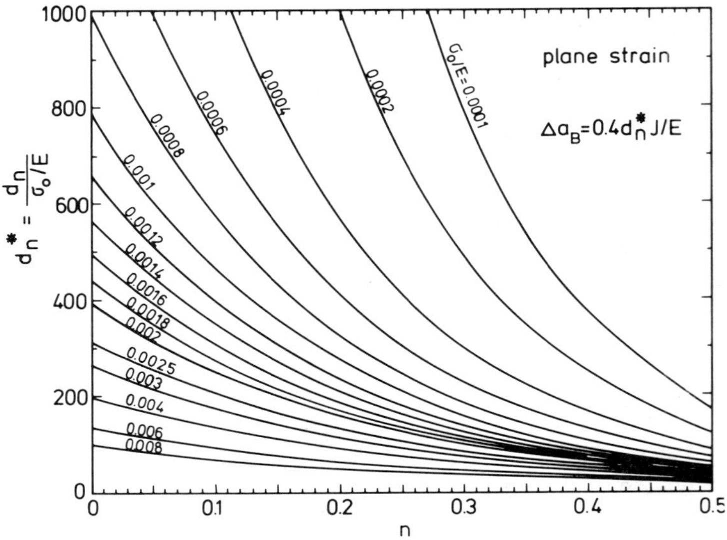

FIG. 14—*Diagram for determining* d_n^* *from* n *and* σ_0/E.

This simple procedure is based on a piece-wise power law which is fixed at the true stresses calculated from $\sigma_{0.2}$ and σ_u, Fig. 13. Now the blunting line

$$\Delta a_B = 0.4\, d_n \frac{J}{\sigma_0}$$

$$= 0.4\, d_n^* \frac{J}{E} \tag{17}$$

can be calculated using the graph for d_n^* in Fig. 14 or Eq 5 for d_n. For more information concerning accuracy and evidence of this method, see Ref *24*.

References

[*1*] Bethmont, M., "French Tentative Test Procedure for the Determination of the Fracture Resistance of Ductile Steels," Committee on the Safety of Nuclear Installations, Workshop on Ductile Fracture Test Methods, Paris, 17–19 April 1985.

[*2*] Neale, B., Curry, K., Green, G., Haigh, J. R., and Akhurst, K. N., "A Procedure for the Determination of the Fracture Resistance of Ductile Steels," Committee on the Safety of Nuclear Installations, Workshop on Ductile Fracture Test Methods, Paris, 17–19 April 1985.

[*3*] Schwalbe, K.-H., "Determination of J_{Ic} and J_R: Present State of Discussion in European Group on Fracture," Committee on the Safety of Nuclear Installations, Workshop on Ductile Fracture Test Methods, Paris, 17–19 April 1985.

[*4*] McCabe, D. E., "Progress Report on the ASTM R Curve Standard," Committee on the Safety of Nuclear Installations, Workshop on Ductile Fracture Test Methods, Paris, 17–19 April 1985.

[*5*] Heerens, J., Cornec, A., and Schwalbe, K.-H., "Determination of J_{Ic} Using New Blunting Line Construction," Committee on the Safety of Nuclear Installations, Workshop on Ductile Fracture Test Methods, Paris, 17–19 April 1985.

[*6*] Cornec, A. and Heerens, J., "Procedure for the Determination of a Theoretical Blunting Line," Committee on the Safety of Nuclear Installations, Workshop on Ductile Fracture Test Methods, Paris, 17–19 April 1985.

[*7*] Schwalbe, K.-H. and Heerens, J., "Suggestion for a Modification of ASTM E 813" in *Elastic-Plastic*

Fracture Test Methods: The User's Experience, ASTM STP 856, American Society for Testing and Materials, Philadelphia, 1985, pp. 411–416.

[*8*] Schwalbe, K.-H. and Heerens, J., "Proposed Modifications of ASTM E 813-81 Standard Test Method for J_{Ic}," 10th MPA Seminar, 10–13 Oct. 1984, Stuttgart, Federal Republic of Germany, *to appear in Nuclear Engineering and Design.*

[*9*] Heerens, J., Schwalbe, K.-H., Hellmann, D., Knaack, J., and Müller-Roos, J., "Application of the DC Potential Drop and the Partial Unloading Methods to Fracture Mechanics Tests," GKSS Report 85/E/16, GKSS Research Center, Geesthacht, Federal Republic of Germany, 1985.

[*10*] Schwalbe, K.-H., "Approximate Calculation of Fatigue Crack Growth," *International Journal of Fracture,* Vol. 9, 1973, pp. 381–395.

[*11*] Schwalbe, K.-H., "Crack Propagation in AlZnMgCu0.5 During Static Loading," *Engineering Fracture Mechanics,* Vol. 6, 1974, pp. 415–434.

[*12*] Schwalbe, K.-H., "Influence of Stress State on Static Crack Growth in AlZnMgCu0.5," *Engineering Fracture Mechanics,* Vol. 9, 1977, pp. 557–583.

[*13*] Schwalbe, K.-H., "Some Properties of Stable Crack Growth," *Engineering Fracture Mechanics,* Vol. 11, 1979, pp. 331–342.

[*14*] Schwalbe, K.-H., "Crack Tip Opening Displacement for Work-Hardening Materials," *International Journal of Fracture,* Vol. 25, 1984, pp. R49–R52.

[*15*] Schwalbe, K.-H., Heerens, J., Hellmann, D., and Cornec, A., "On the Relationships Between Various Definitions of the Crack Tip Opening Displacement," Workshop on the CTOD Methodology, GKSS Research Center, Geesthacht, Federal Republic of Germany, 23–25 April 1985.

[*16*] Heerens, J. and Cornec, A., "Round Robin on Stretch Zone Measurements," to be prepared for publication.

[*17*] Mills, W. J., "On the Relationship Between Stretch Zone Formation and the J-Integral for High Strain Hardening Materials," *Journal of Testing and Evaluation,* Vol. 9, No. 1, Jan. 1981, pp. 56–62.

[*18*] O'Brien, D. M. and Ferguson, W. G., "On the J-Integral Blunting Line for Soft Materials," *International Journal of Fracture,* Vol. 20, 1982, p. R39.

[*19*] Yin, S. W., Gerbrands, R. A., and Hartevelt, M., "An Investigation of the Blunting Line," *Engineering Fracture Mechanics,* Vol. 18, No. 5, 1983, pp. 1025–1036.

[*20*] Shih, C. F., "Relationship Between the J-Integral and Crack Opening Displacement for Stationary and Extending Cracks," Report No. 79CRoo75, General Electric Co., 1979.

[*21*] McMeeking, R. M., "Finite Deformation Analysis of Crack Tip Opening in Elastic-Plastic Materials and Implications for Fracture," *Journal of the Mechanics and Physics of Solids,* 1977, Vol. 25, pp. 357–381.

[*22*] Gibson, G. P. and Druce, S. G., "A Study of Elastic-Plastic Fracture Parameters Through the Thickness of Fracture Toughness Specimens Using Silicone-Rubber Crack Impressions," Report R-10815, Metallurgy Division, Atomic Energy Research Establishment, Harwell, Oxfordshire, U.K., Feb. 1983.

[*23*] Ohji, K., Otsuka, A., and Kobayashi, H., "Evaluation of Several Testing Procedures Recommended in Japan" in *Elastic-Plastic Fracture: Second Symposium. Volume II: Fracture Curves and Engineering Applications, ASTM STP 803,* C. F. Shih and J. P. Gudas, Eds., American Society for Testing and Materials, Philadelphia, 1983, pp. II-398–II-419.

[*24*] Cornec, A., Heerens, J., and Schwalbe, K.-H., "Bestimmung der Rißspitzenaufweitung CTOD und Rißabstumpfung SZW aus dem J-Integral," Report 86/E/15, GKSS-Forschungszentrum Geesthacht, D-2054 Geesthacht, Federal Republic of Germany, 1986.

Fahmy M. Haggag[1] *and Angela K. Richardson*[2]

Precracking and Computerized Single-Specimen J_{Ic} Determination for Irradiated Three-Point Bend Specimens

REFERENCE: Haggag, F. M. and Richardson, A. K., "**Precracking and Computerized Single-Specimen J_{Ic} Determination for Irradiated Three-Point Bend Specimens,**" *Fracture Mechanics: Eighteenth Symposium, ASTM STP 945,* D. T. Read and R. P. Reed, Eds., American Society for Testing and Materials, Philadelphia, 1988, pp. 390–404.

ABSTRACT: An experimental procedure is described for precracking and computerized single-specimen J_{Ic} testing of irradiated three-point bend specimens. The specimens were automatically precracked in cantilever bending at constant deflection. The J_{Ic} values determined by the single-specimen unloading compliance technique were found to be in close agreement with multiple-specimen values. The J_{Ic} testing system comprised a three-point bend fixture and a furnace mounted on a screw-driven Instron testing machine, a Hewlett-Packard computer, and a data acquisition system. This procedure was found to be successful and cost effective in precracking and J_{Ic} testing of the irradiated specimens.

KEY WORDS: automated precracking, cantilever bending, crack propagation, three-point bend specimens, J-integral, unloading compliance, irradiated specimens

The structural integrity of reactor pressure tubes (Type 347/348 stainless steel) depends on the tube material properties of strength, ductility, and fracture toughness, and on the size and orientation of flaws in the tube wall. Service factors that tend to reduce the margin of conservatism in structural integrity are the reduction of material fracture toughness due to fast neutron damage and the growth of flaws due to cyclic stresses associated with plant operation. Hence, the mechanical properties of specimens, either machined from spent tubes or irradiated to similar conditions, at 427°C have been experimentally determined [*1*]. Machined cracks in the in-pile tube specimens were in the longitudinal-circumferential (L-C) orientation. This paper describes the experimental procedure used for precracking and computerized single-specimen J_{Ic} testing of irradiated three-point bend specimens.

Procedure

Flat and curved (having been made from a tube) specimens 5.08 and 10.16 cm long were used in this investigation. Both short and long specimens were tested on the same three-point bend fixture since the span between the two roller pins of the fixture was 4.06 cm. Nominal width and thickness dimensions were the same for both short and long specimens. After fracture toughness testing of each of the 10.16-cm-long specimens, two new specimens 5.08 cm long were machined (and then fatigue precracked) from the broken ends, thus allowing efficient use of the limited irradiated specimens. The geometry of the 5.08-cm-long curved specimens (L-C crack

[1]Development staff member, Oak Ridge National Laboratory, Oak Ridge, TN 37831; previously with the Idaho National Engineering Laboratory.

[2]Engineer, Idaho National Engineering Laboratory, Idaho Falls, ID 83415.

plane orientation) is shown in Fig. 1. The actual notch width of all long and short specimens was taken into account for calculating the vertical displacement at the load-line from measured displacements. The load-line displacement was calculated using the equation shown in Fig. 1.

Machining

The notch in the 5.08-cm-long specimens, shown in Fig. 1, was machined in the hot cell using a slitting saw mounted on the horizontal spindle of a small milling machine. A fixture was designed to machine up to four specimens. Figure 2 shows the notch machining setup. The same milling machine was used for both machining and subsequent precracking operations inside the hot cell.

Fatigue Precracking

Both long and short (10.16 and 5.08 cm) specimens were fatigue precracked in cantilever bending at constant deflection in accordance with ASTM Test Method for J_{Ic}, a Measure of Fracture Toughness (E 813-81). The stress intensity factor range, ΔK, was kept below 23 $MPa \cdot m^{1/2}$ for the final 50% of fatigue precrack extension for all specimens. The calculation of ΔK was based on the formula in Ref *2*. The stress intensity, K, for a beam in cantilever bending is given by [*2*]

$$K = Y \frac{P \cdot L}{BW^{3/2}} \tag{1}$$

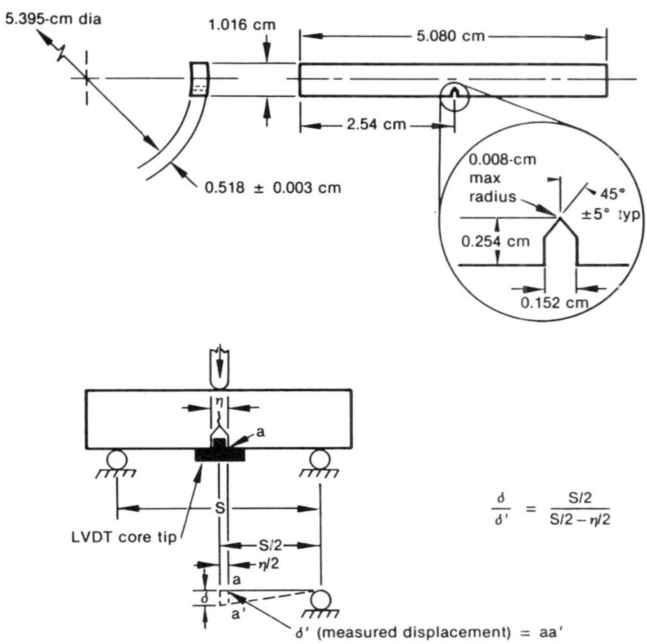

FIG. 1—*Three-point bend specimen geometry and test configuration.*

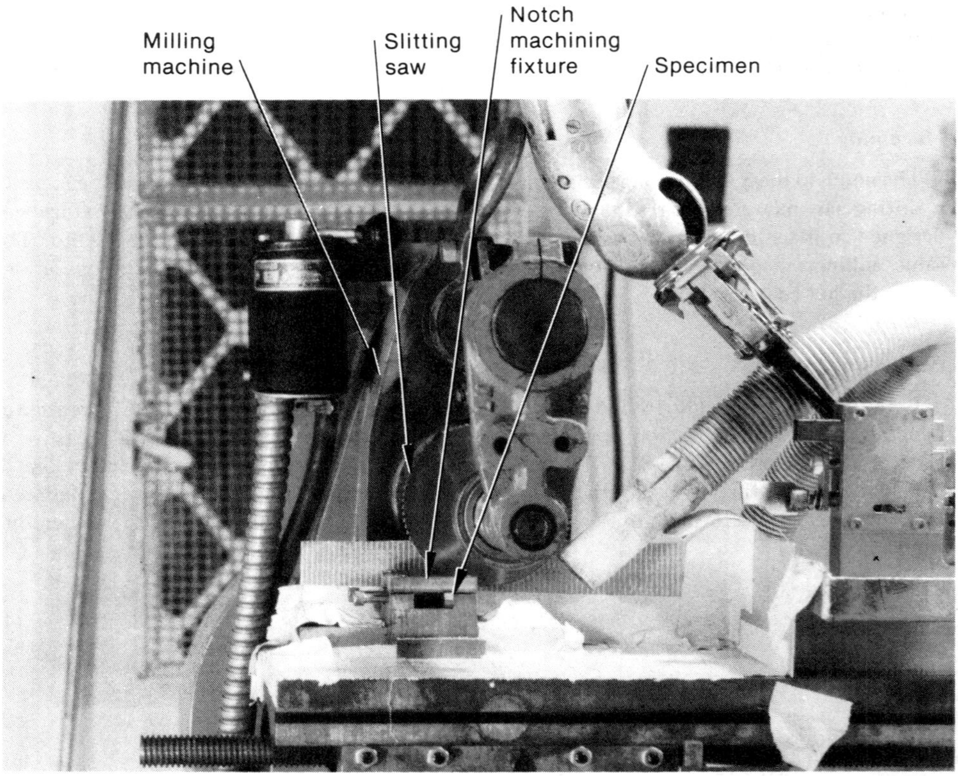

FIG. 2—*Notch machining of bend specimens.*

where

Y = stress-intensity coefficient (obtained from Fig. 4 in Ref *2* as a function of initial crack-to-width ratio),
P = applied load,
L = moment arm,
B = specimen thickness, and
W = specimen width.

The applied load P can be calculated from the measured beam deflection, γ, using the equation [*3*]

$$P = \gamma \frac{3EI}{L^3} \tag{2}$$

where

γ = measured beam deflection,
E = elastic modulus, and
I = moment of inertia of the beam's remaining ligament.

A specially designed fatigue precracking fixture was used in a milling machine inside the hot cell. The moment arm was the same (2.03 cm) for all specimens. Cyclic deflection of the specimens was achieved using a ball bearing mounted eccentrically on the spindle of the milling machine. The amount of deflection, as read on the oscilloscope, while the machine was running, was controlled for each specimen by moving the precracking fixture against the ball bearing until the desired deflection (0.076 to 0.102 mm) was reached. The deflection was measured using a dc-in, dc-out linear variable differential transducer (LVDT) with spring-loaded spindle. Precracking was performed at a cyclic rate of 25 cycles/s (90 000 cycles/h). The cantilever bending precracking setup for single edge-notched (SEN) specimens, as well as the gaged specimen configuration, is shown in Fig. 3. The fatigue crack length was monitored and automatically controlled (that is, the milling machine was stopped when the desired fatigue crack length was reached) through the use of a crack propagation gage mounted on the specimen across from the machined notch with a specially designed electronic circuit (crack propagation detector). The equipment used for fatigue crack monitoring and control is shown in Fig. 4.

The crack propagation gages were supplied by Micro-Measurements. Each gage (5.1 mm wide by 25.4 mm long) consists of 20 resistor strands connected in parallel. The distance between the centerlines of the strands is 0.25 mm. The nominal gage thickness and total resistance are 0.043 mm and 5 Ω, respectively. When the gage is bonded properly to a structure, progression of a surface crack through the gage pattern causes successive open-circuiting of the strands, resulting in a stepped increase of total gage resistance as the strands fracture. If the gage is not bonded properly, the specimen may be precracked without breaking the gage strands or the gage strands may break while the specimen is not cracked. The gages were mounted on the bend specimens across from the notch, as shown in Fig. 3, using an adhesive which required a temperature cure of 1 h at about 177°C. Vacuum mounts (flat, concave, and convex) were used to mount the gages on the specimens using the two locating pins of the precracking fixture. The locating pins were also used to mount the electrical connecting block to apply a d-c voltage through spring-loading pins pushing on the two metallic sheets on the ends of the gage strands. The electrical connection block is shown in Fig. 3 (held by the slave manipulator).

The crack propagation detector (shown on top of the filter and strip chart recorder in Fig. 4) was used to control the fatigue crack length by adjusting the "Failure Level" potentiometer to the setting corresponding to the desired number of broken strands. When the predetermined number of strands was broken, a relay was triggered to stop the milling machine, and the red "Failure" indicator light and an electronic buzzer alarm were turned on, signalling the end of the fatigue precracking operation. A strip chart recorder was used to obtain a step curve of strands broken versus time. It was automatically triggered on and off with the operation of the milling machine by another relay.

Precracking of both flat and curved SEN irradiated specimens in cantilever bending (handled remotely in the hot cell) was performed successfully and efficiently using the techniques described above. The difference between the fatigue crack length at the specimen surfaces and the nine-point average value was less than 7%.

Equipment for Fracture Toughness Testing

Figure 5*a* shows the three-point bend fixture and a furnace mounted on a screw-driven mechanical testing machine. Details of the bend fixture and the tip of the extension rod of the LVDT with a spring-loaded core are shown in Fig. 5*b*. A split-box type electric furnace was used for fracture toughness testing at elevated temperatures. The test temperature was measured by three thermocouples located near the test specimen on both sides of the bend fixture and on the loading ram near its tip, as shown in Fig. 5*a*. For tests conducted at 427°C, the temperature difference among these three locations near the test specimen was less than 6°C. The vertical displacement of the specimen was measured by a dc-dc LVDT placed outside the furnace and slightly above the load cell. The LVDT housing was water-cooled (below 35°C) during testing. The output voltage signal of the LVDT was 10 V at its total range of 1.27 mm.

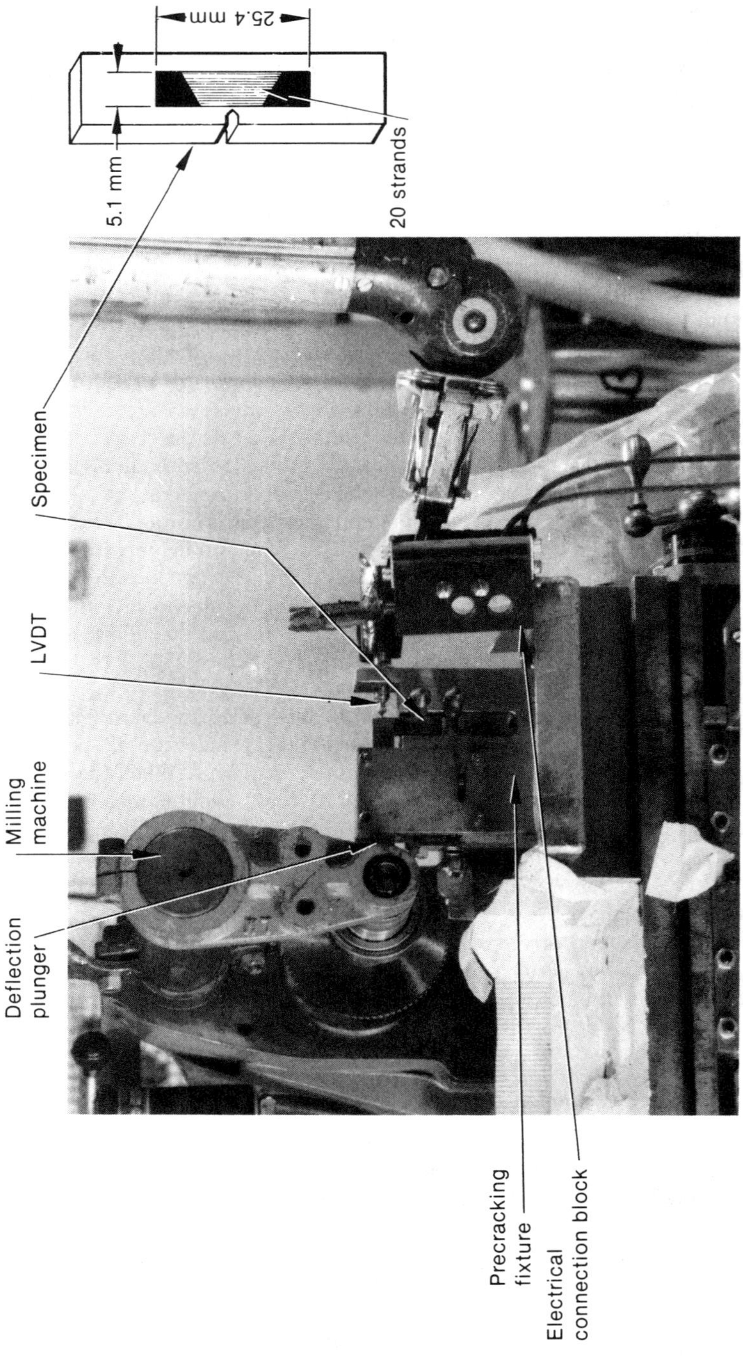

FIG. 3—Cantilever bending precracking of single-edge-notched specimens.

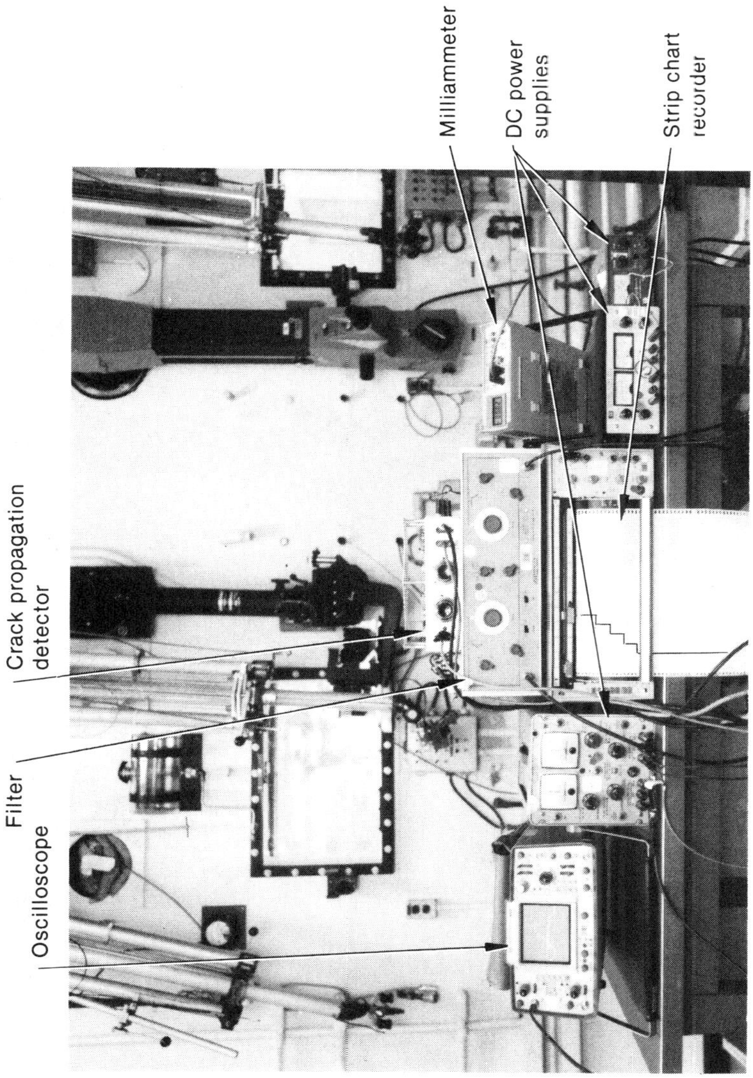

FIG. 4—*Equipment for monitoring and controlling of fatigue precrack extension in bend specimens.*

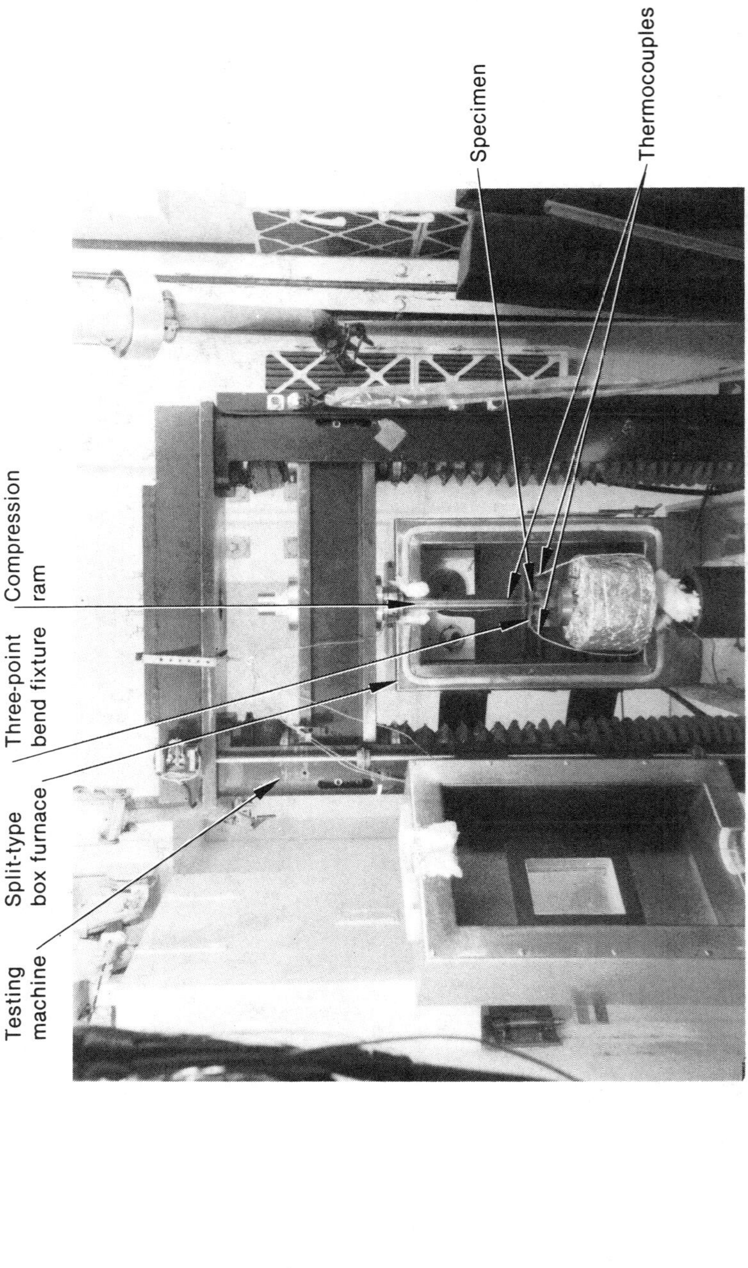
Specimen
Thermocouples
Compression ram
Three-point bend fixture
Split-type box furnace
Testing machine

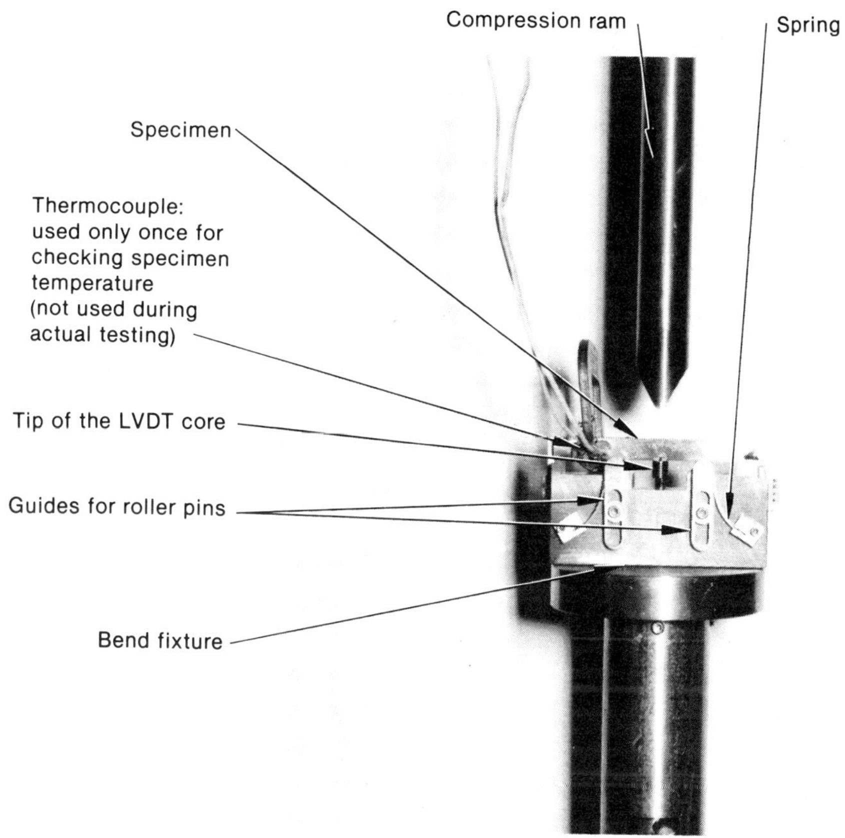

FIG. 5—*High-temperature fracture toughness testing system.* (a) *Three-point bend fixture mounted on testing machine.* (b) *Details of the bend fixture.*

The output voltage from the load cell was filtered and amplified (8.7 V corresponded to full load range).

The test control equipment is shown in Fig. 6. A Hewlett-Packard computer and an automatic data acquisition/control system were used to interactively calculate the values of the J-integral and crack extension (Δa) during the test. The load versus displacement was also recorded autographically on an X-Y recorder. All the voltage signals from the load cell were filtered, amplified, and fed, along with those from the LVDT, to both the data acquisition system (Channels 1 and 2) and the X-Y recorder. These data were later stored on floppy disks for post-test processing as well as to provide a permanent record of the fracture toughness tests. A schematic of the computerized fracture toughness testing system is shown in Fig. 7.

Computer Program

The computer program "UNLOAD" for compact, single-specimen unloading compliance was obtained from Oak Ridge National Laboratory (Pressure Vessel Technology Group).[3] This program has been modified at EG&G Idaho, Inc., for testing three-point bend specimens. The program was also modified to allow manual operation of the Instron mechanical testing machine and data acquisition through a multichannel voltmeter.

This program multiplexes the two channels of the data acquisition system, inputs the voltages to the computer through the digital voltmeter, and then calculates the values of the J-integral and crack extension during the progress of the test. The analysis required that an inverse compliance expression for relating the load-line compliance to a/W (crack length to specimen depth ratio) for three-point bend specimens be determined [*4*]. This inverse compliance expression is given as:

$$a/W = 0.996 - 3.58U - 1.51U^2 - 110U^3 + 1232U^4 - 4400U^5 \tag{3}$$

where

$U = 1/[(EB\delta/P)^{1/2} + 1]$,
E = elastic modulus,
B = specimen thickness,
δ = load-line displacement, and
P = applied load.

Test Procedure

All the fracture toughness tests at 427°C were performed with a cross-head speed of 8.5×10^{-5} cm/s. The initiation J_{Ic} value was determined from multiple loadings-unloadings of a single specimen. Periodically (after loading to certain displacements) during the test, the load was held for about 10 s followed by partial unloading (approximately 14% of the maximum load) and then reloaded. The top 30 to 45% and bottom 7 to 20% of the data (relatively less linear) of each partial unloading were deleted from the linear regression calculation of the slope of this line. (Deletion of these data conformed with recommendations provided by Mills [*5*].[4]) This procedure was used for all specimens for consistency and in order to improve the calculation of the slopes of the unloading lines and hence the crack extension calculations. Longer displacement hold time of 30 s was tried on a few specimens, but the nonlinearity of the unloading slopes was not reduced. The final crack length was marked by heat tinting at 427°C for 10 min. This was achieved simply by leaving each specimen in the furnace 10 min after complete unloading of

[3]Personal communication with R. K. Nanstad, Oak Ridge National Laboratory, 9 Dec. 1983.

[4]Also, personal communication with W. J. Mills, Hanford Engineering Development Laboratory, March 1984.

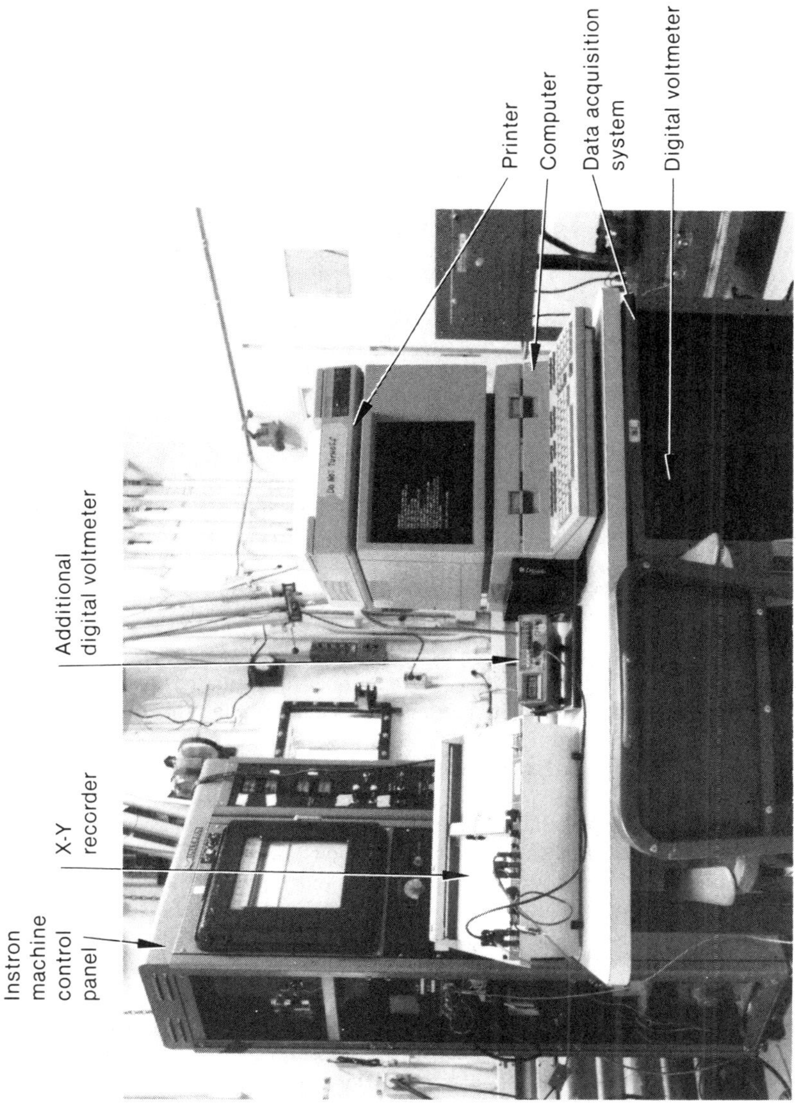

FIG. 6—*Fracture toughness test control equipment (computer, data acquisition system,* X-Y *recorder, and Instron machine control panel).*

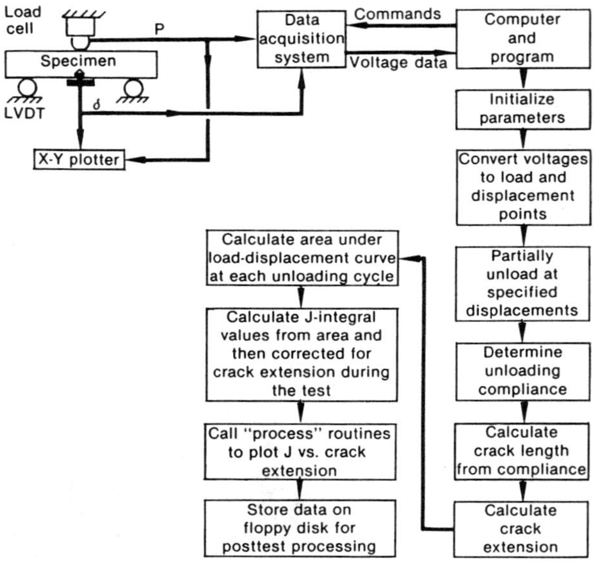

FIG. 7—*Schematic of computerized fracture toughness* (J_{Ic}) *testing system.*

the specimen at the end of the test. After cooling to room temperature, the specimens were broken completely using the fatigue precracking fixture. The initial and final crack lengths were then measured from pictures of the fracture surfaces. Typical initial and final crack lengths were 0.48 and 0.69 cm, respectively.

Results and Discussion

The computerized J_{Ic} test results from a very large number of irradiated specimens are given in Ref *1*; the test results of one specimen are given below as an example. This specimen (No. SS-F-40) was tested in three-point bending at 427°C. The specimen was machined from Type 348 stainless steel and irradiated in the Experimental Breeder Reactor II (EBR-II) to a fluence of 3.4×10^{22} neutrons (n)/cm^2 ($E > 1$ MeV).

The load versus load-line displacement curve for Specimen No. SS-F-40 is shown in Fig. 8. The amplified unloading load-displacement curves are shown in Fig. 9. The specimen was unloaded approximately 14% of maximum load periodically during the test with the first unloading (No. 0) being performed in the initial linear-elastic portion of the load-displacement curve. Using a typical value of 165.5 GPa for the elastic modulus [*6, 7*], the experimental compliance value ($EB\delta/P$) for unloading No. 0 was compared with the theoretical compliance value calculated by Eq 5 of Ref *4*. The difference between the experimental and theoretical compliance values was found to be 4.5% which is less than the 7% criterion required by ASTM E 813-81 (Section 8.7.1). Hence, the equipment was considered to be properly set up. The unloading compliance was computed from a least-squares regression line fitted through the lower part of the unloading curve (excluding the top 45% and bottom 7% data). For each unloading, the instantaneous crack length and the corresponding crack extension were then determined from the unloading compliance using Eq 3, while the J-value was calculated according to Section 9.1.2 of ASTM E 813-81. The J versus crack extension data, using a typical value of the elastic modulus, are shown in Fig. 10. This figure shows that the J_R-curve is shifted to the right of the blunting line.

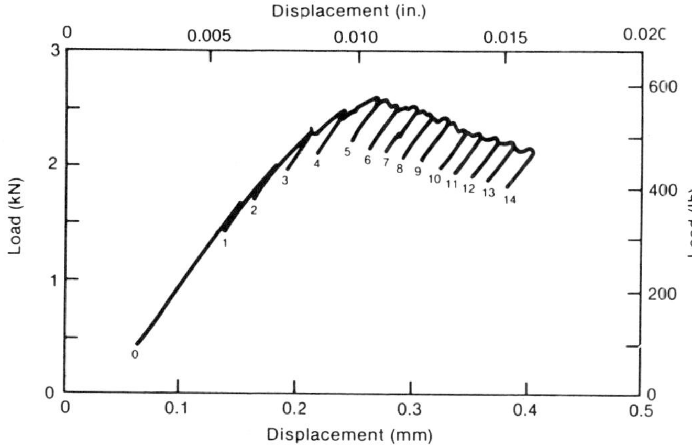

FIG. 8—*Load versus load-line displacement for Specimen No. SS-F-40 tested at 427°C.*

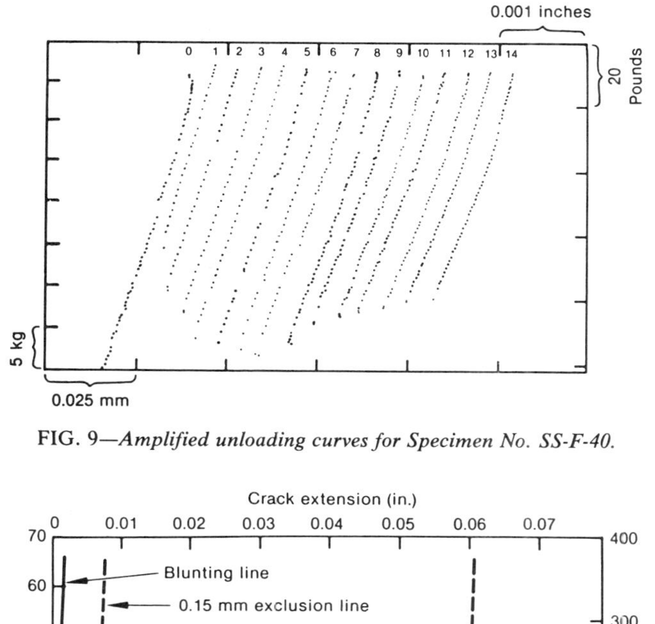

FIG. 9—*Amplified unloading curves for Specimen No. SS-F-40.*

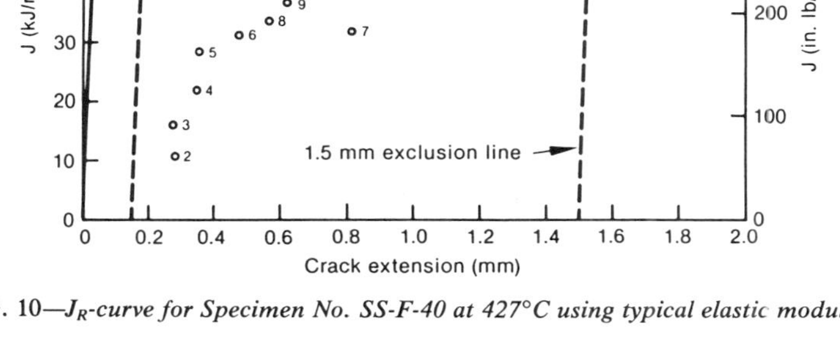

FIG. 10—*J_R-curve for Specimen No. SS-F-40 at 427°C using typical elastic modulus.*

A better J_R-curve (Fig. 11) resulted when an effective elastic modulus calibration was performed at about 80% of maximum load.[5] The effective elastic modulus approach has been used by Nanstad (see footnote 3) and Clarke (footnote 5) because unavoidable test variations would have less effect near maximum load than at points lower on the load-deflection curve. For this specimen (SS-F-40), the effective modulus of 148.2 GPa was calculated at unloading No. 2 (0.77 P_{max}) such that the crack extension, Δa_2, was assumed to be only due to crack front blunting, that is

$$\Delta a_2 = J_2/2\sigma_Y \tag{5}$$

where J_2 is the J-integral at unloading No. 2 and σ_Y is the effective yield strength. In Fig. 11, the data points for unloadings No. 7 and 14 were excluded as outliers since their unloading curves (Fig. 9) were not typically linear. The latter may be due to unfiltered electronic noise or other unavoidable test variations (for example, load cell stability, specimen seating, and friction). Only those data within the offset lines were included in the J_{Ic} calculation, that is, Points Nos. 6, 8, 9, 10, 11, 12, 13. For these points, the linear regression correlation (R^2) was 0.978. The J_{Ic} value determined for this specimen is 19.51 kJ/m^2.

The qualified (per Sections 9.1 to 9.3 of ASTM E 831-81) J-Δa data for three specimens (SS-F-40, SS-F-41, and SS-F-47) are shown in Fig. 12 with a regression line fitted to all data. The solid symbols in this figure represent the measured check extension using the heat tinting method and the final J-integral for each specimen calculated using

$$J = 2A_f/Bb \tag{6}$$

where A_f is the final energy area under the load-deflection curve. The regression line fit through these multiple-specimen data (heat-tinting J_R-curve) indicates a higher J_{Ic} value of 23.68 kJ/m^2 as compared to the unloading compliance value of 20.38 kJ/m^2. Hence, the unloading compliance J_{Ic} value, calculated from the linear regression analysis of 19 points obtained from three specimens, is considered a conservative estimate for the fracture toughness of this irradiated

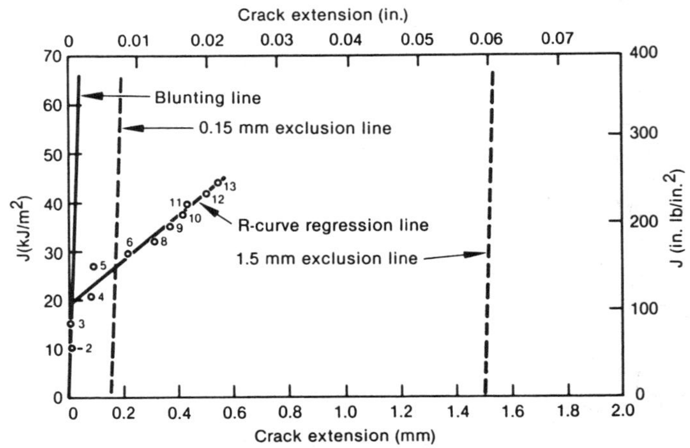

FIG. 11—*J_R-curve for Specimen No. SS-F-40 at 427°C using effective elastic modulus.*

[5]Personal communication with G. A. Clarke, Westinghouse R&D Center, 25 July 1984.

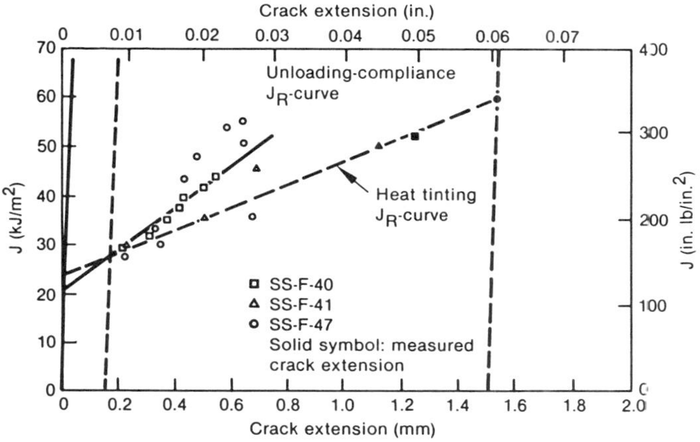

FIG. 12—*J_R-curves for irradiated Type 348 stainless steel (fluence = 3.5 × 10^{22} n/cm^2,* E > *1 MeV).*

material. The linear regression correlation R^2 for the unloading-compliance and the heat-tinting J_R-curves was 0.624 and 0.990, respectively. For these specimens, the unloading-compliance technique underestimated the values of J_{Ic} and K_{Ic} by 16% and 8%, respectively, as compared with those of the multiple-specimen technique. It should be noted here that the heat-tinting J_R-curve (Fig. 12) is not valid per ASTM E 813-81 since the number of available specimens was less than four. The significant crack front bowing of the final crack accounts for the discrepancy between the compliance-calculated and the measured final crack size. Side-grooving of these specimens is not recommended because of the small thickness of these specimens and because its effect on J_{Ic} has not been quantified yet.

Summary and Conclusions

1. Irradiated Type 347/348 stainless steel specimens were precracked successfully in cantilever bending.
2. The J_{Ic} values determined by the single-specimen unloading compliance technique were found to be in close agreement with multiple-specimen values.
3. The automation of the precracking process and the computerized J_{Ic} testing have resulted in considerable saving of the hot cell costs in addition to the efficient use of the limited irradiated material.

Acknowledgment

This work was supported by the U.S. Department of Energy under DOE Contract DE-AC07-75ID01570, Idaho National Engineering Laboratory; and under DOE Contract DE-AC05-84OR21400, Oak Ridge National Laboratory, operated by Martin Marietta Energy Systems, Inc.

References

[1] Haggag, F. M. et al. in *Effects of Radiation on Materials: Twelfth International Symposium, ASTM STP 870,* F. A. Garner and J. S. Perrin, Eds., American Society for Testing and Materials, Philadelphia, 1985, pp. 548–562.

[2] Fisher, D. M. and Repko, A. J., *Materials Research Standards,* Vol. 9, No. 5, May 1969, pp. 28–29.

[3] *Mark's Standard Handbook for Mechanical Engineers,* T. Baumeister, E. A. Avallone, and T. Baumeister III, Eds., McGraw-Hill, New York, 1978, pp. 5-24–5-30.
[4] Haggag, F. M. and Underwood, J. H., *International Journal of Fracture,* Vol. 26, Oct. 1984, pp. R63–R65.
[5] Mills, W. J., "Single Specimen J_{Ic} Testing at Elevated Temperatures," HEDL-TME-81-42, Hanford Engineering Development Laboratory, Hanford, WA, Nov. 1981.
[6] Mills, W. J., "Irradiation Effects on the Fracture Toughness of Austenitic Fe-Cr-Ni Alloys," HEDL-TME-82-117, Hanford Engineering Development Laboratory, Hanford, WA, July 1982.
[7] Little, E. A., "Dynamic J-Integral Toughness and Fractographic Studies of Fast Reactor Irradiated Type 321 Stainless Steel," *Effects of Radiation on Materials: Twelfth International Symposium, ASTM STP 870,* American Society for Testing and Materials, Philadelphia, 1985, pp. 563–579.

G. Mark Tanner[1] *and Walter L. Bradley*[1]

Evaluation of the Fracture Toughness of Ductile Iron Using Fatigue Precracked Charpy, Dynamic Tear, and Compact Tension Specimens

REFERENCE: Tanner, G. M. and Bradley, W. L., **"Evaluation of the Fracture Toughness of Ductile Iron Using Fatigue Precracked Charpy, Dynamic Tear, and Compact Tension Specimens,"** *Fracture Mechanics: Eighteenth Symposium, ASTM STP 945,* D. T. Read and R. P. Reed, Eds., American Society for Testing and Materials, Philadelphia, 1988, pp. 405–418.

ABSTRACT: The fracture toughness of ductile iron has been studied using dynamic tear, fatigue precracked (FPC) Charpy, and compact tension (CT) specimens. Specimens were fractured both quasi-statically and dynamically. Initial crack extension was found to occur at maximum load for both the compact tension specimens and the FPC Charpy specimens. However, initial crack extension occurred prior to maximum load for the dynamic tear specimens. Upper-shelf $K_{J_{Ic}}$ values determined from the compact tension specimens and the FPC Charpy specimens were similar (83 and 76 MPa $\sqrt{m}$) while the J-value at maximum load for the dynamic tear specimens was found to fit nicely on the J-R curve from the CT specimens. The shift in ductile-to-brittle transition temperature with strain rate (comparing quasi-static with impact loading) was similar to what has been observed for structural steel with a similar yield strength.

KEY WORDS: ductile iron, fracture toughness, fatigue precracked Charpy, dynamic tear, $K_{J_{Id}}$, $K_{J_{Ic}}$

A variety of tests has been used during the past ten years to quantify the fracture toughness of ductile cast iron, sometimes called nodular cast iron. These tests, which include standard Charpy tests [*1*], fatigue precracked (FPC) Charpy tests [*2*], dynamic tear tests [*3*], and compact tension (CT) tests [*4*], have given widely varying indications of both the upper-shelf toughness and the ductile-to-brittle transition temperature. This is not surprising since the various tests require specimens with different geometries (including different notch acuity, thickness, and a/W ratio) and are run at different strain rates.

The purpose of this work has been to study the fracture behavior of ductile iron using each of the types of specimens mentioned, but with modifications to see if a self-consistent indication of the fracture toughness of ductile iron could be obtained. To achieve this objective, Charpy specimens with fatigue precracking, dynamic tear specimens with pressed-in notches and with fatigue precracking, and CT specimens with fatigue precracking have all been tested both quasi-statically and dynamically. Thus, the experimental variables in this study are notch acuity, specimen thickness (and thus constraint), ligament width ($W - a$), and strain rate.

Although specimen size, notch acuity, and strain-rate effects on fracture behavior have been established for steels, further study is needed to quantify these effects in ductile cast iron because of its very coarse microstructure and somewhat lower toughness compared with steel. This is particularly important to try to reconcile the discrepancies in indications of upper-shelf

[1]Graduate research assistant and professor, respectively, Mechanical Engineering Department, Texas A&M University, College Station, TX 77843.

toughness and transition temperature from the various fracture toughness tests, as previously noted.

Experimental Procedures and Analysis

In this section specimen preparation, specimen testing, and the approach used to analyze the data are summarized.

Specimen Preparation

A ferritic ductile iron with 2.25% silicon was cast into 2.5-cm-thick (1 in.) plates and subsequently machined into the specimen geometries shown in Table 1. The microstructure of this material is shown in Fig. 1. Tensile properties for this alloy are 310 MPa (45 ksi) yield strength, 450 MPa (65 ksi) tensile strength, and 20% elongation. The compact tension and three-point bend specimens were machined to dimension specifications according to the ASTM Test Method for J_{Ic}, a Measure of Fracture Toughness (E 813-81); while the Charpy and dynamic tear specimens are according to the ASTM Methods for Notched Bar Impact Testing of Metallic Materials (E 23-82) and the ASTM Method for Conducting Drop-Weight Test to Determine Nil-Ductility Transition Temperature of Ferritic Steels (E 208-842). The three-point bend specimens are very similar in size to the dynamic tear specimens. It is worth noting that the standard dynamic tear specimen has a relatively low a/W value of 0.31 after the notch of 0.025-mm (0.000984 in.) root radius has been pressed into the root of the machined in slot. To facilitate comparison between the dynamic tear specimens (which were tested with impact loading) and the three-point bend specimens (which were tested quasi-statically), the initial a/W used was much smaller than that specified by ASTM E 813. We will use the term "dynamic tear specimen" to refer to both the dynamic tear specimen and the three-point bend specimen for the remainder of this paper since they are nearly identical and differ only in the way they were tested in this work—namely, one dynamically and the other quasi-statically.

The fatigue precracking procedures were conducted according to ASTM E 813 except for the Charpy specimens, where fatigue crack initiation in the root of the blunt notch was found to be somewhat difficult to achieve at specified limit loads. It will later be seen that the use of these somewhat higher loads during fatigue precracking of the Charpy specimens may have caused a small underestimate of the upper shelf $K_{J_{Ic}}$ and $K_{J_{Id}}$ values from these particular tests. "Fatigue precracked" will be abbreviated FPC in the remainder of this paper.

Specimen Testing

Charpy Impact Test—The FPC Charpy specimens were fractured dynamically on a Charpy impact machine with an instrumented tup manufacturered by Effects Technology (Model 371)

TABLE 1—*Specimen geometries.*

Type Specimen	Number of Specimens	Dimensions, in.[a]	Type of Notch	$\cong a/w$
compact tension	16	2.5 × 2.4 × 1	fatigue precrack	0.6
Charpy	16	0.394 × 0.394 × 2.165	blunt notch	0.2
Charpy	16	0.394 × 0.394 × 2.165	fatigue precrack	0.5
dynamic tear	8	7.125 × 1.6 × 0.625	pressed-in notch	0.31
dynamic tear	8	7.125 × 1.6 × 0.625	fatigue precrack	0.38
bend	8	7.125 × 1.625 × 0.813	pressed-in notch	0.31
bend	8	7.125 × 1.625 × 0.813	fatigue precrack	0.38

[a]1 in. = 2.54 cm.

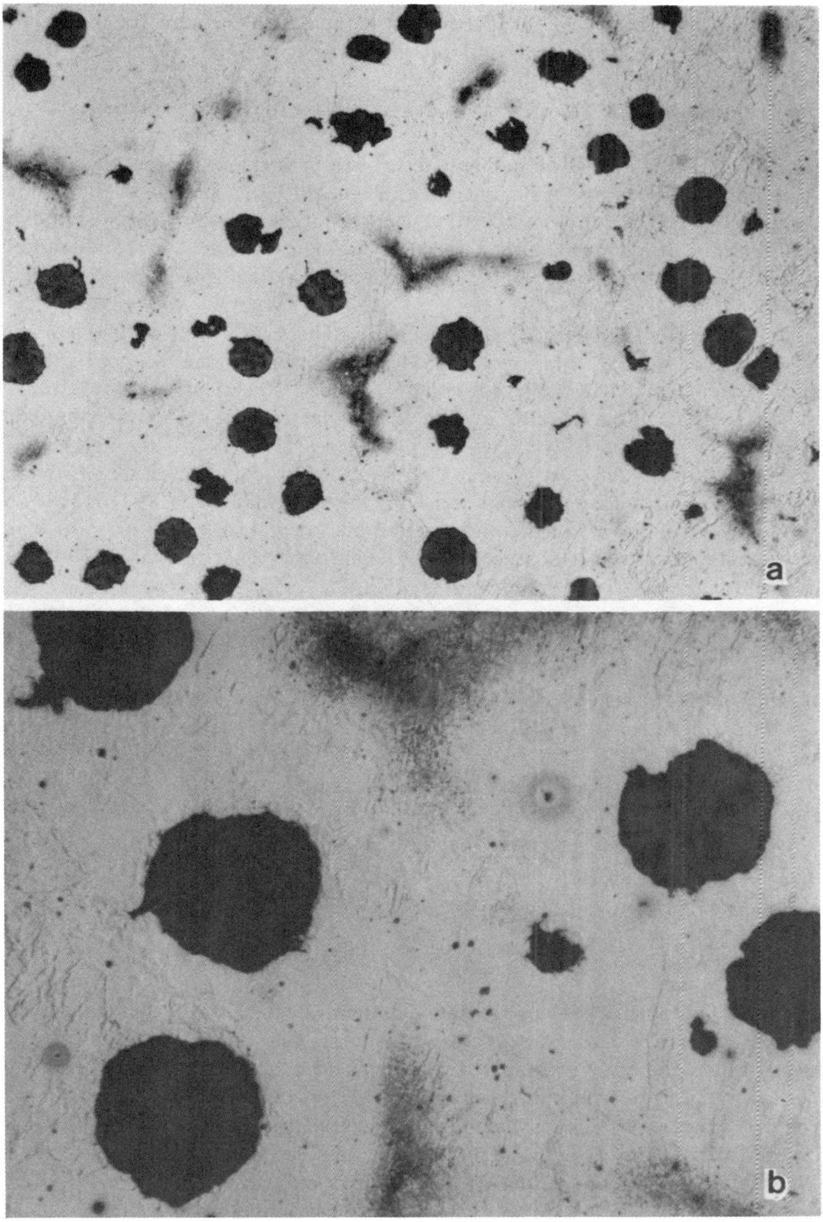

FIG. 1—*Microstructure of ferritic ductile iron. Dark, round regions are graphite nodules; dark peppery regions are residual pearlite; light colored matrix is ferrite.* (a) ×*116;* (b) ×*308.*

with data acquisition made on a digital storage oscilloscope. The data were subsequently transferred to a HP-9816 computer for data reduction.

Quasi-Static Tests of FPC Charpy Specimens—The FPC Charpy specimens were fractured quasi-statically on an MTS materials testing machine with fixtures for three-point bend tests prepared according to the ASTM Test Method for Plane-Strain Fracture Toughness of Metallic Materials (E 399-83). A loading rate suitable for breaking specimens in approximately 900 s was used. The data were recorded directly through a data logger onto an HP-9816 computer. A graphical load-displacement record was also taken to allow the progress of the tests to be monitored.

Dynamic Tear Tests—The dynamic tear tests were performed on a standard drop-weight tower, with the weight dropped from a height of 30 cm (11.8 in.) and the load measured using an instrumented tup. The data were recorded using a storage oscilloscope. The initial height of the tup for both the drop tower and the Charpy test was chosen to avoid the ambiguities in load measurement created by inertial ringing or frequency response of the transducer or both, as described by Server [*5*].

Quasi-Static Testing of Dynamic Tear Specimens—These specimens were tested on an MTS machine with fixtures designed to satisfy ASTM E 399. The data were recorded on an HP-9816 laboratory computer and also displayed visually on a load-displacement chart during the test. The displacement rate of the actuater was adjusted so that the tests would have approximately a 900-s duration.

Quasi-Static Tests of Compact Tension Specimens—The compact tension specimens were tested on an MTS machine with fixtures satisfying the specifications of ASTM E 813. Multiple unloading compliance was used to allow the determination of a J-R curve from which a J_{Ic} value could be determined. All data were recorded through a data logger onto an HP-9816 computer, which also did the necessary data reduction.

Dynamic Test of Compact Tension Specimens—The dynamic tests of compact tension specimens were conducted on an open loop MTS machine capable of displacing 25 m/s. The machine was adjusted so that the displacement rate gave a loading rate and associated stress intensity rate which corresponded to the stress intensity loading rate observed in the Charpy impact tests. The data were recorded on a high-speed data acquisition system supplied by MTS with the machine.

Baths of acetone and dry ice were used to lower the temperature below ambient while a silicon oil bath was used to obtain temperatures above ambient. The temperature was controlled to $\pm 2°C$. All tests were run in stroke or clip gage displacement control to allow for stable crack growth at temperatures where ductile crack propagation was possible.

Data Analysis

It is difficult to determine directly the moment of crack extension for dynamic tests. Thus, it was assumed that crack extension began at maximum load during the tests. This has been demonstrated to be the case for compact tension specimens of ductile iron with $a/W = 0.6$ [*4*] and was also found to be true in this work. Crack extension beginning at maximum load has also been demonstrated in this research for Charpy specimens of ductile iron with an $a/W = 0.5$. However, this has not been found to be the case for dynamic tear specimens with $a/W = 0.31$ or 0.38, which are the values of a/W for the specimens used in this work. As will be shown later, crack growth in these specimens began before maximum load. Therefore the J_{Id} values calculated using the area under the load-displacement curves to maximum load for the dynamic tear specimens are overestimates of the actual values for crack initiation.

Two points require further discussion here. The ASTM E 813 test procedure is for quasi-static testing. There is not a currently accepted or tentative ASTM standard for dynamic elastic-plastic fracture mechanics measurements of fracture toughness. Therefore, we have chosen to follow the procedure of calculating the J-value at maximum load during our tests of FPC Charpy and DT bars loaded in three-point bending (both dynamically and statically). Then if independent testing can verify that crack extension does begin at maximum load, we may consider these values to be critical, or initiation values of J which we will call J_{Id} or J_{Ic}, respectively. We recognize that these are not J-values determined according to ASTM E 813. However, when crack growth does indeed begin at maximum load as assumed in this work, we believe this approach will give a very good estimate of the J-value we would determine from a standard ASTM E 813 test.

Our use of the relationship $J = 2A/Bb$ for a/W values less than 0.5 also needs to be justified (A = area under load-displacement curve; B = specimen thickness; b = uncrack ligament width; W = specimen width). The original derivation by Rice et al. assumed that the plastic zone was restricted to the ligament adjacent to the crack tip, which it was for our specimens. Rice et al. also assumed that the displacement due to elastic deflection of an uncracked specimen must be small relative to the plastic displacements. However, this restriction can be relaxed if one separates the total J into elastic and plastic J components [7] such that

$$J_T = J_e + J_p \tag{1}$$

where $J_e = \eta_e \int P d\delta_e / Bb$ and $J_p = 2\int P d\delta_p / Bb$. The values for $\eta_e(a/W)$ are as follows [8]: $\eta_e(0.3) = 1.67$, $\eta_e(0.35) = 1.76$, and $\eta_e(0.40) = 1.84$. Since the J_p value is generally at least 75% of J_T on the upper shelf for ductile iron and since η_e is only slightly less than the assumed value of 2.0, calculating J_T as $2A/Bb$ will give an estimate of J_T that is within 5% of the actual J_T. Furthermore, the $K_{J_{Ic}}$ values calculated from our approximate J_T values should be within 2.5% of the more exact values.

For the dynamic tear tests and the dynamic Charpy tests, the measured load-time data was converted into load-displacement data using standard conservation of momentum equations [5]. The load-displacement relationships for the dynamically tested compact tension specimens were determined by direct measurement. The kinetic energy of all dynamically tested specimens was calculated and subtracted from the crack initiation energy so that the fracture energy would not be overestimated due to inertial effects.

As has been previously explained, the fracture toughness for ductile fracture was calculated using a J-integral analysis with J_{Ic} and J_{Id} being determined by assuming that crack extension begins at maximum load, except for the CT specimens that were tested quasi-statically, where a J-R curve was actually measured. It should be emphasized that ASTM E 813 requires a J-R curve for a valid J_{Ic} value to be obtained according to the standard test procedure. Disregarding this requirement will only introduce error in the transition zone where a larger specimen size might give a somewhat higher ductile-to-brittle transition temperature range. Otherwise, meeting this requirement is not crucial if one can determine the moment of crack extension in some other way. It should also be noted that one can determine with a fractographic examination in the scanning electron microscope (SEM) whether ductile crack initiation and propagation occurred in a given specimen, thus determining whether one is on the upper shelf or not (that is, if the crack initiation and propagation are 100% void coalescence). In any case, no specimen size dependence of the ductile-to-brittle transition was noted in this study. All calculated values of J_{Ic} and J_{Id} were subsequently converted to $K_{J_{Ic}}$ and $K_{J_{Id}}$ values using the usual relationship $JE = K^2$ where E is the Young's modulus. Once again it should be emphasized that the J_{Ic} and J_{Id} values calculated from our experimental results on DT and FPC Charpy bars are not valid according to ASTM E 813. Nevertheless, for ductile crack initiation at maximum load and ductile crack propagation subsequently, the critical J-values determined should be essentially the same as one would obtain from a valid E 813 test. The good agreement seen in Table 2 between the upper shelf $K_{J_{Ic}}$ values obtained from FPC Charpy bars (where crack growth did begin at maximum load) and the CT specimens results obtained according to ASTM E 813 confirms the validity of the approximation.

TABLE 2—*Comparison of Quasi-static stress intensities* $K_{J_{Ic}}$ *(MPa $\sqrt{m}$).*

Temperature	Charpy (FPC)	Bend (FPC)	Bend (PIN)	Compact Tension
37.8°C	87.1	124.7	122.7	...
23.9°C	...	119.7	...	94.5
10.0°C	76.8	126.4	132.2	...
−3.9°C	79.4	...	...	...
−17.8°C	76.3	107.9	125.3	82.6
−28.9°C	76.9	109.0	106.8	83.0
−40.0°C	55.2	107.1	108.9	81.4
−51.1°C	...	...	...	75.7
−62.2°C	48.8	65.7	83.2	71.6
−73.3°C	...	...	...	37.4

DYNAMIC STRESS INTENSITIES, $K_{J_{Id}}$ (MPa $\sqrt{m}$)

Temperature	Charpy (FPC)	Dynamic Tear (FPC)	Dynamic Tear (PIN)	Compact Tension
65.6°C	86.6	...	162.0	80.9
37.8°C	89.2	147.4	159.9	62.0
23.9°C	80.9	...	...	...
10.0°C	48.6	112.2	127.5	...
−3.9°C	51.1	...	...	...
−17.8°C	41.5	103.3	112.5	...
−28.9°C	39.0	92.5	105.0	...
−40.0°C	30.1	72.7	87.0	...
−73.3°C	33.5	52.9	53.2	...

Experimental Results and Discussion

The values of $K_{J_{Ic}}$ and $K_{J_{Id}}$ determined in this study for ferritic ductile iron are summarized in Table 2. To facilitate comparisons, the dynamic and quasi-static results for FPC Charpy, compact tension, dynamic tear/pressed in notch and dynamic tear/fatigue precracked are all shown in Figs. 2–5. A comparison of the fatigue precracked results to the pressed in notch results (PIN) for dynamic tear type specimens tested dynamically and quasi-statically are seen in Figs. 6 and 7.

The significant observations from these results may be summarized as follows:

1. The effect of strain rate is to give a large shift to the ductile-to-brittle transition temperature, as is clearly seen in the FPC Charpy results (Fig. 2) and the compact tension results (Fig. 3). By contrast, no such significant shift is observed for the dynamic tear specimens, and the transition itself occurs over a very wide temperature range when compared to the Charpy and CT results.
2. The upper-shelf fracture toughness as measured on FPC Charpy and CT specimens is in reasonably good agreement, but the dynamic tear specimen results are much higher than the other two.
3. The notched and fatigue precracked dynamic tear specimens gave essentially identical behavior (see Figs. 6 and 7), indicating that notch acuity effects on the fracture toughness of ductile iron are small for notch root radii below 0.025 mm (0.000984 in.).

The much larger values of $K_{J_{Ic}}$ and $K_{J_{Id}}$ obtained from the dynamic tear specimens compared to the Charpy and CT specimens calls into question the validity of the assumed crack extension at maximum load in the dynamic tear specimens. The lower a/W values used in the dynamic tear specimens make it unlikely that crack extension would occur prior to maximum load

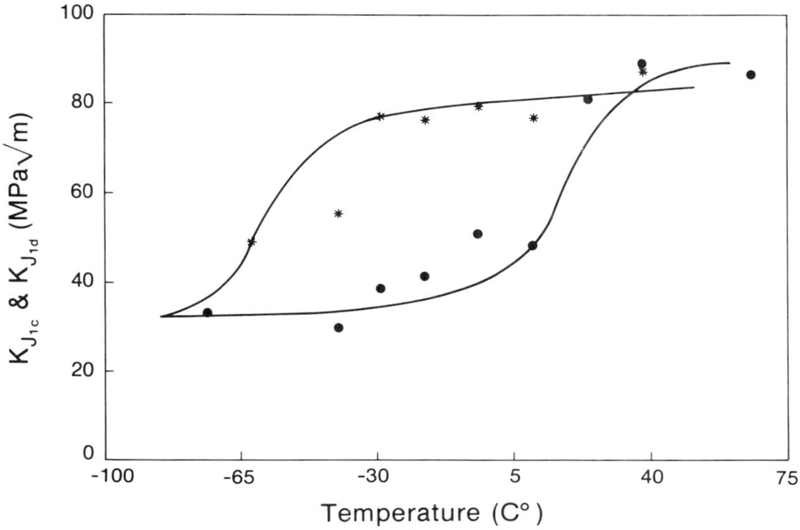

FIG. 2—*Quasi-static and dynamic fracture toughness of ferritic ductile iron, as measured using fatigue precracked Charpy specimens. * quasi-static; • dynamic.*

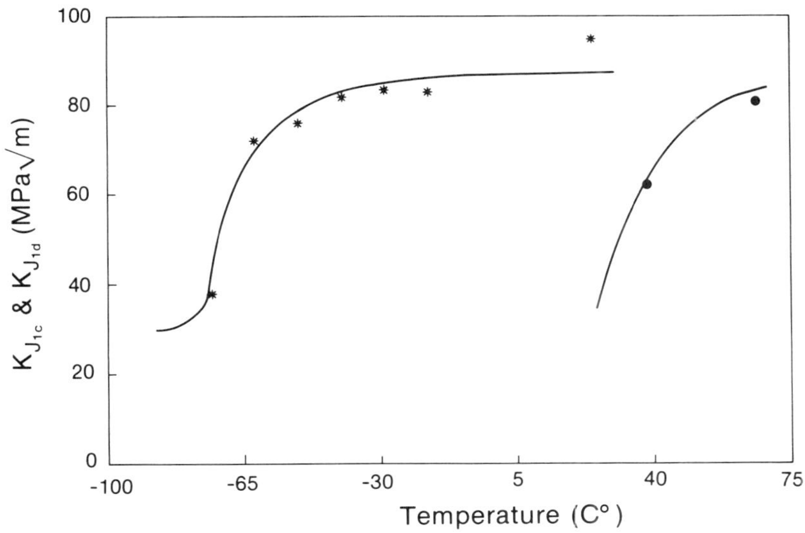

FIG. 3—*Quasi-static and dynamic fracture toughness of ferritic ductile iron, as measured using fatigue precracked compact tension specimens. * quasi-static; • dynamic.*

(rather than at maximum load) in these specimens. To test this hypothesis, a dynamic tear specimen with fatigue precracking to $a/W = 0.38$ was tested quasi-statically until maximum load was reached. Next the test was interrupted and the specimen was fatigue post-cracked and then fractured to determine whether crack extension does indeed begin prior to attainment of maximum load in the test. The results are seen in Fig. 8, which is a macrofractograph of the specimen. It is quite clear that crack extension is well under way by the time maximum load is

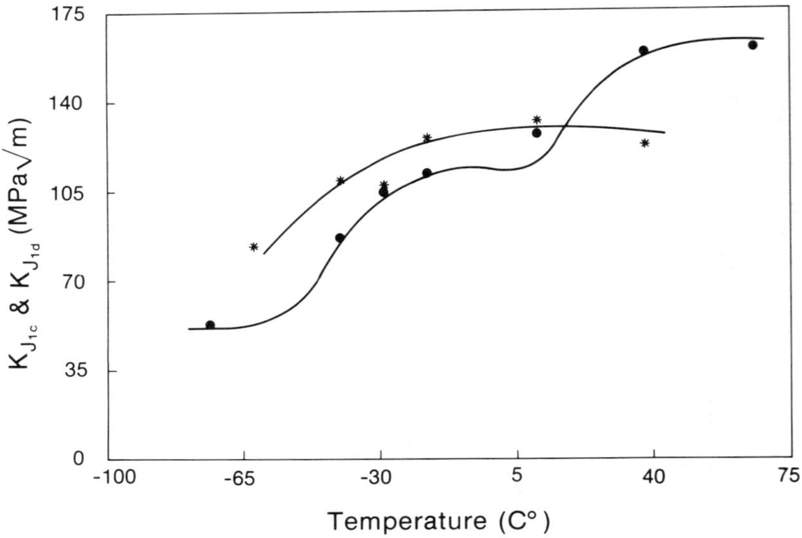

FIG. 4—*Quasi-static and dynamic fracture toughness of ferritic ductile iron, as measured using dynamic tear specimens with a pressed-in notch. * quasi-static; • dynamic. Note values for $K_{J_{Ic}}$ and $K_{J_{Id}}$ are artifically high for reasons explained in text.*

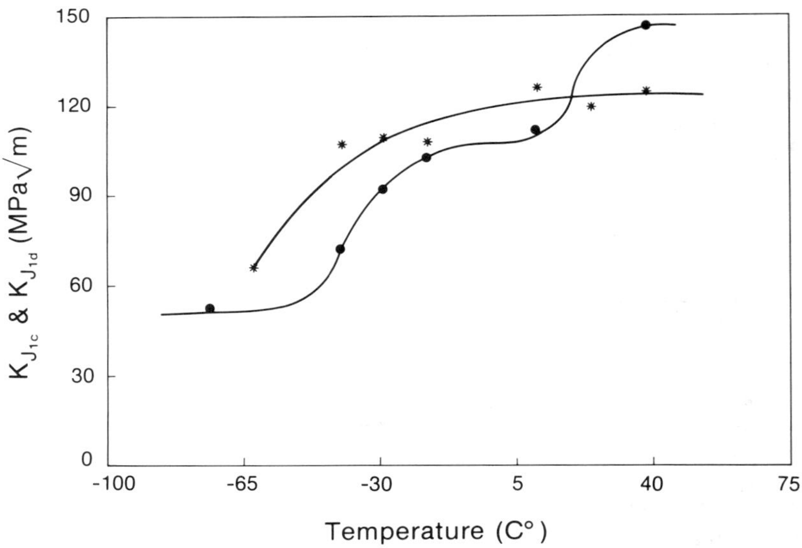

FIG. 5—*Quasi-static and dynamic fracture toughness of ferritic ductile iron, as measured using a dynamic tear specimen with a fatigue precrack. * quasi-static; • dynamic. Note values for $K_{J_{Ic}}$ and $K_{J_{Id}}$ are artifically high for reasons explained in text.*

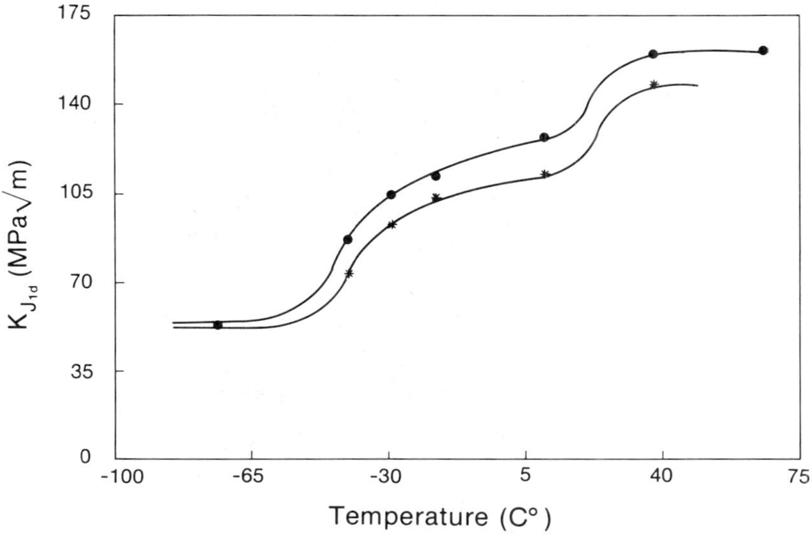

FIG. 6—*Dynamic fracture toughness of ferritic ductile iron as measured using dynamic tear specimens with a pressed-in notch (•) and a fatigue precrack (*). Note values of* $K_{J_{Id}}$ *are artifically high for reasons explained in text.*

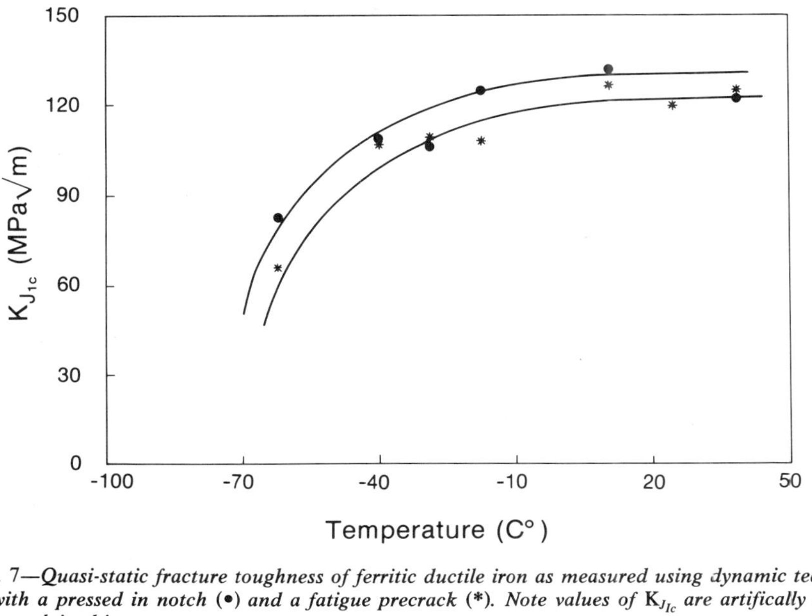

FIG. 7—*Quasi-static fracture toughness of ferritic ductile iron as measured using dynamic tear specimens with a pressed in notch (•) and a fatigue precrack (*). Note values of* $K_{J_{Ic}}$ *are artifically high for reasons explained in text.*

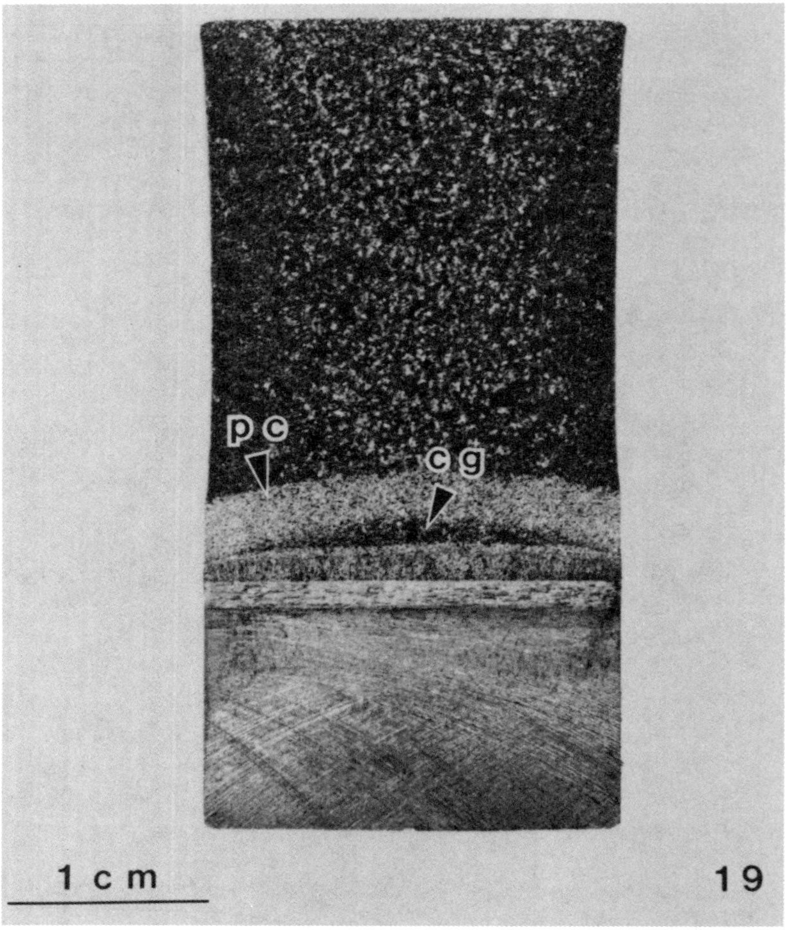

FIG. 8—*Macrofractograph of fatigue precracked, dynamic tear specimen, loaded quasi-statically to maximum load, unloaded and then fatigue post-cracked. Significant crack advance is observed by the time maximum load is reached.*

reached in dynamic tear specimens of ductile iron with $a/W = 0.38$. In fact the average crack extension is in excess of 0.7 mm (0.0275 in.). One may calculate the J-value at maximum load where 0.7 mm (0.0275 in.) of crack extension has already occurred and then place this point on a J-R curve obtained by quasi-statically testing of a CT specimen of the same ductile iron. The results of such an exercise are shown in Fig. 9, where it is seen that the point from the quasi-statically tested dynamic tear specimen fits very nicely on the J-R curve measured on a CT specimen. The actual load-displacement record for this dynamic tear specimen is seen in Fig. 10, with the shaded area indicating the work done to the moment of initial crack extension. It is clear why the assumption of crack extension at maximum load for these specimens gave such an overestimate of the $K_{J_{Ic}}$ and $K_{J_{Id}}$ values.

A similar experiment was performed for two FPC Charpy specimens of ductile iron with $a/W = 0.50$. After the specimens were loaded quasi-statically in displacement control until they reached maximum load, they were unloaded, cooled to $-50°C$, and then broken on a Charpy impact machine to give a very brittle fracture. No indication of ductile crack extension

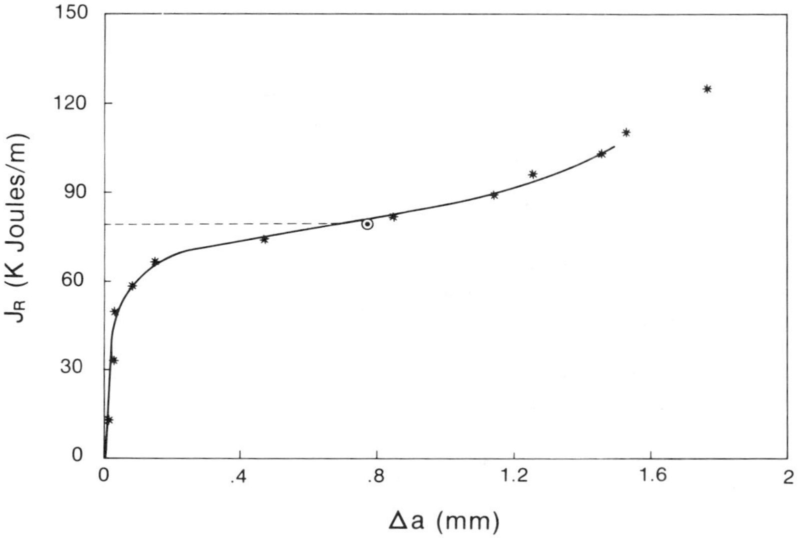

FIG. 9—*J-R curve measured on ferritic ductile iron, compact tension specimen loaded quasi-statically (*). Single point measured at maximum load for quasi-statically loaded dynamic tear specimen (•), with crack advance measured from Fig. 8. Note good agreement.*

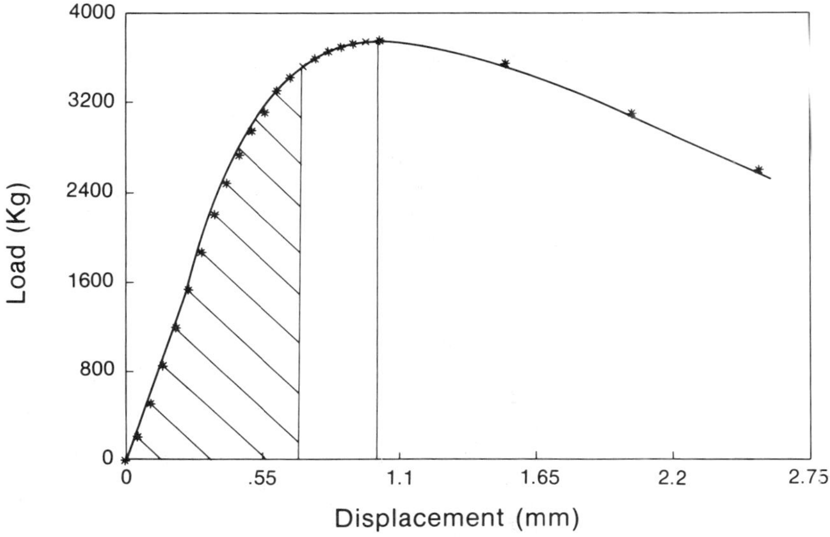

FIG. 10—*Load-displacement record for quasi-statically loaded dynamic tear specimen indicating the moment of crack advance as inferred from Figs. 8 and 9, by assuming that J_{Ic} for this test was same as for CT specimen, and then back-calculating the area under the load-placement curve at crack initiation.*

during the quasi-static loading to maximum load was noted between the fatigue precrack and the subsequent brittle fracture in the Charpy impact tester.

These results indicate that the primary difference between fatigue precracked Charpy results and dynamic tear results is due to the difference in ligament width, not notch acuity or specimen thickness. While the uncracked ligament on both specimens satisfied the requirements for J-controlled growth, the much larger ligament of the dynamic tear specimen gave some stable crack growth prior to maximum load, which makes a precise determination of J_{Id} or $K_{J_{Id}}$ impossible on these specimens.

A second important observation from the results summarized in Table 2 is that the upper shelf values for quasi-static fracture toughness $K_{J_{Ic}}$ determined from the Charpy specimens are consistently 8% lower than similar values determined from CT specimens (which corresponds to a 17% difference in J_{Ic} values). This is opposite of what one might expect based on differences in constraint, since the Charpy specimens [B = 1 cm (0.39 in.), b = 0.5 cm (0.196 in.)] should have less constraint than the CT specimens [B = 2.5 cm (0.98 in.), b = 2.0 cm (0.787 in.)]. We believe that the lower Charpy values are a result of the somewhat excessive loads required to initiate the fatigue precracking in the blunt notch Charpy specimens, as previously mentioned in the experimental procedures section. A similar comparison of the dynamic fracture toughness $K_{J_{Id}}$ from Charpy and CT specimens could not be made since the upper shelf for the CT specimens was not achieved over the temperature range tested.

The lack of a significant notch acuity effect for notch root radii below 0.025 mm (0.000984 in.) may be understood by noting from Fig. 1 that the average spacing between graphite nodules is approximately 0.15 mm (0.0059 in.). Figure 11 shows an SEM fractograph of the fractured

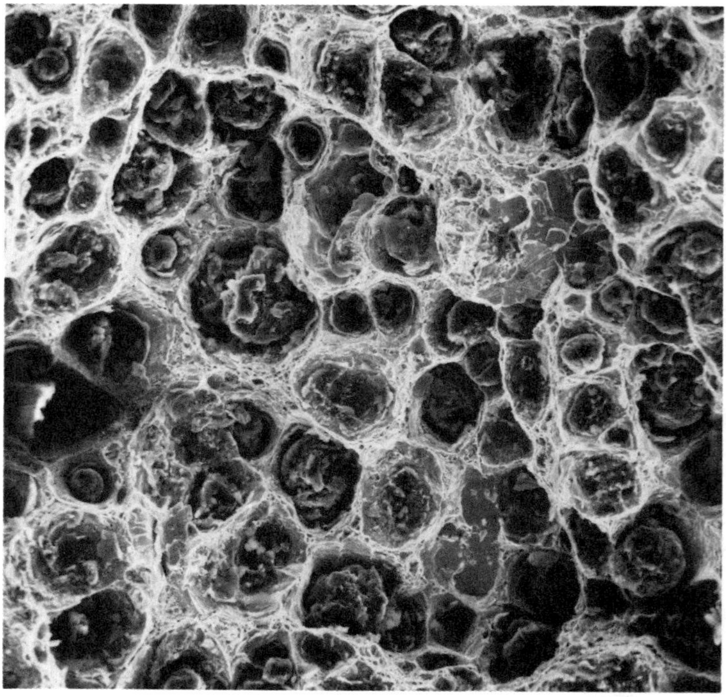

FIG. 11—*SEM fractograph of fractured ferritic ductile iron indicating the prominent role the graphite nodules play in ductile fracture of this material (×100).*

surface and clearly indicates that void formation at nodules dominates the fracture behavior. Thus, the relatively small notch acuity effect, comparing a 0.025-mm (0.000984 in.) pressed in notch root radius to a fatigue precracked specimen, is to be expected when it is recognized that the controlling microstructural feature which is graphite nodule spacing is 6 × coarser than the notch root radius for the pressed-in notch used in this study, namely, 0.025 mm (0.000984 in.). The notch root radius for fatigue precracking is, of course, even smaller than noted for the pressed in notch.

Finally, the ductile-to-brittle transition temperature shift which may be quantified by comparing dynamic to static results for Charpy and CT specimens is similar in magnitude to that observed by Barsom for steels with a similar yield strength [9]. The very diffuse ductile-to-brittle transition temperature noted for the dynamic and quasi-statically tested dynamic tear specimens is a consequence of the fact that the assumed crack initiation energy includes both initiation and propagation energy. Since the transition temperature for crack propagation is somewhat higher than for crack initiation [10] and since it includes both initiation and propagation transitions from ductile-to-brittle fracture, the apparent initiation transition temperature is smeared out over a wider temperature range. Figures 4–6, which include dynamic tear results, have a temperature range sufficient to include the ductile-to-brittle transition for both crack initiation and crack propagation. Thus, the curves have been drawn to show both transitions rather than one smeared-out one.

Summary

1. The upper-shelf quasi-static and dynamic fracture toughness as measured from CT specimens and fatigue precracked Charpy specimens has been shown to be in essential agreement with values of approximately 83 and 76 MPa $\sqrt{m}$.

2. The results from dynamic tear-type specimens tested quasi-statically and dynamically would also apparently be in agreement if the true moment of crack extension were determined; however, the assumption that crack extension begins at maximum load overestimates of the actual values.

3. Notch acuity for notch root radii below 0.25 mm (0.000984 in.) is not important in ferritic ductile iron because of its relatively coarse microstructure.

4. A similar shift in ductile-to-brittle transition temperature for quasi-static as compared to dynamic loading has been observed for crack initiation in Charpy and CT specimen, with the magnitude found to be similar to that for steel with a comparable yield strength. The transition temperature behavior in the dynamic tear specimens was complicated by the fact that it included both crack initiation and crack propagation transitions from ductile-to-brittle behavior.

Acknowledgments

The financial support of this work by the Ductile Iron Society is gratefully acknowledged. A special note of appreciation goes to Mr. Bob Warrick and Mr. Paul Gerhardt, Jr., who supplied the materials and numerous helpful suggestions regarding the work. Finally, our thanks to Mr. P. H. Mani of Wagner Castings, who ran the dynamic tear tests for this research program.

References

[1] Sandoz, G., Bishop, H., and Pellini, W., "Notched Ductility of Nodular Iron," *Transactions of American Society for Metal,* Vol. 46, 1954, p. 418.

[2] McKinney, K. E., Bradley, W. L., and Gerhardt, P. C., Jr., "An Evaluation of the Toughness of Ductile Iron vs. Cast Steel Using Charpy Test Specimens," *Transactions of the American Foundrymen's Society,* 1984, pp. 239–250.

[3] Worzala, F. J., Heine, R. W., and Cheng, Y. W., "Dynamic Tear Characteristics of Ductile Iron," *Transactions of the American Foundrymen's Society,* 1980, pp. 675–686.

[4] Salzbrenner, R. J., Van Den Avyle, J. A., Lutz, J. J., and Bradley, W. L., "Fracture Toughness Testing of Ductile Iron" in *Fracture Mechanics: Sixteenth Symposium, ASTM STP 868,* M. F. Kanninen and A. J. Hopper, Eds., American Society for Testing and Materials, Philadelphia, 1986, pp. 328–345.
[5] Server, W. L., "Impact Three-Point Bend Testing of Notched and Precracked Specimens," *Journal of Testing and Evaluation,* Vol. 6, 1978, pp. 29–34.
[6] Rice, J. R., Paris, P. C., and Makle, J. G., "Some Further Results of J-Integral Analysis and Estimates" in *Progress in Flaw Growth and Fracture Toughness Testing, ASTM STP 536,* American Society for Testing and Materials, Philadelphia, 1973, pp. 231–245.
[7] "An Engineering Approach for Elastic-Plastic Fracture Analysis," Electric Power Research Institute NP-1931, Project 1237l-1, Topical Report, Palo Alto, CA, July 1981.
[8] Towers, O. L., "Stress Intensity Factors, Compliances and Elastic Factors for Six Test Geometrics," Report 136, The Welding Institute, 1981.
[9] Barsom, J. M., "Development of the AASHTO Fracture-Toughness Requirements for Bridge Steels," *Journal of Engineering Fracture Mechanics,* Vol. 7, 1975, p. 605.
[10] Andler, M. R., "The Dynamic Fracture Toughness of Nodular Cast Iron," M.S. thesis, Texas A&M University, College Station, TX, 1983.

C. K. (Peter) Chow[1] *and Leonard A. Simpson*[1]

Determination of the Fracture Toughness of Irradiated Reactor Pressure Tubes Using Curved Compact Specimens

REFERENCE: Chow, C. K. and Simpson, L. A., "**Determination of the Fracture Toughness of Irradiated Reactor Pressure Tubes Using Curved Compact Specimens,**" *Fracture Mechanics: Eighteenth Symposium, ASTM STP 945,* D. T. Read and R. P. Reed, Eds., American Society for Testing and Materials, Philadelphia, 1988, pp. 419–439.

ABSTRACT: This paper describes a case in which nonstandard test configurations had to be used to assess fracture toughness, and an experimental, analytical program to support the validity of the results.

Zirconium alloy pressure tubes (103 mm inside diameter, 5 mm thick) contain the primary coolant in CANDU® reactors. Because irradiated material lacks ductility, the pressure tube cannot be flattened to produce flat specimens to determine fracture toughness. Therefore, a method was devised for spark eroding 17-mm compact specimens from sections of pressure tubing. These specimens retained the normal in-plane dimensions of the standard compact specimens, but possessed the curvature of the tube. By comparing results from flat and curved specimens, this paper shows experimentally that flat-plate equations can be applied to both specimen types to calculate the fracture toughness parameters with little error. A simple analytical calculation estimates that the maximum curvature-induced stress should be less than 10% of the in-plane tensile stress at the crack tip.

KEY WORDS: fracture toughness, *J*-resistance curve, pressure tubes, Zircaloy-2, Zr-2.5Nb, curved specimens

Sometimes the constraints of real-life situations prevent the use of standard test specimen geometries for fracture toughness assessments. This paper describes both a case in which a deviation from a standard procedure was necessary and an experimental program to support the validity of the results obtained from the nonstandard test specimens. A simple analytical calculation is also given in the Appendix to show that the error introduced by the nonstandard method is less than 10%.

Figure 1 is a schematic view of the CANDU reactor, which uses pressure tubes made of either Zircaloy-2 (Zr-2) or Zr-2.5Nb as the primary in-core containment for the pressurized heavy-water coolant. Zr-2 pressure tubes were used in CANDU reactors built before 1971; cold-worked Zr-2.5Nb pressure tubes have been used in all subsequent CANDU reactors. The chemical composition of Zr-2 and Zr-2.5Nb are given in Table 1. To separate the pressure tube from the heavy-water moderator, the pressure tubes are placed concentrically in a zirconium alloy tube called the calandria tube. The pressure tube and calandria tube are separated by a gas annulus. The pressure of the heavy-water coolant is about 10 MPa, and the inlet and outlet temperatures are 240 and 290°C, respectively.

[1]Research scientist and manager, respectively, Materials and Mechanics Branch, Atomic Energy of Canada Limited, Whiteshell Nuclear Research Establishment, Pinawa, MB, R0E 1L0 Canada.

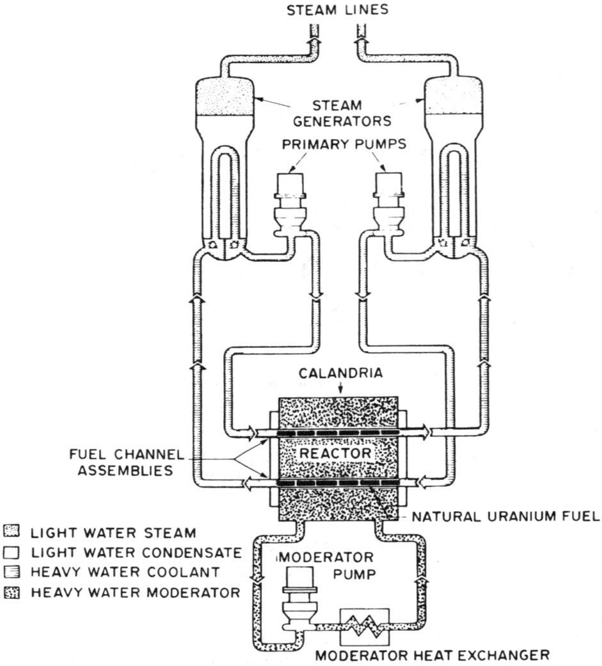

(a)

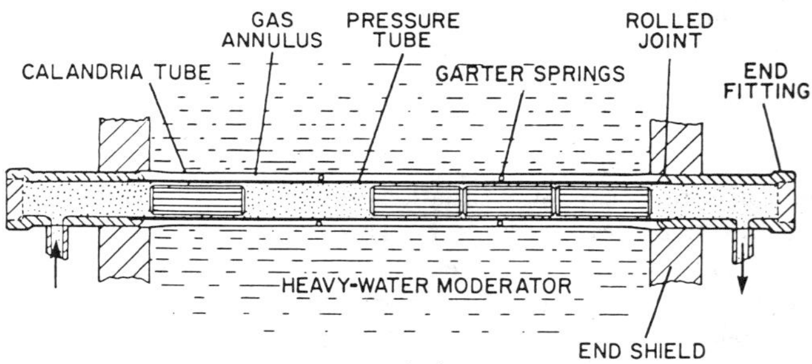

(b)

FIG. 1—(a) *Schematic diagram of CANDU Nuclear Reactor;* (b) *details of fuel channel.*

TABLE 1—*Chemical composition of CANDU reactor pressure tube material (weight%).*

Pressure Tube Material	Sn	Fe	Cr	Nb	O	H	Zr
Zr-2	1.2–1.7	0.07–0.20	0.05–0.15	0.03–0.08	0.1–0.14	<0.0025	balance
Zr-2.5Nb	0.01	0.15	0.02	2.4–2.8	0.09–0.13	<0.002	balance

It is desirable to prevent the unstable failure of pressure tubes. To achieve this, the toughness of the pressure tube and the material conditions that affect the toughness have to be identified. Up to 1983, any pressure tubes that failed had done so in a stable manner known as leak-before-break. In 1974 and 1975, small, axial, through-wall cracks developed in some tubes, near the rolled joints where the pressure tube is attached to the stainless steel end fitting [*1*]. Improper rolling resulted in residual stresses that induced delayed hydride cracking [*2*] from the inside of the tube. The cracks grew through the wall, at which point the pressurized coolant leaked. The leakage was detected, the reactor shut down, and the tubes replaced. The maximum crack length on these failed tubes was 16 mm. In 1982, a similar crack grew in Ontario Hydro's Bruce Unit 2 reactor to a length of 27 mm before leak detection [*3*]. As measured by burst tests, the critical through-wall axial crack length for these pressure tubes ranged from 50 to 102 mm at reactor operating conditions. Thus, the margin for leak-before-break was substantial. However, it was recognized that there was a need to define the material conditions that could reduce the critical crack length (CCL) of the pressure tubes during their service.

This need was made clear on August 1, 1983, when a Zr-2 pressure tube in Pickering Nuclear Generating Station (NGS) Unit 2 failed by unstable crack propagation before any leakage was detected [*4*]. Because of the mislocation of one of the spacers (garter springs) between the hot (240 to 290°C) pressure tube and the externally concentric, cold (75°C) calandria tube, deformation of the pressure tube by creep-sag resulted in contact between the two tubes. The resulting local cooling of the pressure tube caused migration of hydrogen isotopes by thermal diffusion [*5*] to the cold spots, resulting in massive hydride precipitates. Due to the volume expansion associated with the precipitation, the precipitates cracked. Several cracks grew into the zirconium matrix and propagated stably by the delayed hydride mechanism. These cracks linked and grew to a critical configuration before wall penetration occurred.

The analysis of the pressure-tube failure in the Pickering NGS Unit 2 reactor required a test method to map the fracture toughness over the length of the pressure tube. In the past, fracture toughness was characterized by slit burst tests of sections of the pressure tube to estimate the critical crack length [*5–7*]. These tests were very expensive, consumed a lot of material, and obviously could not be used on a burst tube. Therefore, it was desirable to develop small-specimen test methods.

Small specimens produced from flattened pressure tubes have been used to study the effect of microstructure on fracture behavior [*8,9*]. However, these methods require flat-plate specimens, which are not suitable for irradiated pressure tubes. Irradiation hardening renders the pressure tube difficult to flatten. Furthermore, the flattening process might change the properties of the material, especially the unique defect structure produced by irradiation [*10*]. Thus, it was necessary to develop a non-planar specimen geometry.

Small curve specimen methods have been proposed in the literature [*11–13*]. These methods either were not yet well established or could not be instrumented to measure elastic-plastic fracture toughness. They also presented some difficulties in the remote handling of the irradiated materials.

This paper presents a method of testing curved specimens cut directly from the pressure tubes by spark machining. The method is simple enough to enable remote handling in a hot cell. The specimen can be instrumented to measure the *J*-resistance curve using the single specimen method. The chosen geometry is a modification of the ASTM compact specimens. These specimens have the in-plane dimensions proportional to the standard compact specimens but retain the pressure-tube curvature. Since most of the test conditions violate the requirements for plane strain, the thickness is the same as the wall thickness of the tube. Because of the specimen's small size and, hence, limited curvature, flat-plate equations were applied to calculate the fracture-toughness parameter according to established methods [*14*]. The main purpose of this paper is to verify this method experimentally by comparing the results obtained from the curved and flat specimens.

Experimental

Specimen Preparation

Zr-2.5Nb and Zr-2 pressure-tube materials were used in this study. The inside diameters of these pressure tubes were both 103 mm. The nominal wall thicknesses were 5.2 and 4.2 mm for Zr-2 and Zr-2.5Nb tubes, respectively. Curved and flat specimens were produced with both materials.

Both the curved and flat Zr-2.5Nb specimens were obtained from a cold-worked, unirradiated pressure tube. To minimize the metallurgical variation of the flat and curved specimens, they were obtained in pairs from two adjacent rings. One ring was flattened by reverse bending to make flat specimens. The corresponding curved specimens were cut from positions in the remaining ring that were adjacent to the flat ones. The cutting diagram is shown in Fig. 2. The dimensions of the curved specimens are shown in Fig. 3. The flat specimens have the same in-

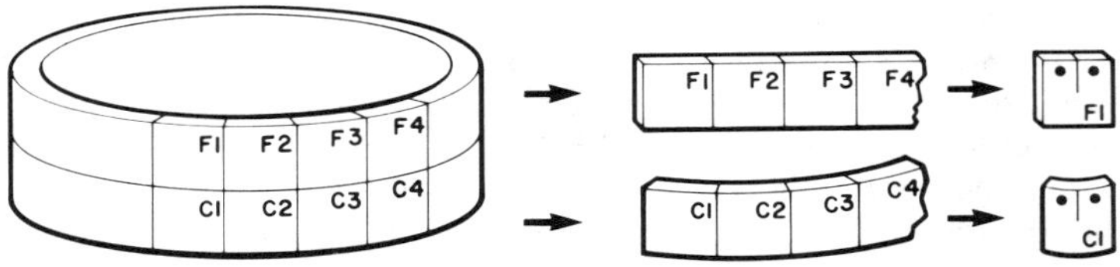

FIG. 2—*Cutting diagram of Zr-2.5Nb compact specimens.*

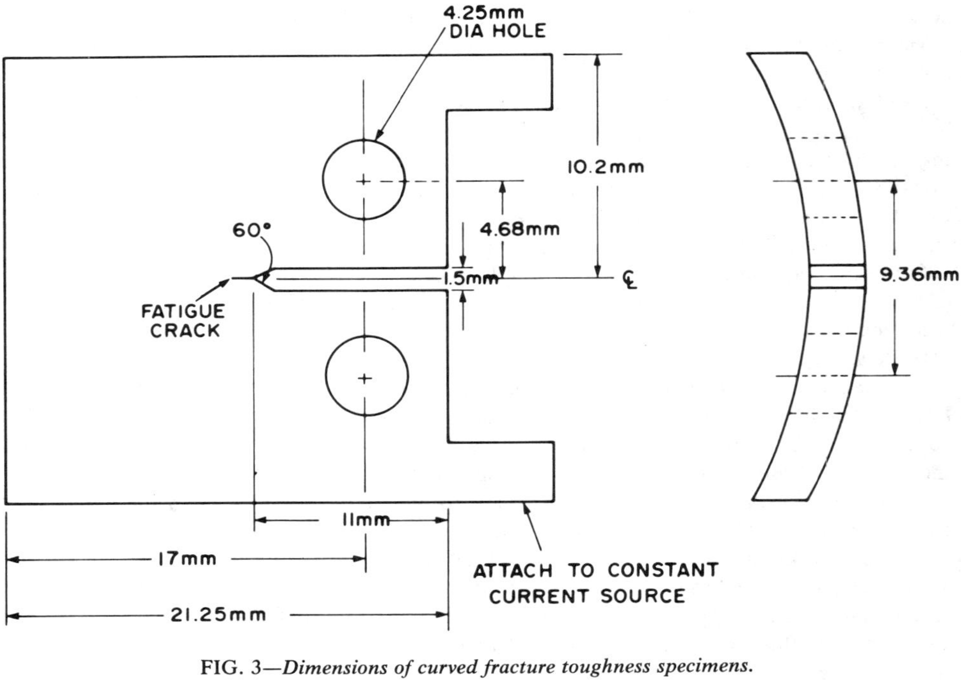

FIG. 3—*Dimensions of curved fracture toughness specimens.*

plane dimensions, which are proportional to the ASTM Test Method for J_{Ic}, A Measure of Fracture Toughness (E 813-81).

Four heat treatments were used to obtain different toughnesses for the Zr-2.5Nb specimens:

(*a*) As-received (~27% cold-worked);
(*b*) 700°C, 7 h, furnace-cooled;
(*c*) 1000°C for 10 min, water-quenched; and
(*d*) 1000 μg/g hydrogen added, then 870°C for 0.5 h, water-quenched.

Hydriding in heat treatment (*d*) was done by heating the specimen to 800°C, admitting a prescribed amount of hydrogen, and holding at temperature until the hydrogen gas was absorbed. Then the specimens were heated to 870°C for 0.5 h. For all four heat treatments, the curved and flat specimen pairs were sealed in the same quartz capsule to ensure that the resulting microstructures were identical.

Transverse Zr-2.5Nb tensile specimens were produced from flattened pressure-tube material. The flattening process added about 2% extra cold-work to the initially 27% cold-worked material. The tensile properties were not changed significantly after flattening. The tensile specimen had a gage length of 25 mm and a nominal size of 4 by 4 mm. They were also heat-treated as previously mentioned.

Zircaloy-2 pressure tubes are 5.2 mm thick, and the 17-mm curved compact specimens can be ground down to flat geometry and still have a thickness of 4.6 mm, 89% of the original thickness. It was decided to compare the fracture-toughness results of the curved and flat-ground specimens from intact sections of a Zr-2 pressure tube. The pressure tube used contained 70 μg/g hydrogen and had been cycled between 350 and 150°C with an internal pressure of 10 MPa. The thermal cycles were introduced to re-orientate the hydride platelets to the radial-axial plane. This orientation encourages brittle behavior [*8, 9*] below a certain temperature, which depends on hydride content. Four specimens with the same dimensions as the Zr-2.5Nb ones were machined from the tube and two of them were ground flat to a final thickness of 4.6 mm. No further heat treatments were performed on the Zr-2 specimens.

Test Procedures

Before testing, the specimens were fatigue-precracked at room temperature. To initiate fatigue cracks evenly on curved specimens, an initial stress intensity factor of about 20 MPa$\sqrt{m}$ was necessary. If the load was too small, the fatigue crack would start only from the inside surface and run unevenly. Once the fatigue crack was started, the load was reduced. The maximum stress intensity factor used for the last 1 mm of fatigue crack was less than 17 MPa$\sqrt{m}$.

The fracture tests were performed in air, and in argon if the test temperature was higher than room temperature. A constant displacement rate of 0.25 mm/min was used. Crack extension was monitored continuously using the d-c potential-drop method [*15*], and load-point displacement was measured by photographing the specimen during the test. The accuracy for the load point displacement measurement was about ±50 μm.

It has been shown [*15*] that for flat compact tension specimens, a close to linear relationship exists between crack length and potential drop. Figure 4 shows that similar relationships exist for curved specimens between a/W = 0.5 to 0.75, where a is the crack length and W is the width of the specimen. For our purpose, we assumed that the change of potential drop was proportional to the change of crack length within this a/W range. The proportionality factors were calculated for each specimen by measuring the total potential change during the test and the total stable crack growth. The total stable crack length was measured by the 9-point average method described in ASTM Standard E 813-81 on the heat-tinted (150°C, 45 min) fracture surface after the test.

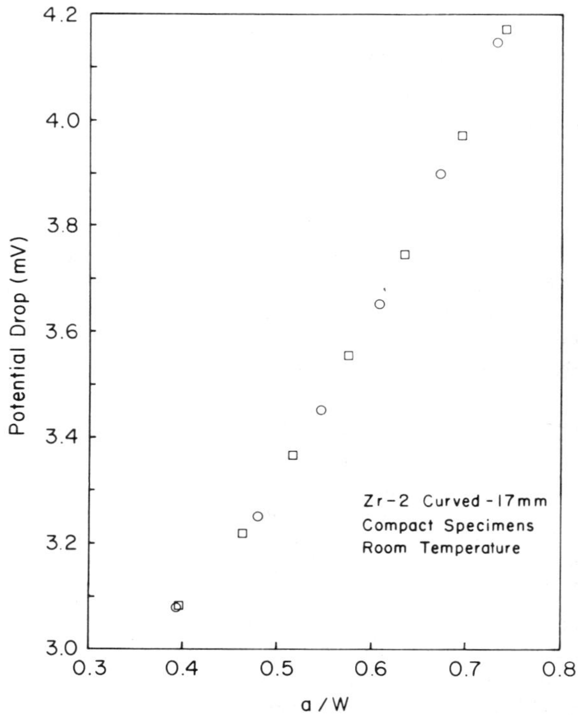

FIG. 4—*D-c potential drop across the crack mouth of curved compact specimens. The two different symbols represent two different specimens.*

Data Analysis

During the test, the load, load-point displacement and potential drop were measured at intervals. These data were used to generate *J*-resistance curves for the ductile specimens. The critical stress intensity factor was used to describe the fracture toughness of brittle specimens, which were those specimens showing no stable crack growth and failing instantaneously at a critical point on a linear load-displacement curve.

Flat-plate equations were used in all the calculations of fracture toughness parameters for both the flat and curved specimens. For brittle specimens, the stress intensity factor, K, was calculated as per ASTM Standard Test Method for Plane Strain Fracture Toughness of Metallic Materials (E 399-83) from the following equation

$$K = \left(\frac{P}{B\sqrt{W}}\right)[(2 + c)/(1 - c)^{3/2}](0.886 + 4.64c - 13.32c^2 + 14.72c^3 - 5.6c^4) \quad (1)$$

where

P = applied load,
B = specimen thickness, and
$c = a/W$.

At unstable crack propagation, the critical stress intensity factor is defined as K_c.

For specimens showing stable crack growth, a crack-growth resistance curve based on the J-integral was calculated for each specimen. The *J*-resistance curves were calculated using the following equation derived by Ernst et al. [*16*] and Clark and Landes [*17*]

$$J_{i+1} = \left[J_i + \left(\frac{f(a_i/W)}{b_i}\right) \cdot \frac{A_{i,i+1}}{B}\right]\left[1 - \left(\frac{\gamma_i}{b_i}\right)(a_{i+1} - a_i)\right] \quad (2)$$

where

$f(a_i/W)$ = tension component correction factor [*16*],
= $2(1 + \alpha)/(1 + \alpha^2)$,
$\alpha = [4(a_i/b_i)^2 + 4(a_i/b_i) + 2]^{1/2} - (1 + 2a_i/b_i)$,
$A_{i,i+1}$ = area under the load, load-point displacement record in energy units from step *i* to $i + 1$,
a_i = crack size at step *i*, and
$b_i = W - a_i$.

J_{Ic} and the initial tearing modulus, T_0, can be derived from the *J*-resistance curves. J_{Ic} was derived as per ASTM Method E 813-81. It is defined as the J-integral at the intersection of the blunting line and the linear regression line for the J-integral data between the exclusion lines. The tearing modulus is defined by [*18*]

$$T_0 = \frac{E}{\bar{\sigma}^2}\frac{dJ}{da} \quad (3)$$

where

E = Young's modulus and
$\bar{\sigma}$ = means of 0.2% yield stress (YS) and the ultimate tensile strength (UTS).

The Young's modulus for zirconium alloys is given by [*19*]

$$E = 95.9 - 0.057T \text{ (GPa)} \quad (4)$$

where T is the temperature (°C).

The initial tearing modulus, T_0, was calculated by equating dJ/da to the slope of the linear regression line in the J_{Ic} calculation. Thus it is the average tearing modulus between the exclusion lines. It has been shown [*8*] that T_0 is a reproducible measure of the ductile tearing resistance of zirconium alloys and is very sensitive to material conditions.

The critical crack length of a through-wall axial crack in a pressure tube was estimated by the tangency condition of the *J*-resistance curve and the crack driving force curve for the pressure tube (see Fig. 5). The crack driving force curve for a tube was calculated as follows

$$J_{\mathrm{cdf}} = \frac{8a\bar{\sigma}^2}{\pi E}\,\ell\mathrm{n}\left(\sec\frac{\pi M\sigma_{\mathrm{H}}}{2\bar{\sigma}}\right) \quad (5)$$

where

σ_H = hoop stress and
M = Folias factor [*20*].

Equation 5 is essentially the Dugdale model with the hoop stress modified by the Folias factor M due to the curvature of the tube. M is a function of the crack length, radius, and wall thickness of the tube. Equation 5 also assumes that

$$J = K^2/E \tag{6}$$

These equations have been used to evaluate the critical crack lengths of zirconium pressure tubes [*21*] and have been shown to yield reasonably conservative predictions.

Results

Tensile Properties

The transverse tensile properties of the Zr-2.5Nb material after the heat treatments are listed in Table 2. Although the flow stresses of these materials are not drastically different, the change in ductility as measured by reduction in area is quite dramatic. For example, the flow stress of as-received Zr-2.5Nb at room temperature is 731 MPa. The specimens hydrided to 1000 μg/g hydrogen and quenched from 870°C had a flow stress of 751 MPa, but the reductions in area at fracture were 60 and 4%, respectively. This change in ductility affects the fracture toughness, as reported in the next section.

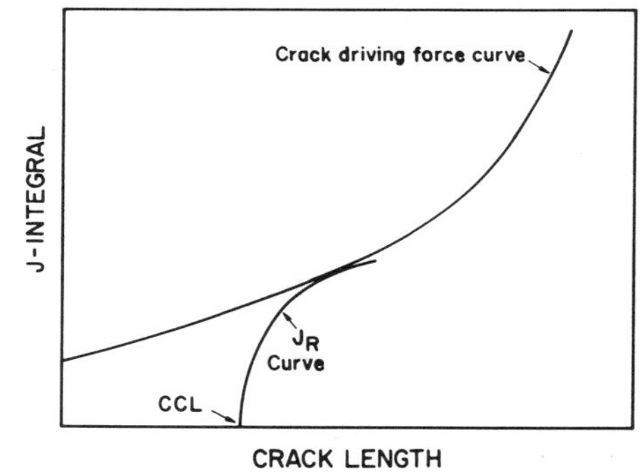

FIG. 5—*Estimation of critical crack length (CCL) by the tangency condition of crack driving force curve of pressure tube and the* J_R *curve.*

TABLE 2—*Tensile properties of Zr-2.5Nb pressure tube material in the transverse direction after heat treatment.*

Heat Treatment	Test Temperature, °C	0.2% Yield Strength, MPa	Ultimate Tensile Strength, MPa	$\bar{\sigma}$, MPa	RA[a], %
(a) as-received	25	690	771	731	60
	250	548	629	589	62
(b) 700°C 7 h, furnace-cooled	25	607	607	607	47
(c) 1000°C 10 min, water-quenched	25	677	752	715	18
(d) [H] = 1000 μg/g 870°C 1/2 h, water-quenched	25	676	825	751	4

[a]RA = reduction in area at fracture.

Fracture Toughness

Zr-2.5Nb Specimens

The fracture toughness of the heat-treated specimens varied widely. Some specimens failed brittlely and others showed ductile behavior. A typical fracture surface for a ductile specimen is shown in Fig. 6, in which three regions can be observed. The fatigue crack fronts were usually quite straight, indicating that the curvature of the curved specimen did not change the stress distribution significantly.

The stable cracks had a thumbnail shape which is quite symmetric with respect to the thickness of the specimen, indicating again that the curvature did not modify the stress field near the crack tip significantly.

A computer program was developed to generate the J-resistance curve using the equations described in earlier. This program also calculates the initial tearing modulus T_0 and the J_{Ic} values. Typical J-resistance curves are given in Fig. 7.

The tearing modulus T_0, initiation toughness J_{Ic}, and critical crack lengths of both the flat (CTFx) and curved (CTCx) specimens are given in Table 3. These results are displayed graphically in Figs. 8 to 10. The solid line represents the case for which both curved and flat specimens give identical results. These figures show that the fracture-toughness parameters of both flat and curved specimens, derived from flat-plate equations, agree very well.

Zr-2 Specimens

The four compact specimens machined from a Zr-2 pressure tube were tested at room temperature. CCT2-5 and CCT2-6 were ground flat, and CCT2-2 and CCT2-9 were left with the curved geometry. Specimens CCT2-6 and CCT2-9 were brittle at room temperature. The K_c values for these two specimens were 35 and 36 $MPa\sqrt{m}$, respectively, in excellent agreement with each other.

The other two Zr-2 specimens, CCT2-2 and CCT2-5, showed ductile behavior. Specimen CCT2-2 was curved and CCT2-5 was ground flat. This variation in behavior resulted from a variation in hydride microstructure with position in the tube. Figure 11 shows that reorientation was not complete through the tube thickness and that the extent of reorientation depended on position. The fracture surfaces of these specimens are shown in Fig. 12. The black lines on the fatigue surface of the curved specimen CCT2-2 were produced by overloading during fatigue cracking to calibrate the potential drop and crack length measurement.

The J-resistance curves for the two ductile specimens are reproduced in Fig. 13. Within the scatter that is usually observed in fracture toughness tests, the two curves are almost identical.

The Zr-2 tests showed no significant difference in fracture toughness between the curved and flat specimens for both brittle and ductile behavior.

Discussion

The J-integral can be used to characterize stable crack growth only when certain conditions are met. Hutchinson and Paris [22] showed that the first of these conditions is that the amount of crack extension Δa, must be a small percentage of the remaining ligament b; that is

$$\Delta a \ll b \tag{7}$$

Exactly how small Δa should be was not specified. For compact specimens, Shih et al. [23] estimated that

$$\Delta a \leq 0.06b \tag{8}$$

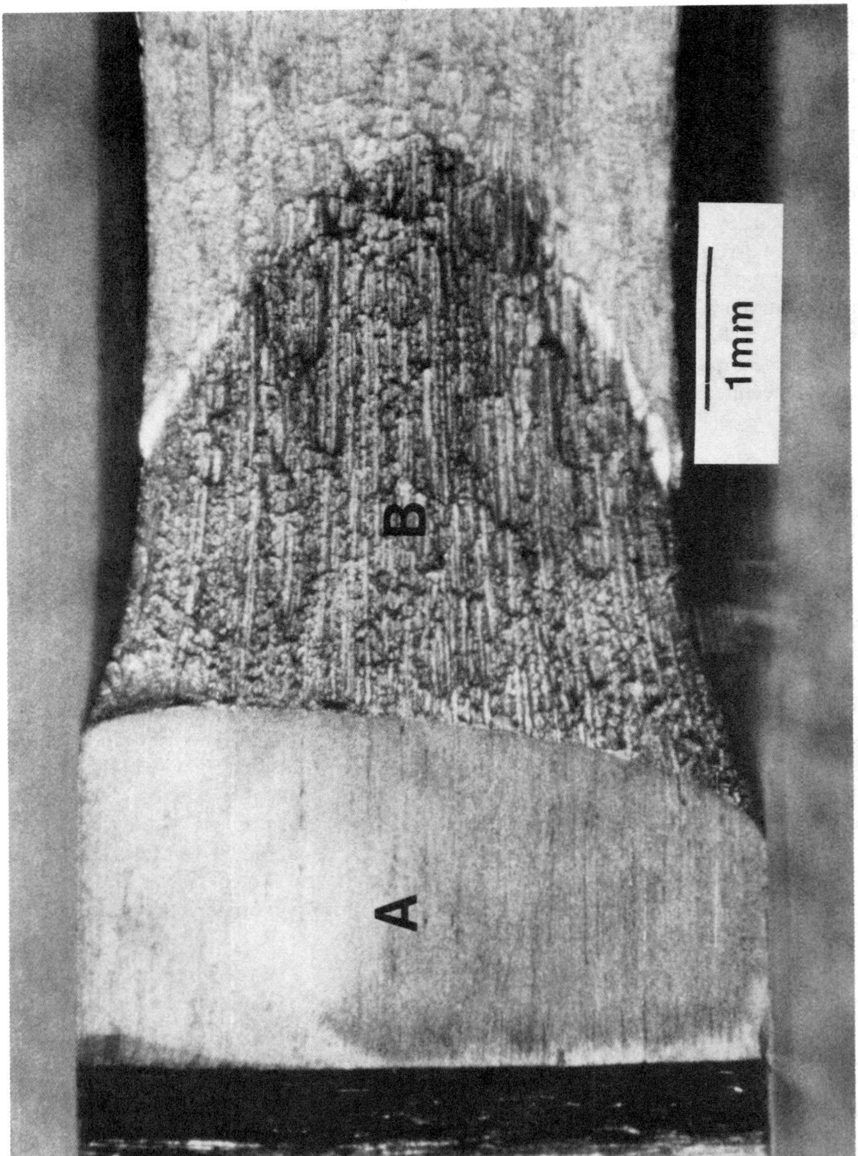

FIG. 6—*Fracture surface of ductile specimen:* (A) *fatigue precrack;* (B) *stable crack.*

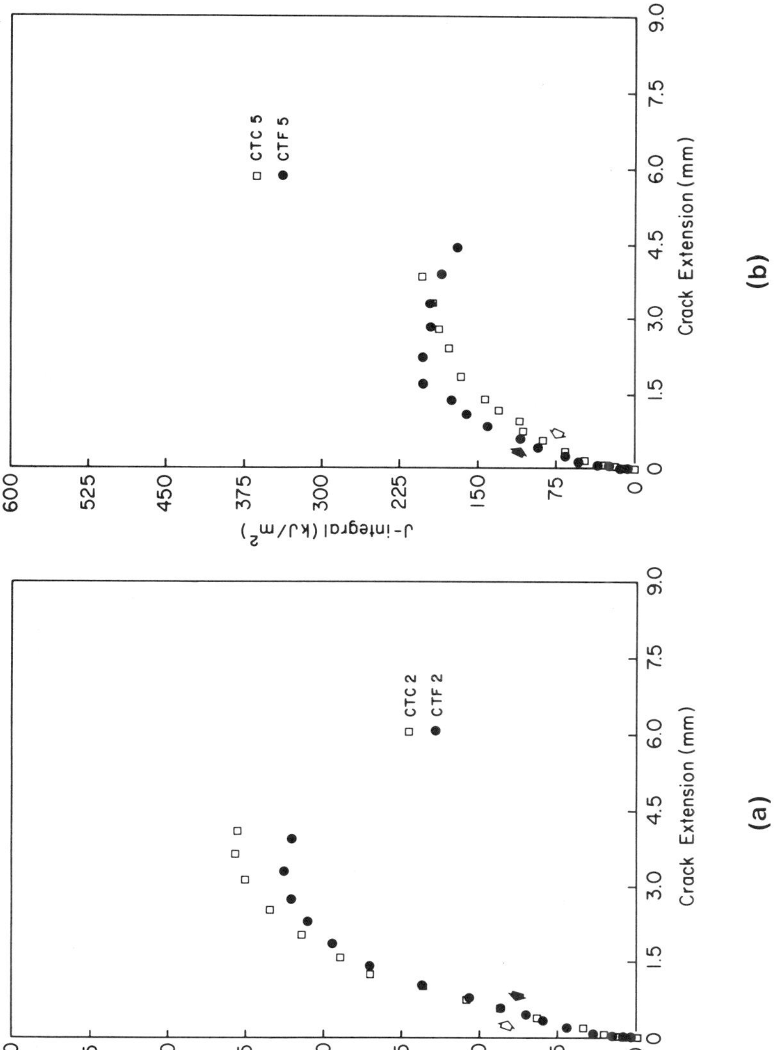

FIG. 7—*J-resistance curves of ductile Zr-2.5Nb specimens. Arrow indicates end of J-controlled crack growth.*

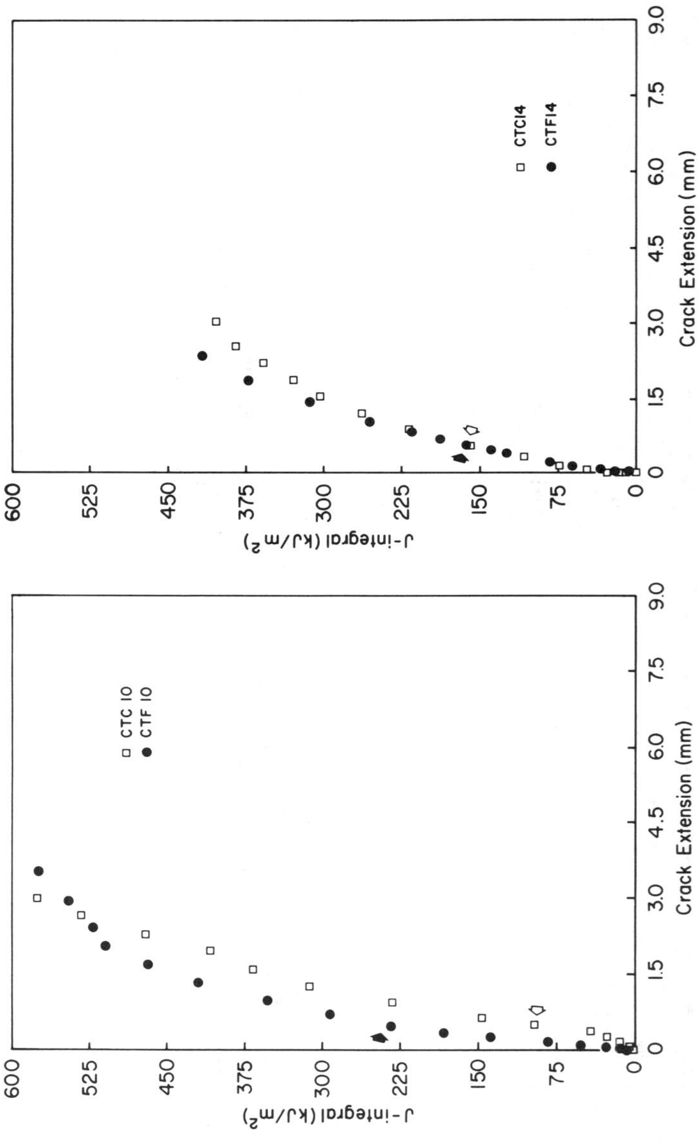
J-integral (kJ/m²)
600
525
450
375
300
225
150
75
0
Crack Extension (mm)
0
1.5
3.0
4.5
6.0
7.5
9.0
CTC 10
CTF 10
(c)
CTC14
CTF14
(d)

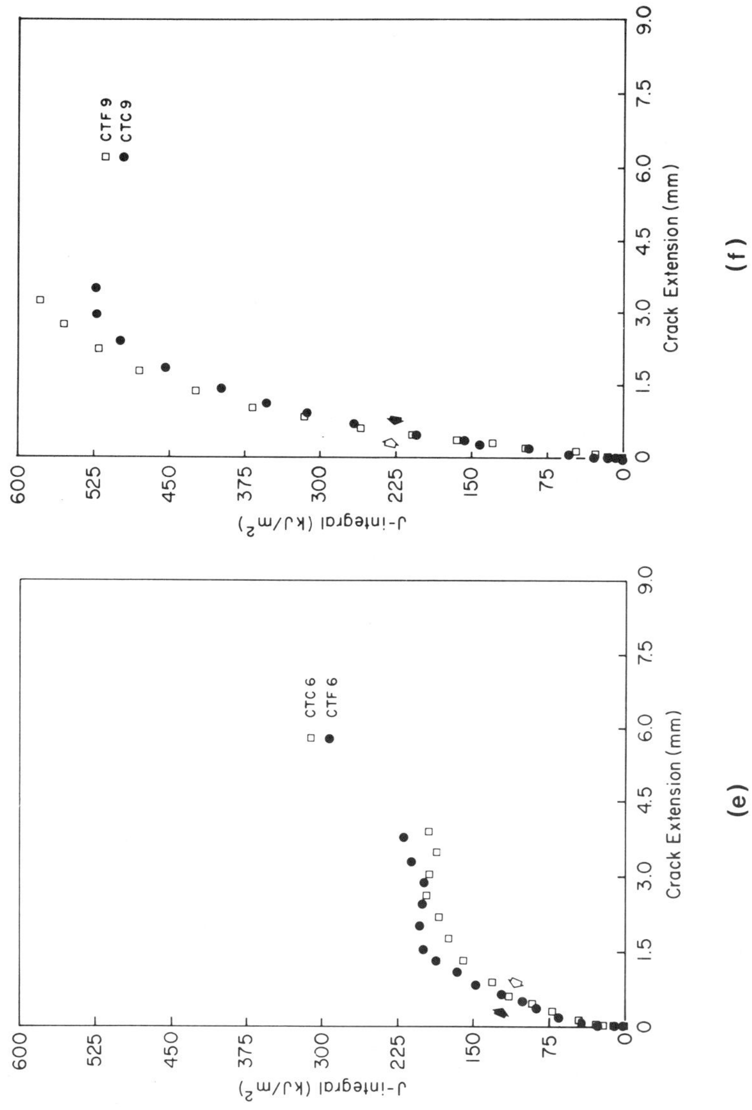

FIG. 7 *Continued.*

TABLE 3—*Fracture toughness parameters for Zr-2.5Nb pressure tube material at room temperature.*

Specimen	Heat Treatment	$\bar{\sigma}$	J_{Ic}, kJ/m²	$J_c{}^a$, kJ/m²	Tearing Modulus	CCL, mm
CTF2	*a*	731	43	...	26	93
CTC2	*a*	731	32	...	29	96
CTF5	*c*	715	53	...	17	79
CTC5	*c*	715	50	...	12	75
CTF6	*c*	715	61	...	17	79
CTC6	*c*	715	58	...	15	76
CTF7	*d*	729	...	5	[b]	15
CTC7	*d*	729	...	7	[b]	19
CTF8	*d*	729	...	7	[b]	18
CTC8	*d*	729	...	6	[b]	17
CTF9	*b*	607	124	...	58	107
CTC9	*b*	607	126	...	50	104
CTF10	*b*	607	162	...	51	106
CTC10	*b*	607	87	...	64	108
CTF14[c]	*a*	589	67	...	47	93
CTC14[c]	*a*	589	76	...	41	90

[a] J_c were calculated by $J_c = K_c^2/E$ for the brittle specimens.
[b] Brittle fracture.
[c] Tested at 250°C.

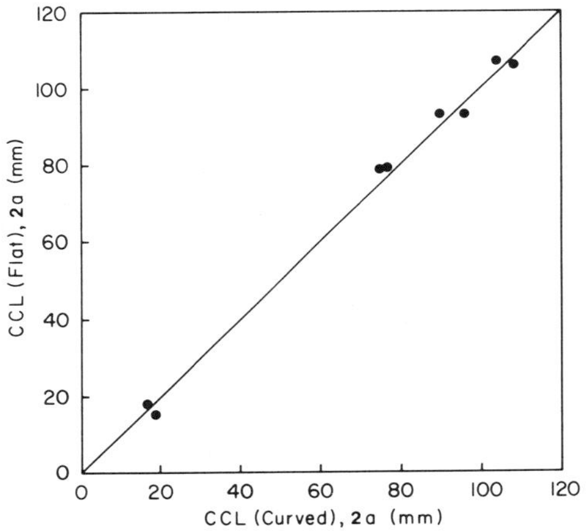

FIG. 8—*Comparison of critical crack lengths derived from flat and curved specimens.*

For the 17-mm compact specimens, $b \simeq 8$ mm, and therefore Δa is limited to about 0.5 mm to ensure J-controlled crack growth. The second condition refers to the ratio of proportional versus non-proportional near-tip strains in the presence of crack growth. The necessary condition for J-controlled growth was given [*22*] as

$$\omega \equiv \frac{b}{J}\frac{dJ}{da} \gg 1 \tag{9}$$

Hutchinson and Paris did not attempt to specify how large ω should be. Current opinions suggest that it should be at least 10. The third condition was given by McMeeking and Parks [*24*].

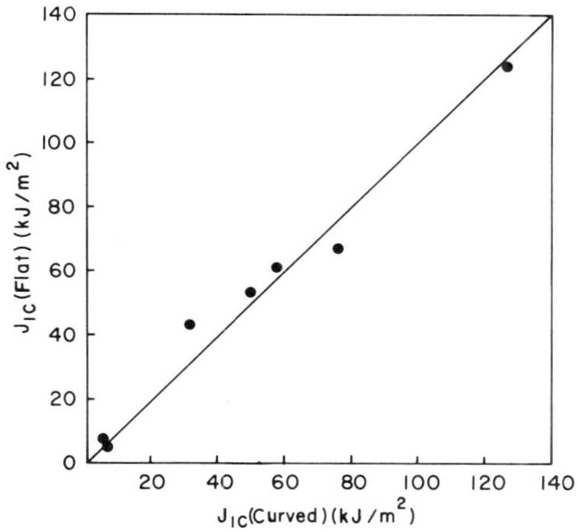

FIG. 9—*Comparison of* J_{Ic} *(J_c for brittle specimens) values obtained from flat and curved specimens.*

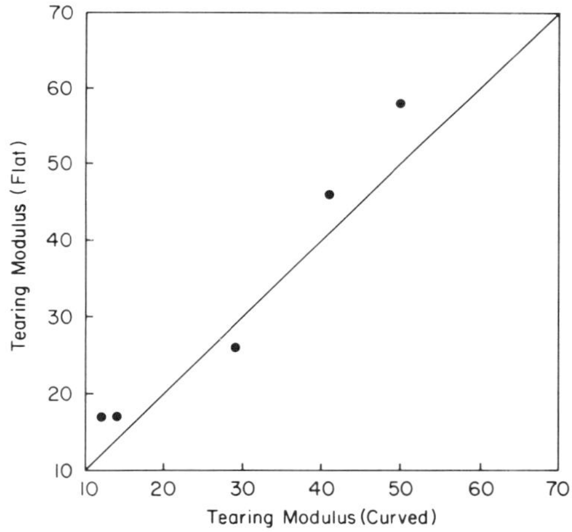

FIG. 10—*Comparison of initial tearing modules* T_0 *obtained from flat and curved specimens.*

In their finite-element studies on the minimum ligament size for which the near-tip fields were *J*-controlled, they concluded that for compact specimens

$$b \geq 25J/\bar{\sigma} \tag{10}$$

According to these authors, all these conditions have to be satisfied to ensure *J*-controlled crack growth. In Fig. 7, the arrows mark the first instance that one of these rules is violated. As indicated in Fig. 7, the *J*-resistance curves of the curved and flat specimens agreed very well with each other, not only when the crack growth was under *J*-control but also beyond the validity limits. Since we expect geometry to affect the *J*-resistance curve shape when these limits are exceeded, these results imply that the influence of specimen curvature is not significant.

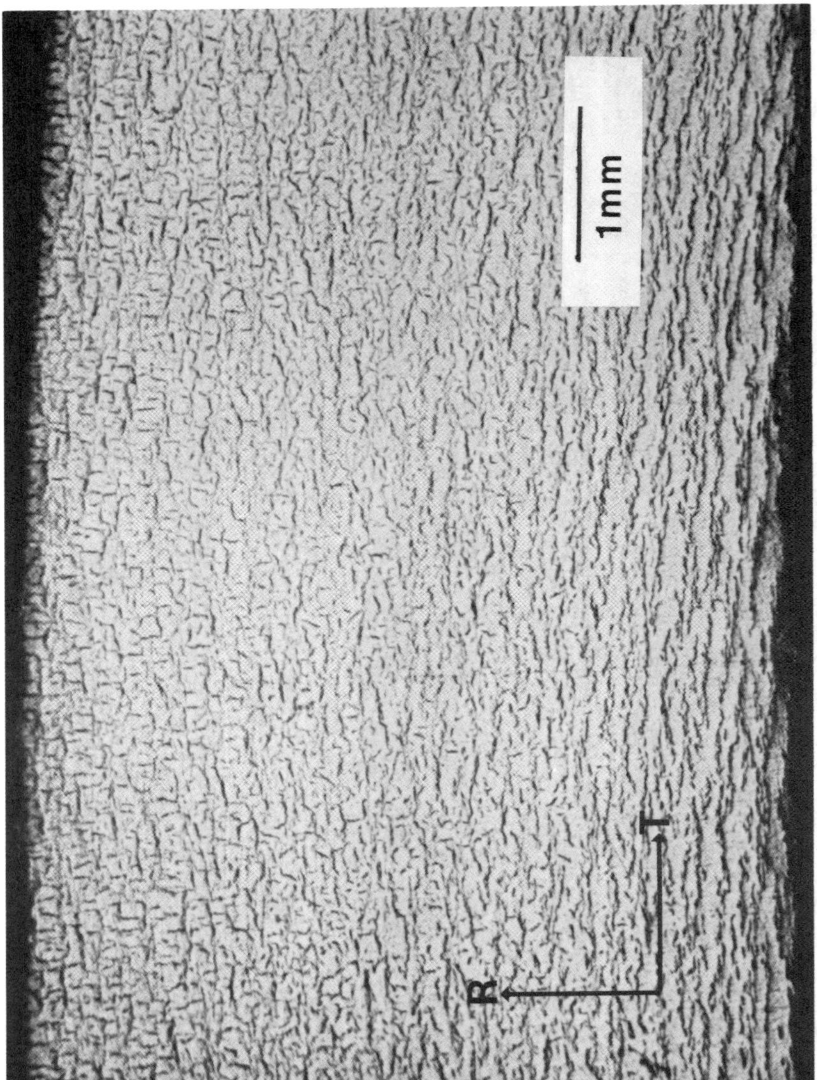

FIG. 11—*Hydride morphology of a Zr-2 specimen.* R *and* T *denote the radial and transverse directions of the tube, respectively.*

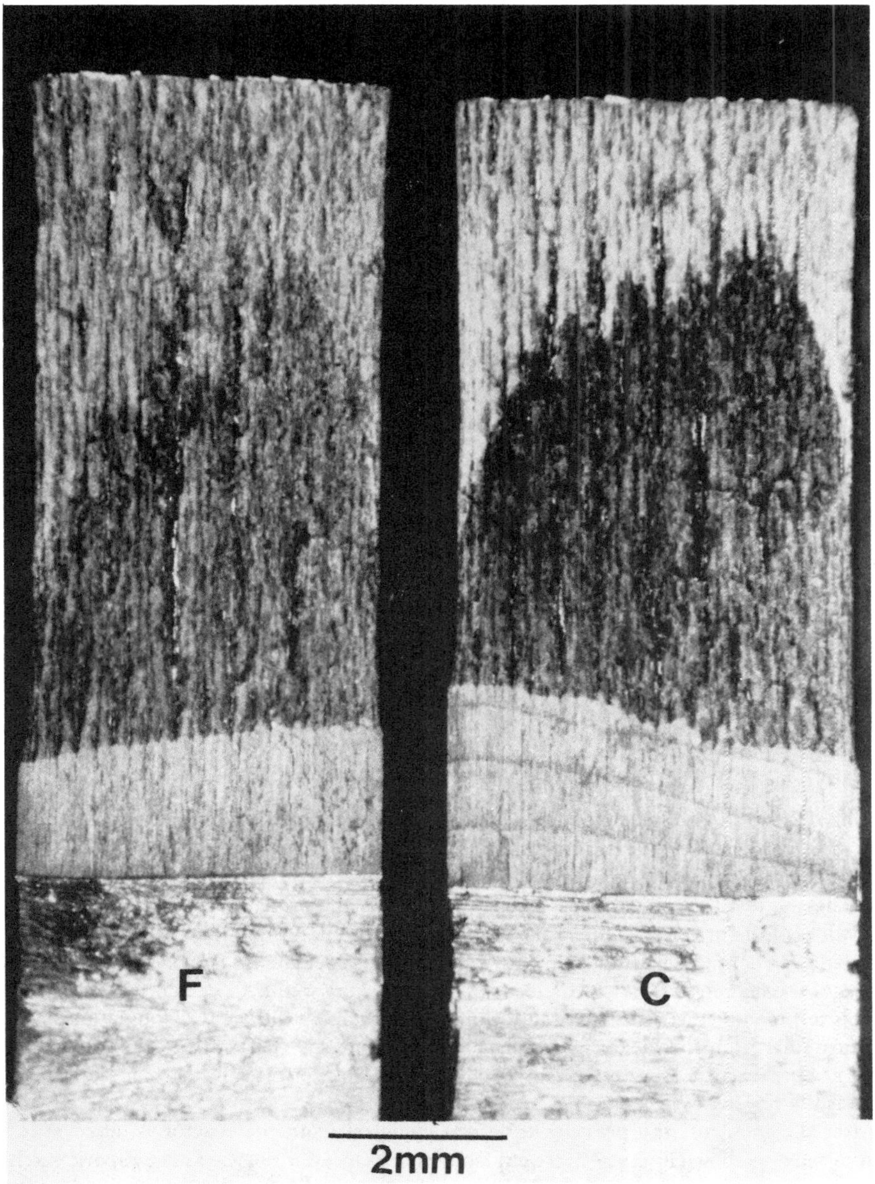

FIG. 12—*Fracture surface of ductile Zr-2 specimens. Specimen marked* F *was ground flat;* C *refers to the curved specimen.*

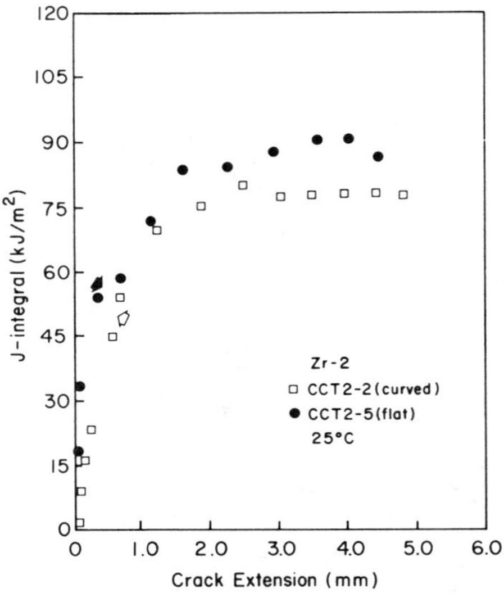

FIG. 13—J_R *curves of the ductile Zr-2 specimens.*

Irrespective of the correctness of the toughness parameter measured in this experiment, our main concern is if the curved specimens give significantly different results than the flat ones: they do not.

For brittle specimens, ASTM Standard ASTM E 399-81 requires that the specimen thickness be

$$B \geq 2.5\left(\frac{K_c}{\bar{\sigma}}\right)^2 \tag{11}$$

so that K_c, the stress intensity factor at fracture, can be qualified as the plane strain fracture toughness K_{Ic}. In this series of tests, none of the specimens satisfied Eq 11. However, the specimen thickness was the full thickness of the pressure tube. Although the K_c may not be the K_{Ic} of the material, it is the critical stress intensity factor for the pressure-tube thickness. The J-integral values of the brittle specimens at fracture were estimated using Eq 6 and tabulated in Table 3 under J_c, to compare with the J_{Ic} values of the ductile specimens. Table 3 indicates that the J_c values of the curved and flat brittle specimens agree very well with each other.

This test method has been successfully applied in mapping the fracture toughness of the burst pressure tube from the Pickering Nuclear Reactor Unit 2 [*14*]. Details of the incident [*4*] and results of the fracture toughness tests [*14*] have been reported. It will suffice to describe the test and results briefly.

After the incident, the pressure tube was removed from the reactor. Curved specimens, 17 mm wide, as shown in Fig. 3, were cut from various locations of the tube by spark machining. These specimens were tested using the methods described here. The results showed that the fracture toughness decreases with deuterium concentration. Near the initiation site, the fracture toughness was found to be between 40 to 50 MPa$\sqrt{\text{m}}$. In some unpublished work, P. H. Davies of Ontario Hydro Research Division applied the finite-element results of Raju and Newman [*25*] to the semi-elliptical boundary of the assumed, initiating crack. She found that the stress intensity factor at the boundary was about 40 to 50 MPa$\sqrt{\text{m}}$, in excellent agreement with the experimental results using small curved specimens.

Conclusions

In this experimental program, the fracture-toughness parameters obtained from curved and flat, 17-mm compact tension specimens were compared. Flat-plate equations were used to calculate the fracture toughness parameters for both specimen types. Heat treatment and hydriding provided a large range of fracture toughness of the specimens. Results show that the fracture toughness parameters of the curved and flat specimens were almost identical through the full range of fracture toughness assessed.

These results show that one can use the curved, 17-mm compact specimens, cut from pressure tubes without flattening, to measure the fracture toughness. This ability was crucial in conducting the failure analysis of the failed pressure tube from the Pickering Reactor Unit 2. Flat-plate equations can be used to calculate the fracture toughness parameters without correction.

Acknowledgment

The authors wish to thank R. B. Ellis and G. R. Brady for technical assistance and M. A. Grant for comments on the manuscript.

APPENDIX

Although the usefulness of the curved specimen method has to be shown experimentally, theoretical estimation of the stresses due to curvature is helpful. In this exercise, an exact analysis has not been carried out because experiments have shown that the fracture parameters obtained for both specimen types, flat or curved, are virtually identical. However, the following estimation shows that the curvature of the specimen does not modify the stress at the crack tip significantly and does serve to support the experiments. We shall compare the stresses due to in-plane loading, (σ_t), to that due to the out-of-plane bending, (σ_c).

For the crack-tip stresses due to in-plane loading, we consider a simple elastic bending analysis of the specimen shown in Fig. 14. The applied bending moment about the neutral axis is

$$P \cdot (a + b/2) \tag{12}$$

where P is the applied load.

The tensile stress, due to simple in-plane bending in the position of the crack tip, is

$$\sigma_t = \frac{P(a + b/2) \cdot b/2}{\dfrac{tb^3}{12}} \tag{13}$$

$$= \left[6\left(\frac{a}{b}\right) + 3\right]\frac{P}{bt} \tag{14}$$

Geometrical analysis shows that the out-of-plane bending moment applied to the specimen ligament is $P \cdot d$, where P is the applied load and d is the displacement of the line of action of the loading forces from the mid-plane of the specimen at the crack plane (Fig. 14*b*). P is assumed to be uniformly distributed on the top of the pin-hole. If we assume simple bending of a rectangular beam (analysis of a curved beam gives almost the same result), the maximum tensile stress due to curvature is at the inner surface of the specimen and is

$$\sigma_c = \frac{MB/2}{I} \tag{15}$$

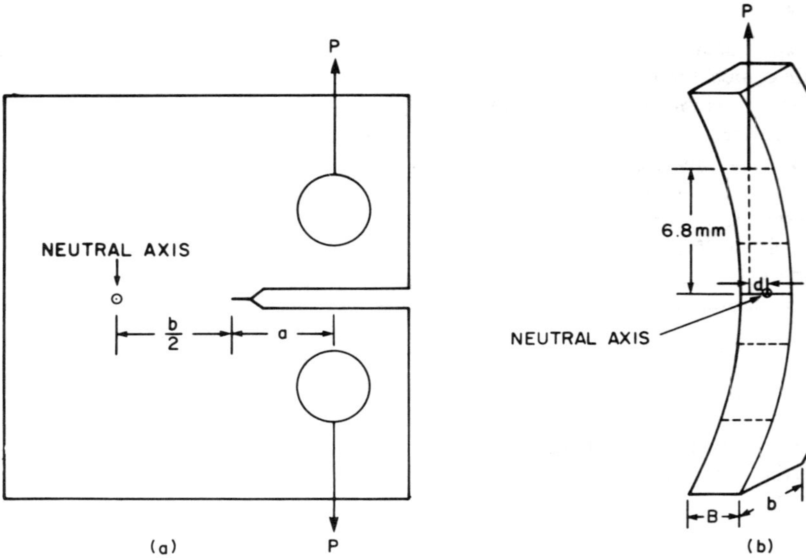

FIG. 14—*Estimation of the tensile stresses due to* (a) *in-plane loading and* (b) *out-of-plane bending due to the curvature.*

where

M = bending moment,
I = moment of inertia of the ligament,
$= bB^3/12$, and (16)

$$\sigma_c = \frac{P \cdot db/2}{bB^3/12} = \frac{6 \cdot P \cdot d}{bB^2} \tag{17}$$

where d is the distance between the neutral axis and the load line.

We assume this stress to be uniformly distributed over the length of the ligament. Thus the relative magnitudes of the stress fields are (for $a = b = 8.5$ mm)

$$\frac{\sigma_c}{\sigma_t} = \frac{6d}{9B} \tag{18}$$

For the Zr-2.5Nb pressure tube geometry, $B = 4.2$ mm, $d = 0.432$ mm

$$\frac{\sigma_c}{\sigma_t} = 6.9\% \qquad \text{(Zr-2.5Nb)} \tag{19}$$

Thus, the maximum deviation in stress due to curvature at the specimen surface is 6.9%. Similar calculations for the Zr-2 geometry show that

$$\frac{\sigma_c}{\sigma_t} = 5.4\% \qquad \text{(Zr-2)} \tag{20}$$

This is not intended to be a rigorous analysis because it ignores stress singularity effects, which would lower the σ_c/σ_t ratio. The estimation above, therefore, provides an upper bound for the σ_c/σ_t ratio.

References

[1] Perryman, E. C. W., *Nuclear Energy,* Vol. 17, 1978, pp. 95-105.
[2] Simpson, L. A. and Puls, M. P., *Metallurgical Transactions,* Vol. 10A, 1979, pp. 1093-1105.
[3] Dunn, J. T. and Jackman, A. H., "Replacement of a Cracked Pressure Tube in Bruce GS Unit 2," Report AECL-8368, Atomic Energy of Canada Limited, 1982.
[4] Field, G. J., Dunn, J. T., and Cheadle, B. A., "Analysis of the Pressure Tube Failure at Pickering NGS 'A' Unit 2," *Canadian Metallurgical Quarterly,* Vol. 24, 1985, pp. 181-188.
[5] Langford, W. J. and Mooder, L. E. J., *International Journal of Pressure Vessels and Piping,* Vol. 6, 1978, pp. 275-310.
[6] Cowan, A. and Langford, W. J., *Journal of Nuclear Materials,* Vol. 30, 1969, pp. 271-281.
[7] Wilkins, B. J. S., Barrie, J. N., and Zink, R. J., "Slit-Burst Testing of Cold-Worked Zr-2.5 wt% Nb Pressure Tubing for CANUD-PHW Reactors," Report AECL-6195, Atomic Energy of Canada Limited, 1978.
[8] Simpson, L. A. and Chow, C. K., "The Effect of Metallurgical Variables and Temperature on the Fracture Toughness of Zirconium Alloy Pressure Tubes," *Zirconium in the Nuclear Industry: Seventh Symposium, ASTM STP 939,* R. B. Adamson and L. F. D. Van Swam, Eds., American Society for Testing and Materials, Philadelphia, 1987, pp. 579-596.
[9] Davies, P. H. and Stearns, C. P., "Fracture Toughness Testing of Zircaloy-2 Pressure Tube Material with Radial Hydrides Using D.C. Potential Drop" in *Fracture Mechanics: Seventeenth Volume, ASTM STP 905,* American Society for Testing and Materials, Philadelphia, 1986, pp. 379-400.
[10] Ondi, T., Kayano, H., and Higashiguchi, Y., *Journal of Nuclear Materials,* Vol. 88, 1980, pp. 226-235.
[11] Honda, S., *Nuclear Engineering and Design,* Vol. 81, 1984, pp. 159-167.
[12] Wallace, A. C., "The Development of Curved Fracture Toughness Specimens for Predicting Crack Growth in CANDU Reactor Pressure Tubes," presented at 11th Canadian Fracture Conference, Ottawa, 14-15 June 1984.
[13] Davies, P. H. and Stearns, C. P., "Curved Double Torsion Fracture Toughness Tests—Round Robin I on Zr-2 Tube 629-15913 (Circumferential Hydrides)," Report M83-196-1C, Ontario Hydro Research Division, 1983.
[14] Chow, C. K. and Simpson, L. A., "An Analysis of the Unstable Fracture of a Reactor Pressure Tube Using Fracture Toughness Mapping" in *Case History Involving Fatigue and Fracture Mechanics, ASTM STP 918,* American Society for Testing and Materials, Philadelphia, 1986, pp. 78-101.
[15] Simpson, L. A. and Clarke, C., "An Elastic-Plastic R-Curve Description of Fracture in Zr-2.5 wt% Nb Pressure Tube Alloy" in *Elastic-Plastic Fracture, ASTM STP 668,* J. D. Landes, J. A. Begley, and G. A. Clarke, Eds., American Society for Testing and Materials, Philadelphia, 1979, pp. 643-662.
[16] Ernst, H. A., Paris, P. C., and Landes, J. D., "Estimations on J-Integral and Tearing Modulus from a Single Specimen Test Record" in *Fracture Mechanics: Thirteenth Conference, ASTM STP 743,* R. Roberts, Ed., American Society for Testing and Materials, Philadelphia, 1981, pp. 476-502.
[17] Clark, G. A. and Landes, J. D., *Journal of Testing and Evaluation,* Vol. 7, No. 5, 1979, pp. 264-269.
[18] Paris, P. C., Tada, H., Zahoor, A., and Ernst, H., "The Theory of Instability of the Tearing Mode of Elastic-Plastic Crack Growth" in *Elastic-Plastic Fracture, ASTM STP 668,* J. D. Landes, J. A. Begley, and G. A. Clarke, Eds., American Society for Testing and Materials, Philadelphia, 1979, pp. 37-64.
[19] Rosinger, H. E., Richie, I. E., and Shillinglaw, A. J., "Young's Modulus of Crystal Bar Zirconium and Zirconium Alloys to 1000 K," Report AECL-5231, Atomic Energy of Canada Limited, 1975.
[20] Folias, E. S., *Engineering Fracture Mechanics,* Vol. 2, 1970, pp. 151-164.
[21] Simpson, L. A. and Wilkins, B. J. S., "Prediction of Fast Fracture in Zr-2.5 wt% Nb Pressure Tubes Using Elastic-Plastic Fracture Mechanics" in *Mechanical Behaviour of Materials,* Miller and Smith, Eds., Vol. 4, Pergamon Press, Oxford, U.K., 1980, pp. 563-572.
[22] Hutchinson, J. W. and Paris, P. C., "Stability Analysis of *J*-Controlled Crack Growth" in *Elastic-Plastic Fracture, ASTM STP 668,* J. D. Landes, J. A. Begley, and G. A. Clarke, Eds., American Society for Testing and Materials, Philadelphia, 1979, pp. 57-64.
[23] Shih, C. F., DeLorenzi, H. G., and Andrew, W. R., "Studies on Crack Initiation and Stable Crack Growth" in *Elastic-Plastic Fracture, ASTM STP 668,* J. D. Landes, J. A. Begley, and G. A. Clarke, Eds., American Society for Testing and Materials, Philadelphia, 1979, pp. 65-120.
[24] McMeeking, R. M. and Parks, D. M., "On Criteria for *J*-Dominance of Crack Tip Fields in Large Scale Yielding" in *Elastic-Plastic Fracture, ASTM STP 668,* J. D. Landes, J. A. Begley, and G. A. Clarke, Eds., American Society for Testing and Materials, Philadelphia, 1979, pp. 175-194.
[25] Raju, I. S. and Newman, J. C., Jr., *Journal of Pressure Vessel Technology,* Vol. 104, 1982, pp. 293-298.

Applications

John M. Barsom[1] *and Stanley T. Rolfe*[2]

Fracture Mechanics in Failure Analysis

REFERENCE: Barsom, J. M. and Rolfe, S. T., "**Fracture Mechanics in Failure Analysis,**" *Fracture Mechanics: Eighteenth Symposium, ASTM STP 945,* D. T. Read and R. P. Reed, Eds., American Society for Testing and Materials, Philadelphia, 1988, pp. 443-467.

ABSTRACT: Fracture mechanics is the best available technology for preventing and analyzing failures of structural and equipment components that are caused by the initiation and propagation of cracks. This paper presents a few observations on the failure and fracture of metals and on the use of correlations to obtain estimates of fracture toughness. The usefulness of these observations in failure analyses is illustrated by analyzing actual structural failures. The paper also presents a brief discussion of some common misconceptions that can lead to incorrect conclusions in a failure investigation.

KEY WORDS: fracture mechanics, failure analysis, fatigue-crack initiation, fatigue-crack propagation, fracture toughness, weldments, strain-rate effects, Charpy V-notch toughness, fracture toughness correlations, brittle fracture, ductile fracture, fracture surfaces

Failure of structural and equipment components may be caused by different mechanisms, including wear, fretting, erosion, corrosion, buckling, stress-corrosion cracking, fatigue, fracture, as well as others. A meaningful failure analysis requires considering all of the possible causes for the failure of a component and isolating the primary contributing factor or factors. In each case the failure analysis must include an investigation of the design (including stresses and details), material, fabrication and assembly, inspection, service conditions, and maintenance. Although general guidelines can be established for investigating any failure, the uniqueness of most failures usually dictates the most relevant steps that should lead to a meaningful analysis.

This paper presents some observations on the fatigue and fracture of metals and on the use of correlations to obtain estimates of fracture toughness. The usefulness of these observations in failure analysis is illustrated by analyzing actual structural failures. Finally, the paper presents a brief discussion of some common misconceptions that can lead to incorrect conclusions in a failure investigation.

Fracture Mechanics

Fracture mechanics is the best available technology for preventing and analyzing failures of structural components that are caused by the initiation and propagation of cracks. This technology has identified and quantified the primary parameters that affect structural integrity, and its application has improved significantly the ability to design safe and reliable structures. Because many failures of structures and equipment are caused by the initiation and the propagation of cracks to a critical size, fracture mechanics technology has become one of the most valuable tools for analyzing failures. The following discussion in this section presents some fracture-mechanics concepts relating to the initiation, propagation, and failure behavior of metals and demonstrates their application to the analysis of failures by analyzing actual structural and equipment failures.

[1]Senior Materials and Product Consultant, USX Corporation, Pittsburgh, PA 15230.
[2]Chairman, Civil Engineering Department, University of Kansas, Lawrence, KS 66045.

Fatigue-Crack Initiation

Initiation of fatigue cracks in structural and equipment components occurs in regions of stress concentration, such as notches, as a result of stress fluctuation. The material element at the tip of a notch in a cyclically loaded component is subjected to the maximum stress range. Consequently, this material element is most susceptible to fatigue damage and, in general, is the site for fatigue-crack initiation [*1*]. The available data show that:

1. The fluctuation in the maximum stress $\Delta\sigma_{max}$ (where $\Delta\sigma_{max} = k_t\Delta\sigma \approx \Delta K/\sqrt{\rho}$ and $\Delta\sigma$ is the fluctuation of the nominal stress, k_t is the stress concentration factor, ρ is the notch-tip radius, and ΔK is the stress-intensity factor calculated by using $\Delta\sigma$ and the notch length) is the primary parameter that controls the initiation of cracks [*1,2*].
2. Compressive stresses are as effective as tensile stresses in initiating cracks. However, once the crack is initiated, the compressive stresses do not contribute significantly to the rate of crack propagation [*2,3*].
3. The fatigue-crack-initiation behavior for steels improves with increased yield strength for the material [*1,2*].
4. A fatigue-crack-initiation threshold exists for steels in a benign environment below which fatigue cracks do not initiate at the root of a notch, Fig. 1 [*1–3*]. The magnitude for the threshold can be predicted from the relationship

$$\frac{\Delta K_{total}}{\sqrt{\rho}} = 10\sqrt{\sigma_{ys}} \quad (1)$$

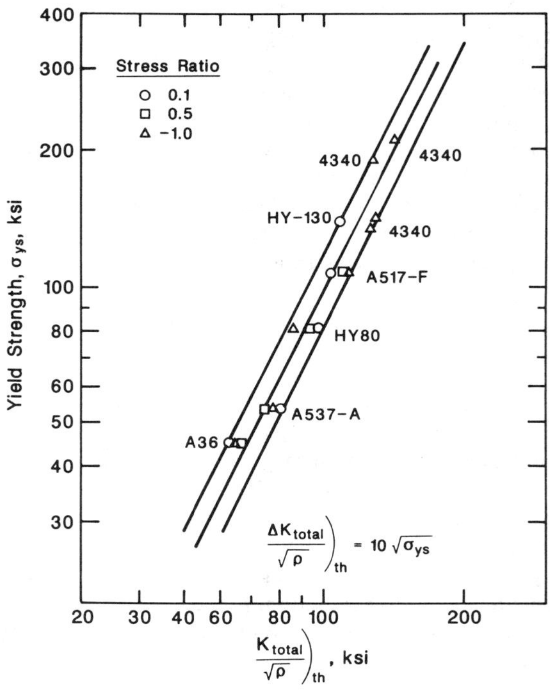

FIG. 1—*Dependence of fatigue-crack-initiation threshold on yield strength.*

where ΔK_{total} is the stress-intensity-factor range calculated by using the tensile-plus-compressive stress range and σ_{ys} is the yield strength for the material.

5. There is a negligible, if any, effect of stress ratio on the fatigue-crack-initiation behavior for steels [*2,3*].

The preceding information should influence material selection and design considerations for structural and equipment components. From a failure analysis viewpoint, the fatigue-crack-initiation behavior suggests that special attention should be placed on verifying the yield strength for the material, the geometry of the component, and the magnitude of the fluctuation for the applied stresses. In many cases the geometry of the component is the most significant contributor to the failure. Figure 2 shows a steel shaft that fractured as a result of a fatigue crack initiating from a keyway having a very sharp corner. Similarly, Fig. 3 is a photograph of an aluminum pipe that fractured because of the excessive weld crown which was intentionally made to decrease the nominal stresses in the weld metal. Unfortunately, the high stress concentration at the weld toe caused the initiation and subsequent propagation of fatigue cracks in the base metal, Fig. 4. The failures for both examples could have been avoided by using less severe, geometrical transitions at the regions of high stress concentrations corresponding to the fatigue-crack initiation sites.

The fatigue-crack-initiation threshold for constructional steels, Fig. 1 and Eq 1, does not exist in distilled water or 3% solution of sodium chloride in distilled water, Fig. 5. Also, the corrosion-fatigue crack-initiation behavior for constructional steels in these environments appears to be independent of material strength, the frequency of the fluctuating stress or the stress ratio, and is governed primarily by the geometry of the component and the magnitude of fluctuation for the applied nominal stresses.

Fatigue-Crack Propagation

The fatigue-crack-propagation behavior for metals in benign environments can be divided into three regions, Fig. 6 [*1-4*]. The behavior in Region 1 exhibits a fatigue-crack-propagation threshold ΔK_{th} that corresponds to the stress-intensity-factor range below which cracks do not propagate under cyclic stresses. The behavior in Region 2 corresponds to a region of crack-propagation behavior that can be represented by a power-law relationship

$$\frac{da}{dN} = A(\Delta K_I)^n \qquad (2)$$

where

a = crack length,
N = number of cycles, and
A, n = constants.

In this region each stress cycle corresponds to a distinct increment of crack extension which usually leaves a fatigue striation on the fracture surface, Fig. 7 [*5*].

The acceleration of fatigue-crack-propagation rates that define the transition from Region 2 to Region 3, Fig. 6, appears to be caused by the superposition of a brittle or a ductile-tearing mechanism onto the mechanism of cyclic-subcritical-crack extension that leaves fatigue striations on the fracture surface [*1*]. These mechanisms occur when the stresses or strains in the vicinity of the crack tip reach critical values.

Failure of structural and equipment components that are caused by the propagation of fatigue cracks are governed by the behavior in Regions 2 and 3, Fig. 6. Most of the useful fatigue life is usually governed by the behavior in Region 2. Extensive fatigue-crack-propagation-rate

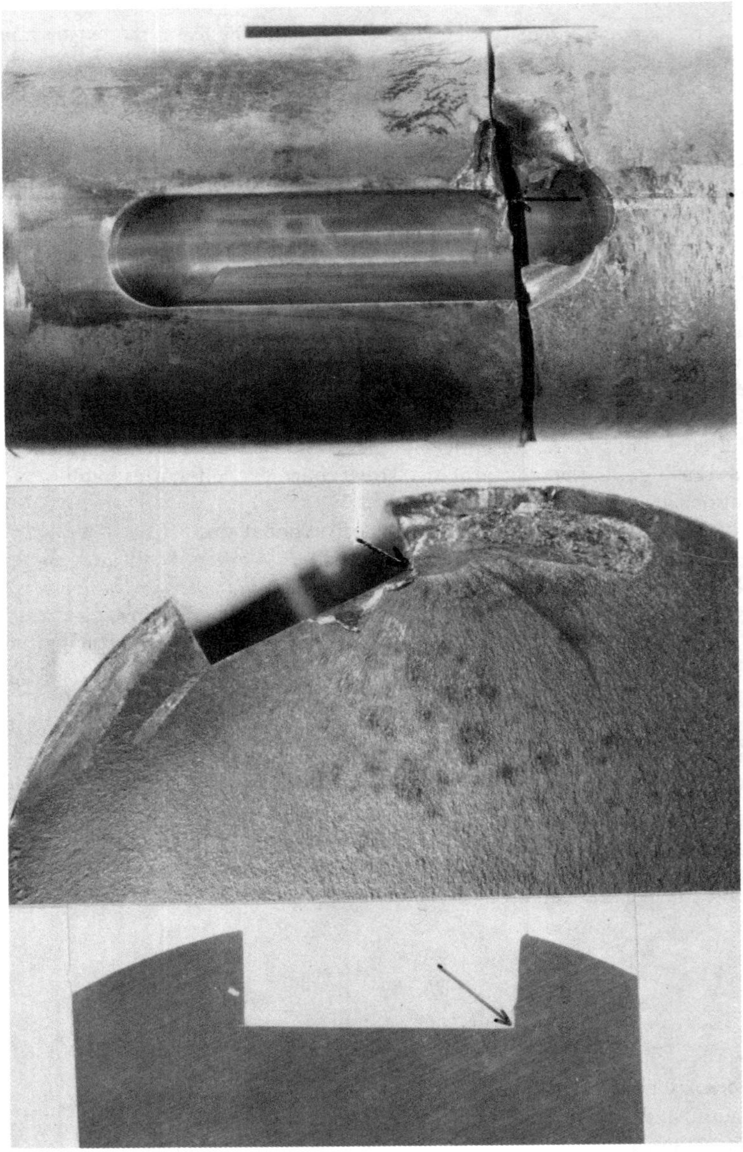

FIG. 2—*Fracture of steel shaft by fatigue crack initiation from very sharp keyway corner (arrows indicate crack initiation site).*

data for various steels show that the primary parameter that affects the propagation rate in Region 2 is the stress-intensity-factor range, ΔK, and that the mechanical and metallurgical properties of the steel have negligible effects on the rate of fatigue-crack propagation [1]. Furthermore, the fatigue-crack-propagation data for metals fall within narrow scatterbands [1]. Consequently, the fatigue-crack-propagation rates for metals can be estimated by using available relationships that have the form given by Eq 2 but with different values for the constants A and n. Furthermore, stress ratio, mean stress, frequency of cyclic loading, and the waveform for

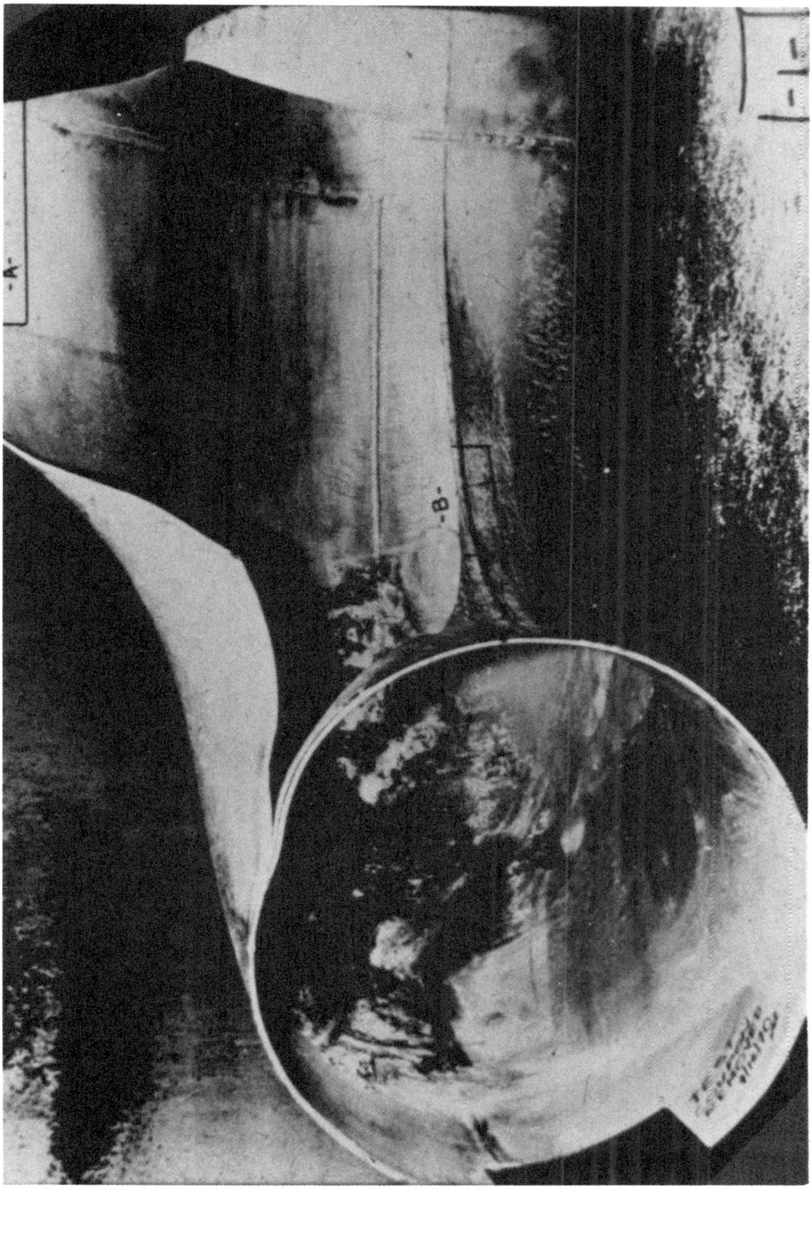

FIG. 3—*Aluminum elbow pipe removed from air separation plant.*

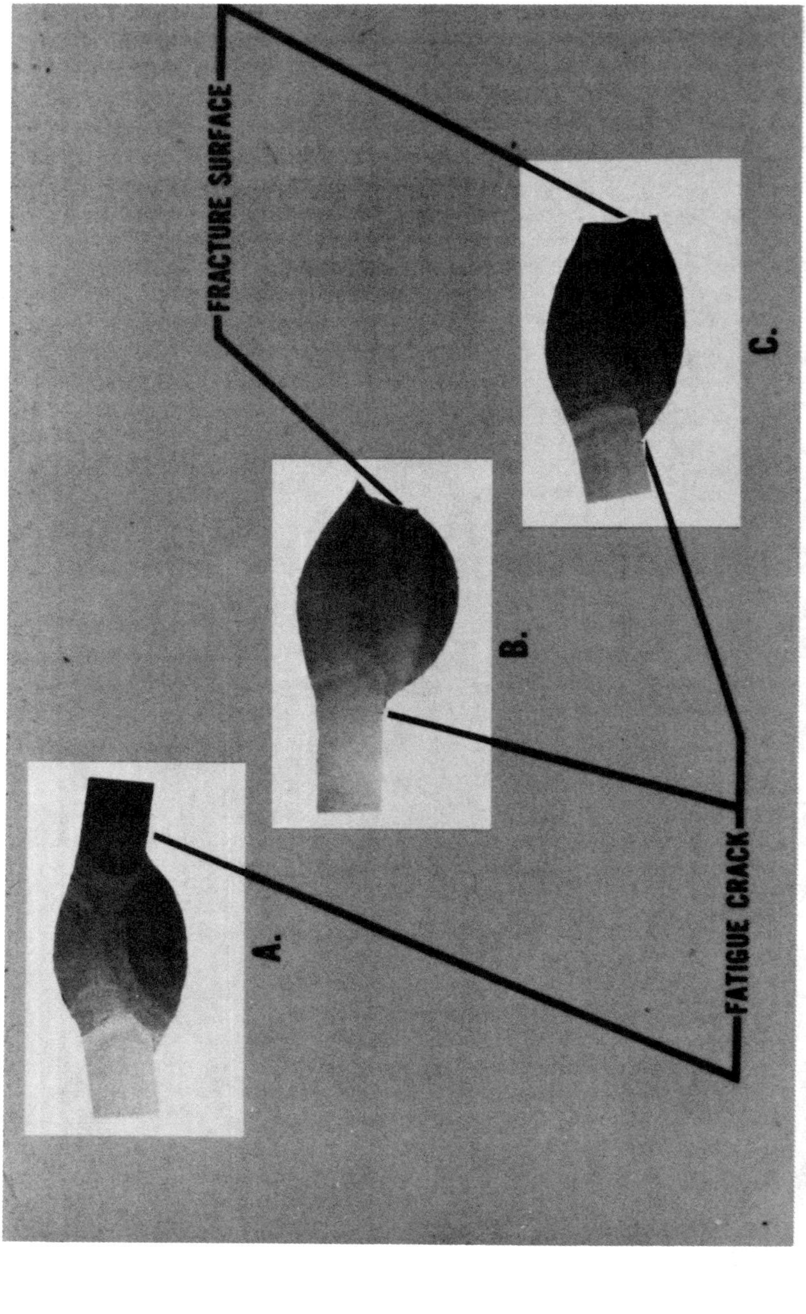

FIG. 4—*Macrographs of cross sections of longitudinal weld in aluminum elbow pipe.*

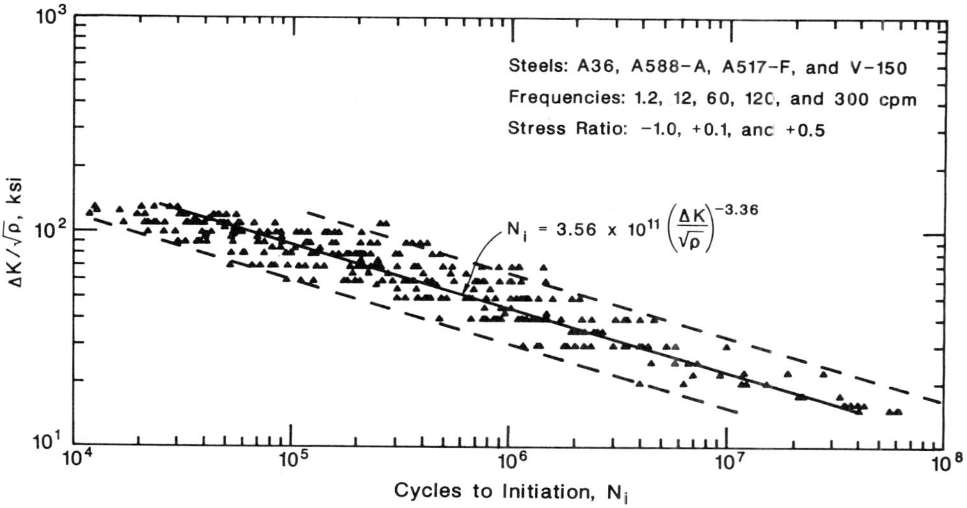

FIG. 5—*Corrosion fatigue crack initiation from notches.*

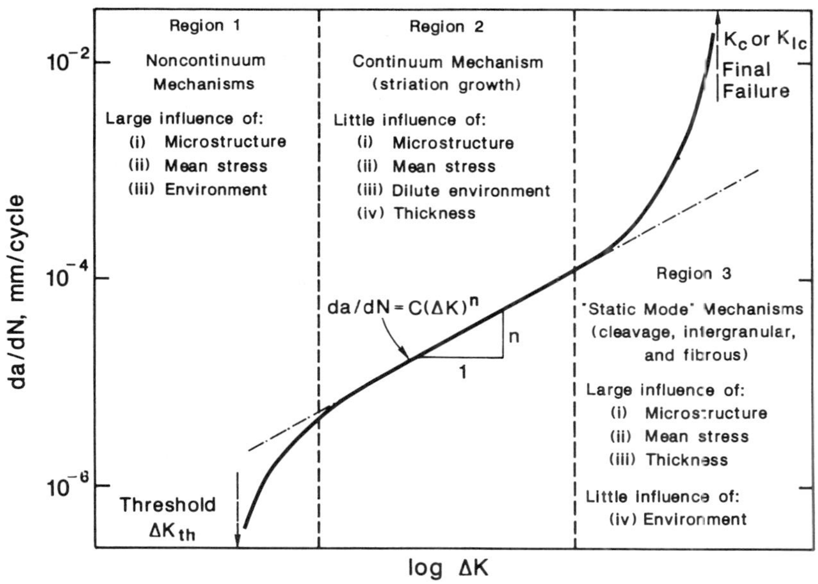

FIG. 6—*Schematic variation of fatigue crack growth rate with stress-intensity-factor fluctuation, showing regions of primary crack growth mechanisms.*

the load (sinusoidal, triangular, square, etc.) have negligible effects on the rate of crack propagation in this region [1]. Thus, the fatigue behavior in Region 2 indicates that analyses of failures that are caused by the propagation of fatigue cracks should place special emphasis on identifying the characteristics (including the size and shape) of the initial and final fatigue crack and on the magnitude of the stress fluctuation. Under ideal conditions, the initial and final cracks and the stress history would be well defined. Figure 8 presents an example of such a case

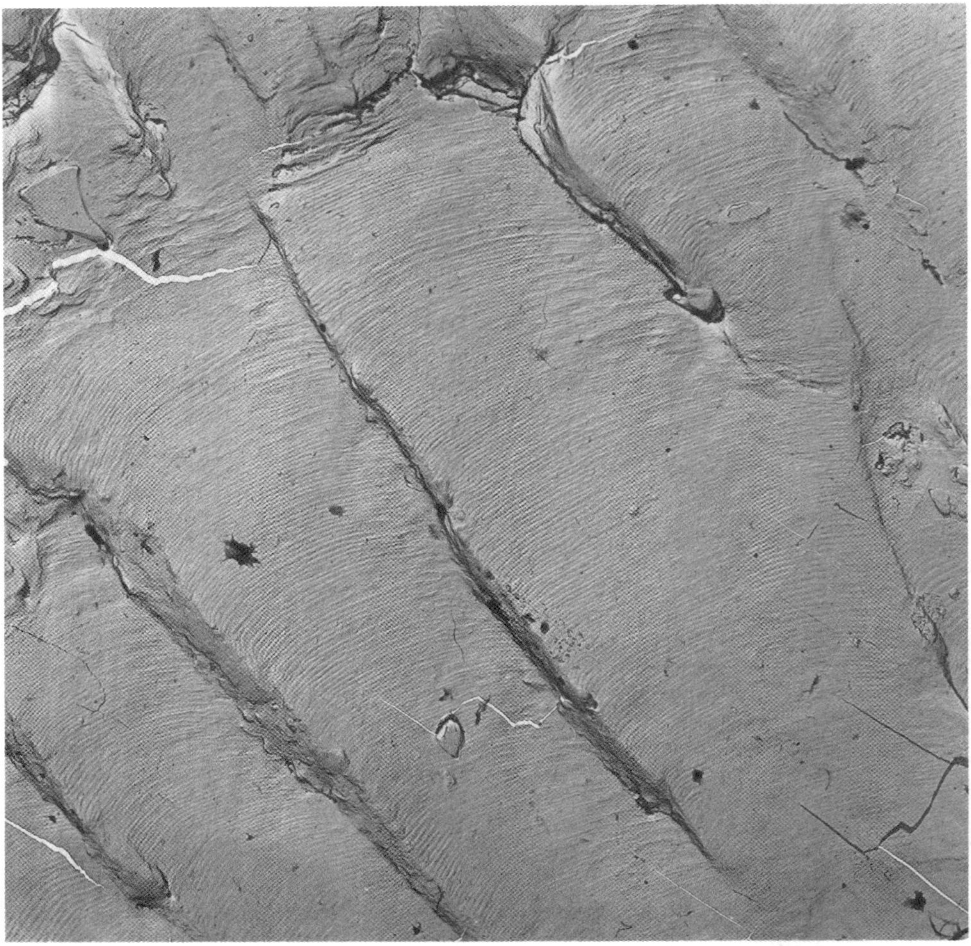

FIG. 7—*Example of ductile fatigue striations in aluminum alloy.*

in which the initial embedded imperfection in a shaft was large, sharp, and penny-shaped, and the final crack was distinct and had a size that was accurately predicted by the plane-strain fracture toughness for the material, K_{Ic}, and the applied stress.

The fatigue life for most welded components is governed by the fatigue-crack-propagation behavior for the weldment. Consequently, the application of fracture-mechanics concepts to analyze failures of welded components by fatigue requires the identification of the size and shape of the initial imperfection and the final crack. Further information, such as the number or the magnitude of the applied cyclic loads, may be necessary to conduct a failure analysis. The specific information that will be necessary depends on the characteristics of the particular failure under consideration. Figure 9 is an example of a fatigue failure of a cover-plate welded detail [*3*]. In principle, analysis of this failure by fracture mechanics is simple; however, in practice it could be very complex. Some of the difficulties encountered in using fracture mechanics to analyze the fatigue failure of weldments have been discussed elsewhere [*6*].

Different environments have different effects on the cyclic behavior of a given material. Similarly, the corrosion-fatigue behavior for different materials is different in the same environ-

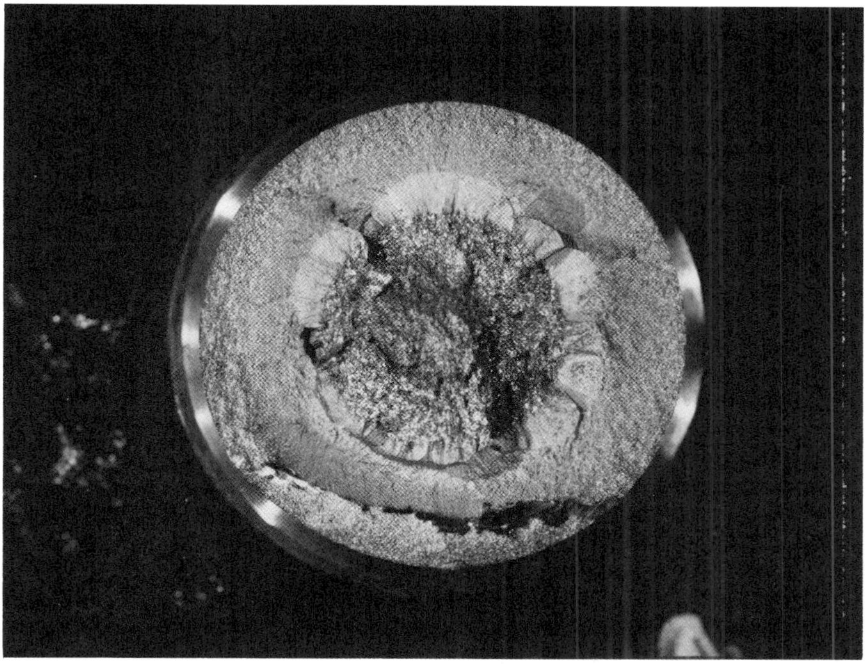

FIG. 8—*Fatigue failure of steel shaft from large embedded imperfection.*

ment. In general, the corrosion-fatigue crack propagation for metals depends on the frequency and on the shape of the cyclic load wave [*1*]. Furthermore, aggressive environments tend to eliminate the striations on a fracture surface, thus eliminating the distinguishing features between corrosion fatigue and stress-corrosion cracking. Also, exposure of fracture surfaces to corrosive environments may mask or eliminate some features that could be significant for the failure analysis and could induce features (for example, pitting) that could lead to erroneous conclusions. These few observations are presented to emphasize that analysis of failures that occur in the presence of aggressive environments can be more complex than in benign environments.

Fracture Toughness

Many tests are available that can be used to measure the fracture toughness for materials. The fracture-mechanics type of tests relate the fracture toughness for the material to the applied stresses and to the size and shape of the crack. Consequently, fracture-toughness characterization of a material by using fracture-mechanics test specimens and test methods is best suited for failure analyses. This characterization can be expressed in terms of a critical stress-intensity factor K [see ASTM Standard Test Method for Plane-Strain Fracture Toughness of Metallic Materials (E 399-83) and ASTM Recommended Practice for R-Curve Determination (E 561-81)], a J-integral [ASTM Test Method for J_{Ic}, a Measure of Fracture Toughness (E 813-81)], or a crack-tip-opening displacement (CTOD) [British Standards Methods for Crack Opening Displacement (COD) Testing (BS 5762)] [*7*]. The values for these parameters depend on the constraint to plastic deformation at the crack tip, the test temperature, and the rate of loading.

The plane-strain fracture toughness, K_{Ic}, for constructional steels under a constant rate of loading increases with increasing temperature [*1,8*]. An example of this behavior is shown in

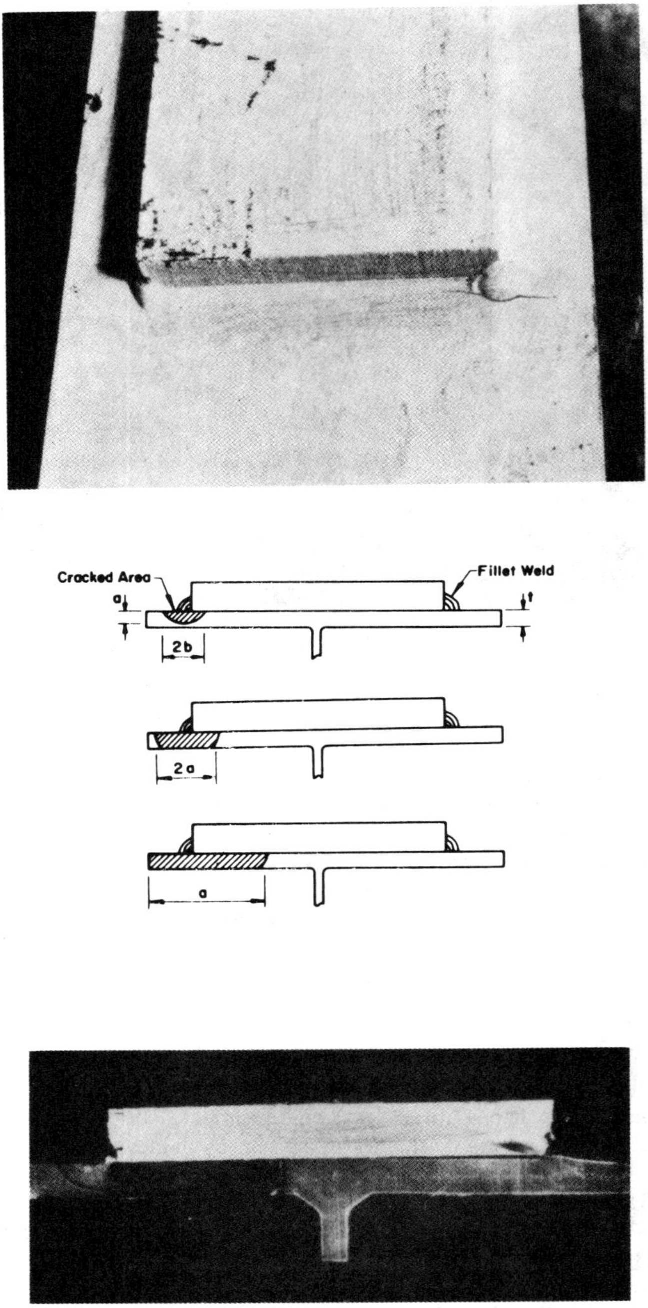

FIG. 9—*Stages of fatigue crack propagation at end of longitudinal welded cover-plate detail.*

Fig. 10 for an A36 steel plate that was tested at three different loading rates. The data also show that the fracture-toughness transition curve is translated (shifted) to higher temperatures as the rate of loading is increased. Thus, at a given temperature, fracture-toughness values measured at high loading rates are generally lower than those measured at slower loading rates. Also, the fracture toughness for constructional steels decreases with decreasing temperature to a minimum K_{Ic} value that is equal to about 27.5 MPa $\sqrt{m}$ (25 ksi $\sqrt{in.}$). This minimum value is independent of the rate of loading used to obtain the fracture toughness transition curve. The available data indicate that a K_{Ic} value of 27.5 MPa $\sqrt{m}$ (25 ksi $\sqrt{in.}$) is an inherent minimum value for constructional steels [*1,8*].

Data for steels having yield strengths between 248 and 1722 MPa (36 and 250 ksi) show that the shift between static and impact plane-strain fracture-toughness curves, Fig. 11 [*1,8*] is given by the relationship

$$T_{shift} = 215 - 1.5\sigma_{ys} \tag{3}$$

for

$$193 \text{ MPa (28 ksi)} < \sigma_{ys} < 896 \text{ MPa (130 ksi)}$$

and

$$T_{shift} = 0$$

for

$$\sigma_{ys} > 896 \text{ MPa (130 ksi)}$$

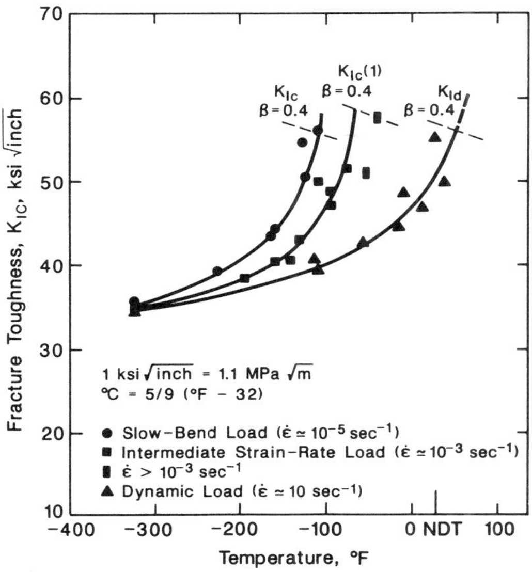

FIG. 10—*Effect of temperature and strain rate on plane-strain fracture toughness behavior of an ASTM A36 steel.*

where T is temperature in °F and σ_{ys} is room-temperature yield strength. The temperature shift between static and any intermediate or impact plane-strain fracture-toughness curves is given by [*8, 9*]

$$T_{\text{shift}} = (150 - \sigma_{ys})\dot{\epsilon}^{0.17} \tag{4}$$

where

T = temperature, °F,
σ_{ys} = room temperature yield strength, ksi, and
$\dot{\epsilon}$ = strain rate, s^{-1}.

The strain rate is calculated for a point on the elastic-plastic boundary for the crack tip according to

$$\dot{\epsilon} = \frac{2\sigma_{ys}}{tE} \tag{5}$$

where t is the loading time and E is the elastic modulus for the material.

Proper application of fracture-mechanics methodology to the analyses of structural and equipment failures requires a knowledge of the temperature and the rate of loading under which the failure occurred.

Correlation of Various Fracture Parameters

Determination of plane-strain fracture toughness K_{Ic} for some failures may not be possible at the failure temperature and loading rate either because sufficient material that possesses the same properties as those at the fracture site is not available or because the material does not exhibit plane-strain behavior under the failure conditions (constraint, temperature, and loading rate). In such cases, the fracture toughness may be estimated by using correlations established for fracture mechanics and non-fracture-mechanics test data or by using elastic-plastic test results [*1, 7, 8*]. The appropriate method (or methods) that should be used for a specific case depends on the characteristics of the particular failure under consideration.

The Charpy V-notch (CVN) impact specimen is the most widely used specimen for failure analysis. The following is a brief discussion of a correlation between plane-strain fracture-toughness test results and Charpy V-notch impact energy absorption curve for constructional steels [*1, 7, 8, 10*].

The Charpy V-notch impact energy absorption curve for constructional steels undergoes a transition in the same temperature zone as the impact plane-strain fracture toughness K_{Id}. Thus, a correlation between these test results has been developed for the transition region and is given by the equation [*1, 8, 10*]

$$\frac{(K_{Id})^2}{E} = 5\ (\text{CVN}) \tag{6}$$

where K_{Id} is in ksi $\sqrt{\text{in.}}$, E is in ksi, and CVN is in ft · lb. The validity of this correlation is apparent from the data presented in Fig. 12 [*8*] for various grades of steels ranging in yield strength from about 248 to about 966 MPa (36 to about 140 ksi) and in Fig. 13 [*8*] for eight heats of SA 533 B, Class 1 steel. Consequently, a given value of CVN impact energy absorption corresponds to a given K_{Id} value (Eq 6) which in turn corresponds to a given toughness behavior at lower rates of loading. The behavior for rates of loading slower than impact can be established by shifting the K_{Id} value to lower temperatures by using Eqs 3 or 4. Conversely, for a

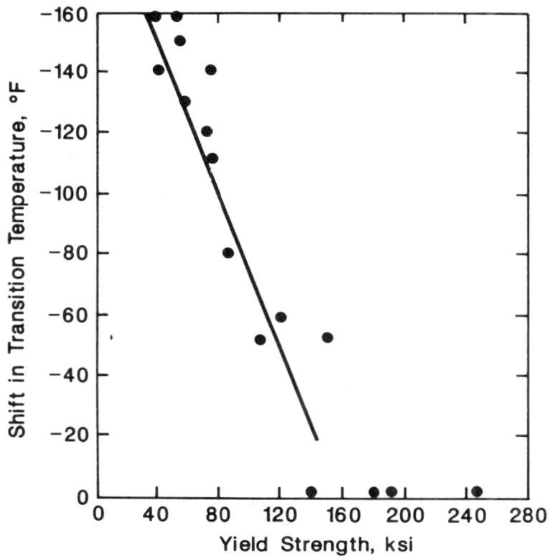

FIG. 11—*Effect of yield strength on shift in transition temperature between impact and static plane-strain fracture-toughness curves.*

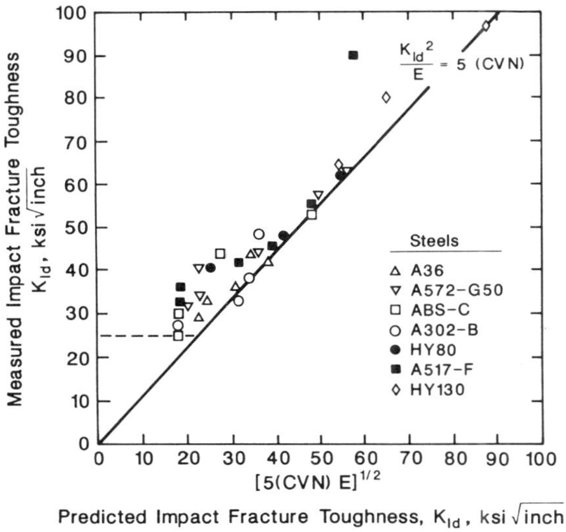

FIG. 12—*Correlation of plane-strain impact fracture toughness and impact Charpy V-notch energy absorption for various steels.*

desired behavior at the minimum operating temperature and maximum in-service loading rate, the corresponding behavior under impact loading can be established by using Eqs 3 or 4, and the equivalent CVN impact value can be established by using Eq 6. Consequently, in many cases, the availability of Charpy V-notch data in the transition zone for a component under investigation may be used, in combination with other information such as temperature and loading rate at failure, to estimate the plane-strain fracture-toughness behavior for the failed component.

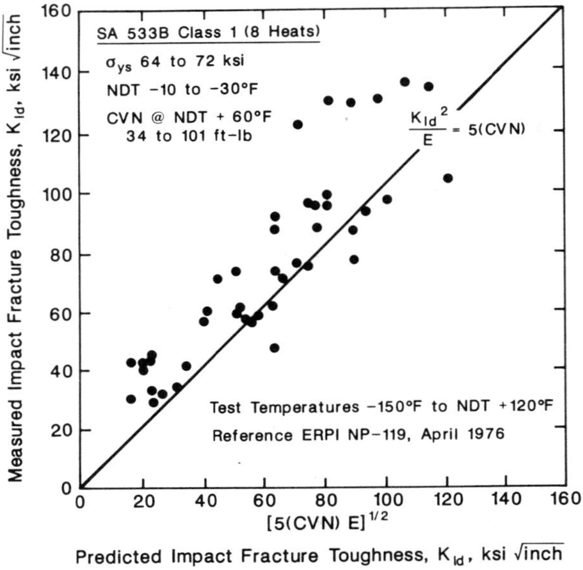

FIG. 13—*Correlation of plane-strain impact fracture toughness and impact Charpy V-notch energy absorption for eight heats of SA533B Class 1 steel.*

Correlation of various fracture parameters for several steels has been discussed by Wellman and Rolfe [*7*]. They investigated the correlation between the stress-intensity factor, K, at fracture and the two commonly used elastic-plastic fracture parameters, J and CTOD. They also discussed correlations with Charpy V-notch test results. Figure 14 is an example of their results and procedure which can be used to establish realistic fracture-toughness values for failures where sufficient material is available to conduct fracture-mechanics type fracture-toughness

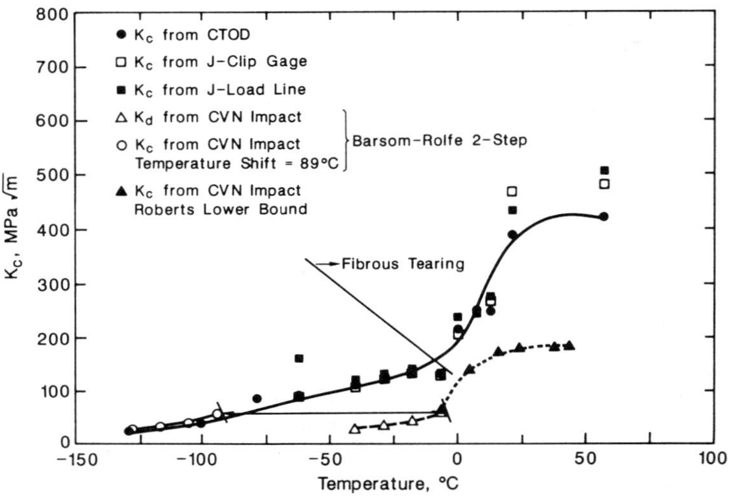

FIG. 14—K_c*-CVN-CTOD-*J *correlations for A131 steel [specimen size 25 by 50 by 200 mm (0.98 by 1.97 by 7.87 in.)].*

tests. Such correlations have been used very effectively in the analysis of failures for engineering structures and equipment.

The fracture event is the final moment in the useful life of a structure. In many cases these events are governed by the fracture toughness of the material which dictates the critical value for the applied stress and crack size and shape that would result in the fracture of the particular component at the temperature and loading rate of interest. Consequently, one can measure or estimate a fracture toughness for any failure that is related to a preexisting crack or crack-like imperfection. However, the significance of the behavior at this final moment in the life of the structure and its relationship, if any, to the primary cause of fracture should be carefully investigated.

The fracture of a flange in the Bryte Bend Bridge represents a simple example for characterizing the fracture event in an engineering structure [3]. The bridge consists of twin parallel structures each carrying three lanes of traffic plus shoulders. The main river section consists of four continuous spans. The superstructure is a trapezoidal steel box supported on reinforced concrete piers. The exterior webs of the box were sloped 1:2, Fig. 15, to reduce the width of the compression flanges in the continuous spans and to improve overall appearance. The boxes

FIG. 15—*Cross section of box girder.*

were designed with the top open. Conventional girder flanges were welded to the tops of the box sides and to a single longitudinal web plate stiffening the center of the box. The bottom plate of the box is longitudinally stiffened with a series of vertical plates.

The bridge was fabricated from plates of A36, A441, and an A517 type steel. The A36 steel was used in all areas of low stress. The A441 steel was used for the maximum positive moment section, and the A517 type steel was used for the maximum negative moment section at the river piers. The steels were not purchased to any toughness requirements, and the bridge specifications in effect at that time did not include fracture toughness requirements. All fabrication was by welding.

In the region of high negative moment over the piers the A517 type steel was used to reduce the size of the members. The A517 type steel flanges were 762 mm (30 in.) wide and 57 mm (2$^{1}/_{4}$ in.) thick.

While the composite concrete deck was being placed in June 1970, a brittle fracture occurred across one of the outer flanges. It initiated at the intersection of a 12.5-mm ($^{1}/_{2}$-in.)-thick lateral attachment welded to the 57-mm (2$^{1}/_{4}$-in.)-thick flange as shown in Fig. 16. The fracture propagated across the entire 762-mm (30-in.)-wide flange and about 101 mm (4 in.) down into the web where it was arrested. The fracture surface was a classic herringbone-type brittle fracture, with the herringbone patterns pointing back toward the crack origin and very small shear lips.

The nominal yield strength of this material was 690 MPa (100 ksi) and the design stress was 310 MPa (45 ksi). At the time the crack propagated, the dead load stress was about 193 MPa (28 ksi) and the ambient temperature about 15.5°C (60°F).

Analysis of the fracture surface indicated that a weld crack about 5 mm (0.2 in.) deep was present in a residual stress field such that sometime during the fabrication or erection, the weld crack propagated out of the residual stress field and arrested at a distance of about 33 mm (1.3 in.) from the edge of the plate. The extent of this 33-mm (1.3-in.) crack is shown as the rusted area in Fig. 17 and is semicircular. During pouring of the concrete deck, as the dead load stress was increased to about 193 MPa (28 ksi), complete fracture of the top flange occurred.

K_{Ic} tests of material from the flange plate that failed showed that at the service temperature of 15.5°C (+60°F), and slow loading rates, the K_{Ic} value was 60.5 MPa $\sqrt{m}$ (55 ksi $\sqrt{in.}$). This was a valid K_{Ic} test that met all the requirements of ASTM E 399. This value agrees very closely with the K_{Ic} value calculated for the applied stress at fracture of 193 MPa (28 ksi) and the 33 mm (1.3 in.) length of the edge crack with a semicircular front that resided in the flange, Fig. 17.

Problems in Failure Analysis Investigations

One of the most frequent errors made in the failure analysis of steel components is caused by the misinterpretation of the visual observations on the fracture surface. Fracture of constructional steels usually exhibit flat cleavage surfaces in the crack propagation region. In many cases, this fracture-surface morphology is erroneously assumed to reflect low fracture-toughness characteristics for the material without regard for the crack-initiation behavior for the material, the temperature at the time of fracture, or the loading rate under which the fracture initiated.

The morphology of fracture surfaces for steel can be understood by considering the fracture-toughness transition behavior under static and impact loading, Fig. 18 [*1,8*]. The static fracture-toughness transition curve depicts the mode of crack initiation at the crack tip. The dynamic fracture-toughness transition curve depicts the mode of crack propagation.

The fracture-toughness curve for either static or dynamic loading can be divided into three regions as shown in Fig. 18. In Region I_s for the static curve, the crack initiates in a cleavage mode from the tip of the fatigue crack. In Region II_s, the fracture toughness to initiate unstable crack propagation increases with increasing temperature. This increase in crack-initiation toughness corresponds to an increase in the size of the plastic zone and in the size of ductile

FIG. 16—*Brittle fracture origin at intersection of lateral attachment and flange plate.*

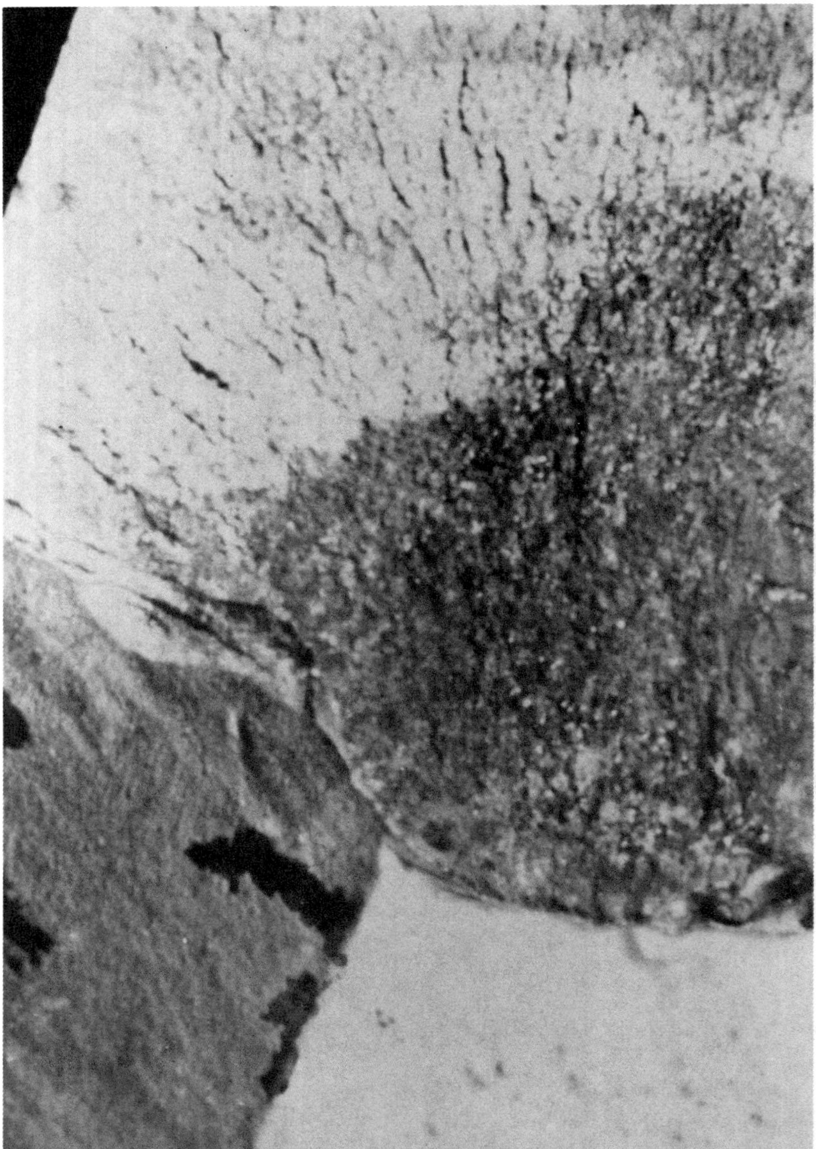

FIG. 17—*Origin of fracture showing 5-mm-deep (0.2 in.) weld crack and 33-mm-deep (1.3 in.) crack.*

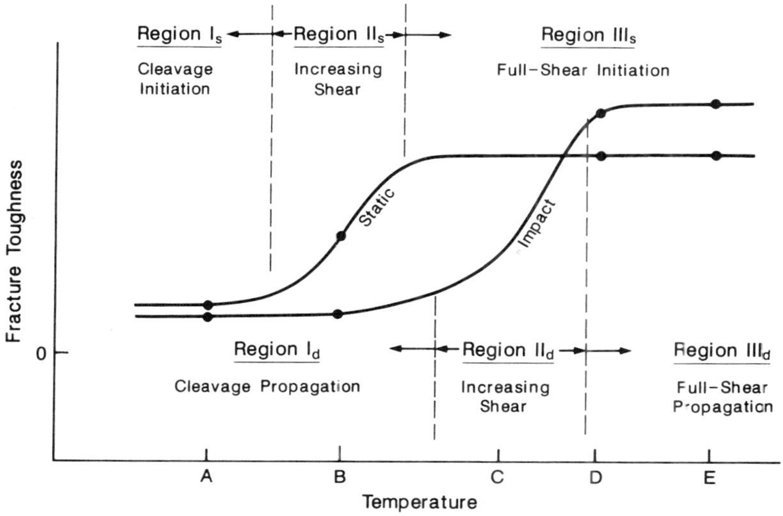

FIG. 18—*Schematic showing relation between static and dynamic fracture toughness.*

tearing (shear) at the tip of the crack prior to unstable crack extension. In this region, the ductile-tearing zone is usually very small and is difficult to delineate by visual examination especially at the lower temperatures. In Region III_s, the static fracture toughness is quite large and somewhat difficult to measure, but the fracture initiates by ductile tearing (shear).

Once a crack initiates under a static load, the morphology (cleavage or shear) of the fracture surface for the propagating crack is determined by the dynamic behavior and degree of plane-strain at the temperature. Regions I_d, II_d, and III_d in Fig. 18 correspond to cleavage, increasing ductile tearing (shear), and full-shear crack propagation, respectively. Thus, at temperature A, the crack initiates and propagates in cleavage. At temperatures B and C, the crack exhibits ductile initiation but propagates in cleavage. The only difference between the behaviors at temperatures B and C is that the ductile-tearing zone for crack initiation is larger at temperature C than at temperature B. At temperature D, cracks initiate and propagate in full shear. Consequently, full shear fracture initiation and propagation occur only at temperatures for which the static and dynamic (impact) fracture behaviors are on the upper shelf.

Most structures such as buildings, bridges, ships, and pressure vessels are subjected to slow or intermediate loading rates under normal operating conditions. Furthermore, these structures are usually subjected to a minimum operating temperature that is lower than that corresponding to temperature D in Fig. 18. Consequently, unless required to exhibit a fracture behavior corresponding to Region D, the crack propagation region for most structural steel failures would exhibit flat cleavage fracture-surface morphology. This crack-propagation morphology cannot be used to predict the crack-initiation morphology or the crack-initiation toughness for the material under normal operating conditions.

The erroneous conclusions derived from the brittle fracture-surface morphology for the crack propagation region are usually justified further by Charpy V-notch test results. In general, when the crack-propagation region exhibits a cleavage surface, Charpy V-notch specimens from the same material exhibit low impact-energy absorption and primarily a cleavage fracture surface at the failure temperature. The low impact fracture toughness for the material may be incorrectly identified as the primary cause for the fracture of a slowly loaded structure. A corollary to this conclusion is to recommend the use of a material with higher fracture toughness to avoid similar future failures. Implementation of such a recommendation develops a false sense of security without eliminating the cause for the failure. Figure 19 [*11*] presents photographs of three 38-mm (1.5-in.)-thick specimens of an A572 Grade 50 steel that were tested at −41°C (−42°F), 3°C (+38°F), and 22°C (+72°F) temperatures. Each of the three specimens exhibited a ductile crack-initiation zone whose size increased with increased test temperature. In each case, the ductile crack-initiation zone was followed by cleavage crack propagation. This behavior occurred even for the specimen that was tested at 22°C (+72°F) which had 67 J (49 ft · lb) Charpy V-notch energy absorption and a K_c of 489.5 MPa $\sqrt{\text{m}}$ (445 ksi $\sqrt{\text{in.}}$).

Flat fractures are usually equated with brittle materials and slant fractures with ductile materials. This observation is fairly accurate for long, fast propagating cracks in plates having large in-plane dimensions. However, brittle materials can exhibit slant fractures and, more important, ductile materials can exhibit flat fractures.

Materials that exhibit brittle flat fracture-surface morphology for fast propagating cracks in thick plates can be made to exhibit ductile slant fracture surfaces by making the plate very thin. This change in fracture-surface morphology is caused by a change in the state of stress at the tip of the propagating crack from plane strain to plane stress. Brittle materials exhibit slant fracture surfaces also when the primary stress component that propagates the crack is inclined at an angle different from that needed to produce a flat fracture. For example, a tensile axial load applied to a bar made of a brittle material results in a flat fracture; however, a torsional load applied to the same bar results in a slant fracture.

Slant fractures in ductile materials usually occur at 45- to 60-deg angle to a plate surface. These slant fracture surfaces are caused by the plastic deformation along shear planes in a ductile material that is subjected to a biaxial stress field. Any loading condition or component geometry that restricts this type of shear deformation can result in a ductile but flat fracture and a decrease or elimination of necking (reduction in thickness or diameter) that usually accompanies such deformations. One method for restricting this type of deformation is by changing the state of stress from biaxial to triaxial. Such a change in the state of stress occurs frequently in engineering structures, in different degrees, especially in regions of high stress concentrations such as along the toe of a transverse weld of a thick cover plate. In these regions the crack initiates and propagates ductilly in a flat mode until it extends beyond the zone of influence of the stress concentration into the region of biaxial stress. Thus, a flat fracture surface is neither a necessary nor a sufficient requirement for identifying a brittle material.

Fracture mechanicians are trained to analyze the behavior of cracks in laboratory specimens or in structures. They also have perpetuated the idea that all structures contain cracks or crack-like imperfections. Consequently, they reason that the fracture of structures must be caused by the action of applied stresses on preexisting cracks or crack-like imperfections. Therefore, the assumption is made that a thorough search of all the fracture surface should reveal an imperfection or crack that is the fracture origin.

Figure 20 is a photograph of the Ingram barge that fractured in Long Island Sound in calm waters in January 1972 [*12*]. This barge was an ocean operating vessel that had been in two hurricanes. Thus it is logical to expect the fracture to have occurred from a fatigue crack that propagated from a weld imperfection or at a region of high stress concentration. Chevron markings on the fracture surface lead to the fracture origin, which was the toe of a weld that attached the kingpost to the deck, Fig. 21. A very thorough examination of the initiation site, Fig. 22, even with scanning electron microscopy, did not reveal any sign of a preexisting crack or a crack-like imperfection. The primary cause for the failure was found to be an abnormally high stress that resulted from improper unloading of the barge. These high stresses combined with a high concentration of stresses at a severe welded detail at the base of the kingpost, Fig. 23, developed stresses equal to the ultimate strength of the material that resulted in failure. Although the crack propagated in clearage, the CVN *impact* toughness of the steel was 75 J (55 ft · lb) at the service temperature of 7°C (45°F).

The Ingram barge fracture demonstrates that the preexistence of cracks or crack-like imperfections is not a necessary requirement for the failure of structures. Moreover, the preexistence

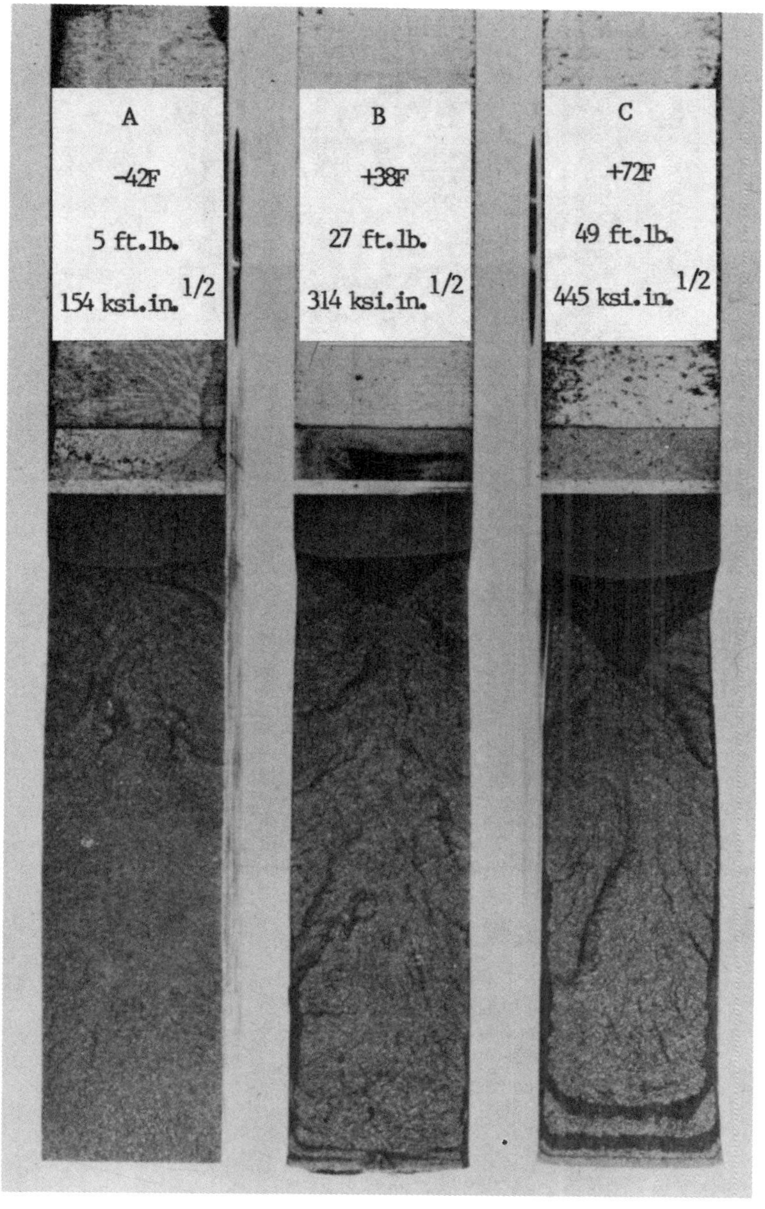

FIG. 19—*Fracture surfaces of full-thickness [*B *= 3.8 cm (1.5 in.)] 4-T compact-tension specimen of A572 Grade 50 steel.*

FIG. 20—*Overall fracture of the Ingram barge.*

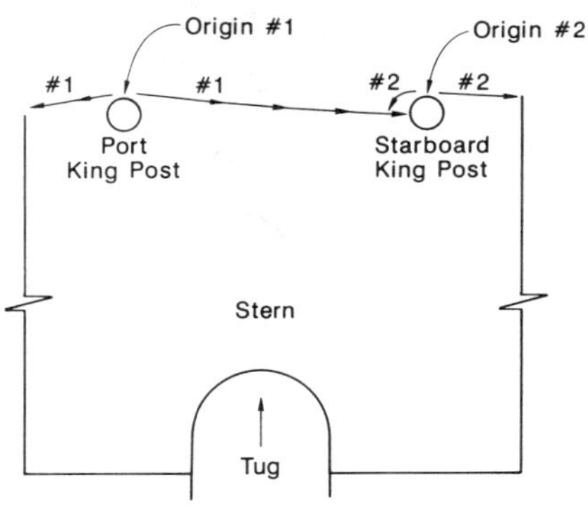

FIG. 21—*Fracture path across the Ingram barge showing two origins.*

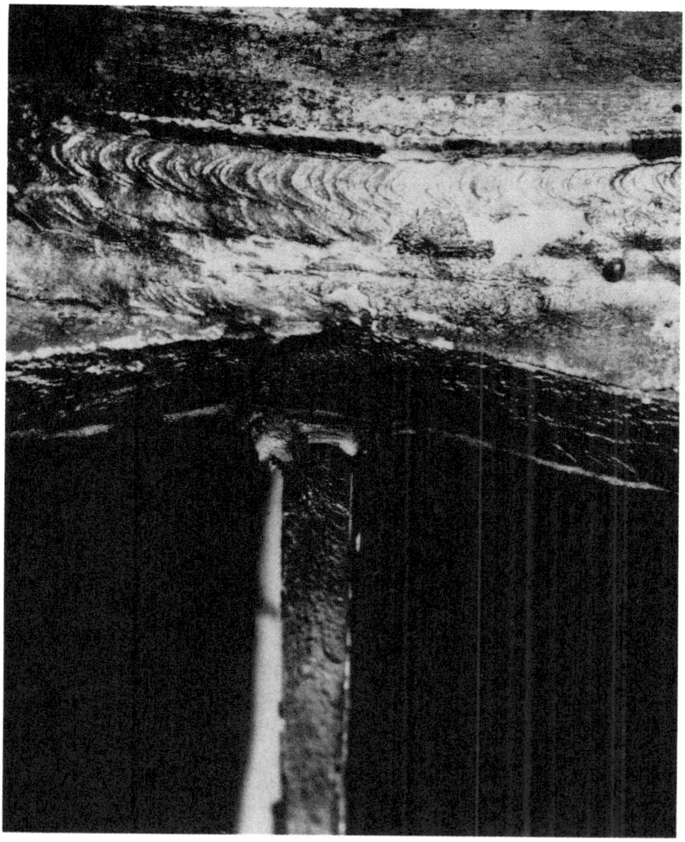

FIG. 22—*Origin of fracture (ring stiffener above deck plate and longitudinal bulkhead below).*

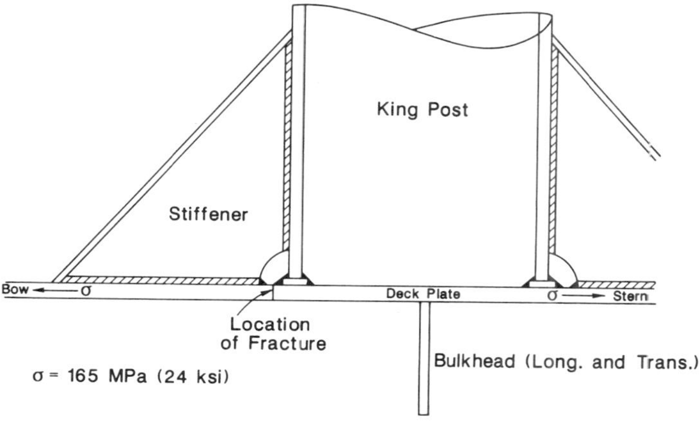

FIG. 23—*Longitudinal section at kingpost showing location of fracture.*

of cracks or crack-like imperfections is not a sufficient requirement to cause the failure of a structure. This observation can be demonstrated by using the following hypothetical example.

Consider a failure analysis of a collapsed steel bridge. A thorough inspection of all the fracture surfaces may reveal the preexistence of a crack or crack-like imperfection. Furthermore, chevron markings may indicate that one of these imperfections was the initiation site for a rapidly propagating crack. However, this initiation site may have been caused by the failure of the structure rather than being the cause of the failure. For example, the collapse of the bridge may have been precipitated by an unexpected movement and displacement of the supports or piers. Such displacements could subject the steel superstructure to abnormally high stresses and loading rates. Under these conditions, crack initiation and propagation can occur at one or more sites as a result of the displacements during the collapse of the structure or due to impact of the structural components against each other, or other objects, or the ground. These sites may or may not have preexisting cracks or crack-like imperfections.

The preceding two examples demonstrate that the existence of cracks or crack-like imperfections is not a necessary or sufficient condition to cause a failure of a structural component. Therefore, it is essential that all failure investigations be based on an objective search for the cause of the failure rather than for cracks or crack-like imperfections.

Summary

Failure of structural and equipment components may be caused by different mechanisms. Fracture mechanics is the best available technology for analyzing failures that are caused by the initiation and propagation of cracks. Fracture-mechanics technology indicates that the stress fluctuation, stress concentration, and the strength of the material are primary parameters that should be studied in failures that are related to fatigue-crack initiation. Fracture-mechanics technology also indicates that the stress fluctuation and the characteristics of the initial and final crack are the primary parameters that should be investigated in failures that are related to fatigue-crack propagation. Finally, the application of fracture-mechanics technology to analyze failures usually requires a knowledge of the temperature and the rate of loading under which the failure occurred.

The crack propagation region for most structural steel failures exhibits flat cleavage fracture-surface morphology. This crack-propagation morphology cannot be used to predict the crack-initiation morphology or the crack-initiation toughness for the material. Furthermore, the existence of cracks or crack-like imperfections is not a necessary or sufficient condition to cause a failure of a structural component. Therefore, it is essential that all failure investigations be based on an objective search for the cause of the failure rather than for cracks or crack-like imperfections.

References

[1] Barsom, J. M. and Rolfe, S. T., *Fracture and Fatigue Control in Structures—Applications of Fracture Mechanics*, Prentice-Hall, Englewood Cliffs, NJ, 1987.

[2] Barsom, J. M., "Fracture Mechanics—Fatigue and Fracture" in *Metals Handbook—Desk Edition*, American Society for Metals, Metals Park, OH, 1985.

[3] Roberts, R., Barsom, J. M., Fisher, J. W., and Rolfe, S. T., "Fracture Mechanics for Bridge Design," FHWA-RD-78-69, Federal Highway Administration, July 1977.

[4] *Application of Fracture Mechanics for Selection of Metallic Structural Materials*, J. E. Campbell, W. W. Gerbrich, and J. H. Underwood, Eds., American Society for Metals, Metals Park, OH, 1982.

[5] Beachem, C. D. and Pelloux, R. M. N., "Electron Fractography—A Tool For The Study of Micromechanisms of Fracturing Processes" in *Fracture Toughness Testing and its Applications, ASTM STP 399*, American Society for Testing and Materials, Philadelphia, 1965.

[6] Barsom, J. M., "Fatigue Consideration for Steel Bridges" in *Fatigue Crack Growth Measurement and Data Analysis, ASTM STP 738*, S. J. Hudak, Jr. and R. J. Bucci, Eds., American Society for Testing and Materials, Philadelphia, 1981.

[7] Wellman, G. W. and Rolfe, S. T., "Engineering Aspects of CTOD Fracture Toughness Testing," Welding Research Council Bulletin 299, Nov. 1984.

[8] Barsom, J. M., "Fracture Toughness" in *Metals Handbook—Desk Edition,* American Society for Metals, Metals Park, OH, 1985.

[9] Barsom, J. M., "Effect of Temperature and Rate of Loading on the Fracture Behavior of Various Steels," *Dynamic Fracture Toughness,* The Welding Institute, Abington, Cambridge, U.K., 1976.

[10] Barsom, J. M., "Development of the AASHTO Fracture-Toughness Requirements for Bridge Steels," *Engineering Fracture Mechanics,* Vol. 7, No. 3, Sept. 1975.

[11] Novak, S. R., *Resistance to Plane-Stress Fracture (R-Curve Behavior) of A572 Structural Steel, ASTM STP 591,* American Society for Testing and Materials, Philadelphia, Feb. 1976.

[12] Rolfe, S. T., "Structural Integrity in Merchant Ships," *Transactions,* American Society of Mechanical Engineers, *Journal of Engineering Materials and Technology,* Vol. 102, Jan. 1980.

John D. Harrison[1] *and Ted L. Anderson*[2]

Developments in the Application of the CTOD Approach to Fracture Assessment of Welded Construction

REFERENCE: Harrison, J. D. and Anderson, T. L., **"Developments in the Application of the CTOD Approach to Fracture Assessment of Welded Construction,"** *Fracture Mechanics: Eighteenth Symposium, ASTM STP 945,* D. T. Read and R. P. Reed, Eds., American Society for Testing and Materials, Philadelphia, 1988, pp. 468–484.

ABSTRACT: The crack-tip opening displacement (CTOD) approach to the assessment of the risk of fracture, which has been incorporated into the British Standard's Published Document PD 6493, has served the United Kingdom welding fabrication industry well for the past 15 years. However, application of the approach and further research have drawn attention to several aspects which need improvement. For example, the PD 6493 document gives no guidance on the derivation of the critical CTOD value to be used in the analysis. Also, the fracture assessment method in PD 6493 suffers from a number of limitations.

This paper describes proposed improvements to the CTOD test and CTOD-based design approaches. A recently drafted appendix for weldment testing, which is proposed to be incorporated into the British CTOD test standard, is discussed briefly. Also, proposals for dealing with stable tearing, pop-in, and scatter in CTOD data are described. Finally, a three-tier fracture assessment procedure is proposed which allows flaws to be assessed at various levels of complexity, depending on the application. The three levels of assessment, in increasing order of complexity and accuracy, are

1. Modified version of the present PD 6493 method.
2. Strip yield model.
3. Reference stress approach.

All three levels utilize CTOD as the fracture toughness input. Some guidelines for choosing the appropriate assessment level are given.

KEY WORDS: fracture mechanics, crack-tip opening displacement, R-curve, plastic collapse, weld imperfections, nondestructive testing

When nondestructive testing (NDT) methods were introduced to inspect welded structures, it was necessary to establish acceptance criteria in order to judge the imperfections that were revealed. Since no rational basis existed for establishing these criteria, they tended to be based on what could be found by radiography, rather than on what might impair structural performance. From an early date [*1*] it was realized that these arbitrary acceptance criteria were unduly conservative and that many of the imperfections that were repaired were innocuous. Nevertheless, vast sums of money are still squandered on removing perfectly acceptable weld metal and heat-affected zone (HAZ) material and replacing it.

[1]Group manager, Engineering and Materials, The Welding Institute, Abington Hall, Abington, Cambridge, CB1 6AL, United Kingdom.
[2]Assistant professor, Mechanical Engineering Department, Texas A & M University, College Station, TX 77843-3123.

The alternative was to consider how the imperfections might lead to failure of the structure. A great deal of research was done in the 1960s and 1970s to establish a rational basis for setting acceptance limits. Initial attempts to apply fracture mechanics to this problem failed because the behavior of most welded structures cannot be adequately described by linear elastic fracture mechanics (LEFM). The crack-tip opening displacement (CTOD) test [*2*] was developed as a measure of elastic-plastic fracture toughness. This was followed by the development of the CTOD design curve [*3,4*], which, in 1980, was adopted in the British Standards Institution (BSI) Published Document PD 6493.

The CTOD design curve approach for assessing the risk of fracture from weld flaws has served the United Kingdom welding fabrication industry well for the past 15 years. Several hundred million dollars have been saved by making realistic assessments and thereby eliminating unnecessary weld repairs. In spite of this generally satisfactory experience, application of the approach and further research have drawn attention to a number of aspects which need to be improved. For example, the British Standard for CTOD testing, BS 5762:1979, does not give adequate guidelines for testing weldments. Also, there are some problems with the interpretation of CTOD test results. In addition, the CTOD design curve approach in PD 6493 suffers from a number of limitations.

This paper describes recent proposals for improving CTOD-based methods. These proposals are concerned with the CTOD test itself as well as with the way in which CTOD data are applied to welded structures. Note that the proposed changes to existing procedures are those recommended by the Welding Institute and have not yet been considered by the appropriate British Standards committees.

Possible improvements to the CTOD test are described next. The topics include guidelines for weldment testing, interpretation of the test results, and the treatment of scatter in the data. This is followed by a discussion of fracture design analyses based on CTOD. A three-tier system is proposed which permits a choice of fracture assessment methods with various levels of complexity and accuracy.

Fracture Toughness

Much of the fracture research in the United States is aimed at avoiding ductile fracture, but this form of failure is relatively rare. The principal objective of fracture control in welded steel construction is the avoidance of brittle cleavage fracture. Although the CTOD test can be applied to upper-shelf behavior, it was developed principally with a view to avoiding brittle fracture, and it has seen its widest application in the brittle-to-ductile transition region.

The CTOD test remains a valuable tool for measuring the fracture resistance of welded structures. However, certain clarifications and improvements are needed.

Weldment Testing

The test method described in BS 5762:1979 and in the current ASTM draft CTOD standard is suitable for testing plain, homogeneous materials. These documents give no guidance on the problems of weldment testing. Weldments generally have highly inhomogeneous microstructures and complex residual stress fields. Also, it is often necessary to machine specimens from configurations other than flat plates (for example, pipes and fillet welds). A suitable appendix [*5*] to the test standard is currently being drafted for discussion by the appropriate subcommittees in ASTM E 24 and in the British Standards Institution. Some of the main points in this draft document are summarized next.

The first step in testing a weld is to choose the specimen geometry and notch orientation. Two types of single-edge notched bend (SENB) specimens are used for the majority of CTOD tests. The so-called preferred geometry specimen has dimensions B by $2B$, where B is the plate thickness, with a through-thickness fatigue crack. The subsidiary geometry specimen has a square cross section (*B* by *B*) and is usually notched perpendicular to the plate surface. The crack length to width ratio (a/W) of the preferred geometry specimen is nominally 0.5, but the a/W of the subsidiary geometry test specimen can vary depending on the application.

The choice of specimen geometry and notch location depends on the objectives of the test. If the purpose is to assess the effect of a particular weld imperfection, then the notch in the CTOD specimen should simulate the size and orientation of the flaw, as illustrated in Fig. 1. If nothing is known about potential or existing weld flaws, the toughness should be evaluated using a notch orientation and specimen geometry which gives a lower-bound measure of toughness. In such cases, the notch should be located in the most brittle region and the crack tip should sample a maximum amount of this low toughness region. Often, a number of tests in different regions are required to locate the microstructure with the lowest toughness.

When test specimens are extracted from a weldment, machining should be kept to a minimum. That is, the test specimen thickness should be representative of the full thickness of the section under consideration, and therefore it is undesirable to remove too much material during machining. When there are slightly curved surfaces, such as those due to distortion or pipe radius, the surfaces should be left unmachined if possible because too much thickness may be

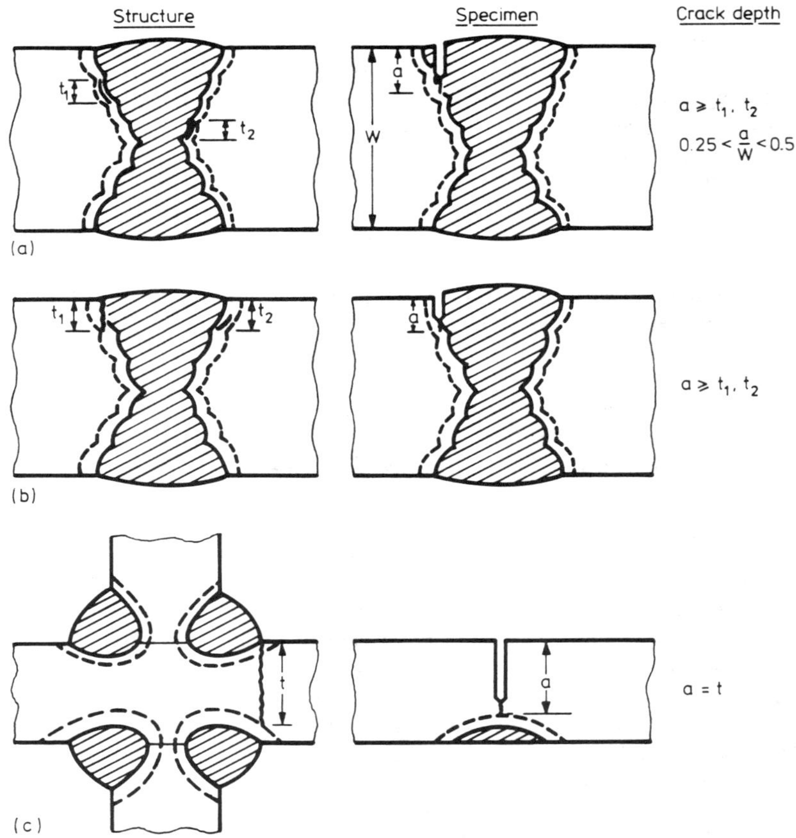

FIG. 1—*Surface notches in specimens to simulate variety of actual or postulated flaws* [8].

lost in machining these surfaces flat. However, severe geometric discontinuities should be removed. Weld overfill is usually removed to facilitate attaching clip-gage knife-edge abutments and applying local compression (see below).

Because of residual stresses, it is very difficult to achieve uniform fatigue cracks in weldment test specimens. Therefore, when testing an as-welded specimen, it is recommended that the specimen be mechanically stress-relieved prior to fatigue cracking. This is usually accomplished by local compression, where an indentation is made on the uncracked ligament such that 1% plastic strain is applied through the thickness. This is normally sufficient to produce acceptable fatigue crack fronts without adversely affecting the material's toughness [*6*].

If the region to be sampled by the fatigue crack is narrow (for example, the coarse-grained HAZ), it is necessary to verify the notch location after the test. First the fracture surface must be examined to identify the initiation site. The broken specimen is then sectioned at the initiation site. The plane perpendicular to the fracture surface is examined metallographically to determine the microstructure from which fracture initiated. This procedure is described in more detail in Refs *7* and *8*.

Interpretation of Test Results

British Standard Document PD 6493 gives little guidance on the derivation of the critical CTOD value used in the assessment. Often, final failure in a CTOD test is preceded by stable tearing. PD 6493 does not specify whether the critical CTOD should be defined by the instability point or the point where tearing initiates.

The method of CTOD test described in BS 5762 calls for the load applied to the SENB specimen to be plotted against the opening at the crack mouth as measured by a clip gage. Examples of load versus clip gage displacement records are given in Fig. 2. The standard identifies four possible measuring points on such records. Where a brittle crack initiates and leads to complete failure of the specimen (Fig. 2, Type 1) or to an arrested brittle crack (pop-in) (Fig. 2, Type 2), the CTOD is defined as δ_c. Brittle initiation of either type may be preceded by stable tearing. In this case, the CTOD at tear initiation is defined as δ_i and that at brittle initiation as δ_u. Finally, for ductile materials the load may pass through a smooth maximum (Fig. 2, Type 3). This is usually preceded by stable tearing and the CTOD at tear initiation is again defined as δ_i, while that at the start of the maximum load plateau is defined as δ_m.

It can be seen that every test will give a value of δ_c, δ_u, or δ_m, these being mutually exclusive, but that every value of δ_u and most values of δ_m would be accompanied by a lower CTOD, δ_i.

The behavior of the specimen after the initiation of tearing ceases to be a function of the material alone, but depends also on the geometry of the test specimen. Thus, in cases where tearing precedes instability, δ_i is the only parameter specific to the material. However, the structural significance of δ_i is not immediately apparent. Certainly, CTOD specimens often fail at values of δ_u or δ_m, which are considerably higher than the associated δ_i. Towers and Garwood [*9*] have thrown light on this aspect by an R-curve approach. They showed that the very highly constrained SENB specimen had both a shallower R-curve and a steeper driving force curve than less constrained but more structurally relevant geometries such as edge, center, or surface-notched tension panels. Thus, since instability is predicted when the driving force curve becomes tangential to the R-curve, this work indicated that δ_m values measured in SENB specimens would underestimate those for structural situations and that δ_u values measured from SENB specimens could be applied to structural situations because there was no danger of the structure failing by tearing instability at a CTOD value less than δ_u. Therefore, it is not necessary to use δ_i measurements in fracture assessments. This is fortunate, because widely used structural steels show quite low values of δ_i. Provisos suggested by Towers and Garwood were that the thickness of the specimen, B, should be greater than $50\delta_m$ (to ensure that adequate constraint was maintained) and that the specimens should comply with BS 5762.

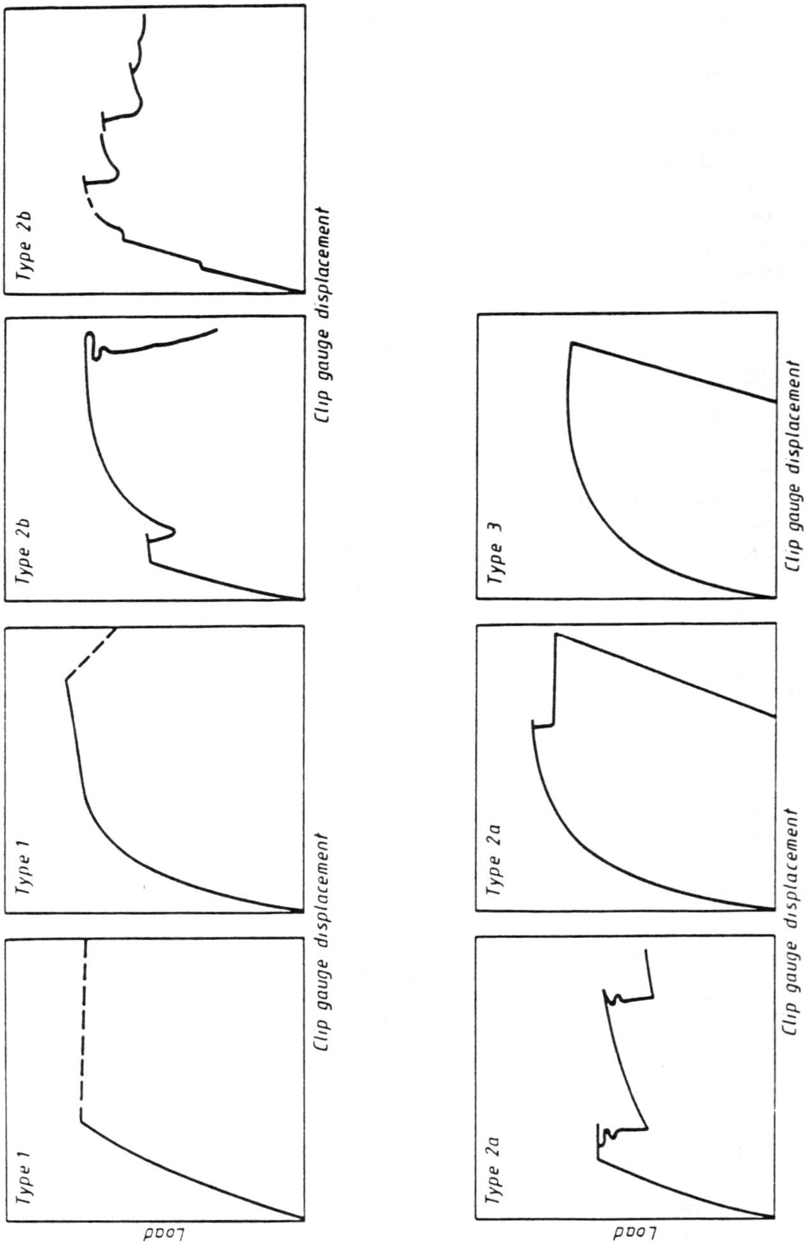

FIG. 2—*Examples of typical load versus clip gage displacement traces observed during CTOD testing.*

Taking account of Towers and Garwood's work, it is concluded that CTOD-based fracture assessment methods such as PD 6493 should recommend the use of δ_c, δ_u, or δ_m, whichever applies.

Pop-in

Traces of Type 2*b* (Fig. 2) indicate that short brittle cracks, or pop-ins, have initiated and arrested and that subsequently the specimen has taken a higher load. (Traces of Type 2*a*, where the load never again reaches that at initiation, are not considered to be significantly different from Type 1.) As in the case of ductile tear initiation, it is not immediately obvious that the initiation of pop-in should be treated as a critical event. Possible causes of pop-in were discussed by Kamath and Gittos [*10*].

The problem concerning its significance is to decide whether the arrest which occurred at the end of the pop-in would also occur in a structure. The arrest in the test may occur because the load drops due to increased compliance, and the concern is that the crack would continue to propagate if the load were maintained as it might be in a structure with much greater stored energy. However, the load drop can be felt by the crack only if a stress wave has propagated from the crack tip to the nearest load point and back to the crack tip. Willoughby and Wood [*11*] use this argument, together with an assumption that stress waves will propagate at speeds not more than 50 times faster than the crack velocity, to show that the brittle crack arrest will not be influenced by reflected stress waves if

$$\Delta a \leq \ell/25 \tag{1}$$

where Δa is the length of crack jump and ℓ is the ligament length.

Willoughby and Wood propose that the fracture surfaces of specimens which have displayed pop-in should be examined. If the crack jump can be identified and its extent is found to be less than $\ell/25$, the ability to arrest the short brittle crack can be assumed to be a property of the material rather than a function of the testing machine/specimen compliance and the pop-in can be ignored. Pop-ins associated with longer crack jumps must be treated as critical.

The study of the fracture surface may reveal other causes of pop-in, such as the linking of two fatigue cracks growing on separate planes, the straightening of a fatigue crack front with a sharply re-entrant angle, or the opening of splits in the rolling plane at right-angles to the fatigue crack. If any of these are identified, it is reasonable to ignore the event.

Scatter in CTOD Data

British Standard Document PD 6493 gives no advice on the statistical treatment of CTOD data which often display considerable scatter, particularly in the transition regime.

The CTOD design curve in PD 6493 is supported by a large number of correlations between CTOD data and the results of wide-plate tests on similar material. Usually there were three CTOD results for each wide-plate test. Kamath [*12*] showed that if the minimum CTOD was used in conjunction with the CTOD design curve, the wide-plate test result would be predicted with 97% confidence in survival. That is, the CTOD design curve was conservative 97% of the time when the minimum of three CTOD values was used.

Obviously, the more CTOD tests are performed, the lower the minimum result is likely to be. Thus if the minimum of, say, ten results is used with the CTOD design curve, the prediction may be unduly pessimistic. However, Williams [*13*] argues that if it is safe to use the minimum of three CTOD values in a fracture assessment, it is also safe to use a value from a larger data set that is equivalent to the minimum of three results. Williams has developed an approach which uses nonparametric statistics to determine this equivalent value.

The statistical approach can be explained as follows. The probability that a random data value will be greater than the true median value is 0.5. (The true median refers to the median value in an infinite population.) If three data values are selected randomly, the probability that the minimum of these three is greater than the true median is $0.5^3 = 0.125$. Thus, if the probability of a given value in a larger data set being greater than the median is also 0.125, then this value is statistically equivalent to the lowest of three random results. Williams used a binomial distribution to obtain the equivalent value

$$p(s) = \sum_{i=0}^{s} \frac{n!}{s!(n-s)!} 0.5^n \leq 0.125 \tag{2}$$

where $p(s)$ is the probability that s results are greater than the median and n is the total number in the sample.

Consider a data set with ten CTOD values. For $s = 2$ (corresponding to the third lowest value in the sample), probability $= 0.117$; for $s = 3$, probability $= 0.205$. Thus, if the third lowest of ten results is used in a fracture assessment, this would be at least as safe as using the minimum of three values.

Towers et al. [*14*] applied the preceding approach to data from a European Coal and Steel Community round-robin program in which over 100 nominally identical CTOD tests and 15 wide-plate tests were performed on the same steel. They showed that if a CTOD value was selected from the 100 results, which was statistically equivalent to the minimum of three results, the CTOD design curve would safely predict the wide plate results within the 97% confidence limit. This corresponds closely with Kamath's conclusions [*12*].

The binomial method is not restricted to determining the minimum of three values. For example, the equivalent minimum of six values can be obtained by setting the probability in Eq 2 equal to $0.5^6 = 0.016$.

The preceding discussion has referred to the confidence in predicting the results of a wide-plate test representing a structural component from the results of a series of CTOD tests. The confidence in predicting the performance of an actual structure is obviously a much wider subject, involving, inter alia, the confidence in the knowledge of the service loads; the accuracy of the assumed minimum working temperature and the interrelationships between temperature and design load (for example, in an offshore platform, what the likelihood is of the 100-year design wave coinciding with the minimum design temperature); the population of weld imperfections and the likelihood that a large imperfection will coincide with a region of high stress; and the probability of locating and the accuracy of sizing imperfections by NDE—all these in addition to material variability. Because designers often assume that each parameter will be at its most pessimistic at one and the same time and location in the structure, it is felt that the probability of the CTOD design curve giving a conservative prediction of structural safety is considerably greater than in the case of the prediction of wide-plate test results. Certainly, as far as the authors are aware, no structural failure has occurred which would not have been predicted by the application of the design curve.

Fracture Assessments with PD 6493

The preceding sections discussed advances in measuring and interpreting CTOD data. Subsequent sections describe new methods for applying CTOD results to fracture control in welded structures. However, before these new analyses are discussed, the existing PD 6493 approach should be mentioned. The PD 6493 fracture assessment method is summarized briefly below. PD 6493 also contains an assessment for fatigue in welded structures, but this is beyond the scope of the present article.

The CTOD design curve forms the basis of the fracture assessment in PD 6493. For ferritic steels and weldments in contained yield situations, where the structure is behaving in a predominantly elastic manner, the design curve can be stated in terms of nondimensional CTOD (Φ) and the pseudo-elastic peak stress, σ_1

$$\Phi = \frac{\delta_{crit}E}{2\pi\sigma_{YS}\bar{a}_m} = \left(\frac{\sigma_1}{\sigma_{YS}}\right)^2 \quad \text{for} \quad \sigma_1 \leq 0.5\sigma_{YS} \tag{3a}$$

$$\Phi = \frac{\sigma_1}{\sigma_{YS}} - 0.25 \quad \text{for} \quad 0.5\sigma_{YS} \tag{3b}$$

where

δ_{crit} = critical CTOD value,
E = Young's modulus,
$\bar{a}_m$ = allowable equivalent flaw size, equal to the half-length of a through-thickness flaw in an infinite plate in tension,
σ_{YS} = yield strength, and
σ_1 = peak stress, defined as

$$\sigma_1 = K_t(\sigma_m + \sigma_b) + \sigma_r$$

where

K_t = elastic stress concentration factor,
σ_m = membrane stress,
σ_b = bending stress, and
σ_r = residual stress.

Equation 3*a* is based on an LEFM analysis with a safety factor of 2; Eq 3*b* was obtained from an empirical correlation between CTOD data and wide-plate tests.

Applied and residual stresses and a critical CTOD value are inserted into Eq 3*a* or Eq 3*b* to obtain an allowable flaw size $\bar{a}_m$. This value can be then converted to allowable dimensions of surface or embedded flaws by means of Figs. 12 and 13 in PD 6493. The inverse of this procedure can be used to assess a known flaw.

The PD 6493 method for avoiding plastic collapse is a simplified and conservative procedure in which a surface flaw is recategorized as a through-wall flaw and an embedded flaw is recategorized as a surface flaw (or in some cases a through-wall flaw) when the material on the remaining ligament exceeds the flow stress. Willoughby [*15*] showed that this approach is satisfactory for long flaws, but is excessively conservative for short flaws because it ignores the support provided by the ligaments on either side of the flaw. He argued that it would be more reasonable to base the collapse check on the loss of cross section due to the flaw.

A Three-Tier Approach to Fracture Assessment

PD 6493 describes a simple and, in most cases, conservative method for specifying flaw acceptance levels. The PD 6493 approach uses the CTOD design curve which is backed by nearly 15 years of service experience. The method does, however, suffer from a number of limitations. In particular:

1. No allowance is made for stress variation through the thickness under post-yield conditions.
2. For stresses greater than $0.5\sigma_{YS}$, the CTOD design curve is empirical.
3. A variable safety factor is inherent in the approach, making probabilistic studies and critical analyses difficult.

4. The CTOD design curve was originally intended to be applied only to cases in which a flaw was small compared to the overall cross section of the structure.
5. Plastic collapse solutions are not an integral part of the analysis.

Various methods for improving the CTOD design curve have been proposed [*16–18*]. For example, the approach can be modified to incorporate stress gradients [*16,17*]. Also, Dawes [*18*] has recently proposed a finite-width correction to the design curve. Unfortunately, these modifications can make the design curve approach considerably more complicated. Even with these improvements, the CTOD design curve still suffers from the weakness of being primarily an empirical approach with a variable safety factor. More accurate and less empirical design equations are available but are generally more complicated than the CTOD design curve.

Anderson et al. [*19*] have proposed a three-tier approach in which flaws can be assessed at a level of complexity appropriate to the problem under consideration. The approach is sufficiently flexible to allow the use of different design equations, with various assumptions for stress distributions, within one overall framework. For example, the CTOD design curve can be used either in its present simple form or with various modifications [*16–18*]. Alternatively, more accurate design equations, corresponding to the second and third tiers, can be used.

The three levels of assessment are summarized next in increasing order of complexity and accuracy. The CTOD design curve occupies the first level. The proposed Level 1 method is similar to the present PD 6493 approach except that additional restrictions have been placed on the analysis. Also, the design equation is expressed in such a way as to allow stress gradients to be incorporated. Level 2 utilizes a strip yield model, similar to the Central Electricity Generating Board (CEGB) R6 assessment method [*20*]. The third and highest level utilizes a reference stress approach [*21*]. The CEGB have included such an approach in the proposals for the latest revision to the R6 method [*22*].

Effective Primary and Secondary Stress

Before the three tiers can be described, it is necessary to define two parameters which are used at each level: the effective primary and secondary stresses [*19*]. These parameters are a convenient way of expressing the stress distributions in the structure under consideration. These definitions allow for some flexibility in the way stress gradients are treated. The effective primary stress is given by

$$\sigma_p = \frac{K_I^p}{\sqrt{\pi a}} \tag{4}$$

where K_I^p is the primary stress intensity factor. The dimension a corresponds to the flaw size of interest, that is, the half-length of a through-thickness flaw, the depth of a surface flaw, or the height of an embedded flaw. σ_p has the units of stress but in general is not equal to the actual primary stress. (The exception is an infinite plate in uniform tension with a through-thickness flaw of length $2a$.) The effective primary stress is simply a convenient parameter that contains the stress intensity and flaw size. Similarly, the effective secondary stress is defined as

$$\sigma_s = \frac{K_I^s}{\sqrt{\pi a}} \tag{5}$$

Secondary stresses, such as thermal and residual stresses, are self-equilibrating in the structure. Primary stresses are the result of externally applied loads and bending moments. A number of different ways of deriving σ_p and σ_s have been proposed by Anderson et al. [*19*].

Level 1: The CTOD Design Curve

For the proposed Level 1 method the CTOD design curve has been reformulated in terms of effective primary and secondary stress

$$\frac{\delta_{\text{crit}} E}{2\pi\sigma_{\text{YS}} a_m} = \left(\frac{\sigma_p + \sigma_s}{\sigma_{\text{YS}}}\right)^2 \quad \text{for} \quad \sigma_p + \sigma_s \leq 0.5\sigma_{\text{YS}} \tag{6a}$$

$$\frac{\delta_{\text{crit}} E}{2\pi\sigma_{\text{YS}} a_m} = \frac{\sigma_p + \sigma_s}{\sigma_{\text{YS}}} - 0.25 \quad \text{for} \quad \sigma_p + \sigma_s \leq 0.5\sigma_{\text{YS}} \tag{6b}$$

The main difference between Eq 6 and the previous version of the CTOD design curve is that the equivalent flaw parameter $\bar{a}_m$ has been replaced by the actual flaw dimension a_m. The LEFM correction factors which were previously contained in $\bar{a}_m$ are now incorporated in σ_p and σ_s.

The advantage of formulating the CTOD design curve in terms of σ_p and σ_s is that stress gradient effects can be taken into account. Also, the form of Eq 6 is consistent with the Level 2 and Level 3 approaches.

Anderson et al. [*19*] have recommended provisionally that the net section stress be limited on 85% of yield. However, this is the subject of further study.

Level 2: Strip Yield Model

The strip yield model [*23–26*] has been used extensively to analyze notched panels. This model assumes plane stress, elastic-perfectly plastic material behavior.

CEGB uses a strip yield based failure assessment diagram in its R6 fracture analysis method [*20*]. The R6 diagram is expressed in terms of the stress intensity (K) ratio versus the stress ratio. Critical K values which are used in the analysis are computed from critical J-integral values measured experimentally. Thus, the R6 method is essentially a J-based fracture analysis. However, the strip yield failure assessment diagram can also be expressed in terms of CTOD [*27*], as illustrated in Fig. 3 (see Eqs 7*a* and 7*b*).

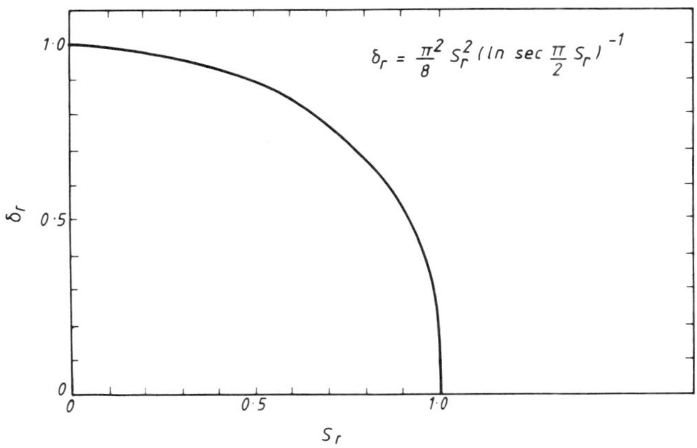

FIG. 3—*Failure assessment diagram expressed in terms of CTOD* [27].

The Level 2 assessment method uses a strip yield equation expressed in terms of effective primary and secondary stress as defined previously. This equation is [*19*]

$$\frac{\delta E}{\pi \sigma_{YS} a} = \left\{ \frac{\sigma_p}{\sigma_n} \left[\frac{8}{\pi^2} \ell n \sec \left(\frac{\pi}{2} \frac{\sigma_n}{\sigma_{YS}} \right) \right]^{1/2} + \frac{\sigma_s}{\sigma_{YS}} \right\}^2 \tag{7a}$$

where σ_n is the net section stress.

This equation can be normalized in terms of CTOD and stress as [*27*]

$$\delta_r = \frac{\pi^2}{8} S_r^2 \left(\ell n \sec \frac{\pi}{2} S_r \right)^{-1} \tag{7b}$$

where δ_r is the ratio of the applied CTOD to the critical CTOD and S_r is the ratio of the applied stress to the stress required to cause plastic collapse of the remaining ligament.

At low stresses, Eq 7 reduces to the LEFM solution, where stress intensities due to primary and secondary stresses are added to obtain the total stress intensity. At higher stresses, there is a plasticity correction on primary stress but not on secondary stress. Consequently, the relative contribution of the secondary stress term decreases with plastic flow. This is consistent with reality, since residual stresses can be relieved by plastic deformation.

The treatment of primary stresses in Eq 7 is essentially identical to the CEGB R6 method [*20*], but the treatment of secondary stresses in the two approaches is slightly different. In Eq 7, the secondary stress term is completely separate from the "ℓn sec" term which defines the collapse limit. This is believed to be a more logical formulation because secondary stresses cannot, by definition, contribute to collapse.

Fracture assessments with Eq 7 are somewhat more complex than the present PD 6493 procedure. It is not possible to solve for critical flaw size explicitly because it appears several places in the equation. In order to estimate critical flaw size from Eq 7, applied CTOD can be plotted for various flaw sizes and the size corresponding to δ_{crit} can then be inferred.

Level 3: The Reference Stress Method

The CTOD design curve (Level 1) is a simple and conservative approach for determining flaw acceptance levels (giving 97% confidence in providing a safe prediction of the behavior of a wide plate test using the minimum of three CTOD results). The strip yield approach (Level 2) provides a higher degree of accuracy and less conservatism. However, both methods rely on simple assumptions which are not valid at high stresses and strains. This is especially true for high work hardening materials. A proposed method for extending fracture analysis beyond net section yield is described below.

A reference stress approach to *J*-based fracture assessments was developed by Ainsworth [*21*]. This approach is based on the *J*-estimation equations developed by Kumar et al. [*28*]. However, unlike the Kumar equations, the Ainsworth approach used a material's stress-strain curve rather than a power law approximation of the flow behavior. Consequently, the reference stress method is more accurate than the Kumar equations when applied to materials whose flow behavior does not obey a simple power law.

Following Ainsworth's *J*-based approach [*19*], Anderson [*29*] has recently developed a reference stress method using CTOD. In Ref *19,* this equation was modified to incorporate secondary stresses and stress gradients and was adopted as the third assessment level. The Level 3 assessment equation is given by

$$\frac{\delta E}{\pi \sigma_{YS} a} = \left[\frac{\sigma_p}{\sigma_{YS}} \left(\frac{a_e}{a} + \frac{E \epsilon_{ref}}{\sigma_{ref}} - 1 \right)^{1/2} + \frac{\sigma_s}{\sigma_{YS}} \right]^2 \tag{8}$$

where

σ_{ref} = reference stress,
ϵ_{ref} = reference strain, and
a_e = effective flaw size, including a plastic zone correction.

The reference stress is equivalent to the effective net section stress defined previously. For a given flaw size and loading condition, σ_{ref} is computed and ϵ_{ref} (at σ_{ref}) is inferred from the material's stress-strain curve. The effective flaw size is given by [28]

$$a_e = a + \frac{1}{2\pi}\frac{(K_I^P)^2}{\sigma_{YS}^2}\frac{1}{1 + \sigma_{ref}/\sigma_{YS}} \tag{9}$$

Equation 8 has a similar form to Eq 7. That is, the equation reduces to the LEFM solution at low stresses and there is a plasticity correction on σ_p. However, the plasticity correction in Eq 8 is based on the actual material flow behavior rather than a strip yield model.

The Level 3 method outlined above is intended to be applied to cleavage fracture analyses, that is, cases where δ_c or δ_u values are measured in fracture toughness tests. If a δ_m value is used in conjunction with Eq 8, the prediction would generally be conservative. If an accurate ductile fracture assessment is required, the crack growth resistance-driving force interaction must be taken into account. A Level 3 approach to ductile fracture which considers this factor has been developed and will be described in another paper [30].

Comparison of Assessment Equations

The CTOD design curve, strip yield model and reference stress approach (Levels 1, 2, and 3, respectively) were compared with experimental data in Ref *19*. For the purpose of this comparison, it should be emphasized that the CTOD design curve is intended to provide a "safe" prediction and, at relatively low stress, it incorporates a factor of safety of ≈ 2. The strip yield and reference stress methods, on the other hand, are intended to predict critical conditions. A structure (that is, a wide-plate specimen) in uniform tension with no secondary stresses was considered for this comparison. This way, the three assessment equations could be compared, independent of considerations of secondary stresses and stress gradients.

The material used for the comparison was 113-mm-thick A533B steel plate which had a yield strength of 460 N/mm^2 and a tensile strength of 595 N/mm^2 at 70°C. The test specimen was 510 mm wide and contained a semi-elliptical surface crack 240 mm long and 38.7 mm deep.

Figure 4 is a plot of applied CTOD versus remote stress (σ_m). The reference stress equation agrees very well with experiment. The strip yield equation gives a reasonable approximation of applied conditions but the CTOD design curve bears little resemblance to experimental data. At high stresses the CTOD design curve appears to be nonconservative. Figure 5 is a plot of the above data in terms of safety factor on applied CTOD. This plot confirms that the CTOD design curve is conservative at low and medium stress levels but is nonconservative at high stress.

A second comparison was made between Level 1 and Level 3 methods for two different values of secondary stress. For this comparison a hypothetical A533B steel structure with a through-thickness flaw was considered. The width of the structure was assumed to be large compared to the flaw size. Equations 6 and 8 were used to compute flaw sizes assuming $\sigma_s = 0$, corresponding to the unwelded parent material, and $\sigma_s = \sigma_{YS}$, a conservative assumption for the residual stresses in an as-welded structure. The ratio of flaw sizes computed from Eqs 6 and 8 are plotted in Fig. 6. If it is assumed that the Level 3 equation is accurate, the ordinate of Fig. 6 can be considered the safety factor of the CTOD design curve.

For $\sigma_s = \sigma_{YS}$, the CTOD design curve appears to be nonconservative at applied stresses above $0.6\sigma_{YS}$, but experimental data [12] and service experience have not revealed a significant num-

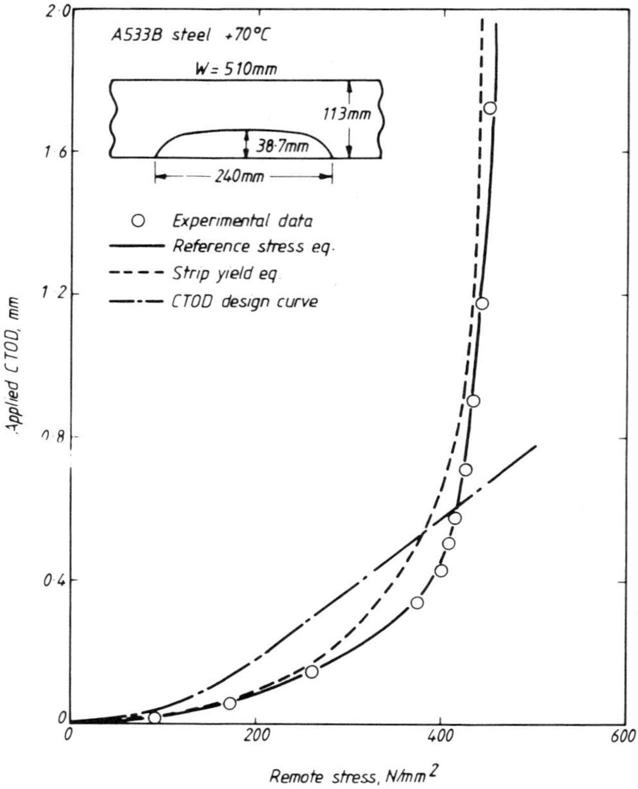

FIG. 4—*Predicted and experimental applied CTOD versus applied stress in A533B steel wide plate* [19].

ber of unsafe predictions at high stresses. There are a number of factors which tend to make the design curve conservative. For example, the minimum of three CTOD values from SENB specimens will tend to underestimate the toughness of a structure or wide-plate specimen [*24*]. Also, the assumption of yield magnitude residual stresses acting uniformly across the section of an as-welded structure is almost certainly conservative in most instances. However, since there is an increasing tendency to use more realistic input data for residual stresses, toughness and stress gradients, it would be unwise to apply the CTOD design curve to high stress situations where it is vulnerable to unsafe predictions.

Discussion

In order to obtain a realistic fracture assessment, two requirements must be satisfied. First, the input data must be accurate and relevant to the structure in question. Second, the fracture design equation, which relates critical flaw size to toughness and applied stress, must accurately describe structural behavior. The proposals in this paper address both of these aspects.

Various aspects of weldment testing were described. The most important (and difficult) task is to locate the fatigue crack tip in the relevant microstructure. Machining and fatigue cracking weldment specimens also create special problems.

It has been proposed that δ_c, δ_u, or δ_m values, but not δ_i, be applied to CTOD-based fracture

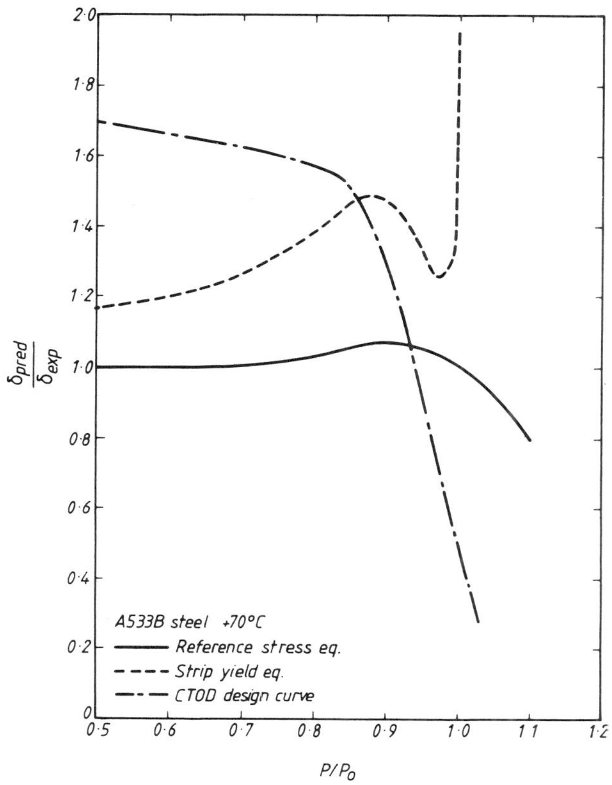

FIG. 5—*Safety factor on applied CTOD in A533B wide plate* [19].

assessments. The exception to this is the Level 3 approach, where only δ_c and δ_u values are used. A Level 3 ductile fracture assessment is described elsewhere [*30*].

A statistical treatment of scatter was described whereby the equivalent minimum of three results is estimated from a larger data set. However, this approach need not be confined to the equivalent minimum of three. Experience has shown that the minimum of three results provides a conservative estimate of toughness when using the CTOD design curve; but a different estimate may prove to be suitable for Level 2 and Level 3 methods. In addition, the treatment of scatter is not limited to the simple nonparametric statistics approach. Other probabilistic methods are currently being investigated at The Welding Institute.

A three-tier approach to fracture assessment has been proposed. The level of complexity which is most appropriate for a given application will depend on a number of factors. The factors which will affect the choice of analysis are summarized in Table 1.

The Level 1 approach is a simple and conservative method for specifying flaw acceptance levels. It is backed by 15 years of service experience. However, it has been suggested that this approach should not be applied at net section stresses above $0.85\sigma_{YS}$. The strip yield model (Level 2) is a relatively simple method that is also reasonably accurate below net section yield. The reference stress method (Level 3) allows fracture analyses to be performed above net section yield. It is valid at all stresses, but its complexity is not warranted at low and medium stresses because it will give results similar to the Level 2 approach at these stress levels.

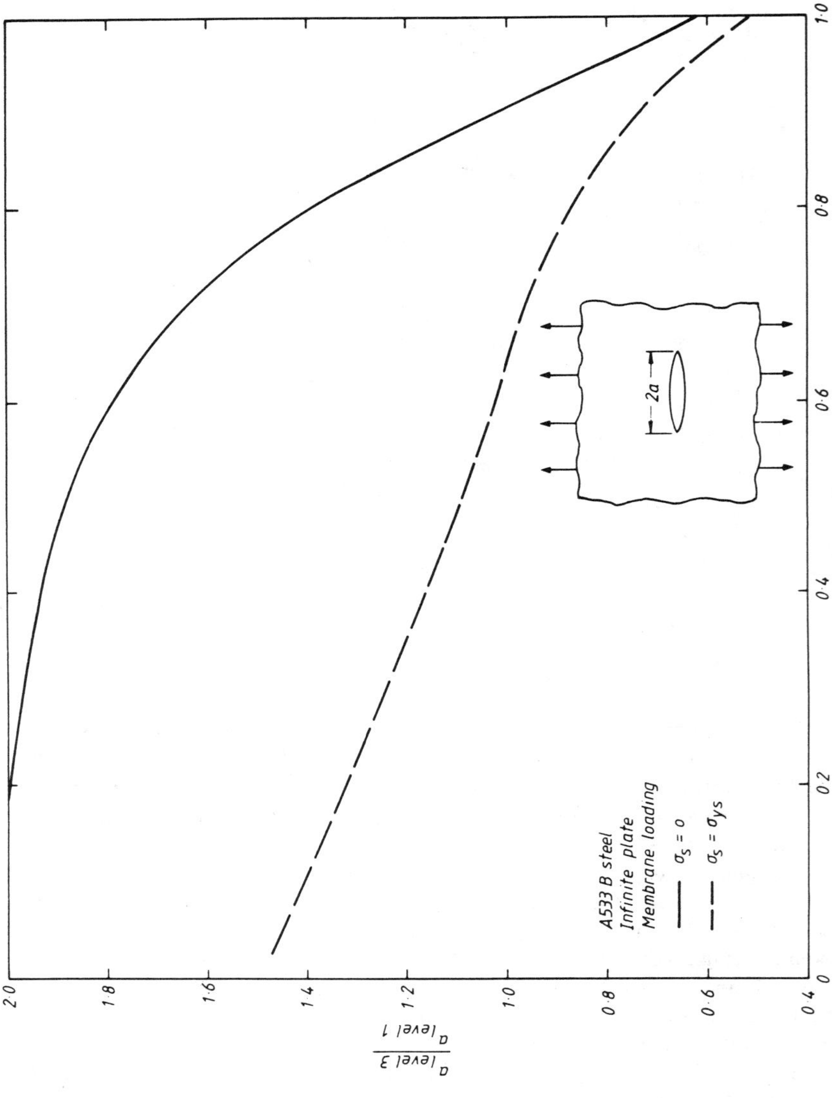

FIG. 6—*Comparison of flaw sizes computed from Eqs 6 and 8 for two assumed secondary stress values.*

TABLE 1—*Summary of the three assessment levels.*

Method	Application	Advantages	Disadvantages
Level 1	low and medium work hardening materials	simple method	empirical approach
		backed by considerable experience	variable safety factor
	$\sigma_n \leq 0.85\sigma_{YS}$		not valid at high stresses
	conservative analysis		
Level 2	low and medium work hardening materials	relatively simple method	invalid above net section yield
	$\sigma_n < \sigma_{YS}$	more accurate than Level 1	inaccurate for high work hardening materials
	critical analysis		
Level 3	medium and high work hardening materials	more accurate than Levels 1 and 2	more complex than Levels 1 and 2
	$\sigma_n > 0.85\sigma_{YS}$[a]	can be applied beyond net section yield	more input data are required
	critical analysis		accuracy of analysis is limited by accuracy of the input data

[a]Although the Level 3 approach is valid at all stresses, its complexity is not warranted at lower stress levels.

Therefore, if a simple analysis with an in-built safety factor is required, Level 1 is suitable provided $\sigma_n \leq 0.85\sigma_{YS}$. Level 2 can be used if a critical analysis is required and primary stresses are below net section yield. Above net section yield, Level 3 must be used.

The way in which stress gradients and secondary stresses are treated has a significant effect on the overall accuracy of a fracture assessment [*19*]. In fact, in some cases these considerations dominate the analysis, and the design equation is of secondary importance.

Conclusions

1. The quality of fracture toughness input to flaw assessment methods can be improved by

(*a*) Measuring weldment toughness properly.
(*b*) Applying δ_c, δ_u, or δ_m (but not δ_i) values to structural assessments.
(*c*) Disregarding minor pop-ins that would be arrested in the structure.
(*d*) Using probabilistic approaches to analyze data scatter.

2. The Level 1 fracture assessment is a simple and conservative method that can be applied at net section stresses less than $0.85\sigma_{YS}$.
3. The Level 2 assessment is a critical analysis below net section yield.
4. If a fracture assessment is to be performed above net section yield, Level 3 should be used.

References

[*1*] Newman, R. P., "The Influence of Weld Faults on Fatigue Strength with Reference to Butt Joints in Pipelines," *Transactions*, Institute of Marine Engineers, Vol. 68, 1956, p. 153.
[*2*] Wells, A. A., "Unstable Crack Propagation in Metals: Cleavage and Fast Fracture" in *Proceedings*, Crack Propagation Symposium, Vol. 1, Cranfield, U.K., 1961.
[*3*] Burdekin, F. M. and Dawes, M. G., "Practical Use of Linear Elastic and Yielding Fracture Mechanics with Particular Reference to Pressure Vessels" in *Proceedings*, Institution of Mechanical Engineers Conference, London, May 1971, p. 28.
[*4*] Harrison, J. D., Dawes, M. G., Archer, G. L., and Kamath, M. S., "The COD Approach and its Application to Welded Structures" in *Elastic-Plastic Fracture, ASTM STP 668*, American Society for Testing and Materials, Philadelphia, 1979, pp. 606–631.

[5] Squirrel, S. J., Pisarski, H. G., and Dawes, M. G., "K_{Ic}, CTOD and J Tests on Weldments," presented at the 3rd ASTM International Symposium on Non-Linear Fracture Mechanics, Knoxville, TN, Oct. 1986.

[6] Towers, O. L. and Dawes, M. G., "Welding Institute Research on the Fatigue Pre-Cracking of Fracture Toughness Specimens" in *Elastic-Plastic Fracture Test Methods: The User's Experience, ASTM STP 856,* American Society for Testing and Materials, Philadelphia, 1985, p. 23.

[7] Pisarski, H. G. and Pargeter, R. J., "Fracture Toughness of HAZs in Steels for Offshore Platforms," *Metal Construction,* Vol. 16, 1984, p. 412.

[8] Pisarski, H. G., Towers, O. L., and Willoughby, A. A., "Guidelines for Fracture Toughness Testing of Heat Affected Zones," *The Welding Institute Research Bulletin,* Vol. 25, 1984, p. 188.

[9] Towers, O. L. and Garwood, S. J., "The Significance of Maximum Load Toughness," *The Welding Institute Research Bulletin,* Vol. 20, 1979, p. 292.

[10] Kamath, M. S. and Gittos, M. F., "The Incidence of Pop-ins in Fracture Toughness Testing," *The Welding Institute Research Bulletin,* Vol. 20, 1979, p. 114.

[11] Willoughby, A. A. and Wood, A. M., "Crack Initiation and Arrest Toughness in 9% Nickel Steel Weldments at Liquefied Natural Gas Temperatures" in *Proceedings,* International Symposium on Storage and Transport of LPG and LNG, Brugges, Belgium, 1984.

[12] Kamath, M. S., "The CTOD Design Curve: An Assessment of Validity Using Wide Plate Tests," Welding Institute Reprint 71/1978/E, Sept. 1979.

[13] Williams, S. and Jutla, T., "Probabilistic Fracture Mechanics—Towards a Practical Approach; Scatter in Fracture Toughness Data," *The Welding Institute Research Bulletin,* Vol. 27, 1986, pp. 20–26.

[14] Towers, O. L., Williams, S., Taylor, A. C., and Harrison, J. D., "European Coal and Steel Community Collaborative Program on Elastic-Plastic Fracture Toughness Testing and Assessment Methods," Welding Institute Report 3571/10M/84 (to be published).

[15] Willoughby, A. A., "Summary of Plastic Collapse Solutions Used in the Failure of Part-wall Flaws," Welding Institute Report 191/1982, Sept. 1982.

[16] Kamath, M. S., "The Crack Tip Opening Displacement (CTOD) Design Curve: Some Proposals for Incorporating Stress Gradient Effects" in *Proceedings,* Conference on Fitness for Purpose Validation of Welded Constructions, The Welding Institute, Paper 23, 1982.

[17] Harrison, J. D., "Fracture Mechanics Developments Related to Weld Defect Acceptance Methods Given in British Standard PD 6493" in *Fracture 84,* 2nd National Conference on Fracture, University of Witswatersrand, Johannesburg, South Africa, Nov. 1984.

[18] Dawes, M. G., "The CTOD Design Curve Approach: Limitations and Finite Size and Application," Welding Institute Report 278/1985, July 1985.

[19] Anderson, T. L., Leggatt, R. H., and Garwood, S. J., "The Use of CTOD Methods in Fitness for Purpose Analysis" in *Proceedings,* Workshop on CTOD Methodology, GKSS, Geesthacht, West Germany, April 1985.

[20] Harrison, R. P., Loosemore, K., Milne, I., and Dowling, A. R., "Assessment of the Integrity of Structures Containing Defects," Central Electricity Generating Board Report R/H/R6-Rev.2, Leatherhead, Surrey, U.K., 1980.

[21] Ainsworth, R. A., "The Assessment of Defects in Structures of Strain Hardening Material," *Engineering Fracture Mechanics,* Vol. 19, 1984, p. 633.

[22] Milne, I., Dowling, A. R., Ainsworth, R. A., and Steward, A. T., "Assessment of the Integrity of Structures Containing Defects—Addendum for Structures Made with a High Capacity for Work Hardening," Central Electricity Generating Board Report R/H/R6-Rev.2, Addendum 1, Leatherhead, Surrey, U.K., Dec. 1984.

[23] Dugdale, D. S., "Yielding in Steel Sheets Containing Slits," *Journal of the Mechanics and Physics of Solids,* Vol. 8, 1960, p. 100.

[24] Burdekin, F. M. and Stone, D. E. W., "The Crack Opening Displacement Approach to Fracture Mechanics in Yielding Materials," *Journal of Strain Analysis,* Vol. 1, No. 2, 1966, p. 144.

[25] Bilby, B. A., Cottrell, A. H., and Swinden, K. H., "The Spread of Plastic Yield from a Notch" in *Proceedings,* Royal Society, Vol. A272, 1963, p. 304.

[26] Heald, P. T., Spink, G. M., and Worthington, P. J., "Post Yield Fracture Mechanics," *Materials Science and Engineering,* Vol. 10, 1972, p. 129.

[27] Garwood, S. J., this publication, pp. 957–985.

[28] Kumar, V., German, M. D., and Shih, C. F., "An Engineering Approach for Elastic Plastic Fracture Analysis," EPRI Report NP 1931, Electric Power Research Institute, Palo Alto, CA, July 1981.

[29] Anderson, T. L., "Elastic-Plastic Fracture Assessments Based on CTOD," Welding Institute Report 276/1985 (to be published).

[30] Anderson, T. L., Gordon, J. R., and Garwood, S. J., "On the Application of R-curves and Maximum Load Toughness to Structures," submitted for presentation at the 19th ASTM Symposium on Fracture Mechanics, 30 June to 2 July 1986, San Antonio, TX.

Rudi M. Denys[1]

The Significance of CTOD in Transversely Loaded Weldments with Weld Metal Overmatching in Strength

REFERENCE: Denys, R. M., **"The Significance of CTOD in Transversely Loaded Weldments with Weld Metal Overmatching in Strength,"** *Fracture Mechanics: Eighteenth Symposium, ASTM STP 945,* D. T. Read and R. P. Reed, Eds., American Society for Testing and Materials, Philadelphia, 1988, pp. 485–501.

ABSTRACT: The purpose of this paper is to review the effects of weld metal overmatching on weld joint behavior. The notch toughness parameters crack-tip opening displacement (CTOD) and crack opening displacement (COD) are used to evaluate the degree of weld metal overmatching on weld joint performance. Exploratory experimental work on wide-plate specimens loaded in tension is presented to support the theoretical observations.

It is concluded that weld metal overmatching has a beneficial effect on the overall performance of transversely loaded welded joints loaded in tension. The results also indicate that the CTOD bend test gives mainly an idea about the local notch tip ductility and that, if a "fitness for purpose" evaluation is to be based upon the small-scale CTOD test, the benefits which accrue from the use of overmatching weld metal on tolerable defect size cannot be described adequately by the design curve approach.

KEY WORDS: overmatching, weld metal, crack-tip opening displacement (CTOD), gross section yielding, CTOD design curve, weld joint

In the assessment of defects entirely located in the weld metal, little consideration is given to the effect of weld metal matching on weld joint performance. In the case of transversely loaded welded joints, a positive difference in yield between the weld metal and the parent plate material will protect the weld metal from being strained plastically. In view of the fact that the fracture behavior in such welded joints will be controlled by the fracture initiation characteristics of the defective weld metal, it is believed that overmatching weld metal will need less notch ductility than commonly required to ascertain the imposed level of overall ductility.

The benefits of using overmatching weld metal are not explicitly formulated in the current defect assessment approaches [see British Standards Guidance on Some Methods for the Derivation of Acceptance Levels for Defects in Fusion Welded Joints (PD 6493: 1980)]. However, the effect of weld metal overmatching can be accounted for in combination with the requirement of gross section yielding [*1*].

The objective of this paper is to extend the knowledge of the influence of the degree of weld metal overmatching on the overall welded joint performance and to determine to what extent weld metal overmatching has a positive effect on the overall ductility.

[1]Lecturer and manager, Fracture Mechanics Division, Laboratory Soete for Strength of Materials and Welding Technology, State University of Gent, Sint Pietersnieuwstraat, 41, 9000 Gent, Belgium.

Extent of Plasticity in Defective Welded Joints

The different ways in which a center-notched welded panel, loaded in tension, may fail are shown schematically in Fig. 1, which puts the various deformation regimes of fracture behavior of welded joints into proper perspective.

In the *weld metal* with a notch, four deformation regimes may be observed, which can be described as (1)(5) elastic, (2)(6) elastic-plastic, (3)(7) net section yielding, and (4)(8) gross section yielding [*1,2*]. For obvious reasons, the deformation mode in the base material is either elastic [Cases (a), (b), (c), and (d)] or plastic [Cases (e), (f), (g), and (h)].

In a defective *welded joint,* the following four deformation modes at fracture can be distinguished: (a) elastic behavior; (b) small scale, contained yield or elastic-plastic behavior; (c)(d) net section yielding; and (e)(f)(g)(h) gross section yielding. Clearly the occurrence of one or the other deformation mode depends on the defect dimensions, the strain hardening properties of the weld metal, and the yield strength properties of the parent plate and the weld metal. The weld joint performance will correspondingly depend on the fracture properties of the weld metal and the relative yield strength of the weld metal and the surrounding parent plate material [*1*].

Weld Metal Behavior in Tension

To establish the interaction between weld metal notch toughness and weld joint performance, the relationship between the crack opening displacement (COD) and the applied gross strain of the weld metal alone should be considered first. [Note: The applied stress or strain will cause a crack opening displacement, and the term COD is used to indicate that quantity as measured at the plate surface of a through-thickness notched plate loaded in tension. This COD is not fully compatible with the commonly encountered definition of crack-tip opening displacement (CTOD), but previous work has shown that the COD is related to CTOD [*3*], so that the use of the term COD will not alter the implications of the general findings which are reported hereinafter.]

The accentuated gross stress-gross strain (or σ-e) and the crack opening displacement-gross strain (or COD-e) curves shown in Fig. 2 model the sequence of events of a weld metal containing a through-thickness defect loaded in tension. The interaction between these curves is as follows [*1,3,4*].

1. Initial loading produces elastic specimen extension, and the relationship between COD (or CTOD) and e can quanitatively be evaluated from the Burdekin-Stone equation [*2*], which was developed for an elastic-plastic material with no work hardening (Fig. 2–OA). A reasonable approximation for the COD is given by

$$\text{COD} = \frac{\pi \sigma^2 a}{E \sigma_{\text{Yw}}} \quad \text{(mm)} \tag{1}$$

where

σ_{Yw} = yield strength of the weld metal,
a = half crack length, and
σ = nominal or gross stress (while the gross strain is defined as $e = \sigma/E$).

It should be noted that in Eq 1, the yield strength relates the yield properties of the material controlling the extent of the plastic zone. Furthermore, the relation in Eq 1 is accurate for only small crack length to plate width ratios and gross stress to yield strength ratios of less than 0.4. For higher stress ratios, a plastic zone ahead of the crack tip will develop, and because of this, Eq 1 provides a too-low estimate of the COD value. However, to facilitate the further assessments and to avoid a complex equation which takes account of a plastic zone adjustment [*5*],

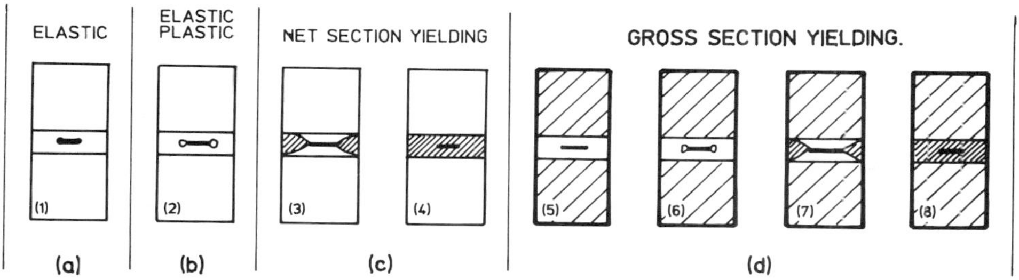

FIG. 1—*Schematic representation of possible deformation modes in defective welded joints. Defect location, weld metal. Hatched areas indicate plastically deformed material.*

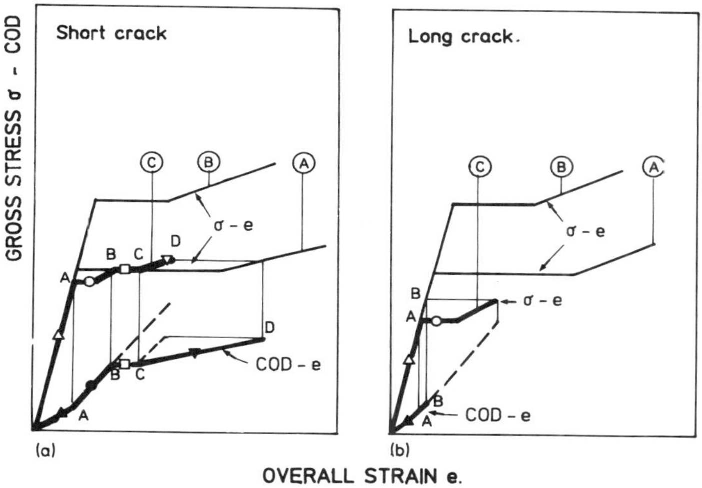

FIG. 2—*Qualitative representation of relationship between gross stress, overall (gross) strain, and crack opening displacement for* (a) *gross section yielding and* (b) *net section yielding.*

the accuracy of Eq 1 is assumed to apply for higher gross stress to yield strength ratios. Also, since this expression has been developed for non-work-hardening material and since the work hardening of the weld metal at the defect tip will contain the deformation more at the crack tip, it is to be noted that real work hardening materials will show a lower COD for gross stresses close to the weld metals yield strength. The effect of work hardening on COD can be accounted for by replacing the yield stress through a flow stress which has a value about halfway between the yield stress and the ultimate tensile strength. Bearing in mind the opposite effects caused by the occurrence of plastic zone at the defect tips and the work hardening on the COD, it is believed that Eq 1 provides an acceptable estimate of the COD for higher stresses: for a through-thickness defect of 10 mm and for a gross stress of 300 and 400 N/mm^2, respectively ($\sigma_{Yw} = 500$ N/mm^2), we find that the amount of COD is 0.027 mm and 0.049 mm, respectively. The conclusion here is that for gross stresses below the weld metal's yield stress the amount of COD will be small.

2. As the stress increases, the plastic zone size ahead of the crack tip enlarges until net section yielding occurs. After the onset of net section yielding, the applied gross strain is entirely

transmitted to the defect tip and a linearity of the COD-versus-*e* relationship becomes noticeable during the Lüders straining phase (Fig. 2–AB). For long cracks, this relationship remains unaltered until fracture occurs (Fig. 2*b*). For short cracks (Fig. 2*a*), a similar behavior is exhibited at the lower gross strain range (Fig. 2–A-AB), but at higher gross strains (Fig. 2–A-BCD), the shape of the COD-versus-*e* curve will be influenced by the strain-hardening characteristics of the weld metal.

3. When strain hardening increases the yield or flow strength in the notch tip area, plastic strains may occur in the gross section; that is, gross section yielding will develop. At the onset of gross section yielding, the remote displacement or the gross strain *e* is absorbed by the defect-free section of the weld metal and COD remains constant during the Lüders straining phase (Fig. 2*a*–BC). As soon as the gross stress is beyond the yield strength of the weld metal, the gross strain is distributed along the entire length of the weld and the COD increases further slowly with *e* (Fig. 2*a*–CD). The COD is no longer proportional to the gross strain *e*, and the slope of this curve depends on the strain hardening properties and the defect size. The bilinear shape of the COD-*e* curve for gross section yielding is shown in Fig. 2*a*. Prior to final fracture, the slope of the COD-versus-*e* curve can gradually increase to that of the first portion of the linear relationship.

The previous observations will not alter when a surface-breaking defect is subjected to tensile loading. However, an even more pronounced effect of defect size on the COD-*e* relationship will result. For a surface defect of the same length, the elastic COD as well as the position of the linear (or bilinear) COD-*e* curve will be lowered by the reduced constraint to plastic flow exhibited by this geometry as well as by the smaller instant defect size. As a consequence, smaller levels of developed COD for comparable applied strain levels will result.

At this stage, it should be pointed out that the assessment of the deformation behavior in the case where the defect is located in the heat-affected zone (HAZ) will be more complicated. There is, however, no reason to disagree with the observation that the same principles will apply.

Weld Joint Behavior in Tension

Having defined the COD-*e* relationship for weld metal only, it is now possible to evaluate the significance of overmatching weld metal yield strength on welded joint behavior, having the knowledge of the degree of weld metal overmatching and the relevant acting gross stress or gross strain level.

A welded joint consists of two different materials with their own stress-strain characteristics (Fig. 3*a*). The stress-strain curve of overmatching weld metal (Curve B) lies above the stress-strain curve of the surrounding parent metal (Curve A). In weld metal containing a defect, yielding will occur at an earlier stage, so that the position of the corresponding stress-strain curve will differ from Curve B. Its position, Curve C, in relation to the weld metal or parent material stress-strain curve, will depend on the defect dimension. For the purpose of the present discussion, Curve C is situated between Curves A and B (Fig. 3*a*).

The actual stress-strain curve of the defective welded joint will lie very close to or coincide with that of the plate material stress-strain Curve A (Fig. 3*b*). As such, only the stress-strain Curves A and C (Fig. 3*b*) are of interest to assess the welded joint performance in the presence of a weld metal crack, defect, or notch.

When analyzing the shape of the COD-*e* curves in Fig. 4, it should be borne in mind that, although the gross stress acting on each cross section of the welded joint is the same, the gross strains in each of these sections may be different. This difference will arise particularly when the applied gross stress reaches the yield stress of a particular region (plate, HAZ, and weld metal) in the welded joint. The use of some typical yield properties would be helpful to illustrate the effects of weld metal overmatching on the relationship between the COD and *e*. The simplified

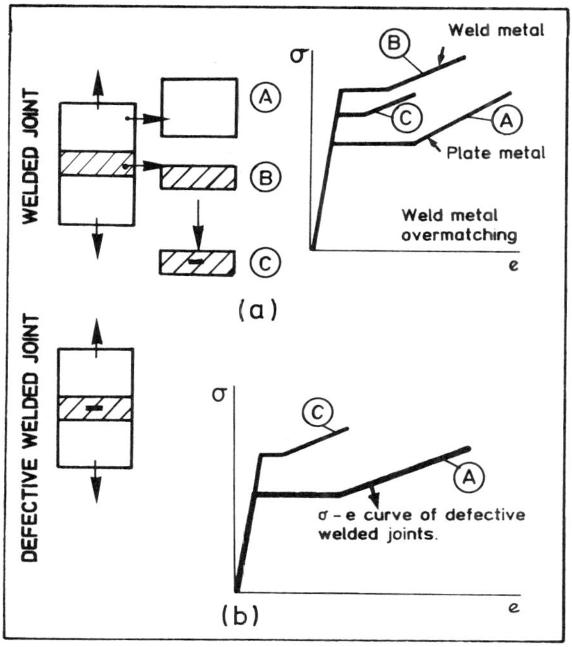

FIG. 3—*Stress-strain curves of* (a) *a welded joint and* (b) *a defective welded joint. Hatched areas represent weld metal.*

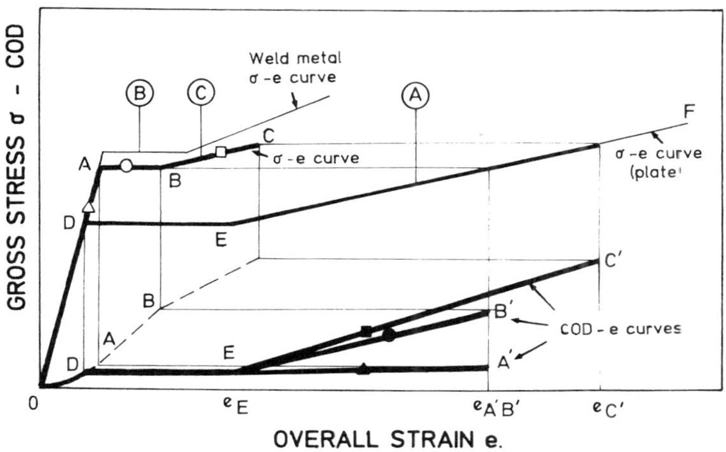

FIG. 4—*Schematic representation of interaction between weld metal stress-strain and the COD gross strain curves.*

stress-strain curves in Fig. 4 model a plate material with $\sigma_{Ypl} = 300$ N/mm^2 (Curve A) and a weld metal with $\sigma_{Yw} = 500$ N/mm^2 (Curve B).

The shape of the COD-versus-*e* curve and thus the amount of COD per unit applied strain can now be evaluated as follows. In the attempt to assess the amount of developed COD, assume first that the crack length is 10 mm and that the yield strength associated with Curve C is 450 N/mm^2 (Fig. 2*a*).

1. For an initial loading up to the yield strength of the parent material ($\sigma = \sigma_{Ypl} = 300$ N/mm^2), the amount of COD (Point D), Eq 1, is 0.027 mm.
2. With further straining, the gross stress-gross strain curve of the whole welded joint (here represented by Curve A) will be controlled by the plate material. As long as the gross strain remains within the Lüders plateau of the parent material (AE), the COD will not increase during the Lüders straining phase and the COD-versus-*e* curve is represented by a horizontal line DE. The amount of gross strain at that very point, e_E, can be as much as 1%, whereas the amount of COD is still 0.027 mm.
3. The subsequent increase in gross stress will provoke a further elastic straining of the weld metal (portion DA on Curve C) and a strain hardening of the parent plate material (portion EF on Curve A). At Point A, the amount of COD is equal to 0.062 mm, and the position of the COD-*e* is represented by the line EA′. By assuming that fracture occurs at Point A (that is, elastic fracture of the weld metal) the gross strain will be as much as $e_{A'}$.
4. If the weld metal toughness permits more COD to develop than the elastic COD, the corresponding COD-versus-*e* relationship will be similar to that of Figs. 2*a* or 2*b* and the weld metal, while the parent material is further plastically strained. The COD-versus-*e* relationship of the welded joint can take one of the following positions:

1. Fracture occurs at B (Curve C—Fig. 4). The COD-overall strain curve is depicted by Curve DEB′. The overall strain at fracture will be slightly larger than at Point A′. In fact the gross strain $e_{B'}$ equals $e_{A'}$ plus the relative increment of COD between AB (=COD_{AB} divided by the gage length).
2. Fracture occurs at C (Curve C—Fig. 4). The position of the COD-gross strain curve is represented by DEC′, and the gross strain is given by $e_{C'}$.

With changing position of Curve C or when the weld metal contains a larger defect which lowers the position of the weld metal stress-strain Curve C (this situation also models reduced overmatching or undermatching if the original length of the defect of Fig. 4 remains unaltered), and by following the foregoing reasoning, it can be verified that the overall welded joint performance can be very poor.

Practical Significance of the Relationship Between COD and Gross Strain

The practical interest in developing the COD-versus-*e* relationship is to be related to the difference between the crack opening displacement as seen in both the bend (CTOD) and the tension (COD) mode of loading [*1,6*]. At low stresses, or for stress levels well below net section yield, and because of the reduced constraint associated with the tension loading mode, the value of the COD is larger than the value of the CTOD. This difference increases sharply with increasing toughness because the effect of work hardening can fully occur in the tension mode of deformation (Fig. 5). In other words, for a (bend) CTOD notch toughness value of 0.25 mm it is not unrealistic to assume that, when the defect is located in overmatched weld metal, fracture in a transversely loaded welded joint will occur at an overall strain level of 2%, 3%, 4%. . . . Since that amount of strain is not necessary it seems reasonable to lower the commonly specified weld metal CTOD requirement and to promote the use of overmatched weld metal.

The amount of weld metal yield strength overmatching is apparently an important factor with regard to the minimum CTOD required to protect the weld metal from being plastically

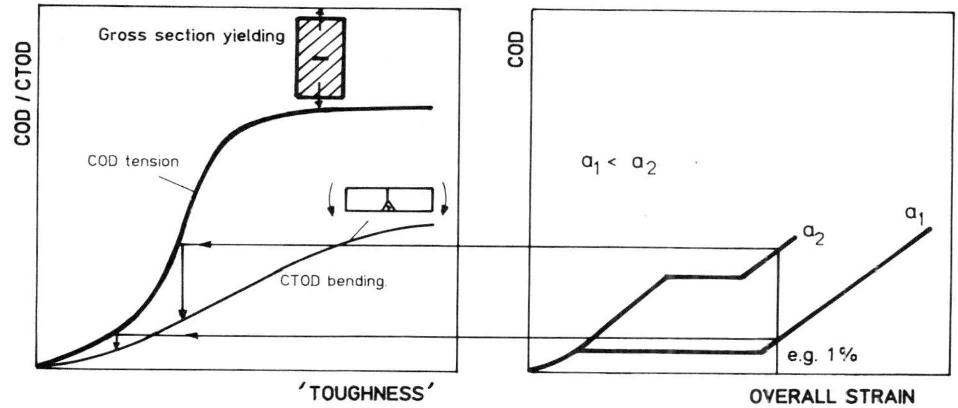

FIG. 5—*Qualitative representation of relationship between COD and CTOD as a function of defect size and notch toughness.*

strained. For example, if the weld metal yield strength of the numerical example given above is lowered from 500 N/mm^2 to 350 N/mm^2, that is, if the amount of overmatching is changed from 67% down to 17%, it would be necessary to increase the required COD from 0.027 mm to 0.039 mm to cause gross section yielding. Accordingly, a larger CTOD value will be required to obtain adequate weld joint performance. Also, an increase of the defect size will provoke a similar effect on the required COD or CTOD value as a reduction of the degree of weld metal overmatch (see also Eq 1).

Verification of Theoretical Observations

To illustrate the theoretical relationship between COD and the gross strain measured in a wide-plate test, a series of flat defective welded joint tension tests has been performed. These tests assessed the effect of weld metal matching on COD, CTOD, and gross strain (or overall wide-plate test performance).

Materials and Weldments

The chemical composition of the base metals and weld deposits is given in Table 1. Two steels (L and M) and two weld consumables (H and S) were used to make four different welded joint combination (Table 2). The welded joints were made by the electrogas welding process. The electrogas welds were made between square cut edges, using a gap of 20 mm. The plate thickness was 22 mm, and no stress relieving was applied after welding.

It can be seen in Table 2 that the base metals had a low (L) and medium (M) yield strength and that the yield strength of one weld deposit was below that of the soft parent plate material (S), whereas the yield strength of the second weld deposit had a much higher value (H). The Charpy V-notch impact energy levels and the CTOD (2B $\times$ B) notch toughness properties, measured at $-30°C$, are given in Table 3.

Specimen Preparation and Testing

The wide-plate tension tests were extracted from 1000-mm-long welded joints. The cross section of the machined specimens, that is, with the weld overfill removed, was 120 by 20/21 mm^2.

TABLE 1—*Chemical composition of plate material and weld deposit.*

Material	C	Mn	Si	P	S	Cu	Ni	Cr	Al	Sn	Mo	Nb
Steel L	0.100	0.56	0.178	0.006	0.028	0.240	0.165	0.075	0.030	0.024	0.012	...
Steel M	0.190	1.25	0.435	0.021	0.022	0.125	0.130	0.105	0.020	0.013	0.024	0.040
Weld L	0.080	0.30	0.120	0.014	0.020	0.040	0.020	0.050	...	...	...	...
Weld H	0.060	0.62	0.050	0.014	0.022	0.080	0.020	0.040	...	...	...	...

TABLE 2—*Tensile properties of plate material and weld deposit at −30°C.*[a]

	Plate		Weld Deposit	
Weldment	σ_{Yp}, N/mm^2	σ_{UTSp}, N/mm^2	σ_{Yw}, N/mm^2	σ_{UTSw}, N/mm^2
LSL	320(1)	462(2)	411(5)	464(6)
LHL	320(1)	462(2)	529(3)	661(4)
MSM	464(7)	630(8)	411(5)	464(6)
MHM	464(7)	630(8)	529(3)	661(4)

[a]The superscript numbers in parentheses were introduced for reference purposes in Figs. 6, 7, and 8.

TABLE 3—*Charpy V-notch and CTOD properties of plate material and weld deposit at −30°C.*

Material	Charpy V[a], J	CTOD[b], mm
Plate L	150	...
Plate M	110	...
Weld S	12	0.030
		0.081
Weld H	37	0.042
		0.065

[a]Average of three specimens: notch orientation T-L; sampling position quarter thickness.
[b]Only two 2B × B specimens were examined; notch orientation T-L.

All specimens were through-thickness notched at the weld metal center. The notches were made up of a drilled hole extended by a 0.10 to 0.15-mm-wide sawcut.

All wide-plate specimens were transversely loaded and the tests were carried out at −30°C. The overall elongation was recorded on a gage length of 200 mm. The local deformation at the very notch tip was monitored by a clip gage, which recorded the relative displacement (COD) of two steel pins straddling the tip of the defect. The distance between these pins was 7 mm. It should be pointed out that this simple COD measuring technique enables a distinction to be made between an overall deformation pattern wherein plastic strains are confined to the net section, or a deformation pattern wherein the plastic deformations are spread out over the entire test specimen [*1*]. During testing, the applied load and the COD were continuously recorded against the overall elongation.

Test Results

The results of all wide-plate tests are graphically depicted in Figs. 6, 7, and 8. Included in these figures are the gross stress at fracture σ, the overall strain at fracture e, and the COD at fracture against the length of the defect for two mutually comparable testing series. The following comparisons in pairs were made: the effect of overmatching in low-strength base metal L (Fig. 6), the effect of undermatching in medium-strength base metal M (Fig. 7), and the effect of increasing overmatching in using the same weld metal H (Fig. 8).

Finally, in Figs. 9 and 10 some typical records of COD versus e are grouped to illustrate the influence of defect length on the COD-e relationship in the LHL weld joints (Fig. 9) and to assess the significance of the degree of weld metal overmatching on the shape of the COD-e curves when the defect length remained unaltered (Fig. 10).

Discussion

The discussion of results focussed on the effect of weld metal matching on tolerable defect size, the amount of gross strain and the COD at fracture, and the shape of the COD versus gross strain records.

Weld Metal Matching and Tolerable Defect Sizes—The effect of weld metal overmatching on tolerable defect size has been assessed with the PD 6493 and gross section yielding methods. These defect assessment methodologies are briefly described in the Appendix.

Calculation of the (CTOD based) tolerable defect length, as described in the Appendix, requires the tensile properties of the weld metal and the parent plate material at the test temperature. First, the yield strength of the material surrounding the notch tip (weld metal) is required for the CTOD design curve. Second, the flow stress of the same region is required for calculation of the conditions giving plastic collapse. Third, the yield strength of the parent material is needed as an input to the gross stress, since this is the condition for gross section yield. Accordingly, Table 2 presents a summary of the required tensile properties of the various weld combinations tested.

With regard to the PD 6493 defect assessment method, the tolerable defect size at gross yield was estimated by assuming an applied stress equal to the yield strength and an applied stress twice the yield strength of the parent material. The latter input value is representative of the situation in which the design stress represents the onset of gross section yielding, whereby a residual stress allowance of yield magnitude is taken into account (as welded). A direct assessment of the wide-plate test results using the gross section stress at fracture or the strain at fracture has not been carried out. Instead, the allowable strain was calculated for the allowable defect length a_{gy}, as obtained from the gross section yielding defect assessment method.

Table 4 gives the results of the CTOD design curve and the plastic collapse calculations. For the former, the residual stress was assumed to be equal to zero because of the reduced plate width. The lower of the two values resulting from the fracture or the plastic collapse (PD 6493) assessment has been underlined. The maximum allowable gross section yielding defect lengths, a_{gy}, which are deduced from the point of intersection of the plate material's yield strength and the average curve fit to the experimental data points of Figs. 6, 7, and 8 as well as the corresponding allowable gross strains (calculated via the CTOD design curve), are also summarized in Table 4 (right-hand side).

A direct comparison between the two methods is not easy. However, bearing in mind that the respective input strains e_Y and $2e_Y$ are much smaller than the actual fracture strains of all combinations tested, it may be seen from Table 4 that the lengths of the allowable gross section yielding defects ($2a_{gy}$) are much larger than the design curve-based tolerable defect lengths. The allowable strains put further emphasis on this observation. Moreover, the effect of weld metal overmatching or the benefit of this on the value of $2a_{gy}$ is much more apparent when these values are compared with the corresponding tolerable defect lengths (design curve).

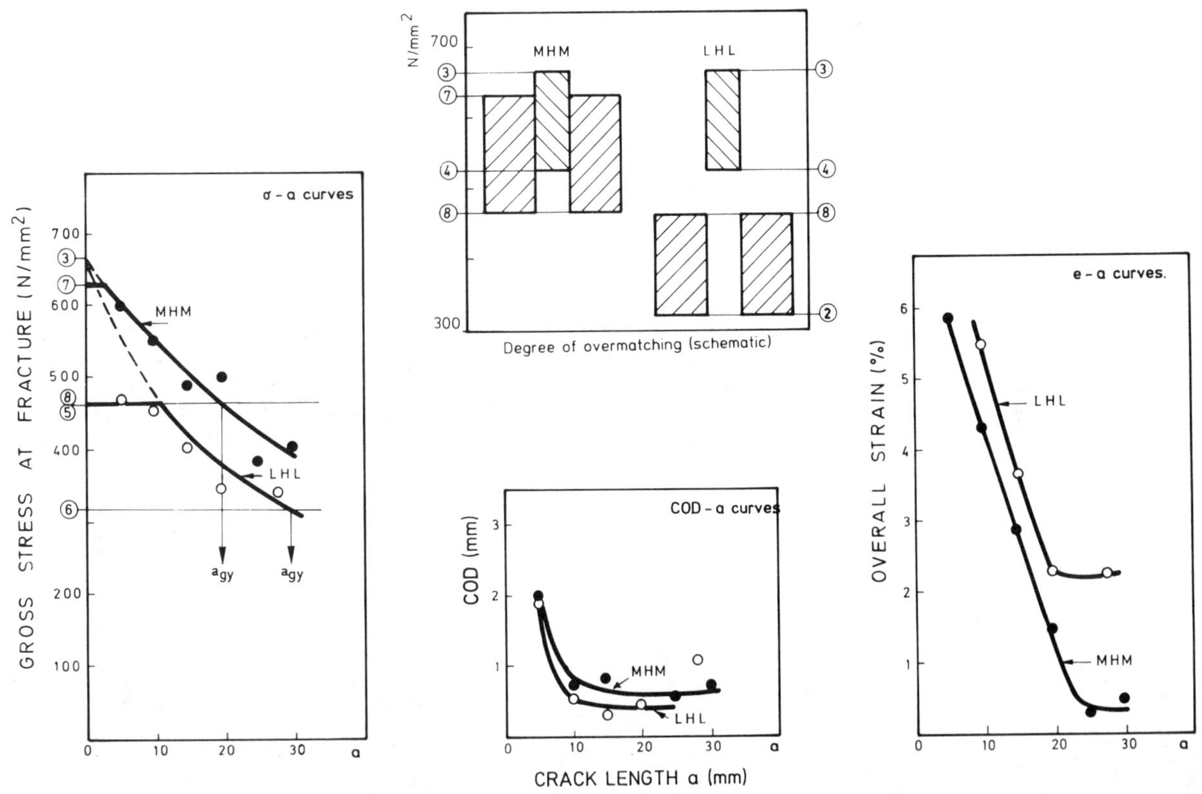

FIG. 6—*Effect of weld metal overmatching in low-strength base metal on gross stress, gross strain, and COD versus notch length (see also Table 4).*

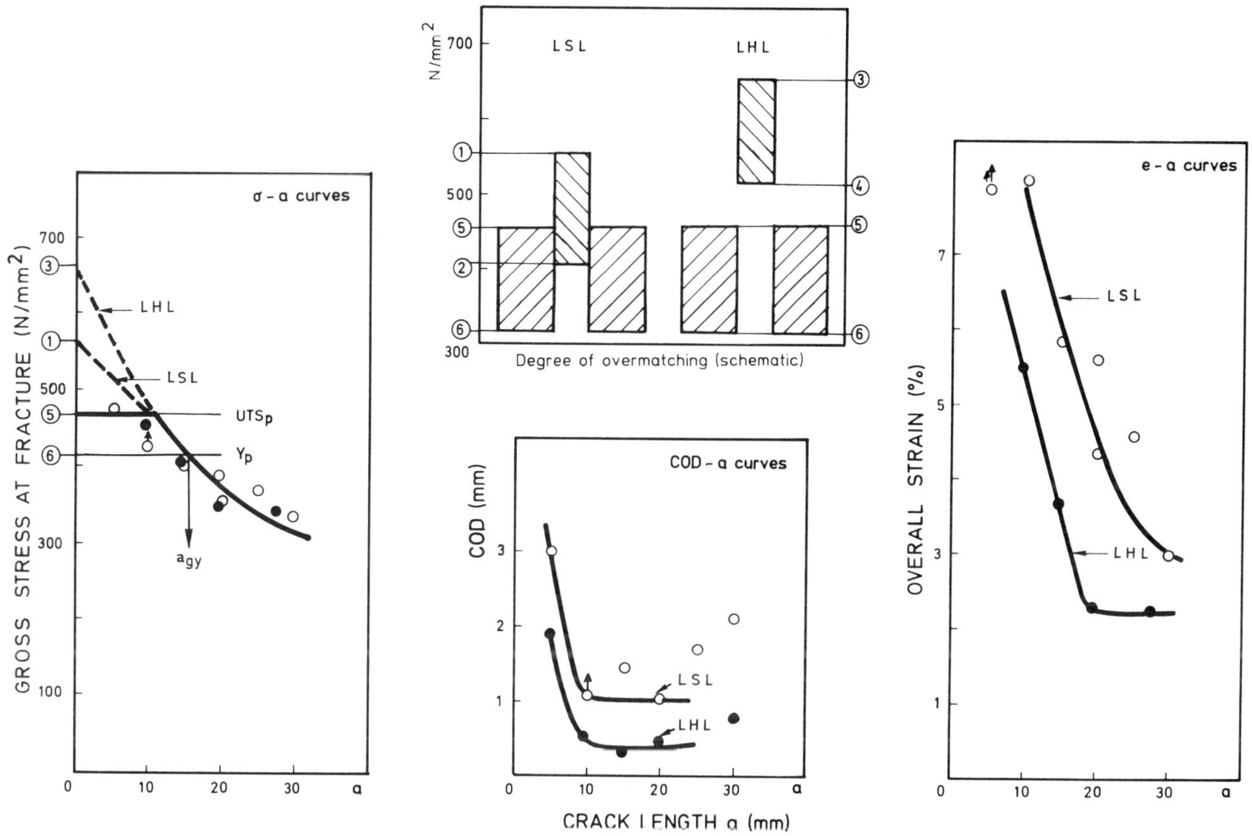

FIG. 7—*Effect of weld metal undermatching in medium-strength base metal on gross stress, gross strain, and COD versus notch length (see also Table 4).*

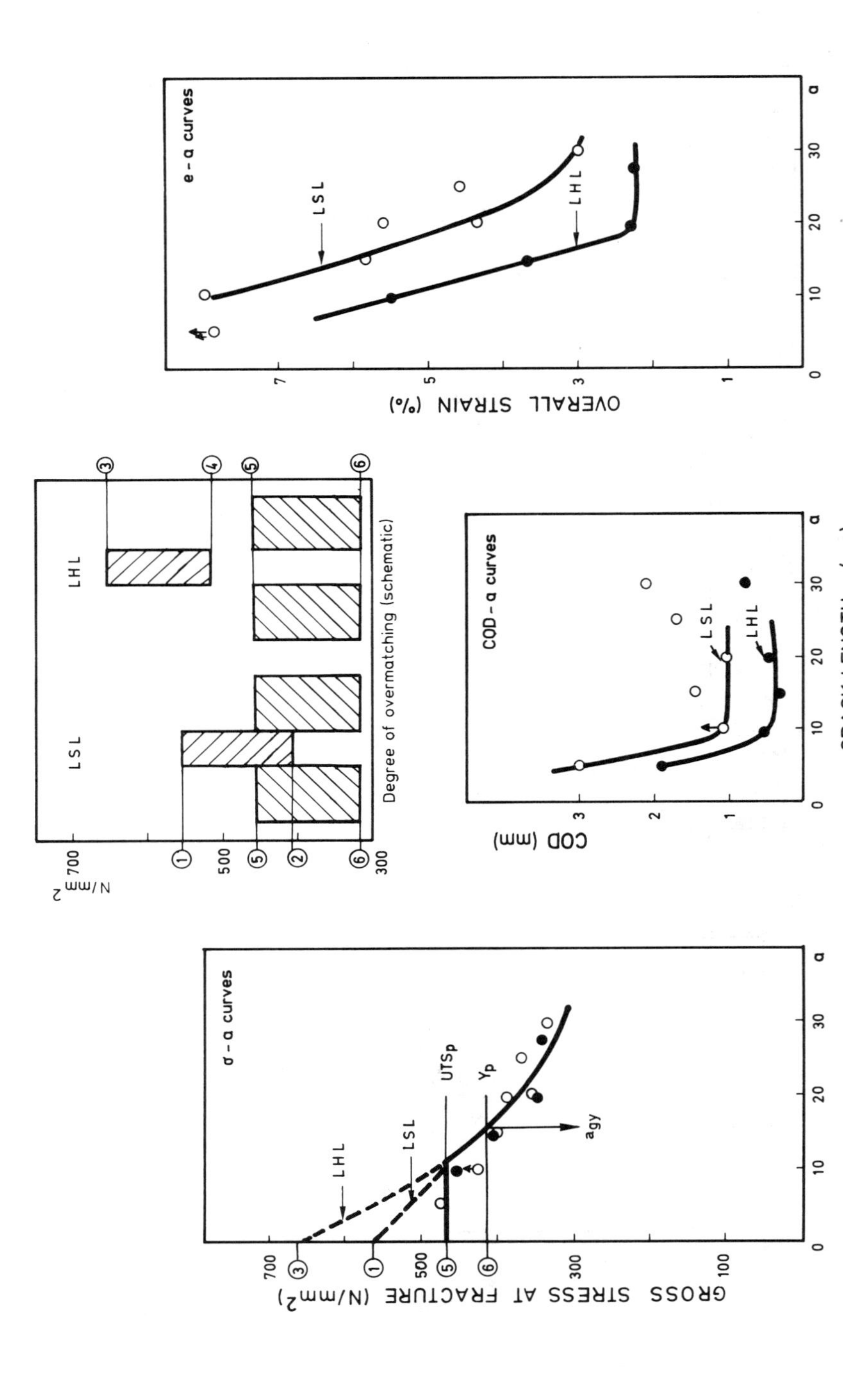

FIG. 8—*Effect of increasing weld metal overmatching on gross stress, gross strain, and COD versus notch length (see also Table 4).*

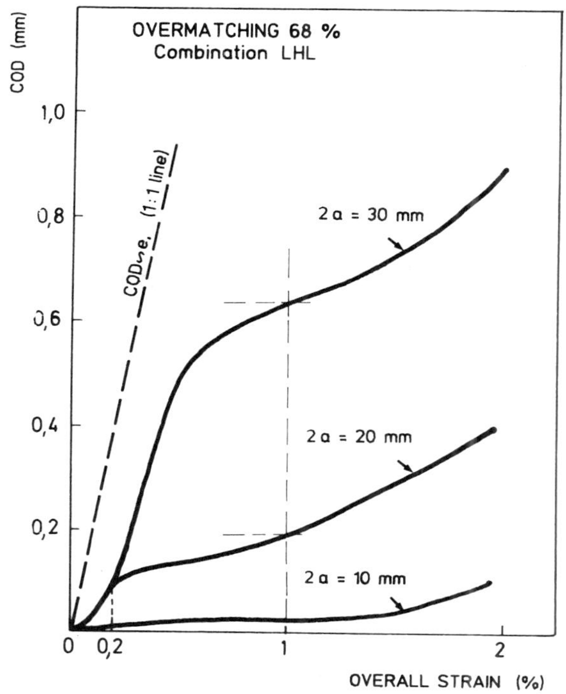

FIG. 9—*Experimental COD-*e *plots in highly overmatched welded joints.*

A further investigation of the $2a_{gy}$ values reveals that the amount of overmatching in low-strength parent metal seems to be less important since the magnitude of $2a_{gy}$ remains unaltered. This behavior is apparently affected by the stress-strain characteristics, that is, the extent of the Lüders plateau and the strain hardening ratio of the plate material.

Gross Strain and COD Data—A steep drop in the gross strain and the COD at fracture (Figs. 6, 7, and 8) is observed at the characteristic defect lengths $2a_{gy}$. This phenomenon is due to the occurrence of plastic straining of the plate material and to the constraint relaxation at the tips of the defect which gives rise to gross section yielding, while in the case of long defects, the plate material is only elastically strained and the amount of COD is also reduced because of the higher degree of constraint exhibited by a long defect.

Shape of COD versus Gross Strain Records—It can be seen in Figs. 9 and 10 that the actual COD-*e* relationship depends on two variables, that is, the length of the defect and the ratio between the weld metal and plate material yield strength. In the case of severe overmatching (LHL)—65%—Fig. 9, gross section yielding was obtained at an early stage, which is illustrated by the nearly horizontal position of the COD-*e* curve for $2a = 10$ mm. Consequently, the applied strain was concentrated mainly in the parent plate material, whereas the weld metal at the tip of the defect did not deform much for a gross strain level which was well above the parent plate materials yield point.

There is no doubt that the slope of the COD-*e* curve is affected by the length of the defect. With increasing length (Fig. 9; $2a = 30$ mm), the benefits of weld metal overmatching gradually decrease, which is emphasized by the linear 1:1 relationship between COD and *e* at the onset of gross section yielding; that is, the applied displacements at the specimen ends are transmitted fully to the defect tips through the slipbands. A further illustration of the interaction

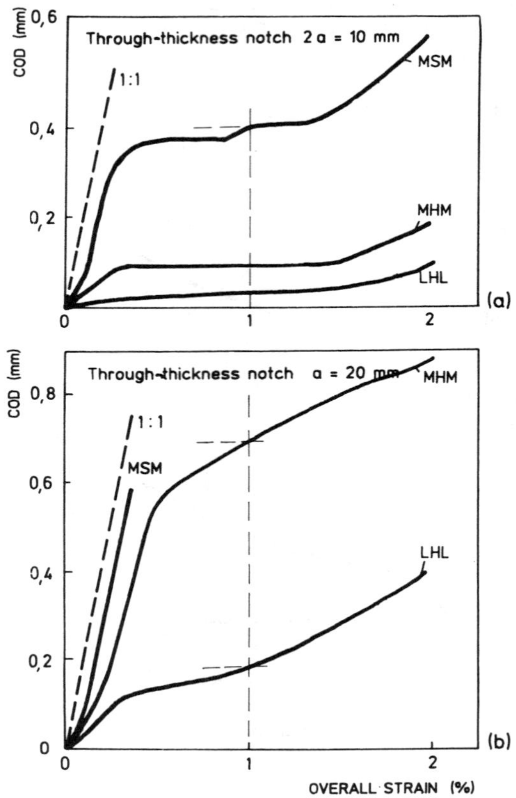

FIG. 10—*Effect of weld metal matching on the COD-*e *relationship for* (a) 2a = *10 mm and* (b) 2a = *20 mm.*

TABLE 4—*Calculated and measured values of tolerable (CTOD design curve) and allowable (gross section yielding) defect lengths.*

	Design Curve		Plastic Collapse	Gross Section Yielding	
	$e_1 = e_{Yp}$	$e_1 = 2e_{Yp}$			
Weldment, %	$2\bar{a}$,[b] mm	$2\bar{a}$, mm	$2\bar{a}_{coll}$,[c] mm	$2a_{gy}$,[d] mm	e_{Yp},[e] %
LSL (+28%)[a]	10	4	27	31	0.52
LHL (+65%)	14	6	46	31	0.48
MSM (−14%)	6	2	0	11	0.88
MHM (+14%)	8	4	22	20	0.58

[a]Degree of weld metal overmatching.
[b]Tolerable defect length (design curve).
[c]Maximum defect length for plastic collapse.
[d]Allowable defect length (gross section yielding).
[e]Allowable strain for zogy (design curve).

between the length of the defect and the degree of weld metal matching is given in Fig. 10. The increase of COD occurs much faster with the decreasing overmatching. A comparison between Figs. 10*a* and 10*b* provides evidence that with increasing notch length, the effects of weld metal overmatching appear to be of second-order interest in the post-gross section yielding phase. The practical engineering significance of this observation can be evaluated from Table 5.

The COD values in Table 5 were obtained by entering the COD-*e* records at 1% gross strain. The required COD values to obtain 1% overall strain vary considerably. These data make it possible to draw some important conclusions. The amount of required COD is undoubtedly smaller with increasing overmatching, and the ratio between weld metal and plate material yield strength seems to be a key factor which affects the amount of COD. Irrespective of the length of the defect, a similar influence of the degree of overmatching on COD was observed. Moreover, the data of Table 5 indicate that the COD and thus CTOD cannot be considered as a material parameter. For instance, it can be seen that for a 10-mm-long defect in weld metal H, the measured values of the COD at a gross strain of 1% were 0.10 mm (Plate M) and 0.03 mm (Plate L), respectively. At this stage, it is of interest to recall the CTOD bend test data of Table 3. The difference between the COD at fracture and the CTOD is quite large. Notch acuity may be one reason among others, but it does not explain these large differences.

Concluding Remarks

As a result of this study, which has been principally concerned with the assessment of the significance of weld metal overmatching in strength on the overall performance of transversely loaded welded joints, the following conclusions can be offered:

1. This study has indicated that it is virtually impossible to obtain realistic information on the overall weld joint performance by testing the plate material, the HAZ, and the weld metal separately. The overall welded joint performance as revealed by wide-plate tension tests indicates that it is the joint action of the parent plate and the weld metal which determines the required local and overall ductility.

2. A comparison of the experimental large-scale tension and small bend test results has indicated that the results of wide-plate test may give a more realistic estimate of the required notch toughness of particular regions in a welded joint. On the basis of wide-plate tension tests, one could accept lower CTOD values which certainly depend on the degree of weld metal matching and defect geometry. Moreover, the critical crack-tip opening displacement may not be considered as a material parameter; in the case of defects located in the weld metal, the amount of crack-tip opening depends on defect size as well as on the plate material properties.

3. The benefits of using overmatching weld metal on overall weld joint performance cannot be described adequately by the design curve fracture analysis. The allowance for the effect of the different yield strengths in a welded joint is given by the factor $e_1 - 0.25e_{Yw}$ (Appendix). For

TABLE 5—*Effect of weld metal matching on the required COD at 1% overall strain for various defect lengths.*

		Required COD at $e = 1\%$, mm Defect Length $2a$, mm				
Weldment	Overmatching, %	10	15	20	25	30
LSL	+28	0.10	0.14	0.19	0.46	0.97
LHL	+65	0.03	...	0.16	...	0.65
MSM	−14	0.40	fract	fract	fract	fract
MHM	14	0.10	0.36	0.70	...	fract

most applications and particularly for as-welded joints, the term 0.25 e_{Yw} as compared to the term e_1 is rather insensitive for changing degrees of weld metal matching.

4. Because the original wide-plate test database used to justify the design curve does not provide wide-plate test results which address directly the effect of weld metal matching, and because for most structures weld metal can be selected which has a higher yield strength than that of the parent plate, it is proposed that serious consideration be given to the development of a small-scale test configuration in which the effect of the differing stress-strain characteristics of the weld metal and parent plate material should be incorporated. In that respect, small-scale notched tension tests could help provide the highly required documentation.

5. In order to establish the significance of weld metal matching on structural performance, it is necessary to have information on (1) the specimen geometry used to assess the relevant notch toughness, (2) the yield strength properties in which the defect lies predominantly, (3) the strain hardening characteristics (or the ratio $V = \sigma_Y/\sigma_{UTS}$) of the plate material, and (4) the size as well as the end shape of the defect.

Acknowledgment

The author wishes to thank A. Vinckier, professor and director of the Laboratory Soete for Strength of Materials and Welding Technology, Gent University, Belgium, for permission to publish this study. The financial support of Nationaal Fonds voor Wetenschappelijk Ondezzoek and Industrieel en Wetenschappelijk Ondezzoek voor Landbouw en Nijverheid is also acknowledged.

APPENDIX

Defect Assessment Routes

PD 6493 CTOD Design Curve Assessment

The PD 6493 CTOD design curve fracture assessment method utilizes CTOD full-thickness data to assess the risk of fracture. The assessment is aimed at determining safe allowable defect sizes rather than critical values. A separate analysis must be carried out to assess the risk of plastic collapse of uncracked ligaments. For the present investigation, the fracture assessment was carried out using the equation

$$\bar{a} = \frac{\delta_c}{2\pi\,(e_1 - 0.25e_y} \qquad \text{for } e_1 > 0.5\,e_y \tag{2}$$

where

δ_c = critical value of CTOD,
a = half-length of an allowable through-thickness defect,
e_Y = yield strain of the defective area, and
e_1 = effective applied strain.

This calculation takes no account of the possibility of plastic collapse of the ligaments surrounding the defect. If this should occur, the fracture assessment based on the CTOD design curve alone is potentially unconservative. To prevent this from occurring, PD 6493 requires that the net section stress on the uncracked ligament be limited to the flow stress (defined as the average of yield and tensile strengths) of the material surrounding the defect. To the procedures of PD 6493, the equation for a through-thickness defect in a tension loaded structure reads

$$\frac{\sigma}{\sigma_f} \leqq \frac{W - 2a}{W} \quad (3)$$

where $2a$ is the total defect length. The maximum tolerable defect parameter $2a$ is then the lesser of the two values calculated by means of the CTOD design curve and the plastic collapse analysis.

Gross Section Yielding

This method is conceptually different from the CTOD design curve analysis in its approach for fracture-safe design. Allowable or critical defect sizes are not calculated on the basis of the maximum applied stress but on the use of the criterion that the cracked structure should be able to withstand yield magnitude stresses in the uncracked regions; that is, gross yielding must occur prior to failures.

The method does not employ a design equation. Instead, a series of specimens containing defects of different sizes is tested. The gross section stress at fracture, σ, for each test is plotted against the length of the defect, and a smooth curve is drawn through these experimental points and the point at $\sigma = \sigma_{UTSW}$ for zero crack length. The allowable gross section yielding defect length a_{gy}, which is defined as the largest defect for which fracture occurs by gross yielding, is obtained from the experimental curve where $\sigma = \sigma_{Yp}$. For cross welded joints it is thus in principle necessary only to exceed the yield stress of the parent plate material to achieve gross section yielding. Alternatively, the gross strain measured over a considerable gage length may be used to estimate a_{gy} [*1*].

Thus, a defect smaller than a_{gy} induces gross section yielding guaranteeing an adequate gross strain and a strength of at least yield point magnitude. Consequently, it can be construed that defects smaller than a_{gy} are *acceptable.* At this stage, three points must be emphasized: (1) the actual material properties are assessed directly; (2) the principles of gross section yielding are straightforward and do not require a subsequent analysis after ensuring that the wide-plate test results obtained meet the gross section criterion; and (3) the requirement of gross section yielding to occur in a wide-plate test specimen is utilized only to define allowable defect dimensions. On the other hand, when the gross section yielding requirement is met, it provides an additional safety margin in case of an accidental overloading.

References

[*1*] Denys, R. M., "The Wide Plate Test and its Application to Acceptable Defect" in *Proceedings,* Welding Institute Conference on Fracture Toughness Testing and Materials, Interpretation and Application, London, June 1982.

[*2*] Burdekin, F. M. and Stone, D. E., "The Crack Opening Displacement Approach to Fracture Mechanics in Yielding," *Journal of Strain Analysis,* Vol. 1, No. 2, 1966, pp. 145–153.

[*3*] Reed, R. P., Kasen, M. B., McHenry, H. I., Fortunko, C. M., and Read, D. T., "Fitness for Service Criteria for Pipeline Girth Weld Quality," Welding Research Bulletin (WRC) No. 296, July 1984.

[*4*] Randall, P. N. and Merkle, J. G., "Gross Strain Crack Tolerance of Steels," *Nuclear Engineering and Design,* No. 17, 1979, pp. 46–63.

[*5*] de Wit, R., "A Review of Generalized Failure Criteria Based on the Plastic Yield Strip Model" in *Fracture Mechanics (Fourteenth Conference), ASTM STP 791,* American Society for Testing and Materials, Philadelphia, 1982, pp. I24–I50.

[*6*] Satoh, K. and Toyoda, M., "Applicability of Bending COD Test of Welds for Securing Required Fracture Performance of Welded Structure," Document 1063-1984, International Institute of Welding, Annual Assembly, Boston, MA, 1984.

Ductile/Brittle Transition

Jon Lereim[1]

Effect of Crack and Plate Geometry on the Fracture Resistance of Wide Plates in Bending

REFERENCE: Lereim, J., **"Effect of Crack and Plate Geometry on the Fracture Resistance of Wide Plates in Bending,"** *Fracture Mechanics: Eighteenth Symposium, ASTM STP 945*, D. T. Read and R. P. Reed, Eds., American Society for Testing and Materials, Philadelphia, 1988, pp. 505–515.

ABSTRACT: The study included examination of the influence of variation in the geometry of the structural component as well as the variation in surface crack geometry. Strong dependency on geometry was experienced on the instability fracture toughness, which is directly a function of the triaxial state of stress, whereas the fracture toughness at initiation of stable crack growth confirmed that the crack-tip opening displacement ($CTOD_i$) essentially is a pure material parameter.

KEY WORDS: wide plates, surface cracks, bending, fracture toughness, crack-tip opening displacement ($CTOD_i$ and $CTOD_{u/c}$), tubular joint

The main support structures of offshore installations are most frequently constructed with joints/connections where large stress concentrations and stress gradients occur. This characteristic leads to extensive local plastic deformation in the joints at the ultimate design loading condition. Furthermore, in the North Sea environment and similar areas offshore, the structures are subject to dynamic loading mainly from the wave actions and from wind on the top structure. Thus, surface fatigue cracks are most likely to appear and develop in the most highly strained areas of the joints. As a consequence, a crack-dependent strength criterion should be included in the assessment of extreme design loading, and the actual fracture resistance of components containing surface cracks should be properly documented. For the actual joints in tubular structures, extensive plastic deformation appears prior to ultimate design load [*1*], and the fracture resistance in the elastic plastic regime for the presence of surface cracks should be requested. An example of a tubular joint with high stress/strain concentration and through-thickness gradient is schematically shown in Fig. 1. In this context, wide plates in bending may reflect the significance of crack geometry, plate thickness, and through-thickness stress/strain gradients typical for tubular joints. These aspects were evaluated in this work through testing of wide plates with different plate thicknesses, widths, crack shapes, and crack depths of structural steel for marine and offshore application.

Materials and Experimental Procedure

Materials

Structural ship hull steel is applied for the study of geometric effects on the fracture behavior of wide plates. This material was selected because cleavage fracture instability was assumed to

[1]Manager structural engineering, Saga Petroleum A. S., N-1322 Hoevik, Oslo, Norway.

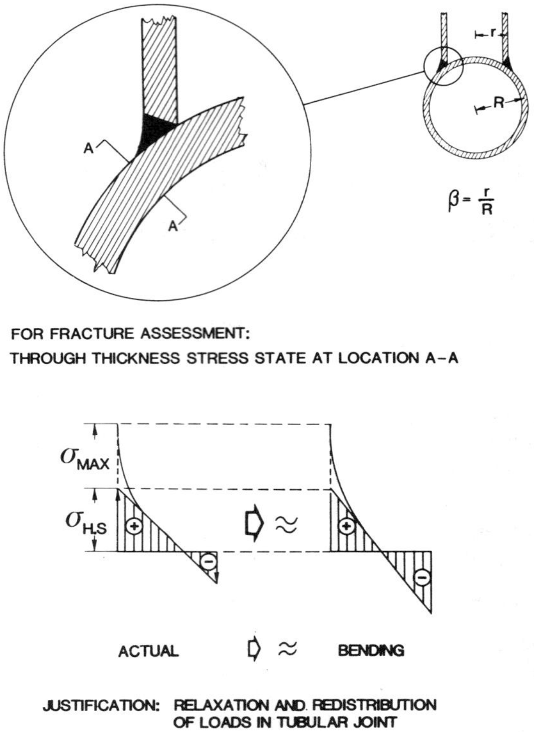

FIG. 1—*Schematic of possible severe stress state in a typical offshore tubular joint.*

occur within realistic test temperature for cooling of the wide plates. The chemical composition, tensile properties, and Pellini/nil ductility test (NDT) temperature are given in Table 1.

Large Wide Plates

Wide plates are tested in three-point bending where the in-plane dimensions are length 1500 mm; span between load points 700 mm; and width 650 mm. The plate thickness varies between 25 and 100 mm where different crack lengths and crack depths are introduced. The cracks were prepared by fatigue precracking of machined notches with different sizes according to the test program. Four of the 50-mm-thick plates also included a local stress concentration in the attempt to simulate the geometry of T-joints and cruciform joints and, indirectly, the local geometry at the highest stressed region in tubular joints.

TABLE 1—*Basic material data, grade NVA [2].*

Plate Thickness, mm	Chemical Composition, Weight %					Tensile Properties, RT			Pellini NDT-Temperature, °C
	C	Mn	Si	S	P	σ_y, MPa	σ_{UTS}, MPa	e_{long}	
25	0.17	1.15	0.22	0.032	0.027	285	457	30	−20
50	0.12	1.15	0.29	0.018	0.028	286	449	26	−20
100	0.11	1.33	0.26	0.021	0.03	276	443	28	−10

The plates are slowly loaded under displacement control by two servo-hydraulic jacks. The following data are recorded: load, load-point displacement, clip-gage displacement across the top of the center of the surface crack, and strain on the top surface to the side of the crack by use of six strain gages (see Fig. 2).

When cooling is required, liquid nitrogen is sprayed on both sides of the plates in the region of the crack. The plates are insulated, and the nitrogen spray is controlled by a thermostat and a thermocouple located in a drilled hole close to the crack surface within the region to the side of the crack surface having essentially zero stress. Thus the 100-mm-thick plates are easily cooled to −40°C, and the 25-mm plates to −60°C, which is sufficient to achieve almost linear elastic behavior of wide plates with semi-elliptic surface notches for the structural steel tested.

For the room temperature wide-plate tests, silicon replicas are made of the crack [*3–5*] at various deformations in order to derive the real crack-tip opening displacement CTOD, for a given clip gage displacement V_g, of a certain crack configuration, plate thickness, and bending strain. Particular CTOD-V_g curves were established and applied in the derivation of CTOD at lower test temperatures. In addition, the initiation value of CTOD, $CTOD_i$, is directly derived from the replicas of the tests performed at room temperature.

Small Wide Plates—From the 100-mm-thick wide plates tested, smaller wide plates were cut and machined. These were machined to thicknesses of 50, 37.5, and 25 mm (see Fig. 3). The width of the small plates was 140 mm. Continuous surface cracks, that is, infinitely long, were introduced where particularly shallow cracks were of main consideration.

In addition, small wide plates containing stress concentrations were made. The same thicknesses were examined as for the ordinary small plates. The crack depth varied from 1.5 to 15 mm in the various thicknesses in order to analyze the possible effect of thickness for a given crack depth.

As for the large wide plates, silicon replicas were made of the crack tip at different deformations for the small wide plates tested at room temperature. At lower test temperatures, silicon replicas were made of both crack surfaces on the broken two halves of the test specimen after reaching ambient temperature. At identical locations, the replicas of the two corresponding fractured surfaces were sectioned, and the initiation value of the crack-tip opening displacement was derived as the sum of the two "half" initiation values from the replicas (see Fig. 4). Twenty parallel sectionings were made on each test specimen within the same region along the crack tip, and the corresponding mean value is the best estimate of the initiation value within the region examined.

Fracture Toughness Tests—The corresponding fracture toughness testing is performed by the three-point bend specimens of the standard 2-by-1 through-thickness notched type [British

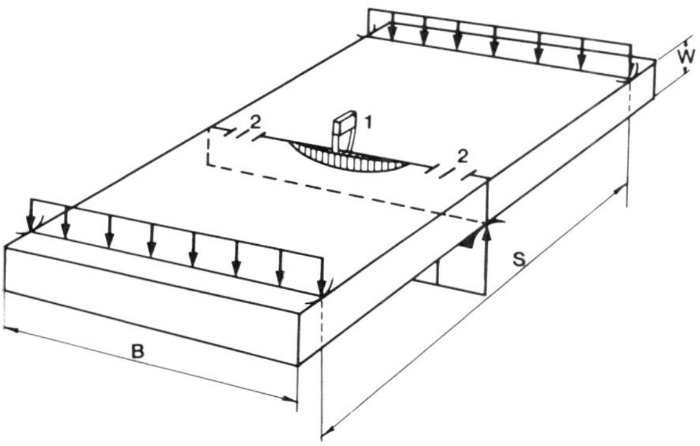

FIG. 2—*Large wide plates containing fatigue precracked surface notches tested in three-point bending. 1 = clip gage; 2 = strain gage; S = 1400 mm; B = 650 mm; and W = 25–100 mm.*

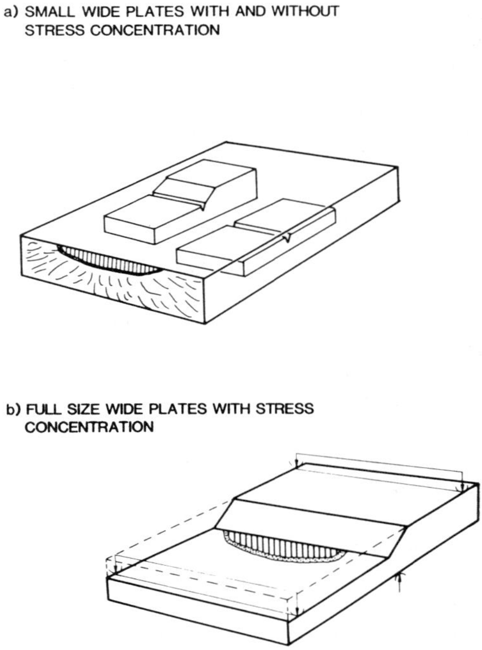

FIG. 3—*Off-sectioning of small wide plates and wide plate containing stress concentration.*

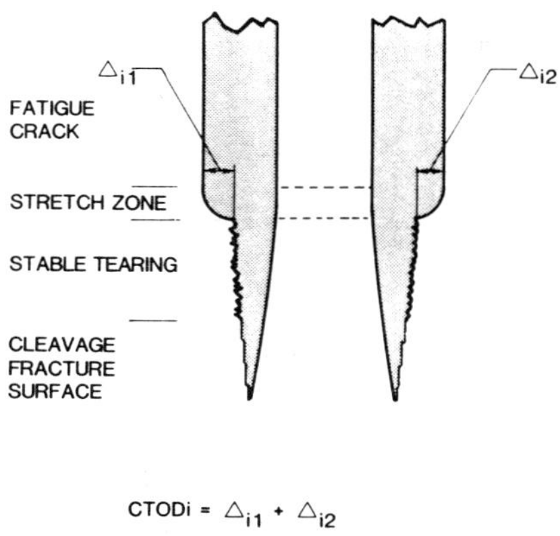

FIG. 4—*Derivation of CTOD$_i$ from replicas of fracture surfaces of the two halves of fractured wide plates tested at low temperatures.*

Standards Methods for Crack Opening Displacement (COD) Testing (BS 5762: 1979)] and of a subsidiary square-sectioned type with surface notches. Also, subsidiary square sectioned specimens with different crack depth to plate thickness ratios were tested.

Results

Effect of Plate Thickness on the Fracture Resistance of Wide Plates

The possible effect of plate thickness on the fracture resistance for a structure containing a surface crack may require full-scale testing when elastic-plastic behavior is expected. This may be performed by doing wide-plate testing, which is applied in the present work for characterization of the fracture resistance of semi-elliptic surface notches subject to three-point bending as shown in Fig. 2. The initial test series included wide plates of 25, 50, and 100-mm-thickness ratio and a/W of 0.3; the wide plates are denoted WP.

The obtained results clearly demonstrate that increasing plate thickness has an unfavorable effect on the transition temperature range between fully ductile and fully brittle behavior, where cleavage instability is preceded by various amounts of plastic deformation and stable ductile crack growth in the transition temperature range for the plates tested. Also, the small wide plate (SWP) series reflected similar behavior and dependence on thickness and temperature (see Fig. 5). Thus, a significant unfavorable thickness effect on the resistance to failure is present at given temperatures and for the fixed crack depth to plate thickness ratio within the transition temperature range, leading to elastic plastic behavior prior to cleavage instability.

Consequently, an unfavorable effect of plate thickness on the fracture resistance is present for surface cracks of fixed ratio of a/W, which may be detrimental with respect to the risk of unstable fracture of structural elements.

Furthermore, the possible effect of plate thickness on the fracture resistance at constant crack depth was briefly examined. The results, summarized in Fig. 6, indicate that no significant effect of thickness can be derived for a given crack depth within the range of crack depths tested. Again, the shallow cracks experience higher toughness levels than the deeper cracks.

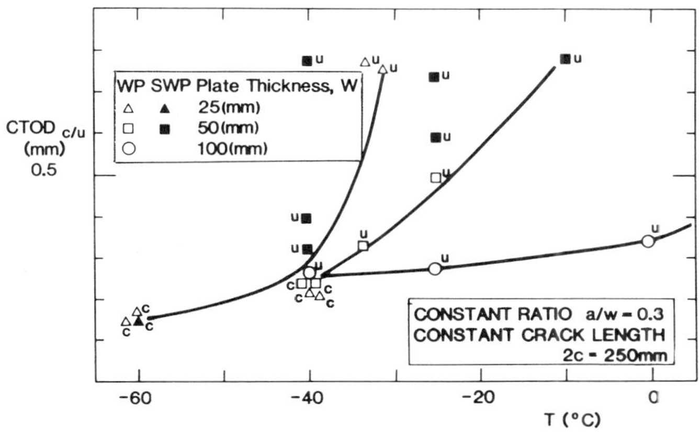

FIG. 5—*CTOD fracture toughness measured on wide plates versus test temperature (WP = large wide plates; SWP = small wide plates). CTOD fracture toughness at ultimate load derived from calibration between clip gage measurement and actual CTOD.*

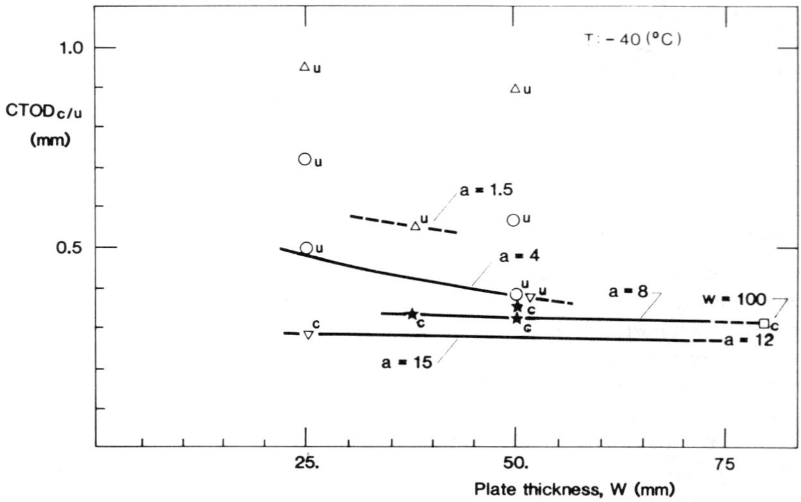

FIG. 6—*Effect of plate thickness for constant crack depths. Tested at* −*40°C. Fracture by cleavage after different amount of stable tearing.*

The practical implications of these results may be as follows:

1. In a fitness-for-purpose condition, no detrimental effect of thickness on the toughness is experienced for *a specific* crack depth, and subsize wide plates with the actual crack depth may simulate the toughness level in the larger plate or tubular connections to be examined.
2. For typical fixed offshore structure loading conditions, fatigue cracking may develop, and fatigue cracks may grow to larger depths in thick connections than in thinner connections at a given number of cycles when having equivalent initial crack sizes. Thus the a/W ratio may become almost constant at a given time regardless of the plate thickness in joints with large strain gradients. Thus reduced toughness levels may still be expected with increasing plate thickness at a certain time.

Effect of Notch Geometry

Linear elastic fracture mechanics (LEFM) describes the relative severity of cracks through the stress intensity factors for a given nominal stress state. Consequently, failure occurs at different nominal stresses, dependent on the crack configuration when assuming a constant fracture toughness. Will a similar behavior be present under elastic-plastic conditions? Parts of this problem are investigated in the present work by bend tests of wide plates containing surface cracks. The results are given in Figs. 7 and 8 for the $\text{CTOD}_{c/u}$ and critical strain, respectively, with variation in the ratio of crack depth to crack length. The crack depth a is 30 mm ($a/W = 0.3$).

The recorded CTOD_c as well as ϵ_c figures increase moderately with increasing ratio of the crack depth a to crack length $2c$ ($a/2c$) up to approximately 0.15, above which a more dramatic increase occurs.

Within the range $0 \leq a/2c \leq 0.15$ for the wide plates tested, the increase in critical strain is similar to the expected variation in nominal stress or equivalent elastic engineering strain (σ/E) under linear elastic conditions according to LEFM theory. This is indicated in Fig. 8, where the dashed line represents the variation in nominal stress or equivalent elastic strain with crack shape for a given stress intensity factor. Therefore, the stress intensity approach according to

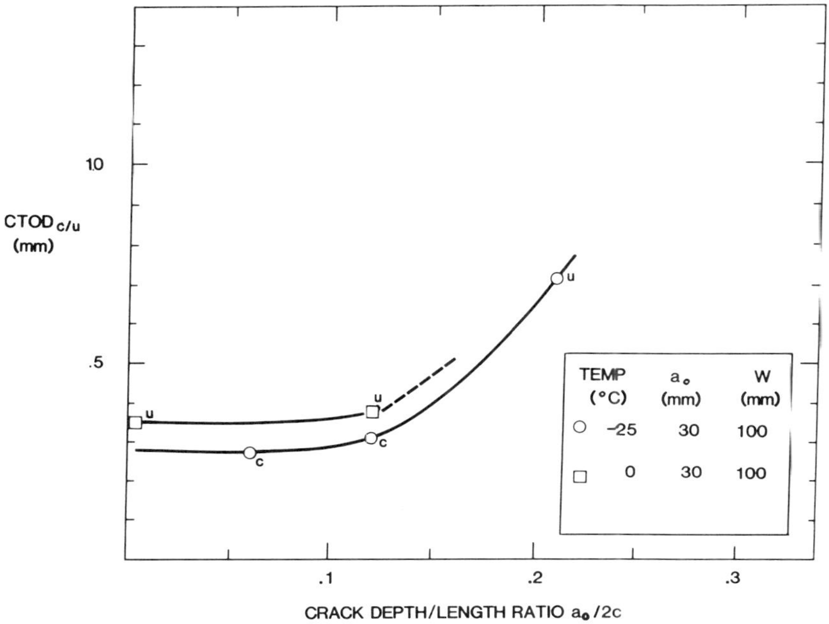

FIG. 7—*Effect of aspect ratio of surface crack on CTOD fracture toughness in wide plates. Final fracture by cleavage.*

LEFM may successfully be used for practical examination of the relative severity of cracks for $a/2c < 0.15$, although a considerable amount of plastic deformation is occurring prior to instability and the condition is far outside the validity range for linear elastic theory.

The present results were obtained under conditions where the failure mechanism terminated with cleavage instability after various amount of stable crack growth. It is, however, highly questionable whether the equivalent stress-intensity factors would reflect the relative deformation capacity for various crack shapes under more extensive plastic deformation. In view of the wide-plate test results, one may limit the acceptable rough application of the equivalency in stress-intensity factors to a critical strain less than five times the engineering yield strain σ_y/E.

Another important geometric variable of the crack configuration is the actual crack depth relative to the plate thickness when referring to the possible variation in fracture resistance during elastic plastic behavior. Seven large wide-plate tests are performed for a plate thickness of 100 mm and variation in crack depth of the surface crack from 12 to 50 mm ($0.12 < a/W \leq 0.5$). In addition, small wide plates of different thickness were tested with variation in the ratio of the crack depth to plate thickness. Although the number of wide-plate tests performed is very limited, there is a significant unfavorable effect of increasing the ratio of crack depth to plate thickness for semi-elliptic surface cracks as well as for infinitely long surface cracks.

See Fig. 9 where the $CTOD_{c/u}$ fracture resistance is plotted versus crack depth ratio for the wide plates. Similar behavior is observed on the corresponding surface-notched subsidiary square-type pure bend fracture toughness specimens having specimen height equal to the plate thickness (see Fig. 10). At a test temperature of $-40°C$ for the steel used, there is a close agreement between the $CTOD_{c/u}$ obtained for the subsidiary fracture toughness specimens and for the wide-plate tests of the 100-mm-thick plates where fracture instability was caused by cleavage with minor or no stable ductile tearing. On the other hand, at higher toughness levels or test temperature, a discrepancy is present between the fracture toughness specimen $CTOD_{c/u}$ and

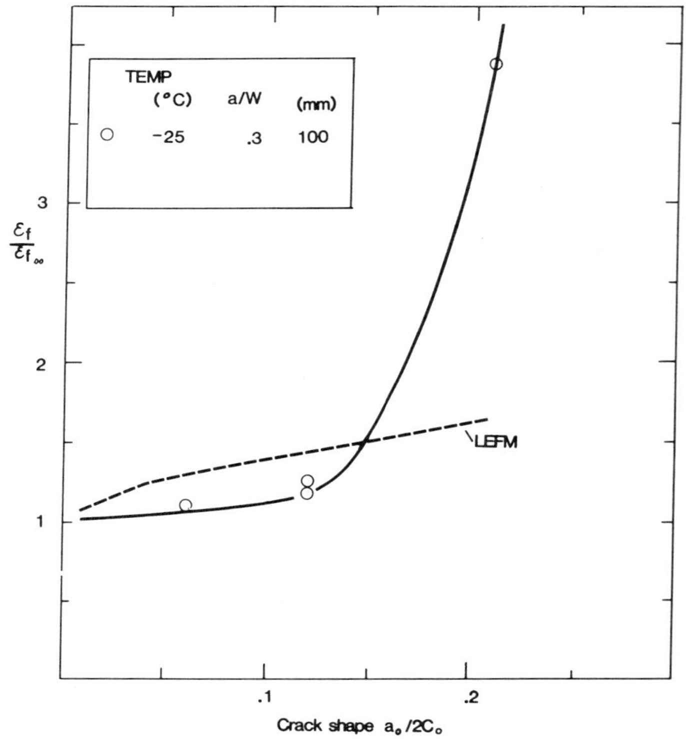

FIG. 8—*Variation in measured critical bending strain ratio with aspect ratio of surface crack. Bending strain measured on top of plastic hinge, developed in the crack cross-sectional plane, but away from crack-tip region.*

the wide-plate test $CTOD_{c/u}$. Although a very limited number of tests are performed and scatter is expected even in parent plate material in the transition temperature range, the surface-notched 1-by-1 subsidiary fracture toughness specimens tend to overestimate the fracture resistance in the wide plates with long surface cracks. A similar difference is recorded for the 50-mm plates and the corresponding 1-by-1 subsidiary-type fracture toughness specimens.

In contrast to the CTOD toughness at cleavage instability, the preceding toughness at initiation of stable ductile crack growth is found independent of crack depth to plate thickness ratio. This is in full agreement with the assertion, well known in the literature, that $CTOD_i$ is a material parameter only (see Fig. 11).

Effect of Stress Concentration

Because cracks are frequently located in regions with local stress concentrations, it is of major importance to examine the possible effect of the local stress concentration on the strength of the structural element of interest.

This study included testing for simulation of a local weld stress concentration which is similar to welded-T and cruciform joints. The fracture strength is characterized through large wide-plate testing of wide plates having a semi-elliptic surface crack at the toe of the stress concentration. The large plates were machined from a 100-mm plate down to 50 mm ahead of the stress concentration as shown in Fig. 3. The in-plane dimensions were identical to the ordinary wide

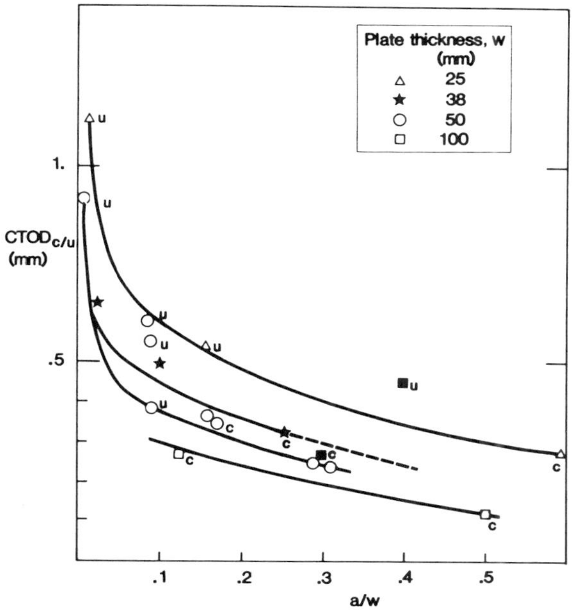

FIG. 9—*Effect of crack depth to plate thickness ratio in small and large wide plates. Tested at −40°C.*

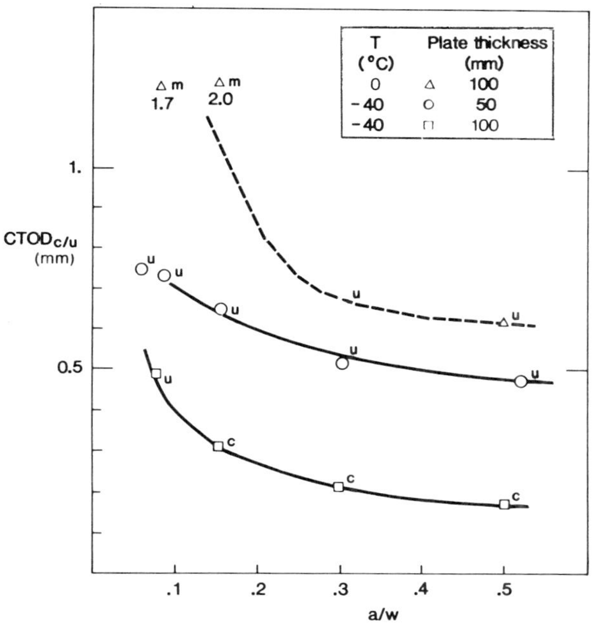

FIG. 10—*Subsidiary 1-by-1 SENB specimens. Effect of crack depth to plate thickness correspondıng to Fig. 9. Tested at −40°C.*

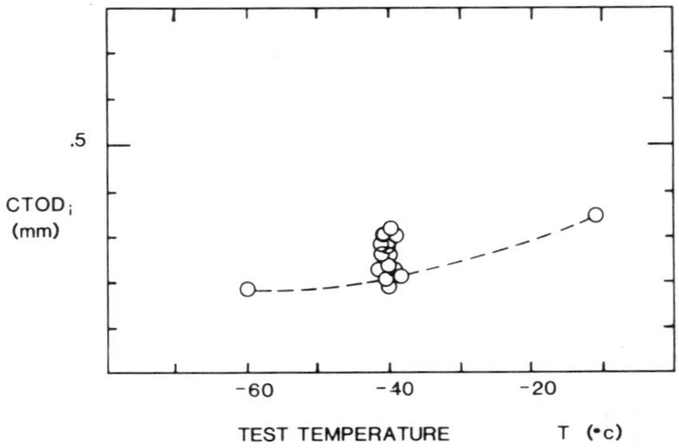

FIG. 11—*Corresponding CTOD fracture toughness at initiation of stable crack growth in the small wide plates tested. Derived from silicon replicas. At room temperature testing, directly on wide plate during testing; at lower-test temperature, according to procedure described in Fig. 4.*

plates used and the plates were loaded in three-point bending. The obtained results are then directly comparable with the corresponding fracture resistance in the ordinary 50-mm-thick wide plates.

Although scatter is present, the results strongly indicate that local stress and strain concentrations have an unfavorable effect on the fracture resistance during elastic-plastic conditions prior to fracture (see Fig. 12). This is, however, partly in disagreement with the expected behavior under linear elastic conditions, since the elastic local stress concentration influences the magnitude of the stress-intensity factor only for very shallow cracks relative to the plate thickness ($a/W < 0.05$) when concerning a bending loading situation. The latter is shown in Fig. 13, where the stress-intensity factor is derived from a finite-element analysis of a plate containing stress concentration [*6*] and by use of the SESAM 69-program system [*7,8*].

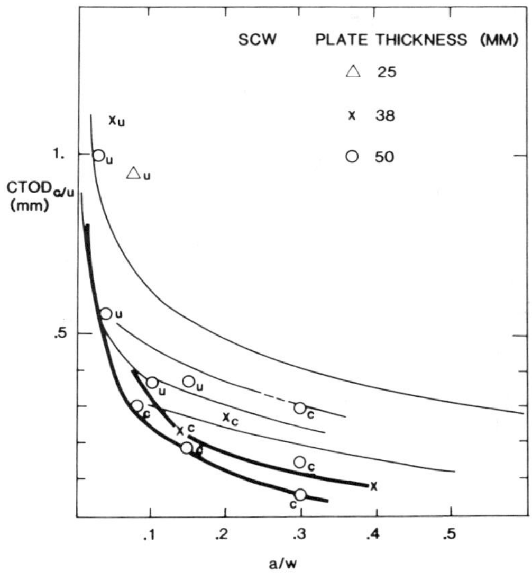

FIG. 12—*Effect of geometric stress concentration on the CTOD fracture resistance in small wide plates. Lower-bound envelopes for ordinary plane plates are included as reference.*

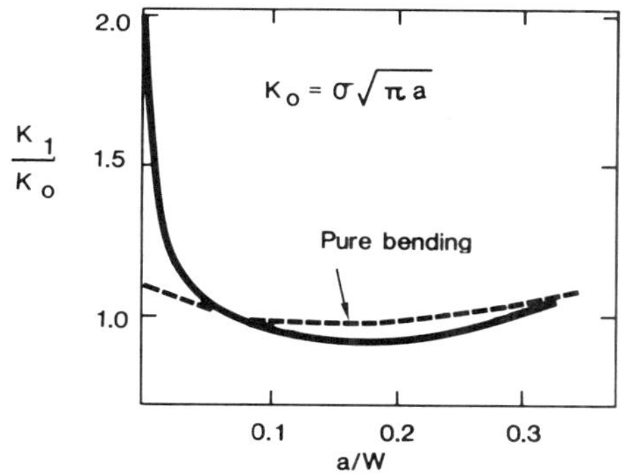

FIG. 13—*Effect of geometric stress concentration on stress intensity factor for application to plate joint analyzed.*

Conclusively, the plates containing localization of stress and strain are more severe than ordinary plane plates, with a ratio of the CTOD toughnesses recorded to be of the order 1:2 for plates containing SCF versus ordinary plane plates. Thus, small wide plates with stress concentration included may be applied as the closest interim-size plate type specimen that may simulate the condition at initiation of unstable fracture in connections containing local stress/strain concentrations.

Concluding Remarks

Based on the assessments and tests performed, the following may be concluded:

1. There is a significant detrimental effect of thickness for a given crack depth to plate thickness ratio in the transition temperature regime.
2. For a given crack size, no detrimental thickness effect can be derived.
3. Wide plates including local geometric stress concentrations gave lower CTOD-toughness values than ordinary plane plates and may be most representative for simplified simulation of tubular joint initiation fracture resistance.

References

[1] *API Recommended Practice for Planning, Designing, and Constructing Fixed Offshore Platforms,* API RP2A, 14th ed., American Petroleum Institute.
[2] *Rules for Classification of Steel Ships,* Materials and Welding Section, Det norske Veritas, Oslo, Norway, Jan. 1980.
[3] Robinson, J. N., *International Journal of Fracture,* Vol. 12, 1976, pp. 723-739.
[4] Fields, B. A. and Miller, K. J., *Engineering Fracture Mechanics,* Vol. 9, No. 137, 1977.
[5] Lereim, J. and Embury, J. D., *Engineering Fracture Mechanics,* Vol. 11, 1979, pp. 161-164.
[6] Lereim, J. in *Proceedings,* International Conference on Offshore Welded Structures, London, Nov. 1982.
[7] General Superelement Program, Computas Report No. 78-926, *Users Manual,* NV 336, SESAM 69, A. S. Computers, Hoevik, Norway.
[8] "Membrane Analysis," *Users Manual,* NV 339, SESAM 69, A. S. Computers, Hoevik, Norway.

Stig Berge,[1] *Fereshteh Ebrahimi,*[2] *and David T. Read*[3]

Wide Plates in Bending: Application of CTOD Design Approach

REFERENCE: Berge, S., Ebrahimi, F., and Read, D. T., **"Wide Plates in Bending: Application of CTOD Design Approach,"** *Fracture Mechanics: Eighteenth Symposium, ASTM STP 945,* D. T. Read and R. P. Reed, Eds., American Society for Testing and Materials, Philadelphia, 1988, pp. 516–534.

ABSTRACT: Wide plates of A572 Grade 50 steel were tested in four-point bending. The plate thickness was 50 mm. The plates were surface cracked, with two different aspect ratios. The test temperature was varied through the transition range. Critical crack-tip opening displacement (CTOD) was measured by scanning electron microscopy (SEM). CTOD at fracture compared well with data from standard bend bar tests for the same materials. Applied strains were measured using three different methods. For low aspect ratios, the measure of applied strain was a major source of uncertainty in the application of the CTOD design curve. Safe predictions from standard bend bar data could be made only if lower-bound CTOD values were assumed. For larger aspect ratios, the design curve was overconservative due to an underestimation of CTOD for a given applied strain. Scatter in CTOD data was due primarily to manganese sulfide stringers which caused delamination at the crack tip under conditions of ductile crack initiation.

KEY WORDS: mechanical properties, fracture test, wide-plate test, temperature transition, crack-tip opening displacement, elastic-plastic fracture, fracture properties, manganese sulfide inclusions, hot-rolled structural steel, fracture design

The crack-tip opening displacement (CTOD) method has been applied extensively in fracture control of welded steel structures, particularly marine structures. The method, which has a semi-empirical basis, has been verified by wide-plate data, with through-thickness cracks in plates loaded in tension [*1*].

In many types of structures, however, the stress tends to be bending, and defects are most likely to be surface fatigue cracks. One typical example is tubular welded connections in offshore structures, which are vulnerable to fatigue damage, and where local stresses in hot spot areas tend to be bending, irrespective of the type of nominal stress. Application of the CTOD method to surface cracks in bending is made by procedures based on shape factors which are derived using linear elastic fracture mechanics (LEFM) [*2*].

The objective of this work was to perform wide-plate testing through the temperature transition range in order to study,

(*a*) correlation between small-scale test data and surface-cracked wide plates in bending and
(*b*) effect of surface crack shape on fracture behavior.

[1]Professor, Division of Marine Structures, The Norwegian Institute of Technology, Trondheim, Norway.
[2]Assistant professor, Department of Materials Science and Engineering, University of Florida, Gainesville, FL 32611.
[3]Physicist, Fracture and Deformation Division, National Bureau of Standards, Boulder, CO 80303.

Material

Several considerations influenced the choice of material for the tests. The main idea was to simulate fracture in welded joints. However, the use of welded material was discouraged, due to the inhomogeneity and inherent scatter in material properties. A hot-rolled steel with a relatively high transition temperature was therefore chosen as a model material, in order to somehow simulate the poor fracture toughness which is generally found in structural steel welds. The chemical composition of the steel and the tensile properties are given in Table 1.

Wide-Plate Tests

Test Procedure

A 1000-kN servohydraulic actuator in a four column test rig was used for the tests. The wide plates were loaded in four-point bending as shown in Fig. 1. Load points and supports were roller bearings, such that tangential stresses were negligible.

The test span between supports was 1300 mm, and the plate width was 500 mm. The span with constant bending moment was 500 mm as shown in Fig. 1. This test rig was also used for fatigue precracking of the notches.

Two crack geometries were chosen for the tests, with aspect ratios 0.1 and 0.2, respectively. Restrictions on the shapes of surface cracks which can be produced by fatigue precracking led to two different crack depths, as shown in Fig. 2.

Fatigue precracking was done in compliance with the British Standards Institution's Methods for Crack Opening Displacement (COD) Testing (BSI 5762: 1969) and the ASTM Standard Test Method for J_{Ic}, a Measure for Fracture Toughness (E 813-81), with regard to stress intensity levels and requirements on fatigue crack extension from the machined notch. Crack growth was monitored by clip gages.

During fracture testing, load and strain were measured. Three different measures of strain were monitored, as shown in Fig. 3.

Local strain was defined by strain-gage readings between the crack tip and the plate edge. Readings from two gages were averaged, at a distance 10 and 125 mm from the edge, respectively. The difference in strain at these two locations was in all the cases within 10%. Furthermore, check measurements at the uncracked (compressive) side of the plate at the edge showed the strains to be very nearly symmetrical through the plate thickness. This is taken as an indication that the readings were not significantly influenced by the crack-tip strain field.

Remote strain was measured at the plate centerline, at both sides of the plate at a distance of 100 mm from the crack plane. These gages were actually the last gages in an array designed in order to measure applied J-integral [*3*]. Measurements along one such array is shown in Fig. 4. As shown, the remote strain was measured outside the strain shadow of the crack, as is evidenced also by the fact that the strain was symmetrical through the crack plane at this location.

Global strain was measured assuming the plate between the load points to deflect with a constant curvature, that is, according to beam theory, with no influence of the crack. The deflection at midpoint was measured relative to the load points, and the curvature was computed on this basis.

CTOD Measurements

CTOD was measured on the wide plates using a scanning electron microscopy (SEM) technique. The model is shown in Fig. 5; actual measurements are depicted in Fig. 6. Figure 6*a*

TABLE 1—*Chemical composition and tensile properties of steel used.*

	Chemical Composition							Tensile Properties			
Specimen	C	Mn	P	S	Si	Cr	V	Temperature, °C	S_y, MPa	S_u, MPa	Reduction in Area, %
Charge	0.09	1.09	0.011	0.009	0.27	...	0.030	22	350	537	74
BB1	0.11	1.06	...	0.012	0.26	0.51	0.028	−3	356	545	72
BB2	0.11	1.03	...	0.012	0.26	0.49	0.028	−26	372	568	73
W1	0.11	1.03	...	0.011	0.26	0.49	0.028	−46	375	579	73
W2	0.11	1.02	...	0.011	0.26	0.50	0.028				

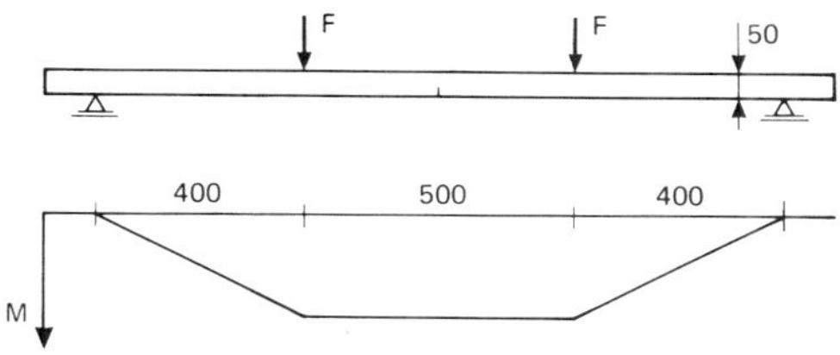

FIG. 1—*Specimen geometry and loading* (top) *and applied bending moment* M (bottom).

shows the crack-tip region at an angle normal to the crack plane. In Fig. 6*b* the specimen is given a 45-deg tilt, and the stretch zone *s* is measured. Critical CTOD is calculated as

$$\delta_{crit} = 2s \cos 45° \tag{1}$$

In CTOD testing, δ_{crit} may be defined in various ways, depending on the mode of fracture. The conventional definitions are listed here according to BSI 5762:

δ_c = unstable fracture or pop-in with no prior stable crack growth,
δ_i = initiation of stable crack growth,
δ_m = attainment of maximum load plateau, and
δ_u = unstable fracture or pop-in after stable crack growth has taken place.

As indicated in Fig. 5, the δ_{crit} measured by SEM was δ_c or δ_i. However, no evidence of ductile tearing was found on the wide plates, and the measured δ_{crit} thus conforms with the definition of δ_c.

The SEM method of measuring CTOD was checked by measurements on three bend bar specimens, where CTOD data measured according to BSI 5762 were available. The data are given in Table 2. As shown, the agreement between the two methods is very good for Specimens BB1 and BB6. For Specimen BB2, the SEM method seems to underpredict the critical CTOD. However, for this specimen, some ductile tearing took place before the onset of unstable fracture. As discussed previously, the SEM method in this case measures δ_i, whereas the clip gage measures δ_u, which explains the discrepancy. For Specimens BB1 and BB6, the extent of ductile tearing was negligible.

CTOD Data from Bend Bar Specimens

CTOD specimens were machined from the plate as shown in Fig. 7. Rolled steel plate has been shown to exhibit large anisotropy with regard to fracture toughness parameters [4]. For this reason, specimens were oriented in the same way as the wide plates with respect to the rolling direction. The reason why a $B \times 2B$ geometry was chosen, and not the $B \times B$, was that J_{Ic} was also measured (not reported here). The test procedure complied with BSI 5762, except that loading was applied in steps, and unloading compliance measurements were performed in accordance with ASTM E 813.

Several specimens showed clear evidence of thumbnail crack extension before instable fracture had been initiated. Following standard procedures, stable crack growth was neglected, and δ_{crit} was reported as δ_c, δ_u, or δ_m, calculated on the basis of the length of the fatigue precrack [2].

Critical CTOD through the temperature transition range is shown in Table 3 and Fig. 8. There is a considerable scatter in the data. The "mean" and "lower bound" temperature transition curves are drawn for later reference.

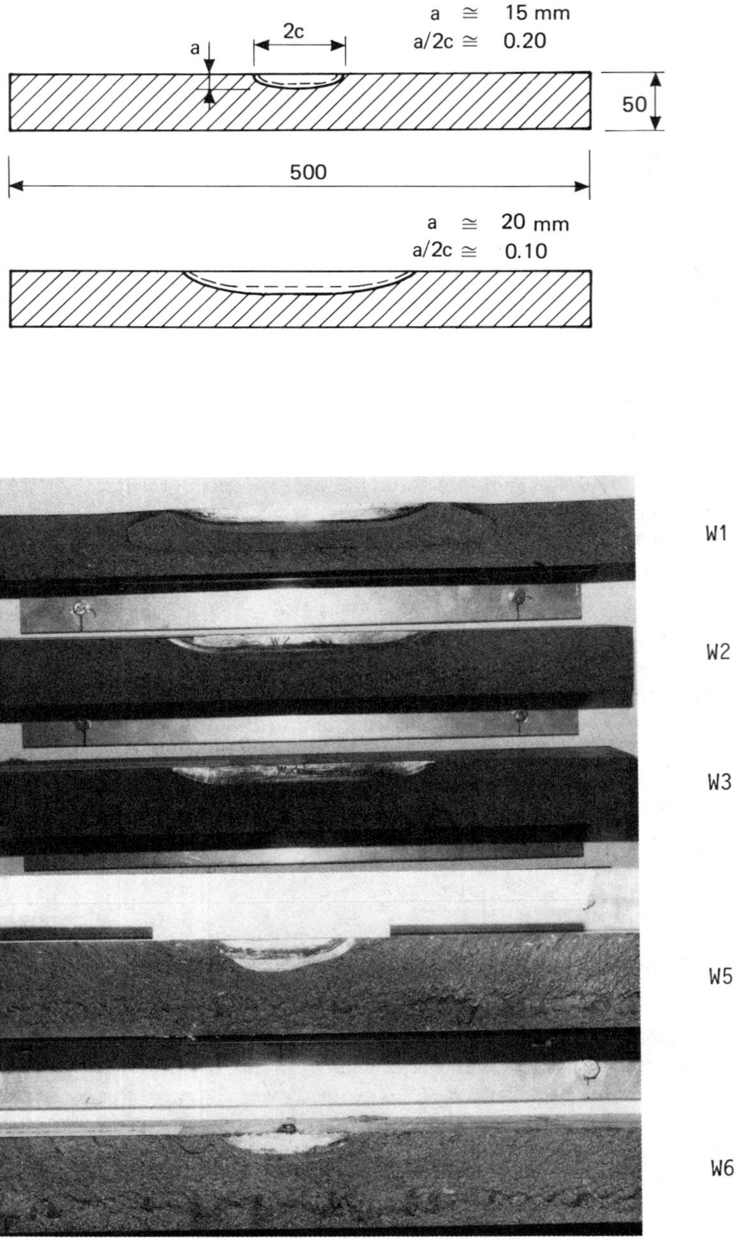

FIG. 2—*Crack geometries showing machined notch* (broken line) *and fatigue pre-crack* (full line). *Photo shows the actual fracture surfaces.*

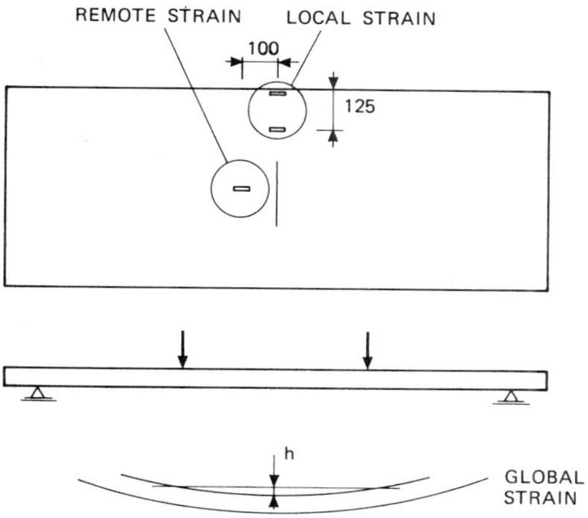

FIG. 3—*Definition and measurement of local, remote, and global strains.*

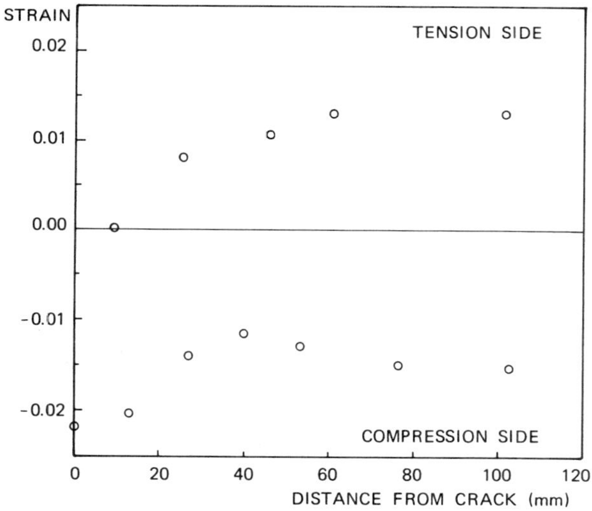

FIG. 4—*Strain measurements along plate centerline for specimen W4 at the end of the test* (no fracture).

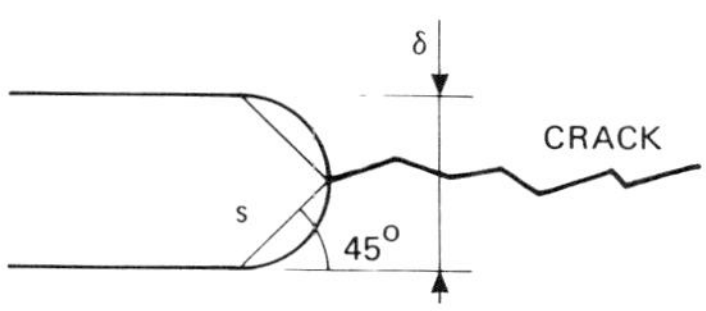

FIG. 5—*The geometric model for measurement of δ_{crit} by SEM.*

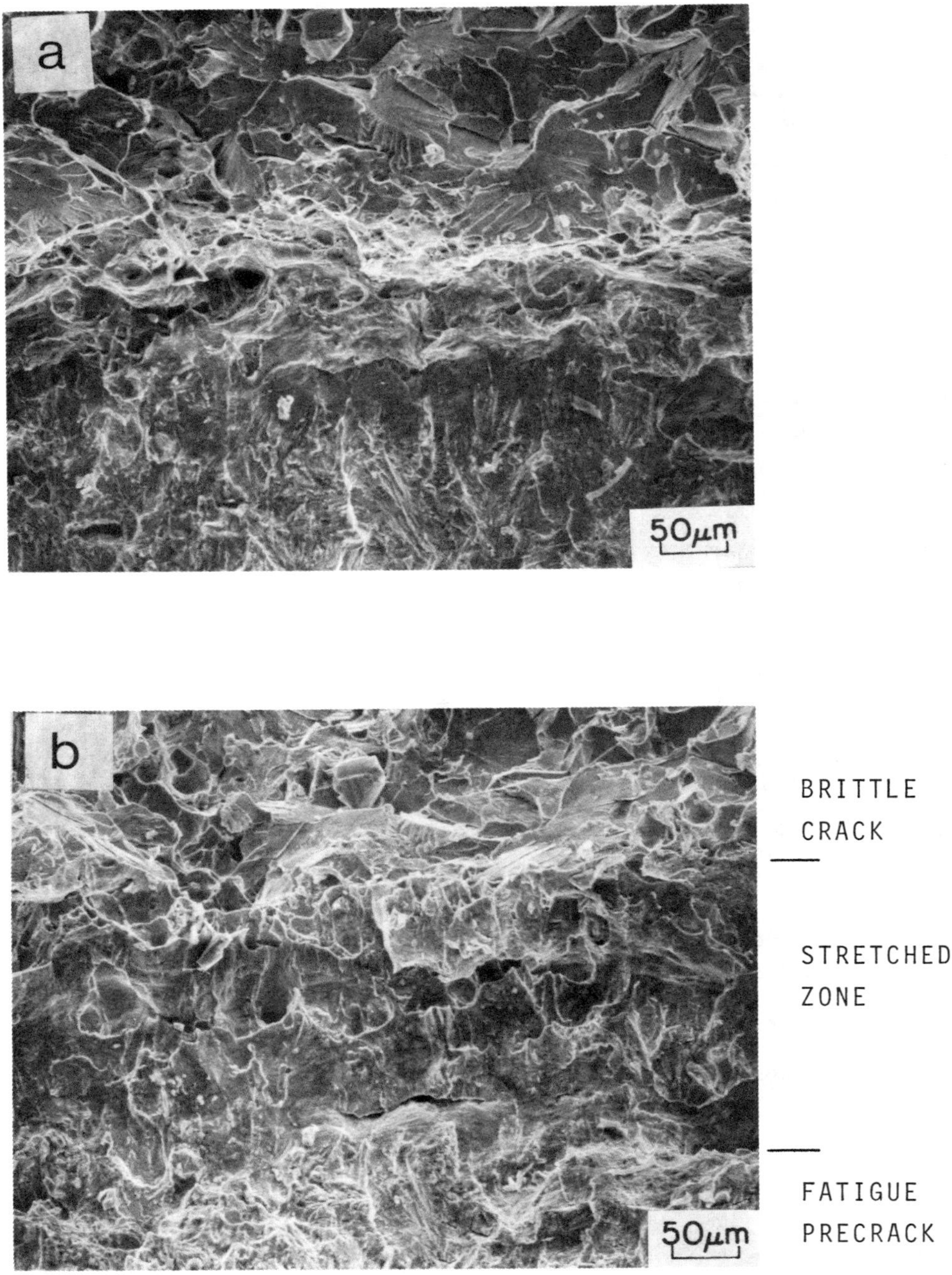

FIG. 6—*SEM fractograph showing the stretched zone.* (a) *0-deg tilt* (b) *45-deg tilt.*

TABLE 2—*CTOD measurement data, bend bar specimens.*

Specimen	Temperature, °C	δ_{crit}, mm		Δa_p,[a] mm
		BSI 5762	SEM	
BB1	23	0.29	0.33 ± 0.07	0.36
BB2	23	0.94	0.40 ± 0.07	1.10
BB6	−29	0.25	0.24 ± 0.05	0.00

[a]Δa_p = ductile crack extension according to ASTM E 813.

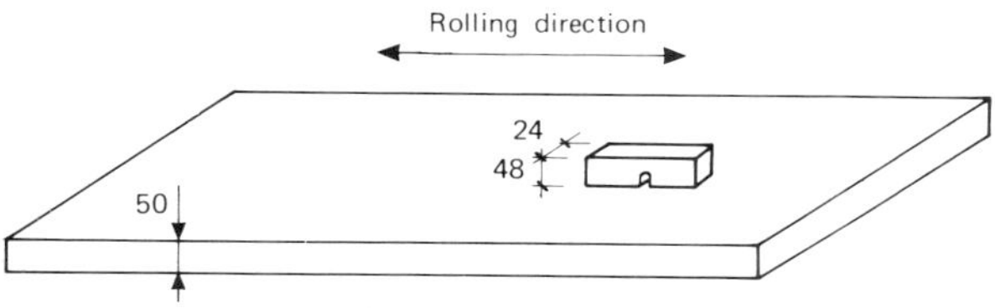

FIG. 7—*Orientation and dimensions of bend bar specimens.*

TABLE 3—*Critical CTOD through-temperature transition range (see also Fig. 8) bend bar specimens.*

Specimen	Temperature, °C	a_i,[a] mm	δ_{crit}, mm	Fracture Mode[b]
BB1	23	26.0	0.29	δ_m
BB2	23	26.4	0.95	δ_m
BB3	51	26.7	1.05	δ_m
BB4	51	26.6	1.15	δ_m
BB5	24	26.8	0.95	δ_m
BB6	−29	27.1	0.25	δ_c
BB7	−10	26.7	0.59	δ_u
BB8	− 5	27.4	1.15	δ_m
BB9	−25	26.8	0.57	δ_u
BB10	−45	27.2	0.09	δ_c
BB11	−58	26.8	0.01	δ_c
BB12	23	27.0	0.60	δ_u
BB13	7	26.7	0.30	δ_u
BB14	2	26.4	0.39	δ_u
BB15	−12	26.5	0.05	δ_c
BB16	−23	26.6	0.40	δ_u

[a]a_i = initial crack length (BS 5762).
[b]Classification according to BS 5762.

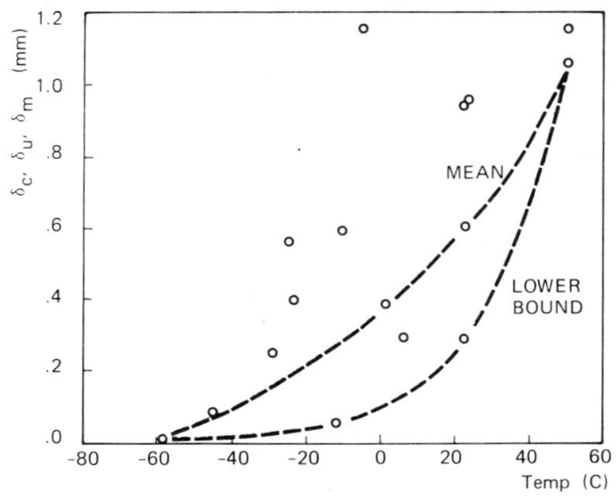

FIG. 8—*Critical CTOD of bend bars, BSI 5762.*

Wide-Plate Test Data

Three wide plates for each crack shape (Fig. 2) were tested at a range of temperatures corresponding to the transition range for the CTOD specimens. Strains were measured as shown in Fig. 3. Critical CTOD was measured using SEM techniques on the basis of the model shown in Fig. 5. Measurements were taken at 50 locations randomly selected in the region around the centerline of each specimen, and data were reported as mean and standard deviations. Test data are given in Table 4.

One plate (W4) did not fail at the full stroke of the test rig. After the test was completed, the plate was broken at liquid nitrogen temperature in order to facilitate measurement of CTOD at maximum strain.

Discussion of the Results

CTOD Design Approach

One main objective of the present work was to study the correlation between CTOD data and wide-plate test data through the temperature transition. This correlation may be influenced by two effects [*5*]:

1. Statistics, due to the fact that in wide-plate tests the crack front samples much more material than in a standard CTOD test.
2. Constraint, which depends on geometry.

The large scatter in the CTOD data (Fig. 8) makes correlations very difficult. Two reference curves are shown in Fig. 8, both of which are based on engineering judgment. The "lower bound" curve should be self explanatory. The "mean" curve is drawn with a view to the probability that the three lowest points are from a poor section of the plate, and not representative of the material properties in general.

The CTOD design approach is based on two assumptions:

1. Fracture in structural components, that is, wide plates, takes place at the same δ_{crit} as in small-scale specimens.
2. CTOD for a given defect and strain can be assessed by the design curve.

TABLE 4—*Wide-plate test data.*

Specimen	Temperature, °C	a, mm	$2c$, mm	$\bar{a}$, mm	e/e_y Global	e/e_y Local	e/e_y Remote	δ_{crit}, mm
W1	23	21.5	197.0	13.1	3.48	5.05	2.13	0.25 ± 0.05
W2	2	20.6	197.0	13.2	5.72	5.98	2.65	0.19 ± 0.04
W3	−25	20.6	199.2	13.4	0.87	0.95	0.77	0.042 ± 0.005
W4	23	16.3	76.5	11.2	11.44	11.32	8.55	0.17 ± 0.04[a]
W5	−30	17.1	78.8	11.1	3.66	3.96	3.12	0.11 ± 0.02
W6	−45	15.8	77.2	11.1	3.09	3.58	2.83	0.11 ± 0.02

[a]No fracture, values at end of test.

The first assumption was checked by comparing δ_{crit} as measured by SEM on the wide-plate specimens, with the δ_{crit} values of the small scale specimens as measured according to BSI 5762 (Fig. 8). The wide-plate data are shown in Fig. 9, with the reference curves from Fig. 8. There seems to be a fair correlation between the wide-plate and the small-scale specimen data. The cracks with aspect ratio $a/2c \simeq 0.1$ seem to follow the lower-bound curve, whereas with aspect ratio $a/2c \simeq 0.2$ two data points fall on the "mean" curve. Bearing in mind that the "mean" curve is actually a 20th percentile curve, it seems justified to state that all the wide plates fractured at a δ_{crit} close to the lower bound of small-scale data. This indicates a size effect in the sense that a large-scale specimen—or a structural component—may exhibit a lower fracture toughness than a standard CTOD specimen. As noted previously, size effects may be caused by constraint and by the so-called statistical effect. Due to the scatter, the data do not justify any discussion on statistical effects versus constraint in this case.

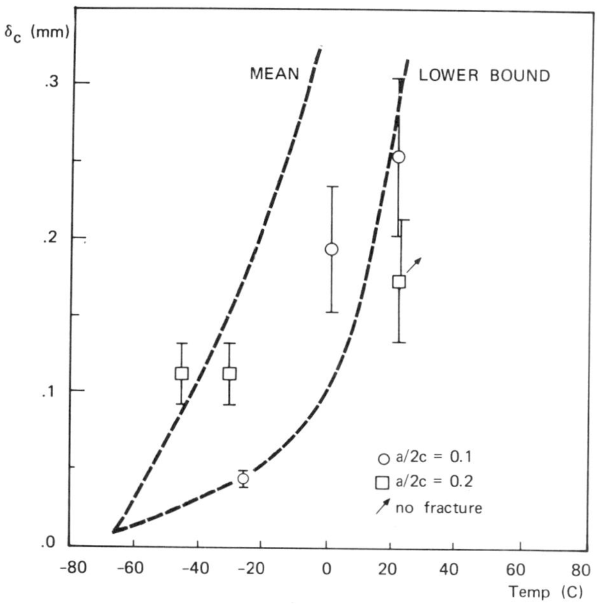

FIG. 9—*Critical CTOD of wide plates as measured by SEM. Reference curves from Fig. 8.*

The second assumption was checked by plotting the wide-plate data on the same format as the design curve, that is

$$\phi = \frac{\delta_{crit}}{2\pi \bar{a} e_y} = \begin{cases} (e/e_y)^2 & e/e_y \leq 0.5 \\ e/e_y - 0.25 & e/e_y \geq 0.5 \end{cases} \tag{2}$$

where

e = applied strain,
e_y = yield strain, and
$\bar{a}$ = equivalent crack length.

Here $\bar{a}$ is the equivalent length of a through crack in a large plate, calculated on the basis that the stress intensity factor of $\bar{a}$ is the same as that of the surface crack. Stress-intensity factors were calculated from Ref *6* (see Table 4).

As shown in Figs. 10 and 11, all the data fall on the conservative side of the design curve. However, the definition of strain is a major source of uncertainty. In the CTOD design approach, strain is to be assessed disregarding the effect of the crack. In elastic-plastic fracture with bending loads, a hinge mechanism is likely to develop at the crack. As shown in Fig. 10, for aspect ratio $a/2c \simeq 0.1$, the strain which was least influenced by the hinge, that is, the remote strain (Fig. 3), gave results which are close to the design curve.

For $a/2c \simeq 0.2$ (Fig. 11), the hinge mechanism was less pronounced, and the variability in strain was less. For this crack geometry, the results are much more on the conservative side, indicating that for large aspect ratios the design curve gives a very conservative prediction of CTOD as a function of applied strain. This tendency for large aspect ratios has been found also in three-point bending [*7*].

In Figs. 12 and 13, the effective crack length $\bar{a}$ is compared with the design critical crack length, using the temperature transition curves from Fig. 8. In this graph, the line 1:1 indicates the design curve to give an accurate prediction of fracture; that is, there is no safety factor. The line 1:3 indicates a safety factor of three on crack length.

Again, the definition of strain introduces a large scatter. On the basis of the "lower bound" temperature transition curve, the predictions based on the design curve are generally conservative, with the exception of one specimen (W1). Using the "mean" temperature transition curve (Fig. 13) gives results which are generally unconservative for crack geometry $a/2c \simeq 0.1$. For $a/2c \simeq 0.2$ the results are conservative also in this case.

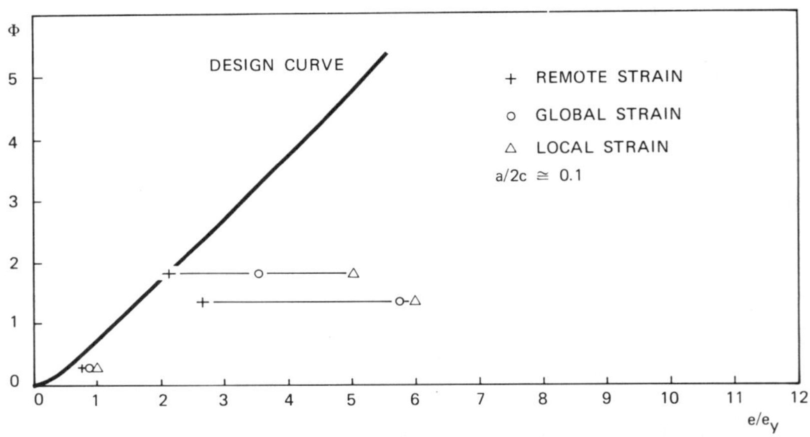

FIG. 10—*Wide plate data plotted on design curve format,* a/2c $\simeq$ *0.1.*

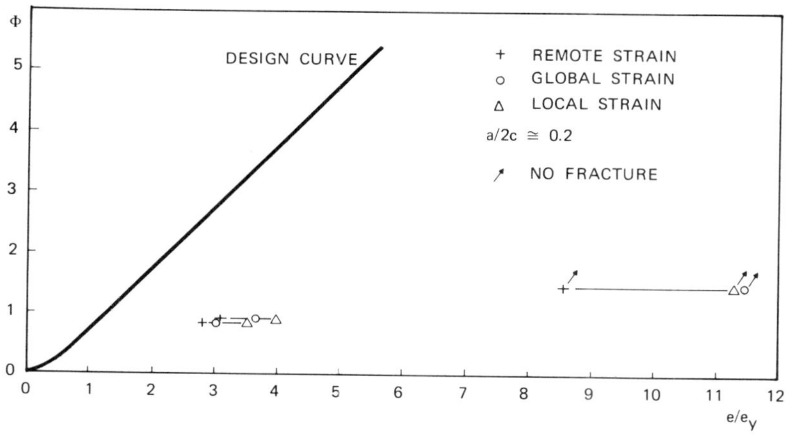

FIG. 11—*Wide plate data plotted on design curve format,* a/2c ≃ *0.2.*

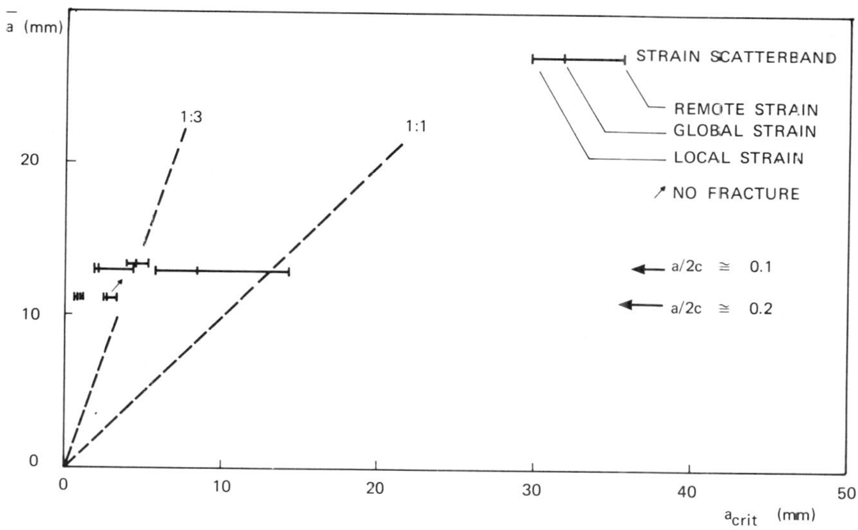

FIG. 12—*Equivalent crack length* $\bar{a}$ *versus critical crack length computed on the basis of the design curve and the lower bound temperature transition curve (Fig. 8).*

Fractography

A fractographic investigation was performed with a particular view to identify the reasons for the large scatter in the CTOD data. Sections of the plate material were studied by optical microscopy. Long stringers of manganese sulfide particles were observed as is shown in Fig. 14 for longitudinal and transverse sections. The microstructure consists of ferrite and pearlite aggregates. Figure 15*a* shows the chemical inhomogeneity of the microstructure as is manifested by the darker etched regions, outlining the original dendritic structure. A higher magnification optical micrograph is presented in Fig. 15*b*. In the dark etched regions, where the concentration of alloying elements is higher, acicular ferrite and extensive subgrain formation can be observed.

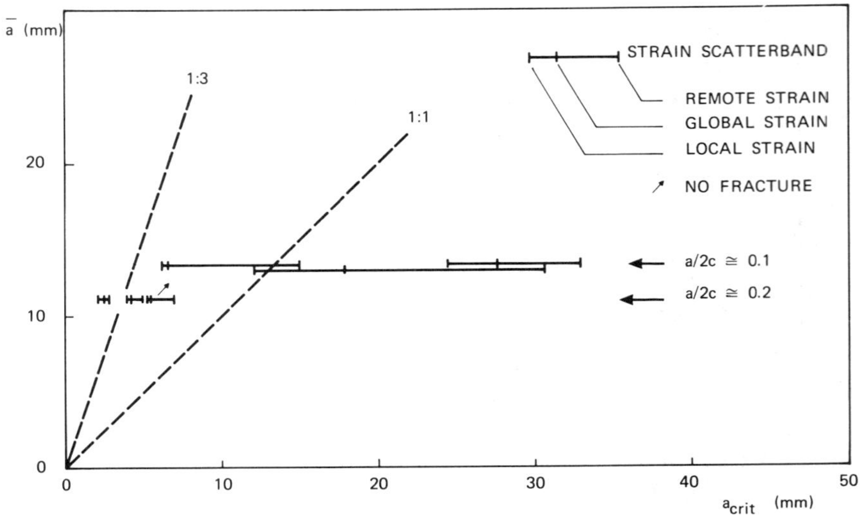

FIG. 13—*Equivalent crack length* $\bar{a}$ *versus critical crack length computed on the basis of the design curve and the "mean" temperature transition curve (Fig. 8).*

Detailed fractography was performed by SEM on all the wide-plate specimens and three bend bar specimens (BB1, BB2, and BB6). A summary of the results is given in Table 5.

All the specimens studied except for W3, W5, and W6 showed a ductile shear zone next to the fatigue precrack, preceding the unstable clevage crack. The mode of crack initiation was classified based on the relative sizes of the ductile shear zone at the midthickness and the edges of a specimen [*8*].

Specimens BB1 and BB2, which were tested at room temperature, showed a significant difference in δ_{crit} as measured by the clip gage. In both specimens, evidence of ductile tearing was found (Table 3). Critical CTOD as measured by SEM was, however, very nearly the same, indicating that ductile tearing was initiated at the same level of CTOD.

Role of Inclusions

Large macroscopic secondary cracks (delamination) were observed on the fracture surface of all specimens that showed some ductility at the crack tip. These secondary cracks, which form due to manganese sulfide stringers, influence the stress state at the crack tip and consequently the competition between ductile and brittle fracture, as demonstrated in Fig. 16. Figure 16*a* shows how the delamination causes a premature brittle cracking. Figure 16*b* shows a secondary crack on the fatigue precrack fracture surface. The manganese sulfide particle had caused a blunting of the tip of the fatigue crack during precracking. Later, when the plate was loaded to the fracture point, a secondary crack was developed and the brittle fracture was joined with the fatigue precrack in a ductile shearing mode. Figure 17 shows the absence of delamination in a specimen tested at a low temperature. In this case adequate stress triaxiality did not develop at the crack tip.

The preceding observations suggest that manganese sulfide stringers modify the stress state at the tip of a crack by a delamination process and consequently influence the measured toughness and the scatter observed in the data. The fact that manganese sulfide particles do not directly initiate cleavage is demonstrated in Fig. 18. Following the river marks, it can be seen that the cleavage facets are stopped at the manganese sulfide particle. However, it was found

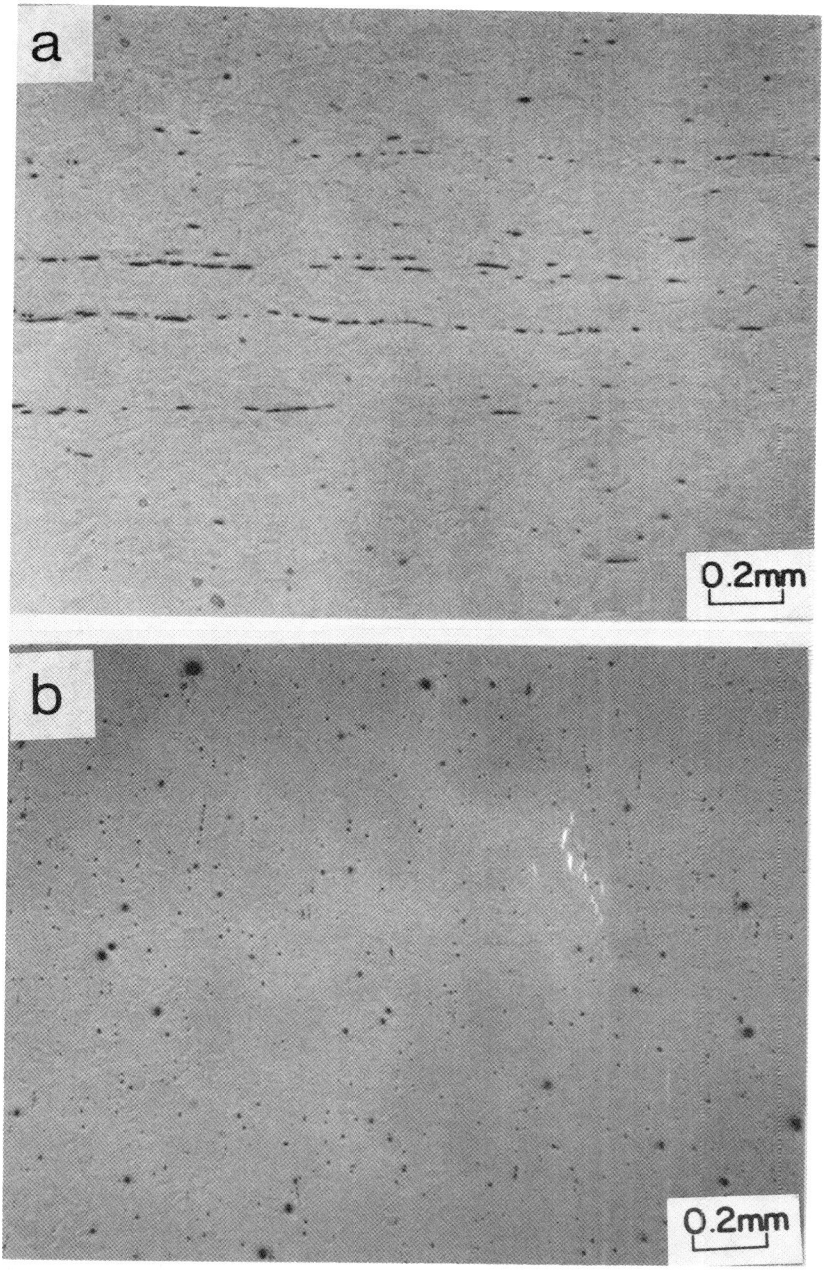

FIG. 14—*Optical micrographs showing inclusions in longitudinal* (a) *and transverse* (b) *sections.*

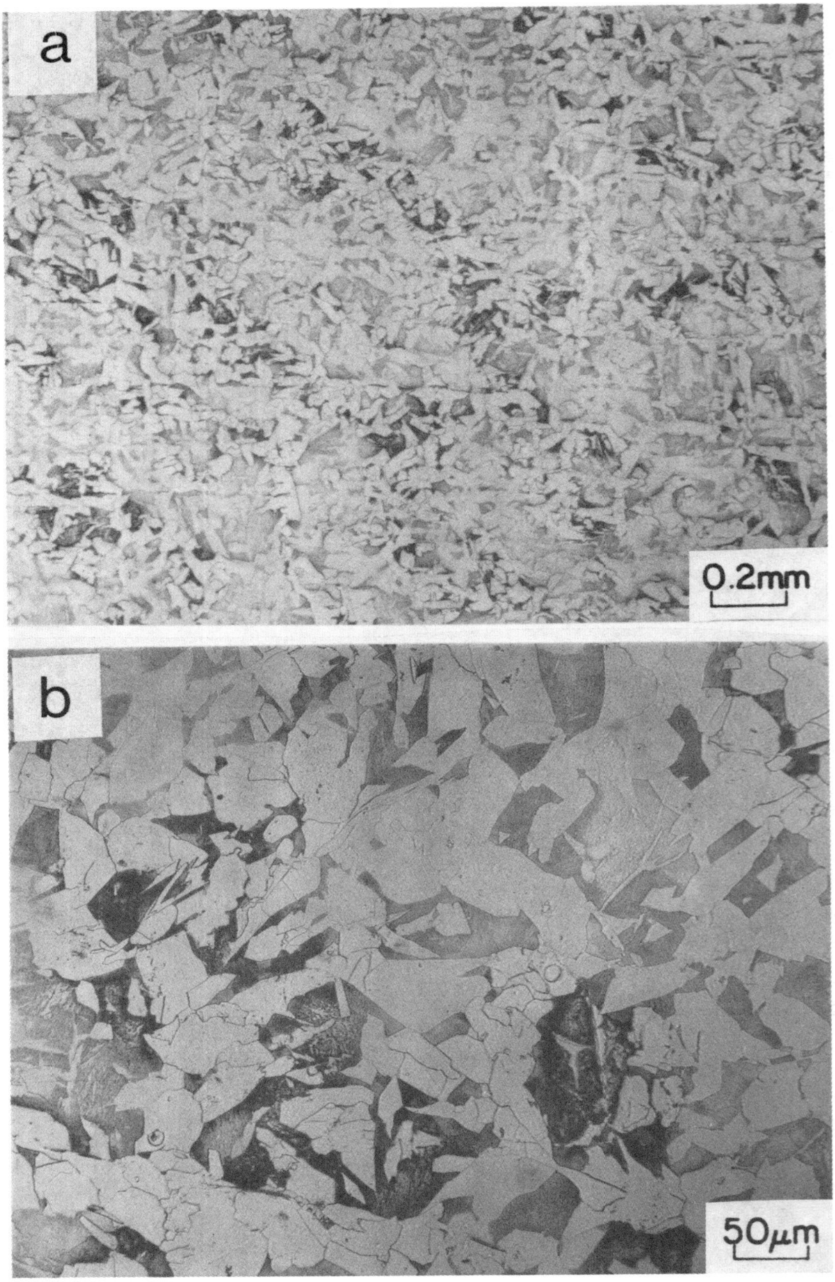

FIG. 15—*Optical micrographs showing the chemical inhomogeneity of the microstructure at low* (a) *and high* (b) *magnification.*

TABLE 5—*Fractography results.*

Specimen	Temperature, °C	δ_{crit}, mm	ℓ_f,[a] mm	Initiation Mode[b]
BB1	23	0.33 ± 0.07	1.32	D
BB2	23	0.40 ± 0.07	1.84	D
BB6	−29	0.24 ± 0.05	0.10	B
W1	23	0.25 ± 0.05	0.92	D
W2	2	0.19 ± 0.04	1.18	D
W3	−25	0.042 ± 0.005	0.0	B
W4	23	0.17 ± 0.04	0–0.6	...
W5	−30	0.11 ± 0.02	0.0	B
W6	−45	0.11 ± 0.02	0.0	B

[a] ℓ_f = shear zone size at midsection [7].
[b] D = ductile, B = brittle [7].

that mainly oxide and silicate particles are responsible for triggering cleavage fracture as shown in Fig. 19.

Role of Chemical Inhomogeneity

It was suspected that the chemical inhomogeneity observed in the microstructure may contribute to the scatter in the data. Semi-quantitative analysis of many cleavage facets along the crack front by EDS for Specimens BB1 and BB2 revealed no significant differences in the amount of chromium and manganese between the two specimens. By the same type of analysis on the polished and etched specimens, no difference could be detected in the amount of alloying elements between the light and dark etched areas shown in Fig. 15. Therefore, at this point no definite conclusion can be made about the effect of chemical inhomogeneity on the probability of brittle crack initiation.

However, the very strong effect of manganese sulfide stringers on the probability of fracture of this particular steel suggests that for a dirty steel the mechanism responsible for the scatter in the toughness data is the delamination process. As with weakest-link models, the probability of fracture depends on the number, size, and geometry of manganese sulfide particles and their proximity to the crack tip. It should be noted that the weakest-link models in the present form assume that weak spots ahead of the crack tip trigger the unstable brittle fracture. However, in this study the effect of manganese sulfide particles is reflected as a change in the stress-state ahead of the crack tip.

Conclusions

CTOD tests were performed on standard bend bar specimens and wide plates in four-point bending with surface cracks. The following conclusions were drawn.

1. CTOD at fracture was of the same magnitude in the two types of specimens.
2. The particular steel employed in the tests showed large scatter in δ_{crit} measured by standard CTOD testing. Only by assuming lower bound values could safe predictions be made with regard to fracture of the wide plates.
3. For surface cracks with aspect ratio $a/2c \simeq 0.1$, the correspondence with the design curve was good. For aspect ratio $a/2c \simeq 0.2$, the design curve was overconservative due to an overestimation of CTOD for a given applied strain.
4. The definition of applied strain in the wide plates was a major source of uncertainty. Due to a hinge mechanism influenced by the crack, different measures of strain could vary by a factor 2.5.

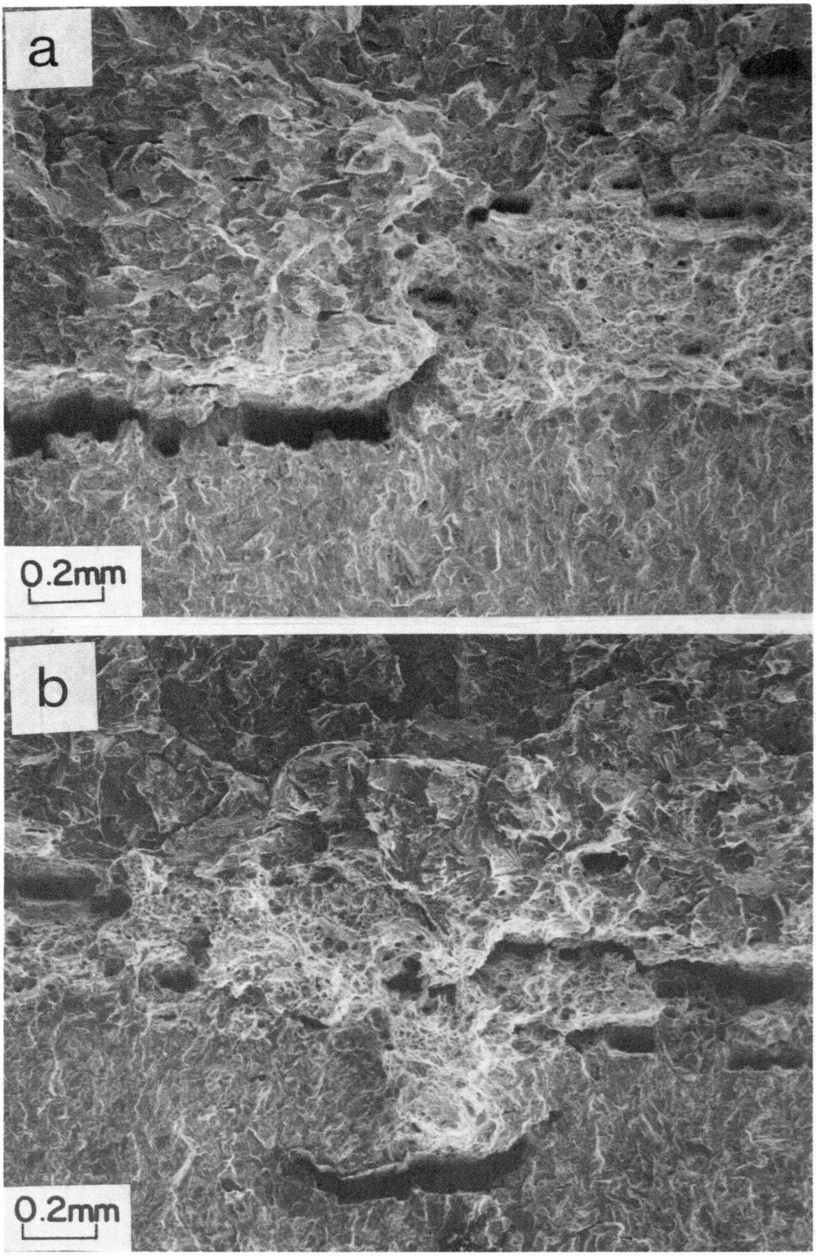

FIG. 16—*SEM fractograph showing effects of delamination on the fracture process at the crack tip. Premature brittle fracture* (a) *and secondary cracking* (b).

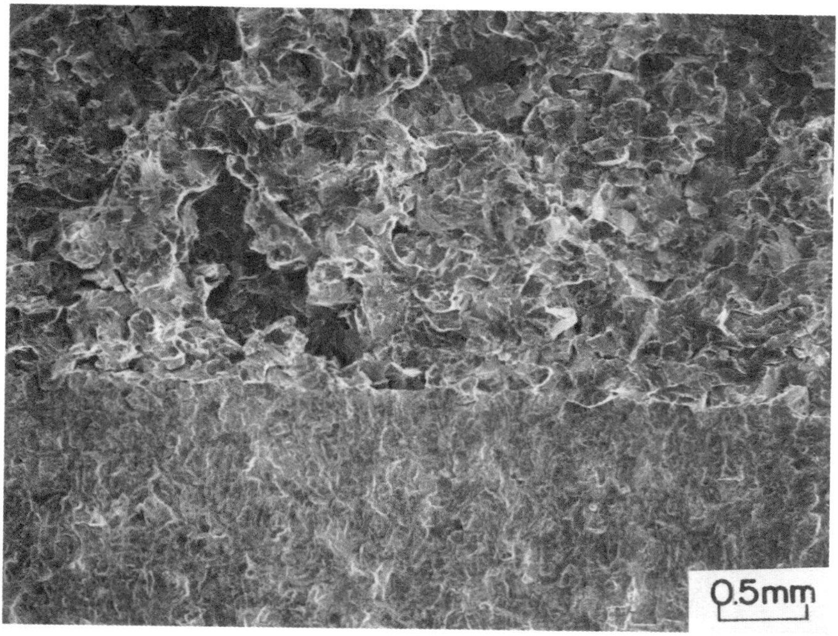

FIG. 17—*SEM fractograph showing the absence of delamination in specimen W3 (δ_{crit} = 0.042 mm).*

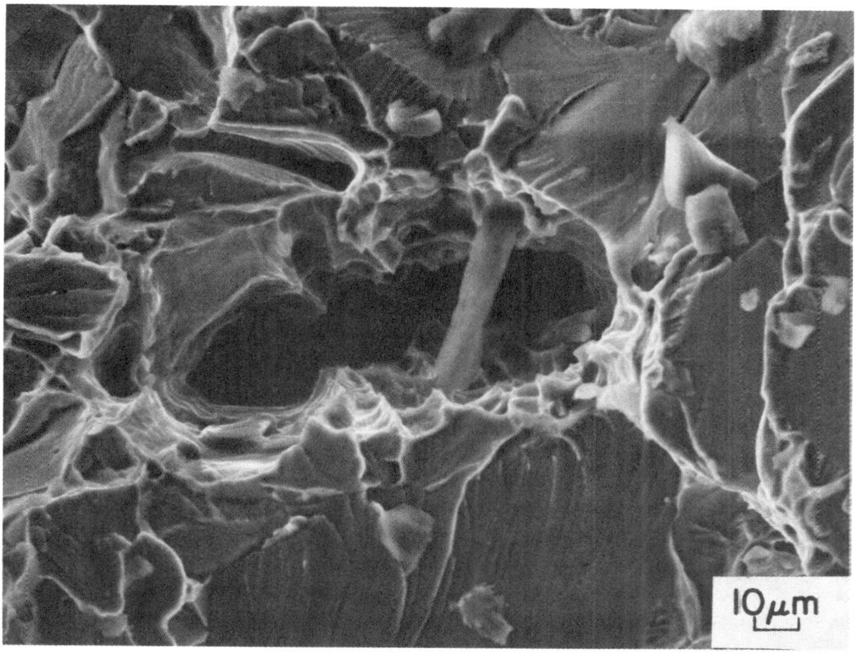

FIG. 18—*SEM fractograph showing that cleavage facets are stopped at large MnS particles.*

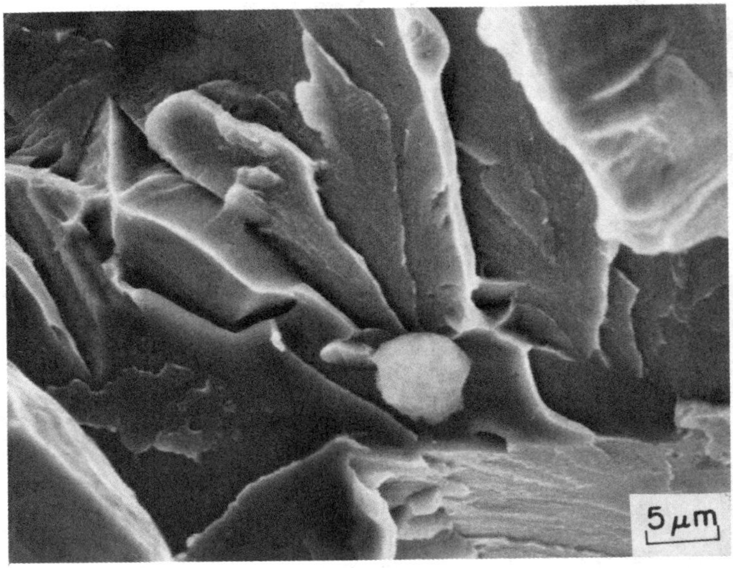

FIG. 19—*SEM fractograph showing the initiation of cleavage facet from a silicate particle.*

5. The main reason for the scatter in the CTOD data was manganese sulfide stringers, which caused delamination at the crack tip under conditions of ductile crack initiation.

Acknowledgments

One of the authors (S.B.) worked on this project while on leave of absence from The Norwegian Institute of Technology. Financial support from the Royal Norwegian Council for Scientific Research is gratefully acknowledged. Financial support from the Minerals Management Service, U.S. Department of the Interior, under Project Monitor Mr. Charles Smith, is greatly appreciated. Mr. David McColskey, of the National Bureau of Standards, Boulder, Colorado, is acknowledged for his excellent assistance to the experimental program.

References

[1] Harrison, J. D. et al., "The COD approach and its application to welded structure," in *Elastic-Plastic Fracture, ASTM STP 668,* American Society for Testing and Materials, Philadelphia, 1979, pp. 606–631.

[2] Harrison, J. D., "The "State-of-the-art" in Crack Tip Opening Displacement (CTOD) Testing and Analysis," Report 7302.10/80/210.3, The Welding Institute, U.K., 1980.

[3] Read, D. T., "Experimental Method for Direct Evaluation of the J-Contour Integral" in *Fracture Mechanics: Fourteenth Symposium, ASTM STP 791.* J. C. Lewis and G. Sines, Eds., American Society for Testing and Materials, Philadelphia, 1983, pp. II-199–II-213.

[4] Lereim, J. and Saether, S., "Fracture Resistance of Three Ship Steels," International Conference on Advances in the Physical Metallurgy and Applications of Steels, Liverpool, U.K., 1981.

[5] Anderson, T. L., "The Effect of Crack-Tip Region Constraint on Fracture in the Ductile-to-Brittle Transition Region, Ph.D. thesis, No. T-2741, Colorado School of Mines, Golden, CO, 1983.

[6] Newman, J. C. and Raju, I. S., "An Empirical Stress-Intensity Factor Equation for the Surface Crack," *Engineering Fracture Mechanics,* Vol. 15, No. 1-2, 1981, pp. 185–192.

[7] Lereim, J., "Elastic-Plastic Fracture Analysis of Heavy Walled Structural Elements," *International Conference on Offshore Welded Structures,* London, U.K., Paper 13, 1983.

[8] Ebrahimi, F., "A Study of Crack Initiation in the Ductile-to-Brittle Transition Region of a Weld," this publication, pp. 555–580.

Gerald W. Wellman,[1] William A. Sorem,[2] Robert H. Dodds, Jr.,[2] and Stanley T. Rolfe[2]

Specimen Thickness Effects for Elastic-Plastic CTOD Toughness of an A36 Steel

REFERENCE: Wellman, G. W., Sorem, W. A., Dodds, R. H., Jr., and Rolfe, S. T., **"Specimen Thickness Effects for Elastic-Plastic CTOD Toughness of an A36 Steel,"** *Fracture Mechanics: Eighteenth Symposium, ASTM STP 945,* D. T. Read and R. P. Reed, Eds., American Society for Testing and Materials, Philadelphia, 1988, pp. 535–554.

ABSTRACT: This paper describes the results of an experimental and analytical study of the effect of specimen size on the fracture toughness behavior of A36 steel, primarily in the lower-transition region. This region is beyond the limits of plane strain as defined by ASTM Method E 399, but in the region where failure occurs prior to any macroscopic fibrous tearing. Experimental results were obtained using crack-tip opening displacement (CTOD) three-point bend specimens, and analytical results were obtained using a 3-D elastic-plastic finite-element analysis. Two specimen sizes, full thickness [31.8 mm (1.25 in.)] and reduced thickness [12.7 mm (0.50 in.)], were investigated using the preferred $t \times 2t \times 8t$ specimen size.

Testing was conducted according to the procedure outlined in BS 5762:1979, Methods for Crack Opening Displacement Testing. CTOD-temperature transition curves were developed for both specimen sizes. A sufficient number of specimens were tested at two temperatures in the lower-transition region to allow statistical characterization of the CTOD results. Both normal and Weibull distributions were used for that characterization.

Three-dimensional elastic-plastic finite-element analyses were conducted to determine the effect of specimen size on the near-crack-tip opening stresses through the specimen thickness. The near-crack-tip stress fields were compared at various CTOD levels well into the elastic-plastic regime, but prior to any stable crack growth. At each of the applied CTOD levels, the maximum near-crack-tip opening stress occurred at the specimen centerplane and was found to be essentially independent of specimen size.

The results of this study showed that the lower bound of the elastic-plastic CTOD toughness values of A36 steel for small specimens is essentially equal to those of larger specimens in the lower-transition region prior to any stable crack growth. Thus it may be possible to characterize the notch toughness of low-strength structural steels, such as A36 steel in the lower-transition region, using small specimens provided that a sufficient number of specimens are tested. However, additional study is required before any recommendations can be made.

KEY WORDS: elastic-plastic fracture mechanics, crack-tip opening displacement (CTOD), size effects, toughness, fracture, finite element, thickness

Early in the evolution of the field of fracture mechanics, the problem of variation of fracture toughness test results with variation of specimen size was recognized [*1–3*]. This variation in test results led to the adoption of a specimen size requirement in the ASTM Standard Test Method for Plane-Strain Fracture Toughness of Metallic Materials (E 399-83) and the British Standards Methods of Test for Plane-Strain Fracture Toughness (K_{Ic}) of Metallic Materials (BS 5447).

[1]Member of the technical staff, Sandia National Laboratories, Albuquerque, NM 87185.

[2]Research assistant, associate professor, and professor, respectively, University of Kansas, Lawrence, KS 66045.

The specimen through-thickness dimension (B) was specified to develop "plane-strain" behavior at the specimen centerplane. The remaining crack ligament ($W - a$) was specified to insure a proportionally small plastic zone size at the crack tip. It is emphasized that this plane-strain behavior is defined in terms of invariant fracture test results, not in the normal manner of zero through-thickness strain.

The development of the J-integral to describe elastic-plastic behavior [*4,5*] allowed relaxation of the small plastic zone size requirement inherent in K_{Ic} testing. However, a plane-strain size requirement similar in intent to that of E-399 was imposed in the ASTM Test Method for J_{Ic}, a Measure of Fracture Toughness (E 813-81). It is interesting to note that for structural steels the specimen size required for plane-strain behavior in K_{Ic} testing is at least one order of magnitude larger than that required in J_{Ic} testing. With the advent of J_{Ic} testing of specimens exhibiting stable crack growth, the effect of specimen size on crack growth resistance (R) curves and subsequently on K_Q values was recognized [*6–8*].

Failures with clearly recognizable instability (predominately cleavage failures with no prior stable crack growth) are unaffected by the R-curve size effects. Two theories describe the effect of specimen size on predominately cleavage failures. The "loss of constraint theory," on which K_{Ic} size requirements are based, requires a certain minimum thickness to limit through-thickness yielding at the crack tip. The loss of constraint prevents the attainment of a state of triaxial tensile stress and thus prevents an elevated opening mode stress at the crack tip which is required for a cleavage failure [*9*]. This theory has been advocated recently by Dawes [*10*], Milne and Chell [*11*], and Pisarski [*12*].

The "weakest link theory" assumes that there is an inherent variability in the material and that both high and low regions of toughness are present within a given specimen. Fracture toughness test results are believed to be controlled by the region of the lowest toughness within each specimen, the "weakest link." The implication is that smaller specimens (less material sampled) will exhibit more scatter but will have the same lower-bound toughness value as that measured with larger specimens. Thus, the results of a single small-specimen test generally should be expected to exhibit a higher toughness value than that of a single large-specimen test since the probability of containing the weakest link is lower. Also, implicit in the weakest link theory is the need for a statistically significant number of tests at each temperature to determine a lower-bound toughness value. The weakest-link theory was proposed by Logsdon [*13*] and later Logsdon and Begley [*14*]. Landes and Shaffer [*15*] provided experimental support through J_c testing of three materials.

Attempts to verify the loss of constraint theory have relied primarily on experimental results. Unfortunately, few data are available from U.S. investigators because testing is routinely performed only to ASTM Methods E 399 or E 813. Neither E 399 nor E 813 test procedures are strictly applicable to the fracture toughness values investigated here. The elastic-plastic specimen behavior and the specimen size invalidate the results according to specification E 399. The brittle fracture (no R-curve development) invalidates the results according to specification E 813. A few U.S. investigators [*5,13–18*] report "invalid" results as K_c or J_{max} values, but the macroscopic failure mechanism is seldom reported. In contrast, the British Methods for Crack Opening Displacement (CTOD) Testing (BS 5762) requires test specimens of the same thickness as the structure of interest. Macroscopic failure mechanisms are routinely reported for these tests. In the absence of an actual structural application, all CTOD results are "valid" for their particular specimen thickness. Therefore, CTOD test results [*10–12,19–23*] are well suited to the investigation of the effects of specimen size on fracture toughness results of brittle failures.

In this paper, the effect of specimen size on fracture toughness test results are investigated both experimentally and analytically. Two sizes, 12.7 by 25.4 by 102 mm (0.5 by 1 by 4 in.) and 31.8 by 63.5 by 127 mm (1.25 by 2.5 by 10 in.), of three-point bend specimens of A36 steel were tested to generate the CTOD-temperature transition curve. In addition, a statistically valid population of both specimen sizes was tested at two temperatures exhibiting cleavage type behavior with no prior stable crack growth. This allowed both a Weibull and normal distribution comparison of the specimen size effect on the fracture toughness test results. Finally, a 3-D elastic-plastic finite-element analysis of both specimen sizes was performed to determine the state of stress near the crack tip. This analysis addresses the degree of "plane-strain" behavior with respect to specimen size for the A36 material. This combined analytical and experimental approach provides insight into the effect of specimen size on elastic-plastic fracture toughness (E-PFM) test results in the region where linear elastic theory does not apply but where failure occurs with no macroscopic fibrous tearing.

Material Properties

A 31.8-mm-thick (1.25-in.) as-rolled plate of A36 steel was used in this study. Tables 1 and 2 give the chemical analysis and mechanical properties of the steel. Table 3 gives the yield strength of this A36 steel plate at various temperatures as determined by Shoemaker and Seeley [*24*]. Charpy-V-Notch impact specimens from this plate were tested at temperatures ranging from −74°C (−100°F) to 82°C (180°F). Figure 1 shows the absorbed energy versus temperature results. The 20.4 J (15 ft · lb) absorbed energy transition occurs at approximately 4°C (40°F) and the 50% shear fracture appearance occurs at 24°C (75°F). The upper shelf region starts at about 49°C (120°F) and a toughness of about 136 J (100 ft · lb).

Metallographic specimens were extracted from the centerplane and the surface of the A36 steel plate to study the microstructure. The ASTM grain size number was approximately six

TABLE 1—*Chemical analysis of A36 steel.*

Element	%
C	0.20
Mn	1.11
P	0.007
S	0.023
Si	0.029

TABLE 2—*Mechanical properties of A36 steel.*

Property	Quantity
Yield strength @ 0.2% offset	248 MPa
Tensile strength	460 MPa
Elongation in 2 in. (50.8 mm)	38%
Reduction in area	67%

TABLE 3—*Yield strength of A36 steel adjusted to temperature.*

Temperature, °C	Yield Strength, MPa
21 to 0	248
−18	262
−43	290
−76	345
−112	462
−195	793

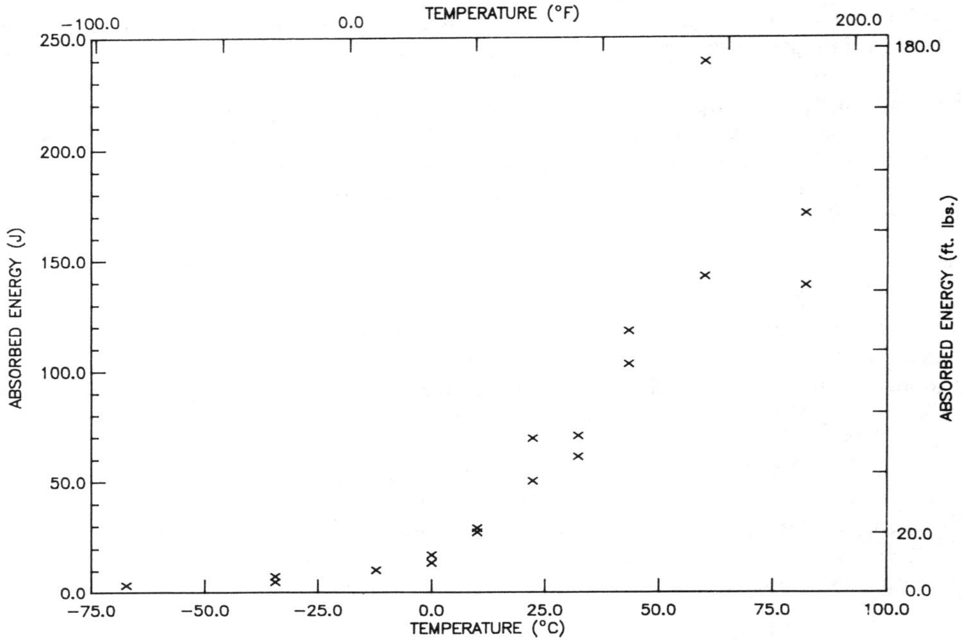

FIG. 1—*CVN impact results for the A36 steel plate used in this study.*

with an average grain diameter of 50 μm (2 mils) was measured from the micrograph. No significant difference was observed between the microstructure at the centerplane and at the surface, thus these values are considered to be representative of both the small and the large specimens.

Experimental Procedure

Preferred type ($t \times 2t$) three-point bend specimens with an a/W ratio of 0.5 were prepared from the A36 steel plate [31.8 by 1372 by 1829 mm (1.25 by 54 by 72 in.)] in its as-rolled condition. Specimen preparation, machining, and fatigue cracking were conducted according to BS 5762 and ASTM E 399. The large three-point bend specimens [31.8 by 63.5 by 254 mm (1.25 by 2.5 by 10 in.)] were prepared with their crack planes perpendicular to the rolling direction of the plate [longitudinal-transverse (L-T) orientation]. The small three-point bend specimens [12.7 by 25.4 by 102 mm (0.5 by 1 by 4 in.)] were machined from the broken ends of the large specimens and followed the same crack plane orientation. These small specimens were extracted from two different locations within the original plate (the 1/4T location and the centerplane location as shown in Fig. 2).

A 200-kN universal closed-loop testing machine was used for both fatigue cracking and the final ramp load to failure. Crack mouth opening displacement was measured by a clip-gage mounted on knife edges machined into the specimen. Both clip-gage displacement and load were transmitted directly from the test machine signal conditioners to an X-Y recorder.

Testing was conducted over temperatures ranging from -195°C (-320°F) to 21°C (70°F). Temperatures below room temperature were attained by submersing the specimen in a heat transfer bath mounted on the test machine actuator. A mixture of ethyl alcohol and solid carbon dioxide provided test temperatures as low as -76°C (-105°F). Liquid nitrogen was circulated through copper radiating coils within the heat transfer bath to lower the ethyl alcohol temperature to -112°C (-170°F). A bath of liquid nitrogen was used to test at -195°C

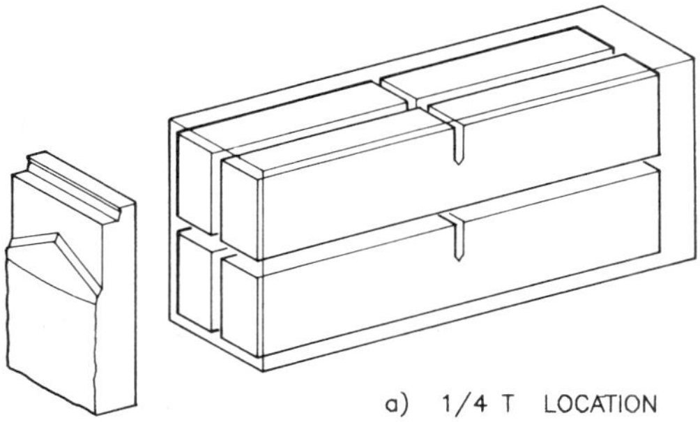

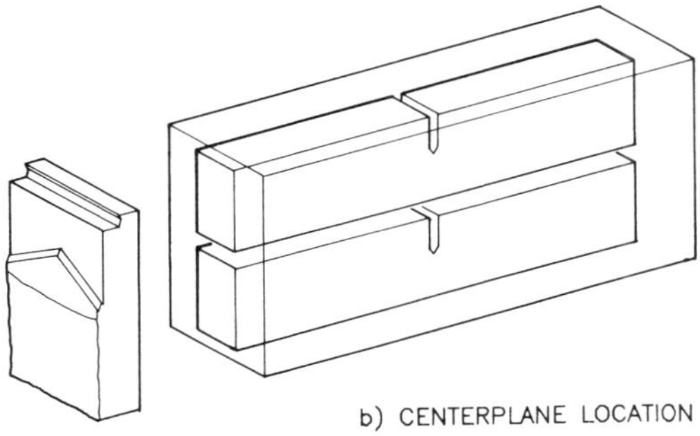

FIG. 2—*12.7-mm specimens as extracted from 31.8-mm specimen broken ends.*

(−320°F). An electric stirrer was used to minimize any thermal gradients within the heat transfer bath. Two bimetal thermometers were used to measure the testing temperatures. These were calibrated with a Leeds and Northrup Potentiometer using a copper and constantan thermocouple. The specimens were held at testing temperature ±2°C (3°F) for approximately 20 min and tested without removing them from the cooling bath.

The first attainment of a local maximum load was used with the equation provided in BS 5762:1979 to determine the critical CTOD.

$$\delta = \frac{K^2(1 - \nu^2)}{2\sigma_{ys}E} + \frac{0.4(W - a)V_p}{0.4W + 0.6a} \tag{1}$$

where

K = stress intensity factor = $YP/B\sqrt{W}$,
Y = stress intensity coefficient for a three-point bend specimen having span (S) = $4W$:

$$= \frac{6(a/W)^{1/2}\{1.99 - a/W[1 - a/W][2.15 - 3.93\,a/W + 2.7\,(a/W)^2]\}}{(1 + 2a/W)(1 - a/W)^{3/2}}$$

P = applied force,
B = specimen depth,
W = specimen width,
ν = Poisson's ratio,
σ_{ys} = 0.2% offset yield strength at temperature of interest,
E = Young's modulus at temperature of interest,
a = effective crack length (initial crack length + stable crack growth), and
V_p = plastic component of clip-gage displacement.

The standard convention was followed in reporting the results of CTOD testing:

(*a*) δ_c for initiation and propagation,
(*b*) δ_u for fibrous initiation followed by brittle propagation (fibrous thumbnail), and
(*c*) δ_m for fibrous tearing initiation and propagation.

Experimental Results

The small specimens were extracted from two different locations within the original plate (the 1/4T location and the centerplane location as shown in Fig. 2), to determine if through-thickness location had any appreciable effect on CTOD results. No difference in CTOD results was observed with variation of through-thickness location. The CTOD-temperatures transition curves for the two specimen sizes are shown in Figs. 3 and 4. Both curves exhibit the typical toughness-temperature relation of low-strength steel. The transition curve for low-strength steel can be divided into four general regions of toughness—lower shelf, lower-transition, upper transition, and upper shelf [*21*]—as shown schematically in Fig. 5.

The lower-shelf region corresponds to linear-elastic plane-strain behavior as defined in ASTM E 399. Thus, the upper boundary of this region is that toughness level or temperature at which "valid K_{Ic}" data can no longer be obtained based on specimen size and yield strength. Table 4 gives the calculated maximum valid K_{Ic} and equivalent CTOD values for A36 steel at various temperatures and specimen sizes according to ASTM E 399. These maximum CTOD values can be compared to the experimental results shown in Figs. 3 and 4. Analysis of the load-displacement records indicate that only the specimens tested at −195°C (−320°F) and two of the large specimens tested at −112°C (−170°F) are "valid" plane-strain tests according to the ASTM E 399 test procedure. The CTOD results in this region are reported as δ_c values.

The lower-transition region corresponds to elastic-plastic behavior between the upper limit of "valid K_{Ic}" behavior and the start of the fibrous tearing mechanism. In this region, the plastic

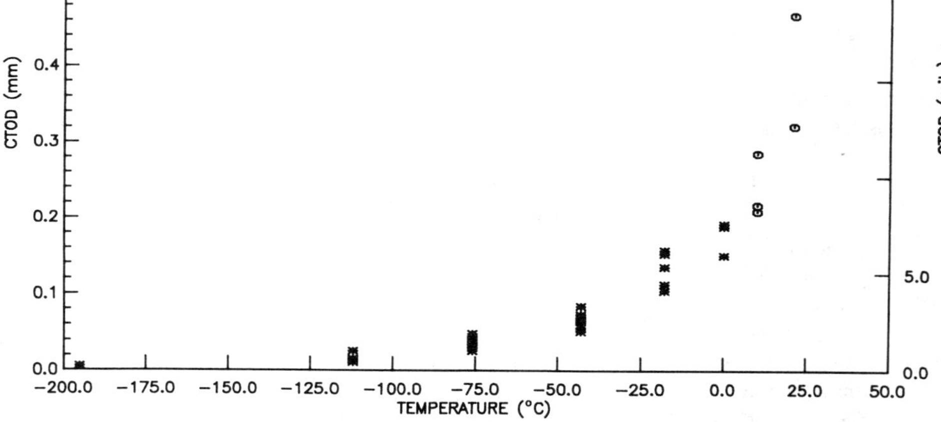

FIG. 3—*CTOD versus temperature test results for 31.8-mm-thick A36 steel specimens.*

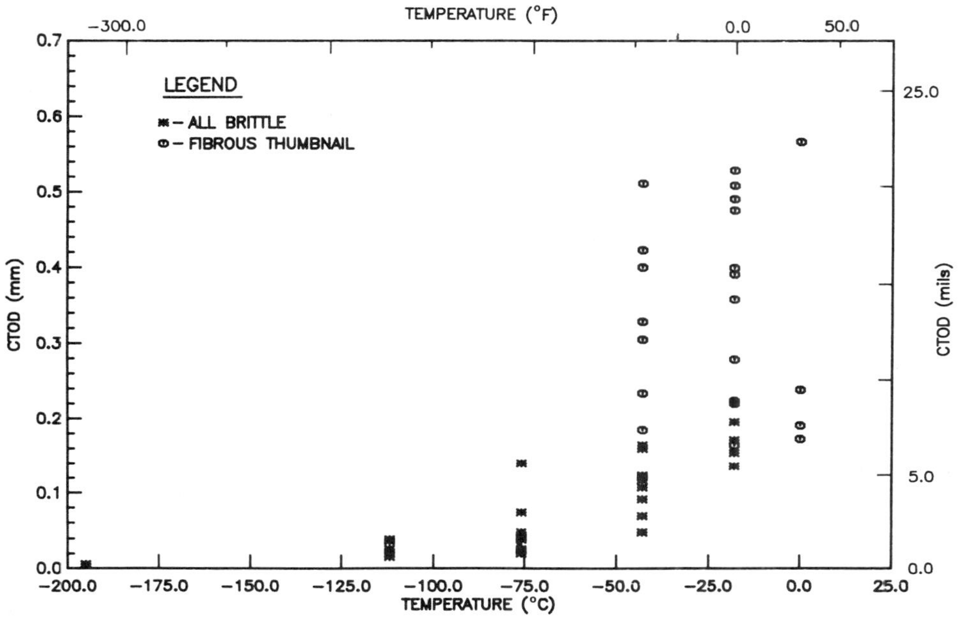

FIG. 4—*CTOD versus temperature test results for 12.7-mm-thick A36 steel specimens.*

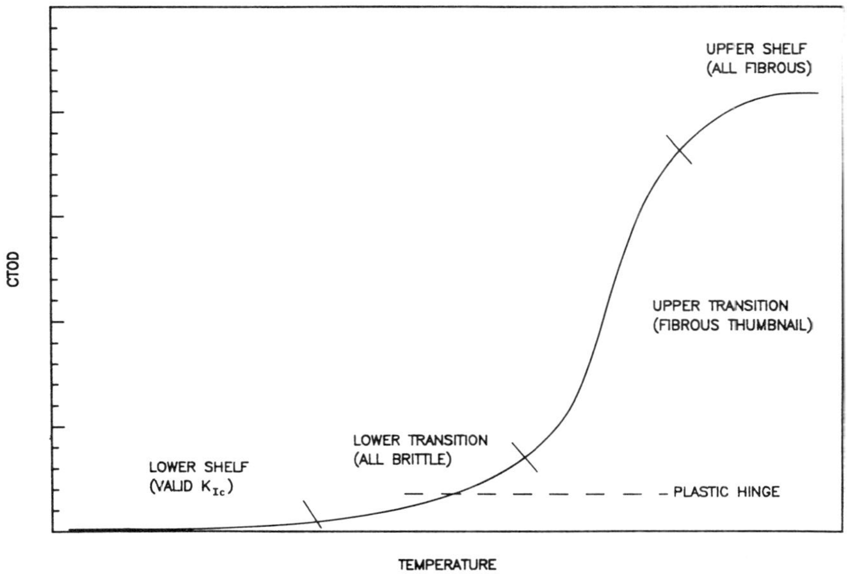

FIG. 5—*Schematic CTOD-temperature transition curve.*

TABLE 4—*Maximum toughness values for "valid" LEFM measurement.*

Temperature, °C	Yield Strength, MPa	Thickness, mm	Maximum K_{Ic}, MPa $\sqrt{m}$	Equivalent Maximum CTOD, mm
−195	793	31.8	89.3	0.071
−112	462	31.8	52.0	0.024
−76	345	31.8	38.9	0.013
−195	793	12.7	56.4	0.028
−112	462	12.7	32.9	0.010

zone size becomes too large for analysis by linear elastic fracture mechanics (LEFM). For the larger specimens, fibrous tearing began at about 10°C (50°F) at a CTOD of 0.208 mm (8.2 mils). The upper boundary for the smaller specimens is not as definitive; a fibrous thumbnail appeared in seven of the eighteen specimens tested at −43°C (−45°F) and four of the fifteen specimens tested at −18°C (0°F) at a CTOD of 0.172 mm (6.8 mil). CTOD results in the lower-transition region are also reported as δ_c values.

The upper transition region is characterized by a very rapid increase in material toughness with increasing temperature. This region starts with a visible change in failure initiation from a brittle cleavage to a fibrous tearing mechanism (fibrous thumbnail). The initiation of stable crack growth is most often studied using crack growth resistance curves. CTOD results in this region can be reported as δ_i if initiation of stable crack growth is measured. Specimens were tested in this region to define the upper boundary of the lower-transition region. No crack growth resistance curves were determined for these tests. Therefore all CTOD results were reported at maximum load attainment as δ_u values.

The upper shelf region is characterized by a "leveling off" of the material toughness with increasing temperature. Fibrous tearing is visible over the entire failure surface. No crack growth instability (cleavage) is experienced. Fracture toughness in this region is most often studied using the J_{Ic} parameter (crack growth resistance). CTOD results in this region are reported as δ_m for maximum attainment of load or δ_i for initiation of stable crack growth. No specimens were tested in the upper shelf region.

At very low temperatures, minimal scatter is observed in the experimental CTOD results. The lowest CTOD values obtained from the small specimens and the large specimens are essentially identical. As testing temperatures increase, the scatter in the CTOD values increases. This increase in data scatter is readily observed in both specimen sizes, but is more pronounced in the smaller specimens. Much of the specimen scatter in the small specimen CTOD results is due to the fact that fibrous tearing precedes failure. Because the CTOD values are based on the load and displacement at failure rather than at initiation of stable crack growth (fibrous tearing), a large variation in results would be expected. However, the lowest values obtained for the two specimen sizes have essentially the same failure mechanism and CTOD value as shown in Fig. 6.

Many more specimens than is normal for elastic-plastic fracture testing were tested at −76°C (−105°F) and at −43°C (−45°F). These temperatures are clearly in the lower transition region for the large specimens, because all the failures occurred by a brittle mechanism. At −76°C (−105°F), all the small specimens also failed in a brittle manner. However, at −43°C (−45°F), seven of the eighteen small specimens exhibited a thumbnail of fibrous tearing prior to final brittle failure. Thus, there is some question as to whether the small specimens at −43°C (−45°F) are in the lower or the upper transition region. For the purposes of this study, the failure mode of the specimen with the lowest CTOD value was considered to be the controlling mechanism. Therefore, these small specimens are also considered to be in the lower transition region.

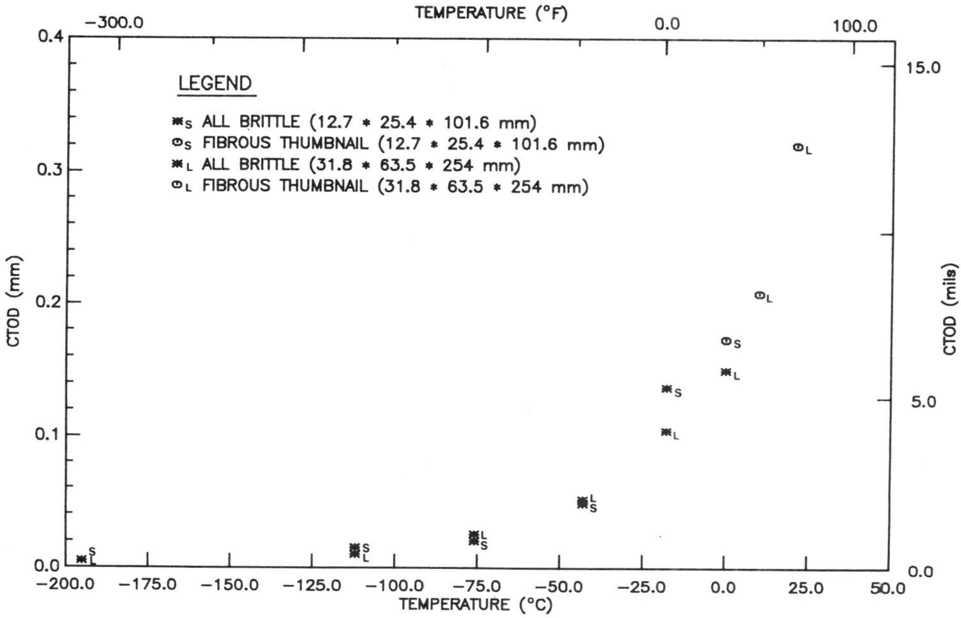

FIG. 6—*Comparison of lowest experimental CTOD values for A36 steel specimens.*

For this plate of A36 steel, there are limits on the CTOD values for specimens exhibiting a brittle failure with no prior fibrous tearing. Because negative CTOD values are physically meaningless (for example, values established from statistical analyses), zero is clearly a lower bound. The limit at which fibrous tearing occurs prior to brittle failure is an upper bound. This upper bound is less accurately established primarily due to the subjective process of identifying fibrous tearing. The data shown in Figs. 3 and 4 suggest that this limit lies in the range of 0.17 to 0.20 mm (0.0068 to 0.008 in.).

Because of these upper and lower limits on the CTOD values of specimens displaying lower transition region behavior, the test results cannot be normally distributed. However, between these limits [0 to 180 mm (0 to 7.2 in.)], a chi-squared test indicates that the data for both sized specimens at both temperatures is closely approximated by a normal distribution. Therefore, because normal distributions are widely used and understood, the results of assuming that the CTOD failure data is normally distributed are given in Table 5. The small specimens at −43°C (−45°F), which exhibited fibrous thumbnails, were excluded from the data analysis. The 95%

TABLE 5—*Normal distribution statistics.*

Specimen Size, mm	Temperature, °C	Median, mm	Mean, mm	Standard Deviation, mm	95% M-1.65 σ, mm	Approximate 95% Confidence Limit on 95% Lower Bound, mm
12.7 by 25.4	−76	0.039	0.046	0.030	−0.004	−0.052 to 0.027
31.8 by 63.5	−76	0.036	0.036	0.007	0.024	0.006 to 0.035
12.7 by 25.4	−43	0.117	0.106	0.039	0.042	−0.041 to 0.085
31.8 by 63.5	−43	0.060	0.066	0.011	0.048	0.016 to 0.064

lower bound (mean − 1.65 standard deviations) was chosen as a basis for comparing lower bounds of the CTOD data. The calculated 95% lower bounds at −76°C (−105°F) are −0.004 mm (−0.16 mil) for the small specimens and 0.024 mm (0.94 mil) for the large specimens. At −43°C (−45°F), these lower bounds are 0.042 mm (1.65 mil) for the small specimens and 0.048 mm (1.89 mil) for the large specimens. The calculated lower-bound value for the small specimens at the lower temperature [−0.004 mm (−0.00016 in.)] is both physically meaningless and outside of the range of the chi-squared test. Therefore, for the small specimens at −76°C (−105°F), the assumption of a normal distribution is not helpful. Despite the relatively large number of test specimens at each temperature compared with typical fracture testing practice, in terms of statistical populations, the CTOD data sets are small. Therefore, the ranges of the 95% confidence limits on the lower bounds in Table 5 are quite wide.

An alternative approach to the assumptions of a normal distribution is available in the Weibull method, which is widely used in the treatment of fatigue failures [*25*]. The Weibull characteristics of the four CTOD data populations discussed above are given in Table 6. The data from the small specimens at −43°C (−45°F) that exhibited a fibrous thumbnail were treated as suspended tests using the Weibull method. That is, they served to adjust the relative placement of the other data, but the actual CTOD values of these data were not used. The 95% lower-bound CTOD values (B5 values in Weibull terminology) at −76°C (−105°F) are 0.018 mm (0.71 mil) for the small specimens and 0.022 mm (0.87 mil) for the large specimens. The 95% lower bounds at −43°C (−45°F) are 0.040 mm (1.57 mil) for the small specimens and 0.043 mm (1.69 mil) for the large specimens. The 95% confidence limits on the 95% lower-bound CTOD values in Table 6 are narrower than those in Table 5, and there are no physically meaningless negative values.

With the exception of the small specimens at −76°C (−105°F), the 95% lower-bound CTOD values are nearly identical using either a normal distribution assumption or the Weibull method. Both methods indicate that there is no effect of specimen size on lower bound CTOD values. Because of the ease of application of the Weibull method, and the problem with the normal distribution assumption for the small specimens at −76°C (−105°F), the Weibull method is recommended for the statistical treatment of CTOD test results.

Finite-Element Analysis Procedure

The purpose of analyzing the bend specimens shown in Fig. 2 was to simulate the deformation up to the point of cleavage type failure with no prior stable crack growth. Therefore no crack growth was modeled in the finite-element analysis. The finite-element solutions employed the conventional, linear strain-displacement relations based on small geometry change assumptions. Numerical computations were performed with the Polo-Finite structural mechanics system [*26,27*].

The 2-D finite-element model for both plane-strain and plane-stress analyses was developed using the 8-node isoparametric element. Due to symmetry of the three-point bend specimen,

TABLE 6—*Weibull distribution statistics.*

Specimen Size, mm	Temperature, °C	Median, mm	Slope	95% Lower Bound, mm	Approximate 95% Confidence Limit on 95% Lower Bound, mm
12.7 by 25.4	−76	0.038	3.4	0.018	0.009 to 0.027
31.8 by 63.5	−76	0.035	5.8	0.022	0.016 to 0.030
12.7 by 25.4	−43	0.148	2.0	0.040	0.008 to 0.096
31.8 by 63.5	−43	0.065	6.3	0.043	0.032 to 0.059

only the right half of the specimen was modeled as shown in Fig. 7. This model was partitioned into linear and nonlinear regions. Static condensation was applied to the linear region to reduce the 19 elements and 78 nodes to just one element having 12 nodes. The condensed substructure was compatible with the 8-node elements in the nonlinear region. The nonlinear region was modeled with 41 elements and 156 nodes including the elements and nodes retained from the condensed linear substructure.

The crack tip was modeled using 8-node isoparametric elements degenerated into triangles as shown in Fig. 8*a*. This is a common technique to develop a singularity of the order $1/\sqrt{r}$ for linear analysis [*28*] and $1/r$ for elastic-plastic analysis [*29*]. Placement of the side nodes at the midpoint rather than the quarterpoint was adopted to reduce the spurious unloading and to allow larger load increments [*30*]. The use of degenerated 8-node elements also provides a convenient means to extract CTOD from the computed displacements of the initially coincident nodes using the Rice [*4*] method. Rice proposed that the CTOD be defined as the separation of the crack flanks at the location where two rays, drawn at right angles to each other from the current crack tip, intercept the crack flanks. The graphical construction in Fig. 9 shows the final displaced locations of the initially coincident nodes and the measuring of CTOD.

The 3-D finite-element model was developed using the 20-node isoparametric element. The finite-element mesh was generated by replicating the 2-D grid pattern in the through-thickness dimension to define three element layers through the half thickness. The element layers varied in thickness with thinner elements near the free surface to accommodate the large strain gradients. Symmetry constraints were applied over the crack plane and over the through-thickness centerplane. Substructuring and condensation similar to that used in the 2-D analysis was employed in the 3-D analysis. Condensation reduced the 57 elements and 402 nodes of the linear region to one element with 69 nodes. This condensed element was retained in the 121 elements and 801 nodes in the nonlinear region.

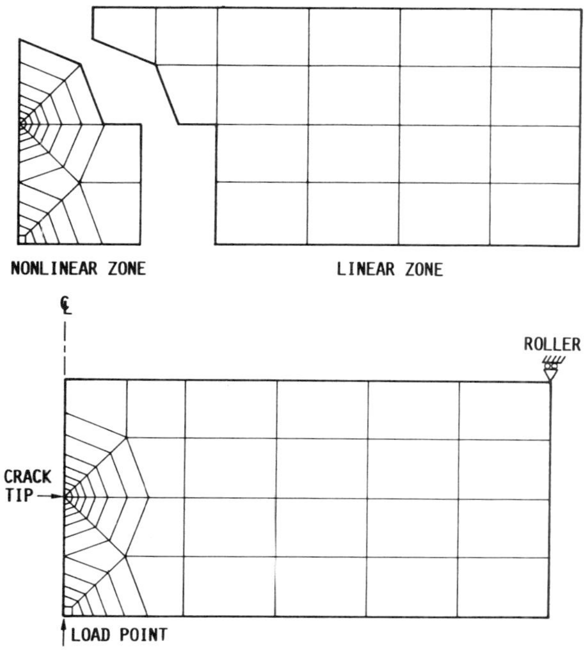

FIG. 7—*2-D finite element analysis mesh for the three-point bend specimen.*

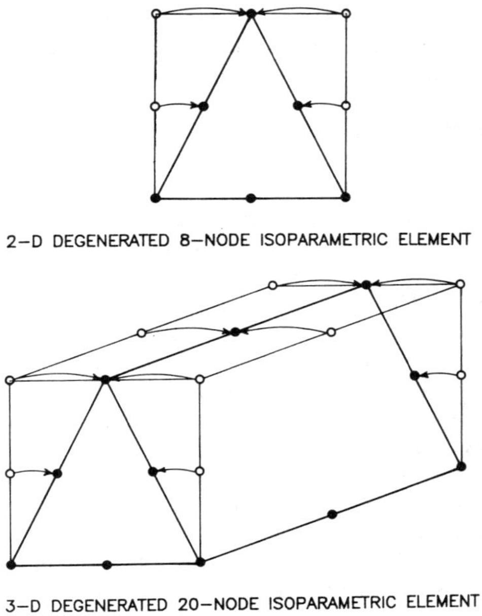

FIG. 8—*Crack tip element geometry for 2-D and 3-D degenerated isoparametric elements.*

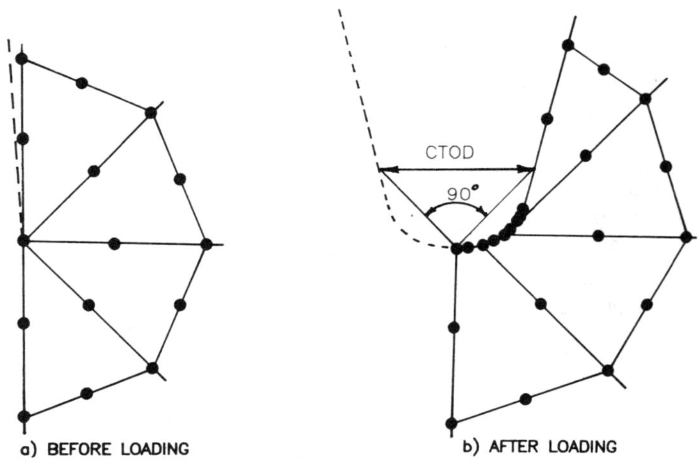

FIG. 9—*90-deg crack flank intercept CTOD as defined by Rice.*

The 3-D, 20-node isoparametric elements were degenerated into triangular prisms at the crack tip as shown in Fig. 8*b*. Their behavior is analogous to that of the degenerated 2-D, 8-node isoparametric element [*31*]. The coincident corner nodes of the degenerated element develop a singularity of the order $1/\sqrt{r}$ for linear analysis and $1/r$ for elastic-plastic analysis. The coincident midside nodes of the degenerated element do not develop an identical singularity. The CTOD was computed from the displacements of the initially coincident corner nodes of the degenerated elements using the same procedure previously outlined for the 2-D analysis. Thus, four CTOD values were generated through the specimen half thickness for the 3-D analysis.

The finite element models for the two specimen sizes studied were obtained simply by scaling the coordinates and elements with the exception of the two rows of elements (8 elements for 2-D and 24 elements for 3-D) nearest the crack tip. These elements were maintained at the same absolute size. Comparisons of stresses and strains were thus made at the same location (radius and angle from the crack tip) independently of the overall specimen size.

Reduced integration (2 by 2 for 2-D and 2 by 2 by 2 for 3-D) was used in the analysis to eliminate artificial locking under the incompressibility conditions imposed by plastic deformation [*32*]. Plasticity was modeled using incremental theory with a von Mises yield surface, associated flow rule, and isotropic hardening [*33*].

The material stress-strain properties for the finite-element analysis were derived from a standard 12.8-mm (0.505-in.)-diameter uniaxial tension test. A piecewise linear approximation provided a better match with the experimental curve than did a power-law approximation. However, due to the Newton-Raphson iteration technique employed to solve equations, a "stiffening" material caused the solution to diverge when large load steps were taken [*30*]. Therefore, the yield plateau was replaced with a straight line from the yield point to a point of tangency on the strain-hardening curve to produce a stress-strain curve with a monotonically decreasing slope.

The finite-element models were loaded by imposing displacements on the bottom of the uncracked ligament. Imposing displacement increments, rather than load increments, made better control of the solution possible, particularly near the limit load. Initially, the displacement was imposed directly below the uncracked ligament at a single node for 2-D analysis and at a line of nodes through the specimen thickness for 3-D analysis. As the size of the plastic zone increased, it was necessary to impose a proportional displacement on the adjacent nodes to be consistent with the local deformation that occurred at the loading roller in the experimental testing.

Finite Element Results

The finite-element results can best be compared with the experimental results using the load versus crack-mouth opening displacement (CMOD) curves. The load and CMOD are the primary experimental measurements and are also readily available from the finite-element analysis. Figure 10 shows the comparison of the analytical plane stress, plane strain, 3-D and experimental load versus CMOD curves for the large specimen. The relationship of the load versus CMOD curves for the small specimen is nearly identical. The experimental curves shown are typical of specimens tested in a temperature range between 0°C (32°F) and 21°C (70°F).

As can be seen in Fig. 10, the plane-strain analysis provides an upper bound, while the plane-stress analysis provides a lower bound to the experimental results. The plane-strain and plane-stress approximations become increasingly inaccurate as the plasticity increases (as a plastic hinge or the limit load is approached). The 3-D analysis provides a good match to the experimental results for both specimen sizes.

Further evidence of the three-dimensional nature of the standard three-point bend specimen is demonstrated by the variation of CTOD through the thickness of the specimen. In Fig. 11, the plane-strain, plane-stress, and 3-D CTOD values are plotted versus CMOD for the large specimen. The CTOD versus CMOD behavior of the small specimen is nearly identical. Using CMOD rather than load is preferable because it permits a more meaningful comparison to CTOD as the limit load is approached. Four values of CTOD were computed through the specimen half-thickness in the 3-D analysis as previously discussed. The CTOD was found to be a maximum at the specimen centerplane and to decrease to a minimum at the specimen free surface. This variation in CTOD is consistent with observations of crack "tunneling" in experimental results and has been described analytically by Robinson and Tetelman [*19*]. No varia-

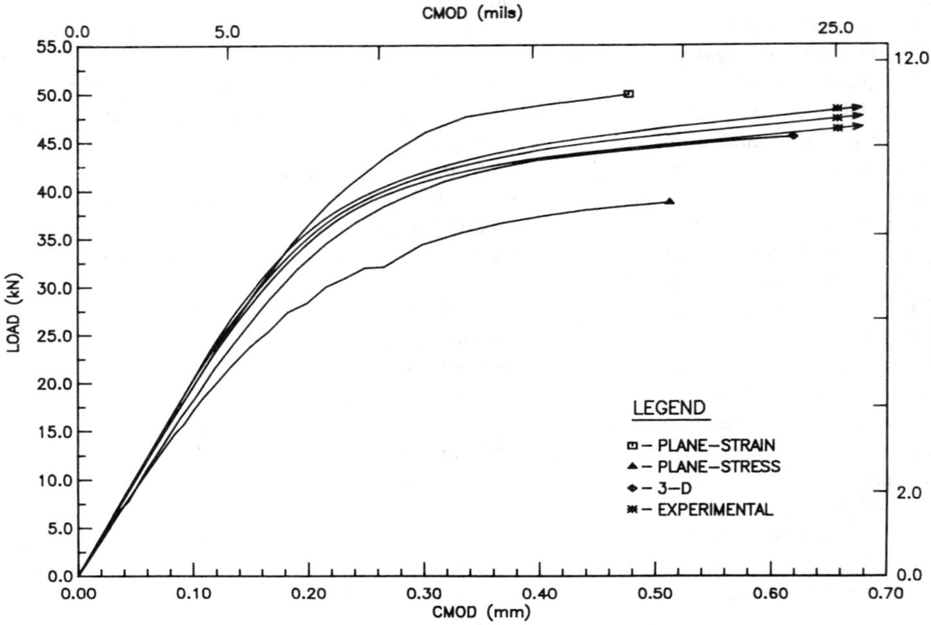

FIG. 10—*Load versus crack mouth opening displacement for 31.8-mm-thick A36 steel specimens.*

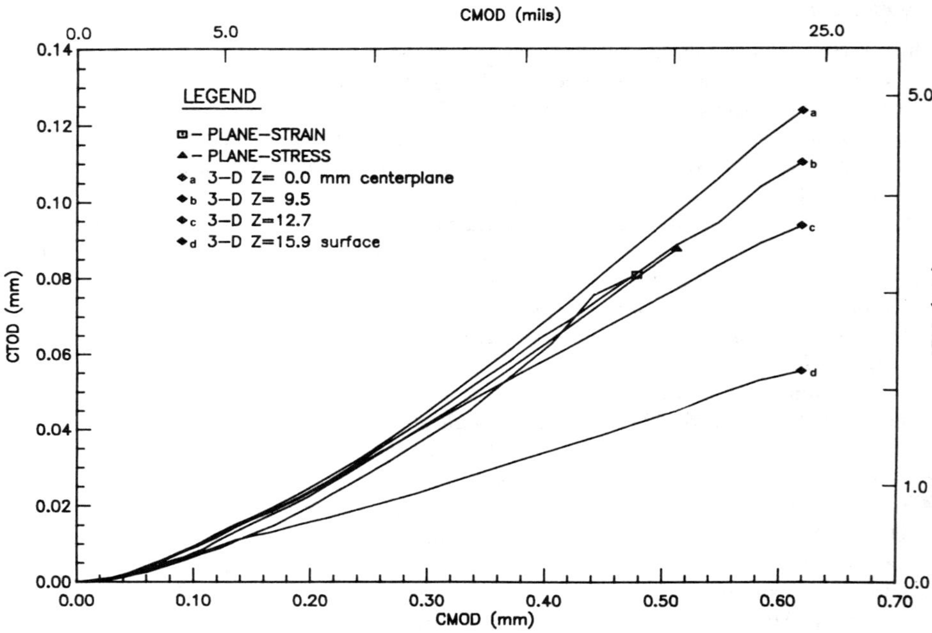

FIG. 11—*CTOD versus crack mouth opening displacement for 31.8-mm-thick A36 steel specimens.*

tion of the analytical CMOD was found through the thickness of the specimen. This implies that the variation in CTOD with through-thickness dimension is due to warpage of the crack face.

The through-thickness variation in crack-opening stress (σ_y) near the crack tip is shown in Fig. 12 for three applied CTOD levels. A comparison of the stresses was conducted at the same location from the crack tip [$r = 0.036$ mm (0.0142 in.) and $\theta = 55$ deg] for the two specimen sizes. This location is the Gauss point nearest the crack tip, but outside the degenerated isoparametric elements. The scale in Fig. 12 is nondimensionalized to accommodate both specimen sizes. The crack opening stress is divided by the room temperature, static yield strength 248 MPa (36 ksi), and the through-thickness dimension is divided by one-half the specimen thickness. This results in a horizontal axis where $z = 0.0$ corresponds to the specimen centerplane and $z = 1.0$ corresponds to the specimen outside surface.

The transverse stress (σ_x) and the through-thickness stress (σ_z) are not shown in these figures. The through-thickness behavior of σ_x is similar to that of σ_y but at approximately one-half the value. The value of σ_z is approximately one half of σ_y at the specimen centerplane but decreases to zero at the specimen free surface.

The two lower curves in Fig. 12 show the crack-opening stress distribution near the crack tip for linear elastic behavior. The CTOD value for the two lower curves corresponds to an approximate stress-intensity factor of 5.5 MPa $\sqrt{\text{m}}$ (5 ksi $\sqrt{\text{in.}}$). The stress distributions for the two specimen sizes are similar with a nearly constant stress (less than 5% variation) over the middle 80% of the specimen. The difference in crack-opening stress between the two specimen sizes is also less than 5%. Therefore, for geometrically similar ASTM E 399 three-point bend specimens exhibiting linear elastic behavior, the stress distribution at a given value of stress-intensity factor is independent of the specimen size. This does not imply that if the ratio of the specimen dimensions is changed, the stress distribution will remain unchanged.

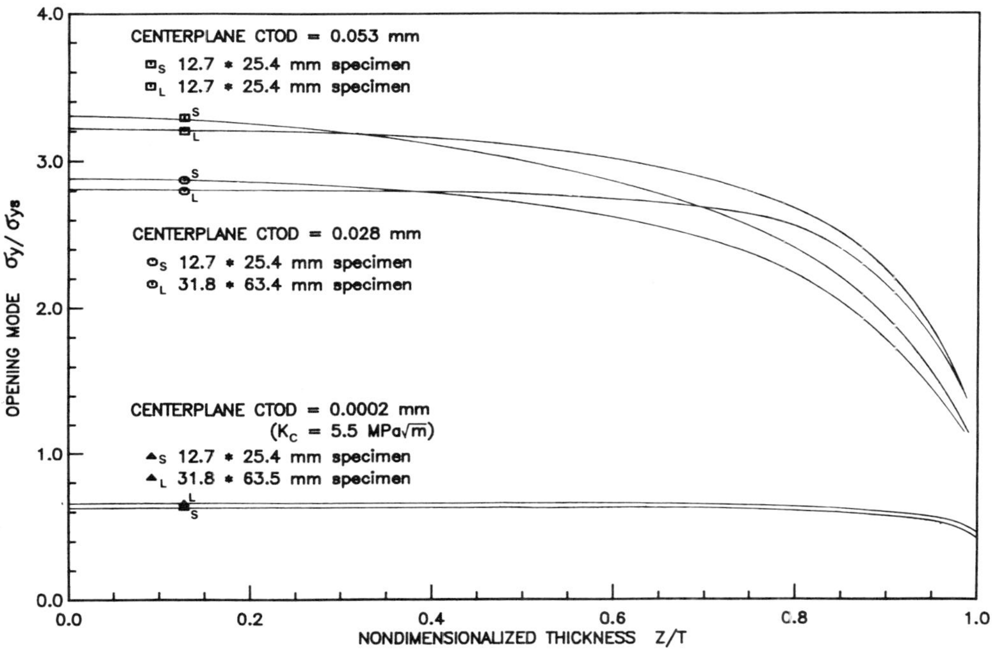

FIG. 12—*Crack opening stress distribution near the crack tip for three levels of CTOD.*

The crack-opening stress distribution for elastic-plastic behavior is shown in the four upper curves in Fig. 12. The two applied CTOD levels studied correspond to the lower-bound CTOD results of the experimental tests conducted at −76°C (−105°F) [CTOD = 0.018 mm (0.7 mil)] and −43°C (−45°F) [CTOD = 0.040 mm (1.6 mil)]. At each of these two applied elastic-plastic CTOD levels, the maximum crack-opening stress occurring at the specimen centerplane is nearly independent of specimen size. Both specimen sizes exhibit a decrease in stress as the outside free surface is approached.

The crack-opening stress near the specimen centerplane and the crack tip is compared with CTOD in Fig. 13 for the two specimen sizes that were modeled. Both curves resemble uniaxial stress-strain curves in shape. The slope is initially steep but undergoes a smooth decrease to an assumed eventual horizontal asymptote. The curves are very similar, but the curve for the smaller specimen lies slightly above the curve for the larger specimen. In this case, the smaller specimen is consistently subjected to a locally more severe stress state than the larger specimen at equivalent CTOD levels. This is supported by the results shown in Table 6. The shape of the σ_y versus CTOD curve has implications concerning experimental testing. For low values of toughness (low CTOD), a large change in σ_y is associated with a small change in CTOD. As the CTOD level increases, small changes in σ_y are associated with large changes in CTOD. Thus, the observation of increased data scatter in the lower-transition region can be explained in terms of Fig. 13.

The size of the plastic zone at a given CTOD value is an important characteristic of structural behavior. The plastic zones near the specimen centerplane and free surface for CTOD values of 0.028 mm (1.1 mil), 0.053 mm (2.1 mil), and 0.074 mm (2.9 mil) are shown in Fig. 14 for the large specimen and 0.028 mm (1.1 mil), 0.053 mm (2.1 mil), and 0.064 mm (2.5 mil) in Fig. 15 for the small specimen. As shown in Figs. 14 and 15, the plastic zone near the free surface is more elongated but only slightly larger than the plastic zone near the specimen centerplane.

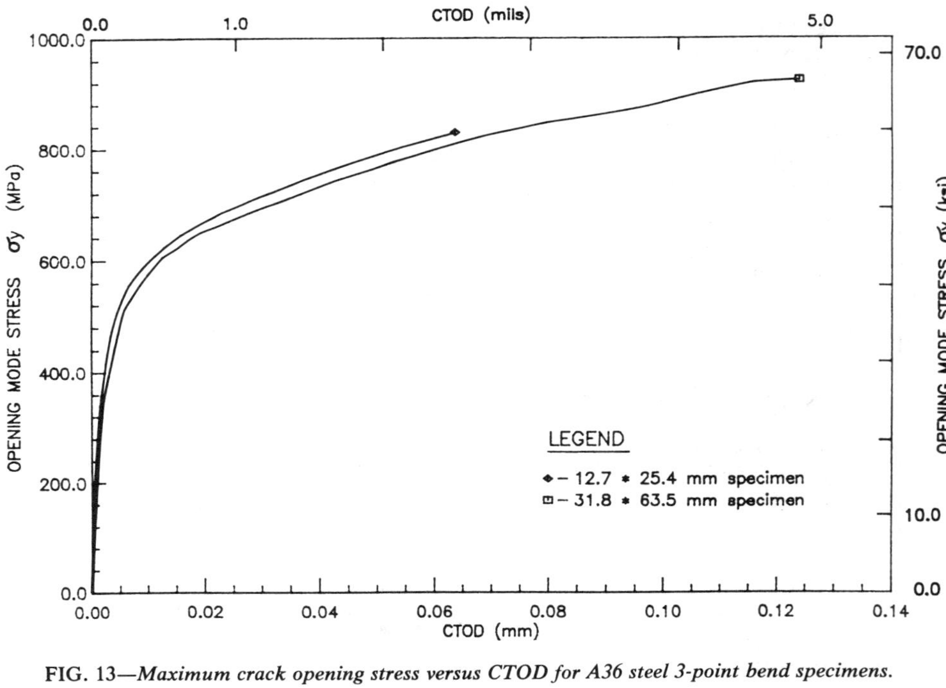

FIG. 13—*Maximum crack opening stress versus CTOD for A36 steel 3-point bend specimens.*

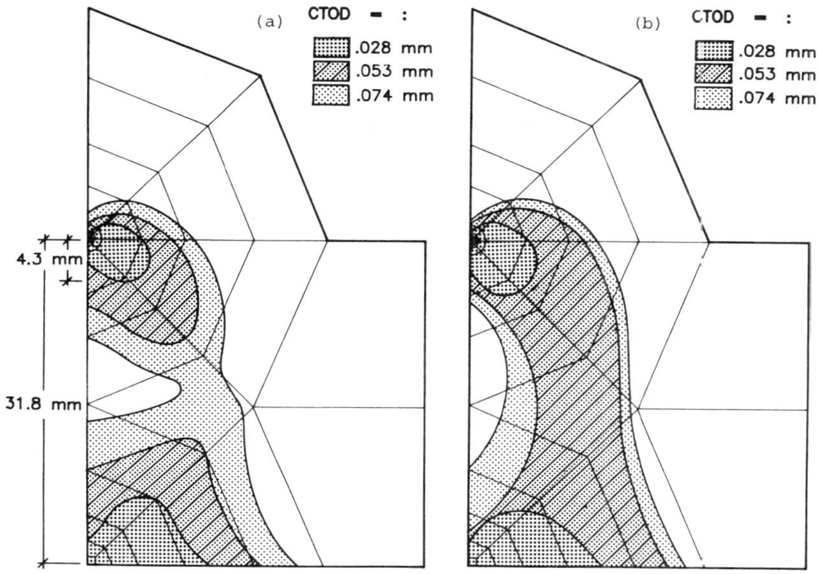

FIG. 14—*Plastic zones for 31.8-mm-thick A36 steel specimens at CTOD levels of 0.028 mm, 0.053 mm, and 0.074 mm:* (a) *centerplane;* (b) *surface.*

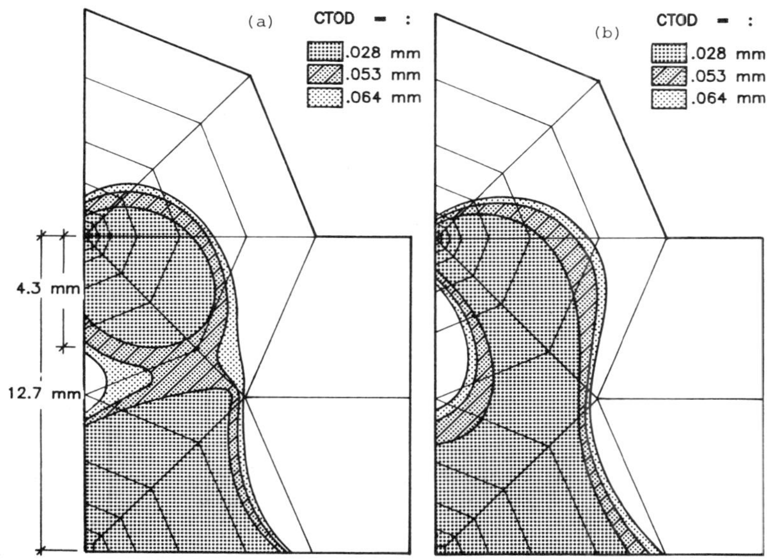

FIG. 15—*Plastic zones for 12.7-mm-thick A36 steel specimens at CTOD levels of 0.028 mm, 0.053 mm, and 0.064 mm:* (a) *centerplane;* (b) *surface.*

When the plastic zone extending from the crack tip intersects the plastic zone extending from the loading point, a plastic hinge is formed. From Figs. 14 and 15, it can be seen that a plastic hinge forms at an approximate CTOD value of 0.028 mm (1.1 mil) in the small specimen and 0.074 mm (2.9 mil) in the large specimen. The formation of a plastic hinge does not correspond to the attainment of a limit load since the low-strength steel undergoes subsequent strain hardening. However, the development of a plastic hinge does produce a significant reduction in specimen stiffness and is an important parameter in the establishment of rational fracture toughness material specifications.

Discussion

A36 steel is not a continuous, homogeneous, isotropic material, but is actually composed of randomly oriented grains. In a body-centered cubic crystal lattice (A36 ferrite grains), cleavage occurs due to a local stress perpendicular to a specific family of crystallographic planes. This leads to one possible explanation of the weakest-link fracture theory. The crack-opening stress profiles presented earlier (Fig. 12) indicated a localized region (process zone) of high crack-opening stress at the crack tip near the specimen centerplane. Brittle fracture of a test specimen occurs when a global crack-opening stress reaches a sufficient magnitude, and cleavage failure occurs rather than slip. If failure occurs when the orientation of this stress coincides with the local normal to a crystallographic plane in one or a series of ferrite grains, then a weakest-link theory can be based on the number of grains within the process zone. Obviously, the wider the specimen, the larger the process zone. Therefore, more ferrite grains are present in the process zone of the large specimen than in the small specimen. Thus a weakest-link theory suggests an increase in data scatter as the specimen size decreases (less of the material is sampled, therefore less chance of the weakest link being present). This theory also suggests that, given a large enough sample (a few large specimens or many smaller specimens), the lower bound of the toughness data will be independent of the specimen size. Both of these explanations are supported by the CTOD test results presented in this paper. In light of this argument, the specimen size requirements of ASTM E 399 can be interpreted as an attempt to achieve the lower-bound fracture toughness value with a single specimen. Given metallurgical and processing differences inherent in producing structural materials of different thicknesses, the ASTM E 399 size requirements appear to be inappropriate. That is, there is an implication that test specimens larger than the actual plate thickness being used in the structure should be tested to obtain "valid" results that satisfy the size requirements. A statistical analysis of test specimens of the same thickness as the structure of interest will give a more realistic estimate of the actual structural performance [*34*]. The results presented in this paper indicate the possibility of using smaller specimens to accurately predict the lower-bound toughness of structures in the lower-transition region. However, additional study is required before any final recommendations can be made.

Conclusions

A combined experimental and 3-D elastic-plastic finite-element analysis investigation of the effect of specimen size on the fracture behavior of A36 steel three-point bend specimens led to the following conclusions:

1. In the lower shelf and lower-transition regions (brittle failure with no prior fibrous tearing) of a low-strength steel, A36, the effect of specimen size on the lower-bound fracture toughness values is very small based on a statistically significant population of failure data.
2. The near-crack-tip opening stress near the specimen centerplane, at a specified CTOD level, is essentially independent of specimen size for preferred type (t by $2t$ by $8t$, $a/W = 0.5$) three-point bend specimens.

3. In the lower-transition region, where brittle fracture exhibits no fibrous crack growth but where large scale or unconfined plasticity occurs, scatter of toughness data is inherent. In this region, a small change in the near-crack-tip opening stress causes a large change in the fracture parameter (CTOD or J). Therefore, a statistical treatment of the fracture data is required. The use of Weibull statistics to obtain the lower-bound toughness value is suggested.

4. The use of ASTM E 399 specimen size requirements to achieve an arbitrarily defined "plane-strain" condition would seem to be too severe. The specimen size requirements can be interpreted as an attempt to achieve the lower-bound fracture toughness value with a single test specimen. That is, the size requirement is an attempt to reduce data scatter and is not related to the through-thickness constraint of geometrically similar specimens. However, the use of specimen size requirements to insure small-scale contained plasticity (necessary to a K_{Ic}-LEFM characterization of the fracture event) is completely appropriate.

5. There does not appear to be a specimen size effect, other than data scatter, in the CTOD characterization of brittle fracture in the lower-shelf and lower-transition regions of low-strength steel (A36). Small specimens exhibit a larger amount of data scatter and therefore require a larger number of test specimens to adequately characterize the lower-bound fracture toughness properties.

The results of this study suggest that specimens smaller than those meeting the various ASTM standard size requirements could be tested to characterize accurately the lower-bound fracture toughness properties of low-strength steel in the lower-shelf and lower-transition regions. However, it is prudent to use specimens of the same thickness as the structure of interest to avoid metallurgical or processing variations. This follows the British Standards BS 5762:1979.

References

[1] Jones, M. H. and Brown, W. F., "The Influence of Crack Length and Thickness in Plane Strain Fracture Toughness Tests" in *Review of Developments in Plane Strain Fracture Toughness Testing, ASTM STP 463*, American Society for Testing and Materials, Philadelphia, 1970, pp. 63–101.

[2] Kaufman, J. G., "Progress in Fracture Testing of Metallic Materials" in *Review of Developments in Plane Strain Fracture Toughness Testing, ASTM STP 463*, American Society for Testing and Materials, Philadelphia, 1970, pp. 3–21.

[3] Brown, W. F. and Srawley, J. E., "Commentary on Present Practice," in *Review of Developments in Plane Strain Fracture Toughness Testing, ASTM STP 463*, American Society for Testing and Materials, Philadelphia, 1970, pp. 216–248.

[4] Rice, J. R., "A Path Independent Integral and the Approximate Analysis of Strain Concentration by Notches and Cracks," *Journal of Applied Mechanics, Transactions*, American Society of Mechanical Engineers, Vol. 35, June 1968.

[5] Begley, J. A. and Landes, J. D., "The J-Integral as a Fracture Criterion," in *Fracture Toughness (Proceedings of the 1971 National Symposium on Fracture Mechanics) Part II, ASTM STP 514*, American Society for Testing and Materials, Philadelphia, 1972, pp. 1–23.

[6] Munz, D. et al., "Effect of Specimen Size on Fracture Toughness of a Titanium Alloy" in *Mechanics of Crack Growth, ASTM STP 590*, American Society for Testing and Materials, Philadelphia, 1976, pp. 219–234.

[7] Munz, D., "Minimum Specimen Size for the Application of Linear Elastic Fracture Mechanics" in *Elastic-Plastic Fracture, ASTM STP 668*, American Society for Testing and Materials, Philadelphia, 1979, pp. 406–425.

[8] Lake, R. L., "What R-Curves Can Tell Us About Specimen Size Effects in the K_{Ic} Test" in *Mechanics of Crack Growth, ASTM STP 590*, American Society for Testing and Materials, Philadelphia, 1976, pp. 208–218.

[9] Ritchie, R. O. et al., "On the Relationship Between Critical Tensile Stress and Fracture Toughness in Mild Steel," *Journal of the Mechanics and Physics of Solids*, Vol. 21, 1973, pp. 395–410.

[10] Dawes, M. G., "Elastic-Plastic Fracture Toughness Based on the COD and J-Contour Integral Concepts" in *Elastic-Plastic Fracture, ASTM STP 668*, American Society for Testing and Materials, Philadelphia, 1979, pp. 307–333.

[11] Milne, I. and Chell, G. C., "Effect of Size on the J Fracture Criterion" in *Elastic-Plastic Fracture, ASTM STP 668*, American Society for Testing and Materials, Philadelphia, 1979, pp. 358–377.

[12] Pisarski, H. G., "Influence of Thickness on Critical Crack Opening Displacement (COD) and J Values," *International Journal of Fracture,* Vol. 17, No. 4, 1981, pp. 427-440.
[13] Logsdon, W. A., "Elastic Plastic (J_{Ic}) Fracture Toughness Values: Their Experimental Determination and Comparison with Conventional Linear Elastic (K_{Ic}) Fracture Toughness Values for Five Materials" in *Mechanics of Crack Growth, ASTM STP 590,* American Society for Testing and Materials, Philadelphia, 1976, pp. 43-60.
[14] Logsdon, W. A. and Begley, J. A., "Dynamic Fracture Toughness of A533 Grade A Class 2 Base Plate and Weldments" in *Flaw Growth and Fracture (Tenth Conference), ASTM STP 631,* American Society for Testing and Materials, Philadelphia, 1977, pp. 477-492.
[15] Landes, J. D. and Shaffer, D. H., "Statistical Characterization of Fracture in the Transition Region" in *Fracture Mechanics (Twelfth Conference), ASTM STP 700,* American Society for Testing and Materials, Philadelphia, 1980, pp. 368-382.
[16] Roberts, R. et al., "Fracture Behavior of Bridge Steels" in *Flaw Growth and Fracture (Tenth Conference), ASTM STP 631,* American Society for Testing and Materials, Philadelphia, 1977, pp. 267-284.
[17] Marandet, B. and Sanz, G., "Experimental Verification of the J_{Ic} and Equivalent Energy Methods for the Evaluation of the Fracture Toughness of Steels" in *Flaw Growth and Fracture (Tenth Conference), ASTM STP 631,* American Society for Testing and Materials, Philadelphia, 1977, pp. 462-576.
[18] Clark, G. A., "Single Specimen Tests for J_{Ic} Determination," *ASTM STP 590,* American Society for Testing and Materials, Philadelphia, 1976, pp. 27-42.
[19] Robinson, J. N. and Tetelman, A. S., "The Relationship Between Crack Tip Opening Displacement, Local Strain and Specimen Geometry," *International Journal of Fracture,* Vol. 11, No. 3, June 1975, pp. 453-468.
[20] Sumpter, J. D. G., "The Prediction of K_{Ic} Using J and COD From Small Specimen Tests," *Metal Science,* Oct. 1976, pp. 354-356.
[21] Wellman, G. W. and Rolfe, S. T., "Engineering Aspects of CTOD Fracture Toughness Testing" in *Elastic-Plastic Failure Test Methods: The User's Experience, ASTM STP 856,* American Society for Testing and Materials, Philadelphia, 1985, pp. 230-262.
[22] Stuber, A. B., "A Study of Crack Growth Behavior and Specimen Size Effects in the Crack Tip Opening Displacement Test," Ph.D. thesis, University of Kansas, Lawrence, KS, Sept. 1980.
[23] Blom, R. C., "Final Report on Testing of the A517 Material," M.S. thesis, University of Kansas, Lawrence, KS, June 1980.
[24] Shoemaker, A. K. and Seeley, R. R., "Summary Report of Round Robin Testing by ASTM Task Group E24.01.06 on Rapid Loading Plain-Strain Fracture Toughness K_{Ic} Testing," *Journal of Testing and Evaluation,* Vol. 11, No. 4, July 1983, pp. 261-272.
[25] Johnson, L. G., *The Statistical Treatment of Fatigue Experiments,* Elsevier, New York, 1964.
[26] Lopez, L. A., "Finite: An Approach to Structural Mechanics Systems," *International Journal for Numerical Methods for Engineering,* Vol. 11, 1977.
[27] Dodds, R. H. and Lopez, L. A., "A Generalized Software System for Nonlinear Analysis," *International Journal for Advances in Engineering Software,* Vol. 2, No. 4, 1980.
[28] Barsoum, R. S., "Application of Quadratic Isoparametric Finite Elements in Linear Fracture Mechanics," *International Journal of Fracture,* No. 10, Dec. 1974.
[29] Barsoum, R. S., "Triangular Quarter-Point Elements as Elastic and Perfectly-Plastic Crack Tip Elements," *International Journal of Numerical Methods in Engineering,* Vol. 11, No. 1, 1977.
[30] Wellman, G. W. et al., "Three-Dimensional Elastic-Plastic Finite Element Analysis of Three-Point Bend Specimen" in *Fracture Mechanics: Sixteenth Symposium, ASTM STP 868,* American Society for Testing and Materials, Philadelphia, 1985, pp. 214-237.
[31] Bloom, J. M. and Van Fossen, D. B., "An Evaluation of the 20-Node Quadratic Isoparametric Singularity Brick Element," *International Journal of Fracture,* No. 12, Feb. 1976.
[32] Dodds, R. H., "Effect of Reduced Integration on the 2-D Quadratic Isoparametric Element in Plane Strain Plasticity," *International Journal of Fracture,* No. 19, 1982.
[33] Nayak, G. C. and Zienkiewicz, O. C., "Elasto-Plastic Stress Analysis: A Generalization for Various Constitutive Relations Including Strain Softening," *International Journal for Numerical Methods in Engineering,* Vol. 5, No. 1, 1972.
[34] Wellman, G. W., Rolfe, S. T., and Dodds, R. H., "Failure Prediction of Notched Pressure Vessels Using the CTOD Approach," presented at the 1984 Pressure Vessel and Piping Conference and Exhibition, San Antonio, TX, June 1984, American Society of Mechanical Engineers, to be published.

Fereshteh Ebrahimi[1]

A Study of Crack Initiation in the Ductile-to-Brittle Transition Region of a Weld

REFERENCE: Ebrahimi, F., **"A Study of Crack Initiation in the Ductile-to-Brittle Transition Region of a Weld,"** *Fracture Mechanics: Eighteenth Symposium, ASTM STP 945,* D. T. Read and R. P. Reed, Eds., American Society for Testing and Materials, Philadelphia, 1988, pp. 555–580.

ABSTRACT: The ductile-to-brittle transition behavior of a shielded metal-arc weld was characterized for various specimen geometries by the crack opening displacement (COD) testing method. The transition curve shifted to higher temperatures with increasing the specimen thickness and the crack length to width ratio (a/W).

Emphasis was placed on mechanisms of crack initiation as characterized by detailed fractography. A ductile shear zone, next to the fatigue precrack, preceding the unstable fracture, was revealed on the fracture surface of all specimens tested, except for specimens that broke under small-scale-yielding conditions. It is shown that brittle or ductile microcracks, depending on the test temperature, initially form at a distance ahead of the tip of a blunted crack. Subsequently, the ligament between these microcracks and the crack tip fails by a shearing mechanism. On the basis of this observation it is proposed that the statistical nature of fracture in the ductile-to-brittle transition regime is related to the probability of propagation of an unstable crack from the brittle or ductile microcracks.

KEY WORDS: crack initiation, crack tip opening, fractography, ductile-to-brittle transition, weldments, specimen size effects

For a particular material, loading mode, and testing setup, the change in failure mode from brittle to ductile can be associated with

(*a*) an increase in test temperature,
(*b*) a decrease in geometric dimensions such as specimen thickness, and
(*c*) an increase in notch root radius.

It has become common to characterize the ductile-to-brittle transition of a material as a function of temperature by the crack opening displacement (COD) testing method. The various regimes of fracture and deformation behavior throughout the ductile-to-brittle transition temperature are summarized schematically in Fig. 1. The fracture behavior can be categorized on the basis of (*a*) yielding, (*b*) stress state, or (*c*) microscopic mode of crack initiation and crack propagation, where (*a*) and (*b*) define the stresses and strains ahead of a crack at the time of fracture event, and (*c*) defines the fracture criterion. It is well known that an increase in thickness and crack length to width ratio (a/W) shifts the ductile-to-brittle transition curve to higher temperatures [*1–5*]. One of the aims of fracture mechanics is to predict this adverse effect.

The crack initiation process at the tip of a crack, either in brittle (cleavage) or in ductile (microvoid coalescence) mode, has been modelled by many investigators [*6–10*]. Unfortunately,

[1]Assistant professor, Department of Materials Science and Engineering, University of Florida, Gainesville, FL 32611.

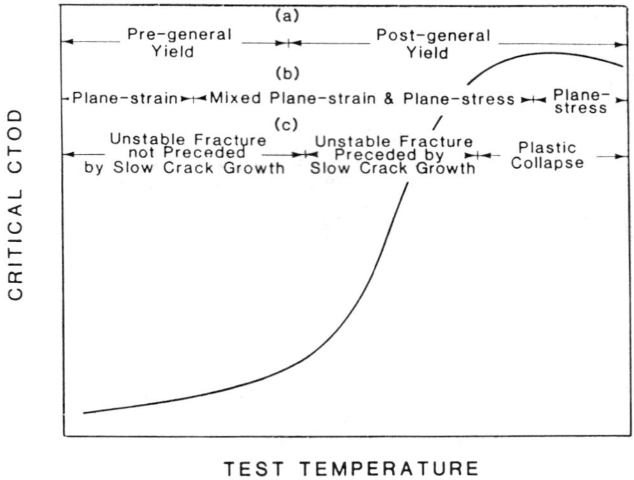

FIG. 1—*Schematic representation of ductile-to-brittle transition curves.*

in the ductile-to-brittle transition regime, where both modes of crack initiation compete with each other, the crack initiation process has not been well defined and there are many questions that need to be answered. For example, does the observation of microvoids in a small region next to the crack tip constitute a ductile fracture criterion? Why does cleavage fracture occur after the extension of a very small ductile crack from the original stationary sharp crack? Therefore, there is a need to clarify the critical fracture processes that establish the criterion for failure of a material in the ductile-to-brittle transition regime.

The purpose of this investigation is to study the ductile-to-brittle transition behavior of a shielded metal-arc weld, a typical weldment in arctic structures. Fracture toughness was measured by the COD testing method at temperatures throughout the ductile-to-brittle transition. The effect of geometric dimensions on fracture toughness was studied by changing the crack length to specimen width ratio (a/W) and the thickness of single-edge notched bend (SENB) specimens. Emphasis was placed on mechanisms of crack initiation as characterized by detailed fractography.

Materials and Experimental Procedures

A 25.4-mm-thick plate of American Bureau of Shipping (ABS) Grade EH36 steel was manually welded by a shielded metal-arc welding process with 8018-C3 electrode. Specimens for tension and COD testing were machined from the welded plate with the length and notch orientation, as shown schematically in Fig. 2.

Round tensile specimens with a reduced cross section were tested at various temperatures ranging from −258°C to +25°C. This specimen geometry was used to enable tensile property measurement of just the weld metal. Tension tests were conducted in an Instron tension testing machine equipped with a fixture designed for cryogenic testing at a crosshead speed of 0.5 mm/min.

Three-point SENB specimens of constant width ($W = 21.8$ mm) with various thicknesses, B, and a/W (see Table 1) were prepared by fatigue precracking at room temperature according to the specifications in the British Standard for COD testing (BS 5762). The SENB specimens were tested in the temperature range −196°C to −45°C, which resulted in critical crack-tip-opening

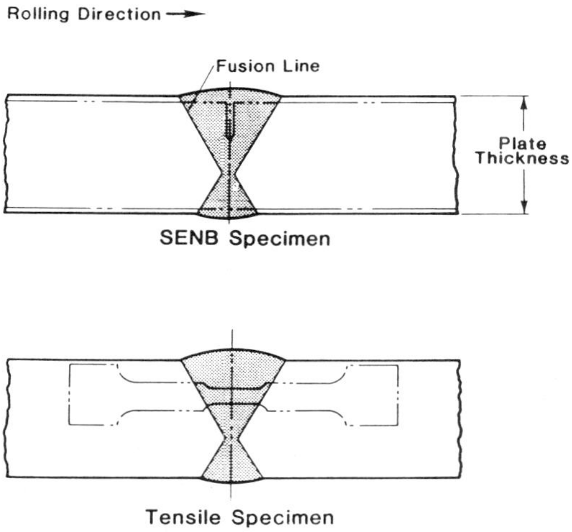

FIG. 2—*Orientation of test specimens relative to steel weldment.*

TABLE 1—*Test matrix for fracture toughness tests of welded SENB specimens (W = 21.8 mm).*

a/W	B = 12.7 mm	B = 25.4 mm
0.25	×	×
0.5		×
0.75		×

displacement (CTOD) values between 0.003 and 0.4 mm for establishing the ductile-to-brittle transition curves.

Both the load-line and mouth-opening displacements were measured during COD testing. A rotational factor, r_p, was calculated based on the relationship between the plastic components of the load-line displacement and the mouth-opening displacement [*11*]. The computed rotational factor was found to change moderately with CTOD, and the average value close to the fracture events was approximately 0.4, the same value as suggested by the British Standard (BS 5762).

The fracture surfaces of the broken specimens were studied using a scanning electron microscope (SEM). Emphasis was placed on characterization of the crack initiation zone in SENB specimens. The microstructures resulting from the multiple-pass welding process and the position of the crack tip for SENB specimens relative to these microstructures were studied by optical microscopy.

Results

Optical Microscopy

Figure 3 shows a low magnification optical micrograph of the weld. Large globular porosities of the order of 0.5 mm in size can be seen at the root of the weld. These defects were large enough to cause a premature fracture of SENB specimens with an $a/W = 0.5$ geometry, which

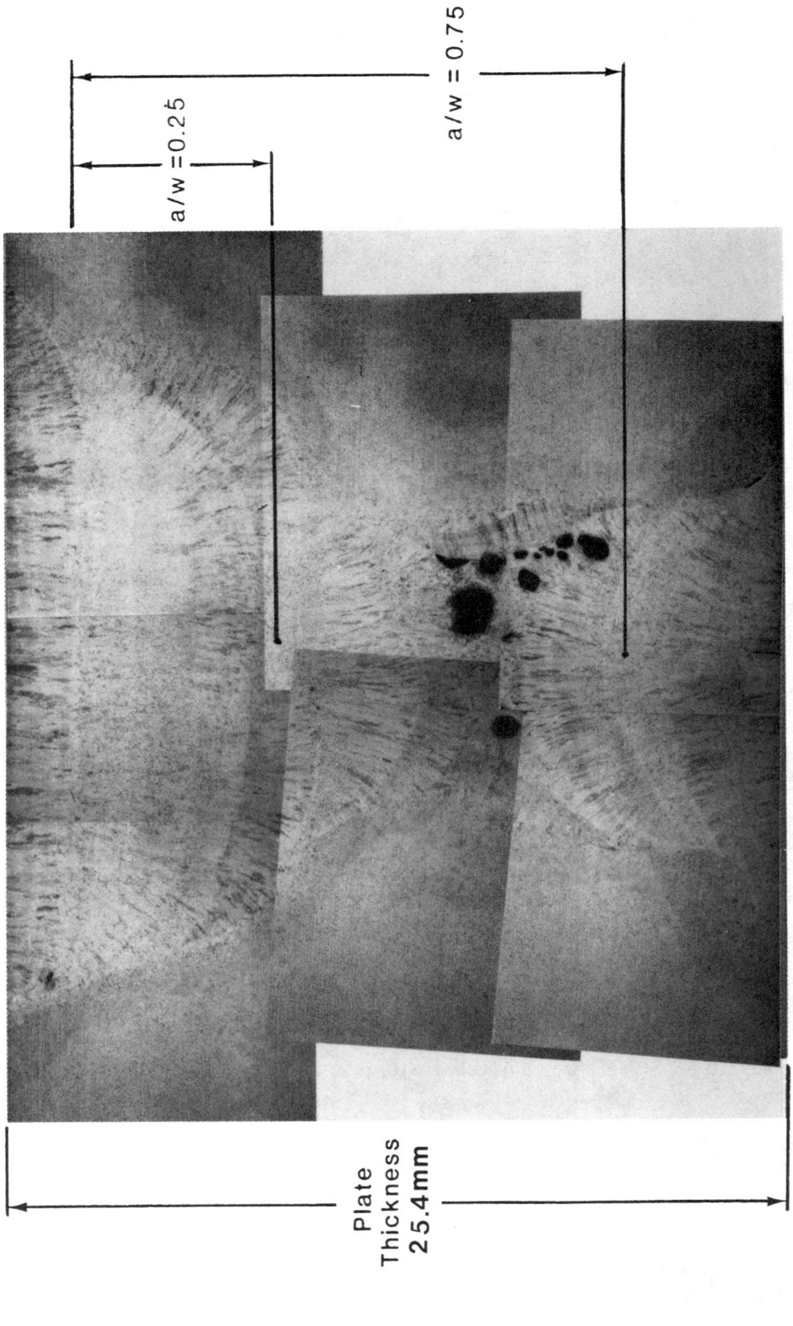

FIG. 3—*Low magnification optical micrograph of weldment. Approximate position of crack tip for a/W values are shown.*

was revealed as a pop-in on the load-displacement curve. Therefore, no meaningful fracture toughness data could be obtained for these specimens.

Different weld passes can be identified in the micrograph shown in Fig. 3. Each inner pass contains two distinctive microstructures. One consists of columnar prior austenite grains delineated by allotriomorph ferrite that resulted from the solidification of the molten weld metal. The other consists of equiaxed prior austenite grains formed by reaustenitizing of the columnar microstructure due to the heat generated from the next weld pass. As shown in Fig. 3, the approximate position of crack tip of specimens with both $a/W = 0.25$ and $a/W = 0.75$ is located in the microstructure with equiaxed grains.

Tensile Properties

The tensile properties of the weld metal are given in Table 2. The ultimate tensile strength and yield strength as a function of test temperature are plotted in Fig. 4. The effect of temperature on strength became more significant for test temperatures below −100°C. The maximum uniform strain did not change considerably with test temperature. The work hardening rate, as reflected by the difference between yield strength and ultimate tensile strength, decreased slightly with decreasing test temperature.

Fracture Toughness and Fractography

The critical CTOD values for the geometries studied are given in Table 3 and are plotted as a function of test temperature in Fig. 5. No macroscopically observable slow crack growth was detected on the fracture surface of specimens tested. Therefore, the critical CTOD values measured are defined as δ_c in accordance to the definitions given in the British Standard 5762.

As mentioned previously, no meaningful data were obtained for specimens with $a/W = 0.5$ geometry, due to weld defects located at the weld root. Figure 5 shows that an increase in the specimen thickness from 12.7 to 25.4 mm raised the ductile-to-brittle transition temperature about 25°C. The effect of an increase in a/W from 0.25 to 0.75 is revealed as a significant change in the shape of ductile-to-brittle transition curve towards a much steeper slope with a slight shift to higher test temperatures.

Figure 6 shows macrographs of fracture surface for $B = 25.4$ mm and $B = 12.7$ mm geometries tested at a temperature corresponding to critical CTOD of 0.35 mm. It can be seen that the

TABLE 2—*Tensile properties of the weld metal.*

Test Temperature, °C	Yield Strength, MPa	Ultimate Tensile Strength, MPa	Uniform Strain, %	Total Strain,[a] %	Fracture Stress, MPa	Reduction in Area, %
25	520	635	9.8	30	. . .	. . .
−31	566	689	11.5	33	1616	75
−67	570	690	11.0	32	1574	72
−83	609	740	12.5	30	1773	73
−118	654	791	12.7	32	1786	71
−143	738	850	13.7	40	1657	57
−173[b]	839	924	8.9	17	1246	27
−196	953	1007	12.0	25	1447	34
−241	. . .	1209	. . .	. . .	1209	. . .
−258	. . .	1129	. . .	. . .	1129	. . .

[a]Gage length = 12 mm.
[b]Premature fracture due to the existence of welding defects in the gage length.

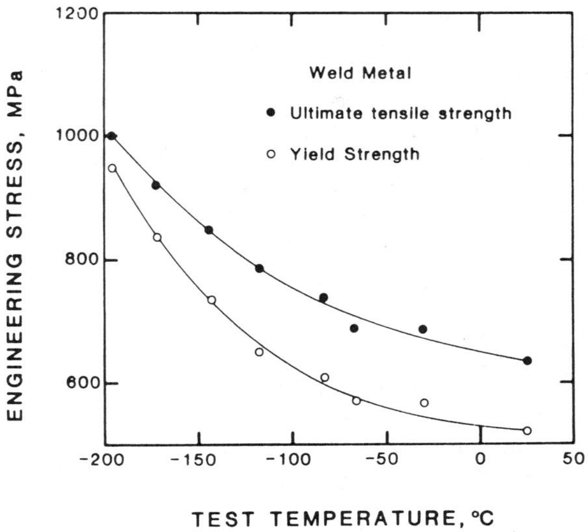

FIG. 4—*Yield and tensile stresses as function of test temperature.*

TABLE 3—*Critical CTOD data for the weld metal.*

B = 12.7 mm, W = 21.8 mm			B = 25.4 mm, W = 21.8 mm			B = 25.4 mm, W = 21.8 mm		
T, °C	a/W	δ_c,[a] mm	T, °C	a/W	δ_c,[a] mm	T, °C	a/W	δ_c,[a] mm
−68	0.245	0.386	−52	0.270	0.359	−48[b]	0.741	0.285
−68[c]	0.241	0.259	−55	0.263	0.297	−50	0.755	0.248
−73	0.241	0.323	−58	0.296	0.270	−53	0.788	0.077
−85	0.231	0.151	−62	0.297	0.149	−55	0.711	0.035
−87	0.238	0.110	−67	0.273	0.109	...	...	...
−92	0.237	0.239	−74	0.273	0.098	...	...	...
−108	0.234	0.078	−83	0.298	0.061	...	...	...
−197[d]	0.242	0.003	−105	0.287	0.022	...	...	...
...	...	...	−196[d]	0.282	0.003	...	...	...

[a] δ_c = critical CTOD.
[b] Initial pop-in due to the differences in microstructure.
[c] Five times higher displacement rate (4 mm/min).
[d] Valid K_{Ic} test; $K_{Ic} = 37$ MPa $\sqrt{m}$.

amount of shear fracture is approximately equal in both specimens and is less than 10% of the total fracture surface.

A microscopic ductile shear zone was observed next to the fatigue precrack and preceding the unstable fracture in all SENB specimens tested, except for specimens tested at liquid nitrogen temperature. This shear zone should not be confused with either the stretch zone, which develops owing to the crack tip blunting process [*12*], or the macroscopic slow crack, which is known as fibrous fracture. In the shear zone the fracture surface is oriented 45 deg from the direction of the maximum tensile stress developed by bending. Figure 7 illustrates the relative orientation of the stretch zone and the shear zone by tilting the specimen in the SEM by 45 deg in two opposite directions.

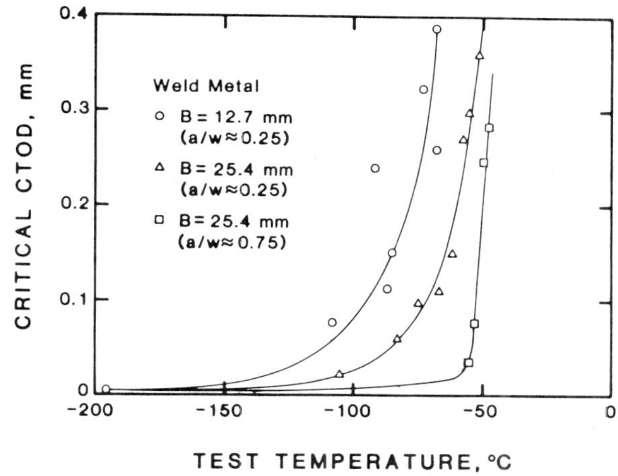

FIG. 5—*Critical CTOD as function of test temperature for three specimen geometries.*

The specimens showed three distinctive types of behaviors with regards to the formation of the shear zone. The specimens broken at very low temperatures showed no ductile shear zone, as has been reported by other investigators, and is shown in Fig. 8. Here, this type of fracture is categorized as Type I. For specimens with critical CTOD of approximately less than 0.15 mm, the shear zone width is the smallest at the midthickness and the largest at edges (Type II), as shown in Fig. 9. At higher temperatures the shear zone has a thumbnail shape (Type III). The larger shear zone width in the midthickness in comparison to the edges for Type III behavior is demonstrated in Fig. 10.

The following observations were made regarding the characteristics of the shear zone and the stretch zone.

1. The ductile crack in the shear zone has a zigzag appearance and follows the planes of maximum shear stress, as shown in Fig. 11. Note the approximately 90 deg angle between the two directions of the ductile crack. The size of these zigzags are much larger than the size of those associated with the propagation of a fibrous crack.
2. Many shear cracks form at the blunted crack tip, as is evident in Fig. 7*c*. The position of these shear microcracks (as marked by arrows on Fig. 7*c*) seems to correspond to the irregularities or the secondary cracks or both that existed on the fatigue precrack fracture surface prior to the blunting process.
3. Across the thickness of the specimen, the ductile crack joins the blunted crack tip through many different paths. Figure 12 shows that at Section A-A, the fatigue precrack is joined by a large single shear crack to the unstable brittle fracture. At Section B-B the ductile crack follows a zigzag path to join the blunted crack tip. At Section C-C the shear crack joins the brittle crack along a plane that makes an angle of 90 deg with the shear crack at Section A-A. As a result of the above behavior, the shear zone intersects with the blunted crack tip surface at various positions with regards to the midsection of the stretch zone; consequently, the width of the stretch zone varies along the crack front.
4. The stretch zone width increases consistently with an increase in the critical CTOD for specimens with Type II behavior ($\delta_c < 0.15$ mm), as shown in Fig. 13. However, the width of the stretch zone does not change significantly among the specimens with Type III behavior, as becomes evident by comparing the width of stretch zones shown in Fig. 7*c* ($\delta_c = 0.297$ mm) and Fig. 13*b* ($\delta_c = 0.149$ mm).

FIG. 6—*Macrographs of fracture surface showing relative amount of shear lips at approximately equal critical CTOD,* δ_c, *values for two different SENB specimen thicknesses.* (a) B = *25.4 mm,* T = *−52°C,* δ_c = *0.359 mm.* (b) B = *12.7 mm,* T = *−68°C,* δ_c = *0.386 mm.*

5. The width of stretch zone for equal critical CTOD values is approximately the same for all the geometries tested.

Discussion

Crack Initiation Process

The results obtained in this study indicate that a shear fracture zone located between the fatigue precrack and the unstable brittle fracture develops in all SENB specimens tested, except for specimens that broke under small-scale-yielding. The formation of this shear zone can be explained by considering microcracking at a distance ahead of the blunted crack tip rather than at the crack tip. Assuming that cleavage is a stress-controlled fracture process, it is well accepted that cleavage microcracks form at a distance ahead of the crack tip [*6,7*]. However, in most models proposed for the ductile fracture process [*8,13,19*], it is assumed that ductile cracks advance from the crack tip in a tearing mode. The results of theoretical models [*14–16*] indicate that the processes of void initiation, growth, and coalescence depend on the stress state as well as the amount of plastic strain. According to Rice and Tracey [*17*], the void growth rate in a strain field with high stress triaxiality is related exponentially to the hydrostatic stress component to yield strength ratio. In precracked specimens the effective strain is maximum directly at the tip of a blunted crack [*8*] and the hydrostatic stress component reaches a maximum value at a distance approximately equal to 2δ [*6*]. Accordingly, the void growth rate is expected to be maximum at a distance between the tip of a crack and the boundary of the large strain region. Knott [*18*] has estimated this distance to be of the order of $1.1\ \delta$. Therefore, conceivably ductile microcracks can form ahead of the crack tip as well as at the crack tip. In previous work [*19,20*] it has been established that, for notched bend specimens, discrete ductile microcracks are indeed formed by a combined tearing and shearing mechanism at a distance removed from the notch surface. These microcracks join small shear cracks, which form simultaneously at the notch tip, by a shearing mechanism. Figure 14 shows the formation of ductile microcracks in an interrupted precracked SENB specimen, which in principle is similar to a notched bend specimen.

The fracture initiation processes for the three types of behavior observed in this study are modeled schematically in Fig. 15, which shows both ductile and cleavage microcracks forming at a distance ahead of the tip of a crack. At very low temperatures (Type I, Fig. 15*a*), where the yield strength is high, the material fails under small-scale-yielding conditions and a small geometry change of the crack tip. In this case, no intense strain concentration develops ahead of the crack tip [*8*]. Characteristic of the microstructure [*21*], cleavage microcracks form at a very small distance ahead of the crack tip, and no ductile fracture develops.

As the temperature is increased within the lower shelf energy range (Type II, Fig. 15*b*), critical fracture stress for cleavage is achieved at larger plastic zones and CTODs, because of a significant drop in yield strength. In this region, cleavage microcracks form only when a certain amount of strain hardening has occurred in the plastic zone. Strain becomes localized ahead of the crack tip, and many small shear cracks develop from the crack tip (see Fig. 7*c*). The process of crack initiation is completed when shear instability is achieved between a propagating cleavage crack and a shear microcrack. As a result, a shear zone is formed which on the fracture surface precedes the unstable cleavage fracture. The shear zone width correlates with the size of plastic zone along the crack front, that is, smallest in the midthickness and largest at the edges (see Fig. 9). As the temperature is raised within this regime (Type II), more deformation is required for the formation of cleavage microcracks, while at the same time shear instability is expected to become easier. These two mechanisms correlate with the increase in the stretch zone width (SZW) and the size of shear fracture zone, respectively. Furthermore, the increase in the SZW with temperature suggests that the formation of cleavage microcracks is the controlling mechanism for Type II behavior. If the shear instability were the controlling mechanism, then

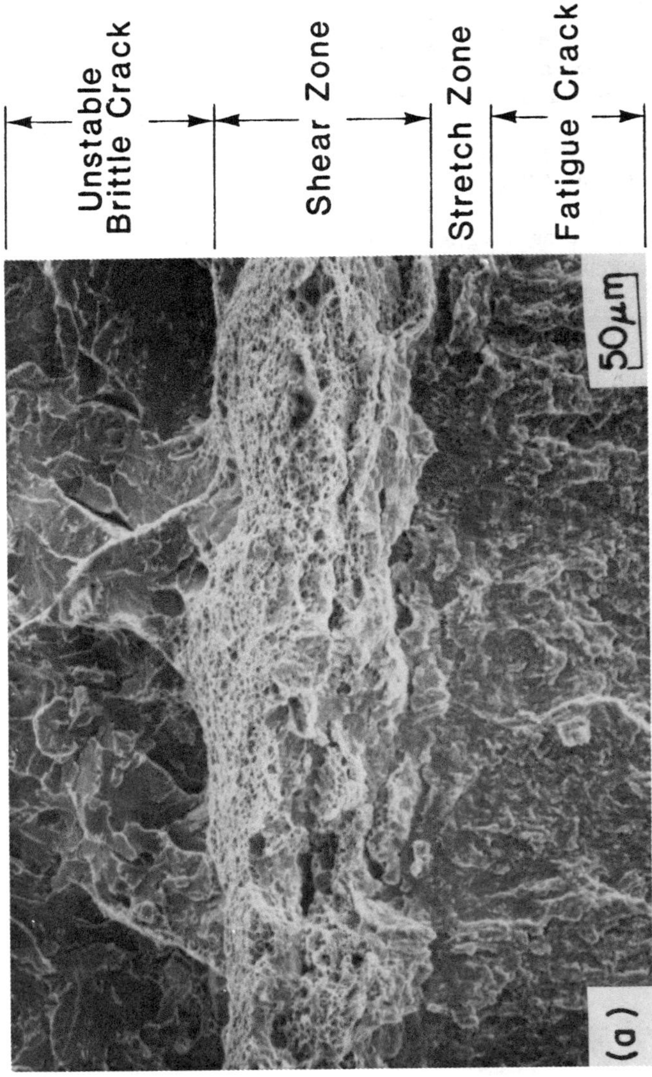
Unstable
Brittle Crack
Shear Zone
Stretch Zone
Fatigue Crack
50μm
(a)

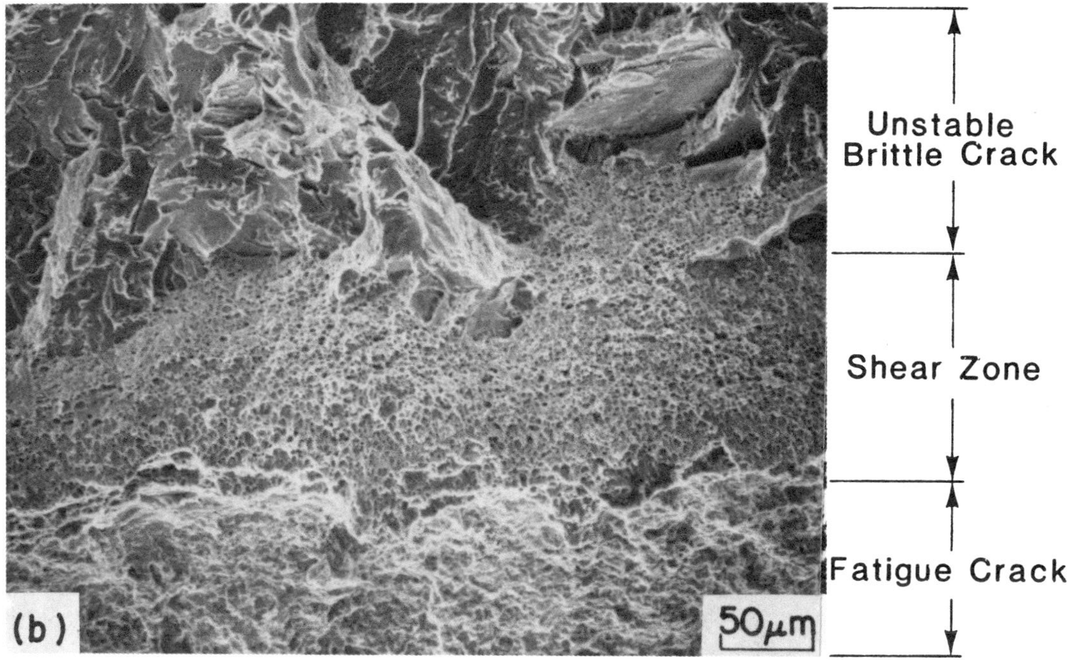

FIG. 7—*SEM fractographs showing relative orientation of stretch zone and shear zone for SENB specimen with* B = *25.4 mm* a/W = *0.263 geometry tested at* $-55°C$ *(*δ_c *= 0.297 mm).* (a) *0 deg tilt;* (b) *−45 deg tilt;* (c) *+45 deg tilt, arrows mark position of shear microcracks;* (d) *schematic showing specimen at +45 deg tilt angle relative to electron beam.*

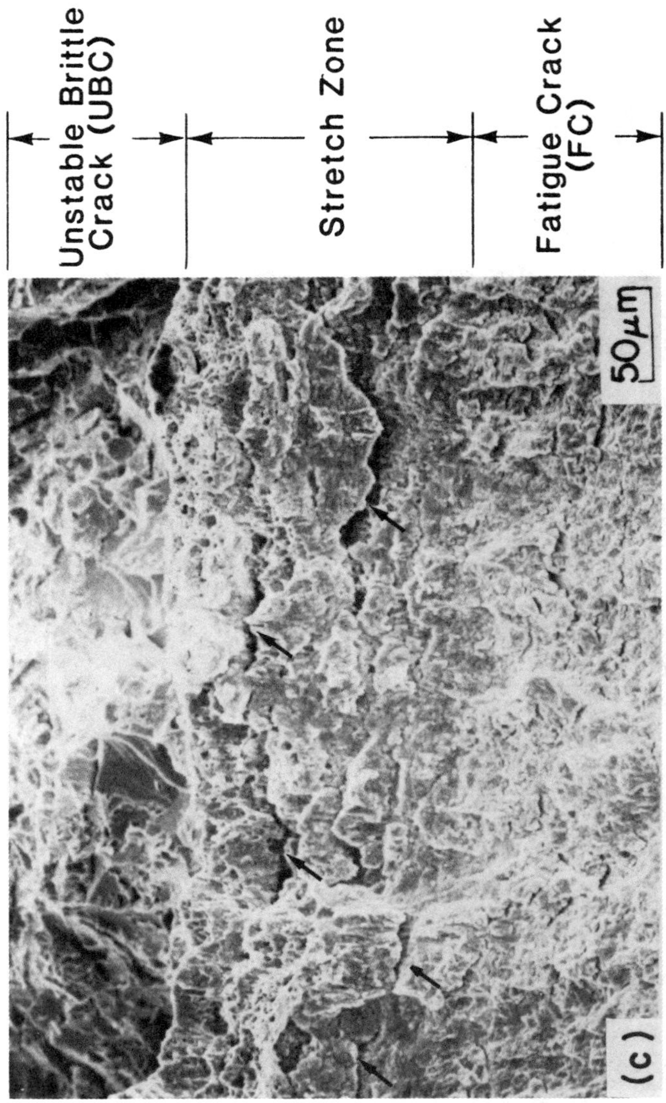

Unstable Brittle Crack (UBC)
Stretch Zone
Fatigue Crack (FC)
50μm
(c)

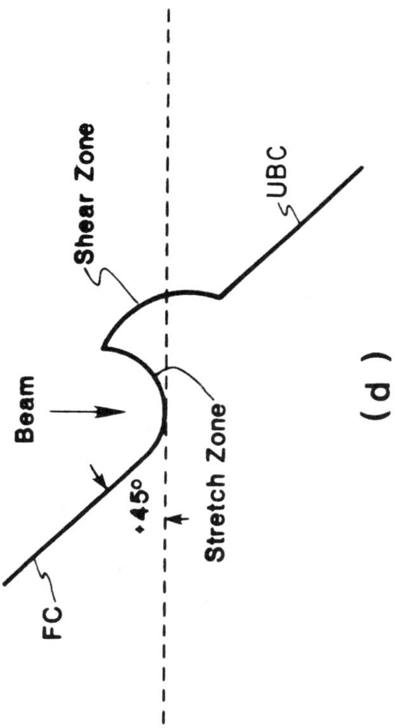

FIG. 7—*Continued.*

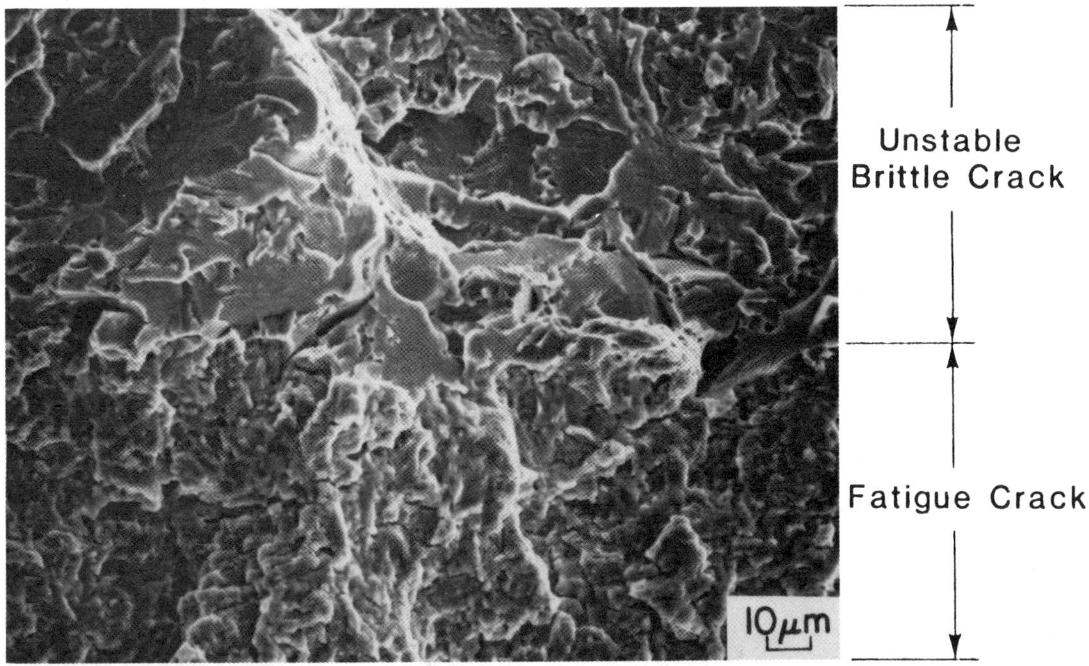

FIG. 8—*SEM fractograph showing the absence of shear zone at crack tip for SENB specimen with* B = *25.4 mm geometry tested at* −*196°C. Type I fracture behavior.*

the SZW would remain constant or slightly decrease with temperature. Therefore, the observation of microvoid coalescence in a small region next to the crack tip does not constitute a ductile fracture criterion.

As the test temperature is raised into the ductile-to-brittle transition temperature range, a point will eventually be reached at which the combination of stresses and strains ahead of the blunted crack favors the formation of ductile microcracks (Type III, Fig. 15*c*). The process of crack initiation is completed when the ductile microcracks, which form ahead of the crack tip, join the shear microcracks at the crack tip. In this temperature regime, the SZW correlates with the critical CTOD for ductile crack initiation, that is, δ_i. The SZW remains relatively constant with temperature, which is consistent with the temperature insensitivity of δ_i [*2*]. For Type III behavior, the size of the shear zone was always less than $2\delta_i$ ($\sim$300 μm, see Figs. 10 and 17), which suggests that cleavage fracture is triggered within the initial strain-hardened volume. Although no macroscopic slow crack growth was observed on the fracture surface of the specimens with Type III behavior, the critical CTOD measured truly should not be classified as δ_c, because ductile crack initiation precedes unstable brittle crack propagation.

In the transition temperature regime, there is a possibility of formation of both ductile and brittle microcracks at a distance ahead of the initial crack tip, as is shown in Fig. 16. Note that the size of a stable cleavage microcrack ($\sim$150 μm) is much larger than the grain size of this material ($\sim$10 μm). This competition between the ductile and brittle crack formation is a manifestation of the relatively high strength and low resistance to ductile fracture of the weldment [*22*]. For such materials, the temperature at which the critical CTOD increases abruptly corresponds to an actual change in the crack initiation mode, that is, from brittle to ductile mode. This point is demonstrated schematically in Fig. 17, where the applied CTOD is plotted as a function of load for a hypothetical specimen geometry and strength versus temperature relationship. For a material with relatively high resistance to ductile crack initiation, that is, with a large δ_i value, the abrupt increase of critical CTOD with temperature occurs in the temperature range where crack is initiated in a brittle mode [Case (a) in Fig. 17]. For a material with the same cleavage fracture stress, specimen geometry and strength versus temperature relationship, but with a lower resistance to ductile crack initiation, the sudden increase of critical CTOD with temperature corresponds to a change in the crack initiation mode [Case (b) in Fig. 17]. This brief discussion shows that a decrease in the ductile crack initiation toughness can raise the ductile-to-brittle transition temperature.

At higher temperatures, where the stress intensification at the tip of ductile microcracks within the initial highly strain hardened plastic zone is not large enough for initiation of a cleavage crack, the crack will extend by fibrous fracture. This slow crack growth may change to an unstable fracture outside of the initial plastic zone. In this case, the critical stress intensification is achieved by an increase in the crack length. This behavior is categorized as Type IV and was not observed for specimens tested in this study.

The crack initiation processes discussed above act as a bridge between the more microscopic sense of initiation, as would be defined by the cracking or the interface decohesion of the first carbide particle, with the more macroscopic sense of crack initiation, as would be defined by J-integral and COD testing methods.

Effect of Specimen Geometry

The effect of thickness on the ductile-to-brittle transition temperature has been explained either by constraint effects [*1,3,5*] or by statistical effects [*23,24*]. The former model attributes the decrease in transition temperature due to a reduction of specimen thickness to a loss in stress triaxiality at the crack tip, which reduces the effective flow stress in the plastic zone. The presence of less than 10% shear lips on the fracture surface of SENB specimens tested in the transition region suggests that crack initiation and propagation occurred under conditions close to plane strain in these specimens. Therefore, it may be concluded that any difference in the

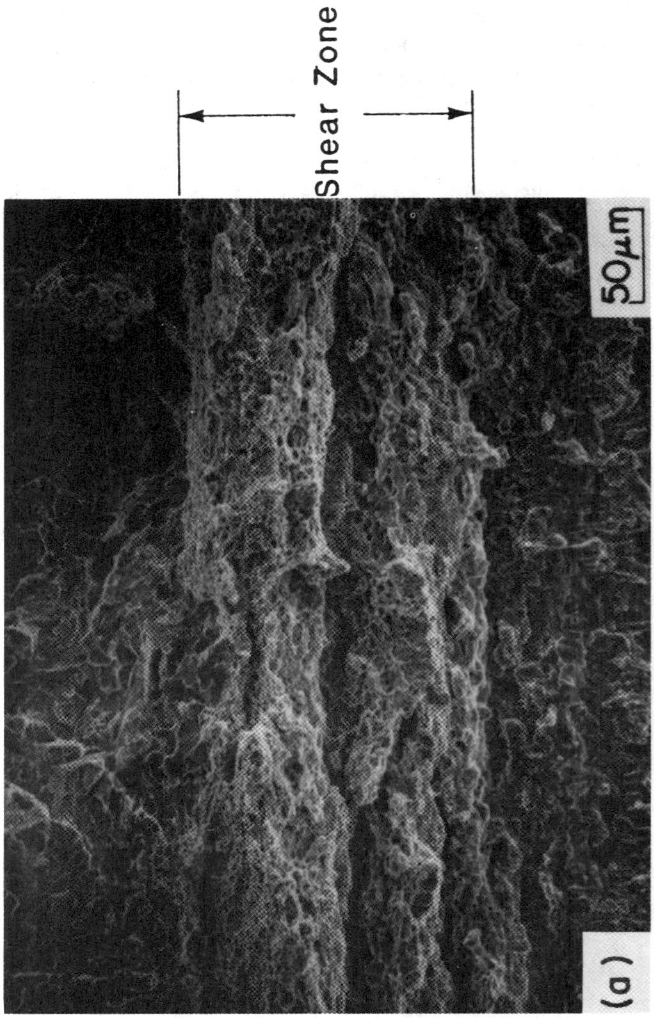
Shear Zone
50µm
(a)

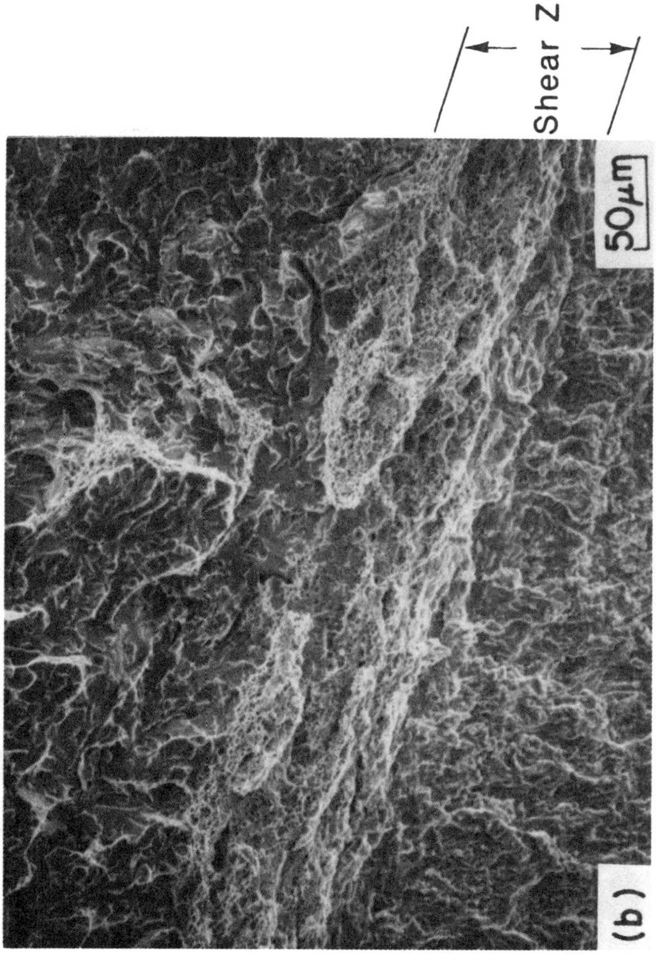

FIG. 9—*SEM fractographs showing difference in size of the shear zone formed at* (a) *midthickness and* (b) *edge of a SENB specimen with* B = 25.4 *mm,* a/W = 0.278 *geometry tested at* −74°C (δ_c = 0.098 *mm*). *Type II fracture behavior.*

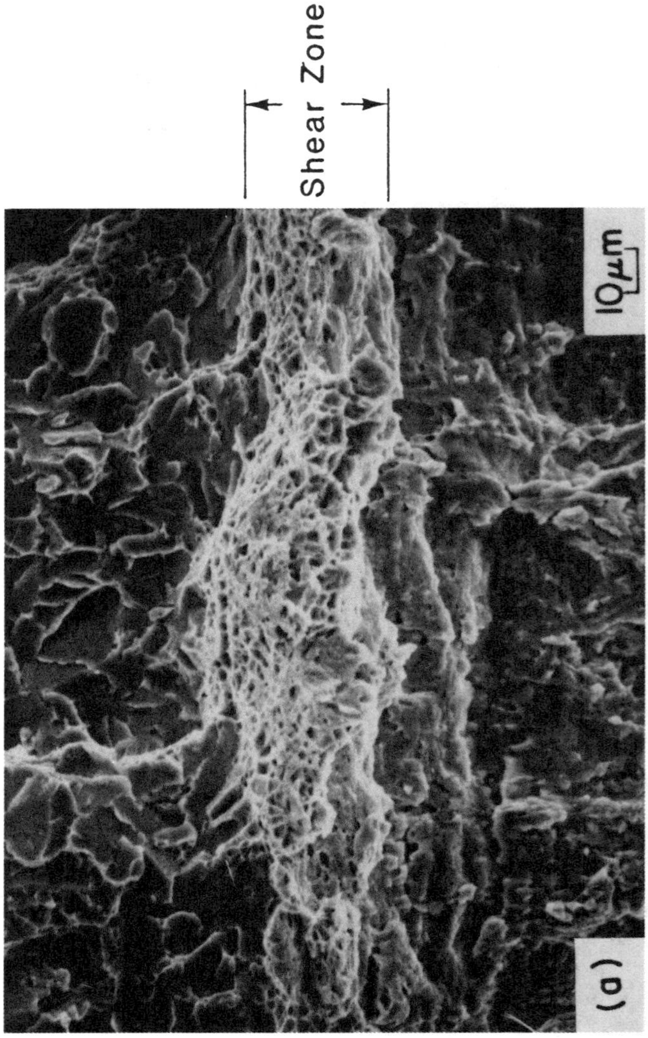
Shear Zone
10μm
(a)

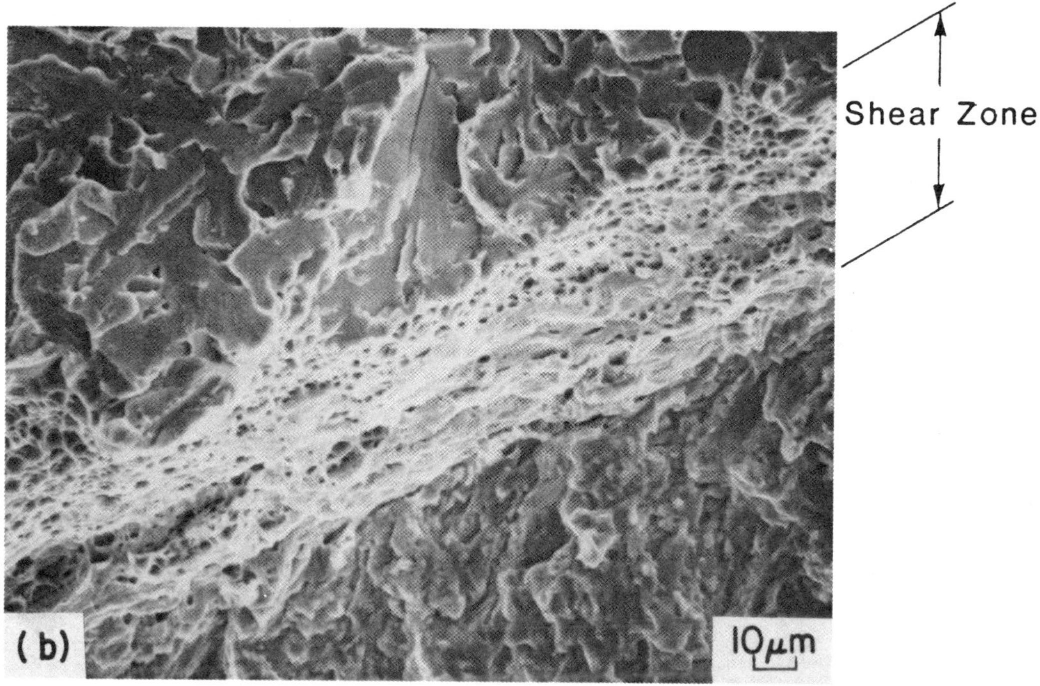

FIG. 10—*SEM fractographs showing difference in size of the shear zone formed at* (a) *midthickness and* (b) *edge of a SENB specimen with* B = *12.7 mm,* a/W = *0.245 geometry tested at −68°C* (δ_c = *0.386 mm).*

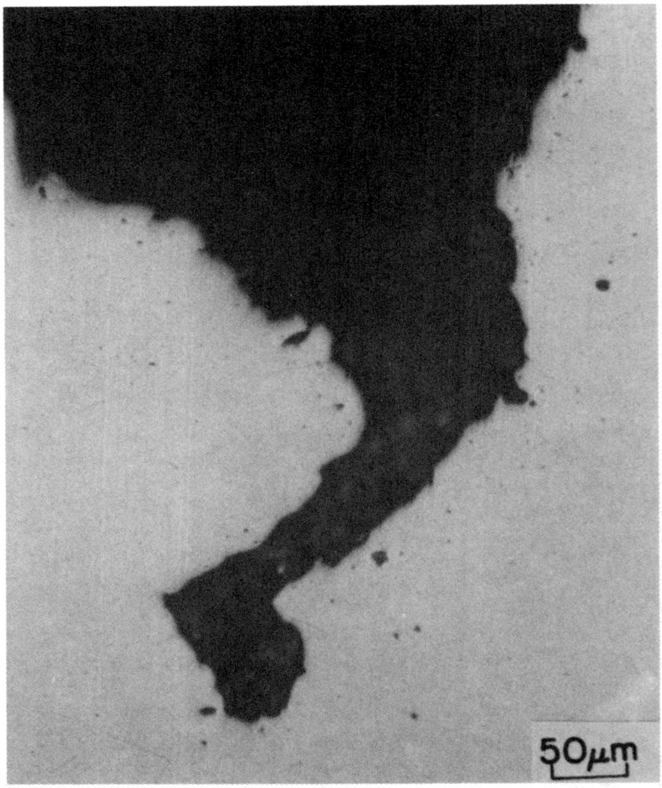

FIG. 11—*Optical micrograph showing the profile of the shear crack in an interrupted SENB specimen tested at −50°C.*

behavior of the edges of the specimen is insignificant in determining failure conditions for the specimens as a whole.

According to statistical models, for thick specimens there is a larger probability of finding weak spots at the tip of a crack than for thin specimens. Consequently, thick specimens should exhibit a lower toughness than thin specimens. The nature of these weak spots has been ill defined other than they may initiate brittle fracture. The statistical nature of cleavage fracture in steels has been explained by considering the probability of the propagation of a carbide microcrack into the ferrite matrix [*21,25*]. The observation of stable cleavage microcracks (see Fig. 16) suggests that the crack propagation from a cracked carbide particle into the ferrite matrix is not the critical fracture event in the ductile-to-brittle transition regime. The size of these microcracks were usually many times the ferrite grain size, which suggests that cleavage microcracks do not necessarily arrest at ferrite grain boundaries. Thus, the arrest of cleavage microcracks can be attributed to either microstructural inhomogeneities or a loss of stress intensification owing to a negative stress gradient in the direction of microcrack propagation.

The statistical nature of fracture and the specimen thickness effect in the ductile-to-brittle transition region can be explained conceptually by considering how a propagating cleavage microcrack may join the crack tip. For Type II fracture behavior, a specimen fractures when the shear instability is reached in the ligament between the stationary blunted crack tip and a prop-

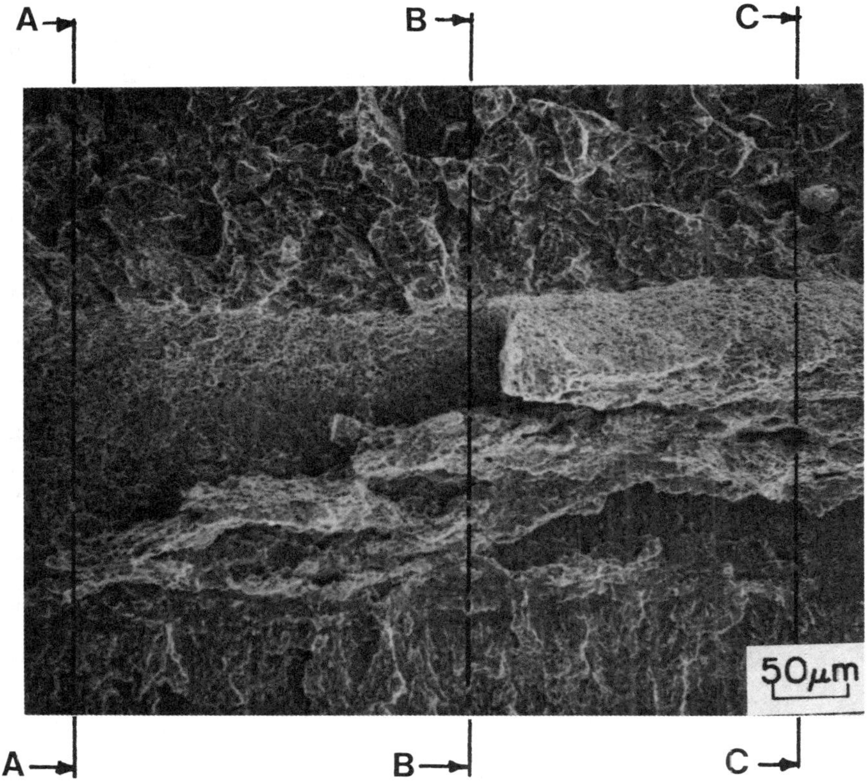

FIG. 12—*SEM fractograph showing different paths that ductile fracture follows to join surface of the stretch zone.*

agating cleavage microcrack, before the cleavage microcrack can arrest. Thus the probability of brittle unstable fracture depends on the position of cleavage microcracks relative to the stress distribution ahead of the stationary crack tip. This position controls how far a cleavage microcrack could grow before it would arrest. If the distance between the possible arrest point of a microcrack and the crack tip is less than the maximum ligament in which shear instability can occur, then the cleavage microcrack would join the blunted crack tip and the specimen as a whole would become unstable. Assuming that cleavage occurs as discrete microcracking across the thickness of the specimen, an increase in the thickness increases the probability of finding a propagating cleavage microcrack that can join the crack tip. In the same manner, increasing specimen thickness raises the brittle-to-ductile transition temperature by shifting the temperature at which fracture behavior changes from Type II to Type III behavior. Within the regime of Type III fracture behavior, the probability of unstable fracture behavior depends on the number of discrete ductile microcracks ahead of the crack tip, which potentially can trigger the initiation of cleavage fracture. Therefore, a specimen thickness effect is expected also when the crack initiation process occurs by ductile microcracking.

The results of this study show that a change in a/W from 0.25 to 0.75 does not affect the transition temperature as much as it influences the slope of the transition curve (see Fig. 5). This observation suggests that an increase in a/W promotes brittle cracking; however, the tem-

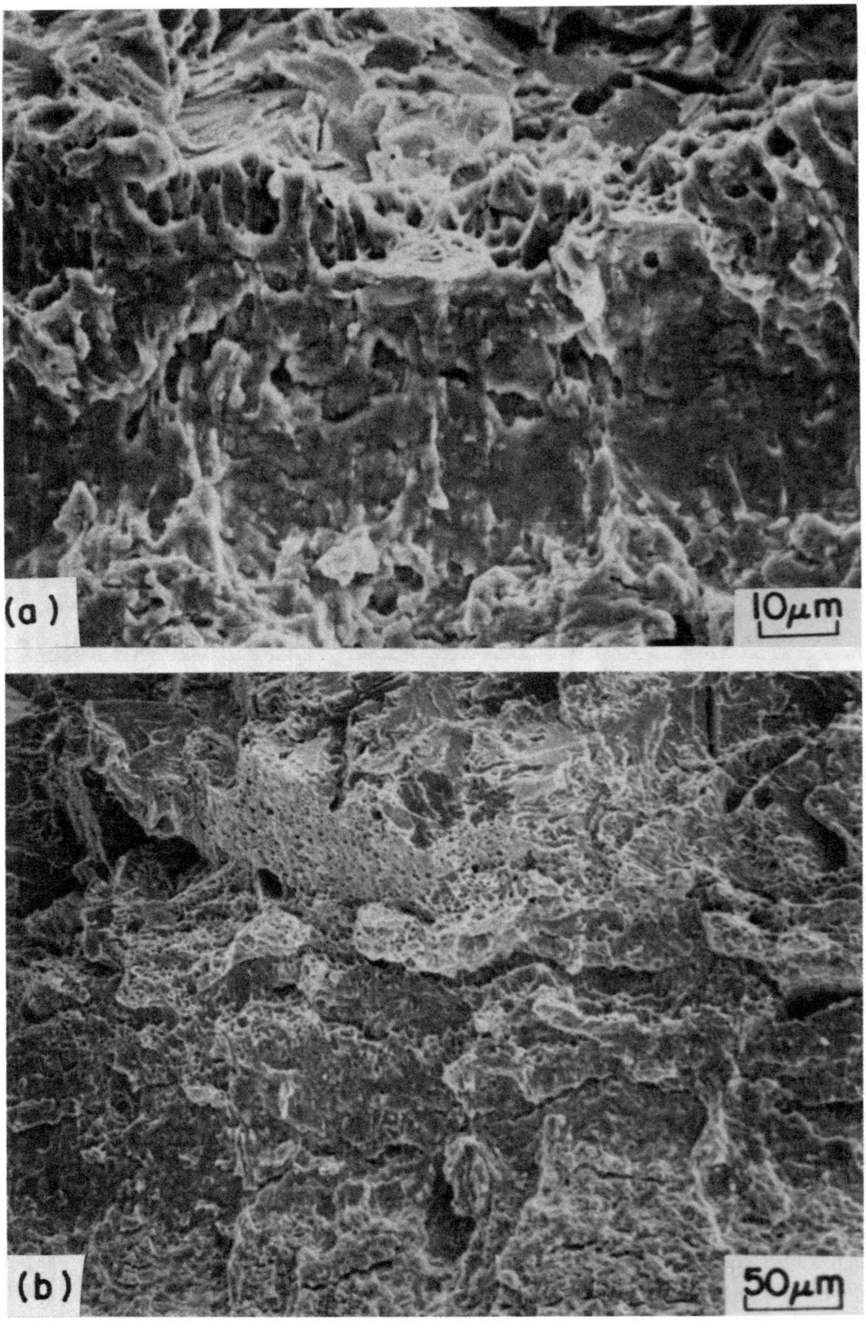

FIG. 13—*SEM fractographs at 45 deg tilt, showing increase in the SZW with an increase in the critical CTOD for Type II fracture behavior. SENB specimen with* B = *25.4 mm geometry.* (a) T = *−83°C,* δ_c = *0.061 mm,* (b) T = *−62°C,* δ_c = *0.149 mm. (Note difference in magnification of the two fractographs.)*

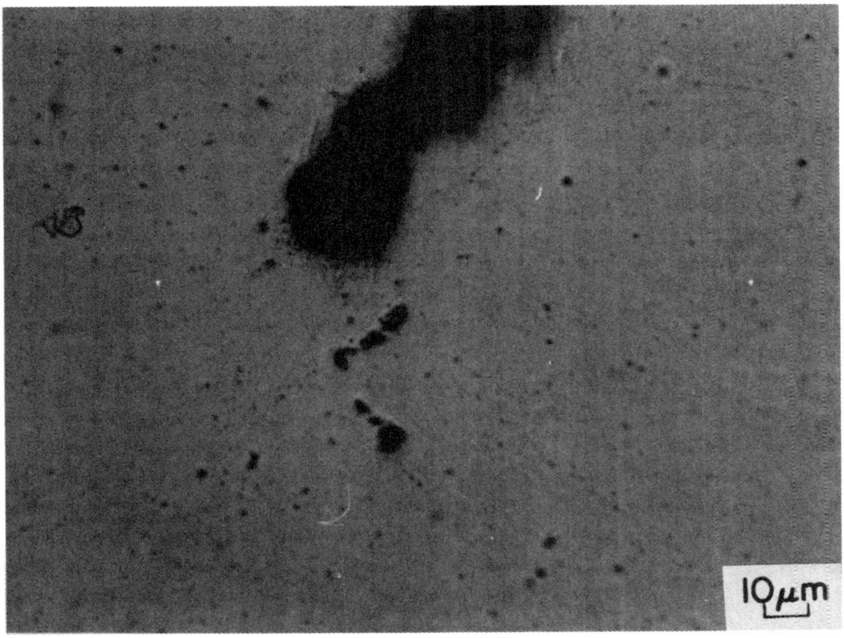

FIG. 14—*Optical micrograph showing formation of microvoids at a distance ahead of crack tip on maximum shear stress planes. An interrupted SENB specimen tested at* $-50°C$.

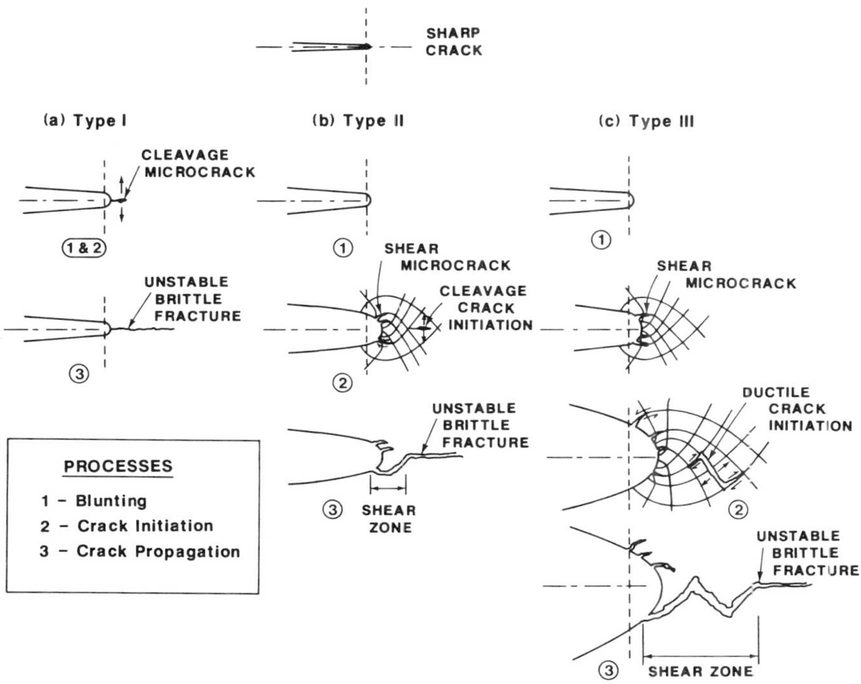

FIG. 15—*Schematic showing step-by-step sequence of crack initiation for* (a) *Type I,* (b) *Type II, and* (c) *Type III fracture behaviors.*

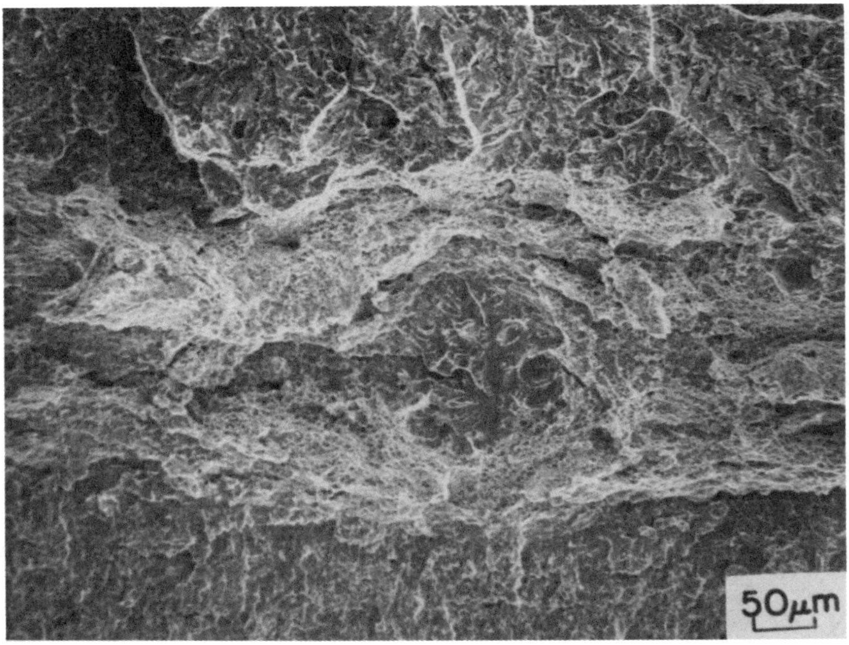

FIG. 16—*SEM fractograph showing a stable cleavage microcrack within the shear zone.* B = *25.4 mm,* T = *−55°C,* δ_c = *0.297 mm.*

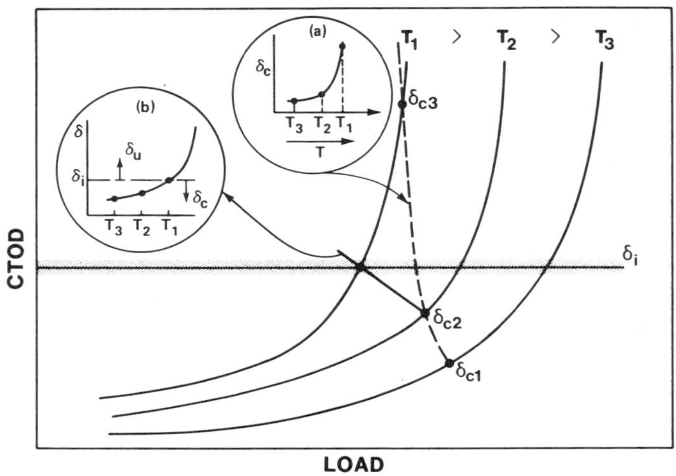

FIG. 17—*Schematic demonstrating effect of resistance to ductile crack initiation on brittle-to-ductile transition temperature:* (a) *crack initiation by cleavage microcracking;* (b) *crack initiation by ductile microcracking.* δ_c = *unstable fracture with no prior ductile cracking;* δ_i = *initiation of ductile fracture;* δ_u = *unstable fracture after ductile crack initiation.*

perature at which ductile cracking occurs is not affected significantly. This is consistent with increasing constraint along the direction perpendicular to the crack front [5].

Conclusions

The ductile-to-brittle transition behavior of a shielded metal-arc weld was studied by the COD testing method and fractography. The following conclusions have been reached.

1. The ductile-to-brittle transition shifts to higher temperatures with increasing specimen thickness and the a/W ratio.
2. In the elastic-plastic regime, brittle or ductile microcracks, or both, depending on the test temperature, develop at a distance ahead of the tip of a blunted crack. These microcracks join the crack tip by a shearing mechanism, which results in the formation of a ductile shear zone next to the fatigue precrack, preceding the unstable fracture.
3. At relatively small critical CTOD values ($\delta_c < 0.15$ mm), the crack is initiated by cleavage and the ductile shear zone width correlates with the plastic zone size; that is, it is smallest at midthickness and the largest at the edges. In this temperature regime, despite the observation of shear fracture at the crack tip, crack initiation is controlled by cleavage microcracking, and the SZW increases with increasing the critical CTOD.
4. At larger critical CTOD values (0.15 to 0.4 mm), the crack is initiated by ductile microcracking and the ductile shear zone has a thumbnail shape with its maximum width located at the midthickness. In this temperature regime, the SZW remains constant and correlates with the critical CTOD for ductile crack initiation, δ_i. The size of shear zone is less than $2\delta_i$, which suggests that the cleavage fracture is triggered within the highly strained region ahead of the blunted crack tip.

Acknowledgments

This work was conducted at the Fracture and Deformation Division of National Bureau of Standards in Boulder, Colorado, under the sponsorship of the Minerals Management Service of the U.S. Department of Interior. Special gratitude is due to the staff members of the Fracture and Deformation Division of National Bureau of Standards.

References

[*1*] Green, G. and Knott, J. F., *Journal of the Mechanics and Physics of Solids*, Vol. 23, 1975, p. 167.

[*2*] Dawes, M. G. in *Elastic-Plastic Fracture, ASTM STP 668*, American Society for Testing and Materials, Philadelphia, 1979, p. 307.

[*3*] Milne, I. and Chell, G. G. in *Elastic-Plastic Fracture, ASTM STP 668*, American Society for Testing and Materials, Philadelphia, 1979, p. 1358.

[*4*] Kotilainen, H. in *Fracture Mechanics, ASTM STP 700*, American Society for Testing and Materials, Philadelphia, 1980, p. 368.

[*5*] Anderson, T. L., "The Effect of Crack-tip Region Constraint on Fracture in the Ductile-to-Brittle Transition Region," Ph.D. thesis No. T-2741, Colorado School of Mines, Golden, CO, 1983.

[*6*] Tetelman, A. S. and McEvily, A. J., Jr., in *Fracture of Structural Materials*, Wiley, New York, 1967.

[*7*] Ritchie, R. O., Knott, J. F., and Rice, J. R., *Journal of the Mechanics and Physics of Solids*, Vol. 21, 1973, p. 395.

[*8*] Rice, J. R. and Johnson, M. A. in *Inelastic Behavior of Solids*, McGraw-Hill, New York, 1970, p. 641.

[*9*] Bates, R. C. in *Fracture: Interactions of Microstructure, Mechanisms Mechanics, Proceedings*, Metallurgical Society of American Institute for Mining, Metallurgical, and Petroleum Engineers, Warrendale, PA, 1984, p. 117.

[*10*] Ritchie, R. O. and Thompson, A. W., *Metallurgical Transactions A.*, Vol. 16, 1985, p. 233.

[*11*] Lin, I.-H., Anderson, T. L., DeWit, R., and Dawes, M. G., *International Journal of Fracture*, Vol. 20, 1982, p. R3.

[*12*] Harrison, J. D., "The State-Of-The-Art In Crack Tip Opening Displacement (CTOD) Testing and

Analysis," Welding Institute Research Report 7302.10/80/210.2, The British Welding Institute, Abington, Cambridgeshire, England, 1980.

[*13*] McClintock, F. A. in *Physics of Strength and Plasticity,* A. S. Argon, Ed., Massachusetts Institute of Technology Press, Cambridge, MA, 1969, p. 307.

[*14*] LeRoy, G., Embury, J. D., Edward, G., and Ashby, M. F., *Acta Metallurgica,* Vol. 29, 1981, p. 1509.

[*15*] Hancock, J. W. and Brown, D. K., *Journal of the Mechanics and Physics of Solids,* Vol. 31, 1983, p. 1.

[*16*] Thomason, P. F., *Acta Metallurgica,* Vol. 33, 1985, p. 1087.

[*17*] Rice, J. R. and Tracey, M. D., *Journal of the Mechanics and Physics of Solids,* Vol. 17, 1969, p. 201.

[*18*] Knott, J. F., *Metal Science,* Vol. 14, 1980, p. 327.

[*19*] Zia-Ebrahimi, F., Matlock, D. K., and Krauss, G., *Scripta Metallurgica,* Vol. 16, 1982, p. 987.

[*20*] Zia-Ebrahimi, F. and Krauss, G., *Acta Metallurgica,* Vol. 32, 1984, p. 1776.

[*21*] Curry, D. A. and Knott, J. F., *Metal Science,* Vol. 12, 1978, p. 511.

[*22*] Zia-Ebrahimi, F., "Ductile-to-Brittle Transition in Steel Weldments for Arctic Structures," NBSIR 85-3020, National Bureau of Standards, U.S. Department of Commerce, Boulder, CO, 1985.

[*23*] Landes, J. D. and Shaffer, D. H. in *Fracture Mechanics, ASTM STP 700,* American Society for Testing and Materials, Philadelphia, 1980, p. 368.

[*24*] Bruckner, A. and Munz, D., *Journal of Fracture Mechanics,* Vol. 18, 1983, p. 359.

[*25*] Wallin, K., Saario, T., and Torronen, K., *Metal Science,* Vol. 18, 1984, p. 13.

J-Integral Test Methods

Paulus A. J. M. Steenkamp[1]

J_R-Curve Testing of Three-Point Bend Specimens by the Unloading Compliance Method

REFERENCE: Steenkamp, P. A. J. M., "**J_R-Curve Testing of Three-Point Bend Specimens by the Unloading Compliance Method,**" *Fracture Mechanics: Eighteenth Symposium, ASTM STP 945,* D. T. Read and R. P. Reed, Eds., American Society for Testing and Materials, Philadelphia, 1988, pp. 583–610.

ABSTRACT: The unloading compliance technique for J_R-curve testing is not well established for single-edge notched specimens loaded in three-point bending (SENB) to date. This paper investigates various effects hampering the successful application of the method to SENB specimens. It is shown that the specimen's compliance is significantly affected by rotation of the specimen halves during deformation and stable crack extension. By contrast, the effects of roller indentation, outer roller movements and friction, and of crack front curvature are relatively minor. Based on these findings, a general procedure is proposed for correcting the experimentally measured compliance of SENB specimens for the above effects.

Crack extensions predicted for three steels from SENB unloading compliance experiments incorporating the proposed corrections are seen to correlate within ±10% with dye penetrant beachmarks made at several points during the tests.

KEY WORDS: J_R-curve, crack growth, unloading compliance, fracture toughness, fracture test, compliance, three-point bend, calibration, elastic-plastic fracture, tunneling

The multiple-specimen technique [*1*] is the simplest and most unambiguous method of determining J_R-curves from laboratory-scale test specimens. Its disadvantages, however, are the large number of specimens needed to establish a complete J_R-curve, and the impossibility to properly correct the determined J-value for crack extension (see Ref *2,* for example). The latter point becomes especially important for larger amounts of crack extension and when testing small-size specimens. These disadvantages have led to the development of single specimen J_R-curve test procedures. Among these, the unloading compliance (UC) method, first described by Clarke et al. [*3,4*], is now widely used for specimens primarily loaded in bending.

A tentative J_R-curve standard, developed and published by a working group of ASTM Committee E24.08.03 [*5*], proposes the use of the UC method for the determination of the amount of crack extension Δa. This tentative standard refers to both compact tension (CT) specimens and single-edge notched specimens, loaded in three-point-bending (SENB). While numerous applications have established its use for the CT geometry [*6–10*], results published to date for SENB specimens (including Charpy-size specimens) were generally unsatisfactory in terms of accuracy and reliability. The UC method tended to significantly underpredict the actual crack extension, and empirical corrections were sometimes needed to force agreement between predicted and actual Δa-values. This has led to the recognition that more effort on and experience with the use of the UC method for SENB specimens is needed before the method may become a reliable procedure for single specimen J_R-curve testing for the SENB geometry.

[1]Mechanical engineer, Shell Eastern Petroleum (Pte) Ltd., Singapore; formerly research associate, Laboratory for Thermal Power Engineering, Delft University of Technology, Delft, the Netherlands.

This paper discusses various impediments to the successful application of the UC test procedure to SENB specimens and proposes corrective measures to overcome these. To this end, effects that influence the specimen's compliance are first analyzed in some detail. Next, UC experiments performed on SENB specimens are described and their results discussed. Finally, recommendations are made for future use of the proposed procedure.

Analyses

The characteristic dimensions of a SENB specimen are given in Fig. 1*a;* Fig. 1*b* schematically depicts the following six areas of uncertainty affecting its elastic compliance:

1. Selection of the proper compliance calibration function from the various analytical expressions relating compliance to the crack length in terms of load-line displacement Δ and of crack mouth opening displacement δ.
2. Effective thickness and Young's modulus. The definition of an effective thickness is of concern for side-grooved specimens, while the effective modulus of elasticity is affected by the stress state (plane stress versus plane strain) in the specimen.
3. Crack front curvature. The specimen's compliance is influenced by the degree of curvature of the crack front.

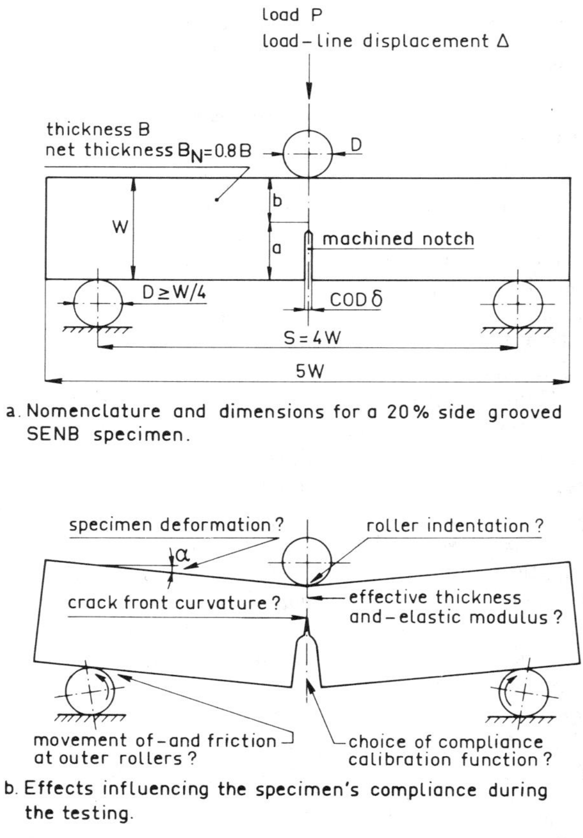

FIG. 1—*SENB specimen: nomenclature, dimensions, and effects influencing its compliance.*

4. Specimen deformation. Analytical calibration functions are valid for straight (undeformed) specimens. Bending of the specimen will influence its compliance, thus invalidating analytical calibrations.
5. Roller motion. The two supporting rollers will move due to specimen deformation, with friction forces occurring between rollers and specimen.
6. Roller indentation. Indentation of the center roller into the specimen causes a deviation from the pure three-point loading situation.

While only the last three factors are specific for SENB specimens, all six will be discussed below in some detail for the sake of completeness. Their quantitative effects on the specimen's elastic compliance will be given where relevant.

Compliance Calibration Functions

The elastic compliance calibration functions published in the literature are usually given in normalized form by polynomial expressions

$$C_\Delta EB = f_\Delta(a/W) \tag{1a}$$

or

$$C_\delta EB = f_\delta(a/W) \tag{1b}$$

where

$C_\Delta = d\Delta/dP$ = load-line compliance, mm/N,
$C_\delta = d\delta/dP$ = crack mouth compliance, mm/N,
P = load, N,
E = Young's modulus, N/mm^2, and
B = specimen thickness, mm.

The load-line compliance C_Δ is a summation of two terms, namely, the compliance of the specimen with no crack present (C_{nc}) and the compliance due to the presence of the crack (C_c). The results of various load-line compliance calibrations found in the literature [*11–14*] differ by up to 5% [*15*]. This is due mainly to uncertainties in C_{nc}, notably regarding the effects of shear stresses, Hertzian stresses at the loading points, and assumed stress state in the uncracked specimen: plane strain or plane stress (see discussion in Ref *14*). It can be shown that for $a/W = 0.5$, C_{nc} accounts for about one third of the total load-line compliance [*15*], implying that uncertainties in the precise value of C_{nc} will have a noticeable effect on the total compliance C_Δ except for very deep cracks ($a/W \rightarrow 1$).

In contrast to the load-line compliance, the crack mouth compliance for the limit case of a perfectly sharp crack (that is, one in which the notch as shown in Fig. 1*a* has zero width) naturally consists of one term only, namely, one that is dependent on the relative crack length a/W. For practical cases (that is, when the COD δ is measured as the increase in distance of two points not initially coinciding), an additional contribution to the crack mouth compliance is made by the deformation of the uncracked specimen. It can be shown, however, that this effect contributes less than about 0.1% to the total compliance, and hence can be neglected. Furthermore, measurement of the load-line displacement of a SENB specimen to a degree of accuracy needed for load-line compliance determination is inherently difficult due to roller indentation effects. These considerations led the present author to the exclusive use of the crack mouth compliance C_δ for the determination of crack length. Accordingly, all further mention of specimen compliance will refer to C_δ.

To the author's knowledge, the only expression for C_δ published in literature is that by Tada et al. [*14*], which reduces for a three-point bend specimen with standard span-to-width ratio $S/W = 4$ to

$$CEB = 24\frac{a}{W}\left[0.76 - 2.28\frac{a}{W} + 3.87\left(\frac{a}{W}\right)^2 - 2.04\left(\frac{a}{W}\right)^3 + \frac{0.66}{\left(1-\frac{a}{W}\right)^2}\right] \quad (2)$$

and is given in inverted form by Ref *13*

$$\frac{a}{W} = 0.999748 - 3.9504U + 2.9821U^2 - 3.21408U^3 + 51.51564U^4 - 113.031U^5 \quad (3)$$

where

$$U = \frac{1}{\sqrt{(CEB)} + 1}.$$

It should be noted that Eq 3 differs from the crack mouth compliance calibration given in the tentative J_R-curve standard [*5*]. As pointed up by Wu [*13*], the recommended crack mouth calibration expression in Ref *5* refers to four-point bending, whereas the specimen geometry is that of a three-point bend bar.

Effective Thickness and Young's Modulus

As a specimen's compliance depends on its thickness, Eq 1, the definition of an effective thickness B_e becomes of concern for side-grooved specimens. Based on elastic three-dimensional finite element calculations, Shih et al. [*16*] found the following approximation for B_e for a compact specimen

$$B_e = B - \frac{(B - B_N)^2}{B} \quad (4)$$

While the tentative J_I-R curve standard [*5*] suggests the use of B_e according to Eq 4 for SENB specimens as well, Neale [*17*] obtained the following approximation for B_e for SENB specimens through three-dimensional finite element computations

$$B_e = \frac{B}{1.41(B/B_N)^2 - 2.89(B/B_N) + 2.48} \quad (5)$$

For a 20% side-grooved specimen ($B_N = 0.8B$), the above equations yield B_e values of $0.96B$ and $0.93B$ for CT and SENB specimens, respectively.

The effective modulus of elasticity E_e is affected by the stress state in the specimen's ligament, that is, its value may vary between the extremes of E (plane stress) and $E' = E/(1-\nu^2)$ (plane strain), where ν is Poisson's ratio. To overcome the resulting uncertainty it has become good practice for one to determine E_e by correlating the compliance C_0 at the first point of partial unloading of the test to the initial crack length a_0, optically measured after termination of the experiment, through

$$E_e = \frac{f(a_0/W)}{C_0 B_e} \tag{6}$$

with B_e obtained from Eq 5.

The determination of an effective Young's modulus from Eq 6 has the important advantage that it serves as a general correction, compensating for all (systematic) factors affecting the value of the initial compliance. However, if one follows this procedure and notes that B and E appear on the same side of the compliance calibration expressions, Eq 1, the question of the exact definition of effective thickness B_e for a side grooved specimen can be avoided by determining an effective EB

$$(EB)_e = \frac{f(a_0/W)}{C_0} \tag{7}$$

with $f(a_0/W)$ according to Eq 2.

Crack Front Curvature

Several studies [*6, 18–20*] have demonstrated that the specimen's compliance is influenced by the degree of crack front curvature. Results of UC experiments on CT specimens show that the final crack length (and hence the amount of crack extension) may be severely underpredicted when crack front tunneling occurs, for example due to the absence of side grooves. For example, De Vries [*18*] reported a 40% underprediction of final crack extension in a UC test of a non-side-grooved A542 1T-CT specimen exhibiting a very pronounced thumbnail-shaped crack front. Elastic three-dimensional finite element computations by Prij [*19*], modeling this geometry, confirmed that this underprediction of Δa could be attributed to the reduction of specimen compliance induced by crack front tunneling.

Although the occurrence of thumbnail-shaped crack front tunneling can be largely avoided by side grooving, such specimens may still display a certain degree of crack front curvature. They are normally fatigue precracked prior to side grooving, which can lead to a curved initial crack front. Moreover, subsequent stable crack extension is not always equally distributed over the specimen's thickness. Elastic three-dimensional finite element computations were performed on meshes with curved crack fronts (but without side grooves) in order to gain insight in the quantitative effects of crack front curvature on the compliance of a SENB specimen. An example of such a mesh given in Fig. 2 shows that, for reasons of symmetry, only one fourth of the specimen was modeled, and that in order to save computer time, only the most highly stressed region (that is, the ligament) was modeled by three-dimensional (20-node brick) elements. The mesh shown in Fig. 2 was originally designed by Bakker [*21*], who used it for an elastic-plastic analysis. A total of seven computations were performed. The variation in crack length over the thickness is given in Table 1. The crack shape of Mesh 3 was modeled after an optically measured fatigue precrack. The other crack shapes were derived from this crack by proportional increase or decrease of its degree of curvature, quantified by the parameter $\delta a/W$

$$\delta a/W = (a_{max} - a_{min})/W = (a_1 - a_5)/W \tag{8}$$

The average crack length $\bar{a}$ was determined according to the nine-point averaging procedure of the ASTM Test Method for J_{Ic}, a Measure of Fracture Toughness (E 813-81).

Series of analyses were performed for meshes with relative average crack length of $\bar{a}/W =$ 0.524 and 0.690. For each series, the computed compliance of a mesh with a straight crack (that is, $\delta a/W = 0$) served as a reference for the results of the analyses with curved cracks. The effect

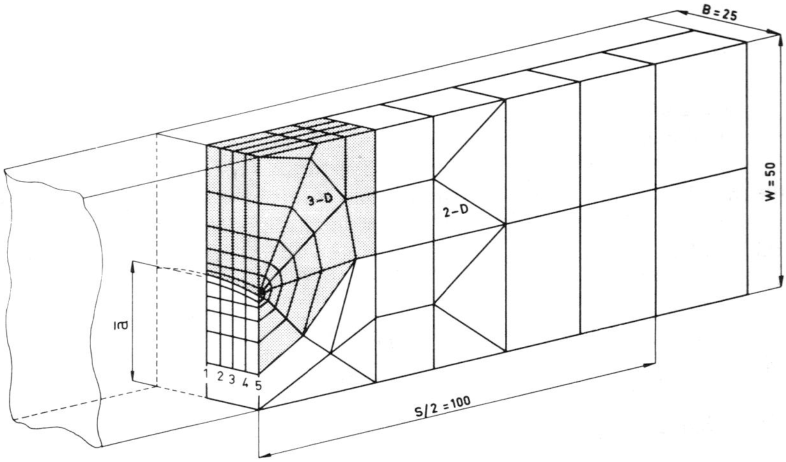

FIG. 2—*Finite element mesh for SENB specimen with a curved crack front. The two-dimensional part consisted of plane-strain elements* [21].

TABLE 1—*Crack front shape parameters for meshes with curved crack fronts.*

			Position Over Half-Thickness (Fig. 2)				
Mesh	$\bar{a}$, mm	$\delta a/W$ (−)	1	2	3	4	5
1	26.2	0.000	26.200	26.200	26.200	26.200	26.200
2	26.2	0.012	26.351	26.327	26.239	26.110	25.743
3	26.2	0.024	26.502	26.455	26.279	26.020	25.288
4	26.2	0.054	26.874	26.767	26.376	25.800	24.167
5	34.5	0.000	34.500	34.500	34.500	34.500	34.500
6	34.5	0.054	35.178	35.071	34.677	34.098	32.456
7	34.5	0.090	35.620	35.443	34.792	33.835	31.120

of crack front curvature on (crack mouth) compliance of an SENB specimen can thus be quantified by the curvature factor FC

$$FC = FC(\bar{a}/W,\ \delta a/W) = \frac{C_{\text{curved}}}{C_{\text{straight}}} \tag{9}$$

The results of the finite element analyses are summarized in Fig. 3. For the ranges in $\bar{a}/W$ and $\delta a/W$ examined, this figure shows that:

(*a*) increasing crack curvature induces a progressive decrease in specimen compliance, and
(*b*) for the same degree of crack curvature, the effect of curvature on compliance becomes more pronounced with increasing crack length.

Both the J_{Ic}-test procedure ASTM E 813, and the tentative J_R-curve standard [*5*] prescribe that none of the individual nine-point measurements for the initial crack a_0 should differ by more than 7% from the average value. For the symmetric crack front shapes of Table 1, this leads to the limits in crack front curvature $\delta a/W$ indicated by the arrows in Fig. 3. It is seen that the above-mentioned restriction by ASTM E 813 and Ref *5* limits the influence of crack curva-

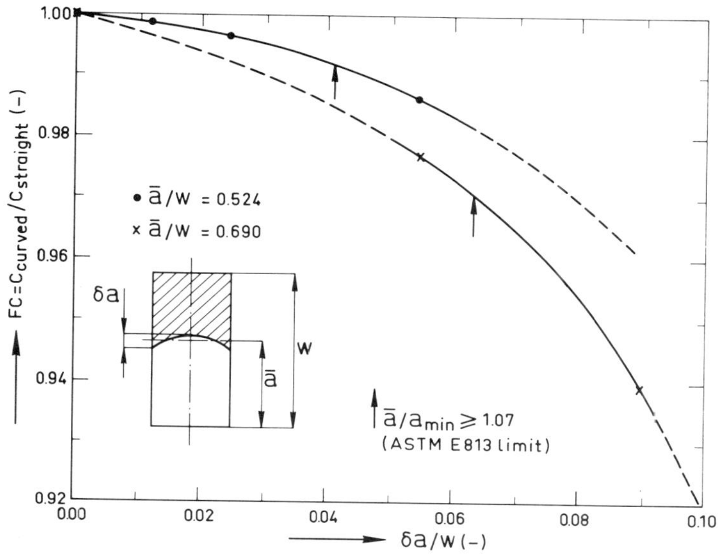

FIG. 3—*Influence of crack front curvature on specimen compliance.*

ture on the initial compliance to less than 1% for moderate values of a_0 (namely, $a_0/W \approx 0.5$), but that this influence on compliance can triple for initial crack lengths up to $a_0/W \approx 0.7$. Furthermore, for subsequent crack lengths, including stable crack extension, the above standards only prescribe that Δa at the edges of the specimen (that is, at the side grooves) should differ from Δa at the specimen's center by no more than $\pm 0.02\ W$. This leaves the possibility for asymmetric crack front shapes or different degrees of crack front curvature or both during the testing of the specimen as compared with the initial crack length, meaning that the influence of crack front shape on the specimen's compliance will change during the experiment.

Specimen Deformation

Elastic compliance calibration functions such as discussed earlier in this paper are based on small (infinitesimal) strain analyses, meaning that load-induced changes in specimen geometry are not taken into account. This will cause deviations where plastic deformation occurs. Such effects will be more pronounced for SENB than for CT specimens because of the well-known fact that for a given material and specimen size W, an SENB specimen requires a larger displacement at lower load than a CT specimen, where the load is applied in tension to reach the same J-level.

Several elastic and elastic-plastic finite element analyses were performed, all under plane-strain conditions, in order to obtain quantitative insight in the influence of specimen deformation on the elastic compliance. For all elastic-plastic computations, the material's stress-strain data used as input to the analyses were those of St.52-3 (Heat 1343), as given in Fig. 14 (see section on Experimental Details).

Figure 4 shows one of the two-dimensional finite element meshes used in this study, with the rectangular eight-node elements at the crack tip collapsed to triangles, each with three independent crack tip nodes (further details are given in Ref *22*). The elastic-plastic finite element computations were performed by incrementally increasing the load-line displacement Δ. At a desired displacement level, say Δ_i, a zero displacement increment (needed for stress redistribution) was applied, followed by a negative increment simulating an unloading of the specimen.

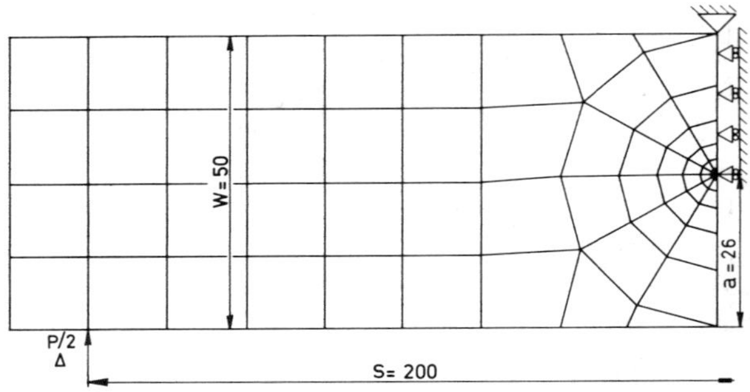

FIG. 4—*Two-dimensional finite element mesh for SENB specimen* [22].

The specimen compliance was subsequently determined from the differences in load (dP_i) and crack opening displacement ($d\delta_i$) between the results of the zero and negative increments at the displacement Δ_i. This procedure was repeated for increasing levels of displacement.

Figure 5 shows how the compliance of the deformed mesh after the ith increment, $C_{D,i}$, is determined. This numerically computed compliance is compared with the theoretical crack mouth compliance C_T given by Eq 2. For large (finite) strain analyses, where the geometry of the mesh is updated after each load increment, C_T must be computed by using the momentary values of crack length a_i and specimen width W_i (as defined in Fig. 5), while C_D is corrected for the increase in span length S_i, thus cancelling the effect of span increase on the compliance, leaving only the true effect of specimen deformation. For small (infinite) strain analyses, these corrections do not apply.

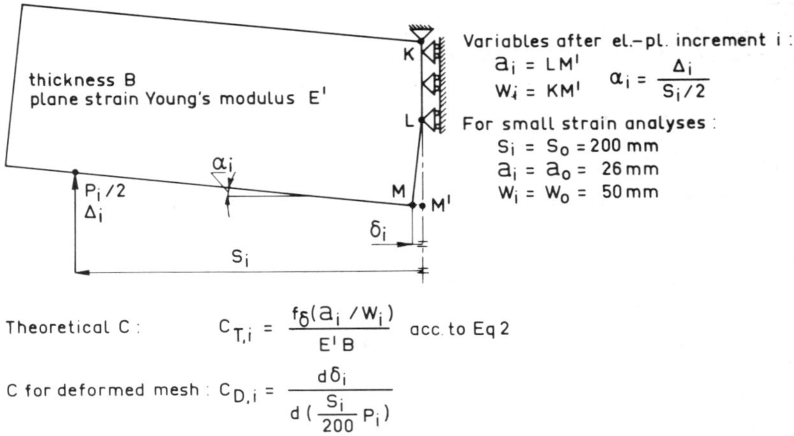

FIG. 5—*Determination of the theoretical and actual compliances of an elastic-plastically deformed SENB specimen.*

The compliances C_0 and $C_{T,0}$, as determined for the first (fully elastic) increment starting from $\alpha = 0$, serve as reference values for the subsequent compliance values. The deformation correction factor FD, quantifying the deviation of the compliance of the deformed SENB specimen from its theoretical value according to Eq 2, can thus be defined as

$$FD = FD(\alpha_i) = \frac{C_{D,i}/C_{DT,i}}{C_0/C_{T,0}} \tag{10}$$

where

$C_{D,i}$ = compliance of deformed specimen, i^{th} increment, mm/N,
$C_{DT,i}$ = theoretical compliance of deformed specimen, i^{th} increment, mm/N,
C_0 = compliance, first increment, mm/N,
$C_{T,0}$ = theoretical compliance, first increment, mm/N, and
α_i = angle of rotation, defined in Fig. 5, rad.

It should be noted that for the small strain analysis, $C_{DT,i} = C_{T,0}$.

For the meshes employed in this study, C_0 was found to differ by less than 1% from $C_{T,0}$ in all cases, indicating that the meshes were sufficiently converged for the present purpose.

With the mesh of Fig. 4, two analyses with constant crack length were performed, the results of which are plotted in Fig. 6 in the form of the deformation correction factor FD (Eq 10) versus the angle of rotation α (chain-dotted curves). Whereas for the small strain analysis the specimen's compliance is hardly affected by the occurrence of plastic deformation, it is seen that for the large strain analysis, the actual specimen compliance decreases with increasing deformation, that is, the deformed specimen acts stiffer than a straight one, which was also observed experimentally by Wallin et al. [23] as well as others.

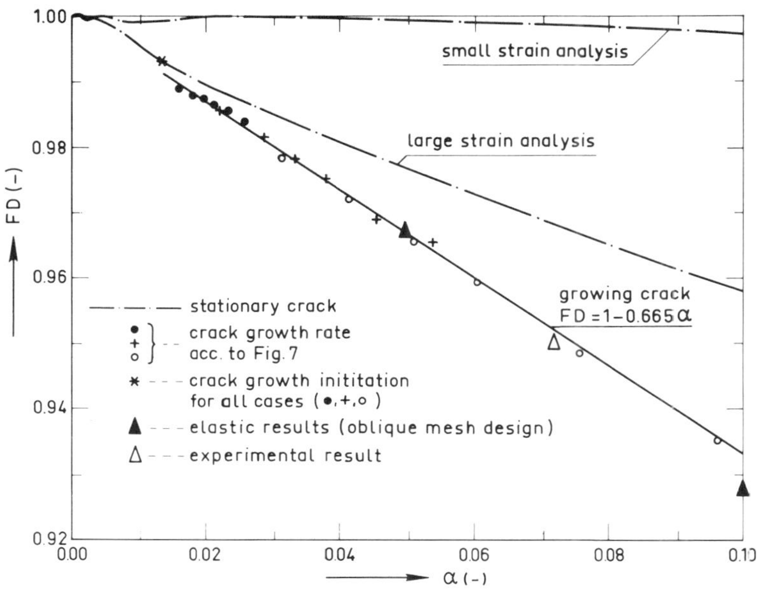

FIG. 6—*Influence of change in specimen geometry (due to rotation) on compliance.*

The large strain analysis curve in Fig. 6 is somewhat at variance with the physical reality, however, as in this computation, the crack continues to blunt without growing (due to the independent nodes at the crack tip). For larger values of imposed Δ (or α), this leads to unrealistically large values of crack tip opening displacement (CTOD). Additional analyses were made with meshes incorporating crack growth in order to account for the possible effect of crack extension on compliance. For this purpose, use was made of earlier work performed in the author's laboratory [*24*].

The mesh used for the crack growth studies was identical to that in Fig. 4 except for the ligament region, where the mesh was more refined (ten rings of elements rather than six around the crack tip). Crack growth was simulated by the node-releasing technique [*24,25*], and the element sizes along the ligament were such that the following values of total crack extension could be successively applied: $\Delta a = 0.6 - 1.2 - 1.8 - 2.4 - 3.3$ and 4.6 mm. Three analyses were made with the three different crack growth rates Δa versus Δ (or α) as given in Fig. 7. These crack growth rates roughly coincide with rates experimentally observed in the author's laboratory for different materials. After each applied crack growth step (each such step involved six elastic-plastic increments), a zero and negative displacement increment were applied and the actual and theoretical compliances determined as described before.

When performing such crack growth simulations in a finite element mesh, it should be noted, however, that the mesh changes after each crack growth step due to the disconnection of nodal points, and that the stress and strain singularities at the crack tip are different for a stationary crack with independent crack tip nodes and a moving crack having only a single crack tip node. Because both factors can be expected to influence the specimen's compliance, two parallel computations, namely a small strain and large strain analysis, incorporating identical crack extension rates, were carried out for each of the three crack growth rates of Fig. 7. By letting the small strain results serve as reference for the large strain results, the above two effects (caused by the finite element modeling) are cancelled, and only the effects of specimen rotation and crack extension remain.

The results of these computations can again be quantified by the deformation correction factor *FD*, now defined as

$$FD = FD(\alpha_i, \Delta a_i) = \frac{C_{D,i}/C_{DT,i}}{C_i/C_{T,i}} \quad (11)$$

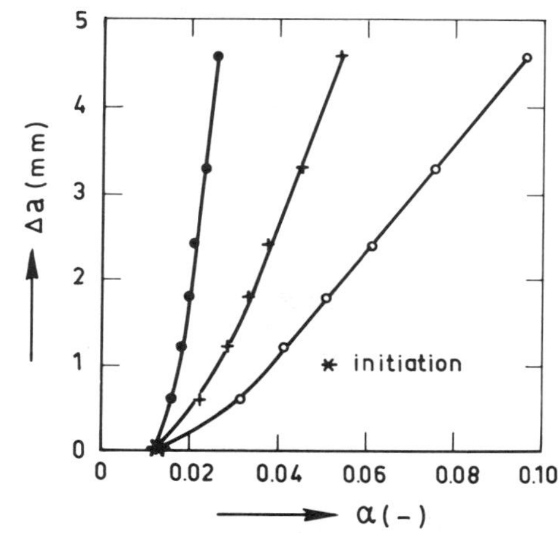

FIG. 7—*Crack growth rates used as input for the crack-growth analyses. The symbols correspond with those of Fig. 6.*

where

$C_{D,i}$ = compliance of deformed specimen, i^{th} increment,
$C_{DT,i}$ = theoretical compliance of deformed specimen, i^{th} increment,
C_i = compliance of specimen, small strain analysis, i^{th} increment, and
$C_{T,i}$ = theoretical compliance, small strain analysis, i^{th} increment.

From result of the above computations, also given in Fig. 6, it is seen that:

1. Up to initiation (denoted by the asterisk in Fig. 6), the results coincide with those of the large strain analysis for constant crack length, as is to be expected.
2. The compliance decreases (as compared with its theoretical value) with increasing deformation.
3. This decrease in specimen compliance appears independent of the crack growth rate.
4. The occurrence of crack growth causes a larger relative decrease in compliance than observed for the case of the stationary crack.

The latter two findings may be explained by the fact that contrary to the large scale blunting occurring at the stationary crack tip, the crack tip is sharp in the crack growth cases, causing stiffer specimen behavior (that is, a lower compliance). Various numerical and experimental studies on the shape of the crack tip profile before and after the onset of stable crack extension have shown that, generally, the growing crack can be considered sharp (that is, having a far lower CTOD) as compared with the crack prior to crack extension initiation, [*26,27*]. This implies that once stable crack growth has commenced, the curve denoted "growing crack" in Fig. 6 is the correct one to assess the influence of deformation on specimen compliance. The observation that this curve appears independent of crack growth rate suggests that the very presence of a sharp crack tip is of more importance than its actual size Δa.

It was stated before that all elastic-plastic finite element computations were performed using the stress-strain data of Fig. 14. Linear-elastic finite element computations were made on obliquely designed meshes such as the specimen shown in Fig. 5 in order to check the dependence of the results just described on the imposed elastic-plastic material behavior. Analyses were made for meshes angled to $\alpha = 0.045$ and 0.10 radians, with a sharp crack in both cases. A straight mesh ($\alpha = 0.00$ rad) again served as reference. Further details are given in Ref *15*. The resulting *FD*-values for these analyses are again given in Fig. 6 (filled triangles), and it is seen that the elastic results agree very well with the results of the growing-crack analyses. Hence it appears possible to draw the important conclusion that the deviation of the actual specimen compliance from its theoretical value, given by the drawn line in Fig. 6, is solely a function of the angle of specimen rotation, independent of material behavior and of crack growth rate.

A final entry to Fig. 6 is made by an experimental result by the present author. A notched SENB specimen ($W = 50$ mm, $B = 25$ mm, and $a_0 = 24$ mm) was loaded in a testing machine to an angular rotation of $\alpha = 0.072$ rad. After unloading, the specimen was side grooved, and a straight and sharp crack (root tip radius 0.05 mm) was extended from the notch by spark erosion to $a_0 = 30$ mm. The compliances of this deformed SENB specimen as well as that of an identical but undeformed specimen were experimentally determined. It was seen that the compliance of the deformed specimen was 5% lower than that of the undeformed one. (The compliances actually differed by more than 5% due to a change in effective span length, but 5 percentage points of the difference could be attributed to specimen deformation). This experimental result is shown in Fig. 6 by the open triangle and correlates well with the numerical findings.

Friction at and Movement of Outer Rollers

When an SENB specimen supported by two outer rollers is centrally loaded, three simultaneous phenomena will occur at these rollers:

(*a*) the rollers move outward;
(*b*) the deforming specimen tilts back on these rollers; and
(*c*) friction forces will occur between rollers and specimen and between the rollers and their supports.

These effects alter the effective bending moment on the specimen with respect to the idealized three-point loading, thus influencing the apparent specimen compliance.

In Fig. 8, it is assumed as a first order approximation that the specimen deforms as two rigid halves rotating around a center of rotation, which is assumed to be located at a distance h from the specimen's lower side

$$h = a + r(W - a) \tag{12}$$

where r is the rotation factor, equal to 0.4 in BS 5762:1979 "Methods for COD Testing." (Other investigators, for example those in Ref *28*, suggest that r be 0.45. The proper value of r obviously depends on the amount of plastic deformation). From Fig. 8 it is seen that although the outer roller moves over a distance e, the effective (half) span length $S'/2$ only increases by e' due to the backward tilt of the specimen half over the roller. With α as the specimen's angle of deflection, e and e' can be determined from geometric considerations [*15*], if it is assumed that the rollers do not slide but only roll

$$e = \frac{(2h + D)\sin\alpha + 4W(\cos\alpha - 1) - \alpha D\cos\alpha}{2(\cos\alpha + 1)} \tag{13a}$$

$$e' = e - \frac{D}{2}\sin\alpha. \tag{13b}$$

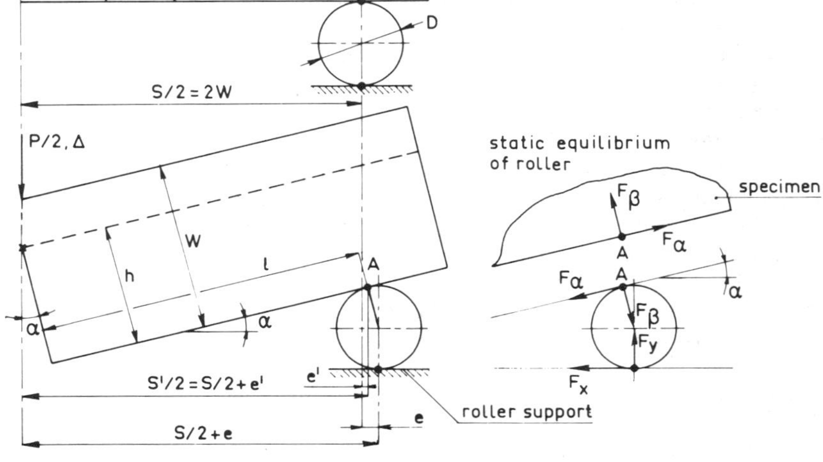

FIG. 8—*Geometric analysis of SENB specimen half, rigidly rotating around its rotation center.*

In order to give an impression of the magnitudes of e and e' for specimen dimensions typical for those used in the author's laboratory, Fig. 9 gives values for e and e' as a function of α for two crack lengths. Some experimentally measured values for e are also included in Fig. 9. The measurements were made on a St.52-3 specimen with initial crack length $a_0 = 26$ mm, which extended to about $a = 33$ mm for $\alpha = 0.1$ rad. The results are seen to correlate well with the values for e as predicted from Eq 13a.

It follows from the preceding discussion and from Fig. 8 that Point A at a distance $S'/2 = 2W + e'$ from the specimen load line serves as the point of contact between specimen and roller. It is shown in Fig. 8 that both a normal force F_β and a friction force F_α act on the specimen in Point A and contribute to the bending moment. It follows from the static equilibrium equations applied to the roller and from the understanding that vertical equilibrium of specimen plus roller requires F_y to be equal to $P/2$ that

$$F_y = F_\beta = \frac{P}{2} \tag{14a}$$

$$F_x = F_\alpha = \frac{\sin \alpha}{1 + \cos \alpha} \frac{P}{2} \tag{14b}$$

The term $(\sin \alpha)/(1 + \cos \alpha)$ in Eq 14b serves as an apparent friction coefficient, and typically reaches a value of 0.05 for $\alpha = 0.1$ rad. From engineering handbooks (for example, Ref *29*), it is seen that this value is easily attainable for metallic contacts, meaning that even without possible indentation of the roller in the specimen and roller supports, sliding should not occur, a fact confirmed by numerous experiments on SENB specimens.

The total effective bending moment M on the specimen is now obtained by multiplying the forces acting on Point A (Fig. 8) by their respective moment arms to the center of rotation, yielding

$$M = F_\alpha h + F_\beta l \tag{15}$$

with l as

$$l = \frac{S/2 + e' - h \sin \alpha}{\cos \alpha} \tag{16}$$

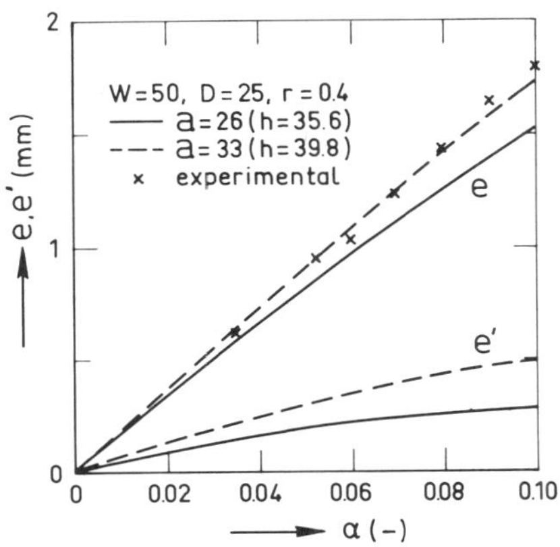

FIG. 9—*Examples of outer roller movement* (e) *and increase in effective span length* (e'). *All dimensions given in mm (*r *is dimensionless).*

Substituting Eqs 13, 14, and 16 into Eq 15 and noting that $S/2 = 2W$, yields, after some algebraic manipulation

$$M = \frac{P}{2}\left(\frac{4W}{\cos\alpha + 1} - \frac{D}{2}\frac{\sin\alpha + \alpha}{\cos\alpha + 1}\right) \tag{17}$$

while for the undeformed specimen ($\alpha = 0$), the bending moment M_0 induced by the three-point load equals

$$M_0 = \frac{P}{2}2W. \tag{18}$$

The combined effect on the bending moment of the above-mentioned phenomena occurring at the outer rollers follows from the comparison of Eqs 17 and 18. It is seen that an identical load increment dP during unloading will induce a different increment in bending moment and hence different displacement increments ($d\Delta$ or $d\delta$) when the specimen and rollers are no longer in their original positions, resulting in different compliance values measured for the same unloading dP. This effect can be described by the roller correction factor FR, defined by

$$FR = FR(\alpha, D/W) = \frac{M}{M_0} = \frac{2}{\cos\alpha + 1} - \frac{D}{4W}\frac{\sin\alpha + 1}{\cos\alpha + 1} \tag{19}$$

with M and M_0 from Eqs 17 and 18, respectively. Note that while the roller movement e and span increase e' are dependent on h and thus on the current value of the crack length, Eq 13, the dependence of M and hence FR on crack length is cancelled by the occurrence of the friction forces in Point A (Fig. 8).

The curves of FR for different values of D/W given in Fig. 10 indicate that, in general, the combined effect on compliance of the phenomena at the outer rollers will be very small. It should be finally noted that the correction of Eq 19, applicable to SENB specimens, is equivalent to the practice followed in CT testing (for example, Ref *4*) to correct the measured specimen compliance for the reduction of the moment arms due to specimen rotation.

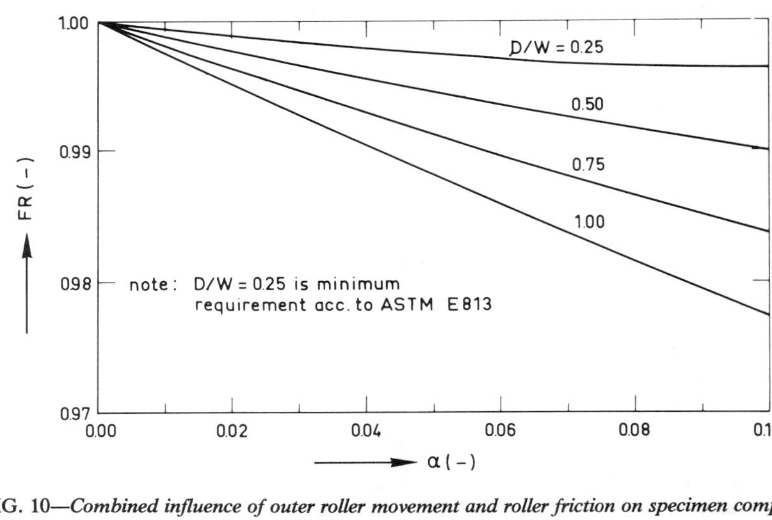

FIG. 10—*Combined influence of outer roller movement and roller friction on specimen compliance.*

Center Roller Indentation

As the load on a SENB specimen increases, the center roller's line of contact with the specimen will grow into a contact area, as depicted in Fig. 11*a*. The stresses in the contact area, initially elastic (Hertzian stresses), will eventually become plastic, thus increasing the size of the contact area. Meanwhile, the specimen halves are rotating, thereby "folding" themselves around the center roller and thus causing a further increase in contact area size (Fig. 11*a*). This indentation effect leads to a departure from the pure three-point bend loading case, thus affecting the specimen's compliance (for example, Ref *30*). Although similar situations occur at the outer rollers, the load at these rollers is only half of that at the center roller. Moreover, the additional increase in contact area at the center roller due to the folding of the specimen halves does not occur at the outer rollers. Therefore only the situation at the center roller will be considered here.

In order to quantify the influence of roller indentation on the compliance, both the stress distribution over the contact area and the size of this area need to be known. Regarding the first point, two elastic solutions [*29,30*] concerning a cylinder and a square stamp as indenters are shown in Fig. 11*b*. However, the extensive plastic deformation generally observed in the present case will level the high peak stresses found for the elastic cases. As a first approximation, a constant stress distribution is therefore assumed. With $2c$ as the width of the contact area, it then follows from Fig. 11*c* that the effective bending moment M due to the distributed load at the center roller reduces to

$$M = \frac{P}{2}\left(2W - \frac{c}{2}\right) \tag{20}$$

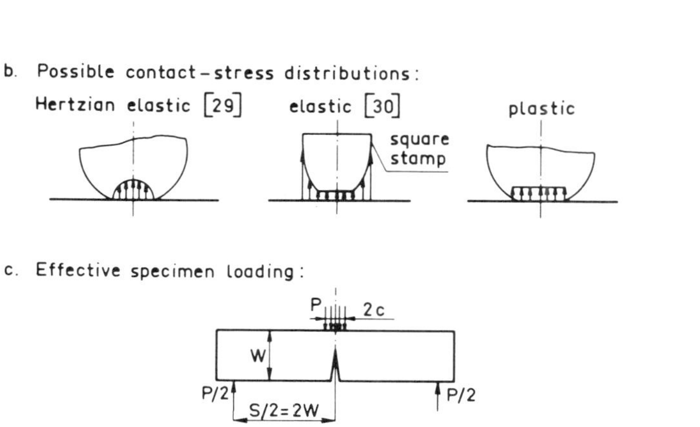

FIG. 11—*Indentation of specimen by center roller.*

Following the same reasoning for the effect of changes in M on compliance as in the preceding section, the influence of the center roller indentation on the specimen's compliance can be described by the indentation correction factor FI

$$FI = FI(\alpha, \text{material, dimensions}) = \frac{M}{M_0} = 1 - \frac{c(\alpha)}{4W} \tag{21}$$

with M_0 according to Eq 18.

The contact area width $2c$ (averaged over the specimen thickness) has to be determined experimentally for the material and specimen dimensions considered. This was done by repeatedly placing a piece of thin aluminum foil between a specimen and the center roller, loading the specimen to a certain deflection, unloading, changing the foil, and reloading to a higher level of displacement. This procedure yielded very clear impressions of the contact area on the foils, the sizes of which could be optically measured by a travelling microscope. This procedure, by avoiding measurements on the specimen itself, permitted the latter to be left in the testing machine and thus to maintain the same specimen alignment throughout this experiment.

The contact area width $2c$ was found to be primarily dependent on the deflection (or angular rotation α), as the value of $2c$ was seen to increase continuously with α even when the load decreased due to extended crack growth. The results of the experiments are therefore presented as curves of FI (according to Eq 21) versus α in Fig. 12, with the indicated specimen types according to Table 3 (see section on Experimental Details). The figure shows that the influence of roller indentation on compliance is small, especially for larger specimen sizes.

Summary

In the preceding paragraphs, an analysis has been made of the factors influencing the compliance of a SENB specimen. The findings of this foregoing analysis are summarized below:

1. The expression for crack mouth compliance, Eq 2 (together with its inverted form, Eq 3), should be used as the basic calibration function relating specimen compliance to crack length.
2. The combined effective Young's modulus and thickness $(EB)_e$ are determined from the initial values of crack length and compliance through Eq 7, thereby circumventing the question of effective thickness for an SENB specimen.

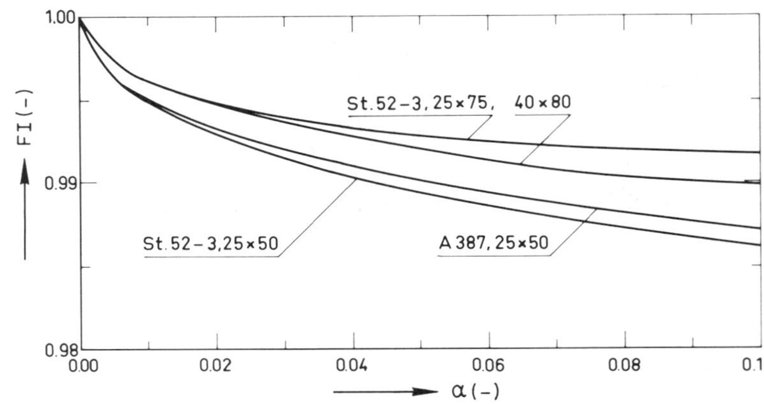

FIG. 12—*Influence of center roller indentation on specimen compliance. The numbers indicate* B × W *(in mm) for the various specimen groups investigated in this paper.*

3. The effects on compliance of specimen deformation, of outer roller movement and friction and of center roller indentation may be expressed as functions of specimen rotation α. The use of α, which is obtained with sufficient accuracy from

$$\alpha = \frac{\Delta}{2W} \tag{22}$$

as independent variable for these functions, generalizes the results for any specimen size.

4. While the effects of specimen deformation and outer roller movement and friction, *FD* and *FR*, respectively, are explicitly quantified, independently of material behaviour (and crack growth rate), the influence of center roller indentation *FI* needs to be determined experimentally for the material and specimen size considered.

5. From the comparison between *FD*, *FR*, and *FI* made in Fig. 13 for some typical cases, it is seen that the influence of specimen deformation becomes predominant at higher α values but that the other two effects are not negligible.

6. The influence of crack front curvature *FC* on specimen compliance increases progressively both with increasing crack curvature and with increasing average crack length, Fig. 3. To determine *FC*, the crack front shape has to be accurately known, which is only possible for the initial and final crack fronts. Linear interpolation with α between the known values of *FC* forms as possible approximation for intermediate compliance values.

The compliance values determined for SENB specimens during UC experiments contain all the effects discussed above. They all lead to a lower compliance than theoretically expected, namely, *FC*, *FD*, *FR*, and *FI* are all smaller than 1. By proper correction, however, a corrected compliance is obtained for which the calibration functions are valid.

Since the definition of the effective $(EB)_e$ through Eq 7 already contains a correction for the above effects at the first unloading, all further corrections of the experimentally measured com-

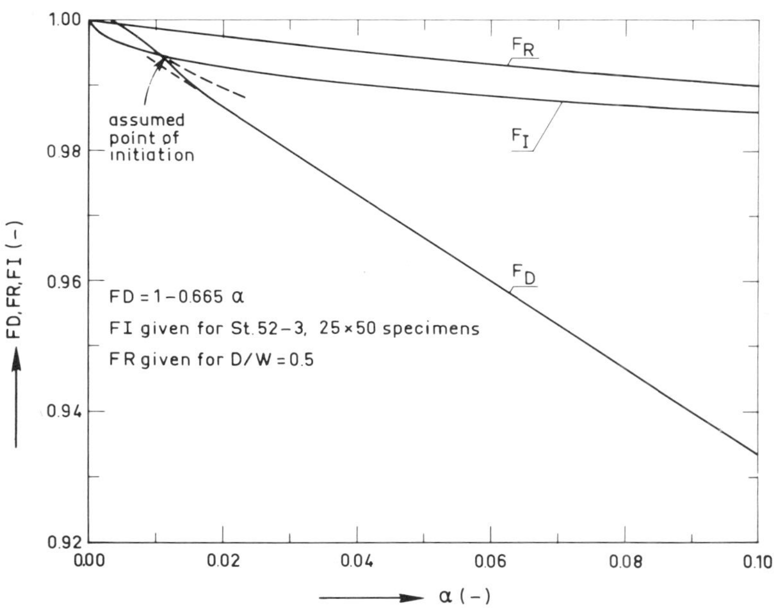

FIG. 13—*Comparison between* FD, FR, *and* FI *for some typical cases.*

pliances for these effects should be applied relative to those of the first unloading; that is, the corrected compliance C_c should be computed from the measured compliance value C by

$$C_c = \frac{C}{FC/FC_0 \cdot FD/FD_0 \cdot FR/FR_0 \cdot FI/FI_0} \tag{23}$$

where

FC = curvature correction factor, Fig. 3,
FD = specimen deformation correction factor, Fig. 6,
FR = outer roller correction factor, Fig. 10 or Eq 19, and
FI = indentation correction factor, Fig. 12 or Eq 21,

and with the subscript "0" referring to the values at the first unloading. C_c is subsequently used to compute the corresponding crack length from Eq 3, with U replaced by U_c

$$U_c = \frac{1}{\sqrt{C_c(EB)_e} + 1} \tag{24}$$

Experimental Details

Material and Specimens

Two heats of mild steel St.52-3 (equivalent to BS 4360 Grade D) and one of A387 Grade D (2¼Cr-1Mo) pressure vessel steel were used. Chemical composition and room temperature stress-strain properties are given in Table 2 and Fig. 14, respectively.

The geometries of the 24 SENB specimens tested are given in Table 3. All specimens were machined according to standard specifications. Although the St.52-3 specimens were fatigue precracked, the A387 specimens were given a spark eroded precrack with a tip radius of 0.05 mm and a crack front shape identical to that of a fatigue precrack. In each case, the initial crack length ratio was $a_0/W = 0.52$, and all specimens were 20% side grooved after precracking.

Test Equipment

The specimens were tested in a servo hydraulic testing machine, using the jig shown schematically in Fig. 15. The load P was recorded as a function of load-line displacement Δ, while the increments in load and crack mouth opening displacement dP and $d\delta$ during unloadings were plotted on a second recorder with expanded scale. The load was measured by the testing machine's load cell (sensitivity 0.04 V/kN), Δ was obtained from a linear variable differential transformer (LVDT) (0.21 V/mm), and δ was monitored by a clip gage (6.06 V/mm). The geometry of the clip gage knife edges shown in Fig. 15 made it possible to measure δ at the

TABLE 2—*Composition of steels in weight percents.*

Material	C	Mn	P	S	Si	Cr	Mo	Ni
A387 Grade D	0.15	0.50	0.018	0.013	0.210	2.40	0.96	0.12
St.52-3, Heat 1343	0.19	1.43	0.040	0.023	0.352	...	...	...
St.52-3, Heat 3297	0.17	1.39	0.032	0.018	0.439	...	...	...

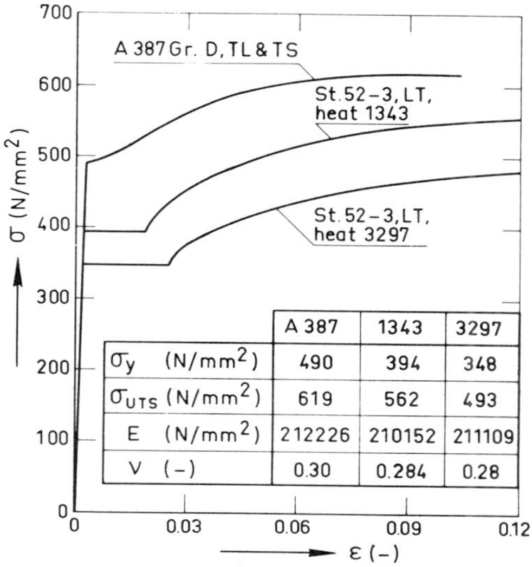

FIG. 14—*Engineering stress-strain data for materials investigated (room temperature).*

TABLE 3—*Test matrix with material, orientation of the crack toward the rolling direction, main dimensions, and number of the specimens.*

Material	Orientation	Dimensions, mm *B* by *W*	Number of Specimens
St.52-3 (Heat 3297)	LT	25 by 50	3
	LT	40 by 80	3
St.52-3 (Heat 1343)	LT	25 by 50	4
	LT	25 by 75	2
A387 Grade D	TL	25 by 50	6
	TS	25 by 50	6

original crack mouth position, that is, the measured δ did not require a correction for the knife edge height.

For larger crack opening displacement values, the knife edges were manually adjusted because the clip gage used had limited working range. Adjusting the knife edges leads to a slight overestimation of the actual δ (or $d\delta$), however, caused by tilting of these knife edges when extended beyond their original position and when the specimen is deformed, (Fig. 15). It can be shown, however, that this effect is very small (about 0.2% for $\alpha = 0.06$ rad) and that its size is determined by the working range of the specific clip gage used. Hence it is sufficient not to further quantify the effect here, but to note only that the measured compliances were duly corrected for this effect.

Test Procedure

The specimens were loaded under displacement control with a loading rate of 0.5 mm/min. The first point of unloading was taken in the initial, linear part of the test. Subsequent unloadings (10 to 15% of current load) were made at equidistant load-line displacement intervals $d\Delta$,

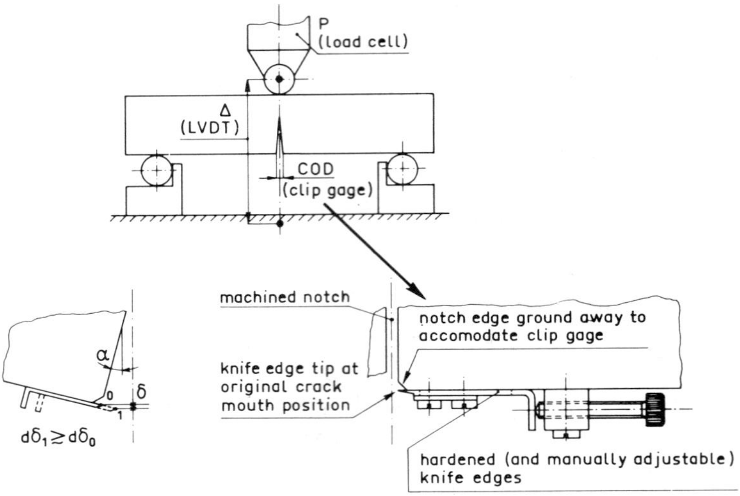

FIG. 15—*Principle of test jig used, with details of the clip gage knife edges.*

typically between 0.1 and 0.5 mm, depending on material and specimen size. At each point of unloading, the displacement was first kept constant for about two or three minutes to allow for load relaxation. Next, four unloadings were made. The first one was intended to ensure full contact between clip gage and knife edges. The slopes of the subsequent three unloadings were averaged to yield the specimen's compliance. For the St.52-3 specimens of both heats, the unloadings were straight, with no appreciable hysteresis effects observed. Hence the measured slopes of the three successive unloadings generally differed by less than 0.5% from each other. For the A387 specimens, however, the initial parts of both unloadings and reloadings were curved, and only the second halves of these lines were straight, even though ample time was taken to allow for load relaxation. Compliance values were determined from the straight parts, and differences up to 2% were observed.

At two or three points of unloading in each test, a dye penetrant beach marking technique [*31*] was used to mark the current crack front. This method extends the possibility to check the accuracy of the UC crack growth prediction against actual Δa at the termination of the test to some intermediate values of Δa as well.

After completion of the test, the crack was extended further by fatigue cycling. The initial crack front and the beach marks were optically measured by means of a travelling microscope, using the nine-point averaging method. These optically measured values of a_0 and the resulting crack front curvatures (Eq 8), together with the values of load, load-line displacement, and compliance at each unloading, were fed into a program, which computed corresponding values of J and Δa, taking the various correction factors into account.

Performing and evaluating the UC experiments proved time-consuming because much of the work was done manually, such as attending the testing machine and measuring the compliance slopes. If UC tests on SENB specimens become routine practice, however, fully computer-controlled procedures may be feasible, which is already the case for CT testing (for example, Refs *9*, *32*).

Determination of J *and* Δa *From Test Results*

For each test, the effective $(EB)_e$ was determined from the first unloading and the optically measured initial crack length a_0 using Eq 7. This $(EB)_e$ was then used throughout this test.

Subsequent compliances were corrected using Eq 23, with the curvature correction factor *FC* estimated by interpolation (linear with specimen rotation α, Eq 22) between the known values for *FC* at the initial and dye-penetrant-marked crack lengths. From this corrected compliance, the corresponding crack length a and hence the corresponding amount of stable crack growth $\Delta a = a - a_0$ were computed from Eq 24.

The J-integral was computed from the increments $A_{i,i+1}$ in area under the specimen's load/load-line displacement record, approximated by trapezoidal rule, through [2]

$$J_{i+1} = \left(J_i + \frac{2A_{i,i+1}}{B_N\, b_i}\right)\frac{b_{i+1}}{b_i} \tag{25}$$

The area increments $A_{i,i+1}$ were duly corrected for extraneous displacements not contributing to J, such as test jig deformation and roller indentations [33].

Results

Prediction of Crack Extension

In Fig. 16 a comparison is made between optically measured dye penetrant beach marks (Δa_{BM}) and crack extension values predicted by the UC method (Δa_{UC}), with the different symbols indicating the different groups of specimens of Table 3. Figure 16*a* gives the uncorrected Δa predictions; Fig. 16*b* shows the predictions obtained by correcting the measured compliance for specimen deformation, roller movement and indentation and crack front curvature. Figures 16*a* and 16*b* give results for $\Delta a_{BM} \leqq 4$ mm, and Figs. 17*a* and 17*b* similarly show the results for $\Delta a_{BM} \geqq 4$ mm. In each figure, $\pm 10\%$ scatter bands are drawn to indicate the quality of crack growth prediction by the UC method. For low values of Δa, however, a relative measure of accuracy becomes unfeasible, and hence it is preferable to use an absolute quantity, for example, ± 0.1 mm [34].

The following four observations can be made from Figs. 16 and 17:

1. In general, the predictions are greatly improved by correcting the measured compliance. The corrected values generally predict the actual crack extension Δa_{BM} to within 10%.
2. For low values of Δa (namely, $\Delta a_{BM} \lesssim 1.2$ mm), the correction procedure tends to cause overprediction of the amount of crack extension. This is probably caused by uncertainty in the precise value of *FD* for small values of displacement around the point of crack growth initiation (Figs. 6 and 13). However, this range of Δa values is of limited importance for J_R-curve testing, where the emphasis is on accurate prediction of more substantial amounts of crack extension, (for example, Ref 35).
3. For all groups of specimens, the effect of the various corrections is seen to be least for the A387 specimens. This is caused by the relatively low resistance to crack extension found for this material, meaning that a certain amount of crack extension was reached at a relatively small value of load-line displacement, thus leading to small correction factors on the compliance (Fig. 13).
4. The predictions for A387 steel appear to be slightly less accurate than for the other specimens. This is attributed to the hysteresis loops that occurred during the unloadings, leading to uncertainty in compliance measurement.

Table 4 gives an impression of the values for the various correction factors applied to the compliance for a typical case. The predictions (uncorrected and corrected) of Δa computed from the compliances made at the beach marks are indicated in Fig. 16. For the specimen considered (specimen 5108), the correction for specimen deformation FD/FD_0 is seen to be largest. Another point worth noting is that the curvature correction factor FC/FC_0 exceeds

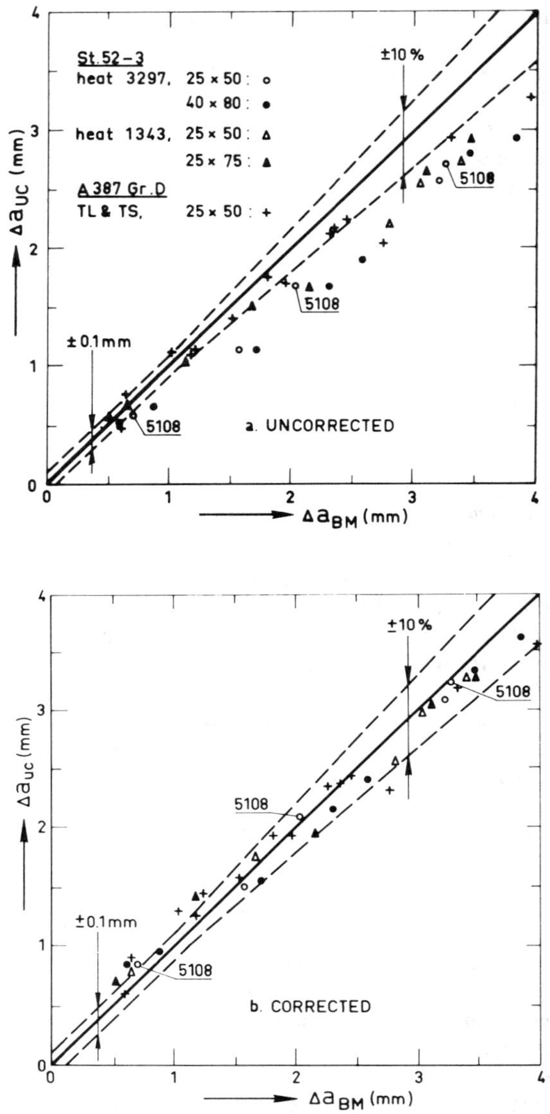

FIG. 16—*Comparison between predicted* (Δa_{UC}) *and measured* (Δa_{BM}) *amounts of crack extension;* $\Delta a \leqq 4$ *mm.*

unity for the first two beach marks of this specimen, indicating a lower degree of crack front curvature for these two beach marks than observed for the initial crack and final beach mark. The table shows that for Specimen 5108, the influence of crack front curvature on the compliance amounted was $\leqq 0.4\%$. This very small influence of crack curvature was typical for most specimens tested (exceptions were some St.52-3, heat 1343 specimens, to be discussed later), which is fortunate because *FC* is the most difficult correction factor to be determined, requiring the actual shape of the crack front to be known.

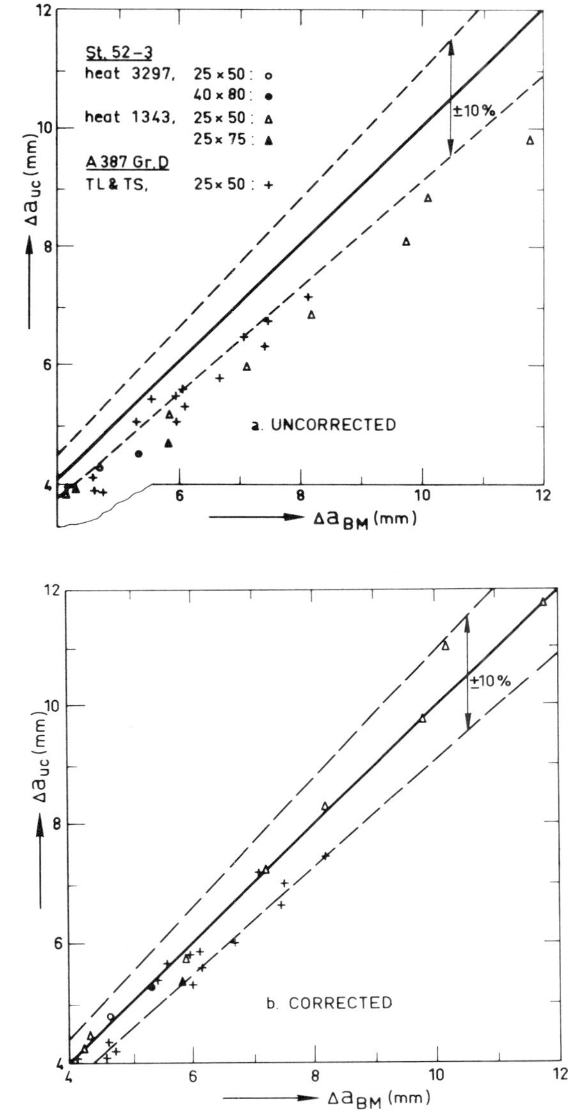

FIG. 17—*Comparison between predicted* (Δa_{UC}) *and measured* (Δa_{BM}) *amounts of crack extension:* $\Delta a \geqq 4$ *mm.*

J_R Curves

Figure 18 shows unloading compliance J_R-curves for two heat 3297 specimens, sized 40 by 80 mm^2 and 25 by 50 mm^2, respectively. Both uncorrected and corrected results are given, and it is seen that the corrections on the compliance result in much improved predictions. The indicated blunting line (with the flow stress $\sigma_f = (\sigma_y + \sigma_{UTS}) / 2$) and the ranges in true-initiation J values J_i for this material heat were determined from stretched zone width and potential-drop measurements, respectively [*21*]. Because the amount of crack extension was relatively small for

TABLE 4—*Correction factors applied for an unloading compliance experiment. Example for a St.52-3, Heat 3297, 25 by 50 specimen, No. 5108. The table lists only the factors for the initial and dye-penetrant beach mark unloadings.*

Unloading	Δ(mm)	α(rad)	FC/FC_0	FD/FD_0	FR/FR_0	FI/FI_0	Total	Remarks
1	0.50	0.005	1.0000	1.0000	1.0000	1.0000	1.0000	a_0
7	3.50	0.035	1.0004	0.9780	0.9965	0.9946	0.9698	beach mark
11	5.50	0.055	1.0002	0.9645	0.9945	0.9927	0.9524	beach mark
14	7.49	0.075	0.9959	0.9537	0.9931	0.9914	0.9352	beach mark = end

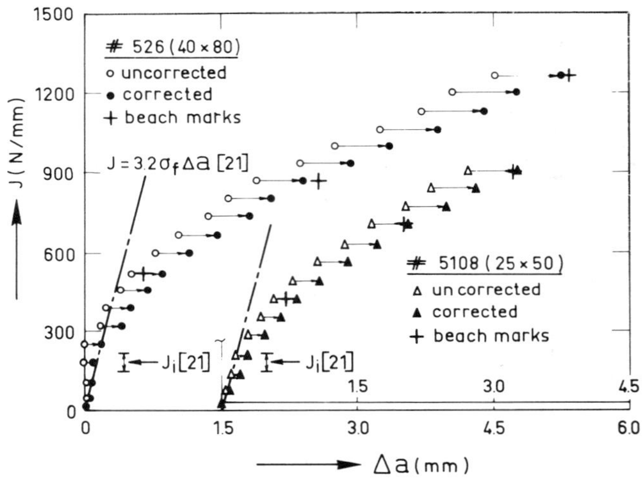

FIG. 18—J_R*-curves for two St.52-3, Heat 3297 SENB specimens. (Note the shifted* Δa*-scale for Specimen 5108.)*

these two specimens, the different Δa prediction did not noticeably influence the corresponding J values (Eq 25).

In Fig. 19, UC J_R-curves are given for a 25 by 50 mm^2 St.52-3 heat 1343 specimen, which was loaded to a large amount of crack extension. The indicated blunting line and true-initiation J-values for this material heat were determined in the manner mentioned above [*36,37*]. Due to the large Δa-values it is seen that the J-values are also affected by the compliance correction procedure. Although the UC method appears to yield very accurate Δa predictions for this specimen (and indeed for the other heat 1343, 25 by 50 mm^2 specimens as well, Fig. 17*b*), it should be noted that large delaminations, typical for this material heat [*22*], developed after a few mm of stable crack extension. This causes a lowering of constraint in the interior of the specimen, leading to a severe degree of reversed tunneling, with subsequent crack growth running ahead at the side grooves, thus yielding crack fronts considered invalid by the ASTM E 813. The compliances for these specimens were corrected by lowering the initially determined effective $(EB)_e$ (Eq 7) by a factor $1 - \nu^2$ (simulating a loss of constraint from plane strain to plane stress conditions) once delamination occurred, and by linearly extrapolating the data in Fig. 3 beyond the known values in order to obtain indicative crack front curvature correction factors. It is obvious that this procedure is fraught with uncertainty, thus the high accuracy of Δa predictions for these specimens may be coincidental.

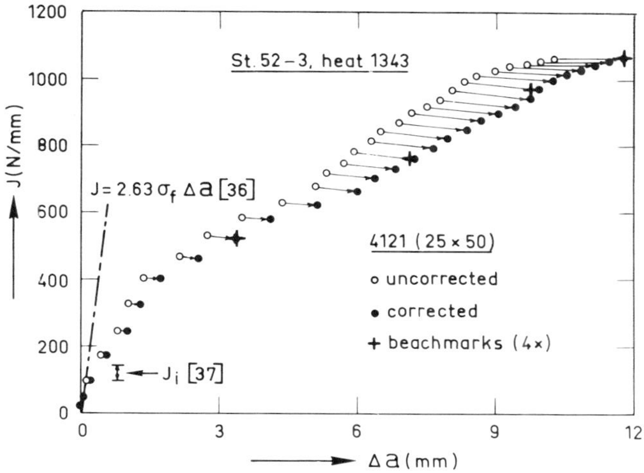

FIG. 19—*J_R-curves for St.52-3, Heat 1343, 25 by 50 SENB specimen.*

In Fig. 20 examples of J_R-curves are given for specimens of A387 Gr.D steel, with cracks in the TL and TS orientation (only the corrected J_R-curves are shown). Although the observed crack fronts were smooth and no delaminations occurred for either the TL or TS specimens, the J_R-curves are somewhat more irregular than those in the preceding figures, also for low Δa-values. This should again be attributed to the hysteresis loops in the unloadings.

In addition to the above discussion of the J_R-curves of Figs. 18 through 20, the following general observations appear in order:

1. The UC experiments, using the test jig of Fig. 15 and applying the compliance correction procedure of Eq 23, yielded no negative crack growth. Exceptions were those cases where hysteresis in the unloadings prevented unambiguous determination of the compliance. These problems are due to specimen material behavior rather than to equipment deficiencies.
2. The results of the initial unloadings are seen to reasonably follow the materials' blunting lines.
3. Appropriately correcting the measured compliances results in shifting the J_R-curve to larger Δa-values, causing a lower slope dJ/da. Not correcting the compliances would therefore lead to overpredicting the material's resistance to fracture.

Conclusions and Recommendations

1. Accurate single-specimen J_R-curves can be obtained from SENB specimens by the unloading compliance technique, provided appropriate corrections are applied to the measured compliance before it is used to compute the corresponding crack length through a theoretical compliance calibration function.
2. The various effects affecting the compliance of an SENB specimen were discussed and quantified. They all lead to apparently stiffer specimen behavior than predicted by compliance calibration functions.
3. The change in specimen geometry due to the rotation of the specimen halves during deformation and stable crack extension was found to have the largest effect on the compliance.
4. Contrary to frequent suggestions, movements of and friction at the outer support rollers were found to have only minor effects on the compliance.

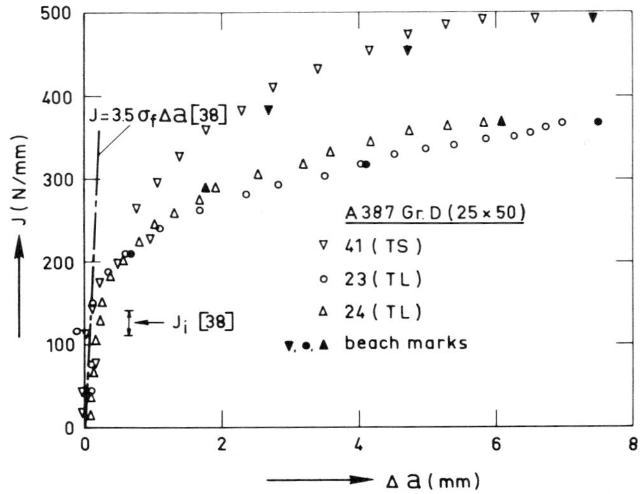

FIG. 20—*J_R-curves for A387 Grade D, TL, and TS SENB specimens, with all corrections applied.*

5. For moderately curved crack fronts, the effect of crack front curvature on the compliance appears small or negligible. However, this effect grows progressively with increasing curvature.
6. Based on the above analysis, a general procedure was proposed for correcting the experimentally measured compliance.

It is recommended that more experiments be performed on SENB specimens of different sizes and materials in order to further investigate the validity of the proposed procedure. This appears particularly desirable for austenitic steels, where large specimen deformations are expected, leading to large corrections on the measured compliance. Interestingly, it has been reported (for example, Ref *39*) that the usual rotation correction for CT specimens [*4*] can be insufficient when testing austenitic steel specimens, and that finite element calculations may be needed to account for the influence of change in CT specimen geometry on the compliance calibration functions. It should be emphasized, however, that for this material, through-thickness contraction may not be negligible, meaning that the effective specimen thickness will change with increasing specimen rotations as well.

Acknowledgments

The work described in this paper was partly sponsored by the Netherlands Ministry of Economic Affairs within the framework of the co-operative research program BROS. This support is greatfully acknowledged.

The author is indebted to Mr. T. C. G. Lo-a-Foe (former graduate student) and to Messrs. R. A. Gerbrands and M. C. de Groot for their contributions to the different aspects of this work, and to Professor D. G. H. Latzko and Dr. A. Bakker for their support and stimulating discussions. Use was made of earlier work by Messrs B. J. Bleeker and G. B. van Nuissenburg (former graduate students).

The author remembers with special gratitude the contributions made by Mr. M. Hartevelt, since deceased.

References

[1] Landes, J. D. and Begley, J. A., "Test Results from J-Integral Studies: An Attempt to Establish a J_{Ic} Testing Procedure" in *Fracture Analysis—Proceedings of the 1973 National Symposium on Fracture Mechanics, Part II, ASTM STP 560,* American Society for Testing and Materials, Philadelphia, 1974, pp. 170–186.

[2] Ernst, H. A., Paris, P. C., and Landes, J. D., "Estimations on J-Integral and Tearing Modulus T from a Single Specimen Test Record" in *Fracture Mechanics—13th Conference, ASTM STP 743,* American Society for Testing and Materials, Philadelphia, 1981, pp. 476–502.

[3] Clarke, G. A., Andrews, W. R., Paris, P. C., and Schmidt, D. W., "Single Specimen Tests for J_{Ic} Determination" in *Mechanics of Crack Growth, ASTM STP 590,* American Society for Testing and Materials, Philadelphia, 1976, pp. 27–42.

[4] Clarke, G. A., "Single-Specimen Tests for J_{Ic} Determination—Revisited" in *Fracture Mechanics—13th Conference, ASTM STP 743,* American Society for Testing and Materials, Philadelphia, 1981, pp. 553–575.

[5] Albrecht, P., Andrews, W. R., Gudas, J. P., Joyce, J. A., Loss, F. J., McCabe, D. E., Schmidt, D. W., and Van Der Sluys, W. A., "Tentative Test Procedure for Determining the Plane Strain J_I-R Curve," *Journal of Testing and Evaluation,* Vol. 10, No. 6, 1982, pp. 245–251.

[6] Carlson, K. W. and Williams, J. A., "The Effect of Crack Length and Side Grooves on the Ductile Fracture Toughness Properties of ASTM A533 Steel," Report CR-1711, Nuclear Regulatory Commission, Washington, DC, 1980.

[7] Davis, D. A., Vassilaros, M. G., and Gudas, J. P., "Specimen Geometry and Extended Crack Growth Effects on J_I-R Curve Characteristics for HY-130 and ASTM A533B Steels" in *Elastic-Plastic Fracture: Second Symposium. Volume II: Fracture Curves and Engineering Applications, ASTM STP 803,* American Society for Testing and Materials, Philadelphia, 1983, pp. II-582–II-610.

[8] Gudas, J. P., Joyce, J. A., and Davis, D. A., "Investigation of Specimen Geometry Modifications to Determine the Conservative J_I-R Curve Tearing Modulus Using the HY-130 Steel System" in *Fracture Mechanics (Eleventh Conference), ASTM STP 677,* American Society for Testing and Materials, Philadelphia, 1979, pp. 474–485.

[9] Voss, B. and Mayville, R. A., "The Use of the Partial Unloading Compliance Method for the Determination of J_I-R Curves and J_{Ic}" in *Elastic-Plastic Fracture Test Methods: The User's Experience, ASTM STP 856,* American Society for Testing and Materials, Philadelphia, 1985, pp. 117–130.

[10] Gudas, J. P. and Davis, D. A., "Evaluation of the Tentative J_I-Curve Testing Procedure by Round Robin Tests of HY 130 Steel," *Journal of Testing and Evaluation,* Vol. 10, No. 6, 1982, pp. 252–262.

[11] Bucci, R. J., Paris, P. C., Landes, J. D., and Rice, J. R., "J-Integral Estimation Procedures" in *Fracture Toughness (Proceedings of the 1971 National Symposium on Fracture Mechanics) Part II, ASTM STP 514,* American Society for Testing and Materials, Philadelphia, 1972, pp. 40–69.

[12] Adams, N. J. and Munro, H. G., "A Single Test Method for Evaluation of the J-Integral as a Fracture Parameter," *Engineering Fracture Mechanics,* Vol. 6, 1974, pp. 119–132.

[13] Wu, S. X., "Crack Length Calculation Formula for Three-Point Bend Specimens," *International Journal of Fracture,* Vol. 24, No. 1, 1984, pp. R33–R35.

[14] Tada, H., Paris, P. C., and Irwin, G. R., *The Stress Analysis of Cracks Handbook,* Del Research Corp., Hellertown, PA, 1973.

[15] Lo-a-Foe, T. C. G., "The Unloading Compliance Method for the Determination of the Amount of Stable Crack Extension," Report MMPP235/EV1338, Delft University of Technology, Laboratory for Thermal Power Engineering, Delft, the Netherlands, 1984 (in Dutch).

[16] Shih, C. F., De Lorenzi, H. G., and Andrews, W. R., "Elastic Compliances and Stress-Intensity Factors for Side-Grooved Compact Specimens," *International Journal of Fracture,* Vol. 13, 1977, pp. 544–548.

[17] Neale, B. K., Curry, D. A., Green, G., Haigh, J. R., and Akhurst, K. N., "A Procedure for the Determination of the Fracture Resistance of Ductile Steels," Report TPRD/B/0495/R84, Central Electricity Generating Board, Leatherhead, Surrey, U.K., 1984.

[18] De Vries, M. I., "Experimental Aspects of Unloading-Compliance J-Tests," Workshop on Ductile Fracture Test Methods, Committee on the Safety of Nuclear Installations, Paris, Dec. 1982, pp. 159–168.

[19] Prij, J., "Some Finite Element Results of CTS Specimen," Workshop on Ductile Fracture Test Methods, Committee on the Safety of Nuclear Installations, Paris, Dec. 1982, pp. 169–180.

[20] Loss, F. J., Menke, B. H., Gray, R. A., Hawthorne, J. R., and Waterson, H. E. in "Structural Integrity of Water Reactor Pressure Vessel Boundary Components," F. J. Loss, Ed., Report CR-1128, Nuclear Regulatory Commission, Washington, DC, 1979.

[21] Bakker, A., "The Three-Dimensional J-Integral: An Investigation into its Use for Post-Yield Fracture Safety Assessment," Ph.D. thesis, Delft University of Technology, Laboratory for Thermal Power Engineering, Delft, the Netherlands, 1984.

[22] Steenkamp, P. A. J. M., *Stable Crack Growth and Instability,* Ph.D thesis, Delft University of Technology, Laboratory for Thermal Power Engineering, Delft, the Netherlands, 1985.
[23] Wallin, K. and Saario, T., "Feasibility of Using Charpy-Size Specimens in J-R Curve Testing," Workshop on Ductile Fracture Test Methods, Committee on the Safety of Nuclear Installations, Paris, April 1985.
[24] Bleeker, B. J., "Numerical Simulation of Stable Crack Growth," Report MMPP233/EV1331, Delft University of Technology, Laboratory for Thermal Power Engineering, Delft, the Netherlands, 1984 (in Dutch).
[25] Kanninen, M. F., Rybicki, E. F., Stonesifer, R. B., Broek, D., Rosenfield, A. R., Marshall, C. W., and Hahn, G. T., "Elastic-Plastic Fracture Mechanics for Two-Dimensional Stable Crack Growth and Instability Problems" in *Elastic-Plastic Fracture, ASTM STP 668,* American Society for Testing and Materials, Philadelphia, 1979, pp. 121–150.
[26] Sham, T. L., "A Finite-Element Study of the Asymptotic Near-Tip Fields for Mode I Plane-Strain Cracks Growing Stably in Elastic-Ideally Plastic Solids" in *Elastic-Plastic Fracture: Second Symposium. Volume I: Inelastic Crack Analysis, ASTM STP 803,* American Society for Testing and Materials, Philadelphia, 1983, pp. I-52–I-79.
[27] Latzko, D. G. H., Turner, C. E., Landes, J. D., McCabe, D. E., and Hellen, T. K., *Post-Yield Fracture Mechanics,* 2nd ed., Elsevier Applied Science Publishers, London, 1984, p. 167.
[28] Wu, S. X., "Plastic Rotational Factor and J-COD Relationship of Three Point Bend Specimen," *Engineering Fracture Mechanics,* Vol. 18, No. 1, 1983, pp. 83–95.
[29] *Dubbel's Handbook for Machine Building,* Part I, 13th ed., F. Sass, C. Bouche, and A. Leitner, Eds., Springer-Verlag, Berlin, 1974, p. 230 and p. 437 (in German).
[30] Yagawa, G. and Hirayama, H., "A Finite Element Method for Contact Problems Related to Fracture Mechanics," *International Journal for Numerical Methods in Engineering,* Vol. 20, 1984, pp. 2175–2195.
[31] Steenkamp, P. A. J. M. and Hartevelt, M., "Crack Front Marking by Dye Penetrants," *International Journal of Fracture,* Vol. 27, 1985, pp. R93–R98.
[32] Van der Sluys, W. A. and Futato, R. J., "Computer-Controlled Single-Specimen J-Test" in *Elastic-Plastic Fracture: Second Symposium. Volume II: Fracture Curves and Engineering Applications, ASTM STP 803,* American Society for Testing and Materials, Philadelphia, 1983, pp. II-464–II-482.
[33] Buzzard, R. J. and Fisher, D. M., "Load-Displacement Measurement and Work Determination in Three-Point Bend Tests of Notched or Precracked Specimens," *Journal of Testing and Evaluation,* Vol. 6, No. 1, 1978, pp. 35–39.
[34] Blauel, J. G. and Schwalbe, K. H., "Minutes of the First Meeting of the European Group on Fracture," Task Group I, Elastic-Plastic Fracture Mechanics, Working Party on Fracture Mechanics Testing Standards, GKSS, Geesthacht, West Germany, July 1984.
[35] McCabe, D. E. and Landes, J. D., "J_R-Curve Testing of Large Compact Specimens" in *Elastic-Plastic Fracture: Second Symposium. Volume II: Fracture Curves and Engineering Applications, ASTM STP 803,* American Society for Testing and Materials, Philadelphia, 1983, pp. II-353–II-371.
[36] Yin, S. W., Gerbrands, R. A., and Hartevelt, M., "An Investigation of the Blunting Line," *Engineering Fracture Mechanics,* Vol. 18, No. 5, 1983, pp. 1025–1036.
[37] Bakker, A., "A DC Potential Drop Procedure for Crack Initiation and R-Curve Measurements During Ductile Fracture Tests" in *Elastic-Plastic Fracture Test Methods: The User's Experience, ASTM STP 856,* American Society for Testing and Materials, Philadelphia, 1985, pp. 394–410.
[38] Steenkamp, P. A. J. M., "Investigation into the Validity of J-Based Methods for the Prediction of Ductile Tearing and Fracture," Ph.D. thesis, Delft University of Technology, Laboratory for Thermal Power Engineering, Delft, the Netherlands, 1986.
[39] Houssin, B., "Rotation Correction for Stainless Steel Specimens in the Unloading Compliance Method," Workshop on Ductile Fracture Test Methods, Paris, Committee for the Safety of Nuclear Installations, 1982, pp. 233–243.

John H. Underwood,[1] Fahmy M. Haggag,[2] and Walter G. Reuter[2]

Unloading-Compliance and Load-Drop Analysis of J_{Ic} Tests of Irradiated 348 Stainless Steel

REFERENCE: Underwood, J. H., Haggag, F. M., and Reuter, W. G., **"Unloading-Compliance and Load-Drop Analysis of J_{Ic} Tests of Irradiated 348 Stainless Steel,"** *Fracture Mechanics: Eighteenth Symposium, ASTM STP 945,* D. T. Read and R. P. Reed, Eds., American Society for Testing and Materials, 1988, pp. 611–621.

ABSTRACT: Procedures for determining J-integral fracture toughness were applied to load versus load-line displacement data from a series of three-point bend specimens of irradiated 348 stainless steel tested at 23 and 427°C. The ASTM Test Method for J_{Ic}, a Measure of Fracture Toughness (E 813-81) was emphasized, using both unloading-compliance and load-drop measures of crack growth.

An effective modulus procedure was included in the unloading—compliance method using recently published load-line displacement results. Two procedures were used for calculating J_{Ic} from the J versus unloading-compliance crack growth plots: the linear fit procedure from ASTM Method E 813 and a power-law fit with a 0.2-mm offset value of J_{Ic}.

Based on the results of these various procedures, conclusions were reached regarding the most consistent measures of fracture toughness from these data. Suggestions were given regarding the use of effective modulus with unloading-compliance measurement of crack growth and the limitations of the load-drop method for measuring crack growth.

KEY WORDS: J_{Ic} tests, unloading compliance, load-line displacement, effective modulus, three-point bend specimen, side grooves, load drop

Fracture toughness testing of ductile materials is a complex procedure in the best of circumstances, and under difficult testing conditions becomes even more complex. One such set of conditions is J_{Ic} testing of irradiated structural materials from nuclear-power-generating equipment [*1*]. Materials which have been irradiated are often hazardous to test using the usual methods, so they must be tested remotely in a closed hot cell. Remote J_{Ic} tests present special problems such as the limitations imposed on specimen displacement measurement, a critical part of the J_{Ic} test procedure. Test results of irradiated 348 stainless steel are described here and analyzed using two methods. One, which has become quite standard in J_{Ic} testing, is the unloading-compliance method [*2*]. The other is the more recently proposed load-drop method [*3*]. Both methods were applied to the same load versus load-line displacement data, P versus δ, from a series of three-point bend specimens of irradiated 348 stainless steel tested at 21 and 427°C, using the ASTM Test Method for J_{Ic}, a Measure of Fracture Toughness (E 813-81) procedures whenever possible.

The objective here is to evaluate the J_{Ic} test procedure under nonideal, remote-testing conditions using two methods of crack growth measurement well-suited for remote testing, the un-

[1]Research engineer, U.S. Army Armament Research Development and Engineering Center, Watervliet, NY 12189.

[2]Engineering specialist and principal engineer, respectively, Idaho National Engineering Laboratory, Idaho Falls, ID 83401. Haggag is currently with Battelle Columbus Laboratories, Columbus, OH 43201.

loading-compliance and load-drop methods. Comparison of the values of fracture toughness with other results and the important relation of toughness to the function of power generation equipment are considered elsewhere [*1,4*]. In this work we describe fracture toughness test procedures which proved useful in the testing of irradiated materials and which may be helpful in other difficult testing conditions and in J_{Ic} testing in general.

Test Procedures

The specimens described here were fabricated from 348 stainless steel sheet, irradiated in an experimental reactor at 427°C to about the same fluence level, and tested in bending at 427 or 23°C in a hot cell. See Table 1 and Fig. 1 for the key test conditions and arrangements. Specimen dimensions were depth, W = 10.0 mm; gross thickness, B = 5.1 mm; span, S = 40.6 mm; and notch width, η = 4.4 mm. Note in Fig. 1 that the displacement gage made contact with the specimen at the edges formed between the notch and the lower specimen surface and that the geometry of this contact was taken into account in calculating the corrected load-line displacement, δ. This calculation of δ does not account for large crack-opening-displacements and specimen rotations, so it may not be adequate for J-R curve determination, for example. Table 1 gives the specimen yield strengths in the temperature and irradiation conditions of the tests. The effective yield strength σ_Y, is the mean value of yield and ultimate strengths, as defined in ASTM Method E 813. The initial crack lengths and the side-groove conditions are given. Specimens 40A and 40B, taken from the same piece as Specimen 40, were tested with 10% side grooves in each side.

Load versus load-line displacement plots were obtained for each specimen, including several partial unloadings to about 86% of the current load at the time of the unloading. Two aspects of the test apparatus and procedure were closely controlled in order to improve the accuracy of the unloading displacement plots. The linear variable differential transformer (LVDT) core was spring-loaded so as to minimize the mechanical hysteresis effect in the measurement of an unloading displacement. Also, the load was allowed to relax at constant displacement for ten seconds before each unloading. Two plots of several unloadings are shown in Fig. 2 for specimens with nearly the same test conditions except for test temperature. The important differences in load-displacement and associated fracture toughness behavior are discussed elsewhere [*1,5*]. The unloading portions of the traces were plotted separately with expanded scales and evaluated for selection of the more linear portions. The top 30 to 45% and the bottom 7 to 20% of the data were eliminated from a linear regression calculation of the unloading slope. This relatively large amount of data elimination was necessary to obtain an accurate slope for this material and for the type of displacement measurement which was dictated by the closed hot cell. The final crack length for each specimen was marked by heat tinting. The unloading slopes were used to calculate crack growth using the procedures in the following discussion of test data analysis.

TABLE 1—*Test conditions for irradiated 348 stainless bend specimens.*

Specimen Number	Test Temperature, °C	Irradiation Fluence, n/cm²	Yield Strength, σ_{ys}, MPa	Effective Yield Strength, σ_Y, MPa	Initial Crack Length, a_o/W	Side Groove, B_{net}/B
40	427	3.4×10^{22}	747	767	0.49	1.0
47	427	3.4×10^{22}	747	767	0.50	1.0
40A	427	3.4×10^{22}	747	767	0.58	0.8
40B	427	3.4×10^{22}	747	767	0.47	0.8
97	23	2.4×10^{22}	1036	1114	0.48	1.0

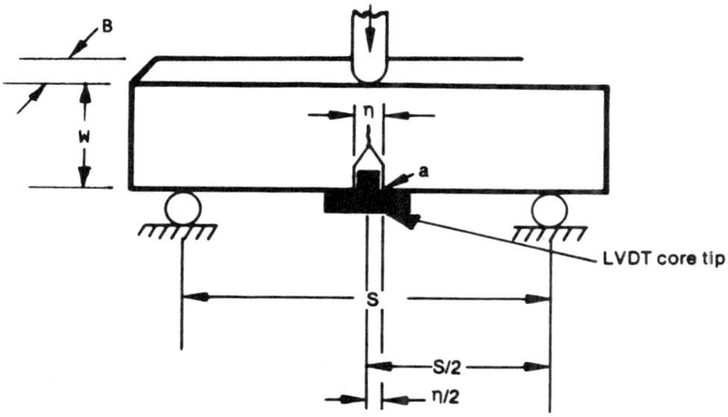

δ = Load-line displacement
δ' = Measured displacement at point a

$$\frac{\delta}{\delta'} = \frac{S/2}{S/2-\eta/2}$$

FIG. 1—*Test arrangement and calculation of load-line displacement for bend tests performed remotely within a hot cell.*

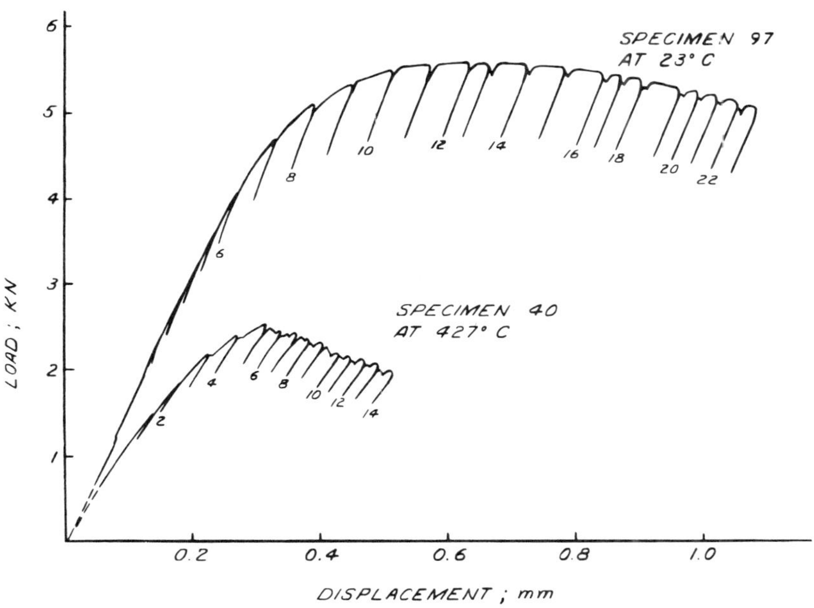

FIG. 2—*Load versus load-line displacement plots from irradiated 348 stainless bend specimens.*

Data Analysis

Unloading-Compliance and Effective Modulus

When J versus Δa data are plotted using Δa obtained directly from an unloading-compliance method, it is often clear that some further analysis is required before J_{Ic} can be determined. This was true for the tests here. Figure 3*a* shows J versus Δa data with J calculated as suggested in ASTM Method E 813 and Δa calculated from unloading compliance using an elastic modulus typical of the irradiated material at 427°C. Note that the first unloading predicts a negative crack growth and that subsequent unloadings result in Δa values which appear to be shifted to the right by about 0.16 mm. This type of zero shift is not uncommon when unloading compliance is used without some sort of calibration process. For the tests here, the shift may be due to unavoidable test variations and to extraneous system and specimen compliance, as well as to uncertainties in the value of elastic modulus. All of these problems are compounded by the testing of an irradiated material.

An effective modulus procedure was used to calibrate the unloading-compliance method and resulted in the plot of Fig. 3*b*. Table 2 outlines the procedure. The slope from an unloading at

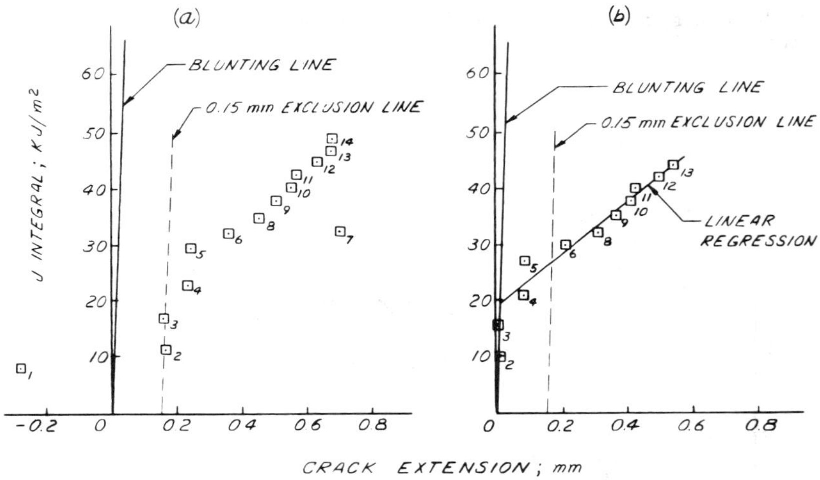

FIG. 3—J *versus* Δa *plots for Specimen 40:* (a) *using typical modulus, 166 GPa;* (b) *using effective modulus from Unloading No. 2, 156 GPa.*

TABLE 2—*Procedure for calculating effective modulus,* E*, *for Specimen 40.*

Using Typical Modulus, E = 166 GPa		Calculate Effective Modulus, E^*				
From Calibration Unloading: No. 2 $EB\delta/P$	Indicated Crack, a/W	Actual Crack, a_o/W	Δa Due to Blunting, $\Delta a_b/W = J/2\sigma_Y W$	Total Crack a_t/W	Effective Compliance, $EB\delta/P)^*$	$E^* = E\ (EB\delta/P)^*/(EB\delta/P)$
56.9	0.504	0.487	0.001	0.488	53.4	156 GPa

about 80% of the maximum load is used with the typical modulus to calculate $EB\delta/P$. An unloading at about 80% of maximum load is used for this material because it is low enough on the curve so that crack growth has not yet begun and, at the same time, it is high enough on the curve that the low-load system irregularities have long since passed. Furthermore, it is advisable to calibrate the unloading-compliance method at a load as close as possible to the loads at which it will be used. The value of $EB\delta/P$ determined as described above from the calibration unloading was used to obtain an indicated value of initial crack length, $a/W = 0.504$, in this case, from the following compliance expressions [6]

$$\frac{EB\delta}{P}\left(\frac{1 - a/W}{S/W}\right)^2 = 1.193 - 1.980\,a/W + 4.478(a/W)^2 - 4.443(a/W)^3 + 1.739(a/W)^4 \quad (1)$$

$$\gamma = \frac{1}{1 + (EB\delta/P)^{1/2}} \quad (2)$$

$$a/W = f(\gamma) = 1 - 3.82\gamma + 7.85\gamma^2 - 384\gamma^3 + 3852\gamma^4 - 12050\gamma^5 \quad (3)$$

Equation 1 is believed to be accurate within 1% for $0 \leq a/W \leq 1.0$ and $S/W = 4$, and Eqs 2 and 3 are believed to be accurate within 2% for $0.4 \leq a/W \leq 1.0$ and $S/W = 4$. Table 3 gives values from Eq 1 for quick reference. The above compliance expressions are used because recent results [6, 7] have shown that they are more accurate than the compliance data in ASTM Method E 813. For example, the value of $EB\delta/P$ for $a/W = 0.5$ from the expressions here is 11.9% higher than the value from ASTM Method E 813. We believe the difference is due to the omission of the no-crack shear displacement in the ASTM E 813 data and to the inaccuracy of the displacement expression which accounts for the presence of the crack in the E 813 data.

Referring again to the calculation of effective modulus, as outlined in Table 2, a total initial crack length, a_t/W, is calculated as the sum of the actual initial crack length from the heat tinted fracture surface and an additional amount due to blunting, $\Delta a_b/W$. The value of effective compliance, $(EB\delta/P)^*$, corresponding to a_t/W, is determined from Table 3 and used as shown to calculate effective modulus, 156 GPa. This value of effective modulus is used to calculate Δa from all the test unloadings, resulting in the plot of Fig. 3*b*. Linear regression was used to fit a line to points from Unloadings 6 and 8 through 13; Unloading 7 was omitted due to excessive nonlinearity, noticeable even in the unexpanded plots of Fig. 2; Unloading 14 was omitted because the automated evaluation of this last unloading may have been interrupted. Linear regression and power law regression fits were performed on the unloading-compliance data for all specimens. Two examples are shown in Figs. 4 and 5. The power law regression was performed as follows

$$\ell n J = \ell n A + n \ell n \Delta a \quad (4)$$

where J was obtained from the crack-growth-corrected procedure of ASTM Method E 813. Linear regression of $\ell n\ J$ on $\ell n\ \Delta a$ was used to obtain an expression in the form

$$J = A\,\Delta a^n \quad (5)$$

The critical value of J is the intersection of the power law curve with a 0.2-mm offset line parallel to the blunting line, as shown in Figs. 4 and 5. This critical value, being considered as a revised J_{Ic} test procedure, will be compared with the current ASTM Method E 813 J_{Ic} value obtained from the intersection of the linear curve with the blunting line.

TABLE 3—*Values of dimensionless load-line displacement,* $EB\delta/P$, *versus crack length,* a/W, *from wide range expression (Ref 5), Eq 1.*

a/W	$EB\delta/P$	*a/W*	$EB\delta/P$	*a/W*	$EB\delta/P$	*a/W*	$EB\delta/P$	*a/W*	$EB\delta/P$
0.400	39.006	0.500	56.052	0.600	88.277	0.700	159.07	0.800	364.95
0.405	39.646	0.505	57.202	0.605	90.576	0.705	164.64	0.805	384.37
0.410	40.305	0.510	58.389	0.610	92.967	0.710	170.51	0.810	405.36
0.415	40.981	0.515	59.615	0.615	95.453	0.715	176.69	0.815	428.10
0.420	41.677	0.520	60.880	0.620	98.040	0.720	183.22	0.820	452.79
0.425	42.393	0.525	62.188	0.625	100.73	0.725	190.11	0.825	479.66
0.430	43.129	0.530	63.538	0.630	103.53	0.730	197.39	0.830	508.98
0.435	43.886	0.535	64.934	0.635	106.46	0.735	205.09	0.835	541.03
0.440	44.666	0.540	66.377	0.640	109.51	0.740	213.25	0.840	576.18
0.445	45.468	0.545	67.870	0.645	112.69	0.745	221.90	0.845	614.83
0.450	46.293	0.550	69.414	0.650	116.01	0.750	231.09	0.850	657.46
0.455	47.144	0.555	71.012	0.655	119.48	0.755	240.85	0.855	704.65
0.460	48.019	0.560	72.666	0.660	123.10	0.760	251.24	0.860	757.04
0.465	48.920	0.565	74.379	0.665	126.89	0.765	262.31	0.865	815.44
0.470	49.849	0.570	76.154	0.670	130.86	0.770	274.12	0.870	880.79
0.475	50.806	0.575	77.994	0.675	135.02	0.775	286.75	0.875	954.24
0.480	51.792	0.580	79.901	0.680	139.37	0.780	300.25	0.880	1037.1
0.485	52.808	0.585	81.879	0.685	143.94	0.785	314.72	0.885	1131.2
0.490	53.856	0.590	83.932	0.690	148.74	0.790	330.26	0.890	1238.6
0.495	54.937	0.595	86.063	0.695	153.78	0.795	346.96	0.895	1361.9

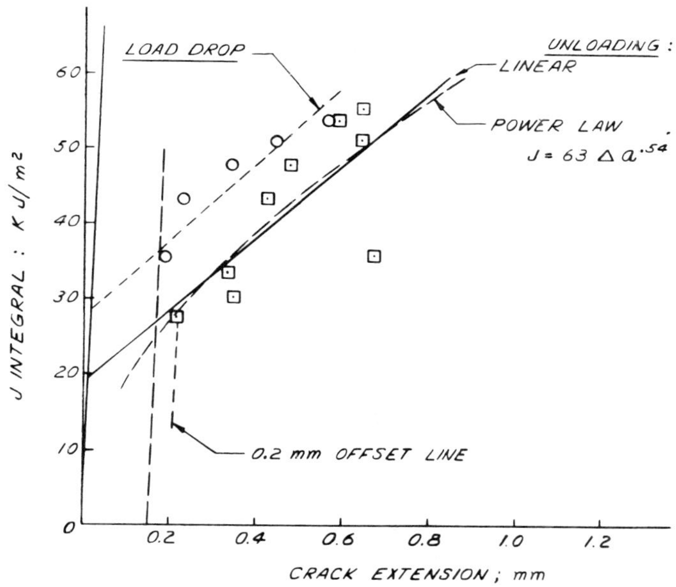

FIG. 4—J *versus* Δa *plots for Specimen 47, using unloading-compliance and load-drop measurement of* Δa.

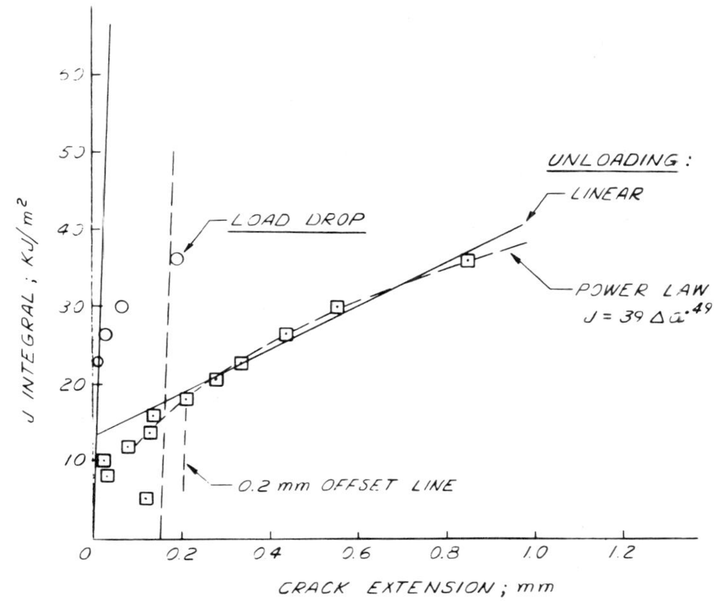

FIG. 5—J *versus* Δa *plots for side-grooved Specimen 40B, using unloading-compliance and load-drop measurement of* Δa.

Load Drop Analysis

Plots of J versus Δa were attempted from the five sets of test results using a load-drop procedure for Δa, where

$$\Delta a = J/2\sigma_y + b_o[1 - (P_{\Delta a}/P_{\max})^{1/2}] \tag{6}$$

In Eq 6 the first term is the blunting contribution to Δa. The second term is the expression [3] for Δa based on the assumptions that no crack growth occurs before maximum load and that all crack growth is described by the bending limit load relation between load and the square of remaining ligament, $P = \text{constant}(b^2)$. Figures 4 and 5 show the type of data obtained. For the nonside-grooved specimens such as in Fig. 4, enough data were obtained to perform a linear regression determination of critical J in the same manner as for a J_{Ic} determination. For the side-grooved specimens, as shown in Fig. 5, little Δa data were obtained above the 0.15-mm exclusion line.

It is worthy of note that the J versus Δa curve obtained from load-drop analysis has essentially the same slope as that obtained from the unloading-compliance method. In Fig. 4, the slope of the load-drop line, in the units shown, is 0.92, compared with 0.88 for the linear unloading-compliance line. A similar slope from both methods would be expected even if only the second of the two assumptions of the analysis were correct. The first assumption, that no crack growth occurred before maximum load, will be proven wrong by subsequent results here. However, this would effect the position of the curve but not its slope. Therefore, there is some indication that the load-drop method is suitable for determining the tearing modulus of the material tested here.

Results

A summary of key results of the tests and analyses is shown in Table 4. Listed first is the load at which the calibration unloading was performed, relative to maximum load P_{calib}/P_{max}. Using a value of about 0.8 for this ratio and applying the effective modulus procedure of Table 2 resulted in the bilinear type of J versus Δa curve which is expected, made up of blunting and crack growth portions. The ratio of effective to typical modulus, E^*/E, which was calculated was generally near unity. The side-grooved specimens, 40A and 40B, gave poorer results in this respect than the nonside-grooved specimens. This may be due to the use of effective specimen thickness in the calculations of Table 2. ASTM Method E 813 calculation of effective thickness was used

$$B_{eff} = B_{gross} - [(B_{gross} - B_{net})^2/B_{gross}] \tag{7}$$

TABLE 4—*Summary of test results.*

Specimen Number	P_{calib}/P_{max}	E^*/E	P_{max}/P_{limit}	$\Delta a_{unload}/\Delta a_{tint}$	J_{Ic} Linear/E 813, kJ/m^2	J_c Power Law, kJ/m^2	J_c Load-Drop, kJ/m^2
40	0.77	0.94	0.71	0.46	19.5	28	28.5
47	0.86	1.01	0.72	0.47	19.5	28	28.4
40A	0.75	0.88	0.60	0.90	8.4	13	...
40B	0.79	0.84	0.64	1.04	13.5	18	31[a]
97	0.85	1.01	0.99	0.69	189.0	285	257

[a]Only one of the four Δa values above the 0.15-mm minimum.

which results in an effective thickness of 4.88 mm for Specimens 40A and 40B. It is interesting to note that if $B_{net} = 4.06$ mm were used in the calculations of Table 2, the resulting values of E^*/E would be 1.05 and 1.01 for Specimens 40A and 40B, respectively. Then all the specimens, side-grooved as well as nonside-grooved, would meet the ASTM E 813 requirement that, essentially, E must equal E^* within 7%.

It is generally of interest to compare the maximum load in a J_{Ic} test with the bending limit load. For the tests here this ratio, shown in Table 4, gives an indication of how well the load-drop method would be expected to work. Load-drop would be expected to give a good measure of Δa only for P_{max}/P_{limit} near unity. This was the case for only one of five tests, so based on this, differences between the load-drop and unloading-compliance results would not be unexpected. It should be noted that the work of Server [8], based on plane-strain slip-line-field analysis, was used to obtain the P_{limit} expression for the tests here

$$P_{limit} = 1.435\,\sigma_Y Bb^2/S \tag{8}$$

in which the constant 1.435 was calculated from Ref *8* for an indentor width of 1 mm and for the assumption of the von Mises' yield criterion. The use of a plane-strain analysis for calculating P_{limit} is believed to be appropriate, because the relatively low strain-hardening of the material encourages slip-line type deformation.

Another requirement of ASTM Method E 813 is that the final measurement of Δa using the unloading-compliance method should agree within 15% of that from heat tinting or other crack length marking after the test. Comparison of these two measures of Δa, listed in Table 4, shows that only the side-grooved tests meet the 15-% requirement. Figure 6 shows the reason. Consid-

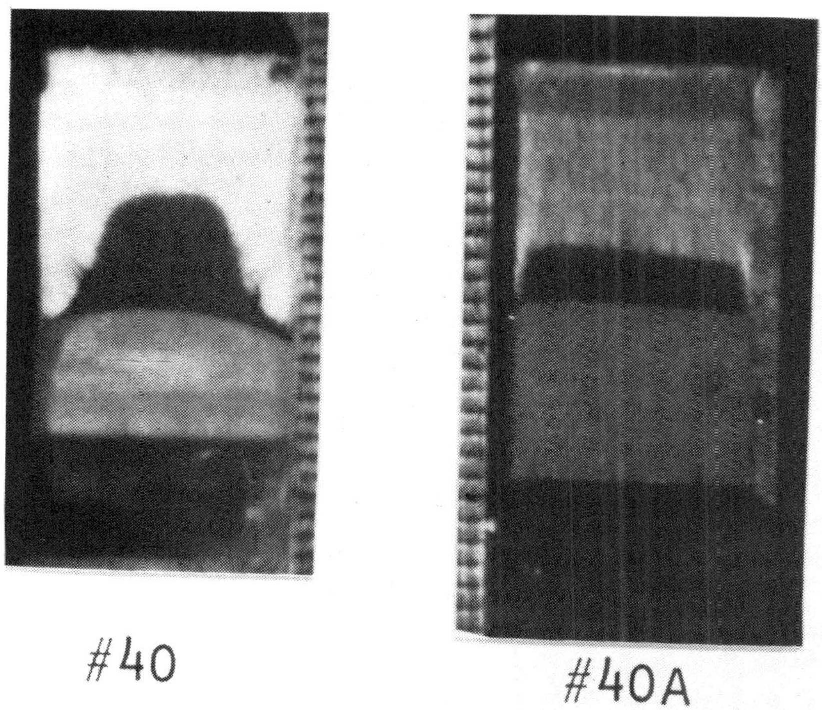

FIG. 6—*Heat-tinted fracture surfaces of two bend specimens.*

erable tunnelling occurred in the nongrooved specimens, whereas relatively straight-fronted crack growth occurred in the side-grooved specimens.

The final three columns in Table 4 list the critical values of J determined using unloading-compliance and linear regression as in Method E 813, unloading-compliance and power law regression, and load-drop and linear regression when there were enough data. One consistent trend is that J from the power law method is significantly above that from the linear ASTM Method E 813, from 33 to 55% above. The critical J from the load-drop method is also above that from the linear method, but not as consistently, from 46% to over 200%. There is no common reason for these two somewhat similar trends. The power law J is above the linear J because the power law method selects a point higher on the J versus Δa curve for these tests. The load-drop J is above the linear J because the load-drop Δa measurements are significantly below those from unloading-compliance; this shifts the J versus Δa curve upward.

An important trend in the linear J_{Ic} results is a significantly lower J_{Ic} value for side-grooved Specimens 40A and 40B, averaging 11 kJ/m^2, compared with nongrooved Specimens 40 and 47, averaging 20 kJ/m^2. The lower side-grooved toughness is clearly caused by the much straighter crack front. However, the choice of which value to use in design is not clear. The side-grooved J_{Ic} is a lower bound, but it may be an overly conservative lower bound because, in practice, no geometry similar to a side-grooved specimen is likely to occur and force such an unnaturally straight-fronted crack and associated low toughness. On the other hand, without knowledge of all possible design and service conditions, there is no guarantee that the lower straight-fronted J_{Ic} value from side-grooved tests is not the appropriate one for a particularly severe set of conditions.

Summary and Conclusions

The inclusion of an effective modulus procedure in the unloading-compliance method used with the tests here significantly improved the self-consistency of the results. The use of effective modulus eliminated large variations in the horizontal position of the J versus Δa curve and the associated variations in the J_{Ic} value. The effective modulus procedure may not be necessary under the best of testing conditions, but in general, it should be a required part of the unloading-compliance J_{Ic} test method. For proper general use of effective-modulus, the load-line compliance data in ASTM Method E 813 should be replaced with more accurate data. The results of Ref *6* are suggested for this purpose. For the tests here, the best results were obtained when calibration unloading and effective modulus procedures were performed at a point approaching maximum load. This is suggested as a general procedure, with care taken to be sure that no crack growth has yet occurred at the calibration unloading point. In general, heat-tinting type tests will be required to positively determine the optimum point for calibration unloading.

The power law regression fit and 0.2-mm blunting line offset approach resulted in critical J values higher by about 40% than those from the linear fit and blunting line approach of ASTM Method E 813. For the tests here, a 0.15 offset would be attractive because this line is already used for data exclusion and it would result in a closer agreement with results from the current ASTM Method E 813.

The load-drop procedure resulted in critical J values considerably above those of the unloading-compliance and linear fit approach of Method E 813. The primary reason for this is believed to be the occurrence of crack growth before maximum load for this material. Such crack growth would not be indicated by the load-drop method, so the J versus Δa curve would be shifted to the left and a higher critical J value would result. The above serves to emphasize that the load-drop method gives a good measure of J_{Ic} only when it is certain that no crack growth has occurred before maximum load. When this is uncertain, the load-drop critical J value must be considered to be unconservative.

References

[1] Haggag, F. M., Server, W. L., Reuter, W. G., and Beeston, J. M., "Effects of Irradiation Fluence and Creep on Fracture Toughness of 347/348 Stainless Steel" in *Effects of Radiation on Materials, ASTM STP 870,* American Society for Testing and Materials, Philadelphia, 1985, pp. 548–562.

[2] Clark, G. A., Andrews, W. R., Paris, P. C., and Schmidt, D. W., "Single Specimen Tests for J_{Ic} Determination" in *Mechanics of Crack Growth, ASTM STP 590,* American Society for Testing and Materials, Philadelphia, 1976, pp. 27–42.

[3] Kapp, J. A. and Underwood, J. H., "Single Specimen *J*-Based Fracture Toughness Test for High-Strength Steels" in *Fracture Mechanics: Fourteenth Symposium—Vol. II: Testing and Applications, ASTM STP 791,* J. C. Lewis and G. Sines, Eds., American Society for Testing and Materials, Philadelphia, 1983, pp. II-402–II-414.

[4] Haggag, F. M. and Richardson, A. K., "Precracking and Computerized Single-Specimen J_{Ic} Determination for Irradiated Three-Point Bend Specimens," presented at Eighteenth National Symposium on Fracture Mechanics, Boulder, CO, June 1985.

[5] Haggag, F. M. and Richardson, A. K., "Fracture Toughness and Stress Relief Response of Irradiated 347/348 Stainless Steel," EG&G Idaho, Inc., PR-T-84-018, Oct. 1984.

[6] Underwood, J. H., Kapp, J. A., and Baratta, F. I., "More on Compliance of the Three-Point Bend Specimen," *International Journal of Fracture,* Vol. 28, 1985, pp. R41–R45.

[7] Haggag, F. M. and Underwood, J. H., "Compliance of a Three-Point Bend Specimen at Load Line," *International Journal of Fracture,* Vol. 26, 1984, pp. R63–R65.

[8] Server, W. L., "General Yielding of Charpy V-Notch and Precracked Charpy Specimens," *Journal of Engineering Materials and Technology,* Vol. 100, 1978, pp. 183–188.

Peter K. Liaw[1] *and John D. Landes*[2]

Effects of Monotonic and Cyclic Prestrain on Fracture Toughness: A Summary

REFERENCE: Liaw, P. K. and Landes, J. D., **"Effects of Monotonic and Cyclic Prestrain on Fracture Toughness: A Summary,"** *Fracture Mechanics: Eighteenth Symposium, ASTM STP 945,* D. T. Read and R. P. Reed, Eds., American Society for Testing and Materials, Philadelphia, 1988, pp. 622–646.

ABSTRACT: The effects of monotonic and cyclic prestrain on fracture toughness properties of 4340 steel, 4335 steel, and 316 stainless steel were investigated. Monotonic prestrain lowered the toughness levels of the J_R-curves and J-T diagrams for these three steels, relative to the virgin material. While cyclic prestrain increased the J_R and J-T toughness of 4340 and 4335 steels, it decreased that of 316 stainless steel. The effects of monotonic and cyclic prestrain on J_R and J-T curves can be rationalized by material strength. Furthermore, the influence of cyclic prestrain was attributed to the cyclic softening or cyclic hardening characteristics of the virgin material. The transition behavior of the 4340 steel was also altered by monotonic and cyclic prestrain. Moreover, material strength appears to provide a rationale to account for the effects of prestrain history on the transition behavior.

KEY WORDS: prestrain, load, history, monotonic, cyclic, fracture toughness, steel, J_R curves, J-T diagrams, material strength, cyclic softening, cyclic hardening, transition behavior, FATT (fracture appearance transition temperature), charpy impact, tensile, upper shelf

In recent years, considerable progress has been made in the area of elastic-plastic fracture mechanics (EPFM) [*1–11*]. The technology of EPFM is now routinely used as a tool to assess the toughness of more ductile engineering materials. In laboratories, most fracture toughness experiments for determining J_{Ic} and J_R-curves are performed on as-processed virgin materials under a simple monotonic loading condition [*12,13*]. In real practice, machine components are, however, subjected to long and complex load histories. Some previous results [*14–22*] indicate that load histories can significantly alter fracture toughness characteristics. Consequently, fracture toughness properties developed in laboratories may not adequately represent the in-service material data. As a result, to ensure structural integrity, it may be important to study the influence of load history on fracture behavior.

Previously, Landes and Leax [*18*] reported that while monotonic prestrain lowered the levels of J_R-curves, relative to the virgin 4335 steel, cyclic prestrain increased fracture toughness properties. It was thought that the effect of cyclic prestrain be related to the cyclic softening characteristics of the 4335 steel. The present paper is an extension of their investigation [*18*]. The effects of prestrain histories on fracture toughness properties were further examined on the cyclic softening 4340 steel and the cyclic hardening 316 stainless steel. Part of these results have been published in a previous paper [*16*], which also contained detailed microstructural and

[1]Fellow engineer, Westinghouse Research and Development Center, Materials Analysis Department, Pittsburgh, PA 15235.

[2]Professor, Engineering Science and Mechanics Department, The University of Tennessee, TN 37996-2030.

fractographical investigations. Since then, more analyses have been conducted on the reported fracture toughness properties [16,18]. This paper provides a summary of these results and the associated analyses. Moreover, results from the recent study of the effect of prestrain history on the transition behavior of the 4340 steel are included.

Experimental Procedure

Material

The materials investigated are 4340 steel, 4335 steel, and 316 stainless steel; their chemical compositions are presented in Table 1. Note that the 4335 steel has a higher level of nickel content and a lower level of carbon content than the 4340 steel. Thus, the 4335 steel was a modified 4340 steel. The 4340 steel was normalized at 954°C, austenitized at 843°C, and water-quenched and tempered at 593°C. The same heat treatment was also given to the 4335 steel. Both 4340 and 4335 steels have a microstructure of tempered martensite.

The 316 stainless steel was annealed at 1054°C. The microstructure of the stainless steel is austenitic with an average grain size of ~25 μm.

Specimen Preparation and Prestrain History

To introduce monotonic and cyclic prestrain, several large tension specimens (Figs. 1 and 2) with a gage section of approximately 124 by 63.5 by 36 mm were manufactured from each steel. A clip-on gage was mounted at the gage section to measure strain levels. Two specimens were subjected to monotonic prestrain at tensile strain levels of 2 and 5%, respectively.

Cyclic prestraining was accomplished by an incremental-step test method [23]; a schematic of this technique is presented in Fig. 3. Cyclic prestraining was performed with a maximum strain amplitude of ±2% under fully reversed conditions. The strain rate was equal to 4×10^{-5} 1/s. The strain levels were consecutively decreased and increased for several strain blocks. Each strain block consisted of 50 strain steps. The specimens were unloaded at three locations: maximum tension and maximum compression positions after two blocks of prestraining, and zero position after two and a half blocks. An electrohydraulic fatigue machine with a load capacity of 364 000 kg was used to introduce cyclic prestraining on tension specimens. The maximum applied load during prestraining equaled 160 000 kg.

As a result, the following six prestrain histories were developed in this investigation: (1) virgin, (2) 2% monotonic prestrain (tension), (3) 5% monotonic prestrain (tension), (4) 2% cyclic prestrain (tension), (5) 2% cyclic prestrain (zero), and (6) 2% cyclic prestrain (compression). Note that in 4335 steel, cyclic prestraining was also conducted at a strain amplitude of 1½% [18].

Two compact tension type (CT) specimens were machined from the gage section of each prestrained tension specimen (Figs. 1 and 2) for fracture toughness testing. Miniature tension specimens were then fabricated from the remainder of the gage section for tension testing. Charpy V-notch specimens were manufactured only from virgin and two prestrained conditions for the

TABLE 1—*Chemical compositions in weight %.*

Material	C	Mn	P	S	Si	Cr	Ni	Mo	V	Co	Cu	Fe
4340 steel	0.42	0.69	0.007	0.011	0.31	0.78	1.85	0.26	0.02	...	...	balance
4335 steel	0.32	0.69	0.007	0.005	0.26	0.91	2.72	0.42	0.09	...	...	balance
316 stainless steel	0.059	1.92	0.017	0.010	0.58	16.70	11.20	2.38	...	0.22	0.36	balance

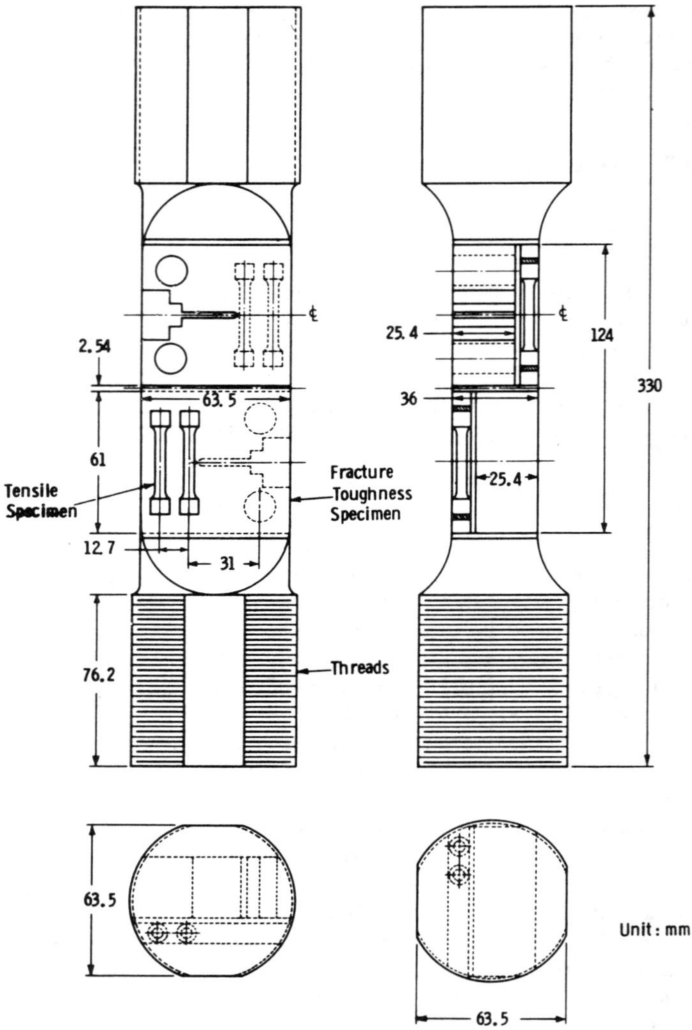

FIG. 1—*Prestrain specimen and layout of fracture toughness and tension specimens.*

4340 steel, namely, (1) virgin, (2) 2% monotonic prestrain (tension), and (3) 2% cyclic prestrain (tension).

Mechanical Testing

To characterize virgin and prestrained 4340 steel, 4335 steel, and 316 stainless steels thoroughly, three kinds of mechanical testing were conducted—incremental step, tension, and Charpy impact.

The incremental-step test [*23*] was performed on each virgin steel. Tension specimens with a gage section of 14-mm length and 6.4-mm diameter were used for incremental-step testing to develop cyclic stress-strain curves (Figs. 1 and 2). The strain rate for incremental-step tests was

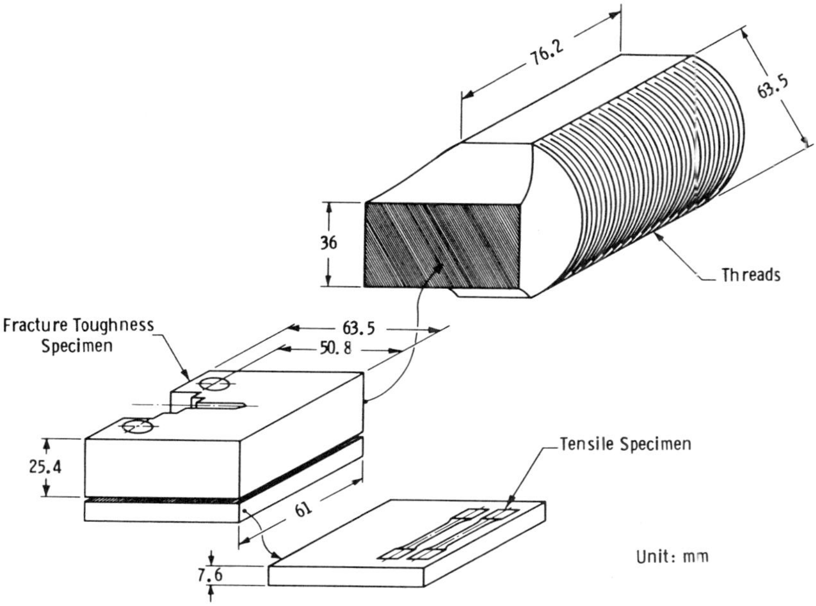

FIG. 2—*Fracture toughness and tension specimens machined from half a prestrained specimen.*

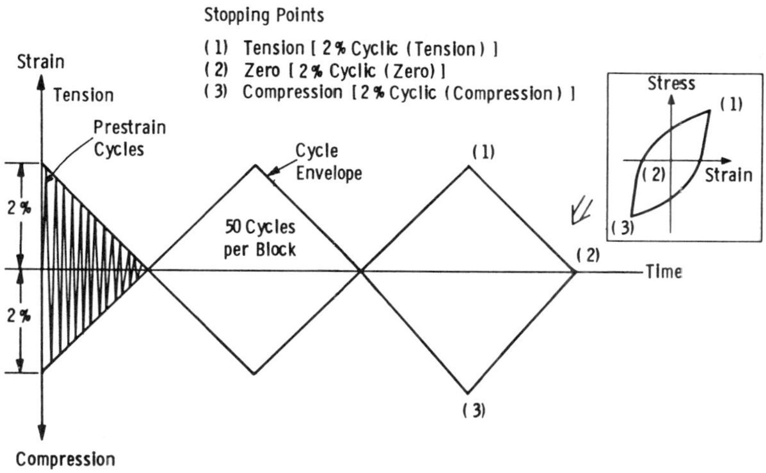

FIG. 3—*Schematic of cyclic prestraining of tension specimens.*

4×10^{-3} 1/s, and the maximum strain amplitude was 2%. The cyclic stress-strain curve was defined at the half-life of each test.

For tension testing, specimens with a gage section of 12.7-mm length and 5.1-mm diameter were machined from the virgin and prestrained specimens (Figs. 1 and 2). Charpy impact tests were performed only for virgin and prestrained 4340 steels. The Charpy V-notch specimens were 55 mm long, 10 mm wide, and 10 mm thick. The notch depth was 2 mm.

Fracture Toughness Testing

As mentioned previously, CT specimens 51 mm wide, 61 mm high, and 26 mm thick were machined from virgin and prestrained steels for developing fracture toughness properties (J_{Ic} and J_R-curves). Prior to testing, specimens were precracked to an a/W ratio of 0.6 (where a is the crack length and W the specimen width) in accordance with the ASTM Standard Test for J_{Ic}, a Measure of Fracture Toughness [*12*]. The detailed procedures for determining J_{Ic} and J_R-curves have been reported in Refs *5, 7, 12,* and *24.* Note that duplicate tests were conducted to ensure the repeatability of fracture toughness results.

In developing J_R-curves, the modified J-integral, J_M, developed by Ernst [*25*] was used

$$J_M = J_D + \int_{a0}^{a} \gamma J_{PL}/b\, da$$

where

J_D = deformation J based on plasticity theory (the detailed computation of J_D can be found in Ref *7*),

$\gamma = 1 + 0.76\, b/W$,

$b = W - a$ (where a is the crack length from the load line of a CT specimen and W is the specimen width from the loadline), and

a_0 = initial crack length.

The computation method for determining J_{PL} has been documented in Ref *25.*

This newly developed J_M was employed in the present investigation because it rationalized the effect of specimen geometry on the fracture toughness properties of structural steels [*25*].

J_R-curves were represented by plotting J_M versus Δa (crack extension). The value of Δa was equal to crack extension beyond the precrack. During fracture toughness testing, crack length was determined by an unloading compliance method [*5,25,26*].

Results

Mechanical Properties

The results of the cyclic stress-strain curves developed by the incremental-step tests are compared with the monotonic stress-strain curves in Figs. 4*a* to 4*c* for virgin 4340 steel, 4335 steel, and 316 stainless steel, respectively. In both 4340 and 4335 steels, the levels of monotonic stress-strain curves are generally higher than those of cyclic stress-strain curves, which indicates the characteristics of cyclic softening. On the other hand, the level of the monotonic stress-strain curve in 316 stainless steel is lower than that of the cyclic stress-strain curve, which demonstrates cyclic hardening.

The monotonic stress-strain curves of virgin and prestrained materials are compared in Fig. 5*a* for 4340 steel. The stress-strain curves of 2 and 5% monotonically prestrained materials are generally higher than those of the virgin material. Increasing the prestrain level from 2 to 5% increases the level of the stress-strain curve. The stress-strain curves of cyclically prestrained specimens are lower than those of the virgin material. This behavior is in agreement with the cyclic softening characteristics of 4340 steel. The positions (tension, zero, and compression) where cyclic prestrain was terminated do not seem to alter the stress-strain curves significantly.

The tension test results of virgin and prestrained 4340 and 4335 steels are presented in Tables 2 and 3, respectively. In 4340 steels, material strengths, yield strength (σ_y), and flow strength [$\sigma_f = 1/2(\sigma_y + \sigma_u)$, where σ_u is the ultimate tensile strength] of monotonically prestrained materials are greater than those of the virgin material. Increasing the prestrain level from 2 to 5% increases material strength. A similar trend was observed in 4335 steel. Note that the values of

σ_u in virgin and monotonically prestrained 4340 steels appear to be comparable and insensitive to prestrain level. In both steels, σ_y, σ_f, and σ_u of cyclically prestrained materials are smaller than those of the virgin materials. Note that in 4335 steel, increasing the cyclic prestrain level from 1.5 to 2% tends to decrease σ_y. In 4340 steel, the values of elongation and reduction in area are essentially identical regardless of prestrain history (Table 2).

Figure 5*b* presents the monotonic stress-strain curves of virgin and prestrained 316 stainless steels. The stress-strain curves of monotonically prestrained specimens are higher than those of the virgin material. Increasing the strain level from 2 to 5% increases the level of the stress-strain curve. The stress-strain curves in cyclically prestrained specimens are higher than those in the virgin material, which is consistent with the observed cyclic hardening behavior of 316 stainless steel. The places where cyclic prestrain was interrupted do not appear to alter the stress-strain response greatly. The level of the stress-strain curve for 5% monotonic prestrain is nearly the same as that for cyclic prestrain.

The tension test results of virgin and prestrained 316 stainless steels are presented in Table 4. Material strengths (σ_y, σ_f, and σ_u) of both monotonically and cyclically prestrained materials are greater than those of the virgin material. The percent of reduction in area is essentially independent of prestrain history, while the elongations of prestrained specimens seem to be smaller than those of the virgin material.

Fracture Toughness Properties

J_R-Curves—The J_R-curves of 4340 steel and 4335 steel are presented in Figs. 6*a* and *b*, respectively. In both steels, monotonic prestrain decreases the level of the J_R-curve relative to the virgin material. The difference in the J_M values of virgin and prestrained steels tends to increase with increasing crack extension (Δa). Increasing the monotonic prestrain level from 2 to 5% decreases the J_R-curve. In monotonically prestrained 4340 steels, cleavage fracture occurred during fracture toughness testing, and thus only an extremely limited amount of stable crack extension was developed in the J_R-curve.

On the contrary, the levels of J_R-curves for cyclically prestrained specimens are higher than those of virgin 4340 and 4335 steels. The difference in the J_M values of virgin and cyclically prestrained specimens increases with increasing Δa. In 4340 steel, the extent of stable crack extension in cyclically prestrained specimens is much greater than that in monotonically prestrained specimens. Thus, monotonic prestrain degrades the ductility of 4340 steels while cyclic prestrain enhances the ductility. A sudden "pop-in" cleavage fracture was observed in some cyclically prestrained 4340 steels, which interrupted the smoothness of the J_R-curves—for example, 2% cyclic (compression) in Fig. 6*a*. It should be noted that in 4335 steel, increasing the cyclic prestrain level from 1.5 to 2% increases the level of the J_R-curve.

The values of J_{Ic}, and J_M corresponding to crack extensions of 0.5 and 1.5 mm are listed in Tables 2 and 3 for 4340 and 4335 steels, respectively. The values of J_{Ic} for monotonically prestrained 4340 steels are approximately 40% smaller than those of the virgin 4340 steel and 10% for 4335 steel. In 4340 steel, increasing the monotonic prestrain level from 2 to 5% decreases J_{Ic} slightly. Relative to the virgin material, cyclic prestrain increases J_{Ic} by approximately 35% for 4340 steel and 50% for 4335 steel. In 4335 steel, decreasing the cyclic prestrain level from 2 to 1.5% (tension) decreases J_{Ic} by about 40%.

Figure 6*c* presents the J_R-curves of 316 stainless steels. The results for the 316 stainless steels show that monotonic prestrain decreases J_R-curves relative to the virgin 316 stainless steel. This trend is in agreement with that observed in 4340 and 4335 steels (Figs. 6*a* and 6*b*). The influence of monotonic prestrain on J_R-curves increases with increasing Δa (Fig. 6*c*). Moreover, the level of the J_R curve in the 2% monotonically prestrained specimen is higher than that in the 5% prestrained specimen. In contrast to the observations in 4340 and 4335 steels, cyclic prestrain decreases the J_R-curve of the virgin material. Thus, the effect of cyclic prestrain on fracture toughness properties appears to be dependent upon test material while that of monotonic pre-

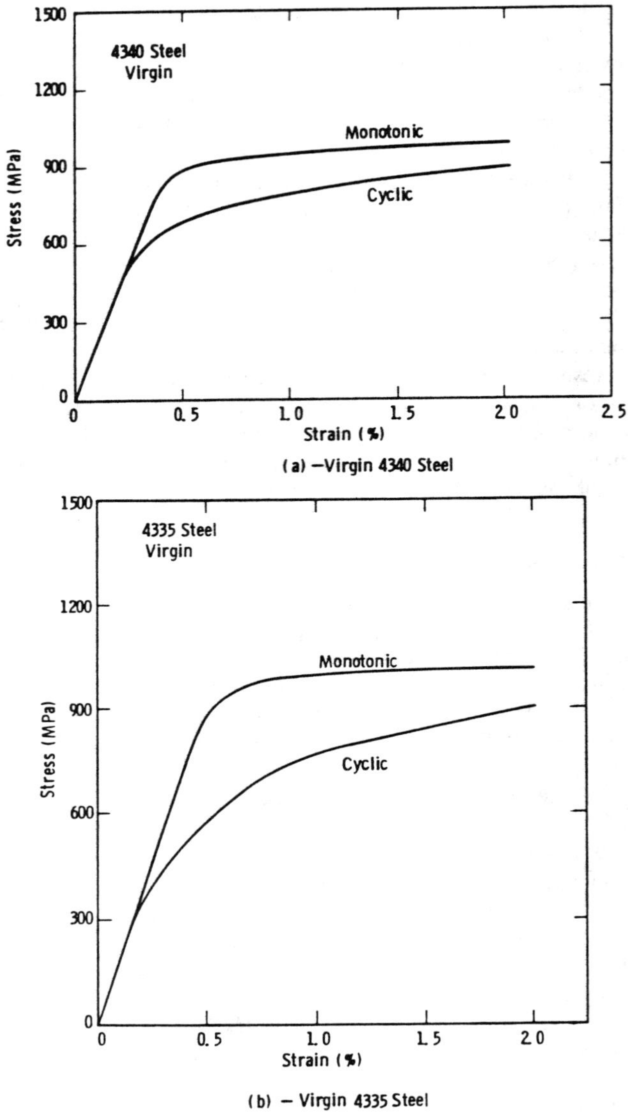

FIG. 4—*Monotonic and cyclic stress-strain curves of virgin materials determined from incremental step tests.*

strain is independent of test material. Note that the cyclic prestrain positions do not significantly affect the J_R-curves of 316 stainless steels. Further, the level of the J_R-curve for the 5% monotonically prestrained 316 stainless steel is nearly the same as that of cyclically prestrained materials.

The values of J_{Ic} and J_M at crack extensions of 2.5 and 3.0 mm are shown in Table 4 for 316 stainless steel. In virgin and 2% monotonically prestrained specimens, J_{Ic} cannot be determined because no intersection between the J_R-curve and the blunting line was observed by following

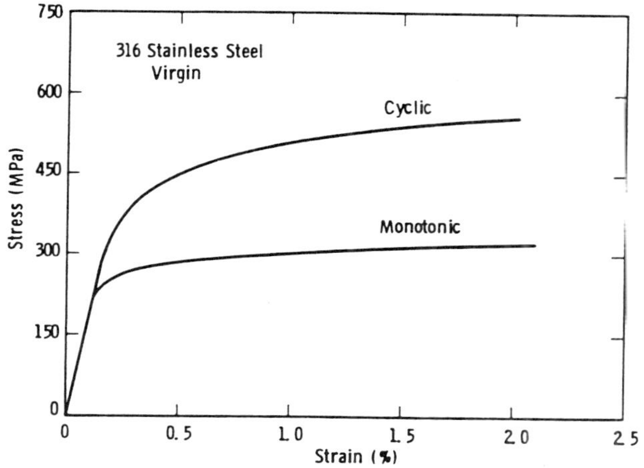

(c) —Virgin 316 Stainless Steel

FIG. 4—*Continued.*

ASTM Method E 813. Note that this behavior indicates that the present ASTM standard for determining J_{Ic} may not be applicable to very high toughness materials such as 316 stainless steel. In Table 4, the value of J_M for the 2% monotonically prestrained specimen is approximately 25% smaller than that of the virgin steel. Increasing the prestrain level from 2 to 5% decreases J_M by about 10%. The values of J_M for cyclically prestrained specimens are nearly 40% smaller than those of the virgin material. The cyclic prestrain positions do not seem to affect the levels of J_{Ic} and J_M.

J-T Diagram—To investigate relative fracture toughness levels in a format used for structural stability evaluation, the J_M values were plotted against the tearing modulus $\{T = (E/\sigma_f^2)\, dJ/da$, where E is Young's modulus [27]$\}$. To determine the value of T, a nonlinear equation was first fitted to the J_R-curve, and T was then calculated by taking the derivative of the equation and using it in the preceding expression.

Figures 7*a* and 7*b* present the J-T plots of 4340 and 4335 steels, respectively. In 4340 steel, the extent of the J_R-curves for monotonically prestrained specimens is extremely limited due to the brittle fracture behavior of these specimens; hence the associated J-T curves are not presented in Fig. 7*a*. In Fig. 7*b*, the level of the J-T curve for the monotonically prestrained 4335 steel is lower than that of the virgin material. In both steels, the J-T curves of cyclically prestrained specimens are higher than those of the virgin materials.

To quantify the effect of prestrain history on J-T curves, the values of J_{50} (J_{50} is defined as the value of J when the line, $J = 50T$ for English unit or $J = 8.8T$ for SI unit, intersects the J-T curve) [27] are listed in Tables 2 and 3 for 4340 and 4335 steels, respectively. In 4335 steel, monotonic prestrain decreases J_{50} by approximately 35% relative to the virgin material. In 4340 steel, cyclic prestrain increases J_{50} by an average of about 100%, while in 4335 steel it is 40%.

The J-T curves of 316 stainless steel are shown in Fig. 7*c*. Both monotonic and cyclic prestrains significantly decrease the levels of J-T curves relative to the virgin steel. Increasing the monotonic prestrain level from 2 to 5% decreases the J-T curve. Note that the J-T curves of the 5% monotonically prestrained specimen are much closer to those of the cyclically prestrained specimens, relative to the 2% monotonically prestrained specimen. As a result, monotonic prestrain decreases the J-T curve of the virgin specimen regardless of test material. On the contrary, cyclic prestrain can either increase or decrease J-T curves depending on the test material.

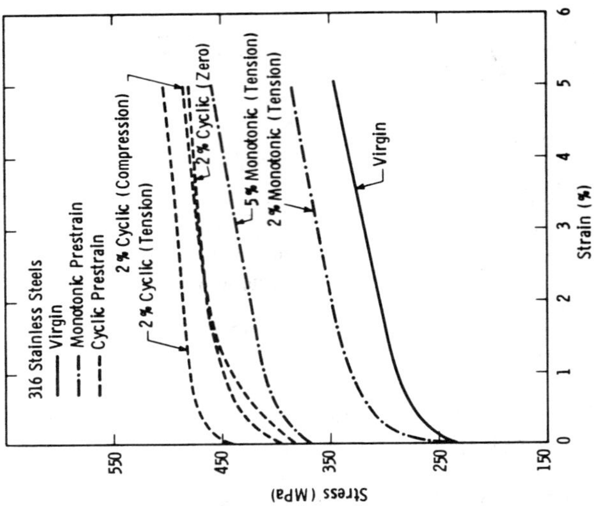

Fig. 5 (b)—316 Stainless Steels

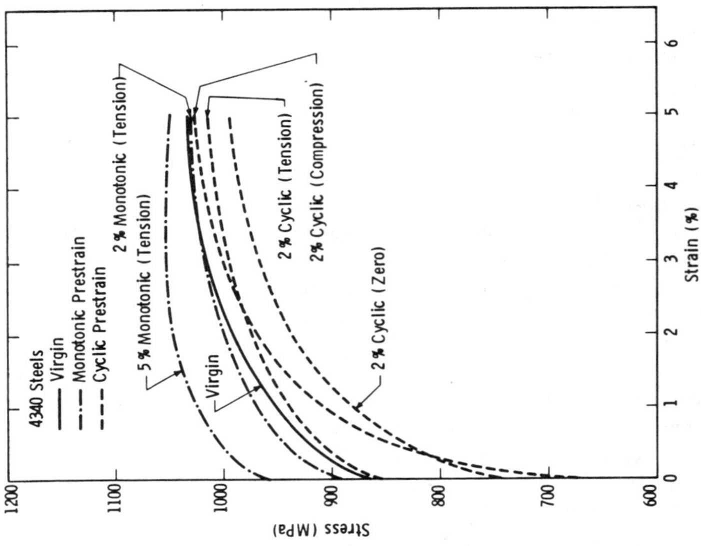

Fig. 5 (a) 4340 Steels

FIG. 5—*Effects of prestrain history on monotonic stress-strain curves.*

TABLE 2—*Mechanical properties and fracture toughness properties of 4340 steels.*

Specimen Designation	Prestrain History	J_{Ic}, kJ/m^2	$J_M/\Delta a$ = 0.5 mm, kJ/m^2	$J_M/\Delta a$ = 1.5 mm, kJ/m^2	J_{50}, kJ/m^2	σ_y,[b] MPa	σ_f,[c] MPa	σ_u,[c] MPa	Elongation,[d] %	Reduction in Area, %
BHFC5-24	virgin	88	123	200	126	893	970	1047	21	48
BHFC5-10	2% monotonic (tension)	52[a]	...	...	...	924	979	1034	20	47
BHFC5-3	5% monotonic (tension)	49[a]	...	...	...	987	1022	1056	19	48
BHFC5-7	2% cyclic (tension)	111	163	276	210	884	951	1017	21	48
BHFC5-6	2% cyclic (zero)	103	200	...	...	779	890	1001	22	48
BHFC5-1	2% cyclic (compression)	142	197	333	307	742	860	977	23	53

[a]Calculated by $J_{Ic} = K_{Ic}^2/E$.
[b]0.2% offset monotonic yield strength.
[c]Flow strength, $\sigma_f = 1/2(\sigma_y + \sigma_u)$ where σ_u is ultimate tensile strength.
[d]Gage length = 12.7 mm.

TABLE 3—*Mechanical properties and fracture toughness properties of 4335 steels.*

Specimen Designation	Prestrain History	J_{Ic}, kJ/m²	$J_M/\Delta a$ = 0.5 mm, kJ/m²	$J_M/\Delta a$ = 1.5 mm, kJ/m²	J_{50}, kJ/m²	$\sigma_y{}^a$, MPa
FC4-CR8	virgin	146	181	264	207	1044
FC4-CR1-1	2% monotonic (tension)	134	159	209	135	1111
FC4-CR3-1	2% cyclic (tension)	250	305	454	297	972
FC4-CR2-2	1½% cyclic (tension)	150	194	297	221	986
FC4-CR4-1	1½% cyclic (zero)	234	270	386	297	...
FC4-CR5-2	1½% cyclic (compression)	246	294	459	319	...

[a]0.2% offset monotonic yield strength.

The values of J_{50} are presented in Table 4 for 316 stainless steel. J_{50} of the 2% monotonically prestrained specimen is approximately 20% smaller than that of the virgin material. Increasing the monotonic prestrain level from 2 to 5% decreases J_{50} by 20%. Cyclic prestrain decreases J_{Ic} of 316 stainless steel by an average of 45%, compared with the virgin material.

Charpy Impact Properties

For the 4340 steels, cleavage fracture was observed in all monotonically prestrained materials and in some of the cyclically prestrained materials (Fig. 6*a*). It appeared that the transition behavior of this material may be affected by prestrain. To further investigate this trend, Charpy impact tests were conducted for three levels of prestrain: virgin, 2% monotonic (tension), and 2% cyclic (tension). The results are presented in Figs. 8*a* through 8*c*. The values of fracture appearance transition temperature (FATT) for these three specimens (virgin, monotonic, and cyclic prestrain) are 66, 77, and 43°C, respectively. Relative to the virgin material, monotonic prestrain increases FATT while cyclic prestrain decreases FATT.

A composite plot of the Charpy impact properties for virgin and monotonically and cyclically prestrained specimens is illustrated in Fig. 9. At a fixed temperature, monotonic prestrain increases the percent of brittle fracture and decreases impact energy relative to the virgin steel, while cyclic prestrain decreases brittle fracture and increases impact energy. Thus, prestrain history affects the Charpy impact transition fracture property. Previously, Troshchenko et al. [22] also reported that load history could alter the transition behavior of a low-carbon steel.

The Charpy impact energy and J_R-curve fracture toughness values follow a similar trend. Both are decreased, relative to the virgin material, following monotonic prestrain, and increased following cyclic prestrain.

Discussion

Tensile and Fracture Toughness Properties

In this investigation, prestrain history was found to alter tensile properties, J_R and J-T curves of 4340 steel, 4335 steel, and 316 stainless steel. In each steel, monotonic prestrain *increases* material strength and *decreases* the levels of J_R and J-T curves, relative to the virgin material, as summarized in Figs. 10 and 11. The effects of cyclic prestrain on fracture toughness properties, however, are dependent upon test material. In 4340 and 4335 steels, cyclic prestrain *decreases*

TABLE 4—*Mechanical properties and fracture toughness properties of 316 stainless steels.*

Specimen Designation	Prestrain History	J_{Ic}, kJ/m²	$J_M/\Delta a$ = 2.5 mm, kJ/m²	$J_M/\Delta a$ = 3.0 mm, kJ/m²	J_{50}, kJ/m²	σ_y,[a] MPa	σ_f,[b] MPa	σ_u,[b] MPa	Elongation,[c] %	Reduction in Area, %
23984-23	virgin	. . .	3748	4177	4921	266	434	602	87	80
PS5-2	2% monotonic (tension)	. . .	2742	3117	4019	309	459	608	80	81
PS2-2	5% monotonic (tension)	2372	2484	2756	3281	390	509	627	74	80
PS3-1	2% cyclic (tension)	1965	2258	2494	2412	473	561	649	73	80
PS4-2	2% cyclic (zero)	2335	2403	2697	2678	417	532	646	71	79
PS1-2	2% cyclic (compression)	2064	2329	2559	2830	397	518	639	74	79

[a] 0.2% offset monotonic yield strength.
[b] Flow strength, $\sigma_f = 1/2(\sigma_y + \sigma_u)$ where σ_u is the ultimate tensile strength.
[c] Gage length = 12.7 mm.

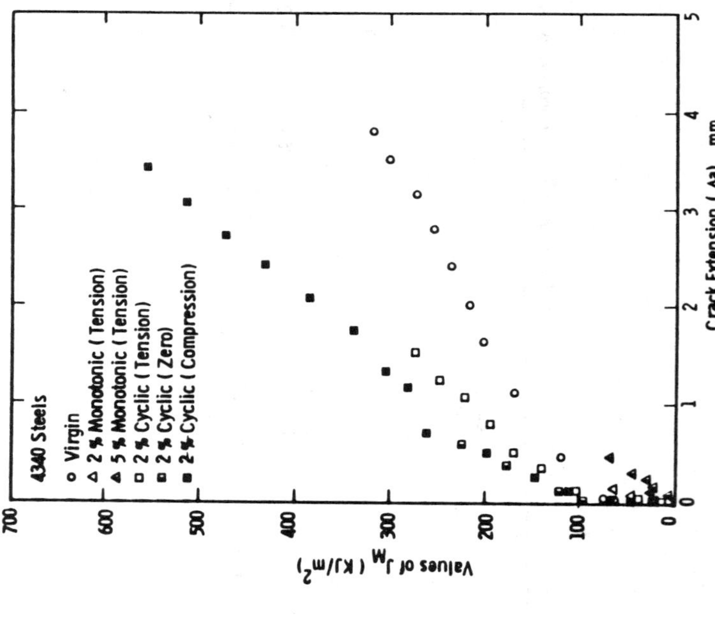

(a) —4340 Steels

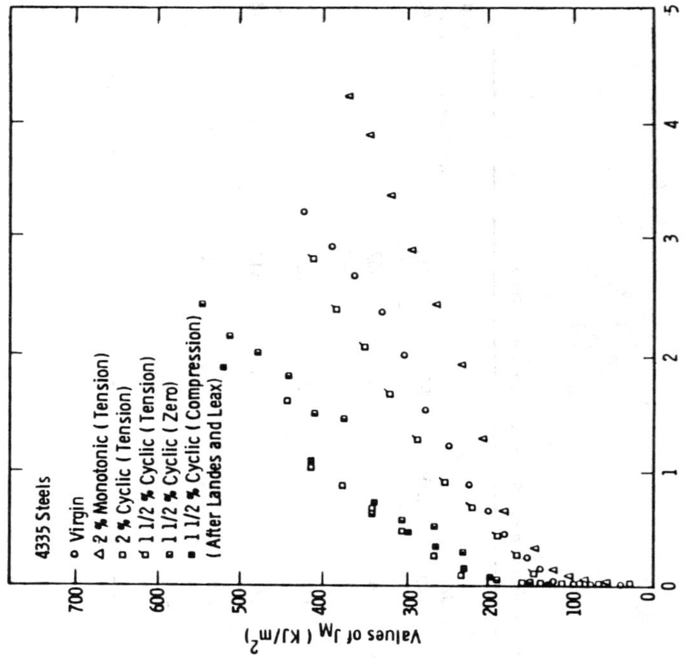

(b) — 4335 Steels (Ref. 18)

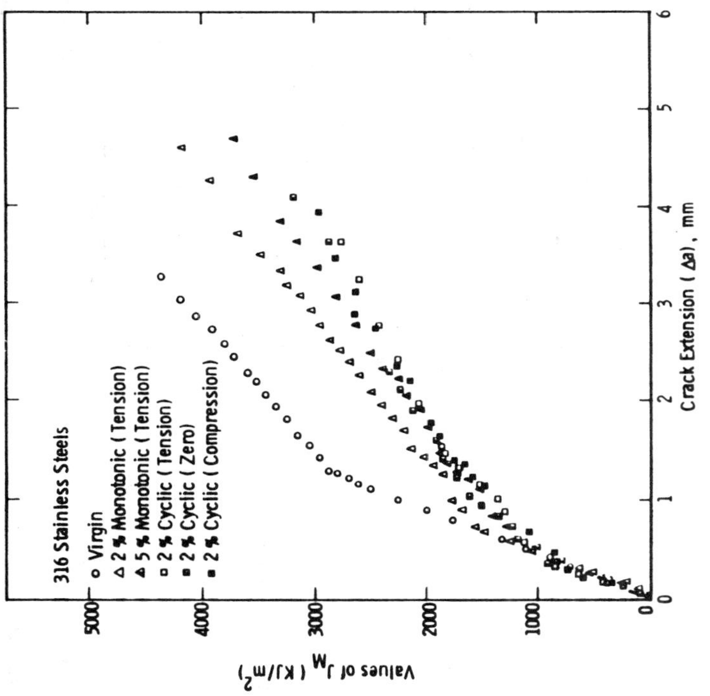

(c) – 316 Stainless Steels

FIG. 6—*Effects of prestrain history on J_R-curves.*

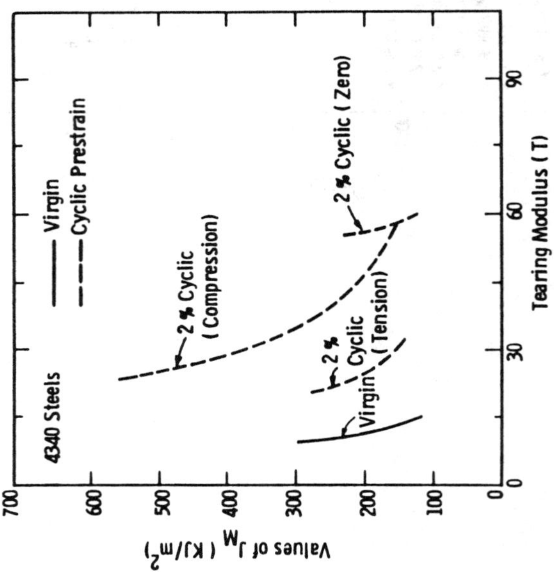

(a) — 4340 Steels

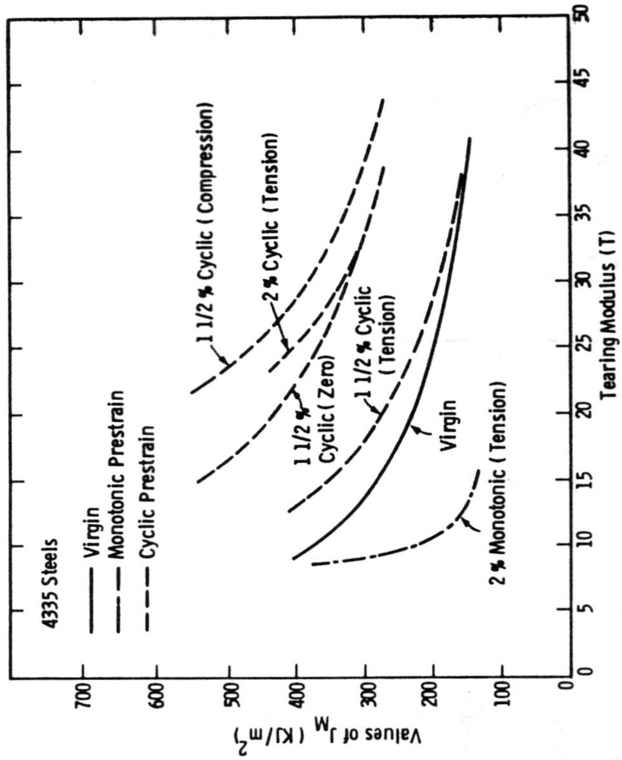

(b) — 4335 Steels

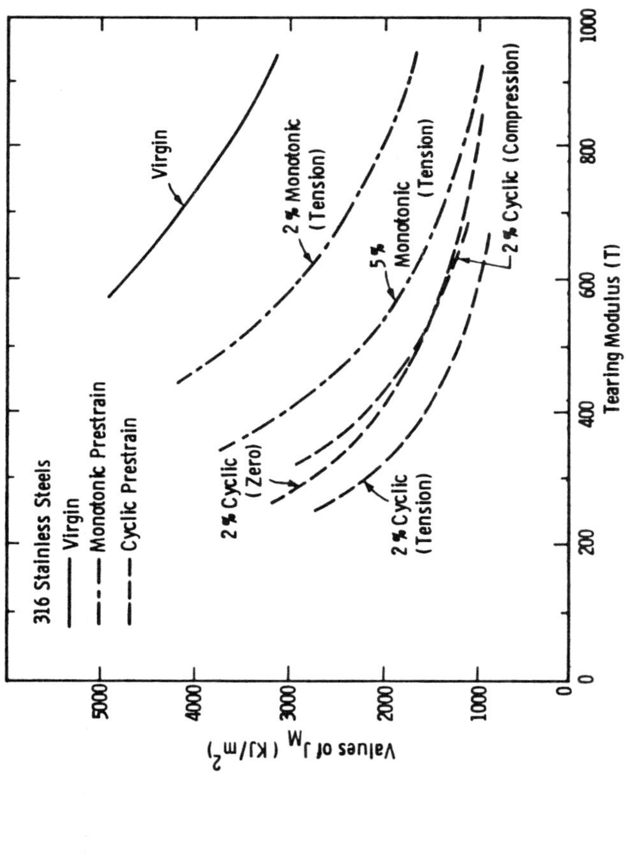

(c) – 316 Stainless Steels

FIG. 7—*Effects of prestrain history on J-T curves.*

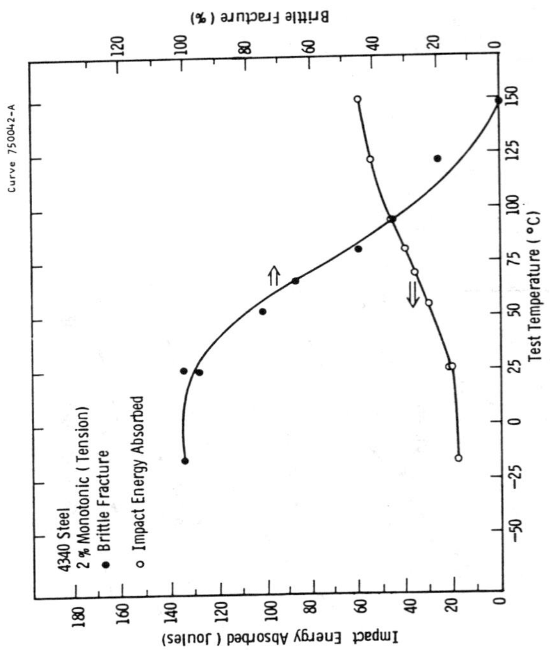

(b)—2 % Monotonic (Tension)

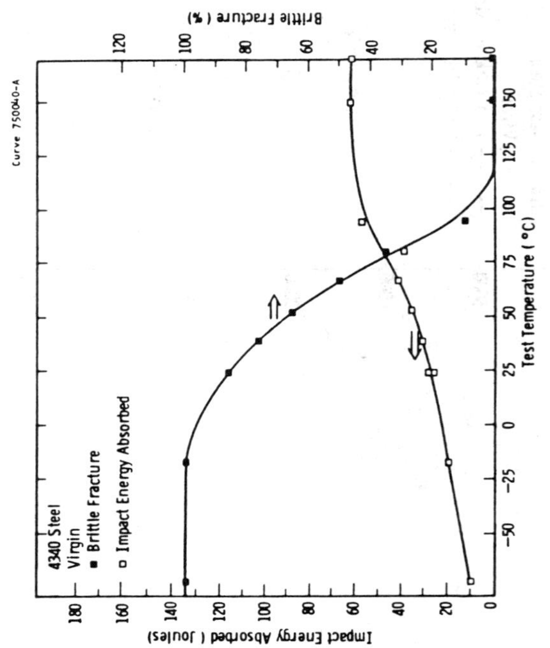

(a) — Virgin

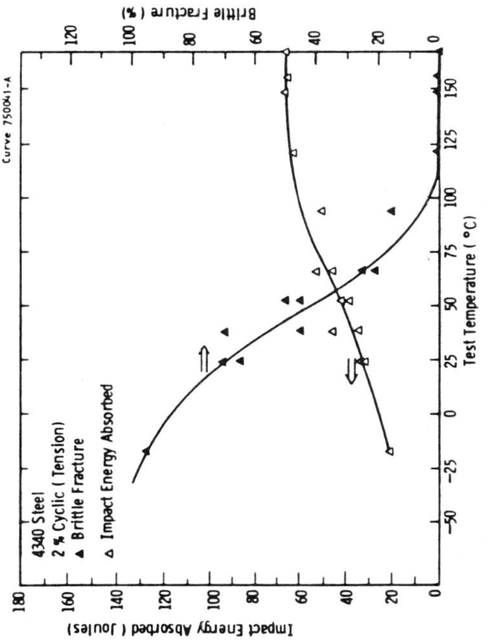

(c) – 2 % Cyclic (Tension)

FIG. 8—*Charpy impact results of 4340 steels.*

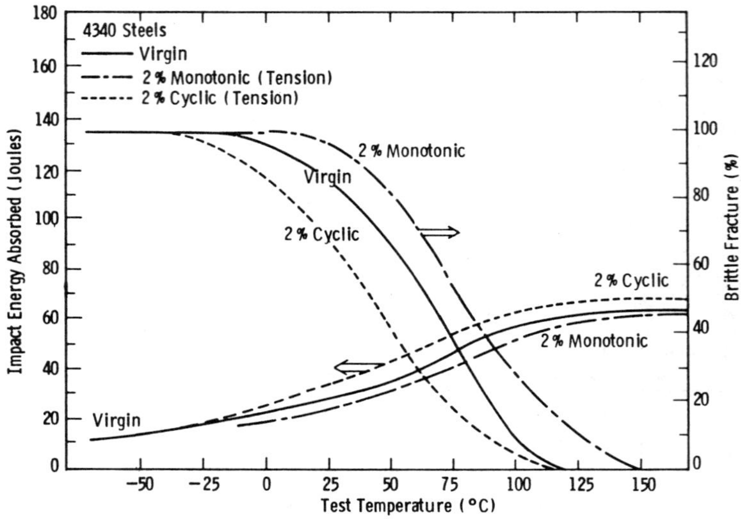

FIG. 9—*Effects of prestrain history on Charpy impact results.*

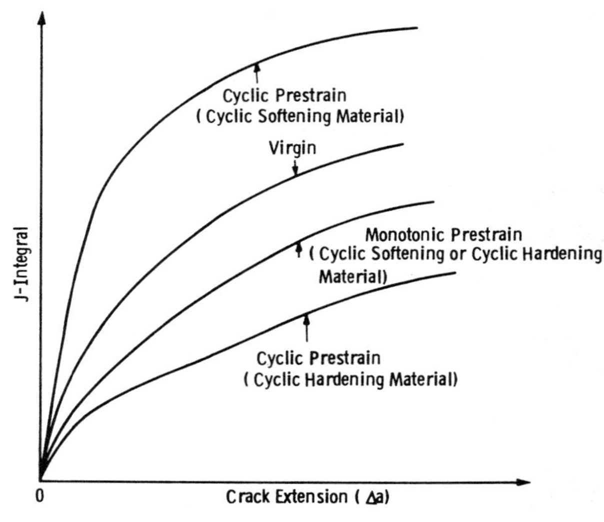

FIG. 10—*Schematic of influence of prestrain history on J_R-curves.*

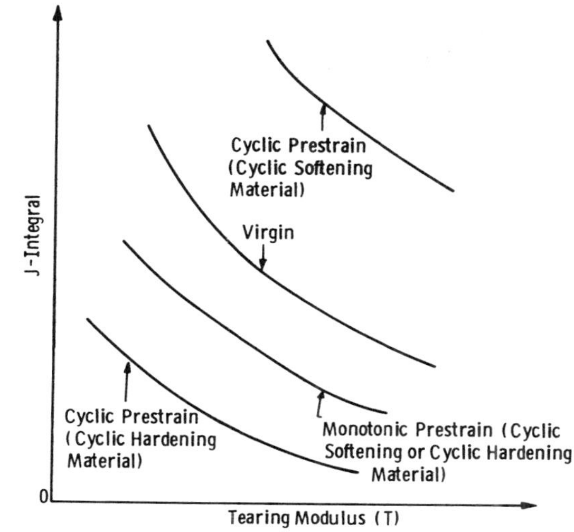

FIG. 11—*Schematic of influence of prestrain history on J-T curves.*

material strength and *increases* the levels of J_R and J-T curves. In 316 stainless steel, cyclic prestrain *increases* material strength and *decreases* J_R and J-T curves.

The preceding results suggest that any prestrain history which increases material strength will degrade fracture toughness properties and vice versa. In other words, an inverse relationship between material strength and fracture toughness properties exists, as typically found in structural materials. This observed relationship is clearly demonstrated by plotting J_{Ic} and J_M values versus material strength (σ_f or σ_y), as shown in Figs. 12*a*, 12*b*, and 12*c* for the three steels, respectively. Note that in 4335 steels, the values of σ_f are not available so that J_{Ic} and J_M are plotted against σ_y (Fig. 12*b*). In each steel, the values of J_{Ic} and J_M generally decrease with increasing σ_f or σ_y regardless of prestrain history. Similarly, an inverse relationship between J_{50} and σ_f or σ_y is presented in Figs. 13*a*, 13*b*, and 13*c* for the three steels. Therefore, the influence of prestrain history on fracture toughness properties can be rationalized by material strength.

The effects of cyclic prestrain on fracture toughness properties can be further related to cyclic softening and cyclic hardening characteristics of test materials, as presented in Figs. 10 and 11. Following cyclic prestrain, the improved fracture toughness properties of 4340 and 4335 steels are due to lower material strengths which result from the characteristics of cyclic softening. In cyclically prestrained 316 stainless steels, the degraded fracture toughness properties are associated with larger material strengths which are derived from cyclic hardening. Thus, cyclic prestrain in cyclic softening materials results in elevated fracture toughness properties. On the other hand, cyclic prestrain in cyclic hardening materials results in lower toughness.

Previously, Amouzouvi and Bassim [*14,15*] reported that increasing monotonic prestrain from 0 to 2% increased fracture toughness of an Atlas SPS steel whose composition was similar to that of a 4340 steel. They attributed the improved fracture toughness properties to the formation of Lüder bands, which produced excessive dislocations and blunt the crack tip. The Lüder band formation behavior was not observed in the present investigation. A further increase in monotonic prestrain from 2 to 15%, however, was found to decrease fracture toughness [*14,15*]. The degraded fracture toughness of these monotonically prestrained specimens can be also related to the increased material strength [*16*]. This trend observed by Amouzouvi and Bassim [*14,15*] is in agreement with the present results.

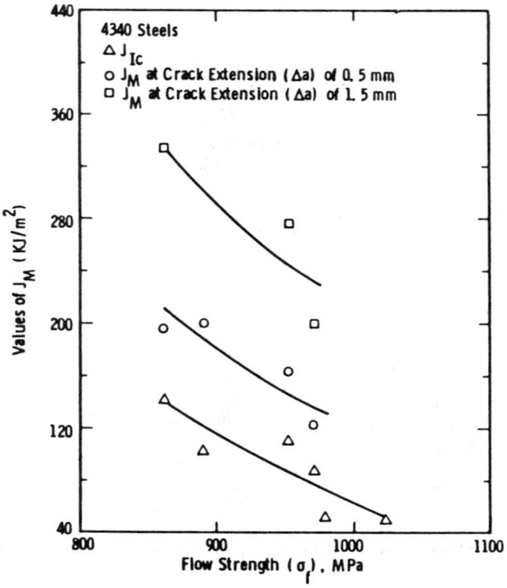

(a) — 4340 Steels

FIG. 12—J_{Ic} *and* J_M *values versus material strength* (σ_f *or* σ_y).

Charpy Impact Properties

Prestrain history was found to alter the transition behavior of Charpy impact properties of 4340 steels (Fig. 9). Monotonic prestrain increases the FATT and slightly decreases upper-shelf impact energy, as compared to the virgin material, while cyclic prestrain decreases the FATT and increases upper shelf impact energy. The effect of prestrain history on the transition behavior is related to material strength. Monotonic prestrain increases material strength, increases FATT, and decreases impact energy, while cyclic prestrain decreases material strength, decreases FATT, and increases impact energy. Thus, a direct relationship between material strength and FATT exists (Fig. 14), while an inverse relationship between material strength and impact energy exists. It is suggested that any prestrain history which increases yield strength will increase FATT and decrease impact energy, and vice versa.

Based on the preceding discussion on tensile, fracture toughness, and Charpy impact properties, prestrain history was found to alter these material properties significantly. Many of in-service machine components have been subjected to long and complex load histories. To ensure structural integrity and to determine the residual lives of these machine components, it is of great importance to investigate the influence of load history on material properties.

Finally, the present results demonstrate that mechanical prestrain can either increase or decrease tensile and fracture properties depending upon prestrain history and test material. Thus, it is further suggested that improved material properties can be achieved by mechanical prestrain if a careful and systematic investigation of load history effects on tensile and fracture behavior of structural materials is conducted.

Conclusions

To determine the structural integrity and the residual life of a structural component, it may be important to investigate the effects of in-service loading histories on material properties. In

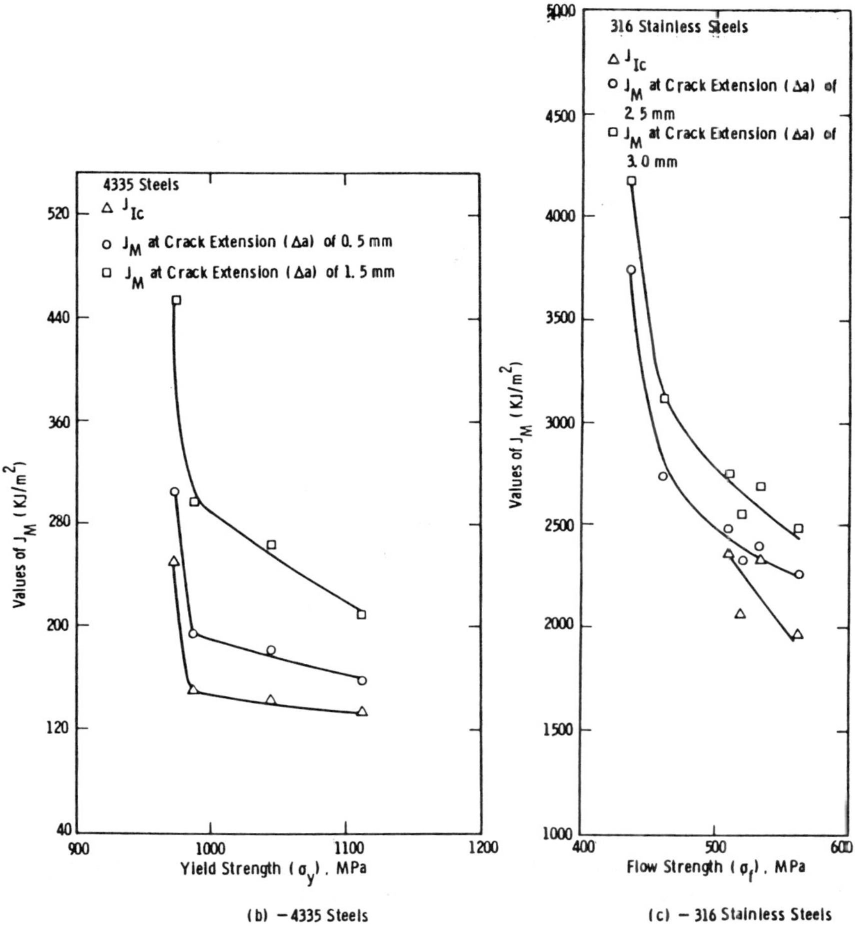

FIG. 12—*Continued.*

this investigation, prestrain histories were found to alter tensile, fracture toughness, and Charpy impact properties of the virgin materials significantly. Pertinent results are summarized as follows:

1. Prestrain histories affected the tensile properties of 4340 steel, 4335 steel, and 316 stainless steel in the following way:

(*a*) Monotonic prestrain increased material strength.
(*b*) Cyclic prestrain decreased material strength in the 4340 and 4335 steels (related to cyclic softening), but increased strength in the 316 stainless steel (related to cyclic hardening).

2. Prestrain histories influenced fracture toughness properties as measured by the J_R curves and the J-T diagrams:

(*a*) Monotonic prestrain decreased toughness in all materials.
(*b*) Cyclic prestrain increased toughness in the 4340 and 4335 steels, and decreased toughness in the 316 stainless steel.

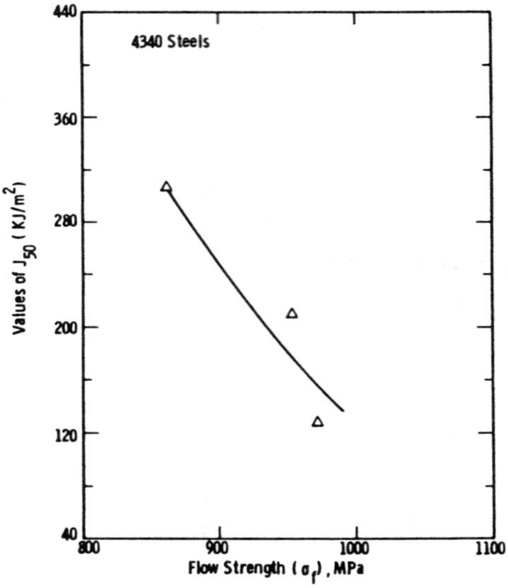

(a) — 4340 Steels

FIG. 13—J_{50} *values versus material strength* (σ_f *or* σ_y).

3. The effect of prestrain on toughness was found to be inversely related to the strength change. Furthermore, the effect of cyclic prestrain on toughness is associated with the cyclic softening or cyclic hardening characteristics of each material.

4. Prestrain histories altered the transition toughness properties of the 4340 steel as measured by the Charpy impact test:

(*a*) Monotonic prestrain increased the transition temperature (as measured by FATT), and slightly decreased upper-shelf impact energy.
(*b*) Cyclic prestrain decreased FATT and increased upper-shelf energy.

5. The transition temperature change because of prestrain is probably related to the strength change.

6. Mechanical prestrain has the potential of improving tensile and fracture properties of structural materials.

Acknowledgments

The authors are very grateful to the late W. H. Pryle for preparing test specimens and for his kind help in many aspects of this work. We would like to thank L. W. Burtner, M. G. Peck, R. B. Hewlett, R. R. Hovan, F. X. Gradich, and P. J. Barsotti for conducting the experiments. This work was supported by Westinghouse Corporate Ladder Funds.

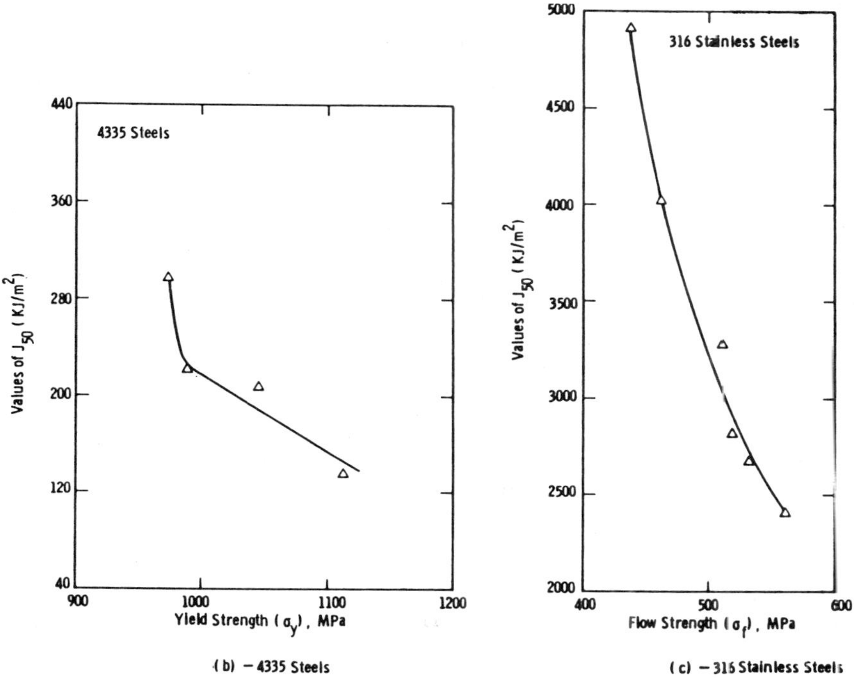

FIG. 13—*Continued.*

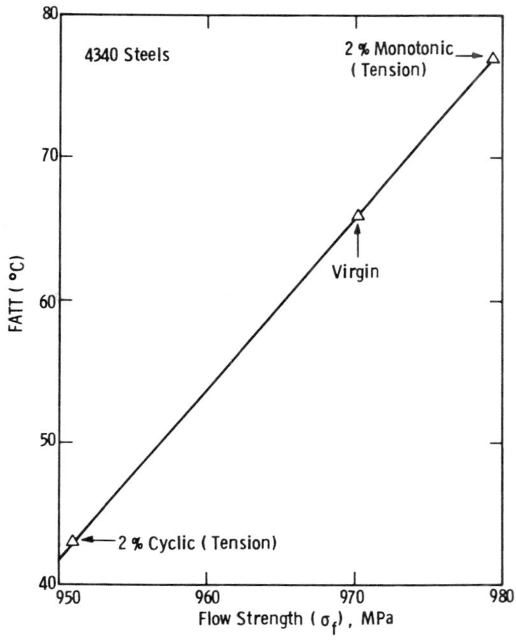

FIG. 14—*FATT versus flow strength* (σ_f) *for 4340 steels.*

References

[1] Landes, J. D. and Begley, J. A. in *Fracture Toughness, ASTM STP 514,* American Society for Testing and Materials, Philadelphia, 1972, p. 24.
[2] Kumur, V., German, M. D., and Shih, C. F., EPRI Topical Report NP-1931 for RP 1237-1, Electric Power Research Institute, Palo Alto, CA, July 1981.
[3] Ritchie, R. O. and Thompson, A. W., *Metallurgical Transactions A,* 1985, Vol. 16A, p. 233.
[4] Liaw, P. K., McCabe, D. E., and Landes, J. D., *International Journal of Fracture,* 1985, Vol. 27, p. R3.
[5] Albrecht, P., Andrews, W. R., Gudas, J. P., Joyce, J. A., Loss, F. J., McCabe, D. E., Schmidt, D. W., and Vandersluys, W. A., *Journal of Testing and Evaluation,* 1982, Vol. 10, p. 245.
[6] Stout, M. G. and Gerberich, W. W., *Closed Loop,* Materials Test System, June 1982, p. 12.
[7] Ernst, H. A., Paris, P. C., and Landes, J. D. in *Fracture Mechanics: Thirteenth Conference, ASTM STP 743,* American Society for Testing and Materials, Philadelphia, 1981, p. 476.
[8] Rice, J. R., Paris, P. C., and Merkle, J. G. in *Fracture Toughness Testing, ASTM STP 536,* American Society for Testing and Materials, Philadelphia, 1973, p. 231.
[9] Bates, R. C. in *Proceedings,* Symposium on Fracture: Interactions of Microstructure, Mechanisms and Mechanics, J. M. Wells and J. D. Landes, Eds., American Institute for Mining, Metallurgical, and Petroleum Engineers, Warrendale, PA, 1984, p. 117.
[10] Gudas, J. P. and Davis, D. A., *Journal of Testing and Evaluation,* 1982, Vol. 10, p. 252.
[11] Sumpter, J. D. G. and Turner, C. E. in *Cracks and Fracture: Ninth Conference, ASTM STP 601,* American Society for Testing and Materials, Philadelphia, 1976, p. 3.
[12] Standard Test for J_{Ic}, A Measure of Fracture Toughness, ASTM E 813-87, *Annual Book of ASTM Standards,* Vol. 03.01, 1987.
[13] Gudas, J. P., "Results of the J-Integral R-Curve Round Robin Program," presented at the 2nd ASTM Symposium on Elastic-Plastic Fracture, Philadelphia, PA, 6–9 Oct. 1981.
[14] Amouzouvi, K. F. and Bassim, M. N., *Material Science and Engineering,* Vol. 55, 1982, p. 257.
[15] Amouzouvi, K. F. and Bassim, M. N., *Material Science and Engineering,* Vol. 60, 1983, p. 1.
[16] Liaw, P. K. and Landes, J. D., *Metallurgical Transactions A,* Vol. 17A, 1986, p. 473.
[17] Juhas, M. C. and Bernstein, I. M., *Metallurgical Transactions A,* Vol. 14A, 1983, p. 1379.
[18] Landes, J. D. and Leax, T. R. in *Fracture Mechanics: Fifteenth Symposium, ASTM STP 833,* American Society for Testing and Materials, Philadelphia, 1984, p. 436.
[19] Landes, J. D. and McCabe, D. E. in *Elastic-Plastic Fracture: Second Symposium, Volume II: Fracture Curves and Engineering Applications, ASTM STP 803,* American Society for Testing and Materials, Philadelphia, 1983, p. II-723–II-738.
[20] Begley, J. A., unpublished results, Westinghouse R&D Center, Pittsburgh, PA, 1987.
[21] Dowling, N. E. in *Flaw Growth and Fracture: Tenth Conference, ASTM STP 631,* American Society for Testing and Materials, Philadelphia, 1977, p. 139.
[22] Troshchenko, V. T., Pokrovsky, V. V., and Prokopenko, A. V., *Fracture 1977,* Vol. 3, ICF4, Waterloo, ON, Canada, June 1977.
[23] Landgraf, R. W., Morrow, J., and Endo, T., *Journal of Materials,* 1969, Vol. 4, p. 176.
[24] Clarke, G. A., Andrews, W. R., Paris, P. C., and Schmidt, D. W. in *Mechanics of Crack Growth, ASTM STP 590,* American Society for Testing and Materials, Philadelphia, 1976, p. 27.
[25] Ernst, H. A. in *Elastic-Plastic Fracture: Second Symposium, Volume I: Inelastic Crack Analysis, ASTM STP 803,* American Society for Testing and Materials, Philadelphia, 1983, p. I-191.
[26] Saxena, A. and Hudak, S. J., Jr., *International Journal of Fracture,* Vol. 14, 1978, p. 453.
[27] Paris, P. C. and Johnson, R. E. in *Elastic-Plastic Fracture: Second Symposium, Volume II: Fracture Curves and Engineering Applications, ASTM STP 803,* American Society for Testing and Materials, Philadelphia, 1983, p. II-5–II-40.

James A. Joyce[1]

Development of a Criterion for the Effect on the J-R Curve of Elastic Unloadings

REFERENCE: Joyce, J. A., "**Development of a Criterion for the Effect on the J-R Curve of Elastic Unloadings,**" *Fracture Mechanics: Eighteenth Symposium, ASTM STP 945,* D. T. Read and R. P. Reed, Eds., American Society for Testing and Materials, Philadelphia, 1988, pp. 647–662.

ABSTRACT: A criterion is proposed which allows estimation of the fraction of cyclic damage which would be expected from a given number of partial or complete unloadings applied during the course of a J-R curve fracture toughness test. The criterion requires both geometry and material inputs, but for the special case of a steel compact specimen, an upper-bound criterion is developed and presented graphically.

Data from a series of compact specimens of a 3% nickel alloy steel are used to verify both the model used and the resulting fractional damage criterion developed.

KEY WORDS: J-R curve testing, crack propagation, cyclic crack growth, elastic-plastic fracture

During the past ten years, partial unloading compliance has become a widely used method to evaluate J-R curves from single-specimen tests. Implicit in this test technique is the assumption that the elastic unloadings do not produce any cyclic crack growth or in any other way influence the specimen toughness properties.

The ASTM Test Method for J_{Ic}, a Measure of Fracture Toughness (ASTM E 813-81) allows the use of unloading compliance, but makes no specification as to the unloading magnitude allowed or to the number of unloadings allowed. The recommended procedure for J-R curve testing [*1*] requires that the unloading do not exceed 0.25 P_L but places no restriction on the number of unloadings used. A recent experimental study by Sutton [*2*] has shown that for cases where 30 unloadings of up to 0.95 P_L are applied, for a single structural steel, no measurable effect was produced on the material J-R curve. On the other hand, work by Kaiser [*3*] has shown that a larger number of 1.0 P_L unloadings can produce a dramatic effect on the apparent material J-R curve, and it appears important that an analytical model be developed to estimate the amount of cyclic crack growth which can occur during unloading compliance tests involving many unloadings so that rational limits can be included in upcoming ASTM J-R curve standards.

In this paper a criterion is proposed to allow estimation of the fraction of cyclic damage which would be expected from a particular geometry, unloading fraction and material so that reasonable restrictions can be included in J-R curve test standards. A limited amount of experimental work is presented to justify the model being used as well as to verify the proposed criterion.

Analysis

In cases where crack extension occurs by a combination of ductile tearing and cyclic crack growth, the simplest possible model is to just superimpose the two effects linearly such that

[1]Associate professor of mechanical engineering, U.S. Naval Academy, Annapolis, MD 21402.

$$\Delta a_{\mathrm{TOT}} = \Delta a_{\mathrm{TEAR}} + \Delta a_{\mathrm{CYCLE}} \tag{1}$$

where

Δa_{TOT} = total crack extension to the present in a J-R curve test,
Δa_{TEAR} = that part due solely to ductile tearing and initial crack tip blunting, and
$\Delta a_{\mathrm{CYCLE}}$ = crack extension resulting from cyclic loading.

The cyclic crack growth component is modeled here by the equation

$$\Delta a_{\mathrm{CYCLE}} = \frac{da}{dN} \Delta N$$

$$= A(\Delta K)^m \Delta N \tag{2}$$

where

da/dN = material cyclic crack growth rate,
ΔN = number of elastic unloadings applied,
ΔK = elastic stress intensity range applied during the unloading, and
A, m = standard Paris [*4*] correlation coefficients obtained, for instance, in accordance with the ASTM Test Method for Constant-Load-Amplitude Fatigue Crack Growth Rates Above 10^{-8} m/Cycle (E 647-83).

This type of direct superposition model has been demonstrated by Kaiser [*3*] to be applicable to the type of load histories used in unloading compliance tests. His results are verified in the experimental work described in the following sections.

The fraction of total crack extension attributable to cyclic loading rather than to ductile tearing is defined here as fractional damage (*FD*) as follows

$$FD = \frac{\Delta a_{\mathrm{CYCLE}}}{\Delta a_{\mathrm{TOT}}} = \frac{\Delta a_{\mathrm{TOT}} - \Delta a_{\mathrm{TEAR}}}{\Delta a_{\mathrm{TOT}}} \tag{3}$$

The objective then is to relate an acceptable level of fractional damage, about $FD = 0.01$, to the unload magnitude ΔP and the number of unloadings applied ΔN. General relationships are thus assumed for the stress-intensity factor range to be of the form

$$\Delta K = \frac{\Delta P}{B\sqrt{W}} F_1(a/W) \tag{4}$$

and for the specimen limit load as

$$P_L = B\sigma_y(W - a)H(a/W) \tag{5}$$

Combining Eqs 2 to 5 gives

$$\frac{\Delta P}{P_L} = \sqrt{W}\, \frac{FD\Delta a_{\mathrm{TOT}}}{A\Delta N} \bigg/ b\sigma_y F_1(a/W)H(a/W) \tag{6}$$

with $b = W - a$, or finally as

$$\frac{\Delta P}{P_L} = \frac{\sqrt{W}}{b\sigma_y} G(a/W)^m \frac{FD\Delta a_{\mathrm{TOT}}}{A\Delta N} \tag{7}$$

where

σ_y = material flow stress,
P_L = material limit load,
ΔP = amount of unload on an individual unloading, and
$G(a/W)$ = geometrical factor depending on the test specimen geometry.

For a particular geometry and material, Eq 7 defines the fraction of cyclic damage expected out of the total crack extension present for a given number of unloadings of magnitude $\Delta P/P_L$.

For the particular case of interest here of the compact specimen, the following relationships were used in Eq 6, that is, that

$$\Delta K = \frac{\Delta P}{B\sqrt{W}} F_1(a/W) \tag{8}$$

[see the ASTM Test Method for Plane-Strain Fracture Toughness of Metallic Materials (E 399-83)] where

$$F_1(a/W) = [(2 + a/W)(0.886 + 4.64a/W - 13.32(a/W)^2 + 14.72(a/W)^3 - 5.6(a/W)^4]/(1 - a/W)^{3/2} \tag{9}$$

and [5]

$$P_L = 1.26\sigma_y Bb\alpha \tag{10}$$

with

$$\alpha = \sqrt{4\left(\frac{a}{b}\right)^2 + 4\left(\frac{a}{b}\right) + 2} - \left(2\frac{a}{b} + 1\right) \tag{11}$$

The result is presented in Fig. 1 for the particular case of a compact specimen where

$a/W = 0.5$,
$W = 50.0$ mm, and
$\Delta a_{TOT} = 2.5$ mm.

The material properties were chosen to represent upper-bound strength and fatigue crack growth rate data expected for structural steels [6] as

$$m = 2.26$$

$$A = 1.35 \times 10^{-10}$$

$$\sigma_y = 960 \text{ MPa}$$

This case is expected to represent a practical limiting case for steels which might be tested by an unloading compliance method and Fig. 1 shows that at the present J-R curve procedure limit of 25% unloading about 70 unloadings would be required before 2.5 mm of crack growth was found to produce even 2% cyclic fractional damage. The same plot shows, however, that five cycles with a 50% unloading could produce 10% cyclic fractional damage. The use of a lower strength material or larger a/W specimens dramatically reduces the predicted fractional damage. Equation 7 is also highly sensitive to the exponent, m, so that for most practical cases, considerably less cyclic damage would be expected than is shown in Fig. 1.

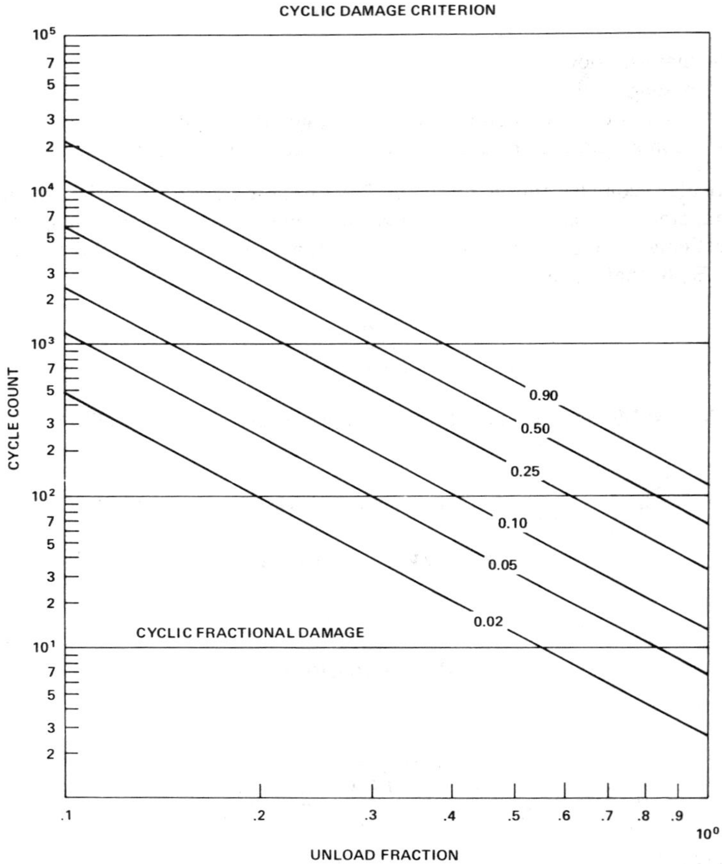

FIG. 1—*Criterion giving number of unloadings required at a given unload fraction to produce a fraction of cyclic crack growth.*

Description of Experimental Work

The objectives of the experimental work undertaken here were:

1. Verify the applicability of the simple superposition rule expressed in Eq 1. This verification has been demonstrated by Kaiser [*3*] for two steels, but it was felt to be such an important step that it was done again here on the structural steel used in this project.
2. Develop some experimental verification of the cyclic damage criterion expressed in Eq 7 by running cyclically loaded specimens with different amounts of unload fraction with various numbers of applied elastic unloadings.
3. Develop an upper-bound form of Eq 7 for structural steels to put some analytical justification into J-R curve standard limitations on unload numbers and intensities to be used in unloading compliance J-R curve methods.

To accomplish this, three types of experimental results have been run. First, unloading compliance J-R curve tests were run using a small number (<25) of unloadings of less than 15% to obtain the J-R curves with negligible cyclic damage. Second, cyclic unloading tests were run with 25, 50, and 100% unloading intensities, including some tests with up to 250 unload cycles.

The more extreme cases produced results in which more than 70% of the total crack length resulted from cycling rather than ductile tearing. The third set of tests was to obtain high cycle fatigue crack growth rate data on the structural steel being used. This data was utilized, as done by Kaiser previously, to verify the superposition statement of Eq 1.

Experimental Testing

Material

The material used for these tests was a 3% nickel structural steel alloy; the chemistry is given in Table 1 and the tensile mechanical properties in Table 2. The material was available as a 32-mm-thick plate and all specimens were tested in the transverse-longitudinal (T-L) orientation. The specimens used were compact specimens as shown in Fig. 2, and 20% side grooves were present on all specimens.

J-R Curve Tests

Standard J-R curve characterization of the above material was run using the recommended procedure of Ref *1* using the computer interactive test method described by Joyce and Gudas [7]. In this method, the spacing of the unloadings and the unloading lengths are chosen by the operator and are not exactly uniform. A typical load-COD record as shown in Fig. 3 demonstrates the relatively small number of unloadings of less than 15% magnitude typically used for this procedure. The scatterband of resulting J-R curves is shown in Fig. 4 for this alloy.

Cyclic J-R Curve Tests

For cyclic J-R curve tests, load-COD histories such as that shown in Fig. 5 were used with unload magnitudes of up to 100% and with unload cycle counts up to about 250. These tests were run on an automated servohydraulic test machine which was programmed to space unloadings at set COD intervals of from 0.005 to 0.05 mm and to control their extent at 25, 50, or 100% of the present maximum load.

Figure 6 shows a plot of the "envelope" load-COD curves for three cases of varying cyclic intensity, showing clearly that application of a large number of extreme unloading cycles effects

TABLE 1—*Chemical composition of steel used for J-R curve testing.*

	Chemical Composition, weight %										
Identification	C	Mn	P	S	Cu	Si	Ni	Cr	Mo	V	Ti
3 Ni steel (FYB)	0.153	0.33	0.012	0.013	0.033	0.18	2.55	1.66	0.37	0.003	<0.001

TABLE 2—*Quasi-static tensile properties of steel used for J-R curve testing [12.8-mm-diameter (0.505 in.) tension specimen].*

Orientation	Ultimate Tensile Strength, MPa (ksi)	Yield Strength 0.2% Offset, MPa (ksi)	Elongation in 2 in., %	Reduction in Area, %
T	731 (106)	614 (89)	23	63
L	724 (105)	607 (88)	26	73

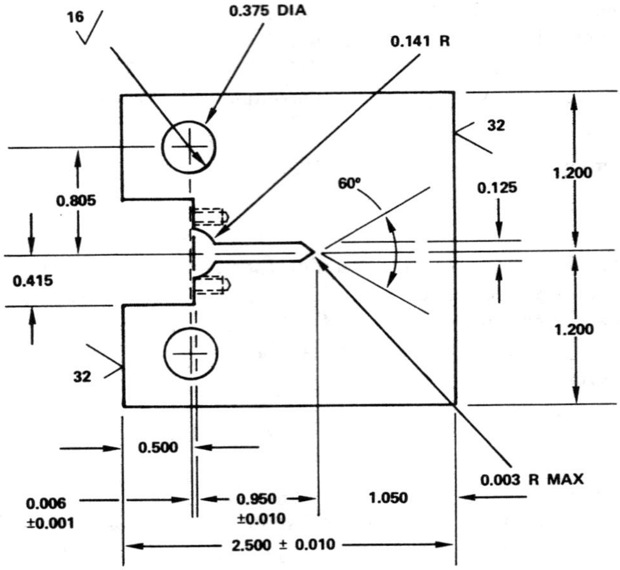

FIG. 2—*Modified compact specimen (1TCT) for J-integral testing [dimensions in inches (1 in. = 25.4 mm)].*

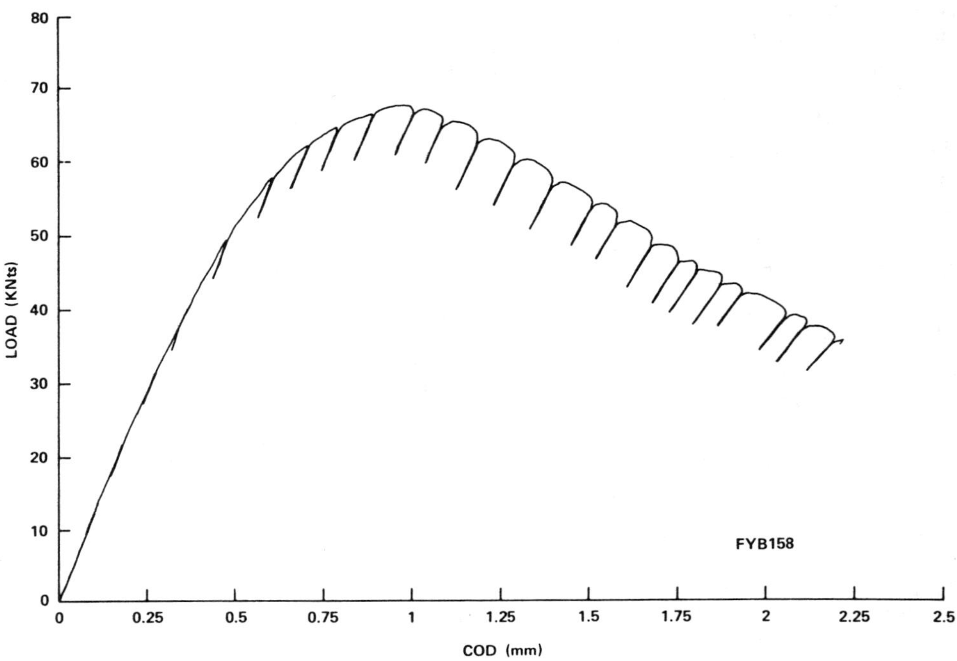

FIG. 3—*Typical unloading compliance load versus load-line crack opening displacement (COD) curve.*

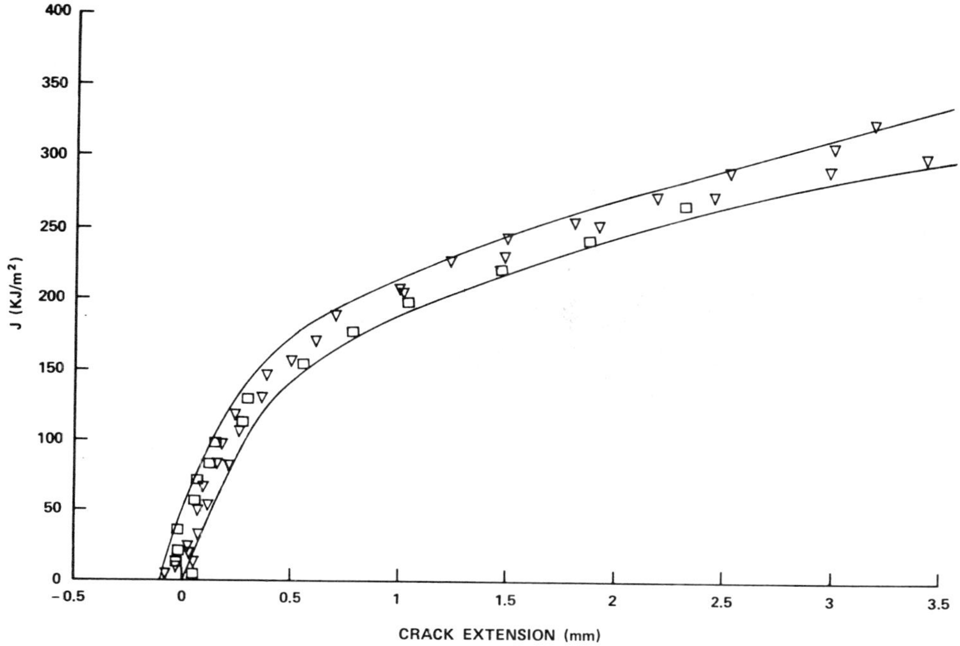

FIG. 4—*J-R curves of baseline unloading compliance data showing scatterband of expected results.*

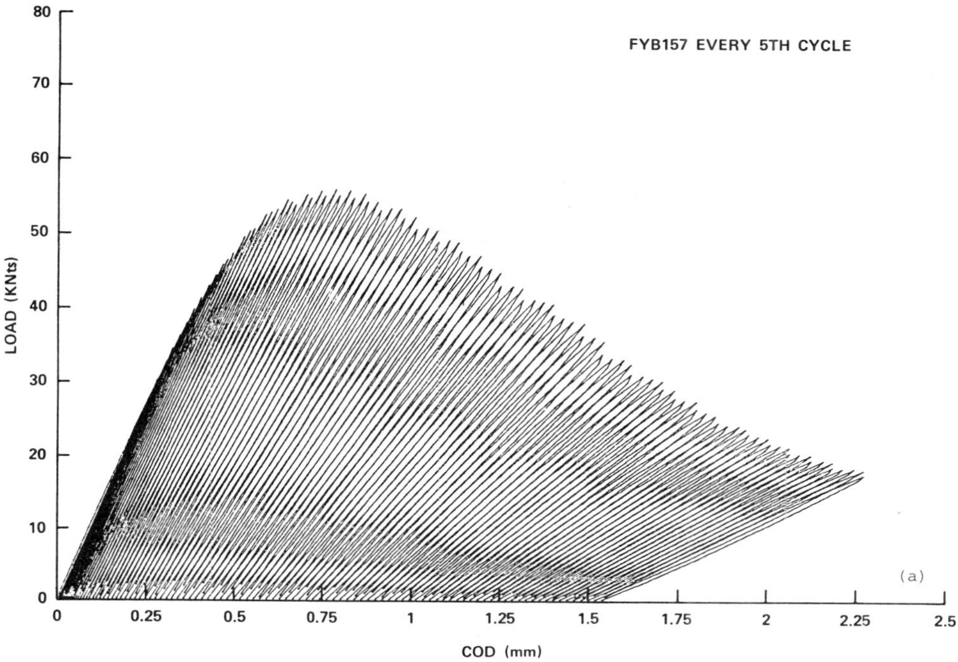

FIG. 5—*Cyclic load-displacement records for three cases with different amounts of cyclic unloading.*

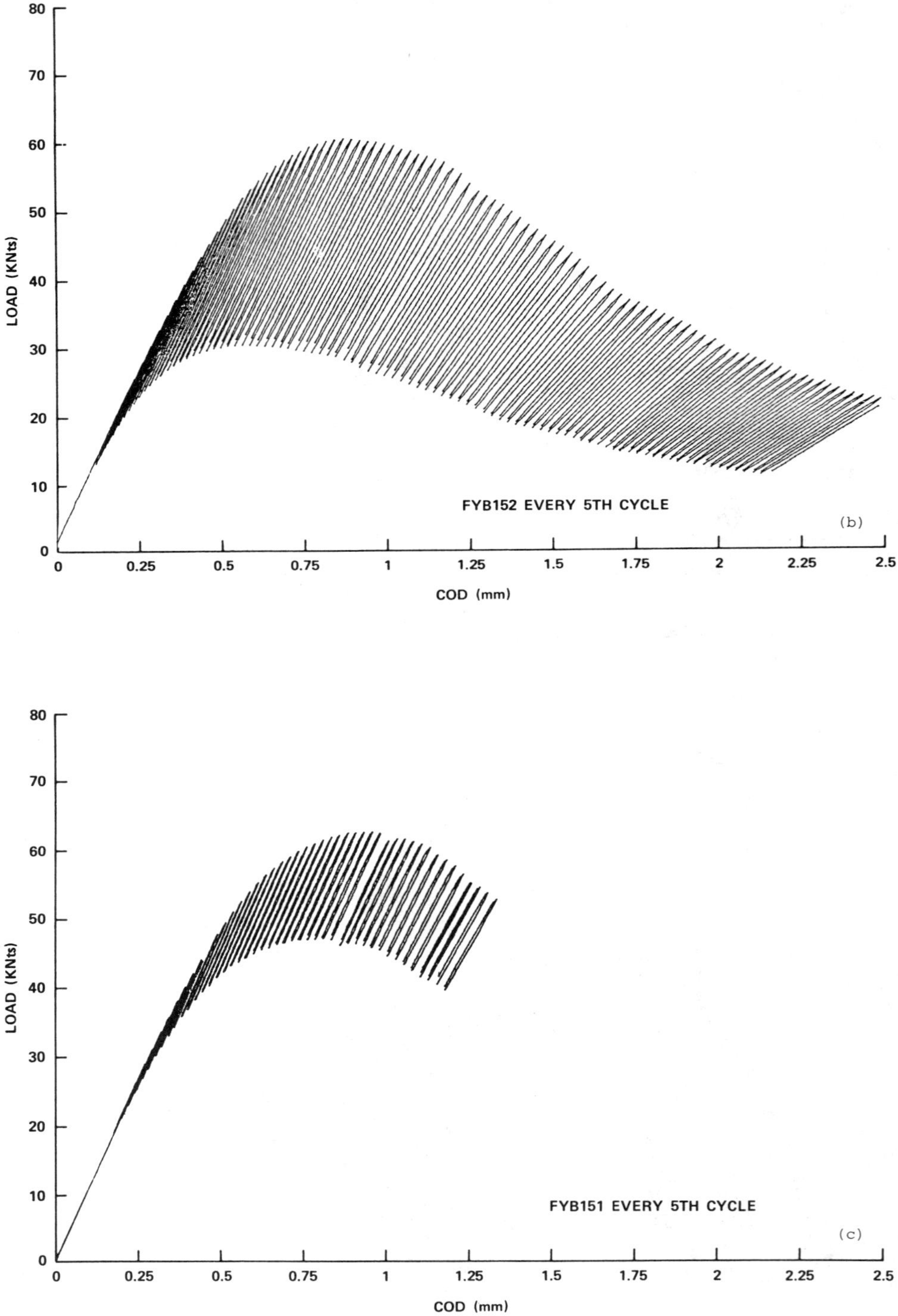

FIG. 5—*Continued.*

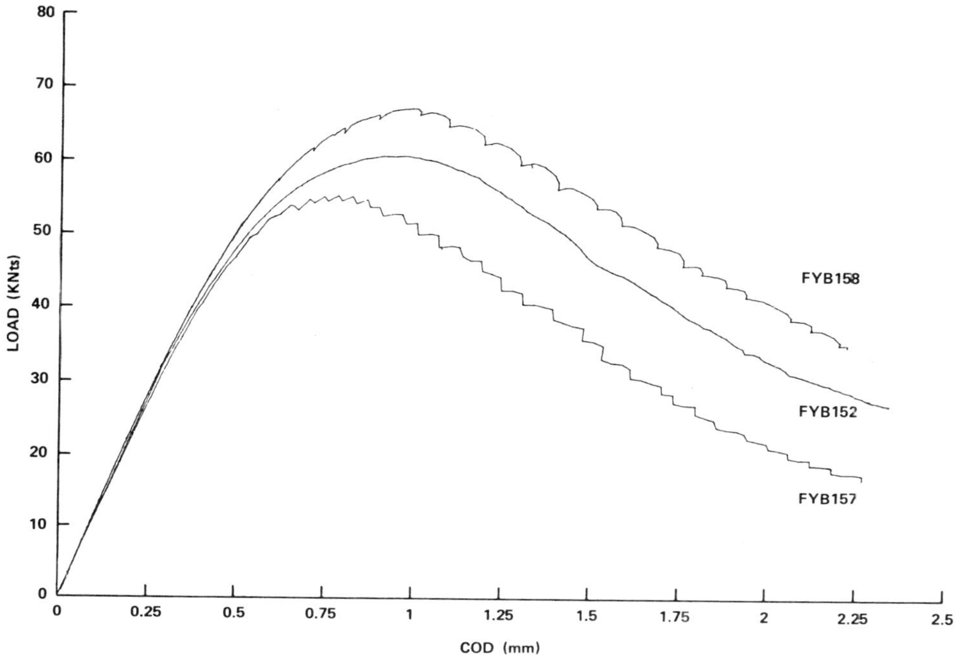

FIG. 6—*Cyclic load-displacement envelops showing effect of cycling on resulting load-displacement record.*

dramatically the measured load-COD record. A comparison of cyclic J-R curves obtained from these specimens shown in Fig. 7 demonstrates the effects of the cycling on the measured material crack growth resistance. For cyclic cases, the crack length was measured by a compliance technique using the unloading slope over the first 30% of the unload, except for the 25% unloadings, in which all 25% obtained was used. *J* was measured from the load displacement curves as shown in Fig. 6 using the equation of Ernst as given by Ref *1*.

The scatterband of unloading compliance data of Fig. 4 is compared in Fig. 7 with cyclic J-R curves for three specimens which have undergone intense cyclic loadings such as those shown in Figs. 5*b* and 5*c*. In all cases cyclic damage has produced a lowering of the J-R curve in comparison with the unloading compliance baseline results, and the more intense the applied cycling, the greater the reduction of the resulting J-R curve. For completeness in testing the fractional damage criterion of Eq 4, some specimens were also tested with an order-of-magnitude fewer 50 or 100% unload cycles as shown in Fig. 8. Unloading cycles were spaced at intervals of COD of approximately 0.05 mm for these specimens. The resulting cyclic J-R curves of two of these specimens are shown in Fig. 9 in comparison with the unloading compliance scatterband. Now, rather surprisingly, an elevation of the cyclic J-R curves is demonstrated, though the effect is small and decreases at larger values of crack extension.

High-Cycle Crack Growth Rate Tests

The high-cycle crack growth rate was measured here only to demonstrate the validity of the superposition model of Eq 1. For this reason, only one specimen was run using a computer-automated methodology, running first with the stress intensity range increasing and then with

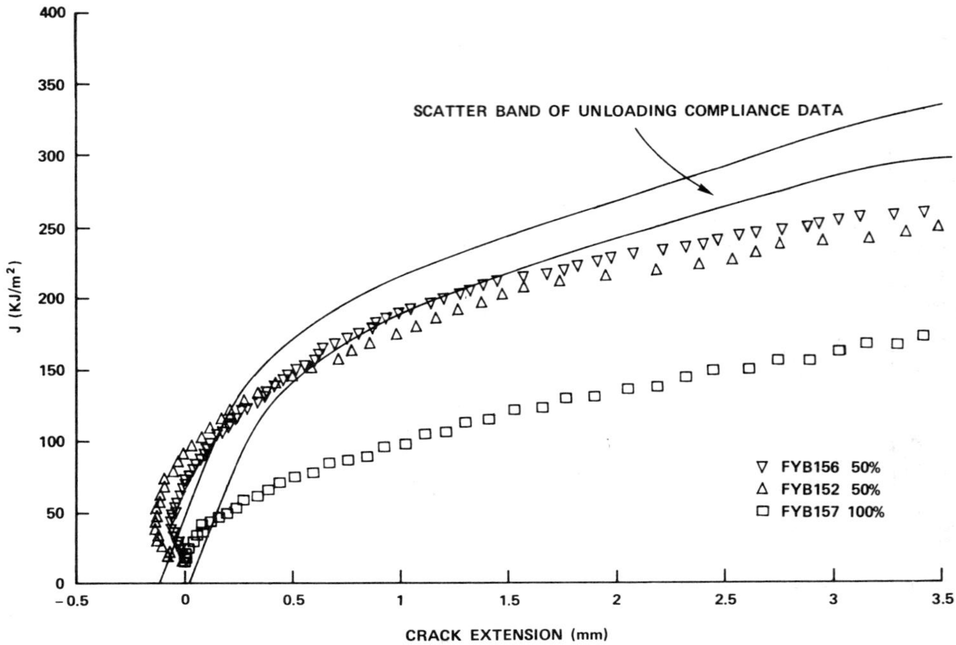

FIG. 7—*Cyclic J-R curves showing effect of cycling in comparison with scatterband of unloading compliance results.*

the stress intensity decreasing. The results are shown in Fig. 10 along with a correlation equation fit to the results. The equation for this alloy steel has the form

$$\frac{da}{dN} = 0.87 \times 10^{-10}(\Delta K)^{3.26} \text{ m/cycle} \tag{12}$$

Verification of the Crack Growth Superposition Model

Equation 1 can now be verified by combining the high-cycle fatigue crack growth rate result of Fig. 10 and Eq 12 and the cyclic J-R curves of the previous section. To evaluate the cyclic crack growth component $\Delta a_{\mathrm{CYCLE}}$ of Eq 1, a standard cycle-by-cycle integration approach was used. The tearing component of crack growth Δa_{TEAR} can be obtained by subtracting the result at each point along the cyclic load-displacement record from the Δa_{TOT} component measured experimentally.

The resulting J-R curves with cyclic damage removed are shown in Fig. 11. At this point a good correspondence exists between the J-R curves obtained by standard unloading compliance techniques and the cyclic J-R curves with the cyclic $\Delta a_{\mathrm{CYCLE}}$ component of crack growth removed. This result promises to be very useful in predicting crack growth rates under elastic plastic cyclic loading conditions, but as shown in Refs *8* and *9,* other types of load histories seem to produce different results and more work needs to be directed to clarifying this area.

For the purposes of this work, however, where the load history expected is just that of an unloading compliance J-R curve test, the results shown in Fig. 11 are taken as validation of the type of superposition assumed in Eq 1. Thus the results also validate the criterion of Eq 7 relat-

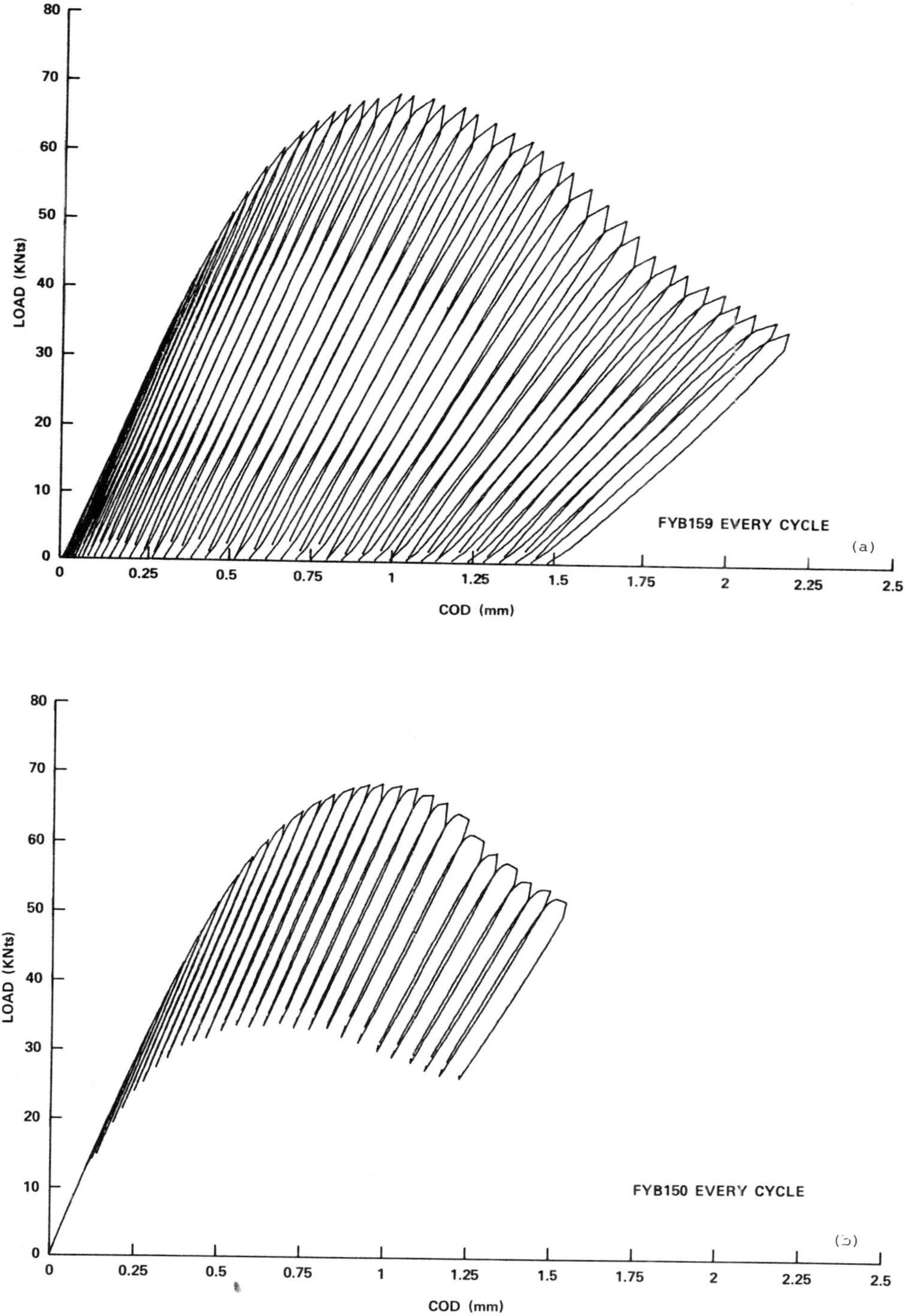

FIG. 8—*Cyclic load-displacement records for two specimens with large* $\Delta a_{TEAR}/\Delta a_{CYCLE}$ *ratios.*

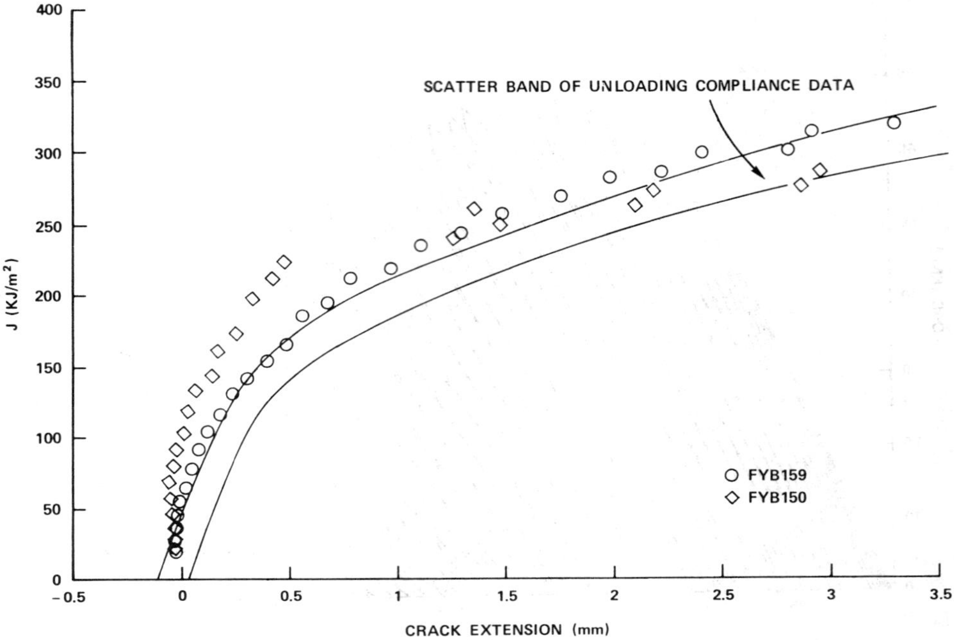

FIG. 9—*Cyclic J-R curves showing cyclic hardening effects of occasional large unloadings in comparison with scatterband of unloading compliance results.*

ing the fractional damage resulting from cyclic unloadings used in the course of standard unloading compliance J-R curve tests.

Discussion

The ductile tearing, cyclic damage superposition model demonstrated in the previous section clearly has useful applications beyond that of developing limits for unloading compliance J-R curve testing, but it is this latter task which is the current objective. A total of eleven specimens were run here, as shown in Table 3.

Although the scatterbands of J-R curve data make comparison difficult in attempting to evaluate the fraction of cyclic damage resulting from a particular combination of unload magnitude and spacing, three columns of fractional damage quantities are presented in Table 3. The information under the column labeled "FD Calculated" was obtained by a cycle-by-cycle calculation of the cyclic crack growth using the 3 nickel steel correlation of Eq 12. Information under the column labeled "FD Upper Bound" was obtained directly from the upper-bound fractional damage criterion of Eq 7 and Fig. 1, while information in the last column was obtained from measurements taken from J-R curves by comparing the cyclically damaged curves with unloading compliance reference curves.

In most cases, the observed damage falls between that calculated using the cycle-by-cycle method and that estimated by the upper-bound criterion. In all cases where small amounts of damage are predicted, small amounts of damage was observed, with Specimen 155 the only case in which the observed cyclic damage appears to exceed the upper-bound criterion prediction. In cases in which a large amount of cyclic damage is predicted, a large amount of damage was observed. The major deviation between measured fractional damage and observed fractional damage occurs in specimens with larger R ratios, that is, with $P_{min}/P_{max} > 0.5$, and corre-

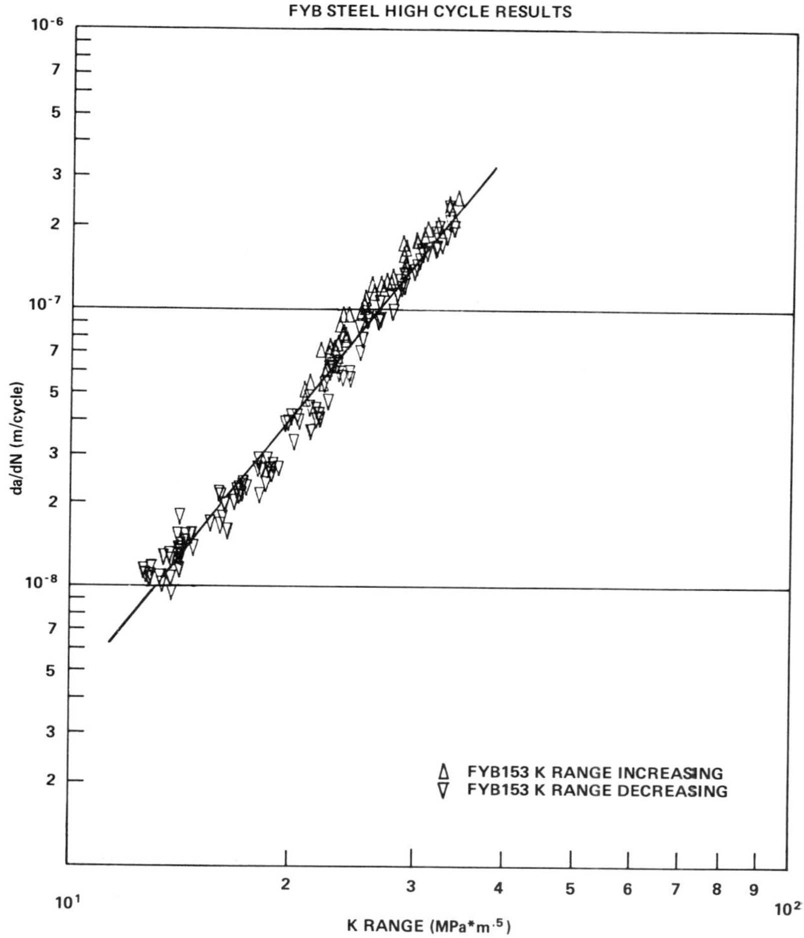

FIG. 10—*High cycle fatigue crack growth properties for 3% nickel alloy steel.*

sponds to a possible *R* ratio dependence or crack closure phenomena which has not been included in the straightforward analysis leading to Eq 7.

These results support the use of the criterion of Eq 7 and Fig. 1 for setting limits for the extent and frequency of unloadings allowed on unloading compliance J-R curve test methods. For the J-R curve standard, the present limitation that the unloadings should not exceed 0.25 P_L seems justified in light of these results since more than 50 unloadings over a crack extension of 2.5 mm (0.1 in.) would be required to produced even 2% cyclic damage by the upper-bound criterion of Fig. 1.

Fractographic Analysis

Scanning electron photomicrographs of the fracture surfaces for three specimens tested with differing mixtures of ductile tearing and fatigue cycling are shown in Fig. 12. The first specimen, FYB158, was tested with only a small number of unloadings of less than 15% and shows no cyclic damage on the fracture surface. The fracture is completely by hole growth and coales-

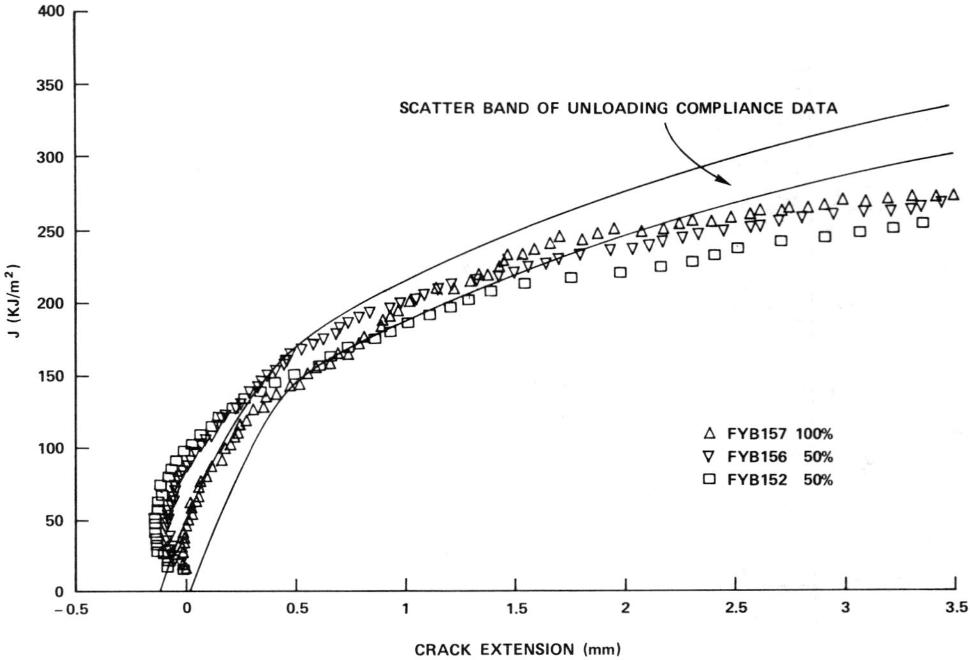

FIG. 11—*J-R curves obtained by correcting cyclic J-R curves to remove the component of cyclic damage.*

TABLE 3—*Experimental results.*

Specimen ID	Unload, %	ΔCOD (mm)/ Cycle	Cycles to Δa = 2.5 mm	FD Calculated	FD Upper Bound	FD Observed
T4T	15	...	16	<0.01	<0.01	...
158	15	...	16	<0.01	<0.01	...
157	100	0.005	145	0.69	0.92	0.75 to 0.85
159	100	0.05	26	0.20	0.20	<0.05
150	50	0.05	26	0.02	0.05	<0.05
152	50	0.005	211	0.12	0.38	0.35 to 0.45
156	50	0.005	220	0.12	0.38	0.15 to 0.25
151	25	0.005	225	0.01	0.06	<0.05
154	50	0.05	24	0.02	0.04	<0.05
155	50	0.0125	71	0.06	0.15	0.25 to 0.35
153		used for high cycle fatigue crack growth				

cence, as would be expected for a structural steel on its upper shelf. The other two specimens were subjected to a much greater number of unloading cycles (211 and 145, respectively) with intensities of 50% unload for Fig. 12*b* and 100% unload for Fig. 12*c*. Both of these fracture surfaces show a mixture of hole growth and fatigue damage. The fatigue damage is demonstrated by striation markings around the perimeter of voids and on other surfaces where rather sharp elevation changes occurred on the fracture surface. Void formation, growth, and coalescence clearly have remained important contributors to crack growth on both of these specimens, and it is also clear that the fewer number of 100% unloading cycles produced much more fa-

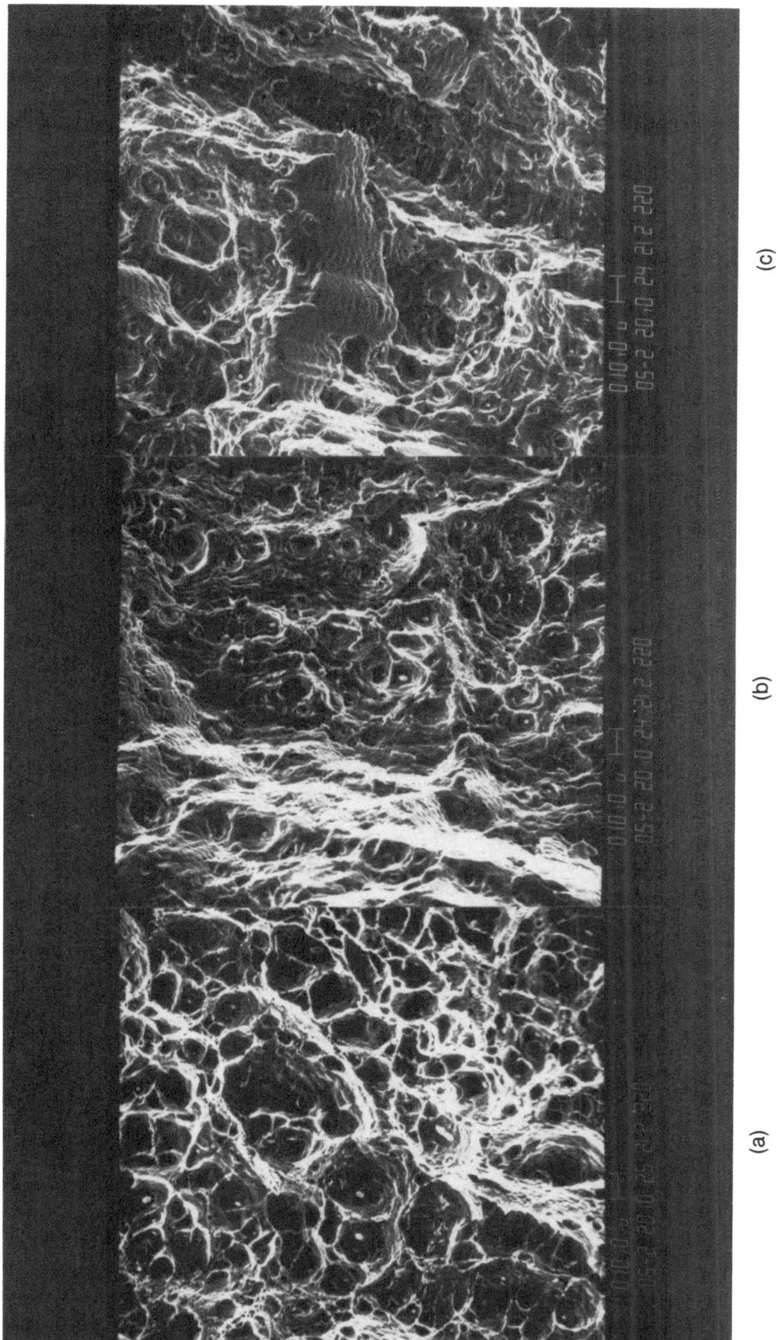

FIG. 12—*Photomicrographs of fracture surfaces showing combination of fatigue damage and ductile tearing found on these specimens.*

tigue damage than the greater number of 50% unloading cycles in accordance with the damage criterion presented above and the experimental results discussed in the previous section.

Conclusions

1. For a load history typical of a standard unloading compliance J-R curve test, a straightforward superposition of crack growth due to ductile tearing and crack growth due to the cyclic unloadings seems to correspond well with measured total crack extension.
2. This simple model thus can be used to develop a criterion to limit cyclic damage encountered during unloading compliance J-R curve tests, and the criterion developed herein was found to correspond well with experimental results obtained on one structural steel alloy.
3. The present limitation of 25% P_L used by the recommended ASTM J-R curve test procedure appears adequate to keep cyclic effects from entering the resulting J-R curve.

References

[1] Albrecht, P. et al., "Tentative Test Procedure for Determining the Plane Strain J_I-R Curve," *Journal of Testing and Evaluation,* Vol. 10, No. 6, Nov. 1982, pp. 245-251.

[2] Sutton, G. S. and Vassilaros, M. G., "Study of Affects of Elastic Unloadings on the *J-R* Curves from Compact Specimens," presented at the 17th National Symposium on Fracture Mechanics, Albany, NY, July 1984.

[3] Kaiser, S., "On the Relation Between Stable Crack Growth and Fatigue," *Fatigue of Engineering Materials and Structures,* Vol. 6, No. 1, 1983.

[4] Paris, P. C., "Fatigue—An Interdisciplinary Approach" in *Proceedings,* 10th Sagamore Conference, Syracuse University Press, Syracuse, NY, 1964, p. 107.

[5] Merkle, J. G. and Corten, H. T., "A J-Integral Analysis for the Compact Specimen, Considering Axial Force as well as Bending Effects," *Journal of Pressure Vessel Technology,* Nov. 1974, pp. 286-292.

[6] Rolfe, S. T. and Barsoum, J. M., *Fracture and Fatigue Control in Structures,* Prentice Hall, Englewood Cliffs, NJ, pp. 237-243.

[7] Joyce, J. A. and Gudas, J. P., "Computer Interactive J_{Ic} Testing of Navy Alloys" in *Elastic-Plastic Fracture, ASTM STP 668,* American Society for Testing and Materials, Philadelphia, 1979, pp. 451-468.

[8] Landes, J. D. and McCabe, D. E. in *Elastic-Plastic Fracture: Second Symposium, Vol. II—Fracture Resistance Curves and Engineering Applications, ASTM STP 803,* C. F. Shih and J. P. Gudas, Eds., American Society for Testing and Materials, Philadelphia, 1983, pp. II 723-738.

[9] Landes, J. D. and Leax, T. R. in *Fracture Mechanics: Fifteenth Symposium, ASTM STP 833,* R. J. Sanford, Ed., American Society for Testing and Materials, Philadelphia, 1984, pp. 436-448.

Analysis

J. C. Newman, Jr.,[1] *B. C. Booth,*[2] *and K. N. Shivakumar*[3]

An Elastic-Plastic Finite-Element Analysis of the *J*-Resistance Curve Using a CTOD Criterion

REFERENCE: Newman, J. C., Jr., Booth, B. C., and Shivakumar, K. N., **"An Elastic-Plastic Finite-Element Analysis of the *J*-Resistance Curve Using a CTOD Criterion,"** *Fracture Mechanics: Eighteenth Symposium, ASTM STP 945,* D. T. Read and R. P. Reed, Eds., American Society for Testing and Materials, Philadelphia, 1988, pp. 665–685.

ABSTRACT: The tentative test procedure for determining J-R curves (crack-growth resistance curves based on the J-integral) on compact and three-point bend specimens showed a discrepancy between specimen types: the bend specimen gave a higher J-R curve than the compact specimen. The purpose of this paper is to investigate this discrepancy, numerically, by simulating the fracture process on compact and bend specimens using a two-dimensional finite-element analysis and a critical crack-tip opening displacement (CTOD) criterion.

An elastic-plastic (incremental and small strain) finite-element analysis was used with the critical CTOD criterion to model stable crack growth in compact specimens made of HY-130 steel. A combined analysis between plane-stress and plane-strain conditions was found to simulate the fracture process quite well. In the combined analysis, plane-strain conditions were imposed around the crack tip while plane-stress conditions were imposed elsewhere. The same analysis was then used to predict stable crack growth behavior of bend specimens. J-R curves were then calculated from the numerical results for each specimen type using several different methods. In general, the J-R curves obtained from the bend specimens were higher than those obtained from the compact specimens, especially beyond maximum load. Up to maximum load, however, the modified deformation theory of plasticity J_D^* and the contour-integral J_Γ methods gave essentially the same J-R curves for both specimen types.

KEY WORDS: fracture, cracks, plastic deformation, stress analysis, finite-element method, J-integral, crack-tip opening displacement

Nomenclature

- a Crack length, m
- a_0 Initial crack length, m
- B Specimen thickness, m
- B_n Specimen thickness minus side-groove depths, m
- b Uncracked ligament length (w-a), m
- b_0 Initial uncracked ligament length (w-a_0), m
- d Minimum element size along crack line, m
- E Young's modulus, N/m^2
- J Path independent integral, N/m
- J_D J computed from deformation theory of plasticity, N/m
- J_D^* J computed from modified deformation theory of plasticity, N/m

[1]Senior scientist, NASA Langley Research Center, Hampton, VA 23665.
[2]Former graduate student, The Johns Hopkins University, Baltimore, MD 21218.
[3]Senior scientist, Analytical Services and Materials, Inc., Hampton, VA 23666.

J_e Elastic component of J, N/m
J_p Plastic component of J, N/m
$J_{\Delta A}$ J computed from change in total potential energy, N/m
J_Γ J computed from contour in finite-element analysis, N/m
K Stress-intensity factor, N/m$^{3/2}$
n Ramberg-Osgood strain-hardening power
P Applied load, N
P_f Maximum load, N
r_m Mean radius of contour, m
V Load-point displacement, m
V^p Plastic component of load-line displacement, m
V_{LL} Load-line displacement, m
w Specimen width, m
Γ Contour for J-integral
Δa Physical crack extension, m
δ_c Critical crack-tip opening displacement, m
ϵ Uniaxial strain
κ Ramberg-Osgood strain-hardening coefficient, N/m^2
ν Poisson's ratio
σ Uniaxial stress, N/m^2
σ_u Uniaxial tensile strength, N/m^2
σ_{ys} Uniaxial yield stress (0.2% offset), N/m^2

One of the primary goals of fracture mechanics is to develop methodologies to predict crack growth and failure of flawed structural components under given operating conditions. The stress-intensity factor [*1*] has provided an accurate method for evaluating fracture toughness for brittle materials. However, for metallic materials exhibiting large amounts of plastic deformation and stable crack growth during fracture, the use of the stress-intensity factor is limited. For these materials, the J-integral [*2–4*] provides an analogous method of measuring fracture toughness.

Currently, the plane-strain J-integral resistance (J-R) curve is used to assess the fracture safety of nuclear pressure vessels and piping [*5*]. To evaluate the current procedure and to develop an improved method, a task group of ASTM Committee E 24 initiated the development of a test procedure for determining the plane-strain J-R curve. This test procedure was developed to characterize the slow-stable crack growth resistance of bend-type specimens in such a manner that the J-R curve is configuration insensitive under conditions of full plane-strain constraint [*6*].

In 1980, a round-robin test program was conducted on HY-130 steel plate [*7*] to measure J-R curves. This program, conducted at 19 laboratories in the United States, Canada, and Europe, tested both compact and three-point bend specimens each with side grooves. The configurations are shown in Fig. 1. The results from the various laboratories showed that the compact specimen provided closely grouped J-R curves. The three-point bend specimen results, however, were not as clearly defined as those of the compact specimen. Of major concern was the apparent difference between the average J-R curves of the two specimens. Figure 2 shows J-R curves on a compact and bend specimen tested at the same laboratory. (The bend test results from E. Hackett, David W. Taylor Naval Ship Research and Development Center, are unpublished.) When the deformation theory of plasticity J_D formulation was used, the bend specimen gave a resistance curve that was higher than the results from the compact specimen [*7*]. This lack of uniqueness questioned the adequacy of the test procedure and the J calculation method in measuring a configuration insensitive J-R curve.

The purpose of this study was to analyze numerically the discrepancy between the compact and bend specimen J-R curves. The approach was to accurately model the fracture process in the two specimens and then to assess the various methods of calculating the J-integral.

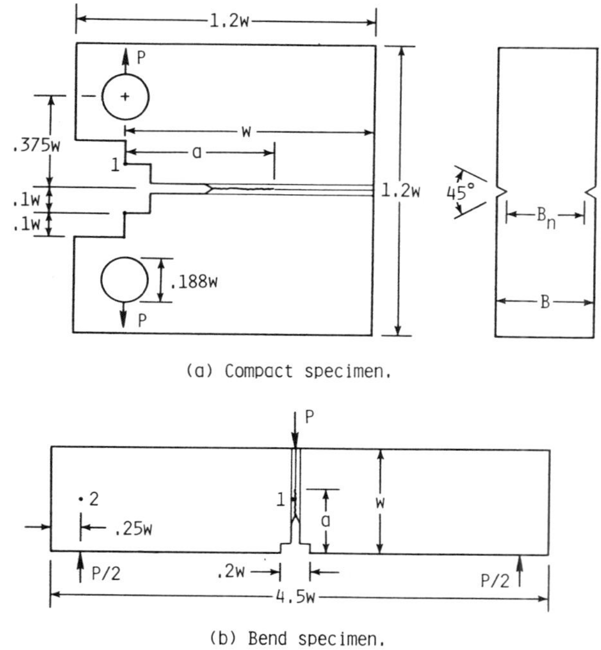

FIG. 1—*Compact and three-point bend specimens.*

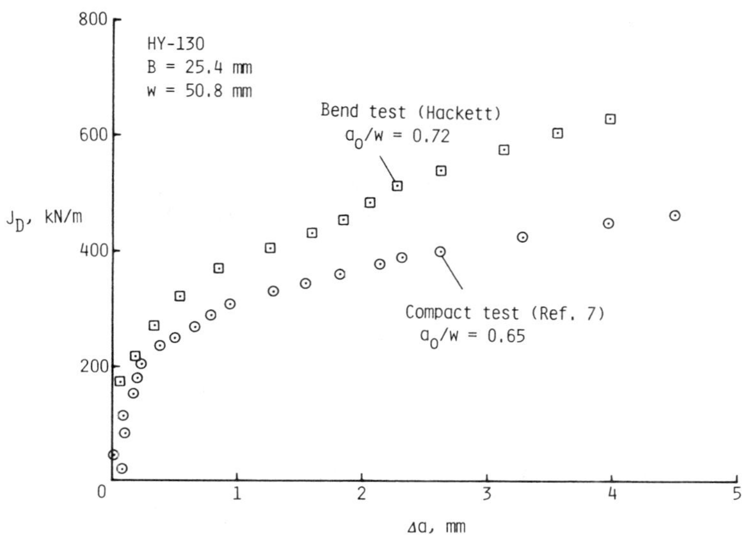

FIG. 2—J_D *resistance data for compact and bend specimens.*

The stable crack growth behavior of the two specimens were analyzed using a two-dimensional, elastic-plastic, finite-element computer program developed at the NASA Langley Research Center [8]. A critical crack-tip opening displacement (CTOD) criterion [9] was used in the analysis to model stable crack growth behavior in compact specimens made of HY-130 steel. The same analysis was then used to predict stable crack growth behavior in bend specimens. Comparisons were made between experimental and predicted load-against-crack-extension and load-against-displacement behavior. Comparison were also made between experimental and predicted maximum loads on both specimen types.

J-R curves were then calculated from the numerical results for each specimen type using three different methods. They were the deformation theory of plasticity formulation J_D, a modified deformation theory of plasticity formulation, J_D^*, and the contour integral J_Γ. Comparisons were made among these J-evaluation methods for both specimen types. The variation in total potential energy was also used to calculate the J-integral for stationary cracks to assess the accuracy of the J-integral calculations.

Analysis

Finite-Element Method

The elastic-plastic analysis of the two specimen types (Fig. 1) employed the finite-element method and the initial-stress concept [9,10]. The elastic-plastic analysis was based on incremental flow theory with a small strain assumption. The finite-element models for these specimens were composed of two-dimensional, constant strain, triangular elements under either plane-stress or plane-strain conditions (Fig. 3). Mesh patterns were selected so that the minimum element size along the line of crack extension would be the same in all specimens. From previous studies [9], a mesh size d of 0.4 mm was selected. Fictitious springs were used to change

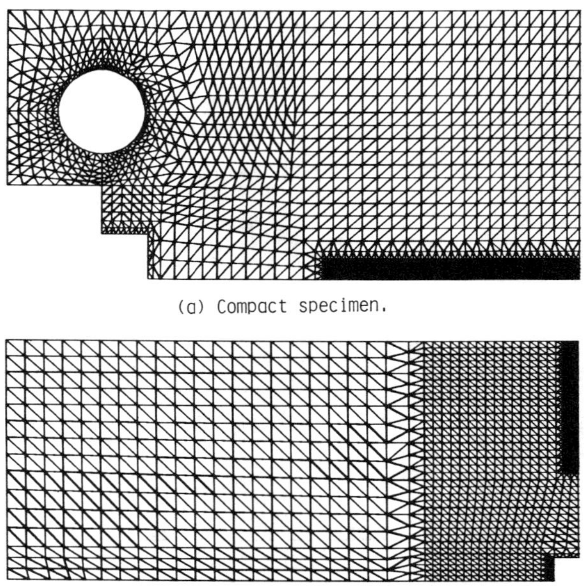

(a) Compact specimen.

(b) Bend specimen.

FIG. 3—*Finite-element models for compact and bend specimens (Mesh I).*

boundary conditions associated with crack extension. For free nodes along the crack line, the spring stiffnesses were set equal to zero, and, for fixed nodes, the stiffnesses were assigned extremely large values. (See Refs *8* or *9* for details in the elastic-plastic finite-element analysis with crack extension.)

Because the specimens tested were side grooved by 20% ($B_n = 0.8B$), the thicknesses of the elements in the side-groove region were reduced by an amount equal to the side grooving in the actual specimen. This was done by changing the volume integration in the element stiffness matrix calculation. A further modification was made in the analysis program to specify plane-stress or plane-strain conditions for individual elements. This allowed for plane-strain conditions to be imposed around the crack tip while elements remote from the crack tip were under plane-stress conditions. Herein, this type of analysis is referred to as a "mixed state-of-stress formulation" and is denoted by the acronym (P_σ/P_ϵ).

The crack-growth criterion used was a critical crack-tip-opening displacement δ_c at a specified distance d from the crack tip. The distance d was the element size along the crack line—in other words, d was the distance between the crack tip and the first free node along the crack surface. [The critical CTOD criterion is also equivalent to a critical crack-tip opening angle (CTOA) criterion, since CTOA $= 2\tan^{-1}(\delta_c/2d)$.] During incremental loading under displacement-control, whenever the CTOD equaled or exceeded a preset critical value δ_c, the crack-tip node was released (see Ref *9* for details) and the crack advanced to the next fixed node. This process was repeated until the desired crack length was reached. The use of the CTOD (or CTOA) criterion does require that the absolute size d and arrangement of elements in the crack-tip region and along the line of crack extension be the same in all crack configurations considered.

The average material stress-strain curve used in the finite-element analysis was approximated by the Ramberg-Osgood equation [*11*] as

$$\epsilon = \frac{\sigma}{E} + \left(\frac{\sigma}{\kappa}\right)^n \quad (1)$$

where κ and n are the strain-hardening coefficient and power, respectively. For the HY-130 steel plate, κ was 1200 MPa and n was 21. The yield stress (σ_{ys}) was 945 MPa, the ultimate tensile strength (σ_u) was 970 MPa, and Young's modulus (E) was 200 GPa.

J-Integral

Three methods were used to calculate the J-integral from the results of the finite-element simulation of fracture, namely, the contour integral J_Γ [*2–4*], the deformation theory of plasticity J_D formulation [*12*], and a modified deformation theory of plasticity J_D^* formulation [*12*]. Mathematical expressions for the three methods are briefly described in Appendix I. An evaluation of the elastic and elastic-plastic calculation of the J-integrals from the finite-element analyses is also given in Appendix II.

Finite-Element Simulation of Fracture

The intent of this section was to model the load-against-crack extension and load-against-load-line displacement behavior of the compact and bend specimens. The elastic-plastic finite-element analysis and the critical CTOD criterion were used to simulate the fracture process. The compact specimen was first analyzed to establish the critical CTOD for the material. The same critical CTOD value was then used to predict the fracture behavior of bend specimens. Both experiments and analyses were performed under displacement-control conditions. Comparisons are made with experimental data.

The compact specimen was initially analyzed under plane-strain conditions using Mesh I, as shown in Fig. 3*a*. A critical CTOD value ($\delta_c = 0.048$ mm) was selected by trial-and-error to match the maximum load of the experiment. The load plotted against physical crack extension from the experiment and analysis is shown in Fig. 4. The open symbols show crack extension values measured by the unloading compliance method, and the solid symbol shows the average crack extension value measured on the broken specimen. Beyond maximum load, the plane-strain analysis gave more crack extension than was found in the experiment. Using a more refined model (Mesh II) under plane-strain conditions gave essentially the same results beyond maximum load (dashed curve). The load at initiation of crack growth was lower using Mesh II ($d = 0.2$ mm) than that obtained using Mesh I ($d = 0.4$ mm). (Mesh II was obtained from Mesh I by subdividing each element into four elements by connecting midsides, see Appendix II.) The δ_c value for Mesh II was also selected by matching maximum load. Under plane-stress conditions, the results from an analysis using Mesh I ($\delta_c = 0.128$ mm) agreed well with the experimental results. However, the load at initiation was higher than the experimental load and was very close to the maximum load.

Comparisons among the experimental and numerical load-against-load-line displacement records are shown in Fig. 5. Calculations were made under plane-stress or plane-strain conditions using Mesh I or II. The solid curves show results from Mesh I and the dashed curves from Mesh II. Plane-stress results using Mesh I agreed well with the experimental data up to maximum load. Beyond maximum load, however, the plane-stress calculations gave much larger displacements than the experiment. Plane-stress results using Mesh II also gave larger displacements than the experiment. Under plane-strain conditions, the results from Mesh I and II gave much lower displacements than the experiment. These results suggest that the fracture process is three-dimensional in nature and could possibly be simulated by a "mixed state-of-stress" formulation between plane-stress and plane-strain conditions. In subsequent simulations of fracture, only Mesh I will be used.

The plane-strain plastic-zone regions for a compact specimen at maximum load are shown in Fig. 6. This figure shows the plastic yielding at the crack tip and the compressive yielding at the

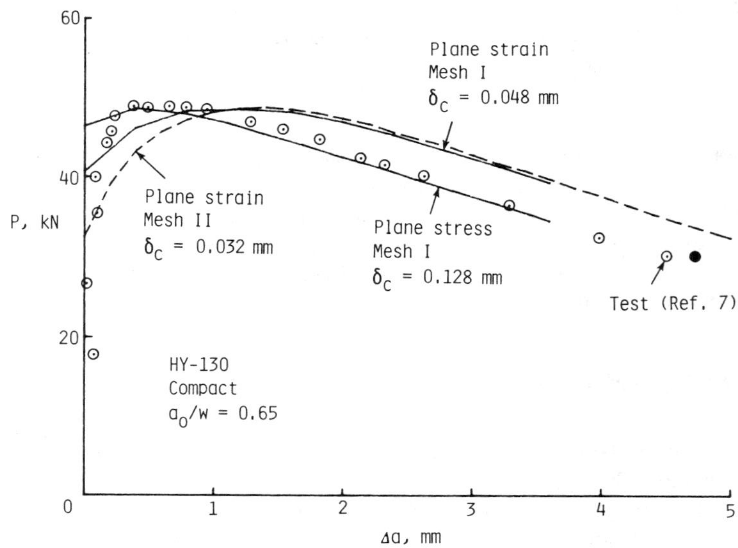

FIG. 4—*Comparisons of experimental and finite-element analyses of crack growth in a compact specimen.*

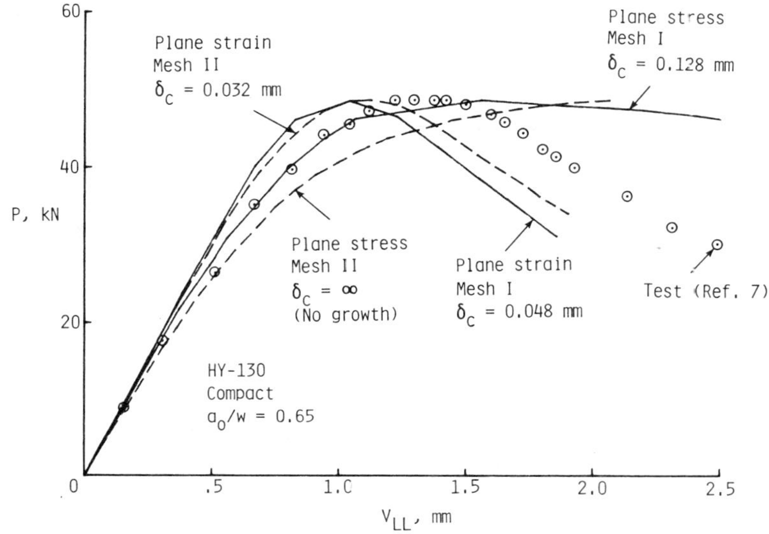

FIG. 5—*Comparisons of experimental and finite-element analyses of load-line displacement records on a compact specimen.*

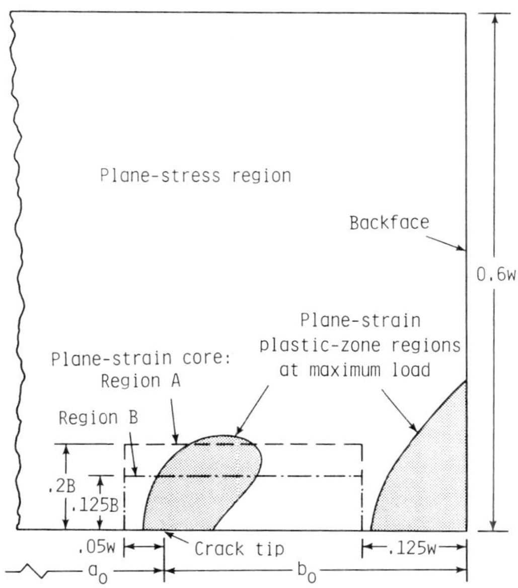

FIG. 6—*Definitions of mixed state-of-stress simulations.*

backface. In the proposed mixed state-of-stress formulation, plane-stress conditions will be assumed everywhere except for a plane-strain core around the crack tip. The dashed and dash-dot lines show the two assumed plane-strain cores, Regions A and B, respectively. Elements which have their centroid within these core regions will have plane-strain conditions. Region A was selected to enclose most of the plane-strain yield zone as the crack grows. Region B was selected because 0.125B corresponds to the definition of the plastic-zone size for plane-strain fracture conditions [ASTM Test Method for Plane-Strain Fracture Toughness of Metallic Materials (E 399-83)]. Both of these mixed state-of-stress formulations, denoted as $(P_\sigma/P_\epsilon)_A$ and $(P_\sigma/P_\epsilon)_B$, will be used to simulate fracture on the compact and bend specimens.

Figure 7 shows the experimental load-against-crack-extension results (symbols) measured on two compact specimens with the same initial crack length ($a_0/w = 0.65$). These tests, conducted by two different laboratories, were part of the round-robin test program [7]. The open symbols show results obtained by unloading compliance, and the solid symbols show the average crack extension values measured from the broken specimens. The solid and dashed curves show the calculations using Mesh I with the mixed state-of-stress (P_σ/P_ϵ) Conditions A and B, respectively. Both calculations agreed well with the experimental data except at the initiation load.

Comparisons among the experimental [7] and calculated load-against-load-line displacement records for the compact specimen are shown in Fig. 8. The load-line displacement (V_{LL}) was measured (symbols) and calculated (curves) at Point 1 in Fig. 1*a*. In the elastic regime, both analyses gave slightly lower displacements (about 5%) than the tests because the finite-element method used gives a lower bound on displacements. For all load levels, the analysis using Condition A (solid curve) gave lower displacements than the tests. Although calculations using Condition B were in better agreement with the test at high values of displacement, the shape of the load-displacement record was not as good as that shown by Condition A. In comparison with the load-displacement calculations shown in Fig. 5, these calculations were considered adequate.

In the round-robin test program [7], the initial-crack-length-to-width (a_0/w) ratios ranged from 0.62 to 0.7 for the compact specimens. Figure 9 shows the maximum load P_f plotted

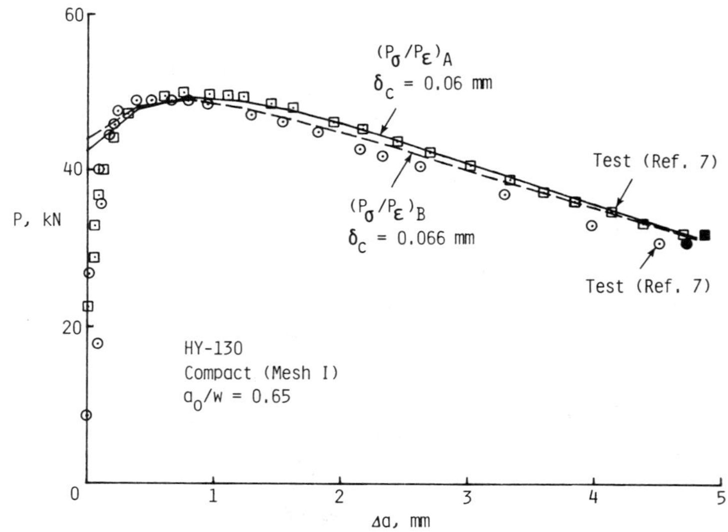

FIG. 7—*Comparisons of experimental and mixed state-of-stress analyses of crack growth in compact specimens.*

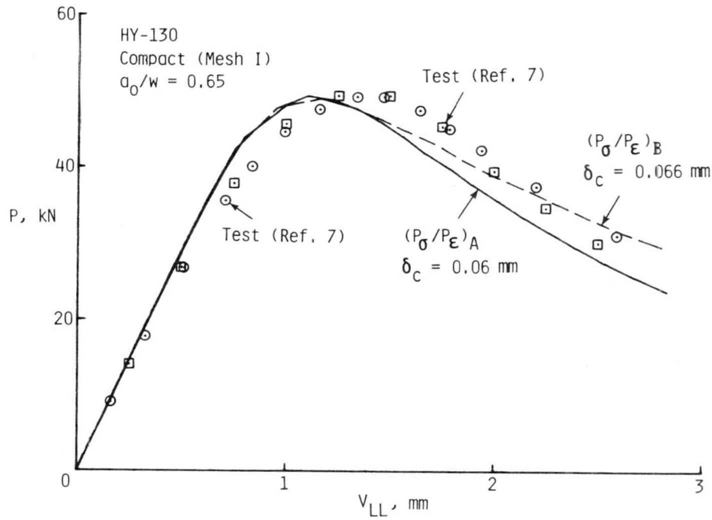

FIG. 8—*Comparisons of experimental and mixed state-of-stress analyses of load-line displacements on compact specimens.*

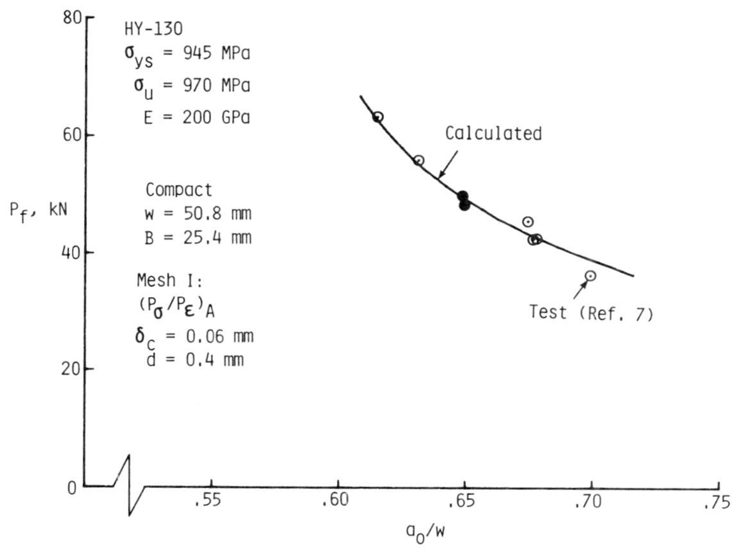

FIG. 9—*Comparison of experimental and calculated maximum loads on compact specimens as a function of* a_0/w.

against a_0/w for these tests. The solid symbols show the two test results that were used to obtain δ_c. The open symbols show the results from the other tests. The curve shows the calculations from the mixed state-of-stress formulation (Condition A). The calculations agreed well with the experimental data.

Because the critical CTOD value was determined from compact specimens, the good maximum load predictions on other compact specimens may have been fortuitous. To test the CTOD criterion for specimen type dependence, bend specimens were also analyzed.

Figure 10 shows comparisons between experimental and predicted load-against-crack-extension behavior on bend specimens. The symbols show test data on two bend specimens (from the same laboratory) with nearly the same initial-crack-length-to-width ratio (see Ref 7). Predictions made using the mixed state-of-stress Conditions A and B are shown by the solid and dashed curves, respectively. The analyses tended to predict higher loads (about 10%) than the tests for a given amount of crack extension.

A comparison between the experimental and predicted load-against-displacement behavior of the bend specimen is shown in Fig. 11. The load-line displacement V_{LL} was measured (symbols) and predicted (curves) from the difference in vertical displacements at Points 1 and 2 in Fig. 1*b*. In the elastic range, the measured displacements were considerably less than those predicted by the analysis. The measured displacements, however, should have been slightly greater (about 5%) than the predictions. Beyond maximum load, the predicted results were in fair agreement with the test data.

Figure 12 shows experimental and predicted maximum loads on bend specimens tested in the round robin test program [*7*]. The predicted results agreed well with the test data. This shows that the CTOD criterion can be used to predict maximum loads on one specimen type using results obtained from another type. Results shown in this section also demonstrate that the CTOD criterion and the mixed state-of-stress formulation accurately model the fracture behavior of side-grooved HY-130 steel specimens. These results provide a basis for evaluating the various methods of calculating J-R curves.

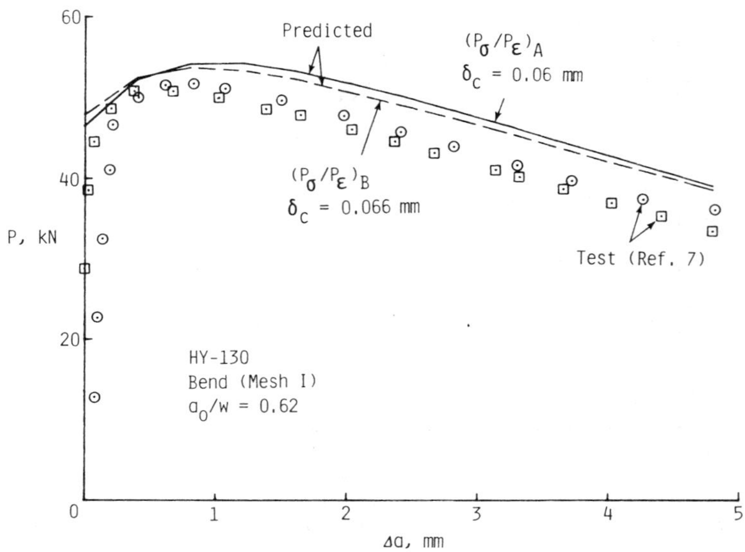

FIG. 10—*Comparisons of experimental and predicted crack growth in bend specimens using the mixed state-of-stress analyses.*

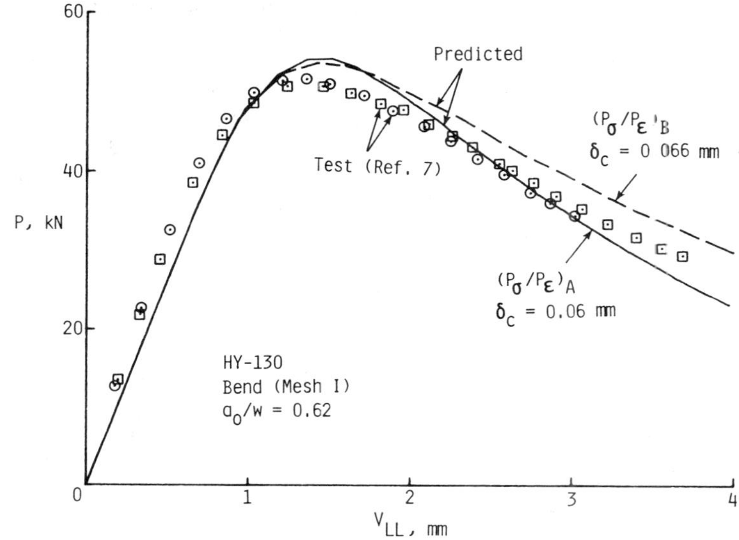

FIG. 11—*Comparisons of experimental and predicted load-line displacements on bend specimens using the mixed state-of-stress analyses.*

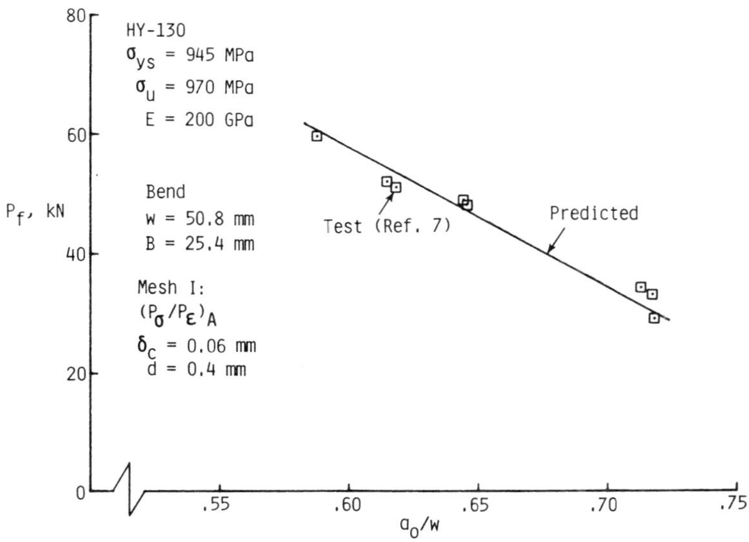

FIG. 12—*Comparison of experimental and predicted maximum loads on bend specimens as a function of* a_0/w.

J-R Curve Evaluations

In the following, the predicted load-against-crack-extension and load-against-displacement behavior of compact and bend specimens were used to calculate *J*-resistance curves. Three different methods of calculating the J-integral were investigated. They were the deformation theory of plasticity J_D, a modified deformation theory of plasticity J_D^*, and the contour integral J_Γ. The tentative test procedure for determining J-R curves developed by Albrecht et al. [*6*] and used in the round robin test program [*7*] was based on the J_D formulation. J_D^* was developed because J_D did not reduce to the elastic solution. The contour integral J_Γ was evaluated on Contour 6, as shown in Fig. 13. Contour 6 is identical in both the compact and bend specimens except for the exclusion of a small region near the load application point in the bend specimen. Appendix II develops the rational for selecting Contour 6 and verifies the use of J_Γ for growing cracks.

From the finite-element simulation of fracture using the mixed state-of-stress formulation, *J*-resistance curves based on J_D were calculated from Eq 4 in Appendix I and are shown in Fig. 14. The solid and dashed curves show J_D plotted against Δa for the compact and bend specimens, respectively. These results show that the bend specimen gives a higher resistance curve than the compact specimen, similar to that shown by the experimental data in Fig. 2. The magnitude of the computed resistance curves, however, were lower than the experimental curves (Fig. 2) because the finite-element method underestimates the load-line displacements (see Fig. 8). The relative difference between the two specimens, however, is expected to be correct because the element sizes and arrangements in the cracked region of both specimens are identical. The arrows on Fig. 14 indicate the amount of crack extension at maximum load and at 10% of the initial uncracked ligament length (b_0).

The modified deformation theory J_D^* formulation (Eq 7 in Appendix I) was used to compute the *J*-resistance curves for the compact and bend specimens, shown in Fig. 15. Here again, the bend specimen gives a higher resistance curve than the compact specimen. But the resistance curves were in fair agreement with each other for small amounts of crack extension. The slight

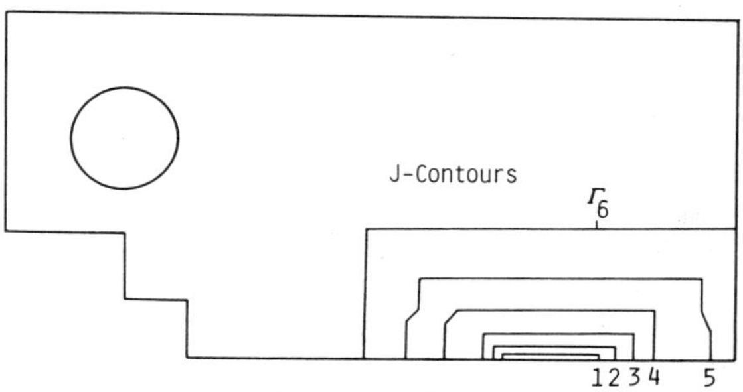

(a) Compact specimen.

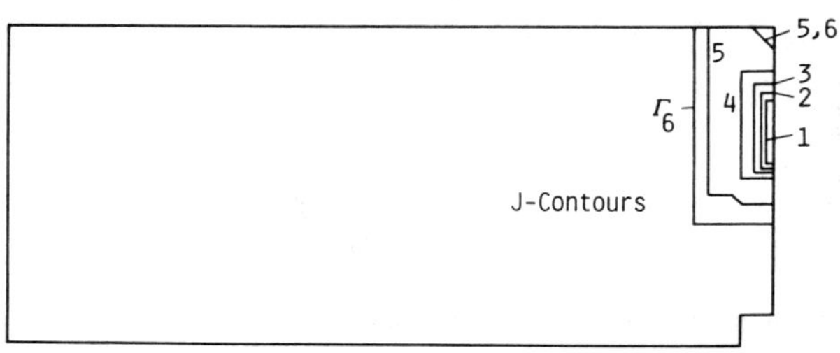

(b) Bend specimen.

FIG. 13—*Contours for J-integral evaluations on compact and bend specimens.*

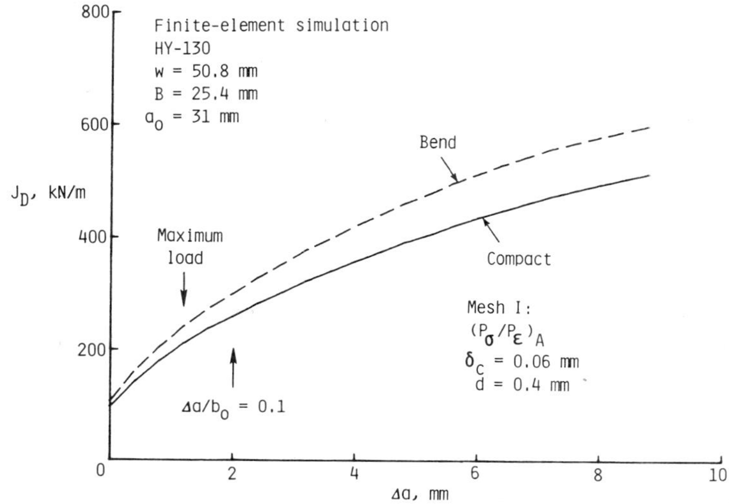

FIG. 14—*Computed* J_D *resistance curves for compact and bend specimens using deformation theory.*

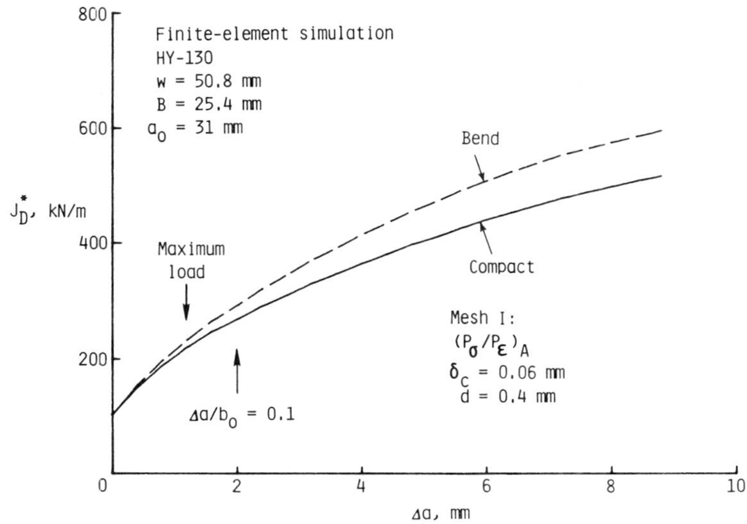

FIG. 15—*Computed* J_D^* *resistance curves for compact and bend specimens using modified deformation theory.*

improvement in correlation between resistance curves, over that shown for the J_D formulation, was due to expressing J in terms of the elastic and plastic components.

A comparison between the resistance curves computed from remote contour integrals are shown in Fig. 16 for both specimens. Up to maximum load, the J_Γ curves computed from Contour 6 (see Fig. 13) agreed well. Beyond maximum load, the bend specimen, again, gives a higher resistance curve than the compact specimen. The contour-integral method also gave higher values of J than either J_D or J_D^* for large amounts of crack extension ($\Delta a/b_0 > 0.1$). Also, beyond maximum load, the relationship between J_Γ and Δa was almost linear.

The results shown in Fig. 16 indicate that the η-factors used in J_D and J_D^* calculations need to be changed to give better correlation between specimen types, at least, up to maximum load.

Conclusions

The simulation of fracture using the finite-element method and the crack-tip opening displacement criterion for compact and bend specimens made of HY-130 steel support the following conclusions:

1. Plane-stress and plane-strain analyses bound the experimental load-against-load-line displacement results on compact specimens.
2. An analysis with a mixed state-of-stress formulation combining plane-stress and plane-strain conditions accurately models the fracture behavior of compact and bend specimens.
3. The CTOD fracture criterion correlates and accurately predicts the maximum load behavior of compact and bend specimens.
4. The J-resistance curves based on deformation theory J_D are specimen-type dependent.
5. The J-resistance curves based on modified deformation theory J_D^* and on the contour integral J_Γ are insensitive to specimen type up to maximum load.
6. Beyond maximum load, the J-resistance curves based on deformation theory J_D, modified deformation theory J_D^*, and the contour integral J_Γ are specimen-type dependent.

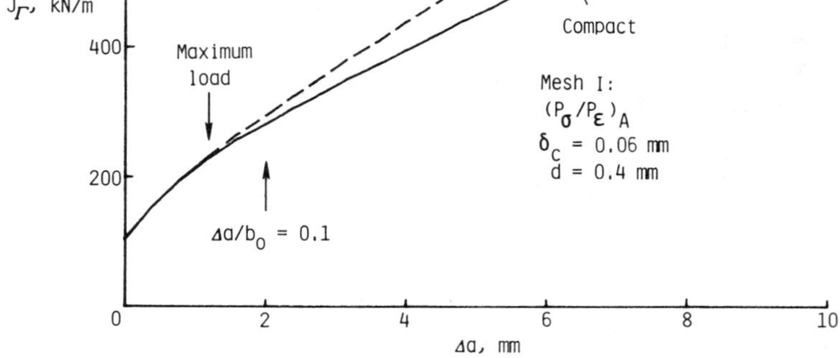

FIG. 16—*Computed* J_Γ *resistance curves for compact and bend specimens using the contour integral.*

Acknowledgments

The authors would like to thank Dr. John P. Gudas, David W. Taylor Naval Ship Research and Development Center, for his encouragement and financial support to undertake this project, and the second author would like to especially thank Dr. Andrew Douglas, The Johns Hopkins University, for his guidance in conducting this research.

APPENDIX I

J-Integral Calculation Methods

The three methods used to calculate the J-resistance curves from the finite-element simulations of fracture are briefly described herein. The method used to calculate J from the variation in total potential energy is also presented.

Contour Integral J_Γ

Rice [2], Cherepanov [3], and Eshelby [4] derived a path independent integral as

$$J_\Gamma = \int_\Gamma \left\{ W dy - T \frac{\partial u}{\partial x} ds \right\} \tag{2}$$

where

W = strain energy density,
T = surface traction on contour Γ,
u = displacement vector,
s = arc length along the contour, and
x, y = cartesian coordinate system with the crack along the x-axis.

J_Γ is a measure of the deformation field around the crack tip and is path independent within the "proportional" loading region [2]. Shih et al. [13] have shown that the path-independent integral is also valid for a growing crack if the contour is in the nearly proportional loading region.

J_D and J_D^* Integrals

Ernst et al. [12] developed a simple analytical expression to calculate the J-value using the load, load-line displacement, and the current crack length data from a single specimen test. These equations were developed by using elastic-plastic finite-element results from Kumar et al. [14]. The first equation, based on the deformation theory of plasticity, gives J_D as

$$J_D = \frac{1}{B} \int_0^V \frac{\gamma}{b} P dV \tag{3}$$

where V is the total load-line displacement and γ is a dimensionless coefficient which depends upon specimen type. For an incremental analysis, Eq 3 was written as

$$J_{D_{i+1}} = \left[J_{Di} + \left(\frac{\eta}{b}\right)_i \frac{A_{i,i+1}}{B} \right] \left[1 - \left(\frac{\gamma}{b}\right)_i (a_{i+1} - a_i) \right] \tag{4}$$

where

$\left.\begin{array}{l}\eta = 2 + 0.522\ b/w \\ \gamma = 1 + 0.766\ b/w\end{array}\right\}$ for compact specimen,

$\left.\begin{array}{l}\eta = 2 \\ \gamma = 1\end{array}\right\}$ for three-point bend specimen, and

$A_{i,i+1}$ = area under the load-against-load-line displacement record between displacements V_i and V_{i+1}.

The last term in brackets in Eq 4 represents the correction for crack growth in the ith interval. The J_D equation, however, fails to reduce to the elastic solution because of the use of the plastic η-factors.

Ernst et al. [12] also modified Eq 4 by expressing J as the sum of the elastic and plastic components. The modified J_D^* expression is

$$J_D^* = J_e + \frac{1}{B} \int_0^{V_p} \frac{\gamma}{b} P\, dV^p \tag{5}$$

where

$$J_e = \frac{K^2}{E}(1 - \nu^2) \tag{6}$$

for plane-strain conditions and V^p is the plastic component of the load-line displacement. For an incremental analysis, Eq 5 becomes

$$J_{D_{i+1}}^* = J_{e_{i+1}} + \sum_{j=1}^{i} \left[\left(\frac{\eta}{b}\right)_j \frac{A_{j,j+1}^p}{B}\right]\left[1 - \left(\frac{\gamma}{b}\right)_j (a_{j+1} - a_j)\right] \tag{7}$$

where $A_{j,j+1}^p$ is the area under the load against load-line plastic displacement record between displacements V_j^p and V_{j+1}^p.

Variation in Total Potential Energy $J_{\Delta A}$

For a stationary crack, Rice [2] showed that the path integral (Eq 2) reduces to the variation of total potential energy (Π) for a cracked plate. Therefore, J is the area between the load-against-load-point displacement records for two different crack lengths differing by a small increment. This method of evaluating J is

$$J_{\Delta A} = \frac{\partial \Pi}{B \partial a} = \frac{\partial}{B \partial a} \int_0^V P dV \tag{8}$$

where P is the load and V is total load-point displacement. In the present numerical analysis, Eq 8 is written as

$$J_{\Delta A} = \sum_{i=1}^{N} \{(P_a^{i+1} + P_a^{i+1})\Delta V_a^i - (P_{a'}^i + P_{a'}^{i+1})\Delta V_{a'}^i\}/(2B\Delta a) \tag{9}$$

where

P^i and P^{i+1} = loads at the beginning of intervals i and $i + 1$, respectively,
ΔV_a^i = incremental displacement of interval i, and
N = total number of intervals to reach a fixed value of displacement.

The subscripts a and a' refer to different crack lengths. The incremental crack length Δa ($= a' - a$) in the present analysis was 0.8 mm (two element sizes).

APPENDIX II

Elastic and Elastic-Plastic Evaluation of J-Integral

The purpose of this appendix is to establish the accuracy of the finite-element meshes, to select contours for J_Γ calculations, and to evaluate J-integrals from various methods using stationary and growing cracks. The contour integral J_Γ and the deformation theory integrals J_D and J_D^* are evaluated and compared with $J_{\Delta A}$ calculated from the variation in total potential energy.

Plane-strain finite-element analyses of a compact specimen made of HY-130 steel were made using two different meshes (denoted I and II). Constant strain triangular elements were used in both meshes. Mesh I is shown in Fig. 3*a*; Mesh II was obtained by subdividing each element in Mesh I into four elements by connecting midpoints of the three sides. The number of elements and nodes in Mesh I is 2688 and 1462, respectively, and in Mesh II is 10752 and 5612, respectively. Six contours were selected in Mesh I (see Fig. 13) and seven contours in Mesh II for calculating J_Γ. The first five contours in Mesh I and II were the same. The sixth and seventh contours had a mean radius (r) 50 and 100% larger than that for Contour 5, respectively. A displacement-controlled analysis was simulated by imposing an applied displacement at the load position P shown in Fig. 1*a*. Elastic nodal displacements, element stresses, and strains were calculated for a unit applied displacement. J_Γ, J_D, and J_D^* were calculated using Eqs 2, 4, and 7, respectively. Load-point displacements were applied incrementally, and an elastic-plastic analysis was performed at each increment. When the CTOD at the first free node from the crack tip equalled or exceeded the critical CTOD (δ_c), the J-values were computed. The crack was then extended by releasing the crack-tip node to advance the crack one element size to the next node. For the new crack-tip location, the applied displacement was again incremented until the CTOD became critical. Again, the J-values were computed. This process was repeated until the desired crack length was reached. A similar analysis was also made for the bend specimen using the same critical CTOD value. Results are discussed in the following sections.

Elastic Analyses

Figure 17 shows elastic J-integral (J_e values plotted against mean-radius-to-initial-crack-length ratio (r_m/a_0) for Meshes I and II. J-values from the three methods (J_Γ, J_D, and J_D^*) are

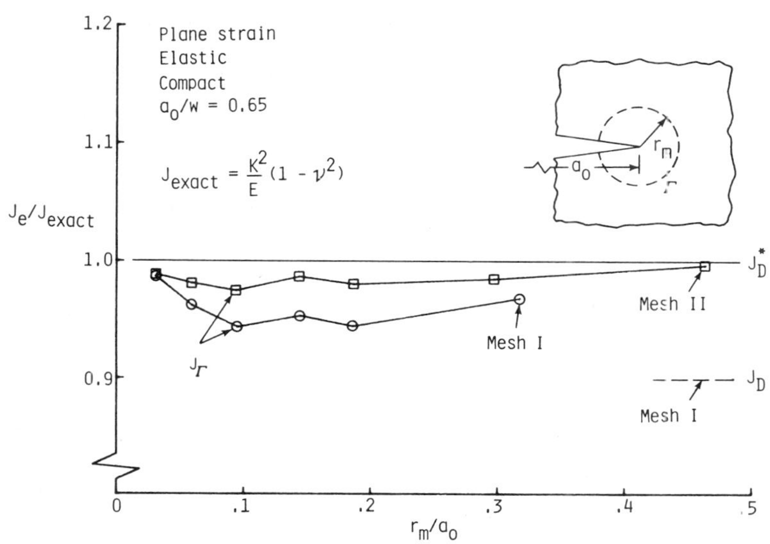

FIG. 17—*Evaluation of* J *calculation methods under elastic conditions for a compact specimen.*

normalized with respect to the exact elastic J-value from Eq 6 using the K expression in Ref *15*. Because the cracked plate was elastic, the J_D^* value was equal to J_{exact}, whereas J_D was about 10% less than J_{exact}. Five percent of this difference was due to the finite-element (Mesh II) being too stiff and the other 5% was due to differences in the elastic and plastic η-factors (see Ref *12*). Except for the first few contours, the J_Γ values become more accurate for larger r_m/a_0 ratios. The sixth contour of Mesh I ($r_m/a_0 = 0.32$) gave a J_Γ value 3% lower than J_{exact}. As expected, Mesh II gave more accurate J_Γ values than those from Mesh I. But Mesh II was much more expensive to run on the computer than Mesh I.

Comparisons among the various J-values computed for compact and bend specimens are shown in Fig. 18 for Mesh I. The analyses were conducted under plane-strain conditions and at an applied load such that the highest stressed element at the crack tip was just at the yield stress of the material for both specimens. The solid and dashed lines show results for the compact and bend specimens, respectively. Because the specimens were elastic, J_D^* is equal to J_e and is given by Eq 6, see upper solid and dashed lines. The K-values were computed from well-known solutions for these specimens [*15*]. The contour integrals J_Γ are shown as a function of r_m/a_0 for values of r_m/a_0 from 0.03 to 0.32. For contours around the crack tip with low r_m/a_0 ratios, the J_Γ values from both specimens agreed well. But J_Γ values were 1 to 5% lower than the exact values for the compact specimen and were 2 to 10% lower for the bend specimen. The J_D value for the deeply-cracked compact specimen was also 10% lower than that for the deeply-cracked bend specimen, as shown on the extreme right side of the figure.

Because J_D was evaluated from load-displacement records, several observations about the J_D formulation for elastic bodies can be made. First, the calculated load-displacement records for both the compact and bend specimens are about 5% too stiff. Thus, the area under the load-displacement records are also 5% too low, and consequently, the J_D calculation would be about 5% lower than the exact values. Because the bend specimen results for J_D are in good agreement with the exact value, this means that the J_D equations for the bend specimen overestimate J_D by about 3%. In contrast, the J_D equations for the compact specimen underestimate J_D by about 7%. Thus, the difference in J_D between the two specimens was 10%.

Elastic-Plastic and Crack-Extension Analyses

The following section evaluates the various methods of calculating J under elastic-plastic conditions with crack extension. Figure 19 shows J plotted against normalized mean radius (r_m/a_0) for a compact specimen under plane-strain conditions for various amounts of crack extension ($\Delta a/b_0$). J is shown by symbols. J_Γ results for $\Delta a/b_0 = 0$ represent the J-values just before crack initiation ($\delta = \delta_c$). Here all contour integrals gave essentially the same value of J_Γ (within 2%). For large $\Delta a/b_0$ values, the contours closest to the crack surfaces (low r_m/a_0 ratios) gave very low J-values because they lie in the region of nonproportional loading and of elastic unloading. These two conditions make the calculation of J_Γ invalid. J_Γ values calculated on contours with $r_m/a_0 \geq 0.3$ (Contour 6) were not adversely affected by crack extensions up to $0.3 b_0$. The contours with $r_m/a_0 > 0.3$ will be used in the text to calculate the J-integral during crack extension.

In Fig. 19, J_D and J_D^* are shown by the solid and dashed lines, respectively, on the extreme right side of the figure. Each pair of lines (solid and dashed) correspond to the indicated value of $\Delta a/b_0$. J_D was consistently lower (5 to 10%) than J_D^*. J_D^* agreed well with J_Γ calculated for $r_m/a_0 > 0.1$ and $\Delta a/b_0 < 0.1$. For larger values of $\Delta a/b_0$, J_D^* was lower than J_Γ (10% at $\Delta a/b_0 = 0.3$) for the remote contour.

Figure 20 shows calculated J resistance curves for an HY-130 steel compact specimen under plane-strain conditions using a critical CTOD criterion ($\delta_c = 0.032$ mm). J was calculated using four different methods and was plotted against crack extension (Δa). The maximum load in the simulated test occurred at $\Delta a/b_0$ of about 0.09. The contour integral (J_Γ) was calculated on Contour 6, and the results are shown as the solid curve. J_D and J_D^* were calculated from the simulated load-against-crack-extension and load-against-displacement records using Eqs 4 and 7, respectively. The J_D curve (dashed) was consistently lower (5 to 10%) than the J_Γ curve, whereas the J_D^* curve (dashed-dot) agreed well with the J_Γ curve up to maximum load. Beyond maximum load, the J_D and J_D^* curves were also lower than the J_Γ curve. At crack extension values of 10 and 25% of the initial ligament length (b_0, $J_{\Delta A}$ values were calculated from the change in potential energy. The $J_{\Delta A}$ values are shown by the symbols. They were obtained by

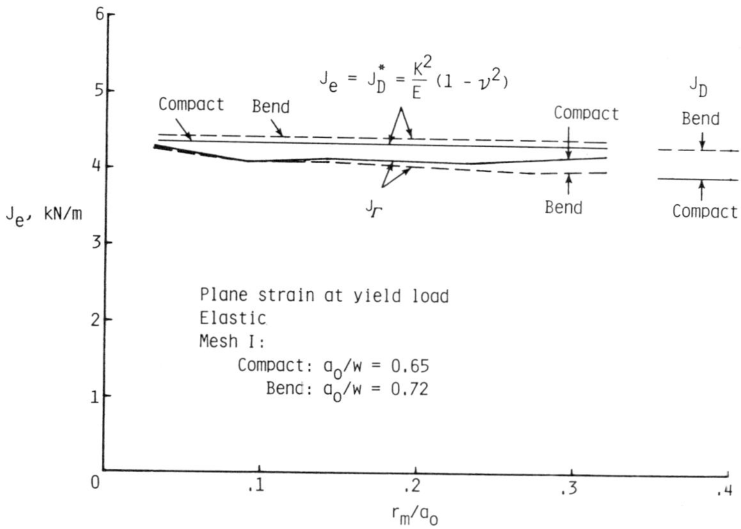

FIG. 18—*Comparison of* J *calculation methods under elastic conditions for compact and bend specimens.*

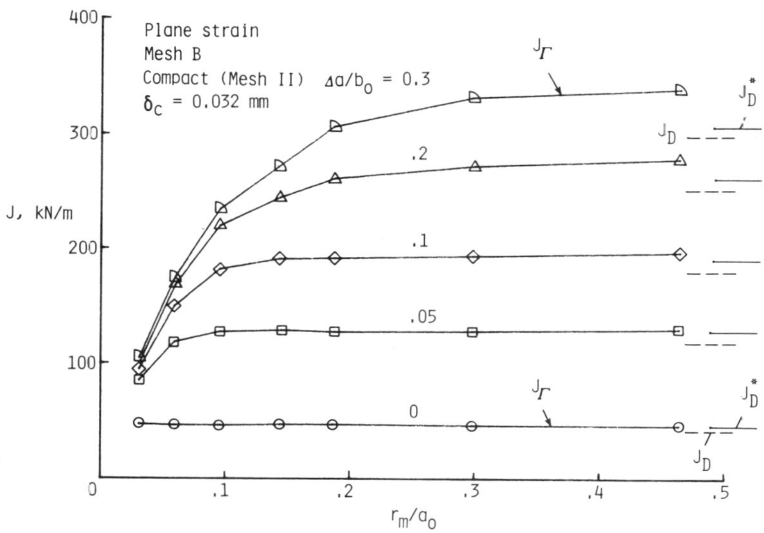

FIG. 19—*Evaluation of* J *calculation methods under elastic-plastic conditions with crack extension for a compact specimen.*

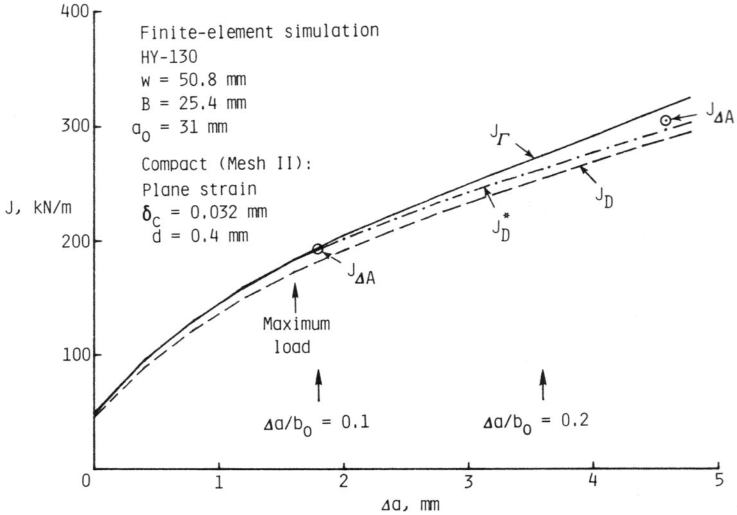

FIG. 20—*Finite-element simulation of various* J*-resistance curves under plane-strain conditions for a compact specimen.*

conducting an elastic-plastic finite-element analysis with a stationary crack of length $a - d$ and a stationary crack of length $a + d$ where d is the element size at the crack tip. The total potential energy was calculated for each stationary crack length. The $J_{\Delta A}$ value was then calculated from Eq 9 in Appendix I. $J_{\Delta A}$ agreed well with J_Γ and J_D^* for $\Delta a/b_0 = 0.1$, but was between J_Γ and J_D^* for $\Delta a/b_0 = 0.25$. These results show that J evaluated from a remote contour gives a good estimate for the J-integral even with crack extensions up to 25% of the initial ligament length (b_0).

References

[1] Irwin, G. R., "Fracture" in *Encyclopedia of Physics,* S. Flugge, Ed., Springer, Berlin, Vol. 6, 1958, pp. 551-590.

[2] Rice, J. R., "A Path Independent Integral and the Approximate Analysis of Strain Concentrations by Notches and Cracks," *Transactions,* American Society of Mechanical Engineers, *Journal of Applied Mechanics,* Vol. 90, Series E, No. 2, 1968, pp. 379-386.

[3] Cherpanov, G. P., "Crack Propagation in Continuous Media," *USSR Journal of Applied Mathematics and Mechanics,* Translation 31, 1967, p. 504.

[4] Eshelby, J. D., "The Calculation of Energy Release Rates" in *Prospects of Fracture Mechanics,* G. C. Sih, H. C. van Elst, and D. Brock, Eds., Noordhoff International Publishing, Leyden, the Netherlands, 1974, pp. 69-84.

[5] "Resolution of the Reactor Vessel Materials Toughness Safety Issue," NUREG-0744, Nuclear Regulatory Commission, Washington, DC, Vols. 1 and 2, Sept. 1981.

[6] Albrecht, P., Andrews, W. R., Gudas, J. P., Joyce, J. A., Loss, F. J., McCabe, D. E., Schmidt, D. W., and VanDerSlugs, W. A., "Tentative Test Procedure for Determining the Plane Strain J(I)-R Curve," *Journal of Testing and Evaluation,* Vol. 10, No. 6, Nov. 1982, pp. 245-251.

[7] Gudas, J. P. and Davis, D. A., "Evaluation of the Tentative J(I)-R Curve Testing Procedure by Round Robin Tests of HY-130 Steel," *Journal of Testing and Evaluation,* Vol. 10, No. 6, Nov. 1982, pp. 252-262.

[8] Newman, J. C., Jr., "Finite-Element Analysis of Fatigue Crack Propagation—Including the Effects of Crack Closure," Ph.D. thesis, Virginia Polytechnic Institute and State University, Blacksburg, VA, May 1974.

[9] Newman, J. C., Jr., "Finite-Element Analysis of Initiation, Stable Crack Growth, and Instability Using a Crack-Tip-Opening Displacement Criterion," NASA TM-84564, National Aeronautics and Space Administration, Washington, DC, Oct. 1982.

[10] Zienkiewicz, O. C., Valliappan, S., and King, I. P., "Elasto-Plastic Solutions of Engineering Problems, Initial Stress, Finite Element Approach," *International Journal for Numerical Methods in Engineering,* Vol. 1, 1969, p. 75–100.

[11] Ramberg, W. and Osgood, W. R., "Description of Stress-Strain Curves by Three Parameters," NACA TN-902, National Advisory Committee for Aeronautics, Washington, DC, 1943.

[12] Ernst, H. A., Paris, P. C., and Landes, J. D., "Estimations of J-Integral and Tearing Modulus T from a Single Specimens Test Record" in *Fracture Mechanics, ASTM STP 743,* R. Roberts, Ed., American Society for Testing and Materials, Philadelphia, 1981, pp. 476–506.

[13] Shih, C. F. and German, M. D., "Requirements for a One Parameter Characterization of Crack Tip Fields by the HRR Singularity," *International Journal of Fracture,* Vol. 17, 1981, pp. 27–43.

[14] Kumar, V., German, M. D., and Shih, C. F., "An Engineering Approach for Elastic-Plastic Fracture Analysis," EPRI-NP-1931, Elastic Power Research Institute, Palo Alto, CA, July 1981.

[15] Tada, H., Paris, P. C., and Irwin, G. R., *The Stress Analysis of Cracks Handbook,* Del Research Corp., Hellertown, PA, 1973.

Ronald A. Mayville[1] and Peter D. Hilton[1]

A Study of Plasticity Effects on the Crack Geometry Correction Factors Used with the COD-Design Curve

REFERENCE: Mayville, R. A. and Hilton, P. D., **"A Study of Plasticity Effects on the Crack Geometry Correction Factors Used with the COD-Design Curve,"** *Fracture Mechanics: Eighteenth Symposium, ASTM STP 945,* D. T. Read and R. P. Reed, Eds., American Society for Testing and Materials, Philadelphia, 1988, pp. 686–698.

ABSTRACT: A recent application of the COD-Design Curve to fracture data for surface defects in pipeline girth welds, revealed that use of the elastic defect geometry correction factors leads to a nonuniform factor of safety with respect to surface defect aspect ratio. Calculations were performed in this paper which show that the correction factors should be different in the presence of elastic-plastic deformation. This is a possible explanation for the observed nonuniform safety factor. Tabulated solutions based on the elastic-plastic estimation approach for the center cracked panel, single edge notched specimen, and hollow cylinder with internal, complete circumferential crack were used to study the influence of strain on the crack geometry correction factors for long defects. The calculations show that the correction factors must be greater the greater the amount of plasticity. Elastic-plastic finite-element solutions from the literature were used to study the influence of plasticity on the COD-Design Curve correction factors for shorter surface defects.

KEY WORDS: elastic-plastic fracture mechanics, COD-Design Curve, correction factors, pipeline girth weld, surface defects

The COD-Design Curve is a semi-empirical approach for determining allowable defect sizes and applied stresses and strains for engineering structures. As originally conceived, the Design Curve applied to wide plates containing small through cracks and loaded in remote tension [*1*]. The fundamental aspects of the Design Curve are based on this geometry. The COD-Design Curve, because of its derivation and based on experience, has been found to have a factor of safety of approximately two. Applications of the Design Curve, carried out mainly through the British PD6493 [*2*], now include edge cracks in wide plates and surface and buried defects. The half crack length $\bar{a}$, known as the effective defect parameter, used in the COD Design Curve is equated to the crack depth for surface cracks and to one half the crack height for buried defects. Such applications of the Design Curve for these defect geometries seem logical when sections of the surface and buried defects are considered, as in Fig. 1.

It was recognized in applying the COD-Design Curve to geometries other than the wide plate, that some correction was needed to account for the crack shape and the proximity of the crack tip to the front and back surfaces. The corrections used in the British document PD6493 are based on linear elastic fracture mechanics (LEFM). Given the geometry of the defect and the thickness of the structure, one finds the crack length for the infinite plate containing a through crack which will give the same crack tip opening displacement as the crack in the structure of interest. This procedure retains the same factor of safety that is in the basic COD-Design Curve provided the structure is elastic. In the presence of plastic deformation it is not clear how the use of corrections based on linear elasticity will affect the factor of safety.

[1]Arthur D. Little, Inc., Cambridge, MA 02140.

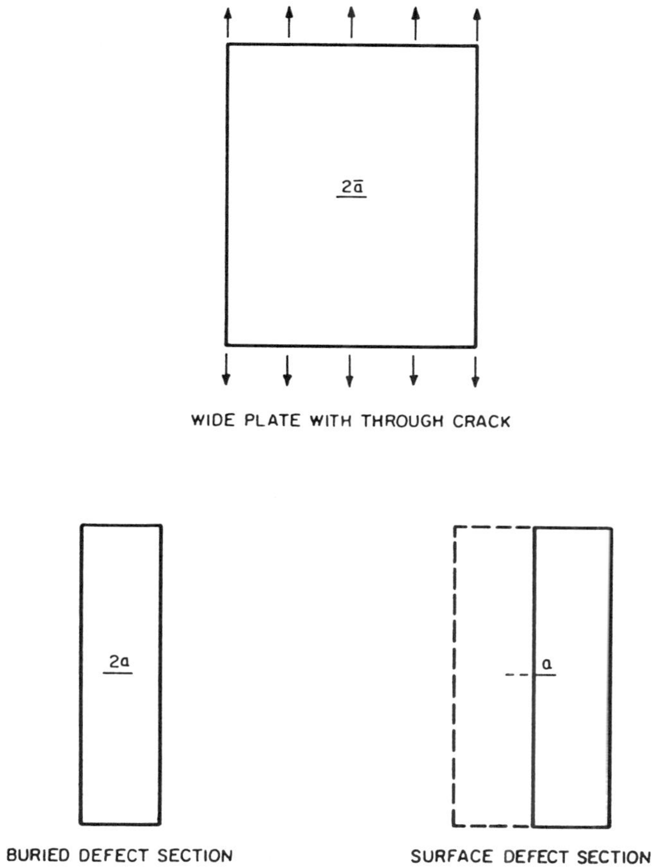

FIG. 1—*Similarities between crack geometries for which the COD-Design Curve is used.*

Recently, in applying the COD-Design Curve to pipeline girth welds, Glover and Coote [*3*] found that the factor of safety, based on strain, was nonuniform with respect to defect aspect ratio; the factor was smaller for crack depth-to-length ratios (a/ℓ) near zero. They recommended that the defect geometry correction factors be changed so that a factor of safety of two would be retained for long defects. A large amount of data were used to empirically generate the new corrections.

Failure strains in the tests considered by Glover and Coote exceeded 80% of the yield strain, suggesting that plasticity effects may have contributed to the nonuniformity of safety factor in applying elastic correction factors to the data.

This paper presents an analysis of the effects of plasticity on the crack geometry correction factors used with the COD-Design Curve. The elastic-plastic estimation approach [*4*] is used to show that a nonuniform safety factor is anticipated when the current crack geometry corrections are used with the COD-Design Curve in the presence of plasticity. The resulting curves are found to be in reasonable agreement with the empirical curves recommended by Glover and Coote. Although this provides evidence that the nonuniform safety factor observed by Glover and Coote is due solely to plasticity, it is important to acknowledge the possible influence of other factors. These include ductile tearing, residual stresses and values of stress and strain at fracture for which plastic collapse may be a better description of failure.

COD-Design Curve

The COD-Design Curve is given by

$$\begin{aligned} \delta &= 2\pi e_0 \bar{a}(e/e_0)^2; \qquad e/e_0 \leq 0.5 \\ \delta &= 2\pi e_0 \bar{a}(e/e_0 - 0.25); \qquad e/e_0 \geq 0.5 \end{aligned} \tag{1}$$

where

δ = CTOD toughness,
e_0 = yield strain,
e = applied strain, and
$\bar{a}$ = equivalent half crack length.

The value of $\bar{a}$ is obtained by equating the elastic crack tip opening displacement (CTOD) for the crack geometry of interest to the CTOD for a through crack in an infinite plate under uniform tension; this is equivalent to equating the linear elastic stress intensity factors. Thus, we have

$$\sigma\sqrt{\pi\bar{a}} = M\sigma\sqrt{\pi a}$$

$$\bar{a} = M^2 a$$

and normalizing by the plate thickness

$$\frac{\bar{a}}{t} = M^2 \frac{a}{t} \tag{2}$$

Values of M are obtained from analyses similar to that of Newman and Raju [5]. The resulting correction factors for surface defects expressed in terms of the equivalent half crack length are shown graphically in Fig. 2 as they appear in PD6493. The curve for $a/\ell = 0$ corresponds to the single edge notched specimen in tension. We note that for deep, relatively long defects the equivalent crack length a can be larger than the plate thickness. Also shown in Fig. 2 are the modified correction curves suggested by Glover and Coote ($a/\ell = 0.01$) and correction curves recommended in a recent draft of the Canadian welding standard CSA Z184 [6] for $a/\ell = 0.02$ and 0.05. Some of the data used to generate these curves are included in Fig. 2. The modifications apply only for relatively long defects; there are no suggested modifications for $a/\ell \geq 0.1$.

Analysis for Long Defects

Calculation of correction factors that include the effects of plastic deformation requires elastic-plastic CTOD solutions for the geometry of interest—in this case, surface defects in plates—and the center cracked plate in tension. The elastic-plastic estimation approach is used for this purpose.

The elastic-plastic estimation procedure separates the CTOD, and other crack parameters, into elastic and plastic components. The elastic component is calculated using LEFM with an Irwin plastic zone correction. The plastic component is obtained from detailed finite element calculations. Elastic solutions are available for many geometries, but the fully plastic solutions are tabulated for only a limited set of geometries [4]. These include, among others, the center

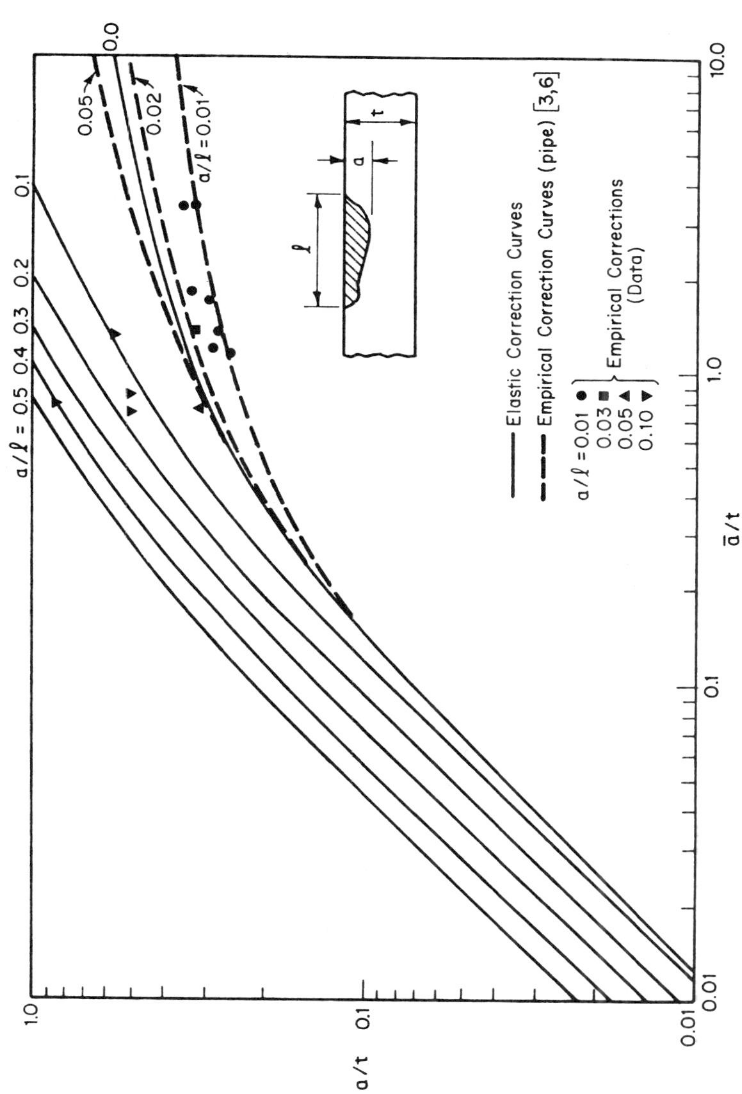

FIG. 2—*The elastic correction curves used with the COD-Design Curve in PD6493* [2] *and the suggested modifications of Glover and Coote* [3].

cracked panel (CCP), the single edge notched (SEN) specimen, and the hollow cylinder containing an internal, full circumferential crack, all loaded in tension.

The CCP solutions can be used to investigate the factor of safety on the COD-Design Curve. We use for this purpose the elastic-plastic estimation formula for a crack length-to-width ratio of 0.125 to model an infinite plate; this is the smallest a/W ratio tabulated in the handbook.

The value of the CTOD given by the elastic-plastic estimation formula is

$$\delta = \delta_e + \delta_p = f(a_e)\frac{P^2}{E'\sigma_0} + \alpha e_0 a h\left(\frac{a}{W}, n\right)\left(\frac{P}{P_0}\right)^{n+1} \tag{3}$$

where

f = function of crack and component geometry,
a_e = plastically adjusted crack length,
E' = Young's modulus; $E' = E$ for plane stress, $E' = E/(1 - \nu^2)$ for plane strain; υ = Poisson's ratio,
σ_0 = yield strength; $\sigma_0 = Ee_0$,
P = applied load,
P_0 = reference load; approximately the plastic limit load,
h = function of crack geometry and hardening exponent n, and
W = component width.

The values of α and n describe the stress-strain relation for the material according to

$$\frac{e}{e_0} = \frac{\sigma}{\sigma_0} + \alpha\left(\frac{\sigma}{\sigma_0}\right)^n \tag{4}$$

For a pipeline steel base metal with $\sigma_0 = 469$ MPa (68 ksi) we find that $\alpha = 1.55$, $n = 20$ [*7*]. Figure 3 shows a plot of Eqs 1 and 3 for a material with: $\sigma_0 = 469$ MPa (68 ksi), $E = 207 \times 10^3$ MPa (30×10^6 psi), $\alpha = 1.55$, $n = 20$, $a/W = 0.125$, and $a = \bar{a} = 3.18$ mm (0.125 in.). The COD-Design Curve is seen to overestimate the CTOD by a factor of about two for the range of strains shown.

The single edge notched panel geometry is used to study the effect of crack depth on the COD-Design Curve correction factors in the presence of plasticity. This geometry is equivalent to a surface crack with $a/\ell = 0$, $a > 0$. Recall that the elastic SEN solution is the one used to generate the correction curve in Fig. 2 for $a/\ell = 0$. The elastic-plastic estimation formula for this geometry is very similar to Eq 3. Figure 4 shows a plot of CTOD for the SEN geometry, normalized by the CTOD for the CCP geometry with $a/W = 0.125$, as a function of strain; as before, $\sigma_0 = 469$ MPa (68 ksi), $E = 207 \times 10^3$ MPa (30×10^6 psi), $\alpha = 1.55$, $n = 20$. Curves are shown for two values of a/W for the SEN geometry: $a/W = 0.125$, 0.25.

Curves such as those shown in Fig. 4 can be used to calculate crack geometry correction curves as in Fig. 2 for various levels of strain. The elastic-plastic estimation Eq 3 can be written as the product of the crack length times a function of the ratio a/W for the different geometries under consideration here [*8*]; that is

$$\delta = aF(a/W, \ldots)$$

Therefore, the crack length for the wide plate containing a small through crack—modeled by the CCP geometry with $a/W = 0.125$—required to give the same CTOD as the SEN geometry is given by

$$\bar{a} = a\,\frac{F_{\mathrm{SEN}}(a/W)}{F_{\mathrm{CCP}}(a/W = 0.125)}. \tag{5}$$

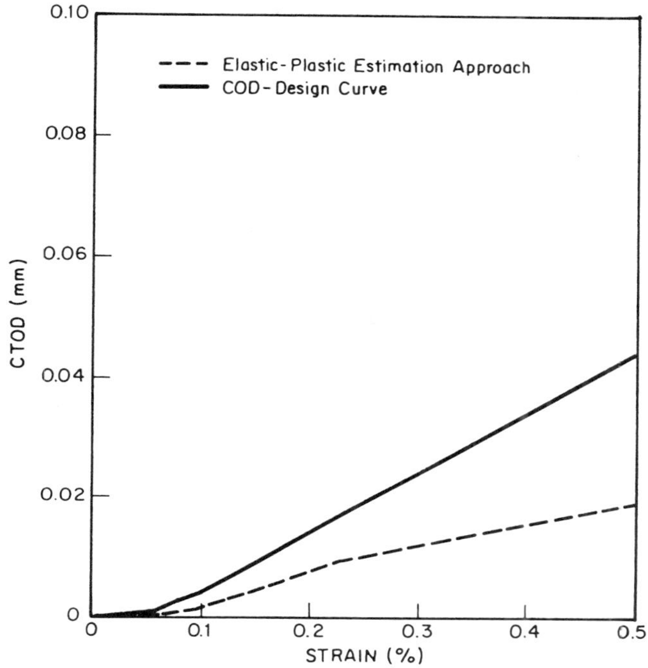

FIG. 3—*Comparison of the CTOD toughness given by the COD-Design Curve and the CTOD given by the elastic plastic estimation approach for* $\sigma_0 = 469$ *MPa,* $E = 207 \times 10^3$ *MPa,* $\alpha = 1.55$, $n = 20$, $a/W = 0.125$ *and* $a = \bar{a} = 3.18$ *mm.*

This equation is equivalent to the elastic Eq 2. One can also see that the ratio of F-factors on the right side of Eq 5 is the same ratio plotted on the ordinate of Fig. 4; that is, it is also equal to the ratio of the CTOD values for equal crack lengths. Using data of the type shown in Fig. 4 for various ratios of a/W, one can plot a curve of $\bar{a}/W$ versus a/W similar to those shown in Fig. 2 for various levels of strain. This has been done in Fig. 5, but with W replaced by t to represent the plate thickness. The figure shows that for nominally elastic strains, $e = 0.10\%$, the curve follows the elastic curve, as it should, except for large values of a/t. At large values of a/t the SEN geometry is close to its plastic limit load so that, even though far field stresses are below yield, significant yielding is occurring in the cracked cross section. At higher levels of strain, the deviation of the plastic correction curves from the elastic correction curves occurs at lower crack depths. Also shown in Fig. 5 is the shifted correction curve recommended by Glover and Coote for $a/\ell = 0.01$.

Use of the SEN model to determine crack geometry correction factors for circumferential cracks in pipelines may not be appropriate. In an SEN specimen loaded in remote tension, some rotation will occur at the crack. This results in a larger CTOD than is expected in a pipe, since the pipe will tend to resist this rotation due to its cylindrical geometry. This can be investigated by calculating correction factors from solutions for the hollow cylinder containing an internal, complete circumferential crack. Elastic-plastic estimation formulas also exist for this geometry, but are tabulated only for hardening exponents of 10 and below. Figure 6 shows a plot of CTOD for the cylinder and SEN geometries normalized by the CTOD for the CCP geometry with $a/W = 0.125$, as a function of strain; $\sigma_0 = 469$ MPa (68 ksi), $E = 207 \times 10^3$ MPa (30 $\times$ 10^6 psi), $\alpha = 1.0$, $n = 10$, R_i (internal radius) = 254 mm (10.0 in.), and R_0 (outer radius) =

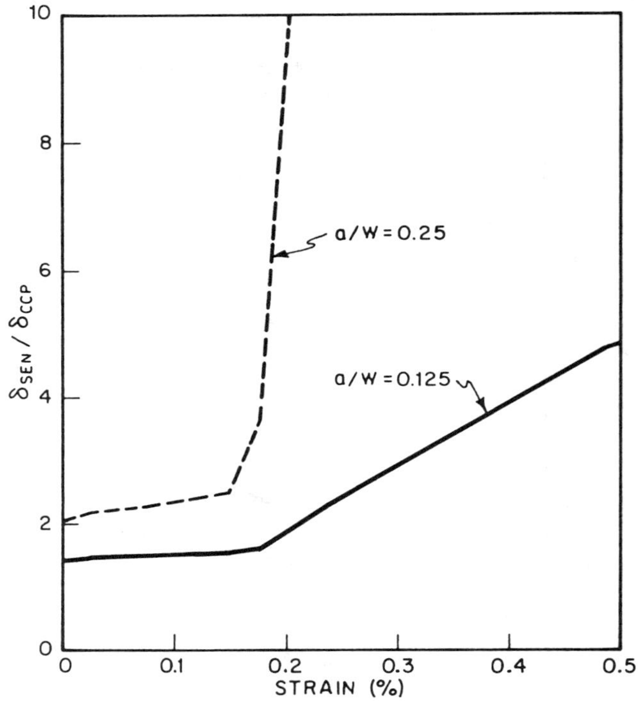

FIG. 4—*The ratio of CTOD for the SEN geometry to the CTOD for the center cracked panel geometry for two values of SEN* a/W; σ_0 = *469 MPa,* E = *207* × *10*^{3} *MPa,* α = *1.55,* n = *20.*

267 mm (10.5 in.). The correction factors based on the cylinder geometry are not as great as those based on the SEN geometry, but are still greater than the elastic correction factors.

Short Defects

The analysis of the plasticity effects on geometry corrections for long surface flaws was possible because of the plain strain approximation via the SEN geometry. Shorter surface cracks cannot be analyzed as easily because there are currently no analytic or tabulated solutions which provide a method to calculate CTOD in the presence of plasticity for this geometry. Some numerical results are available and one set of these will be used to study the plasticity correction for short defects. Wilkening, deLorenzi, and Barishpolsky [*9*] have conducted three-dimensional elastic-plastic finite element analyses for axial surface cracks in a pressure vessel. They considered a cylinder with internal radius R_i = 2.3 m (90 in.) and wall thickness t = 229 mm (9 in.), which gives R_0/R_i = 1.10. Two defect depths and three defect aspect ratios were treated; a/t = 0.25, 0.75, and ℓ/a = 6, 4, and 3. The internal pressures studied caused elastic stresses and stresses that exceeded the uniaxial yield strength, and crack face pressure was modeled. We consider here, only applied stresses below the yield strength.

Equation 4 was used in Ref *9* to model the stress-strain behavior with: σ_0 = 414 MPa (60 ksi), e_0 = 0.002, α = 1.4, and n = 8.6. A partial list of their results for CTOD and J at the deepest point in the crack are given in Table 1.

In order to estimate the geometry correction factors for these defects, calculations were made, as before, using the elastic-plastic estimation procedure with the center cracked panel geometry

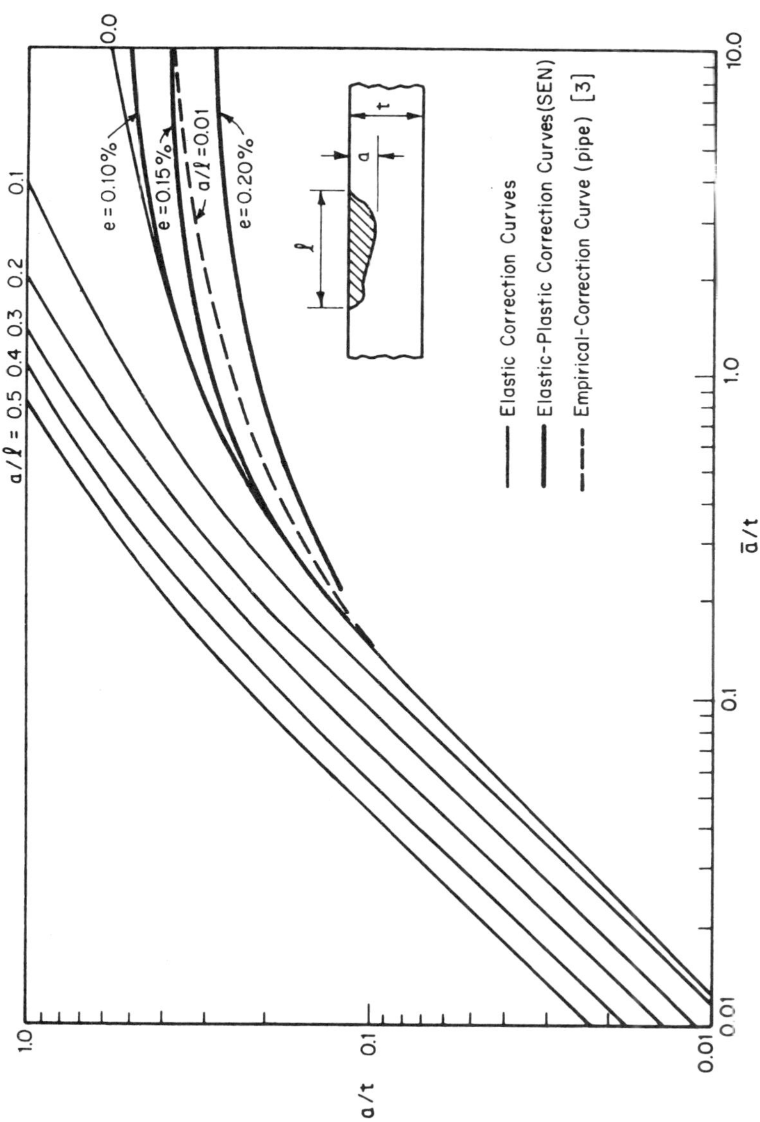

FIG. 5—*Calculated elastic-plastic correction curves in comparison to the elastic correction curves and the curve suggested in* [3]; $\sigma_0 = 469$ *MPa*, E = 207×10^3 *MPa*, $\alpha = 1.55$, n = 20.

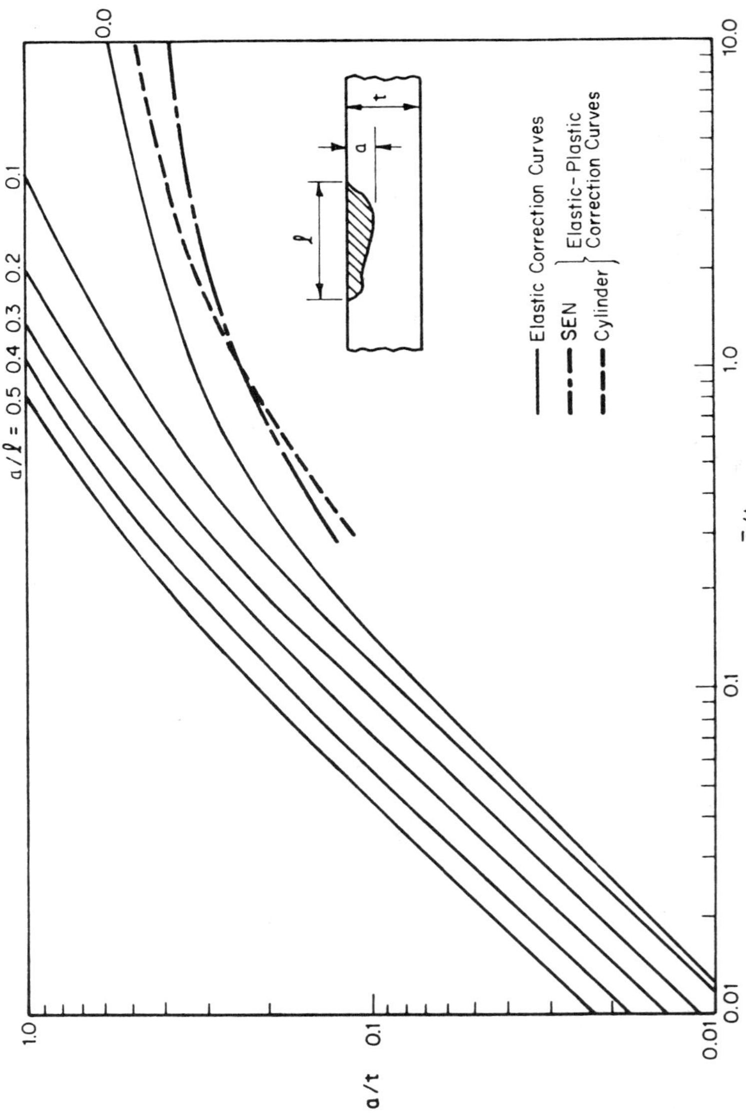

FIG. 6—*Comparison of elastic-plastic correction curves calculated using the SEN and cylinder geometries;* $\sigma_0 = 469$ *MPa,* E = 207×10^3 *MPa,* $\alpha = 1.0$, n = 10.

TABLE 1—*Values of J and CTOD at the deepest crack extent for axial surface cracks in a pressure vessel* [9]; a/t = *0.25 for all flaws;* σ_0 = *414 MPa.*

Internal Pressure, MPa	a/ℓ	CTOD, mm	J, kJ/m^2	$d = \delta(\sigma_0/J)$
17.2	0.167	0.045	32.5	0.57
	0.250	0.037	23.2	0.66
	0.333	0.030	18.3	0.68
31.0	0.167	0.171	129	0.55
	0.250	0.134	91	0.61
	0.333	0.109	70	0.64

in plane-strain tension and Eq 5, except with δ_{SEN} replaced by the CTOD-values obtained by Wilkening et al. [*9*].

Some difficulty arises in comparing CTOD-values from a cylinder to a plate because of curvature effects and the variation of circumferential stress through the thickness. The average of the circumferential stresses at the inside and outside radii of the pressure vessel was used as input to the center cracked plate equations; the stresses at the inside and outside radii differ by about 10%. In addition, a stress equal to the internal pressure was added to the far field circumferential stress to account for crack face pressure.

The same material properties as used by Wilkening et al. were used in the elastic-plastic estimation approach. The elastic-plastic estimation approach requires a value of d to calculate CTOD, where

$$\delta = d\left(\frac{J}{\sigma_0}\right)$$

For $n = 8.6$ and in plane strain, it is recommended [*4*] that $d = 0.63$. However, as Table 1 shows, d varies from 0.51 to 0.68 for the particular range of conditions reported here. We have chosen a value of $d = 0.60$ in calculating δ for the center cracked panel geometry, recognizing that we are studying plasticity effects as opposed to obtaining precise correction factors.

Table 2 shows the results of the calculated equivalent crack sizes. The stresses corresponding to an internal pressure of 17.2 MPa are essentially elastic, and the equivalent crack sizes are seen to agree reasonably well with the equivalent crack sizes currently used with the COD-Design Curve which are based on linear elasticity. For the high pressure, or applied strain, the elastic-plastic values of $\bar{a}/t$ are seen to decrease with respect to the elastic values in contrast to the behavior for the long surface defects.

Discussion

Glover and Coote's observation of a nonuniform factor of safety with defect aspect ratio in the application of the COD-Design Curve to pipeline girth weld defects can be explained by the results of our analysis. They found that the factor of safety for small defect aspect ratios—short defects—was three to five times that for large defect aspect ratios. Our results show that the equivalent crack sizes for long defects should be greater than those based on LEFM for strains on the order of the yield strain. The results also indicate that the opposite is true for short defects. This effect is anticipated to occur for all cases in which the COD-Design-Curve is applied with current correction factors to surface defects under elastic-plastic conditions. Accounting for this difference by the correction curves recommended by Glover and Coote should result in a more uniform factor of safety.

TABLE 2—*Elastic-plastic equivalent crack sizes for short surface defects in comparison to elastic equivalent crack sizes;* a/t = *0.25 in all cases.*

Internal Pressure, MPa	Average Stress, MPa	Strain, %	a/ℓ	δ_{ccp}, mm	δ(FEM), mm	$\delta(FEM)/\delta_{ccp}$	Elastic-Plastic, $\bar{a}/t$	Elastic, $\bar{a}/t$
17.2	190	0.092	0.167	0.044	0.045	1.02	0.25	0.27
31.0	341	0.218	0.167	0.201	0.171	0.85	0.21	0.27
17.2	190	0.092	0.250	0.044	0.037	0.85	0.21	0.22
31.0	341	0.218	0.250	0.201	0.134	0.67	0.17	0.22
17.2	190	0.092	0.333	0.044	0.030	0.69	0.17	0.18
31.0	341	0.218	0.333	0.201	0.109	0.54	0.14	0.18

There are other possible explanations for the observations of Glover and Coote. Fracture in the pipe tests occurred by ductile tearing; a mode of failure the COD-Design-Curve is not meant to treat. Fracture by tearing is governed by the driving force and resistance to ductile crack growth. Irwin's model for the CTOD of a deep surface crack [8] can be used to show that the increase of CTOD with crack length is given by

$$\frac{d\delta}{da} = \frac{2\sigma_0}{E}\frac{\ell}{t}$$

Thus, the driving force for tearing increases with surface crack length. It is also possible that differences in constraint associated with short and long defects will affect the material resistance to tearing, although the way in which it is affected is not clear.

We also note that because the failure strains for the pipes were in excess of 80% of the uniaxial yield strain some of the data used to generate the modified correction curves are probably better characterized by plastic collapse criteria.

Because of these relatively large strains, it is unlikely that residual stresses played a significant role in causing the wide variation in safety factor. This is supported by a test reported by Glover and Coote [3] who observed that stress relief of a girth weld containing a long surface defect did not alter the failure strain significantly.

Conclusions

Glover and Coote [3] have observed a nonuniform factor of safety on strain with surface defect aspect ratio when the COD-Design-Curve is applied to pipe bending fracture data using linear elastic correction factors for defect geometry. The factor of safety is relatively small for long flaws and relatively large for short flaws. This nonuniformity is expected for all surface defects experiencing elastic-plastic deformation and can be explained by plasticity effects on the defect geometry correction factors. The elastic-plastic estimation approach was used to show that for long flaws, modeled by the SEN geometry, the elastic-plastic correction factors can be substantially greater than those given by elastic analysis. Consideration of solutions for internal, circumferentially cracked cylinders in tension indicates that the actual corrections are not as great as predicted by consideration of the SEN geometry, but that they are still greater than the elastic corrections. A set of numerical data for CTOD of surface defects in pressure vessels was used to assess the elastic-plastic correction factors for short defects. These results suggest that the elastic-plastic correction factors are smaller than the elastic correction factors.

Though plasticity accounts for the trend of the nonuniform safety factor, there may be other explanations. In particular, since the pipes failed by ductile tearing, there is a higher driving force for tearing for long defects. Plastic collapse as a mode of fracture may also influence the computed factors of safety.

Acknowledgments

The authors would like to acknowledge the support for this work by Douglas Chisholm and the Research and Special Projects Administration of the U.S. Department of Transportation under Contract DTRS-57-83-C-00078.

References

[1] Kamath, M. S., "The COD Design Curve: An Assessment of Validity Using Wide Plate Tests," *International Journal of Pressure Vessel and Piping,* Vol. 9, 1981, pp. 79–105.
[2] "Guidance on Some Methods for the Derivation of Acceptance Levels for Defects in Fusion Welded Joints," PD6493:1980, British Standards Institution, London.

[3] Glover, A. G. and Coote, R. I., "Full-Scale Fracture Tests of Pipeline Girth Welds" in *Circumferential Cracks in Pressure Vessels and Piping—Vol. II.*, PVP, Vol. 95, G. Wilkowski, Ed., American Society of Mechanical Engineers, New York, 1984, pp. 107–121.
[4] Kumar, V., German, M. D., and Shih, C. F., "An Engineering Approach for Elastic-Plastic Fracture Analysis," Report NP-1931, Electric Power Research Institute, Palo Alto, CA, 1981.
[5] Newman, J. C., Jr., and Raju, I. S., "Analyses of Surface Cracks in Finite Plates Under Tension and Bending Loads," NASA Technical Paper No. 1578, 1979, Washington, DC.
[6] Draft modification to Canadian Standards Association Z184, Gas Pipeline Systems, Pipeline Systems and Materials, 1984.
[7] Erdogan, F., "Theoretical and Experimental Study of Fracture in Pipelines Containing Circumferential Flaws," Report DOT-RC-82007, Lehigh University, Bethlehem, PA, Aug. 1982.
[8] Paris, P. C., "A Method of Application of Elastic-Plastic Fracture Mechanics to Nuclear Vessel Analysis," Appendix B, NUREG-0744, Vol. 1, Revision 1, Oct. 1981, B1–B48.
[9] Wilkening, W. W., deLorenzi, H. G., and Barishpolsky, M., "Elastic-Plastic Analyses of Surface Flaws in a Reactor Vessel," *Transactions, Journal of Pressure Vessel Technology,* American Society of Mechanical Engineers, Vol. 106, Aug. 1984, pp. 247–254.

C. William Smith,[1] *W. Randolph Lloyd,*[1] *and Olekan Olaosebikan*[1]

An Assessment of the Corresponding Stress-Intensity Factor for Accounting for Boundary Effects in Cracked Bodies

REFERENCE: Smith, C. W., Lloyd, W. R., and Olaosebikan, O., **"An Assessment of the Corresponding Stress-Intensity Factor for Accounting for Boundary Effects in Cracked Bodies,"** *Fracture Mechanics: Eighteenth Symposium, ASTM STP 945,* D. T. Read and R. P. Reed, Eds., American Society for Testing and Materials, Philadelphia, 1988, pp. 699–710.

ABSTRACT: An integrated frozen stress photoelastic and moiré interferometric method is briefly described and the results of applying the method to study boundary effects for several crack geometries in finite-thickness nearly incompressible bodies are discussed. Results confirm Benthem's analysis at the free surface but suggest a thick transitional zone through the thickness. A linear elastic fracture mechanics quantity called the corresponding stress-intensity factor is found to be a conservative way of interpreting test data near crack tips in elastic nearly incompressible materials.

KEY WORDS: stress intensity factors, boundary effects, Eigenfunction problems, photoelastic and moiré analysis, linear elastic fracture mechanics

When a straight front crack intersects a free surface at right angles, it has been shown analytically for a quarter infinite crack in a semi-infinite body [*1,2*], and experimentally [*3,4*] for a compact bending specimen, that the inverse square-root singularity in the local elastic stresses associated with classical linear elastic fracture mechanics (LEFM) is lost. This effect is especially significant in nearly incompressible materials and may affect stress intensity results by as much as a factor of 2 for such materials. Recently it has been proposed [*5,6*] that one may account for this effect through the use of a "corresponding" stress intensity factor (SIF) which may be obtained from a simple mapping function and which accounts for the effect of the boundary layer in a classical LEFM approach. Preliminary studies suggest that the corresponding SIF can be obtained for crack border intersection angles other than 90 deg and for curved crack fronts. Using an integrated frozen stress-moiré interferometric technique combined with special computerized algorithms for converting optical data into corresponding SIF values, classical and corresponding SIF distributions have been obtained for a variety of problems. In this paper, the features and limitations of this method are briefly described and results of its application to several different three-dimensional cracked body problem geometries are presented. Implications of the results are summarized and the significance of the corresponding SIF is assessed.

Method of Analysis

As noted above, the method of analysis employed to study boundary effects in cracked body problems involves the tandem application of two well-known experimental methods: frozen

[1]Alumni professor, research assistant, and assistant professor, respectively, Department of Engineering Science and Mechanics, Virginia Polytechnic Institute and State University, Blacksburg, VA 24061.

stress photoelasticity and moiré interferometry. However, certain refinements are necessary in both methods in order to achieve desired measurement precision close to boundaries.

Moiré Analysis

Conventional coarse moiré analysis can be described as the mechanical interference between two superimposed gratings: for in-plane displacements one grating (the master grating) is undeformed, and the second grating is deformed by a deformation field. The interference results in an optical fringe pattern where fringes are proportional to the in-plane displacement normal to the master grating direction. However, this system rarely exceeds a grating line density of 40 lines/mm. For near tip measurements in the present application, it was estimated that an effective grating line density of 1200 lines/mm would be required. In order to achieve this, a "virtual" grating [7] was made. In this approach, a laser beam is split into two coherent beams which then are recombined for form-layered regions of constructive and destructive interference. These regions form an optical or "virtual" grating in space. By viewing the deformed grating on the specimen surface through the virtual grating, one obtains optical fringes proportional to the specimen in-plane surface displacements as before. Line densities of 1200 to 2400 lines/mm can be achieved by this method. Figure 1 shows such a pattern near the tip of a sharp vee-notch under Mode I loading. A detailed description of this approach is found in Ref *7*.

Frozen Stress Analysis

This method capitalizes on the fact that the transparent model material exhibits diphase mechanical and optical behavior. At room temperature, it responds to load in a viscoelastic Kelvin-like manner (Fig. 2). However, when the model's temperature is raised to its "critical" value, the viscous coefficient vanishes and the mechanical response becomes linearly elastic with

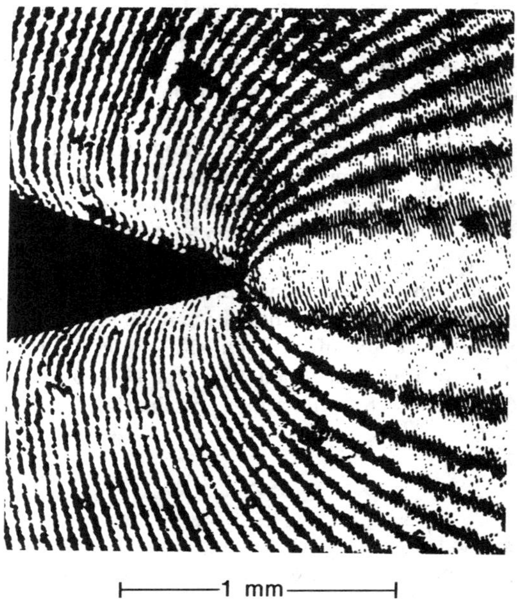

FIG. 1—*Moiré interferometric pattern for near-tip displacement normal to crack plane for Mode I loading (effective frequency 1200 ℓ/mm).*

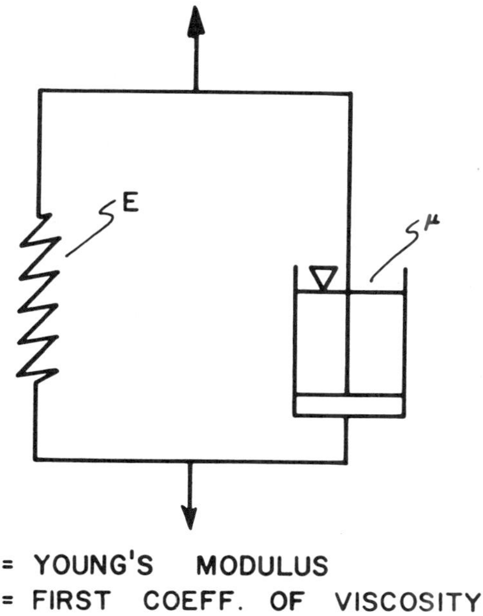

FIG. 2—*Viscoelastic (Kelvin) model.*

a corresponding reduction in material modulus of some two and a half orders of magnitude and with an increase in optical sensitivity to stress fringes of about 25 times the room temperature sensitivity. To take advantage of this behavior, we heat the cracked model to critical temperature, load, and cool under load. Upon load removal, and even after slicing through the body after load removal, the deformations and stress fringes produced above critical temperature are retained and can be analyzed photoelastically in the usual way. The photoelastic fringes are proportional to the maximum shearing stress in the plane of the slice, and the algorithms for converting the data into stress intensity information are referred to a local cartesian coordinate system *tnz* (Fig. 3) with slices always taken parallel to the *nz* plane for Mode I loading [*8, 9*].

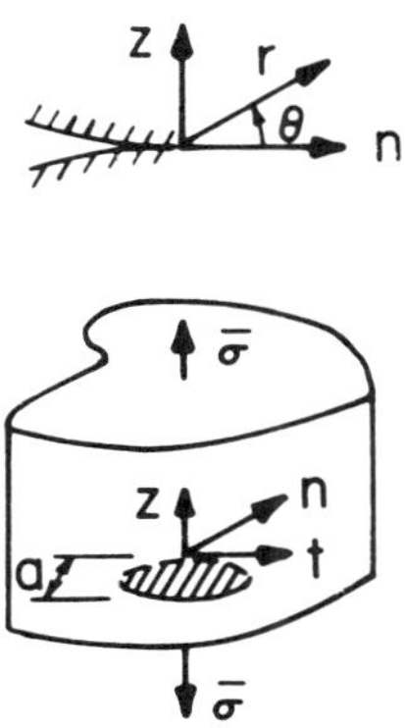

FIG. 3—*General problem geometry and notation.*

In stress freezing work the loads must be kept small to avoid finite deformations. Moreover, in studying boundary effects, it is desirable to use very thin slices (≈ 0.5 mm) in order to define the stress field near the boundary. However, load and thickness reduction reduce the number of stress fringes. Consequently, a system for permitting the combined application of partial mirror fringe multiplication [*10*] and Tardy compensation [*11*] was constructed; a sketch of the system is shown in Fig. 4. In Fig. 4, a laser light source enters at the base of the structure and is reflected upward through two partial mirrors which form a sandwich around the frozen slice, through a focusing lens and an index card for selecting the appropriate fringe multiple. It is then reflected onto a screen for viewing. Using Tardy compensation on the fringe pattern shown, one-tenth fringe order increments are readily produced. The system gives readable patterns up to the seventeenth multiple before applying Tardy compensation. Figure 5 shows a typical near-tip magnified photoelastic fringe pattern (about $\times 19$) before and after fringe multiplication. Details of the fringe multiplication system are found in Ref *12*.

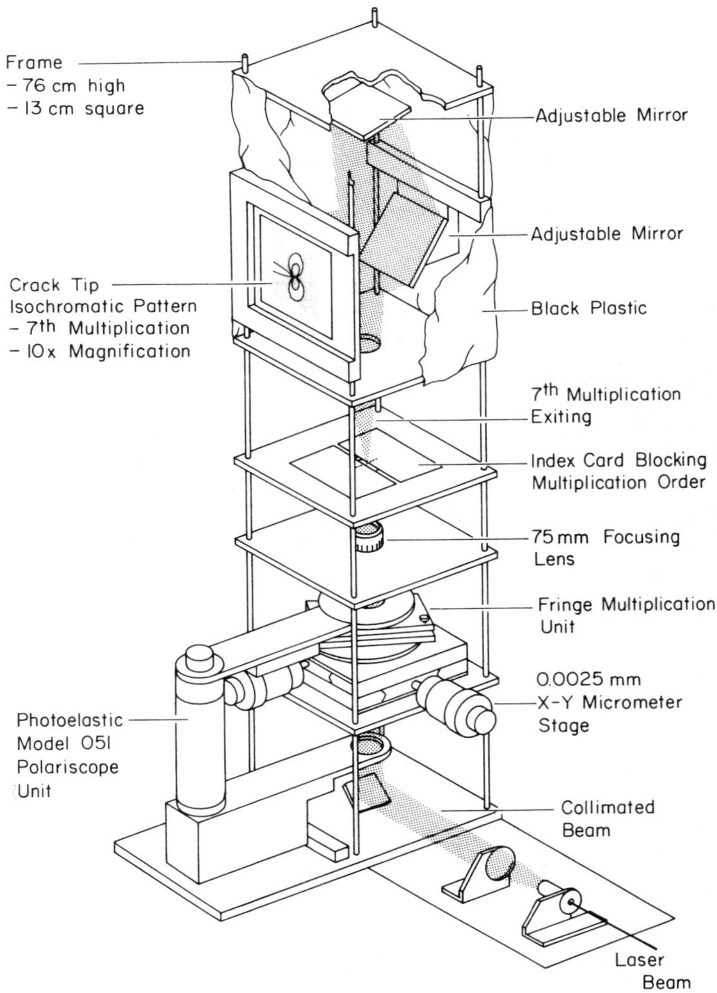

FIG. 4—*Stress fringe multiplication system.*

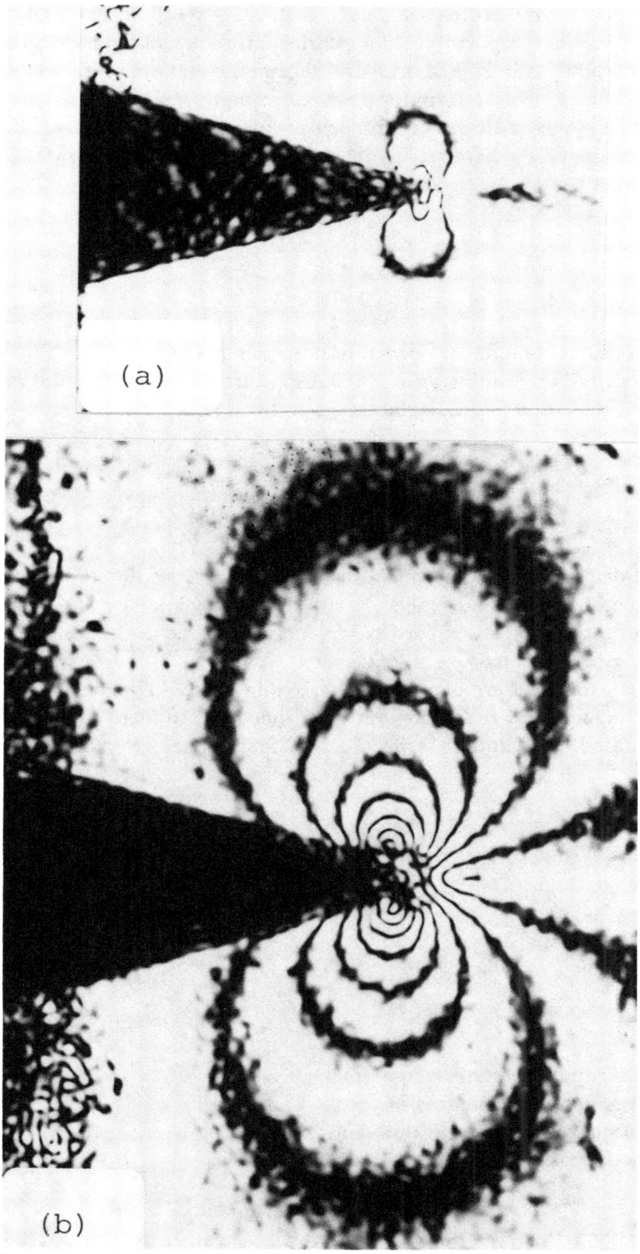

FIG. 5—(a) *Unmultiplied photoelastic near tip Mode I fringe pattern.* (b) *Fifth multiple pattern all at about ×19 for slice from an artificial crack.*

Algorithms for Computing Fracture Parameters

Once data have been obtained from the methods described previously, it is necessary to convert the optical data into fracture parameters. Benthem's solution [2] revealed that the order of the stress singularity at the point where a quarter infinite crack intersected a free surface was not $-1/2$ as it normally is in LEFM problems. Using the method of separation of variables, he assumed a series eigenfunction form for the stress components. By focusing on the most dominant eigenvalue term as is customary in such problems, and noting that, for Mode I, moiré displacement fringes (Fig. 1) are readily measurable along $\theta = \pm\pi/2$, we may write, near the crack tip

$$U_z = Cr^{\lambda_u} \tag{1}$$

where

U_z = displacement component normal to the crack plane,
C = coefficient of the leading term in the displacement series, constant for $\theta = \pm\pi/2$, and
λ_u = first, or most dominant displacement eigenvalue.

Therefore

$$\log U_z = \log C + \lambda_u \log r \tag{2}$$

and λ_u may be determined in the proper measurement zone as the slope of a log-log plot of U_z versus r. The measurement zone is the smallest linear range in log r and should be nearly common to all slices for a given crack border. In general the zone lies between $r = 0.5$ mm and 1.5 mm from the crack tip along $\theta = \pm\pi/2$.

When photoelastic data are extracted from a thin slice, they have been averaged optically over the slice thickness. By restricting our data zone so as to focus upon the in-plane singular effect, we can describe the data in terms of a two-dimensional stress function which leads to [5]

$$\tau_{nz}^{\max} = \frac{(K_\lambda)_{AP}}{r^{\lambda_\sigma}} f(\lambda_\sigma) \qquad \text{AP} = \text{apparent} \tag{3}$$

whence[2]

$$\log \tau_{nz}^{\max} = \log\{(K_\lambda)_{AP} f(\lambda_\sigma)\} - \lambda_\sigma \log r \tag{4}$$

where

$\pi_{nz}^{\max}$ = maximum shear stress in nz plane,
$(K_\lambda)_{AP}$ = an apparent eigenstress factor,
λ_σ = dominant stress eigenvalue, and
$f(\lambda_\sigma)$ = coefficient of $(K_\lambda)_{AP}$ for $\theta = \pi/2$.

By plotting $\log \tau_{nz}^{\max}$ versus $\log r$, λ_σ can be determined. A typical set of data from a slice is shown in Fig. 6. Note that $\lambda_\sigma = 1 - \lambda_u$. However, each can be determined independently of the other.

[2]Although adequate for showing the relative values of K_I and K_{cor} as discussed here, a more accurate algorithm for curved crack fronts requires replacing the left side of Eq 4 with $\log(\tau_{nz}^{\max} - \tau_0)$ [13]. When applied, this refinement will lift the semi-elliptic crack data in Fig. 8 to near the straight front data.

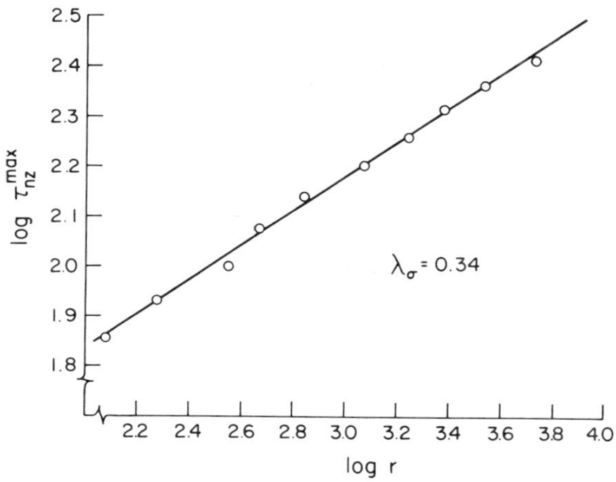

FIG. 6—*Determination of λ_σ from photoelastic data for surface flaw surface slice.*

Results and Discussion: Concept of K_{cor}

The methods described above have been applied to several different problem geometries over the past two years. The standard procedure involves stress freezing the cracked model, removing thin slices along the crack border parallel to the *nz* plane (Fig. 3) analyzing photoelastically, digitizing the data, applying a moiré grating to one face of each slice, annealing the slices to deform the grating, and performing a moiré analysis, again digitizing the data. In some cases, the moiré analysis was omitted.

We shall focus upon three geometries here as pictured in Fig. 7. All of the cracks had about the same relative maximum depth but widths varied due to the nature of the crack shapes. The straight front crack and surface flaw preserved the crack border free surface intersection angle of 90 deg while the intersection angle for the part-through crack did not.

Figure 8 shows the distribution of λ_σ for the three crack geometries. Moiré gratings were used to obtain λ_σ values at the free surface for the straight front and semi-elliptic flaws, but not for the part circular crack. However, progressive thinning of the slice nearest the surface of the part circular crack produced a rapidly declining value of K_1 as computed from LEFM.

It is clear from Fig. 8 that the value of λ_σ at the free surface is not very sensitive to the shape of the crack and this is not totally unexpected. However, the relative insensitivity of λ_σ, albeit an extrapolated value, to the angle change produced in the part circular crack, seems surprising and suggests a need for further study.

A second point which is made clear by Fig. 8 is that the variation in λ_σ from a value of one half is not confined to a thin boundary layer in any of the examples studied, but rather extends quite some distance into the body thickness.

Due to the well-established variation of λ_σ from the classical LEFM value at the free surface, it would appear that test measurements made on the boundary surface near the crack tip will require an interpretation other than that of classical LEFM. In this connection, the authors and their associates have developed the following approach.

In photoelastic analysis the optical fringe order is proportional to the maximum in-plane shearing stress. In LEFM, we relate this quantity to an apparent stress-intensity factor through the near tip field equations as [*8*]

$$\tau_{nz}^{max} = \frac{K_{AP}}{r^{1/2}} \tag{5}$$

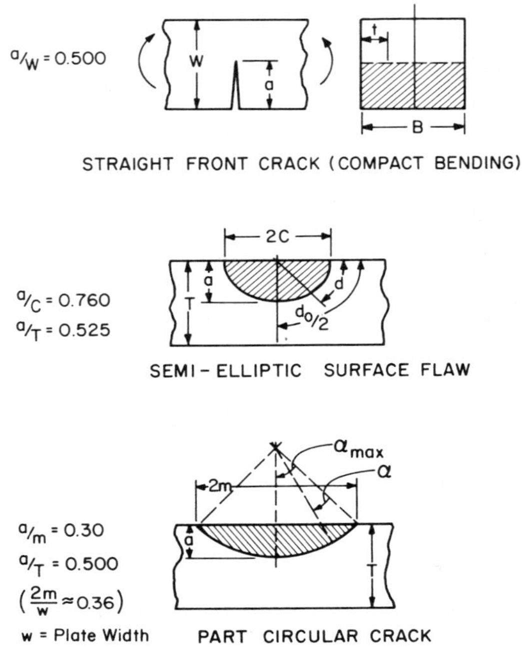

FIG. 7—*Crack geometries studied.*

But from Eq 3

$$\tau_{nz}^{\max} = \frac{(K_\lambda)_{AP}}{r^{\lambda_\sigma}} f(\lambda_\sigma)$$

Thus

$$(K_{cor})_{AP} = \left\{ \frac{(K_\lambda)_{AP}}{r^{\lambda_\sigma}} f(\lambda_\sigma) \right\} r^{1/2} \tag{6}$$

where the quantity on the left side is called the corresponding apparent stress intensity factor and represents the classical LEFM equivalent of $(K_\lambda)_{AP} f(\lambda_\sigma)$. By converting $(K_\lambda)_{AP} f(\lambda_\sigma)$ values to $(K_{cor})_{AP}$ values through Eq 6, we may then proceed to extrapolate across a near-tip nonlinear zone to obtain a value for K_{cor} as is done in LEFM [*8*]. An example of this last step is shown in Fig. 9.

In Fig. 10, we compare the results obtained by the above method with those obtained from LEFM for the three crack geometries studied. The comparison shows that, for each of the cases considered, the K_{cor} values are higher than the K_1 values from LEFM except where $\lambda_\sigma \approx 0.5$ where K_{cor} coalesces with the value of K_1 from LEFM. The straight front crack result suggests that use of Eq 6 in the manner described may eliminate the boundary effect.

Summary and Conclusions

An integrated frozen stress-moiré interferometric method for studying near tip three-dimensional effects in cracked bodies was briefly reviewed and results obtained from applying the

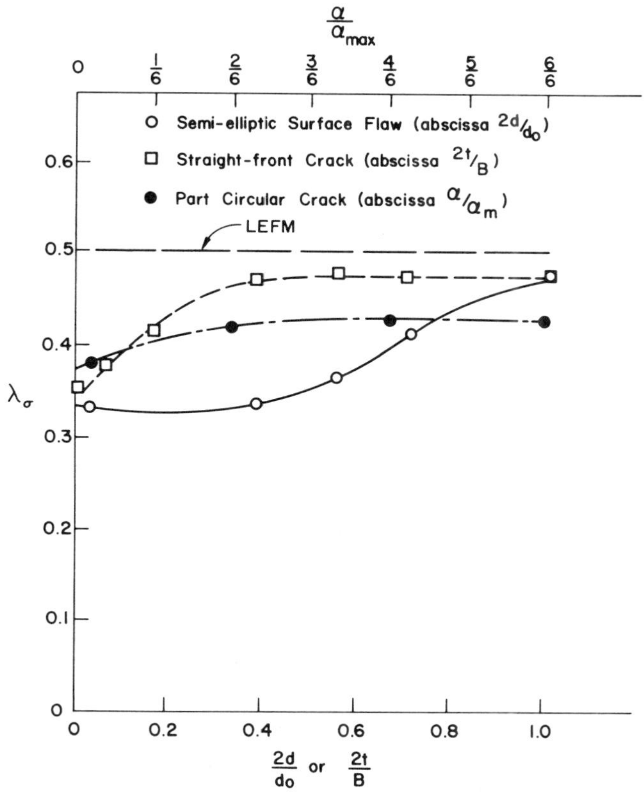

FIG. 8—*Distribution of λ_σ along crack borders.*

method to a variety of crack shapes in nearly incompressible finite thickness bodies were presented. Results suggest that the order of the stress singularity is different from 1/2 not only at the point of intersection of the crack border with the free surface, but also in a thick transitional zone extending inward from the boundary. It may be conjectured that the thickness dimension of the transition zone may be due to the high value of Poisson's ratio ($\nu \approx 0.48$).

It was shown that a conservative interpretation of the data in terms of conventional K_1 of LEFM could be achieved through the use of a corresponding stress intensity factor computed from the data. In view of the conservative nature of the preliminary results presented here, it seems reasonable—subject to further confirmation—to suggest the use of K_{cor} when interpreting test measurements on the surface of nearly incompressible, elastic materials such as rubbers, certain plastics, or solid rocket propellants.

No experimental method of analysis is without some data scatter. Experience with the present method suggests that a scatterband of ±5% is quite adequate for the present work.

Acknowledgments

The authors wish to acknowledge the contributions of Dr. J. S. Epstein, whose research and dissertation provided the foundations from which much of the present results were obtained, and to gratefully acknowledge the support the National Science Foundation under Grant No. MEA 8320252.

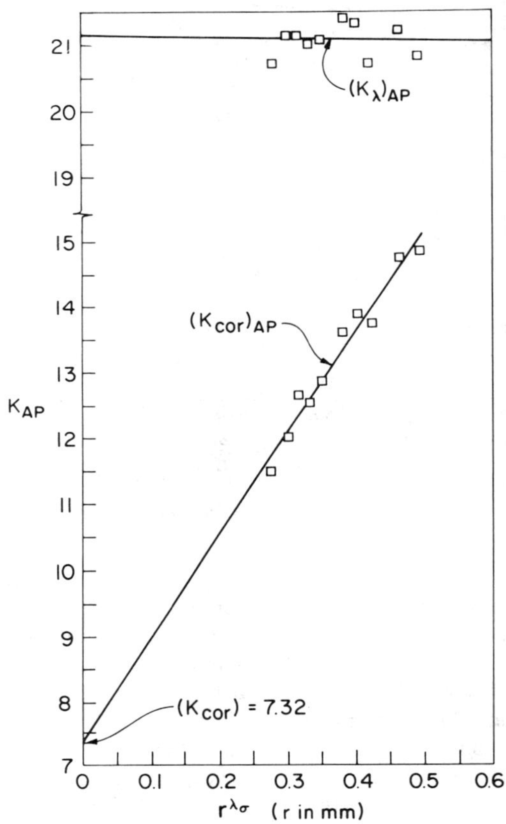

FIG. 9—*Determination of* K_{cor} *for same slice as in Fig. 6.*

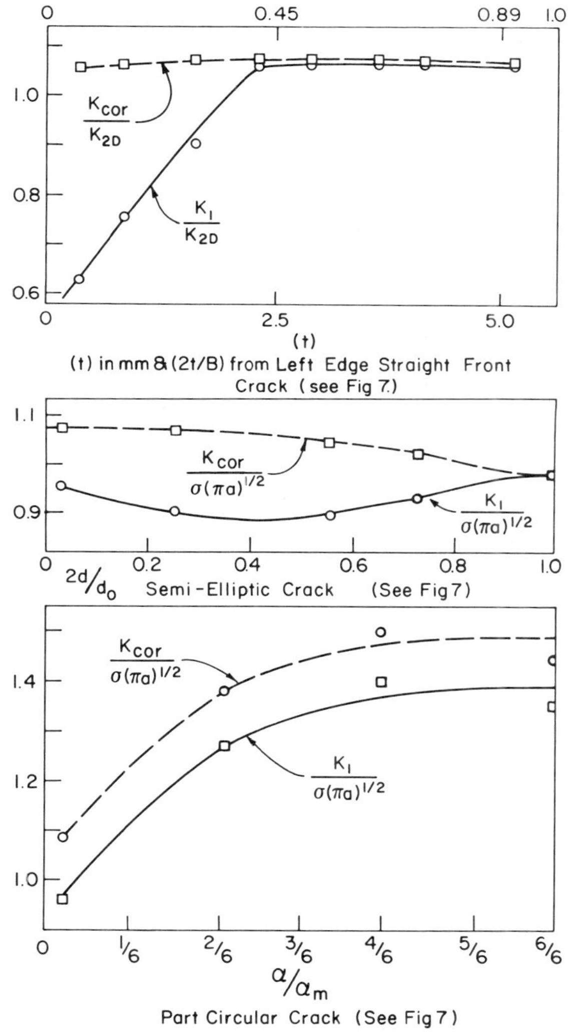

FIG. 10—K_{cor} *versus* K_I *distributions.*

References

[1] Benthem, J. P., "On an Inversion Theorem for Conical Regions in Elasticity Theory," *Journal of Elasticity,* Vol. 9, No. 2, 1979, pp. 159–169.

[2] Benthem, J. P., "The Quarter Infinite Crack in a Half Space: Alternative and Additional Solutions," *International Journal of Solids and Structures,* Vol. 16, 1980, pp. 119–130.

[3] Smith, C. W. and Epstein, J. S., "Experimental Boundary Layer Studies in Three Dimensional Fracture Problems," *Advances in Aerospace Structures, Materials and Dynamics,* ASME-A0-06, American Society of Mechanical Engineers, Nov. 1983, pp. 119–126.

[4] Smith, C. W. and Epstein, J. S., "Boundary Layer Measurements in Cracked Body Problems" in *Proceedings,* 12th Southeastern Conference on Theoretical and Applied Mechanics, Vol. I, May 1984, pp. 355–360.

[5] Smith, C. W. and Epstein, J. S., "Measurement of Three Dimensional Effects in Cracked Bodies" in *Proceedings,* 5th International Conference on Experimental Mechanics, June 1984, pp. 109–110.

[6] Smith, C. W., Olaosebikan, O., and Epstein, J. S., "A Proposed Rationale for Accounting for Boundary Layer Effects in Designing Against Fracture in Three Dimensional Problems" in *Proceedings,* 11th Canadian Fracture Conference, June 1984, pp. 209–219.

[7] Nicoletto, G., Post, D., and Smith, C. W., "Moiré Interferometry for High Sensitivity Measurements in Fracture Mechanics" in *Proceedings,* Joint Meeting of Society for Experimental Stress Analysis and Japan Society for Mechanical Engineers, May 1982, pp. 266–270.

[8] Smith, C. W., "Use of Three Dimensional Photoelasticity and Progress in Related Areas," *Experimental Techniques in Fracture Mechanics 2,* SESA Monograph No. 2, A. S. Kobayashi, Ed., Chapter 1, Society for Experimental Stress Analysis, 1975, pp. 3–58.

[9] Smith, C. W., "Use of Photoelasticity in Fracture Mechanics," *Experimental Evaluation of Stress Concentration and Intensity Factors, Mechanics of Fracture,* G. C. Sih, Ed., Vol. 7, Martinus-Nijhoff, The Hague, The Netherlands, Chapter 2, 1981, pp. 162–187.

[10] Post, D., "Isochromatic Fringe Sharpening and Fringe Multiplication in Photoelasticity" in *Proceedings,* Society for Experimental Stress Analysis, Vol. 12, No. 2, 1955, pp. 143–156.

[11] Tardy, M. H. L., "Methode Pratique d'Examen de Mesure de la Birefringeneces des Verres d'Optique," *Optics Review,* Vol. 8, 1929, pp. 59–69.

[12] Epstein, J. S., Post, D., and Smith, C. W., "Three Dimensional Photoelastic Measurements with Very Thin Slices," *Experimental Techniques,* Dec. 1984, pp. 34–37.

[13] Smith, C. W., "Boundary Effects in Cracked Bodies" in *Proceedings,* Euromechanics Colloquium on Physical-Numerical Modelling in Nonlinear Fracture Mechanics, Stuttgart, West Germany, Sept. 1985.

Poster Session—Elastic-Plastic Fracture Mechanics II

Kwang S. Kim[1] *and Thomas W. Orange*[2]

A Review of Path-Independent Integrals in Elastic-Plastic Fracture Mechanics

REFERENCE: Kim, K. S. and Orange, T. W., "**A Review of Path-Independent Integrals in Elastic-Plastic Fracture Mechanics,**" *Fracture Mechanics: Eighteenth Symposium, ASTM STP 945,* D. T. Read and R. P. Reed, Eds., American Society for Testing and Materials, Philadelphia, 1988, pp. 713–729.

ABSTRACT: The objective of this paper is to review the path-independent (P-I) integrals in elastic-plastic fracture mechanics which have been proposed in recent years to overcome the limitations imposed on the J-integral. The P-I integrals considered herein are the J-integral by Rice, the thermo-elastic P-I integrals by Wilson and Yu, and by Gurtin, the J*-integral by Blackburn, the J_Θ-integral by Ainsworth et al., the $\hat{J}$-integral by Kishimoto et al., and the ΔT_p- and ΔT_p^*-integrals by Atluri et al. The theoretical foundation of these P-I integrals is examined with an emphasis on whether or not the path independence is maintained in the presence of nonproportional loading and unloading in the plastic regime, thermal gradients, and material inhomogeneities. The similarities, differences, salient features, and limitations of these P-I integrals are discussed. Comments are also made with regard to the physical meaning, the possibility of experimental measurement, and computational aspects.

KEY WORDS: elastic-plastic fracture mechanics, path-independent integral

Nomenclature

In most papers reviewed in this study the index notation has been used. Accordingly, we will use the index notation for all the P-I integrals. The variables to be used in the test are described below. Unless stated otherwise, the definitions of variables are as given here. For the nomenclature related to integration paths and areas, the reader is also referred to Fig. 1.

- a Crack length
- A $\lim_{\epsilon \to 0}(A_\Gamma - A_\epsilon)$
- A_Γ Area surrounded by Γ and the crack surface
- A_ϵ Area surrounded by Γ_ϵ and the crack surface
- E Young's modulus
- J The J-integral
- J_G Gurtin's P-I integral
- J_W Wilson and Yu's P-I integral
- J_Θ Ainsworth et al.'s P-I integral
- J^* Blackburn's P-I integral
- $\hat{J}$ Kishimoto et al.'s P-I integral
- K_I Mode I stress intensity factor
- n_i Outward unit normal vector
- P Potential energy
- s Arc length along the contour

[1]Engineer, General Electric Company, Cincinnati, OH 45215.
[2]Research engineer, NASA Lewis Research Center, Cleveland, OH 44135.

S	Total external boundary
S_t	Part of S where traction is given
S_u	Part of S where displacement is given
t_i	Traction vector
u_i	Displacement vector
W	Strain energy density
x_i	The Cartesian coordinates
α	Thermal expansion coefficient
Γ	A counterclockwise path surrounding the crack tip, starting on the lower crack surface, and ending on the upper crack surface
Γ_c	$\Gamma_c^+ + \Gamma_c^-$
Γ_c^+	Upper crack surface from Γ to Γ_ϵ
Γ_c^-	Lower crack surface from Γ_ϵ to Γ
Γ_ϵ	Γ with radius ϵ and with its origin at the crack tip; $\epsilon \rightarrow 0$ is implied without limit notation
δ_{ij}	Kronecker delta
Δ	Increment of the subsequent variable from time τ to $\tau + \Delta\tau$
ΔT	Atluri's P-I integral
ΔT_p	Atluri et al.'s P-I integral
ΔT_p^*	Atluri et al.'s P-I integral
ϵ_{ij}	Strain tensor
Θ	Relative temperature
λ, μ	Lame's constants
ν	Poisson's ratio
σ_{ij}	Stress tensor
τ	Time
Φ	Global thermodynamic potential
Ψ	Helmholz free energy
Ω	Specific internal energy
$(\cdot)^e$	Elastic component of $(\cdot)$, for example, u_i^e, ϵ_{ij}^e, W^e
$(\cdot)^p$	Plastic component of $(\cdot)$
$(\cdot)^\Theta$	Thermal component of $(\cdot)$, for example, ϵ_{ij}^Θ

The J-integral has received much attention among researchers since its introduction to fracture mechanics. As a result, a wide body of literature exists in this connection. The J-integral is a measure of severity of the deformation field at the crack tip and has proven to be a viable parameter for prediction of ductile crack initiation and growth under monotonic loading in the nonlinear regime. Since it can be determined from the far-field numerical solution, which is usually more accurate than the near-field solution, the J-integral has been quite popular in the computational fracture mechanics. Unfortunately, however, the theoretical foundation which the J-integral is based on does not permit the extension of its utility to more practical engineering problems. The path-independence of the J-integral is valid only within the deformation theory of plasticity; hence, it cannot be used in the presence of substantially nonproportional loading and unloading after plastic deformation. The J-integral cannot be used in the presence of a temperature gradient and material inhomogeneity. Other factors such as body force and crack surface loading,[3] though not as frequently encountered in applications as the above, are also excluded in the J-integral.

In recent years there have been considerable efforts to modify or reformulate the path-independent (P-I) integral such that these limitations are removed. Accordingly, there emerged a

[3]The crack surface traction can be included in the J-integral simply by adding the crack surface to the integration path.

number of new P-I integrals[4] in the literature. Some of them are merely slight modifications of the original J-integral and others were formulated on different theoretical bases. None of the new P-I integrals has yet been used enough so that a consensus on the utility in fracture mechanics is reached.

The purpose of this study is to review the new P-I integrals from the standpoint of theoretical bases of their formulation. We will restrict our attention here to rate-independent materials and will not consider the effects of inertia, body force,[5] and large deformation. The P-I integrals in this category and to be reviewed in the text are the J-integral [*1*], Wilson and Yu's integral [*20*], Gurtin's integral [*24*], the J_Θ-integral by Ainsworth et al. [*26*], the J*-integral by Blackburn [*27*], the $\hat{J}$-integral by Kishimoto et al. [*29*], and the ΔT_p- and ΔT_p^*-integrals by Atluri et al. [*32*]. These P-I integrals will be recast in the same nomenclature, and the similarities, differences, salient features and limitations will be examined. Comments will be also made with regard to the physical meaning, the possibility of experimental measurement, and the ease of computation.

The J-Integral

The introduction of the J-integral to fracture mechanics is due to Rice [*1*], although some related earlier studies can be found in Eshelby [*2*], Sanders [*3*], and Cherepanov [*4*]. The J-integral was later extended to large elastic deformation by other researchers [*5–7*]. In the following, we will review the concept of the J-integral.

Let us consider a cracked body shown in Fig. 1. We assume that the crack is straight and

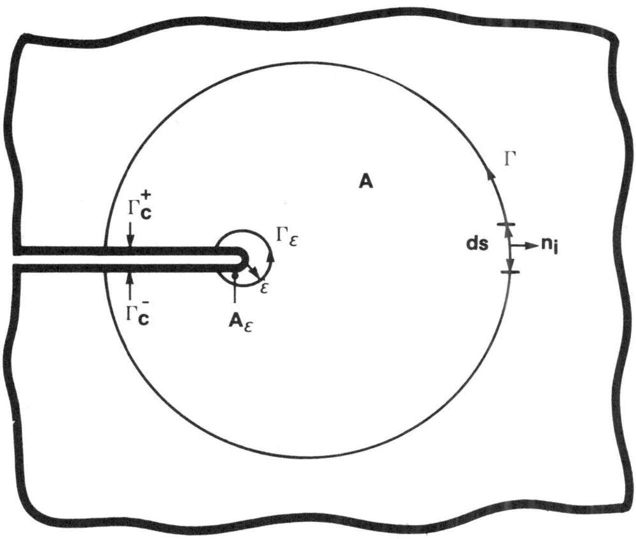

FIG. 1—*Integration paths and areas.*

[4]In the classical sense, the path-independent integral is a line integral performed along an open or closed curve. Here, the notion of "path independence" is somewhat different. As we will see, most path-independent integrals contain not only line integrals, but also area integrals.

[5]The body force can be included easily in any P-I integral. Usually it generates an area integral term.

oriented in the x_1-direction. We will also assume that the body force is negligible and the crack surface is traction-free. Then, the following integral denoted by J is independent of the path

$$J = \int_{\Gamma} (n_1 W - t_i u_{i,1}) ds \tag{1}$$

where the strain energy density function W is given by

$$W = \int_0^{\epsilon_{ij}} \sigma_{ij} d\epsilon_{ij} \tag{2}$$

Because W is a uniquely defined function of the current strain, the J-integral is considered to be valid for elastic materials.

The path-independence of J is easily proved by use of the equilibrium equation

$$\sigma_{ij,j} = 0 \tag{3}$$

and the divergence theorem

$$\int_C n_i f ds = \int_A f_{,i} dA \tag{4}$$

where

f = piecewise continuously differentiable function,
C = closed contour, and
A = area surrounded by C.

The homogeneity of the material, at least in the x_1-direction, is also assumed to prove the path-independence of J. In this connection, we note the work by Smelser and Gurtin [8], who showed that the J-integral can be extended to bimaterials without change, provided the bond line is straight and parallel to the crack.

The J-integral is a measure of the severity of crack-tip deformation. For linearly elastic materials, J can be related to the stress-intensity factor by direct substitution of the singular solution [21]. For instance, the Mode I solution yields

$$J = \frac{\kappa}{E} K_{\mathrm{I}}^2 \tag{5}$$

where

$\kappa = 1$ for plane-stress problems and
$\kappa = 1 - \nu^2$ for plane-strain problems.

Equation 5 has been widely used to determine K_{I} from J computed from far-field data.

The J-integral is also valid for elastic-plastic materials within the deformation theory of plasticity. For power-law hardening materials the crack-tip field is characterized by the Hutchinson-Rice-Rosengren (HRR) singularity [9,10]. The J-integral is again related to the strength of the singular field by

$$\sigma_{ij} = \sigma_0 \left(\frac{EJ}{\sigma_0^2 I_n r} \right)^{\frac{1}{n+1}} \tilde{\sigma}_{ij}(\theta, n) \tag{6}$$

$$\epsilon_{ij} = \frac{\sigma_0}{E}\left(\frac{EJ}{\sigma_0^2 I_n r}\right)^{\frac{n}{n+1}} \tilde{\epsilon}_{ij}(\theta, n) \tag{7}$$

where

σ_0 = yield stress,
n = hardening exponent,
r, θ = polar coordinates with the origin at the crack tip,
I_n = constant which is a function of n [*9,10*], and
$\tilde{\sigma}_{ij}, \tilde{\epsilon}_{ij}$ = dimensionless functions of θ and n.

A study by Kumar et al. [*11*], among other papers, revealed that J is a viable parameter for prediction of ductile crack extension under monotonic loading.

The J-integral computed from the numerical results of finite-element analyses, based on the incremental theory of plasticity, is essentially path independent provided that the degree of nonproportionality in loading is not severe. It is also worth noting McMeeking's work [*12*], which shows the path dependence of J in the very vicinity of the crack tip where the loading is significantly nonproportional. One of the limitations on the utility of the J-integral is that it cannot be used in the circumstances where unloading takes place after plastic deformation. This carries much significance since most crack propagation problems involve unloading at either global or local level. Some researchers [*13,14*] correlated the crack growth data to ΔJ defined operationally using the approximation formula for J in Refs *15* and *16*. This type of approach, despite a lack of theoretical rigor, may be worthwhile insofar as no other parameters are available for crack growth prediction in the elastic-plastic regime.

Rice [*17*] has shown that the J-integral can be interpreted as the rate of potential energy decrease for two bodies differing in crack length by an infinitesimal amount, namely

$$J = -\frac{dP}{da} \tag{8}$$

where

$$P = \int_V W(\epsilon_{ij})dV - \int_{S_t} t_i u_i dA \tag{9}$$

is the potential energy, V is the volume of the body, and S_t is the boundary where traction is prescribed. In view of Eq 8, J is often referred to as the "crack driving force" or the "energy release rate." This interpretation is, of course, valid within the framework of nonlinear elasticity. The relation (Eq 8) renders the experimental measurement of J possible as discussed by Dowling and Begley [*18*], and Rice et al. [*19*].

Wilson and Yu's Integral

Wilson and Yu [*20*] modified the J-integral to include the thermal strain. Their integral is given by (for convenience, we will use J_W for their integral)

$$J_W = \int_\Gamma (n_1 W - t_i u_{i,1})ds - \alpha(3\lambda + 2\mu)\int_A [{}^1/_2(\Theta\epsilon_{ii})_{,1} - \epsilon_{ii}\Theta_{,1}]dA \tag{10}$$

where

$$W = {}^1/_2 \sigma_{ij}\epsilon_{ij} \tag{11}$$

and ϵ_{ij} includes the thermal strain. The stress component σ_{ij} is related to ϵ_{ij} by

$$\sigma_{ij} = \lambda\epsilon_{kk}\delta_{ij} + 2\mu\epsilon_{ij} - \alpha(3\lambda + 2\mu)\delta_{ij}\Theta \tag{12}$$

Notice that the modification was made only for the case of homogeneous, isotropic, linearly elastic materials. Alternative forms of J_W were also suggested in Ref *20* using different definitions of W

$$J_W = \int_\Gamma (n_1 W^* - t_i u_{i,1})ds + \alpha(3\lambda + 2\mu)\int_A \epsilon_{ii}\Theta_{,1}dA \tag{13}$$

$$J_W = \int_\Gamma (n_1 \bar{W} - t_i u_{i,1})ds - \alpha(3\lambda + 2\mu)\int_A \Theta\epsilon_{ii,1}dA \tag{14}$$

where

$$W^* = {}^1\!/\!{}_2\sigma_{ij}\epsilon_{ij} - \alpha(3\lambda + 2\mu)\Theta\epsilon_{ii} \tag{15}$$

and

$$\bar{W} = {}^1\!/\!{}_2\sigma_{ij}\epsilon_{ij} + \alpha(3\lambda + 2\mu)\Theta\epsilon_{ii} \tag{16}$$

The path Γ must be replaced by $\Gamma + \Gamma_c$ if the crack surface is not free of traction. The near field expression for J_W can be written as

$$J_W = \int_{\Gamma_\epsilon} (n_1 W - t_i u_{i,1})ds \tag{17}$$

Using the singular solutions [*21,22*], J_W can be related to the stress intensity factors for a bounded crack-tip temperature.

Noted that the material inhomogeneity can be included in J_W in a manner similar to Atluri [*32*]; that is, the additional term in $\partial W/\partial x_1$ due to material inhomogeneity can be expressed in terms of stress, strain, temperature, and their derivatives. The resulting equation is, however, identical to the J*-integral, Eq 31, to be discussed later.

The singularity of the temperature field at the crack tip, if not bounded, must be weaker than the square-root singularity for the area integral to exist. The possibility of an unbounded temperature at the crack tip was discussed in Nguyen [*35*] and Bui [*36*]. Notice also that for a bounded Θ the integrand of the area integral in Eqs 10 and 14 behaves like $r^{-3/2}$ near the crack tip while the integrand in Eq 13 behaves like $r^{-1/2}$. In this sense, Eq 13 would be somewhat better for the computational purpose.

The Wilson and Yu's integral was reconsidered by McCartney [*23*] with a thermodynamic background. He showed that

$$J_W = \int_\Gamma (n_1 \rho\Psi - t_i u_{i,1})ds + \int_A \rho\eta\Theta_{,1}dA \tag{18}$$

where

ρ = density of the material,
Ψ = Helmholz free energy,
η = entropy, and
Θ = absolute temperature.

For the adiabatic process (η = constant), Eq 18 reduces to

$$J_W = \int_\Gamma (n_1 \rho \Omega - t_i u_{i,1}) dS \tag{19}$$

where Ω is the specific internal energy. It is worth noting that Eq 18 can be written as

$$J_W = \int_\Gamma (n_1 \rho \Omega - t_i u_{i,1}) ds - \int_A \rho \Theta \eta_{,1} dA \tag{20}$$

Here, $\Psi = \Omega - \Theta\eta$ has been used. Then, replacing Γ with $\Gamma + \Gamma_c$ and using the divergence theorem, the near-field expression is obtained as

$$J_W = \int_{\Gamma_\epsilon} (n_1 \rho \Omega - t_i u_{i,1}) ds \tag{21}$$

It is of interest to note the work of Nguyen [*35*] and Germain et al. [*37*] in which they showed that Eq 21 can be expressed as

$$J_W = -\frac{\partial \Phi}{\partial a} \tag{22}$$

where Φ is a global thermodynamic potential with arguments such as strain, temperature, crack length, and other internal parameters. Equation 22 is equivalent to Eq 8 for J and, as such, carries a significant meaning from the standpoint of experimental measurement. Notice also that Eq 21 is equivalent to Eq 17 if Ω is identical to W at the crack tip. This condition is satisfied if the entropy at the crack tip is maintained constant. For a stationary crack this implies a finite crack-tip temperature.

Gurtin's Integral

Gurtin [*24*] proved the following conservation law for the thermoelastic field satisfying Eqs 3 and 12 and the steady-state temperature condition, $\Theta_{,ii} = 0$

$$\int_S \left[n_i W - t_k u_{k,i} - \frac{\alpha^2 (3\lambda + 2\mu)^2}{2(\lambda + \mu)} \Theta^2 n_i + \frac{\alpha\mu(3\lambda + 2\mu)}{(\lambda + \mu)} \left(\Theta \frac{\partial u_i}{\partial n} - u_i \frac{\partial \Theta}{\partial n} \right) \right] dA = 0 \tag{23}$$

where S is the boundary of the volume,

$$\frac{\partial}{\partial n} = n_j \frac{\partial}{\partial x_j}$$

and W is the strain energy density defined by

$$W = \mu \epsilon_{ij} \epsilon_{ij} + \frac{\lambda}{2} (\epsilon_{kk})^2 \tag{24}$$

Equation 23 is valid in two or three dimensions. In particular, for the former with traction free crack surface

$$J_G = \int_\Gamma \left[n_1 W - t_k u_{k,1} - \frac{\alpha^2(3\lambda + 2\mu)^2}{2(\lambda + \mu)} \Theta^2 n_1 + \frac{\alpha\mu(3\lambda + 2\mu)}{(\lambda + \mu)} \left(\Theta \frac{\partial u_1}{\partial n} - u_1 \frac{\partial \Theta}{\partial n} \right) \right] ds \tag{25}$$

is path independent provided that $\Theta u_{1,2}$ and $u_1 \Theta_{,2}$ are continuous across the crack. This condition is a severe restriction from the application standpoint. For instance, the continuity of the two quantities cannot be satisfied for Mode I crack problems. It would be more practical to change the integration path from Γ to $\Gamma + \Gamma_c$, thereby eliminating the restrictions. With this modification and with the assumption that the singularity of the temperature at the crack tip is weaker than $r^{-1/2}$, the near-field expression for J_G can be written as

$$J_G = \int_{\Gamma_\epsilon} (n_1 W - t_i u_{i,1}) ds \tag{26}$$

This equation and Eq 17 yield the same results, although different definitions of W were used.

It is seen that Eq 25 has clearly a computational advantage over J_W because it does not contain area integrals. It would be also possible to determine J_G experimentally by measuring the variables in the integrand along the boundary. This type of measurement was performed by Read and McHenry [*25*] to determine J for a single-edge-notch specimen. Notice, however, that the J_W-integral does not require the steady-state temperature condition, while J_G does.

The J_Θ-Integral by Ainsworth et al.

The J_Θ-integral introduced by Ainsworth et al. [*26*] is given by

$$J_\Theta = \int_\Gamma (n_1 W - t_i u_{i,1}) ds + \int_A \sigma_{ij} \epsilon^\Theta_{ij,1} dA \tag{27}$$

where ϵ^Θ_{ij} is the thermal strain

$$W(\epsilon'_{ij}) = \int_0^{\epsilon'_{ij}} \sigma_{ij} d\epsilon'_{ij} \tag{28}$$

and

$$\epsilon'_{ij} = \epsilon_{ij} - \epsilon^\Theta_{ij}$$

Notice that ϵ'_{ij} includes the elastic and plastic strain due to mechanical loads. For elastic deformation it can be easily shown that J_Θ is identical to J_W. The J_Θ-integral is valid only within the deformation theory of plasticity, and it cannot be used with unloading after plastic deformation and material inhomogeneity. Ainsworth et al. [*26*] assert that J_Θ is related to the potential energy by

$$J_\Theta = \lim_{\Delta a \to 0} \frac{1}{\Delta a} \left[\int_\Gamma n_j \sigma_{ij} \Delta u_i ds - \int_A \sigma_{ij} \Delta \epsilon_{ij} dA \right] \tag{29}$$

where Δ denotes the increment as the crack length changes from a to $a + \Delta a$. In fact, Ainsworth et al. started from Eq 29 and arrived at Eq 27 in their derivation. It is noted, however, that J_Θ given by Eq A7 in Ref *26* is not identical to Eq 27 unless the integration of $n_1 W$ along Γ_ϵ vanishes. Notice also that their Eq A7 is the same as the $\hat{J}$-integral given by Eq 36.

The J*-Integral by Blackburn

Blackburn [27] proposed the J*-integral defined by

$$J^* = \int_{\Gamma_\epsilon} (1/2\,\sigma_{ij}u_{i,j}dx_2 - t_iu_{i,1}ds) \tag{30}$$

or alternatively

$$J^* = \int_{\Gamma+\Gamma_c} (1/2\,\sigma_{ij}u_{i,j}dx_2 - t_iu_{i,1}ds) + \int_A (1/2\,\sigma_{ij}u_{i,j1} - 1/2\,\sigma_{ij,1}u_{ij})dA \tag{31}$$

It can be shown that the second term in Eq 31 vanishes in the elastic case; thus, J^* is identical to J. In the thermoelastic case, it is identical to J_W, J_G, and J_Θ in view of the near-field expression of these P-I integrals. Furthermore, the J*-integral given by Eq 31 is path independent even in the presence of material inhomogeneity since only the equilibrium and the divergence theorem were used to obtain Eq 31 from Eq 30. For power-law hardening materials, direct substitution of the HRR field furnishes the relation between J^* and J

$$J^* = g(n)J \tag{32}$$

where n is the hardening coefficient and $g(n)$ can be computed once the angular variation of field variables is known; see the Appendix. For general elastic-plastic materials with smooth stress-strain curves, the integrands in Eq 31 are continuous everywhere in the body if the material undergoes either loading or unloading in the global sense. For loading conditions that give rise to loading and unloading zones in the body at the same time and for materials with kinks in the stress-strain curves (for example, a bilinear material), the continuity of the stresses and the derivatives of displacement must be guaranteed across the loading-unloading boundary or the elastic-plastic boundary for the path independence of J^*. The continuity of the stresses can be proven from the equilibrium of a thin slice of the body containing the boundary and from the continuity of the effective stress. It is also believed that the strain will be continuous across the boundary (excluding the elastic-perfectly plastic material) considering that the stress continuity will ensure the continuity of the elastic strain and the plastic strain will change continuously across the boundary (but the plastic strain rate is discontinuous). The continuity of strain will then guarantee the continuity of the displacement gradient; hence, use of the divergence theorem, Eq 4, and consequently the path independence of J^* is justified also in these circumstances. Unfortunately, no numerical results for J^* are found for problems of this sort.

The major difference of J^* from other P-I integrals is that the strain energy density function does not appear in the integral. Instead, an explicit expression $1/2\,\sigma_{ij}u_{i,j}$ was used. This type of approach will expand the scope of the applicability of the integral, yet it makes the physical meaning obscure. In fact, the physical meaning of J^* in the plastic regime is not known and may be difficult to find. The determination of J^* through experiment is not likely unless J^* is related to a potential in the manner analogous to Eq 8. Notice that Eq 30 requires accurate measurement of near-field variables and that Eq 31 has an area integral which can be determined only by measuring the field data everywhere in A. Neither of these is expected to be feasible. Other P-I integrals containing area integrals have the same problem.

The computation of Eq 31 requires caution in view that the integrand behaves like r^{-2} in the vicinity of the crack tip. The second term can be written near the crack tip as

$$\int_0^r \int_{-\pi}^{\pi} [r^{-2}f(\theta) + \text{less singular terms}]r\,d\theta\,dr \tag{33}$$

Thus, for J^* to exist (this is guaranteed by Eq 30), we must have

$$\int_{-\pi}^{\pi} f(\theta)d\theta = 0 \tag{34}$$

This can be readily verified for linearly elastic materials, as noted by Stonesifer et al. [*34*]. For other types of materials the angular distribution of the field variables is not explicitly known.

Blackburn et al. [*28*] presented some numerical results of J^* for specimens subjected to loading, unloading, and temperature gradient.

The Ĵ-Integral by Kishimoto et al.

The Ĵ-integral proposed by Kishimoto et al. [*29*] is given by

$$\hat{J} = -\int_{\Gamma_{end}} t_i u_{i,1} ds \tag{35}$$

where Γ_{end} is the path surrounding the so-called "fracture process zone" in which the continuum mechanics fails to work effectively. In the process to arrive at Eq 35, the equilibrium equation and the divergence theorem were used. Also, it was assumed that the displacement field on Γ_{end} is maintained constant in the crack propagation, namely, $\partial u_i/\partial a = 0$ on Γ_{end}. With this assumption, $\hat{J}$ represents the rate of work done on the fracture process zone by the surrounding medium. Then they convert Eq 35 into the following by assuming that Γ_{end} shrinks and vanishes

$$\hat{J} = -\int_{\Gamma+\Gamma_c} t_i u_{i,1} ds + \int_A \sigma_{ij}\epsilon_{ij,1} dA \tag{36}$$

In a paper published later [*38*], Γ_{end} was taken as a rectangular path of width 2ϵ and height 2δ with the center at the crack tip. This path was shrunk to the crack tip in the order of δ and ϵ. Equation 36 is now changed to

$$\hat{J} = -\int_{\Gamma+\Gamma_c} t_i u_{i,1} ds + \lim_{\epsilon\to 0}\lim_{\delta\to 0}\int_A \sigma_{ij}\epsilon_{ij,1} dA \tag{37}$$

Then, Ref *38* showed that

$$\lim_{\epsilon\to 0}\lim_{\delta\to 0}\int_{\Gamma_{end}} W^e n_1 ds = 0 \tag{38}$$

and obtained

$$\hat{J} = \int_{\Gamma+\Gamma_c} (n_1 W^e - t_i u_{i,1}) ds + \lim_{\epsilon\to 0}\lim_{\delta\to 0}\int_A \sigma_{ij}\epsilon^*_{ij,1} dA \tag{39}$$

where

$$\epsilon^*_{ij} = \epsilon_{ij} - \epsilon^e_{ij} = \epsilon^p_{ij} + \epsilon^\Theta_{ij} \tag{40}$$

and

$$W^e = \int_0^{\epsilon_{ij}^e} \sigma_{ij} d\epsilon_{ij}^e \tag{41}$$

For the elastic case, they also showed that Eq 39 is equivalent to

$$\hat{J} = \int_{\Gamma_\epsilon} (n_1 W^e - t_i u_{i,1}) ds \tag{42}$$

which is the usual J-integral. For thermoelastic problems it can be shown that $\hat{J}$ given by Eq 39 is identical to J_Θ, J_W, J_G, and J^* provided that the material is homogeneous and the area integral is independent of the limiting process. It is also noted that the physical meaning they claim is valid if there is a finite rigid fracture process zone at the crack tip independent of the crack size. It is unclear, however, whether or not the singular crack-tip field is recovered when $\Gamma_{\epsilon r d}$ shrinks and vanishes, and if it is recovered, whether or not the physical meaning remains intact.

For elastic-plastic problems, the crack-tip field is dominated by the plastic deformation, as verified by Hutchinson [9] for power-law hardening materials. Equation 38, therefore, is acceptable for a vanishingly small fracture process zone regardless of its shape and the limiting process, and Eq 39 would be identical to Eq 36.

The Ĵ-integral can be used for elastic-plastic materials subjected to loading and unloading. Aoki et al. [30] demonstrated the path independence of $\hat{J}$ for a specimen subjected to global loading and then unloading. For problems undergoing local unloading, no numerical results are available. However, it is expected that $\hat{J}$ will be again path-independent in view that the integrand of the line integral will be continuous across the loading-unloading boundary as discussed in the J* integral. $\hat{J}$ can be also used with a temperature gradient and material inhomogeneity since no constitutive relations were assumed in the derivation.

The computation of the area integral in $\hat{J}$ is similar to that of the J*-integral. The experimental measurement of $\hat{J}$ would be as difficult as J^*.

The ΔT-Integrals by Atluri et al.

Atluri [31] proposed a path-independent integral given in the incremental form which takes into account the finite strain, body force, inertia, and crack surface traction. The temperature gradient and material inhomogeneity were not included in the formulation. The small deformation version of the ΔT-integral is given by

$$\begin{aligned} \Delta T &= \int_{\Gamma+\Gamma_c} [n_1 \Delta W - n_j(\sigma_{jk} + \Delta\sigma_{jk}) \Delta u_{k,1}] ds \\ &\quad - \int_A \sigma_{jk,1} \Delta u_{j,k} dA \\ &= R + \int_{\Gamma_\epsilon} [n_1 \Delta W - n_j(\sigma_{ij} + \Delta\sigma_{ij}) \Delta u_{i,1}] ds \end{aligned} \tag{43}$$

where

$$\Delta W = (\sigma_{ij} + 1/2 \Delta\sigma_{ij}) \Delta u_{i,j} \tag{44}$$

and R is the contribution of the discontinuities in the material response along the loading-unloading boundary. The incremental stress $\Delta\sigma_{ij}$ is related to the incremental strain by the Prandtl-Reuss equation

$$\Delta\sigma_{ij} = 2\mu\Delta\epsilon_{ij} + \lambda\delta_{ij}\Delta\epsilon_{kk} - \frac{12\alpha\mu^2\Delta\epsilon_{kl}\sigma'_{kl}\sigma'_{ij}}{\sigma'_{mn}\sigma'_{mn}(6\mu + 2\partial F/\partial W^P)} \tag{45}$$

where

$$\sigma'_{ij} = \sigma_{ij} - 1/3\,\delta_{ij}\sigma_{kk} \tag{46}$$

$$F = \left(\frac{3}{2}\,\sigma'_{ij}\sigma'_{ij}\right)^2 - F_0(W^P) = 0 \tag{47}$$

and

$$W^P = \int^{\epsilon_{ij}^P} \sigma_{ij}\,d\epsilon_{ij}^P \tag{48}$$

The ΔT-integral defined by Eq 43 includes not only the measure of the severity of conditions near the crack tip, but also the effect of the transition from loading to unloading zones in A. The ΔT-integral was proven to be path independent for the paths either in the entirely loading zone or in the entirely unloading zone. Atluri [*31*] also showed for the case of nonsingular crack-tip field that the ΔT-integral implies the rate of change in potential energy (per unit crack length) in the time interval τ to $\tau + \Delta\tau$ of two bodies which are identical in shape and load history but have an infinitesimal difference in crack length.

Later, Atluri, Nishioka, and Nakagaki [*32*] presented two modified versions of the ΔT-integral. In the new integrals, denoted by ΔT_p and ΔT_p^*, the contribution of the elastic-plastic boundary was expressed in terms of field variables. The ΔT_p^* and ΔT_p integrals are given by

$$\begin{aligned}\Delta T_p^* &= \int_{\Gamma_\epsilon} [n_1\Delta W - (t_i + \Delta t_i)\Delta u_{i,1} - \Delta t_i u_{i,1}]ds \\ &= \int_{\Gamma+\Gamma_c} [n_1\Delta W - (t_i + \Delta t_i)\Delta u_{i,1} - \Delta t_i u_{i,1}]ds \\ &\quad + \int_A [\Delta\sigma_{ij}(\epsilon_{ij,1} + 1/2\,\Delta\epsilon_{ij,1}) - \Delta\epsilon_{ij}(\sigma_{ij,1} + 1/2\,\Delta\sigma_{ij,1})]dA\end{aligned} \tag{49}$$

$$\begin{aligned}\Delta T_p &= \int_S [n_1\Delta W - (t_i + \Delta t_i)\Delta u_{i,1} - \Delta t_i u_{i,1}]ds \\ &= \int_{\Gamma+\Gamma_c} [n_1\Delta W - (t_i + \Delta t_i)\Delta u_{i,1} - \Delta t_i u_{i,1}]ds \\ &\quad + \int_{A_s - A_\Gamma} [(\sigma_{ij,1} + 1/2\,\Delta\sigma_{ij,1})\Delta\epsilon_{ij} - (\epsilon_{ij,1} + 1/2\,\Delta\epsilon_{ij,1})\Delta\sigma_{ij}]dA\end{aligned} \tag{50}$$

In the ΔT_p-integral, S is the external boundary including the crack surface, A_Γ is the area enclosed by Γ, and A_s is the total area. The area integrals on the extreme right-hand sides of Eqs

49 and 50 vanish when the loading is proportional. Thus, inasmuch as the deformation theory of plasticity is applicable, the following identity holds

$$\Delta T_p^* = \Delta T_p = \Delta J \tag{51}$$

where

$$\Delta J = \int_{\Gamma_\epsilon} [n_1 \Delta W - (t_i + \Delta t_i)\Delta u_{i,1} - \Delta t_i u_{i,1}]ds \tag{52}$$

is considered the increment of the J-integral due to the increment of the external loading in the time interval τ to $\tau + \Delta\tau$.

It is seen from Eq 49 that ΔT_p^* is a direct measure of the severity of deformation at the crack tip, but it is not experimentally measurable. On the other hand, in view of Eq 50, ΔT_p is not related to the immediate crack-tip field unless the loading is proportional. Note, however, that ΔT_p can be determined by measuring field data along the external boundary. Nakagaki et al. [*33*] presented some numerical data to demonstrate the path independence of ΔT_p^* and ΔT_p for a compact tension specimen which was plastically loaded, unloaded, and then reloaded. They also computed ΔJ at the peaks using Eq 52 and showed that it is no longer path independent after the specimen is unloaded.

A salient feature of the ΔT_p^* and ΔT_p-integrals is that they are based on the incremental theory of plasticity, which enables these P-I integrals to be used even in nonproportional loading and elastic unloading following plastic deformation. It can be also verified by adding $-(2\mu + 3\lambda)\alpha\sigma_{ij}\Delta\Theta$ to Eq 45 and following the procedure in Ref *32*, that Eqs 49 and 50 can be used without change for thermomechanical loading. The material inhomogeneity is also taken into account in Eqs 49 and 50.

The computation of ΔT_p^* would be similar to other P-I integrals involving area integrals. The behavior of the integrand of the area integral of ΔT_p^* in the vicinity of the crack tip is similar to the area integral of Eq 31.

Atluri et al. [*32*] showed that ΔT_p is related to the rate of an incremental energy ($\Delta\Pi$) with respect to crack length if the loading is proportional. For nonproportional loading the physical meaning of ΔT_p is not clear, nor is that of ΔT_p^*.

Engineering Applications

In this paper, several P-I integrals have been examined on a theoretical basis. To be useful for engineering applications, a P-I integral must satisfy each of three conditions. First, the integral must be path independent for realistic loading and temperature conditions. The integral must be calculable without difficulties and must be reasonably path independent.

Second, the integral has to be determined (or closely approximated) from measurements made on laboratory test specimens at various stages of loading. If the integral can be expressed as the rate of change of a potential with crack length, in principle it can be determined from a load-displacement record. If the integral consists only of a line integral applicable over the specimen boundaries, it may be measured using strain gages and extensometers as in Ref *25*. Area integrals would require a whole-field stress and strain analysis. While such may be possible, they are tedious and results are not immediately available.

Third, the integral must correlate various types of crack propagation behavior. That is, crack propagation must depend on the imposed value of the integral and be independent of crack size and specimen or component geometry. This condition is most likely to be satisfied if the integral is a measure of the crack-tip severity and if a sound physical meaning is associated with the P-I integral in relation to crack propagation.

The preceding three aspects are summarized in Table 1 in conjunction with the deformation and temperature conditions under which the integrals are path-independent.

Conclusions

We have reviewed a number of P-I integrals proposed for use in elastic-plastic fracture mechanics. The results are summarized in Table 1. It is seen that there exists as yet no P-I integral with well-established physical background and all the desirable features encountered in applications. The J-integral and J_Θ-integral may be used only for monotonic loading without substantial nonproportionality. For cyclic loading, utilization of these P-I integrals may be possible only by use of appropriate operational definitions of the range of these quantities. The J_W and J_G integrals are usable only for thermoelastic problems with homogeneous material properties. These integrals may be useful for prediction of crack growth in a rather small temperature gradient field and under small-scale yielding conditions. The path independence of the J*, $\hat{J}$, ΔT_p, and ΔT_p^* integrals is maintained for more general elastic-plastic problems, including nonproportional loading, unloading, temperature gradients, and material inhomogeneities. This is clearly a salient feature upon application to crack growth analyses under cyclic and thermomechanical loading. However, the physical meaning of these P-I integrals needs to be further investigated. Finally, it is noted that no significant effort has yet been made to utilize these P-I integrals to consolidate crack growth data under mechanical or thermomechanical load cycling. Experimental efforts along with further analytical studies would be necessary to evaluate the utility of these P-I integrals in fracture mechanics.

Acknowledgment

This work was supported by the NASA Lewis Research Center under Contract No. NAS3-23940.

APPENDIX

Substitution of the HRR singular solution into the near-field expression of the J*-integral furnishes

$$\int_{\Gamma_\epsilon} ({}^1\!/\!_2\,\sigma_{ij}\epsilon_{ij}n_1 - \sigma_{ij}n_ju_{i,1})ds = \alpha K^{n+1}\epsilon^{(n+1)(s-2)+1}I^* \tag{53}$$

where

$s = (2n + 1)/(n + 1)$,
ϵ = radius of Γ_ϵ,
K = constant related to the boundary condition [9], and

$$I^* = \int_{-\pi}^{\pi}\left\{{}^1\!/\!_2\,\tilde{\sigma}_e^{n+1}\cos\Theta - \left[\sin\Theta\left(\tilde{\sigma}_r\left(\tilde{u}_\Theta - \frac{\partial\tilde{u}_r}{\partial\Theta}\right) - \tilde{\sigma}_{r\Theta}\left(\tilde{u}_r + \frac{\partial\tilde{u}_\Theta}{\partial\Theta}\right)\right)\right.\right.$$
$$\left.\left. + [n(s-2)+1]\cos\Theta(\tilde{\sigma}_r\tilde{u}_r + \tilde{\sigma}_{r\Theta}\tilde{u}_\Theta)\right]\right\}d\Theta \tag{54}$$

TABLE 1—*Summary of P-I integrals.*

		Physical Meaning[b]			Capable of Handling					
P-I Integral[a]	Measure of Crack-Tip Severity	Elastic	Thermo-elastic	Plastic	Proportional Loading (Plastic Regime)	Nonproportional Loading and Unloading (Plastic Regime)	Thermal Strain	Material Inhomogenity	Computation (Integrals Involved)	Experimental Measurement[c]
J	yes	$-\frac{\partial P}{\partial a}$	. . .	$-\frac{\partial P}{\partial a}$	yes	no	no	no	line	yes
J_w	yes	$-\frac{\partial P}{\partial a}$	$-\frac{\partial \Phi}{\partial a}$[d]	. . .	no	no	yes	no	line + area	yes
J_G	yes	$-\frac{\partial P}{\partial a}$	$-\frac{\partial \Phi}{\partial a}$	. . .	no	no	yes	no	line	yes
J_Θ	yes	$-\frac{\partial P}{\partial a}$	$-\frac{\partial \Phi}{\partial a}$	$-\frac{\partial P}{\partial a}$	yes	no	yes	no	line + area	yes
J^*	yes	$-\frac{\partial P}{\partial a}$	$-\frac{\partial \Phi}{\partial a}$	unknown	yes	yes	yes	yes	line + area	no
$\hat{J}$	yes	rate of work done to crack tip by surrounding material[e]			yes	yes	yes	yes	line + area	no
ΔT_p^*	yes	$-\frac{\Delta \Pi}{da}$ for proportional loading			yes	yes	yes	yes	line + area	no[f]
ΔT_p	no	Unknown for nonproportional loading			yes	yes	yes	yes	line + area	yes

[a] $J = J_W = J_G = J_\Theta = J^*$ for elastic and thermoelastic deformation of homogeneous materials, $J = \Sigma \Delta T_p^* = \Sigma \Delta T_p$ for monotonic proportional loading.
[b] P = Potential energy; Φ = global thermodynamic potential, $\Delta \Pi$ = incremental energy.
[c] Yes, if it can be expressed as the rate of a potential, or if it can be expressed as a line integral along the boundary.
[d] Provided that the internal energy is identical to the strain energy at the crack tip.
[e] With the assumption of a rigid fracture process zone at the crack tip independent of the crack size.
[f] Yes, if the loading is proportional.

Here, $\tilde{\sigma}_e$, $\tilde{\sigma}_r$, $\tilde{\sigma}_\Theta$, $\tilde{\sigma}_{r\Theta}$, $\tilde{u}_r$, and $\tilde{u}_\Theta$ are functions of Θ only [*9*] associated with the angular distribution of the corresponding stress or displacement components in the vicinity of the crack tip. Note that I^* is a function of n only. Equation 53 can be written as

$$J^* = g(n)J$$

where $g(n) = I^*/I$, and I is defined by Eq 24 of Ref *9*.

References

[*1*] Rice, J. R., *Journal of Applied Mechanics,* Vol. 35, 1968, pp. 379-386.
[*2*] Eshelby, J. D. in *Solid State Physics,* Vol. 3, Academic Press, New York, 1956, pp. 79-144.
[*3*] Sanders, J. L., Jr., *Journal of Applied Mechanics,* Vol. 27, 1960, pp. 352-353.
[*4*] Cherepanov, G. P., *Journal of Applied Mathematics and Mechanics,* Vol. 31, No. 3, 1967, pp. 476-488.
[*5*] Knowles, J. K. and Sternberg, E., *Archive for Rational Mechanics and Analysis,* Vol. 44, 1972, pp. 187-211.
[*6*] Green, A. E., *Archive for Rational Mechanics and Analysis,* Vol. 50, 1973, pp. 73-80.
[*7*] Chen, F. H. K. and Shield, R. T., *Journal of Applied Mathematics and Physics,* Vol. 28, 1977, pp. 1-22.
[*8*] Smelser, R. E. and Gurtin, M. E., *International Journal of Fracture,* Vol. 13, 1977, pp. 382-384.
[*9*] Hutchinson, J. W., *Journal of the Mechanics and Physics of Solids,* Vol. 16, 1968, pp. 13-31.
[*10*] Rice, J. R. and Rosengren, G. F., *Journal of the Mechanics and Physics of Solids,* Vol. 16, 1968, pp. 1-12.
[*11*] Kumar, V., German, M. D., and Shih, C. F., "An Engineering Approach to Elastic-Plastic Fracture," Report NP-1931, prepared by General Electric, Electric Power Research Institute, Palo Alto, CA, 1981.
[*12*] McMeeking, R. M., *Journal of the Mechanics and Physics of Solids,* Vol. 25, 1977, pp. 357-381.
[*13*] Dowling, N. E. in *Cyclic Stress-Strain and Plastic Deformation Aspects of Fatigue Crack Growth, ASTM STP 637,* American Society for Testing and Materials, Philadelphia, 1977, pp. 97-121.
[*14*] Kaisand, L. R. and Mowbray, D. F., *Journal of Testing and Evaluation,* Vol. 7, No. 5, 1979, pp. 270-280.
[*15*] Shih, C. F. in *Mechanics of Crack Growth, ASTM STP 590,* American Society for Testing and Materials, Philadelphia, 1976, pp. 3-26.
[*16*] Shih, C. F. and Hutchinson, J. W., "Fully Plastic Solutions and Large Scale Yielding Estimating for Plane Stress Crack Problems," Report No. DEAP-S-14, Division of Engineering and Applied Physics, Harvard University, Cambridge, MA, 1975.
[*17*] Rice, J. R. in *Fracture,* Vol. 2, H. Liebowitz, Ed., Academic Press, New York, 1971, pp. 191-311.
[*18*] Dowling, N. E. and Begley, J. A. in *Mechanics of Crack Growth, ASTM STP 590,* American Society for Testing and Materials, Philadelphia, 1976, pp. 82-103.
[*19*] Rice, J. R., Paris, P. C., and Merkle, J. G. in *Flow Growth and Fracture Toughness Testing, ASTM STP 536,* American Society for Testing and Materials, Philadelphia, 1973, pp. 231-245.
[*20*] Wilson, W. K. and Yu, I. W., *International Journal of Fracture,* Vol. 15, 1979, pp. 377-387.
[*21*] Irwin, G. R. in *Hanbuch der Physik,* Vol. 6, Springer, Berlin, 1958.
[*22*] Sih, G., *Journal of Applied Mechanics,* Vol. 29, 1962, pp. 587-589.
[*23*] McCartney, L. N., *International Journal of Fracture,* Vol. 15, 1979, pp. R217-R221.
[*24*] Gurtin, M. E., *International Journal of Fracture,* Vol. 15, 1979, pp. R169-R170.
[*25*] Read, D. T. and McHenry, H. I. in *Proceedings,* 5th International Conference on Fracture, Cannes, France, 1981, pp. 1715-1722.
[*26*] Ainsworth, R. A., Neale, B. K., and Price, R. H. in *Proceedings,* Institute of Mechanical Engineers' Conference on Tolerance of Flaws in Pressurized Components, London, 1978, pp. 171-178.
[*27*] Blackburn, W. S., *International Journal of Fracture Mechanics,* Vol. 8, 1972, pp. 343-346.
[*28*] Blackburn, W. S. and Jackson, A. D., *International Journal of Fracture,* Vol. 13, 1977, pp. 183-200.
[*29*] Kishimoto, K., Aoki, S., and Sakata, M., *Engineering Fracture Mechanics,* Vol. 13, 1980, pp. 841-850.
[*30*] Aoki, S., Kishimoto, K., and Sakata, M., *Engineering Fracture Mechanics,* Vol. 16, 1982, pp. 405-413.
[*31*] Atluri, S. N., *Engineering Fracture Mechanics,* Vol. 16, 1982, pp. 341-364.
[*32*] Atluri, S. N., Nishioka, T., and Nakagaki, M., "Incremental Path-Independent Integrals in Inelastic and Dynamic Fracture Mechanics," Report No. GIT-CACM-SNA-83-27, Georgia Institute of Technology, Atlanta, GA, May 1983.
[*33*] Nakagaki, M., Atluri, S. N., and Nishioka, T., "On the Path Independent Integral, ΔT_p, in Elastic-

Plastic Fracture Mechanics," presented at the SECTAM XII Conference, Pine Mountain, GA, May 1984.

[*34*] Stonesifer, R. B. and Atluri, S. N., "Creep Crack-Growth: A New Path-Independent Integral (Tc) and Computational Studies," NASA Report 1678917, National Aeronautics and Space Administration, Washington, DC, July 1982.

[*35*] Nguyen, Q. S. in *Proceedings,* IUTAM Symposium on Three-Dimensional Constitutive Relations and Ductile Fracture, North-Holland Publishing Company, Amsterdam, The Netherlands, 1981, pp. 315–330.

[*36*] Bui, H. D. in *Proceedings,* IUTAM Symposium on Three-Dimensional Constitutive Relations and Ductile Fracture, North-Holland Publishing Company, Amsterdam, The Netherlands, 1981, pp. 357–362.

[*37*] Germain, P., Nguyen, Q. S., and Suquet, P., *Journal of Applied Mechanics,* Vol. 50, 1983, pp. 1010–1020.

[*38*] Aoki, S., Kishimoto, K., and Sakata, M., "Energy Flux into Process Region in Elastic-Plastic Fracture Problems," *Engineering Fracture Mechanics,* Vol. 19, No. 5, 1984, pp. 827–836.

Stojan Sedmak[1] *and Blagoj Petrovski*[1]

Application of Direct Measurement of J-Integral on a Pressure Vessel with Axial Notch

REFERENCE: Sedmak, S. and Petrovski, B., **"Application of Direct Measurement of J-Integral on a Pressure Vessel with Axial Notch,"** *Fracture Mechanics: Eighteenth Symposium, ASTM STP 945,* D. T. Read and R. P. Reed, Eds., American Society for Testing and Materials, Philadelphia, 1988, pp. 730–740.

ABSTRACT: The *J*-contour integral value was determined by the method of direct evaluation on a pressure vessel with an axial crack in a welded specimen. The specimen, cut from a weldment made by manual-arc welding, was machine-notched and fatigue precracked. It was mounted by welding on the pressure vessel, positioned in the opening cut on the cylindrical shell. The test was performed by gradually increasing water pressure. Crack mouth opening displacement (CMOD) was measured by clip gage, and strains were measured by strain gages, positioned along the J-integral path. J-integral, CMOD, and hoop stress are presented as a function of applied pressure. The fatigue crack length was extended by approximately 10 mm during pressurization. The development of an experiment to determine the J_R curve in a full-scale test and to predict the residual strength of a cracked pressure vessel is considered.

KEY WORDS: contour integral, direct evaluation, experimental full-scale test, pressure vessel, fracture mechanics, crack, strain, stress, displacement, weldment

The operational safety of pressurized structures (such as pressure vessels, penstocks, and pipelines) primarily depends on the behavior of the welded joint under loading. The possibility of the existence of defects and cracks in welded structures must be recognized, warranting a fracture mechanics approach to investigating the behavior of weld defects.

The choice of Japanese HSLA steel SM 80P for the construction of the underground penstock for the Bajina Bašta Hydroelectric Pumping-up Power Plant, recently constructed in Yugoslavia, and the high load level required a detailed analysis of weldment behavior. Full-scale prototypes of the penstock in the form of pressure vessels (with 47-mm wall thickness) were made and tested. The crack arrest concept was the basis for the evaluation of penstock weldments in the first experiment [*1*], and the crack initiation concept was accepted as the basis for the second experiment [*2*]. The results of these experiments led to the conclusion that the Bajina Bašta penstock welding technology was suitably defined and the operational safety of the penstock is satisfactory up to the load level of the test conditions.

The problem of stable crack growth in a welded structure was not considered in the two experiments performed. In order for us to gain insight into possible crack propagation, it was necessary to extend the experiments to stable crack growth behavior. The basis of the accepted approach to calculations of crack extension is to compare the crack driving "force" for the structure to the material crack resistance properties [*3*], which can be defined by fracture mechanics tests on specimens. The next step could be to evaluate crack behavior in full-scale tests. The opportunity to implement this approach was offered by newly developed experimental

[1] Professors, Faculty of Technology and Metallurgy, University of Belgrade, Belgrade, Yugoslavia.

method for direct evaluation of J-integral [4]. With this method, J-integral was found to be a convenient fracture mechanics parameter for crack behavior during and after stable growth. On this basis it was possible to develop a technique for life prediction of a cracked structure [5]. For the safe operation of a cracked structure, the J-integral measured on the pressure vessel material has to be of higher value than the J-integral calculated from the crack driving force under the same stress and strain condition.

The full-scale prototype tests were extended to one additional pressure vessel (Fig. 1) made of SM 80P steel by manual-arc welding (MAW) and submerged-arc welding (SAW). The tensile properties of SM 80P steel and its weld metals are given in Table 1.

The pressure vessel in the uncracked condition (as-welded) was pressurized up to 130 bar (10 bar = 1 MPa) in order to verify its general behavior under loading. The values of strains on base and weld metal, measured by strain gages (SG), were in good agreement with the calculated values.

Application of J-Integral to Residual Strength Prediction of Pressure Vessel

The residual strength of a cracked pressure vessel can be predicted by comparing the crack driving force generated in the pressurized vessel and the crack resistance of material used for pressure vessel weldments. Crack driving force can be expressed by the J-integral value and

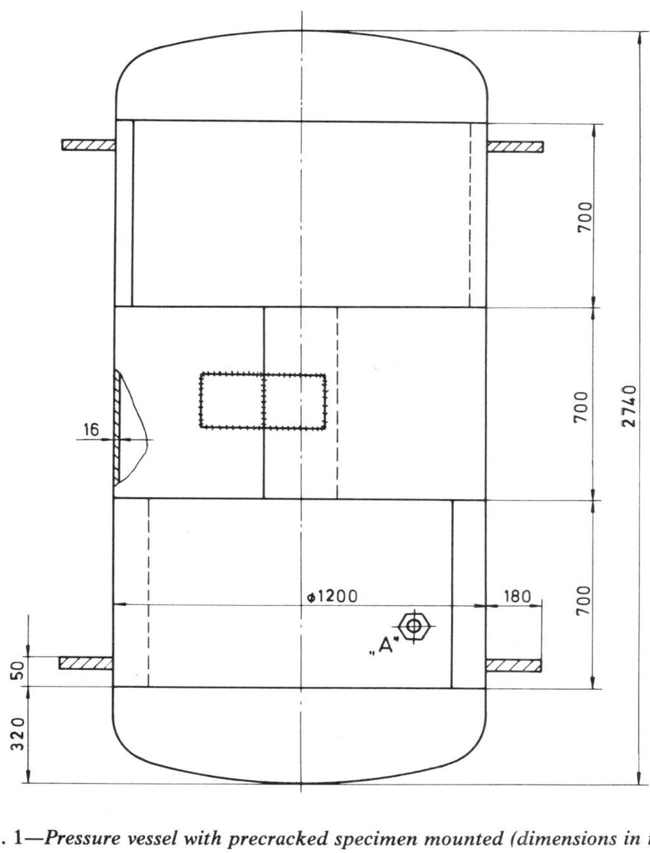

FIG. 1—*Pressure vessel with precracked specimen mounted (dimensions in mm).*

TABLE 1—*Tensile properties of SM 80P steel and its manual arc welding (MAW) and submerged arc welding (SAW) weld metals.*

		Tensile Properties		
	Specimen Position	Yield Point, MPa	Ultimate Tensile Strength, MPa	Elongation, δ, %
SM 80P plate	rolling direction	755 to 794	804 to 834	24 to 29
	cross rolling direction	755 to 794	795 to 834	22 to 23
MAW	weld metal	722	810	22
SAW	weld metal	687	804	23

calculated for specified pressure vessel geometry and loading conditions. Crack-resistance properties of structural material can be experimentally determined from the fracture mechanics tests and expressed by the J-integral value as well.

Following this approach, the analysis of crack growth behavior in both the stable and unstable regimes can be performed [*5,6*]. For the pressure vessel of Fig. 1, the shell parameter λ can be obtained as

$$\lambda = [12(1 - \nu^2)]^{1/4} \frac{a}{\sqrt{Rt}} = 0.6 \tag{1}$$

for Poisson's ratio $\nu = 1/3$, nominal shell radius $R = 592$ mm, wall thickness $t = 16$ mm, and observed crack length $2a = 65$ mm (Fig. 2).

Using the procedure described in Ref *5,* a set of curves $\sqrt{J}$ versus d/t (d being crack depth, Fig. 2) could be drawn for specified values of pressure levels (Fig. 3). These curves represent the crack driving force, expressed by J-integral, for shell parameter $\lambda = 0.6$ and for different (crack depth/wall thickness) ratios.

Crack resistance curves for the specimens with the crack positioned in base metal (BM) and submerged-arc weld metal (WM) are superimposed in Fig. 3. Curves 80-1 (BM) and N-2 (WM) are obtained by three-point bending, using the specimens made in accordance with the ASTM Test Method for J_{Ic}, a Measure of Fracture Toughness (E 813-81). Curves BM-2 (BM) and 83 (WM) are obtained by tension testing using the procedure for J-integral direct evaluation [*4*] applied on the specimens made in the form of tension panels with the part-through surface crack [*6*].

Some information on crack behavior in the pressure vessel considered can be obtained from Fig. 3. The weld metal exhibits lower crack resistance compared to the base metal, and Point A of the critical pressure level can be clearly defined for weld metal. The stable crack growth for the initial crack depth 9.1 mm (Specimen N-2) can be expected up to the pressure level 150 bar (Point A), and crack extension in a stable manner to 11.5 mm can be predicted. This is an additional confirmation of operational safety of the Bajina Bašta penstock, since the working pressure in it is as high as 90 bar and the waterhammer effect can raise the pressure up to 120 bar.

The J-integral values for weld metal (Specimens N-2 and 83) are of the same order of magnitude as in other tests (V-8A welds in Ref *7*).

It is to be noted that J_R-curves were obtained by testing the specimens, and crack driving force was calculated for the idealized pressure vessel. The loading conditions in specimens and in the pressure vessel are different and, accordingly, the stress and strain states have to be different. The crack resistance behavior in bend and tension tests could be evaluated as similar

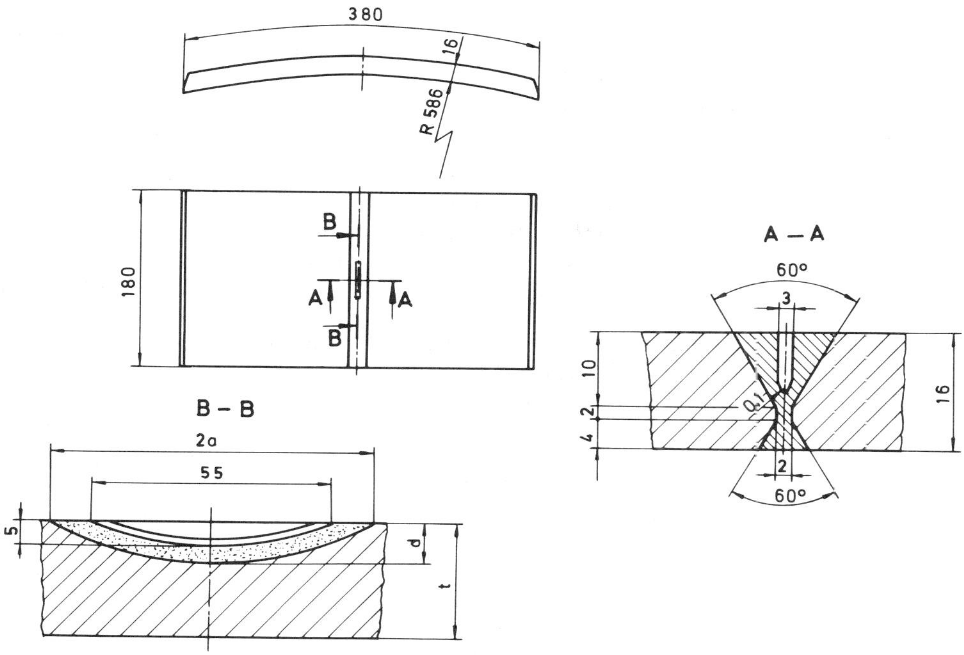

FIG. 2—*Specimen prepared for mounting on the pressure vessel (dimensions in mm).*

for similar (crack depth/wall thickness) ratios [*6*], but it is not quite clear if this will be the case when biaxial loading is applied, as in the pressure vessel. Additional analysis of real crack behavior on the pressure vessel can be performed by measuring crack resistance parameters in full-scale tests.

Experiment with Precracked Pressure Vessel

From a theoretical background and for experimental convenience it was concluded that the J-integral could be selected as the fracture mechanics parameter for full-scale testing. The theoretical background is described elsewhere (for example, Ref *8*). The method of direct evaluation of the J-integral [*4*] has been found promising. To apply this experimental method to a real full-scale structure, a precracked pressure vessel was tested.

Two essential experimental problems had to be solved in comparison with the uncracked pressure vessel test:

1. Preparing a cracked area on the pressure vessel.
2. Sealing cable feedthroughs for SG positioned on the inside of the vessel.

The part-through surface crack is accepted as typical in the real operational situation in the case of pressure vessels. The selected technique for preparation of such a crack involved five steps:

1. Fabricating cold bent and welded specimens at the same time that the pressure vessel was made and under the same conditions;
2. Machining the specimen to 380 by 180 by 16 mm with a notch of circular arc shape in MAW weld metal (Fig. 2). The machined notch was 55 mm long and 5 mm deep;

3. Fatigue precracking the specimen by three-point bending, with maximum force F_{max} = 100 kN and loading ratio $F_{min}/F_{max} = 0.1$, with a span of L = 200 mm for 350 800 cycles. The final length of the crack was $2a$ = 65 mm;
4. Cutting the opening for the specimen in the pressure vessel wall;
5. Mounting the specimen on the pressure vessel by MAW welding (Fig. 1).

The specimen welded by MAW procedure was selected in this experiment, since this welding procedure is accepted as inevitable for repair welding, and the weld defects are found mostly in such weldments of Bajina Bašta penstock. However, the crack behavior of base metal and SAW weld metal are required for complete evaluation of penstock safety. And finally, crack behavior of heat-affected zone (HAZ) is probably of the greatest importance, but the solution of this problem requires further investigation.

In addition to strain gages positioned for the testing of the uncracked pressure vessel, a new series of strain gages was positioned along the path selected for J-integral evaluation (Fig. 4), and a clip gage was provided for measuring of crack mouth opening displacement (CMOD). The problem of sealing cables was solved following the design developed by JRC (Ispra-Italy) (private communication) and presented in Fig. 5.

The pressure vessel was loaded by gradually rising pressure, obtained by an electrically powered pump controlled manually. The scanning of the instrumentation was made at predetermined pressure levels. The results obtained in this experiment for certain pressure levels are given in Fig. 6.

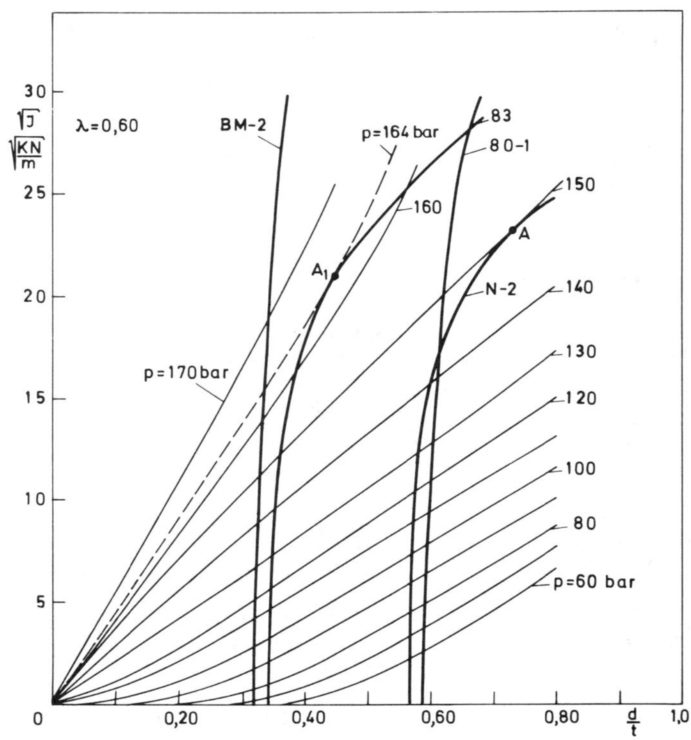

FIG. 3—*Scheme for residual strength prediction.*

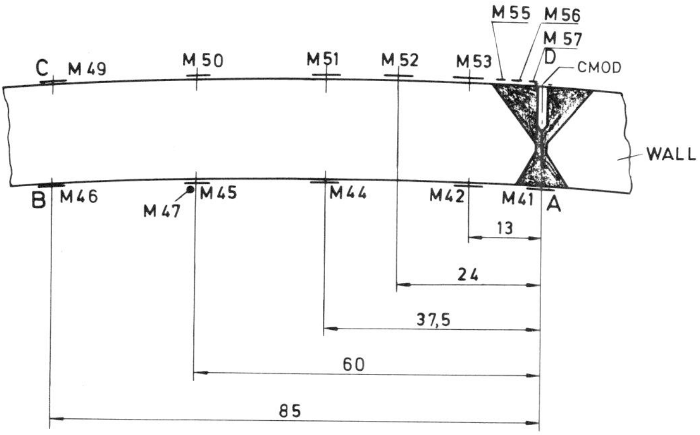

FIG. 4—*J-integral path and the instrumentation applied for the direct evaluation of J-integral in cracked pressure vessel full-scale test.*

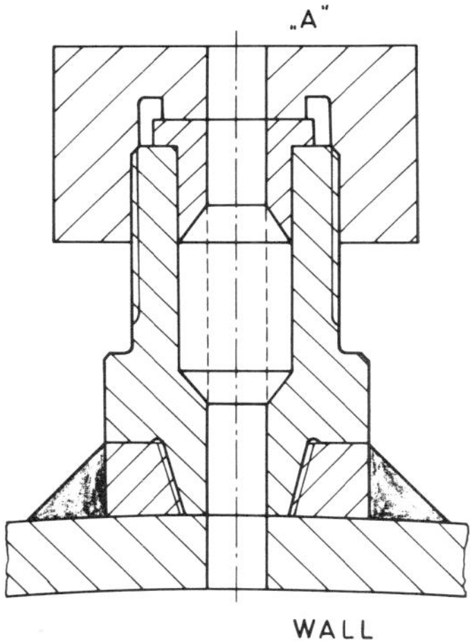

FIG. 5—*Design of cables sealing.*

Application of Method for Direct Evaluation of J-Contour Integral to the Pressure Vessel

Following the procedure described in Ref *4* and also applied on tension panel with a part-through surface flaw [*6*], J-contour integral values for an artificial axial crack on the tested pressure vessel could be evaluated.

Strain distribution along J-integral path, given in Fig. 6, together with CMOD from Fig. 7, represents the basis for J-integral evaluation.

The fundamental expression for Rice's J-integral [*8*] is

$$J = \int_{\Gamma} \left(W dy - \vec{T} \frac{\partial \vec{u}}{\partial x} ds \right) \qquad (2)$$

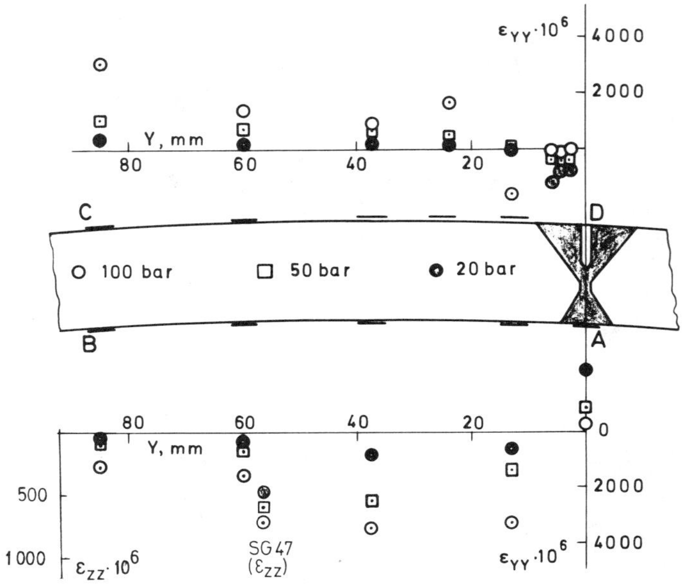

FIG. 6—*Strain distribution along uncracked and cracked side of pressure vessel wall for different pressure levels.*

where

W = strain work density,
$\vec{T}$ = traction vector,
$\vec{u}$ = displacement vector,
x and y = rectangular coordinates, and
s = arc length along the contour.

For the path ABCD defined in Fig. 4, the J-integral reduces to the form

$$J = 2SW_{\mathrm{UCR}} + 2SW_{\mathrm{CR}} - ST \tag{3}$$

where SW_{UCR} and SW_{CR} stand for strain work density contribution of uncracked and cracked side, respectively, and ST is the contribution from the traction-bending term $\vec{T} \cdot (\partial \vec{u}/\partial x) ds$.

Assuming elastic-perfectly plastic behavior of material (for SM 80P steel and its MAW weld metal the difference between yield stress and ultimate tensile strength is small, Table 1, and their tensile behavior is close to this assumption), the strain work density can be obtained from

$$W = \frac{E \cdot \epsilon_{yy}^2}{2} \quad \text{for} \quad \epsilon_{yy} < \frac{\sigma_Y}{E}$$

$$W = \frac{\sigma_Y^2}{2E} + \sigma_Y\left(\epsilon_{yy} - \frac{\sigma_Y}{E}\right) \quad \text{for} \quad \epsilon_{yy} > \frac{\sigma_Y}{E} \tag{4}$$

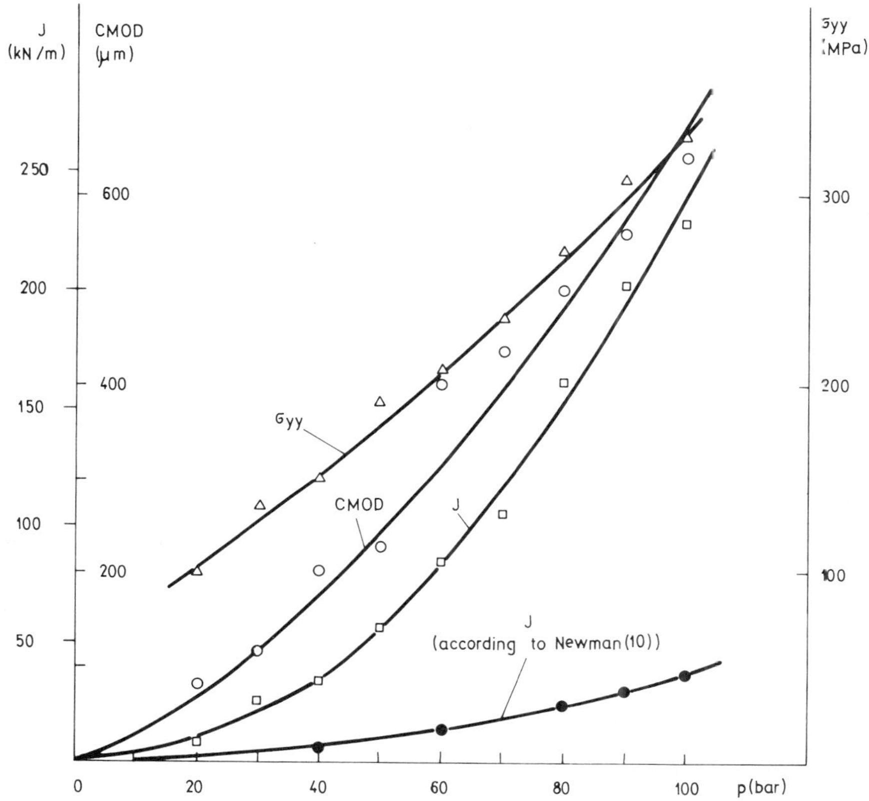

FIG. 7—*Hoop stress σ_{yy}, CMOD and J-integral versus pressure; J-integral versus pressure according to Newman's solution [9] for an axial crack in cylindrical pressure vessel.*

and the first term in Eq 2 is obtained by numerical integration along the path contour. The traction vector $\vec{T}$ reduces to its component $T_Y = \sigma_{yy}$ [4]. The acting bending term $\partial u_y/\partial x$ across the thickness $\overline{BC} = t = 16$ mm is approximated by

$$\frac{\partial u_y}{\partial x} = \frac{u_y(C) - u_y(B)}{x(C) - x(B)} = \frac{u_y(C) - u_y(B)}{16} \tag{5}$$

and

$$u(C) = 2\int_D^C \epsilon_{yy}\,dy + \text{CMOD}$$
$$u(B) = 2\int_A^B \epsilon_{yy}\,dy \tag{6}$$

The integrals (Eq 6) had been calculated numerically following the strain distribution from Fig. 6. The terms from the right side of Eq 5 and the evaluated J-integrals are given in Table 2.

TABLE 2—*Values of J-integral for different pressure levels.*

Pressure p, bar[a]	Contributing Parts			
	SW_{UCR}, kN/m	SW_{CR}, kN/m	ST, kN/m	J-integral, kN/m
20	4.68	−0.96	4.07	7.79
30	24.32	−2.75	3.58	25.15
40	15.96	−5.19	22.64	33.41
50	40.85	−9.18	24.00	55.67
60	51.28	−18.72	52.18	81.75
70	60.93	−25.64	70.34	105.62
80	106.83	−32.50	86.82	161.18
90	127.79	−40.70	115.19	202.28
100	115.72	−39.84	152.36	228.24

[a]10 bar = 1 MPa.

Results and Discussion

The results of the experiment are given in Fig. 7 in graphic form of J-integral, CMOD and σ_{yy} dependance on pressure p. The hoop stress σ_{yy} was calculated from

$$\sigma_{yy} = \frac{E}{1 - \nu^2} (\epsilon_{yy} + \nu \cdot \epsilon_{zz})$$

with the value ϵ_{yy} measured on SG45 and ϵ_{zz} measured on SG47, positioned in the axial direction, near SG45 (Fig. 6).

The relationships in Fig. 7 show the regular behavior of pressure vessel deformation during elastic loading and stable crack growth. It is to be noted that the crack propagated not only in depth, but also in length. The length extension during fatigue was about 10 mm, and during pressurization another 10 mm, and could be followed by the naked eye.

The results, presented in Figs. 3 and 7, can be compared with the elastic solution for a part-through crack in a pressure vessel, given by Newman [*9*]. For this, the initial crack depth $d_0 = 9$ mm was assumed, and the initial crack length $2d_0 = 65$ mm could be accepted from the experiment. The J-integral values obtained for different pressure levels are superimposed on Fig. 7.

For $d_0/t = 9/16$, similar to the N-2 curve in Fig. 3, the $\sqrt{J}$-integral ($\simeq 8\sqrt{kN/m}$) at 100-bar pressure level can be compared to $J = 35.8$ kN/m from the elastic solution (Fig. 7). However, the J-integral value at 100 bar in the full-scale test was found to be $J = 228.2$ kN/m, indicating higher crack resistance of the structure (pressure vessel) compared with the specimen. This means that the prediction from Fig. 3 could be taken as on the safe side despite the fact that the exact value of d is still unknown in the full-scale test. This value will be available for analysis after the final fracturing of the welded specimen (Fig. 2).

It was shown during this experiment that it is possible to evaluate directly the J-contour integral on the full-scale pressure vessel. However, many aspects of direct J-integral evaluation also have to be considered, both from the experimental and theoretical points of view.

The first question is how to evaluate the residual stresses induced by specimen welding, or how to avoid them. The ASTM Standard Method for Determining Residual Stresses by the Hole-Drilling Strain-Gage Methods (E 837-81), could be used for the evaluation of residual stresses induced by welding. On the other hand, the residual stresses induced by specimen welding could be avoided by machining the notch in the original pressure vessel wall. In such case, the machined notch or crack produced by controlled explosion [*10*] could replace the fatigue precracking.

The second question is connected with the further development of the experiment in order to obtain the J_R-curve for the full-scale pressure vessel. In this way it would be possible to predict the pressure vessel residual strength when existing crack develops in a stable manner. Such a procedure was performed for the pressure vessel used for this experiment, but the J_R-curves in Fig. 3 were obtained by testing the tension panels with part-through surface crack and three-point bend specimens [*6*].

Applying the partially unloading method and compliance technique described in ASTM E 813 to the CMOD-pressure relationship, it might be possible to obtain the J_R-curve for the full-scale pressure vessel. The comparison of the J_R-curve for the full-scale pressure vessel with the curves obtained for the specimens could help in understanding the size and stress state effects on J-integral behavior.

At least two problems have to be considered from a theoretical point of view. The first is the biaxial stress effect on the J-contour integral, and the second is in connection with the shell curvature effect on the fracture mechanics parameters. For the final solution of the cracked pressure vessel strength prediction these two effects have to be taken into account.

Conclusions

1. It was shown by the experiment performed on the full-scale pressure vessel with an axial crack in MAW weld metal that the experimental method for direct evaluation of the J-integral is applicable not only for specimens, but also for full-scale structures.
2. The sealing of SG cables using the design developed by JRC-Ispra was satisfactory for this experiment.
3. Further development of J-integral evaluation in the full-scale test could be achieved through J_R-curve determination. The compliance technique with CMOD-pressure dependance and partially unloading test procedure could be applied, together with the method for direct evaluation of the J-integral, for J_R-curve determination on a pressure vessel with an existing crack.
4. Additional experiments and theoretical analyses are needed for understanding of the biaxial stress effect and shell curvature effect on the J-integral.

Acknowledgments

The authors are indebted to the GOŠA Industry for permission to publish the experimental results. They wish to thank Dr. D. Drenić for his help in experiment organization. They also acknowledge the cooperation of R. P. Reed, H. I. McHenry, and D. T. Read from the Fracture and Deformation Division of the National Bureau of Standards, who have contributed to the general development of the project "Fracture Mechanics of Weldments," for which this experiment was organized.

References

[*1*] Rak, I., Gliha, V., and Kuder, J., *Metalna Strokovni Bilten '84 (Metalna Technical Journal),* Vol. 1, No. 1, Sept. 1984, pp. 15–23.

[*2*] Sedmak, S., Radović, A., and Nedeljković, Lj. in *Proceedings,* 3rd International Conference on Mechanical Behavior of Materials, Cambridge, England, K. J. Miller and R. F. Smith, Eds., Aug. 1979, Pergamon Press, Oxford, U.K., Vol. 3, pp. 435–446.

[*3*] Ratwani, M. M. in *Introduction to Fracture Mechanics and Fracture-Safe Design,* S. Sedmak, Ed., TMF-Goša, Belgrade, Yugoslavia, 1980, pp. 169–198, 550–556.

[*4*] Read, D. T. in *Fracture Mechanics: Fourteenth Symposium, Volume II: Testing and Applications, ASTM STP 791,* American Society for Testing and Materials, Philadelphia, 1983, pp. II-199–II-213.

[*5*] Ratwani, M. M., Petrovski, B., and Sedmak, S., "Primena krivih otpornosti u predvidjanju loma suda pod pritiskom sa površinskom prslinom" ("Resistance Curve Approach to Failure Prediction in

Pressure Vessel with Surface Flaw") in *Mehanika loma zavarenih spojeva (Fracture Mechanics of Weldments),* TMF-Goša, Belgrade, Yugoslavia, 1985.

[*6*] Petrovski, B., "Odredjivanje preostale nosivosti suda pod pritiskom sa površinskom prslinom" ("Residual Strength Prediction in Pressure Vessel with Surface Crack"), Ph.D. thesis, Faculty of Technology and Metallurgy University of Belgrade, Belgrade, Yugoslavia, 1985.

[*7*] Bryan, R. H., "Results of Low Ductile Shelf Intermediate Vessel Test V-8A*," Oak Ridge National Laboratory, Oak Ridge, TN, 1982, pp. 289-300.

[*8*] Rice, J. R., *Journal of Applied Mechanics,* Vol. 35, 1968, pp. 379-386.

[*9*] Newman, J. C., Jr., "Fracture Analysis of Surface and Through Cracks in Cylindrical Pressure Vessels," TN D-8325, National Aeronautics and Space Administration, Washington, DC, 1976.

[*10*] Kuder, J., "Making of Crack-Starters by Explosion," Document-X-916-78, International Institute of Welding, 1978.

Ted L. Anderson[1]

A Comparison of J-Integral and CTOD as Fracture Toughness Parameters

REFERENCE: Anderson, T. L., **"A Comparison of J-Integral and CTOD as Fracture Toughness Parameters,"** *Fracture Mechanics: Eighteenth Symposium, ASTM STP 945,* D. T. Read and R. P. Reed, Eds., American Society for Testing and Materials, Philadelphia, 1988, pp. 741–753.

ABSTRACT: The J-integral and the crack-tip opening displacement (CTOD) are widely used throughout the world to measure the fracture toughness of structural materials. There are many similarities between these two parameters, as well as a number of important differences. Because of these differences, the CTOD parameter tends to be more conservative when applied to structures.

The J-integral and CTOD were compared qualitatively and analytically. This work differs from previous comparisons in that the present study compared engineering estimates of J and CTOD, which are not necessarily equal to the true values of J and CTOD. However, the fracture toughness of a material is nearly always characterized in terms of the engineering approximations of J or CTOD.

The main difference between J and CTOD as measures of toughness is that J is directly proportional to the area under the load-displacement curve whereas CTOD is proportional to the clip gage displacement. This means that the relationship between the two parameters depends on strain hardening and constraint.

KEY WORDS: crack-tip opening displacement, elastic-plastic fracture mechanics, J-integral, fitness for purpose

Nomenclature

a	Crack length
B	Specimen thickness
E	Young's modulus
G	Crack extension force
G_Y	G at yield
J	J-integral
J_c	Critical J
J_{Ic}	Critical J at initiation when ASTM Test Method for J_{Ic}, a Measure of Fracture Toughness (E 813-81) criteria are satisfied
J_p	Plastic component of J
K	Stress intensity factor
K_{Ic}	Critical K under plane-strain conditions
L	Notch constraint factor
m	Dimensionless constant which relates J to CTOD
m_e	Elastic m
m_0	Overall m
m_p	Plastic m
N	Hardening exponent

[1]Assistant professor of mechanical engineering, Texas A & M University, College Station, TX.

n Ramberg-Osgood strain hardening exponent ($= 1/N$)
P Load
P_Y Load at net section yield
q Load line displacement
q_p Plastic component of q
q_Y Elastic q at yield
r_p Plastic rotational factor
U Energy absorbed by specimen
U_p Plastic component of energy
V_p Plastic clip gage displacement
W Specimen width
Y Linear elastic fracture mechanics (LEFM) correction factor
Z Nondimensional compliance
z Knife edge height
δ Crack-tip opening displacement (CTOD)
δ_c Critical CTOD
δ_e Elastic CTOD
δ_p Plastic CTOD
σ_Y Flow stress
σ_{YS} Yield strength

Soon after the development of fracture toughness (K_{Ic}) testing based on linear elastic fracture mechanics (LEFM), it was recognized that LEFM could not be applied to all engineering materials; high toughness materials, for instance, often undergo extensive plastic deformation prior to fracture initiation. For this reason the concepts of fracture mechanics were extended to account for elastic-plastic behavior. The two most commonly used elastic-plastic toughness parameters are the crack-tip opening displacement (CTOD) and the J-integral.

The CTOD parameter was originally proposed by Wells [*1*] in 1961. Wells observed that when a fracture specimen was loaded beyond its elastic limit, the crack tip blunted and the faces of the crack moved apart. It was hypothesized that fracture would occur when the CTOD reached a critical value. Since that time, CTOD has been widely used as a fracture toughness parameter. In 1979, the British Standards Institution published Methods for Crack Opening Displacement (CTOD) Testing (BS 5762: 1979); an ASTM standard for CTOD testing is currently in draft form.

The J contour integral, which was developed by Rice [*2*] in 1968, is a line integral which characterizes the stress and strain fields near the crack tip. For linear and nonlinear elastic materials, the J integral is path independent; that is, the value of J is the same regardless of the path of integration around the crack tip. J is nearly path-independent for elastic-plastic materials that are loaded monotonically.

In the early 1970s Landes and Begley [*3–5*] proposed J as a measure of fracture toughness. Since it was not practical to calculate J by performing line integrations on laboratory specimens, simplified J estimation procedures were devised. These procedures allowed J values to be estimated from load displacement records. For bend specimens, J is given by

$$J = \frac{2U}{B(W - a)} \tag{1}$$

where

U = energy absorbed by the specimen (defined as the area under the load versus load line displacement curve),

B = specimen thickness,
W = specimen width, and
a = crack length.

A similar expression has been derived for compact tension (CT) specimens [6]. Equation 1 was originally derived for deep-notched specimens ($a/W \geq 0.5$). However, recent results [7] indicate that this equation can be applied down to a/W ratios of 0.2.

In the 1970's, J was primarily used in the United States while CTOD was used primarily in Britain. However, in the past few years both parameters have been used throughout the world. Investigators with backgrounds in solid mechanics or numerical analysis tend to favor J over CTOD because J can be derived from nonlinear elasticity theory. Also, J values can be obtained numerically by integrating around a contour in a finite element mesh, whereas a numerical analysis of CTOD requires special elements which accurately model crack blunting. Many engineers, however, prefer CTOD as a measure of toughness because it is a physical parameter that can be understood intuitively.

There are a number of widely held misconceptions about J and CTOD. Because J was originally defined by a complex analytical expression, many people have conferred a high status upon J as a fracture parameter. Conversely, since CTOD is a physical parameter, it is often regarded as a simple, qualitative measure of material toughness, similar to the Charpy or drop-weight tests. However, as shown later in this paper, J and CTOD are closely related. It is therefore misleading to state categorically that one parameter is "better" or "more scientific" than the other.

One advantage of the J test is that a critical J value can sometimes be used to calculate the K_{Ic} of a material. This is a highly desirable quality because prohibitively large specimens are often required for valid K_{Ic} tests. However, only a K_{Ic} value for ductile fracture can be obtained from a J_{Ic} value, that is, the value of J (in a valid J_{Ic} test) at the onset of ductile crack growth. Upper-shelf initiation toughness, whether quantified by J or CTOD, is not a strong function of specimen size [4, 7]. For this reason, specimen size requirements for valid J_{Ic} tests are much less stringent than the K_{Ic} validity requirements. However, fracture toughness in the ductile-to-brittle transition region, where fracture occurs by a cleavage mechanism, is highly dependent on the size and geometry of the test specimen [7–11]. The magnitude of these size effects is roughly the same for critical J and CTOD values [7]. Therefore, K-values computed from critical J-values in the transition region are not necessarily equal to K_{Ic} for cleavage fracture.

The CTOD method has the advantage of being the only standardized fracture toughness test that can be applied to the entire toughness range of steels, from the lower shelf to the upper shelf. Because of size restrictions, valid K_{Ic} results can usually be obtained only on the lower shelf. The J_{Ic} test, as defined in ASTM E 813, is intended exclusively for the upper shelf. According to BS 5762:1979, CTOD tests can be performed on the upper and lower shelves, as well as in the transition region of steels. Critical J values can be measured in the transition region [7,10–12], but these tests do not conform to any existing standard. Dawes [13] has recently proposed a J equivalent of the CTOD test. This would give the same continuity in J-based tests as in CTOD-based tests.

The differences and similarities between J and CTOD are explored in more detail next. The ramifications with respect to design are also investigated.

Relationship Between J and CTOD

Previous Work

Wells [1] derived a relationship between CTOD and crack extension force under conditions of small-scale yielding at the crack tip. This relationship has the form

$$\delta = \frac{G}{m\sigma_{YS}} = \frac{K^2}{m\sigma_{YS}E} \tag{2}$$

where

δ = CTOD,
G = crack extension force,
σ_{YS} = yield strength,
E = Young's modulus,
K = stress intensity, and
m = dimensionless constant.

For the linear elastic case, $J = G$. Therefore

$$J = m\sigma_{YS}\delta \tag{3}$$

A number of investigators [*7,10,14–16*] have applied Eq 3 to the elastic-plastic regime. Experimental values of m generally were found to lie between 1.0 and 2.0. Recent revisions to ASTM E 813 and ASTM Terminology Relating to Fracture Testing (E 616-82) define the J-CTOD relationship as follows

$$J = m'\sigma_Y\delta \tag{4}$$

where σ_Y is the flow stress, defined as the average between σ_{YS} and the tensile strength. Equation 4 provides a means for taking strain hardening effects into account. In the present paper, the original definition of the m factor (Eq 3) is used because hardening effects are taken into account by means of a Ramberg-Osgood stress-strain relationship (see Appendix).

Shih [*14*] derived analytical equations that relate J to CTOD. These relationships, which have a similar form to Eq 3, are based on the Hutchinson, Rice, and Rosengren [*17,18*] stress-strain singularity at the crack tip. Shih conducted numerical studies that showed that his equations work well for small scale yielding but break down under large-scale yielding conditions. Unfortunately, many have ignored these results and applied the Shih equations to cases where large scale yielding occurs.

Present Work

Unlike the Shih equations [*14*], the relationships derived in this paper are valid for large-scale yielding. Another important difference is that the Shih relationships and a subsequent analysis by Schindler [*19*] are based on the fundamental definitions of J and CTOD, whereas the relationships derived here are based on the engineering equations that are applied to laboratory specimens.

There is some doubt as to whether the J estimations in ASTM E 813 are valid in many cases [*20*]. That is, the laboratory J test might not measure the true J as defined by Rice [*2*]. The same applies to the CTOD test. Since the CTOD equation in BS 5762:1979 contains several simple assumptions about the deformation behavior of test specimens, a calculated CTOD value is not necessarily equivalent to the actual displacement at the crack tip.

That engineering estimates of J and CTOD may not correspond to "true" values is not important in the present analyses. It can be argued that it is more important to understand the relationship between experimental estimates of J and CTOD because the fracture toughness of engineering materials is almost always quantified in terms of these estimates rather than "true" J and CTOD values. Tests performed according to a J or CTOD standard provide a self-consistent measure of toughness, that is, the ability of the material to withstand elastic and plastic deformation prior to fracture.

Qualitative Comparisons

Both J and CTOD are usually obtained from load-displacement curves. According to Eq 1, J is proportional to the area under the load versus load line displacement curve. In the plastic regime, CTOD is proportional to both load line and mouth opening displacements. This is because the CTOD equation in BS 5762:1979 is based on the assumption that the bend specimen used for the test rotates about a plastic hinge point. If a material work hardens after yielding, the area under the load-displacement curve will increase at a faster rate than displacement, and the m factor in Eq 3 will increase with displacement. This is illustrated schematically in Fig. 1. For a nonhardening material, the ratio of area to displacement is constant in the plastic region. Therefore, the ratio of J to CTOD is not a function of displacement. However, the J/CTOD ratio increases with displacement in the hardening material.

The m factor is also a function of constraint, which is dependent on test specimen geometry. The constraint effect is illustrated in Fig. 2. An increase in constraint elevates the load at yield [*7*], thereby increasing the area under the curve at a given displacement. Therefore, m increases with constraint.

Analytical Comparisons

The relationship between J and CTOD has been derived for the case where both parameters are measured simultaneously on the same specimen. Analytical solutions were obtained (see Appendix) for single-edge notched bend (SENB) specimens and notched tensile panels. The solution for tensile loading is applicable to center-cracked tension (CCT) panels and single-edge notched tension (SENT) specimens with restrained bending. A J-CTOD relationship was not obtained for compact tension (CT) specimens because currently there is not a generally accepted equation for CTOD in this geometry.

These analytical solutions are merely algebraic relationships between engineering estimates of J and CTOD. The only assumption that has been made is that the plastic portion of the load-displacement curve fits a power-law expression. Such an assumption can be shown to be valid if the material obeys a Ramberg-Osgood stress-strain law and if crack growth is negligible [*21*]. Because of this assumption, the equations derived in this article can only be applied below maximum load.

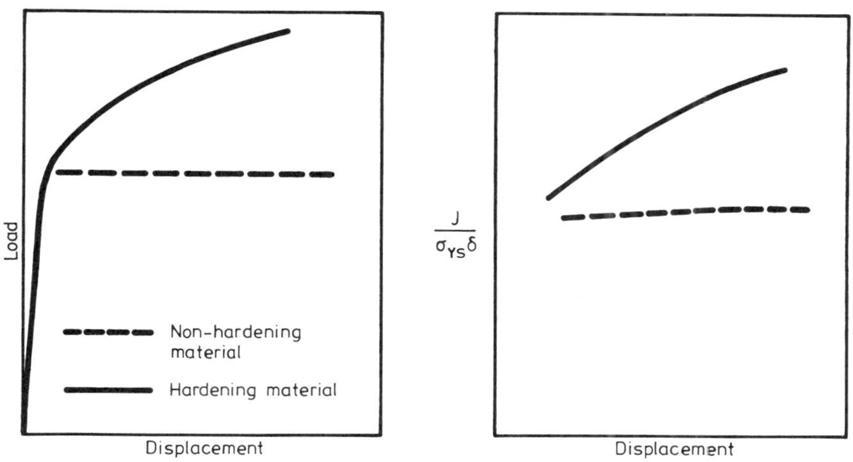

FIG. 1—*Comparison of the* m *factor* ($= J/\sigma_{YS}\delta$) *for a hardening and a nonhardening material.*

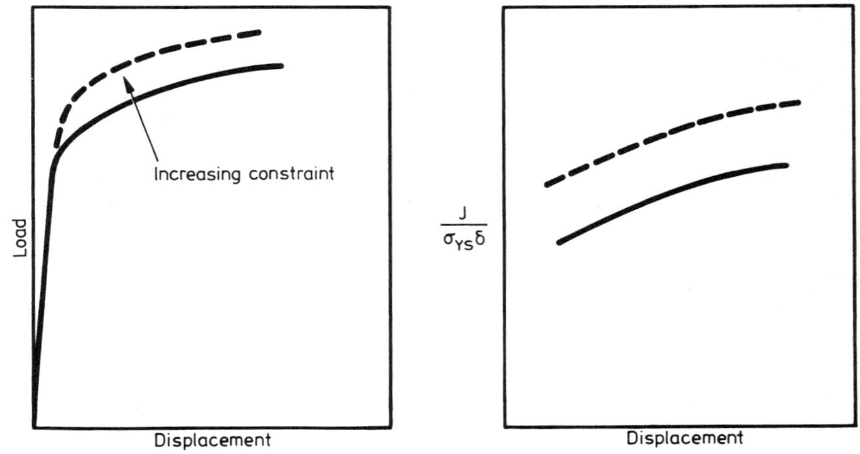

FIG. 2—*Effect of specimen constraint on the* m *factor.*

The derivations of the *J*-CTOD relationships are given in the Appendix. In each case, the equations for *J* and CTOD were separated into elastic and plastic components. For example, Eq 3 can be rewritten as

$$J = G + J_p = m_e \sigma_{YS} \delta_e + m_p \sigma_{YS} \delta_p \tag{5}$$

where the subscripts *e* and *p* denote elastic and plastic behavior, respectively. An inherent assumption of the CTOD equation is that $m_e = 2.0$. Therefore, the overall *m* factor is given by

$$m_0 = \frac{J}{\sigma_{YS}\delta_c} = \frac{2\delta_e + m_p\delta_p}{\delta_c} \tag{6}$$

Figure 3 is a plot of m_0 as a function of critical CTOD for three configurations of test specimen. The analytical expressions are compared with experimental data for a 25.4-mm (1-in.)-thick plate of ABS EH36 steel, a normalized carbon-manganese (C-Mn) steel with a yield strength of 350 MPa at 25°C. The fracture toughness data used to compute experimental m_0 values were originally published by Anderson et al. [*8*]. At low CTOD values, m_0 approaches 2 because the elastic term in Eq 6 dominates $m_0 \simeq m_e$. At high CTOD values, the plastic term in Eq 6 dominates.

The factor m_0 is highly dependent on specimen configuration. As illustrated in Fig. 2, m_p is a function of notch constraint. Increasing the constraint by changing the loading from tensile to bending or by increasing the crack length to width ratio, a/W, results in an elevation of m_0.

Much of the experimental data in Fig. 3 do not fall on the corresponding curves. The scatter in the data is probably caused by variations in material flow properties. Also, the power law expression for the load-displacement curves (Eq 9) may be slightly inaccurate, particularly at low displacements.

At CTOD values greater than 0.1 mm, there appears to be only a slight dependence of m_0 on toughness. There are a number of reasons for this. The curves and data are plotted over a fairly narrow range of CTOD values; the abscissa is expanded relative to the ordinate, resulting in a low apparent slope in the curves. In addition, the experimental data are somewhat scattered because of variation in material properties between specimens; this masks any tendency for *m* to increase with toughness. These difficulties are alleviated in Fig. 4, which was taken from

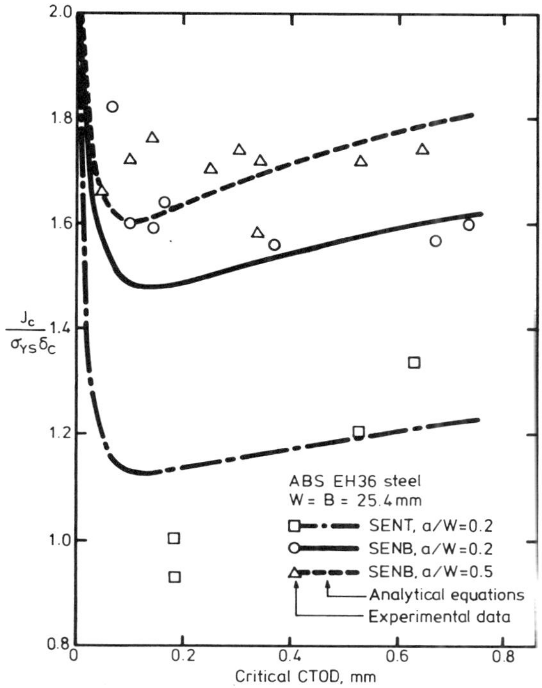

FIG. 3—m_0 *as a function of CTOD for three test specimen configurations.*

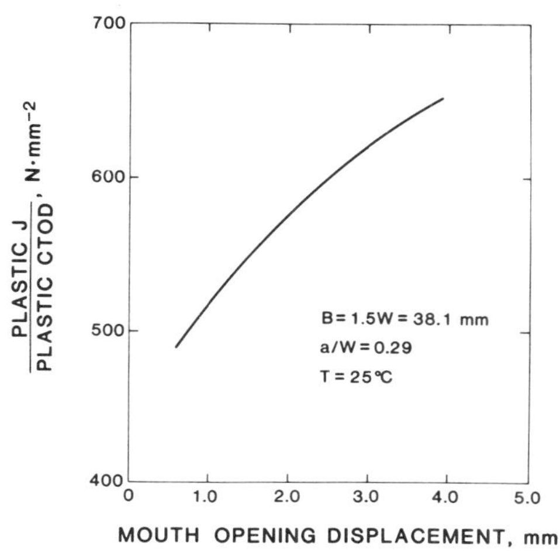

FIG. 4—*Experimental ratio of plastic* J *to plastic CTOD for a single SENB specimen of ABS EH36 steel* [7].

Ref 7. Here the plastic component of the J/CTOD ratio (determined experimentally) for a single specimen is plotted against displacement. At the test temperature of 25°C, $\sigma_{YS} = 350$ MPa. Thus, in the range displacements plotted, m_p ranges from 1.43 to 1.86 due to strain hardening effects (Fig. 1).

Another reason why the experimental data and computed curves in Fig. 3 appear to be insensitive to CTOD is that the strain hardening rate in the ABS EH36 steel is relatively low. As defined in Fig. 5, the hardening exponent (N) is only 0.1. However, the CTOD dependence of m_0 would be very pronounced in materials with high strain hardening rates. This is illustrated in Fig. 5, in which m_0 is plotted against CTOD for $N = 0.1$, 0.2, and 0.3. An N value as high as 0.3 is probably unusual in metals. An N value of 0.2 (that is, twice the strain hardening rate of the C-Mn steel used in this study) corresponds approximately to an austenitic stainless steel.

Figure 6 compares predicted J_c values with experimental J_c values for the three configurations of ABS EH36 steel specimens. The predicted values were obtained by multiplying δ_c, σ_{YS}, and the predicted m_0 values (Fig. 3). Aside from the scatter mentioned previously, the agreement between the predictions and experimental data was quite good. This agreement is not surprising since the predictions were based on the algebraic relationship between experimental estimates of J and CTOD, as discussed previously.

Application to Design

In design, the fracture behavior of a structure is usually inferred from a laboratory toughness test on a CT or SENB specimen. It is generally assumed that the fracture toughness of the test specimen is equivalent to that of the structure. However, this is not always the case, as illustrated next.

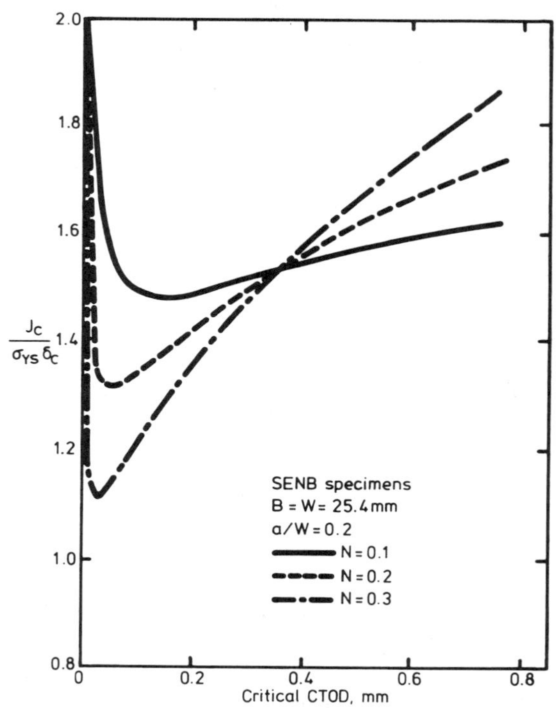

FIG. 5—m_0 *as a function of CTOD for three strain hardening exponents.*

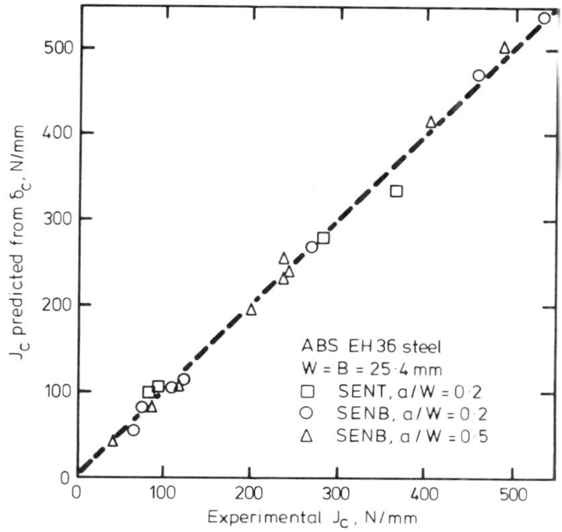

FIG. 6—*Comparison of predicted and experimental* J_c *values for three test specimen configurations.*

Consider a structure, loaded in tension, with a long surface crack. This hypothetical structure can be simulated by the SENT configuration that was used to obtain the data given in Figs. 3 and 5. Assume that a fracture toughness test is performed on a 25 by 25 mm SENB specimen with $a/W = 0.5$ and that a critical CTOD of 0.2 mm is measured. According to Fig. 3, m_0 for the test specimen is 1.63. Thus, assuming $\sigma_{YS} = 400$ N/mm^2, a critical J value of 130 N/mm would have been measured from the same test.

If it is assumed that the critical CTOD of the test specimen and the structure are equal, the critical J of the structure would be 90 N/mm because $m_0 = 1.13$ (see Fig. 3). Thus, a J test on a laboratory specimen would overestimate the critical J of the structure.

Conversely, if it is assumed that the critical J values of the structure and test specimens are both 130 N/mm, the critical CTOD of the structure would be approximately 0.28 mm, or 0.08 mm greater than the CTOD of the test specimen.

Thus if critical J and CTOD values are measured on a test specimen, these parameters cannot both be equal to the corresponding values in a structure. This is because the J-CTOD relationship is a function of the configuration of the specimen or structure. Conservative results are more likely to be obtained if toughness is quantified in terms of CTOD.

The fact that the J-CTOD relationship is geometry-dependent leads to the question: Which fracture toughness parameter is the least sensitive to configuration? That is, if critical J and CTOD values both are obtained from a given test specimen, which parameter will best describe the critical conditions for fracture in a structure? The answer to this question is unclear at present, but it may depend on the micromechanism of fracture. For cleavage fracture, which is controlled by local stress at the crack tip, the J-integral may be the more appropriate parameter. At a constant CTOD, measured J values will increase with constraint, as discussed previously. However, cleavage fracture toughness tends to decrease with constraint [*7–11*]. Thus, if cleavage toughness is quantified in terms of J, these two effects may tend to cancel one another, resulting in a lower geometry dependence of toughness. On the other hand, CTOD might be more appropriate for defining the onset of stable crack growth in ductile materials. Microvoid coalescence is controlled by the local strain at the crack tip and is less sensitive to constraint than cleavage. Since CTOD is related to the shape of the deformed crack tip, it may be the better parameter for characterizing the local strain field.

Existing experimental data do not shed much light on the subject of the relative geometry dependence of critical J and CTOD values. In Ref *7*, J and CTOD were measured simultaneously on various specimen configurations over a wide range of temperatures. (Some of these data were used in Fig. 3). Cleavage fracture toughness in the transition region was geometry dependent, regardless of whether toughness was quantified in terms of J or CTOD. The upward shifts in transition with increasing thickness and a/W was roughly the same for both parameters. The scatter was too great to determine conclusively if one parameter exhibited slightly less geometry dependence than the other. Critical J and CTOD values at the onset of stable tearing both appeared to be geometry independent for the range of specimens considered. Based on the above discussion one might expect J values at initiation to be slightly geometry dependent but the data set (5 geometries, 18 tests total) was not large enough to reveal any such effects.

The results in Ref *7* imply that the geometry dependence of the J-CTOD relationship may be overwhelmed by other factors, such as material variability and constraint effects on toughness. Thus, the choice of fracture toughness parameter for a given situation is probably of secondary importance. However, as discussed previously, when material toughness is measured using highly constrained bend specimens, CTOD is inherently more conservative when applied to structures loaded in tension.

Summary

The engineering equations for J and CTOD were compared qualitatively and analytically. The ratio of these two parameters depends on yield strength, toughness, strain hardening rate, and notch constraint. Although J and CTOD are closely related, they are not identical measures of toughness. The J-integral is proportional to the energy absorbed by the specimen, whereas CTOD is roughly proportional to the clip gauge displacement. Because of this difference, the CTOD test tends to give a more conservative estimate of structural toughness than does J.

Acknowledgments

This work was funded by the Materials Chemicals and Vehicles Requirements Board and the Research Members of The Welding Institute. The data in Fig. 3 were obtained from fracture toughness tests performed at the National Bureau of Standards in Boulder, Colorado. These tests were sponsored by the Minerals Management Service of the U.S. Department of Interior.

APPENDIX

Analytical Relationships Between J and CTOD

Bend Specimens

The equation for CTOD in BS 5762:1979 is given by

$$\delta = \frac{K^2}{2\sigma_{YS}E} + \frac{r_p(W - a)V_p}{r_p(W - a) + a + z} \tag{7}$$

where

r_p = plastic rotational factor,
V_p = plastic component of mouth opening displacement, and
z = knife edge height.

In BS 5762:1979, r_p is assumed to equal 0.4. However, r_p has been left as a variable in order to make the equations which follow more general.

Equation 7 expresses CTOD in terms of elastic and plastic components. Similarly, Eq 1 for J can be separated in such a manner [12]. The plastic component of J is given by

$$J_p = \frac{2U_p}{B(W-a)} \tag{8}$$

The plastic component of Eq 7 can be expressed in terms of the plastic load line displacement, q_p:

$$\delta_p = \frac{r_p(W-a)q_p}{W} \tag{9}$$

Solving for J_p in terms of δ_p leads to

$$J_p = \frac{2U_p W\delta_p}{r_p(W-a)^2 Bq_p} \tag{10}$$

However

$$U_p = \int_0^P P dq_p \tag{11}$$

where P is load.

It was assumed that the load-displacement curve obeys a power-law function

$$\frac{P}{P_Y} = \alpha\left(\frac{q_p}{q_Y}\right)^N \tag{12}$$

where P_Y is the load at net section yield. This assumption is valid [21] as long as crack growth is small and the material obeys a Ramberg-Osgood strain-strain law

$$\frac{\epsilon}{\epsilon_0} = \frac{\sigma}{\sigma_0} + \alpha'\left(\frac{\sigma}{\sigma_0}\right)^n \tag{13}$$

where

σ = true stress,
ϵ = strain,
σ_0 = reference value of stress,
$\epsilon_0 = \sigma_0/E$, and
$n = 1/N$.

For SENB specimens

$$P_Y = \frac{L\sigma_{YS}(W-a)^2 B}{4W} \tag{14}$$

where L is the notch constraint factor, a dimensionless constant that is a measure of the elevation of P_Y due to the presence of a notch. The elastic load point displacement at yield, q_Y, is given by

$$q_Y = \frac{L(W-a)^2 e_Y Z}{4W} \tag{15}$$

where Z is the nondimensional compliance [22] given by

$$Z = \frac{qBE}{P} = f\left(\frac{a}{W}\right) \tag{16}$$

Combining Eqs 9–15 leads to

$$J_p = m_p \sigma_{YS} \delta_p \tag{17}$$

where

$$m_p = \frac{L\alpha}{2r_p(N+1)} \left[\frac{4W^2\delta_p}{Lr_p e_Y Z(W-a)}\right]^N \tag{18}$$

The overall relationship between J and CTOD can be obtained by combining elastic and plastic components

$$m_0 = \frac{J}{\delta \sigma_{YS}} = \frac{m_e \delta_e + m_p \delta_p}{\delta} \tag{19}$$

In BS 5762:1979, m_e is assumed to equal 2.0.

Notched Tensile Panels

Rice et al. [23] have derived an equation for J in edge-cracked and center-cracked panels in tension. In the edge-cracked case, the panel is assumed to be restrained against bending. The J estimation equation is given by

$$J = \frac{K^2}{E} + \frac{1}{B(W-a)}\left(2\int_0^{V_p} P\,dV_p - PV_p\right) \tag{20}$$

where V_p is the plastic component of the clip gage displacement.

For pure tension, that is, no rotation of the crack faces during plastic deformation

$$\delta_p = V_p \tag{21}$$

The power-law expression (Eq 12) was used to define the load-displacement curve. The load at net section yield is given by

$$P_Y = L\sigma_{YS} B(W-a) \tag{22}$$

Substituting Eqs 12, 21, and 22 into Eq 20 and integrating with respect to δ_p results in an expression in the form of Eq 17. For pure tension

$$m_p = L\alpha\left[\frac{\delta_p}{Le_Y(W-a)}\right]^N \left[\frac{2}{N+1} - 1\right] \tag{23}$$

Elastic and Plastic CTOD

To use the equations derived in this paper it is sometimes necessary to separate a CTOD value into its elastic and plastic components. This is accomplished by calculating δ_e and subtracting it from the total CTOD. The elastic CTOD at yield is given by

$$\delta_Y = \frac{L^2(W-a)^4 Y^2 e_Y}{32W} \tag{24}$$

At loads greater than yield, the elastic CTOD is greater than δ_Y

$$\delta_e = \alpha^2 \delta_Y \left(\frac{q_p}{q_Y} \right)^{2N} \tag{25}$$

Since q_p is usually not known in advance, an iterative calculation with Eqs 9 and 25 must be performed to determine δ_e and δ_p for a given δ_c.

References

[1] Wells, A. A., "Unstable Crack Propagation in Metals: Cleavage and Fast Fracture" in *Proceedings*, Crack Propagation Symposium, Vol. 1, Paper 84, Cranfield, U.K., 1961.

[2] Rice, J. R., "A Path Independent Integral and the Approximate Analysis of Strain Concentration by Notches and Cracks," *Journal of Applied Mechanics*, June 1968, pp. 379–386.

[3] Begley, R. A. and Landes, J. D., "The J-Integral as a Fracture Criterion" in *Fracture Toughness, ASTM STP 514*, American Society for Testing and Materials, Philadelphia, 1972, pp. 1–23.

[4] Landes, J. D. and Begley, J. A., "The Effect of Specimen Geometry on J_{Ic}" in *Fracture Toughness, ASTM STP 514*, American Society for Testing and Materials, Philadelphia, 1972, pp. 24–39.

[5] Landes, J. D. and Begley, J. A., "Test Results from J-Integral Studies: An Attempt to Establish a J_{Ic} Testing Procedure" in *Fracture Analysis: Eighth Conference, ASTM STP 560*, American Society for Testing and Materials, Philadelphia, 1974, p. 170.

[6] Merkle, J. G. and Corten, H. T., "A *J*-Integral Analysis for the Compact Specimen, Considering Axial Force as Well as Bending Effects," Paper No. 74-PVP-33, American Society of Mechanical Engineers.

[7] Anderson, T. L., McHenry, H. I., and Dawes, M. G., "Elastic-Plastic Fracture Toughness Tests with Single Edge Notched Bend Specimens," in *Fatigue and Fracture Toughness—Cryogenic Behavior, ASTM STP 856*, American Society for Testing and Materials, Philadelphia, 1985, p. 210.

[8] Anderson, T. L. and Williams, S., "Assessing the Dominant Mechanism for Size Effects in the Ductile-to-Brittle Transition Region" in *Fracture Mechanics: Seventeenth Volume, ASTM STP 905*, American Society for Testing and Materials, Philadelphia, 1986, pp. 715–740.

[9] Anderson, T. L., "A Constraint-Based Model for Fracture in the Ductile-to-Brittle Transition Region," Report 248/1984, The Welding Institute, Cambridge, U.K., Aug. 1984.

[10] Dawes, M. G., "Elastic-Plastic Fracture Toughness Based on the COD and *J*-Contour Integral Concepts" in *Elastic-Plastic Fracture, ASTM STP 668*, American Society for Testing and Materials, Philadelphia, 1979, p. 307.

[11] Iwadate, T., Tanaka, Y., Ono, S., and Watanabe, J., "An Analysis of Elastic-Plastic Fracture Toughness Behaviour for J_{Ic} Measurement in the Transition Region" in *Elastic-Plastic Fracture, Second Symposium, ASTM STP 803*, American Society for Testing and Materials, Philadelphia, 1983, p. II-531.

[12] Sumpter, J. D. G. and Turner, C. E., "Method for Laboratory Determination of J_c" in *Cracks and Fracture: Ninth Volume, ASTM STP 601*, American Society for Testing and Materials, Philadelphia, 1976, p. 3.

[13] Dawes, M. G., "A J_I Equivalent of the CTOD Test—A Proposal," *Welding Institute Research Bulletin*, Vol. 23, No. 2, 1982, p. 50.

[14] Shih, C. F., "Relationship Between the J-Integral and the Crack Opening Displacement for Stationary and Extending Cracks," *Journal of the Mechanics and Physics of Solids*, Vol. 29, 1981, p. 305.

[15] Andrews, W. R., "An Investigation of the Relationships Between *J* and CTOD," Report No. 82TTL227, General Electric Inc., Schenectady, NY, April 1983.

[16] Dawes, M. G., "The Application of Fracture Mechanics to Brittle Fracture in Steel Weld Metals," Ph.D. thesis, Council for National Academic Awards, The Welding Institute, Cambridge, U.K., Dec. 1976.

[17] Hutchinson, J. W., "Singular Behaviour at the End of a Tensile Crack in a Hardening Material," *Journal of the Mechanics and Physics of Solids*, Vol. 16, 1968, p. 13.

[18] Rice, J. R. and Rosengren, G. F., "Plane Strain Deformation Near a Crack Tip in a Power-Law Hardening Material," *Journal of the Mechanics and Physics of Solids*, pp. 1–12.

[19] Schindler, H. J., "On the Relationship Between J-Integral and Crack Opening Displacement," *Engineering Fracture Mechanics*, Vol. 20, No. 2, 1984, p. 281.

[20] Schroedl, M. A. and Jolles, M. I., "Crack Tip Parameters for Elastic-Plastic Fracture," presented at the ASTM 17th National Symposium on Fracture Mechanics, Albany, NY, Aug. 1984.

[21] Kumar, V., German, M. D., and Shih, C. F., "An Engineering Approach for Elastic-Plastic Fracture Analysis," EPRI Report NP-1931, Electric Power Research Institute, Palo Alto, CA, July 1981.

[22] Towers, O. L., "Stress Intensity Factors, Compliances and Elastic η Factors for Six Test Geometries," Report 136/1981, The Welding Institute, Cambridge, U.K., March 1981.

[23] Rice, J. R., Paris, P. C., and Merkle, J. G., "Some Further Results on J-Integral Analysis and Estimates" in *Fracture Toughness Testing, ASTM STP 536*, American Society for Testing and Materials, Philadelphia, 1973, pp. 231–245.

Poster Session—Fatigue

Srivathsan Venkataraman,[1] *Theodore Nicolas,*[1] *and Mohammad Khobaib*[2]

Elevated Temperature Fatigue Crack Growth of Single Crystal René N4 Superalloy

REFERENCE: Venkataraman, S., Nicolas, T., and Khobaib, M., **"Elevated Temperature Fatigue Crack Growth of Single Crystal René N4 Superalloy,"** *Fracture Mechanics: Eighteenth Symposium, ASTM STP 945,* D. T. Read and R. P. Reed, Eds., American Society for Testing and Materials, Philadelphia, 1988, pp. 757–764.

ABSTRACT: Constant-load-amplitude fatigue crack growth tests were conducted at 0.33 Hz over a temperature range of 538 to 871°C for single crystal René N4 nickel base superalloy. Crack growth behavior was found to be similar in the Paris law regime for the entire temperature range tested except for a slight increase in growth rate at 871°C at higher levels of the stress intensity range. Mainly, two distinct modes of fatigue fracture were observed, namely, a flat, featureless, Stage II fracture morphology and a faceted, crystallographic, Stage I cracking mode. The Stage II fatigue fracture occurred over the entire range of ΔK at the three highest temperatures. At 538°C, however, a transition from Stage II to Stage I fatigue fracture occurred with increase in ΔK. The higher crack growth rates at higher temperatures in the high-stress-intensity range were attributed to casting defects present in the material.

KEY WORDS: fatigue crack growth, single crystal, nickel base superalloy, elevated temperature, fractography, fracture mechanism

Improvements in the fuel efficiency and performance of modern day aircraft gas turbine engines require turbine airfoil materials to operate at increasingly higher temperatures over longer periods of time while still retaining adequate mechanical and corrosion properties. In the seventies, conventionally cast alloys were replaced by directionally solidified (DS) nickel base superalloys which were developed to meet this need [*1*]. Lately, however, maximum operating temperature requirements have generally increased beyond the capabilities of even the DS alloys. The solution to these requirements has been the recent development of a class of single crystal (SC) nickel-base superalloys, such as René N4, which possess superior creep rupture life and elevated temperature fatigue strength along with improved oxidation resistance over cast and DS alloys. These properties make them attractive candidates as future turbine blade materials [*2,3*]. These improvements in SC superalloys have been achieved through the elimination of transverse grain boundaries and by selection of preferred crystallographic orientations for optimum design. However, there are other potentially detrimental features present in SC superalloys such as inclusions, porosity, dendrites, and subgrains, as well as anisotropic properties along various crystallographic orientations. Understanding of their impact on the material behavior is quite important. Detailed information on the mechanical properties and fatigue and fracture behavior of these alloys is prerequisite for their reliable application in future turbine engines as well as for any life prediction modelling.

[1]Resident NRC research associate and senior scientist, respectfully, AFWAL MLLN, Wright-Patterson Air Force Base, OH 45433.

[2]Research scientist, University of Dayton Research Institute, Dayton, OH 45429.

Research is currently underway in both government and industrial laboratories to evaluate the creep-fatigue-environmental interactions and to understand the underlying crack growth mechanisms in the newly developed SC nickel base superalloys. This paper deals with the fatigue crack growth rate characteristics of SC René N4 alloy and the associated crack growth micromechanisms.

Experimental Procedures

A series of isothermal fatigue crack growth tests were conducted on SC René N4 superalloy. The composition of this alloy is shown in Table 1. The cast SC slabs of nominal dimensions 115 by 50 by 13 mm were solution treated at 1270°C for 2 h followed by aging at 1080°C for 4 h and 900°C for 16 h in protective atmosphere. Mini compact tension specimens of dimensions shown in Fig. 1 were then machined out of the heat treated slabs. The loading line axis of the specimens was oriented along [001] while the crack growth direction was along [100] as shown in Fig. 1.

Tests were carried out in an automated servo-hydraulic testing machine under computer control. The specimen was maintained at constant temperature in a resistance heated oven having quartz windows through which optical surface crack length could be periodically monitored with the aid of a travelling microscope. The primary crack length data were obtained from

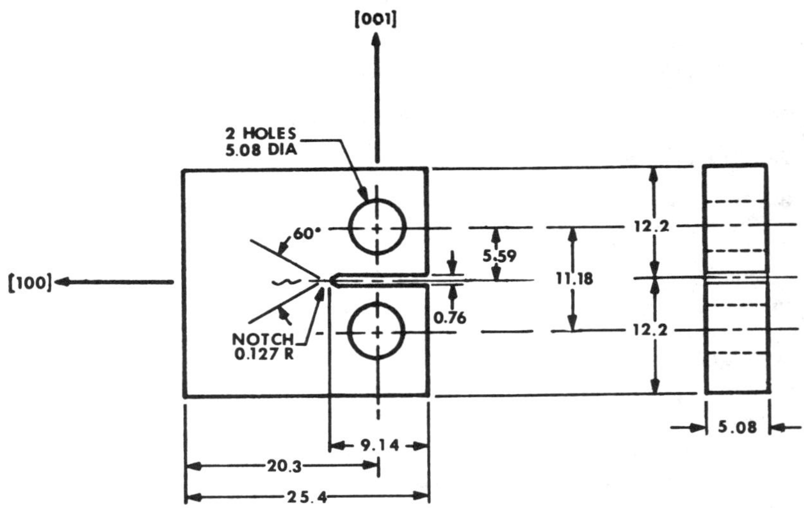

FIG. 1—*Schematic of minicompact tension specimen. Dimensions are in millimeters.*

TABLE 1—*Chemical composition of single crystal René N4 nickel base superalloy.*

Element	weight %	Element	weight %
Cr	9.25	W	6.0
Co	7.7	Mo	1.5
Al	3.7	Cb	0.5
Ta	4.0	Ni	balance
Ti	4.2		

compliance measurements from load-displacement data. Crack mouth opening displacements were obtained with the aid of an extensometer with quartz rod extension arms. The air-cooled extensometer was mounted outside the oven. Compliance was determined from a least squares linear regression to digital load-displacement data over a range of 95% maximum load to a lower window (generally around 50% maximum load) which is above the closure load. The compliance determined crack lengths were verified through optical surface crack length measurements.

Precracking was conducted following the ASTM standard procedure. For precracking these single crystals the temperature and the stress intensity level were found to be critical. Lower temperature and higher ΔK levels led to crystallographic shear cracking deviating from normal to the loading axis. Preliminary testing was conducted at several temperatures and ΔK levels to establish precracking procedures. A temperature of 538°C and a narrow band of initial ΔK level between 20 and 25 MPa $\sqrt{\text{m}}$ were found to be optimum for precracking. A sinusoidal loading waveform with a frequency of 0.33 Hz and minimum to maximum load ratio (R) of 0.1 were utilized for both precrack and crack growth tests. Constant-load-amplitude fatigue crack growth tests were then carried out over a temperature range of 538 to 871°C.

The fatigue crack growth data were reduced to the form of crack growth rate per cycle (da/dN) versus stress intensity range (ΔK) using a modified incremental polynomial method [4]. In this procedure, all crack length data falling within a crack length interval (Δa_{reg}) are fitted with a second degree polynomial, and da/dN and ΔK are calculated at the midpoint of the interval. This process is repeated with the sliding regression being successively incremented by an amount of Δa_{inc}. Values of 1.524 and 0.254 mm were used for Δa_{reg} and Δa_{inc}, respectively. This procedure limits the scatter normally obtained using an ASTM standard seven point polynomial, especially below the "knee" portion of da/dN versus ΔK plots.

To understand the mechanism of fatigue crack growth, detailed fractographic analyses were carried out on all fractured test specimens using a scanning electron microscope. The solidification substructures as well as post heat treatment microstructure were also studied.

Results and Discussion

The fatigue crack growth rate data of SC René N4 superalloy obtained for the temperature range of 538 to 871°C are given in Fig. 2 in the form of da/dN versus ΔK relationship. There is not much variation in the growth rate with increase in temperature in the Paris law regime above the knee in the curves except for a slight increase at 871°C at the higher ΔK levels (above 40 MPa $\sqrt{\text{m}}$). However, in the near threshold to knee region certain abnormal behavior is observed. Although threshold has not been reached at the lowest growth rates of approximately 2×10^{-8} m/s, the threshold at 538°C appears to be lower than that at the other three higher temperatures. For a given ΔK, therefore, the growth rate at 538°C is higher in this region as compared to other higher temperatures. Similar results have also been reported by other investigators [2] who observe an increase in threshold and growth rate with temperatures up to approximately 760°C. No mechanistic differences could be traced through the fracture surface analysis. The faster oxidation kinetics at higher temperatures may have resulted in reasonable oxide induced crack closure leading to a lower effective ΔK causing such abnormal behavior. In fact considerable oxidation of the fracture surface was noticed at temperatures of 649°C and above. No closure studies have yet been made in this investigation to confirm this suggestion.

Fractographic examination of the failed specimen at 538°C revealed two types of crack propagation mechanisms. First, in the lower ΔK region (below 25 MPa $\sqrt{\text{m}}$), a flat, featureless fracture surface was observed, as shown in Fig. 3. On the other hand, at higher ΔK regions a faceted, crystallographic fracture mode was observed as indicated in Fig. 4. The facets were aligned parallel to available {111} planes. In addition, the crack surface also deviated from normal to the principal stress axis. Measurement of interfacet spacing matched closely with the growth rate at the corresponding ΔK levels. This transition in fracture morphology was not

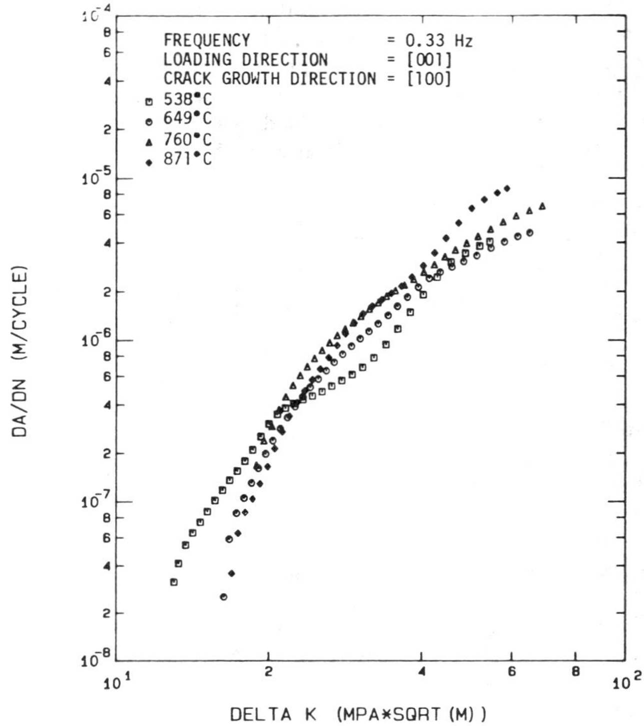

FIG. 2—*Fatigue crack growth rate* (da/dn) *versus stress-intensity range* (ΔK) *plots for single crystal René N4 superalloy tested between 538 and 871°C.*

observed at temperatures of 649°C and above. In these cases the entire fracture surface was flat and featureless. However, an increasing amount of secondary cracks were observed in and around casting defects such as inclusions and porosities which became more dominant at higher ΔK levels and temperatures as shown in Fig. 5. This increase in the secondary cracking should have resulted in a slight increase in the growth rate at 871°C and higher ΔK levels. Further, at 871°C the fracture surface was considerably oxidized as shown in Fig. 6.

Earlier investigations on fatigue fracture of nickel base superalloy single crystals have shown a tendency for low temperature and high strain rates to favor planar, heterogeneous slip, and crystallographic fatigue crack growth along {111} slip planes (Stage I) while high temperatures and low strain rates led to wavy, homogeneous slip and fatigue fracture (Stage II) [*5–8*]. This reasoning applies to the fracture behavior of SC René N4 observed in this investigation also.

A thermally activated deformation process must be involved in determining the fracture mode. It is likely, for example, that planar bands of dislocations are generated at the tip of stress concentrations, but are dispersed by thermally activated recovery processes such as dislocation climb or cross slip. At 538°C and lower ΔK regions (below 25 MPa $\sqrt{m}$) this phenomenon leads to cause Stage II crack growth. With increasing ΔK at this temperature a larger number of such bands form on selected {111} slip planes (easy glide). Higher growth rate limits the probability of the dislocation dispersion, leading to Stage I fracture. With increasing temperatures, more thermal energy becomes available to aid the dispersion of dislocations resulting in homogeneous slip and Stage II fatigue fracture for the entire ΔK region tested.

At temperature above 649°C and higher ΔK levels, extensive secondary cracking was observed at casting defects such as porosities. Upon studying the solidification substructures of

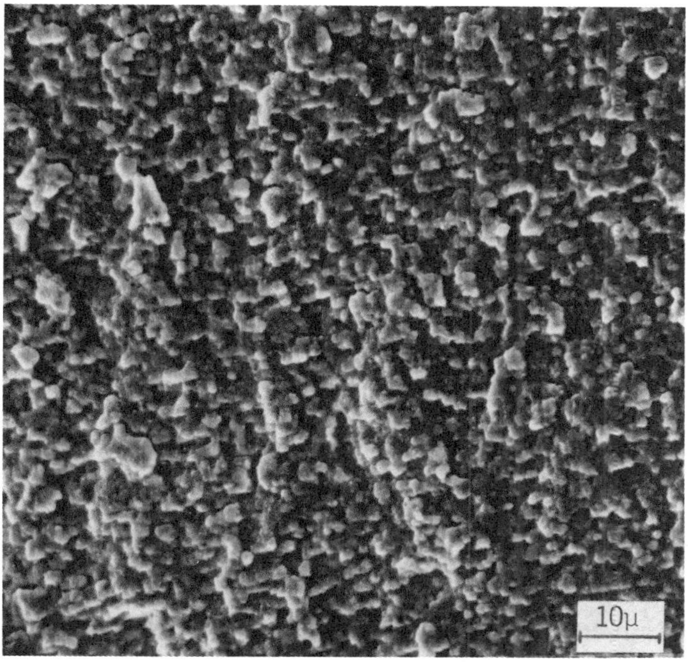

FIG. 3—*Stage II: Noncrystallographic fatigue fracture observed at low ΔK regime at the test temperature of 538°C.*

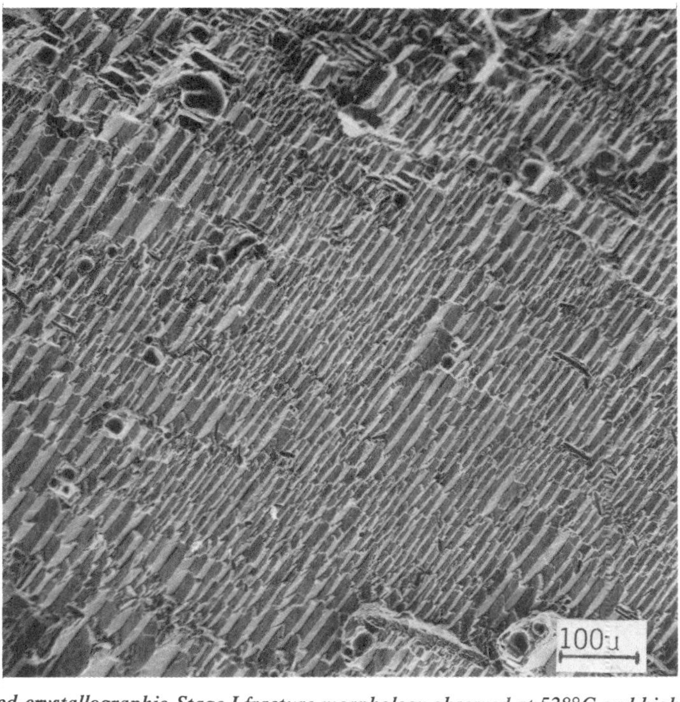

FIG. 4—*Faceted-crystallographic-Stage I fracture morphology observed at 538°C and higher ΔK levels.*

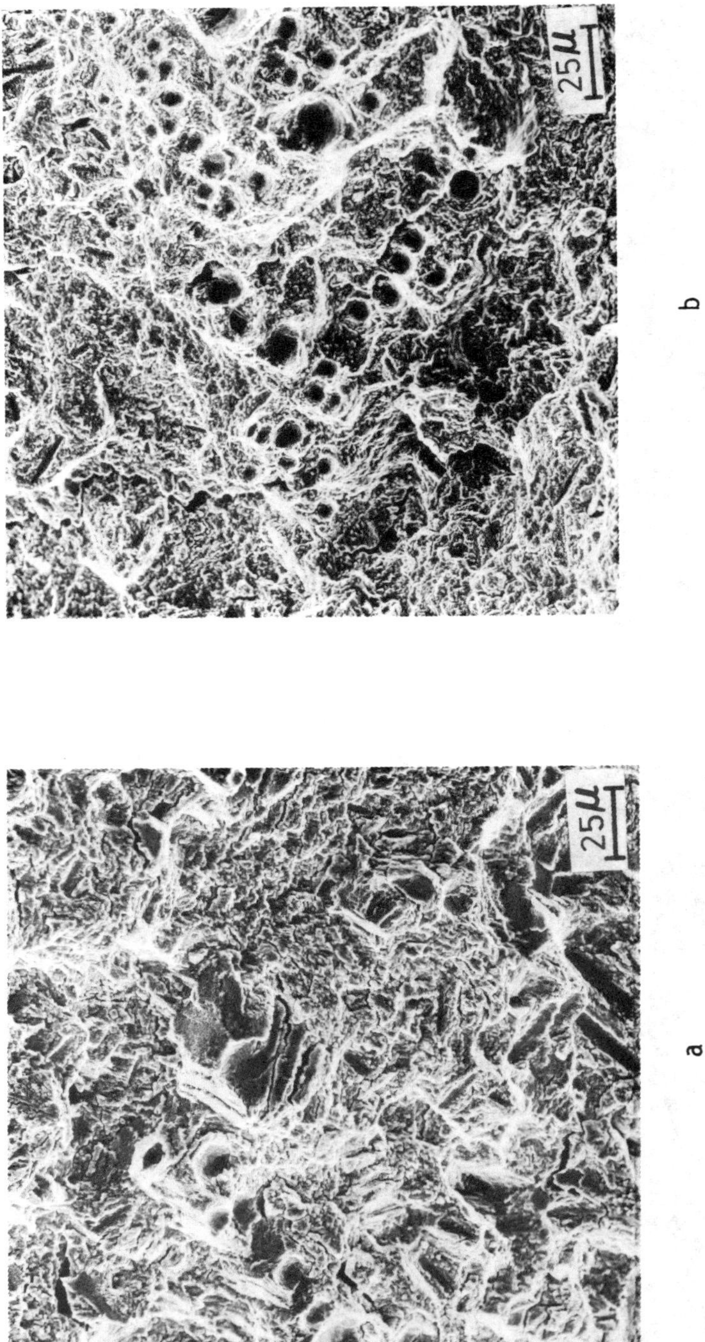

FIG. 5—*Secondary cracking observed at casting defects such as* (a) *inclusions and* (b) *pores. These defects provide the weak sites for accelerated fatigue crack growth. Test temperatures are* (a) *649°C and* (b) *760°C.*

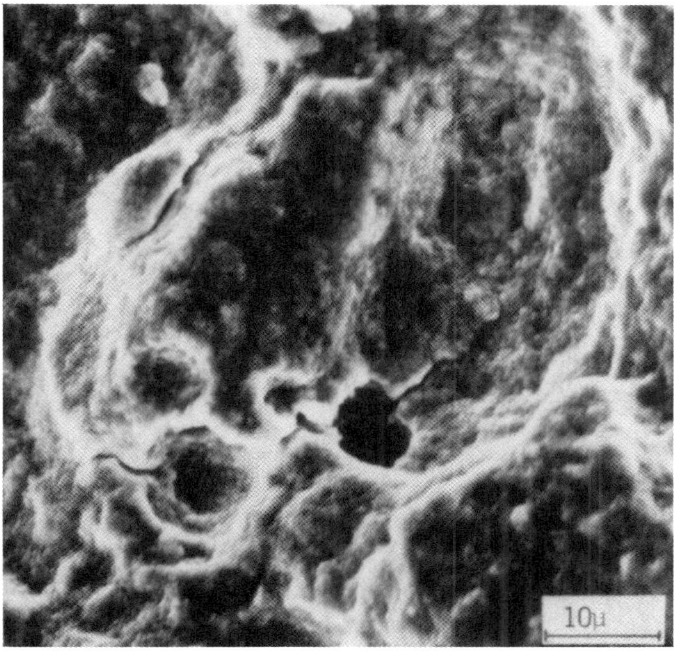

FIG. 6—*Severe oxidation of fracture surface at 871°C. Note the presence of casting pores.*

this SC material, most of the defects were found to be present between the dendrite arms as shown in Fig. 7. These defects act as nucleation sites as well as weak locations causing accelerated fatigue crack growth.

Conclusions

1. In the temperature range of 538 to 871°C the fatigue crack growth rates for SC René N4 superalloy were comparable in the Paris law regime except for slight increase at 871°C under higher ΔK levels. This illustrates the excellent elevated temperature fatigue properties of the material.

2. Two distinct fatigue fracture modes were observed. At lower ΔK levels, a flat, featureless, Stage II fracture occurred, while a faceted, crystallographic, Stage I cracking mode was observed with increasing ΔK levels. However, this transition was not noticed at temperatures of 649°C and above.

3. Casting defects present in the interdendritic locations of the SC alloy weakened the material at higher temperatures (649°C and above) and caused accelerated growth at higher ΔK levels. Improvements in the casting technology to reduce the concentration of such defects should certainly improve the elevated temperature fatigue life of this superalloy.

Acknowledgments

The first author would like to thank the National Research Council for sponsoring his research work. The authors would like to express their appreciation to Dr. P. K. Wright of General Electric Company, Evendale, Ohio, for providing the SC René N4 material and for his helpful comments. This work was supported by the US Air Force under project 2302P1.

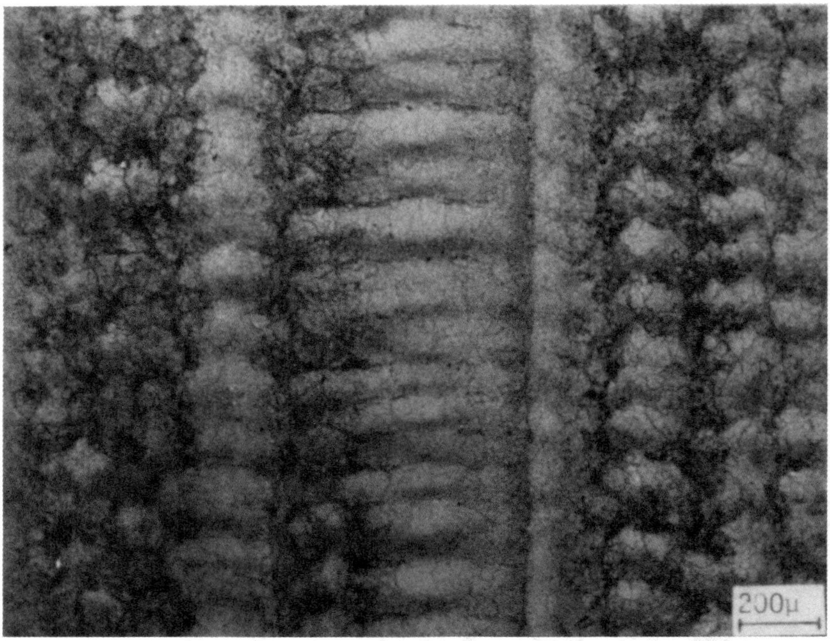

FIG. 7—*Solidification substructure of single crystal René N4. Large number of pores are present along dendrite boundaries.*

References

[1] VernSnyder, F. L. and Shank, M. E., *Materials Science and Engineering,* Vol. 6, No. 4, 1970, p. 321.

[2] Wright, P. K., Jang, H., and Popp, H. G., "Fatigue and Fracture of Advanced Blade Materials," AFWAL-TR-84-4166, Wright-Patterson AFB, Ohio, Feb. 1985.

[3] Gell, M., Duhl, D. N., and Giamei, A. F., *Superalloys 1980,* American Society for Metals, Metals Park, OH, 1980, p. 205.

[4] Larsen, J. M. in *Fracture Mechanics: Seventeenth Volume, ASTM STP 905,* American Society for Testing and Materials, Philadelphia, 1986, pp. 226–238.

[5] Gell, M. and Leverant, G. R., *Fracture 1969,* Chapman and Hall Ltd., London, 1969.

[6] Gell, M., Leverant, G. R., and Wells, C. H., *Achievement of High Fatigue Resistance in Metals and Alloys, ASTM STP 467,* American Society for Testing and Materials, Philadelphia, 1970, pp. 113–153.

[7] Leverant, G. R. and Gell, M., *Transactions,* The Metallurgical Society of the American Institute of Mining, Metallurgical, and Petroleum Engineers, 1969, Vol. 245, p. 1167.

[8] Leverant, G. R. and Gell, M., *Metallurgical Transactions,* Vol. 6A, Feb. 1975, p. 367.

Mehan M. Ratwani,[1] *Han-Pin Kan,*[1] *and Sheng-Ming Hsu*[1]

Fatigue Crack Growth at Root Radii in Sheets with Variable Thickness

REFERENCE: Ratwani, M. M., Kan, H.-P., and Hsu, S. M., **"Fatigue Crack Growth at Root Radii in Sheets with Variable Thickness,"** *Fracture Mechanics: Eighteenth Symposium, ASTM STP 945,* D. T. Read and R. P. Reed, Eds., American Society for Testing and Materials, Philadelphia, 1988, pp. 765–780.

ABSTRACT: Fatigue crack growth data have been obtained for cracks at root radii in sheets with variable thickness. These test data were obtained under constant-amplitude as well as spectrum loading. Analytical stress-intensity factors were developed using mathematical techniques. The Mode I stress-intensity factors were approximately 10% lower than that of a plate with uniform thickness. Crack growth predictions for the test specimens were made using the analytical stress-intensity factors and a crack growth equation for a skin with uniform thickness. A good correlation is shown between observed and predicted crack growth behavior.

KEY WORDS: crack growth, stress intensity factors, spectrum loading, land, root radii, integral equation

Nomenclature

u	Displacement in the x-direction
v	Displacement in the y-direction
σ_{yy}	Normal stress
σ_{xy}	Shear stress
ν	Poisson's ratio
E	Young's modulus
μ	Shear modulus
c	Half-crack length
t_1	Thickness of land
t_2	Thickness of skin
A, B, C, D	Constants
f_1, f_2	Unknown functions
h	Half-land width
h_1	Distance from edge of land to crack
da/dN	Crack growth rate per cycle
R	Stress ratio
i	Subscript or superscript denoting different regions shown in Fig. 6; $i = 1, 2, 3, 4$

Crack growth in simple structures has been evaluated by several investigators, and it has been shown that linear elastic fracture mechanics can be used to predict crack growth in these structures with a high degree of confidence. However, structural details in aircraft construction are complex, and cracks initiate and propagate at these details under flight loads. In order to pre-

[1]Manager, Structures Research, senior technical specialist, Strength and Life Assurance Research, and senior engineer, Specimen Preparation and Mechanical Test, M and P Research and Technology, Northrop Corporation, Aircraft Division, One Northrop Ave., Hawthorne, CA 90250.

dict crack growth behavior at these complex structural details, it is necessary to know the stress intensity factor so that crack growth models for life prediction may be used.

The majority of the analytical studies reported in the literature are on sheets of constant thickness. The stress-intensity solution for stiffened panels may be found in Refs *1* and *2*. Analysis of cracks in skins with lands is discussed in Ref *3*. Crack growth behavior in skins with lands and repaired with composite patches has been discussed in Refs *4* and *5*.

The objective of this study is to experimentally investigate crack growth behavior at root radii in sheets of variable thickness, develop stress intensity factors for these complex geometries, and verify that the crack growth behavior can be predicted using analytical stress intensity factors.

Analytical techniques of Refs *5* and *6* have been extended to obtain stress-intensity factor for cracks at root radii in skins with variable thicknesses. The Mode I stress-intensity factors thus computed are approximately 10% lower than cracks in an infinite plate with uniform thickness. The Mode II stress-intensity factors are found to be small. The reduced Mode I stress-intensity factor will cause approximately 30% reduction in crack growth rate for typical aluminum alloys.

Experimental Program

An experimental program was implemented to obtain crack growth data at complex structural details such as cracks at root radii to verify the accuracy of analysis in predicting crack growth behavior.

Test Specimens

All test specimens were fabricated from 7075-T6 aluminum sheets 3.2 mm (0.125 in.) thick. The test specimen used to obtain baseline crack growth data for 3.2-mm (0.125-in.)-thick material is shown in Fig. 1.

The test specimens with lands were chem-milled from 7075-T6 aluminum 3.2-mm (0.125 in.)-thick material. The test specimen geometry for cracks at a root radii is shown in Fig. 2. Two cracks were made in each specimen so that two sets of data were obtained. The cracks were spaced sufficiently far apart so that there was no significant interference between the cracks. Through cracks were started with a jeweler's saw and then precracked to the desired length under constant amplitude loading before crack growth measurements were taken.

Specimen Testing

All tests were conducted in a 45 000-kg (100 kip) MTS testing machine at a frequency of 10 Hz under room temperature ambient environment. All constant-amplitude tests were conducted at R-ratio (minimum stress to maximum stress) of 0.1. Crack growth measurements were taken at regular intervals. The cracks at a root radii specimen configuration were also tested under spectrum loading.

The fatigue load spectrum selected for the program is shown in Fig. 3. The maximum stress in the spectrum was 270-MPa (39.2 ksi) tension and minimum stress was 69-MPa (10 ksi) compression. The selected spectrum is considered to be a severe example of fatigue loading for the tension skin of a fighter aircraft wing near the root and is therefore more severe than the actual spectrum at other locations.

To prevent buckling under the compression loading part of the fatigue spectrum, stabilization plates were placed on each side of the test specimen as shown in Fig. 4. Two slits in the stabilization plate at the same locations as the cracks were used to observe the crack growth.

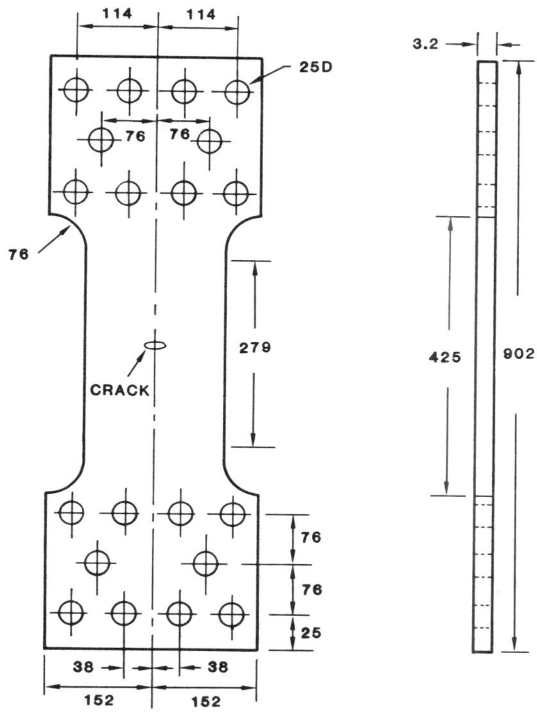

FIG. 1—*General test specimen (dimensions in mm).*

Analytical Stress Intensity Factors

Consider the skin with lands with a crack at the root radii as shown in Fig. 5. The analysis of the problem shown in the figure is beyond the present state of analytical developments. For the purposes of present analytical developments, the influence of the radius of curvature at the land will be neglected, and it will be assumed that the land is symmetrically placed about the centerline of the skin thickness as shown in Fig. 6. Thus, the geometry of Fig. 5 will be replaced by that of Fig. 6 for developing analytical techniques.

Formulation

The analysis of this problem is based on the following assumptions:

1. The skin and the land are either under generalized plane-stress or plane-strain condition.
2. Displacements in x- and y-directions are continuous at the junctions of the skin and the land.
3. Stresses in the skin and the land at the junctions are discontinuous. However, the total force in the loading direction and total shear force at the junctions are continuous.

The displacements and stresses for various regions of the symmetric problem (about the Y-axis) [*6, 7*] shown in Fig. 6, are given by

$$u_i = \frac{2}{\pi}\int_0^\infty [(A_i + B_i y)e^{-\alpha y} + (C_i + D_i y)e^{\alpha y}] \sin \alpha x d\alpha$$

$$v_i = \frac{2}{\pi}\int_0^\infty \left\{\left[A_i + \left(\frac{\kappa_i}{\alpha} + y\right)B_i\right]e^{-\alpha y} + \left[-C_i + \left(\frac{\kappa_i}{\alpha} - y\right)D_i\right]e^{\alpha y}\right\} \cos \alpha x d\alpha$$

$$\frac{1}{2\mu_i}\sigma^i_{yy} = \frac{2}{\pi}\int_0^\infty \left\{-\left[\alpha(A_i + B_i y) + \frac{\kappa_i + 1}{2}B_i\right]e^{-\alpha y} + \left[-\alpha(C_i + D_i y) + \frac{\kappa_i + 1}{2}D_i\right]e^{\alpha y}\right\} \cos \alpha x d\alpha$$

$$\frac{1}{2\mu_i}\sigma^i_{xy} = \frac{2}{\pi}\int_0^\infty -\left[\alpha(A_i + B_i y) + \frac{\kappa_i - 1}{2}B_i\right]e^{-\alpha y} + \left[\alpha(C_i + D_i y) - \frac{\kappa_i - 1}{2}D_i\right]e^{\alpha y}\Big\} \sin \alpha x d\alpha \tag{1}$$

where $\kappa_i = (3 - \nu_i)/(1 + \nu_i)$ for generalized plane stress and $(3 - 4\nu_i)$ for plane strain.

Continuity conditions for the problem under consideration are given by

$$\begin{aligned} &\text{At } y = 2h,\ u_1 = u_2 \\ &v_1 = v_2 \qquad \text{(Assumption 2)} \\ &t_1\sigma^1_{yy} = t_2\sigma^2_{yy} \\ &t_1\sigma^1_{xy} = t_2\sigma^2_{xy} \qquad \text{(Assumption 3)} \end{aligned} \tag{2}$$

$$\begin{aligned} &\text{At } y = 0,\ u_1 = u_3 \\ &v_1 = v_3 \qquad \text{(Assumption 2)} \\ &t_1\sigma^1_{yy} = t_2\sigma^3_{yy} \\ &t_1\sigma^1_{xy} = t_2\sigma^3_{xy} \qquad \text{(Assumption 3)} \end{aligned} \tag{3}$$

$$\begin{aligned} &\text{At } y_1 = -h_1,\ \sigma^3_{yy} = \sigma^4_{yy} \\ &\sigma^3_{xy} = \sigma^4_{xy} \end{aligned} \tag{4}$$

where $u_i(x, y)$ and $v_i(x, y)$ are the displacement fields in the respective regions shown in Fig. 6, and $\sigma^i_{yy}(x, y)$ and $\sigma^i_{xy}(x, y)$ are the stress in y-direction and shear stress, respectively, in various regions shown in Fig. 6.

The problem is formulated in terms of two unknowns defined by

$$\frac{\partial}{\partial x}(u_3 - u_4) = f_1(x)$$

$$\frac{\partial}{\partial x}(v_3 - v_4) = f_2(x), \qquad -a < x < +a \tag{5}$$

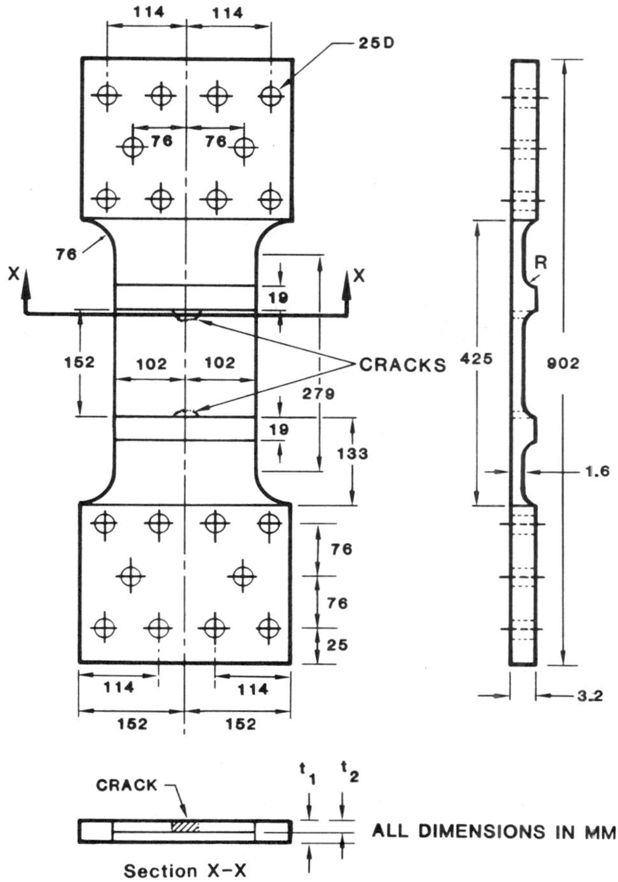

FIG. 2—*Test specimen for cracks at root radii.*

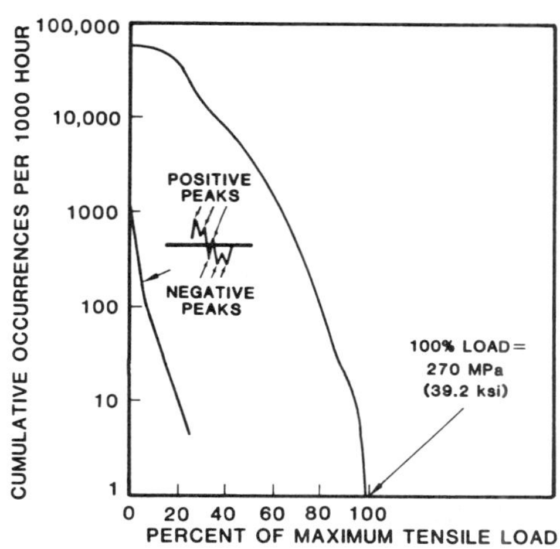

FIG. 3—*Fatigue spectrum used in testing.*

FIG. 4—*Fatigue test arrangement.*

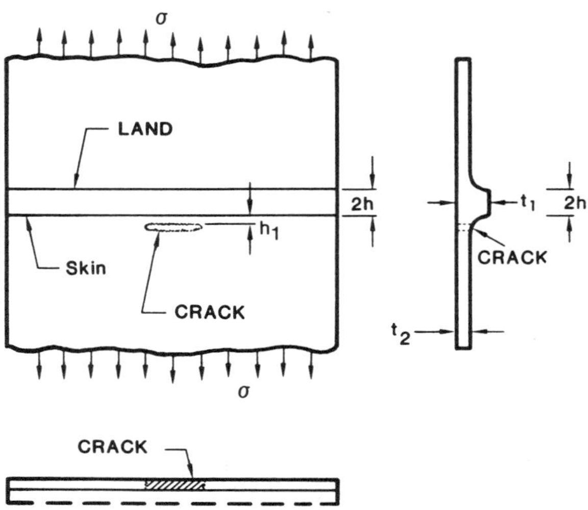

FIG. 5—*Crack at a land.*

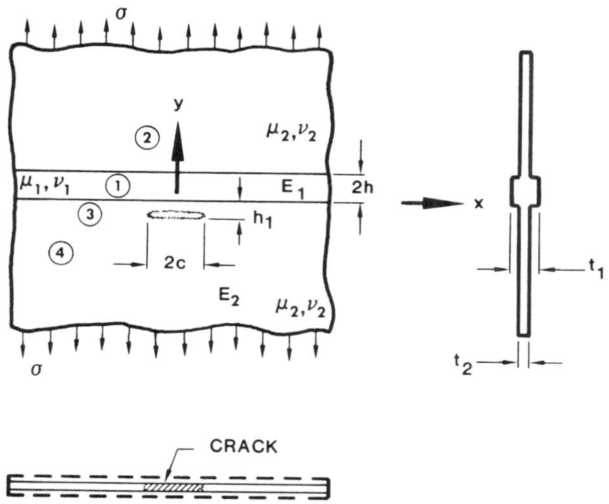

FIG. 6—*Idealized geometry of skin with lands.*

where $u_i(x, y)$ and $v_i(x, y)$ $(i = 3, 4)$ are the displacement fields in the regions 3 and 4, respectively, shown in Fig. 6.

The crack surface is subjected to the following stress state

$$\sigma_{yy}^{3}(x, -h_1) = \sigma_{yy}^{4}(x, -h_1) = p_1(x)$$

$$\sigma_{xy}^{3}(x, -h_1) = -\sigma_{xy}^{4}(x, -h_1) = p_2(x), \qquad -a < x < +a \tag{6}$$

Using the boundary conditions given by Eqs 1 through 4 and Eq 5, the constants A_i, B_i, C_i, and D_i are determined in terms of $f_1(x)$ and $f_2(x)$. Using these constants in terms of $f_1(x)$, the problem reduces to the solution of simultaneous equations given by

$$\int_{-a}^{+a}\left[\frac{f_1(t)}{t-x} + M_{11}(x, t)f_1(t) + M_{12}(x, t)f_2(t)\right]dt = \frac{1+\kappa_2}{2\mu_2}\,p_1(x)$$

$$\int_{-a}^{+a}\left[\frac{f_2(t)}{t-x} + M_{22}(x, t)f_2(t) + M_{21}(x, t)f_1(t)\right]dt = \frac{1+\kappa_2}{2\mu_2}\,p_2(x) \tag{7}$$

The kernels $M_{ij}(i, j = 1, 2)$ are given by

$$M_{11}(x, t) = -\frac{1}{2}\int_0^{\infty}\left[\lambda_3 + S_8(\alpha) + \frac{(\lambda_1 - \lambda_2)}{D_1(\alpha)D_2(\alpha)}\{\lambda_3 S_7(\alpha) + [S_7(\alpha)S_8(\alpha) - 8\alpha h(1 - 2\alpha h_1)]e^{-4\alpha h}\}\right]e^{-2\alpha h_1}\sin\alpha(t-x)\,d\alpha$$

$$M_{12}(x, t) = -\frac{1}{2}\int_0^{\infty}\left[\lambda_3 + S_6(\alpha) + \frac{(\lambda_1 - \lambda_2)}{D_1(\alpha)D_2(\alpha)}\{\lambda_3 S_7(\alpha) + [16\alpha^2 h h_1 - S_6(\alpha)S_7(\alpha)]e^{-4\alpha h}\}\right]e^{-2\alpha h_1}\cos\alpha(t-x)\,d\alpha$$

$$M_{21}(x, t) = M_{12}(x, t)$$

$$M_{22}(x, t) = -\frac{1}{2}\int_0^\infty \left[\lambda_3 + S_9(\alpha) + \frac{(\lambda_1 - \lambda_2)}{D_1(\alpha)D_2(\alpha)}\{\lambda_3 S_7(\alpha) - [S_7(\alpha)S_9(\alpha) - 8\alpha h(1 + 2\alpha h_1)]e^{-4\alpha h}\}\right]e^{-2\alpha h_1}\sin\alpha(t - x)d\alpha \tag{8}$$

where

$$D_1(\alpha) = -\lambda_2 + (4\alpha h + \lambda_1 e^{-2\alpha h})e^{-2\alpha h},$$

$$D_2(\alpha) = \lambda_2 + (4\alpha h - \lambda_1 e^{-2\alpha h})e^{-2\alpha h},$$

$$S_6(\alpha) = (1 - 4\alpha^2 h_1^2)/\lambda_3,$$

$$S_7(\alpha) = -\lambda_2 + \lambda_1 e^{-4\alpha h},$$

$$S_8(\alpha) = (1 - 2\alpha h_1)^2/\lambda_3,$$

$$S_9(\alpha) = (1 + 2\alpha h_1^2)/\lambda_3,$$

$$\lambda_1 = \frac{t_1\mu_1\kappa_2 - t_2\mu_2\kappa_1}{t_2\mu_2 + t_1\mu_1\kappa_2},$$

$$\lambda_2 = \frac{t_1\mu_1 + t_1\mu_1\kappa_1}{t_1\mu_1 - t_2\mu_2},$$

$$\lambda_3 = \frac{t_2\mu_2 + t_1\mu_1\kappa_2}{t_2\mu_2 - t_1\mu_1},$$

t_1 = land thickness, and

t_2 = skin thickness.

Solution of Integral Equations

The unknown functions f_i in Eq 7 have integrable singularities at the end points; therefore, Eq 7 must be solved subject to the single valued conditions given by

$$\int_{-a}^{+a} f_i(t)dt = 0, \qquad i = 1, 2 \tag{9}$$

The singular integral equations (Eqs 7) are solved simultaneously by using the numerical techniques described in Ref *8*. Using the technique described in the reference, we reduce the problem to a system of simultaneous algebraic equations for unknown functions. A computer program is written to solve the set of simultaneous equations and compute the stress intensity factors.

Stress-Intensity Factors

Stress-intensity factors have been obtained for several crack and structural configurations using the techniques discussed earlier. In obtaining these stress-intensity factors, we assumed that the skin material and land material have the same Poisson ratio and elastic moduli. The ratio of land thickness and skin thickness is varied in the calculations. The computer program that was developed can be used for cases in which skin and land have different Poisson ratios and elastic moduli.

Typical variations of stress-intensity factors with the a/h_1 (ratio of crack length to distance from the edge of the land) are shown in Figs. 7 through 10. Figure 7 shows the stress-intensity factors for h/a (ratio of land width to crack length) of 0.25 and land-to-skin thickness ratio of 2.0. It is seen that the Mode I stress-intensity factor decreases as the crack approaches the land due to the stiffening effect of the land. When the crack is far away from the land, the normalized stress-intensity factor increases to unity (stress-intensity factor for a crack in an infinite plate). It is seen that in addition to Mode I, stress-intensity factor K_I, Mode II stress-intensity factor K_{II} increases as the crack approaches the land. The magnitude of the Mode II stress-intensity factor K_{II} is much smaller than that for Mode I, K_I. Similar results for $h/a = 0.5$ and 1.0 are shown in Figs. 8 and 9, respectively. Figure 10 shows the results for $h/a = 0.25$ and $t_1/t_2 = 3.0$.

The stress-intensity factors for other land to skin thickness ratios and crack lengths showed the magnitude of K_{II} to be less than 10% of the Mode I stress-intensity factor.

Correlation Between Predictions and Test Data

Fatigue crack growth predictions were made for the crack at a root radius specimen using the analytical stress-intensity factors and the Forman crack growth equation [*9*]. The results of predicted crack growth were compared with test results.

Baseline Crack Growth Data

Crack growth data, obtained from a 7075-T6 center-cracked specimen with 3.2 mm (0.125 in.) thickness, were used to obtain constants in the Forman equation. The equation is given by

$$\frac{da}{dN} = \frac{C(\Delta K)^n}{(1 - R)K_c - \Delta K} \tag{10}$$

where

da/dN = crack growth rate,
ΔK = stress-intensity factor range,
K_c = fracture toughness,
R = ratio of minimum stress to maximum stress, and
C, n = constants.

The value of K_c for the material was found to be 1669 MPa$\sqrt{\text{mm}}$ (48 000 psi$\sqrt{\text{in.}}$). For 7075-T6 aluminum, the Forman equation constants were found to be $C = 2.3953 \times 10^{-10}$ and $n = 2.3412$.

Constant-Amplitude Fatigue Tests

The analytical stress-intensity factors for cracks at root radii were used in the Forman equation (Eq 10) to predict crack growth behavior at root radii. Crack growth constants for 3.2 mm

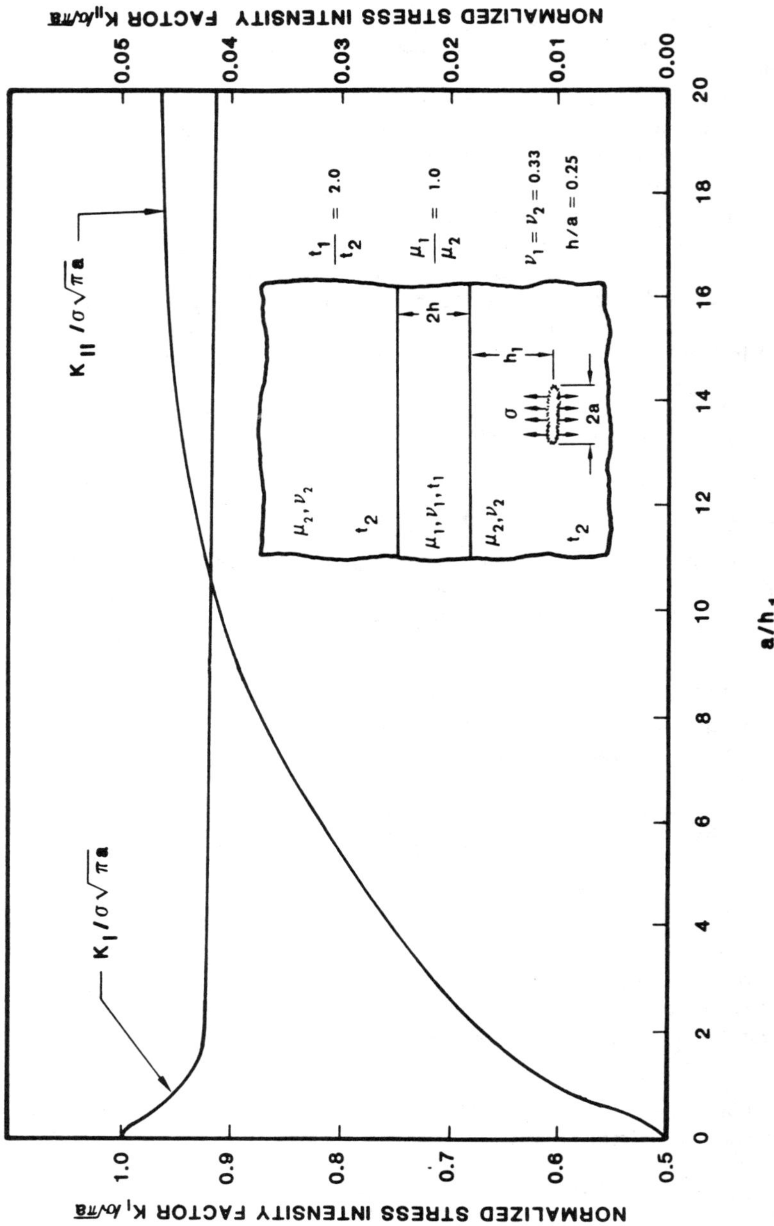

FIG. 7—*Variation of stress intensity factors with distance from the edge of the land* ($h/a = 0.25$; $t_1/t_2 = 2.0$).

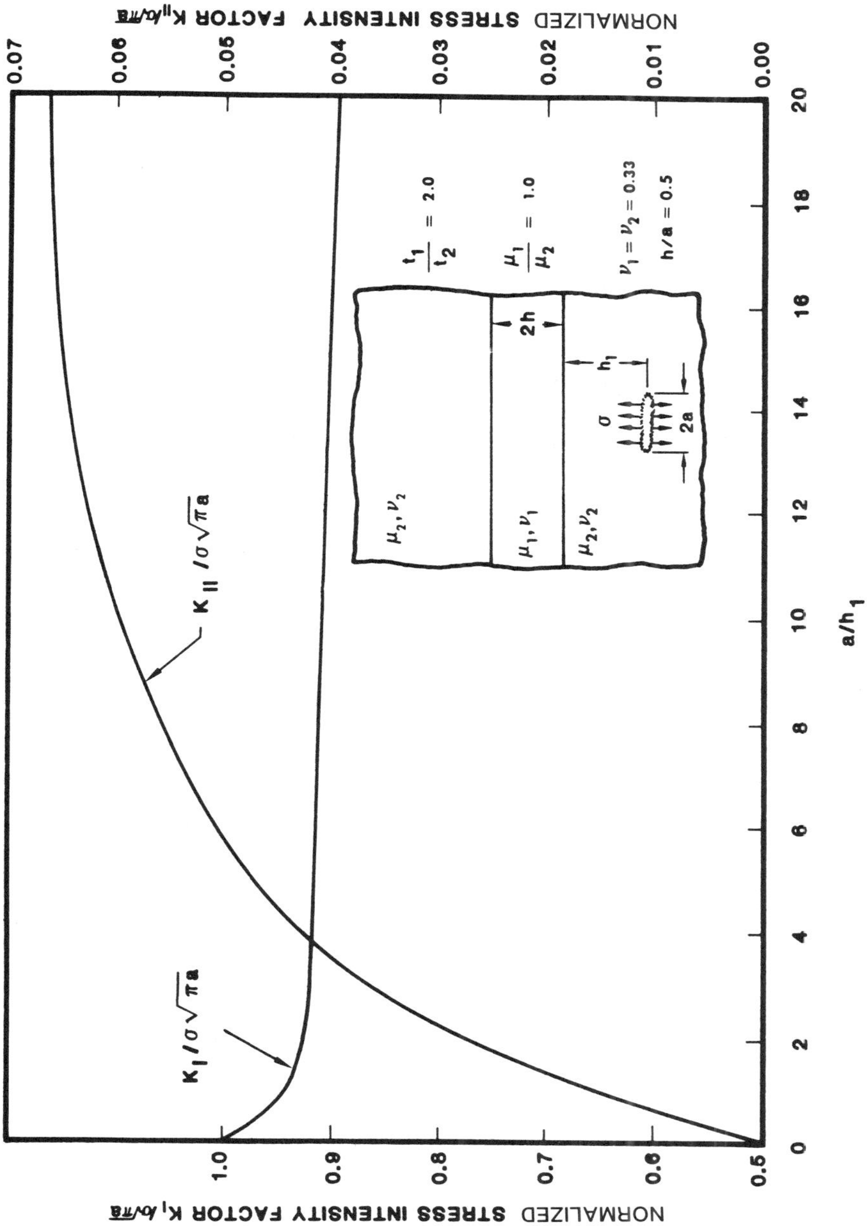

FIG. 8—*Variation of stress intensity factors with distance from the edge of the land* (h/a = 0.5; $t_1/t_2 = 2.0$).

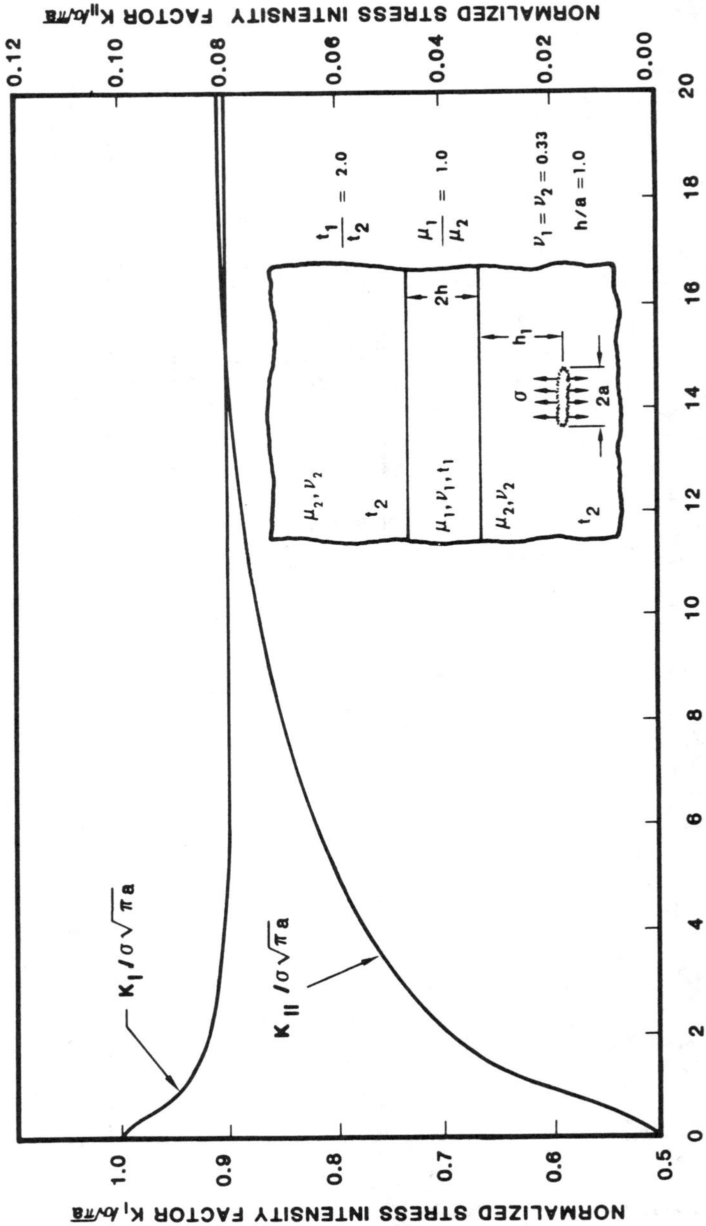

FIG. 9—*Variation of stress intensity factors with distance from the edge of the land* (h/a = 1.0; t_1/t_2 = 2.0).

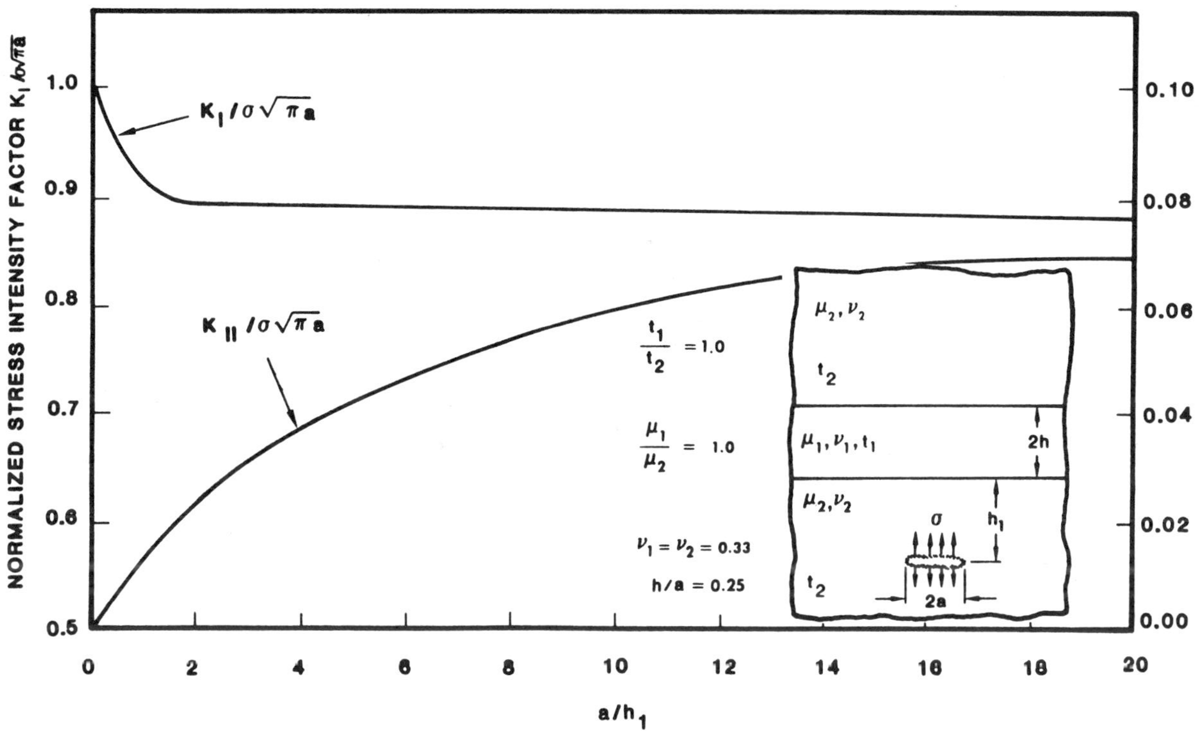

FIG. 10—*Variation of stress intensity factors with distance from the edge of the land* (h/a = *0.25*; t_1/t_2 = *3.0*).

(0.125 in.) thickness were used in making these predictions because the skin thickness at root radius is between 1.6 mm (0.063 in.) and 3.2 mm (0.125 in.) and also because the specimen was chem-milled from 3.2-mm (0.125-in.) thick material.

A comparison between predicted and observed crack growth behavior is shown in Fig. 11. Only the Mode I stress-intensity factor was used in predictions, even though both Mode I and Mode II stress-intensity factors are present due to the fact that the magnitude of Mode II stress-intensity factor is small compared with Mode I stress-intensity factor. Also, the crack growth rate equation used in predictions is based on Mode I crack growth only. The agreement between predicted and observed behavior is good.

Spectrum Loading Fatigue Tests

A comparison of predicted and observed crack growth behavior in a specimen with a crack at a root radius and tested under spectrum loading is shown in Fig. 12. The Northrop retardation model [*10*] was used in the prediction; this model is a modification of the Willenborg model in which the effects of compressive stress are taken into consideration. The comparison between predicted and observed crack growth is good.

Conclusions

The studies undertaken in this research program have resulted in the following conclusions:

1. Crack growth behavior at complex structural details can be reliably predicted using analytical stress-intensity factors and the crack growth rate obtained from a simple specimen.
2. Crack at a root radius propagates under Mode I and Mode II stress-intensity factors. However, the Mode II stress-intensity factor is small compared with Mode I. The crack growth behavior can be predicted with reasonable accuracy using Mode I stress-intensity factors.

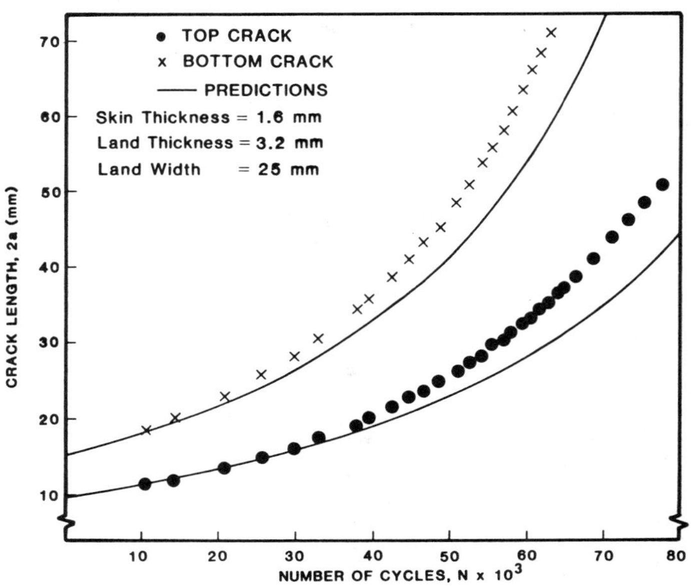

FIG. 11—*Comparison of observed and predicted crack growth behavior for cracks at root radii, tested under constant amplitude landing.*

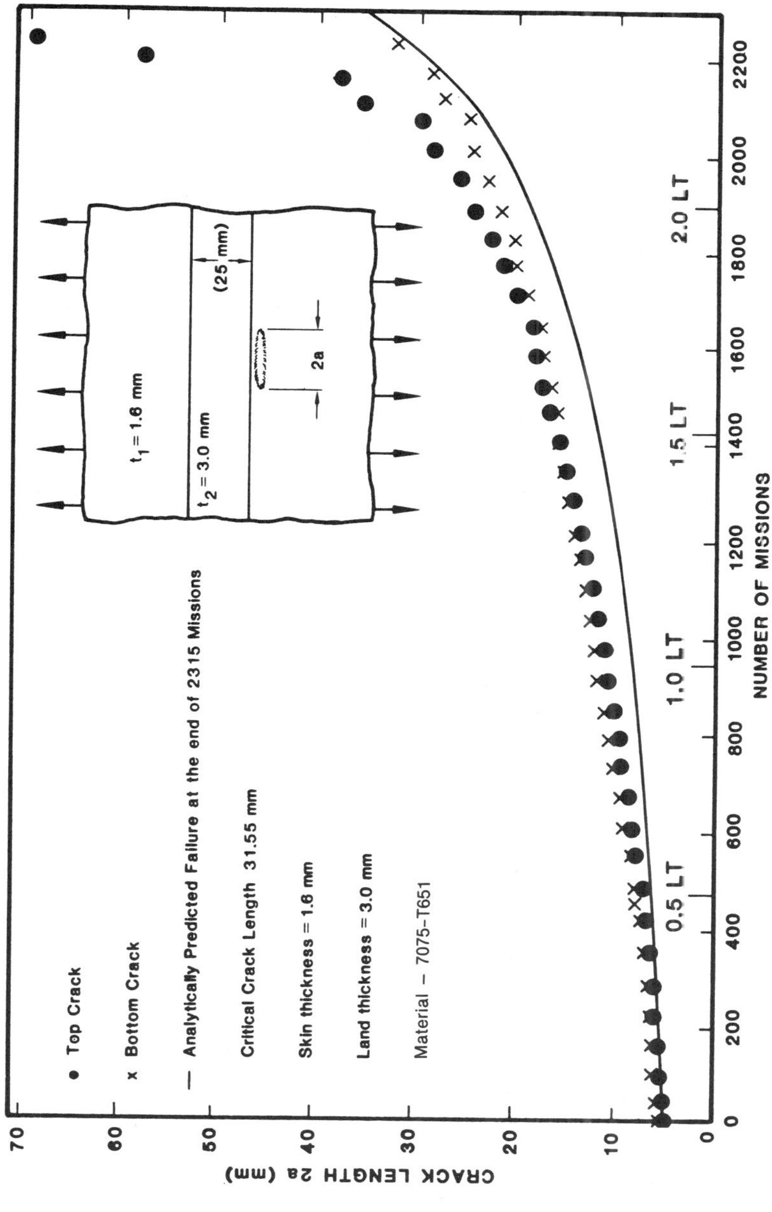

FIG. 12—*Comparison of observed and predicted crack growth in a crack at a root radius specimen, tested under spectrum loading.*

Acknowledgment

The work described herein was performed as part of Northrop's Independent Research and Development program.

References

[*1*] Isida, M., "On the Determination of Stress Intensity Factors for Some Common Structural Problems," *Engineering Fracture Mechanics,* Vol. 2, 1970, p. 61.

[*2*] Tada, M., Paris, P. E., and Irwin, G. R., "The Stress Analysis of Cracks Handbook," Del Research Corp., Hellertown, PA, 1973.

[*3*] Ratwani, M. M. and Kan, H. P., "Analysis and Growth of Cracks in Skins with Variable Thickness," *Fracture Mechanics: Fifteenth Symposium, ASTM STP 833,* R. J. Sanford, Ed., American Society for Testing and Materials, Philadelphia, 1984, pp. 44–56.

[*4*] Ratwani, M. M. and Kan, H. P., "Development of Composite Patches to Repair Complex Cracked Metallic Structures," Report No. NADC-80161-60, Naval Air Development Center, Vol. I and II, Jan. 1982.

[*5*] Ratwani, M. M., Kan, H. P., and Fitzgerald, J. H., "Experimental Investigation of Fiber Composite Reinforcement of Cracks in Complex Metallic Structures," presented at the Joint Society for Experimental Stress Analysis/Japan Society of Mechanical Engineers Conference, Hawaii, May 1982.

[*6*] Erdogan, F. and Gupta, G. D., "The Stress Analysis of Multilayered Composites with a Flaw," *International Journal of Solids and Structures,* Vol. 7, 1971, pp. 39–61.

[*7*] Erdogan, F. and Gupta, G. D., "Layered Composites with an Interface Flaw," *International Journal of Solids and Structures,* Vol. 7, 1971, pp. 1089–1107.

[*8*] Erdogan, F. and Gupta, G. D., "On the Numerical Solution of a Singular Integral Equations," *Quarterly Journal of Applied Mathematics,* Vol. 30, 1972, pp. 525–534.

[*9*] Forman, R. G., Kearney, V. E., and Engle, R. M., "Numerical Analysis of Crack Propagation in Cyclically Load Structures," *Journal of Basic Engineering, Transactions,* American Society of Mechanical Engineers, Vol. 89, Series D, Sept. 1967, p. 459.

[*10*] Chanani, G. R., "Fundamental Investigation of Fatigue Crack Growth Retardation in Aluminum Alloys," AFML-TR-76-156, Air Force Materials Laboratory, Wright-Patterson AFB, OH, 1976.

Royce G. Forman,[1] *Vankataraman Shivakumar,*[2] *James C. Newman, Jr.*[3], *Susan M. Piotrowski,*[2] *and Leonard C. Williams*[2]

Development of the NASA/FLAGRO Computer Program

REFERENCE: Forman, R. G., Shivakumar, V., Newman, J. C., Jr., Piotrowski, S. M., and Williams, L. C., **"Development of the NASA/FLAGRO Computer Program,"** *Fracture Mechanics: Eighteenth Symposium, ASTM STP 945,* D. T. Read and R. P. Reed, Eds., American Society for Testing and Materials, Philadelphia, 1988, pp. 781-803.

ABSTRACT: A preliminary version of a new computer program has been developed for calculating crack propagation in cyclically loaded structures. The new program, titled NASA/FLAGRO, has numerous enhanced features over previous programs used for safe life analysis on space systems. The essential features of the new program consist of modular design, promptive input using a cathode ray tube terminal, improved crack growth rate analytical methods, many new crack case solutions, and incorporation of user-defined files for material properties and fatigue spectrums. Also, five options made available in the program are to compute (1) safe life of a part with a preexisting crack, (2) critical crack size for a specified stress level, (3) stress-intensity factor numerical values for making comparisons or checks of solutions, (4) least-squares fit of growth rate equations to crack growth rate data, and (5) to convert *a* versus *N* data to ΔK versus da/dN.

KEY WORDS: cracks, crack propagation, fracture mechanics, damage tolerance, fatigue (materials)

Nomenclature

a	Depth, length, or half-length of crack
c	Width or half width of crack, distance from neutral axis to outer fibers for flexure stress problem
d	Exponent in Eq 5
m, n, p, q	Exponents in Eq 1
m_w	Exponent in Walker's growth rate equation
t	Thickness of plate or shell
t_0	Thickness to meet plane-strain condition
A_K, B_K	Fit parameters in Eq 6
A_n	Net area
B	Geometric dimension or biaxial stress ratio
C	Growth rate coefficient
C_0	Fit parameter in Eq 5
D	Diameter of hole or cylinder or other specified dimension
D_1, D_2	User-defined dimensions
F_0, F_1, F_2, F_3	Stress-intensity magnification factors

[1]Materials engineer, NASA, Lyndon B. Johnson Space Center, Houston, TX 77058.
[2]Advanced systems engineering specialist, computer programmer, and scientific programming specialist, respectively, Lockheed-EMSCO, 2400 NASA Rd. 1, Houston, TX 77258.
[3]Senior scientist, NASA, Langley Research Center, Hampton, VA 22665.

I_n Moment of inertia of a net section
K Stress-intensity factor
K_c Critical stress-intensity factor for fracture
K_{max} Maximum stress-intensity factor in fatigue cycle
K_{min} Minimum stress-intensity factor in fatigue cycle
K_{Ic} Plane-strain fracture toughness
K_{Ie} Effective fracture toughness for surface or elliptically shaped crack
K_{Iscc} Environmental threshold stress-intensity factor
K_R Stress-intensity factor for resistance curve
ΔK Stress-intensity factor range
ΔK_0 Fatigue threshold stress-intensity factor range at $R = 0$
ΔK_{th} Fatigue threshold stress-intensity factor range
M Bending moment
N Number of fatigue cycles
P Applied force
R Cycle ratio, K_{min}/K_{max}, or radius dimension
S_{max} Maximum applied stress
S_n Nominal failure stress on net section
S_u Stress required to produce full plastic region
S_0, S_1, S_2, S_3 Applied stresses
W Plate or specimen width
X Hole offset in arc-shaped specimen
α Enclosed angle of crack in cylinder, three-dimensional constraint factor on tensile yielding
β_R Crack closure factor
η Exponent in polynomial series equations for F_0, F_1, F_2, F_3
ρ Crack-tip yield zone size
σ_0 Flow stress
σ_{ys} Tensile yield strength
ω Exponent in Eq 6

The NASA/FLAGRO computer program has been developed to provide a user-friendly, state-of-the-art software package for fracture control analysis. This work grew, in part, from the need for a reliable and consistent analysis methodology for Space Shuttle payloads and, in the coming years, for the Space Station. The primary feature of the computer program is to provide an automated procedure for calculating the fatigue life of cyclically loaded structures with initial crack-like defects. The earlier version of the program, FLAGRO4, was developed by Rockwell International for fracture control analysis on the Space Shuttle and was the primary analysis tool used by Rockwell and the Shuttle subcontractors for this purpose. A review of the fracture control analysis on the Space Shuttle using FLAGRO4, and the capabilities and limitations of the computer program, are given in Ref *1*.

The new version of the program, NASA/FLAGRO, described herein, is a differently designed and coded program, but which retains a number of the useful features of FLAGRO4. The most important features retained are:

1. The numerical integration technique of varying the fatigue cycle increment size by specifying the increment size of crack growth.
2. The two-dimensional crack-growth model that predicts flaw growth independently at the major and minor axes of elliptically shaped cracks.
3. A check for failure of a part-through crack in the mode of a through crack when net ligament yielding occurs.
4. The storage and automatic call-up of different crack case solutions and material properties.

The enhanced features of NASA/FLAGRO are based on more than ten years' experience using FLAGRO4, and on the state-of-the-art improvements both in fracture mechanics and computer technology. Essentially, the new features incorporated in NASA/FLAGRO are listed as follows:

1. Modular design to allow for easier revisions and operation on mini-computers.
2. Promptive input from CRT terminals.
3. Many new or more accurate crack case solutions.
4. Simple input format for combined load application problems.
5. Proof test flaw screening analysis.
6. Table look-up using nonlinear interpolation for both ΔK versus da/dN and crack case K_I solutions.
7. User-defined files for both load spectrums and material properties.
8. Least squares fit of growth rate equations to crack growth rate data.
9. More accurate crack growth rate equations.
10. A limit-load fracture criterion routine.
11. Option of using either U.S. customary units or SI units.

The following section describes the basic theoretical background and assumptions made in the program. After that, descriptions are given of special subroutines and files developed for numerous capabilities. In the last section, both short-range and long-range plans for future enhancements and research areas are discussed. Lastly a description is given in Appendix I of the several different nonlinear interpolation routines used for the table lookup capability, and in Appendix II an example problem is presented.

Theoretical Background

The NASA/FLAGRO computer program provides the option in safe-life analysis of using either a crack growth rate equation or a linear interpolation routine operating on tabular data of ΔK versus da/dN. The growth rate equation incorporated into NASA/FLAGRO is a modified Forman equation expressed as

$$\frac{da}{dN} = \frac{C(1 - R)^m \Delta K^n [\Delta K - \Delta K_{\text{th}}]^p}{[(1 - R)K_c - \Delta K]^q} \tag{1}$$

where ΔK is assumed to be the full range of ΔK, including R values less than zero. The form of Eq 1 gives improved accuracy, versatility, and can be fit to data using standard least-squares methods. More specifically, Eq 1 can be converted to the following commonly used equations by assigning values for the exponents m, p, and q, as given in Table 1.

The listed equation options are available in both the safe-life analysis routine and the least-squares equation fit routine. Also, for the safe-life analysis routine, if $p = 0$ is entered, the computer program still calculates zero crack growth rate for the threshold condition $\Delta K - \Delta K_{\text{th}} \leq 0$. Similarly, for $q = 0$, crack instability is still predicted for the condition $(1 - R)K_c - \Delta K \leq 0$.

The computer program also incorporates the capability to use crack closure analysis to calcu-

TABLE 1—*Conversion of Eq 1.*

Exponent Values	Equation Form
$m = p = q = 0$	Paris
$m = p = 0, q = 1$	Forman
$p = q = 0, m = (m_w - 1)n$	Walker

late the effect of R on crack growth rate under constant amplitude loading. The analysis is based on Newman's [2] equation expressed in the form

$$\Delta K_2 = \left[\frac{1 - (S_0/S_{max})_1}{1 - (S_0/S_{max})_2}\left(\frac{1 - R_2}{1 - R_1}\right)\right] \Delta K_1 \tag{2}$$

In Eq 2, ΔK_1 is a baseline or known ΔK value corresponding to da/dN for $R = R_1$ ($R_1 = 0$ in NASA/FLAGRO). The ΔK_2 variable is the ΔK value that gives the same da/dN at a different R value (that is, R_2).

The solution to Eq 2 was derived numerically by Newman for a center crack tension specimen and is a function of a "constraint" factor α on tensile yielding, the stress ratio R, and the stress level ratio S_0/S_{max}. Equations for S_0/S_{max} for various R ratios, stress levels (S_{max}), and constraint (α) values are also given in Ref *2*.

The method used to incorporate crack closure into Eq 1 is to calculate the value of m which gives the needed shift in ΔK from ΔK_1 to ΔK_2 for a given R_2 and specified constant da/dN. The value of da/dN for this calculation is set at approximately the midrange of the linear part of the da/dN curve (for most materials) as shown in Fig. 1. The exponent values of $p = q \leq 0.50$ were found to give a satisfactory linear behavior in the central part of the curve and also give a good fit to experimental results in the upper and lower parts which are representative of threshold and instability behavior.

For constant-amplitude fatigue loading, Newman and Raju [3] have shown that multiplication of ΔK by a crack-closure factor β_R, gives more accurate crack growth predictions for semi-elliptical surface cracks and quarter-elliptical corner cracks. The β_R factor is applied at points where a crack intersects a free surface, and is a function of the stress ratio R. For $R > 0$, β_R is given by

$$\beta_R = 0.9 + 0.2R^2 - 0.1R^4 \tag{3}$$

For $R < 0$, β_R is assumed to have the value 0.9. The closure factor is applied in NASA/FLAGRO only to the cases of a surface crack in a plate, a surface crack in a hollow cylinder, and all corner crack problems.

The interpolation routine for computing da/dN is coded for use with either one-dimensional or two-dimensional tables. The two-dimensional tables are applicable to fatigue spectrums with multiple R values, particularly for problems relating to environmentally accelerated crack growth that cannot be analyzed by Eq 1. The form of the tabular input (storable in user-defined files) is shown in Fig. 2.

The $\Delta K_c(R)$ values are the critical stress-intensity factor ranges that cause fracture. For computational purposes, these values are assumed to correspond to data points having a growth rate of 25.4 E-2 mm/cycle (E-2 in./cycle), or any other growth rate value entered in the first row of the table. In a similar manner, the $\Delta K_{th}(R)$ values for no growth correspond to points with a growth rate of 25.4 E-8 mm/cycle (E-8 in./cycle), or also any value entered in the last row of the table.

To analyze problems with combined loading, which is often the case, the stress-intensity factor is expressed in the additive form

$$K = [S_0F_0 + S_1F_1 + S_2F_2 + S_3F_3]\sqrt{\pi a} \tag{4}$$

The terms S_0, S_1, S_2, and S_3 are different possible applied stresses, consisting of tension, compression, bending, or pin bearing pressure, and the F values are magnification factors applicable to each specific crack problem and type of applied stress. Since $\Delta K = K_{max} - K_{min}$ and $R = K_{min}/K_{max}$, the fatigue stress spectrum requires input of the individual minimum and

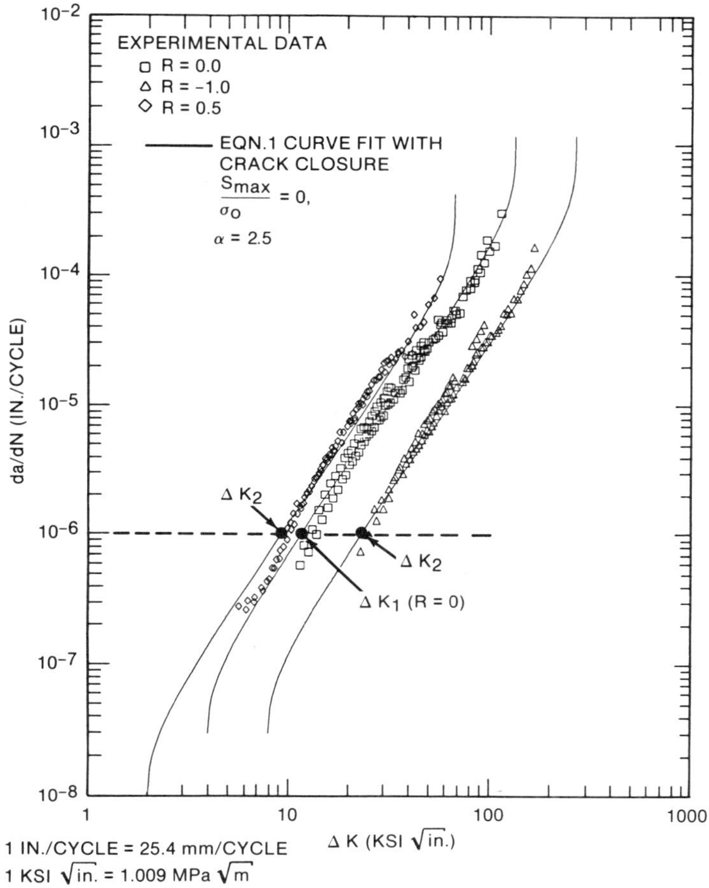

FIG. 1—*Curve fit of Eq 1 with crack closure to crack-growth rate data (from Ref 15) for 300 M steel.*

da/dN, mm/cycle (in./cycle)	ΔK			
	R→	⋮	⋮	⋮
25.4 E-2 (E-2)	$\Delta K_c(R)$ ⟶		⋮	⋮
—		⋮	⋮	⋮
—	⟵ 4 to 8 columns ⟶			
—	(All elements must be filled)			
—		⋮	⋮	⋮
25.4 E-8 (E-8)	$\Delta K_{th}(R)$ ⟶		⋮	⋮

(6 to 25 rows)

FIG. 2—*Interpolation routine for computing* da/dN *in tabular form.*

maximum values for the types of stresses, and these are used to calculate K_{min} and K_{max}. The maximum number of combined stresses currently allowed is four, as shown in Eq 4.

When using growth rate Eq 1, the fatigue threshold, ΔK_{th}, is calculated using the equation

$$\Delta K_{th} = (1 - C_0 R)^d \Delta K_0 \tag{5}$$

Equation 5 is a combination of two proposed threshold equations for which the fit parameters C_0 and d have been derived. By letting $C_0 = 1$, the Klesnil and Lukas [*4*] equation is obtained; by letting $d = 1$, the Barsom [*5*] equation is obtained. Values of ΔK_{th} or ΔK_0, C_0, and d for numerous alloys are given in Refs *6–9*. In cases where insufficient fatigue threshold data exist for different R values, the assumption that $C_0 = d = 1$ gives conservative results for $R \geq 0$ type load spectrums. Also, in relation to fatigue threshold analysis, the assumption is made in the computer code that $\Delta K_0 = 0$ for very small or shallow cracks. This occurs for Eq 5 when the dimension a of a through crack or dimensions a or c of an embedded, surface, or corner crack are equal to or less than 0.635 mm (0.025 in.).

The computer program also requires the input of the plane-strain fracture toughness, K_{Ic}, the fracture toughness for an elliptical crack, K_{Ie}, and fit parameters A_k and B_k, in the following expression for the variation of K_c with thickness

$$K_c/K_{Ic} = 1 + B_k e^{-\omega} \tag{6}$$

where

$$\omega = (A_k t/t_0)^2$$

and

$$t_0 = 2.5(K_{Ic}/\sigma_{ys})^2$$

Equation 6 is a modification of the relationship proposed by Vroman [*10*] which can be derived by letting $A_k = 5$ and $B_k = 1$. A typical curve fit of Eq 6 to experimental data is shown in Fig. 2. The data and plot routine to generate Fig. 3 are from a database program which will be discussed in the next section.

Both K_c and K_{Ie} are required for most of the part-through-crack cases because of checks made for failure as a through crack when net ligament yielding occurs [*11*]. The ligament yielding check is made with the criterion

$$a + \rho \geq t \tag{7}$$

where the yield zone dimension, ρ, is given by

$$\rho = (1/2\pi)(K_{max}/\sigma_{ys})^2 \tag{8}$$

The final version of the program will use two criteria for fracture. First, a K_R resistance curve routine will be used to determine whether the applied loading and crack length are critical or how much crack extension has occurred during proof testing. Second, a limit-load criterion, which is already developed and incorporated in the program, will be also used. The limit-load criterion is needed because the K_R resistance curve does not apply for very ductile materials. For these materials, the limit-load criterion does apply. The combination of these two criteria can be used to predict failure of a wide variety of crack configurations and material fracture behavior [*12*].

The limit-load criterion assumes that the crack configuration is under plane-stress conditions

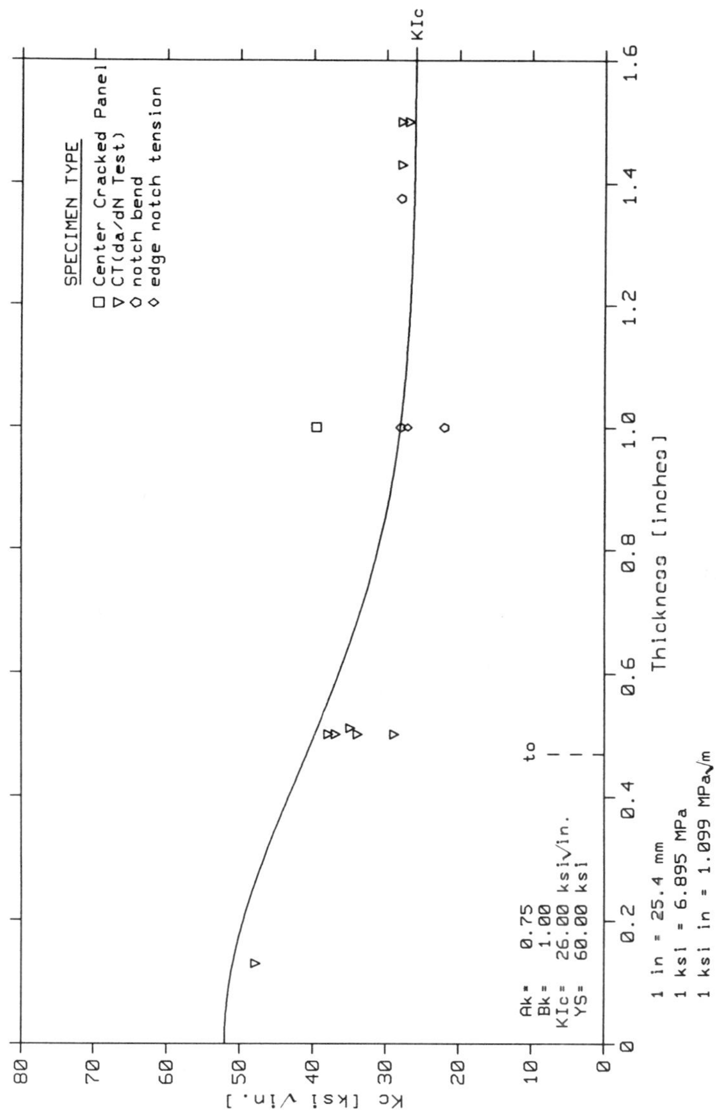

FIG. 3—*Curve fit of Eq 6 with fracture toughness data for 7075-T7351 aluminum.*

and that limit-load is defined as the load (or loads) required to produce a plastic hinge on the net section. The criterion is expressed in terms of nominal (or net section) stress as

$$S_n = P/A_n + Mc/I_n \tag{9}$$

where

A_n = net area,
P = resultant force,
M = resultant moment, and
I_n = moment of inertia on the net section.

Fracture occurs when $S_n \geq S_u$ where S_u is the nominal stress at the limit-load criterion.

The final input required for fracture mechanics material properties is the value of the environmental crack growth threshold K_{Iscc}. The purpose for this input is to make sure that K_{max} does not exceed K_{Iscc} for material and environmental combinations that could cause sustained load or environmentally induced crack growth. This is applicable to pressure vessels or other metallic components that are exposed to propellant fluids, gaseous hydrogen, high temperature air, or other incompatible fluid-material combinations. Since many parts are exposed to environments with low K_{Iscc} values for only a portion of their load spectrum, the application of the K_{Iscc} check along with the numerical value is entered during the fatigue spectrum input, and $K_{max} < K_{Iscc}$ is checked only for the steps or load applications which are selected.

Finally, three additional capabilities provided by NASA/FLAGRO are to calculate the critical crack size for a specified stress level, to calculate the stress-intensity factor for a given stress level and crack size, and to perform a least-squares fit of growth rate equations to *da/dN* data. The critical crack size calculation, which also includes the limit-load criterion check, is useful for the application of proof test flaw screening or, possibly, as a screening criterion to determine if a full safe-life analysis is required. Full safe-life analysis would not be required for many spectrums when the critical crack size exceeds the nondestructive examination (NDE) crack size by a large order of magnitude. The calculation of K_I for a given stress and flaw size is useful for checks or comparisons of K_I solutions in the program with other numerical solutions. The least-squares fit routine is included so that the program will incorporate mostly all of the essential computational procedures often required in a fracture control analysis.

Special Subroutines and Files

The NASA/FLAGRO program has a number of special subroutines and files which provide enhanced capabilities and easy entry of data. The most important of these are:

1. Crack case solutions.
2. Cyclic load spectrums.
3. Nondestructive examination (NDE) initial flaw sizes.
4. Table interpolation.
5. Material properties.

The crack case solution subroutines incorporated in the program are listed in Table 2, and the crack model geometries are shown in Figs. 4 through 7. For the loads spectrum file, an example of a spectrum provided for use on space transportation system (STS) payloads is shown in the example problem in Appendix II. Other spectrums can be entered or deleted by using procedures given by promptive input instructions. The NDE flaw size subroutine was also developed for STS payloads analysis, and the sizes for the different crack cases are listed in Table 3. The table interpolation subroutines are used for numerous applications in the program, and their descriptions and use are given in Appendix I.

TABLE 2—*Description of crack cases.*

THROUGH CRACKS
TC01 through crack in center of plate
TC02 single edge through crack
TC03 through crack from hole in plate
TC04 through crack from hole in lug
TC05 through crack from hole in row of holes
TC06 through crack in sphere
TC07 through crack in cylinder in longitudinal direction
TC08 through crack in cylinder in circumferential direction
TC09 through crack from hole in a plate under combined loading
CORNER CRACKS
CC01 corner crack in plate or bar (2D)[a]
CC02 corner crack from hole in plate (2D)
CC03 corner crack from hole in lug (2D)
EMBEDDED CRACKS
EC01 embedded crack in plate (2D)
SURFACE CRACKS
SC01 surface crack in finite width plate (2D)
SC02 surface crack in solid cylinder
SC03 surface crack in pipe, longitudinal on inner or outer diameter (2D)
SC04 surface crack in hollow cylinder, circular on outer diameter (2D)
SC05 surface crack in bolt thread (circumferential crack)
SC06 surface crack in threaded pipe
SC07 surface crack in bolt thread (thumbnail crack)
DATA TABLES
DT01 one-dimensional data table for through crack
DT02 two-dimensional data table for through crack
DT03 two-dimensional data table for elliptical crack (2D)
POLYNOMIAL SERIES
PS01 $F_0, F_1, F_2, F_3 = A_0 + A_1(a/D) + \ldots + A_5(a/D)^{5\eta}$
STANDARD SPECIMENS
SS01 standard center-cracked tension specimen
SS02 standard compact-type specimen
SS03 standard round compact-type specimen
SS04 standard arc-shaped specimen
SS05 standard bend specimen

[a]Two-dimensional type growth of elliptically shaped cracks.

A description of the materials properties files will be given in more detail because of their particularly special development. These files are divided into two types, a user-defined file and a permanent or fixed file. For the user-defined file, data are entered and stored during program execution, similar to the operation used for the loads spectrum file. This file is originally empty and the entry of data is left to the discretion of the program user. Conversely, the fixed file contains already coded-in property value data for many different materials. This file cannot be changed during program execution by the user. An essential purpose for providing the fixed materials file is to have property values that are consistent and acceptable for fracture control analysis on NASA space hardware.

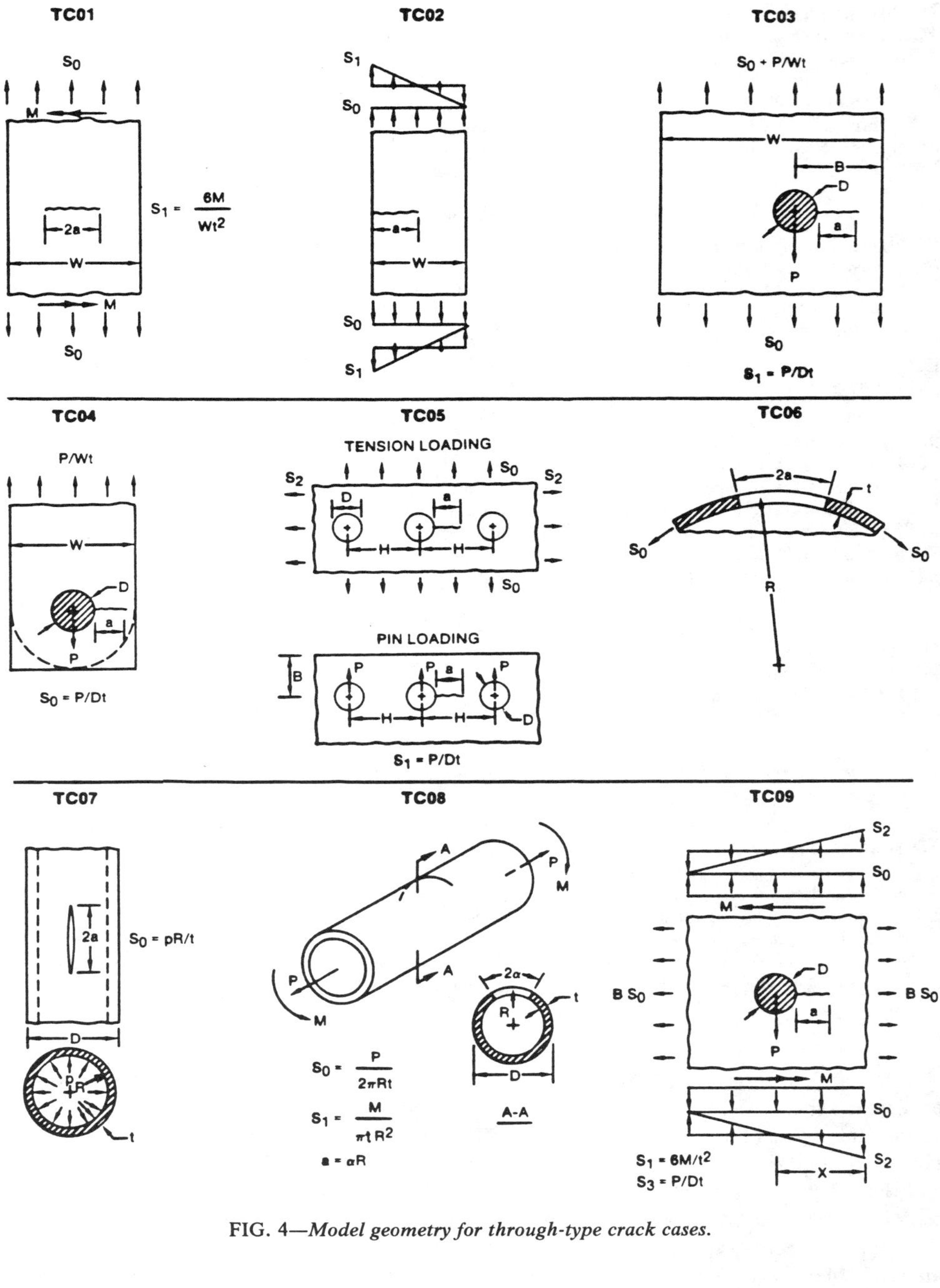

FIG. 4—*Model geometry for through-type crack cases.*

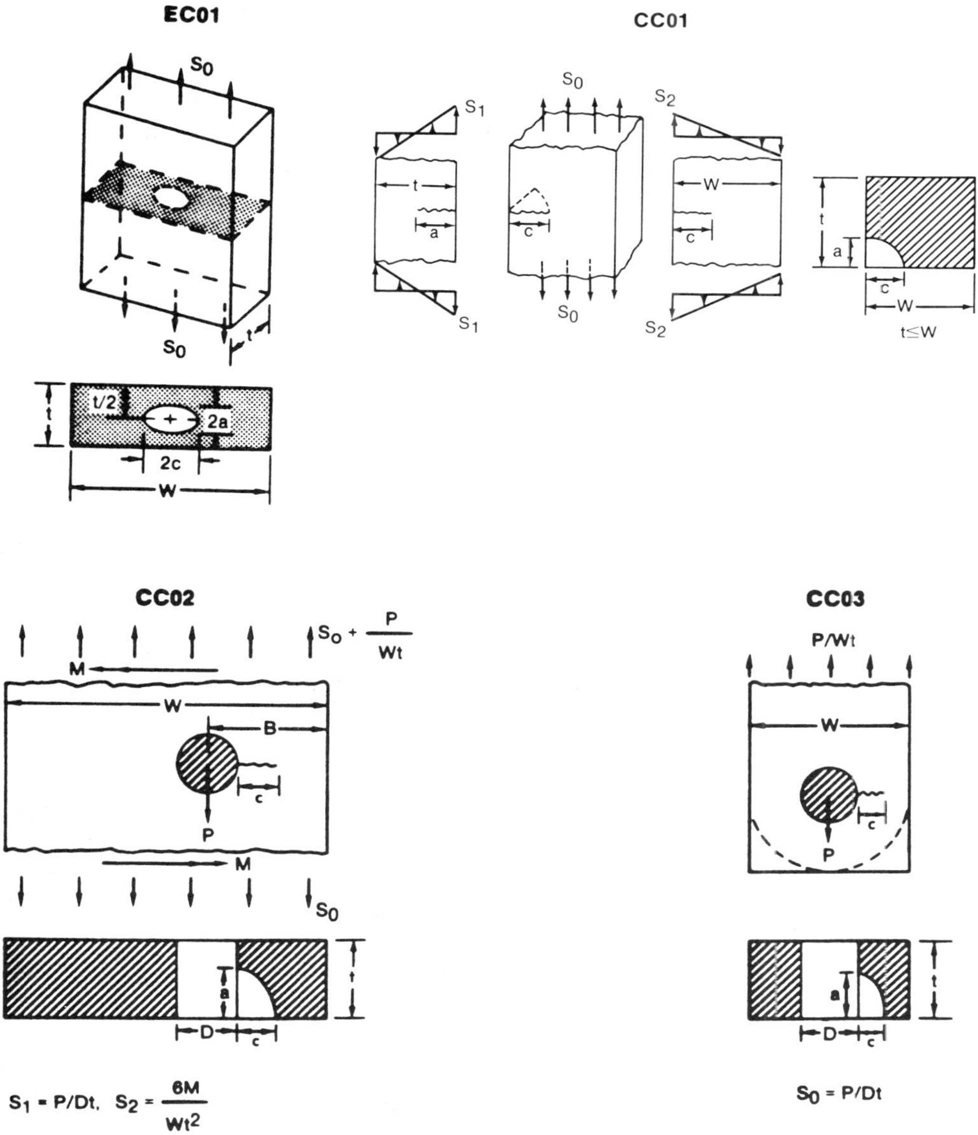

FIG. 5—*Model geometry for embedded- and corner-type crack cases.*

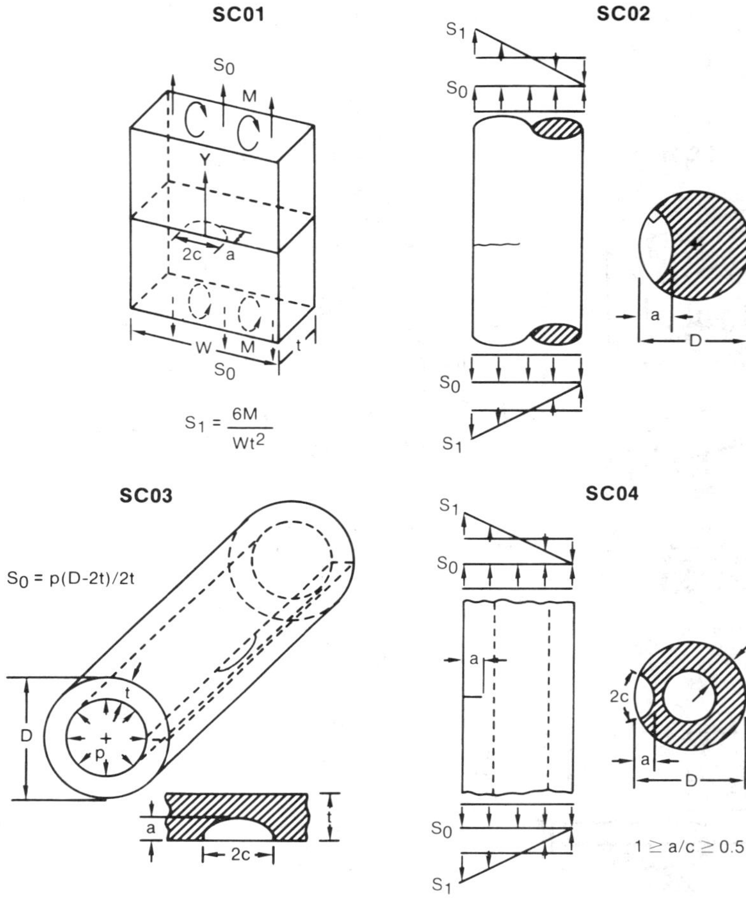

FIG. 6—*Model geometry for surface-type crack cases.*

The code system developed for the materials identification and promptive input is unique and designed to be both simple and highly expandable. Basically, the code system and order of promptive material selection is designed for three entry modes:

1. Letter entry from menu to select material category.
2. Number entry from menu to select material type.
3. Letter and number entry from menu to select material condition/heat treatment and environmental condition.

A scan option for the two materials files is also provided so that the program user can easily search for material data availability.

The NASA/FLAGRO materials file was generated by use of a separately developed database program with a code system compatible with NASA/FLAGRO. There were six sources of the basic data which include *da/dN* for both through cracks and surface cracks and fracture toughness values of K_c, K_{Ic}, and K_{Ie}:

1. References in Hudson's compendium of references (1184 listed) of published *da/dN* and fracture toughness data [*13,14*].

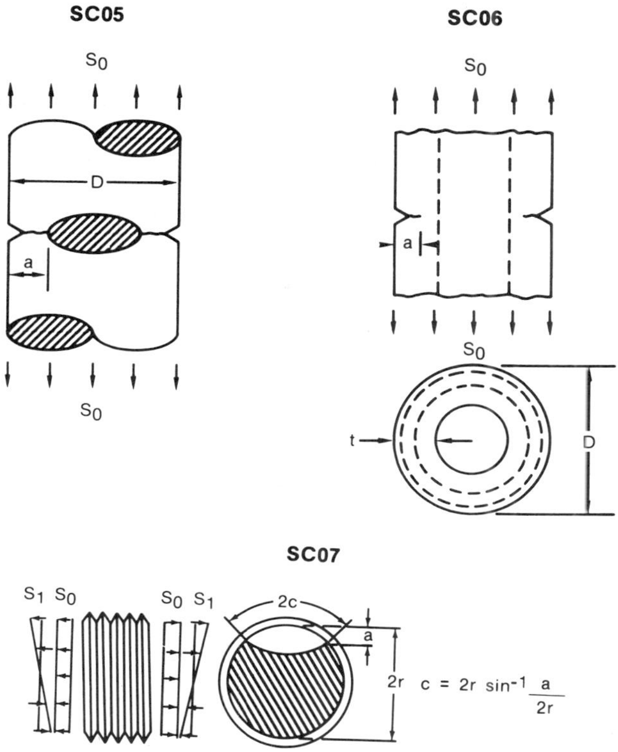

FIG. 6—*Continued.*

2. Data tape of fracture mechanics data used for generating the *Air Force Damage Tolerant Design Handbook* [*15*].
3. Contractor fracture mechanics test reports (25 total) written for the space shuttle program.
4. NASA contractor reports related to the Apollo program.
5. L. B. Johnson Space Center internal test reports and data sheets of unreported data.
6. Open literature references not in Hudson's compendiums (for example, references later than 1980).

Even though an extensive amount of data from the above references was collected, only data of primary interest for space applications have at this writing been entered into the database system. All data entered were screened for acceptability in meeting ASTM standards where they applied such as those for compact tension type specimens. The important criteria for all data, however, for entry into the database consisted of having sufficient information specifying:

1. Specimen type.
2. Maximum net section stresses.
3. Grain direction.
4. Specimen thickness.
5. Material strength or condition/heat treatment.

Least-squares methods and plotting routines were used to derive the appropriate constants in Eqs 1, 5 and 6 (that is, the da/dN, ΔK_{th}, and K_c equations). The da/dN curve fits were espe-

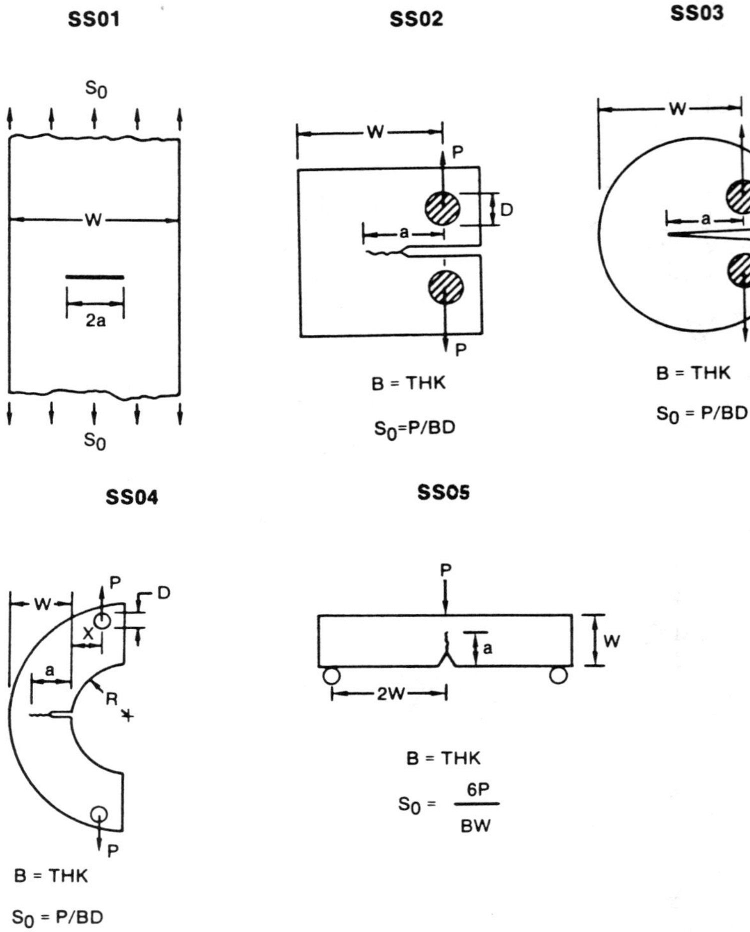

FIG. 7—*Model geometry for standard specimens.*

cially difficult because most of the available data were for single values of the stress ratio R. The problem was solved by using a computer analysis of the influence of R on da/dN based on Newman's crack closure equation. As stated previously, for this analysis, the controlling parameters for the spread of da/dN for various R ratios were (1) the three-dimensional constraint factor, α, on tensile yielding, and (2) the ratio of maximum stress to flow stress, S_{max}/σ_0. (The term σ_0 is assumed to be the average between uniaxial yield strength and uniaxial ultimate tensile strength of the material.)

Typical α values were determined for different categories of materials by fitting available da/dN data that had variable R values. The ratio S_{max}/σ_0 was assumed to be zero, and the fits were mostly performed on moderately thick specimens [for example, between 6 and 18 mm (0.234 and 0.702 in.)]. From the numerous fits, the values of α to be used for different material categories were determined as given in Table 4.

When fitting Eq 1, the exponents p and q were assumed in most cases to have values ≤ 0.50. This gave generally satisfactory curve fits and allowed for variations in K_c and ΔK_0 (for example, when applying Eqs 5 and 6) without introducing unwanted effects in the da/dN curves. For example, a change in K_c should not introduce a significant change in da/dN at the lower growth

TABLE 3—*Standard NDE flaw sizes for STS payloads.*

Crack Model No.	Type of Inspection	Crack Dimensions (in.)[a]	
TC01	dye penetrant	$a = 0.010 \geq t$	
TC02	dye penetrant	$a = 0.010 \geq t$	
TC03	(5)	(1)	
TC04	(5)	(1)	
TC05	(5)	(1)	
TC06	dye penetrant	$a = 0.010 \geq t$	
TC07	dye penetrant	$a = 0.010 \geq t$	
TC08	dye penetrant	$a = 0.010 \geq t$	
TC09	(5)	(1)	
EC01	radiographic	$a = 0.35t$, $c = 0.075$	for $t \leq 0.214$
		$a = c = 0.35t$	for $t > 0.214$
	ultrasonic	$a = c = 0.05$	
CC01	dye penetrant	$a = 0.010$	
CC02	(5)	(2) or (3)	
CC03	(5)	(2) or (3)	
SC01	radiographic	$a = 0.7t$, $c = 0.075$	for $t \leq 0.107$
		$a = c = 0.7t$	for $t > 0.107$
	dye penetrant	$a = c = 0.075$ or (4)	
SC02	dye penetrant	$a = 0.075$	
SC03-exterior flaw	dye penetrant	$a = c = 0.075$ or (4)	
SC03-interior flaw	ultrasonic	$a = c = 0.075$ or (4)	
SC04	dye penetrant	$a = c = 0.075$ or (4)	
SC05	(5)	$a = 0.005 +$ thd. depth	
SC06	(5)	$a = 0.005 +$ thd. depth	
SC07	none specified	none specified	

[a]1 in. = 25.4 mm.

NOTES: (1) $a = 0.1 \geq t$ for drilled hole, $a = 0.05 \geq t$ for reamed hole.
(2) $a = c = 0.1 < t$ for drilled hole, $a = c = 0.05 < t$ for reamed hole.
(3) $a = c = 0.005$ for cold-worked hole or hole with driven rivet.
(4) Equivalent area with minimum dimension ≥ 0.020 in. (0.508 mm) $0.1 \leq a/c \leq 1$.
(5) Flaw size based on assumed maximum possible manufacturing defect size.

TABLE 4—*α-Values for different materials categories.*

Material Category	α
Ultra-high-strength steel	2.5
Precipitation hardenable steel	2.0
Superalloys	2.0
Titanium alloys	1.75 to 2.0
Low yield strength stainless steel	1.75
Aluminum alloy	1.75

rate or threshold part of the curve. Similarly, a change in ΔK_0 should not cause a significant change in da/dN in the middle and upper part of the growth rate curve.

Future Enhancements

Future versions of NASA/FLAGRO will be released as major enhancements are completed. The next version to be released will incorporate improvements that are currently under development. Following this, subsequent releases will depend on progress made in relevant research studies in fracture mechanics and the completion of future identified improvements.

Seven improvements are currently under development and will be included in the next version of the program:

1. A boundary integral solution routine for calculating K_I values for through-crack problems.
2. A crack closure analysis routine for variable amplitude cyclic loading.
3. A weight function analysis routine for calculating surface crack K_I solutions in parts with nonlinear stress gradients.
4. Additional and more accurate crack case solutions.
5. Revised NDE flaw assumptions (for example, for threaded parts, new crack cases, and additional NDE methods).
6. Expanded materials file, including both aerospace and nonaerospace type alloys.
7. Addition of a transportable graphics package to be used with the plot routines.

For later versions of NASA/FLAGRO, the most important capabilities that are needed are for the analyses of environmentally accelerated and random load-induced fatigue crack growth. The methodology for these analyses problems has not yet been adequately developed, and further research on these problems is needed. The current development work on crack closure, however, may be significant for the analysis of both environmentally accelerated and random load-induced fatigue crack growth. Incorporation of crack closure theory into the research on these problems may help provide the improvement in the methodology needed for their solution.

Acknowledgments

The task of developing the NASA/FLAGRO computer program included a combined effort by many individuals working for NASA, the European Space Agency, U.S. aerospace contractors, and European aerospace contractors. The development and technical evaluation of the program was both monitored and assisted by the NASA Fracture Control Analysis Methodology Panel (formed in March 1985). The members and invited observers on the panel are:

Panel Members
- R. H. Weinstein, NASA Headquarters
- R. G. Forman, L. B. Johnson Space Center
- J. C. Newman, Jr., Langley Research Center
- T. W. Orange, Lewis Research Center
- G. C. Faile, Marshall Space Flight Center
- S. J. Brodeur, Goddard Space Flight Center
- R. W. Hampton, Ames Research Center
- S. Murray, Kennedy Space Center
- M. C. Lou, Jet Propulsion Laboratory

Invited observers:
- J. B. Chang, Aerospace Corp.
- G. G. Reibaldi, European Space Agency

Several fracture mechanics specialists with aerospace companies were also requested to assist in the technical evaluation. Their contributions and comments were particularly beneficial. Their names and companies are:

- C. R. Saff, McDonnell-Douglas Corp.
- G. A. Vroman, TRW

L. R. Hall, Boeing Aerospace Co.
T. Hu, Rockwell International
H. L. Eidinoff, Grumman Aerospace Corp.

Special mention should be made of several additional individuals for their work on the NASA/FLAGRO computer program. At Lockheed/EMSCO, M. F. Trznadel, H. L. Nguy, D. Goodrich, C. Sapp, and J. Beek contributed to the development of the fracture mechanics properties materials data base and computer code. In the technical evaluation and checking of NASA/FLAGRO, special mention should be made of the contributions of T. K. Henriksen at the European Space Agency and of G. J. Pattison at the Goddard Space Flight Center. At the National Aerospace Laboratory, The Netherlands, A. U. de Koning is assisting in the design and coding for the SI units conversion and in the design of the crack closure subroutine for the later version of the program.

APPENDIX I

Interpolation Routines

Five main subroutines (abbreviated here as S1, S2, S3, S4, and S5) perform interpolation in NASA/FLAGRO. The first (S1) is the controlling subroutine for interpolation in a table of crack length versus stress-intensity factor. The second (S2) controls interpolation in a table of stress-intensity factors with two apparently independent variables, the nondimensional crack lengths $a/D1$ and $a/D2$. The third (S3) performs the same function for the case of two-dimensional flaws with two truly independent variables, the nondimensional crack length a/D, and the crack shape a/c. The fourth (S4) handles interpolation in a table of stress intensity range ΔK versus crack growth rate da/dN, the calculations being performed on the logarithms of the data supplied. The fifth (S5) controls interpolation in the bivariate case obtaining da/dN for given values of ΔK and the stress ratio R, working again in the logarithmic domain (except for R).

Cubic spline interpolation methods are used in subroutines S1, S2, S3, and in the first-stage fitting of ΔK versus R (for constant da/dN) in S5. The formulation and coding of these subroutines are modifications of subroutines found in de Boor's tract [*16*].

The crack growth interpolating subroutines, S4 and S5, use a different curve-fitting technique. Since the spline fits are sensitive (unacceptably in the case of crack growth data, as shown by the solid line in Fig. 8) to the number of data points, an alternative method using piecewise hermite polynomial fits is employed. This method uses piecewise cubics, the interval endpoint values, and slopes of which are made to assume the spline-fitted values. The slope at each endpoint of the interval is taken to be the mean of the slopes of the straight lines joining it and the two points adjacent to it. For instance, the slope at an interval endpoint, P_i, is set to be the mean of the slopes of the straight lines segments, $\overline{P_{i-1}P_i}$ and $\overline{P_iP_{i+1}}$. At the end points of the table, P_1 and P_N, a different method is used since they each lack, on one side, a point adjacent to them. Here, a constant rate of change of slope is assumed. Thus, denoting slope by S_i, S_1 is set equal to $(2S_2 - S_3)$ and S_N equal to $(2S_{N-1} - S_{N-2})$. This improved hybrid fit ensures functional and slope continuity throughout while avoiding the problem of excessive flexibility that arises with splines. The dashed line in Fig. 7 shows the improvement achieved by this method, even when used on sparse data.

In S5, first-stage cubic spline fits are used to fit $\ln \Delta K$ to R for each value of da/dN. In the calculational mode, the first step consists of using the spline fits to obtain (using a search procedure that minimizes the number of spline evaluations) a sequence of $\ln \Delta K$'s corresponding to different da/dNs for the given value of R. A running fit of $\ln \Delta K$ is made, using the piecewise polynomials of the preceding paragraph, wherefrom da/dN for the given value of ΔK is evaluated. This hybrid method ensures that functional and slope continuity of da/dN (as a function of either independent variable) are both preserved.

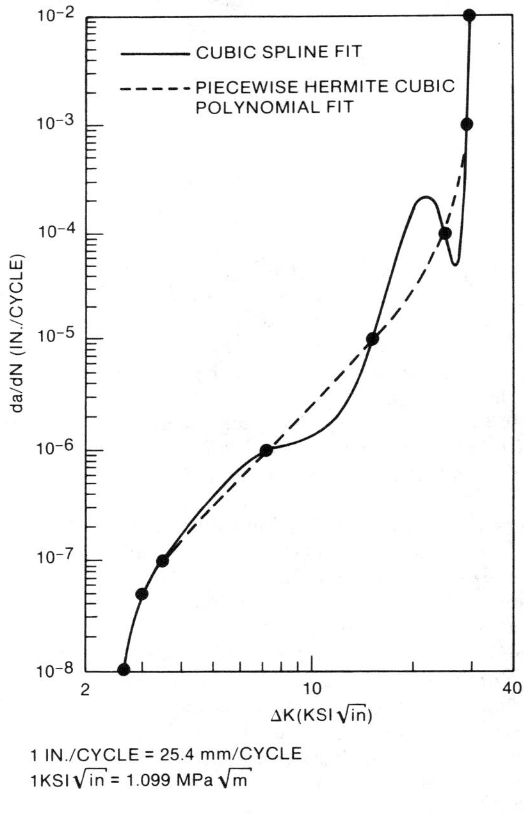

FIG. 8—*Comparison of cubic spline fit with piecewise hermite cubic polynomial fit for sparse data.*

APPENDIX II

Example Problem Input Sequence

```
                    ************************************
                    *       FATIGUE CRACK GROWTH       *
                    *         COMPUTER PROGRAM         *
                    *      " N A S A / F L A G R O "   *
                    ************************************
                  VERSION: MAY 1986, FORTRAN 77 LANGUAGE

   DEVELOPED BY:                         DEVELOPMENT SUPPORT BY:
   R. G. FORMAN,   NASA-JSC              DR. A. U. DE KONING,  NLR,  NETH.
   DR. V. SHIVAKUMAR,  L/E*              T. K. HENRIKSEN,  ESA/ESTEC
   DR. J. C. NEWMAN,  NASA-LANGLEY       J. B. CHANG,  AEROSPACE CORP.
     *L/E: LOCKHEED/EMSCO                DR. M. C. LOU,  JPL

   PROGRAMMED BY:                        ENGINEERING SUPPORT BY:
   L. C. WILLIAMS,  L/E                  J. M. BEEK,  L/E
   S. M. PIOTROWSKI,  L/E                M. F. TRZNADEL,  L/E
   C. A. SAPP,  L/E                      H. L. NGUY,  L/E

     MONITORED BY: NASA FRACTURE CONTROL ANALYTICAL METHODOLOGY PANEL

                    ************************************
                    *    Hit RETURN to start program   *
                    ************************************
```

NOTE: -> indicates keyed input values.

```
  Enter type of units:
  1    for U.S. Customary Units
  2    for S.I. Units
->1

  Enter one of the following options:
  1    to compute Safe Life
  2    to    "     critical crack size
  3    to    "     stress-intensity solutions
  4    to    "     growth rate constants
  5    to convert a vs N data to da/dN vs Delta K
  0    to terminate session
  NOTES:
  1. Enter all values in U.S. customary units [inches, ksi, ksi sqrt(in)]
  2. Always hit RTN to complete entry.
  3. For hard copies, print the contents of File 8 after terminating session.
     e.g. on the VAX 780 system, enter "PRINT FOR008.DAT".
->1

                          ****************************
                          * COMPUTATION OF SAFE LIFE *
                          ****************************

                 U.S. customary units [inches, ksi, ksi sqrt(in)]

  Enter Computation Title (up to 60 characters):
->EXAMPLE PROBLEM 1 - SAFE LIFE COMPUTATION

                          * CRACK MODEL DATA *
            U.S. customary units [inches, ksi, ksi sqrt(in)]

  Enter Model type from the following :
  TC   for Through Crack
  CC   for Corner Crack
  EC   for Embedded Crack
  SC   for Surface Crack
  DT   for Data Table
  PS   for Polynomial Series
  BI   for Boundary Integral
  SS   for Standard Specimen
->SC

  Enter Model designation from the following SURFACE crack cases:
  SC01  -  in finite width plate
  SC02  -  in solid cylinder (circumferential plane)
  SC03  -  in hollow cylinder (longitudinal at ID or OD)
  SC04  -  in hollow cylinder, circular on outer dia.
  SC05  -  in bolt thread (circumferential crack)
  SC06  -  in threaded pipe
  SC07  -  in bolt thread (thumbnail crack)
->SC01
```

FIG. 9—*Appendix II.*

```
  Enter Thickness, t:
->.5

  Enter Width, W:
->20

  Enter initial flaw size option:
  U   for User-specified flaw size
  S   for  Standard NDE  flaw size
->S

  Enter Radiographic (R) or Dye Penetrant (D):
->D

  Enter initial a/c:
->1

  Standard NDE.  Initial a   =    0.0750
                 Initial a/c =    1.0000

  Hit RTN (or C and RTN to change/rescan Crack Model data
          or S and RTN to start over again)

                         * MATERIAL PROPERTY DATA *
               U.S. customary units [inches, ksi, ksi sqrt(in)]

  Enter Total Number of Materials:
->1

  Enter option for Material No. 1:
  1   for manual input & (if reqd) save on User Defined materials file
  2   for  input from    NASA/FLAGRO materials file
  3   to examine data on         "            "       "
  4   for  input from    User Defined materials file (UDMF)
  5   to examine data on    "     "          "        "
  6   for deleting entries from UDMF (option is unrelated to this computation)
  0   to terminate materials entry
->2

  Material  1 Input Option: File

  Enter one of the codes in brackets [ ] to display a detailed menu
      or 0 to exit

                        NASA/FLAGRO MATERIALS FILE

             FERROUS ALLOYS                       NON-FERROUS ALLOYS

   Iron, Alloy or Cast              [A]*  2000-7000 Series Aluminum     [M]
   Low Alloy or Carbon Steel        [B]*  Foreign Aluminum Alloys       [N]*
   Ultra High Strength Steel        [C]   Misc. and Cast Aluminum       [O]
   Stainless Steel                  [D]   Titanium Alloys               [P]
   Age Hard Steel                   [E]   Nickel Base Alloys            [Q]
   Ni Cr Steel                      [F]   Ni Co Base Alloys             [R]
   Tool Steel                       [G]*  Cobalt Base Alloys            [S]*
   Miscellaneous Ferrous            [H]*  Miscellaneous Non-ferrous     [T]
                    NOTE: WELDS ARE NOT HEAT TREATED AND
                   ENVIRONMENT IS RT AIR UNLESS SPECIFIED.
  * - No data in files
->M

  Enter ID Number in brackets to choose a Material Type
      or 0 to return to main material menu.

                     2000-7000 SERIES ALUMINUM  (CODE M)

     2014                      [1]*    5000 Series          [8]*
     2024                    --[2]     6000 Series        --[9]
     2124                    --[3]     7050                 [10]*
     2219                    --[4]     7075               --[11]
     Misc. 2000 Series         [5]*    7178                 [12]*
     3000 Series               [6]*    Misc. 7000 Series    [13]*
     4000 Series               [7]*
  * - No data in files
->11
```

FIG. 10—*Appendix II continued.*

```
  Enter Code in brackets to choose Material Condition.
  Note: Y is Eqn Constants  (closure option)      1 is 1-d table, DK    vs  da/dN
        N is  "     "     (non-closure option)    2 is 2-d   "  , DK(R) vs  da/dN
                                                                   :Y:N:1:2:
  [A1]    7075-T6 AL, L-T                                          :+:*:*:*:
  [A5]    7075-T6 AL, L-T, HHA (0-9Hz)                             :+:*:*:*:
  [A6]    7075-T6 AL, T-L, HHA (0-9Hz)                             :+:*:*:*:
  [B1]    7075-T651 AL, L-T                                        :+:*:*:*:
  [B3]    7075-T651 AL, S-T                                        :+:*:*:*:
  [B5]    7075-T651 AL, L-T, HHA (0-9Hz)                           :+:*:*:*:
  [B6]    7075-T651 AL, T-L, HHA (0-9Hz)                           :+:*:*:*:
  [F1]    7075-T73 AL, L-T, LHA & HHA                              :+:*:*:*:
  [F2]    7075-T73 AL, T-L, LHA & HHA                              :+:*:*:+:
  [G1]    7075-T7351 AL, L-T, LHA & HHA                            :+:*:*:*:
  [G2]    7075-T7351 AL, T-L, LHA & HHA                            :+:*:*:*:
  [L2]    7075-T7651 AL, T-L                                       :+:*:*:*:
                                           (+ Available    * Not available)
->F2

  Enter type of data desired if available:
  Y   Crack Growth Eqn. Constants (closure option)
  N   Crack Growth Eqn. Constants (no closure option)
  1   1-D table - Delta K vs. da/dN
  2   2-D table - Delta K ( R ) vs. da/dN
->Y

  Matl No. 1: 7075-T73 AL, T-L, LHA & HHA

  Material Properties:
```

Matl No.	YS	K1e	K1c	Ak	Bk	Thk	Kc	KIscc
1	62.00	41.0	26.0	0.75	1.25	0.500	41.7	

Matl No.	Crack Growth Eqn Constants (closure) C	n	p	q	DKo	Co	d	DK1	Alpha	Smax/SIGo
1	0.173D-08	3.643	0.50	0.50	2.5	1.0	1.0	7.33	1.75	0.0

```
  Hit RTN (or C and RTN to change/rescan Material Property data
           or S and RTN to start over again)

                         * LOAD SPECTRUM DATA *
              U.S. customary units [inches, ksi, ksi sqrt(in)]

  Enter Spectrum Heading (up to 60 characters):
->NASA GODDARD PAYLOAD LAUNCH & LANDING SPECTRUM

  Enter No. of Load Steps / Spectrum Block:
->13

  Enter Total No. of Blocks:
->10

  Enter Stress Scaling Factor (SSF) for S0:    (S0 to be entered later),
  where SSF * S0 = Remote Uniform Tension Stress
->.25

  Enter Stress Scaling Factor (SSF) for S1:    (S1 to be entered later),
  where SSF * S1 = Plate Bending Stress
->.25

  Enter one of the following options:
  1   Display Menu of Spectra on File
  2   Enter a Spectrum from File
  3   Enter/Edit a Spectrum Manually
  4   Delete a Spectrum from file
  0   Continue (if Spectrum Input complete)
->2

  Enter Spectrum name:
->GSFC

  Enter Material Number:
->1

  Spectrum will be added beginning with Load Step No.    1

  Hit RTN (or C and RTN to Change above number)
```

FIG. 11—*Appendix II continued.*

```
  Enter one of the following options:
  1   Display Menu of Spectra on File
  2   Enter a Spectrum from File
  3   Enter/Edit a Spectrum Manually
  4   Delete a Spectrum from file
  0   Continue (if Spectrum Input complete)
->0

  Spectrum Table:
  [Note : Input S(i) = Actual Stress / Stress Scaling Factor]

   S: M: NUMBER :       S0        :       S1        :
   T: A:   OF   :                 :                 :
   E: T: FATIGUE:                 :                 :
   P: L: CYCLES :  min   :  max   :  min   :  max   :
   --:--:-------:--------:--------:--------:--------:
   1: 1:       2:-100.00: 100.00:-100.00: 100.00:
   2: 1:       4: -90.00:  90.00: -90.00:  90.00:
   3: 1:       8: -80.00:  80.00: -80.00:  80.00:
   4: 1:      15: -70.00:  70.00: -70.00:  70.00:
   5: 1:      49: -60.00:  60.00: -60.00:  60.00:
   6: 1:      81: -50.00:  50.00: -50.00:  50.00:
   7: 1:     178: -40.00:  40.00: -40.00:  40.00:
   8: 1:     641: -30.00:  30.00: -30.00:  30.00:
   9: 1:    3120: -20.00:  20.00: -20.00:  20.00:
  10: 1:    3405: -10.00:  10.00: -10.00:  10.00:
  11: 1:    5019:  -7.00:   7.00:  -7.00:   7.00:
  12: 1:   28853:  -5.00:   5.00:  -5.00:   5.00:
  13: 1:   91655:  -3.00:   3.00:  -3.00:   3.00:

  Check of Kmax < KIscc for sustained stress: NOT SET

  Hit RTN (or S and RTN to set/change above check & enter KIscc)

  Hit RTN (or C and RTN to change/rescan Load Spectrum data
           or S and RTN to start over again)

  Enter Block Interval for printing results - at end of a block
  (0 = final results only):
->0

  Enter Block Interval for printing results - at each step in block
  (0 = none):
->0

  Hit RTN (or C and RTN to change)

                       FATIGUE CRACK GROWTH ANALYSIS RESULTS

                  U.S. customary units [inches, ksi, ksi sqrt(in)]

           FINAL RESULTS:
           Critical Crack Size has NOT been reached.
           at Cycle No.          91655.
           of Load Step No.         13
           of Block No.             10
           Crack Size     a =  0.812243E-01     ,   a/c =  0.973355

  Hit RTN (or C and RTN to continue growing the present crack)

  Hit RTN to continue (or P and RTN to flag for printing)

  Enter one of the following options:
  1 Another Safe Life Analysis
  0 Return to Master menu
->0

  Enter one of the following options:
  1   to compute Safe Life
  2   to    "     critical crack size
  3   to    "     stress-intensity solutions
  4   to    "     growth rate constants
  5   to convert a vs N data to da/dN vs Delta K
  0   to terminate session
  NOTES:
  1. Enter all values in U.S. customary units [inches, ksi, ksi sqrt(in)]
  2. Always hit RTN to complete entry.
  3. For hard copies, print the contents of File 8 after terminating session.
     e.g. on the VAX 780 system, enter "PRINT FOR008.DAT".
->0

                          ***********************
                          * SESSION TERMINATED *
                          ***********************
```

FIG. 12—*Appendix II continued.*

References

[1] Forman, R. G. and Hu, T., "Application of Fracture Mechanics on the Space Shuttle" in *Damage Tolerance of Metallic Structures, ASTM STP 842,* American Society for Testing and Materials, Philadelphia, 1984, pp. 108–133.

[2] Newman, J. C., Jr., "A Crack Opening Stress Equation for Fatigue Crack Growth," *International Journal of Fracture,* Vol. 24, No. 3, March 1984, pp. R131–R135.

[3] Newman, J. C., Jr., and Raju, I. S., "Prediction of Fatigue Crack-Growth Patterns and Lives in Three-Dimensional Cracked Bodies," presented at Sixth International Conference on Fracture, New Delhi, India, Dec. 1984.

[4] Klesnil, M. and Lukus, P., "Effect of Stress Cycle Asymmetry on Fatigue Crack Growth," *Material Science and Engineering,* Vol. 9, 1972, pp. 231–240.

[5] Barsom, J. M., "Fatigue Behavior of Pressure-Vessel Steels," WRC Bulletin 194, Welding Research Council, New York, May 1964.

[6] Hertzberg, R. W., *Deformation and Fracture Mechanics of Engineering Materials,* 2nd ed., Wiley, New York, 1983.

[7] Rolfe, S. T. and Barsom, J. M., *Fracture and Fatigue Control in Structures, Applications of Fracture Mechanics,* Prentice-Hall, Englewood Cliffs, NJ, 1977.

[8] Backlund, J., Blom, A. F., and Beevers, C. J. in *Fatigue Threshold, Fundamentals and Engineering Applications, Proceedings,* International Conference, sponsored by the Aeronautical Research Institute of Sweden and The Swedish Plastic and Rubber Institute, Stockholm, 1–3 June 1981.

[9] Beevers, C. J., "A Review of Fatigue Crack Growth in Metallics at Low Stress Intensities, Final Scientific Report," Finmet Consultants, prepared for European Office of Aerospace Research and Development, Jan. 1983.

[10] Vroman, G. A., "Material Thickness Effect on Critical Stress Intensity," Monograph No. 106, TRW Space and Technology Group, Redondo Beach, CA, Feb. 1983.

[11] Orange, T. W., Sullivan, T. L., and Calfo, F. D., "Fracture of Thin Sections Containing Through and Part-through Cracks" in *Fracture Toughness Testing at Cryogenic Temperatures, ASTM STP 496,* American Society for Testing and Materials, Philadelphia, 1970, pp. 61–81.

[12] *Elastic-Plastic Fracture Mechanics Technology, ASTM STP 896,* J. C. Newman, Jr., and F. J. Loss, Eds., American Society for Testing and Materials, Philadelphia, 1985.

[13] Hudson, M. C. and Seward, S. K., "Compendium of Sources of Fracture Toughness and Fatigue Crack Growth Data for Metallic Alloys," *International Journal of Fracture,* Vol. 14, 1978, pp. R151–R184.

[14] Hudson, C. M. and Seward, S. K., "A Compendium of Sources of Fracture Toughness and Fatigue Crack Growth Data for Metallic Alloys—Part II," *International Journal of Fracture,* Vol. 20, 1982, pp. R59–R117.

[15] *Damage Tolerant Design Handbook,* MC1C-HB-01R, Metals and Ceramics Information Center, Battelle Columbus Laboratories, Columbus, OH, Dec. 1983.

[16] de Boor, C., *A Practical Guide to Splines,* Springer-Verlag, New York, 1978.

Jon Dag Evensen,[1] *Per A. Gradin,*[2] *and Jon Lereim*[3]

The Influence of Cracks on the Buckling Capacity of Axially Compressed Columns

REFERENCE: Evensen, J. D., Gradin, P. A., and Lereim, J., **"The Influence of Cracks on the Buckling Capacity of Axially Compressed Columns,"** *Fracture Mechanics: Eighteenth Symposium, ASTM STP 945,* D. T. Read and R. P. Reed, Eds., American Society for Testing and Materials, Philadelphia, 1988, pp. 804–818.

ABSTRACT: When using slender beams as compression load-carrying members, one has to consider the risk of buckling failure. Apart from cross-sectional properties, length of the beam, and elastic properties, the initial lateral deflection and the boundary conditions will be important in determining the critical load, or buckling capacity. The presence of cracks may release the end constraints and drastically change the boundary conditions.

A growing fatigue crack may then lead to a gradual change from a rigid joint to a more flexible one with its lower extreme as moment-free. Thus it may be expected that the buckling strength will decrease with increasing fatigue crack size and therefore be time-dependent. Such a behavior would be of great concern in dynamically loaded structural elements containing growing fatigue cracks. This situation is highly realistic for a variety of offshore structures which are constructed as complex welded frames where high local stress levels in the nodal points in combination with dynamic loads due to current, wind, and waves may cause fatigue cracks to grow.

In order to provide some further insight into this problem, experiments and corresponding numerical analyses were performed with a Euler-III column of rectangular cross section and with surface cracks of varying length at the clamped end. The numerical results were combined with possible crack length versus load cycle relations to show the time dependence of the buckling load.

From the experiments it was concluded that the calculated elastic buckling load versus crack length relation (based on linear material behavior) represents an upper bound for the buckling load in the crack length interval considered.

KEY WORDS: buckling, cracks, fatigue, numerical analysis

When designing load-carrying frame work structures—for example, certain types of offshore installations—one has to consider several possible collapse mechanisms such as plastic collapse, fatigue failure, local and global buckling, and brittle fracture. Considering global buckling, the codes regarding design and construction of steel structures supply procedures for determining the buckling strength of each member in the frame structure in the light of the geometry and joint flexibility. Although a rough estimate of the influence of cracks is given in the British Standards Guidance on Some Methods for the Derivation of Acceptance Levels for Defects in Fusion Welded Joints CBS PD 6493-807, the effect of the release of end constraint caused by growing cracks is in general not treated by the codes.

In the case of a crack growing incrementally in, for example, a welded joint, the buckling strength will be time-dependent, and the safety factor against this type of failure will decrease

[1]Manager, Research and Development, Hydro Aluminum, P. O. Box 222, N-3081, Holmestrand, Norway.

[2]Senior research engineer, The Swedish Plastics and Rubber Institute, S-85171 Sundsvall, Sweden.

[3]Manager, Structural Engineering, Saga Petroleum, A.S., P. O. Box 9, N-1322 Hoevik, Norway.

during service. This time dependence of strength criteria was pointed out by Lereim [1,2] who also gave simple expressions for possible relations between buckling load and crack depth.

In order to more thoroughly investigate the influence of cracks on the elastic buckling capacity, we considered some numerical examples, the results of which are given in the first part of the paper. In the numerical analysis, a simple torsional spring model was employed and the resulting stability problem was solved by use of the so-called Berry's functions [3]. To show the time dependence of the buckling strength, the calculated buckling load versus crack length relations were combined with possible crack length versus number of cycle relations.

The last part of the paper gives the results from buckling experiments performed on aluminum beam columns. These results are compared with both the numerical results obtained from the spring model and with those obtained from the approach proposed in BS PD 6493.

Spring Model

Consider the case shown in Fig. 1, that is, a straight edge-cracked beam (crack length a) with the cracked end clamped and the other end loaded with a load P having an eccentricity ϵ with respect to the central axis of the beam. (This was the case considered in the experiments.) The same figure also shows a possible deformation mode. Assuming linear elastic conditions during every instant of the loading, any initial lateral deflection (or load eccentricity) will not affect the buckling strength. This is not true when the material is elasto-plastic.

Real-life structures always contain initial lateral deflections and the present design codes allow this within certain limits. The applied load point eccentricity ϵ simulates the lateral deflection because there is a direct correspondence between the influence of the initial lateral deflection and the load eccentricity, on the load bearing capacity, so that it should be possible, by adjusting the load eccentricity in Fig. 1, to model the influence of a given initial lateral deformation.

It is now assumed (see Fig. 2) that the stiffness k of the torsional spring and the distance e can be chosen so as to give the same influence as the crack on the beam, in points away from the cracked end.

The stiffness k is further on assumed to depend on the crack length only, whereas e will depend on both load and crack length. To determine the buckling load we consider a small

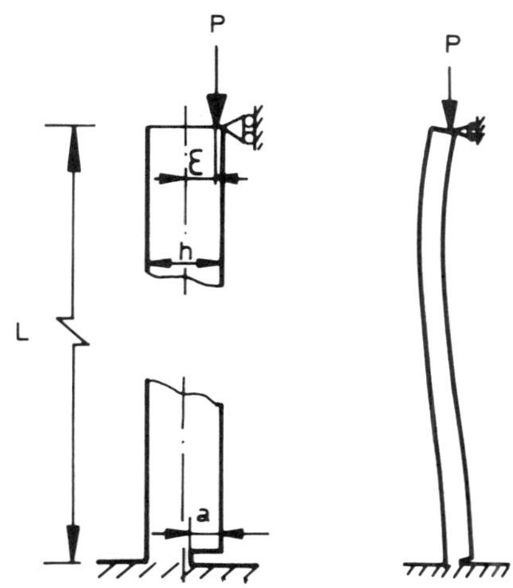

FIG. 1—*Cracked beam column under axial load.*

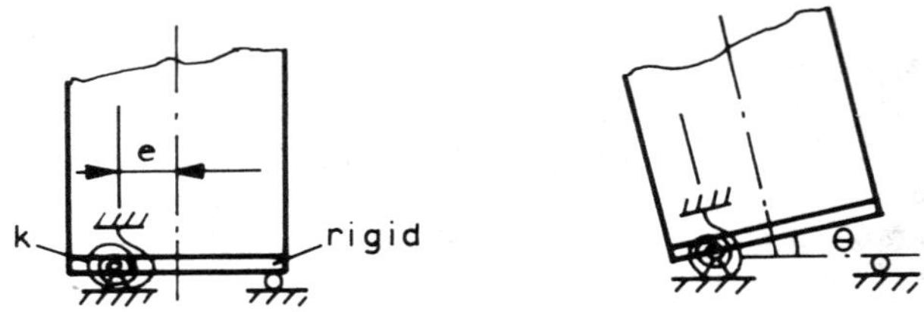

FIG. 2—*End conditions.*

angular deflection θ of the joint in Fig. 2. The load P with an eccentricity ϵ in the upper end of the beam column (see Fig. 1) can be replaced by a centrally applied load P and a moment $P\epsilon$ acting clockwise. In the lower end of the column there will be a reaction force acting of magnitude P and eccentricity e. This reaction force is now replaced by a load P acting along the center of the column, and a moment Pe acting clockwise. In addition we will have a moment $k\theta$, due to rotation, acting in the same direction. According to Pfluger [4], one will in this case have the following reaction between θ and P

$$\theta = -\frac{(Pe + k\theta)L\phi(\gamma)}{3EI} + \frac{P\epsilon L}{6EI}f(\gamma) \tag{1a}$$

where EI and L are the bending stiffness and length of the beam, respectively, and f and ϕ are two of Berry's functions defined through

$$f = \frac{6}{\gamma}\left(\frac{1}{\sin\gamma} - \frac{1}{\tan\gamma}\right) \tag{2}$$

$$\phi = \frac{3}{\gamma}\left(\frac{1}{\gamma} - \frac{1}{\tan\gamma}\right) \tag{3}$$

where γ is given by

$$\gamma^2 = L^2\frac{P}{EI} \tag{4}$$

Solving Eq 1 for θ, one obtains

$$\theta = \frac{P(\epsilon f - 2e\phi)}{2\left(\dfrac{3EI}{L} + k\phi\right)} \tag{1b}$$

Assuming e to be bounded, the buckling load is obtained from Eq 1 as the root to

$$\phi = -\frac{3EI}{kL} \tag{5}$$

The stiffness k versus crack length was obtained through comparison of the displacements obtained from a linear finite-element (FE) analysis performed on the cracked beam with those obtained from technical beam theory applied to the spring model, the loading in both cases being an end moment. In all calculations of torsional stiffness, compliance, etc., the FE code ADINA was used [5].

Numerical Examples

Three different cases were studied numerically to determine the relation between crack length and buckling load. In the first case (Fig. 3) the particular geometry used in the experiments was considered, that is, an aluminum beam with rectangular cross section with end conditions as in Fig. 1. This particular geometry is very close to the condition for a cross section in the midplane between stiffeners in a stiffened plate structure. The second case was a tubular steel member with cracks in both ends, growing equally with respect to each other. In the third and last case, the effect of crack growth in a tubular steel T-joint was studied. Figure 3 shows the geometry of the first case.

The Young's modulus of the aluminum alloy was determined through experiments to be 72 000 MPa, and Poisson's ratio to be 0.31. Figure 4 gives the calculated torsional stiffness versus length relation for the aluminum beam. When k versus a from Fig. 4 is used together with Eq 5, the buckling load P_c can be calculated as a function of the crack length. The result of this is shown in Fig. 5. Note that P_c is normalized with respect to P_{E3}, the elastic buckling load for a Euler 3 beam column (one end free to rotate, the other end clamped).

The second geometry is shown in Fig. 6, that is, a pipe with cracks extending over an angle 2ψ, in both ends. The pipe material was assumed to be steel with Young's modulus equal to 210 000 MPa and Poisson's ratio equal to 0.3. Torsional stiffness and critical load versus crack length for the cracked pipe are shown in Figs. 7 and 8, respectively. Shell elements were used in the numerical analysis. Note that in this case the buckling load P_c normalized with respect to P_{E4}, denotes the elastic buckling load for a Euler 4 beam column (both ends clamped). An attempt to model the influence of the buckling load of cracks penetrating just a part of the wall thickness was made simply by decreasing the shell element thickness in the nodes at the clamped ends. This was done for different angles 2ψ, and in all cases it seemed to have a very small influence on the elastic buckling load of cracks not penetrating the whole wall thickness.

To give an idea of the time dependence of the elastic buckling strength, an axial load varying sinusodically between −100 and 100 kN was applied to the geometry in Fig. 6, and the stress intensity factor range ΔK_I (in this case equal to the stress intensity factor value at the maximum load 100 kN) was determined by use of the compliance method [6], where required data on energy versus crack length were obtained from the FEM analyses. For determining the crack length (or crack angle rather) the Paris law was used [6]; that is, the following relation between

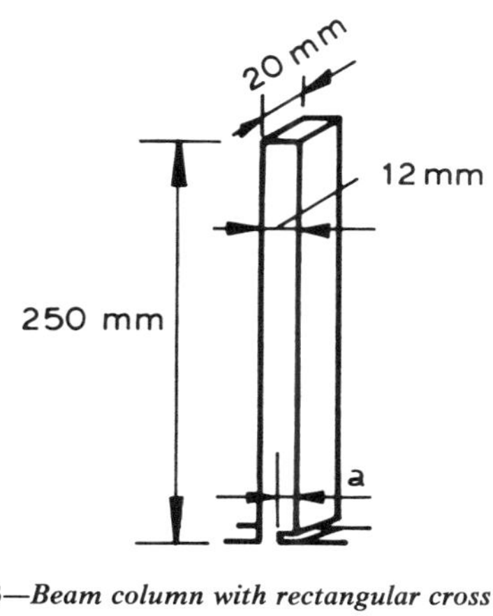

FIG. 3—*Beam column with rectangular cross section.*

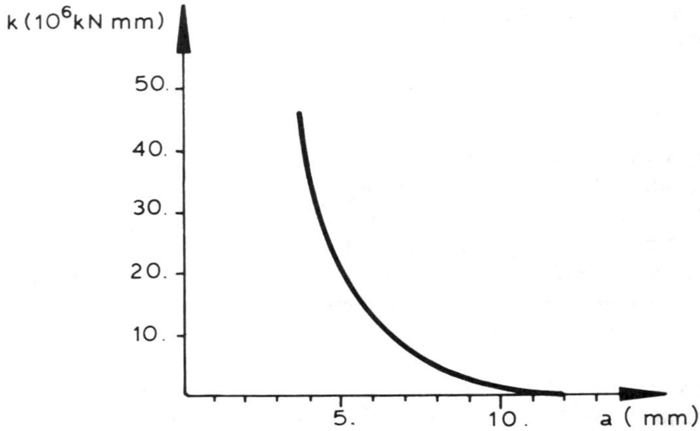

FIG. 4—*Torsional stiffness versus crack length.*

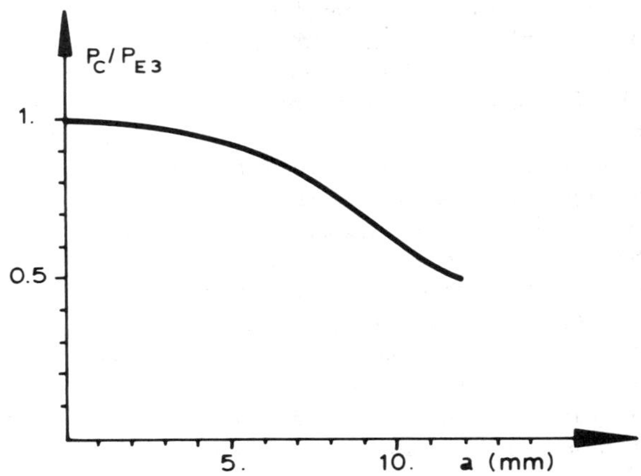

FIG. 5—*Elastic buckling strength versus crack length.*

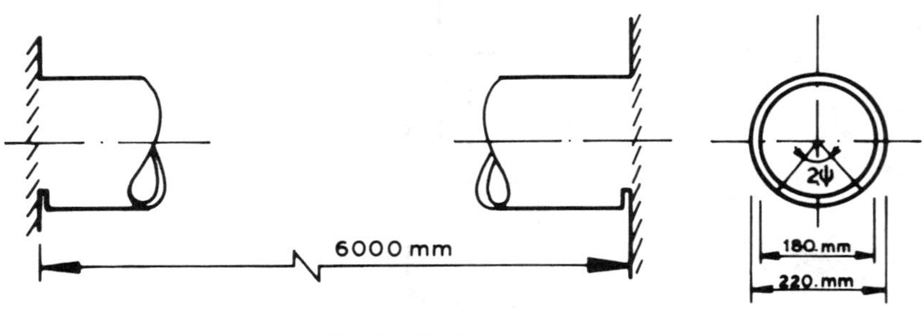

FIG. 6—*Cracked steel pipe.*

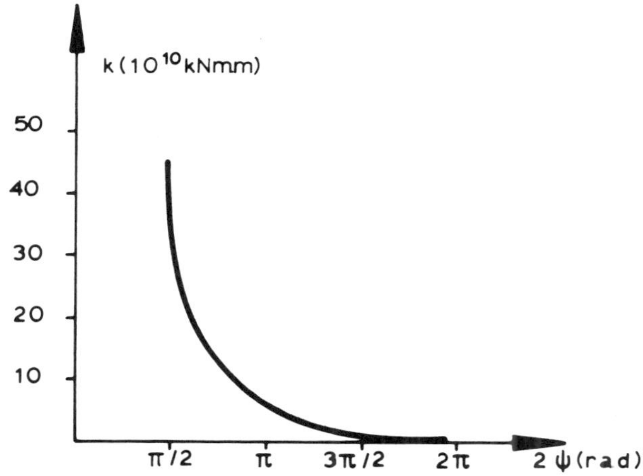

FIG. 7—*Torsional stiffness versus crack angle.*

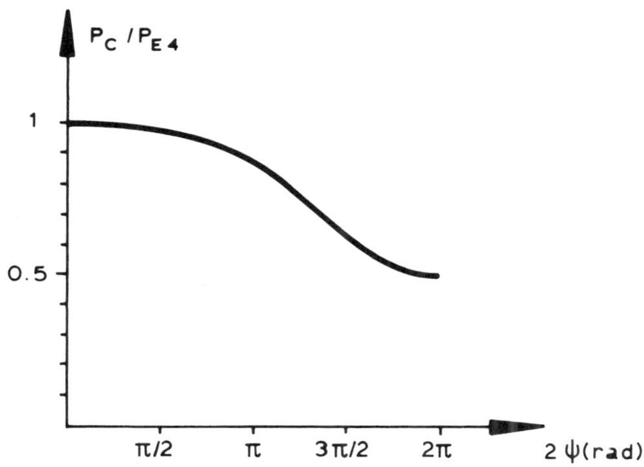

FIG. 8—*Elastic buckling strength versus crack angle.*

crack length $2\psi R$ (R is the middle radius of the tube cross section) and number of cycles N was assumed

$$\frac{d(2\psi R)}{dN} = C(\Delta K_I)^m \tag{6}$$

C and m were taken to be 3.2 10^{-12} $mm^{5.5}N^{-3}$ and 3, respectively. The crack angle was obtained by numerical integration of Eq 6 and is shown versus the number of load cycles N in Fig. 9. An initial crack angle 2ψ equal to 0.45π was assumed. When combining Figs. 8 and 9, a curve showing the load cycle (or time) dependence of the buckling load can easily be generated as shown in Fig. 10. It can be seen that in this particular case, the decrease in buckling load will appear rather late in the lifetime of the tubular member.

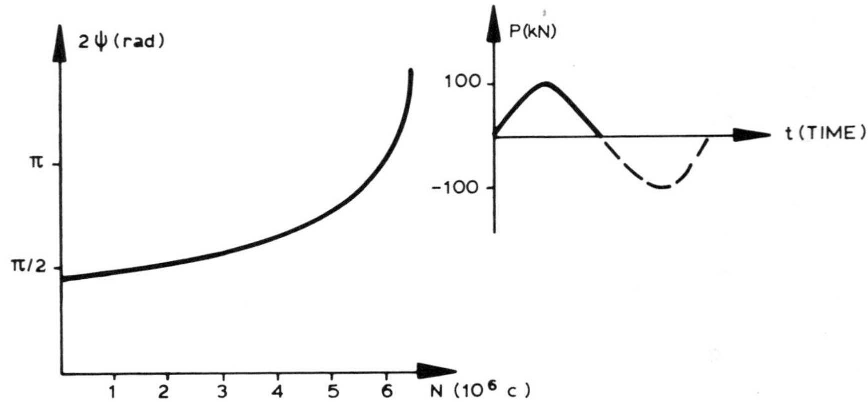

FIG. 9—*Crack angle versus number of load cycles.*

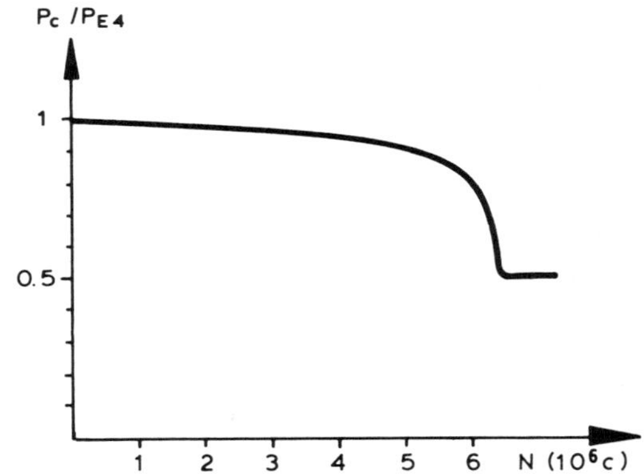

FIG. 10—*Elastic buckling strength versus number of load cycles.*

The third and last case considered was the structure shown in Fig. 11. The material considered was steel with Young's modulus and Poisson's ratio equal to 210 000 MPa and 0.3, respectively. Point A in Fig. 11 was assumed to be the point where the crack growth along the intersection between the brace and chord was initiated, which may be characteristic for in-plane bending of real tubular joints in, for example, conductor frames in offshore structures. Symmetric crack growth with respect to A was assumed. The calculations of buckling load versus crack length relation and buckling load versus number of load cycles were based on data found in Ref *7*. The same values for the constants in Paris law C and m as in the second case were assumed, and further on the loading also was assumed to be the same. Figure 12 shows the buckling load P_c (normalized with respect to P_{E4}) versus the crack length for two cases, one case where the chord has the same Young's modulus as the brace and one where the chord is rigid. The length of the intersection line between the chord and brace was about 640 mm. Exactly in the same way as before, the crack length versus number of load cycles, and thereafter the buckling load versus number of load cycles, was determined. These two relations are shown in Figs. 13 and 14, respectively. When the relation shown in Fig. 13 was calculated, an initial crack of 100 mm was assumed. Also, in this case it can be seen that the decrease in buckling load will appear rather late in the joint lifetime. However, this conclusion is based upon the assumption

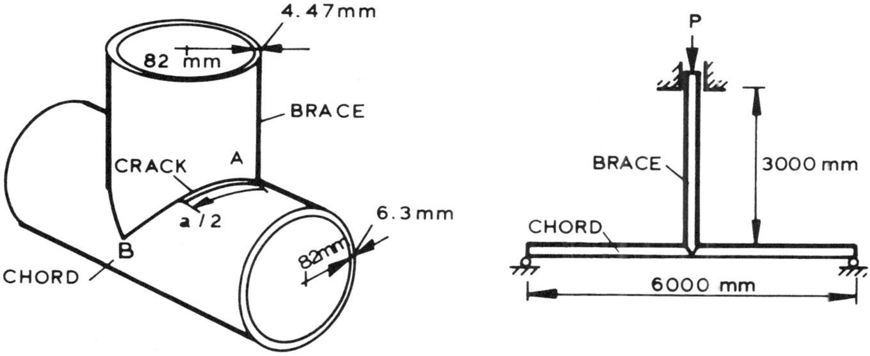

FIG. 11—*Tubular* T-*joint.*

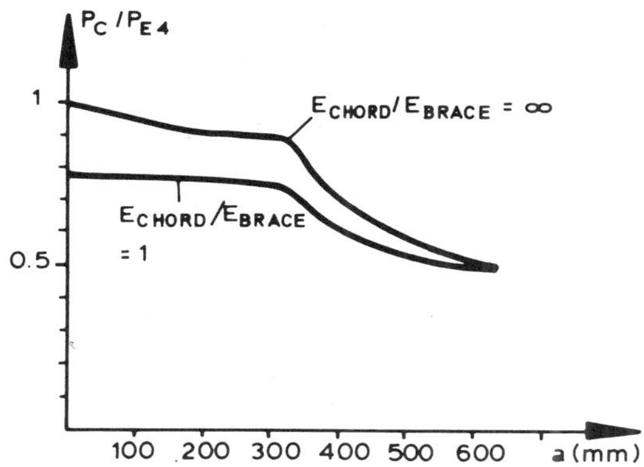

FIG. 12—*Elastic buckling strength versus crack length.*

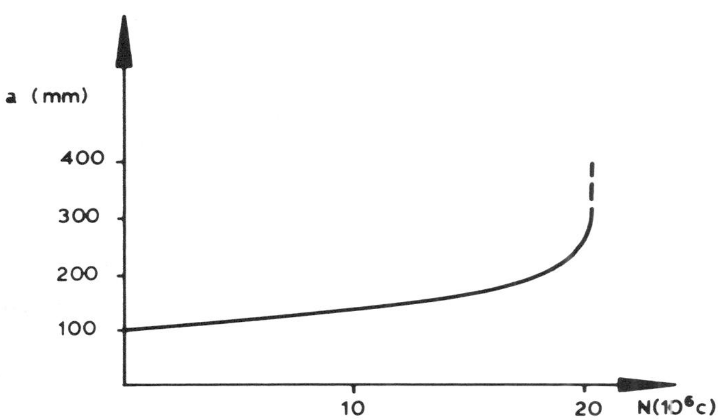

FIG. 13—*Crack length versus number of load cycles.*

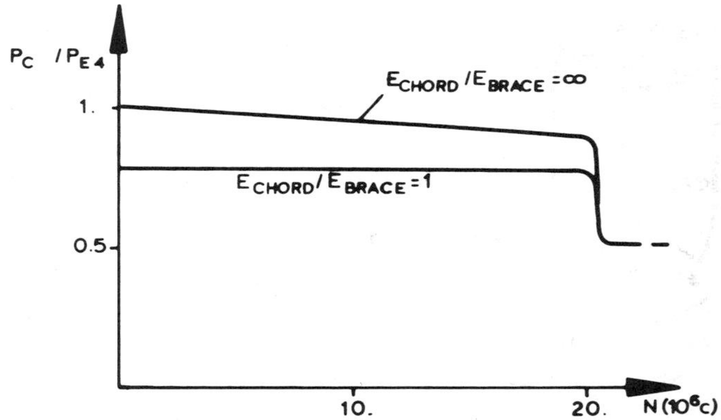

FIG. 14—*Elastic buckling strength versus number of load cycles.*

that the load is of constant-amplitude type. Situations could well be imagined where the loading will be such that the crack length versus number of load cycles relation becomes for example, convex upwards, which will make the decrease in buckling load appear earlier. Another assumption is that the brace buckles in the plane of the paper. One situation where this might not be true is when the crack grows from Point B in Fig. 11, hot spot for axial loading and out-of-plane bending, and this could possibly affect the torsional stiffness of the joint in such a way so as to invalidate the above conclusion.

Experiments

The specimens were manufactured in the following way: First, aluminum rods with length 350 mm and cross-sectional dimensions 20 by 20 mm were notched to a depth of 5 mm and then loaded by a fatigue load in three-point bending so as to produce a fatigue crack of a specified length. Then the notch was machined away, and the side of the specimen opposite to the notched side was also machined in order to get a straight specimen with cross-sectional dimensions 12 by 20 mm. When the specimen was inserted into the testing machine (MTS Servo Hydraulic Testing Machine), 100 mm of the specimen below the crack was fixed in the hydraulic grip, so that the free length of the specimen was 250 mm (see Fig. 3). A wedge was used to apply the load to the upper end of the specimen. It was possible to move this wedge with respect to the centerline of the testing machine; that is, it was possible to vary the load eccentricity. Two strain gages were glued to each specimen, one on the cracked side and one on the opposite side and approximately at the specimen midpoint. These strain gages were used to obtain a load eccentricity as small as possible, but still sufficient to make the specimen to deflect in such a way as to make a deflection mode according to Fig. 1 possible. This means that when the specimen was loaded with a compressive load, the load eccentricity was adjusted so that the strain on the cracked side was numerically somewhat larger than on the opposite side. The material used was an aluminum alloy with Young's modulus 72 000 MPa, Poisson's ratio 0.31, yield strength (stress at 0.2% plastic strain) 564 MPa, and ultimate tensile strength 592 MPa. During the loading, which was performed in controlled displacement mode, the load-point displacement and lateral deflection at a point 140 mm from the cracked end were recorded. Figure 15 shows a photo of the experimental setup.

An experimentally obtained load-deflection ($P - \delta$) relation is shown in Fig. 16 for a nominal (as measured on the surface of the specimen) crack length of 4 mm (all other cases were similar to Fig. 19). Figure 17 shows the measured maximum load P_{max} as a function of the crack length. Due to the curved crack front, the crack length was defined as the average of the crack lengths

FIG. 15—*Experimental setup.*

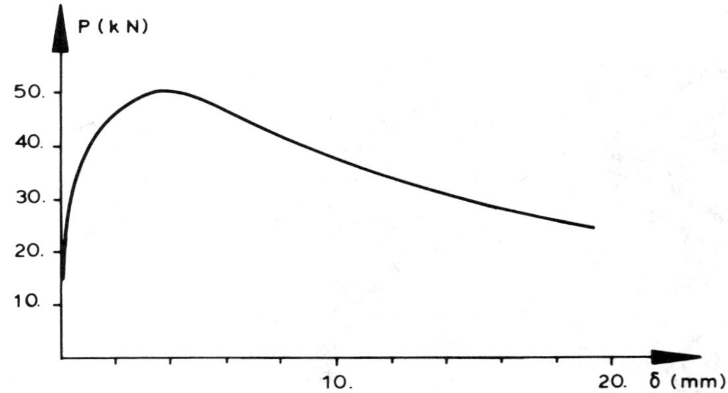

FIG. 16—*Load versus lateral deflection.*

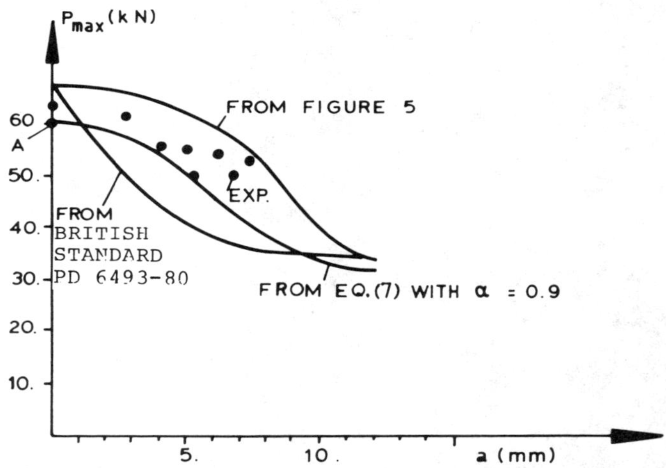

FIG. 17—*Maximum load versus crack length.*

in three points located at 3, 6, and 9 mm along the crack front, in accordance with the ASTM Test Method for Plane-Strain Fracture Toughness of Metallic Materials (E 399-81). Also shown along with the experimental results is the relation from Fig. 5 (with $P_c = P_{max}$), a relation based on BS PD 6493, that is, a relation that assumes that the elastic buckling load varies between the two extremes P_{E3} and P_{E2} (the buckling load when both ends are moment free) as does the moment of inertia for the unbroken ligament. Finally a relation proposed by the authors, namely

$$P_{max} = P_{E3}\frac{\alpha}{2}\left(1 + \cos\frac{\pi a}{h}\right) + P_{E2}\frac{\alpha}{2}\left(1 - \cos\frac{\pi a}{h}\right) \tag{7}$$

is shown where α is a positive constant less then unity and h is the height of the cross section. Point A in Fig. 17 is the buckling load for an uncracked column with an initial midpoint deflection of 0.22 mm.

A parameter that most certainly will affect the maximum load is the load eccentricity. To obtain the eccentricity for each test case, it was assumed that the specimen at the beginning of the loading deflected as a Euler 3 beam column, and by adjusting the load eccentricity ϵ so that the deflection obtained, assuming Euler 3 conditions, fitted the first part of the experimental curve, ϵ could be determined. Figure 18 shows a typical case, both the experimentally obtained and Euler 3 lateral deflections, for a nominal crack length of 6 mm, and ϵ equal to 0.45 mm. Also shown in the same figure is the lateral deflections obtained by assuming e (see Fig. 2) to be constant and equal to (see Eq 1)

$$e = \frac{\epsilon f}{2\phi} \tag{8}$$

where f and ϕ are calculated for the load in the branch Point B. Table 1 gives the load eccentricity for the crack lengths considered.

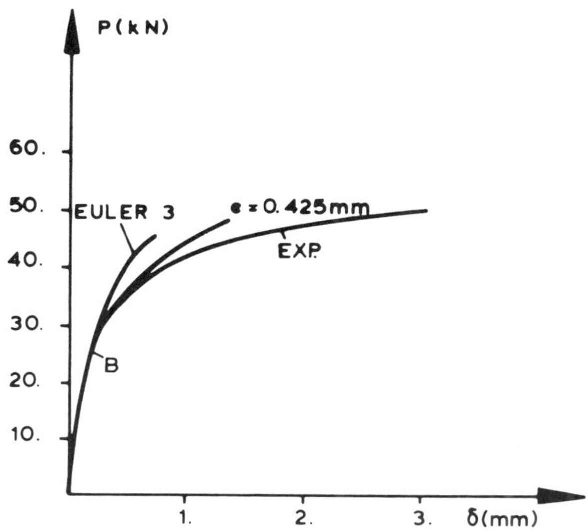

FIG. 18—*Determination of load eccentricity.*

TABLE 1—*Load eccentricity for crack lengths considered.*

Crack Length, mm		
Nominal	Average	Load Eccentricity, mm
1.0	2.7	0.012
2.0	3.9	0.195
3.0	4.9	0.267
4.0	5.1	0.550
5.0	6.1	0.064
6.0	6.8	0.450
7.0	7.4	0.092

Table 1 together with Fig. 17 indicates, by comparing the maximum load for an average crack length 4.9 mm with that for 5.1 mm and the maximum load for 6.8 mm with that for 7.4 mm, that the load eccentricity will cause larger deviation from the elastic solution for shorter than for longer cracks.

A very interesting observation is that for lateral deflections larger than about 5 mm and for nominal crack lengths 4 to 7 mm, the various load-lateral deflection relations were almost exactly the same, indicating that for sufficiently large deflections the curves converge and the contribution to the stiffness from the cracked end will be the same irrespective of the initial crack length. This is shown in Fig. 19, where the load-lateral deflection relations for nominal crack lengths 4 to 7 mm are compared with those for a nominal crack length equal to zero, and with no midpoint deflection. It was observed that the bending deformation of all cracked specimen was concentrated to a point at about 140 mm from the cracked end (see Fig. 15); this was in fact almost exactly the point where the lateral deflection was measured. A simple plastic hinge model is shown in Fig. 20, where it has been assumed that at the cracked end the force is applied with an eccentricity equal to 6 mm or half the cross-section height (which is equal in order to assume that the length of the ligament is equal to zero) and that the plastic hinge is located at a distance 140 mm from the cracked end. P and δ in Fig. 20 are axial load and lateral deflection, respectively. Figure 21, finally, shows that the model according to Fig. 20 describes the real situation quite well. The points are experimental results for nominal crack lengths 4 to 7 mm.

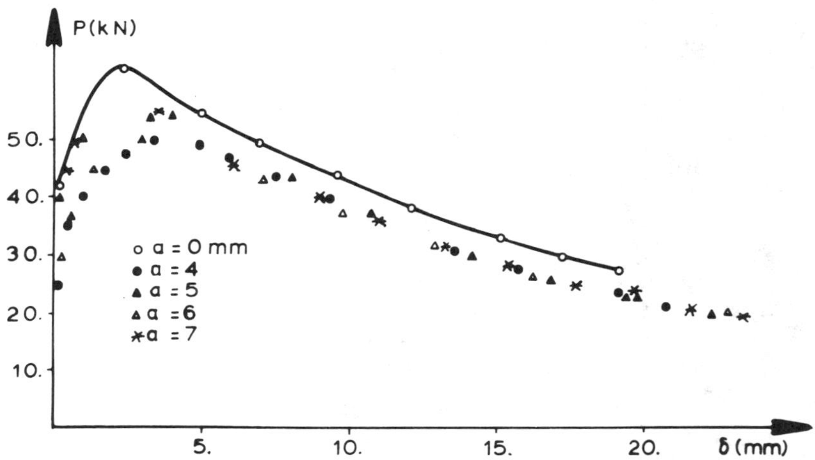

FIG. 19—*Load-lateral deflection relations for different nominal crack lengths.*

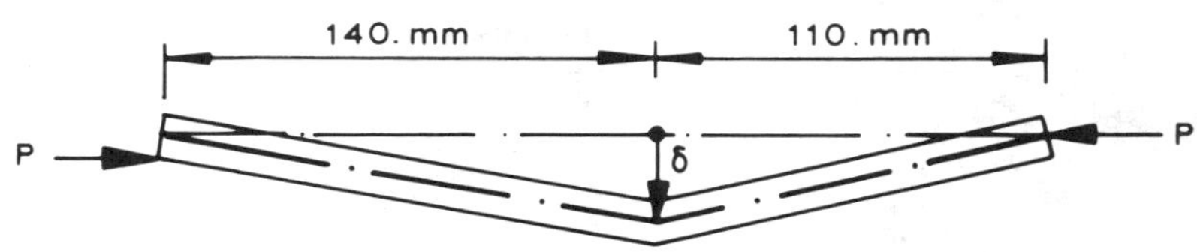

FIG. 20—*Plastic model.*

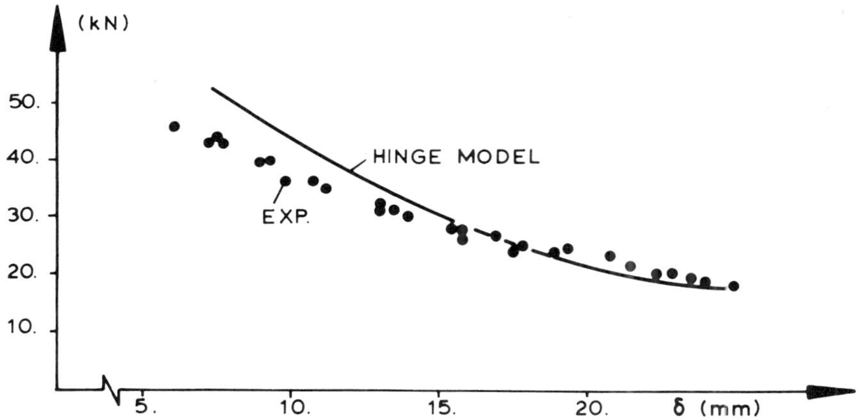

FIG. 21—*Load versus lateral deflection.*

Summary and Conclusions

A method for calculating the elastic buckling load for cracked beam columns has been presented. This method was used to calculate for three different geometries the influence of a growing crack on the buckling load. Experiments were performed on one of the geometries, and the results were compared with those obtained from an elastic analysis which was shown to be an upper bound for the experimental results. An expression for assigning buckling strength to a beam column in terms of a safety factor and the buckling loads for the two extreme cases—no crack and a fully penetrating crack—was suggested. Finally, it was concluded from the experiments that for large lateral deflections, the influence of the cracked end can be neglected regardless of the size of the initial crack.

For future research it is proposed that cracked members with circular cross sections be tested and also that combined material and geometrical nonlinear problems be treated numerically.

Acknowledgment

Mr. S. Rystad of Alprofil is gratefully acknowledged for providing the high-strength aluminum alloy.

The views expressed in this paper represent those of the authors and do not necessarily coincide with the policy of Det norske Veritas, which we acknowledge here for allowing the publication of this work.

References

[1] Lereim, J., "Influence of Cracks on Structural Design" in *Proceedings,* Application of Fracture Mechanics to Materials and Structures, GCSG, E. Sommer and W. Dahl, Eds., M. N. Nijkoff Publishers, Freiburg, West Germany, June 1984, pp. 999–1011.

[2] Lereim, J., "Significance of Fracture Mechanics on Design of Offshore Structures," paper presented at the International Conference on Fracture Prevention in Energy and Transport Systems, Rio de Janeiro, Brazil, Nov. 1983.

[3] "Formelsamling i Haallfasthetslaera," Publication number 104, Institute foer haallf. KTH, Stockholm, Sweden, 1978 (in Swedish).

[4] Pfluger, A., *Stabilitetsprobleme der Elastostatik,* 2nd ed., Springer, Berlin, 1964.
[5] "Automatic Dynamic Incremental Nonlinear Analysis," ADINA Engineering, Report 81.1.
[6] Hellan, K., Introduction to Fracture Mechanics, McGraw-Hill, New York.
[7] Brown, D. K., Hancock, I. W., and Green, D. R., "On the Analysis of a Welded Tubular T-joint With a Through Thickness Crack" in *Proceedings,* 3rd International Conference on Numerical Methods in Fracture Mechanics, Swansea, Wales, U.K., March 1984.

Subcritical Crack Growth

Robert O. Ritchie[1]

A Comparison of Fatigue Crack Propagation in Modes I and III

REFERENCE: Ritchie, R. O., **"A Comparison of Fatigue Crack Propagation in Modes I and III,"** *Fracture Mechanics: Eighteenth Symposium, ASTM STP 945,* D. T. Read and R. P. Reed, Eds., American Society for Testing and Materials, Philadelphia, 1988, pp. 821-842.

ABSTRACT: The propagation behavior of fatigue cracks in Mode III (anti-plane shear), measured under cyclic torsion, is described and compared with more commonly encountered behavior under Mode I (tensile opening) loads. It is shown that a unique, global characterization of Mode III growth rates, akin to the Paris "law" in Mode I, is only possible if characterizing parameters appropriate to large-scale yielding are employed and allowance is made for crack tip shielding from sliding crack surface interference (that is, friction and abrasion) between mating fracture surfaces. Based on the crack tip stress and deformation fields for Mode III stationary cracks, the cyclic crack tip displacement, $(\Delta CTD)_{III}$, and plastic strain intensity range, $\Delta\Gamma_{III}$, have been proposed and are found to provide an adequate description of behavior in a range of steels, provided crack surface interference is minimized. The magnitude of this interference, which is somewhat analogous to crack closure in Mode I, is further examined in the light of the complex fractography of torsional fatigue failures and the question of a "fatigue threshold" for Mode III crack growth. Finally, micromechanical models for cyclic crack extension in anti-plane shear are briefly described, and the contrasting behavior between Mode III and Mode I cracks subjected to simple variable amplitude spectra is examined in terms of the differing role of crack tip blunting and closure in influencing shear, as opposed to tensile opening, modes of crack growth.

KEY WORDS: fatigue (materials), cracking (fracturing), crack propagation, torsion, Mode I (tensile opening), Mode III (anti-plane shear), crack closure and sliding interference, fatigue thresholds, variable-amplitude loading

Over the past 30 years, there have been numerous analytical models proposed for cyclic crack extension in Mode III (anti-plane shear) [*1-9*], presumably because of the mathematically convenient form of continuum solutions describing the local Mode III crack tip stress and deformation fields [*2,10,11*]. In contrast, experimental studies of Mode III crack *growth* behavior are rare [*3,7-9,12-21*], particularly when compared to the extensive literature on the propagation behavior of fatigue cracks in Mode I (tensile opening loading) (reviewed, for example, in Ref *22*). This is somewhat surprising, since torsional fatigue fractures are a common mode of engineering service failures [*23*], particularly in the automotive industry with springs and transmission components, such as drive and propeller shafts, and in the power generation industry with large turbo-generator rotors. In addition to the paucity of fracture mechanics data on Mode III crack growth rates, mechanistic aspects are also relatively unknown. As metallurgical consultant D. J. Wulpi writes in a recent paper [*23*], ". . . after many years of studying and analyzing fractures, it has become obvious that the various types of torsional fractures are not well understood. . . ."

The intent of this paper is to review what is known about the fracture mechanics and micromechanistic aspects of Mode III fatigue crack growth and to compare such behavior with that

[1]Professor materials science, Materials and Chemical Sciences Division, Lawrence Berkeley Laboratory, and Department of Materials Science and Mineral Engineering, University of California, Berkeley, CA 94720.

more commonly encountered in Mode I. We examine first the anti-plane strain crack tip fields and the choice of appropriate characterizing parameters for crack advance, under both small-scale yielding and elastic-plastic conditions, and apply these to steady-state Mode III fatigue crack propagation. It is shown that a global characterization of growth rates in small- or large-scale yielding can be achieved in terms of the range of crack tip displacement (ΔCTD), similar to that in Mode I, although the Mode III crack extends over a far smaller proportion of the ΔCTD each cycle. That crack surfaces remain in sliding contact is attributed in part to the occurrence of "sliding crack surface interference" [*14–16*] where, analogous to crack closure in Mode I, abrasion and interlocking between asperities on mating fracture surfaces provide a potent mechanism for "shielding" the crack tip from the full applied "driving force." The significance of this interference to both the fractography of Mode III failures and the relative rates of crack growth compared to Mode I behavior is discussed in some detail. In addition, several recent models [*8*] for Mode III crack extension are examined and compared to torsional crack growth data in a range of steels. Finally, the sharp contrast between Mode I and Mode III crack growth behavior under variable amplitude cyclic loading is described in the light of current ideas on "damage accumulation" and fatigue crack closure.

Crack Tip Fields

For both Mode I and Mode III cracks, the stress and strain fields in the vicinity of the crack tip have been determined using asymptotic continuum mechanics analysis. For linear elastic solids in Mode I, the local singular stress field at distance r, polar angle θ, from the tip of a stationary crack is given in terms of the Mode I stress intensity factor K_I, in the limit of $r \to 0$, as [*24*]

$$\sigma_{ij} \to \frac{K_\text{I}}{\sqrt{2\pi r}} \cdot f_{ij}(\theta) \tag{1}$$

whereas for elastic-plastic solids (actually nonlinear elastic solids where $\bar{\epsilon}/\epsilon_0 = \alpha(\bar{\sigma}/\sigma_0)^n$), the local Mode I singular stress, strain and displacement fields are given in terms of the path-independent integral J [*25*], as $r \to 0$ [*26,27*]

$$\frac{\sigma_{ij}(r, \theta)}{\sigma_0} \to \left(\frac{J}{\alpha\sigma_0\epsilon_0 I_n r}\right)^{1/(n+1)} \tilde{\sigma}_{ij}(\theta, n) \tag{2a}$$

$$\frac{\epsilon_{ij}(r, \theta)}{\epsilon_0} \to \alpha\left(\frac{J}{\alpha\sigma_0\epsilon_0 I_n r}\right)^{n/(n+1)} \tilde{\epsilon}_{ij}(\theta, n) \tag{2b}$$

$$\frac{u_i(r, \theta)}{r} \to \alpha\epsilon_0\left(\frac{J}{\alpha\sigma_0\epsilon_0 I_n r}\right)^{n/(n+1)} \tilde{u}_i(\theta, n) \tag{2c}$$

where

σ_0 = yield stress,
ϵ_0 = yield strain,
α = material constant of order unity,
f_{ij} = dimensionless function of θ,
$\tilde{\sigma}_{ij}(\theta)$ = normalized stress function of θ,
$\tilde{\epsilon}_{ij}(\theta)$ = normalized strain function of θ,
$\tilde{u}_i(\theta)$ = normalized displacement function of θ, and
I_n = numerical constant weakly dependent upon the strain hardening exponent n.

For Mode III anti-plane shear, the corresponding solutions for the linear elastic stationary crack are given, in the limit of $r \to 0$, by [28]

$$\sigma_{xy} = \frac{-K_{III}}{\sqrt{2\pi r}} \sin\frac{\theta}{2} \tag{3a}$$

$$\sigma_{yz} = \frac{K_{III}}{\sqrt{2\pi r}} \cos\frac{\theta}{2} \tag{3b}$$

where K_{III} is the stress intensity factor in Mode III. For the nonlinear elastic crack, the singularity field can be written as $\tau \to 0$ [2]

$$\tau_p(r, \theta) \to \tau_0\left(\frac{r_y}{r}\right)^{1/(n+1)} \tilde{\tau}_p(\theta) \tag{4a}$$

$$\gamma_p(r, \theta) \to \gamma_0\left(\frac{r_y}{r}\right)^{n/(n+1)} \tilde{\gamma}_p(\theta) \tag{4b}$$

where

τ_0 = shear yield stress,
γ_0 = shear yield strain,
r_y = distance from the crack tip to the elastic-plastic boundary, and
$\tilde{\tau}_p(\theta)$ = normalized shear stress function of θ, and
$\tilde{\gamma}_p(\theta)$ = normalized shear strain function of θ.

For elastic-perfectly plastic stationary cracks, the field (which exactly satisfies incremental plasticity theory) can be expressed in simpler form as [29]

$$\gamma_p = \gamma_0\left(\frac{r_y}{r}\right) \tag{5}$$

Such solutions are used to define appropriate characterizing parameters for crack extension. In Mode I, where the plastic zone size under plane-strain conditions is fabiform in shape and given approximately by [30]

$$r_y \sim \frac{1}{2\pi}\left(\frac{K_I}{\sigma_0}\right)^2 \tag{6}$$

the stress intensity K_I provides characterization of the unique linear elastic crack tip field when the plastic zone is small compared with crack size. Under such small-scale yielding conditions, a crack tip (opening) displacement $(CTD)_I$ can be estimated as [30]

$$(CTD)_I \sim \frac{1}{2}\frac{K_I^2}{\sigma_0 E} \tag{7}$$

where E is the elastic (Young's) modulus. In the presence of more extensive plasticity, the nonlinear elastic crack tip field can be characterized by the J-integral or the $(CTD)_I$.

In Mode III, where the plastic zone is circular in shape, with a diameter given by [2,29]

$$r_y \sim \frac{1}{\pi}\left(\frac{K_{\text{III}}}{\tau_0}\right)^2 \tag{8}$$

linear elastic fields can be characterized by K_{III}, with the Mode III crack tip displacement $(\text{CTD})_{\text{III}}$ under small-scale yielding given by [*2,29*]

$$(\text{CTD})_{\text{III}} \sim \frac{2}{\pi}\frac{K_{\text{III}}^2}{\tau_0 G} \tag{9}$$

where G is the shear modulus. With large scale plasticity, the cyclic crack tip displacement and the *plastic strain intensity range* $\Delta\Gamma_{\text{III}}$ have been suggested as appropriate characterizing parameters for Mode III crack extension [*6*]. Γ_{III} is defined in terms of the radius of the plastic zone r_y, and the Walsh and Mackenzie solution [*10*] for the shear strain distribution in a circumferentially notched cylinder of elastic-rigidly plastic material subjected to a torque M

$$\gamma = \frac{\tau_0}{G}\left(\frac{r'}{r_p}\right)^2\left(\frac{r_N - r_p}{r_N - r'}\right) \tag{10}$$

where the radii r', r_N, and r_p are defined in Fig. 1. Although the plastic strain is infinite at the crack tip ($r' = r_N$), the parameter $(r_N - r')\gamma$ clearly is finite and defines the intensity of plastic shear strain in the crack plane. For cyclic loading, this gives [*6*]

$$\Delta\Gamma_{\text{III}} = \lim_{r_N - r' \to 0} [(r_N - r')\Delta\gamma] = \frac{2\tau_0}{G}\Delta\left[\left(\frac{r_N}{r_p}\right)^2 r_y\right] \tag{11}$$

Note that the plastic strain intensity so defined ($\Gamma_{\text{III}} = \lim[(r_N - r')\gamma]$) is analogous to the linear elastic stress intensity where $K_{\text{I}} = \lim[\sigma(r)\sqrt{2\pi r}]$. In the limit of small-scale yielding, where $r_N \to r_p$, the plastic strain intensity range can be related to the Mode III cyclic crack tip displacement simply by the expression

$$(\Delta\text{CTD})_{\text{III}} = 2\Delta\Gamma_{\text{III}} \tag{12}$$

For large-scale yielding, Eq 12 is only approximately correct. The computation of Γ_{III} values under these conditions specifically involves determination of analytical expressions for r_y, and has been discussed in detail elsewhere [*14,20,31*].

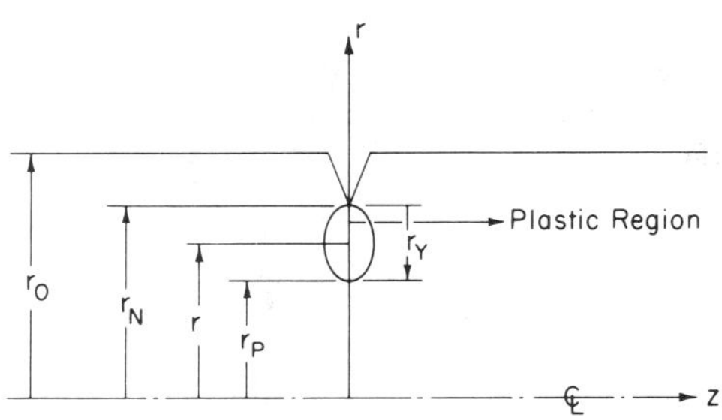

FIG. 1—*Nomenclature for plastic zone and respective radii in circumferentially notched cylindrical torsion specimen.*

Mode III Fatigue Crack Growth Behavior

The large majority of experimental research on the propagation behavior of fatigue cracks has been conducted under Mode I loading. For behavior in other modes, there have been several studies (for example, Refs *32–37*) on the "slant" crack loaded in tension, which experiences both Mode I and either Mode II or Mode III displacements. In general, it has been found that such cracks are unstable with respect to Mode I and hence tend to deviate to the Mode I plane [*37*]. To obtain stable non-Mode I growth, the tensile mode must be suppressed. This has been demonstrated by Smith [*36*], who obtained stable Mode II crack extension of a slant crack through the application of cyclic compression, which suppresses the tensile mode by closing the crack surfaces together.

Unlike behavior in Mode II, with sufficient plasticity stable crack advance in Mode III can be readily attained. Whereas Mode I behavior commonly is measured in bending or tension, with rare exception [*12*] fatigue crack propagation rates in pure Mode III are determined on cylindrical bars or tubes tested in cyclic torsion (for example, Refs *7* to *9*). Because the surfaces of maximum shear are perpendicular to and parallel to the axis of the cylinder, Mode III cracking can occur both longitudinally and radially (Fig. 2). For this reason, a circumferential notch generally is employed such that, provided extreme care is taken to minimize misalignment which induces undesired bending moments and hence asymmetric cracks, radially inward concentric crack growth is Mode III in torsion and Mode I in tension. Shear crack growth, however, is promoted by extensive plasticity [*7*], as can be demonstrated by twisting a brittle material like a piece of chalk, which will fail by a helical fracture along plane of maximum tensile stress, compared with a ductile material like plasticene, which simply will shear off. Accordingly, at low stresses, that is, less than 70% of yield [*7*], Mode III crack extension becomes unstable to Mode I, with the result that torsional failures occur either macroscopically in Mode I with a

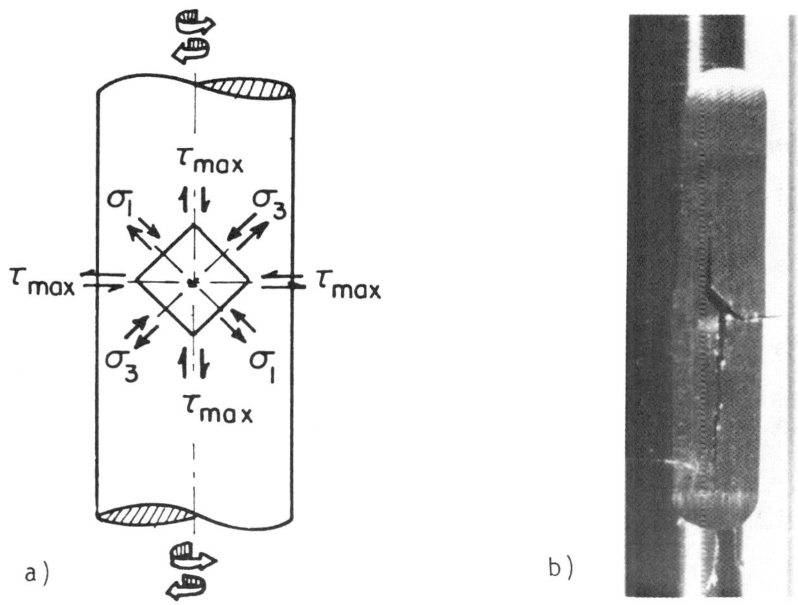

FIG. 2—(a) *State of stress and* (b) *longitudinal and radial shear cracks in a cylindrical bar subjected to cyclic torsion. Shear cracks in A469 rotor steel are Mode II on the surface and Mode III in the interior (courtesy of H. Nayeb-Hashemi).*

helical fracture surface, locally in Mode I with 45 deg branch cracks, or by a combination of modes [*7*].

Typical crack growth data [*7,8,14-20*], determined by such procedures using electrical potential methods to monitor crack extension [*38*], are shown in Figs. 3-5. In Fig. 3, Mode III crack growth rates $(dc/dN)_{\text{III}}$ are plotted as a function of the cyclic crack tip displacement $(\Delta\text{CTD})_{\text{III}}$ and plastic strain intensity range $(\Delta\Gamma_{\text{III}})$, under cyclic conditions of zero mean stress ($R = -1$), for several steels, ranging in tensile yield strength (σ_y) from 260 MPa for AISI 1018 to 956 MPa for quenched and 650°C tempered AISI 4340 steel. Although plastic zone sizes can approach 50% of the section diameter at the higher crack tip displacements, that is, conditions are well beyond small-scale yielding, Mode III growth rates clearly are well characterized by $(\Delta\text{CTD})_{\text{III}}$, or $\Delta\Gamma_{\text{III}}$. Variations in behavior between the different alloys are not large, apart from the earliest published data (for example, Refs *7* and *13*) where sliding crack surface interference was not accounted for (see below).

In Fig. 4, Mode III growth rates for an A469 rotor steel (yield strength 621 MPa) are compared with corresponding Mode I rates at equivalent cyclic crack tip displacements (ΔCTD) [*8*]. As several authors have noted previously [*7-9,13,34*], Mode I growth rates exceed those in Mode III when compared at equivalent ΔCTD values or stress intensity ranges. In fact, the crack growth increment in Mode III is only a small fraction ($<1/100$) of the crack tip displacement per cycle, consistent with micro-mechanical models [*7,8*]. Tschegg [*15,18,20*] has shown that where procedures are used to remove *completely* crack surface interference, Mode III cracks actually may grow faster than in Mode I, at least for higher growth rates above 5×10^{-3} mm/cycle, when compared at equivalent ΔCTD values. This, however, may be a somewhat inappropriate comparison since the corresponding interference between crack surfaces in Mode I, that is, crack closure, was not similarly accounted for.

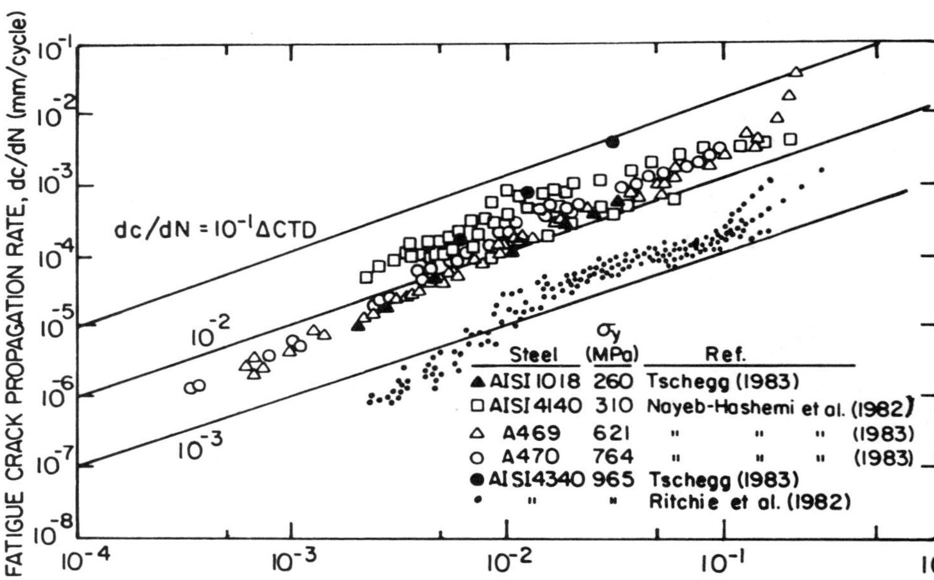

FIG. 3—*Mode III fatigue crack growth data* $(dc/cN)_{III}$ *plotted as a function of the cyclic crack tip displacement,* $(\Delta CTD)_{III}$ *and* $\Delta\Gamma_{III}$, *for AISI 1018, AISI 4140, AISI 4340, A469, and A470 steels, after Refs* 7,8,14-20.

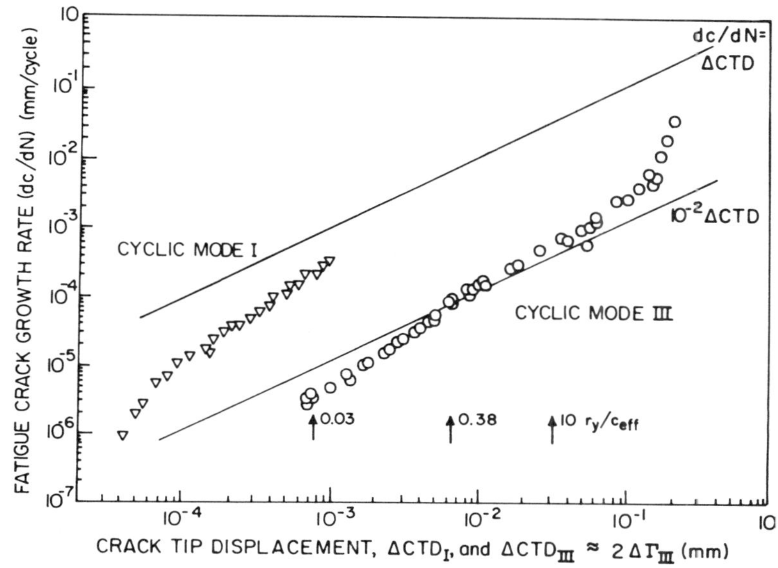

FIG. 4—*Comparison of Mode III and Mode I fatigue crack growth rates, as a function of their cyclic crack tip displacements,* $(\Delta CTD)_{III}$ *and* $(\Delta CTD)_I$ *respectively, for A469 rotor steel with* $\sigma_y = 621$ *MPa. Mode III results were measured in cyclic torsion with a small superimposed axial load* [8].

As data are scarce for Mode III crack extension, the influence of various mechanical, microstructural, and environmental variables are largely unknown. For example, Mode III corrosion fatigue crack growth behavior has not been characterized to date, although certain authors have examined the problem of corrosion fatigue life in torsion using *S/N* tests on unnotched specimens (for example, Ref *39*). In view of the widely differing hydrostatic stress states between tension and anti-plane shear, this would seem to be an important area for research, especially for environments which induce hydrogen embrittlement. Microstructural effects on Mode III crack growth also are essentially unexplored, although it is apparent that material strength has little influence on crack extension rates (Fig. 3). Limited data [*8,13*] are available on the role of mean stress, or load ratio *R* (that is, ratio of minimum to maximum load or stress). Unlike Mode I behavior where increasing the load ratio can markedly increase growth rates due primarily to a reduced influence of crack closure at high *R* (40), in Mode III where there is no closure as such, load ratios effects appear insignificant, as illustrated by the $R = -1$ and -0.5 results [*8*] for the A469 rotor steel in Fig. 5. It has been noted, however, that at higher *R* values, for example, ~0.1, Mode III cracks show a tendency to become destabilized and to form Mode I branch cracks more readily, although in many cases the effect on growth rates is not that significant [*13*]. The influence of other mechanical variables, such as cyclic frequency, wave form, and geometry, has not been examined.

Fractography

The fractography of torsional failures is complex and is associated with several distinct fracture morphologies, summarized in Fig. 6 [*21*]. Pure Mode III cracks have flat, featureless fracture surfaces with clear signs of abrasion, that is, evidence of fretting oxide debris and fractographic details obscured by friction and rubbing (Fig. 7*a*). In fact, Tschegg [*18*] reports the

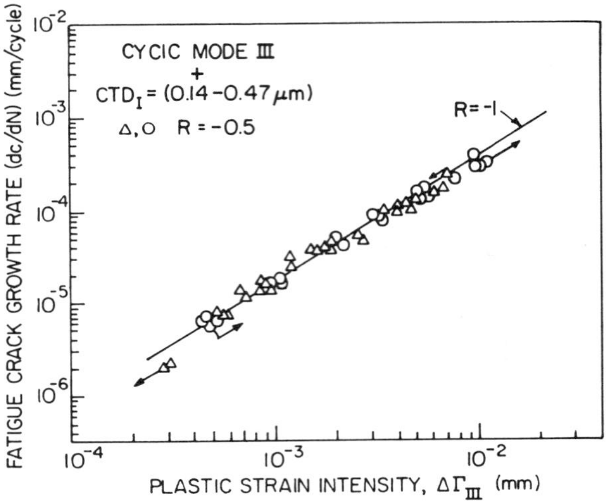

FIG. 5—*Influence of load (or stress) ratio* R *on Mode III fatigue crack growth rates in A469 rotor steel. Data for* R = *−1* (solid line) *and* R = *−0.5* (data points) *were obtained in cyclic torsion with a small superimposed axial load. Arrows indicate direction of crack growth measurements, with* Δ *denoting decreasing* $\Delta\Gamma_{III}$ *and* ○ *increasing* $\Delta\Gamma_{III}$ [8].

presence of black spherical Fe_3O_4 particles on torsional fracture surfaces in 4340 steel after testing and red Fe_2O_3 powder squeezing out of the crack during the test. The presence of Mode III fatigue striations also has been claimed for torsional failures in mild steel [*20*], although their origin is uncertain.

At lower growth rates, typically below 10^{-4} to 10^{-5} mm/cycle [*7,13–16*], or with the superposition of small tensile loads [*7*], there is an increasing tendency for secondary or branch cracks to form and deviate off the macroscopic radial Mode III plane (Fig. 7*b*). In certain materials, these secondary cracks are 90 deg longitudinal Mode III cracks which intersect the radial plane (Fig. 6*d*) [*14*]. However, more commonly they are primarily intersecting Mode I branch cracks which form on complementary planes at ±45 deg to the radial Mode III surface, as shown in Fig. 8. This mode of failure, which is locally pure Mode I and only "nominally" Mode III in the macroscopic sense, has been termed "factory-roof" fracture [*41,7*] and is a common type of torsional fracture found in service [*23*].

Factors responsible for the transition from pure radial Mode III to local Mode I factory-roof fractures (for example, Fig. 6*c*) are far from understood. However, the transition is promoted at lower growth rates [*7,14–16*], with increasing crack length (*15,16*) and in larger specimen diameters [*18*]. This can be explained qualitatively in terms of the lower "effective" ΔK_{III} or $(\Delta CTD)_{III}$ experienced at the tip, favoring Mode I separation, because the latter two factors have been shown to enhance crack tip shielding due to sliding crack surface interference from the increased fracture surface area [*14–16,18*].

It has been suggested that "true" Mode III crack growth, as evidenced by the flat, featureless surfaces in Figs. 6*a* and 7*a*, is nothing more than a local Mode I factory-roof fracture with the Mode I facets destroyed by the rubbing [*34*]. This is unlikely, however, because "true" Mode III fractures are only observed at higher $(\Delta CTD)_{III}$ or ΔK_{III} values, where the larger plastic zones promote sufficient crack opening to minimize the abrasion between sliding crack surfaces, as discussed below.

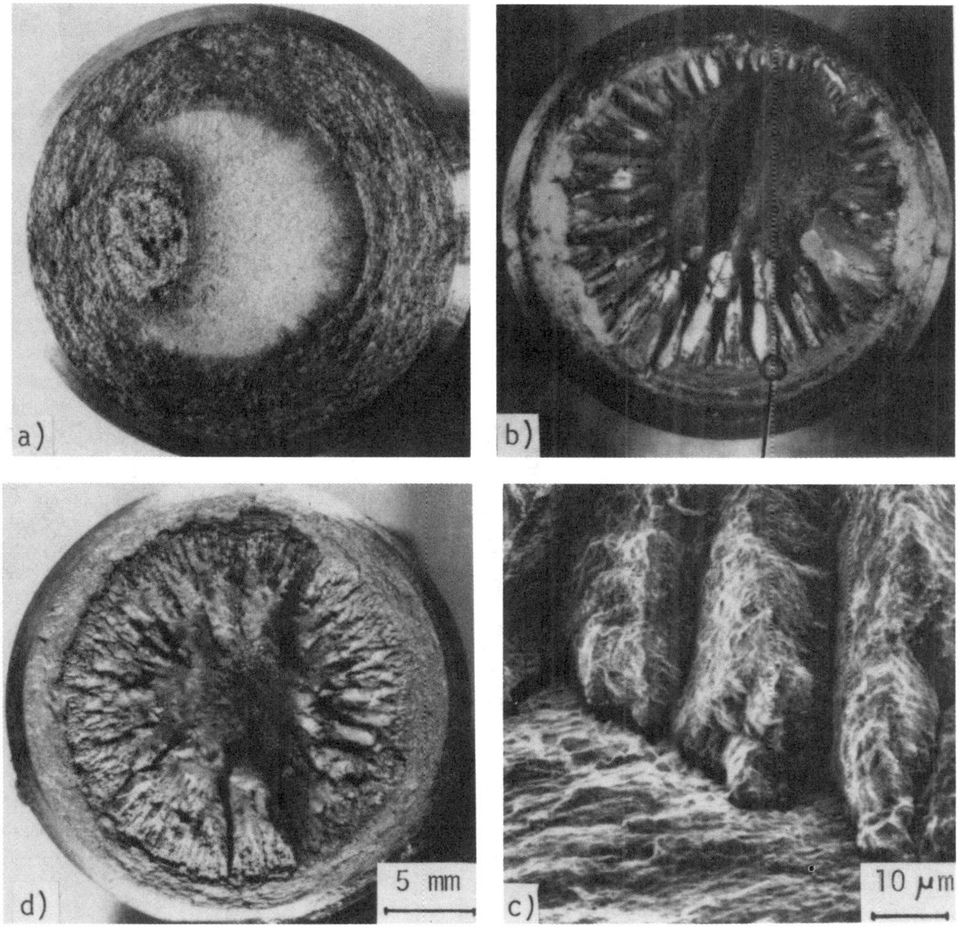

FIG. 6—*Survey of torsional fatigue failures* [21] *showing* (a) *macroscopically flat fracture (true radial Mode III) in 4140 steel at* $(\Delta CTD)_{III} \approx$ *100–200 μm* [14], (b) *"factory-roof" fracture (radial Mode III with 45-deg Mode I branch cracks) in 4340 steel,* (c) *enlargement of* (b) *at* $(\Delta CTD)_{III} \approx$ *12 μm* [7], *and* (d) *"factory-roof" fracture (radial Mode III with longitudinal Mode III branch cracks) in 4140 steel at* $(\Delta CTD)_{III} \approx$ *6–120 μm* [14].

Crack Surface Interference

As noted above, a prominent characteristic of Mode III crack extension, in fact of all non-Mode I cracking, is the rubbing together of sliding crack surfaces during cyclic loading [*14–16*]. Aside from causing heating of the specimen and abrasion of the fracture surfaces, which obscures fractographic details and promotes fretting wear debris, this phenomenon leads problems with electrical shorting in electrical potential crack monitoring [*38*] and most importantly to a marked reduction in the "effective crack tip driving force" for Mode III crack extension. The latter effect is known as *sliding crack surface interference* [*14–16*], or sometimes torsional or Mode III crack closure due to its somewhat analogous effect to Mode I crack closure (for example, Ref *40*) in shielding the crack tip from the full applied loads. Probable mechanisms

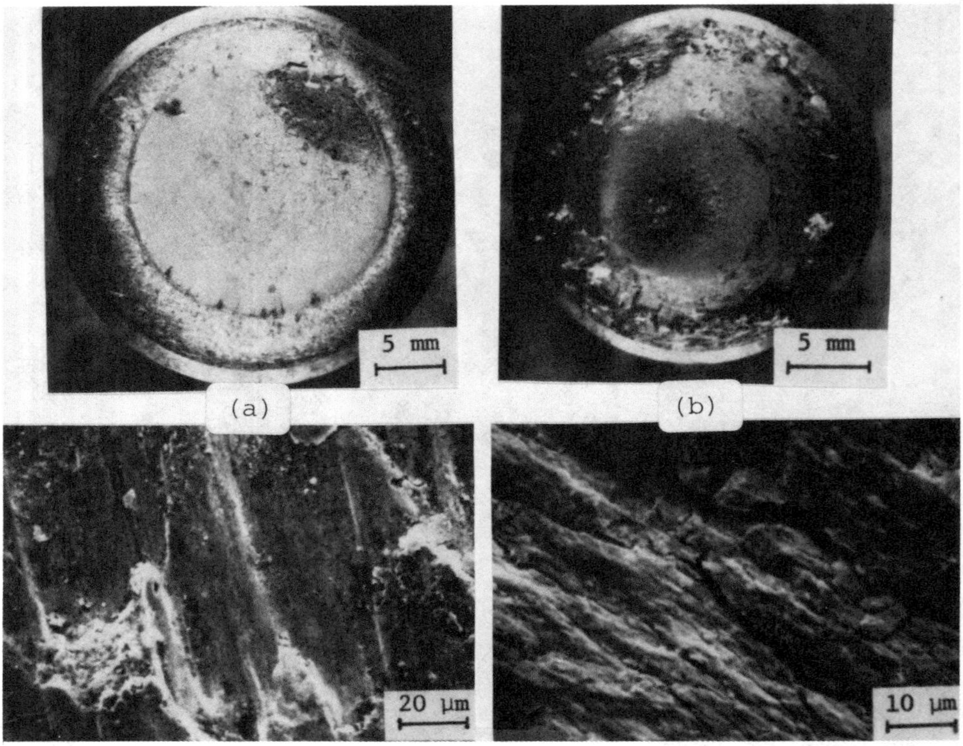

FIG. 7—*Macroscopically flat true radial Mode III failures in 4340 steel (σ_y = 956 MPa) showing* (a) *abrasion of fracture surface at ΔK_{III} = 25 MPa $\sqrt{m}$ with no superimposed axial load (K_I = 0), and* (b) *decreased abrasion and evidence of secondary branch cracks at ΔK_{III} = 30 MPa $\sqrt{m}$ with superimposed axial load (K_I = 10 MPa $\sqrt{m}$). Arrows indicate general direction of radial crack extension* [7].

for such interference, involving friction, abrasion, debris formation, and interlocking, are illustrated in Fig. 9 and clearly enable the broken portion of the test piece to carry a portion of the applied torque. The analogy to Mode I crack closure is reasonable, particularly in the case of the oxide-induced and roughness-induced mechanisms [*40*], although in contrast to behavior in Mode III, Mode I closure does not act over the entire fatigue cycle (Fig. 10).

The effect of crack surface interference on Mode III growth rates can be dramatic. When cracks have been cycled at constant ΔK_{III} [*14–16*] or at constant $\Delta\Gamma_{III}$ [*20*], where fracture mechanics similitude would normally dictate constant rates of crack advance independent of crack size, growth rates are essentially nonunique and decelerate progressively by several orders of magnitude with increasing crack size. This is illustrated in Fig. 11 by torsional results [*20*] on mild steel (σ_y = 260 MPa). The effect can be seen to be enhanced at larger crack lengths, particularly after the transition to factory-roof fractures, but is diminished at higher cyclic torques where the larger plastic zones permit larger crack openings.

Providing a fracture mechanics characterization of Mode III fatigue crack extension without accounting for sliding crack surface interference simply is not feasible, as demonstrated for example in Ref *14*. This is because a unique growth rate cannot be associated with any *nominal* value of ΔK_{III}, $(\Delta CTD)_{III}$, or $\Delta\Gamma_{III}$. Two experimental techniques have been used to overcome this problem, both intended to yield upper bound measurements of Mode III growth rates compared to service behavior. The first of these involves extrapolation of the $(dc/dN)_{III}$ versus crack

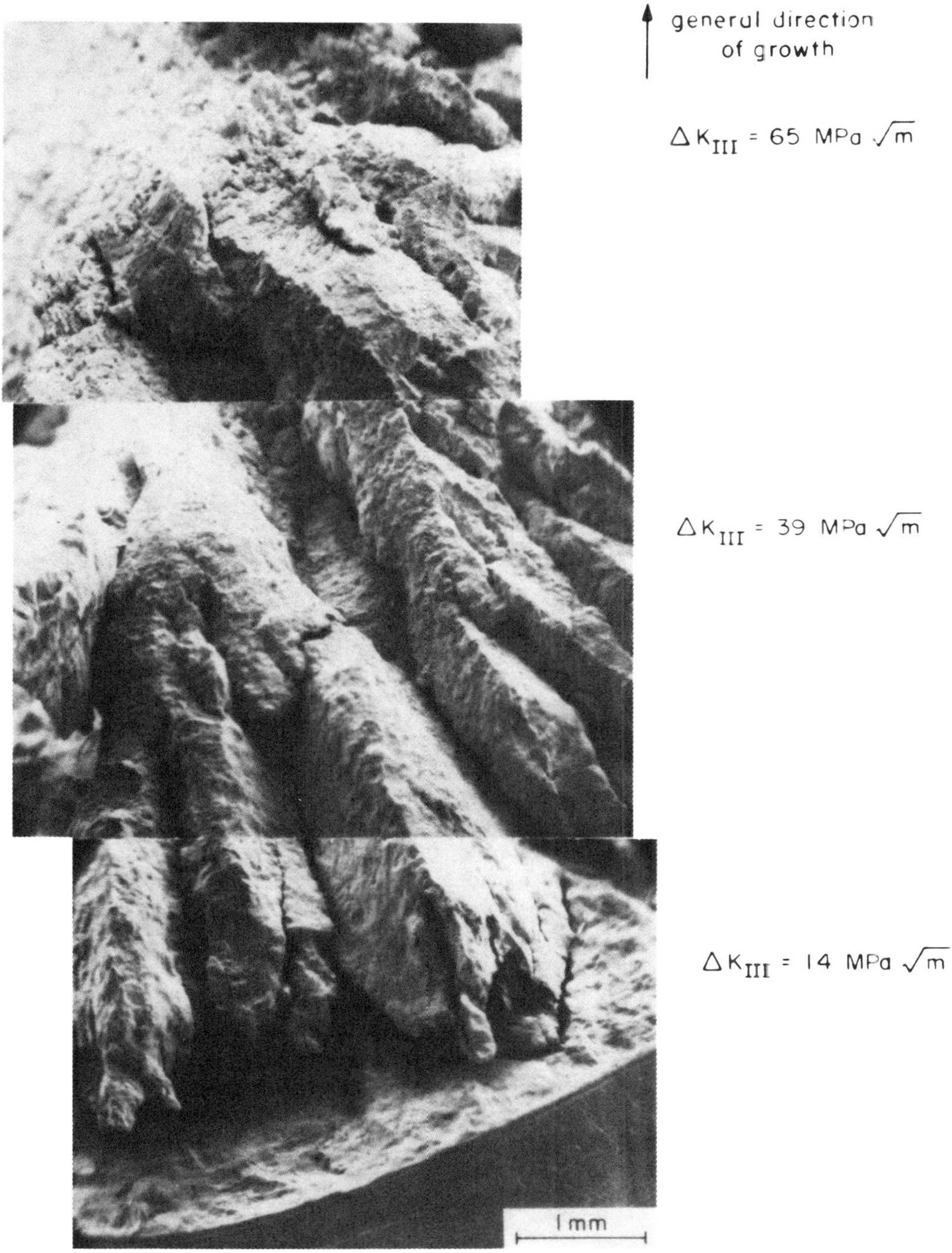

FIG. 8—*"Factory-roof" fracture (radial Mode III with 45-deg Mode I branch cracks) in 4340 steel. Note transition at* $\Delta K_{III} = 14$ *MPa* $\sqrt{m}$ *from macroscopically flat pure radial Mode III fracture (courtesy of H. Nayeb-Hashemi).*

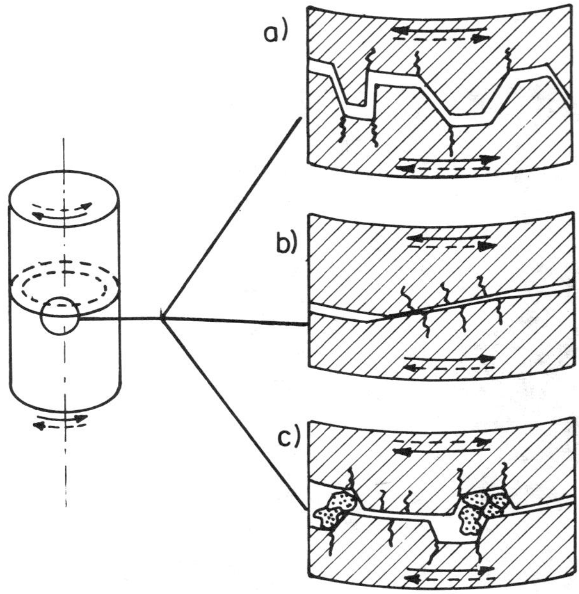

FIG. 9—*Schematic illustration of mechanisms of sliding crack surface interference for Mode III cracks showing* (a) *interlocking of asperities,* (b) *friction and abrasion, and* (c) *fretting debris formation* [19].

length c curves, measured at constant nominal "driving force" (for example, Fig. 11), to a fictitious zero crack length where no interference can be present [*15,18–20*]. The method is somewhat artificial and involves taking several measurements at different crack sizes to obtain one $(dc/dN)_{\text{III}}$ data point, but does yield upper bound results, as shown by results [*20*] for mild steel in Fig. 12. The more common technique [*8,9,14,21,38*], however, is to superimpose a small tensile mean load onto the cyclic torsion, thereby minimizing interference by the increased crack opening. This technique also yields upper bound results (Fig. 13) and been shown to produce unique $(dc/cN)_{\text{III}}$ data over a wide range of testing conditions [*8,14*]. Predictably, it works well where crack surfaces are not overly rough where the required Mode I mean loads are small enough such as not interfer with crack growth behavior [*8,9*]. This required mean load, however, will change at different growth rates [*19*], due to differences in crack path morphology, and can lead to an increase in Mode I branch cracks [*7*]. Such problems are reflected in the data on mild steel shown in Fig. 14 where increasing the magnitude of the superimposed K_{I} during cyclic torsion at two $\Delta\Gamma_{\text{III}}$ levels does not completely remove crack surface interference [*19*]. Moreover, at high levels of K_{I}, for example, with $K_{\text{I}} = 10.5$ MPa $\sqrt{\text{m}}$ at $\Delta\Gamma_{\text{III}} = 0.002$ mm, the trend in $(dc/dN)_{\text{III}}$ with increasing c can be seen to be changed abruptly, presumably arising from enhanced surface interference from an increased propensity of Mode I branch cracks.

The mechanistic details of sliding crack surface interference have not been quantified in detail, although one simple analytical solution has been reported. By assuming that the fracture surface asperities compel the crack faces to separate, thereby causing a normal force which in turn generates a frictional force that acts in opposition to the applied torque, Gross [*42*] has recently estimated the torque lost to friction M_f, in terms of friction coefficient μ, asperity height t, crack size c, and ligament size r_N, namely

$$M_f = \frac{\mu G t}{(1 - \nu)} \left[\frac{c^2}{2} - 2cr_N + r^2 \ell n \left(\frac{c}{t} \right) \right] \quad (13)$$

where G is the shear modulus and ν is Poisson's ratio. By reducing the applied torque by M_f, computed using asperity heights between 100 and 200 μm, Eq 13 has been shown to provide reasonable predictions of the progressive decay in growth rates, as depicted in Fig. 11 [*42*].

Behavior under Cyclic Mode I with Steady Mode III

Several authors [*41,43*] have examined the influence of steady Mode III superimposed on cyclic Mode I, producing studies analogous to those described previously on the influence of steady Mode I loading superimposed on cyclic Mode III [*7,21*]. Whereas in the former instance the superimposed Mode I load can lead to faster Mode III growth rates by limiting crack surface interference by opening the crack, results for superimposing a Mode III load on cyclic Mode I invariably show Mode I growth rates to be decreased progressively with increasing K_{III}. This is shown for a Ti-5Al-2.5Sn alloy (σ_y = 760 MPa) in Fig. 15 [*41*]. Similar behavior has been reported in steels [*43*].

In all cases, the decrease in growth rates (typically by two orders of magnitude) is concurrent with a marked change in crack path morphology from being relative linear to a zigzag profile characteristic of a factory-roof fracture. Accordingly, one would reason that the reduction in Mode I growth rates with superimposed Mode III results principally from a reduced "effective driving force" actually experienced at the crack tip, that is, a lower local stress intensity range, due to crack deflection from the Mode I plane and the resulting roughness-induced crack closure. Similar effects are seen under (nominal) pure Mode I loading where the promotion of crack path meandering, either by crystallographic deflection [*44–46*] or deflection at hard phases [*47*], can lead to major reductions in crack growth rates.

"Thresholds" for Crack Growth

Extensive information is now available, for a wide range of materials and testing conditions, on values of Mode I fatigue thresholds ΔK_{Ith}, below which tensile cracks are assumed to remain dormant (for example, Refs *22,40*). Furthermore, the existence of a Mode I threshold has been related mechanistically to the magnitude of crack closure generated along the crack flanks (for example, Refs *48,49*). Corresponding understanding of the Mode III threshold ΔK_{IIIth} is far less advanced. The majority of data has been reported by Pook and co-workers [*34,35,37*], who define the value of ΔK_{IIIth} in terms of the stress intensity to initiate crack growth from a torsionally loaded circumferential crack. This can be questioned as a true Mode III threshold for crack advance in shear, however, because cracking does not commence in Mode III. The measured threshold instead represents the value of ΔK_{III} at which a Mode I branch crack initiates from the Mode III pre-crack, consistent with crack deflection calculations which accurately predict that the measured Mode III thresholds should be a factor of 1.35 larger than corresponding values in Mode I [*35*].

Ideally, a threshold for crack growth should be defined under conditions of decreasing "driving force" until crack growth arrests, as is commonly employed in Mode I testing [*22*]. If this is adopted in Mode III, at low values of ΔK_{III} the Mode III cracking mode changes to a local Mode I (factory-roof) fracture long before crack arrest can occur (for example, Refs *7,13*). Accordingly, several authors [*13,18*] have defined the Mode III threshold in terms of this Mode III/Mode I transition rather than in terms of the cessation of cracking. Values of ΔK_{IIIth} defined in this way, however, are not comparable with crack arrest thresholds in Mode I, and in fact are considerably larger in magnitude.

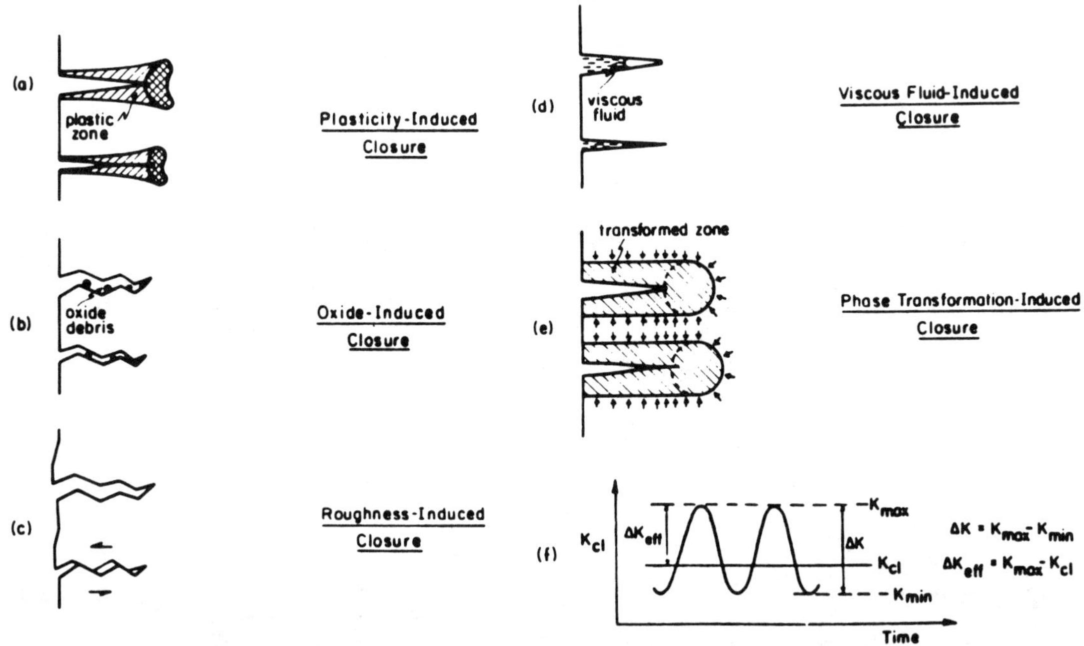

FIG. 10—*Schematic illustration of primary mechanisms of crack closure during Mode I fatigue crack propagation* [40].

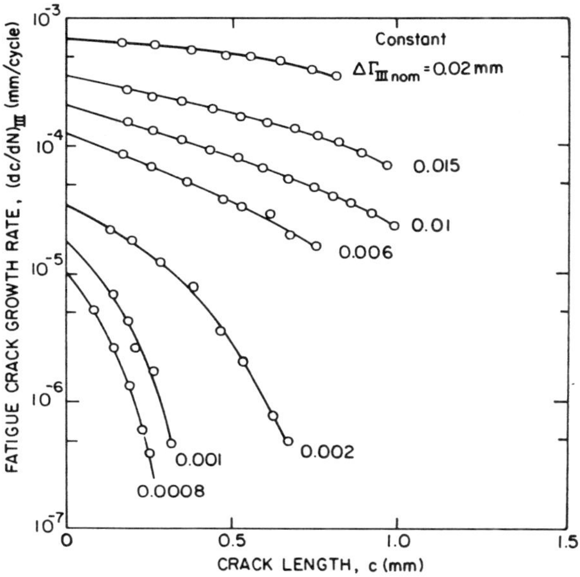

FIG. 11—*Mode III fatigue crack growth rates in 1018 steel (σ_y = 260 MPa) as a function of crack length, measured in cyclic torsion at constant nominal $\Delta\Gamma_{III}$ values. Growth rates are reduced with increasing crack length, particularly at low $\Delta\Gamma_{III}$ levels, due to crack surface interference (after Tschegg* [20]*).*

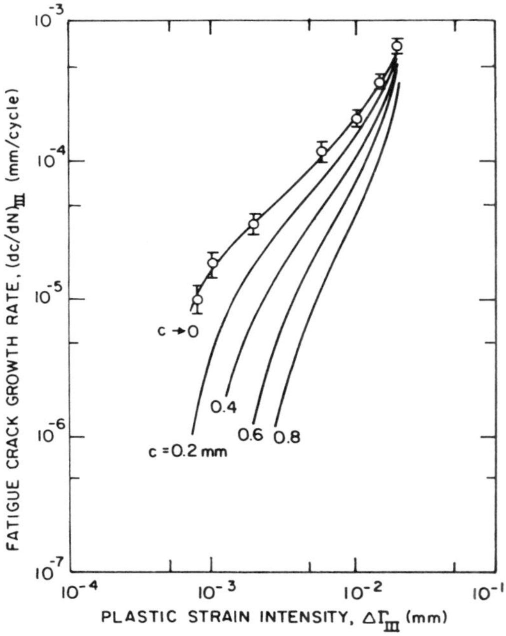

FIG. 12—*Variation in Mode III fatigue crack growth rates in 1018 steel with crack length,* c, *measured from notch. The curve for* c → 0 *is derived by extrapolating the data in Fig. 11 and is taken to represent behavior in the absence of crack surface interference (after Tschegg* [20]*).*

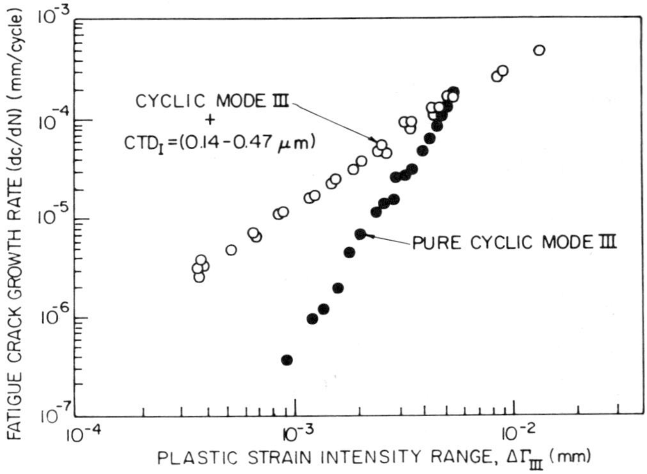

FIG. 13—*Comparison of Mode III fatigue crack propagation rates in A469 rotor steel measured in cyclic torsion with and without a small superimposed axial load. Data obtained with superimposed axial load are taken to represent behavior in the absence of crack surface interference* [8].

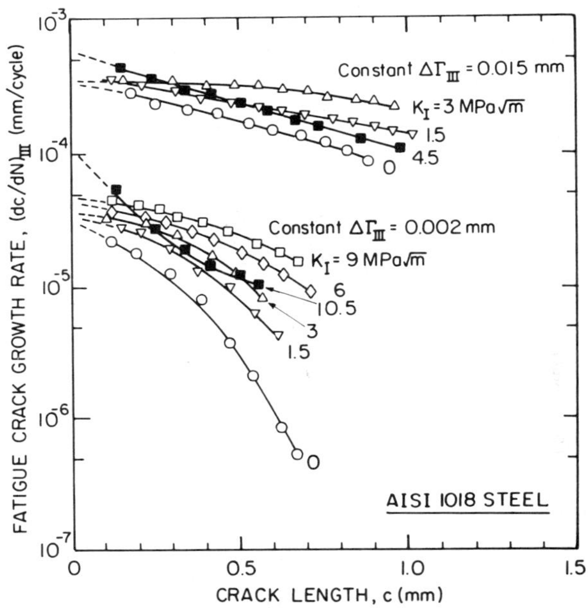

FIG. 14—*Influence of superimposed axial load (*K_I *varying from 0 to 10.5 MPa* $\sqrt{m}$*) on Mode III fatigue crack growth rates in 1018 steel measured in cyclic torsion at constant nominal* $\Delta\Gamma_{III}$ *(after Tschegg* [19]*).*

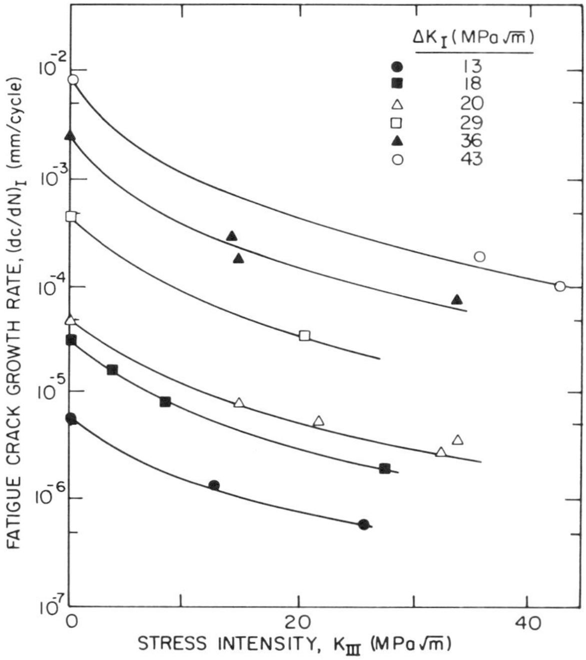

FIG. 15—*Influence of superimposed torsional load (*K_{III} *varying from 0 to 40 MPa* $\sqrt{m}$*) on Mode I fatigue crack growth rates in Ti-5Al-2.5Sn alloy with* $\sigma_y = 760$ *MPa (after Hourlier et al.* [41]*).*

Crack Propagation Models

Recent micromechanical modeling of Mode III fatigue crack growth has been formulated on the premise that Mode III crack advance occurs via the Mode II shear linkage of voids parallel to, and in the immediate vicinity of, the main Mode III crack front [*7–9*]. Although unlikely to be a universal mechanism, this model is consistent with the early work of Tipnis and Cook [*50*] and with more recent observations on Mode III cracks broken open in liquid nitrogen, where the coalescence of elongated voids can be clearly seen to be confined locally along the main crack front [*8*].

The proposed mechanism is illustrated schematically in Fig. 16 [*7*]. Micromechanical models for this mechanism have been developed based either on considerations of instantaneous crack tip displacement [*7,8*] or on damage accumulation arguments [*8,9*]. Since for a Mode III crack tip displacement along the length of the void, the Mode III crack will only advance a distance equal to the void width, the models in general predict $(dc/dN)_{III}$ to be a small fraction of the $(\Delta CTD)_{III}$ each cycle, consistent with the experimental results shown in Fig. 3, for example.

Since the two models have been described in detail in Ref *8*, only a very brief summary is included here. In the crack tip displacement model, growth rates in anti-plane shear are related geometrically to rates of Mode II shear coalescence along the crack front, assuming that crack extension is proportional to the instantaneous ΔCTD. Accordingly, similar to equivalent models for Mode I growth [*51*], Mode III growth rates are predicted to be dependent upon the first power of the crack tip displacement range, consistent with experimental results in certain materials such as 4340 and 4140 [*7,14*]. Conversely, in the damage accumulation model, the relationship between Mode III crack advance and Mode II shear along the crack front is achieved by assuming that accumulated damage, defined by the Coffin-Manson expression, governs the rate of shear coalescence. Here, Mode III growth rates are predicted to depend

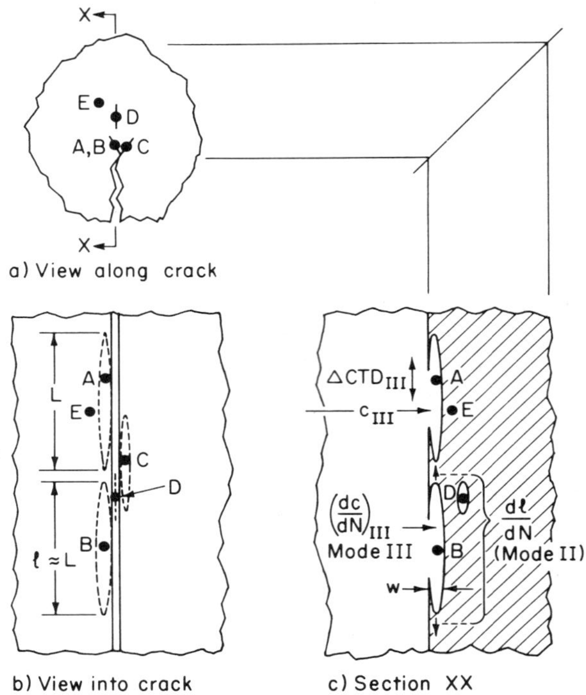

FIG. 16—*Schematic illustration of micromechanical model for fatigue crack propagation in anti-plane shear showing orthographic views of Mode III crack advance occurring by microscopic Mode II shear along the main Mode III crack front* [7].

upon the crack tip displacement range, raised to the power 1.5, consistent with experimental results in several rotor steels [*8, 9*].

Behavior Under Variable-Amplitude Loading

The transient growth rate behavior of Mode I cracks subjected to certain variable-amplitude loading spectra is now well-documented in the literature [*52*]. For example, the transient retardations observed following single tensile overloads or high-low block loading sequences have been attributed to crack tip-yield zone interactions, crack tip blunting, crack branching, and principally to fatigue crack closure. In Mode III, however, where crack tip blunting and closure as such do not exist, behavior under variable amplitude loads is very different [*9, 17*].

As illustrated in Fig. 17 for torsional results on A469 rotor steel [*9*], an initial acceleration is observed for Mode III cracks above a 50% overtorque,[2] whereas a single positive overload results in a *retardation* in Mode I growth rates. Similarly, for single fully reversed overloads, Mode III cracks tend to *accelerate* on application of overtorques of 50% or larger, again in contrast to Mode I where a *deceleration* occurs. The response to high-low block loading sequences also is entirely opposite in Mode III compared to Mode I.

As suggested previously, these differences have been attributed to the lack of crack closure mechanisms during Mode III crack extension under variable amplitude loads [*9*]. Whereas fol-

[2]The terms overtorque and overtwist are defined as $(\Delta M_0 - \Delta M_B)/\Delta M_B$ and $(\Delta\varphi_0 - \Delta\varphi_B)/\Delta\varphi_B$, respectively, where ΔM_0 and $\Delta\varphi_0$ are the torque and crack mouth rotation at the overload, and ΔM_B and $\Delta\varphi_B$ are the corresponding baseline values.

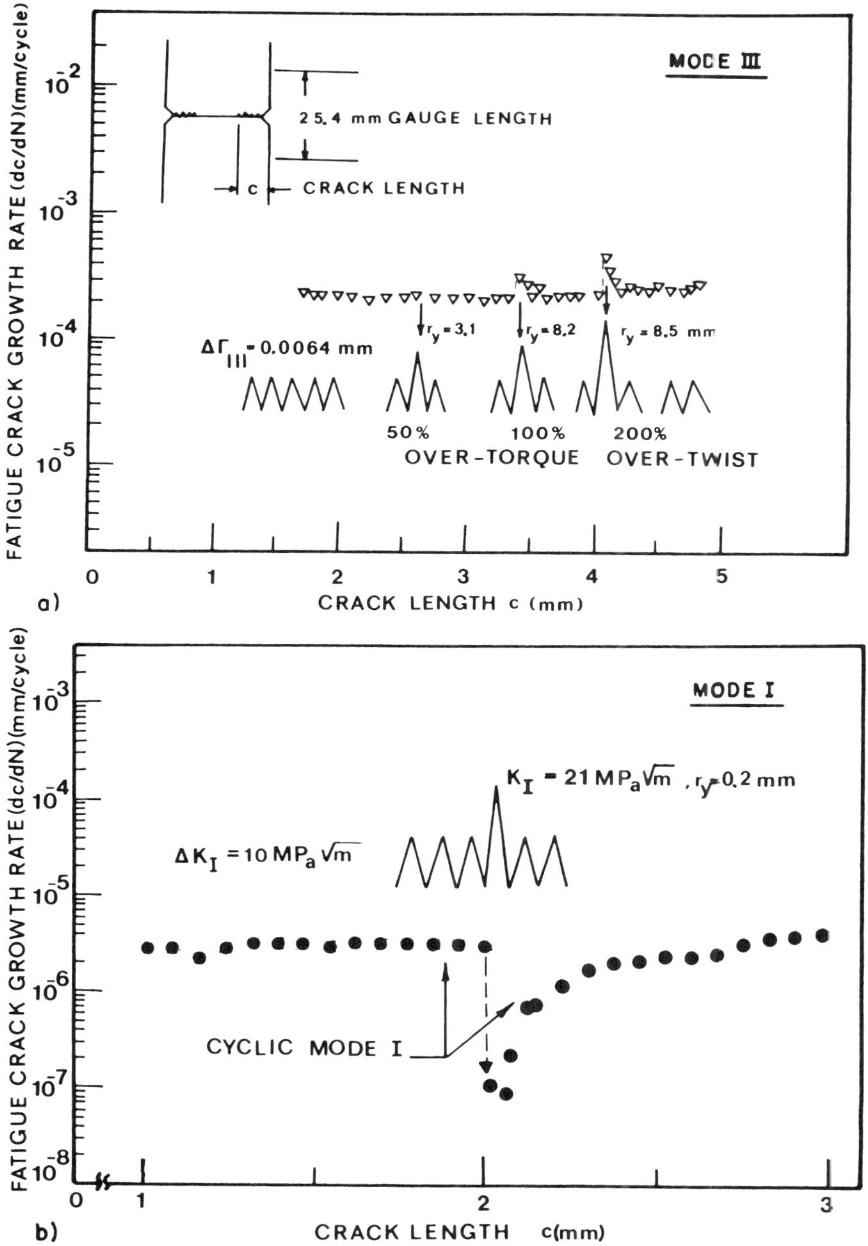

FIG. 17—*Comparison of the effect of single (spike) positive overloads on Mode III and Mode I fatigue crack propagation in A469 rotor steel. Note how the transient post-overload response is an acceleration in Mode III compared with a retardation in Mode I* [9].

lowing a single overload on a Mode I crack, the "damage" caused by the higher load excursion is more than compensated by concomitant increases in crack tip blunting and crack closure, these mechanisms have little relevance for Mode III cracks. Correspondingly, the "damage" resulting from single overloads simply accelerates Mode III cracks in accordance with the classical predictions of Miner's rule, without competition from crack-shielding mechanisms.

Summary and Conclusions

Fracture mechanics characterizations of fatigue crack propagation behavior in Mode III (anti-plane shear), specific to damage-tolerant design and lifetime prediction, have important applications in engineering service, although clearly they are less frequently used compared to corresponding methodologies for Mode I. One such application is in the lifetime prediction of large turbo-generator shafts in electrical power generation and transmission systems, where following particular electrical fault or line switching events, complex high-amplitude torsional oscillations can be set up which approach full-scale yielding of the shaft [*53*]. Lifetime prediction procedures have been developed using both fracture mechanics (damage-tolerant) analyses of crack growth [*54*] and classical stress/life (S/N) curves [*55*], although in view of the high cost of failure, the inherently conservative nature of the damage-tolerant approach clearly is preferable in this application.

As a basis for such a damage-tolerant analysis of Mode III fatigue, the present article attempts to describe what is currently understood about the mechanics and mechanisms of crack growth in anti-plane shear. It is apparent that to provide a global characterization of Mode III growth rates, similar to the well-known Paris "law" for Mode I fatigue [*56*], methodology must be developed which both utilizes a characterizing parameter (that is, "crack driving force") appropriate to large-scale yielding (since Mode III cracking is promoted by extensive crack tip plasticity) and which accounts for the problem of sliding crack surface interference. In the present approach, the Mode III crack tip displacement $(\Delta CTD)_{III}$ and plastic strain intensity range $\Delta\Gamma_{III}$ are proposed as characterizing parameters for Mode III crack extension under large-scale yielding, and experimentally measured growth rates on a range of steels are shown to be proportional to $(\Delta CTD)_{III}$ and $\Delta\Gamma_{III}$ raised to power 1 to 1.5. However, to obtain upper bound data uniquely relating $(dc/dN)_{III}$ to these parameters, allowance must be made for crack surface interference, either by superimposing a small mean tensile load on the cyclic torsional loads or by extrapolating $(dc/dN)_{III}$ versus crack length data to zero crack length.

Micromechanical models for cyclic crack advance in anti-plane shear have been described and are found to be consistent with the fractography of Mode III torsional failures. Such models have been applied [*9*] to Mode III crack growth under variable-amplitude loading, where the transient growth rate behavior following single tensile and fully reversed overloads and high-low block loading sequences is completely opposite to that observed for Mode I cracks. Specifically, behavior in Mode III is attributed to the lack of crack tip shielding from blunting and crack closure mechanisms, which results in transient *accelerations* in growth rates, compared to the commonly reported transient *retardations* primarily induced by such shielding mechanisms following overload sequences in Mode I.

Acknowledgments

This paper was supported by the director, Office of Energy Research, Office of Basic Energy Sciences, Materials Science Division of the U.S. Department of Energy under Contract No. DE-AC03-76SF00098. In addition, early Mode III fatigue work was funded by the Division of Energy Conservation, under Contract No. W-7405-ENG-26, Subcontract 9062, from Oak Ridge National Laboratory. While the opinions expressed in the manuscript are my own, they result largely from collaborative studies with my colleagues, Professor F. A. McClintock and Dr. E.

K. Tschegg, and former students, of whom Professors H. Nayeb-Hashemi and S. Suresh deserve special mention.

References

[1] McClintock, F. A. in *Proceedings,* International Conference on Fatigue of Metals, Institution of Mechanical Engineers, London, 1956, p. 538.
[2] Hult, J. A. and McClintock, F. A. in *Proceedings,* 9th International Congress of Applied Mechanics, University of Brussels, Belgium, Vol. 8, 1956, p. 51.
[3] Hult, J. A., *Transactions of the Royal Institute of Technology,* Stockholm, Sweden, No. 119, 1958.
[4] Kayan, I., *Bulletin of the Technical University of Istanbul,* Vol. 14, 1961, p. 45.
[5] McClintock, F. A. in *Fracture of Solids,* D. C. Drucker and J. J. Gilman, Eds., Interscience, New York, The Metallurgical Society of the American Institute of Mining, Metallurgical, and Petroleum Engineers' Conference Series 20, 1963, p. 65.
[6] McClintock, F. A. and Ritchie, R. O. in *Mechanics of Fatigue,* T. Mura, Ed., American Society for Mechanical Engineers, New York, AMD-Vol. 47, 1981, p. 1.
[7] Ritchie, R. O., McClintock, F. A., Nayeb-Hashemi, H., and Ritter, M. A., *Metallurgical Transactions A,* Vol. 13A, 1982, p. 101.
[8] Nayeb-Hashemi, H., McClintock, F. A., and Ritchie, R. O., *International Journal of Fracture,* Vol. 23, 1983, p. 163.
[9] Nayeb-Hashemi, H., McClintock, F. A., and Ritchie, R. O., *Engineering Fracture Mechanics,* Vol. 18, 1983, p. 736.
[10] Walsh, J. B. and Mackenzie, A. C., *Journal of the Mechanics and Physics of Solids,* Vol. 7, 1959, p. 247.
[11] Rice, J. R., *International Journal of Fracture Mechanics,* Vol. 2, 1966, p. 426.
[12] Wright, R. P. and Queeney, R. A., *International Journal of Fatigue,* Vol. 4, 1982, p. 27.
[13] Hurd, N. J. and Irving, P. E. in *Design of Fatigue and Fracture Resistant Structures, ASTM STP 761,* American Society of Testing and Materials, Philadelphia, 1982, p. 212.
[14] Nayeb-Hashemi, H., McClintock, F. A., and Ritchie, R. O., *Metallurgical Transactions A,* Vol. 13A, 1982, p. 1860.
[15] Tschegg, E. K., *Materials Science and Engineering,* Vol. 54, 1982, p. 127.
[16] Tschegg, E. K., Ritchie, R. O., and McClintock, F. A., *International Journal of Fatigue,* Vol. 5, 1983, p. 29.
[17] Nayeb-Hashemi, H., Suresh, S., and Ritchie, R. O., *Materials Science and Engineering,* Vol. 59, 1983, p. L1.
[18] Tschegg, E. K., *Journal of Materials Science,* Vol. 18, 1983, p. 1604.
[19] Tschegg, E. K., *Materials Science and Engineering,* Vol. 59, 1983, p. 127.
[20] Tschegg, E. K., *Acta Metallurgica,* Vol. 9, 1983, p. 1323.
[21] Ritchie, R. O., McClintock, F. A., Tschegg, E. K., and Nayeb-Hashemi, H. in *Multiaxial Fatigue, ASTM STP 853,* K. J. Miller and M. W. Brown, Eds., American Society of Testing and Materials, Philadelphia, 1985, p. 203.
[22] Ritchie, R. O., *International Metals Reviews,* Vol. 20, 1979, p. 205.
[23] Wulpi, D. J., *Metals Progress,* Vol. 123, 1981, p. 25.
[24] Williams, M. L., *Journal of Applied Mechanics, Transactions,* American Society of Mechanical Engineers, Vol. 24, 1957, p. 109.
[25] Rice, J. R., *Journal of Applied Mechanics, Transactions,* American Society of Mechanical Engineers, Vol. 35, 1968, p. 379.
[26] Hutchinson, J. W., *Journal of the Mechanics and Physics of Solids,* Vol. 16, 1968, p. 13.
[27] Rice, J. R. and Rosengren, G. R., *Journal of the Mechanics and Physics of Solids,* Vol. 16, 1968, p. 1.
[28] Tada, H., Paris, P. C., and Irwin, G. R. in *The Stress Analysis of Cracks Handbook,* Del Research Corp., Hellertown, PA, 1973.
[29] Chitaley, A. D. and McClintock, F. A., *Journal of the Mechanics and Physics of Solids,* Vol. 19, 1971, p. 147.
[30] Shih, C. F., *Journal of the Mechanics and Physics of Solids,* Vol. 29, 1981, p. 305.
[31] McClintock, F. A. and Tschegg, E. K., *International Journal of Fracture,* Vol. 26, 1984, p. R11.
[32] Erdogan, F. and Sih, G. C., *Journal of Basic Engineering, Transactions,* American Society of Mechanical Engineers, Vol. 85D, 1963, p. 519.
[33] Gao, H., Alagok, N., Brown, M. W., and Miller, K. J. in *Multiaxial Fatigue, ASTM STP 853,* K. J. Miller and M. W. Brown, Eds., American Society for Testing and Materials, Philadelphia, 1985, p. 184.
[34] Pook, L. P. in *Multiaxial Fatigue, ASTM STP 853,* K. J. Miller and M. W. Brown, Eds., American Society for Testing and Materials, Philadelphia, 1985, p. 249.

[35] Pook, L. P. and Sharples, J. K., *International Journal of Fracture*, Vol. 15, 1979, p. R223.
[36] Smith, R. A. in *Fundamentals of Tribology*, N. P. Suh and N. Saka, Eds., Massachusetts Institute of Technology Press, Cambridge, MA, 1980, p. 605.
[37] Pook, L. P., *International Journal of Fatigue*, Vol. 7, 1985, p. 21.
[38] Ritter, M. A. and Ritchie, R. O., Fatigue of Engineering Materials and Structures, Vol. 5, 1982, p. 91.
[39] Sudarshan, T. S., Louthan, M. R., and Place, T. A., *Materials Science and Engineering*, Vol. 66, 1984, p. L27.
[40] Suresh, S. and Ritchie, R. O. in *Fatigue Crack Growth Threshold Concepts*, D. L. Davidson and S. Suresh, Eds., The Metallurgical Society of the American Institute of Mining, Metallurgical, and Petroleum Engineers, Warrendale, PA, 1984, p. 227.
[41] Hourlier, F., McLean, D., and Pineau, A., *Metals Technology*, Vol. 5, 1978, p. 154.
[42] Gross, T. S., *Scripta Metallurgica*, Vol. 19, 1985, p. 1185.
[43] Akhurst, K. N., Lindley, T. C., and Nix, K. J., *Fatigue of Engineering Materials and Structures*, Vol. 6, 1983, p. 345.
[44] Suresh, S., *Metallurgical Transactions A*, Vol. 14A, 1983, p. 2375.
[45] Suresh, S., Vasudévan, A. K., and Bretz, P. E., *Metallurgical Transactions A*, Vol. 15A, 1984, p. 369.
[46] Zaiken, E. and Ritchie, R. O., *Materials Science and Engineering*, Vol. 70, 1985, p. 151.
[47] Dutta, V. B., Suresh, S., and Ritchie, R. O., *Metallurgical Transactions A*, Vol. 15A, 1984, p. 1193.
[48] Minakawa, K., Newman, J. C., and McEvily, A. J., *Fatigue of Engineering Materials and Structures*, Vol. 6, 1983, p. 359.
[49] Zaiken, E. and Ritchie, R. O., *Metallurgical Transactions A*, Vol. 16A, 1985, p. 1467.
[50] Tipnis, V. A. and Cook, N. H., *Journal of Basic Engineering, Transactions*, American Society of Mechanical Engineers, Vol. 89, 1967, p. 533.
[51] Irving, P. E. and McCartney, L. N., *Metal Science*, Vol. 11, 1977, p. 351.
[52] Suresh, S., *Engineering Fracture Mechanics*, Vol. 18, 1983, p. 577.
[53] Jackson, M. C., Umans, S. D., Dunlop, R. D., Horowitz, S. H., and Parikh, A. C., *Institute of Electrical and Electronics Engineers Transactions for Power Apparatus and Systems*, Vol. 98, 1979, p. 2299.
[54] McClintock, F. A., Nayeb-Hashemi, H., Ritchie, R. O., Wu, E., and Wood, W. T., "Assessing Expired Fatigue Life in Large Turbine Shafts," Oak Ridge National Laboratory Report ONRL/Sub/83-9062/1, Oak Ridge, TN, July 1984.
[55] Tanaka, K., Matsuoka, S., Kouzu, F., and Nagata, K., *Fatigue of Engineering Materials and Structures*, Vol. 6, 1983, p. 103.
[56] Paris, P. C. and Erdogan, F., *Journal of Basic Engineering, Transactions*, American Society of Mechanical Engineers, Vol. 85, 1963, p. 528.

Takeo Yokobori,[1] *Juro Watanabe,*[2] *Takao Aoki,*[3] *and Tadao Iwadate*[4]

Evaluation of the K_{ISCC} Testing Procedure by Round Robin Tests on Steels

REFERENCE: Yokobori, T., Watanabe, J., Aoki, T., and Iwadate, T., **"Evaluation of the K_{ISCC} Testing Procedure by Round Robin Tests on Steels,"** *Fracture Mechanics: Eighteenth Symposium, ASTM STP 945,* D. T. Read and R. P. Reed, Eds., American Society for Testing and Materials, Philadelphia, 1988, pp. 843–866.

ABSTRACT: Round robin test programs of the tentative test procedure for the determination of the threshold stress-intensity factor K_{ISCC} were conducted under the auspices of the 129th Committee, Japan Society for the Promotion of Science. The test programs used JIS SNCM 439 (AISI 4340) steels. Thirty-one laboratories participated in these programs. Tests were carried out with both 25.4-mm (1-in.)-thick bolt loaded wedge-opening-loading (WOL) specimens and 25.4 by 25.4-mm cantilever beam (CB) specimens in a 3.5% sodium chloride solution of distilled water.

The results of the round robin test programs showed some informations to standardize the K_{ISCC} testing procedure. For valid K_{ISCC} values to be obtained, the test must last 7000 h for a WOL specimen and 4000 h for a CB specimen. In the WOL tests, the final crack length and width ratio a_f/W of a specimen must be smaller than 0.8. An elastic bolt analysis of WOL specimens was recommended to obtain the valid K_{ISCC} values and the good relationship between da/dt and K_I. In the CB tests, the estimated K_{ISCC} value coincided well with that in the WOL tests when at least six CB specimens were used. Slight effect of crack front curvature, test temperature, and pH of the solution on the variance of the K_{If} values was observed.

With these test results, the K_{ISCC} testing procedure was standardized by the Committee and entitled "Standard Test Method for Threshold Stress Intensity Factor K_{ISCC} in Stress Corrosion Cracking."

KEY WORDS: environmental testing, fracture mechanics, stress corrosion cracking, stress intensity factor, subcritical flaw growth, test technique, evaluation

Since 1975, the 129th Committee (Professor T. Yokobori, Committee Chairman), Japan Society for the Promotion of Science have planned an interlaboratory test program to evaluate the testing procedure for the determination of the threshold stress-intensity factor K_{ISCC} for stress corrosion cracking. The overall objective of this test program was to develop a standard test procedure for measuring K_{ISCC} and the crack growth rate da/dt in metallic materials subjected to static loading in deleterious environments. In 1974, ASTM Joint Task Group E24.04.02/G01.06.04 initiated the same interlaboratory test program on test methods for measurement of K_{ISCC} and SCC growth rate, in which one of the authors participated [*1*].

[1]Professor Emeritus of Mechanical Engineering, Tohoku University, Aramaki, Sendai, Japan.

[2]Director, Research and Development Division, The Japan Steel Works, Ltd., 1-1-2, Yurakucho, Chiyodaku, Tokyo, Japan.

[3]Manager, National Research Institute for Metals, 2-3-12, Nakameguro, Meguroku, Tokyo, Japan.

[4]General manager, Research Laboratory, The Japan Steel Works, Ltd., 4 Chatsu-machi, Muroran, Japan.

The Japanese interlaboratory round robin test program was initiated in 1977 using JIS SNCM 439 (AISI 4340) steel with yield strength of 1376 MPa in 3.5% sodium chloride solution. Since this stage, the correspondence and the cooperation given by E24.04.02 Task Group, ASTM have helped to form the test program on K_{ISCC} and SCC growth rate measurement. Testing was completed in 1979, and the results were analyzed for the modification of the testing procedure. From June 1981 through April 1983, the second round robin test program was performed using JIS SNCM 439 steels with yield strength levels of 1168 and 1241 MPa. Tests were carried out with both 25.4-mm (1-in.)-thick bolt loaded wedge-opening-loading (WOL) specimens and 25.4 by 25.4-mm single-edge-notched cantilever beam (CB) specimens in a 3.5% sodium chloride solution of distilled water. The 31 laboratories shown in Table 1 participated in these test programs.

In this paper, the tentative test procedure will first be reviewed, and the details of the recently completed second interlaboratory round robin test program will be presented. The results of the round robin test program which were analyzed by the Committee to develop a standard test procedure entitled "Standard Test Method for Threshold Stress Intensity Factor K_{ISCC} in Stress Corrosion Cracking" will be then reported.

TABLE 1—*K_{ISCC} interlaboratory round robin test participants.*

Participant	Affiliation
Hironobu Nishitani	Faculty of Engineering, Kyushu University, Fukuoka
Kanetoshi Iwamoto	Department of Mechanical Engineering, Kurume College of Technology, Kurume
Akio Otsuka	Department of Iron and Steel Engineering, Nagoya University, Nagoya
Hisashi Izumi	Faculty of Engineering, Himeji Institute of Technology, Himeji
Masateru Ohnami	Faculty of Science and Engineering, Ritsumeikan University, Kyoto
Keizo Fujitani	Faculty of Science and Engineering, Ritsumeikan University, Kyoto
Shozo Matsuda	Technical Institute, Kawasaki Heavy Industries, Ltd., Akashi
Hisashi Takada	Central Research Laboratories, Kobe Steel, Ltd., Kobe
Kunio Nishioka	Central Research Laboratories, Sumitomo Metal Industries, Ltd., Amagasaki
Shigeki Sawa	Central Research Laboratory, Daido Steel Company, Ltd., Nagoya
Tadayoshi Endo	Takasago Technical Institute, Mitsubishi Heavy Industries, Ltd., Takasago
Ryuichiro Ebara	Hiroshima Technical Institute, Mitsubishi Heavy Industries, Ltd., Hiroshima
Masao Kinoshita	Technical Research Institute, Technology & Development Headquarters, Hitachi Zosen Corporation, Osaka
Naritaka Yamamoto	Faculty of Engineering, Kanagawa University, Yokohama
Hiroyuki Okamura	Faculty of Engineering, University of Tokyo, Tokyo
Hideo Kitagawa	Institute of Industrial Science, University of Tokyo, Tokyo
Hajime Nakazawa	Faculty of Engineering, Tokyo Institute of Technology, Tokyo
Shigemi Sasaki	University of Electro-Communications, Tokyo
Yuichi Kawada	The Metropolitan College of Technology, Tokyo
Takao Aoki	National Research Institute for Metals, Tokyo
Shinsaku Matsuyama	Railway Technical Research Institute, Tokyo
Jiro Hoshino	Nippon Kaiji Kyokai (NK Register of Shipping), Tokyo
Munemitsu Fukagawa	Research Institute, Ishikawajima-Harima Heavy Industries, Corporation, Ltd., Tokyo
Yoshifumi Nakano	Research Laboratories, Kawasaki Steel Corporation, Chiba
Tomomi Murata	R & D Laboratories 1, Central R & D Bureau, Nippon Steel Corporation, Kawasaki
Masayuki Tanimura	Technical Research Center, Nippon Kokan K.K., Kawasaki
Mitsuru Arii	Toshiba Corporation, Tokyo
Takeo Yokobori	Faculty of Engineering, Tohoku University, Sendai
Tadashi Kawasaki	Faculty of Engineering, Tohoku University, Sendai
Hiraku Yamazaki	Government Industrial Research Institute, Tohoku, Sendai
Juro Watanabe	Research & Development Division, Japan Steel Works, Ltd., Tokyo

Tentative Test Procedure

Scope

The method covers the determination of the threshold stress-intensity factor K_{ISCC} and the crack growth rate da/dt for metallic materials subjected to static loading in deleterious environments. The testing procedure in a 3.5% sodium chloride solution with WOL type compact tension specimens and CB specimens are detailed herein, but the testing procedures in other environments will be referred to this document.

Test Environment

The 3.5% sodium chloride solution will be used as the test environment. The solution shall be prepared by dissolving reagent-grade sodium chloride in ASTM D-1193-70, Type II distilled water to a concentration of 36.3 g of sodium chloride in each 1.0 L of distilled water.

In WOL tests, the 1 to 2 L solution per 1T-WOL specimen is required where the solution level shall be maintained to at least 6 mm above the tip of the fatigue crack, whereas in CB tests, the 0.5 L solution per 25.4 by 25.4-mm CB specimen is required where the full immersion of the test section should be insured. The solution temperature shall be aimed at 25 ± 3°C, and the solution pH shall be 5.0 to 7.0 during the test.

Specimen Configuration, Dimensions, and Preparation

The recommended WOL and CB specimens are shown in Figs. 1 and 2. If the validity criteria are met, the other alternate specimens may be used. The use of side grooves to ensure straight

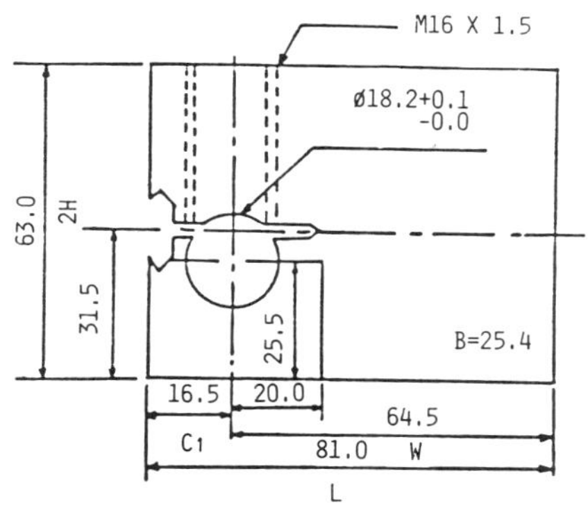

FIG. 1—*Bolt-loaded WOL specimen.*

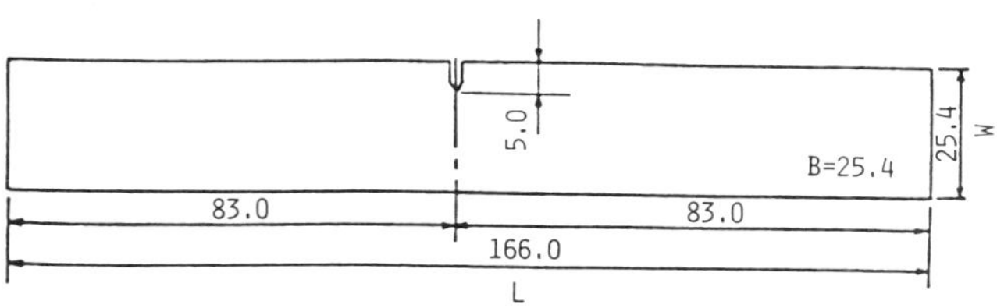

FIG. 2—*Cantilever beam specimen.*

crack fronts during crack growth is allowed. The 10 to 20% semicircular side grooves to original thickness are recommended which should be machined after fatigue precracking.

All specimens shall be precracked in fatigue under cyclic loading. The maximum stress-intensity factor K_{max} for the final 1.3 mm of fatigue crack extension shall not exceed 0.8 K_{IO} with $0.05 \leqq R \leqq 0.20$, where K_{IO} is the stress-intensity factor at the start of the test and R is the stress-intensity factor ratio (K_{min}/K_{max}). The stress-intensity factor K of a WOL specimen can be calculated from Eq 1

$$K = \frac{P\sqrt{a}}{BW}[30.96 - 195.8(a/W) + 730.6(a/W)^2 - 1186.3(a/W)^3 + 754.6(a/W)^4] \quad (1)$$

where

P = applied load,
a = crack length,
B = specimen thickness (for side grooved specimens, B is replaced with $\sqrt{BB_N}$, where B_N = net specimen thickness), and
W = specimen width.

For cantilever beam specimens, the stress-intensity factor K is given by Eq 2

$$K = \frac{6M\sqrt{a}}{BW^2}[1.99 - 2.47(a/W) + 12.97(a/W)^2 - 23.17(a/W)^3 + 24.80(a/W)^4] \quad (2)$$

where M is bending moment at the notched section.

The initial crack length a_0 must be between 0.35 and 0.37 W for a WOL specimen and 0.35 and 0.40 W for a CB specimen. The difference between the fatigue cracks on both surfaces of a specimen shall not exceed 0.035 W.

Procedure

WOL Test

It is recommended that at least three specimens be tested at any different K_{IO} values because the final crack length a_f must lie between 0.45 and 0.80 W.

1. Using the average crack length a_0 and specimen width W, the initial crack mouth opening displacement V_0 corresponding to the specified K_{IO} value is calculated from

$$V_0 = \left(\frac{K_{IO}}{E}\right)\sqrt{a_0}\left[\frac{C_6(a_0/W)}{C_3(a_0/W)}\right] \quad (3)$$

$$C_3(a/W) = 30.96(a/W) - 195.8(a/W)^2 + 730.6(a/W)^3 - 1186.3(a/W)^4 + 754.6(a/W)^5$$

$$C_6(a/W) = \exp[4.495 - 16.130(a/W) + 63.838(a/W)^2 - 89.125(a/W)^3 + 46.815(a/W)^4]$$

2. The plastic tape is affixed to the back face of the specimen to protect the reference surface from corrosion.
3. The specimen is bolt-loaded to the calculated crack mouth opening displacement V_0 using a calibrated clip gage and then placed into the test solution.
4. The visual measurements of crack length shall be conducted at any time intervals during the test. The average crack length a_i is to be determined from measurements of the uncracked ligament at each specimen surface.

5. Overall test duration shall be 4000 to 7000 h. The tests shall be terminated when the crack growth rate becomes less than 10^{-4} mm/h.
6. At the test termination, the change of crack mouth opening displacement ΔV_t from the loaded to the unloaded states is measured with a clip gage by unloading the specimen in a vise. The applied bolt load at test termination P_f is determined by reloading the test specimen up to the displacement equal to the measured ΔV_t with a tension test machine, where the load versus crack mouth opening displacement curve must be recorded.
7. The final average crack length a_f of five measurements is determined from the fracture surface.

CB Test

In the cantilever-beam technique, at least six specimens are required to determine the valid K_{ISCC} below which specimens do not exhibit any subcritical crack extension in a given period of time.

1. Using the average crack length a_0, thickness B, specimen width W, and moment M, the load required to produce the desired K_{IO} is calculated from Eq 2.
2. Test specimen is placed into the test solution so that the test section is fully immersed.
3. The test specimens are deadweight loaded by gradually lowering the load pan with a jack, where the loading rate is limited within the range of 31 to 155 MPa$\sqrt{\text{m}}$/min in terms of $\dot{K}$.
4. Each specimen shall be tested to failure or to 1000 to 4000 h. If failure does not occur after 1000 to 4000 h, the specimen may be retested at higher K_{IO} levels after resharpening the crack by fatigue.
5. At the test termination, the fatigue precrack length a_0 and the final stress corrosion crack length a_f are measured as the average of five measurements from the fracture surface.

Data Analysis

In WOL specimens, the applied bolt load P_i decreases as the crack length a_i increases. In a rigid-bolt analysis which is based on the assumption that a constant crack opening displacement at the bolt loading line is maintained throughout the test, the applied load P_i can be calculated as

$$P_i = \left[\left(\frac{a_0}{a_i}\right)\left(\frac{a_i + C_1}{a_0 + C_1}\right)\right] \frac{EBV_0}{C_6\left(\dfrac{a_i}{W}\right)} \tag{4}$$

where C_1 is distance defined as in Fig. 1, 16.5 mm. When an elastic-bolt analysis considering the elastic relaxation of the bolt with increasing crack length is used, the true applied bolt load P_{EB} in the specimen at any time can be calculated using Eq 5 [2].

$$\frac{P_{EB}}{P_i} = \frac{1 + \dfrac{a_0 + C_1}{a_0}\dfrac{L_B B}{A_b}\dfrac{1}{C_{60}}}{1 + \dfrac{a_i + C_1}{a_i}\dfrac{L_B B}{A_b}\dfrac{1}{C_{6i}}} \tag{5}$$

where

L_B = effective free length of loading bolt and tup (16.5 mm for 1T-WOL specimens) and,
A_b = cross-sectional area of the bolt.

The stress-intensity factor K based on the elastic-bolt analysis can be obtained by substituting P_{EB} into Eq 1. The crack growth rate da/dt is to be reported along with the corresponding value of K obtained by this elastic bolt analysis, where the crack growth rate shall be determined from a curve obtained by plotting the average crack length a_i versus time relation on a square type graph paper.

The stress-intensity factor K_{If} at test termination can be obtained by substituting the measured bolt load P_f or the calculated bolt load P_{EB} into Eq 1. When the K_{If} value is satisfied with the following specimen size and crack length criteria

$$B, a \geqq 2.5\left(\frac{K_{If}}{\sigma y}\right)^2 \tag{6}$$

$$0.45 \leqq a_f/W \leqq 0.80 \tag{7}$$

$$a_{max} - a_{min} \leqq 0.07a_f \tag{8}$$

the K_{If} value is defined as the threshold stress-intensity factor K_{ISCC} of the material.

In CB specimens, the initial stress-intensity factor K_{IO} will be calculated from Eq 2 using the average initial crack length a_0 measured from the fracture surface. From the K_{IO} versus time to failure relation plotted on a graph paper, the threshold stress-intensity factor K_{ISCC} shall be estimated. When the estimated K_{ISCC} value is satisfied with the above Eqs 6 through 8, the estimated K_{ISCC} value is judged to be valid.

Round Robin Test Program

In June 1981, the second round robin test program was initiated after the analysis of the results obtained by the first round robin test program. Twenty-four laboratories participated in this program, where twenty-two laboratories carried out WOL tests and six laboratories performed CB tests.

Material and Test Procedure

The material used in the round robin test program is a 28-mm-thick and 85-mm-wide plate of JIS SNCM 439 (AISI 4340) steel. To provide the materials with two strength levels, the different heat treatments were conducted. Tables 2 and 3 present the chemical composition and the tensile and fracture toughness properties of the materials, respectively. The yield strength of the materials is 1168 MPa for Steel A and 1241 MPa for Steel B.

TABLE 2—*Chemical composition (%) of the materials (check analysis).*

Material	Composition, %
Carbon	0.40
Silicon	0.25
Manganese	0.66
Phosphorus	0.009
Sulfur	0.002
Copper	0.02
Nickel	1.82
Chromium	0.82
Molybdenum	0.25

TABLE 3—*Tensile and fracture toughness properties of the materials.*

Material	Yield Strength, MPa	Tensile Strength, MPa	Elongation, %	Reduction of Area, %	Fracture Toughness K_{Ic}, MPa$\sqrt{m}$ (24°C)
Steel A	1168	1260	18.0	56.1	127
Steel B	1241	1339	16.9	54.6	116

The bolt loaded WOL and CB specimens with no side grooves were machined according to Figs. 1 and 2, where the specimen thickness is 25.4 mm and the specimen orientation was L-T per the ASTM Test Method for Plane-Strain Fracture Toughness of Metallic Materials (E 399-83). Twenty-two laboratories were provided with four WOL specimens, and six laboratories were provided with five CB specimens for each material.

Table 4 shows the initial stress-intensity factor K_{IO} in WOL and CB test programs which were determined after the preliminary tests by the Committee. The WOL tests were performed with two specimens for each K_{IO} where the two levels of K_{IO} were selected for each materials, whereas the CB tests were carried out with five specimens which were tested at the two specified K_{IO} and the three suitable K_{IO} selected by each laboratory.

Each participant was required to perform the following steps in accordance with the tentative K_{ISCC} test procedure:

1. specimen machining,
2. specimen precracking,
3. side grooving,
4. specimen measurement,
5. solution preparation,
6. test,
7. solution temperature and pH measurements,
8. crack length measurement,
9. opening of tested specimen,
10. final crack length measurement,
11. data analysis, and
12. report of individual measurements and calculations.

TABLE 4—*Initial stress intensity factor* K_{IO}.

		WOL Test		CB Test	
Material	Yield Strength, MPa	K_{IO}, MPa$\sqrt{m}$	Number of Specimen	K_{IO}, MPa$\sqrt{m}$	Number of Specimen
Steel A	1168	68.2 83.7	2 2	43.4 52.7 <65.1	1 1 3
Steel B	1241	37.2 49.6	2 2	27.9 37.2 <49.6	1 1 3

Experimental Results and Discussion

Results of WOL Test

Twenty-two participating laboratories reported the results of WOL tests. Figures 3 and 4 show the crack length versus time records for each initial K_{IO} of Steel A and B materials. At the elapsed time of 4000 h, the test time was extended from 4000 h to more than 7000 h, because the majority of the specimens showed no crack arrest, resulting in the test duration ranging from 4200 to 15 000 h. These figures demonstrate that the incubation time of each specimen is greatly different, which may lead to the different crack growth behavior, although a little difference between the actual and specified K_{IO} was observed. Test results also indicate that the data

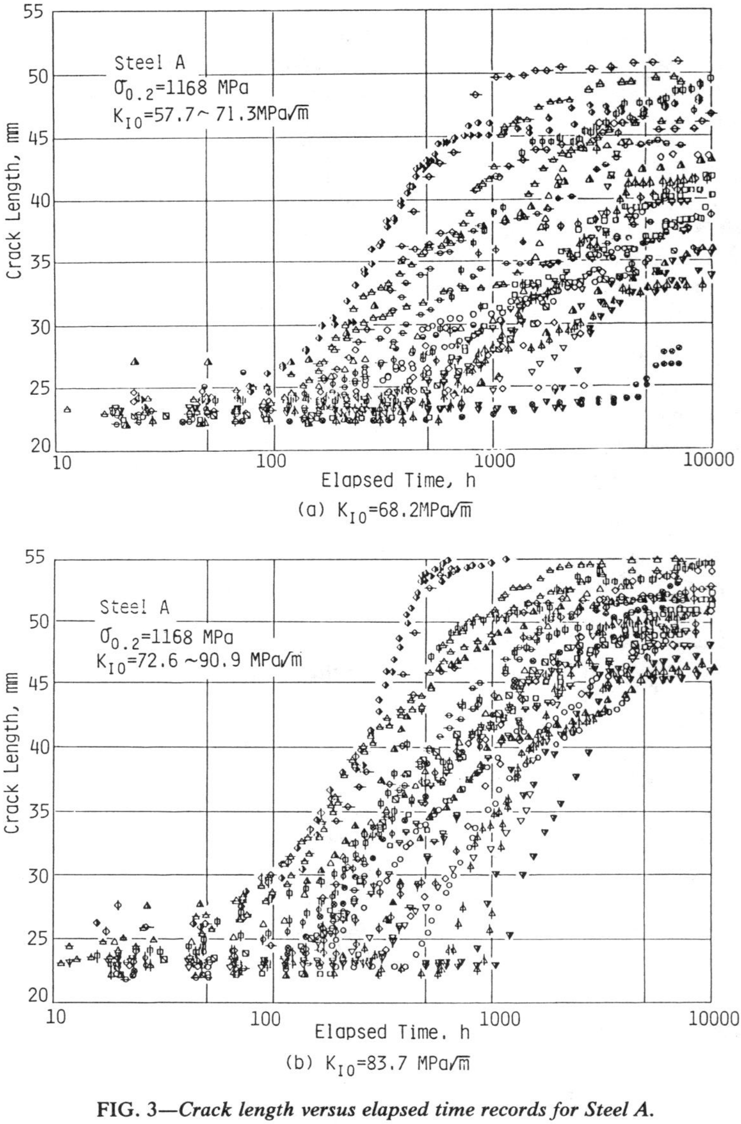

FIG. 3—*Crack length versus elapsed time records for Steel A.*

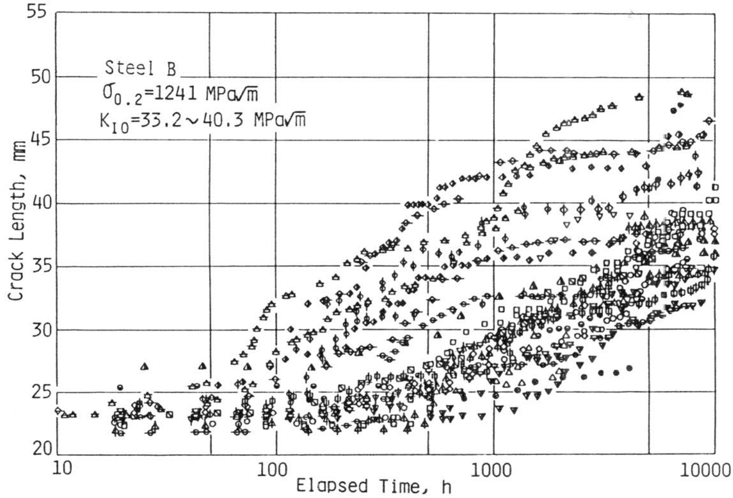

(a) K_{IO}=37.2 MPa√m

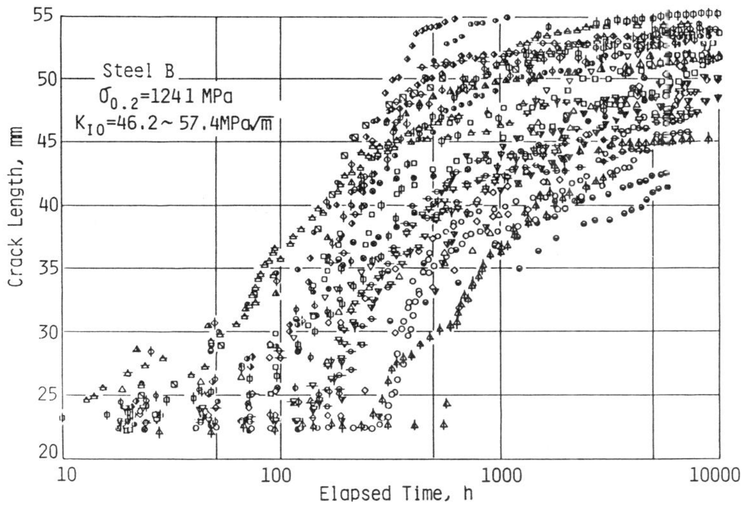

(b) K_{IO}=49.6 MPa√m

FIG. 4—*Crack length versus elapsed time records for Steel B.*

reported by some of the laboratories show the long or the short periods of incubation experienced for all specimens tested within a laboratory, where the same symbols show the data obtained by the same laboratories.

Figure 5 shows the stress corrosion crack growth rate da/dt obtained with the data shown in Figs. 3 and 4 as a function of stress-intensity factor K based on a rigid-bolt analysis. A great deal of scatter of crack growth rate is clearly observed, in particular for the Steel A material with lower yield strength which may be caused by that the K_{ISCC} value is situated at the transition region of the K_{ISCC} versus yield strength relation [3]. The higher or the lower crack growth rate for all specimens tested within a laboratory was also observed.

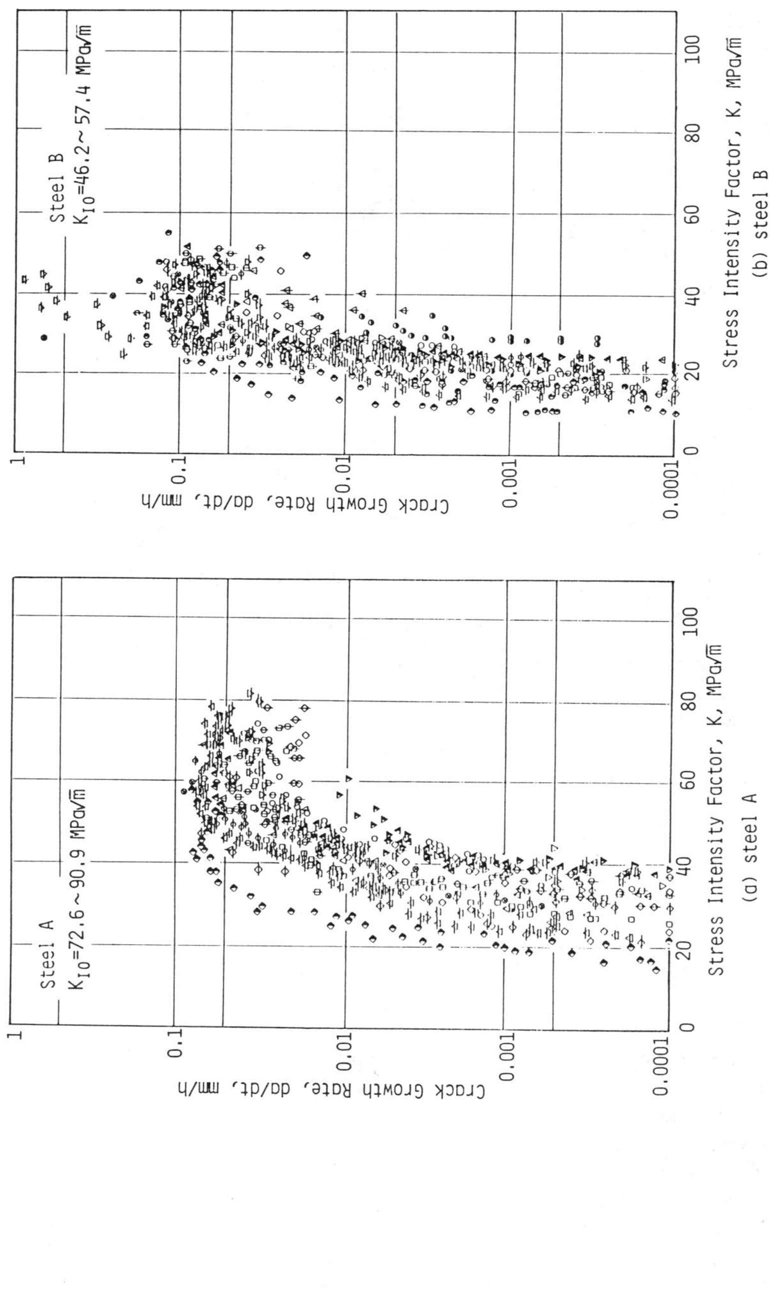

FIG. 5—*Crack growth rate as a function of* K *from a rigid-bolt analysis.*

Figure 6 presents the bar charts of final stress-intensity factor K_{If} at 7000 h for Steel A obtained from both of a rigid-bolt analysis and an elastic-bolt analysis. The K_{If} values of a rigid-bolt analysis are about 3 MPa$\sqrt{m}$ lower than those of an elastic-bolt analysis, which can be explained by that the value of P_{EB}/P_i in Eq 5 is generally larger than one because $a_i > a_0$ and $C_{6i} > C_{60}$. The difference of the K_{If} values from both analyses becomes larger as the crack length increases. This suggests that the elastic-bolt analysis considering the elastic bolt relaxation with increasing crack length is recommended for the evaluation of K_{If} and the da/dt versus K relation.

Figure 7 shows the K_{If} values at the elapsed time of 4000, 7000, and 10 000 h obtained from

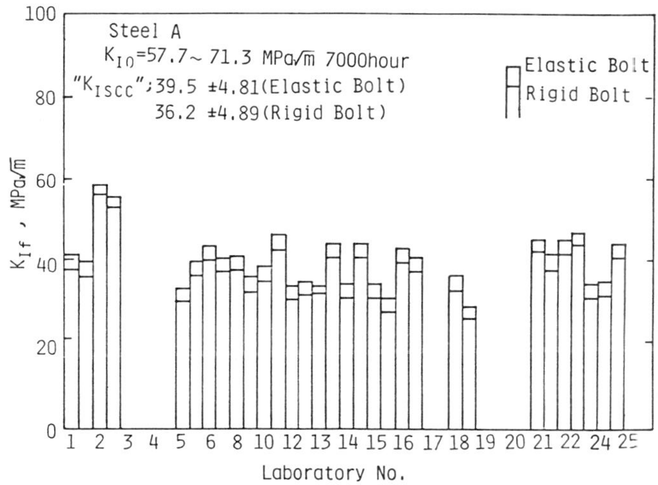

FIG. 6—K_{If} *at 7000 h for Steel A obtained from both of a rigid-bolt analysis and an elastic-bolt analysis.*

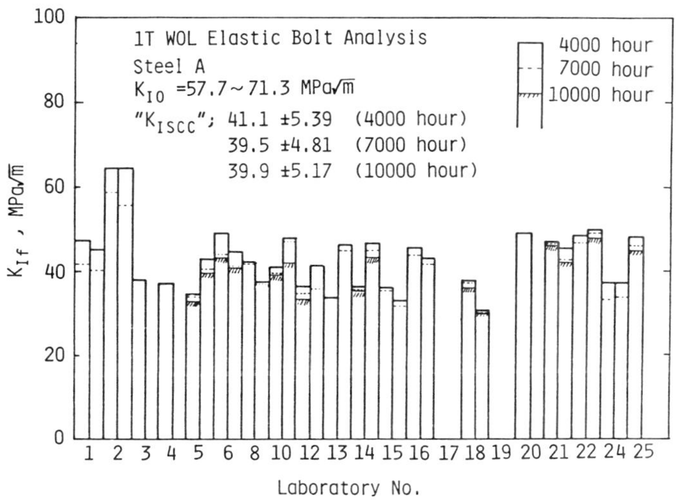

FIG. 7—K_{If} *at the elapsed time of 4000, 7000, and 10 000 h obtained from an elastic-bolt analysis for Steel A.*

an elastic-bolt analysis for Steel A. Even after 4000 h specified in the original program instructions, the crack growth is clearly observed in all specimens. This means that test duration much longer than 4000 h is required to obtain the valid K_{ISCC} values, especially in the lower K_{IO} levels, because the lower K_{IO} level results in longer incubation time and lower crack growth rate.

Figure 8 shows the comparison of the K_{If} values calculated by the elastic-bolt analysis using both of the average crack length of five measurements and the surface crack length. Some of the specimens indicate the large difference between the K_{If} values obtained using both type of crack length, which may be caused by the crack front curvature of a specimen as shown in Fig. 9.

Figures 10 and 11 present the comparison of the K_{If} values obtained from both of the elastic-bolt analysis and the final-load analysis using the average crack length at the test termination for all specimens. Some of the specimens show a large difference between the K_{If} values, although the K_{If} values from both analyses should be the same when the accuracy of initial stress-intensity factor K_{IO} and final load P_f are assured. The large variance of K_{If} is also observed for those obtained from the final-load analysis. These suggest that the much more requirements should be provided for the initial bolt loading procedure and the final load measurement at test termination. These figures also show that the specimens with $a_f/W > 0.85$ give a significant decrease of K_{If} and that the smaller K_{IO} leads to a little larger K_{If}, in particular for the Steel A with lower yield strength.

Variance of K_{If} *for WOL Specimens*

The sources for variance of K_{If} values for WOL specimens were examined using the data obtained from the elastic-bolt analysis with the average crack length at test termination.

Figure 12 shows the relationship between the K_{If} values and the yield strength measured with the 5-mm-diameter tension specimens sampled from its corresponding WOL specimens. The plate-to-plate variations in yield strength is small, and to find the effect of yield strength variation on the variance of K_{If} is very difficult.

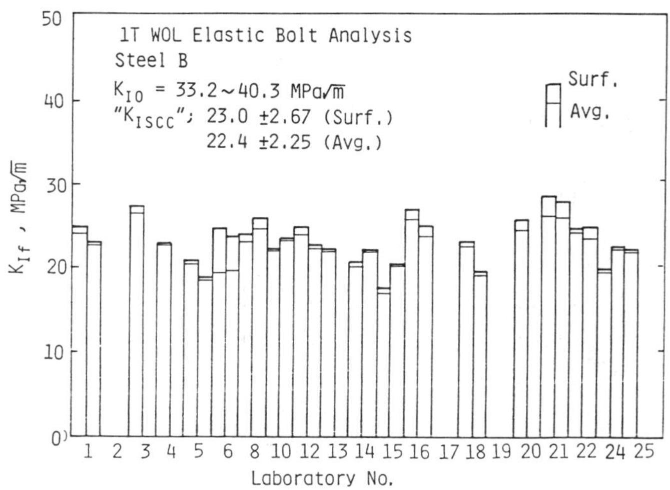

FIG. 8—*Comparison of* K_{If} *values calculated by an elastic-bolt analysis using average crack length and surface crack length.*

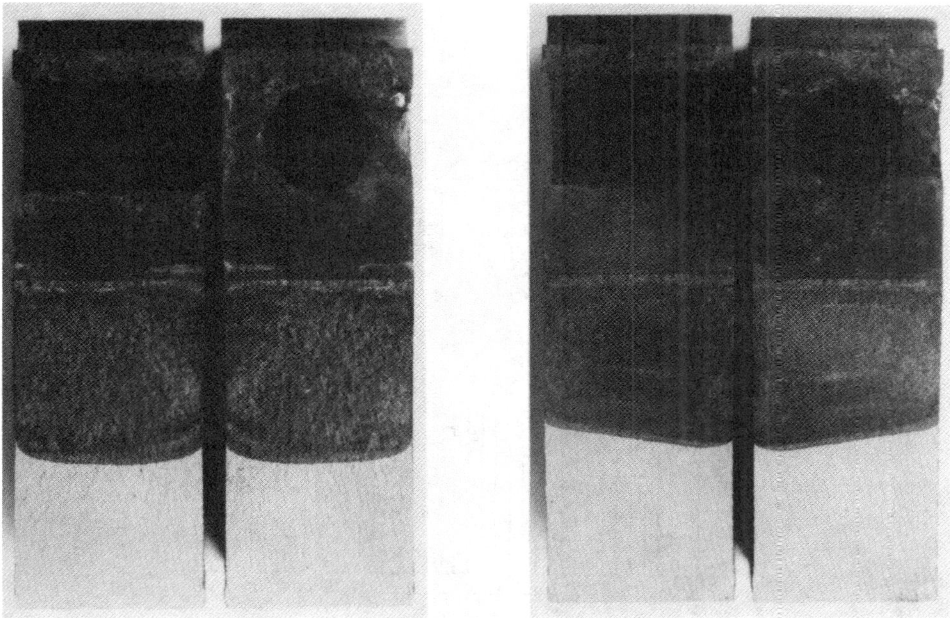

FIG. 9—*Fracture surfaces of WOL specimens.*

Figure 13 shows the effect of crack front curvature, where the K_{If} values were plotted as a function of $(a_{max} - a_{min})/B$ obtained from five measurements. A slight increase of K_{If} with increasing $(a_{max} - a_{min})/B$ is observed, which suggests that the validity criteria on crack front curvature may be needed.

Figures 14 and 15 present the effect of solution temperature and pH on the variance of K_{If}. The solution temperature ranges from 7 to 29°C, and the solution pH ranges from 3.8 to 9.5 for all specimens tested. From the figures, it is very difficult to find the effect of solution temperature and pH on the variance of K_{If}.

Figure 16 shows an example of the load versus displacement record at the final load measurement. The deviation from a linearity of the load versus displacement curve is clearly observed which is caused by the wedge effect of corrosion products in the specimen. Figure 17 presents the relationship between the reduction of specimen thickness and the elapsed time. Even after the same elapsed time, a wide scatter of the thickness reduction is observed. This means that the solution environment may be slightly different among the participating laboratories.

In conclusion, it is expected that the sources for the variance of K_{If} for WOL specimens may be based on the combined effect of the factors such as material properties, crack front curvature, solution environment, test duration, and measurement accuracy of K_{IO} and P_f.

Statistical Analysis of K_{If} *for WOL Specimens*

Using each K_{If} value from the rigid-bolt analysis, elastic-bolt analysis and final-load analysis at test termination, the variance of the mean for k laboratories (S_m^2), the within-laboratory variance (or repeatability) (S_w^2), and the between-laboratory variance (or reproducibility) (S_b^2) that have been defined by Mandel [4] were calculated using the following equations

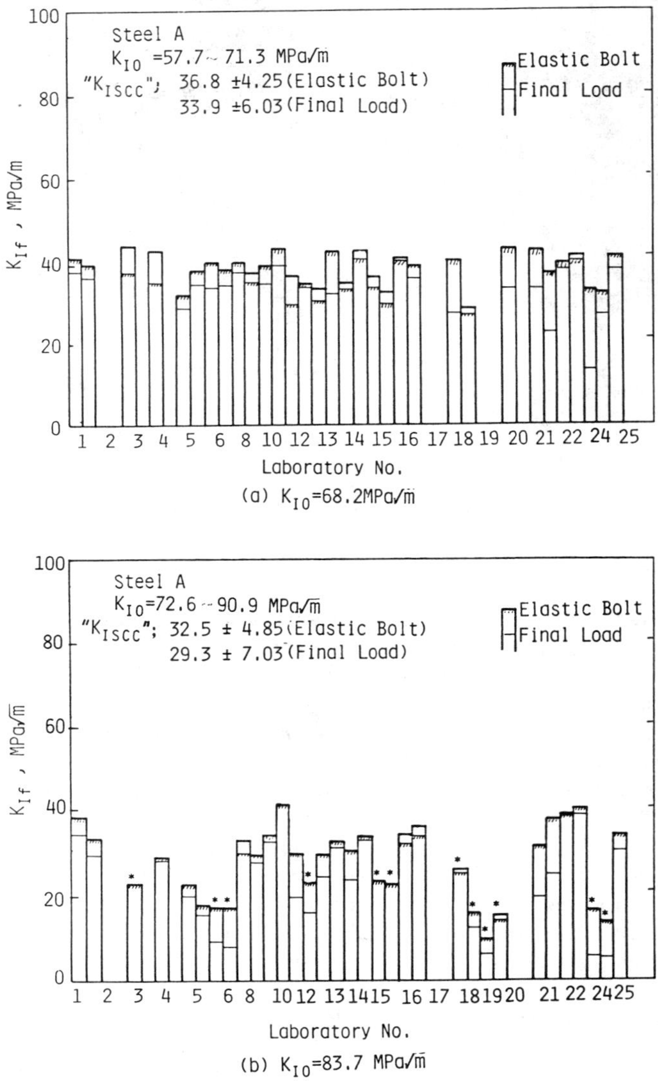

FIG. 10—*Comparison of* K_{If} *values obtained from an elastic-bolt analysis and a final-load analysis for Steel A.*

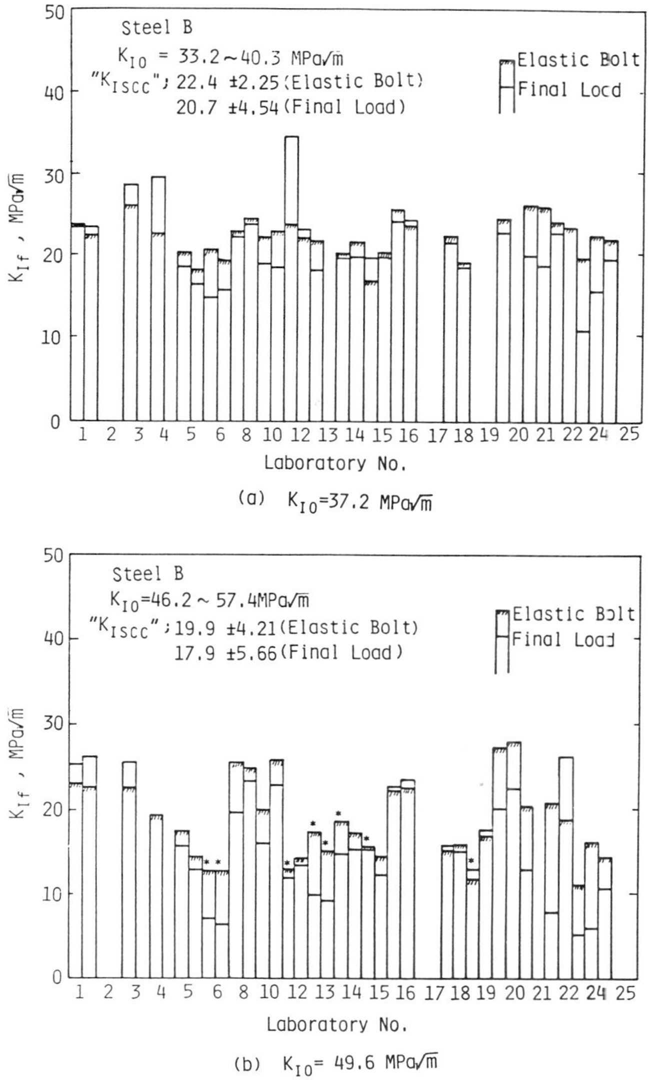

FIG. 11—*Comparison of* K_{If} *values obtained from an elastic-bolt analysis and a final-load analysis for Steel B.*

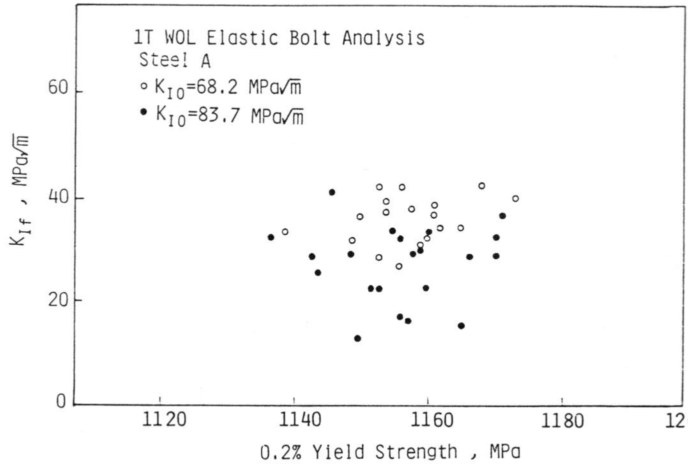

FIG. 12—*Relationship between the* K_{If} *values and the yield strength measured with the tension specimens sampled from its corresponding WOL specimens.*

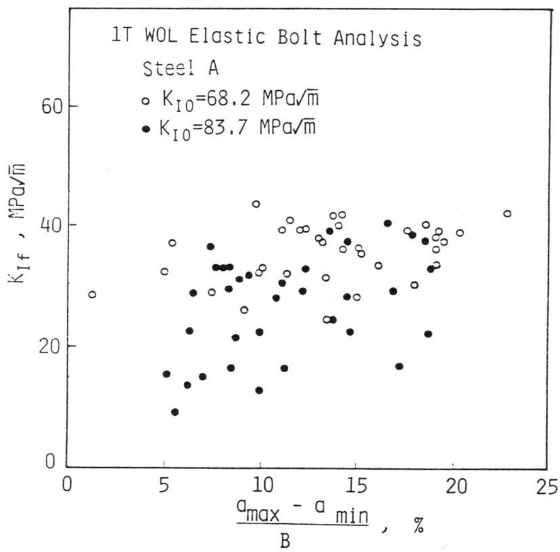

FIG. 13—*Relationship between the* K_{If} *values and the crack front curvature* $(a_{max} - a_{min})$/B *of WOL specimens.*

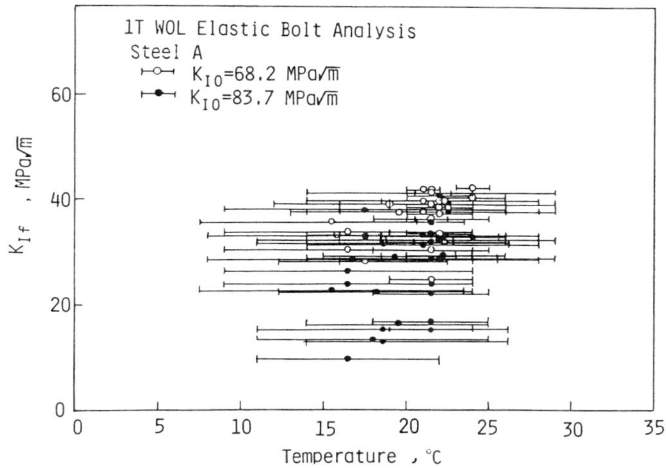

FIG. 14—*Effect of solution temperature on the variance of* K_{If} *of WOL specimens.*

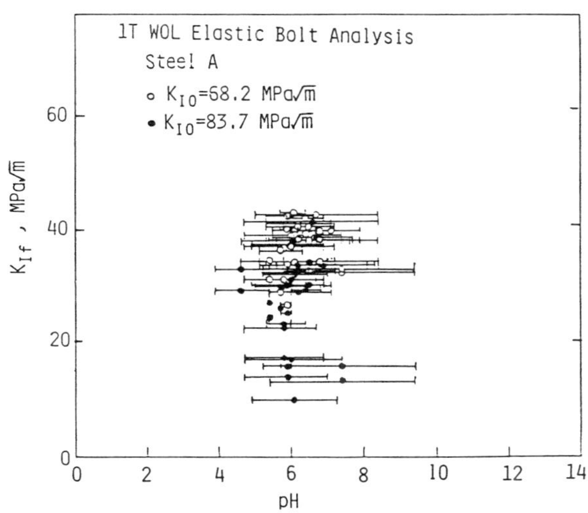

FIG. 15—*Effect of solution pH on the variance of* K_{If} *of WOL specimens.*

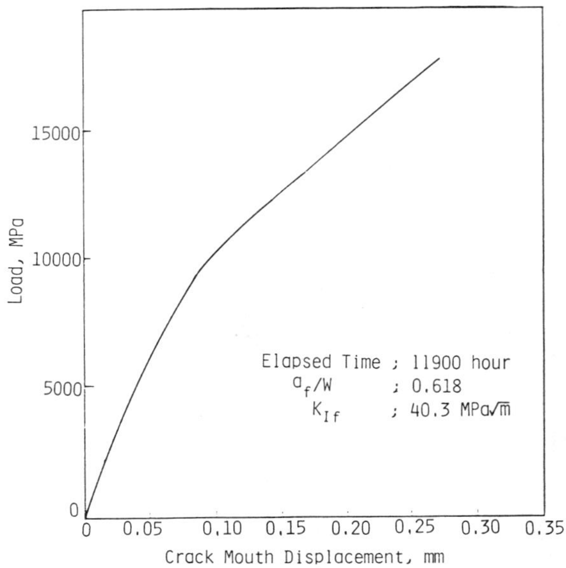

FIG. 16—*An example of the load versus displacement record at the final load measurement of a WOL specimen.*

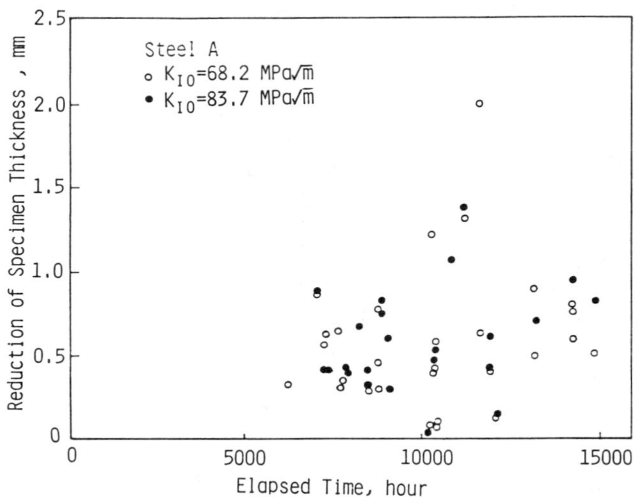

FIG. 17—*Relationship between the reduction of specimen thickness and the elapsed time for WOL specimens.*

TABLE 5—*Statistical analysis of K_{If} values obtained with WOL specimens.*

Material	Yield Strength, MPa	K_{IO}, MPa$\sqrt{m}$	Test Duration, h	Analysis Method	Number of Laboratory	Number of Specimen	$a_{surface}$: K_{If}, MPa$\sqrt{m}$	$a_{surface}$: Standard Deviation, MPa$\sqrt{m}$: Within *Sw*	Between *Sn*	Total *Sm*	a_{avg}: K_{If}, MPa$\sqrt{m}$	a_{avg}: Standard Deviation, MPa$\sqrt{m}$: Within *Sw*	Between *Sn*	Total *Sm*
Steel A	1168	68.2	4 200	rigid bolt	12	24	34.9	3.34	2.90	3.92	33.0	2.13	3.98	3.53
			↓	elastic bolt	12	24	38.1	3.35	2.82	3.90	36.3	2.09	3.82	3.42
			15 000	final bolt	12	24	28.5	5.21	3.70	3.97	33.3	4.69	4.18	5.55
		83.7	4 200	rigid bolt	10	20	33.3	3.19	3.20	3.92	29.7	3.03	2.80	3.62
			↓	elastic bolt	10	20	37.0	3.41	3.51	4.22	33.2	3.30	3.25	4.02
			15 000	final bolt	9	18	25.2	6.20	3.51	6.68	29.8	6.17	4.19	6.83
Steel B	1241	37.2	4 200	rigid bolt	13	26	21.2	2.35	1.41	2.56	20.2	2.04	1.36	2.25
			↓	elastic bolt	13	26	23.0	2.34	1.38	2.53	22.2	1.67	1.64	2.03
			15 000	final bolt	13	26	18.5	4.28	2.55	4.65	20.2	3.43	2.71	3.93
		49.6	4 200	rigid bolt	9	18	20.1	3.56	1.70	3.76	19.3	5.51	1.63	4.13
			↓	elastic bolt	9	18	22.5	3.76	1.84	3.98	21.4	4.02	1.75	4.20
			15 000	final bolt	8	16	17.9	0.70	4.78	3.45	19.6	1.52	5.05	3.88

$$
\begin{aligned}
S_i^2 &= \frac{1}{(m-1)} \sum_{j=1}^{m} (Y_j - \bar{Y})^2 \\
S_m^2 &= \frac{1}{(k-1)} \sum_{i=1}^{k} (Y_i - \bar{\bar{Y}})^2 \\
S_w^2 &= S_{\text{Rep}}^2 = \frac{1}{k} \sum_{i=1}^{k} S_i^2 \\
S_B^2 &= S_m^2 - \frac{1}{m} S_w^2
\end{aligned}
\tag{9}
$$

where

m = number of replicas,
k = number of laboratories,
Y_j = K_{If} of any specimen,
$\bar{Y}$ = average of K_{If} within-laboratory,
Y_i = average of K_{If} within-laboratory, and
$\bar{\bar{Y}}$ = average of K_{If} between-laboratory.

Table 5 presents the standard deviations for each analysis, where the data with $a/W > 0.8$ and that the actual K_{IO} is larger than $\pm 20\%$ of the specified K_{IO} are eliminated for the calculations. In the case of elastic-bolt analysis using the average crack length, the standard deviation of within-laboratory variance S_w is 2.2 to 3.4 MPa$\sqrt{\text{m}}$ for Steel A and 1.5 to 4.0 MPa$\sqrt{\text{m}}$ for Steel B, and also the standard deviation of between-laboratory variance S_B is 3.1 to 3.7 MPa$\sqrt{\text{m}}$ for Steel A and about 1.5 MPa$\sqrt{\text{m}}$ for Steel B. In both of the within-laboratory variance and between-laboratory variance, that the standard deviation of K_{If} from the final-load analysis is larger than that from the elastic-bolt analysis are demonstrated.

Results of CB Test

Six participating laboratories reported the results of CB tests. Figure 18 shows the relationship between the initial K_{IO} and the elapsed time to failure for Steels A and B. After some of tests, the test duration was extended from 1000 h to more than 4000 h because of the long periods of incubation for the steels. Apparent K_{If} values at 4000 h estimated on the basis of data developed in each laboratory are shown as a bar graph in Fig. 19. The standard deviation of K_{If} at 4000 h is 4.3 MPa$\sqrt{\text{m}}$ for Steel A and 5.5 MPa$\sqrt{\text{m}}$ for Steel B.

Table 6 shows the comparison of K_{If} values of the WOL and CB test specimens, where in the WOL specimens the K_{If} values obtained from the elastic-bolt analysis with the average crack length are used. The K_{If} values obtained with the CB specimens are 7.7 to 9.3 MPa$\sqrt{\text{m}}$ higher than those with the WOL specimens. It is noted that in Fig. 18 many CB specimens produce the lowest value of K_{If} which is very close to the K_{If} values obtained with the WOL specimens. This suggests that at least six specimens are required to obtain the valid K_{If} values in the cantilever beam (CB) tests.

Summary and Conclusions

The objective of the 129th Committee, Japan Society for the Promotion of Science, was to develop a standard test procedure for measuring the threshold stress-intensity factor K_{ISCC} for stress corrosion cracking and the crack growth rate da/dt in metallic materials. The interlaboratory round robin test programs were conducted at first 1977 through 1979 and the second

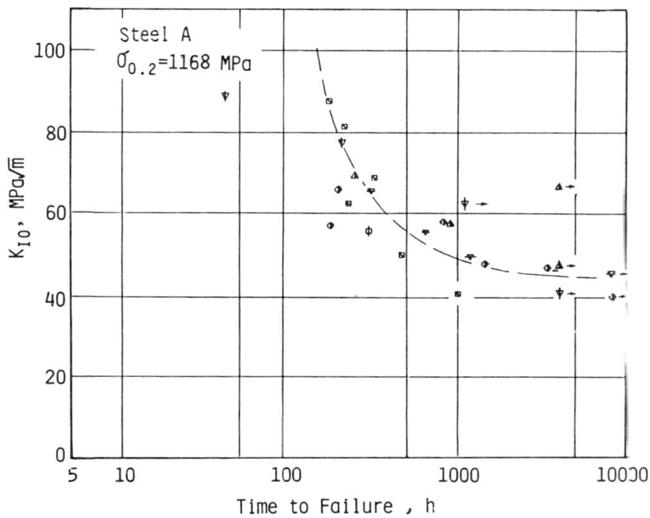

(a) steel A

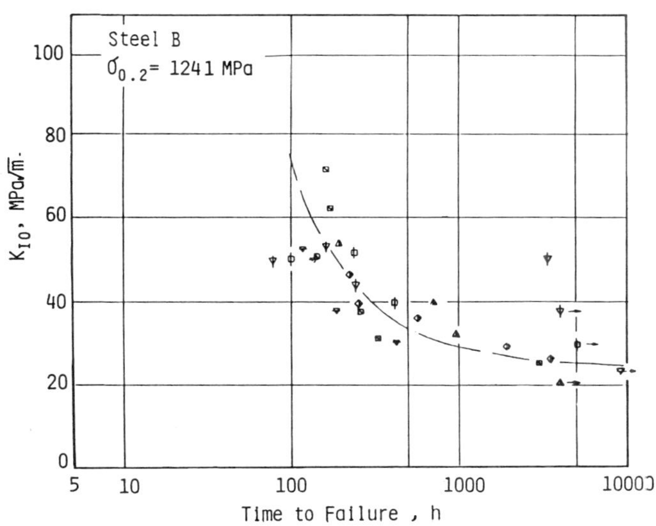

(b) steel B

FIG. 18—*Relationship between the initial* K_{IO} *and the elapsed time to failure for cantilever beam specimens.*

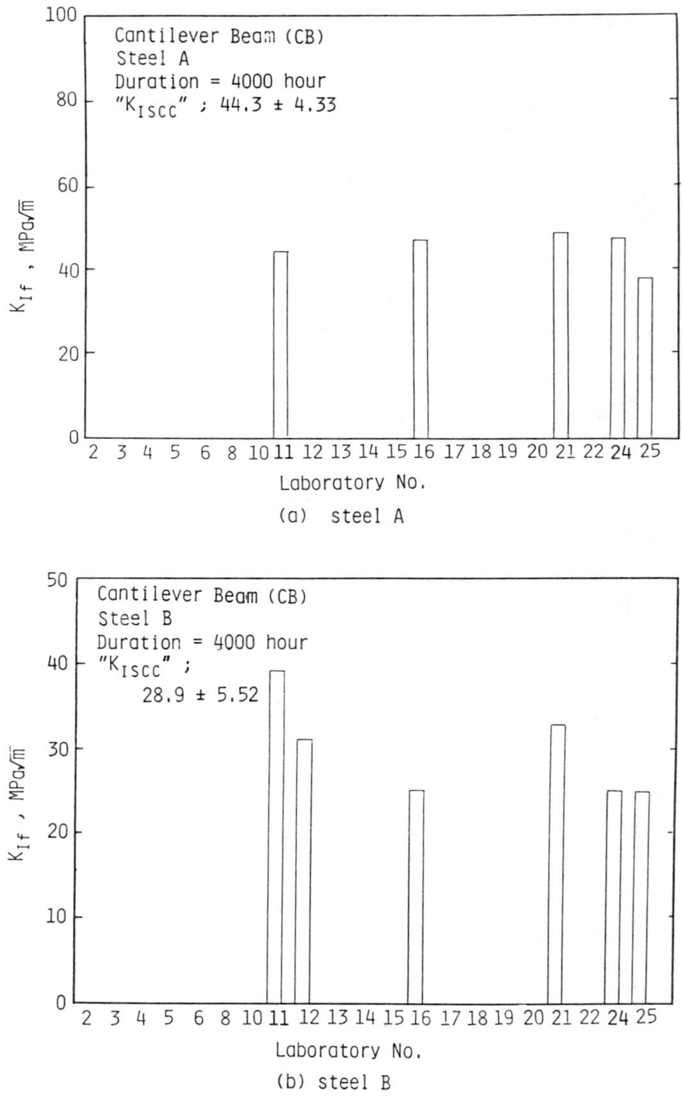

FIG. 19—K_{If} *values at 4000 h estimated on the basis of data in each laboratory for cantilever beam specimens.*

TABLE 6—*Comparison of* K_{If} *values of the WOL and CB test specimens.*

Material	Specimen	K_{IO}, $MPa\sqrt{m}$	Number of Laboratory	Number of Specimen	Test Duration, h	K_{If},[a] $MPa\sqrt{m}$	Standard Deviation,[a] $MPa\sqrt{m}$
Steel A	WOL	68.2	18	30	4 000 ↓ 15 000	36.8	4.2
		83.7	14	23	4 200 ↓ 15 000	32.5	4.9
	CB	—	5	25	4 000	44.3	4.3
Steel B	WOL	37.2	19	32	4 200 ↓ 15 000	22.4	2.3
		49.6	18	27	4 200 ↓ 15 000	20.4	4.2
	CB	—	6	29	4000	28.9	5.5

[a]Elastic-bolt analysis using the average crack length at test termination.

1981 through 1983 using JIS SNCM 439 (AISI 4340) steels. Thirty-one laboratories participated in these programs. The tests were carried out with both 25.4-mm (1-in.)-thick bolt loaded WOL specimens and 25.4 by 25.4-mm CB specimens in a 3.5% sodium chloride solution.

The analysis of the test results showed that the tentative test procedure is an adequate document to guide the K_{ISCC} testing, although some of the revisions and additions such as specimen preparation and testing criteria were conducted in the original document.

The results obtained in the recently completed second interlaboratory round robin test program to develop a standard test procedure are summarized as follows:

1. The test duration is an important valuable because the incubation time up to the onset of crack growth is greatly dependent on the initial K_{IO} and the materials.
2. The test duration of more than 7000 h for WOL specimens and 4000 h for CB specimens were recommended.
3. The elastic-bolt analysis was recommended for the evaluation of K_{If} and the da/dt versus K relation of WOL specimens because the K values from a rigid-bolt analysis give the lower values than those from an elastic-bolt analysis.
4. A slight effect of the initial K_{IO} on the apparent K_{If} of WOL specimens was observed.
5. Many more requirements or limits for the initial bolt loading procedure and the final load measurement of WOL specimens are needed to reduce the variance of K_{If} at test termination.
6. The validity criteria of the final crack length $a_f/W < 0.80$ is required for WOL specimens.
7. The sources for the variance of K_{If} for WOL specimens may be based on the combined effect of the factors such as material properties, crack front curvature, solution environment, test duration, and measurement accuracy of K_{IO} and P_f.
8. In the CB tests, at least six specimens are required to obtain the valid K_{If} values.

The results mentioned above support the findings given in the ASTM report [*1*] prepared by Wei and Novak for E24/G1 Task Group. The objective of the round robin test program was clearly met. Based upon the results of the interlaboratory round robin test program, a standard test procedure of the 129th Committee, Japan Society for the Promotion of Science entitled "Standard Test Method for Threshold Stress Intensity Factor K_{ISCC} in Stress Corrosion Cracking" was established.

Acknowledgments

On behalf of the 129th Committee, Japan Society for the Promotion of Science, one of the authors (Prof. T. Yokobori) expresses sincere thanks to the kind cooperation on this work given by E24.04.02 Task Group, ASTM. The work in the present paper is patterned after the E24/G1 Round Robin Test Program in which the Japanese participated.

References

[1] Wei, R. P. and Novak, S. R., in *Environment-Sensitive Fracture: Evaluation and Comparisons of Test Methods, ASTM STP 821,* S. W. Dean, E. N. Pugh, and G. M. Ugiansky, Eds., American Society for Testing and Materials, Philadelphia, 1984, pp. 75–79.

[2] Novak, S. R. and Rolfe, S. T., *Journal of Materials,* Vol. 4, No. 3, 1969, pp. 701–728.

[3] Iwadate, T., Tanaka, Y., Kabutomori, T., Kanno, N., and Nakamura, M., Preprint, 16th Annual Offshore Technology Conference, Houston, TX, May 1984, pp. 21–30.

[4] Mandel, J., *Standardization News,* American Society for Testing and Materials, Philadelphia, Vol. 5, No. 3, 1977, p. 17.

Carl E. Jaske[1]

Long-Term Creep-Crack Growth Behavior of Type 316 Stainless Steel

REFERENCE: Jaske, C. E., "**Long-Term Creep-Crack Growth Behavior of Type 316 Stainless Steel,**" *Fracture Mechanics: Eighteenth Symposium, ASTM STP 945,* D. T. Read and R. P. Reed, Eds., American Society for Testing and Materials, Philadelphia, 1988, pp. 867–877.

ABSTRACT: For most engineering applications, crack-growth rates of 10^{-9} m/s or less are of practical interest, but this is the regime where little creep-crack growth data currently are available. Long-term creep-crack growth experiments of about 1000 h or more duration are required to develop such data. In the present study, creep-crack growth data were developed at rates between 10^{-11} and 10^{-7} m/s for Type 316 stainless steel. Three specimen configurations were used: compact type (CT); center-cracked tension (CCT); and multiple edge-notch tension (MENT). The CT and CCT specimens had long (25 to 35 mm) and the MENT specimens had short (100 to 500 μm) cracks. Creep-crack growth rate was evaluated both in terms of the linear elastic stress intensity factor K and the C*-integral. It was concluded that the C*-integral is useful for characterizing long-term as well as short-term creep-crack growth rate data for both short and long cracks on a consistent basis.

KEY WORDS: creep, fracture mechanics, crack growth, stainless steel, testing, temperature, evaluation

Engineering equipment that operates at elevated temperatures is subjected to loadings that may cause creep damage. Examples of such equipment are steam turbines, steam generators, gas turbines, nuclear power plant structural components, rocket engines, and systems for chemical and petroleum processing. This equipment is usually fabricated from metals, and its structural integrity depends on the ability of the material to resist creep damage from loadings experienced during service operations.

Because typical engineering components often contain small flaws or cracks when they are put into service, the development of procedures for quantifying creep-crack propagation is an important consideration. Dealing directly with crack growth is a powerful approach that enables service life to be computed from some initial flaw size to some final critical crack size by integration of the appropriate crack-growth rate relationships. This type of approach involves the use of fracture mechanics. In general, the deformations associated with creep are inelastic; thus, inelastic fracture mechanics (IFM), rather than linear elastic fracture mechanics (LEFM), should be applied to creep-crack growth problems.

For most engineering applications where the possibility of creep damage is of concern, crack-growth rates of 10^{-9} m/s or less are of practical interest. Although a number of recent studies [*1–12*] have dealt with creep-crack growth, except for the work by Shih on Type 304 stainless steel [*11*] and by Saxena et al. on chromium-molybdenum-vanadium (Cr-Mo-V) steel [*12*], most of the data developed in those studies have been limited to growth rates above 10^{-9} m/s. None of those studies investigated the creep-crack growth of short cracks (less than 1 mm). The objective of this study was to develop methods for characterizing long-term creep-crack growth

[1]Senior research scientist, Battelle's Columbus Division, Columbus, OH 43201-2693.

behavior of both long (25 to 35 mm) and short (100 to 500 μm) cracks at rates between 10^{-11} and 10^{-7} m/s under inelastic straining.

Procedure

The experimental work has concentrated on evaluating the creep-crack growth behavior of Type 316 stainless steel in air at 593 and 649°C. The material, specimen preparation techniques, equipment, and crack-growth measurement methods used in this study are described in the following sections.

Material

The material used in this study was a 0.61-m by 0.61-m by 12.7-mm sample from a plate of solution annealed Type 316 stainless steel, Reference Heat 8092297. The processing history and properties of that heat have been documented extensively and characterized by Beaver and Martin [*13*]. A small (about 38 by 22 mm) specimen was given a laboratory solution anneal of 0.5 h at 1066°C, followed by a rapid quench (at a rate greater than 82°C per min) to room temperature. The hardness of 71 Rockwell B, grain size of ASTM 3, microstructure, and chemical composition of the laboratory-annealed specimen were typical of those expected for Type 316 stainless steel [*14*]. The creep and stress-rupture behavior of this heat of Type 316 stainless steel also have been characterized [*14*].

Specimen Preparation

The three specimen configurations shown in Figs. 1, 2, and 3—(1) compact type (CT), (2) center-cracked tension (CCT), and (3) multiple edge-notch tension (MENT)—were used in this study. All specimens were oriented so that the loading axis was parallel to the final rolling direction of the plate; thus, the direction of crack propagation was perpendicular to the final rolling direction. After machining, all specimens were given the laboratory solution annealing heat treatment described previously. An argon atmosphere was used to protect the specimen surface from oxidation.

Equipment

The CT specimens were tested at a constant displacement rate using an electrohydraulic system operated in closed-loop, load-line displacement control. The applied load was measured using a strain-gage-equipped load cell in series with the loading train. The CCT and MENT specimens were tested using a standard creep frame with deadweight loading. The specimens were heated to temperature using a standard laboratory resistance heated furnace. Load-line displacement was measured on the CT and CCT specimens using a high-temperature extensometer equipped with a linear variable differential transformer (LVDT). Axial strain was measured over a 38.1-mm gage length on the MENT specimens using an optical platinum-strip extensometer.

Crack-Growth Measurement

The d-c potential-drop method has been successfully used in measuring creep-crack growth in other studies [*1,12,15–17*], so it was chosen for use with the CT and CCT specimens in the current study. The experimental setup was patterned after that used by Vosikovsky [*18*] because of its simplicity. A constant d-c current of 10 A was supplied to the CT specimens and of

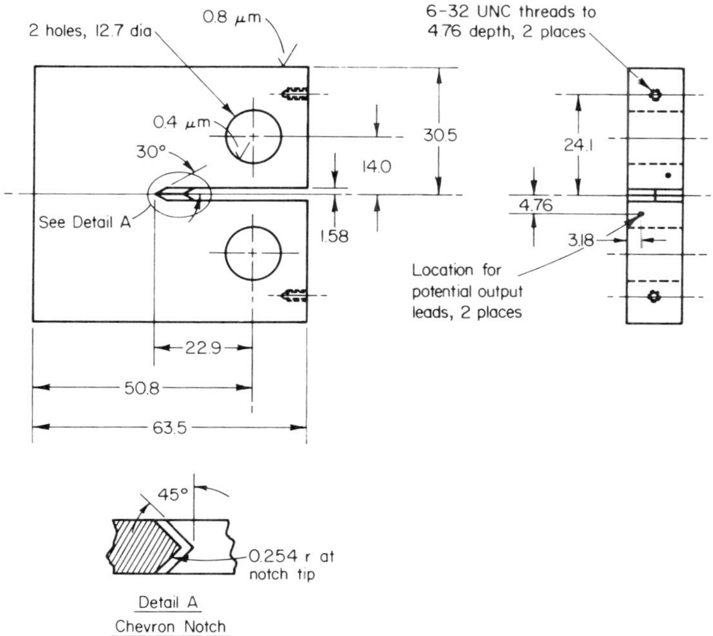

FIG. 1—*Configuration of compact-type (CT) specimen (all dimensions in mm).*

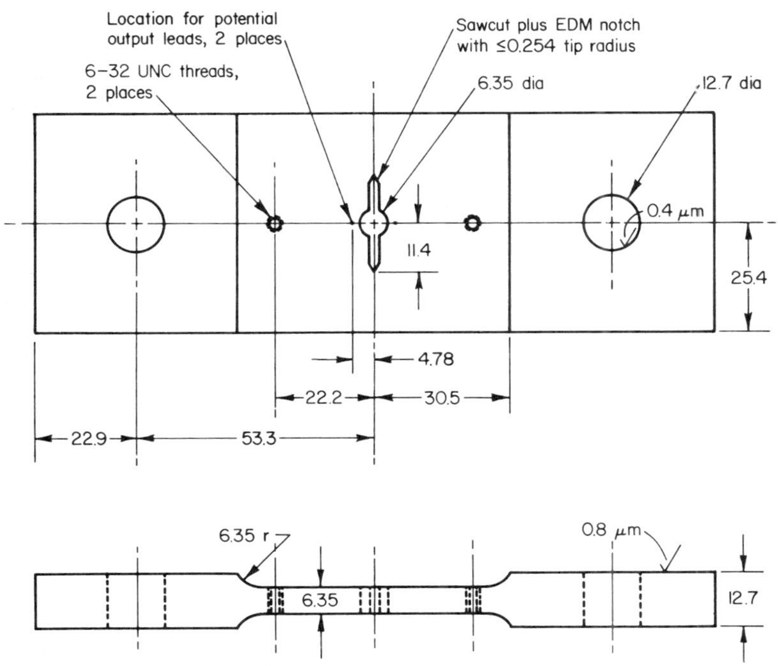

FIG. 2—*Configuration of center-cracked tension (CCT) specimen (all dimensions in mm).*

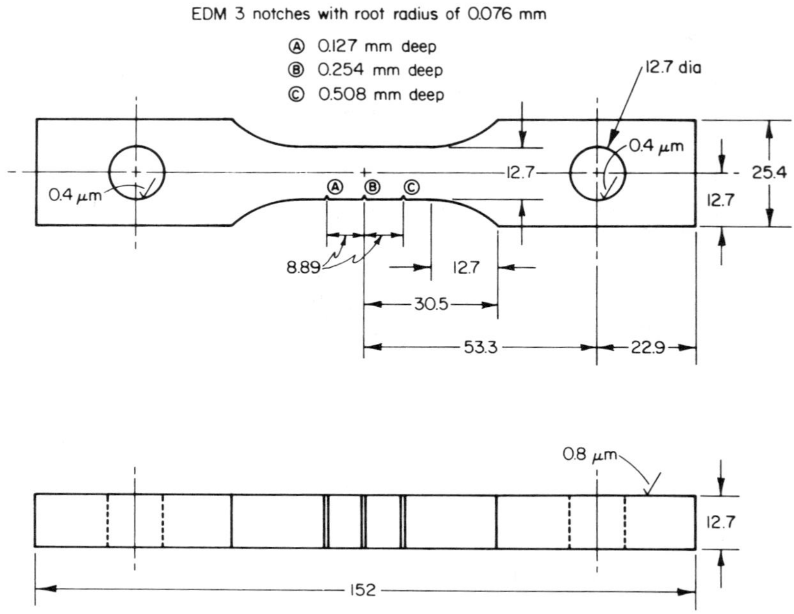

FIG. 3—*Configuration of multiple-edge-notch tension (MENT) specimen (all dimensions in mm).*

20 A to the CCT specimens using a stabilized power supply and 2.0-mm-diameter nickel-chromium alloy lead wires attached directly to the specimens. The potential change was monitored using a commercial nanovoltmeter with 0.5-mm-diameter nickel-chromium alloy lead wires spot welded to the specimens near the crack tip. The system is capable of detecting crack-size changes with a sensitivity of 25 μm and an accuracy of ±50 μm. The change in d-c potential was continuously monitored and recorded using a standard laboratory strip-chart recorder.

For the CT specimens, the potential-drop measurements were calibrated using fractographic measurements of crack length at crack-front-profile "benchmarks" on the specimen fracture surface. For the CCT specimens, calibration was achieved by periodically interrupting the experiment, cooling the specimen to room temperature, and optically measuring the surface-crack length using a traversing microscope. The calibration data for the CT and CCT specimens are presented in Figs. 4 and 5, respectively. The normalized change in crack length for the CT specimens is $\Delta a/(a_0 + 0.25w)$, where Δa is the change in crack length, a_0 is the initial crack length, and w is the specimen width [*16*]. The normalized change in crack length for the CCT specimens is $\Delta a/a_0$. In both cases, $\Delta V/V_0$ is the normalized change in d-c potential, where ΔV is the measured d-c potential change and V_0 is the initial reference d-c potential, given by

$$V_0 = V_T - V_{th} \tag{1}$$

and where V_T and V_{th} are the d-c potential and thermal electromotive force (emf), respectively, at the start of a test.

As is shown in Figs. 4 and 5, the calibration data are quite consistent, reproducible, and in good agreement with Saxena's [*16*] data for Type 304 stainless steel at 593°C. For $\Delta V/V_0$ less than 0.4, Johnson's analytical solution of Ref *19* agreed well with the data for the CCT speci-

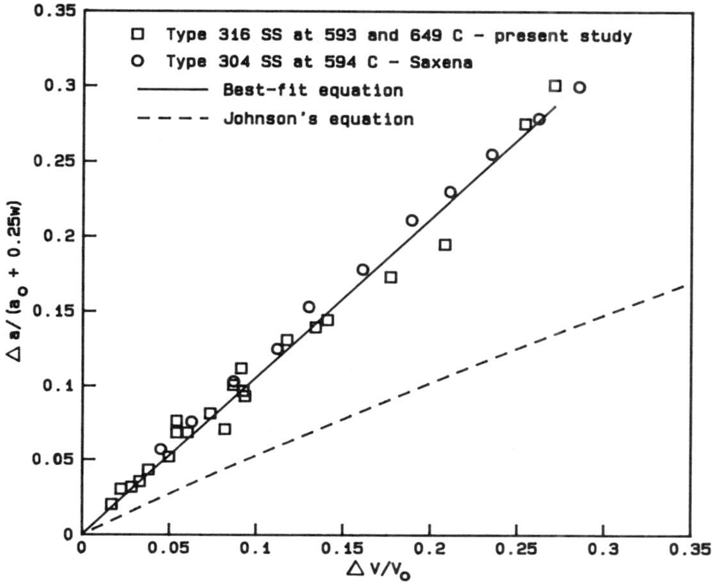

FIG. 4—*D-c potential-drop calibration data for CT specimen.*

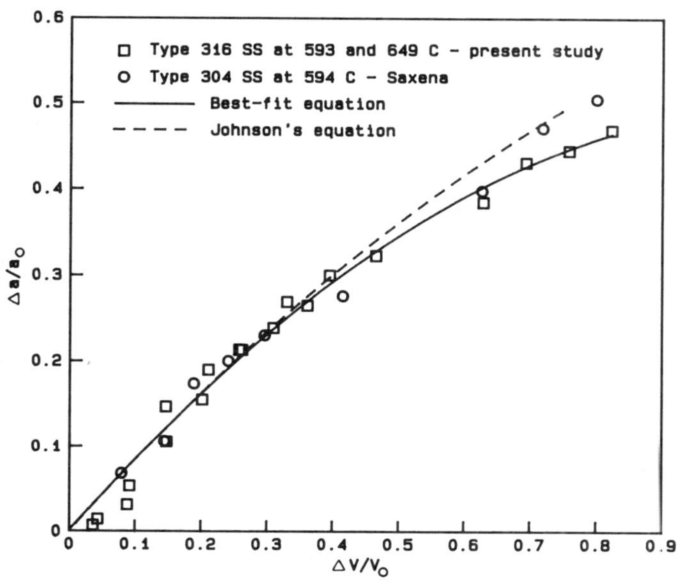

FIG. 5—*D-c potential-drop calibration data for CCT specimen.*

mens, but it fell well below all of the data for the CT specimens. These results for the CT specimens contrast with those reported by Schwalbe and Hellman [20], which showed good agreement between Johnson's equation and their calibration data, because those investigators attached the current leads at the midpoints of the top and bottom faces, rather than to the end-face locations (see Fig. 1) used in the current study. Thus, if Johnson's equation is used to compute crack length, the current leads must be attached at the proper locations.

To facilitate the computation of crack length from d-c potential data, the best-fit calibration curves shown in Figs. 4 and 5 were obtained by linear least-squares regression analysis. The best-fit equation for the CT specimens is

$$\Delta a/(a_0 + 0.25w) = 1.059(\Delta V/V_0) \quad (2)$$

and the best-fit equation for the CCT specimens is

$$\Delta a/a_0 = 0.8908(\Delta V/V_0) - 0.3970(\Delta V/V_0)^2 \quad (3)$$

In all cases, the final crack length for which data were obtained had a $\Delta V/V_0$ value that fell reasonably close to the best-fit calibration curve. Once the initial values of a_0 and V_0 were measured, it was a straightforward process to calculate values of Δa corresponding to the measured values of ΔV using Eqs 2 and 3.

For the MENT specimens, the amount of overall creep-crack growth at each of the three notches was determined by post-test fractography. No crack length measurements were made during the test. The test was stopped when the creep strain began to accelerate before failure occurred. The specimen then was fatigue-cycled at room temperature to help mark the crack fronts and to fracture the specimen at the location of the largest crack. The two smaller cracks were opened by carefully sawing the unbroken ligament. Figure 6 shows an example of the fracture surface of each notch for one of the MENT specimens evaluated at 649°C. The average value of Δa was determined from crack-length measurements made at nine equally spaced points along the crack front.

Experimental Results

The data obtained from the tests of CT and CCT specimens consisted of corresponding sets of load, load-line displacement, and crack length as a function of time. The load-line displacement and crack-length data were analyzed using the procedures described elsewhere [*21*] to obtain values of displacement rate $d\delta/dt$ and crack-growth rate da/dt by either the secant method or the seven-point incremental polynomial method of the ASTM Test Method for Constant-Load-Amplitude Fatigue Crack Growth Rates Above 10^{-8} m/Cycle (E 647-83). Values of C^* were computed using the relationships derived by Jaske [*14,22*]. Values of the linear elastic stress intensity factor K were computed using the relationships given in ASTM E 647.

The data obtained from the tests of the MENT specimens consisted of load, axial strain as a function of time, and the total change in crack length at each notch. Figure 7 shows a representative plot of strain versus time for a MENT specimen; this specimen was tested at 121 MPa and 649°C for 1006.8 h. The time from the beginning of steady state creep (about 180 h in Fig. 7) to the end of the test was used to calculate the average creep rate de_c/dt, that is, the creep strain during that time period divided by the length of that period. The average values of da/dt for each notch were calculated by dividing the amount of creep-crack growth at each notch by the length of that same time period. Thus, three values of da/dt were obtained for each MENT specimen that was tested. Corresponding values of C^* were computed using the formula derived by Jaske [*14,22*]

$$C^* = 1.2544P(de_c/dt)a[3.85n^{1/2}(1 - (1/n)) + \pi/n]/Bwg \quad (4)$$

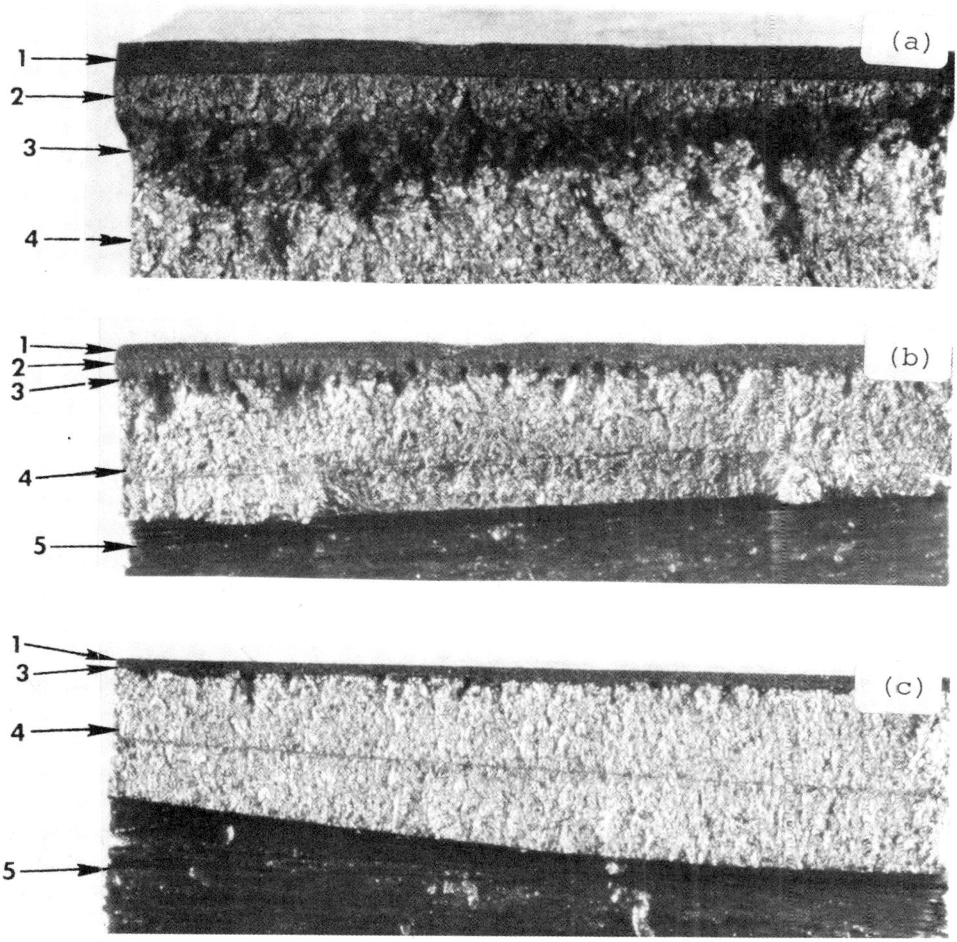

FIG. 6—*Fracture surfaces at the notches of an MENT specimen tested 1006.8 h at 121 MPa and 649°C: EDM notch (1); fatigue precrack (2); creep crack (3); fatigue after test (4); and sawcut (5).* (a) *Crack at 508-μm-deep notch (×10, Neg. No. 7L754).* (b) *Crack at 254-μm-deep notch (×10, Neg. No. 7L755).* (c) *Crack at 127-μm-deep notch (×10, Neg. No. 7L756).*

where

P = applied load,
a = crack length,
n = power-law creep exponent,
B = specimen thickness, and
g = gage length.

Values of K were computed from the relationship

$$K = 1.12P(\pi a)^{1/2}/Bw \tag{5}$$

Therefore, analysis of the data from the experiments on MENT specimens is straightforward.

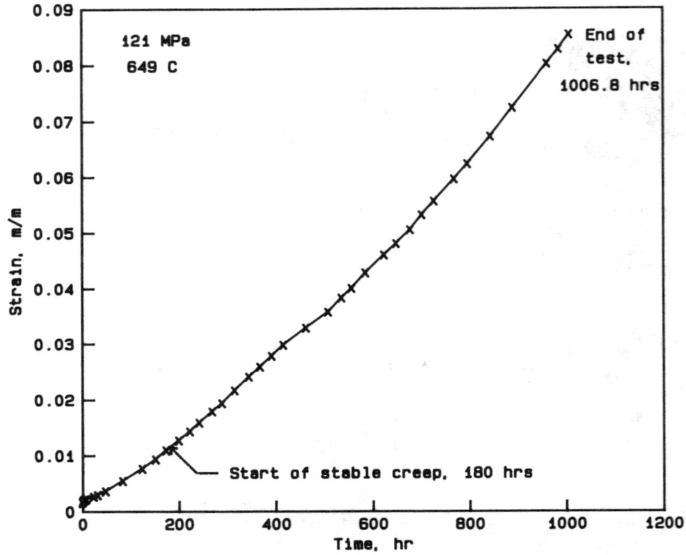

FIG. 7—*Creep behavior of a MENT specimen of Type 316 stainless steel.*

For each set of data points, Riedel and Rice's [*23*] transition time, t_1, from K to C^* controlled stress-strain fields for stationary cracks was computed from the following equation

$$t_1 = (K)^2(1 - \mu^2)/(n + 1)EC^* \tag{6}$$

where E is the elastic modulus and μ is Poisson's ratio. In all cases where crack-growth rates were calculated in this study, the actual test time corresponding to that crack-growth rate was greater than the calculated value of t_1. This result indicated that C^* was the appropriate parameter to use. Two characteristic crack-tip zone sizes for growing creep cracks, r_1 defined by Hui and Riedel [*24*] and r_2 defined by Riedel and Wagner [*25*], were calculated from the following equations

$$r_1 = [AK^{n-1}/(da/dt)]^{2/(n-3)} \tag{7}$$

$$r_2 = [(da/dt)/A]^{(n+1)/2}(A/EC^*)^{(n-1)/2} \tag{8}$$

where A is the power-law creep coefficient. The values of r_1 and r_2 were compared with the crack length, the specimen thickness, and the uncracked ligament size (that is, with B, a, and w-a). The values of r_1 were similar to or greater than a, B, and w-a, indicating that K is not the appropriate parameter, because r_1 must be small for valid use of K. The values of r_2 were much smaller than B, a, and w-a, indicating that C^* is the appropriate parameter, because r_2 must be small for the use of C^* to be valid. Thus, both the transition-time (Eq 6) and the crack-tip-zone-size (Eqs 7 and 8) criteria indicated that C^*, not K, was the parameter to use in characterizing the creep-crack growth data developed in this study.

The experimental results are summarized in Fig. 8. The creep-crack growth rate is shown both as a function of K in Fig. 8*a* and as a function of C^* in Fig. 8*b*.

As expected, the results were not well correlated by K, and there appeared to be a separate crack-growth rate curve for each independent test condition. The K parameter provided espe-

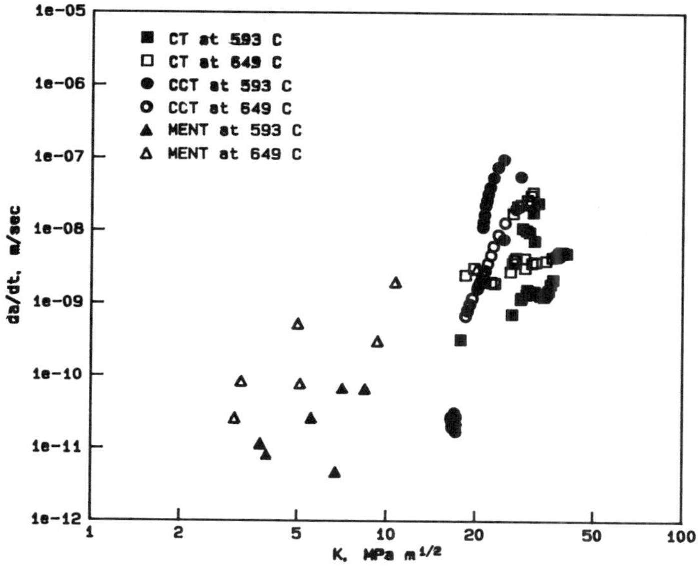

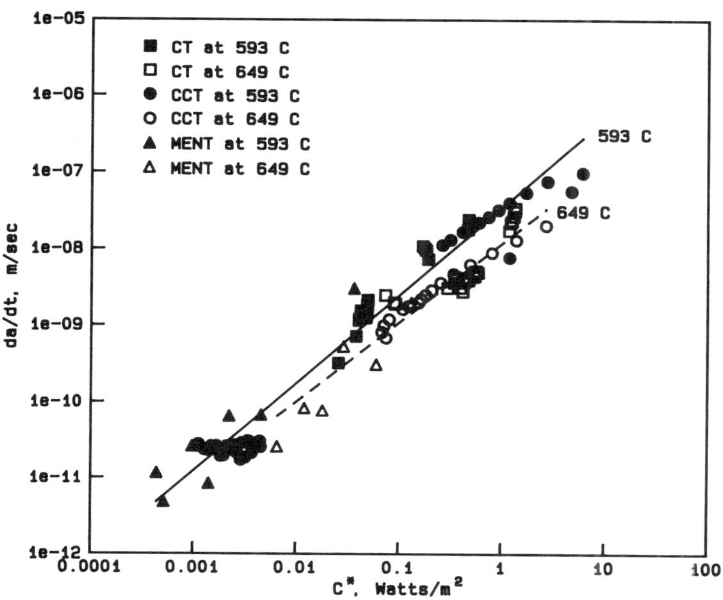

FIG. 8—*Creep-crack growth of Type 316 stainless steel.* (a) *Correlated using* K. (b) *Correlated using* C*.

cially poor correlation for the MENT specimens because they contained short cracks with significant creep deformation in uncracked ligament.

The C^* parameter provided an excellent correlation of the crack-growth data. As is shown in Fig. 8*b*, the data for tests at 593°C fell slightly above those for comparable tests at 649°C. At each temperature, the data for different specimen configurations gave the same da/dt versus C^* relationship.

Discussion

Under creep conditions in ductile, low-strength engineering alloys, the basic criteria of linear elastic fracture mechanics are violated and the linear elastic stress intensity factor is not expected to be the appropriate parameter for characterizing creep-crack growth behavior. This was found to be the case for Type 316 stainless steel in the present study. For this reason, the C*-integral was proposed as a possible fracture-mechanics parameter for characterizing creep-crack growth behavior [*1*].

The results of this study show that the C*-integral is a useful parameter for characterizing the creep-crack growth response of Type 316 stainless steel at 593°C and 649°C. It can be used with different specimen configurations and with both short and long cracks. Creep-crack growth rates as low as 10^{-11} m/s or 0.315 mm/year were measured reliably. To measure rates this low, it was necessary to conduct experiments for durations as long as 5000 h. Using d-c electric potential apparatus for these long periods significantly adds to the cost of creep-crack growth testing. The fractographic crack-measurement procedure used with the MENT specimens provided an economical alternative to the d-c electric potential method at low crack-growth rates. The data obtained using fractography were in good agreement with those obtained using d-c electric potential drop. Thus, it is possible to develop, in a reasonably economical way, the long-term data needed for most engineering applications.

It is encouraging that the long-term crack-growth rates could have been estimated reasonably well by extrapolating the trend of the higher growth rate data in Fig. 8*b*. Similar trends between long-term and short-term growth rates have been observed for Type 304 stainless steel [*11*] and Cr-Mo-V steel [*12*]. Therefore, when a limited amount of time is available for testing similar alloys, it would be reasonable to characterize the short-term creep-crack growth behavior and extrapolate the results to longer times. However, in establishing values for use in engineering design, it is always best to have data in the relevant regime of growth rates.

Once long-term creep-crack growth data have been developed, those data need to be applied to actual engineering components. Doing so requires the calculation of C^* for typical crack configurations and orientations in various structural components. Performing these calculations rigorously requires detailed finite element creep calculations [*26*]. However, it is likely that such calculations can be performed for a particular component configuration using the approximation and estimation techniques that have been developed for application of the J integral in the same manner that those techniques have been used to obtain estimates of C^* [*14,22*] for laboratory specimens. Thus, although they need further refinement and development, the tools required to deal with creep-crack growth in engineering applications are available.

Conclusions

It is concluded that the C^* parameter is a useful tool for characterizing the creep-crack growth behavior of Type 316 stainless steel at 593 and 649°C. Long-term creep-crack growth behavior at rates as low as 10^{-11} m/s can be characterized reliably using either conventional CT and CCT fracture-mechanics specimens or the MENT specimen. The behavior of both long (25 to 35 mm) and short (100 to 500 μm) cracks can be characterized by the same C^* versus da/dt

relationship. The information and analytical methods required to deal with creep-crack growth are becoming available for everyday engineering application.

Acknowledgments

This work was performed under the sponsorship of the U.S. Department of Energy, Office of Basic Energy Sciences, Division of Mathematical, Engineering, and Geosciences through Contract W-7405-ENG-91-131, BCL NO. G5051-3100.

References

[*1*] Landes, J. D. and Begley, J. A. in *Mechanics of Crack Growth, ASTM STP 590,* American Society for Testing and Materials, Philadelphia, 1976, pp. 128–148.
[*2*] Sadananda, K. and Shahinian, P. in *Fracture Mechanics,* University of Virginia Press, Charlottesville, VA, 1978, pp. 685–703.
[*3*] Ellison, E. G. and Harper, M. P., *Journal of Strain Analysis,* Vol. 13, No. 1, 1978, pp. 35–51.
[*4*] Fu, L. S., *Engineering Fracture Mechanics,* Vol. 13, 1980, pp. 307–330.
[*5*] Sadananda, K. and Shahinian, P., *Engineering Fracture Mechanics,* Vol. 15, Nos. 3–4, 1981, pp. 327–342.
[*6*] Christian, E. M., Smith, D. J., Webster, G. A., and Ellison, E. G. in *Advances in Fracture Research,* Pergamon Press, Oxford, U.K., 1980, pp. 1295–1302.
[*7*] Taira, S., Ohtani, R., and Kitamura, T., *Journal of Engineering Materials and Technology,* Vol. 101, 1979, pp. 154–161.
[*8*] Harper, M. P. and Ellison, E. G., *Journal of Strain Analysis,* Vol. 12, No. 3, 1977, pp. 167–179.
[*9*] Nibkin, K. M., Webster, G. A., and Turner, C. E. in *Fracture 1977,* Vol. 2, ICF4 (Fourth International Conference on Fracture), Pergamon Press, Oxford, U.K., 1977, pp. 627–634.
[*10*] Saxena, A. in *Fracture Mechanics: Twelfth Conference, ASTM STP 700,* American Society for Testing and Materials, Philadelphia, 1980, pp. 131–151.
[*11*] Shih, T. T. in *Fracture Mechanics: Fourteenth Symposium—Volume II: Testing and Applications, ASTM STP 791,* American Society for Testing and Materials, Philadelphia, 1983, pp. II-232–II-247.
[*12*] Saxena, A., Shih, T. T., and Ernst, H. A. in *Fracture Mechanics: Fifteenth Symposium, ASTM STP 833,* American Society for Testing and Materials, Philadelphia, 1984, pp. 516–531.
[*13*] Beaver, R. J. and Martin, W. R., "Procurement of Type 316 Stainless Steel Reference Heat for LMFBR Research and Development Programs," Report No. ORNL/TM-5196, Oak Ridge National Laboratory, Oak Ridge, TN, Jan. 1976.
[*14*] Jaske, C. E., "Damage Accumulation by Crack Growth Under Creep and Fatigue," Topical Report No. BMI-2116, Battelle's Columbus Laboratories, Columbus, OH, 1984.
[*15*] McIntyre, P. and Priest, A. H., "Measurement of Sub-Critical Flaw Growth in Stress Corrosion, Cyclic Loading and High Temperature Creep by the DC Electrical Resistance Technique," BISRA Report No. MG/54/71, British Steel Corp., London, 1971.
[*16*] Saxena, A., *Engineering Fracture Mechanics,* Vol. 13, 1980, pp. 741–750.
[*17*] Sooley, P., "An Investigation into the Fatigue/Creep Crack Growth Characteristics of 2.25Cr-1Mo Steel," M.S. thesis, University of Toronto, Toronto, Ont., Canada, 1981.
[*18*] Vosikovsky, O., "Determination of Fatigue Crack Growth Rate by the Potential Drop Method—Set-Up and Calibration," Report No. EPR/PMRL-75-15(R), Physical Metallurgy Research Laboratories, CANMET, Ottawa, Ont., Canada, 1975.
[*19*] Johnson, H. H., *Materials Research and Standards,* Vol. 5, 1965, pp. 442–445.
[*20*] Schwalbe, K.-H. and Hellmann, D., *Journal of Testing and Evaluation,* Vol. 9, No. 3, 1981, pp. 218–221.
[*21*] Jaske, C. E., "An Automated Systematic Procedure for Evaluating Creep-Crack-Growth Data," paper presented at 19th National Symposium on Fracture Mechanics, San Antonio, TX, 30 June–2 July 1986.
[*22*] Jaske, C. E. in *Proceedings,* International Conference and Exposition on Fatigue, Corrosion Cracking, Fracture Mechanics and Failure Analysis, American Society for Metals, Metals Park, OH, 1986.
[*23*] Riedel, H. and Rice, J. R. in *Fracture Mechanics: Twelfth Conference, ASTM STP 700,* American Society for Testing and Materials, Philadelphia, 1980, pp. 112–130.
[*24*] Hui, C. Y. and Riedel, H., *International Journal of Fracture,* Vol. 17, No. 4, 1981, pp. 409–425.
[*25*] Riedel, H. and Wagner, W. in *Advances in Fracture Mechanics,* ICF5 (Fifth International Conference on Fracture), Vol. 2, Pergamon Press, Oxford, U.K., 1981, pp. 691–698.
[*26*] Wu, D., Christian, E. M., and Ellison, E. G., *Journal of Strain Analysis,* Vol. 19, No. 3, 1984, pp. 185–195.

Fatigue I

Ralph I. Stephens,[1] Hyun W. Lee,[2] Ren Bu,[3] and Gregory K. Werner[4]

The Behavior of Short and Long Fatigue Cracks at Threshold and Near-Threshold Levels

REFERENCE: Stephens, R. I., Lee, H. W., Bu, R., and Werner, G. K., **"The Behavior of Short and Long Fatigue Cracks at Threshold and Near-Threshold Levels,"** *Fracture Mechanics: Eighteenth Symposium, ASTM STP 945,* D. T. Read and R. P. Reed, Eds., American Society for Testing and Materials, Philadelphia, 1988, pp. 881–895.

ABSTRACT: The fatigue crack growth behavior of physically short cracks ($0.2 \leq a \leq 1$ mm), intermediate cracks ($1 < a \leq 2$ mm), and long cracks ($2 < a < 25$ mm) at near-threshold and threshold levels was investigated with single-edge-notched-type specimens using four materials with stress ratios R of 0.05, −1, and 0.5. The materials were SAE 0030 cast steel, AISI 1020 and 1045 hot-rolled steels, and 7075-T6 aluminum. Short cracks grew much faster than long cracks based upon applied ΔK. The transition crack lengths at which there was similitude with ΔK were between 1 and 2 mm. Mean stress effects were significant at these low da/dN values for both long and short cracks. Crack closure played an important role in this behavior, but appears not to be the entire cause. Scanning electron microscopic fractographic analysis did not indicate a difference in fatigue crack growth morphology between the long and short cracks.

KEY WORDS: fatigue (materials), crack growth, short cracks, long cracks, steels (cast and wrought), aluminum alloy, crack closure, mean stress, thresholds

Similitude in fatigue crack growth rate behavior using the stress-intensity factor as the driving force has been the key to successful damage-tolerant fatigue crack growth life predictions. Anomalous nonconservative behavior with short cracks, however, has been a recent key research topic [*1–6*]. This research deals principally with through-the-thickness physically short cracks between 0.2 and 1 mm in length. The purpose of this research is to compare room temperature constant amplitude threshold and near-threshold fatigue crack growth behavior of physically short cracks with that of intermediate cracks ($1 < a \leq 2$ mm) and of long cracks ($2 < a < 25$ mm) using single-edge-notch (SEN) type specimens. The significance of crack closure is also investigated, and estimates are made of a transition crack length above which no apparent fatigue crack growth differences between short and long cracks occur. Load ratios R equal to 0.05 (~0), −1, and 0.5 were used. A multiple-specimen approach was used for short cracks; the more conventional ASTM standard procedures were used for long cracks.

[1]Professor, Department of Mechanical Engineering, The University of Iowa, Iowa City, IA 52242.
[2]Assistant professor, Mechanical Engineering Department, Pusan National University, Pusan, Korea.
[3]Head, Fatigue and Fracture Section, Shenyang Aircraft Corporation, Shenyang, China.
[4]Senior engineer, John Deere, Dubuque, IA, 52001.

Test Materials and Procedures

The materials used in this research were SAE 0030 cast steel, AISI 1020 hot-rolled steel, AISI 1045 hot-rolled steel, and 7075-T6 aluminum. 0030 cast steel blocks 117 by 102 by 76 mm were poured in conventional green sand molds, normalized at 900°C and tempered at 627°C. The 1020 steel was hot-rolled to 9.5-mm-thick plate, and the 1045 steel was hot-rolled to a diameter of 63.5 mm and then normalized. The 7075-T6 was hot-rolled to 9.5-mm-thick plate. The 0030 cast steel contained 25% pearlite and 75% ferrite with an average nominal grain diameter of 0.016 mm. The 1020 steel was 20% pearlite and 80% ferrite with an average nominal grain size of 0.060 mm. The 1045 steel consisted of 70% pearlite and 30% ferrite with an average nominal grain diameter of 0.010 mm. The 7075-T6 aluminum alloy had an average nominal grain diameter of 0.025 mm.

The chemical compositions are given in Table 1; the monotonic tensile properties along with the average nominal grain diameters, which ranged from 0.010 to 0.060 mm for the four materials, are summarized in Table 2. SEN specimens with a net section thickness of 4.1 mm were used for both long and short crack growth tests. The specimen width for long crack tests was either 25 or 50 mm. For short cracks, the specimen width ranged from 8 to 18 mm. This range existed because the short crack specimens were made by precracking 25-mm-width specimens and then symmetrically machining away two sides to form initial crack lengths between 0.2 and 0.5 mm. This resulted in the initial physically short crack lengths to be from 4 to 50 times longer than the average nominal grain diameters of the four materials. Specimens were gripped by either a single pin connection or a three-hole end connected to a spherical monoball. The monoball assured axial alignment in either tension or compression. Tests were performed using an 89-kN closed-loop electrohydraulic test system. Frequencies between 10 and 50 Hz were used for both precracking and actual testing, depending upon crack length and R ratio. The crack length was measured with a $\times 33$ magnification traveling telescope which has a least reading of 0.01 mm. The reading accuracy was ± 0.02 mm.

Precracking for all specimens was done using a load shedding technique with $R \approx 0$. Long crack da/dN versus ΔK data were obtained using ASTM Test Method for Constant-Load-Amplitude Fatigue Crack Growth Rates Above 10^{-8} m/Cycle (E 647-83), along with the ASTM proposed recommendation for near-threshold and threshold conditions (10^{-8} to 10^{-11} m/cycle) [7]. For the long crack tests, between two and four specimens were used to obtain the complete da/dN versus ΔK data for each R ratio. A multiple-specimen approach was used for short and intermediate cracks to obtain data. Between three and ten specimens were used for each R ratio for each material. For the three steels, the short cracks were formed by incremental load shedding precracking followed by machining of both edges and then vacuum stress relieving below recrystallization temperatures. Because 7075-T6 aluminum cannot be stress relieved by heat treatment without altering its mechanical properties and microstructure, a more tedious multi-machining and load shedding technique was employed. Two or three sequences of load shedding and machining were needed to obtain initial crack lengths between 0.2 and 0.4 mm with the precrack plane stress plastic zones less than 0.01 mm and ΔK very near the threshold region. The preceding precracking techniques were incorporated to minimize residual stress effects.

Short and intermediate near-threshold and threshold crack growth data were obtained from polished test specimens under constant amplitude conditions with crack growth increments between about 0.1 and 0.3 mm. These data were converted to da/dN versus ΔK using a secant method between adjacent data points. Substantial data were invalid due to difficulties with crack tip eccentricity in these physically short cracks. Cracks were often longer at one free surface than at the other free surface. Crack lengths a were obtained by measuring at three to five locations on the fracture surface. The centerline, quarter points, and free surfaces were used for the measurements. The average of these measurements was taken as the nominal crack length correction to be used in obtaining ΔK. The applied stress-intensity factor range ΔK, which is

TABLE 1—*Chemical composition by weight percent.*

Material	C	Mn	Si	S	P	Cr	Ni	Cu	Mo	Ti	Cb	V	Mg	Zn
0030 Cast	0.24	0.71	0.44	0.026	0.015	0.10	0.10	0.05	0.08	0.013	...	...	...	...
1020 Hot Rolled	0.21	0.44	0.018	0.015	0.002	0.01	0.017	0.004	...	...	0.004	0.001	...	...
1045 Hot Rolled	0.44	0.70	0.23	0.046	0.019	0.05	0.03	0.03	...	0.002	0.008	0.002	...	...
7075-T6	...	0.06	0.10	...	...	0.19	...	1.50	...	0.03	...	...	2.48	5.93

TABLE 2—*Average monotonic tensile properties and grain size.*

Property	0030 Cast	1020 Hot Rolled	1045 Hot Rolled	7075-T6
S_Y: 0.2% yield strength, MPa	303	262	382	503
S_u: ultimate strength, MPa	496	413	622	572
σ_f: true fracture stress, MPa	750	999	985	...
ϵ_f: true fracture strain	0.62	0.87	0.71	...
E: Young's modulus, GPa	207	203	202	70
Average nominal grain diameter, mm	0.016	0.060	0.010	0.025

referred to as the applied positive stress-intensity factor range in this research, was calculated from [*8*]

$$\Delta K = \frac{\Delta P}{BW}\sqrt{a}\left[1.99 - 0.41\left(\frac{a}{w}\right) + 1.87\left(\frac{a}{w}\right)^2 - 38.48\left(\frac{a}{w}\right)^3 + 53.85\left(\frac{a}{w}\right)^4\right] \quad (1)$$

where ΔP is the applied positive load range. That is

$$\Delta P = P_{\max} - P_{\min} \quad (2)$$

where for $R = -1$, $P_{\min}$ is considered equal to zero instead of its actual negative value. B is the specimen thickness, W is the specimen width, and a is the mean crack length.

Crack closure measurements were carried out at the near-threshold and threshold regions for all long crack tests and for many of the short and intermediate crack tests by using a clip-gage with a 12.7 mm gage length. The gage was mounted to the specimen at the edge containing the crack by springs or rubber bands. Curves of load P versus crack opening displacement (COD) were obtained at a reduced test frequency of about 1 Hz.

Results

The reduced data of crack growth rate da/dN versus the applied positive stress-intensity factor range ΔK, for R ratios of 0.05, -1, and 0.5, are shown in Figs. 1 to 4 for the four different materials. The short, intermediate, and long crack data were plotted together on the same scale for each of the different R ratios for better comparison. The solid lines were drawn through the median of the long crack data and the dashed lines were drawn through the median of the short crack data. No line was drawn for the intermediate crack length data. More scatter occurred with the short and intermediate crack data than with the long crack data due to the experimental complexity with the physically short cracks. For the long cracks, four to five orders of da/dN magnitude were obtained for each R ratio and material, whereas only the threshold and near threshold regions were obtained for most of the short crack tests. The threshold stress-intensity factor ranges, ΔK_{th}, were taken at 10^{-10} m/cycle. These values are given in Table 3 for both the long and the short cracks (row two and row four, respectively).

Figures 5 to 8 show the superimposed median curves of da/dN versus ΔK for both short and long cracks for the different R ratios in order to obtain (along with Table 3) a better indication of the influence of mean stress on the fatigue crack growth behavior for both short and long cracks.

Figures 9–11 show representative data of applied stress-intensity factor range ΔK versus crack length for different ranges of crack growth rates da/dN with $R = 0.05$. Data similar to Figs. 9–11 also were determined for the R ratios of -1 and 0.05. For better distinction, a scatterband encompasses each da/dN range data set in Fig. 11. The purpose of Figs. 9–11 is to better determine a transition from similitude for the crack lengths studied.

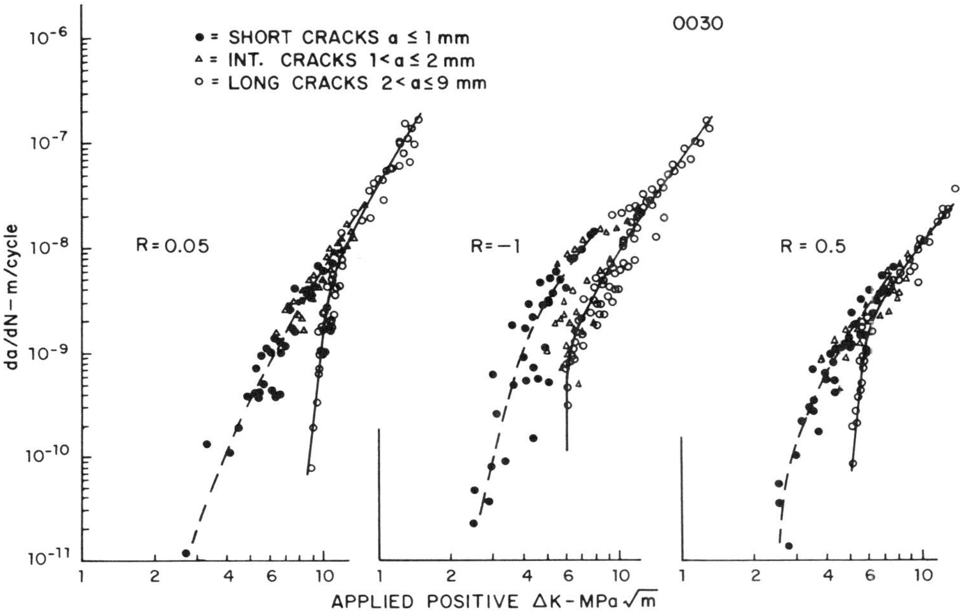

FIG. 1—da/dN *versus applied positive* ΔK, *0030 cast steel.*

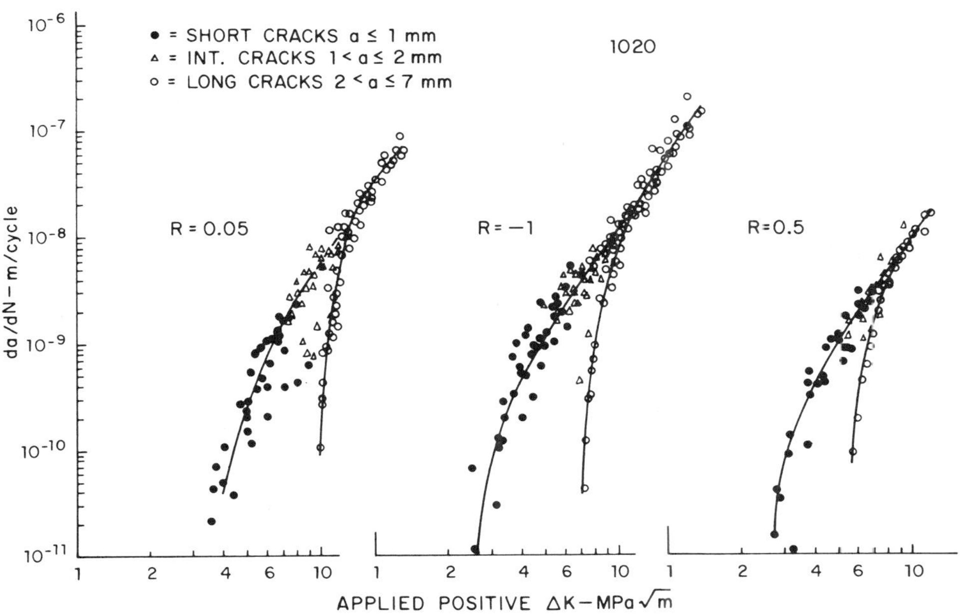

FIG. 2—da/dN *versus applied positive* ΔK, *1020 hot-rolled steel.*

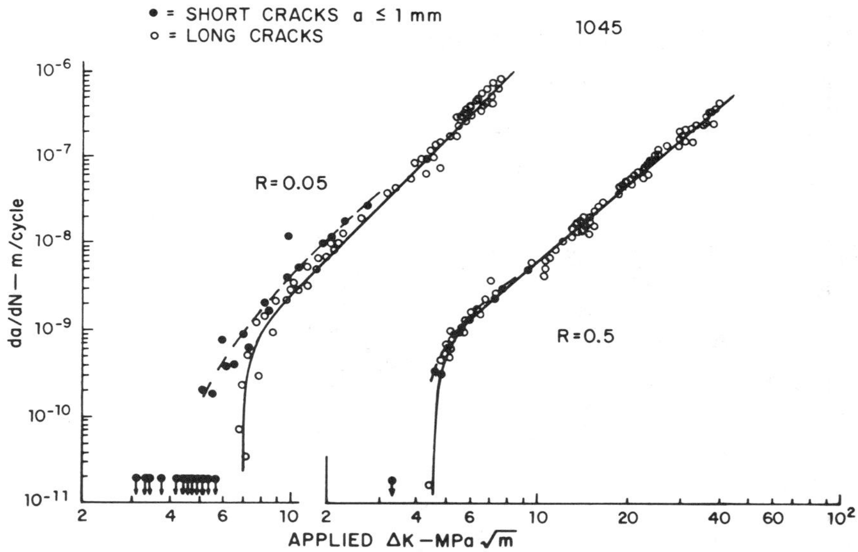

FIG. 3—da/dN *versus applied positive* ΔK, *1045 hot-rolled steel.*

Load-versus-COD curves obtained for the short cracks were essentially single straight lines. Thus sufficient change in compliance did not occur with the short cracks to obtain a crack opening load P_{op} or a crack closure load $P_{c\ell}$. Two key reasons can be presented for this lack of compliance change with the short cracks: first, because the crack closure phenomenon predominates only in the wake of the crack and the short cracks have a very limited extent of this wake; and second, because of possible insensitivity of a crack mouth COD gage which tends to provide more of a bulk compliance rather than the local crack-tip compliance for these short cracks.

For the long cracks, the loading and unloading P-versus-COD curves were essentially coincident with rather distinct changes in compliance. The difference between crack opening load and crack closure load, however, could not be distinguished. Typical recorded plots of load versus COD at near threshold regions are shown in Fig. 12 for the long crack specimens. For these long cracks, P_{op} was taken at the intersection of two straight lines drawn through the median of the two rather distinct slope regions. Values of P_{op}/P_{max} versus ΔK for the long cracks are shown in Figs. 13 and 14 for $R = 0.05$ and -1, respectively. The $R = 0.5$ data were not included since in most cases P_{op}/P_{max} was less than 0.5 except at essentially threshold levels where they ranged from 0.5 to 0.55, which shows that crack closure was essentially missing for $R = 0.5$ tests.

Discussion

The elastic stress-intensity factor K is sufficiently accurate in describing the crack tip stress-strain state if the plastic zone radius r_y is small compared to the crack length, thickness, and width and if the grain size is small relative to the crack dimensions. All the short crack threshold data satisfied the above criteria, because initial short crack lengths were 4 to 50 times greater than average nominal grain sizes and at the threshold levels the plane stress plastic zone radius, r_y, where

$$r_y = \frac{1}{2\pi}\left(\frac{K_{max}}{S_Y}\right)^2 \tag{3}$$

TABLE 3—ΔK_{th} *and effective* ΔK_{th} *at* 10^{-10} *m/cycle for long and short cracks,* $MPa\sqrt{m}$.

	0030 Cast			1020 Hot Rolled			1045 Hot Rolled		7075-T6		
R ratio	0.05	−1	0.5	0.05	−1	0.5	0.05	0.5	0.05	−1	0.5
Long crack ΔK_{th}	9.0	6.0	5.1	10.0	7.3	5.8	7.1	4.6	2.5	2.3	1.7
Long crack effective ΔK_{th}	4.8	3.0	4.7	5.1	3.8	5.4	3.5	4.6	1.1	1.0	1.5
Short crack ΔK_{th}	4.0	3.0	3.0	4.4	3.1	3.1	4.8	3.7	1.8	1.6	1.4
$\dfrac{\Delta K_{th}\text{ long}}{\Delta K_{th}\text{ short}}$	2.2	2.0	1.7	2.3	2.3	1.9	1.5	1.2	1.4	1.4	1.2
$\dfrac{\text{Effective }\Delta K_{th}\text{ long}}{\Delta K_{th}\text{ short}}$	1.2	1.0	1.6	1.3	1.2	1.7	0.7	1.2	0.6	0.6	1.1

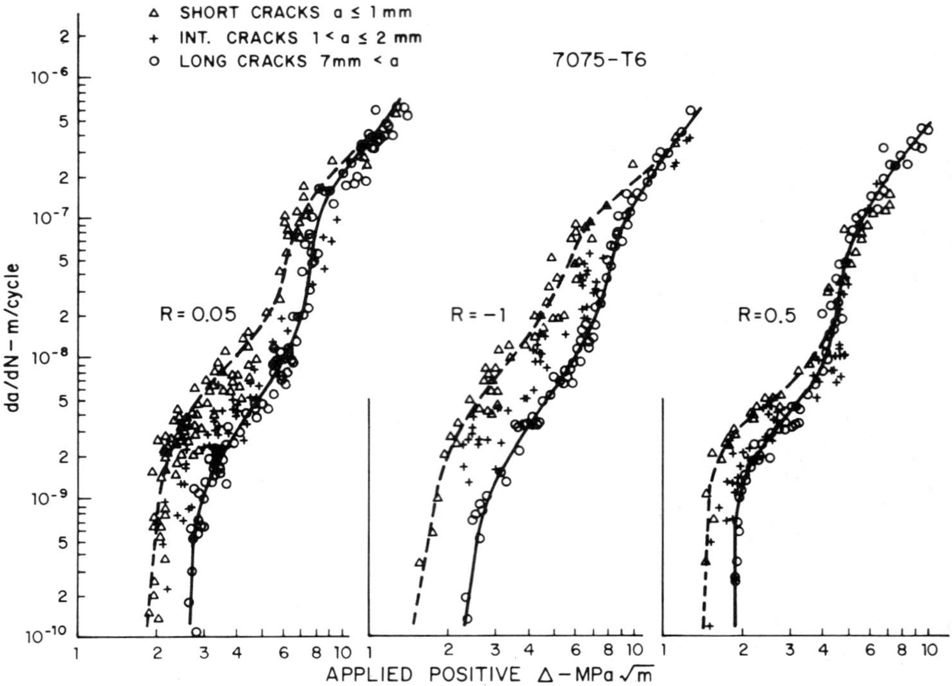

FIG. 4—da/dN *versus applied positive* ΔK, *7076-T6 aluminum.*

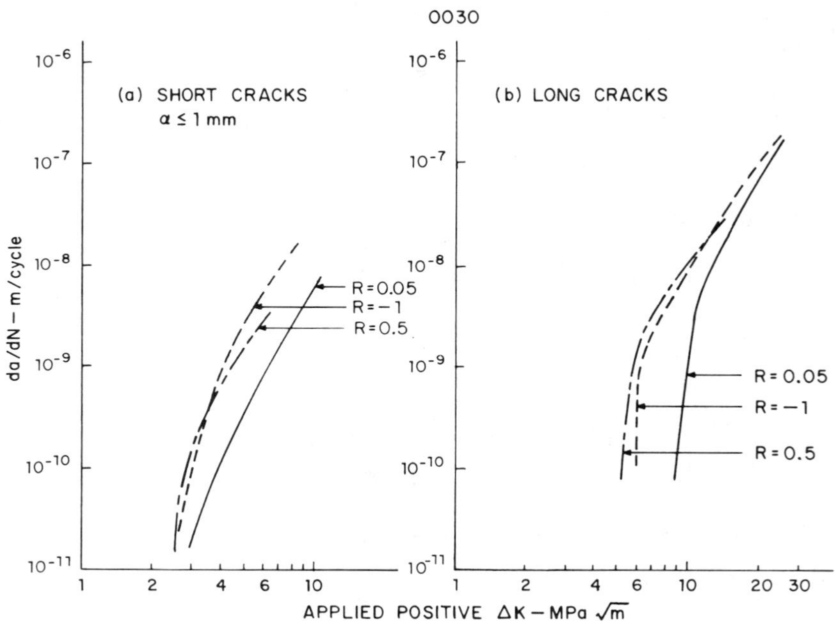

FIG. 5—R-*ratio effects for both short and long cracks, 0030 cast steel.*

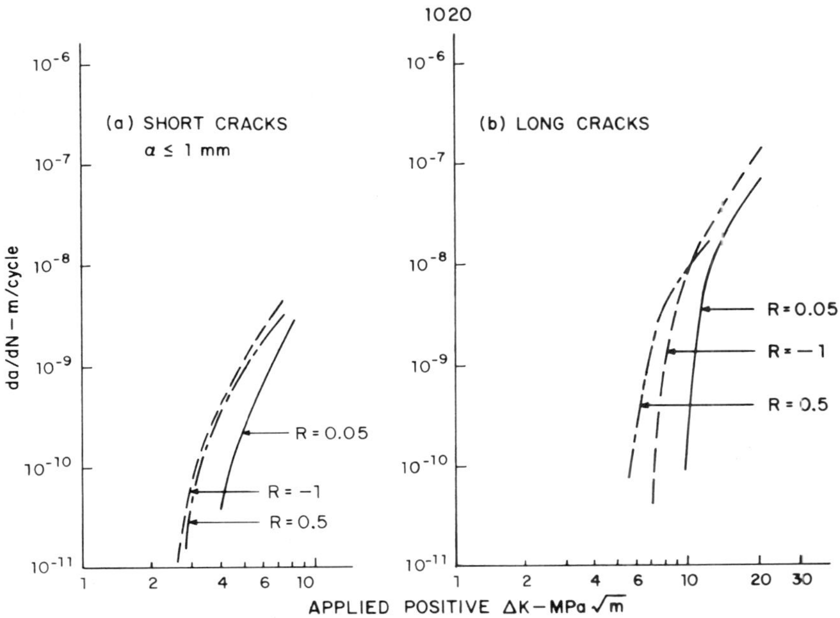

FIG. 6—*R-ratio effects for both short and long cracks, 1020 hot-rolled steel.*

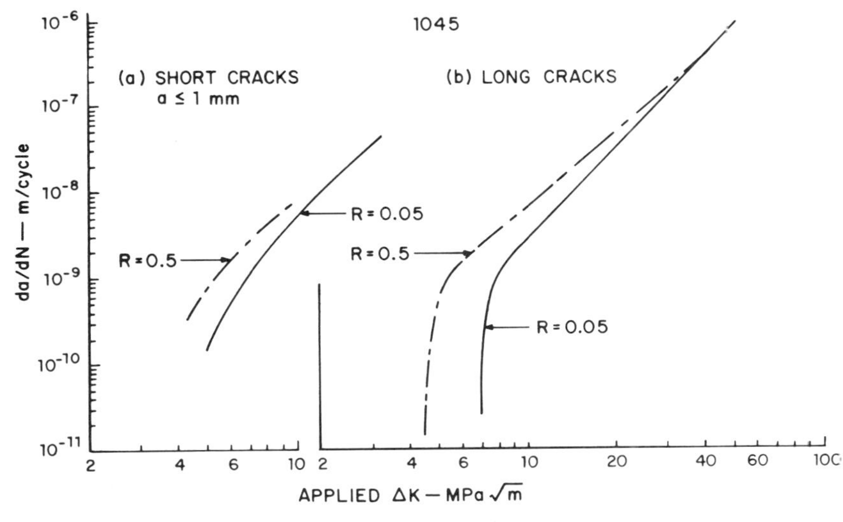

FIG. 7—*R-ratio effects for both short and long cracks, 1045 hot-rolled steel.*

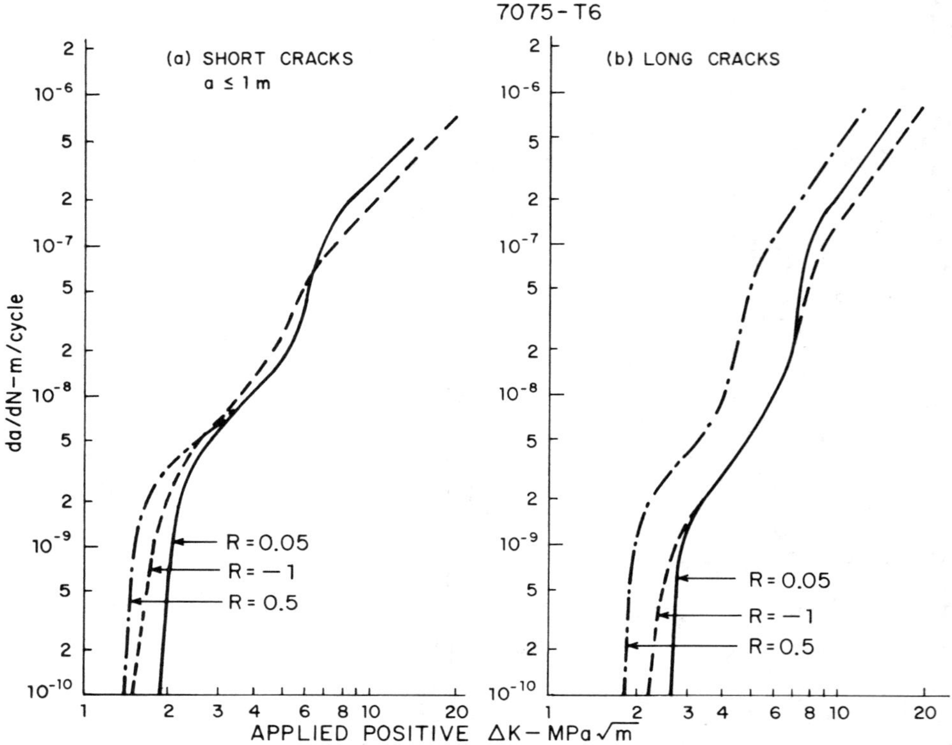

FIG. 8—R-*ratio effects for both short and long cracks, 7075-T6 aluminum.*

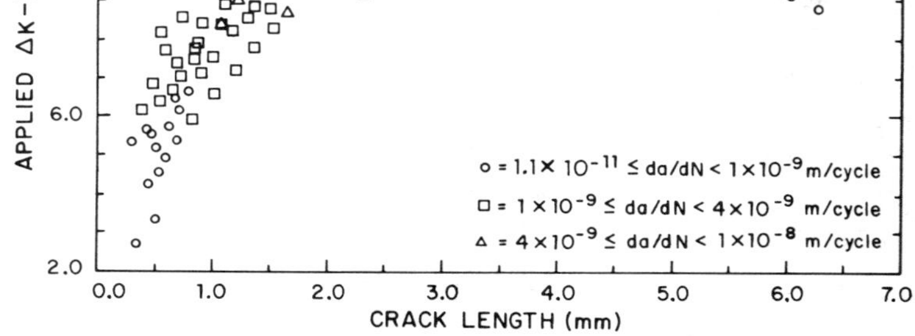

FIG. 9—*Applied* ΔK *versus crack length for different* da/dN, R = *0.05, 0030 cast steel.*

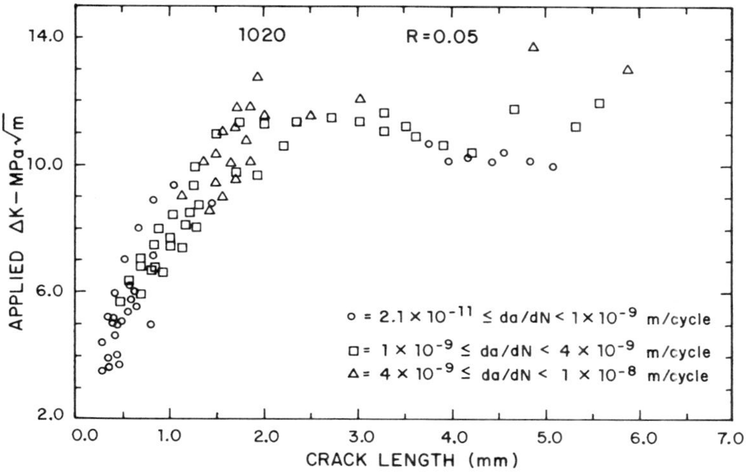

FIG. 10—*Applied* ΔK *versus crack length for different* da/dN, R = *0.05, 1020 hot-rolled steel.*

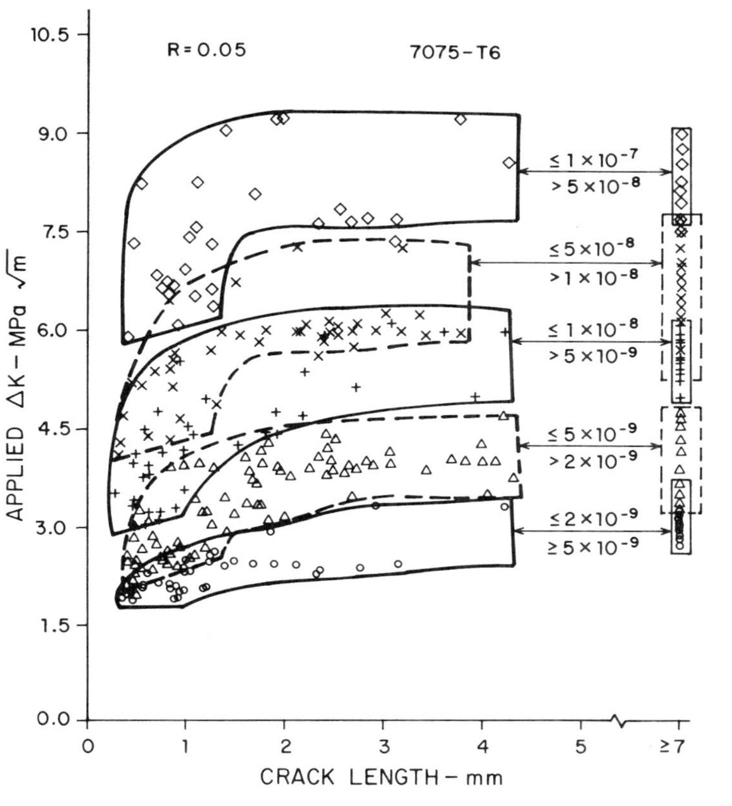

FIG. 11—*Applied* ΔK *versus crack length for different* da/dN, R = *0.05, 7075-T6 aluminum.*

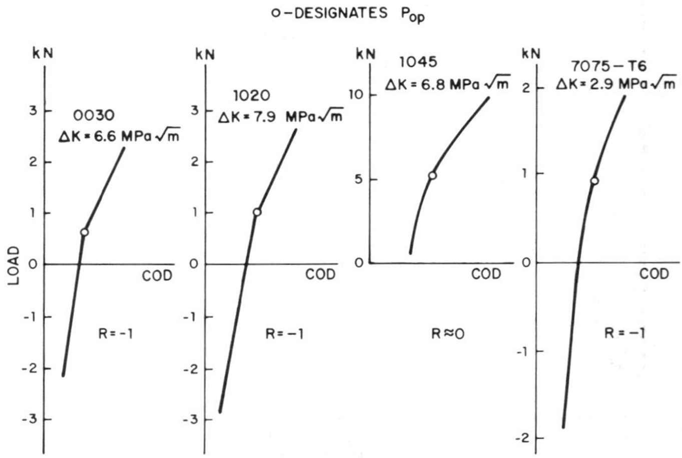

FIG. 12—*Typical load vs. COD for crack opening determination of long cracks at near threshold.*

was less than 0.01 mm for all tests. At the highest values of ΔK for the short cracks ($5 \times 10^{-9} \leq da/dN \leq 5 \times 10^{-7}$ m/cycle), r_y ranged from 0.02 to 0.45 mm. Thus a few of the short crack data points at the highest crack growth rates may stretch the validity of using ΔK as the crack driving force. However, these are less than a few percent of the data points shown in Figs. 1-4. Thus the use of ΔK in this research for the short and the long cracks is applicable.

The effective stress-intensity factor range was obtained from $K_{max} - K_{op}$ where K_{op} ($\geq K_{min}$) is the crack opening stress-intensity factor calculated using P_{op}. Effective threshold stress-intensity factors for long cracks have been included in Table 3 along with applied positive threshold stress-intensity factors. Also included in Table 3 are ratios of threshold levels for long and short cracks. From Figs. 1-4 and Table 3, it is evident that for a given material and R ratio, near threshold and threshold short crack behavior based upon applied positive stress-intensity factor range is poorer for the short cracks. At threshold levels, the ratio of long crack to short crack

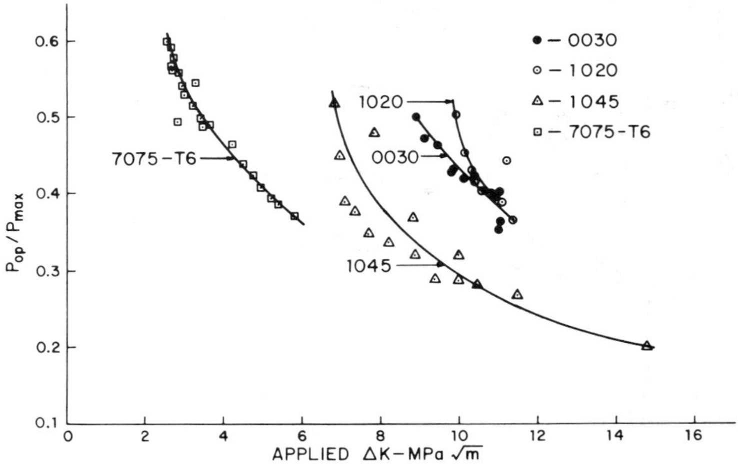

FIG. 13—P_{op}/P_{max} *versus applied positive* ΔK, R = *0.05.*

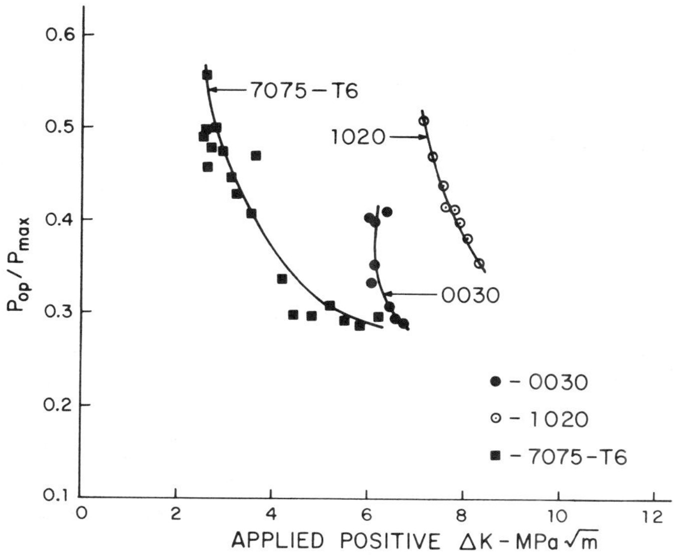

FIG. 14—P_{op}/P_{max} *versus applied positive* ΔK, R = *−1.*

applied positive ΔK_{th} values, given in Table 3, ranged from 1.2 to 2.3. At higher crack growth rates, however, the long and short crack data tend to converge for all R ratios as shown in Figs. 1-4. Since no crack opening/closure could be determined with the crack mouth clip gage for the short cracks, the applied positive ΔK short crack values also represent effective ΔK values. However, more sensitive COD measurements may show closure effects with the short cracks [*9*]. Based upon the obtained effective ΔK_{th} values, the threshold ratios for long/short cracks were substantially reduced except at $R = 0.5$ where very little long crack closure is involved. From Figs. 12 and 13, it is seen that P_{op}/P_{max} for long cracks increases sharply at the lower ΔK levels. At threshold levels the crack opening load ratios are high with an average value of about 1/2. At higher ΔK ranges this ratio dropped to an average of about 1/3. Thus crack closure is still significant at these long crack higher ΔK values. However, at all ΔK levels, crack closure does not completely eliminate discrepancies between long and short cracks.

The effect of R ratio, or mean stress, on da/dN versus ΔK is best seen in the superimposed median curves of Figs. 5-8 and Table 3. At threshold and near-threshold levels, compression was detrimental for both the long and the short cracks based upon either applied positive ΔK or effective ΔK. The threshold and near-threshold fatigue crack growth resistance for $R = 0.5$ was poorer than for $R = 0.05$ or -1 when based upon ΔK, but mixed results occurred when based upon effective ΔK. Thus, crack closure has a substantial influence on long crack mean stress effects, but is not the only controlling factor.

The applied ΔK versus crack length data for different $R = 0.05$ low crack growth rate ranges shown in Figs. 9-11 indicate that a transitional behavior exists with the long and short crack growth between 1 and 2 mm of crack length. This same transitional behavior occurred for both $R = -1$ and 0.5 in most of the tests. Thus, under these experimental conditions, crack lengths of 2 mm or more show reasonable similitude with applied ΔK.

Both a macro and micro fractographic analysis was made on the long and short crack tests. In general, greater macrofretting damage was found with the $R = -1$ tests due to the extensive compression. The $R = 0.05$ tests also showed macrofretting damage whereas the $R = 0.5$ tests had very little. Typical SEM fractographs for near threshold growth rates of the four materials are shown in Fig. 15 for $R = 0.05$. Striations were generally not apparent for all four materials

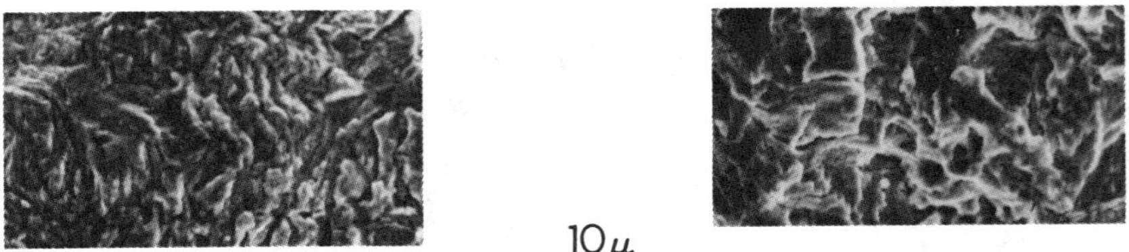

10μ

SHORT CRACK LONG CRACK

(a) 0030

10μ

SHORT CRACK LONG CRACK

(b) 1020

33μ

SHORT CRACK LONG CRACK

(c) 1045

20 μ

SHORT CRACK LONG CRACK

(d) 7075-T6

FIG. 15—*SEM fractographs in the near-threshold region,* R = *0.05.*

at these low crack growth rates. The extensive pearlite colonies of the 1045 steel, however, produced features which had similarities to striations. Secondary cracks were often evident particularly at $R = -1$. In general, the fracture surfaces for a given material appear to be very similar for both short and long cracks and for all R ratios. Thus from both the macro and micro examination, the evidence does not clearly indicate significant morphological differences as a function of short crack length versus long crack length, or of stress ratio.

Conclusions

1. Since plastic zone sizes and grain sizes were sufficiently small compared with the physically short through cracks ($a \geq 0.2$ mm), the use of the stress-intensity factor with these four materials at threshold and near threshold levels was applicable.

2. Similitude based upon the applied stress-intensity factor range ΔK existed in these experiments for a given R ratio at crack lengths greater than about 2 mm. With $da/dN \leq 10^{-7}$ m/cycle, a transition from similitude occurred between 1 and 2 mm for all four materials. Thus at lower fatigue crack growth rates, the physically short cracks grew faster than the long cracks based upon ΔK.

3. Ratios of P_{op}/P_{max} obtained from long crack tests increased steeply with decreasing ΔK for all three R ratios. At threshold levels P_{op} was about one half of P_{max}.

4. Crack closure effects are partially responsible for the discrepancies between physically short crack and long crack behavior in this study.

5. The fatigue crack growth resistance for $R = 0.5$ was poorer compared with that of $R = 0.05$ at near-threshold and threshold levels for both long and short cracks. Crack closure is also partially responsible for the mean stress long crack influence.

6. The scanning electron microscopic (SEM) fractographs did not clearly show the evidence of significant morphological differences as a function of short cracks versus long cracks as well as stress ratio R at near-threshold and threshold levels.

References

[1] El Hadded, M., Smith, K. N., and Topper, T. H., "Fatigue Crack Propagation of Short Cracks," *Journal of Engineering Materials and Technology,* Vol. 101, 1979, p. 42.

[2] Leis, B. and Forte, T., "Fatigue Growth of Initially Physically Short Cracks in Notched Aluminum and Steel Plates" in *Fracture Mechanics, ASTM STP 743,* R. Roberts, Ed., American Society for Testing and Materials, Philadelphia, 1981, p. 100.

[3] Lankford, J., "The Growth of Small Fatigue Cracks in 7075-T6 Aluminum," *Fatigue of Engineering Materials and Structures,* Vol. 5, 1982, p. 233.

[4] Ritchie, R. O. and Suresh, S., "Mechanics and Physics of the Growth of Small Cracks" in *Behavior of Short Cracks in Airframe Components,* Advisory Group for Aerospace Research and Development, Proceedings No. 328, 1982, p. 1.

[5] Zurek, A. K., James, M. R., and Morris, W. L., "The Effect of Grain Size on Fatigue Growth of Short Crack," *Metallurgical Transactions A,* Vol. 14A, 1983, p. 1697.

[6] De Los Rios, E. R., Tang, Z., and Miller, K. J., "Short Crack Fatigue Behavior in a Medium Carbon Steel," *Fatigue of Engineering Materials and Structures,* Vol. 7, 1984, p. 97.

[7] Bucci, R. J., "Development of a Proposed ASTM Standard Method for Near-Threshold Fatigue Crack Growth Rate Measurement" in *Fatigue Crack Growth Measurement and Data Analysis, ASTM STP 738,* S. J. Hudak, Jr. and R. J. Bucci, Eds., American Society for Testing and Materials, Philadelphia, 1981, p. 5.

[8] Fuchs, H. O. and Stephens, R. I., *Metal Fatigue in Engineering,* Wiley Interscience, New York, 1980.

[9] Larsen, J. M., Jira, J. R., and Weerasooriya, T., this publication, pp. 896–912.

James M. Larsen,[1] *Jay R. Jira,*[1] *and Tusit Weerasooriya*[2]

Crack Opening Displacement Measurements on Small Cracks in Fatigue

REFERENCE: Larsen, J. M., Jira, J. R., and Weerasooriya, T., **"Crack Opening Displacement Measurements on Small Cracks in Fatigue,"** *Fracture Mechanics: Eighteenth Symposium, ASTM STP 945,* D. T. Read and R. P. Reed, Eds., American Society for Testing and Materials, Philadelphia, 1988, pp. 896–912.

ABSTRACT: A computerized interferometric displacement gage was used to monitor the crack mouth opening behavior of naturally initiated small fatigue cracks in the high-strength alloy Ti-6Al-2Sn-4Zr-6Mo. Complete load versus crack opening displacement curves were obtained for naturally initiated semicircular cracks of surface length as small as 50 μm. For a crack of this size, the maximum crack mouth opening displacement was less than 0.5 μm, and the precision of the individual displacement measurements was approximately 0.01 μm. An excellent correlation between crack mouth opening compliance and surface crack length was demonstrated, and the data were shown to agree with an analytical expression for part-through surface cracks. A matrix of experiments was performed for a range of loading conditions, and crack growth data were plotted versus the nominal ΔK and ΔK_{eff}, which was defined using crack closure measurements. Experimental measurements obtained for a small crack indicated that closure often develops discontinuously as the crack initially extends. This behavior was speculated to result from the formation of individual asperities on the fracture surface. In general, the interferometric method proved to be very effective and demonstrated a unique capability to monitor the behavior of very small fatigue cracks.

KEY WORDS: automated, crack propagation, fatigue (materials), fracture mechanics, mechanical properties, minicomputer, microcrack, short crack, small crack, test methods

Over the past decade a number of investigators have observed that, compared with long cracks, very small fatigue cracks may grow anomalously fast in conventional specimens subjected to a nominally equivalent crack driving force. Additionally, small cracks have often been observed to propagate under loading conditions that were well below the threshold stress intensity factor range (ΔK_{th}) that exists for long cracks. This apparent failure of conventional linear elastic fracture mechanics (LEFM) methods to predict the growth of very small cracks has important implications. Life prediction methodologies for fatigue crack growth, such as those being implemented by the U.S. Air Force for some advanced gas turbine engines [*1–4*], may seriously underestimate the actual growth rate of small cracks. Such methodologies predict the growth of a fatigue crack from a minimum detectable crack size to failure. As long as the minimum detectable crack size is larger than that corresponding to the small-crack regime, conventional fracture mechanics methods will be adequate. For some materials and initial crack sizes, however, the growth rates of small cracks may be seriously underpredicted. Until a fundamental understanding of the mechanics and microstructural aspects of small-crack behavior has

[1]Materials research engineer and materials engineer, respectively, Materials Laboratory Air Force Wright Aeronautical Laboratories, AFWAL/MLLN, Wright-Patterson Air Force Base, OH 45433.
[2]Research scientist, University of Dayton Research Institute, Dayton, OH 45469.

been attained, the practical crack size limit for the valid application of available fracture mechanics methods will be uncertain.

The inability of fracture mechanics methods to correlate the behavior of both large and very small cracks, sometimes termed the "small-crack effect," has been extensively reviewed [*5–8*]. The anomalous behavior of small cracks has been attributed to a number of factors that include (1) violation of the continuum assumption of fracture mechanics (a microstructural effect), (2) violations of the assumptions of LEFM (for example, that small-scale yielding exists at the crack tip), or more generally, (3) a failure to maintain similitude between the processes that control the propagation of small and large fatigue cracks.

It has often been suggested that the anomalous behavior of small fatigue cracks may be due, at least in part, to differences in the crack closure response for small and large cracks. Essentially, the phenomenon of crack closure (recently reviewed in Refs *9* and *10*) may be described as the premature contact of the crack's fracture surfaces during the unloading portion of a fatigue cycle. Crack closure has been shown to be produced by three primary mechanisms associated with fracture surface plasticity [*11,12*], oxides [*13*], and fracture surface roughness [*14*]. As a result of crack closure, the tip of a crack that is subjected to a nominal applied stress intensity factor range of $\Delta K = K_{max} - K_{min}$ may actually experience a lesser effective stress intensity factor range, $\Delta K_{eff} = K_{max} - K_{cl}$, where K_{cl} = closure stress-intensity factor. Thus, variations in closure level may alter the effective driving force for fatigue crack propagation. This concept is particularly applicable to small cracks. A crack of zero length cannot experience closure, and the full development of closure is argued to require some finite crack extension. The result is a lack of similitude between the closure levels of small and large cracks, and this lack of similitude corresponds to a difference in effective crack driving force for the two nominally equivalent cracks. Newman [*15*] has used the finite-element method to show that the development of closure can account for the anomalous behavior of small cracks under conditions for which plasticity induced closure is dominant. However, experimental measurement of the closure level of small cracks is extremely difficult, and few data exist in the literature.

The experimental determination of crack closure level is generally accomplished by measuring crack opening displacement as a function of applied load. Although this is a relatively simple procedure for large cracks in conventional specimens, similar measurements for small cracks are much more difficult for two primary reasons. The magnitude of crack mouth opening displacement (CMOD) that must be measured for small cracks is very small, and CMOD must be measured at locations very near the crack mouth in order to attain sufficient sensitivity to the local displacement field of the crack. The following discussion describes a unique experimental method that overcomes these difficulties. Crack length and closure behavior of very small cracks are determined from very precise measurements of crack opening displacement made very near the crack mouth. Data are presented for a matrix of tests of a high-strength titanium alloy tested over a range of loading conditions.

Experimental Approach

The general approach used during this investigation followed a number of guidelines. The guidelines were established in order to provide an opportunity to make crack closure measurements on physically small cracks without the complication of having clearly violated the founding assumptions of linear elastic fracture mechanics. The small-crack growth tests were performed on smooth specimens under nominally elastic loading, and the scale of alloy microstructure was kept intentionally small. Aggressive environments (other than air) were avoided. To avoid complications associated with the use of an artificial crack starter, only naturally initiated small cracks were examined. Finally, crack growth data generally were taken only from small cracks that were sufficiently separated to insure noninteracting stress fields.

Material

All testing was performed on the alloy Ti-6Al-2Sn-4Zr-6Mo, whose specific chemical composition is given in Table 1. This material was selected due to its extensive use for high-performance turbine engine components whose lives are calculated using fracture mechanics methods to predict the growth of initially small fatigue cracks. The material as acquired had been cast and isothermally forged at 913°C to the shape of circular disks approximately 50 mm thick and 500 mm in diameter and heat-treated as follows: 921°C, 2 h, fan air-cooled; 824°C, 2 h, water-quenched; 593°C, 8 h, and air-cooled. This heat treatment produced a fine duplex microstructure of equiaxed primary α-phase (hexagonal close-packed) in a matrix of Widmanstätten $\alpha + \beta$ (body-centered cubic) phase. The material had a high strength (σ_y = yield strength = 1160 MPa; σ_u = ultimate strength = 1230 MPa) and was essentially isotropic as determined by sonic and mechanical testing.

All small crack testing was performed using the specimen shown in Fig. 1 [*16*]. This simple geometry employed a mild notch to initiate surface cracks naturally in a localized field. The elastic stress concentration factor (K_t) due to the notches was determined by a two-dimensional elastic finite-element stress analysis to be $K_t = 1.037$. Thus, the through-thickness stress was essentially uniform, while the reduced gage section effectively localized crack initiation within a small region. In order to eliminate surface residual stresses and roughness produced during machining, the gage sections of all specimens were carefully electropolished to a depth of at least 0.20 mm. The electropolishing also produced a highly reflective surface that enhanced the detection and resolution of small surface cracks while highlighting the material's microstructure.

The general procedure for small crack testing was as follows. A small precrack was initiated under fully reversed axial loading ($\sigma_{min}/\sigma_{max} = R = -1.0$; $\sigma_{max} = 0.6\sigma_y$; frequency = 20 Hz) and detected visually. The specimen was removed from the testing machine in order to place Vickers hardness indentations above and below the crack mouth for use with crack opening displacement measurements. Subsequently, the specimen was replaced in the testing machine, and crack opening displacement measurements were obtained periodically as the crack was extended under constant load amplitude fatigue at a selected stress level and stress ratio.

A scanning electron micrograph of a small crack is presented in Fig. 2, and the fine microstructure of the Ti-6Al-2Sn-4Zr-6Mo alloy is evident. The crack shown represents the approximate lower limit in crack size ($2c \approx 10\ \mu m$) that could be detected by optical inspection. The photograph also shows a pair of microhardness indentations for use in making crack opening displacements. The crack opening displacement measurements were effective only for cracks of surface length greater than approximately 40 μm. The axis of loading was parallel to a line through the indentations. Although some of the cracks observed in this investigation initiated on planes that were inclined to the axis of loading, most of the small cracks quickly took on a macro-orientation that was principally normal to the axis of loading (Mode I). The shapes of the small surface cracks were found in a previous investigation to remain very nearly semicircular over a broad range in crack sizes [*16*].

TABLE 1—*Chemical composition of Ti-6Al-2Sn-4Zr-6Mo.*

Element	Weight %	Element	Weight %
Al	6.33	N	0.01
B	<0.01	O	0.10
C	0.02	Si	0.09
Cu	<0.01	Sn	2.07
Fe	0.12	Ti	remainder
H	0.0017	Y	0.0010
Mo	5.76	Zr	3.93

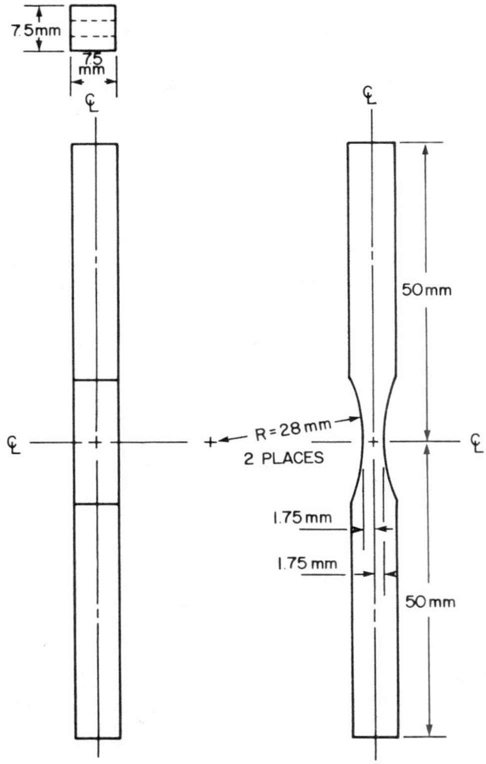

FIG. 1—*The small crack fatigue specimen.*

Interferometric Displacement Gage

The determination of mouth opening compliance and closure of small fatigue cracks was accomplished using a unique laser interferometric displacement gage (IDG) originally developed by Sharpe (see, for example, Refs *17–19*). In the present application, the method utilizes the coherent light of a helium-neon (He-Ne) laser to make very precise measurements of mouth opening displacement across a small surface crack. Figure 3 illustrates the basic principle schematically. A beam of laser light illuminates a pair of Vickers hardness indentations, placed approximately 50 μm apart, that span the crack mouth. Each indentation has the shape of a regular four-sided pyramid, and the four pairs of parallel faces of the indentations reflect the laser light to four positions in space. Of the four reflections, the two shown in the figure are used for CMOD measurement. Each of the pairs of reflections interferes optically to produce observable fringe patterns as illustrated. The relationship between a fringe pattern and the gage length d separating the indentations is given by

$$n\lambda = d \sin(\alpha_0) \tag{1}$$

where

n = order of the interference fringe,
λ = wavelength of the laser light (0.6328 μm), and
α_0 (~44 deg) = reflection angle shown in the figure.

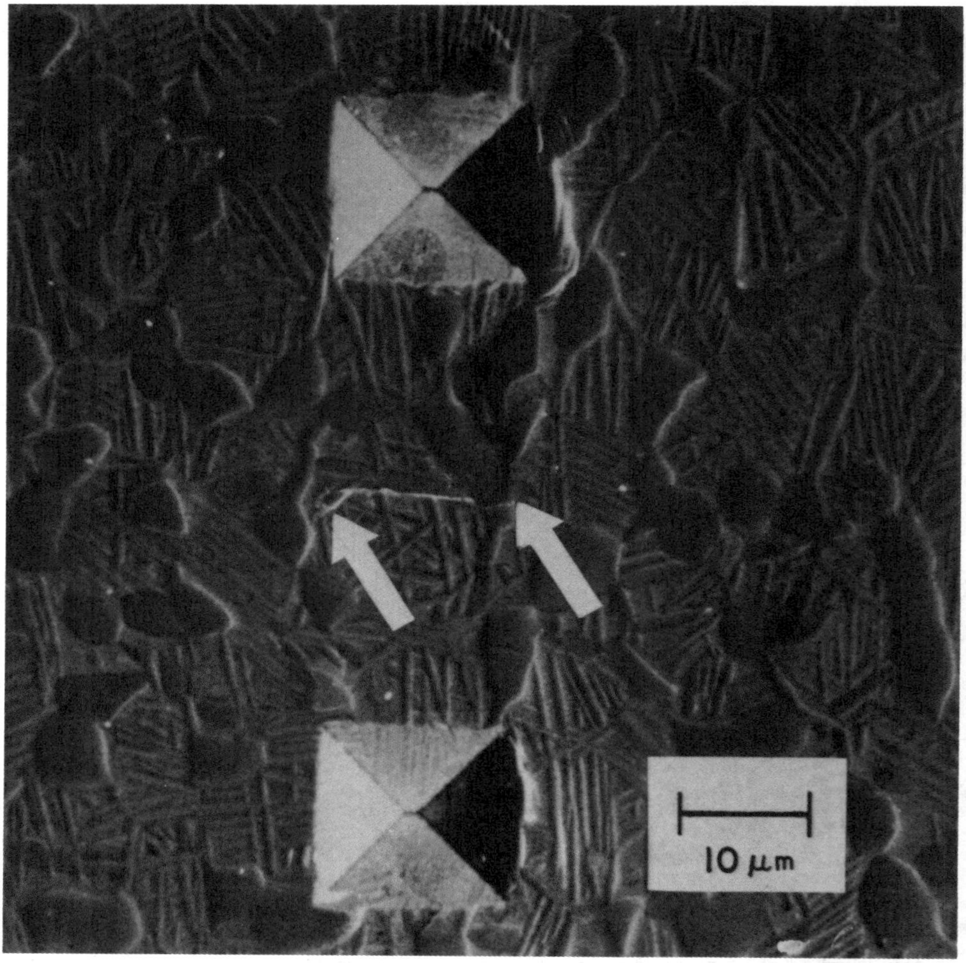

FIG. 2—*Scanning electron micrograph showing a small crack in the alloy Ti-6Al-2Sn-4Zr-6Mo. The Vickers hardness indentations are used in making CMOD measurements.*

Solving for d and differentiating gives

$$\delta d = \delta n \lambda / \sin(\alpha_0) \approx 0.9 \ \mu\text{m} \tag{2}$$

for this geometry and $\delta n = 1$. Thus, by simply counting the fringes that pass a fixed point in space, displacements between the indentations can be resolved to approximately 0.9 μm.

Improved displacement resolution was attained by monitoring fractional fringe motion. This was accomplished using the system illustrated schematically in Fig. 4. Servocontrolled rotating mirrors were used to sweep the fringe patterns across the photodetectors located at α_0. This allowed the system to monitor the fringe pattern as a function of time (indentation displacement), and real time numerical determination of the fringe position was performed utilizing a Digital Equipment Corp. 1124 minicomputer. This procedure could locate interference fringe position to within approximately 0.01 of the fringe wavelength, and this corresponded to a measurement precision for displacement between the indentations of $\delta d \approx 0.009$ μm (90 Å). This

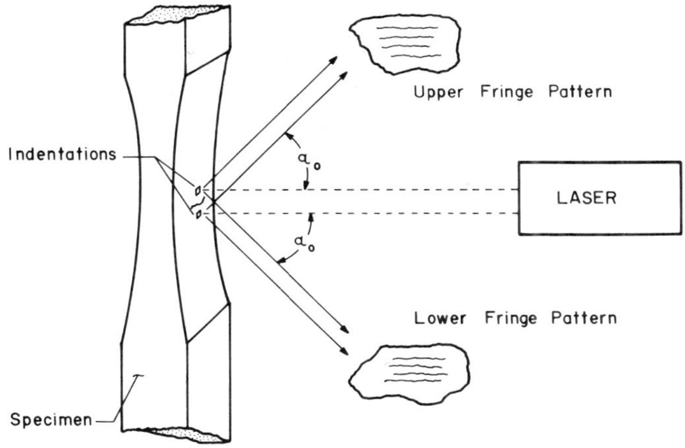

FIG. 3—*Schematic illustrating the principle of operation of the interferometric displacement gage.*

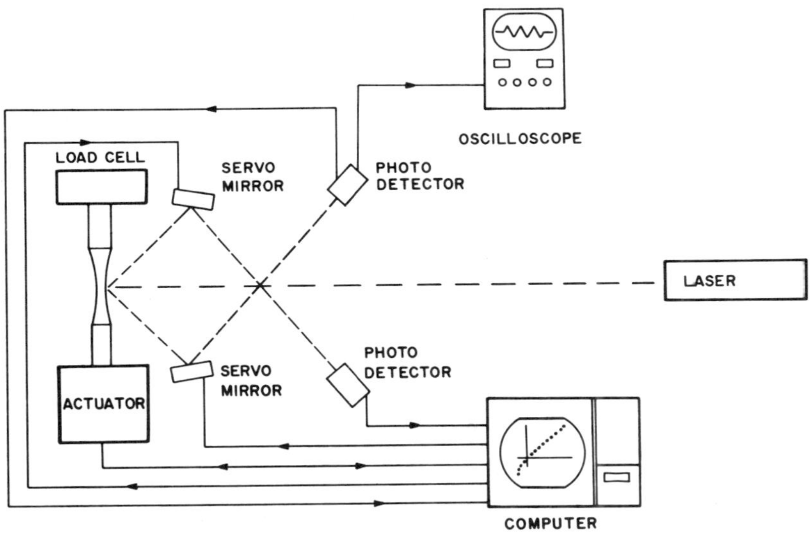

FIG. 4—*Schematic illustration of the computerized interferometric system.*

precision is approximately equivalent to the resolution of scanning electron microscopy (SEM), and the numerical nature of the data produced by the interferometric system made this method more efficient and potentially much more accurate than methods that use SEM for the measurement of CMOD.

Experimental Results and Discussion

Compliance and Closure Measurements on Small Cracks

Crack opening displacement measurements were obtained periodically by interrupting constant load amplitude sinusoidal fatigue cycling to apply a low-frequency (0.025 Hz) loading cycle having a triangular waveform. During this cycle, 280 load versus displacement measurements were made, and data were obtained such as those presented in Fig. 5 for a small crack of surface length $2c = 74$ μm. For this case, the total CMOD measured under maximum load was approximately 0.5 μm. The extremely high precision of the individual load-displacement measurements is evident in the figure. The data corresponded to a complete fatigue cycle, and essentially no load-displacement hysteresis occurred in this case. The effective size of the small crack was calculated from the elastic compliance determined from the slope of the linear region of the data. The crack closure load (P_{cl}) was determined from the break in the load versus differential displacement curve as illustrated. Here, the differential load-displacement data were obtained by fitting a line to the upper linear portion of the load-displacement data and subtracting the resulting linear elastic component of crack opening from the actual measured displacement given by individual data points. The differential load-displacement data were easier to interpret and simplified the determination of crack closure load. For the data shown ($R =$ 0.1), crack closure occurred at approximately 57% of the maximum load.

As discussed, the crack mouth opening compliance was measured at periodic intervals throughout the test. Typical compliance data are plotted against optically measured surface crack half-length in Fig. 6. This particular crack was grown under $R = 0.1$ constant load amplitude fatigue with $\sigma_{max} = 0.6\sigma_y$, and crack length ranged from $2c = 58$ μm to over 1 mm at failure. The experimental data corresponded closely to the solid line which represents an analytical relationship for a semicircular surface crack [20]. The excellent correlation between the experimental data and the analytical expression allowed the compliance measurements to be

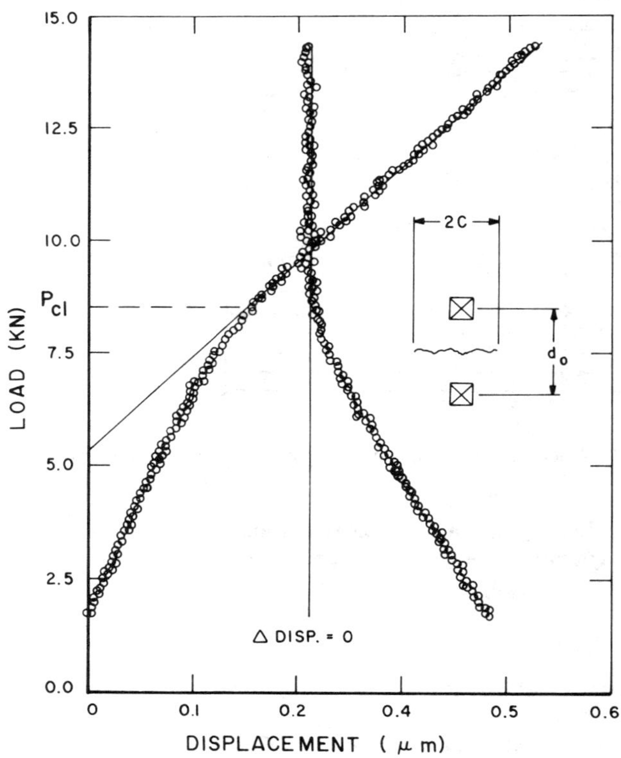

FIG. 5—*Typical load versus CMOD data for a small surface crack. Differential compliance data are also shown.*

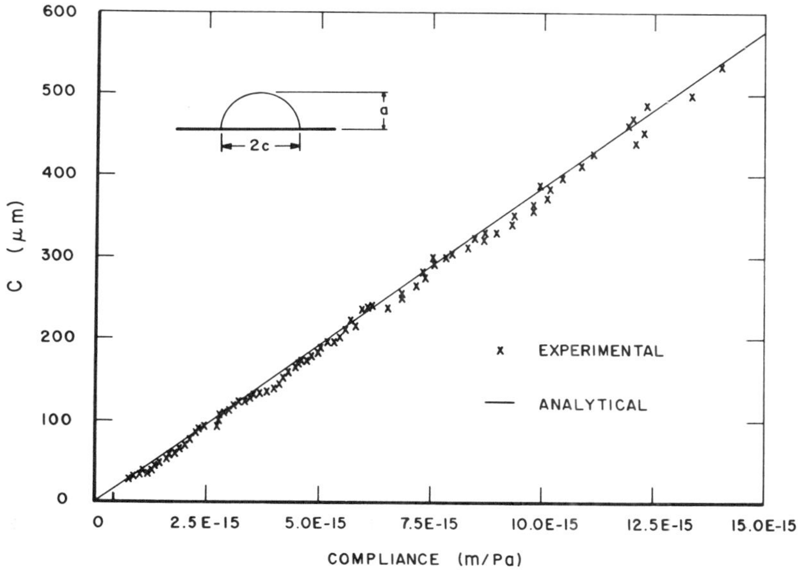

FIG. 6—*Crack length versus compliance data for a small surface crack.*

used to determine the effective length of small surface cracks. In fact, most of the variability in the data presented was associated with the optical crack length measurements which were made using a long focusing microscope mounted with a carefully calibrated filar eyepiece. Typically, the standard deviation of repeated crack length calculations determined from compliance measurements was less than 3 μm.

Corresponding to the results shown in Fig. 6, data showing crack closure level as a function of crack length are illustrated in Fig. 7. Over the range in crack size shown, the crack closure loads were greatest for the shortest crack lengths and diminished as the crack extended. The closure measurements varied significantly for $2c \leq 200$ μm and became more consistent as the crack propagated farther. The crack closure level changed abruptly at the point indicated by the arrow. The discontinuity in the measured closure load corresponded precisely to the bifurcation of one of the visible tips of the surface crack. During this same period, however, the actual crack length increased in a continuous manner as determined from compliance and as measured optically. It is speculated that the increase in closure load was produced by the abrupt formation of a significant fracture surface asperity and that the subsequent decline in closure load resulted as the crack tip moved away from the newly formed asperity. However, it has been reported that crack closure measurements performed using elastic compliance techniques depend on measurement location [*21*]. It is possible that some of the closure observations reported here were influenced by the nearness of the measurement location to the crack tip. However, considering all of the crack closure data collected in this investigation, no consistent pattern of observations was found to indicate that the measurements were biased by measurement location. Additional data are required to assess fully the importance of measurement location with respect to closure measurements on small cracks.

A number of experiments were performed to investigate the nature of crack growth and closure development for small cracks propagated under a range of loading conditions. The preceding figures typify data generated during testing outlined in the matrix given in Table 2. All tests were performed in room temperature laboratory air using naturally initiated surface cracks. Fatigue conditions ranged from fully reversed cycling ($R = -1.0$), where the effects of closure

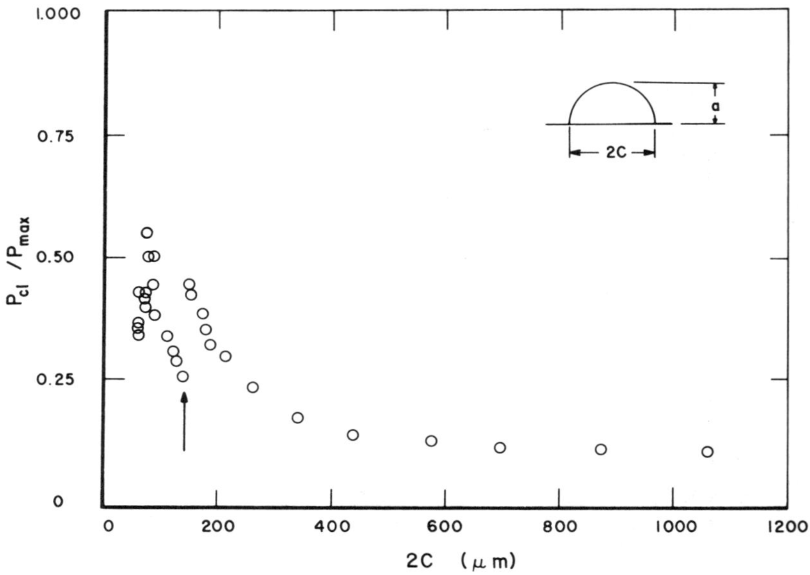

FIG. 7—*Data of normalized crack closure load versus length of a small crack.*

TABLE 2—*Small-crack testing conditions.*

σ_{max}/σ_y	$R = -1.0$	$R = 0.1$	$R = 0.5$
0.6	X	X	X
0.9	X	X	X

were expected to be most accentuated, to an $R = 0.5$ condition for which the effect of closure was expected to be minimal. Although all testing was performed under nominally elastic conditions, maximum stress levels for the tests ranged from 0.6 to 0.9 times the material's yield strength in order to significantly influence the extent of crack tip plasticity. The crack propagation data were reduced to the form of crack growth rate (da/dN) versus nominal applied stress intensity factor range (ΔK) using the stress intensity factor solution of Newman and Raju [*22*] and an incremental polynomial regression [*16*].

Figures 8–10 present crack growth rate data plotted as a function of nominally applied ΔK for the full matrix of tests. Figure 8 shows data for the growth of small cracks under the $R = 0.5$ test condition and the two stress levels. The data fall slightly above the solid line which represents $R = 0.5$ data for "large" cracks in conventional compact type (CT) specimens of 40-mm width and 10-mm thickness. Figure 9 illustrates similar data for testing at a stress ratio of $R = 0.1$, and the small crack data fall slightly below the long crack trend curve. The open symbols in this figure are from the same test as the data presented in Figs. 5–7. Figure 10 presents the data from three tests under $R = -1.0$ loading (two tests had $\sigma_{max}/\sigma_y = 0.9$). The small crack ΔK values were calculated using only the positive portion of the loading range, while the long crack data are from $R = 0.1$ testing. The small crack data from the high stress testing lie above the

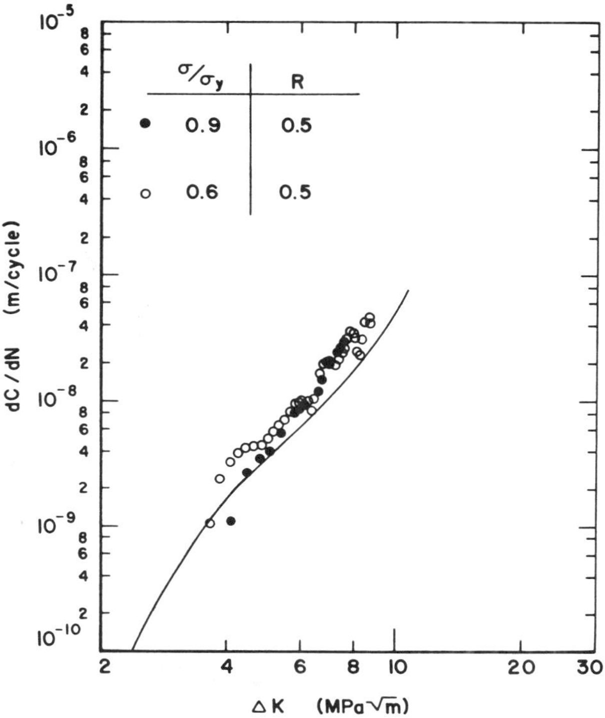

FIG. 8—*Crack growth rate data plotted versus* ΔK *for small surface cracks as compared with the trend line for long cracks in conventional CT specimens.*

long crack trend, while the small and large cracks grow at approximately equal rates at the lower stress.

The crack growth rate data just discussed were also plotted versus ΔK_{eff} (defined as $K_{max} - K_{cl}$). The values of K_{cl} were determined from interferometric displacement gage measurements for the small cracks, while K_{cl} for the large cracks in CT specimens was determined from crack opening displacement measurements obtained using a conventional extensometer. Since K_{min} under $R = 0.5$ fatigue was consistently greater than K_{cl}, the ΔK_{eff} data for this loading condition were identical to those of Fig. 8. The ΔK_{eff} data for loading under $R = 0.1$ and $R = -1.0$ are presented in Figs. 11 and 12, respectively. The $R = 0.1$ data generally agree well with the long-crack trend when plotted against ΔK_{eff}. However, at the smallest values of ΔK_{eff}, the small-crack data fall above the long-crack trend. The data for the $R = -1.0$ tests (Fig. 12) are even more unusual. At this stress ratio, the data for the low stress condition were less effectively correlated by ΔK_{eff} than by the nominal ΔK. Furthermore, some of the data of Fig. 12 display a decreasing growth rate with an increase in ΔK_{eff}. The remainder of the data follow the long-crack trend.

The ability of ΔK (and ΔK_{eff}) to correlate the data of large and small cracks was quite remarkable. This was particularly true for the testing under conditions for which $R = -1.0$ and $\sigma_{max} = 0.9\sigma_y$. Figure 13 presents a photograph of a length of a crack tested under these conditions (Fig. 12 shows the associated crack propagation data). As was generally the case, the crack path often followed the habit plane of the Widmanstätten α-platelets and proceeded transgranularly across the equiaxed primary α-grains. The photograph of this unloaded specimen also reveals some secondary cracking, and a residual CMOD is evident. The residual CMOD is

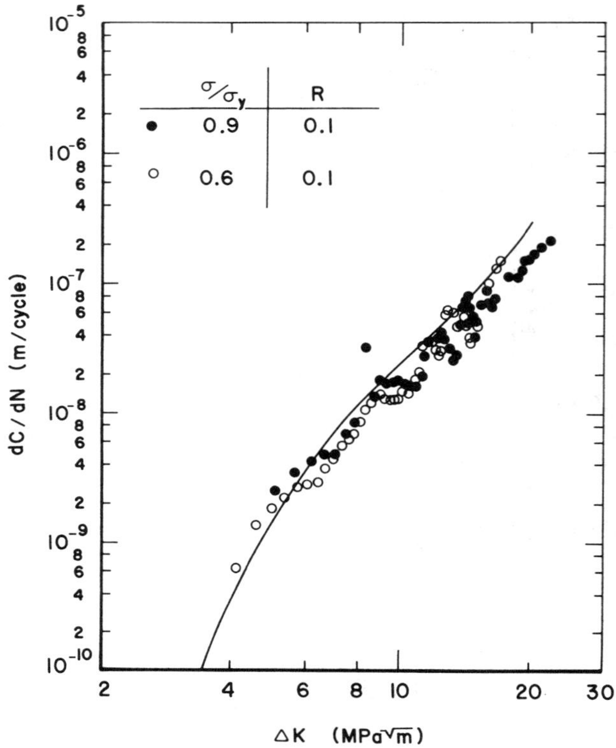

FIG. 9—*Crack growth rate data plotted versus* ΔK *for small surface cracks as compared with the trend line for long cracks in conventional CT specimens.*

consistent with the observation of negative crack closure loads throughout this test. Apparently, the extensive crack tip plasticity that developed under the tensile portion of the fully reversed cycle produced a residual crack opening that was closed only when the specimen was sufficiently loaded into compression. Thus, K_{cl} was negative, and the nominal applied ΔK (calculated for tensile loading only) was less than ΔK_{eff}, which more effectively correlated these particular small and large crack data. The utility of ΔK and ΔK_{eff} becomes even more impressive when one inspects the same crack at a lower magnification, Fig. 14. The photograph shows the crack to have followed a tortuous path, often deflecting at microstructural features, and extensive secondary cracking is evident. In view of the very small initial crack sizes tested and violations of the basic assumptions of linear elastic fracture mechanics, the data from the entire matrix of tests conducted represent a remarkable endorsement of LEFM.

As stated earlier, crack closure was expected to be developed fully only after some finite crack extension. During the initial transient period, the driving force for growth of the small crack (estimated by ΔK_{eff}) was expected to exceed that of a large crack subjected to a nominally equivalent ΔK. However, except for an occasional minor exception, the closure behavior of small and large cracks was equivalent and the associated crack growth rates were similar. The similarity between the propagation behavior of small and large cracks was expected to occur for small cracks that were larger than a critical size. The approximate size below which a small crack was

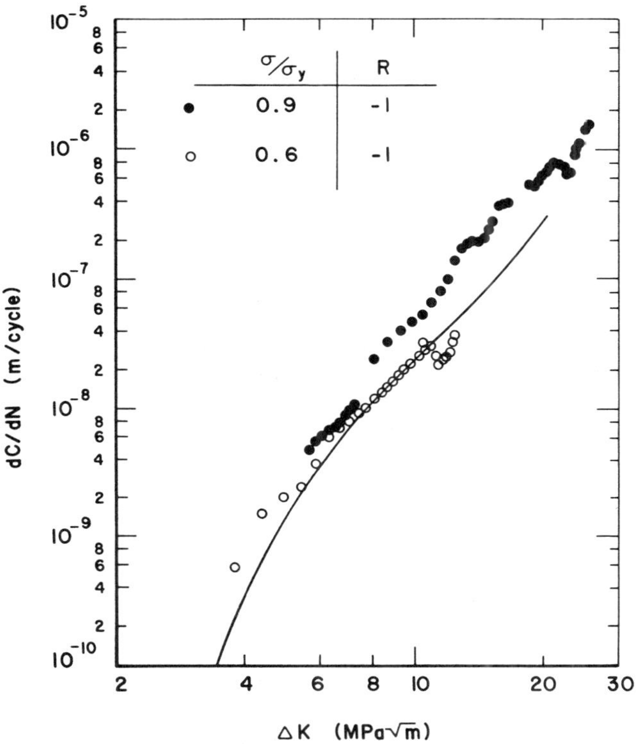

FIG. 10—*Crack growth rate data plotted versus* ΔK *for small surface cracks as compared with the trend line for long cracks in conventional CT specimens.*

expected to exhibit unusually rapid rates of fatigue crack propagation was estimated by the equation $a_0 = (\Delta K_{th}/\Delta\sigma_e)^2/\pi$ [*23*], where ΔK_{th} is the material's threshold stress-intensity factor range, and $\Delta\sigma_e$ is the fatigue limit of the material. Since both ΔK_{th} and $\Delta\sigma_e$ depend on stress ratio, a_0 was calculated to range from approximately 55 to 75 μm for fatigue under the three stress ratios ($R = -1.0$, 0.1, and 0.5).

In a recent review of carefully selected experimental data, Lankford [*24*] concluded that necessary conditions for the anomalously rapid growth of small fatigue cracks were that $r_p/a \geq 0.05$ and $r_p/M < 1$, where r_p is the plastic zone size ($\sim 0.15(K_{max}/\sigma_y)^2$), a is the crack depth, and M is the relevant microstructural element size. For semicircular surface cracks that are small with respect to the specimen dimensions, the stress-intensity factor solution of Newman and Raju [*22*] simplifies to

$$K = 1.174\sigma\sqrt{a} \tag{3}$$

and substitution into the expression for r_p/a indicates that an expression that is equivalent to the first criterion is

$$\sigma_{max}/\sigma_y \geq 0.49 \tag{4}$$

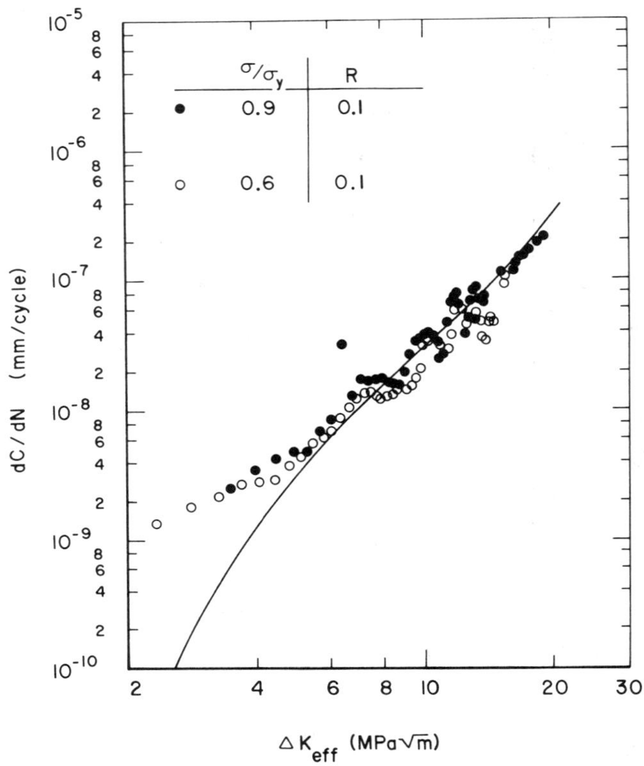

FIG. 11—*Crack growth rate data plotted versus* ΔK_{eff} *for small surface cracks as compared with the trend line for long cracks in conventional CT specimens.*

All of the present experiments satisfy this condition. Substituting Eq 3 into the second criterion indicates that rapid growth of small cracks may occur when

$$a \leq 4.84M\ (\sigma_y/\sigma_{max})^2 \tag{5}$$

For the current material, the critical microstructural element is probably of a size ranging from that of the primary α-grains ($\sim$5 μm) to the size of the prior β-grains ($\sim$15 μm). Using these two dimensions, the second criterion therefore predicts the rapid growth of small cracks for crack depths below approximately 30 and 90 μm, respectively, for the $\sigma_{max}/\sigma_y = 0.9$ experiments and 67 and 200 μm for the $\sigma_{max}/\sigma_y = 0.6$ tests. Thus, at least the low stress experiments should have demonstrated the rapid growth of small cracks.

Although the data presented did not display anomalous small crack behavior, the experiments provided an optimum test of the capabilities of the interferometric displacement gage. The interferometric system demonstrated a unique capability for making crack opening displacement measurements of small fatigue cracks. The very short gage length and extreme measurement precision that are characteristic of the interferometric technique offer potential for novel research of the mechanisms that control the growth of small cracks. Crack length and crack closure data can be acquired as a small crack extends, and the natural development of the steady-state crack growth

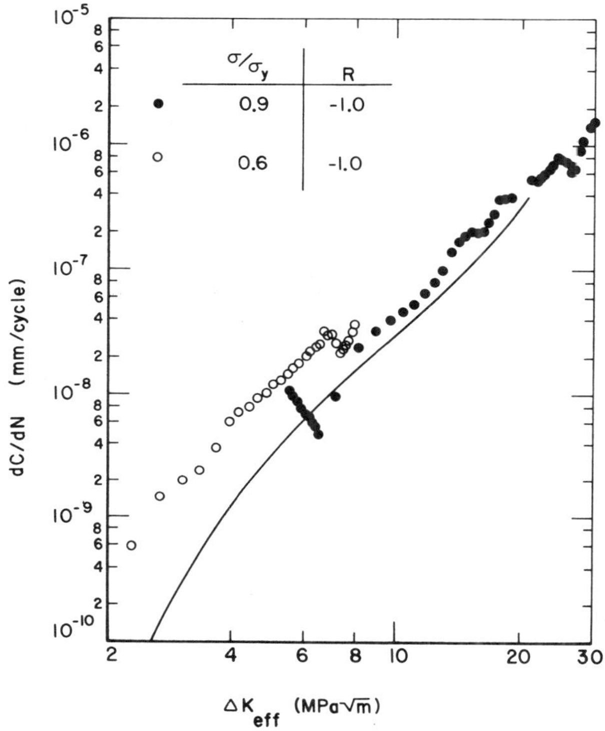

FIG. 12—*Crack growth rate data plotted versus* ΔK_{eff} *for small surface cracks as compared with the trend line for long cracks in conventional CT specimens.*

process can be monitored. Crack closure data such as those presented have previously been difficult or impossible to obtain. The computerized interferometric system can acquire and analyze these data extremely efficiently and quickly.

Summary and Conclusions

A matrix of tests was performed to investigate the behavior of small, naturally initiated surface cracks. A real-time computerized interferometric system was used to monitor CMOD as a function of applied load. These data were used to determine both effective crack length from compliance and the crack closure load. Data were obtained for semicircular cracks of length ranging from approximately 50 μm to 2 mm at fracture. For a 50-μm crack, the maximum CMOD was approximately 0.5 μm and the precision of the individual displacement measurements was of the order of 0.01 μm. An excellent correlation between crack mouth opening compliance and surface crack length was demonstrated, and the data were shown to agree with an analytical expression for part-through surface cracks. Crack closure measurements were obtained for tests performed at three stress ratios ($R = -1.0$, 0.1, and 0.5) and two stress levels ($\sigma_{max}/\sigma_y = 0.6$ and 0.9). It was found that closure often develops discontinuously as the crack initially extends. This behavior was speculated to result from the formation of individual asperi-

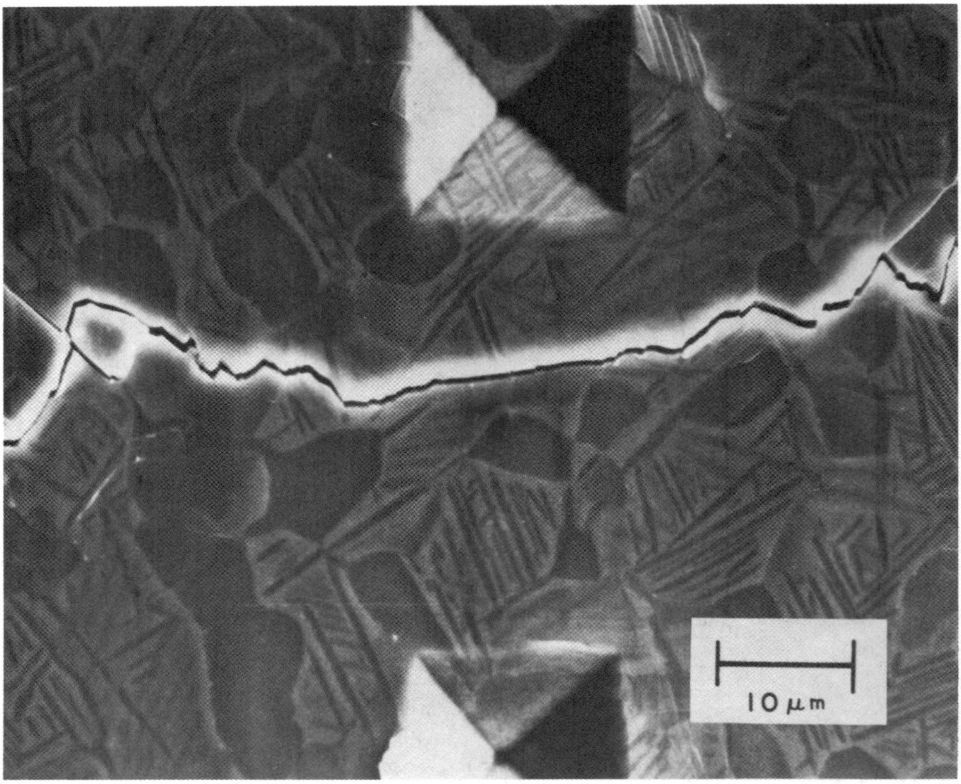

FIG. 13—*Photograph taken at the end of a small-crack test (*R $= -1.0$; $\sigma_{max} = 0.9\sigma_y$*), showing a residual crack opening under zero load.*

ties on the fracture surface. Crack growth rate data from the matrix of tests were plotted versus ΔK and ΔK_{eff} and compared with equivalent data from conventional tests of specimens containing large cracks. Although various estimates predicted otherwise, the small- and large-crack data were both well consolidated by ΔK and ΔK_{eff}.

Acknowledgment

This research was performed under Materials Laboratory in-house research project No. 2302P101, Air Force Wright Aeronautical Laboratories (AFWAL), Wright-Patterson Air Force Base, OH. The authors wish to acknowledge the helpful discussions of Dr. T. Nicholas of the Materials Laboratory and R. P. Sincavage of the Air Force Aeronautical Systems Division, who was principally responsible for the early development of the interferometric system. One of the authors (J. M. Larsen) also wishes to acknowledge helpful discussions with Dean J. C. Williams and Professor A. W. Thompson of Carnegie-Mellon University, Pittsburgh, PA. J. M. Larsen's contribution to the research discussed in this paper was performed as partial fulfillment of the requirements for the degree Doctor of Philosophy in Metallurgical Engineering and Materials Science at Carnegie-Mellon University.

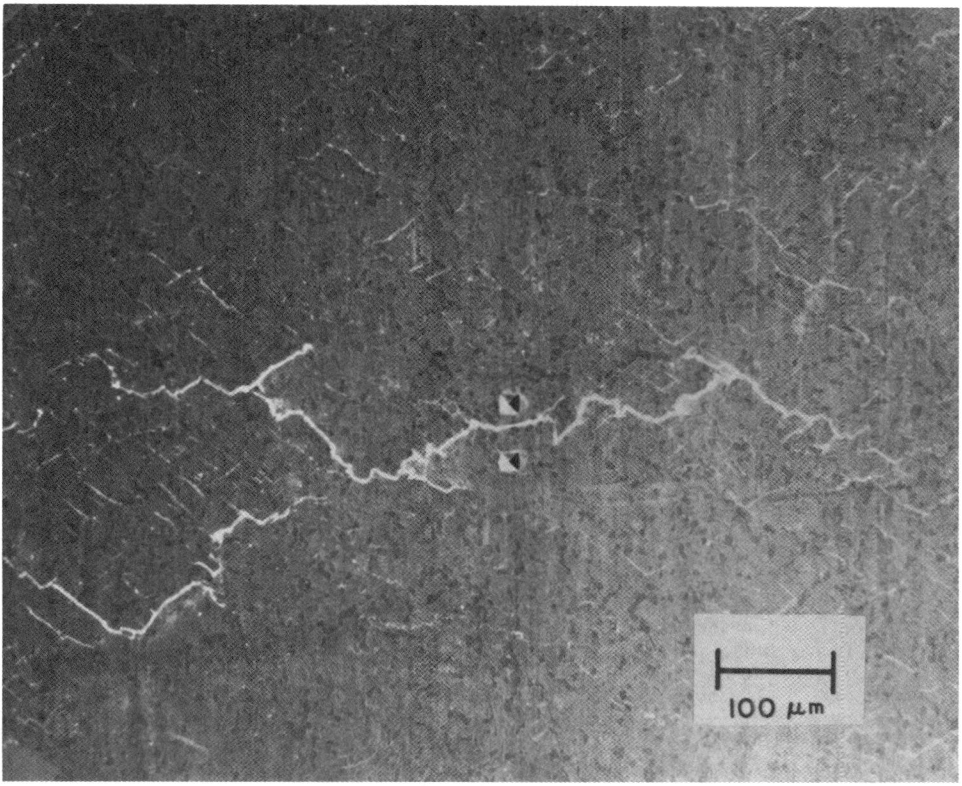

FIG. 14—*A lower-magnification photograph of the crack pictured in Fig. 13, showing a tortuous crack path and many secondary cracks.*

References

[1] Harris, J. A., Jr., Sims, D. L., and Annis, C. G., Jr., "Concept Definition: Retirement for Cause of F100 Rotor Components," Air Force Wright Aeronautical Laboratories Report AFWAL-TR-80-4118, Wright-Patterson Air Force Base, OH, 1980.

[2] "Engine Structural Integrity Program," U.S. Air Force, Military Standard 1783, Aeronautical Systems Division, Wright-Patterson Air Force Base, OH, 1984.

[3] King, T. T., Cowie, W. D., and Reimann, W. H., "Damage Tolerant Design Concepts for Military Engines," *Damage Tolerance Concepts for Critical Engine Components,* AGARD Conference Proceedings No. 393, NATO Advisory Group for Aerospace Research and Development, 1985, pp. 3.1–3.9.

[4] King, T. T., Hurchalla, J., and Nethaway, D., "United States Air Force Engine Damage Tolerance Requirements," presented at the 21st American Institute of Aeronautics and Astronautics/American Society of Mechanical Engineers/Society of Automotive Engineers Joint Propulsion Conference, Monterey, CA, July 1985.

[5] Hudak, S. J., Jr., "Small Crack Behavior and the Prediction of Fatigue Life," *Journal of Engineering Materials and Technology, Transactions,* American Society of Mechanical Engineers, Vol. 10, 1981, pp. 26–35.

[6] Suresh, S. and Ritchie, R. O., "Propagation of Short Fatigue Cracks," *International Metals Reviews,* Vol. 29, 1984, pp. 445–476.

[7] Leis, B. N., Kanninen, M. F., Hopper, A. T., Ahmad, J., and Broek, D., "A Critical Review of the

Short Crack Problem in Fatigue," Air Force Wright Aeronautical Laboratories Report AFWAL-TR-83-4019, Wright-Patterson Air Force Base, OH, 1983.

[8] Gangloff, R. P. and Ritchie, R. O., "Environmental Effects Novel to the Propagation of Short Fatigue Cracks" *Fundamentals of Deformation and Fracture,* B. A. Bilby, K. J. Miller, and J. R. Willis, Eds., Cambridge University Press, Cambridge, U.K., 1984, pp. 531-558.

[9] Banerjee, S., "A Review of Crack Closure," Air Force Wright Aeronautical Laboratories Report AFWAL-TR-84-4031, Wright-Patterson Air Force Base, OH, 1984.

[10] Allison, J. E., "The Measurement of Crack Closure During Fatigue Crack Growth," *Fracture Mechanics: Eighteenth Symposium, ASTM STP 945,* American Society for Testing and Materials, Philadelphia, 1987, pp. 913-933.

[11] Elber, W., "Fatigue Crack Closure Under Cyclic Tension," *Engineering Fracture Mechanics,* Vol. 2, 1970, pp. 37-45.

[12] Elber, W., "The Significance of Fatigue Crack Closure," *Damage Tolerance in Aircraft Structures, ASTM STP 486,* American Society for Testing and Materials, Philadelphia, 1971, pp. 230-242.

[13] Ritchie, R. O., Suresh, S., and Moss, C. M., "Near-Threshold Fatigue Crack Growth in 2¼Cr-1Mo Pressure Vessel Steel in Air and Hydrogen," *Journal of Engineering Materials and Technology, Transactions,* American Society of Mechanical Engineers, Vol. 102, 1980, pp. 293-299.

[14] Walker, N. and Beevers, C. J., "A Fatigue Crack Closure Mechanism in Titanium," *Fatigue of Engineering Materials and Structures,* Vol. 1, 1979, pp. 135-148.

[15] Newman, J. C., Jr., "A Nonlinear Fracture Mechanics Approach to the Growth of Small Cracks," *Behavior of Short Cracks in Airframe Components,* AGARD Conference Proceedings No. 328, NATO Advisory Group for Aerospace Research and Development, 1982, pp. 6.1-6.26.

[16] Larsen, J. M., "An Automated Photomicroscopic System for Monitoring the Growth of Small Fatigue Cracks," *Fracture Mechanics: Seventeenth Volume, ASTM STP 905,* American Society for Testing and Materials, Philadelphia, 1986, pp. 226-238.

[17] Sharpe, W. N., Jr., "Interferometric Surface Strain Measurement," *International Journal of Nondestructive Testing,* Vol. 3, 1971, pp. 59-76.

[18] Sharpe, W. N., Jr. and Martin, D. R., "An Optical Gage for Strain/Displacement Measurement at High Temperature Near Fatigue Crack Tips," Air Force Materials Laboratory Report AFML-TR-77-153, Wright-Patterson Air Force Base, OH, 1977.

[19] Sharpe, W. N., Jr., and Ward, M., "Benchmark Cyclic Plastic Notch Strain Measurements," *Journal of Engineering Materials and Technology, Transactions,* American Society of Mechanical Engineers, Vol. 105, 1983, pp. 235-241.

[20] Mattheck, C., Morawietz, P., and Munz, D., "Stress Intensity Factor at the Surface and at the Deepest Point of a Semi-Elliptical Surface Crack in Plates Under Stress Gradients," *International Journal of Fracture,* Vol. 23, 1983, pp. 201-212.

[21] D. E. Macha, D. M. Corbly, and J. W. Jones, "On the Variation of Fatigue-Crack-Opening Load with Measurement Location," *Experimental Mechanics,* 1979, pp. 207-213.

[22] Newman, J. C., Jr., and Raju, I. S., "Stress-Intensity Factor Equations for Cracks in Three-Dimensional Finite Bodies," *Fracture Mechanics: Fourteenth Symposium—Volume I: Theory and Analysis, ASTM STP 791,* J. C. Lewis and G. Sines, Eds., American Society for Testing and Materials, Philadelphia, 1983, pp. I-238-I-265.

[23] El Haddad, M. H., Smith, K. N., and Topper, T. H., "Fatigue Crack Propagation of Short Cracks," *Journal of Engineering Materials and Technology, Transactions,* American Society of Mechanical Engineers, Vol. 101, 1979, pp. 42-46.

[24] Lankford, J., "The Influence of Microstructure on the Growth of Small Fatigue Cracks," *Fatigue of Engineering Materials and Structures,* Vol. 8, 1985, pp. 161-175.

John E. Allison[1]

The Measurement of Crack Closure During Fatigue Crack Growth

REFERENCE: Allison, J. E., **"The Measurement of Crack Closure During Fatigue Crack Growth,"** *Fracture Mechanics: Eighteenth Symposium, ASTM STP 945,* D. T. Read and R. P. Reed, Eds., American Society for Testing and Materials, Philadelphia, 1988, pp. 913–933.

ABSTRACT: In recent years, the dominant role of fatigue crack closure in determining fatigue crack growth behavior in metals has been recognized. Attention is now turning to more precise quantification of this process and to the application of standardized measurement techniques. This paper reviews the various techniques which have been used to measure the crack opening stress intensity K_{OP}. These include thickness-averaging compliance methods, indirect methods, various techniques which measure crack closure on the surface near the crack tip, and techniques for measuring crack closure in the interior of thick specimens. The practical and fundamental advantages and limitations of these methods are discussed. Of these techniques, the thickness-averaging compliance techniques (back-face strain and crack mouth opening displacement methods) appear to provide reproducible and practical global measurements of the average closure response and are the most logical choices for standardization. The interferometric displacement gage and direct scanning electron microscope measurements although difficult to standardize, allow more detailed studies of local crack-tip closure processes. The increased use of techniques which measure closure in the specimen interior is important for understanding the three-dimensional nature of crack closure. Finally, this paper reviews selected crack closure data measured using thickness-averaging compliance techniques in several different alloy systems. Even when such "standard" techniques are used, significant variations in crack closure are observed, and general patterns for crack closure are difficult to discern. This points to the need for further standardization of measurement techniques and numerical data analysis methods.

KEY WORDS: fatigue (materials), fatigue crack growth, crack closure, metals and alloys, measurement techniques, crack-tip opening displacement (CTOD)

The existence of some form of crack closure during fatigue crack growth has long been postulated [*1,2*]; experimental indication of its existence were observed as early as 1962 [*3–6*]. The first attempts to quantify crack closure using a fracture mechanics framework were made by Elber [*5,7*], who defined two important parameters: the opening stress intensity K_{OP} and an effective ΔK given by the now well-known relationship

$$\Delta K^{EFF} = K_{max} - K_{OP} \tag{1}$$

Elber also introduced the concept that fatigue crack growth rates were a function of ΔK^{EFF} rather than the remote ΔK [*7*]. From this point an extensive literature on the subject evolved, and the decade after Elber's measurements was marked by significant controversy over the existence of crack closure and its importance [*8–17*].

It is now widely recognized that fatigue crack closure plays a dominant role in determining the fatigue crack growth (FCG) rates of metals. In many cases, the quite significant influences of such variables as microstructure, slip character, residual stresses, and load ratio have been

[1]Research staff, Metallurgy Department, Ford Motor Company, P.O. Box 2053, Dearborn, MI 48121-2053.

shown to be manifested largely through the crack closure process. A number of different closure mechanisms have been reported and reviewed in detail elsewhere [*18–22*]. Two such mechanisms have been widely observed, namely, plasticity-induced crack closure and asperity-induced crack closure. Plasticity-induced crack closure arises from elastic constraint acting on the plastically stretched material in the wake of the crack tip. As originally suggested by Elber [*7*], this leads to interference between mating fracture surfaces and thus crack closure. Asperity-induced crack closure (frequently called roughness-induced crack closure) arises due to the rough nature of fracture surfaces coupled with some mismatch of these asperities as the crack closes. The level of crack closure has been shown to increase with increasing fracture surface asperity height [*19,23*] and in principle should also increase with increasing asperity mismatch [*24*]. Another important crack closure mechanism is crack closure due to the build up of oxide debris in the crack-tip region, or oxide-induced crack closure [*25–27*]. Appreciable accumulation of oxide debris occurs only in certain specific alloy/microstructure/environment combinations and generally only at near-threshold growth rates.

Although crack closure effects are most pronounced at low ΔK, they can occur at higher ΔK as well, and in some cases their influence at high ΔK is quite substantial. An example of this is demonstrated in Fig. 1, using results from an investigation of FCG in the titanium-aluminum system [*19,23*]. In this system slip character is altered from wavy to planar slip by increasing aluminum content from 4 to 8 weight%; aging the Ti-8Al leads to coarse, planar slip (strain localization). As can be seen in Fig. 1*a*, the influence of slip character on FCG is quite large when crack growth rates are plotted as a function of the remote ΔK, with the threshold ΔK varying from 4 to 17 MPa $\sqrt{m}$. Moreover, the influence of slip character is evident over several decades of FCG rate and at relatively high levels of ΔK. When crack growth rates are plotted as a function of ΔK^{EFF}, as shown in Fig. 1*b*, the data fall in one central band, indicative of the fact that the influence of slip character shown in Fig. 1*a* is actually manifested through a crack

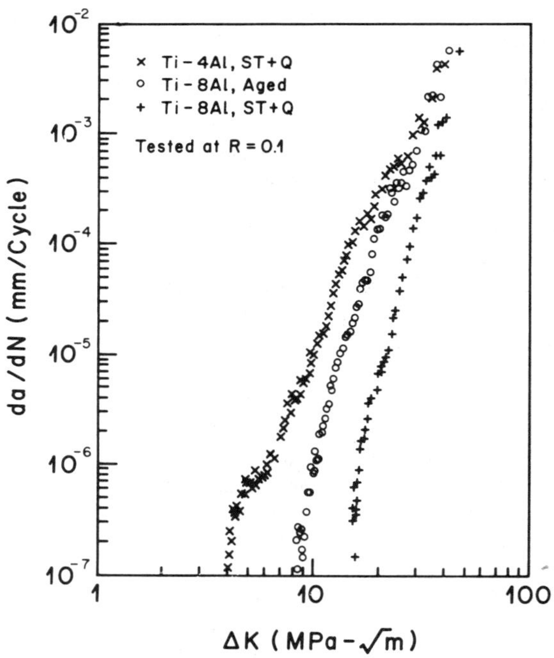

FIG. 1*a*—*Fatigue crack growth rates versus remote* ΔK *in titanium-aluminum alloys, showing the pronounced influence of slip character.*

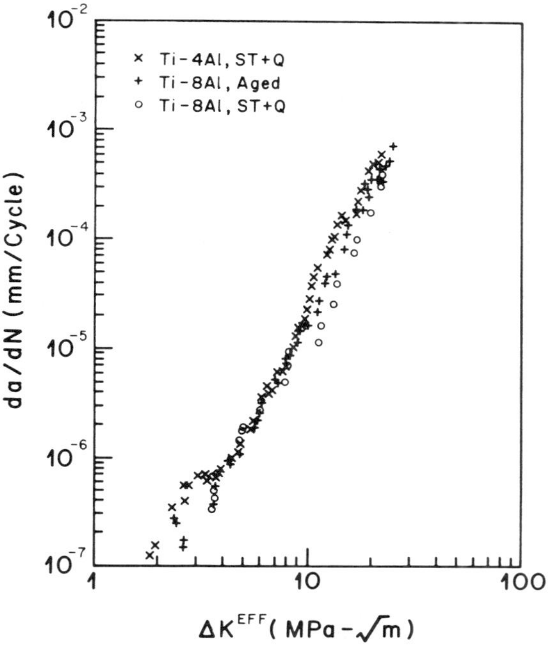

FIG. 1*b*—*Fatigue crack growth rates versus* ΔK^{EFF} *in titanium-aluminum alloys, showing little influence of slip character (Ref* 23*).*

closure process. Similar effects of crack closure over a broad range of ΔK and at relatively high crack growth rates have been observed in ferrous alloys [*28–30*].

With the increased awareness of the importance of crack closure comes a need for reliable determination of K_{OP}. Although analytical models for predicting K_{OP} have been developed [*24,31–36*], there remains a need for accurate and quantitative measurements of crack closure. The purpose of this paper is to review the various techniques which have been used to measure K_{OP} and to discuss their advantages and limitations. Important issues concerning the measurement and characterization of crack closure are discussed. For the sake of brevity, this paper is restricted to the measurement and characteristics of crack closure in standard, long crack specimens under constant cyclic amplitude (load or stress intensity) conditions.

Measurement Techniques

The various techniques which have been applied to the quantitative measurement of crack closure can be divided into four categories: global or thickness-averaging compliance techniques, indirect thickness-averaging techniques, surface techniques used primarily to measure closure near the crack tip, and techniques which measure closure in the specimen interior.

Thickness-Averaging Compliance Techniques

Crack Mouth Opening Displacement

Measurement of crack mouth opening displacement (CMOD) is historically the method used most often for quantifying a thickness-averaged (also called global or bulk) value of K_{OP}

[*12,20,22,37–60*]. It is accomplished most often using a clip gage mounted across the notch mouth and located either on the specimen load line or at the specimen edge. Alternates such as miniature clip gages or strain gages mounted on the specimen face across the crack plane and located well behind the crack tip have been also used [*22,39*] to give similar measurements. K_{OP} is determined by plotting load versus displacement and noting the deflection point as shown schematically in Fig. 2. The reciprocal of the slope of the upper portion of the load-displacement curve is a measure of the compliance of the specimen with the crack fully open. The reciprocal slope of the lower portion of the curve reflects the compliance of the specimen with the crack fully closed and is an indication of the distance behind the crack tip over which the crack is closed. In practice this deflection point is not always as distinct as shown in Fig. 2, and some interpretation is required as will be discussed in a subsequent section.

The attachment of clip gages must be accomplished with care because improper attachment or gage misalignment can lead to nonlinear or noisy load displacement curves. Friction between the loading pin and the specimen reportedly leads to similar effects and to rather pronounced hysteresis between the loading and unloading portions of the load-displacement curve [*42*]. Some researchers have found it necessary to eliminate loading pin/specimen friction by using specially designed low-bending clevis grips [*42*] or grips with roller bearings [*47,61*], although others have found that such procedures are not required [*62,63*]. At high stress intensities some hysteresis occurs due to significant crack-tip plasticity. Standard clip gages are limited to use at frequencies below 100 Hz due to their long attachment arms; however, special displacement gages have been used at considerably higher frequencies. When crack lengths are measured using electric potential techniques, the clip gage attachment points must be electrically insulated from the specimen. Clip gages equipped with quartz extension rods have been used for measuring crack closure at elevated temperature [*62,64*]; however, the increased length of these attachment arms causes a decrease in the maximum permissible operating frequency to approximately 5 Hz.

Even with proper attachment and specialized gripping, the deflection in the load-displacement curve due to crack closure is often indistinct and difficult to quantify using graphical techniques. A number of researchers have adopted a signal processing technique to enhance the sensitivity of the clip gage [*22,42,43,47,49,50,53,56,57*]. This technique, commonly called the "differential compliance" or "unloading compliance" technique [*65*], uses a plot of load versus differential displacement to exaggerate the deflection point in the load-displacement curve. In the calculation of the differential displacement, the reciprocal slope (that is, compliance, $C = \Delta\delta/\Delta P$) of the upper portion of the load-displacement curve is calculated during the unloading portion of the cycle. This value of the unloading compliance is then used in the following relationship to calculate the differential displacement, $\Delta\delta$

$$\Delta\delta = k(\delta - CP) \tag{2}$$

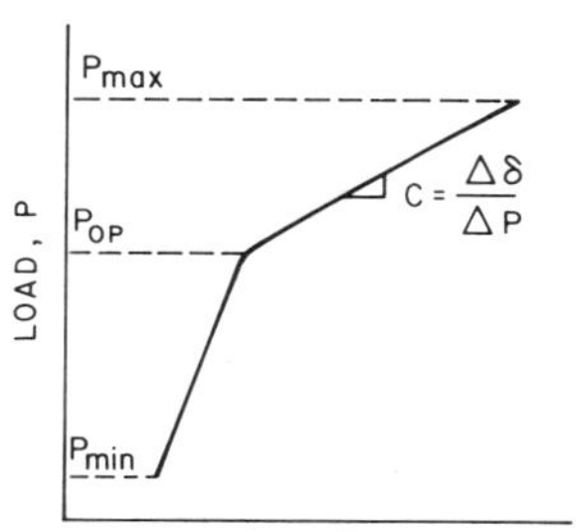

FIG. 2—*Schematic load-displacement or load-strain curve, showing deflection in curve which is indicative of crack closure.*

in which k is used as a "gain" factor. A plot of load versus $\Delta\delta$ can be constructed, as shown in Fig. 3. At loads above the deflection point, where the crack is fully open, $\Delta\delta = 0$ and the plot of load versus $\Delta\delta$ is a vertical line. At the onset of crack closure, $\Delta\delta$ becomes nonzero and the use of a high gain factor (for example, $k = 10$–20) enables the deflection point to be quantified with high sensitivity and a minimum of subjectivity. This signal processing can be accomplished using either analog circuitry [*22,42*] or digital signal processing.

Back-Face Strain

The back-face strain was originally developed as an alternative to the clip gage compliance technique for monitoring fatigue crack growth [*66*]. The technique uses a standard strain gage bonded to the back edge (face) of a fatigue crack growth specimen as shown in Fig. 4, with either digital or analog recording of the load-strain curve. A plot of load versus back-face strain is made, similar to the load versus displacement curve shown in Fig. 2. This technique was first used by Kikukawa et al. [*65*] to measure crack closure in steels. Its use has increased considerably in recent years [*19,22,23,29,30,51,53,60,67–70*], probably due to its low cost, relatively high sensitivity, and ease of use. Hysteresis in the load-strain curve is essentially nonexistent except when significant crack-tip plasticity has occurred. This technique is most effective on specimens which are loaded in bending, such as compact specimens or notched bend bars. In these situations the strain on the specimen back face is compressive during remote tensile load-

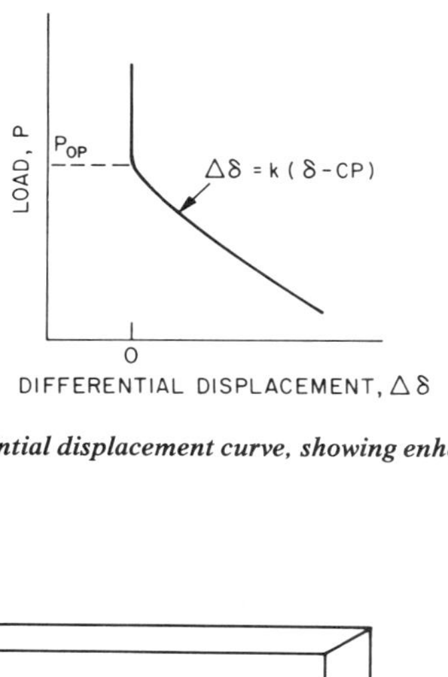

FIG. 3—*Schematic load-differential displacement curve, showing enhancement of the deflection point of Fig. 2.*

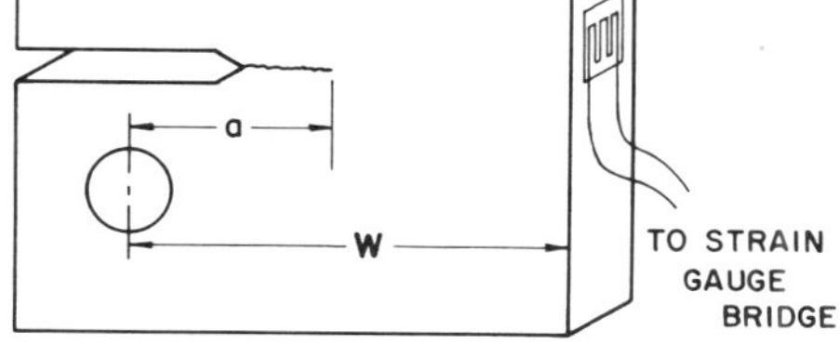

FIG. 4—*Compact specimen showing location of back-face strain gage.*

ing, which, because of the hinging action at the neutral stress axis, tends to magnify the specimen deflections. The back-face strain technique can be used in corrosive environments if adequate protection of the gage is provided and, applying advanced strain gage technology, it can be also used at elevated temperatures [*62*].

In addition to quantifying K_{OP}, the back-face strain technique can be used to determine the bulk residual strain due to crack closure [*22,67*]. For accurate measurement of these residual strains, good strain-gage practice requires the use of strain gages which are matched to the coefficient of thermal expansion of the specimen material. To preclude thermal drift, a second gage can be used in a half-bridge configuration, with the gage mounted either in a location of the specimen experiencing zero strain or on a dummy specimen of the same material located adjacent to the test specimen. If residual strain values are not of interest these restrictions on strain-gage circuitry are not as severe since the quantification of K_{OP} depends only on accurate quantification of the load deflection point for which a relative value of strain is sufficient.

When used on compact specimens, the back-face strain technique is sufficiently sensitive that signal processing is not generally required to discern the deflection point in the load-strain curve. Some investigators, however, have used the previously described differential compliance technique with back-face strain when investigating nonbending specimens or when very high sensitivity was desired [*30,53,65,68*].

Indirect (Noncompliance) Thickness-Averaging Techniques

Effective Threshold Method

A recently developed technique, the effective threshold method [*71*], requires no special measurement apparatus, but instead is based on the use of "nontraditional" load histories and automated fatigue crack growth technology. Investigations of the fatigue threshold behavior of many different materials indicate that an effective threshold, defined by $\Delta K_{TH}^{EFF} = K_{max}^{TH} - K_{OP}^{TH}$, is independent of load ratio and can be determined by conducting fatigue crack growth tests in which K_{max} is held constant and K_{min} is increased [*72–74*]. In such an experiment, the ΔK at which fatigue crack growth ceases is defined as ΔK_{TH}^{EFF} and is assumed to be an intrinsic material property [*74*]. Subsequent to determination of this effective threshold, a traditional K-increasing or K-decreasing experiment is conducted; however, such a test is periodically interrupted to determine K_{OP}. This is accomplished by stopping the test and applying to the specimen a constant ΔK slightly greater (10%) than ΔK_{TH}^{EFF} and slowly increasing the load ratio until the crack begins to grow. At the load ratio at which the crack just begins to grow, K_{min} is assumed to be equivalent to K_{OP}, thus providing a measurement of this quantity. Aside from the complexity of this procedure, the primary concern would appear to be validating the assumption that ΔK_{TH}^{EFF} in fact is independent of load history, R ratio, etc. This technique, however, is intriguing in that it uses the actual da/dN behavior of a material to determine K_{OP} and thus circumvents the concerns associated with relating a global compliance measurement with local crack-tip closure processes.

Electric Potential

Although the d-c electric potential technique is an effective method for monitoring the length of a propagating crack (for example, Ref *75*), it has not proven to be a reliable technique for quantifying K_{OP}. Some of the early controversy concerning crack closure was based on results obtained using the electric potential technique [*9,11,12,16,17*]. These investigators related deflections in the load-versus-potential curve to crack closure similar to the procedure described for compliance techniques. When tests are conducted in air, little change is observed in the potential as load is applied, and these early researchers took this as an indication that crack closure was not occurring. In contrast, when tests were conducted in inert environment, rather pronounced increases in potential were observed as the load was increased, and thus high values of K_{OP} were reported.

Studies in which both electric potential and compliance techniques were compared have shown that when testing is conducted in air, even though compliance techniques indicate substantial load transfer across the crack faces, electric potential shows none [*15,22,39,45,67*]. This is due to the formation of insulating oxides on the fracture surfaces [*39*], which causes the electric potential to indicate a fully or almost fully opened crack even when the crack is completely closed. When tests are conducted in inert environment, the formation of these oxides is circumvented. However, in experiments in Ti-6Al-4V in inert environment, Bachman and Munz [*39*] have shown conclusively that non-load-bearing asperity contact can occur and can lead to a decrease in the electric potential without causing crack closure; that is, the crack is mechanically open but electrically closed. On the basis of these studies it has been concluded that the use of the electric potential technique for quantifying crack closure must be used with extreme caution [*76*]; past and present [*56,77*] investigations in which the electric potential technique has been used to quantify crack closure must be viewed with this in mind.

Ultrasonic and Acoustic Techniques

Ultrasonic techniques have been applied in only a limited way to the measurement of crack closure in large crack specimens [*78–84*]. For such specimens, longitudinal bulk waves with nominal frequencies of 2 to 14 MHz have been employed in either a reflected or through-transmission mode. Ultrasonic surface wave techniques have been used for determining crack closure characteristics of surface flaws [*85–87*]. Similar to compliance or electric potential measurements, a deflection in the load versus ultrasonic intensity occurs due to sound transmission through fracture surfaces in contact. Attempts have been made to relate this deflection point to crack closure, and while the results are subject to interpretation, it appears as if the value of K_{OP} determined via ultrasound is higher than that measured using a CMOD gage [*74*]. This may be indicative of sound transmission through asperities which are in contact but are not load-bearing and thus, in a manner similar to electric potential, the use of ultrasonic techniques for determining K_{OP} is questionable. When properly focused transducers are coupled with a traversing device or detailed diffraction analysis, ultrasonic techniques may be, however, effective for indicating the extent of fracture surface contact in the specimen interior [*83*].

Acoustic emission techniques have been also applied to this problem [*48*]; however, no consistent correlation has been found between acoustic emission and crack closure. The emission of acoustic stress waves appears to be related more to mechanical abrasion of the fracture surface, which is not necessarily indicative of the amount of load transfer across the fracture surface.

Near-Crack-Tip Surface Measurement Techniques

Interferometric Techniques

Several interferometric techniques have been used to measure crack closure [*48,88–92*]. Of these, the most promising in terms of quantifying crack closure has been the interferometric displacement gage (IDG) originally developed by Sharpe et al. [*88,93*]. The IDG employs laser interferometry to measure the relative displacement between shallow reflecting indentations located 0.05 to 1 mm apart across the crack plane [*48,88*]. Since the laser radiation is monochromatic and coherent, interference fringes form in space; the motion of these fringes is a measure of the displacement of the indentations. The motion of the fringes can either be detected using photographic film or photodetectors. Using the IDG technique, crack closure measurements can be made near the crack tip with very high resolution. Since it is a noncontacting method, attachment problems are avoided; as long as the indentations remain clear, the IDG can be used at elevated temperatures or in aggressive environments. Significant limitations are the cost and complexity of the apparatus. In addition, the use of the IDG typically requires that the FCG experiment be slowed to approximately 0.02 Hz during closure measurements. Recent developments making use of photodetector arrays show promise for increasing the response time of the IDG and reducing its complexity [*94*].

One notable finding which has been made using the IDG is the observation that when K_{OP} is measured adjacent to the crack tip, the measured value is strongly dependent on the measurement location and significantly higher than that measured in the bulk. As shown in Fig. 5, using a series of indents spaced in roughly 0.5 mm intervals behind a crack tip, Macha, Corbly, and Jones [*48*] observed that immediately adjacent to the crack tip a K_{OP} value of 25 MPa $\sqrt{m}$ was measured; values of K_{OP} decreased continuously as the distance behind the crack tip increased. At distances greater than 5 mm, a K_{OP} value of 9 MPa $\sqrt{m}$ was measured independent of measurement location. This was equivalent to a thickness-averaged value of K_{OP} equal to 9 MPa $\sqrt{m}$ as measured by a CMOD gage.

Direct Optical, SEM, and TEM Measurements

A number of investigators have used the optical microscope [*8,95*], the scanning electron microscope (SEM), or the transmission electron microscope (TEM) to make direct measurements of the crack opening displacements as a function of load. In this way an estimate of K_{OP} can be made which represents the stress intensity required for complete separation of the crack tip at the surface. Figure 6 shows an example of such a measurement made on cracks in 1070 steel, indicating that K_{OP} for complete crack-tip separation is between 10 and 15 MPa $\sqrt{m}$ [*96*]. Although this measurement was made for small cracks, similar measurements are possible on large crack specimens. To facilitate these measurements, some investigators have equipped the SEM with a loading platform to view the crack-tip displacements *in situ* [*97–100*] while others have used two-stage replicas viewed in either SEM [*22,40,63,96,101,102*] or a TEM [*103*]. Skill and considerable diligence are required in producing and quantifying these direct measurements, and thus the use of these techniques has been limited. Estimates of crack closure have been also determined from SEM measurements of striation spacings [*104–105*] by relating these spacings to an average *da/dN* and remote ΔK^{EFF} and calculating K_{OP} using Eq 1.

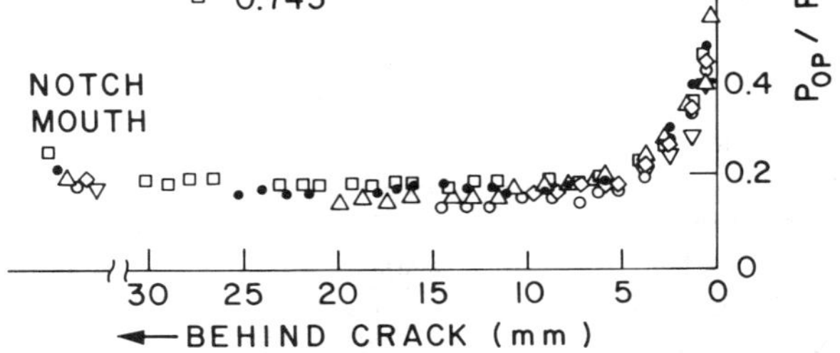

FIG. 5—*Normalized crack opening load versus distance behind the crack tip as measured on the specimen surface using the interferometric displacement gage (IN 100,* K_{max} *= 35 MPa* $\sqrt{m}$*) (Ref* 48*).*

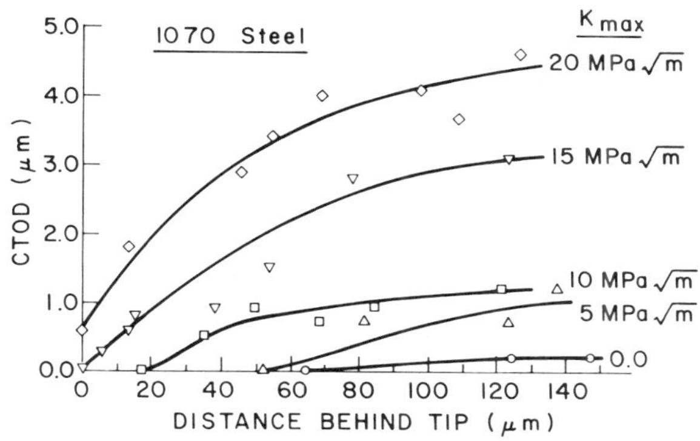

FIG. 6—*Direct SEM observation of crack-tip opening displacement versus distance behind the crack tip as measured on the specimen surface at various static stress intensities (1070 steel,* $K_{max} = 20$ *MPa* $\sqrt{m}$*) (Ref* 96*).*

Crack Tip Compliance Gages

At least three specialized measurement techniques can be placed in this category: crack-tip opening displacement (CTOD) gages, crack-tip strain gages, and near-crack-tip strain gages. The first such gages were single-cantilever CTOD gages developed by Elber [*7*] and Nowack [*106*] and first used to quantify crack closure by Elber [*7*]. Subsequent to Elber's measurements, a number of researchers have used [*10,17,22,39,102,107–110*] and in some cases improved upon [*22*] such gages to measure near-crack-tip closure. In general, CTOD gages have gage lengths of 1 to 3 mm, although gage lengths as large as 9 mm have been used successfully by at least one investigator [*22*]. CTOD gages are typically placed at an arbitrary location approximately 1 mm behind the crack tip. Some difficulty occurs when attempting to affix CTOD gages to the specimens, and in general their mechanical stability is poor. Strain gages bonded across the crack tip have been used in a manner similar to the CTOD gage [*22,111,112*] and yield similar results. Strain gages have been also bonded near but not across the crack tip, either in front of or behind it [*8,22,113–117*]. The deflection point in the load-displacement curve is more pronounced when measured near the crack tip, and thus these gages are very sensitive. Where higher sensitivity is required, it is possible to apply the differential compliance technique to these gages.

Using crack-tip compliance techniques, several investigators [*22,102,103,107,113*] have corroborated the observation of Macha et al. [*48*] that close to the crack tip the value of K_{OP} is a strong function of measurement location. From a reproducibility and standardization viewpoint this is a serious limitation for crack-tip compliance techniques. The reproducibility of the measurement is directly related to the accuracy with which the gage can be located. An equally serious limitation is the fact that when K_{OP} is measured at a location behind the crack, it is a measure neither of bulk closure nor of complete crack-tip separation. It is rather merely a measure of the load at which some arbitrary point behind the crack tip opens. Extrapolation of this value to predict the point of complete crack-tip separation may be limited by a dependence on material and loading conditions. In addition, other localized effects such as local crack contact can mask the measurement of crack closure when using a crack-tip compliance technique. Local crack contact can lead to high local values of K_{OP} and increased data scatter [*114*]. Thus, because of the ambiguous nature and inherent lack of reproducibility of crack-tip compliance techniques, much of the early controversy regarding crack closure would appear to be related to the use of these methods.

Interior Measurement Techniques

Push Rod Compliance Technique

A new type of gage has been developed by Fleck to measure K_{OP} in the interior of thick compact specimens [*22,118*] or on the interior of part-through, elliptical cracks [*119*]. This gage has provided the only quantitative measurement of crack closure in the interior of thick metal specimens. For the gage to be used, the fatigue test is interrupted and two small, parallel holes are drilled just behind the crack front. One hole is drilled to a depth slightly below the crack plane, the other to a depth slightly above the crack plane. A push-rod assembly is then mounted on the specimen, and the relative displacement of the hole bottoms is measured using a clip gage. The point of crack closure can be determined by measuring the deflection point in the load-displacement curve or in the load-differential compliance curve. To date this technique has been applied only to the study of crack closure in a low-strength steel (BS4360) where it was observed that K_{OP} measured in the interior of a 24-mm-thick specimen was equivalent to the thickness-averaged value of K_{OP} and significantly lower than that measured on the surface near the crack tip.

Interferometry

Optical interferometry has been used on transparent polymers to determine the three-dimensional aspects of crack closure [*120,121*]. Interference fringe patterns are obtained by shining a monochromatic light source through the polished top surfaces of either compact or notched bend (edge crack) specimens. Fringe patterns can be photographed as a function of applied load, thus providing a three-dimensional topographic map of the crack-tip region as a function of applied load. The opening stress intensity is defined as the point at which complete crack-tip separation occurs. Results from both polycarbonate and polymethylmethacrylate indicate that K_{OP} measured in the center of the specimen is finite but considerably lower than that measured on the specimen surface. Moreover, at zero load a much smaller region near the crack tip is closed in the specimen interior than along the free surface. Such measurements are obviously not possible on metals, and the extent to which these results can be extrapolated to metals remains an open question. Nevertheless, this technique gives results which are consistent with observations made in metals and three-dimensional numerical predictions of plasticity-induced crack closure [*122*], as well as an interesting three-dimensional picture of the crack closure process.

Discussion

The Three-Dimensional Nature of Crack Closure

The results of several investigators indicate that when plasticity-induced crack closure is the dominant crack closure mechanism, crack closure is three-dimensional in nature. In the surface regions, where plane stress conditions exist, the extent of the plastic zone is large and thus high closure stress intensities are expected relative to the interior of thick specimens. In the center of thick specimens, plane-strain conditions prevail and the plastic zone is small, leading to lower opening stress intensities. That crack closure occurs in the interior of thick specimens but at levels lower than at the surface has been confirmed in a low-strength, carbon-manganese steel [*22*] and in polymers [*120,121*]. Using a vacuum infiltration technique, Bowles has shown that significant fracture surface contact occurs in the interior of aluminum specimens [*123*]. In addition, crack closure measurements have been made on aluminum and titanium alloys in which surface layers were removed followed by a remeasuring of K_{OP} using crack-tip compliance techniques [*124–127*]. In these investigations, as the surface layers were removed the level of K_{OP}

decreased; however, even at the center of 10-mm specimens [*125*] significant levels of closure were measured. For 2024-T3 aluminum tested at intermediate ΔK levels, the decrease in K_{OP} was most pronounced after the first surface layer removal, that is, within the first 1 mm of the specimen thickness [*125*]. Although these results are clouded by the inability to reproducibly place the CTOD gage at a precise location behind the crack tip, they do provide a qualitative indication that the level of crack closure is influenced by the state of stress. As a further indication of the influence of state-of-stress on crack closure, several investigators have observed that crack closure levels are higher in thin specimens than in thick specimens [*22, 76*].

Fleck [*22*] has made the important observation that regardless of the differences in K_{OP} between the surface and interior, the thickness-averaged value of K_{OP} is equal to that measured in the interior. This implies that in thick specimens the higher value of crack closure evident on the surface plays an insignificant role in determining the bulk closure response of the specimen. In addition, Fleck has observed that in thin specimens where plane-stress conditions prevail through the entire thickness, the bulk crack closure is equivalent to that measured on the surface. Finally, Fleck observed that crack growth rates correlated well with a ΔK^{EFF} calculated using the K_{OP} measured globally for both thick and thin specimens. This indicates that regardless of closure levels measured on the surface, it is the global (bulk) closure values which are reflected by the average crack growth rates.

Observations of the three-dimensional nature of crack closure have generally been determined on specimens which were tested under conditions which favor plasticity-induced crack closure, that is, at relatively high ΔK. Similar experiments under conditions which favor other forms of crack closure, for example, low ΔK, have not been conducted. For alloys or microstructural conditions in which asperity-induced crack closure predominates, crack closure might be expected to be independent of thickness and through-thickness variations will be minimized. For conditions which promote oxide-induced crack closure, oxide buildup is more pronounced in the specimen interior and one might speculate that crack closure levels will be higher in these interior regions. In both situations, it has been shown that thickness-averaged crack closure levels correlated well with crack growth rates.

Interpretation of Compliance Curves

The interpretation of load-displacement or load-strain curves is central to precise determination of the crack closure load or stress intensity. A problem for all compliance techniques is defining the point on the load-displacement curve which best quantifies the crack opening or closure load, since, as shown in Fig. 7, the deflection in the load-displacement curve is not

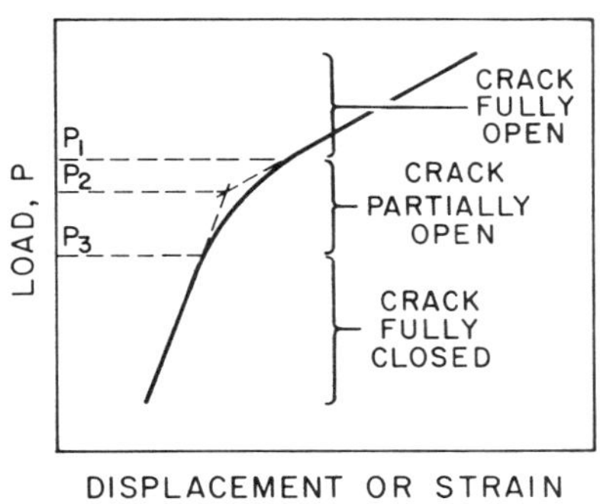

FIG. 7—*Load-displacement or load-strain curve, showing three different points reflecting crack closure.*

always a sharp one. That the deflection point is not a distinct point when measured via thickness-averaging compliance techniques is probably related to the fact that closure is a three-dimensional process occurring at progressively lower loads at different points through the specimen thickness and along the crack length. There are a number of points on the load-displacement curve which might be used as a measurement of the opening load, three of which are shown in Fig. 7. Point P1 is the point which represents the fully opened crack, that is, the point at which the first load-bearing contact between fracture surfaces occurs on the unloading portion of the cycle. Point P3 is a measure of the load at which the crack is fully closed. Presumably any load above this point causes stress intensification at some location on the crack front. Point P2 represents an average opening load where a significant portion of the crack is open, but where portions of the crack surface remain in contact. It is typically determined at the intersection of the extrapolated tangents of the upper and lower portions of the load-displacement curve. More detailed investigations are required to determine which of these points most accurately reflects the crack closure process; however, there are indications [*19,20,127*] that the use of a point nearer to P2 gives superior consolidation of da/dN versus ΔK^{EFF} data.

In addition to the differential compliance method, with the advent of computerized fatigue crack growth testing, a number of numerical procedures have been developed for extracting some measure of K_{OP} from digitally stored load-displacement or load-strain curves [*62,127–129*]. These procedures reduce the amount of subjectivity involved in determining K_{OP} and allow the automated analysis of large amounts of crack closure data. To date, no comparative study has been undertaken to assess the relative abilities of these various methods to accurately quantify the crack closure process.

A further consideration is whether crack closure is measured on the load-increasing (K_{OP}) or load-decreasing (K_{CL}) portion of the load cycle. For global measurements there is little distinction between the two, and the term K_{OP} is often applied as a general term to cover both. For near-crack-tip compliance methods, there is some experimental evidence [*130*] supporting analytical predictions [*33,131*] that K_{OP} is greater than K_{CL}.

Standardization of Measurement Techniques

With the increased awareness of the importance of crack closure comes a need for more widespread use of standard techniques for quantifying crack closure. This is required to allow comparisons between investigators, to develop a clearer understanding of basic crack closure phenomena, to augment and accelerate the development of analytical crack closure models, and to develop a crack closure database for use in design. Of the previously described techniques, the thickness-averaging compliance techniques based on back-face strain and CMOD are the most appropriate for standardization. In principle they yield a reproducible measurement of crack closure (although this has not been experimentally proven) and have the required sensitivity. In addition, they are inexpensive and relatively simple to use for most types of fatigue crack growth experiments. Because indirect techniques such as electric potential and ultrasound have led to erroneous indications of crack closure, their use for measuring K_{OP} should be avoided. Likewise, the use of crack-tip compliance techniques has led to confusion because of problems associated with the influence of measurement location on the measured value. In addition to standardizing measurement techniques, there is a need to standardize numerical procedures for extracting K_{OP} from load-displacement data.

Although the use of back-face strain and CMOD gages is recommended for routine use, there remains a need for additional fundamental studies of the crack closure process. Detailed two- and three-dimensional measurements of crack-tip opening displacements using SEM and interferometric techniques are important for this purpose. Additional measurements of crack closure in the interior of thick specimens using techniques such as the push-rod displacement gage are clearly warranted.

Mechanisms and Crack Closure Variability

Although three primary forms of crack closure have been identified, it is difficult to predict how these mechanisms interact when present together. One way in which these interactions can be viewed is by investigating the dependence of K_{OP} on K_{max}. Figure 8 shows qualitatively the manner in which K_{max} will influence K_{OP} for the three primary forms of crack closure. For plasticity-induced crack closure analytical and numerical models indicate that K_{OP} increases linearly with K_{max} [*32–34*], as shown in Fig. 8. In those alloy/microstructural/environment combinations where it occurs, oxide-induced crack closure is generally restricted to the low ΔK regime because it is there that the crack opening displacements are small and lead to enhanced fretting-assisted oxidation of the fracture surfaces. Crack opening displacements are proportional to the square of K_{max}, and thus K_{OP} can be expected to decrease rapidly as K_{max} increases when the dominant closure mechanism is oxide-induced crack closure.

The influence of K_{max} on crack closure due to asperity-induced crack closure is not well documented. K_{OP} has been shown to increase as the height of fracture surface asperities increases [*19,23*], and simple models indicate that the level of closure should increase as the degree of asperity mismatch increases [*24,35,36*]. Thus the influence of K_{max} on both fracture surface topography and asperity mismatch, as well as the combination of the two, are important. In many materials, the topography of the surface becomes smoother at high ΔK, which would lead to the speculation that K_{OP} would decrease with increasing K_{max} as indicated in Fig. 8. However, this is not the case for titanium alloys [*19*] and some aluminum alloys [*132*], where for most microstructural and alloy conditions the roughness of the fracture surface is independent of K_{max}. In these alloys, K_{OP} due to asperity-induced crack closure would be independent of K_{max} as indicated in Fig. 8. Because the underlying causes of asperity mismatch have not been clearly identified, predicting the dependence of asperity mismatch on K_{max} is difficult. Popular descriptions of asperity-induced crack closure suggest that asperity mismatch is a result of Mode II crack-tip displacements [*18,24,43,44*]. Since it is generally thought that Mode II displacements are more pronounced at low ΔK, asperity-induced crack closure should be most potent at low values of K_{max}. It is likely, however, that the mismatch of asperities is due in large part to crack branching which leads to localized asperity pullout and interaction at many loca-

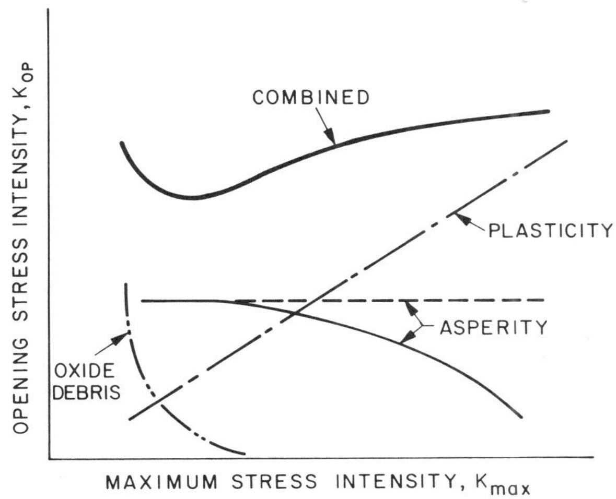

FIG. 8—*Influence of* K_{max} *on crack closure for three forms of crack closure—plasticity-induced crack closure, asperity-induced crack closure, and oxide-induced crack closure—and their combined influence.*

tions behind the crack tip. The propensity for crack branching as a function of K_{max} has not received widespread attention, and generalizations are difficult because crack branching is highly dependent on the alloy and microstructure in question.

Crack closure has been studied in a number of alloys, in a variety of microstructural conditions, and even with the increased use of "standard" thickness-averaging measurement techniques, no general pattern has emerged. To illustrate this point, crack closure data from a number of different investigators using either back-face strain or CMOD techniques is presented as a function of K_{max} in Figs. 9 to 11. As is evident from these data, the influence of K_{max} on K_{OP} is complex and depends on both alloy and microstructure. These variations may be due to variations in the dominant crack closure mechanism. They may be also indicative of differ-

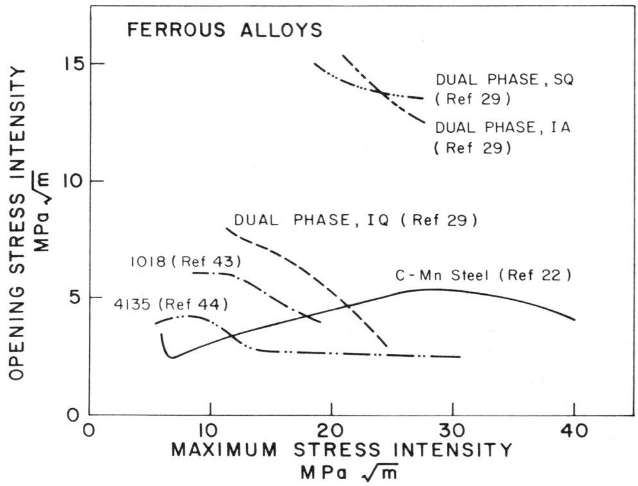

FIG. 9—*Crack closure behavior for several ferrous alloys.*

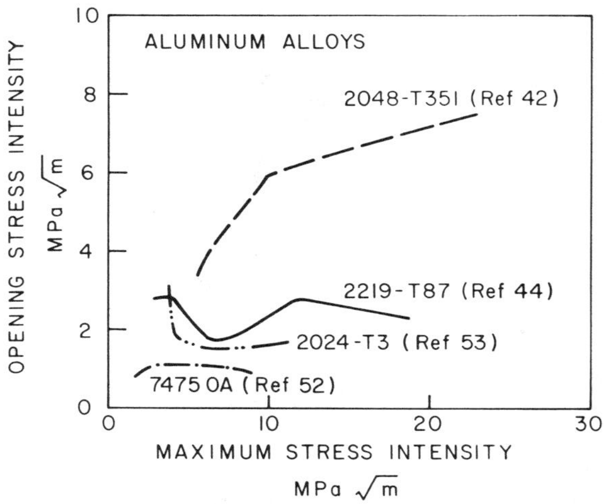

FIG. 10—*Crack closure behavior for several aluminum alloys.*

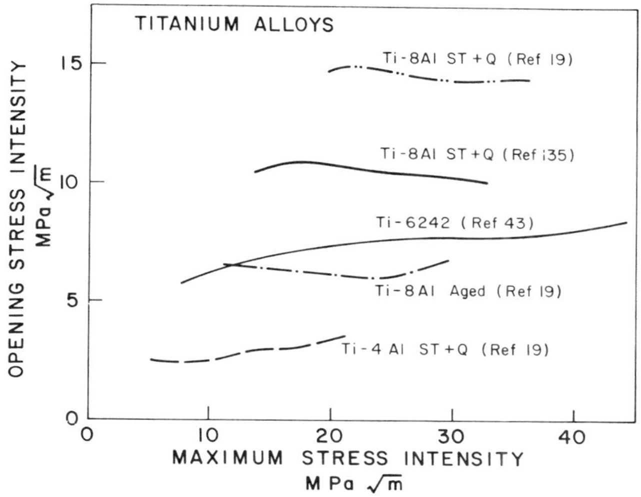

FIG. 11—*Crack closure behavior for several titanium alloys.*

ences in test technique (that is, precracking load history or rate of load shedding), which also can have a significant effect on crack closure [*22,133*].

In ferrous alloys (Fig. 9) the dependence of K_{OP} on K_{max} follows no general pattern. In 1018 steel [*43*] and in three different microstructural conditions of a dual-phase steel [*29*], K_{OP} was observed to decrease with increasing K_{max}, whereas in 4135 steel K_{OP} initially decreased and then became independent of K_{max} [*44*]. Both of these behaviors would appear to indicate the dominance of asperity-induced crack closure over plasticity-induced crack closure in these alloys. This is most apparent in the dual-phase steel, where microstructural variations strongly influenced crack closure. Although these three microstructures had equivalent yield strengths, the level of crack closure was significantly higher in microstructures produced by step-quenching (SQ) and intercritical annealing (IA) than in a microstructure produced by an intermediate quenching (IQ) procedure. In contrast, results in a low strength, carbon-manganese steel (BS4360) show that K_{OP} initially decreases sharply with increasing K_{max}, possibly indicative of oxide-induced crack closure, and then increases with K_{max} [*22*]. The dependence of K_{OP} on K_{max} for the carbon-manganese steel is roughly in agreement with numerical predictions developed by Newman [*34*] for plasticity-induced crack closure (using a plane-strain constraint factor of three). Results similar to those observed in the carbon-manganese steel have been observed in 17-4 PH steel [*53*] and a tool steel [*22*].

The crack closure behavior of aluminum alloys (Fig. 10), like that of steels, follows no general pattern. The crack closure values reported for aluminum alloys are much lower than those measured in steels or titanium alloys, normally not reaching above 3 MPa $\sqrt{m}$. Even at these low absolute values, K_{OP} is generally a significant fraction of K_{max} at near-threshold growth rates, and closure effects are significant. An alloy showing exceptionally high levels of crack closure is 2048 aluminum, where values of K_{OP} as high as 7.5 MPa $\sqrt{m}$ have been deduced from the literature [*42*]. In this alloy, K_{OP} appears to increase with increasing K_{max}, perhaps indicative of plasticity-induced crack closure. In 2048-T351 aluminum, the dependence of K_{OP} on K_{max} is equivalent to that predicted by Newman's model (using a plane-strain constraint factor of three); however, the absolute values of K_{OP} are considerably higher than those predicted by this model [*34*]. In both 2219-T87 [*44*] and 2024-T3 [*53*] aluminum, the opening stress intensity initially decreases and then either becomes constant (2024) or increases (2219) with further increases in K_{max}. The initial decrease in K_{OP} may be attributed to oxide-induced crack closure, although at least one study has suggested that this mechanism should not be pronounced in 2xxx series aluminum alloys [120]. A transition in fracture surface topography from rough to smooth has been observed in 2024-T3 at low ΔK [*42,44*], and this may also cause such a sharp drop in crack closure. In a study of 7475 aluminum, Carter et al. [*52*] have observed levels of crack closure which are essentially constant over a range of values of K_{max}, perhaps indicative of asperity-induced crack closure. An example of this is shown for an overaged (OA) microstructure in Fig. 10.

For titanium alloys (Fig. 11), the opening stress intensity is either independent of K_{max}, as is the case for Ti-8Al [*19,70,135*], or increases slightly with increasing K_{max}, as is the case for Ti-4Al [*19*], Ti-6242 [*43*], and Ti-6-4 [*70*]. This would indicate that oxide-induced crack closure does not play a major role in determining closure levels in titanium alloys, a fact substantiated by the lack of significant accumulation of oxide debris on titanium fracture surfaces [*19*]. The independence or weak dependence of K_{OP} on K_{max} indicates that the predominant mechanism of crack closure in titanium alloys is asperity-induced crack closure, with some influence of plasticity-induced crack closure possible in Ti-6-4, Ti-6242, and Ti-4Al in certain (fine grain size) microstructural conditions. Two sets of crack closure data are shown in Fig. 11 for Ti-8Al alloys from two different studies [*19,135*]. Although these were from nominally identical plates of Ti-8A1, the large difference between the levels of crack closure indicates the high degree of variability that exists in crack closure measurements, due to slight differences in microstructural details, fatigue crack growth test procedures, or crack closure measurement techniques.

Summary

The significant influence of crack closure on constant-amplitude fatigue crack growth rates is now widely recognized. More precise and quantitative measurements of crack closure are required to facilitate a more in-depth understanding of the crack closure process. There is a requirement for increased understanding of the three-dimensional nature of crack closure especially at near-threshold crack growth rates. To facilitate comparisons between different investigators there is a need for standardization of measurement techniques. In this review, an attempt has been made to describe the various techniques used to quantify crack closure, and to offer opinions regarding their practical and fundamental advantages or limitations. Of these techniques, thickness-averaging compliance techniques such as back-face strain and crack mouth opening displacement appear to be the most appropriate for standardization and for widespread use. Surface measurement techniques such as the interferometric displacement gage and direct SEM observations of crack-tip opening displacements give valuable information concerning the closure behavior of the surface regions of cracks. It is not yet clear under which conditions crack closure measured on the surface accurately reflects the bulk closure behavior, and thus it is recommended that thickness-averaging compliance methods be used in conjunction with measurements made with such surface crack-tip methods. The use of crack tip compliance techniques such as the crack-tip opening displacement gage is widespread; however, in addition to the limitations inherent in other surface closure measurements, these techniques are further limited by their sensitivity to measurement location. This leads to problems with reproducibility and to an arbitrary definition of crack closure. Measurement techniques which measure crack closure in the specimen interior have been developed and can provide valuable information concerning the three-dimensional nature of fatigue crack closure. Because indirect measurement techniques such as electric potential and ultrasonic methods can produce erroneous indications of crack closure, their use in measuring K_{OP} should be avoided.

Even when "standard" thickness-averaging compliance techniques are used, significant variations in crack closure are observed. For similar alloys, crack closure has been observed to increase, decrease, or remain essentially constant with increasing K_{max}. Thus it is important to develop a knowledge of the influence of fatigue crack growth testing procedures on K_{OP} and of the influence of the various methods used to define K_{OP} from load-displacement data. Such understanding can lead to the development of fatigue-resistant materials, to increased confidence in crack closure models, and to the development of crack closure data for use in the design of fatigue-resistant structures.

Acknowledgments

The author would like to thank N. A. Fleck and M. J. Couper for providing copies of their Ph.D. thesis prior to publication. Helpful discussions with several colleagues, most notably S. Suresh, J. M. Larsen, and D. L. Davidson, are gratefully acknowledged.

References

[*1*] Christensen, R. H., *Applied Mechanics Reviews,* Oct. 1963, pp. 207-210.
[*2*] Endo, K., Komai, K., and Ohnishi, K., "Effects of Stress History and Corrosive Environment on Fatigue Crack Propagation," *Memoirs of the Faculty of Engineering,* Kyoto University, Japan, Vol. 31, 1969, pp. 25-46.
[*3*] Steigerwald, E. A. and Hanna, G. L. in *Proceedings,* American Society for Testing and Materials, Philadelphia, Vol. 62, 1962, pp. 885-913.
[*4*] Anctil, A. A., Kula, E. B., and Dicesare, E. in *Proceedings,* American Society for Testing and Materials, Philadelphia, Vol. 63, 1963, pp. 799-808.
[*5*] Elber, W., *Engineering Fracture Mechanics,* Vol. 2, 1970, pp. 37-45.
[*6*] Cheng, Y. F. and Brunner, H., *International Journal of Fracture,* Vol. 6, 1970, pp. 431-434.
[*7*] Elber, W. in *Damage Tolerance in Aircraft Structures, ASTM STP 486,* American Society for Testing and Materials, Philadelphia, 1971, pp. 230-242.
[*8*] Adams, N. J. I., *Engineering Fracture Mechanics,* Vol. 4, 1972, pp. 543-554.
[*9*] Shih, T. T. and Wei, R. P., *Engineering Fracture Mechanics,* Vol. 6, 1974, pp. 19-32.
[*10*] Katcher, M. and Kaplan, M. in *Fracture Toughness and Slow-Stable Cracking, ASTM STP 559,* American Society for Testing and Materials, Philadelphia, 1974, pp. 264-282.
[*11*] Lindley, T. C. and Richards, C. E., *Materials Science and Engineering,* Vol. 14, 1974, pp. 281-293.
[*12*] Irving, P. E., Robinson, J. L., and Beevers, C. J., *Engineering Fracture Mechanics,* Vol. 7, 1975, pp. 619-630.
[*13*] Bachmann, V. and Munz, D., *International Journal of Fracture,* Vol. 11, 1975, pp. 1075-1056.
[*14*] Irving, P. E., Robinson, J. L., and Beevers, C. J., *International Journal of Fracture,* Vol. 11, 1975, pp. 713-716.
[*15*] Bachmann, V. and Munz, D. in *Fatigue Testing and Design,* Society of Environmental Engineers, Vol. 2, 1976, pp. 35.1-35.32.
[*16*] Unangst, K. D., Shih, T. T., and Wei, R. P., *Engineering Fracture Mechanics,* Vol. 9, 1977, pp. 725-734.
[*17*] Clarke, C. K. and Cassatt, G. C., *Engineering Fracture Mechanics,* Vol. 9, 1977, pp. 675-688.
[*18*] Suresh, S. and Ritchie, R. O. in *Fatigue Crack Growth Threshold Concepts,* D. Davidson and S. Suresh, Eds., The Metallurgical Society of the American Institute of Mining, Metallurgical, and Petroleum Engineers, Warrendale, PA, 1984, pp. 227-261.
[*19*] Allison, J. E., "The Influence of Slip Character and Microstructure on Fatigue Crack Growth in Titanium Alloys," Ph.D. thesis, Carnegie-Mellon University, Pittsburgh, PA, 1982.
[*20*] Couper, M. J., "Fatigue of an Al-Mg-Si Casting Alloy," Ph.D. thesis, Monash University, Clayton, Victoria, Australia, 1982.
[*21*] Banerjee, S., "A Review of Crack Closure," Technical Report AFWAL-TR-84-4031, AFWAL, Wright-Patterson Air Force Base, OH, 1984.
[*22*] Fleck, N. A., "An Investigation of Fatigue Crack Closure," Ph.D. thesis, Cambridge University, Cambridge, U.K., 1984.
[*23*] Allison, J. E. and Williams, J. C. in *Titanium, Science and Technology,* Vol. 4, G. Luetjering, U. Zwicker, and W. Bunk, Eds., DGM Publications, Oberufsel, West Germany, 1985, pp. 2243-2250.
[*24*] Suresh, S. and Ritchie, R. O., *Metallurgical Transactions A,* Vol. 13A, 1982, pp. 1627-1631.
[*25*] Suresh, S., Zamiski, G. F., and Ritchie, R. O., *Metallurgical Transactions A,* Vol. 12A, 1981, pp. 1435-1443.
[*26*] Benoit, D., Namdar-Irani, R., and Tixier, R., *Materials Science and Engineering,* Vol. 45, 1980, pp. 1-7.
[*27*] Freeman, B. L., Smith, D., and Stewart, A. T. in *Fatigue Thresholds,* J. Backlund, A. Blom, and C. J. Beevers, Eds., EMAS Publications Ltd., Warley, U.K., 1981, pp. 547-562.
[*28*] Minakawa, K., Matsuo, Y., and McEvily, A. J., *Metallurgical Transactions A,* Vol. 13A, 1982, pp. 439-445.

[29] Dutta, V. B., Suresh, S., and Ritchie, R. O., *Metallurgical Transactions A*, Vol. 15A, 1984, pp. 1193–1207.
[30] Tzou, J.-L. and Ritchie, R. O., *Scripta Metallurgica*, Vol. 19, 1985, pp. 751–755.
[31] Newman, J. C., Jr. in *Mechanics of Crack Growth, ASTM STP 509*, American Society for Testing and Materials, Philadelphia, 1976, pp. 281–301.
[32] Ogura, K. and Ohji, K., *Engineering Fracture Mechanics*, Vol. 9, 1977, pp. 471–480.
[33] Budiansky, B. and Hutchinson, J. W., *Transactions*, American Society of Mechanical Engineers, Vol. 45, 1978, pp. 267–276.
[34] Newman, J. C., Jr. in *Methods and Models for Predicting Fatigue Crack Growth Under Random Loading, ASTM STP 748*, American Society for Testing and Materials, Philadelphia, 1981, pp. 53–84.
[35] Beevers, C. J., Carlson, R. L., Bell, K. B., and Starke, E. A., *Engineering Fracture Mechanics*, Vol. 19, 1984, pp. 93–100.
[36] Carlson, R. L. and Beevers, C. J., *Engineering Fracture Mechanics*, Vol. 20, 1984, pp. 687–690.
[37] Robin, C., Dominiak, S., and Pluvinage, G., *Materials Science and Engineering*, Vol. 29, 1977, pp. 145–150.
[38] Brown, R. D. and Weertman, J., *Engineering Fracture Mechanics*, Vol. 10, 1978, pp. 757–771.
[39] Bachmann, V. and Munz., D., *Engineering Fracture Mechanics*, Vol. 11, 1979, pp. 61–71.
[40] Walker, N. and Beevers, C. J., *Fatigue of Engineering Materials and Structures*, Vol. 1, 1979, pp. 135–148.
[41] Clerivet, A. and Bathias, C., *Engineering Fracture Mechanics*, Vol. 12, 1979, pp. 599–611.
[42] Asaro, R. J., Hermann, L., and Baik, J. M., *Metallurgical Transactions A*, Vol. 12A, 1981, pp. 1133–1135.
[43] Minakawa, K. and McEvily, A. J., *Scripta Metallurgica*, Vol. 15, 1981, pp. 633–636.
[44] Minakawa, K. and McEvily, A. J., *Fatigue Thresholds*, J. Backlund, A. Blom, and C. J. Beevers, Eds., EMAS Publications Ltd., Warley, U.K., 1981, pp. 35.1–35.17.
[45] Halliday, M. D. and Beevers, C. J., *Journal of Testing and Evaluation*, Vol. 9, 1981, pp. 195–201.
[46] Nakai, Y., Tanaka, K., and Nakanishi, T., *Engineering Fracture Mechanics*, Vol. 15, 1981, pp. 291–302.
[47] Paris, P. C. and Hermann, L. in *Fatigue Thresholds*, J. Backlund, A. Blom, and C. J. Beevers, Eds., EMAS Publications Ltd., Warley, U.K., 1981, pp. 3–32.
[48] Macha, D. E., Corbly, D. M., and Jones, J. W., *Experimental Mechanics*, Vol. 19, 1979, pp. 207–213.
[49] Liaw, P. K., Hudak, S. J., Jr., and Donald, J. K., *Metallurgical Transactions A*, Vol. 13A, 1982, pp. 1633–1645.
[50] Liaw, P. K., Swaminathan, V. P., Leax, T. R., and Donald, J. K., *Scripta Metallurgica*, Vol. 16, 1982, pp. 871–876.
[51] LaFarie-Frenot, M. C. and Gasc, C., *Fatigue of Engineering Materials and Structures*, Vol. 6, 1983, pp. 329–344.
[52] Carter, R. D., Lee, E. W., Starke, E. A., Jr., and Beevers, C. J., *Metallurgical Transactions A*, Vol. 15A, 1984, pp. 555–563.
[53] Blom, A. F. in *Fatigue Crack Growth Threshold Concepts*, D. Davidson and S. Suresh, Eds., The Metallurgical Society of the American Institute of Mining, Metallurgical, and Petroleum Engineers, Warrendale, PA, 1984, pp. 263–279.
[54] Venables, R. A., Hicks, M. A., and King, J. E. in *Fatigue Crack Growth Threshold Concepts*, D. Davidson and S. Suresh, Eds., The Metallurgical Society of the American Institute of Mining, Metallurgical, and Petroleum Engineers, Warrendale, PA, 1984, pp. 341–357.
[55] Esaklul, K. A., Wright, A. G., and Gerberich, W. W. in *Fatigue Crack Growth Threshold Concepts*, D. Davidson and S. Suresh, Eds., The Metallurgical Society of the American Institute of Mining, Metallurgical, and Petroleum Engineers, Warrendale, PA, 1984, pp. 299–326.
[56] Vincent, J. N. and Remy, L. in *Fatigue 84*, Vol. II, C. J. Beevers, Ed., EMAS Publications Ltd., Warley, U.K., 1984, pp. 697–706.
[57] Jono, M. and Song, J. in *Fatigue 84*, Vol. II, C. J. Beevers, Ed., EMAS Publications Ltd., Warley, U.K., 1984, pp. 717–726.
[58] Dickson, J. I., Handfield, L., Bailon, J.-P., and Sahoo, M. in *Fatigue 84*, Vol. III, C. J. Beevers, Ed., EMAS Publications Ltd., Warley, U.K., 1984, pp. 191–200.
[59] Vecchio, R. S., Crompton, J. S., and Hertzberg, R. W. in *Fatigue 84*, Vol. II, C. J. Beevers, Ed., EMAS Publications Ltd., Warley, U.K., 1984, pp. 1379–1388.
[60] Byrne, J. and Duggan, T. V. in *Fatigue Thresholds*, J. Backlund, A. Blom, and C. J. Beevers, Eds., EMAS Publications Ltd., Warley, U.K., 1981, pp. 759–775.
[61] Saxena, A., Hudak, S. J., Jr., Donald, J. K., and Schmidt, D. W., *Journal of Testing and Evaluation*, Vol. 6, 1978, pp. 167–174.
[62] Ashbaugh, N. E., "Time Dependent and Cycle Dependent Threshold Evaluation in a Nickel-Base

Superalloy," presented at ASTM 18th National Symposium on Fracture Mechanics, Boulder, CO, June 1985.

[63] Allison, J. E. in *Fatigue Crack Closure,* American Society for Testing and Materials, Philadelphia, 1986, to be published.

[64] Zawada, L. P. and Nicholas, T., this publication, pp. 192-205.

[65] Kikukawa, M., Jono, M., and Tanaka, K. in *Proceedings,* Second International Conference on Mechanical Behavior of Materials, Boston, American Society for Metals, 1976, pp. 254-277.

[66] Deans, W. F. and Richards, C. E., *Journal of Testing and Evaluation,* Vol. 7, 1979, pp. 147-154.

[67] Mayes, I. C. and Baker, T. J., *Fatigue of Engineering Materials and Structures,* Vol. 4, 1981, pp. 79-96.

[68] Fujitani, K., Sakai, T., Nakagawa, A., and Tanaka, T., *Bulletin of the Japan Society of Mechanical Engineers,* Vol. 25, 1982, pp. 1195-1201.

[69] Zaiken, E. and Ritchie, R. O., *Materials Science and Engineering,* Vol. 70, 1985, pp. 151-160.

[70] Gray, G. T. and Luetjering, G. in *Fatigue 84,* Vol. II, C. J. Beevers, Ed., EMAS Publications Ltd., Warley, U.K., 1984, pp. 707-716.

[71] Doeker, H. and Bachmann, V. in *Fatigue Crack Closure,* American Society for Testing and Materials, Philadelphia, 1986, to be published.

[72] Doeker, H., Bachmann, V., and Marci, G. in *Fatigue Thresholds,* J. Backlund, A. Blom, and C. J. Beevers, Eds., EMAS Publications Ltd., Warley, U.K., 1981, pp. 45-57.

[73] Radon, C. J. in *Fatigue Thresholds,* J. Backlund, A. Blom, and C. J. Beevers, Eds., EMAS Publications Ltd., Warley, U.K., 1981, pp. 113-124.

[74] Doeker, H. and Peters, M. in *Fatigue 84,* Vol. I, C. J. Beevers, Ed., EMAS Publications Ltd., Warley, U.K., 1984, pp. 275-285.

[75] Aronson, A. and Ritchie, R. O., *Journal of Testing and Evaluation,* Vol. 7, 1979, pp. 208-215.

[76] Schijve, J., *Engineering Fracture Mechanics,* Vol. 11, 1979, pp. 167-221.

[77] Petit, J., *Fatigue Crack Growth Threshold Concepts,* D. Davidson and S. Suresh, Eds., The Metallurgical Society of the American Institute of Mining, Metallurgical, and Petroleum Engineers, Warrendale, PA, 1984, pp. 3-24.

[78] Frandsen, J. D., Inman, R. V., and Buck, O., *International Journal of Fracture,* Vol. 11, 1975, pp. 345-349.

[79] Sharpe, W. N., Jr., Corbly, D. M., and Grandt, A. F. in *Fatigue Crack Growth Under Spectrum Loading, ASTM STP 595,* American Society for Testing and Materials, Philadelphia, 1976, pp. 61-77.

[80] Sewell, G. and Marcus, H. L., *Scripta Metallurgica,* Vol. 11, 1977, pp. 521-524.

[81] Mahulikar, D. S. and Marcus, H. L., *Fatigue of Engineering Materials and Structures,* Vol. 3, 1981, pp. 257-264.

[82] Suresh, S., Parks, D. M., and Ritchie, R. O. in *Fatigue Thresholds,* J. Backlund, A. Blom, and C. J. Beevers, Eds., EMAS Publications Ltd., Warley, U.K., 1981, p. 391.

[83] Buck, O., Thompson, R. B., and Rehbein, D. K. in *Fatigue Crack Closure,* American Society for Testing and Materials, Philadelphia, 1986, to be published.

[84] Bouami, D. and De Vadder, D., *Engineering Fracture Mechanics,* Vol. 23, No. 5, 1986, pp. 913-920.

[85] Buck, O., Ho, C. L., and Marcus, H. L., *Engineering Fracture Mechanics,* Vol. 5, 1973, pp. 23-34.

[86] Buck, O., Frandsen, J. D., and Marcus, H. L., *Engineering Fracture Mechanics,* Vol. 7, 1975, pp. 167-171.

[87] Resch, M. T., London, B. D., Yuce, H. H., and Nelson, D. V. in *Proceedings,* ASTM Symposium on Fundamental Questions and Critical Experiments in Fatigue, Dallas, TX, Oct. 1984, American Society for Testing and Materials, Philadelphia.

[88] Sharpe, W. N. and Grandt, A. F. in *Mechanics of Crack Growth, ASTM STP 590,* American Society for Testing and Materials, Philadelphia, 1976, pp. 302-320.

[89] Rueping, J. E. and Hillberry, B. M., "Fatigue Crack Closure Behavior: A Comparative Study," Technical Report AFOSR-TR-76-1090, AFOSR, Bolling Air Force Base, Washington, DC, Aug. 1976.

[90] Jones, J. W., "The Interferometric Measurement of Near Crack Tip Displacements in a Nickel Base Superalloy at Ambient and Elevated Temperature," Technical Report AFML-TR-78-159, Air Force Materials Laboratory, Wright Patterson Air Force Base, OH, 1979.

[91] Ohta, A., Kosuge, M., and Sasaki, E., *International Journal of Fracture,* Vol. 14, 1978, pp. 251-264.

[92] Marci, G. and Packman, P. F., *International Journal of Fracture,* Vol. 16, 1980, pp. 133-153.

[93] Macha, D. E., Sharpe, W. N., Jr., and Grandt, A. F., Jr. in *Cracks and Fracture, ASTM STP 601,* American Society for Testing and Materials, Philadelphia, 1976, pp. 490-505.

[94] Hartman, G., private communication, 1985.

[95] Staal, H. U. and Elen, J. D., *Engineering Fracture Mechanics,* Vol. 11, 1979, pp. 275-283.

[96] Sehitoglu, H. in *Fracture Mechanics: Sixteenth Symposium, ASTM STP 868,* American Society for Testing and Materials, Philadelphia, 1985.
[97] Davidson, D. L. and Nagy, A., *Journal of Physics E: Scientific Instruments,* Vol. 11, 1978, pp. 207-210.
[98] Lankford, J., Davidson, D., and Chan, K. S., *Metallurgical Transactions A,* Vol. 15A, 1984, pp. 1579-1588.
[99] Davidson, D. L. and Lankford, J., *Metallurgical Transactions A,* Vol. 15A, 1984, pp. 1931-1940.
[100] Davidson, D. L., Hudak, S. J., Jr., and Dexter, R. J., "Measurement and Analysis of Critical Crack Tip Processes During Fatigue Crack Growth," Contractor Report NASA-172597, National Aeronautics and Space Administration, Hampton, VA, June 1985.
[101] Nisitani, H. and Kage, M. in *Fracture 1977, ICF4,* D. M. R. Taplin, Ed., Waterloo University Press, Waterloo, Ont., Canada, 1977, pp. 1099-1108.
[102] Ohta, A., Kosuge, M., and Sasaki, E., *International Journal of Fracture,* Vol. 15, pp. R53-R57.
[103] Shaw, W. J. D. and LeMay, I. in *Fracture Mechanics: Eleventh Symposium, ASTM STP 677,* American Society for Testing and Materials, Philadelphia, 1979, pp. 233-246.
[104] Pelloux, R. M., Faral, M., and McGee, W. M., *Fatigue of Engineering Materials and Structures,* Vol. 1, 1979, pp. 21-35.
[105] Sunder, R. and Dash, P. K., *International Journal of Fatigue,* April 1982, pp. 97-105.
[106] Nowack, H., "Ein Beitrag zur Untersuchung der Shadensakkumulation auf der Grundlage biharmonischer Belastungsablaeufe," Technical Report DLR-FB-71-23, Deutsche Luft- und Raumfahrt, Cologne, West Germany, 1971.
[107] Romaniv, O. N., Nikiforchin, G. N., and Andrusiv, B. N., *Soviet Materials Science,* Vol. 19, 1983, pp. 212-225.
[108] Romaniv, O. N., Nikiforchin, G. N., and Andrusiv, B. N., *Soviet Materials Science,* Vol. 20, 1984, pp. 62-67.
[109] Vazquez, J. A., Morrone, J. A., and Ernst, H., *Engineering Fracture Mechanics,* Vol. 12, 1979, pp. 231-240.
[110] Chand, S. and Garg, S. B. L., *Engineering Fracture Mechanics,* Vol. 18, 1983.
[111] Roberts, R. and Schmidt, R. A., *International Journal of Fracture,* Vol. 8, 1972, pp. 469-471.
[112] Schmidt, R. A. and Paris, P. C. in *Progress in Flaw Growth and Fracture Toughness Testing, ASTM STP 536,* American Society for Testing and Materials, Philadelphia, 1973, pp. 79-94.
[113] Gan, D. and Weertman, J., *Engineering Fracture Mechanics,* Vol. 15, 1981, pp. 87-106.
[114] Haenny, L. and Dickson, J. I., *International Journal of Fracture,* Vol. 16, 1980, pp. R121-R125.
[115] Liaw, P. K., Leax, T. R., Williams, R. S., and Peck, M. G., *Metallurgical Transactions A,* Vol. 13A, 1982, pp. 1607, 1617.
[116] Kawai, S., *Engineering Fracture Mechanics,* Vol. 16, 1982, pp. 857-870.
[117] Park, D. H. and Fine, M. E., *Fatigue Crack Growth Threshold Concepts,* D. Davidson and S. Suresh, Eds., The Metallurgical Society of the American Institute of Mining, Metallurgical, and Petroleum Engineers, Warrendale, PA, 1984, pp. 145-161.
[118] Fleck, N. A. and Smith, R. A., *International Journal of Fatigue,* Vol. 4, 1982, pp. 157-160.
[119] Fleck, N. A., Smith, I. C. F., and Smith, R. A., *Fatigue of Engineering Materials and Structures,* Vol. 6, 1983, pp. 225-239.
[120] Pitoniak, F. J., Grandt, A. F., Jr., Montulli, L. T., and Packman, P. F., *Engineering Fracture Mechanics,* Vol. 6, 1974, pp. 663-670.
[121] Ray, S., Grandt, A. F., Jr., and Andrew, S. in *Proceedings,* Symposium on Fundamental Questions and Critical Experiments in Fatigue, Dallas, TX, Oct. 1984, American Society for Testing and Materials, Philadelphia.
[122] Chermahini, R. G. and Newman, J. C., Jr. in *Fatigue Crack Closure,* American Society for Testing and Materials, Philadelphia, 1986, to be published.
[123] Bowles, C. Q. and Schijve, J. in *Fatigue Mechanisms: Quantitative Measurements of Physical Damage, ASTM STP 811,* American Society for Testing and Materials, Philadelphia, 1983, pp. 400-426.
[124] McEvily, A. J., *Metal Science Journal,* Aug./Sept. 1977, pp. 274-284.
[125] Ewalds, H. L. and Furnee, R. T., *International Journal of Fracture,* Vol. 14, 1978, pp. R53-R55.
[126] Ewalds, H. L., *Engineering Fracture Mechanics,* Vol. 13, 1980, pp. 1001-1007.
[127] Donald, J. K., *Fatigue Crack Closure,* American Society for Testing and Materials, Philadelphia, 1986, to be published.
[128] Ritchie, R. O. and Yu, W. in *Small Fatigue Cracks,* R. O. Ritchie and J. Lankford, Eds., The Metallurgical Society of the American Institute of Mining, Metallurgical, and Petroleum Engineers, Warrendale, PA, 1986.
[129] Roberson, L. J. and Kirk, M. T. in *Fatigue Crack Closure,* American Society for Testing and Materials, Philadelphia, 1986, to be published.
[130] Sehitoglu, H., *Engineering Fracture Mechanics,* Vol. 21, No. 2, 1985, pp. 329-339.
[131] Newman, J. C., Jr. in *Behavior of Short Cracks in Airframe Components,* Proceedings No. 328,

North Atlantic Treaty Organization-Advisory Group for Aeronautical Research and Development, 1982, pp. 6.1–6.26.
[*132*] Bretz, P. E., Petit, J. I., and Vasudevan, A. K., *Fatigue Crack Growth Threshold Concepts,* D. Davidson and S. Suresh, Eds., The Metallurgical Society of the American Institute of Mining, Metallurgical, and Petroleum Engineers, Warrendale, PA, 1984, pp. 163–183.
[*133*] Kikukawa, M., Jono, J., and Kondo, J., *Journal of the Society of Materials Science,* Japan, Vol. 30, 1981, pp. 56–62.
[*134*] Vasudevan, A. K. and Suresh, S., *Metallurgical Transactions A,* Vol. 13A, 1982, pp. 2271–2280.
[*135*] Larsen, J., private communication, 1986.

David L. Davidson[1] *and Stephen J. Hudak, Jr.*[1]

Alterations in Crack-Tip Deformation During Variable-Amplitude Fatigue Crack Growth

REFERENCE: Davidson, D. L. and Hudak, S. J., Jr., **"Alterations in Crack-Tip Deformation During Variable-Amplitude Fatigue Crack Growth,"** *Fracture Mechanics: Eighteenth Symposium, ASTM STP 945,* D. T. Read and R. P. Reed, Eds., American Society for Testing and Materials, Philadelphia, 1988, pp. 934-954.

ABSTRACT: Fatigue crack growth rate changes caused by single overloads and an overload followed by an underload have been studied in the high-strength aluminum alloy 7091. Crack-tip plasticity parameters were measured at each step in the loading sequence using the stereoimaging technique. Effective stress-intensity factor was measured with high resolution immediately at the crack tip both before an overload and during the subsequent growth rate retardation period. Crack-tip opening displacement and strain were altered in the same way as growth rate following the overload/underload events, and similitude with respect to crack-tip plasticity was preserved during the growth retardation period. Effective stress-intensity factor was found to correlate well with crack growth rate. Strains within the plastic zone were converted to stress at each step in load change. These stresses were then summed to determine the zone of residual stress caused by the overload. The region of compressive residual stress ahead of the crack tip was found to increase with overload severity but was limited to yield stress in magnitude. An underload shrank the spatial extent of the residual stresses caused by the overload.

KEY WORDS: fatigue crack growth, variable amplitude loading, similitude, crack-tip plasticity, plastic zone size, crack tip opening displacement, aluminum alloy, overload, crack closure, effective stress-intensity factor, strain

Research conducted over the past two decades has clearly shown that intermittent overloads, or high-low block loading, can cause significant crack growth retardation relative to the steady-state rate which occurs under constant-amplitude loading. The reverse effect—crack growth acceleration due to underloads, or low-high block loading—can also occur; however, this phenomenon is much less pronounced than is crack growth retardation. Consequently, neglecting loading interaction effects and using linear damage accumulation is likely to result in conservative life predictions. Depending on the load history and operational requirements of the structure, this conservatism may not be tolerable, for example, in aircraft or aerospace structures for which weight savings is a primary design consideration.

The concept of an effective stress-intensity factor has been useful in attempting to understand and model load interaction effects. This local alteration of the crack "driving force" is believed to arise from a variety of processes: crack-tip blunting [*1,2*], crack closure due to compressive residual stresses generated ahead of the crack tip [*3–7*], crack closure due to contact along the crack flanks arising from the plastically deformed wake [*8–10*], and microscopic crack branching/asperity contact [*11,12*]. Effective stress-intensity factor is only likely to be of value for describing crack growth rate if similitude is maintained, as has been emphasized by Broek [*13*].

[1]Institute scientist, staff scientist Department of Materials Sciences, Southwest Research Institute, 6220 Culebra Road, San Antonio, TX 78284.

An alternative—or perhaps supplemental—view is that the overload alters the materials intrinsic resistance to crack growth. For example, it has been suggested that load interaction effects are related to an increase in fatigue resistance due to strain hardening of the material within the overload plastic zone [*14,15*]. Thus, similitude would be violated and an effective stress-intensity factor could not be used to describe the crack growth rate.

The contribution of the above processes to crack growth, under either constant-amplitude or variable-amplitude loading, is difficult to isolate since most of these processes are inextricably related to the plastic deformation at the advancing crack tip.

The current study focuses on the deformation response of the tip of a growing fatigue crack since this process is central to understanding the interrelationships governing the other elemental processes believed to contribute to load interaction effects. Information on the crack-tip displacements, opening loads, and stress-strain states have been obtained on the fine-grained powder metallurgy aluminum alloy 7091-T7E69 using the stereoimaging technique [*16*]. This alloy has mechanical properties similar to 7075-T651. These measurements of crack-tip deformation are compared for constant- and variable-amplitude loading and their correlations with the transient crack growth responses following simple load excursions are also examined. This information is used to assess the state of similitude during variable-amplitude loading, as well as the viability of the effective stress-intensity approach to describing crack growth rate.

Experimental

Several different definitions have evolved to describe the magnitude of load range variations. An overload cycle may be referred to a baseline of zero load or to the range of cyclic stress intensity experienced by the crack prior to the variable load. The two resulting terms for overload severity, OLR and OLR*, are defined in Fig. 1. The term OLR*, used in this paper, is related to OLR by

$$\text{OLR*} = (\text{OLR} - R)/(1 - R) \tag{1}$$

A similar definition of underload severity, ULR*, is shown in Fig. 1. Three levels of overload severity, as well as the effect of baseline ΔK and R, were experimentally examined.

Single-edge-notched specimens with gage section dimensions of 20 by 3 mm were precracked in a laboratory servohydraulic loading frame and then transferred to a similar loading stage constructed to fit within a scanning electron microscope (SEM) [*17*]. The high resolution and depth of field of the SEM were useful especially for making photographs of the crack-tip region. The stereoimaging technique compares two photographs made under different conditions of

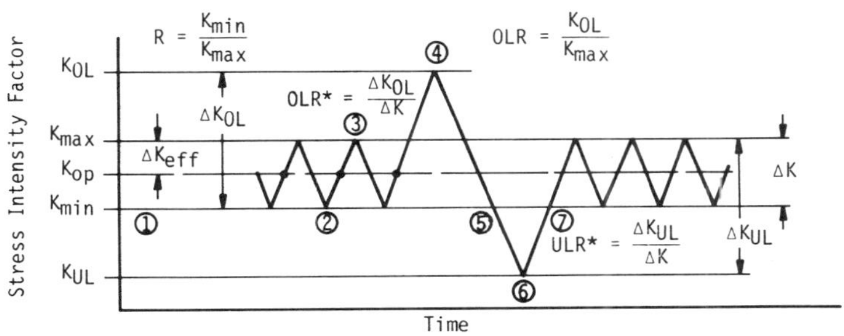

FIG. 1—*Definitions of terms describing fatigue crack growth and load range variation severity. Circled numbers designate photographs made of the crack-tip region during the experiment that were subsequently compared using stereoimaging.*

loading. Displacements resulting from the action of the load are measured using the techniques of photogrammetry. The fatigue crack grown under constant-amplitude loading was examined prior to the application of the load range variations by comparison of photographs made at minimum and maximum load. These are labeled as Photographs 2 and 3 in Fig. 1. Stereoimaging also allows very accurate, local determination of the crack opening load to be made by comparing photographs taken at minimum load with those made with increasing levels of loading. Crack opening load was carefully determined both prior to and after the load variations.

After the overload cycle was applied, Photograph 4 was taken at the maximum load, and 5 taken afterwards at the minimum load. Comparison of Photographs 2 and 4 defines the displacements due to the loading portion of the overload, while the reversibility due to subsequent unloading was shown by comparison of Photographs 4 and 5. Strains determined at each step in this loading sequence were converted to stresses, using previously developed methods [*18*], so that the residual stresses caused by the load excursion could be computed.

In addition to the analyses made for the overload cycle, two additional photographs were made during an underload. Stresses caused by strains which occurred during the underload cycle were then added to those due to the overload.

Subsequent to the cycles of load variation, the specimen was removed from the SEM and cycled in the laboratory loading frame at the cyclic stress intensity used previously. Since these experiments were at constant ΔK, it was necessary periodically to reduce load as the crack lengthened. Load reduction was done in very small increments to avoid this as an experimental variable. Periodically, the specimen was transferred to the SEM and reloaded for photographing and determination of the crack opening load. Crack growth rates were determined from examination of the specimen by both optical microscopy and SEM.

Numerical values of the parameters defined in Fig. 1 that were actually used in the experiments are indicated in Fig. 2. Three levels of overload, one followed by an underload, were examined.

Results

Fatigue Crack Growth Rate and Opening Load

The number of cycles during which growth rate is retarded is defined as the number of delay cycles, N_D. This may be determined from a plot of crack length versus cycles or crack growth rate. Crack growth rate, normalized to the constant-amplitude rate preceding the load range variation, is shown plotted versus crack length in Fig. 3*a* for the smallest overload studied, OLR* = 2.0. The number of cycles corresponding to the excursion in crack growth rate after the overload N_D is shown on this figure and on Figs. 3*b* and 3*c*, which have comparable information for the other load range variations studied. Also shown in the figures is the effective stress-intensity factor, ΔK_{eff}, as derived from local measurements of crack opening load.

The number of delay cycles, minimum crack growth rate, and the spatial extent of the delay region increase with severity of the overload, in agreement with results from numerous other investigators. Change in effective cyclic stress-intensity factor parallels fatigue crack growth rate, with the minimum occurring at the same location (Figs. 3*a* and 3*b*), except for the overload/underload, which shows a difference between the minimum in crack growth rate and ΔK_{eff} (Fig. 3*c*).

Displacements—Application of Cyclic Load

An example of the displacements caused by loading the crack to the maximum constant-amplitude load is shown in Fig. 4. Note that the scale of displacements on the right has been made different from the spatial scale on the left in order for the displacements to be seen. The

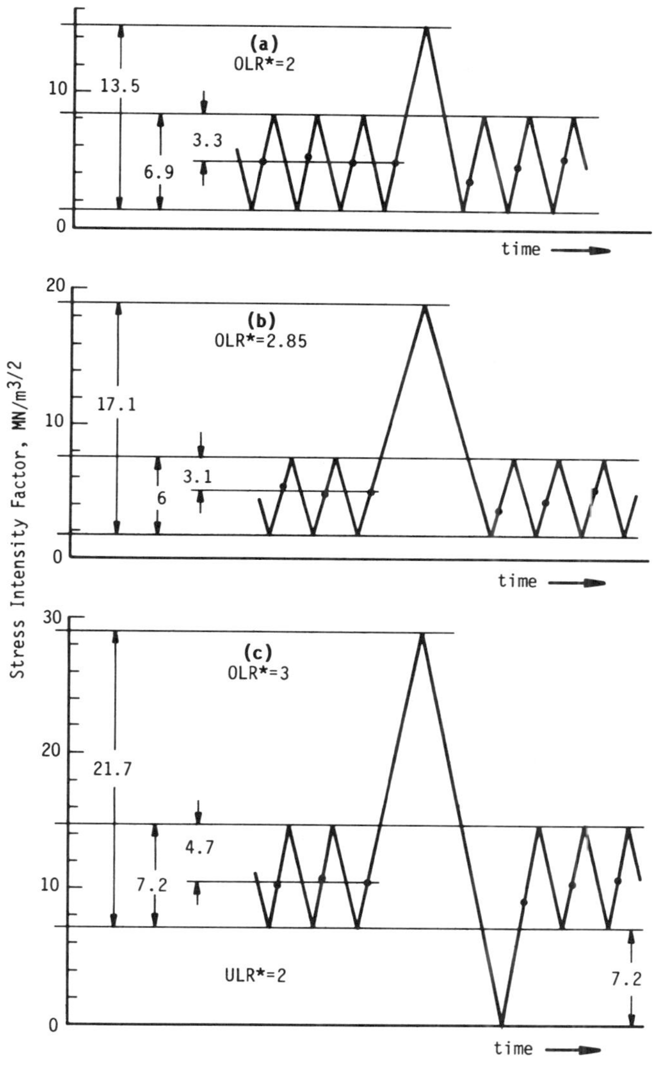

FIG. 2—*Schedules of load application for the three load range variations examined. Value of* ΔK_{eff} *shown is that measured just prior to the load range variation.*

crack opening and displacements ahead of the tip are asymmetric in agreement with previous observations in this [*19*] and other alloys [*20,21*].

Displacements—Overload and Underload Cycles

On the overload cycle, the crack opens to a much larger extent than occurred during cyclic loading, and the displacements are much larger (Fig. 5). The field is also much more symmetric than for cyclic loading and more in the direction of loading. On the unloading portion of the overload cycle, considerable reversibility occurs. The net result of the overload cycle is a small

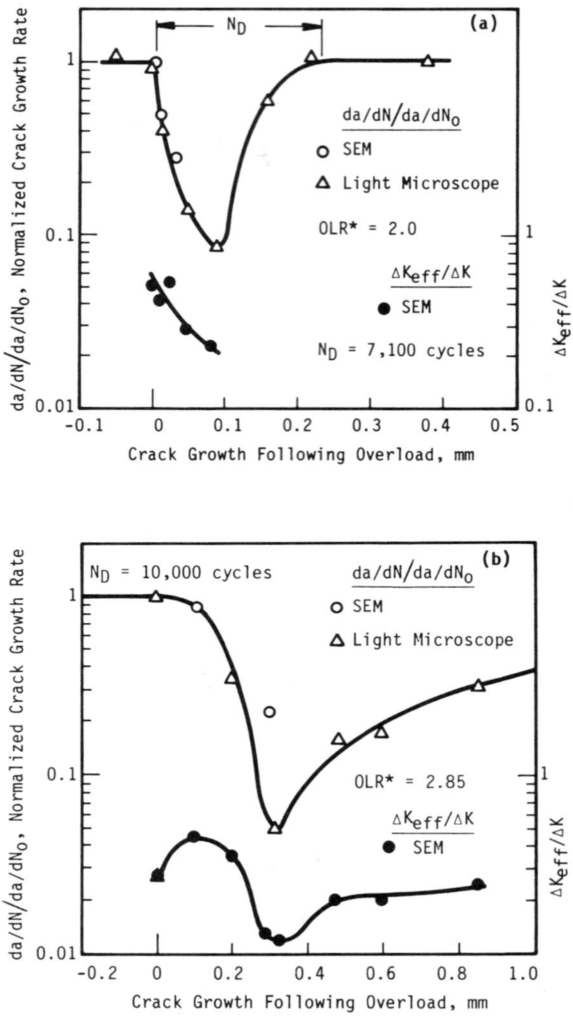

FIG. 3—*Measured crack growth rate response and* ΔK_{eff} (a) *following a single overload of OLR** $= 2.0$, R $= 0.22$. *Plastic zone size* $= 165$ μm. (b) *Following a single overload of OLR** $= 2.85$, R $= 0.22$. *Plastic zone size* $= 172$ μm. (c) *For a single overload followed by an underload to near zero (*R $= 0.5$, *OLR** $= 3.0$, *ULR** $= 2.0$*). Plastic zone size* $= 1290$ μm.

displacement of generally tensile nature, accompanied by a crack which remains open for a distance of up to several millimetres behind the crack tip (the entire length of the crack in these experiments). An underload cycle increases the reversibility of displacements caused by unloading from the maximum of the overload cycle, and decreases the residual crack opening displacement.

Crack Opening Displacement

Displacements behind the crack tip were measured parallel to (COD_x) and perpendicular to (COD_y) the direction of loading. Figure 6 shows COD in both directions before application of

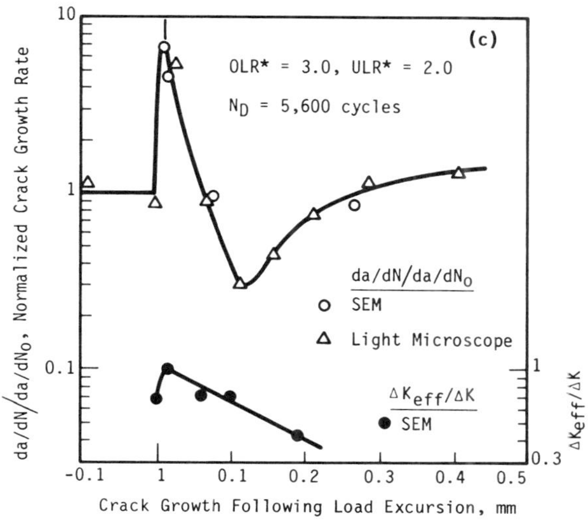

FIG. 3—*Continued.*

the load excursion and for the crack at two locations after the overload and near the point of minimum crack growth rate. A similar result was found for all the load range variations, namely, COD_x decreased after the overload, in a manner similar to the crack growth rate, but COD_y was changed very little. The exception is Fig. 6*c*, the overload/underload, which was cycled at $R = 0.5$ and produced a different combination of COD_x and COD_y than the other two cracks, which were cycled from a lower R. In other words, the differences in COD exhibited by the pre-overload cracks are due to the R ratio, rather than effects caused by the load range variation.

The changes in COD resulting from the variable-amplitude loading are very similar to those caused by decreasing ΔK into the near-threshold crack growth region [*22*]. Also note that at low R there are negligible changes in slopes of these lines before and after the overload.

If the resultant COD $= (COD_x^2 + COD_y^2)^{1/2}$ is plotted similarly to Fig. 6, a straight line may be fit to the data, giving the equation

$$COD = C|y|^p \tag{2}$$

Crack-tip opening displacement (CTOD) is defined as that value of COD at distance of 1 μm from the crack tip (C in Eq 2).

Strains Within the Plastic Zone

Crack Tip Strains—Strains at the crack tip due to both cyclic and overload cycles are shown versus the applied stress-intensity factor in Fig. 7. The good correlation indicates that similitude (in terms of deformation) is preserved before and during the overload cycle. It should be noted, however, that this correlation differs from that derived for a similar experimental alloy MA-87 [*19*].

Strain Distribution—Overload Cycle—The overall distribution of strain caused by the overload (Fig. 8), is generally similar to the distribution caused by constant-amplitude cyclic loading [*23*]. However, a detailed comparison of the distributions of strain ahead of the crack tip shows differences. Effective strain ahead of the crack tip for the overload cycles is shown in Fig. 9.

FIG. 4—*Displacements near the tip of a crack grown under constant-amplitude loading at* $\Delta K = 6.9$ $MN/m^{3/2}$. *Displacements* (right scale) *are shown on a grid 10 μm in spacing* (left scale). *Loading axis was horizontal. Note the asymmetry of the displacements relative to the direction of crack growth.*

These strains are normalized by strain at the crack tip for comparison. The equation which describes the strain distribution has the general form

$$\epsilon' = m \ln (A/(r + B)) \tag{3}$$

where r is the distance ahead of the crack tip in micrometres and the other factors are fitting parameters. This equation fits strain data within the plastic zone reasonably well for cyclic loading, and for an overload cycle, although the fitting parameters change for each type of loading (Table 1).

Strain Distribution—Cyclic Loading—Distributions of effective strain within the plastic zone of cyclically loaded cracks, both pre- and post-overload, are compared in Fig. 10. Although scatter is found in the data, there does not appear to be any systematic deviation of these strain distributions from that described by Eq 3. This result implies that similitude is maintained during the period of slow growth following a load excursion. The fitting parameters for Eq 3 are listed in Table 1, and are different from those for single overloads.

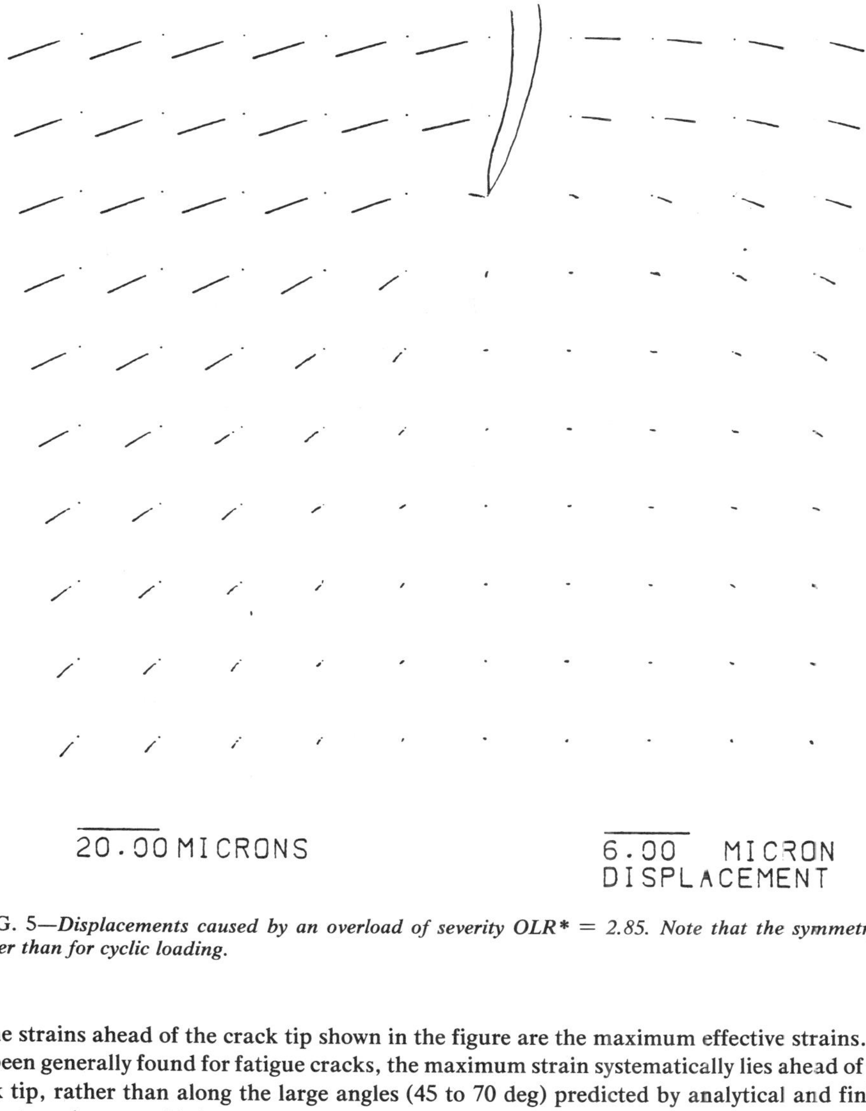

FIG. 5—*Displacements caused by an overload of severity OLR* = 2.85. Note that the symmetry is greater than for cyclic loading.*

The strains ahead of the crack tip shown in the figure are the maximum effective strains. As has been generally found for fatigue cracks, the maximum strain systematically lies ahead of the crack tip, rather than along the large angles (45 to 70 deg) predicted by analytical and finite-element analyses, and is in agreement with the large Mode II component found in crack opening displacement.

Plastic Zone Sizes—The size of the plastic zone directly ahead of the crack tip was computed from the equations of Table 1 by extrapolation to a value of elastic strain. The computed values are listed in Table 2. As expected, plastic zone sizes for the overload cycle are larger than for either pre- or post-overload cycles, and the plastic zone sizes are the smallest for the slowest growing cracks.

Correlations Between Crack-Tip Strain, CTOD, and Crack Growth Rates

There are two additional results which indicate the maintenance of similitude for cracks in the post-overload growth period: There is a correlation between crack tip strain and CTOD (Fig. 11), and between crack tip strain and crack growth rate (Fig. 12).

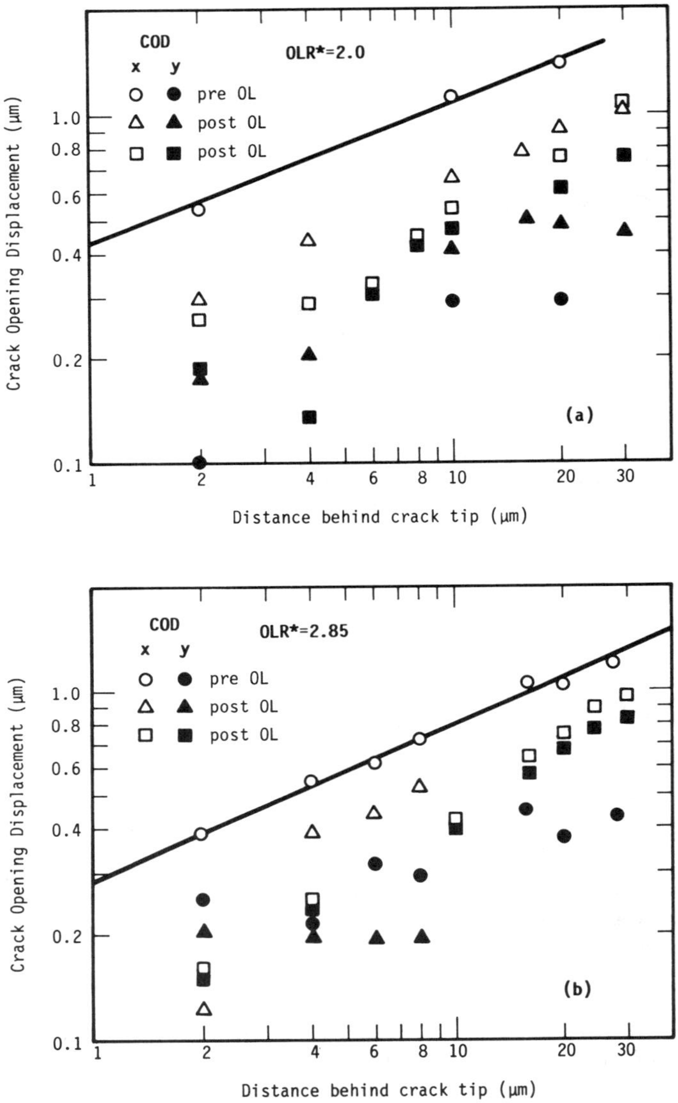

FIG. 6—*Crack opening displacements parallel* (x) *and perpendicular* (y) *to the direction of loading for cracks before and after the load range variation. For both single overloads,* (a) *and* (b), *crack opening after the overload is lowered in the direction of loading (Mode I), while that in Mode II is not measurably altered. However, in* (c), *the overload/underload from* R = *0.5, the crack opening is altered in both directions.*

There is a good correlation between strain and CTOD for constant-amplitude cyclic loading before and after the load variations, but the correlation apparently changes on overload cycles. This is another indication that similitude is maintained during constant-amplitude loading, but it calls into question the existence of similitude during overload cycles. Note that there is no data point for OLR* = 3 because COD was difficult to define due to tearing which took place at this high level of stress intensity. These results, indicating a loss of similitude on the overload cycle, are contrary to those of Fig. 7.

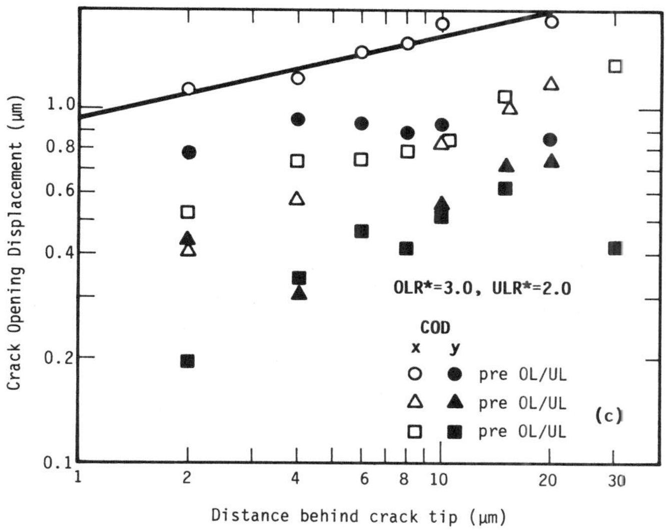

FIG. 6—*Continued.*

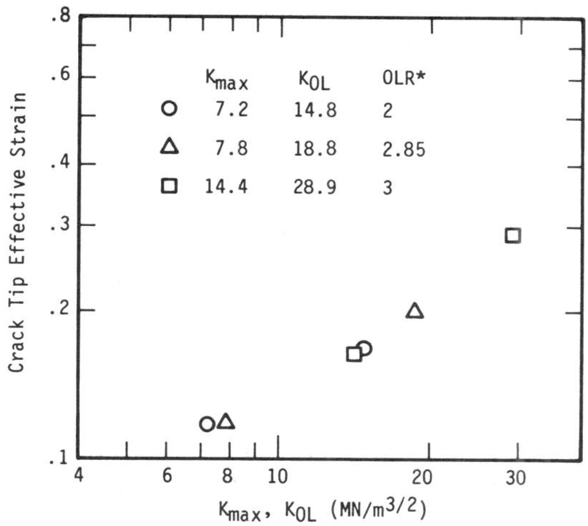

FIG. 7—*Crack-tip strains as a function of stress-intensity factor for constant amplitude and overloads.*

Crack-tip strains also correlate well with crack growth rates (Fig. 12). This is another indication that the growth mechanism has not been altered by the load excursions, and further indication that crack growth is a strain-controlled process with crack advance being controlled by cumulative strain [*19–24*].

Residual Stress Fields

The change in strain which occurred during each step in the load range variation was determined by comparison of photographs made at the loads depicted in Fig. 1. The stress change

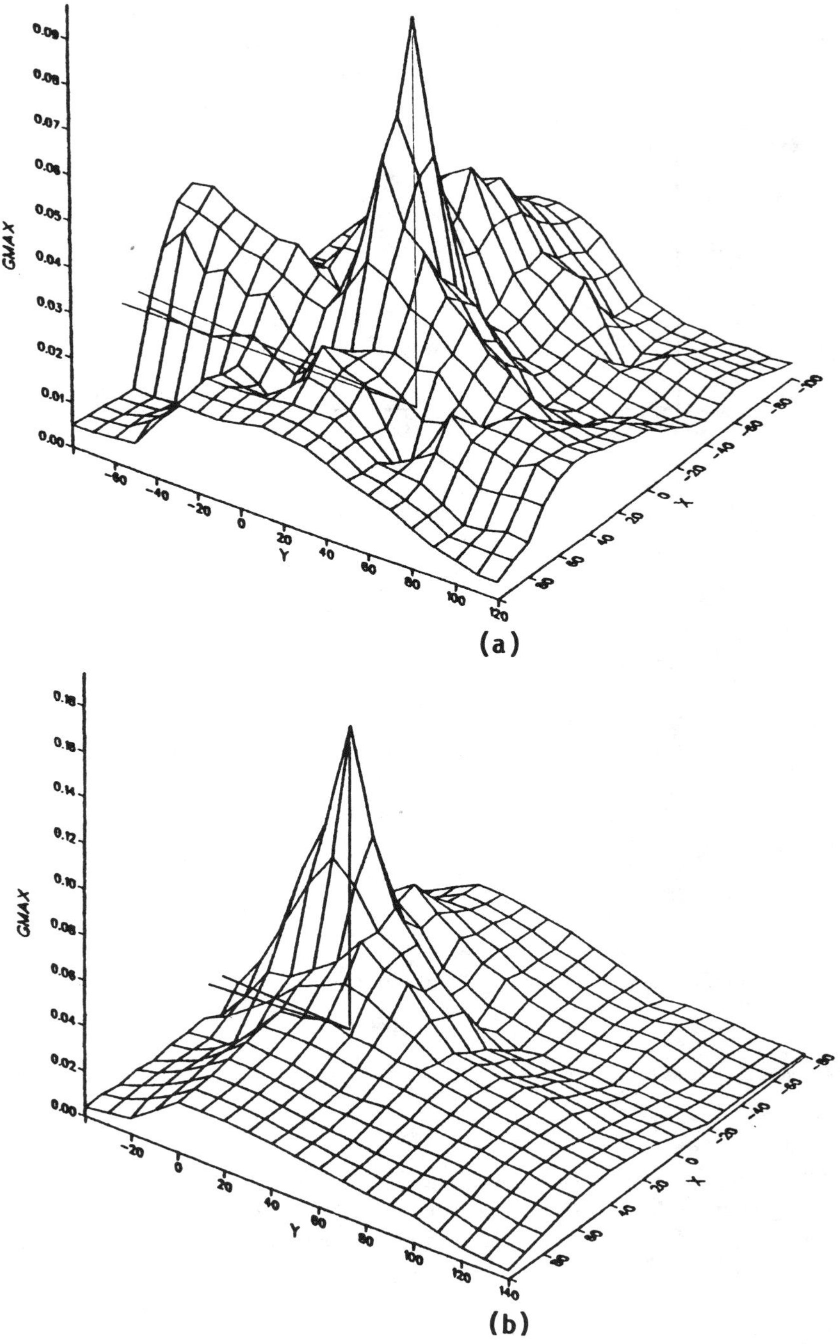

FIG. 8—*Comparison of the maximum shear strains resulting from each of the overload cycles analyzed.* (a) *OLR* = 2.0,* (b) *OLR* = 2.85, and* (c) *OLR* = 3.0.*

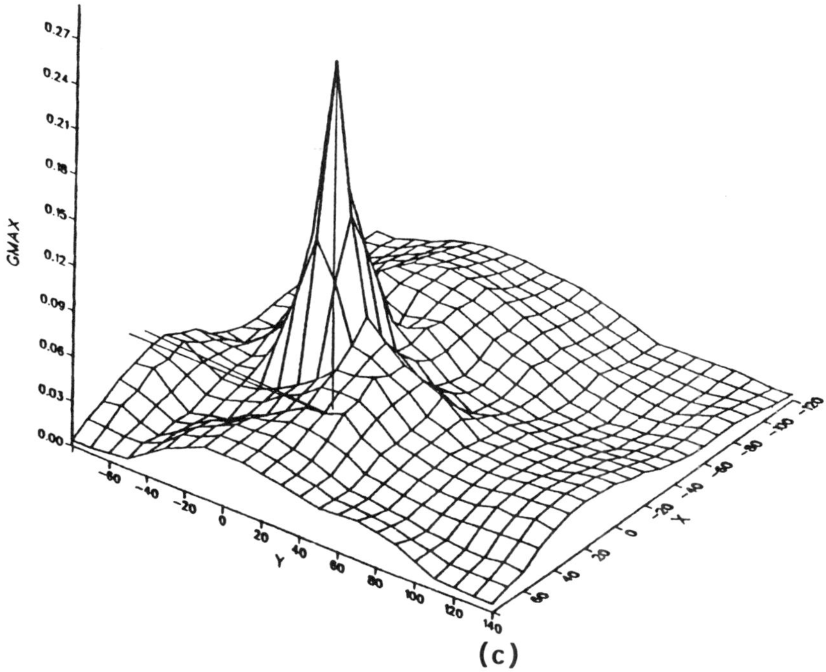

FIG. 8—*Continued.*

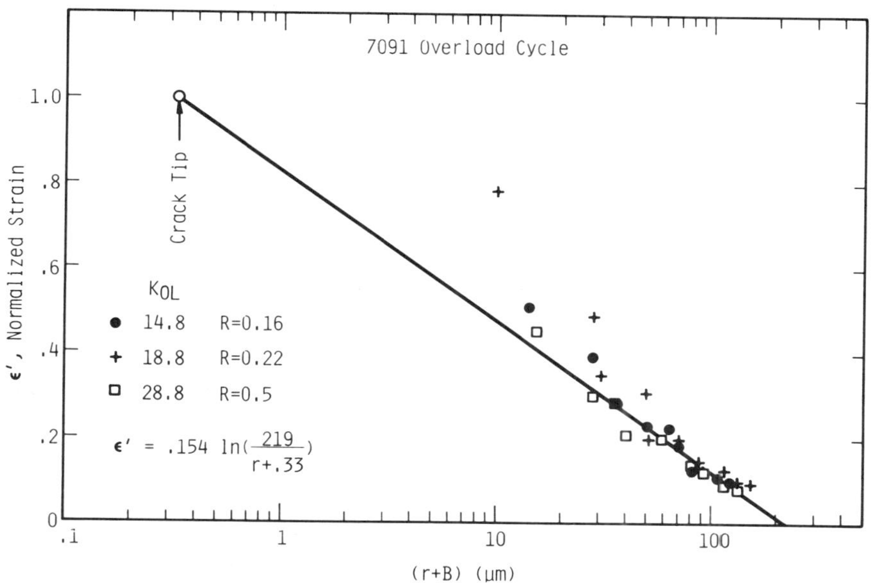

FIG. 9—*Normalized effective strain ahead of the crack tip on the overload cycle.*

TABLE 1—*Equations for normalized strain ahead of crack tip,* r *in* μm.

Loading	Equation
cyclic	$\epsilon' = 0.22 \ln\left(\frac{57.7}{r + 0.612}\right)$
overload cycle	$\epsilon' = 0.154 \ln\left(\frac{219}{r + 0.33}\right)$
monotonic	$\epsilon' = 0.180 \ln\left(\frac{64.7}{r + 0.32}\right)$

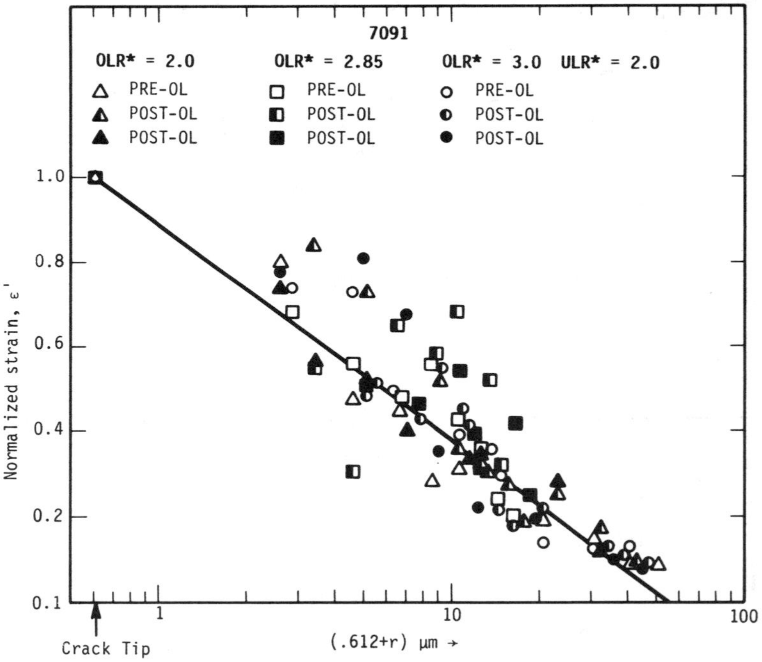

FIG. 10—*Normalized effective strain ahead at the tips of cracks grown under constant-amplitude loading before and after the load range variation.*

accompanying each change in load was computed using the procedure previously derived [*18*]. This computation requires knowledge of the appropriate stress-strain curves. Consequently, the monotonic and cyclic stress-strain curves were measured for this material. The cyclic and monotonic curves were nearly equal, and there is essentially no difference between the tensile and compressive monotonic curves. Cyclic hysteresis loops were very nearly symmetrical, and stabilized very rapidly after a step change in strain range.

Values of the stress tensor were computed at points on a grid within the plastic zone for each step in the load change, and then added together to produce a net residual stress. For each load change, the result of the stress computation was six components of stress at each point of the

TABLE 2—*Distance ahead of crack to elastic strain boundary (plastic zone size).*

Crack	ΔK	$r(\mu m)$
	OLR* = 2, ULR* = 1, R = 0.16	
Pre	6.2	42.8
OL	14.8	164
Post	6.2	29.3
Post	6.2	30.0
	OLR* = 2.85, ULR* = 1, R = 0.22	
Pre	6	55.5
OL	17.1	172
Post	6	23.7
Post	6	15.5
	OLR* = 3, ULR* = 2, R = 0.5	
Pre	7.2	46.6
OL	21.7	190
Post	7.2	35.8
Post	7.2	41.9

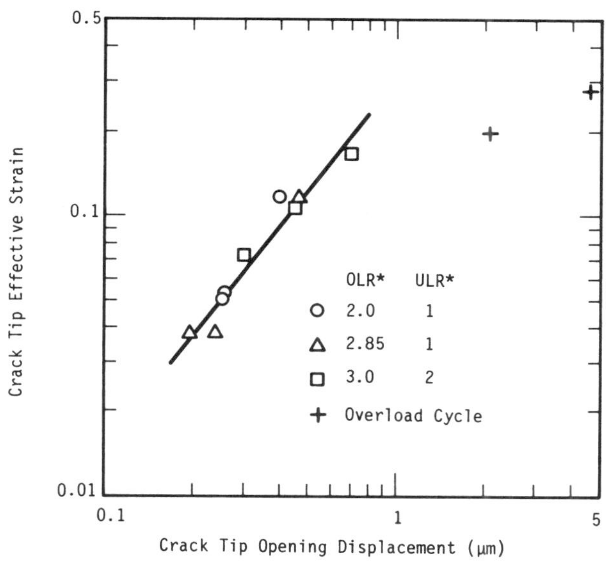

FIG. 11—*Correlation between crack-tip strain and CTOD. The lack of correlation on the overload cycles shows that the crack-tip shape has changed from that due to constant-amplitude loading.*

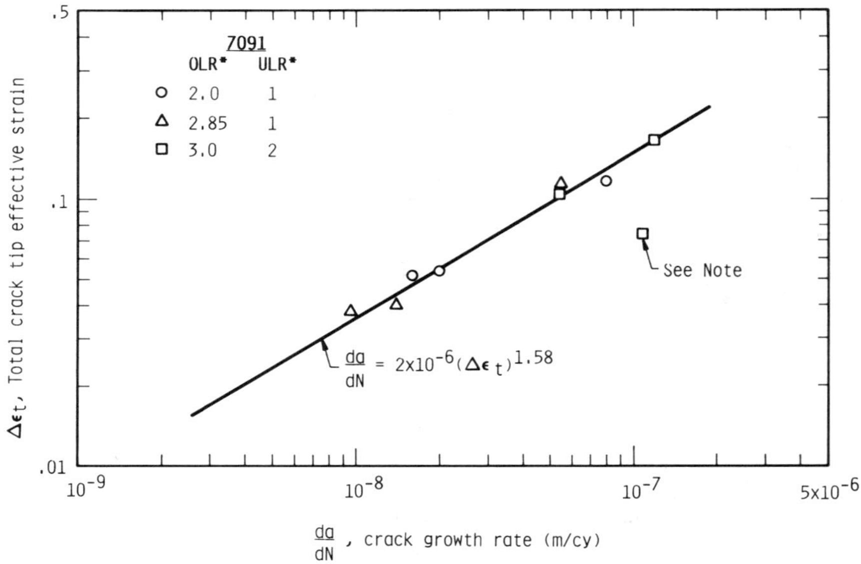

FIG. 12—*Correlation between crack-tip strain and growth rate. Crack growth rate for the point lying off the correlation line was averaged over 100 μm, giving a value which is probably much larger than at the time crack-tip strain was measured.*

analysis. It is not known which element of the stress tensor, or which combination of stresses, is the most important in retarding fatigue crack growth, but the stress in the loading axis (σ_x) and the total effective stress (σ_{eff}) were chosen as most likely to influence crack growth. Stress in the loading axis could be expected to exert a direct clamping influence on the crack, and the effective stress provides a method for including both the normal and shear stresses. Stresses in the loading axis caused by the loading portion of an overload are shown in Fig. 13*a*, and those due to the unloading part of the overload are shown in Fig. 13*b*.

Summing the stresses shown in Fig. 13 results in small tensile values of stress. Strains on the loading part of the overload are somewhat larger than those on the unloading part of the cycle, indicating flow irreversibility. However, these resultant stresses are referenced to the value of stress at the beginning of the overload cycle, which is not zero because of the compressive residual stresses present at the tip of the crack loaded under constant amplitude. To reference the residual stresses caused by the load excursion to zero, these compressive stresses must be added to the stresses which occurred at each step in the load excursion.

For the constant-amplitude cyclic crack, compressive stresses within the plastic zone at minimum load were assumed to be proportional to that part of the loading cycle during which the crack remained closed. Thus, the computed stresses due to constant-amplitude cycling were multipled by $[(\Delta K_{eff}/\Delta K) - 1]$. When these stresses are added to those of Fig. 13, the resulting stress distribution is as shown in Fig. 14. Qualitatively, the residual stress distribution is somewhat bowl-shaped, with the deepest part of the bowl ahead of the crack tip. The maximum value of residual stress computed for all the overload values studied was approximately that of yield. The residual effective stresses were computed similarly and are shown in Fig. 15. The result is the same qualitatively as for stresses in the loading axis.

A result similar to that shown in Fig. 14, but more pronounced and well formed due to the increased severity of the overload, is shown in Fig. 16. Again the maximum value of compressive stress is approximately equal to the yield stress.

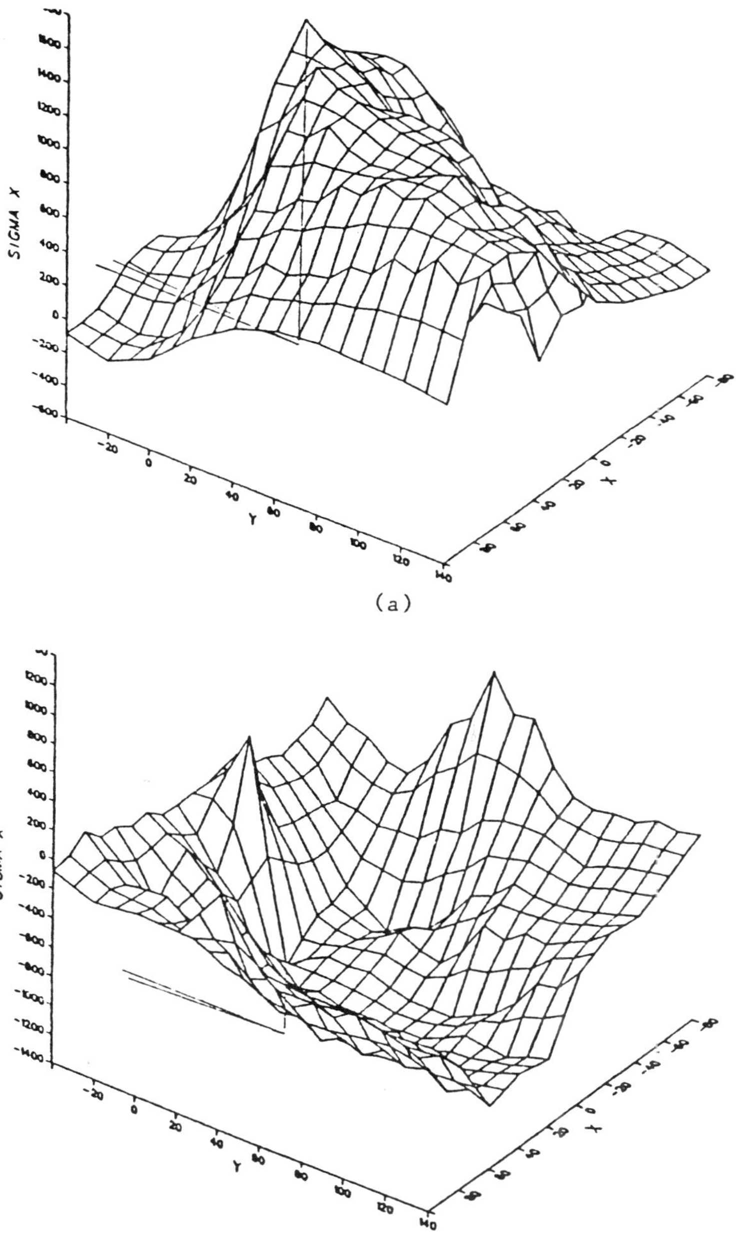

FIG. 13—(a) *Stresses on the loading portion of the overload cycle and* (b) *on the unloading portion of the overload cycle in the axis of the load application, OLR* = 2.85.*

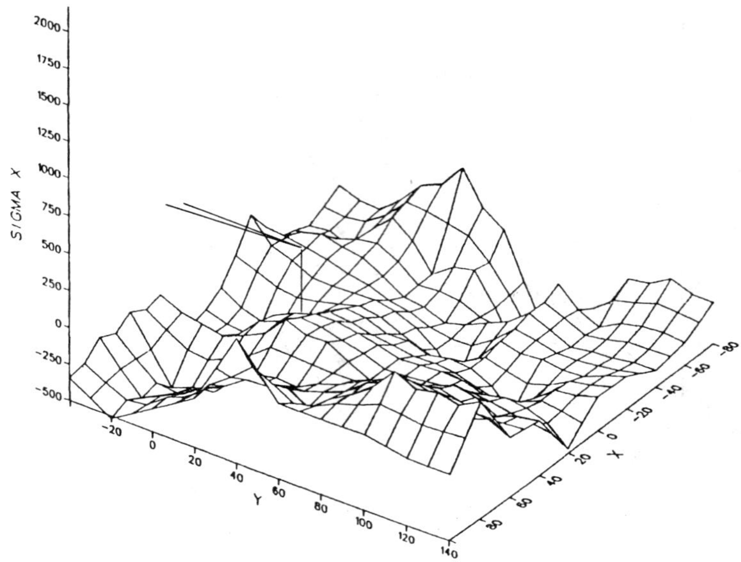

FIG. 14—*Residual stress field in the direction of loading caused by the OLR* = 2.85.*

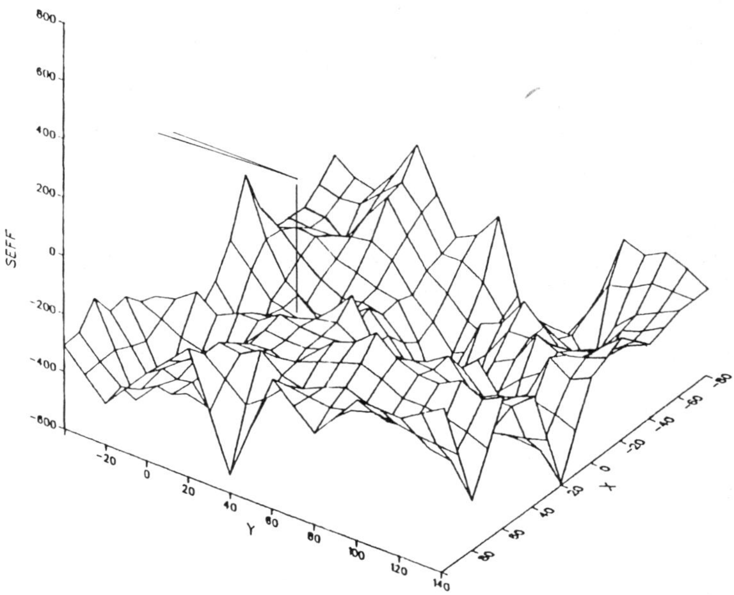

FIG. 15—*Residual effective stress field for OLR* = 2.85.*

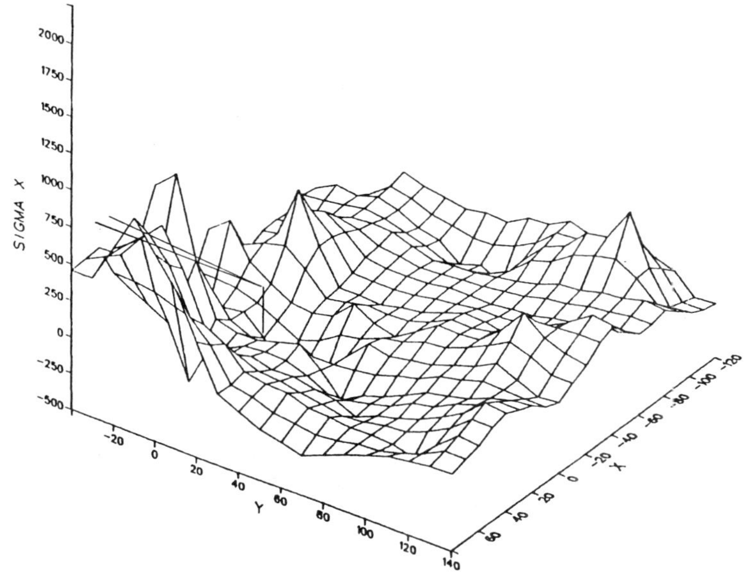

FIG. 16—*Residual stress field in the direction of loading caused by OLR* = 3.0. Note that the crack was not subsequently grown through this condition, but was followed by an underload.*

Discussion

One of the main findings of this research has been that similitude in fatigue crack plasticity is maintained during periods of crack velocity change resulting from variable-amplitude loading. Similitude, unaltered by load range variations, is manifested in three ways:

(*a*) the distribution of strain ahead of the crack tip within the plastic zone,
(*b*) the relationship between crack-tip opening displacement and strain, and
(*c*) the correlation between crack growth rate and crack-tip strain.

It follows from the above findings that fatigue crack growth rate is not easily determined during variable amplitude loading because the driving force for crack growth is not known. Thus, the problem of predicting crack growth rate is that of determining the correct driving force.

The measurements of ΔK_{eff} following an overload showed a decrease which paralleled crack growth rate (Fig. 3). This result suggests that ΔK_{eff} could be used to predict crack growth rate following a load range variation, if some means could be determined for its computation from constant-amplitude crack growth data. Thus, the constant-amplitude data given in Fig. 7 were used to derive a correlation between crack tip strain, $\Delta\epsilon(0)$, and ΔK^*_{eff}

$$\Delta\epsilon(0) = 4.13 \times 10^{-2} \Delta K^{*\,0.56}_{eff} \tag{4}$$

where $\Delta K^*_{eff} = \Delta K - \Delta K_{TH}$ and where $\Delta K_{TH} = 1.6$ MN/m$^{3/2}$ at $R = 0.1$. Values of crack-tip strain derived after the load range variations allow values of ΔK^*_{eff} to be computed from Eq 4; those values are given in Table 3.

These results indicate reasonable agreement in about half the cases, with the least agreement after the overload/underload. Differences in these values could arise either from measurement of ΔK_{eff} or its computation. The error in measurement of opening load could have been easily

TABLE 3—*Comparison between derived and measured effective stress-intensity factors.*

ΔK	OLR*/ULR*	$\Delta\epsilon(0)$ After OL	ΔK^*_{eff}, MN/M$^{3/2}$	$\Delta K^*_{eff}/\Delta K$	$\Delta K_{eff}/\Delta K$
6.9	2/1	0.0505	2.18	0.31	0.56
		0.0518	2.26	0.32	0.31
6.0	2.85/1	0.038	1.4	0.23	0.12
		0.038	1.4	0.23	0.20
7.2	3/2	0.072	3.75	0.52	0.62
		0.109	7.1	0.99	0.38

$\Delta K^*_{eff} = \Delta K - \Delta K_{TH}$.
$\Delta K_{TH} = 1.6$ MN/m$^{3/2}$.

± 100 N which translates roughly into a measurement variation of $\Delta K_{eff}/\Delta K$ of ± 0.05, for OLR* = 2 and 2.85, and ± 0.1 for OLR* = 3/ULR* = 2. An estimate of the errors likely in calculating ΔK^*_{eff} is more difficult to make, but the error would be reduced by increasing the data base. It is likely that the error could be as much as $\pm 20\%$. Thus, there is reasonably good consistency between the values of ΔK^*_{eff} and ΔK_{eff}, except for the last case in Table 3, for which ΔK^*_{eff} surely is too large.

That it is possible to derive approximate values of $\Delta K^*_{eff} = \Delta K - \Delta K_{TH}$ indicates that this concept is reasonably correct. This assumption implies that when $\Delta K = \Delta K_{TH}$ there is zero driving force for crack growth, that is, $\Delta K_{eff} = 0$; thus, it is compatible with the usual definition of $\Delta K_{eff} = K_{max} - K_{open}$.

Comparison of the crack opening displacements of Fig. 6 with previous results [*19,22*] indicates that the effect of an overload is to shift fatigue crack-tip parameters to those found closer to the threshold for crack growth. This is also consistent with the lowering of crack-tip strain (Fig. 12) and decrease in ΔK_{eff} (Table 3). When the load excursion is applied from a high R, the effect is to shift the crack-tip opening displacements to values more like those found at low R (Fig. 6*c*).

Results of the computation of residual stresses after the load range variations (Figs. 14 to 16), indicate that the magnitude and extent of compressive residual stresses at the crack tip are increased by the overload. The trends implied by the results are summarized schematically in Fig. 17, which also shows the effect of the overload/underload combination. Magnitude of the residual stresses is limited to yield, probably because of the low work-hardening rate of this alloy. The increase in size of the residual stress region undoubtedly causes a decrease in driving force, even though we do not yet know how to compute ΔK^*_{eff} from the combination of applied and residual stresses, and that remains as a challenge in predicting crack growth under variable-amplitude loading from constant-amplitude tests.

Summary and Conclusions

1. The magnitude of the overload/underload cycle controls the spatial extent of the compressive residual stress field both parallel and perpendicular to the direction of crack growth, whereas the magnitude of this compressive stress field is limited to the flow stress of the material. These results are consistent with the amount of crack growth retardation following the overload/underload cycle and thereby attest to the importance of crack-tip residual stress during variable-amplitude fatigue crack growth.

2. A correlation exists between the fatigue crack growth rate and the crack-tip strain measured before and after the overload, thus indicating that the intrinsic mechanism of crack growth is strain controlled and is the same as that for constant-amplitude fatigue crack growth.

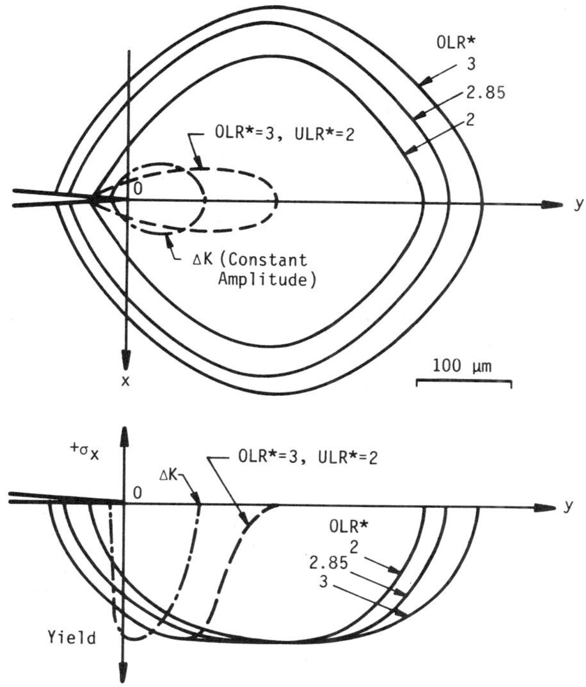

FIG. 17—*Idealized representation of the magnitude and spatial extent of the region of compressive stress in the loading direction. The effect of increasing the overload severity is to enlarge the region of compression, compared with that for constant-amplitude loading, and the effect of an underload following an overload is to shrink the size of the compressive zone, with the largest contraction lateral to the crack growth direction.*

3. Similitude in crack-tip plasticity is maintained during variable-amplitude fatigue crack growth. This similitude is manifested by a uniqueness in (*a*) strain distribution within the crack-tip plastic zone, and in the relationship between (*b*) crack-tip strain and crack-tip opening displacement, and (*c*) crack-tip strain and crack growth rate.

4. It follows from Conclusions 2 and 3 that the problem of predicting fatigue crack growth rate during variable-amplitude loading is one of determining the proper "driving force" for crack growth. Incorporating crack closure concepts into an effective stress-intensity factor appears to be a viable approach to characterizing the crack "driving force."

5. After an overload, the effective stress-intensity factor, as determined from locally measured crack opening load, is decreased. The magnitude is consistent with values derived from a correlation between effective ΔK and crack growth rate using constant-amplitude loading.

6. Crack opening displacement is shifted toward a larger proportion of Mode II after a load range variation, similar to the shift which has been found upon approach to the threshold during constant-amplitude crack growth rate tests.

Acknowledgment

This work was supported by NASA Langley Research Center, Virginia. The careful experimental work of Jim Spencer and John Campbell is greatly appreciated.

References

[*1*] Christensen, R. H., *Metal Fatigue*, McGraw-Hill, New York, 1959.
[*2*] Rice, J. R., "Mechanics of Crack Tip Deformation and Extension by Fatigue" in *Fatigue Crack Propagation, ASTM STP 415*, American Society for Testing and Materials, Philadelphia, 1967, pp. 247–311.
[*3*] Hudson, C. M. and Hardrath, H. F., "Effects of Changing Stress Amplitude on the Rate of Fatigue Crack Propagation of Two Aluminum Alloys," NASA Technical Note D-960, National Aeronautics and Space Administration, Washington, DC, 1961.
[*4*] Lankford, J. and Davidson, D. L., *Journal of Engineering Materials Technology, Transactions*, American Society of Mechanical Engineers, Vol. 98, 1976, pp. 17–23.
[*5*] Davidson, D. L. and Lankford, J., *Fatigue of Engineering Materials and Structures*, Vol. 1, 1979, pp. 439–446.
[*6*] Lankford, J. and Davidson, D. L. in *Advances in Fracture Research*, D. Francois et al., Eds., Vol. 2, Pergamon Press, New York, 1981, pp. 899–906.
[*7*] Bathias, C. and Vancon, M., *Engineering Fracture Mechanics*, Vol. 10, 1978, pp. 409–421.
[*8*] Elber, W., *Engineering Fracture Mechanics*, Vol. 2, 1970, pp. 37–45.
[*9*] Elber, W. in *Damage Tolerance of Aircraft Structures, ASTM STP 486*, American Society for Testing and Materials, Philadelphia, 1971, pp. 230–242.
[*10*] von Euw, E. F. J., Hertzberg, R. W., and Roberts, R. in *Stress Analysis and Growth of Cracks, ASTM STP 513*, American Society for Testing and Materials, Philadelphia, 1972, pp. 230–259.
[*11*] Suresh, S., *Scripta Metallurgica*, Vol. 16, No. 8, 1982, pp. 995–999.
[*12*] Suresh, S., *Engineering Fracture Mechanics*, Vol. 18, No. 3, 1983, pp. 577–593.
[*13*] Broek, D., *Elementary Engineering Fracture Mechanics*, Martinus Nijhoff, Boston, 1982.
[*14*] Jones, R. E., *Engineering Fracture Mechanics*, Vol. 5, 1973, pp. 585–604.
[*15*] Knott, J. F. and Pickard, A. C., *Metal Science*, Vol. 11, 1977, pp. 399–404.
[*16*] Williams, D. R., Davidson, D. L., and Lankford, J., *Experimental Mechanics*, Vol. 20, 1980, pp. 134–139.
[*17*] Davidson, D. L. and Nagy, A., *Journal of Physics E*, Vol. 11, 1978, pp. 207–210.
[*18*] Davidson, D. L., Williams, D. R., and Buckingham, J., *Experimental Mechanics*, Vol. 23, 1983, pp. 242–248.
[*19*] Davidson, D. L. and Lankford, J., *Fatigue of Engineering Materials and Structures*, Vol. 7, 1984, pp. 29–39.
[*20*] Davidson, D. L. and Lankford, J., *Fatigue of Engineering Materials and Structures*, Vol. 6, No. 3, 1983, pp. 287–290.
[*21*] Davidson, D. L. and Lankford, J., *Metallurgical Transactions A*, Vol. 15A, 1984, pp. 1931–1940.
[*22*] Davidson, D. L. and Lankford, J., *Materials Science and Engineering*, Vol. 60, 1983, pp. 225–229.
[*23*] Davidson, D. L. and Lankford, J., "Fatigue Crack Tip Strains in 7075-T6 by Stereoimaging and Their Use in Crack Growth Models" *Fatigue Mechanisms: Advances in Quantitative Measurement of Physical Damage, ASTM STP 811*, American Society for Testing and Materials, Philadelphia, 1983, pp. 371–399.
[*24*] Davidson, D. L., *Acta Metallurgica*, Vol. 32, 1984, pp. 707–714.

Elastic-Plastic Fracture Mechanics II

Stephen J. Garwood[1]

A Crack-Tip Opening Displacement Method for the Analysis of Ductile Materials

REFERENCE: Garwood, S. J., **"A Crack-Tip Opening Displacement Method for the Analysis of Ductile Materials,"** *Fracture Mechanics: Eighteenth Symposium, ASTM STP 945,* D. T. Read and R. P. Reed, Eds., American Society for Testing and Materials, Philadelphia, 1988, pp. 957–985.

ABSTRACT: Although crack-tip opening displacement (CTOD) analysis procedures are well established for the prediction of tolerable crack sizes in ferritic steels, no specific procedures exist which incorporate a resistance (R) curve to facilitate estimation of the ductile instability conditions for a fully ductile material.

This paper discusses a modification to the CTOD procedure, based on the strip yield model, to enable the prediction of maximum nominal stress and the extent of stable ductile crack growth, using a CTOD R-curve as input.

The effectiveness of this procedure is assessed by predicting the behavior of wide plate tension tests and burst tests in API 5LX56 and A533B Class I ferritic parent steels and of wide plates in stainless steel plate and weldments. The predictions are then compared with physical data obtained from the tests.

Reasonable predictions of maximum applied stress are achieved. However, the accuracy is very dependent on the plastic collapse estimations. This is particularly true for the high work-hardening stainless steel plate where, to avoid an overestimate of collapse, the effective flow stress must be taken to be the stress at 3% strain in a uniaxial tension test.

Predictions of the test results are also made using maximum load toughness values, measured using CTOD test specimens complying with BS 5762, with either the CTOD design curve or the modified strip yield model approach. Predictions give safe estimates of nominal stress.

It is concluded that maximum load values provide the most practical general approach when detailed resistance curve information is not available.

KEY WORDS: elastoplastic analysis, fracture toughness, fracture mechanics, critical values, ductile fracture, low-alloy steels, austenitic stainless steels, welded joints, pressure tests, static fracture tests

Nomenclature

- a Half-length of a through-thickness crack in a tension plate; total length of a crack in a bend specimen
- $\bar{a}$ Defect parameter, equal to the half-length of an equivalent through-thickness crack
- $\bar{a}_m$ Tolerable defect parameter
- B Specimen thickness
- CTOD Crack-tip opening displacement
- E Elastic modulus
- e Plate thickness
- FAD Failure assessment diagram
- J J-integral

[1]Head of fracture, The Welding Institute, Abington, Cambridge CB1 6AL, United Kingdom.

K Stress intensity factor
ℓ Total length of a surface crack
LEFM Linear elastic fracture mechanics
R Crack growth resistance
S_r Stress ratio: σ/σ_{pc}
t Depth of a surface crack
W Specimen width
δ CTOD
δ_i δ at initiation of tearing
δ_m δ at the first attainment of maximum load in a CTOD bend test
δ_0 δ referred to the original crack-tip position when ductile tearing occurs
δ_r CTOD ratio: applied CTOD ($\delta_{applied}$)/critical CTOD ($\delta_{critical}$)
δ_u δ associated with an unstable fracture by cleavage after the initiation of tearing
Δa Incremental ductile crack extension
σ Nominal stress
σ_{pc} Plastic collapse stress
$\bar{\sigma}$ Flow stress
σ_{YS} Yield strength, defined as the stress at 0.2% strain
σ_{TS} Tensile strength
Φ Nondimensional CTOD

The crack-tip opening displacement (CTOD) design curve procedure [*1*] which forms the basis of the assessment route in BS PD6493 "Guidance on Some Methods for the Derivation of Acceptance Levels for Defects in Fusion Welded Joints" has a number of drawbacks, particularly when used to assess ductile materials. The major ones are:

1. At primary stress to yield strength ratios above 0.5 the assessment is empirical and has a variable safety factor. This makes probabilistic studies and critical assessments very difficult.
2. The empirical CTOD design curve has only been validated on ferritic steels [*2*]. In PD 6493, other materials have a more conservative assessment line based on linear elastic fracture mechanics (LEFM) for the entire range of primary stress to yield strength ratios.
3. There is no mechanism for incorporating a representative crack growth resistance curve in the assessment for ductile materials. The fracture toughness input is single valued. However, no guidance is given on how to determine this value.
4. Plastic collapse solutions are not an integral part of the PD 6493 analysis. Collapse predictions are used as a separate assessment to indicate when the CTOD design curve is inappropriate, or in defect recategorisation procedures, to allow conservative predictions when local collapse occurs.

In complementary studies aimed at addressing these drawbacks, Dawes [*3*] outlines the limitations of use of the CTOD design curve and discusses simple finite width corrections, while Anderson [*4*] considers complex assessment procedures for accurate predictions.

This paper presents a collapse-modified strip yield model as an alternative assessment route to the CTOD design curve for general assessments of ductile materials. This model can:

(*a*) provide predictions of critical conditions;
(*b*) be used on other materials;
(*c*) accommodate a resistance curve as input; and
(*d*) incorporate collapse solutions directly.

The accuracy of this model is evaluated from the assessment of various wide plates and pipeline burst tests on fully ductile materials carried out at the Welding Institute. The use of maximum load toughness with this model and the CTOD design curve is also assessed, based on the results of the experimental tests.

Modified Strip Yield Model/R-Curve Approach

Derivation of the Collapse Strip Yield Model

Procedures for incorporating a representative material resistance (R) curve in the fracture assessment of fully ductile materials have been developed by Milne [5] to allow modifications of the Central Electricity Generating Board's R6 procedure [6] using J-R curve input data. A J-analysis route has also been adopted by Kumar et al. [7] to complement LEFM with a plasticity correction, to provide procedures covering linear elastic to fully plastic behavior.

At present, no specific analysis route is available using CTOD procedures where R-curve data are employed to predict ductile instability or maximum stress and the extent of stable ductile tearing. However, Tanaka and Harrison [8] did propose a modification to the CTOD design curve procedure to estimate CTOD driving force curves.

In an earlier study [9], modifications to the CTOD design curve and to the Dugdale (strip yield) model [10], as applied by Burdekin and Stone [11] and Heald, Spink, and Worthington [12], were considered, to allow the incorporation of an R-curve as input. Schwalbe [13] and Erdogan and Bakioglu [14] have also suggested the use of a modified strip yield model for critical predictions.

Predictions obtained using the derived models were assessed with the results of wide plates in the nuclear pressure vessel steel A533B Class I. The assessments indicated that a strip yield-based model could provide reasonable results if linked to an accurate collapse solution.

The strip yield model for the CTOD (δ) at the tip of a through-thickness crack of length $2a$ in an infinite sheet (which gives an identical solution to the Bilby, Cottrell, and Swinden [15] solution for Mode III deformation) can be written in the form

$$\delta = \frac{8\sigma_{YS}a}{\pi E} \ell n \sec \left(\frac{\pi\sigma}{2\sigma_{YS}}\right) \tag{1}$$

To predict nominal stress, Eq 1 can be rewritten as

$$\sigma = \frac{2\sigma_{YS}}{\pi} \cos^{-1}\left[\exp\left(-\frac{\pi E\delta}{8\sigma_{YS}a}\right)\right] \tag{2}$$

For tough materials this equation predicts stresses asymptotic to the yield strength as the $\cos^{-1}$ term tends to $\pi/2$.

To allow for the presence of a defect which is a significant proportion of the plate width, Eq 2 can be modified to be asymptotic to the collapse stress, σ_{pc}, that is

$$\sigma = \frac{2\sigma_{pc}}{\pi} \cos^{-1}\left[\exp\left(-\frac{\pi E\delta}{8\sigma_{YS}a}\right)\right] \tag{3}$$

In Ref *9*, the best predictions of nominal stress (σ) were obtained from a collapse-modified Heald-Spink-Worthington model of the form

$$\sigma = \frac{2\sigma_{pc}}{\pi} \cos^{-1}\left[\exp\left(-\frac{\pi E\sigma_{YS}\delta}{8\sigma_{TS}^2\bar{a}}\right)\right] \tag{4}$$

In this model, the use of σ_{TS} allows for work hardening. However, the disadvantage of this model, which can be rewritten as

$$\delta = \frac{8\sigma_{TS}^2 \bar{a}}{\pi E \sigma_{YS}} \ell n \sec\left(\frac{\pi\sigma}{2\sigma_{pc}}\right) \tag{5}$$

is that it does not reduce to the linear elastic solution at very low σ/σ_{pc} ratios. To do so, the σ_{TS} term in Eq 5 must be replaced by the collapse stress, σ_{pc}. This results in the *collapse-modified strip yield model* which is studied in this report, that is

$$\delta = \frac{8\sigma_{pc}^2 \bar{a}}{\pi E \sigma_{YS}} \ell n \sec\left(\frac{\pi\sigma}{2\sigma_{pc}}\right) \tag{6}$$

Work hardening can be still incorporated in this model by the adoption of a flow stress, $\bar{\sigma}$, in the prediction of σ_{pc}, the collapse stress.

Another advantage of the adoption of this equation is that, if plane stress conditions are assumed, then the stress intensity factor K is given by

$$\frac{K^2}{E} = \sigma_{YS}\delta \tag{7}$$

Substituting for δ from Eq 7, Eq 6 becomes

$$K^2 = \frac{8\sigma_{pc}^2 \bar{a}}{\pi} \ell n \sec\left(\frac{\pi\sigma}{2\sigma_{pc}}\right) \tag{8}$$

which is the basis of the Central Electricity Generating Board (CEGB) R6 procedure, giving direct equivalence between the CTOD-based procedure used here and the K- or J-based analysis of R6.

A failure analysis diagram (FAD) can also be derived equivalent in form to that adopted in R6 by defining

$$\delta_r = \frac{\delta_{applied}}{\delta_{critical}} \quad \text{and} \quad S_r = \frac{\sigma}{\sigma_{pc}}$$

Then, assuming linear elastic conditions for a crack in an infinite plate, the stress intensity factor is given by

$$K = \sigma\sqrt{\pi\bar{a}}$$

Substituting in Eq 7 for K gives

$$\delta_{applied} = \frac{\sigma^2 \pi \bar{a}}{E\sigma_{YS}} \tag{9}$$

Then, using Eq 6 to predict $\delta_{critical}$, and substituting for $\bar{a}$ gives

$$\delta_{critical} = \frac{8\sigma_{pc}^2 \delta_{applied}}{\pi^2 \sigma^2} \ell n \sec \frac{\pi}{2} \frac{\sigma}{\sigma_{pc}} \tag{10}$$

Rewriting the equation in the form of δ_r and S_r gives

$$\delta_r = \frac{\pi^2}{8} S_r^2 \left(\ell\text{n} \sec \frac{\pi}{2} S_r \right)^{-1} \tag{11}$$

The resulting form of FAD using Eq 14 is shown in Fig. 1.

Incorporation of the Resistance Curve

In terms of CTOD, a representative resistance curve, as defined by Appendix A of BS 5762:1979 "Methods for Crack Opening Displacement (COD) Testing," is measured by estimating the CTOD, at the position of the original crack tip (before stretch zone development and ductile tearing commences), δ_0, and plotting this displacement against ductile tearing, excluding stretch zone. However, it is often not possible to distinguish between the stretch zone and tearing region experimentally, particularly when heat tinting is employed. Thus, it is now becoming more common to plot δ_0 against total crack extension (Δa).

Thus, rewriting Eq 6 as

$$\sigma = \frac{2\sigma_{pc}}{\pi} \cos^{-1}\left[\exp\left(-\frac{\pi E \sigma_{YS} \delta_0}{8\sigma_{pc}^2 \bar{a}}\right)\right] \tag{12}$$

the variation of nominal stress σ can be evaluated as the crack extends by increments of Δa by insertion of the relevant δ_0 value.

The σ_{pc} value in Eq 12 is the relevant nominal stress to cause plastic collapse for the geometry under consideration and will, of course, decrease with increasing Δa.

Conceptually, the derivation of the half length of the equivalent through-thickness crack $\bar{a}$ should not be affected by crack growth, as the CTOD (δ_0) refers to the position of the original crack tip. However, in Ref *9* it was shown that not only is it more conservative to allow for crack extension in $\bar{a}$, but it is also more realistic to do so. Milne's modification to the R6 procedure [*5*] also alters $\bar{a}$ to allow for crack extension. Thus, to maintain consistency with the R6 method, and to provide better predictions, $\bar{a}$ has been modified to allow for crack growth in the assessments carried out in this paper.

Therefore, with a knowledge of a representative resistance curve, the crack size, material yield strength, and an appropriate collapse solution, a locus of the variation of nominal stress

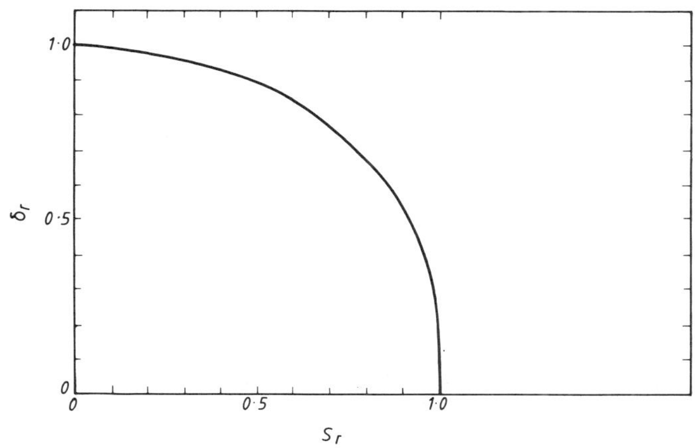

FIG. 1—*CTOD failure analysis diagram derived using the strip yield model.*

with crack extension can be derived from Eq 12. This curve can be plotted with the collapse prediction, as shown in Fig. 2, and the maximum stress and crack growth predictions read off. In addition, the form of the curve illustrates structural performance. For example, Fig. 2 illustrates initiation of tearing under contained yielding conditions, with nominal stress and plasticity increasing with crack extension until collapse is achieved. This behavior is expected when a steep resistance curve, that is, if high tearing resistance is measured.

With a low tearing resistance, maximum nominal stress could be experienced under fracture-controlled conditions (also illustrated in Fig. 2).

In the extreme, the material may be very tough or have a very small flaw, in which case collapse occurs at or prior to initiation of tearing. Alternatively, the material may have such a low tearing resistance or large flaw that maximum load occurs at initiation of tearing under fracture-controlled conditions. This latter case defines the most unstable ductile condition.

Use of Maximum Load Toughness

The use of maximum load toughness CTOD (δ_m) values, determined following BS 5762 procedures, as input to the CTOD design curve has been fully discussed elsewhere [*16*].

To ensure safe predictions, the choice of geometry for δ_m determinations must be treated with care, because maximum load toughness is geometry-dependent. For through-thickness cracks, full thickness (B) specimens of the preferred geometry (that is, $B \times 2B$) of BS 5762 are usually specified. For part-wall flaws, $B \times B$ specimens would normally be selected and the value should be conservative, provided the following criterion is satisfied

$$B > 50\delta_m$$

In all cases, the crack length to width (a/W) ratio of the specimens should exceed the structural case. For situations where full-section thickness tests cannot be adopted, side-grooves in subthickness specimens may be used, or the thickness criterion implemented. However, in these cases it must be emphasized that the thickness criterion is relevant only to fully ductile behavior and does not cover the constraint which may be necessary to assess the risk of cleavage.

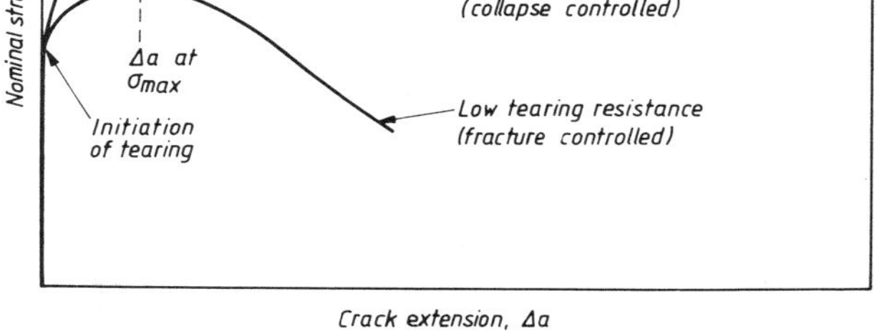

FIG. 2—*Maximum stress assessment derived from the locus of the predicted nominal stress with increasing crack extension.*

In traditional assessments, the δ_m value and nominal stress condition would be used as input in the CTOD design curve equations [1]

$$\bar{a}_m = \frac{\delta E \sigma_{YS}}{2\pi\sigma^2} \qquad \text{for} \quad \frac{\sigma}{\sigma_{YS}} \leq 0.5$$

$$\bar{a}_m = \frac{\delta E}{2\pi(\sigma - 0.25\sigma_{YS})} \qquad \text{for} \quad 1.0 \geq \frac{\sigma}{\sigma_{YS}} > 0.5 \tag{13}$$

to predict the tolerable defect parameter $\bar{a}_m$. This value is then compared with physical flaw dimensions using the recommendations of PD 6493 to assess fitness-for-purpose. No allowance is normally made for crack extension when determining $\bar{a}$.

In the form presented here, the CTOD design curve equations apply only at stress levels below net section collapse and do not incorporate an allowance for residual stress or a stress concentration factor (SCF).

For the tests to be analyzed in this report, σ/σ_{YS} ratios are greater than 0.5 at maximum stress. Thus, rewriting the relevant part of Eq 13 gives

$$\sigma = \frac{\delta E}{2\pi\bar{a}} + 0.25\sigma_{YS} \tag{14}$$

Predictions can then be made using this equation, with δ_m as input, which will have an upper limit of the plastic collapse stress σ_{pc} for validity.

As recommended by Garwood and Willoughby [17], the collapse solution should be adjusted for the amount of stable crack extension experienced in the maximum load bend tests. In addition, for consistency, $\bar{a}$ should be adjusted to allow for tearing.

Maximum load toughness values can also be used as the critical value input to the collapse-modified strip yield approach to estimate the maximum nominal stress conditions. Predictions of nominal stresses using both R-curve and maximum load toughness inputs are compared below.

Description of Experimental Tests

Materials

To assess the collapse-modified strip yield model/R-curve approach and the use of maximum load toughness CTOD values, a number of experimental tests carried out at the Welding Institute for various sponsors have been reanalyzed. The materials chosen had similar flow stresses, $\bar{\sigma}$, where

$$\bar{\sigma} = \frac{\sigma_{YS} + \sigma_{TS}}{2} \tag{15}$$

but different work-hardening characteristics. All materials were fully ductile and covered a range of thicknesses between 13 and 110 mm:

110-mm-thick A533B Class I plate,
90-mm-thick A533B Class I plate,
50-mm-thick A533B Class I plate,
13-mm-thick API 5LX56 pipe,
50-mm-thick 316 stainless plate, and
50-mm-thick 316 stainless submerged-arc weld.

The chemical analyses of these materials are given in Table 1. Table 2 indicates their tensile properties and Charpy V energies. Relevant fracture toughness (CTOD) data obtained using three-point bend specimens tested using BS 5762 procedures are quoted in Table 3. These data were interpreted from the results of batches of specimens tested using a multiple-specimen resistance curve technique [*17*].

The stress-strain behavior of the different materials is given in Fig. 3. (Note that in the fracture toughness program, the A533B plates were tested in the longitudinal direction only, whereas the API 5LX56 and 316 plates were tested only transversely). The 90-mm-thick A533B plate behaved differently than the 50- and 110-mm plates, which suggests that it was cross-rolled. The 316 stainless plate is dramatically different from the ferritic steels, whereas the submerged arc weld metal indicates similar tensile properties.

Wide Plate and Burst Tests

110-mm-Thick A533B Wide Plates—Nine 500-mm-wide plates in 110-mm-thick A533B Class I were tested in tension at +70°C. Three had through-thickness cracks, three long surface defects, and three semi-elliptical surface cracks [*18*]. These three geometries are described as center-cracked tension (CCT), single-edge notched tension (SENT), and semi-elliptical surface notched tension (SESNT), respectively, in Table 4*a,* which gives the plate and flaw dimensions.

90-mm-Thick A533B Wide Plates—Four 500-mm-wide plates were tested in 90-mm-thick A533B Class I at +70°C. Two tests were CCT in the T-L orientation and two were SESNT tests in the L-S orientation. These tests were used to study time-dependent crack growth [*19*]; however, the maximum stress behavior of these tests is still relevant. The dimensions of these tests are also given in Table 4*a*.

50-mm-Thick A533B Wide Plates—One semi-elliptical surface cracked, 500-mm wide plate test was carried out in 50-mm-thick A533B. This test was also carried out at +70°C in the L-S orientation, and its dimensions are set out in Table 4*a*.

Wide Plates and Pipe Burst Tests in 13-mm-Thick API 5LX56—Six 900-mm-wide plates and two burst tests in 13-mm-thick API 5LX56 were reported in Ref *20*. All tests were carried out at room temperature and with either through-thickness cracks or surface cracks, as indicated in Table 4*b* and *c*. The CCT geometries were notched in the T-L orientation, with the surface cracks in the T-S orientation (see Fig. 4). The surface-cracked wide plate geometries indicated unstable ductile behavior (one experiencing complete fracture), while the other four geometries were stable, the semi-elliptical surface cracks breaking through to become CCT geometries. Each burst test had two axial cracks, one on each side of the pipe, as indicated in Table 4*c*. In the first test, unstable ductile failure occurred from the surface crack, while in the second (which had a smaller surface crack but identical through-thickness crack), failure occurred from the through-thickness crack.

Wide Plates in 50-mm-Thick 316 Stainless Plates—Two wide plates of center-cracked tension configuration in the T-L orientation in 316 stainless steel plate were carried out with the crack dimensions quoted in Table 4*d*. Specimen WP4 was entirely parent plate, while WP3 contained a longitudinal submerged arc weld in the center of the plate. Thus, although the crack tips were positioned in base metal, those in WP3 would have been affected by the residual stress field of the weld. For this assessment, however, the contribution of the residual stresses has been ignored. (In fact, the two plates exhibited very similar load-displacement records.)

Wide Plates in 50-mm-Thick 316 Stainless Submerged-Arc Welds—Two wide plates with surface notches into the weld center line of a transverse 316 stainless steel submerged-arc weld were also tested. WP1 was in the as-welded condition, whereas WP2 received a mechanical stress relief from the application of a nominal stress equal to parent material yield transverse to the weld prior to notching. Unlike all other tests described here, which had fatigue cracks, these two plates had 0.015-mm-wide saw-cuts as starter notches. The two tests showed virtually iden-

TABLE 1—*Chemical composition of the materials studied.*

Material	Element, weight %															
	C	S	P	Si	Mn	Ni	Cr	Mo	V	Cu	Nb	Ti	Al	B	Sn	Co
A533B																
110 mm	0.20	<0.005	0.006	0.20	1.39	0.59	0.12	0.50	<0.01	0.10	0.005	<0.01	0.023	<0.01	0.01	0.01
90 mm	0.19	0.005	0.005	0.17	1.36	0.61	0.18	0.50	<0.01	0.13	<0.002	<0.01	0.024	<0.001	<0.01	<0.01
50 mm	0.19	0.004	0.010	0.23	1.42	0.57	0.20	0.48	0.002	0.14	<0.002	0.003	0.027	<0.001	0.01	0.02
API 5LX56	0.18	0.02	0.01	0.27	1.20	0.05	0.03	<0.01	0.14	<0.005	<0.003	<0.065	<0.001	0.01	<0.01	<0.01
316 Plate[a]	0.06	0.022	0.031	0.42	1.62	11.21	17.18	2.51	...	...	...	...	...	...	...	...
316 Submerged-arc weld metal[b]	0.06	0.02	0.02	0.50	1.0	12.6	21.6	2.09	...	...	...	...	...	...	...	...

[a]Supplier's figures.
[b]Manufacturer's figures for Bostrand 316 wire used in conjunction with saturn arc BX600 flux.

TABLE 2—*Mechanical properties of the materials studied.*

	Tensile Properties				Upper Shelf Charpy Data		
Material	Orientation to Rolling Direction[a]	σ_{YS} Yield Strength,[b] N/mm^2	σ_{TS} Tensile Strength, N/mm^2	Reduction of Area, %	Temperature, °C	Orientation	Average Charpy V Energy, J
A533B							
110 mm	L	461	600	65	+70	L-S	194
	T	467	605	63	+70	L-T	197
90 mm	L	439	567	65	+70	L-S	216
	T	427	559	62	+70	L-T	244
50 mm	L	471	591	70	+70	L-S	287
	T	475	597	66	+70	L-T	293
API 5LX56	L	362	561	59	+22	T-S	55
	T	438	584	45	+22	T-L	36
316							
Plate	L	249/347	580/593	61/70	+20	L-T	145
	T	250	573	69	+20	T-L	127
Submerged-arc weld metal	weld metal	410	600	50	+20	weld metal	75

[a] L = longitudinal; T = transverse.
[b] 0.2% proof.

TABLE 3—*Fracture toughness data.*

Material	Orientation	Specimen Dimensions $W \times B$, mm	Temperature, °C	Initiation of Tearing Toughness, δ_i, mm	Minimum Recorded Maximum Load Toughness, δ_m, mm	Crack Growth at Maximum Load Toughness, Δa, mm	Initial Slope of the R-Curve, $d\delta/da$
A533B							
110-mm		range of sizes,					
	L-S	thicknesses of	+70	0.23	2.0	3	0.72
	L-T	55 + 110 mm	+70	0.23	2.1	3	0.72
90-mm	L-S	45 × 90	+70	0.55	2.2	4	0.78
	L-T	45 × 90	+70	0.59	2.3	3	0.58
50-mm	L-S	50 × 50	+70	0.39	1.3	2	0.56
API 5LX56	T-L	26 × 13	+22	0.075	0.11	...	0.11
	T-S	13 × 13	+22	0.19	0.25	...	0.24
316							
Plate	T-L	50 × 100	+20	0.46	2.55	4	0.72
Submerged-arc WM	surface-notched	50 × 50	+20	~0	0.45	2.5	0.33

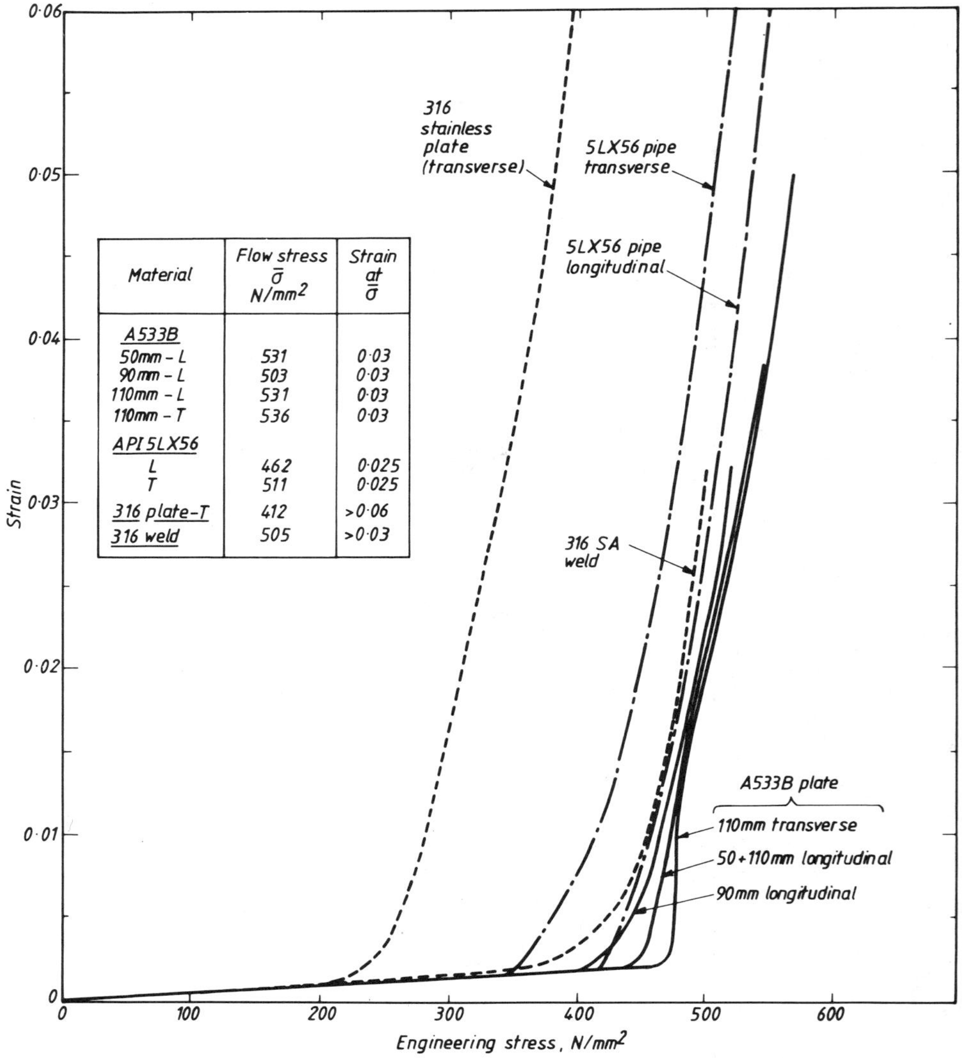

FIG. 3—*Typical stress-strain relationships derived from wide plate test materials using Hounsfield tension specimens.*

tical load-displacement records, so the effect of the residual stresses has also been ignored for the analysis of WP2.

Prediction of Test Results

Collapse-Modified Strip Yield Model/R-Curve Approach

To assess the accuracy of the collapse-modified strip yield model for the prediction of the behavior of the wide plates and burst tests, the most relevant comparison is on the basis of predicted to measured maximum nominal stress. Thus, the strip yield model was employed

TABLE 4—*Test specimen dimensions.*

Specimen Identification	Crack Orientation	Type	Plate Width, W, mm	Plate Thickness, e, mm	Half Through-Thickness Length, a, mm	Crack Depth, t, mm	Crack Length, t, mm
(a) WIDE PLATE TEST DATA—A533B MATERIAL							
3545							
1008	L-T	CCT	487.4	112.5	102.35	...	...
1042	L-T	CCT	491.3	110.0	127.75	...	...
1043	L-T	CCT	486.0	112.0	65.6	...	...
1009	L-S	SENT	508.0	110.0	...	37.4	$= W$
1045	L-S	SENT	509.0	109.9	...	33.0	$= W$
1044	L-S	SENT	508.0	108.7	...	51.1	$= W$
1006	L-S	SESNT	508.0	110.0	...	22.8	152.0
1007	L-S	SESNT	510.0	112.9	...	38.7	240.0
1046	L-S	SESNT	508.0	113.1	...	56.2	328.0
3679							
WP1	L-T	CCT	499.0	88.2	48.6	...	...
WP3	L-T	CCT	495.0	87.3	47.2	...	...
WP2	L-S	SESNT	495.0	87.7	...	29.1	206.0
WP4	L-S	SESNT	483.0	86.6	...	19.7	105.7
3740							
WP1	L-S	SESNT	496.0	51.5	...	24.9	145.8
(b) WIDE PLATE TEST DATA—API 5LX56 MATERIAL							
3591							
1176	T-S	SESNT	858	13	...	2.7	42
1177	T-S	SESNT	899	13	...	4.8	90
1178	T-L	CCT	898	13	43	...	...
1179	T-L	CCT	892	13	132.5	...	...
1180	T-S	SENT	885	13	...	2.1 (avg) 3.5 (max)	$= W$
1181	T-S	SENT	898	13	...	4.6 (avg) 5.5 (max)	$= W$
(c) BURST TEST DATA API 5LX56 MATERIAL							
3591							
PV1	T-L	CCT	[a]	13	19	...	...
PV1	T-S	SESNT	[a]	13	...	5.4	234

TABLE 4—(*Continued*)

Specimen Identification	Crack Orientation	Type	Plate Width, W, mm	Plate Thickness, e, mm	Half Through-Thickness Length, a, mm	Crack Depth, t, mm	Crack Length, t, mm
PV2	T-L	CCT	[a]	13	19	...	...
PV2	T-S	SESNT	[a]	13	...	4.5	87
		(*d*) Wide Plate Test Data—316 material					
7801							
WP1	weld centerline	SENT submerged-arc weld	475.0	48.0	...	12.5	$= W$
WP2	weld centerline	SENT submerged-arc weld	475.0	49.0	...	11.0	$= W$
WP3	T-L	CCT plate	477.0	51.1	31.0	...	...
WP4	T-L	CCT plate	480.0	51.07	37.9	...	...

[a]The pressure vessels had a diameter of 900 mm—all cracks were axial.

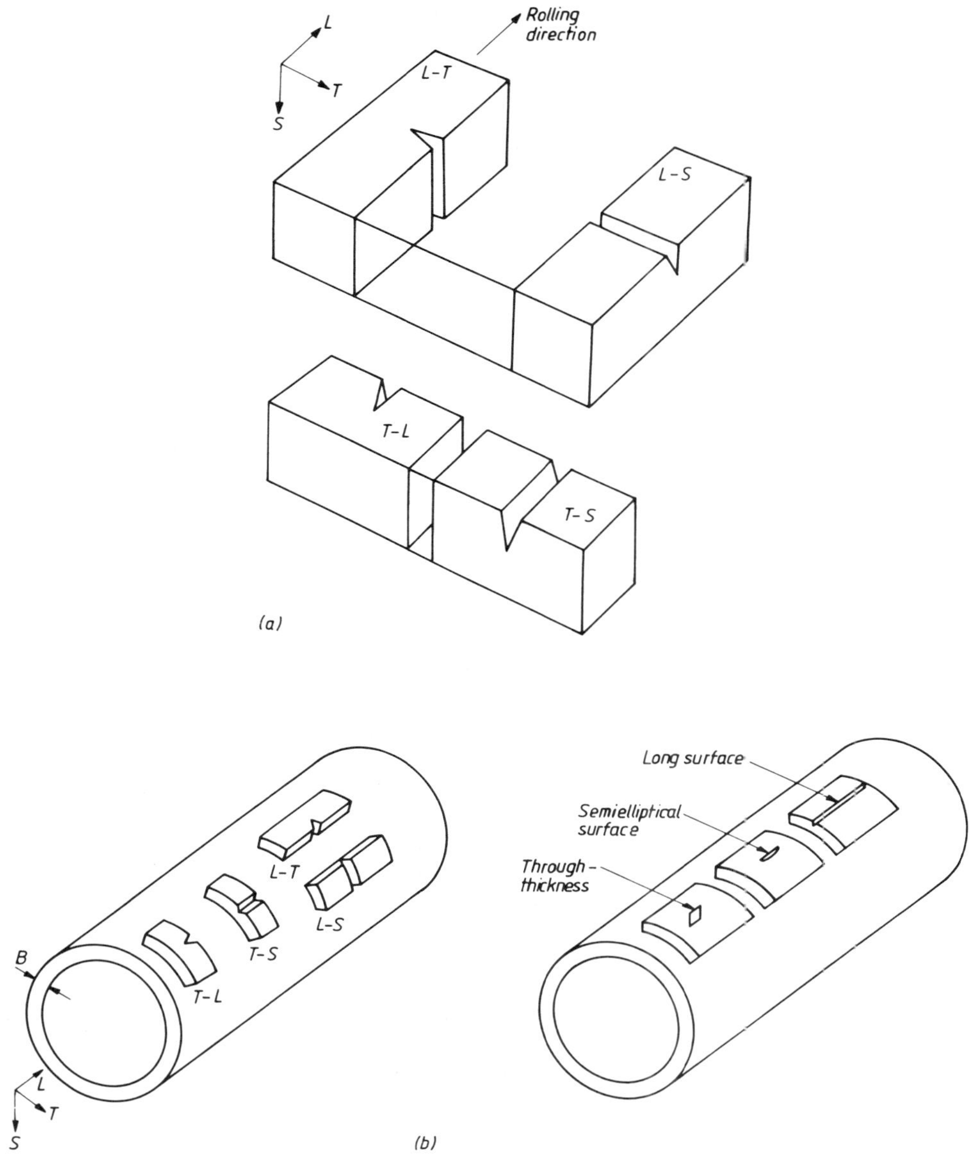

FIG. 4—*Specimen orientations:* (a) *plate specimens;* (b) *pipe tests.*

using the relevant R-curve input from Table 3. (Note, however, that these relate to bend tests and will be conservative estimates of the behavior in tension [*18*].) Because this assessment was intended to be a critical one, plastic collapse solutions representative of average rather than bound behavior were adopted.

For consistency, flow stress was calculated using Eq 15 for all materials. The elastic modulus was taken as 207 000 N/mm^2 for the ferritic materials and 190 000 N/mm^2 for the 316 stainless steel plate and welds.

The plastic collapse solutions adopted for the wide plate geometries were as follows:

CCT

$$\sigma_{pc} = \bar{\sigma}\left(1 - \frac{2a}{W}\right) \tag{16}$$

SENT

$$\sigma_{pc} = \bar{\sigma}\left(1 - \frac{t}{e}\right) \tag{17}$$

SESNT

$$\sigma_{pc} = \bar{\sigma}\left(1 - \frac{\pi \ell t}{4eW}\right) \tag{18}$$

For realistic estimates of the SENT geometry, the depth t of the surface crack was taken as the average value across the plate width. For the SESNT geometry, t is the maximum depth of the crack.

For the burst tests, the plastic collapse criteria derived by Kiefner et al. [*21*] were adopted as described in Ref *20*. For the center-cracked wide plate geometries, the half-length through-thickness cracks set out in Table 4 were taken directly as the initial $\bar{a}$ inputs to Eq 12, no allowance for finite width being made (this would have only a very small effect in any case).

For the SENT and SESNT geometries, the equivalent crack length $\bar{a}$ was calculated using the relationships used to define Fig. 12 of PD 6493 (see Refs *22,23*), which converts surface defect dimensions to equivalent through-thickness crack dimensions. For the burst tests, the bulging correction factor of Fig. 15 of PD 6493 was applied to the $\bar{a}$ value, calculated using the same procedure adopted for the wide plate tests.

For all crack shapes, $\bar{a}$ values were recalculated incrementally to allow for crack extension. For the CCT geometry, Δa was added to a to calculate the new $\bar{a}$. For the SENT geometry, Δa was added to t prior to the calculation of $\bar{a}$. The value of t was also incremented for the SESNT geometry, but ℓ was kept constant. (This models the experiments of Ref *18*.)

Thus, using Eq 12 with the relevant R-curve information from Table 3, the appropriate $\bar{a}$ values, and the plastic collapse solutions (adjusted for incremental crack extension, calculated using the specimen dimensions set out in Table 4), a table of nominal stress predictions for increasing values of crack extension (Δa) was produced for each test geometry which was used to draw assessment loci of the form shown in Figs. 2*a* and 2*b* for each test geometry, to facilitate the predictions of maximum nominal stress.

Predictions of the maximum nominal stress and the crack extension at maximum stress have been interpolated from the figures and are set out in Table 5, with the actual maximum nominal stresses measured on the specimens. Where possible, the actual extension at maximum stress is also included in the table. As described earlier, the stress versus Δa loci can be used to indicate the controlling failure mechanism, and indications from the predicted curves are also set out in Table 5. The nominal collapse stresses (based on initial flaw dimensions) are also included in Table 5 to indicate which specimens failed at stresses below nominal collapse.

Maximum Load Toughness Approach

The collapse-modified strip yield model of Eq 12 can be also used with a single value input, such as maximum load toughness, δ_m. Such predictions have been made for all the tests consid-

TABLE 5—*Summary of predictions for the collapse-modified strip yield model/R-curve procedure.*

		Predicted Data			Recorded Data		
Specimen Number	Type	Predicted Maximum Nominal Stress, N/mm^2	Predicted Crack Extension at Maximum Stress, mm	Controlling Failure Mechanism	Nominal Collapse Stress,[a] N/mm^2	Recorded Maximum Nominal Stress, N/mm^2	Actual Crack Extension at Maximum Stress, mm
				(*a*) A533B Wide Plates			
3545							
1008	CCT	303	1.3	collapse	307	366	7.0
1042	CCT	250	1.2	collapse	254	281	7.0
1043	CCT	383	2.0	collapse	387	455	7.0
1009	SENT	335	2.5	near collapse	350	394	8.0
1045	SENT	360	1.8	near collapse	371	425	7.5
1044	SENT	257	3.5	fracture controlled	281	329	6.5
1006	SESNT	503	1.3	collapse	504	609	9.0
1007	SESNT	458	1.8	collapse	463	503	11.0
1046	SESNT	390	2.0	collapse	396	415	9.0
3679							
WP1	CCT	402	1.0	collapse	405	474	6.5
WP3	CCT	404	1.0	collapse	407	499	10.6
WP2	SESNT	443	2.0	collapse	448	452	5.0
WP4	SESNT	482	1.0	collapse	483	614	8.5
3740							
WP1	SESNT	467	1.0	collapse	472	476	6.0
				(*b*) API 5LX56 Wide Plates			
3591							
1178	CCT	445	12.0	near collapse	462	403	17
1179	CCT	335	15.0	near collapse	359	316	18
1180	SENT	425	0	near collapse	428	414	...
1181	SENT	301	0	fracture controlled	330	351	...
1176	SESNT	506	0	collapse	507	483	...
1177	SESNT	490	2.0	collapse	496	451	...
				(*c*) API 5LX56 Pressure Vessels			
3591							
PV1	CCT	473	5.0	near collapse	491	>370	...

TABLE 5—(*Continued*

		Predicted Data			Recorded Data		
Specimen Number	Type	Predicted Maximum Nominal Stress, N/mm²	Predicted Crack Extension at Maximum Stress, mm	Controlling Failure Mechanism	Nominal Collapse Stress,[a] N/mm²	Recorded Maximum Nominal Stress, N/mm²	Actual Crack Extension at Maximum Stress, mm
PV2	CCT	473	5.0	near collapse	491	457	...
PV1	SESNT	338	0.25	near collapse	391	370	...
PV2	SESNT	458	0.5	near collapse	470	>457	...
			(*d*) 316 STAINLESS WIDE PLATES				
7801							
WP1	SENT	348	1.5	near collapse	374	340	...
WP2	SENT	371	1.5	near collapse	392	348	...
WP3	CCT	356	1.0	collapse	359	311	14
WP4	CCT	344	1.0	collapse	347	337	13

[a]Based on initial flaw size.

ered here, using δ_m values from Table 3, with appropriate adjustments to the collapse stress and $\bar{a}$ values to allow for the crack extension experienced at δ_m, as indicated in Table 3.

As this assessment route is meant to provide conservative assessments, slight adjustments were made to the collapse predictions employed for two cases as follows:

1. For the surface cracks in thin-walled pipe material, maximum depths of crack were used in Eq 17 for both SENT and SESNT geometries. This follows the recommendations of Ref *24* for $t/\ell < 0.1$. This was not necessary for the thicker wall geometries, which are much less sensitive to small variations in crack depth, and had t/ℓ ratios > 0.1.
2. From Fig. 3 it can be seen that the flow stress assumption is equivalent to assuming a 3% strain on the stress-strain curve for all materials considered except the 316 stainless plate. Thus, to provide a more conservative estimation of flow stress in this case, a value of $\bar{\sigma} = 340$ N/mm^2 (that is, the stress at 3% strain) was assumed for the 316 stainless steel plate.

Collapse stress estimates and predictions using the collapse-modified strip yield model, based on δ_m values, are set out in Table 6. This table also contains predictions of nominal stress, using the CTOD design curve approach of Eq 14. In this approach, a separate check is needed to prevent collapse, and the solutions used are mostly very conservative. Therefore, in this column, if the predicted collapse stress (using the procedure adopted for the strip yield-model) is exceeded, this is noted and the predicted stress recorded in brackets. If this predicted stress exceeds the material flow stress, the value is not relevant and collapse is the dominant mechanism.

To allow comparison between the use of δ_m and initiation of tearing, δ_i, the calculations have been repeated for the A533B specimens using δ_i values from Table 3, and initial flaw dimensions for the σ_{pc} and $\bar{a}$ estimations. These values are set out in Table 7. Since δ_m and δ_i are similar for the pipeline material, and the initiation of tearing is difficult to define in the 316 stainless tests, this exercise was limited to the A533B data.

Discussion of Predicted Results

Collapse-Modified Strip Yield Model/R-Curve Approach

Figure 5 compares LEFM methods, the CTOD design curve, and the standard strip yield model. The LEFM curve, with a safety factor of two on flaw size, is equivalent to the "other materials" curve which is adopted in Fig. 14 of PD 6493. The strip yield model plotted is asymptotic to σ_{YS}, whereas the collapse-modified model would be asymptotic to collapse, which could be lower or higher than the yield strength. However, Fig. 5 can still be used to illustrate trends between the collapse-modified model considered here and the other assessment lines.

Below $\sigma/\sigma_{YS} = 0.5$, the strip yield model, LEFM with plastic zone size correction and LEFM, are coincident. The design curve and the "other materials" curve give exactly double the value of nondimensional CTOD, Φ, as they contain a safety factor of two on $\bar{a}$.

As σ/σ_{YS} increases, LEFM gives the most optimistic predictions. LEFM with plastic zone correction predicts the same value of Φ as the strip yield model until $\sigma/\sigma_{YS} \approx 0.8$, after which, as collapse is approached, the strip yield model becomes more conservative (that is, a higher value of Φ is predicted for a given value of σ). The design curve and "other materials" plot are the most conservative until near collapse, when the strip yield model crosses them.

Thus, if the strip yield model were represented giving exact material behavior, the PD 6493 approach would tend to be unconservative very near collapse. In fact, this never happens in practice owing to the use of conservative collapse solutions.

It can be postulated that the true shape of the assessment curve should reflect the stress-strain behavior; Ainsworth [*25*] and Akhurst and Milne [*26*] have proposed methods of producing a FAD using material data. The stress-strain curves relevant to the materials tested here are plotted in Fig. 3 in a form which is compatible with the Φ-versus-σ/σ_{YS} plot of Fig. 5. Thus, it may

TABLE 6—*Predictions of nominal stress using maximum load toughness.*

Specimen Number	Type	δ_m, Maximum Load CTOD, mm	Δa, Allowance, mm	$\bar{a}$, mm	Predictions of Nominal Stress: σ_{pc}, Collapse Stress, N/mm^2	CTOD Design Curve, N/mm^2	Collapse Modified Strip Yield Model, N/mm^2	Actual Maximum Stress, N/mm^2
				(*a*) A533B Wide Plates				
3545								
1008	CCT	2.1	3.0	105	301	collapse	301	366
1042	CCT	2.1	3.0	131	248	collapse	248	281
1043	CCT	2.1	3.0	69	381	collapse	381	455
1009	SENT	2.0	3.0	152	335	collapse	332	394
1045	SENT	2.0	3.0	112	357	collapse	356	425
1044	SENT	2.0	3.0	423	267	(271)	253	329
1006	SESNT	2.1	3.0	28	499	collapse	499	609
1007	SESNT	2.1	3.0	50	456	collapse	456	503
1046	SESNT	2.1	3.0	81	387	collapse	387	415
3679								
WP1	CCT	2.3	3.0	52	399	collapse	399	474
WP3	CCT	2.3	3.0	50	401	collapse	401	499
WP2	SESNT	2.2	4.0	42	441	collapse	441	452
WP4	SESNT	2.2	4.0	23	479	collapse	479	614
3740								
WP1	SESNT	1.3	2.0	36	467	collapse	467	476

				(b) API 5LX56 Wide Plates and Burst Tests				
3591								
1176	SENT	0.25	0.5	4.8	385	collapse	385	483
1177	SENT	0.25	0.5	11.2	303	collapse	303	451
1178	CCT	0.11	0	43.0	462	194	253	403
1179	CCT	0.11	0	133.0	359	110	149	316
1180	SENT	0.25	0.5	11.4	353	collapse	353	414
1181	SENT	0.25	0.5	37.5	275	collapse (329)	267	351
PV1	SENT	0.25	0.5	23.9	360	collapse (454)	347	370
PV1	CCT	0.11	0	20.8	491	284	343	>370
PV2	SENT	0.25	0.5	10.6	463	collapse	457	>457
PV2	CCT	0.11	0	20.8	491	284	343	457
				(c) 316 Stainless Steel Wide Plates				
7801								
WP1	SENT	0.45	2.5	43.7	347	collapse (414)	331	340
WP2	SENT	0.45	2.5	33.7	366	collapse (506)	355	348
WP3	CCT	2.55	4.0	35.0	290[a]	collapse	290	311
WP4	CCT	2.55	4.0	41.9	281[a]	collapse	281	337

[a]Flow stress at 3% strain assumed, that is, $\bar{\sigma} = 340$ N/mm^2.

TABLE 7—*Predictions of nominal stress using initiation toughness for A533B wide plates.*

				Predicted Nominal Stress			
Specimen Number	Type	δ_i, Initiation CTOD, mm	$\bar{a}$, mm	σ_{pc}, Collapse Stress, N/mm^2	CTOD Design Curve, N/mm^2	Collapse Modified Strip Yield Model, N/mm^2	Actual Maximum Stress, N/mm^2
3545							
1008	CCT	0.23	102	307	192	227	366
1042	CCT	0.23	128	254	177	199	281
1043	CCT	0.23	66	387	236	286	455
1009	SENT	0.23	123	350	179	220	394
1045	SENT	0.23	91	371	202	250	425
1044	SENT	0.23	336	281	139	139	329
1006	SESNT	0.23	25	504	428	424	609
1007	SESNT	0.23	47	463	283	339	503
1046	SESNT	0.23	78	396	217	266	415
3679							
WP1	CCT	0.59	47	407	collapse (510)	389	499
WP3	CCT	0.59	49	405	collapse (522)	387	474
WP2	SESNT	0.55	37	448	collapse (507)	431	452
WP4	SESNT	0.55	20	483	collapse	480	614
3740							
WP1	SESNT	0.39	34	472	collapse (497)	430	476

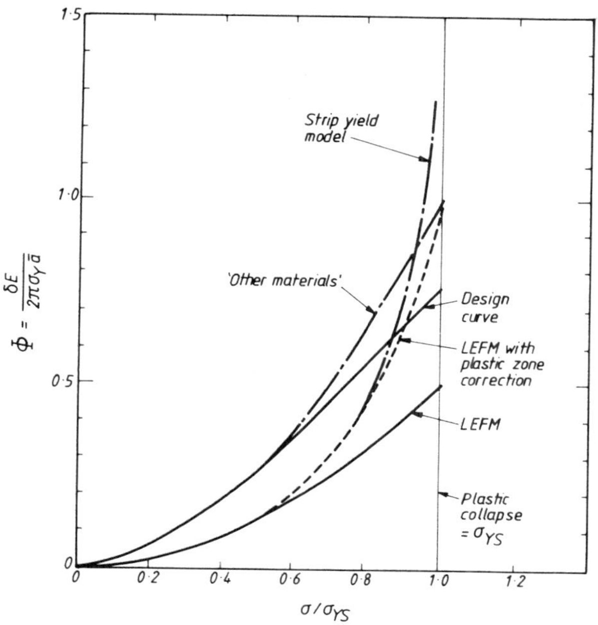

Non-dimensional CTOD $\Phi = \frac{\delta E}{2\pi \sigma_{YS} \bar{a}}$

LEFM

$$\Phi = \frac{1}{2}\left(\frac{\sigma}{\sigma_{YS}}\right)^2$$

LEFM + plastic zone correction

$$\Phi = (\sigma/\sigma_{YS})^2 / (2 - (\sigma/\sigma_{YS})^2)$$

Design curve

$$\Phi = \left(\frac{\sigma}{\sigma_{YS}}\right)^2 \quad \text{for } \frac{\sigma}{\sigma_Y} \leqslant 0.5$$

$$\Phi = \frac{\sigma}{\sigma_{YS}} - 0.25 \quad \text{for } \frac{\sigma}{\sigma_Y} > 0.5$$

'Other materials'

$$\Phi = \left(\frac{\sigma}{\sigma_{YS}}\right)^2$$

Strip yield model

$$\Phi = \frac{4}{\pi^2} \ln \sec \left(\frac{\pi \sigma}{2\sigma_{YS}}\right)$$

FIG. 5—*Comparison of elastic-plastic models.*

be expected that the stainless plate, pipeline steel, and A533B will give quite different predictions using the strip yield model type approach.

The accuracy of the nominal stress predictions, using the collapse-modified strip yield model/R-curve approach can be assessed from Table 5. Most specimens fail at or near collapse; however, these are the most relevant tests to analyze since, theoretically, all the design models converge at lower stress ratios. This is discussed further by Anderson [*4*].

All A533B specimens (Table 5*a*) recorded failure stresses above nominal collapse. Thus the accuracy of the nominal stress prediction is very dependent on the collapse analysis and relatively insensitive to the resistance curve adopted. A more indicative measure of the model is the Δa predictions at maximum stress of Table 5*a*, which underestimate actual measurements. Examples of the predicted loci, compared with physical data points for the CCT geometries in A533B, are given in Fig. 6.

The pipeline material tests (Tables 5*b* and 5*c*), indicate entirely different behavior; most specimens failed just below the nominal collapse stress. In this situation, reasonable estimates of crack extension are predicted for the CCT wide plates (Table 5*b*) with slight overestimates of the maximum nominal stress, as illustrated in Fig. 7. In fact, reducing the collapse stress by

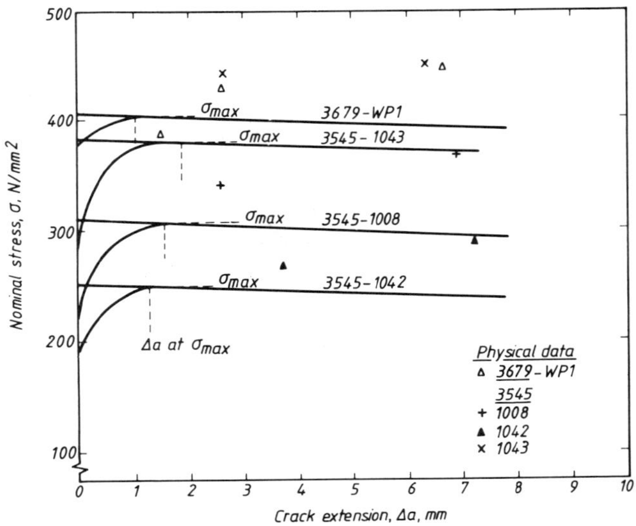

FIG. 6—*Predictions compared with test results for the center-cracked tension wide plates in A533B material.*

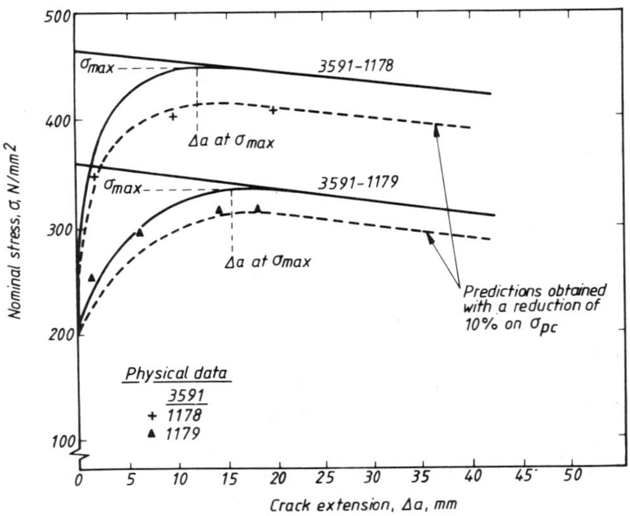

FIG. 7—*Predictions compared with test results for the center-cracked tension wide plates and the through-cracked burst test in the 5LX56 pipeline material.*

10% produces an accurate model of the behavior of the CCT plates (see Fig. 7). For the surface notched geometries, little or no crack extension is predicted, and this was evident in practice; the SENT plates experienced unstable ductile fracture just after initiation of tearing.

The model gives reasonable predictions for the SENT type 316 stainless steel submerged arc weld tests. However, it grossly underestimates the amount of tearing occurring in the CCT type 316 base metal tests (Table 5*d*), as illustrated in Fig. 8, which in turn causes an overestimate of the maximum load achieved.

In summary, very good predictions were obtained for the pipeline material, which shows a relatively low initial work hardening capacity, but, as the work hardening capability of the materials increases, the predictions become less accurate. This is in complete agreement with the findings of Akhurst and Milne [*26*] and Anderson [*4*].

To make accurate assessments of the behavior of high work hardening materials, a procedure based on actual stress-strain data (for example, Ainsworth [*25*]), would be necessary. However, the necessity for such a level of complex assessments must be questionable. Although not modelling material behavior near collapse, the design curve and strip yield models are accurate at lower stress ratios, even for these high work hardening materials. Conservative estimates, near collapse, can be ensured by conservative collapse formulae.

Maximum Load Toughness Approach

The complexity and cost of producing an R-curve and a stress-strain curve for the majority of structural predictions is unwarranted, particularly because:

(*a*) material variations (for example, longitudinal and transverse orientation, or plate-to-weld variations evident from Fig. 3), could dominate the choice of failure assessment line, and

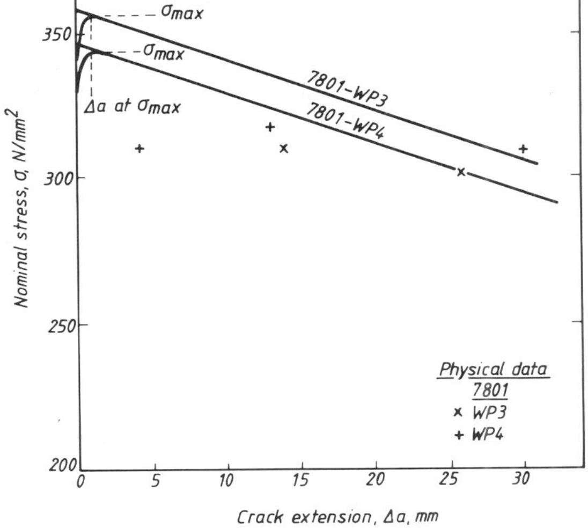

FIG. 8—*Predictions compared with test results for the center-cracked tension wide plates in 316 stainless plate.*

(*b*) cracks large enough to cause plastic collapse problems are unlikely to be left in structures.

A more practical assessment route is to adopt the maximum load toughness when fully ductile behavior is exhibited [*16*] and to use conservative collapse solutions [*24*].

The results of assessments using δ_m in the collapse-modified strip yield model and the design curve shown in Table 6, indicate conservative estimations of nominal stress for all test geometries.

Initiation of tearing toughness values have been also used as input for the A533B material, as shown in Table 7. Using maximum load values, collapse is predicted in all cases for the A533B material with safety factors on stress in the range 1.0 to 1.3. Using initiation values, fracture-dominated failures are predicted in certain cases (these did not occur in practice) with safety factors on nominal stress in the range 1.0 to 2.4.

As initiation values range from a third to a ninth of the maximum load CTOD; the use of initiation values causes the safety factor on flaw size to be increased by a factor of 3 to 9 in the fracture-dominated region. Although many of the A533B tests initiated in the fracture-dominated region, collapse had been achieved at the attainment of maximum nominal stress. Thus, the use of initiation values for A533B leads to misleading conclusions of structural behavior and, in certain cases, to considerable underestimates of failure stresses.

Of all the tests analyzed here, only three geometries apparently achieved a maximum nominal stress in the fracture-controlled regime on the basis of maximum load toughness predictions. These were the through-thickness cracks in the pipeline steel (Specimens 3591-1178, 1179, and PV2). For this material, δ_m is approximately 1.5 times the CTOD at the initiation of tearing (that is, 0.11 compared to 0.075 mm). Table 8 indicates the safety factors on crack size, using initiation and maximum load toughness values, for both the design curve and the collapse-modified strip yield model. Although the strip yield model is intended to be nearly critical, the minimum factor of safety, using δ_m, was 2.6. The use of δ_i and the design curve resulted in safety factors up to 11.

Recommended Ductile Fracture Assessment Procedure

Except in very specialized circumstances, the use of a procedure involving analysis of the material's stress-strain curve is unwarranted. The use of maximum load toughness, with the CTOD design curve procedure, has been argued as being safe on theoretical grounds in Ref *16*. However, the empirical nature of the method, together with the conservatisms of using δ_m estimated from bend specimens for tension geometries, can give very variable safety factors. In addition, a separate check for plastic collapse must be employed.

TABLE 8—*Safety factors on defect parameter* ($\bar{a}$) *for the fracture-dominated failures in the tests analyzed.*

Specimen Number	Actual a, mm	Safety Factor on $\bar{a}$: Design Curve, Using δ_i	Design Curve, Using δ_m	Collapse-Modified Strip Yield Model, Using δ_i	Collapse-Modified Strip Yield Model, Using δ_m
3591					
1178	43	5.1	3.5	4.8	3.3
1179	133	11.1	7.6	7.6	5.2
PV2	20.8	2.9	2.0	3.8	2.6

A generalized approach, which can be also used on other materials as well as ferritic steels, is to use the collapse-modified strip yield model, given by Eq 6 with maximum load toughness as the fracture toughness input.

Anderson [4] and Akhurst and Milne [26] have indicated that a strip yield-type approach might be optimistic for high work-hardening materials. This problem can be avoided by the use of a conservative value of flow stress. Conservative predictions, using the modified strip yield model of the behavior of the 316 stainless plate tested here, resulted when the stress associated with 3% strain was used in place of the traditional flow stress. This is in agreement with the findings of Akhurst and Milne, who suggested that a 3% strain criterion would always provide safe answers using the R6 procedure.

Implications to General Analysis Procedures

In this paper, four drawbacks of the CTOD design curve procedure, as used in PD 6493, are highlighted. The use of a collapse-modified strip yield model is proposed to overcome these deficiencies. A further repercussion of the use of this model is that it is a CTOD equivalent to the R6 assessment method which has been adopted by the Central Electricity Generating Board and used successfully in the Power Generation Industry.

Thus, adoption of the collapse-modified strip yield model does provide the potential of incorporating an R6-type analysis within PD 6493 while basing the procedure on CTOD values. The method could be used as an alternative to the CTOD design curve, of particular use for critical assessments, and for the analysis of other materials than ferritic steels and for fully ductile materials.

A major advantage of the method is that the strip yield equation is descriptive of failure over the complete range of behavior, from linear-elastic to collapse.

The study carried out here indicates that the method works well for fully ductile materials and can be used with either an R-curve or, for safe predictions, a maximum load toughness as input. The incorporation of residual stresses, stress gradients, and SCFs is not examined here but is discussed in Ref 27. The use of the model for materials failing by cleavage will almost certainly require additional considerations.

Conclusions

1. A modified strip yield model, incorporating a collapse analysis, is proposed for the assessment of fully ductile materials, in place of the current procedures in BS PD 6493.
2. Reasonable predictions of the behavior of wide plates in various ferritic and stainless steels and burst tests in a pipeline steel are obtained using the modified strip yield model with a representative resistance curve as input.
3. It is recommended that, for high work-hardening materials, a flow stress relating to 3% strain be adopted to ensure safe collapse predictions.
4. Conservative predictions of the ductile instability condition are obtained using a maximum load CTOD as input to either the modified strip yield model or the CTOD design curve.
5. The use of initiation of tearing toughness measurements in ductile analysis procedures can give unduly conservative estimates of ductile instability and misleading predictions of the controlling failure mechanism.

Acknowledgments

The author wishes to acknowledge the financial support of the Research Members of the Welding Institute and of the Materials, Chemical and Vehicles Requirement's Board. Thanks are also due to the Nuclear Installations Inspectorate and NV Gasunie for sponsoring the test programs referred to in this report. The author is also grateful for the assistance of Dr. A. A.

Willoughby with the pipeline data and for the helpful discussions with colleagues in the Fracture Department.

References

[1] Dawes, M. G., "The COD design curve," *Advances in Elasto-Plastic Fracture Mechanics* (2nd Advanced Seminar on Fracture Mechanics, Ispra, Italy, April 1979) L. H. Larsson, Ed., Applied Science Publishers, London, 1979, pp. 279-300.

[2] Kamath, M. S., "The COD Design Curve: An Assessment of Validity Using Wide Plate Tests," *International Journal of Pressure Vessels and Piping,* Vol. 9, 1981, pp. 79-105.

[3] Dawes, M. G., "The Crack Tip Opening Displacement (CTOD) Design Curve Approach: Limitations, Finite Size and Application," Welding Institute Research Report 278/1985, Cambridge, U.K., July 1985.

[4] Anderson, T. L., "Elastic-Plastic Fracture Assessment Based on CTOD," Welding Institute Research Report 276/1985, Cambridge, U.K., June 1985.

[5] Milne, I., "Failure Analysis in the Presence of Ductile Crack Growth," Central Electricity Research Laboratory Note No. RD/L/N/179/78, Surrey, U.K., Oct. 1978.

[6] Harrison, R. P., Loosemore, K., Milne, I., and Dowling, A. R., "Assessment of the Integrity of Structures Containing Defects," Central Electricity Generating Board R/H/R6, Revision 2, Surrey, U.K., April 1980.

[7] Kumar, V., German, M. D., and Shih, C. F., "An Engineering Approach for Elastic-Plastic Fracture Analysis," Electric Power Research Institute Report No. NP1931, Project 1237-1, Palo Alto, CA, July 1981.

[8] Tanaka, K. and Harrison, J. D., "An R-Curve Approach to COD and J for an Austenitic Steel," *International Journal of Pressure Vessels and Piping,* Vol. 6, No. 3, May 1978, pp. 177-202.

[9] Garwood, S. J., "The Use of Elastic-Plastic Fracture Analysis in the Assessment of Pressure Vessel Integrity," Welding Institute Final Conference Report No. 3645/2/83, Cambridge, U.K., June 1983.

[10] Dugdale, D. S., "Yielding in Steel Sheets Containing Slits," *Journal of the Mechanics and Physics of Solids,* Vol. 8, 1960, p. 100.

[11] Burdekin, F. M. and Stone, D. E. W., "The Crack Opening and Displacement Approach to Fracture Mechanics in Yielding Materials," *Journal of Strain Analysis,* Vol. 1, No. 2, 1966, pp. 145-153.

[12] Heald, P. T., Spink, G. M., and Worthington, P. J., "Post Yield Fracture Mechanics," *Materials Science and Engineering,* Vol. 10, 1972, pp. 129-137.

[13] Schwalbe, K.-H., "A Modification of the COD Concept and Its Tentative Application to the Residual Strength of Center Cracked Panels," *Fracture Mechanics (Twelfth Conference), ASTM STP 700,* American Society for Testing and Materials, Philadelphia, 1980, pp. 500-512.

[14] Erdogan, F. and Bakioglu, M., "Crack Opening Stretch in a Plate of Finite Width," *International Journal of Fracture,* Vol. 11, No. 6, Dec. 1975, pp. 1031-1039.

[15] Bilby, B. A., Cottrell, A. H., and Swinden, K. H., "The Spread of Plastic Yield from a Notch," *Proceedings of the Royal Society of London,* Vol. A272, 1963, p. 304.

[16] Towers, O. L. and Garwood, S. J., "The Use of Maximum Load Toughness for Ductile Fracture Assessments," *Advances in Fracture Research, Proceedings,* Fifth International Conference on Fracture, Cannes, France, Vol. 4, April 1981, pp. 1731-1740.

[17] Garwood, S. J. and Willoughby, A. A., "Fracture Toughness Measurements on Materials Exhibiting Stable Ductile Crack Extension," *Proceedings,* International Conference on Fracture Toughness Testing—Methods, Interpretation and Applications, London, June 1982, Paper No. 32.

[18] Garwood, S. J., "Crack Propagation Toughness of Structural Steels at Temperatures Above the Brittle-Ductile Transition Temperature," *Fracture Mechanics (Twelfth Conference), ASTM STP 700,* American Society for Testing and Materials, Philadelphia, 1980, pp. 271-295.

[19] Garwood, S. J., "Time-Dependent Ductile Crack Extension of Reactor Pressure Vessel Steels" in *Proceedings,* 7th International Conference on Structural Mechanics in Reactor Technology, Chicago, Aug. 1983.

[20] Willoughby, A. A. and Garwood, S. J., "The Application of Maximum Load Toughness to Defect Assessments in a Ductile Pipeline Steel," *Fracture Mechanics, Sixteenth Symposium, ASTM STP 868,* American Society for Testing and Materials, Philadelphia, 1984, pp. 632-655.

[21] Kiefner, J. F., Maxey, W. A., Eiber, R. J., and Duffy, A. R., "Failure Stress Levels of Flaws in Pressurized Cylinders" in *Progress in Flaw Growth and Fracture Toughness Testing, ASTM STP 536,* American Society for Testing and Materials, Philadelphia, 1974, pp. 461-481.

[22] Newman, J. C., "A Review and Assessment of the Stress Intensity Factors for Surface Cracks," *Part-Through Crack Fatigue Life Prediction, ASTM STP 687,* American Society for Testing and Materials, Philadelphia, 1979, pp. 16-42.

[23] Tada, H., Paris, P. C., and Irwin, G. R., *The Stress Analysis of Cracks Handbook,* Del Research Corporation, Hellertown, PA, 1973.
[24] Willoughby, A. A., "A Survey of Plastic Collapse Solutions Used for the Assessment of Part Wall Defects," Welding Institute Research Report No. 191/82, Sept. 1982.
[25] Ainsworth, R. A., "The Assessment of Defects in Structures of Strain Hardening Material," Central Electricity Generating Board Report No. TPRD/B/0175/N83, Surrey, U.K., Jan. 1983.
[26] Akhurst, K. N. and Milne, I., "Failure Assessment Diagrams and J Estimates: Validation for an Austenitic Steel," *Proceedings,* Conference on Application of Fracture Mechanics to Materials and Structures, Freiburg, West Germany, June 1983, pp. 401-404.
[27] Anderson, T. L., Leggatt, R. H., and Garwood, S. J., "The Use of CTOD Methods in Fitness for Purpose Analysis," *Crack Tip Opening Displacement in Elastic Plastic Fracture Mechanics,* K. H. Schwalbe, Ed., Springer Verlag, Berlin, Heidelburg, 1986, pp. 281-313.

Mohammad R. Etemad,[1] *Sabu J. John,*[1] *and Cedric E. Turner*[1]

Elastic-Plastic R-Curves for Large Amounts of Crack Growth

REFERENCE: Etemad, M. R., John, S. J., and Turner, C. E., **"Elastic-Plastic R-Curves for Large Amounts of Crack Growth,"** *Fracture Mechanics: Eighteenth Symposium, ASTM STP 945,* D. T. Read and R. P. Reed, Eds., American Society for Testing and Materials, Philadelphia, 1988, pp. 986–1004.

ABSTRACT: Several J-estimation procedures for use in the presence of stable crack growth are reviewed to show the distinction between those based on internal energy or on work done and on the choice of "incremental" or "sequential" analysis. These are then related and compared theoretically and experimentally to show very little differences between them for the J-controlled regime with small amounts of growth. For large amounts of growth, rising and drooping R-curves are discussed using data from open literature. A deformation theory J is shown to give a drooping R-curve shortly after $\omega < f(\eta)$. Two recent expressions for J, J_m, and J_a, formulated to satisfy the Rice, Drugan, and Sham criterion of $\dot{J}_{pl} \neq f(\dot{a})$, are discussed in light of these arguments. Finally, a reinterpretation of R-curves is discussed for large amounts of growth suggesting $\Delta a/c$ for the R-curve abscissa instead of Δa with the view to obtaining a geometry-independent R-curve.

KEY WORDS: materials, fracture mechanics, R-curves, J, crack opening displacement (COD), tearing, crack growth, normalized R-curves

Nomenclature

a	Crack length (mm)
B	Thickness (mm)
b	Ligament length (mm)
d	Deformation theory value of J (subscript)
el	Elastic component (subscript)
pl	Plastic component (subscript)
J	J contour integral (kN/mm)
m	Modified value of J
R	Resistance
ss	Steady-state value of J (subscript)
TPT	Three-parameter technique (subscript)
U	Work done (kN mm)
w	Integral energy (kN mm)
δ	Crack opening displacement (mm)
η	Geometry dependent constant

It is generally recognized that toughness of a material, ductile on the micromode, is not a unique value, but can be presented as a so-called resistance curve or R-curve of toughness versus crack growth.

[1]Post-doctoral fellow, graduate assistant, and professor of Materials, respectively, Mechanical Engineering Department, Imperial College, London SW7, United Kingdom.

It would be helpful if this resistance to ductile tearing could be expressed as an R-curve that was independent of geometry, since the main purpose of fracture mechanics is to allow the transfer of data from a test specimen to a component of a different configuration. Thickness, which has a strong and direct effect on constraint, may remain an exception to this rule, thereby requiring tests at the thickness of interest. But, if feasible, other elements of geometry, including crack size and component width and length, should not affect the R-curve for a given temperature and strain rate if an R-curve is to be readily usable. However, so far, R-curves have been found to be not only size-dependent by any definition, but also dependent on the particular definition adopted.

The J-integral has been used for R-curve determination more extensively than other toughness parameters (such as crack opening displacement, COD), primarily due to its mathematically rigorous derivation, albeit for deformation rather than incremental plasticity theory or with no unloading. An overall summary of J, COD, and their applications to R-curves has already been given [*1,2*]. Inadequacies regarding the use of J-integral for crack growth have been treated to some extent by restricting the amount of crack growth to a so-called J-controlled regime [*3*]. Attention is given here to J-estimation procedures used with crack growth in excess of that prescribed by using data from the literature and from tests on HY 130 steel and a titanium alloy [*13*]. Two candidates for a J-like term based on the understanding which has emerged for the limit plastic range behavior described in Ref *4*, J_m and J_a, are examined and compared. For large amounts of crack growth, the choice of abscissa for the R-curve also emerges as an important variable with the term conventionally used since 1965 [*5*], the absolute value of crack advance Δa, seeming not always to be the most appropriate one [*6*]. Only behavior ductile on the microscale is discussed here, although it should be recognized clearly that in some materials and circumstances (notably steels at rather low temperature) a brittle fracture may follow, and even ductile tearing can lead to unstable behavior under appropriate circumstances [*1–3,7,8*].

J and COD Evaluation with Crack Growth

A number of definitions of J, for use after initiation, have arisen. There are two main choices. First, if J is formulated in terms of the area of the loading diagram

$$J = \eta(\text{area})/Bb \tag{1}$$

where

B = thickness,
b = ligament width, and
η = a geometry dependent term,

there is the choice of actual area as a measure of work done, $U + dU$, or of internal energy, $w + dw$. Prior to growth leaving aside the question of whether the internal (strain) energy is or is not wholly recoverable, $U = w$, so the choice does not matter. After growth these terms are related, for deformation theory

$$dw = dU - BJda \tag{2}$$

leading to a definition of J as in Ref *9*. Second, there is the question of size of increment of crack growth and convenience of usage. If each increment is small and a single test specimen method is used (for example, unloading compliance), it is simple to use a truly incremental formulation. If the steps are large or the results scattered, as in a multispecimen method, it is easier to treat each result in relation to the total growth Δa and obtain a sequence of points, each independent of the intermediate steps, here called a sequential method. These two differences, coupled with whether or not the geometric terms η and b are updated, have given rise to the several formulations that now exist. Note all early tests used a multispecimen method. Incremental analysis was advocated in Ref *9,* but the simply performed sequential analysis is often used in practice, for multispecimen data.

The two J-estimation procedures used most frequently with crack growth [*9,10*] are "incremental." This refers to analysis in a multi- or single-specimen procedure where test data are spaced at small intervals of Δa and J evaluated always from the previous step and not from the original or first data point. The first is the "three parameter technique" of Garwood et al., J_{TPT} [*9*], stated here in its general form as opposed to the simplified deep notch three point bend version given originally (see Appendix I)

$$J_{TPT(n)} = J_{(n-1)}\left\{1 + \frac{\Delta a_{n,(n-1)} g(\eta)_n}{(W - a_n)}\right\} + \frac{\eta_n \Delta U_{n,(n-1)}}{B(W - a_n)} \tag{3a}$$

where suffix n refers to the nth step and

$$g(\eta) = 1 + \frac{b}{\eta}\frac{d\eta}{da} - \eta$$

Note that the "three-parameter technique" was originally devised for a multispecimen method, and the variability in the original test specimen dimensions, B_0 and b_0, is accounted for by scaling $J_{(n-1)}$ for the nth step, that is

$$\text{effective } J_{(n-1)} = \text{actual } J_{(n-1)}(Bb/\eta)_n/(Bb/\eta)_{n-1} \tag{3b}$$

in order that the increment $\Delta U_{n,n-1}$, required in Eq *3a* is not unduly influenced by random differences in the intended dimensions of successive test specimens. The second is the procedure currently adopted as the ASTM Test Method for J_{Ic}, a Measure of Fracture Toughness (E 813-81) and derived by Ernst et al. [*10*] from the concept of variables separable (that is, load is a separable function of displacement and crack length), $Q = Q(q, a)$

$$J_{(j+1)} = \left[J_j + \frac{\eta}{Bb_j}(A_{(j+1)} - A_j)\right]\left[1 - \left(\frac{\gamma}{b_j}\right)(a_{(j+1)} - a_j)\right] \tag{4}$$

where subscript j refers to the jth step, and $\gamma = -g(\eta)$, A is the area under the load-displacement diagram, that is, $U + dU$ at the $j + 1$ step. Since, in Fig. 1, Eq 4 is applied to multitest specimen data, the results have been corrected for variability of initial dimensions according to Eq 3*b* although that is not part of the normal single test specimen procedure.

Both equations can be stated in a sequential manner, and Eq 4 was originally stated [*10*] as

$$J_R = \left(J + \eta\frac{\Delta U}{Bb}\right)\left(1 - \gamma\frac{\Delta a}{b}\right) = J_{RO} - J_{RO}\gamma\frac{\Delta a}{b} \tag{5a, b}$$

where J_{RO} is the simplest and earliest definition for $J + \Delta J$, that is

$$J_{RO} = J + \Delta J = \frac{\eta(U + \Delta U)}{Bb} \tag{6a, b}$$

then for deep-notch three-point bending, where $\gamma = +1$, $J_R = J_{RO}[1 - (\Delta a/b)]$. One set of data from Ref *11* is shown, Fig. 1, using these four methods: J_{TPT}, J_R, $J_{(j+1)}$, and J_{RO}. The first two are shown with a "best fit" curve according to a second-degree polynomial as well as a "by

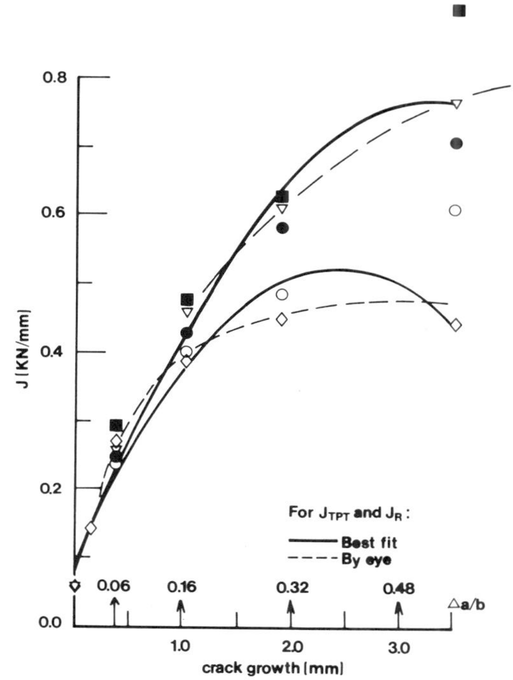

FIG. 1—*Effect of method of analysis for HY130; SENB,* S = 4W, *4* W = B = *50 mm,* a_0/W = *0.5, TL crack.* ■ J_{RO}, ◇ J_R, ▽ J_{TPT}, ● *(corrected), and* ○ *(uncorrected)* $J_{(j+1)}$.

eye" fit. For more than, for example, 15% growth, the sequential treatment by J_R falls clearly below the two incremental analyses, J_{TPT} and $J_{(j+1)}$, which differ little between themselves when both are corrected for variations in size of test specimens (Eq 3*b*). A rising or falling curve as $\Delta a/b$ becomes large here seems to depend on choice of curve fitting, incremental or sequential formulation and definition of *J*. A further discussion on effect of choice of *dU* or *dw* is given later.

To define initiation or to evaluate tearing resistance or both for small growth in terms of the COD concept, the British Standards Institutions Methods for Crack Opening Displacement (COD) Testing (BS 5762, 1979) was used

$$\delta = \delta_{el} + \delta_{pl} = \frac{K^2}{2\sigma_y E} + \frac{r(W - a)}{rW + (1 - r)a + z} Vg_{pl} \tag{7}$$

where the rotational factor *r* that defines the center of hinge rotation as a fraction of ligament was taken as 0.4. Vg_{pl} is the plastic component of the clip gage reading taken a distance *z* above the surface of the test specimen, *E* is the tensile modulus, and σ_y is the yield stress. In this method, COD is split into elastic and plastic components and the elastic component expressed in terms of *K*, itself derived from load as in a conventional linear elastic fracture mechanics (LEFM) test. The conversion from *K* to δ_{el} is made using

$$J = \frac{K^2}{E} = m\sigma_y \delta \tag{8a,b}$$

with $m = 2$, that is,

$$\delta = K^2/2\sigma_y E \tag{8c}$$

Relationship Between *J* and COD

The qualitative relationship between J and COD, Eq 8*b*, was investigated using Eqs 3*a* and 7 for the data detailed in Fig. 2 [*11*], with $\sigma_y = 0.90$ kN/mm^2. It was shown in Ref *11* that the data were predominantly in plane strain. The value of m at initiation was found to be 1.61, and at $\Delta a = 3.57$ mm was 1.66, varying between 1.87 to 1.55 with a slight tendency to decrease with growth, but with a mean value of 1.69 ± 0.18. This is some 15% lower than the often used value of 2 for plane strain, and indeed close to the value of $\sqrt{3}$ implied by the LEFM treatment of plastic zone size. The value of $m = 1.69 \pm 0.18$ quoted above is in general agreement with results of Ref *12* using an Hutchinson-Rice-Rosengren (HRR) type solution. It also falls within the range of values ($m = 1.32$ to 1.89) mentioned by Ref *4* from the computational studies (expressed there as a $1/m$) that led to the present understanding of the stress fields ahead of an advancing crack tip (that is, maintenance of a quasi-Pradtl field even for growing cracks), so that there is indirect support for the use of these J and COD values. Note that use of flow stress σ_{fl} (the mean of yield and tensile strength) instead of σ_y in the discussion of m makes little difference for HY130 (about 5%) since the tensile-to-yield ratio is only 1.09, but clearly might be relevant for steels with a higher degree of work-hardening.

Similar results have been obtained from a titanium alloy [*13*] using a multispecimen method with three-point bend specimens of $2B = W = 35$ mm and $S = 4W$, $a/W = 0.5$. Results show $m = 1.80$ at initiation, $J_{j+1} = 0.075$ kN/mm, and $m = 1.37$ at $\Delta a = 4.9$ mm ($\Delta a/b_0 = 0.28$), $J_{j+1} = 0.22$ kN/mm. A decreasing trend for m is again observed with a mean value $m = 1.57 \pm 0.19$. The general reasonableness of these m values adds support to the use of these J and COD equations as well as Eq 8*b* relating them. Moreover, it shows that the methods are not restricted to one steel, although the materials discussed here have rather similar properties of yield, toughness and work-hardening. However, similar conclusions had emerged earlier when two steels with quite different properties were examined [*14*].

Beyond *J*-Controlled Crack Growth

Although within the range of data in Fig. 2 there seems no a prior reason to prefer J to COD or vice versa, the more extensive analysis is offered in terms of J, reflecting other aspects of its use, notably ease of estimation, compatibility with LEFM, and relation to other parameters for a range of circumstances beyond the scope of these test. The near constant value of m (= 1.69 or 1.57) relating J to COD ($J = m\sigma_y\delta$) during crack growth, and the agreement of that value with values suggested elsewhere by computational studies, tends to support the use of J or COD as closely linked alternatives. However, beyond this restrictive region, the COD approach with its constant r assumption is not considered here.

For J, Eqs 3*a* and 4 are not numerically equivalent for large growth; see Figs. 1 and 2 for $\Delta a/b >$ about 15%. This is because the common correction term, $1 + g(\eta)\Delta a/b$, is used more severely in Eq 4 where it acts on the total area $U + dU$ rather than only the term U as in Eq 3*a*, thus resulting in larger values of J_{TPT} than for J_{j+1}. Equation 4 or its sequential form, Eq 5, may yield a negative dJ/da, since for $\Delta a \rightarrow b$ and deep notch bending, that is, $g(\eta) = -1$, $J \rightarrow 0$. Similar but less severe results may be obtained from Eq 3*a* where the correction term does not act on ΔU. The negative dJ/da is possible because a J-R curve as defined above is a measure of the change in internal energy, dw, defined by Eq 2, which can decrease. However, for crack growth in an elastic-plastic body an increasing J will result if a work-based definition of J is adopted. This would be so even in plane-stress LEFM R-curves where a constant dJ/da is main-

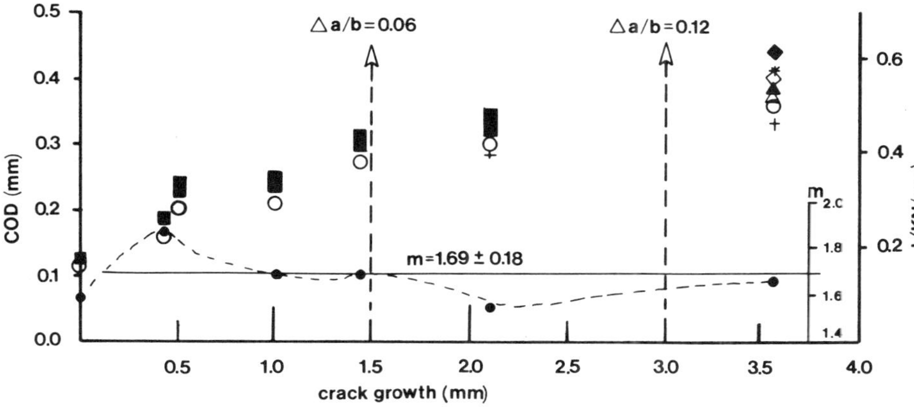

FIG. 2—*Comparison of COD and* J *for HY130; SENB,* S = 4W, 2B = W = 50 *mm,* a_0/W = *0.5, TS crack. Left hand scale:* ○ *BS5762 for COD. Right hand scale:* * J_{RO} † J_R △ $J_{(j+1)}$ ▲ J_m ◇ J_{TPT} ◆ J_a ■ *by all definitions. Inner right hand scale* ●--- *m from Eq.* 3a.

tained with a linearly increasing J (when based on work), as the maximum plastic zone size is developed and carried ahead of the advancing crack tip by the imposed deformation.

A steady-state condition giving a constant J, $dJ/da = 0$, has been addressed by Rice, Drugan, and Sham [*4*]. This special case is possible with a nonlinear elastic (NLE) definition (where deformation is recoverable contrary to elastic-plastic behavior with unrecoverable work stemming from plasticity) giving the deformation theory J, J_d, for a constant near-tip-crack profile assumption, as discussed fully in Ref *4*. Note that both J_{TPT} and J_{j+1} (or any other definition of J based on internal energy) using Eq 1 are conceptually identical with J_d, all relating to a NLE definition for J. Some related theoretical and experimental work by Rice and Hermann [*15*] gives steady-state values for J_d. Moreover, the experimental values for J_d were found to decrease with further crack growth, giving a negative dJ/da. Other data detailing "drooping" R-curves have been reported [*16*], and their numerical and physical significance interpreted [*17*]; see Appendix II. A further set of data from Ref *16* is shown in Fig. 3.

As argued in Ref *17*, any R-curve defined in terms of internal energy w denoted $R_{(w)}$, may droop. Clearly (Eq 27) dw/da can become zero when $\omega = f(\eta)$ and thereafter becomes negative, and dJ/da is likely to become zero and then negative, shortly afterwards when $dw/da < -(w/b)f(\eta)$. Conversely, if dw/da remains greater than this value, then the R-curve will continue to rise. Note for all cases of remote loading so far known, $f(\eta)$ is positive. These expressions are not so much for numerical use since dw/da is only known by a circuitous argument using Eq 2 but offer some physical insight into the meaning of a rising or drooping R-curve. If, however, the R-curve arises from a work based definition of J (for example, Eq 6), denoted $R_{(u)}$, then a rising curve will always be found since stable growth requires a positive increment of work.

Both J_{TPT} and J_{j+1} give $R_{(w)}$ because although they use the actual area under the Q-q diagram, that is, $(U + \Delta U)$, the $(w + dw)$ is retrieved by correcting for crack growth using Eq 2. Figure 3 shows both equations giving drooping R-curves, whereas one example of $R_{(u)}$, given by J_{RO} (Eq 6), gives a rising R-curve. Furthermore, a third condition was stated in Ref *17* relating to a constant $R_{(w)}$, that is, $dJ_d/da = 0$. This was thought similar if not identical to the steady-state deformation theory condition, $J_d|_{ss}$, in Ref *4*. Using Eq 29 in Appendix II, an expression for $J_d|_{ss}$ at $dJ_d/da = 0$ may be obtained, that is

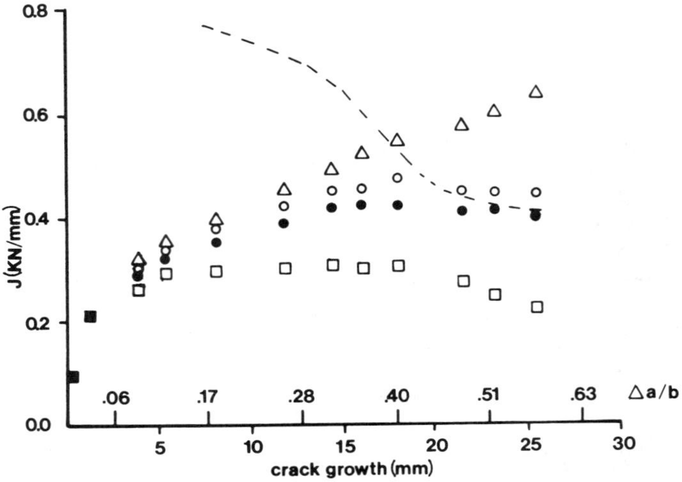

FIG. 3—*Drooping R-curves for HY130; CT,* 4B = W = *100 mm,* a_0/W = *0.6.* Δ J_{RO} □ J_R ● $J_{(j+1)}$ ○ J_{TPT} ■ *by all definitions. Broken line showing the trend of* $J_d|_{ss}$.

$$J_d|_{ss} = -\frac{\eta}{g(\eta)}\frac{\Delta U}{B\Delta a} \tag{11}$$

This is identical to Eq 15 of Ref *17* if $g(\eta) = -1$ (that is, deep notch three-point bend). Values of $J_d|_{ss}$ calculated from Eq 11 are also shown (the decreasing curve) in Fig. 3. Clearly, a "plateau" of values is emerging for large crack growth compatible with J_d (that is, J_{TPT} or $J_{(j+1)}$) rather than J_{RO}.

Further corroborating evidence from the titanium alloy mentioned earlier is shown in Fig. 4, where drooping is observed at much smaller absolute values of crack growth. However, both titanium and HY130 start to droop, $\Delta a/b \approx 0.32$ or 0.34. Note this set of titanium data was generated using a computer-interactive single-specimen partial unloading compliance method [*13*], and compact tension specimens, whereas the HY130 data, Fig. 3, are multitest specimens.

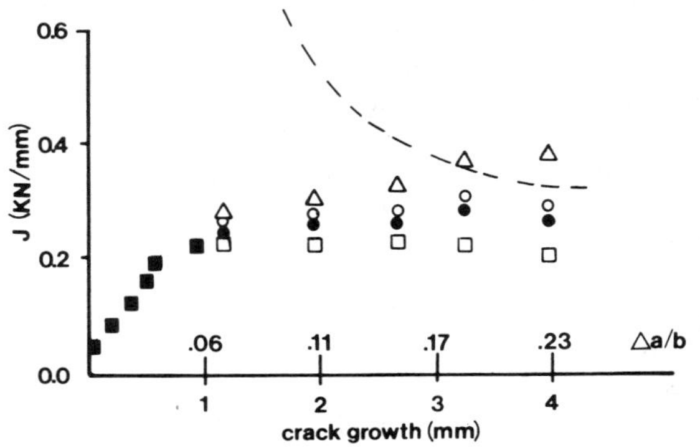

FIG. 4—*Drooping R-curves for a titanium alloy;* CT, *2*B = W = *35 mm,* a_0/W = *0.5. Symbols same as Fig. 3.*

It is noted that the steady-state J_d is similar in value to the maximum on the $R_{(w)}$ curve in Figs. 3 and 4 where η is independent of both geometry and degree of deformation (that is, a constant such as 2 for deeply cracked bend specimens). It is not clear whether this will always be so, since as commented earlier, the maximum value of J reached in a fully yield condition is not truly steady state in the sense used in Ref *4* and for elastic-plastic material may well be a geometry-dependent combination of tearing and work-hardening.

A *J* Term Independent of Crack Growth

Choice of Definition

A theoretical guide on the elastic-plastic behavior of growing cracks is given by Rice, Drugan, and Sham [*4*]. The criterion for crack growth was taken as "the crack maintaining a steady near-tip geometrical profile." This is interpreted here to be similar to criteria such as "constant crack opening angle" (COA) and "constant advancing crack tip opening displacement" (ACTOD) as apart from the original tip COD. Numerical results of Shih et al. [*12*] confirm COA as a likely candidate, while experimental results of Garwood et al. [*18*] have shown ATCOD to be constant during crack growth, and hence similarly appropriate. Thus the physical modeling by Ref *4* appears reasonable, although the implication of its restriction to rigid plastic materials (that is, $\sigma_0/E \rightarrow 0$) and to steady-state behavior is not fully clear. However, in that regime the analysis shows that the appropriate J must satisfy the requirement (called the RDS requirement)

$$\dot{J}_{\text{pl}} \neq f(\dot{a}) \tag{12}$$

This prompted McCabe and Ernst [*20*] to introduce a "modified" formula for satisfying Eq 12

$$J_m = J_d + \int_{a_0}^{a} m \frac{J_{\text{pl}}}{b} \Delta a \tag{13}$$

where

$$m = \left(-\frac{b}{J_{\text{pl}}}\right) \frac{\partial J_{\text{pl}}}{\partial a}\bigg|_q$$

$$J_d = J_R$$

and

$$J_{\text{pl}} = J_d - G$$

J_m may be expressed in the familiar η and $d\eta/da$ terms by differentiating $J_{\text{pl}} = \eta_{\text{pl}} U_{\text{pl}}/Bb$ where $U_{\text{pl}} = Q_{\text{pl}} q_{\text{pl}}$ and recalling

$$J_{\text{pl}} = -\frac{1}{B} \frac{\partial U}{\partial a}\bigg|_q$$

$$\frac{\partial J_{\text{pl}}}{\partial a}\bigg|_q = \frac{J_{\text{pl}}}{b}\left(1 + \frac{b}{\eta_{\text{pl}}} \frac{d\eta_{\text{pl}}}{da} - \eta_{\text{pl}}\right) \tag{14}$$

and hence

$$m = \eta_{pl} - 1 - \frac{b}{\eta}\frac{d\eta_{pl}}{da} = \eta_{pl} - f(\eta_{pl}) = -g(\eta_{pl}) \tag{15}$$

which for deep notch bending gives $m = 1$. Furthermore, Eq 13 may be evaluated for $b \gg \Delta a$ using $\delta = rb\theta$ giving $d\delta = rbd\Theta$ [that is, neglecting $d(rb)$], and with $d\delta = \alpha da$ where α (or $\tan\alpha$) is the crack opening angle

$$J_m = J_d + \frac{1}{b}\int_{a_0}^{a} (J_{pli} + \Delta J_{pl})\Delta a$$

$$= J_d + \frac{1}{b}\left[\int_{a_0}^{a} J_{pli}\Delta a + \int_{a_0}^{a} \frac{\eta_{pl} Q_{pl} \Delta q_{pl}}{Bb}\Delta a\right] \tag{16a}$$

The term J_{pli} refers to the contribution of J_{pl} at initiation and for a given geometry is assumed to be a constant (for HY130, $J_{pli} = 0.04$ kN/mm). The term Δq_{pl} can be evaluated after initiation using the bend angle, Θ

$$\Delta q_{pl} = \frac{S}{2}\frac{d\Theta}{2} = \frac{S}{2}\frac{d\delta}{2rb} = \frac{S}{2}\frac{\alpha}{2rb}da \tag{16b}$$

Using Eq 5 for J_d and $J_{RO} = J_{el} + J_{pl}$ simplifies Eq 16*a* to

$$J_m = (J_{el} + J_{pl})\left(1 - \frac{\Delta a}{b}\right) + J_{pli}\frac{\Delta a}{b} + \frac{\eta Q}{Bb}\frac{S}{2}\frac{\alpha}{2rb}\frac{\Delta a^2}{b} \tag{16c}$$

or

$$J_m = J_{pl}\frac{\Delta a}{b}\left(\alpha\frac{\Delta a}{\delta} - 1\right) + J_{RO} - G\frac{\Delta a}{b} + J_{pli}\frac{\Delta a}{b} \tag{16d}$$

using $\alpha = (\delta - \delta_i)/\Delta a$ gives

$$J_m = J_{RO} - \frac{G\Delta a}{b} + J_{pl}\left(\frac{\delta_i}{\delta}\right) + J_{pli}\left(\frac{\Delta a}{b}\right) \tag{16e}$$

Assuming $(\delta_i/\delta) = (J_i/J)$ with J tending to J_{pl}

$$J_m = J_{RO} - G\frac{\Delta a}{b} \tag{16f}$$

Note that assuming $m = 1$, $b \gg a$, constant J_{pl}, and using Eq 5 for J_d gives

$$J_m = J_{RO}\left(1 - \frac{\Delta a}{b}\right) + \frac{\Delta a}{b}\left[J_{RO}\left(1 - \frac{\Delta a}{b}\right) - G\right] \tag{17a}$$

or

$$J_m = J_{RO} - G\frac{\Delta a}{b} - J_{RO}\left(\frac{\Delta a}{b}\right)^2 \tag{17b}$$

which differ by the small second-order term $J_{RO}(\Delta a/b)^2$ from Eq 16*f*. These latter approximations become ever poorer as $\Delta a/b$ becomes significant but seem adequate for discussion of a J-controlled regime, conventionally taken up to $\Delta a \leqq 0.06b$. In these cases for small Δa, $J_m \rightarrow J_d$, for intermediate Δa, $J_m = J_{RO} - G\Delta a/b$. More recently Turner [2] observed Eq 12 to write J_a for an advancing crack

$$\Delta J_a = \Delta J_d - J_d g(\eta) \frac{\Delta a}{b}(1 - n) = \Delta J_{RO} + nJ_d \frac{\Delta a}{b} \quad (18)$$

where n is the hardening exponent $0 \leq n \leq 1$. For $n = 0$ (that is, $\sigma_0/E \rightarrow 0$), $G \rightarrow 0$, this gives $J_a = J_{RO} = \eta(U + \Delta U)/Bb$ and for LEFM with $n = 1$ it gives $\Delta J_a = \Delta J_d = \Delta G$. In short, Eq 18 simply interpolates between the conditions for LEFM and those set out by Eq 12 for steady-state advance with plasticity. Moreover, in these two limits Eqs 16 and 18 are identical, that is, LEFM $J_a = J_m = G$ and rigid plastic $J_a = J_m = J_{RO}$. For deep notch three-point bending with $g(\eta) = -1$, Eq 18 gives

$$_{(n)}J_a = {}_{(n-1)}J_a + \Delta J_d + J_d \frac{\Delta a}{b}(1 - n) \quad (19)$$

For small amounts of growth $J_a \rightarrow J_d$; for intermediate values it rises above J_d depending on the value of n. Note that in Eqs 13, 16, 18, and 19, J_d or ΔJ_d may be treated incrementally as in Eqs 3*a* and 4 or sequentially as in Eq 5. In either case, J_d and ΔJ_d imply a definition based on internal energy. The leading term in ΔJ_a, Eq 18, is ΔJ_{RO}, which, as seen Eq 6, is based on increment of work at fixed (original) values of b and η. For that term a sequential treatment would be adequate, but unless n is very small the last term in Eq 18 contributes a substantial amount and, for that term, just as for J_m, the incremental treatment seems preferable. In short, it seems that choice of incremental or sequential is based on the general arguments of plasticity theory rather than some specific argument about R-curves, except for the limiting case of using J_{RO} to model the RDS state.

It is also noted that Eq 12 is for a rigid ideally plastic material which also, however, tends to a steady-state J-value, thus specifying an R-curve where $dJ/da = 0$ in the plastic limit rather than the more or less parabolic curve familiar in small-scale testing. The question therefore seems to be, "Should test data tend to a steady state?" This surely implies either $\Delta a \ll b$ and a quasi-LEFM condition of advance into virgin (elastic) material. Or should it be recognized that since all components are of finite size, then if there is appreciable plasticity a steady-state tearing is by definition not attainable?

Some Experimental Evidence

Data presented in Figs. 1 to 4 from Refs *11, 13,* and *16* are employed for a comparative analysis of the equations outlined previously. As shown, there is very little difference due to the estimation procedure when initiation toughness (or for that matter a little crack growth, say less than 6% b) is required. This is because the equations presented here offer an estimate of toughness in the presence of restricted crack growth that uses the area under the $Q - q$ diagram, normalized for size and shape by Eq 1, as a measure of toughness. Even at double the prescribed datum of $\Delta a/b = 0.06$, the data show differences that are quite tolerable. However, it is clear that in this range (6 to 12% b), the choice of sequential or incremental analysis starts to influence the slope of the R-curve, with the sequential R-curve being the first to fall below the others and even perhaps droop. Originally the sequential analysis was encouraged by the simplified formulation of $J_R = J_{RO}(1 - \Delta a/b)$ for deeply cracked bend bars. It is still useful when testing by the multispecimen method and sustaining small crack growths, $\Delta a/b < 0.10$. Use of incremental analysis, $J_{(j+1)}$ and J_{TPT}, is clearly more meaningful and not much more difficult than sequential analysis to perform when single-specimen methods and, in particular, computer-interactive partial unloading compliance are used. For intermediate amounts of growth ($0.1 < \Delta a/b < 0.25$), the differences between J estimation procedures may not be tolerable, but there is no marked change in the shape of the R-curve, that is, all rising with slopes depending upon the definition of J. For large growth, $\Delta a/b > 0.25$, drooping R-curves are unacceptable if J is viewed as a measure of work dissipation, $R_{(u)}$, Figs. 3 and 4, while acceptable if J is viewed as a measure of change of internal energy, $R_{(w)}$.

For large growth in finite size test specimens, $R_{(u)}$ has been seen to satisfy the limiting RDS requirement (for example, via J_{RO}), while to satisfy LEFM deformation theory J must be regained, $R_{(w)}$. The two new candidates detailed here, Eqs 13 and 18, when based on incremental analysis give the results marked on Fig. 2, both exceeding J_R. The distinction between J_m being obtained incrementally ($j + 1$), or from a sequential type analysis, R, is shown in Fig. 5, using another set of data from Ref *11*. J_m values from the incremental analysis are larger since $J_{(j+1)} > J_R$. Using a sequential analysis for J_a will result in marked numerical differences (similar to J_m); for example, the sequential analysis gives a value 13% smaller for J_a for the last data point on Fig. 2. Hence two different types of R-curves seem to have emerged, both starting from initiation, itself not clearly defined:

1. What is termed $R_{(u)}$ [*17*], based on $J_{(u)} = \eta(U + \Delta U)/Bb$ with η and b referring to original test specimen dimensions. This curve will always rise and may achieve a steady state with constant (nonzero) slope. The J giving this curve has sometimes been referred to as "far field J," J_{ff}, and can be made to satisfy the requirement of RDS [*4*] for the limiting rigid plastic condition.
2. What was termed $R_{(w)}$ [*17*], based on $J_{(w)} = \eta\omega/Bb$ where w is internal energy. This J refers to deformation theory, J_d, and has been evaluated by correcting $J_{(u)}$ for crack growth. Prior to growth $J_{(u)} = J_{(w)}$. For growing cracks in finite-size test specimens, $R_{(w)}$ will give a rising and then drooping curve.

Either $R_{(u)}$ or $R_{(w)}$ can be treated incrementally or sequentially (in essence, updating or not updating the dimensions of the component as it tears). While conventional arguments for small increments and the preferred incremental theory of plasticity imply use of an incremental analysis, it is the sequential use of $J_{(u)}$ that is consistent with the RDS requirement of Eq 12. For $R_{(w)}$, chosen either because internal energy is seen as physically desirable here, or as that part of either J_m or J_a that satisfies the LEFM limit, then an incremental analysis that updates b and η seems essential. By casting J_m and J_a in the form of Eqs 13 and 19, an incremental analysis is both appropriate and convenient, but the underlying point remains that it is a term such as J_{RO}, Eq 6, that is satisfying the RDS limit. Whether that is indeed desirable, because the practical

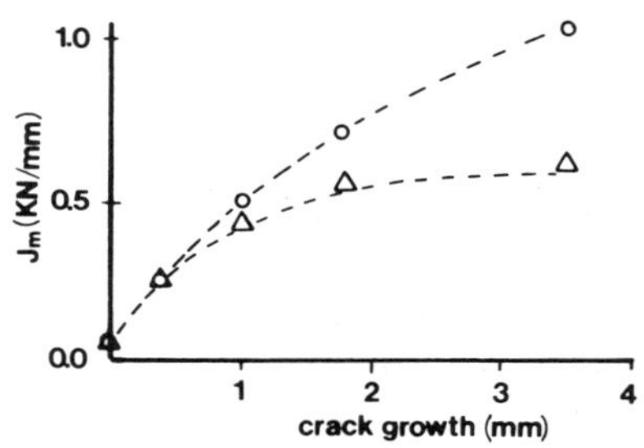

FIG. 5—*Definition dependence of* J_m *for HY130; SENB,* S = 4W, 4W = B = *50 mm,* a_0/W = *0.5, TL crack.* ○ *using* $J_{(j+1)}$, Δ *using* J_R.

limit is surely elastic-plastic rather than rigid plastic behavior (that is, it is only the elastic component that provides a driving force), seems open to question. Moreover, it is not clear how the limiting value can be achieved in finite size test specimens unless either extensive yielding or no yielding has occurred prior to tearing, giving essentially a flat R-curve with $J = 2\gamma_{eff}$. The steady-state condition specified in Ref *4* may be numerically the same as the maximum on the $R_{(w)}$ if the deformation pattern is self-similar, that is, in cases where there is a constant η factor, as discussed earlier.

The main objective in discussing these various curves and definitions has been to establish the most likely combination to give a unique R-curve. It was suggested in Ref *19* that J_m produced a unique R-curve. Clearly, thickness as a geometric variable is an exception due to its strong effect on constraint as indeed, in Ref *20,* R-curves were shown to be independent of plan view size only. Results of the two sets of data from Ref *11* which gave geometry-dependent R-curves were reanalyzed according to J_m, Eq 16, Fig. 6. It is seen that a unique R-curve was not obtained. It may be argued that this failure of J_m in producing a unique R-curve was due to the smaller-sized test specimens not adequately fulfilling the size requirements; that is, based on a J-value at initiation of about 0.15 kN/mm and a yield strength of 0.90 kN/mm^2 the series $4W = B = 50$ mm is judged invalid ($50\ J/\sigma_y = 8.3$ mm, whereas $b = 6$ mm). Hence, other specimens were selected from Ref *11* which satisfied $50\ J/\sigma_y$ and were treated using J_m analysis. Results are shown in Fig. 7 and, for the range of crack growth studied, these show such small differences between the two sizes that for all practical purposes it is a unique R-curve. The same observa-

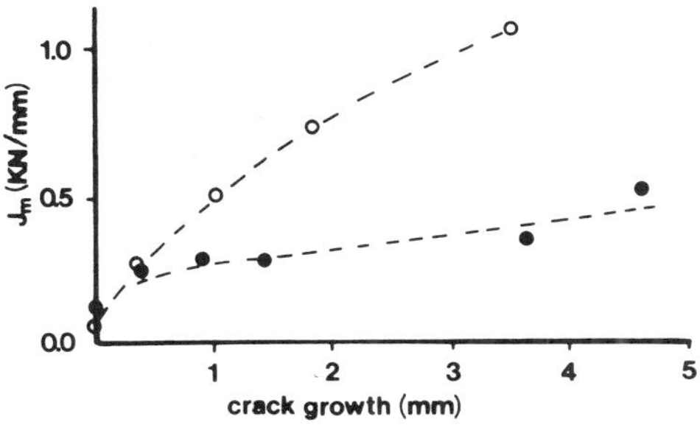

FIG. 6—*Geometry dependence of* J_m *for HY130; SENB,* S = 4W, a_0/W = *0.5,* B = *50 mm, TL crack.* ○ W = *12.5 mm,* ● W = *50 mm.*

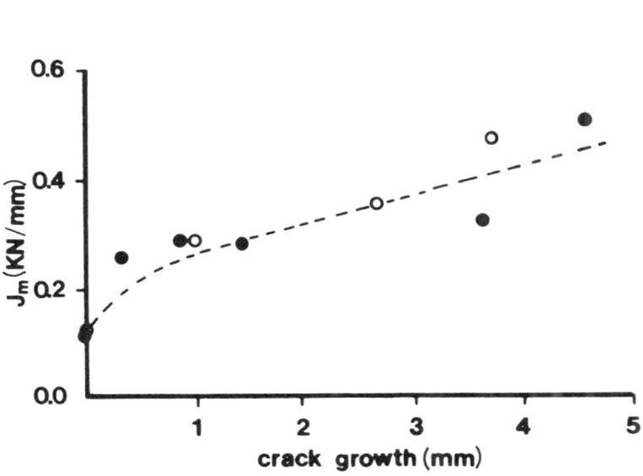

FIG. 7—*Geometry independence of* J_m *for HY130; SENB,* S = 4W, a_0/W = *0.5,* B = *50 mm, TL crack.* ○ W = *66 mm,* ● W = *50 mm.*

tion was made when the data were tested with all of the other equations outlined here, showing insensitivity toward the method of analysis. But this was expected in any case; R-curves obtained from the highly constrained SENB geometry would be only slightly different from each other if strict size requirements are complied with. However, the two curves will diverge at values of Δa, giving $\Delta a/b$ for the smaller-sized series approaching unity, while for the larger it would only be $\Delta a/b = 0.76$.

A Unique R-curve

The data of Fig. 6 reanalyzed according to all of the equations presented here is given in Fig. 8*a*. The smaller-sized series shows more sensitivity toward the method of analysis, since for the same absolute value of Δa, it has a larger $\Delta a/b$. Moreover, it is seen that a term such as J_R appears to produce a geometrically independent R-curve with $dJ/da = 0$, whereas the other definitions of J do not. However, J_R is here rejected as a large growth toughness parameter since it can give drooping R-curves, although that condition may not prove to be as unacceptable as it at first seemed, as already discussed in terms of internal energy. Furthermore, as outlined earlier, effects of constraint due to thickness can not be accounted for by the method of analysis, be it J_R or some other J capable of accounting for large growth. Indeed, R-curve variation due to geometry (constraint) and large growth are not one and the same.

This can be seen more clearly in other data from Ref *11* which are presented in Fig. 9*a*, where variation due to constraint is not accounted for by any definition. A fundamental treatment suggested by Garwood [*21*] was to divide the R-curve into flat fracture and oblique (shear lip) components. However, it seems that in some cases an alternative treatment may be possible. This was detailed in "A Reinterpretation of R-curves beyond J-controlled Crack Growth for Plane Stress and Fully Plastic Plane Strain" [*6*]. Briefly, using the data already presented here, the diverging R-curves of Fig. 8*a* can be brought onto a single R-curve by altering the abscissa to $\Delta a/b$, Fig. 8*b*. The same is true for Fig. 9*a*, which can give a near single R-curve by plotting toughness versus $\Delta a/B$, Fig. 9*b*. The definition of J is not seen as crucial to this discussion except insofar as it no doubt affects the slope of the R-curve, and with the definitions allowing

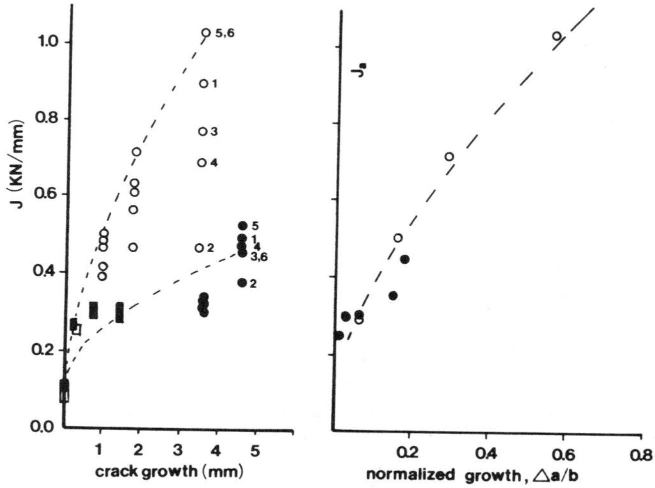

FIG. 8—*Effect of width for HY130; SENB, S = 4W, $a_0/W = 0.5$, B = 50 mm, TL crack.* ○ W = 12.5 *mm,* ● W = 50 *mm. 1* J_{RO} *2* J_R *4* $J_{(j+1)}$ *5* J_m *3* J_{TPT} *6* J_a *and square symbols for all definitions.*

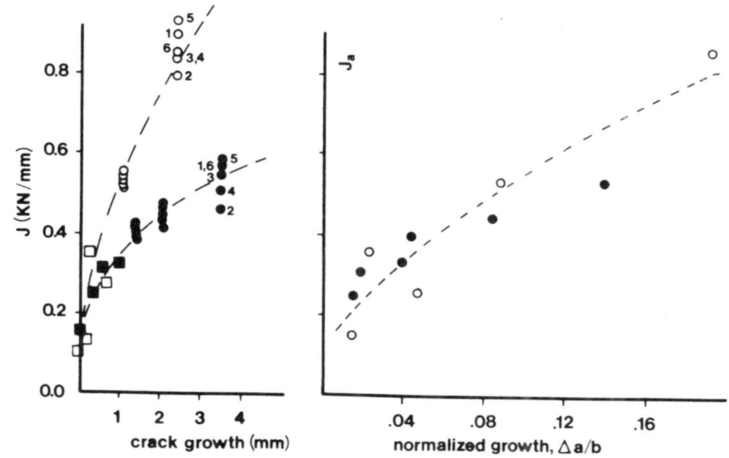

FIG. 9—*Effect of thickness for HY130; SENB,* S = 4W, a_0/W = *0.5,* W = *50 mm, TS crack.* ○ B = *12.5 mm,* ● B = *25 mm.* J *values as in Fig. 8.*

drooping R-curves being unacceptable. Further evidence is cited here from the recently published data of Davis et al. [*16*]. A first set shown in Fig. 10 normalizes with $\Delta a/b$, giving a unique R-curve. Note the thickness is of valid size; B_n = 40 mm and is larger than 50 J/σ_y = 22 mm, and is kept larger than the ligament, which does not always satisfy the minimum size requirements for plane-strain fracture. A second set of data from Ref *16*, Fig. 11, normalizes with $\Delta a/B$, giving a unique R-curve with the exception for the smaller of the three thicknesses. The latter at B_n = 9.5 mm is clearly invalid according to 50 J/σ_y = 22 mm, but the heavy side grooves have forced an R-curve similar to the data at B_n = 40 mm. For this set to be successfully normalized, Δa must be divided by the equivalent plane-strain section size, that is, 40 mm, and not the unrealistic but side-grooved 9.5 mm. Other sets of data for HY130 steel may be detailed from Davis et al. [*16*], which are actually not useful because both B and W were kept constant while varying a/W, hence B/b, but since heavily side-grooved all data are forced into plane strain. The band of results obtained may be presented by $J = 0.18 + 0.002\Delta a$.

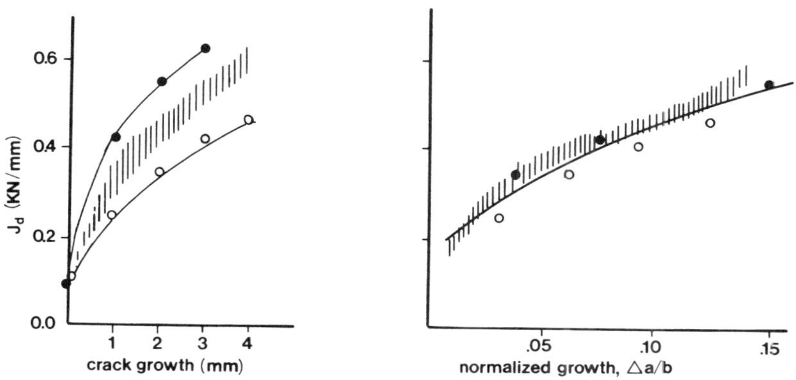

FIG. 10—*Effect of width for A533B; CT,* B_n = *40 mm.* ○ a_0/W = *0.66,* b = *32 mm.* ● a_0/W = *0.86,* b = *13 mm. Shaded;* a_0/W = *0.77 and 0.82 where* b = *23 and 18 mm, respectively.*

It is clear from the discussion above and Figs. 8 and 10 that results in or near plane-strain regime with large growth are functions of $\Delta a/b$ for a given thickness. Equally from Figs. 9 and 11, results in or near plane-stress regime with large growth are functions of $\Delta a/B$ for a given ligament. A third regime identified in Ref *6* was related to geometrically similar specimens with $\Delta a/\bar{r}$ where

$$\bar{r} = \frac{1}{2\pi}\frac{JE}{\sigma_y^2}$$

Thus, generally the abscissa is taken as $\Delta a/c$, where c is that factor (B, b, $\bar{r}$, or other) which limits the zone of plastic work dissipation, of which J is some measure.

The implication of the preceding discussion is far-reaching when considering use of R-curve methodology in design. The most immediate point is that since the R-curve data from small (laboratory) and large (real structure) test specimens are not the same, that is, geometric variability in R-curves; therefore, use of R-curves from small test specimens in design of large components must be made carefully. Use of normalization as detailed above may be one solution. Use of specimens that model the structure closely may be another, subject still to scale effects. Moreover, preliminary analysis shows numerical differences in the prediction of allowable crack growth, when a unique R-curve is not available (due to large growth perhaps), while leaving estimates of toughness unchanged. Further analysis is in progress.

Conclusions

1. Leaving aside the definition implied by the standard procedures, there must conceptually be agreement in the value of initiation toughness in terms of J or COD where both types of estimation procedures use a similar definition of initiation (that is, the same value of crack opening stretch). If J and COD are related through $J = m\sigma_y\delta$, the resulting value of m was 1.69 for data of HY130 steel and 1.57 for data of a titanium alloy. These did not vary beyond 18 and 19% for $\Delta a/b = 0.14$ and 0.28 for HY130 and the titanium alloy, respectively.

2. If strict size requirements are adhered to in order to maintain the characteristic meaning of J, then a unique R-curve is identifiable, within the scatter of experimental data and theoretical models for the range of bending-type geometries examined here.

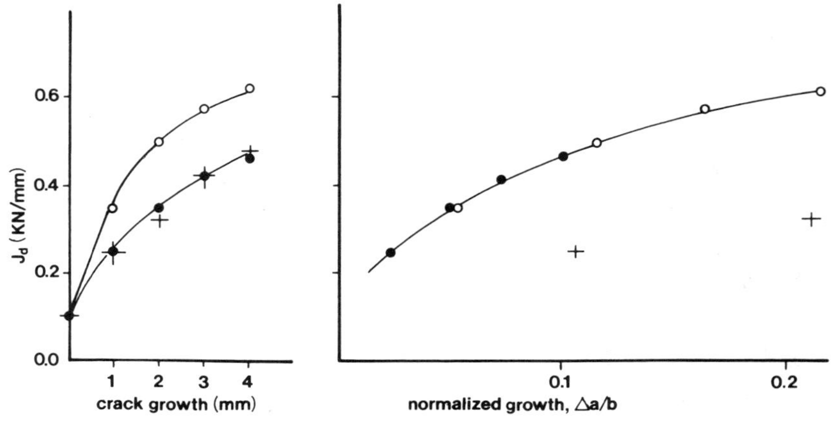

FIG. 11—*Effect of thickness for A533B; CT,* a_0/W *= 0.64 to 0.67 where* b *= 30 to 34 mm.* ○ B_n *= 18.5 mm,* ● B_n *= 40 mm* † B_n *= 9.5 mm.*

3. For intermediate and large amounts of crack growth (for example, $\Delta a/b > 0.2$) there is a divergence of R-curve data according to the method of analysis used, curves based on a truly incremental formulation or on work (rather than internal energy) rising above other definitions or both.

4. For very large amounts of crack growth or for large departures from conventional J-controlled conditions, a unique R-curve was obtained by using the concept of a normalized abscissa $\Delta a/c$, where c is that dimension which limits the amount of the plastic work dissipation.

APPENDIX I

A General Derivation for J_{TPT}

Using the fictitious loading diagram of Garwood et al. [*9*], Fig. 12, it was shown that

$$U_1 + U_4 = U_2 + U_3$$

and given $J_1 = \eta_1 U_1/Bb$, $J_2 = \eta_2 U_2/Bb_2$, and assuming J varies linearly during a small incremental growth of Δa, hence $U_3 = B(J_1 + J_2)\Delta a/2$, leading to

$$(J_1 Bb_1/\eta_1) + U_4 = (J_2 Bb_2/\eta_2) + B(J_1 + J_2)\Delta a/2$$

and therefore

$$J_2 = J_1 \frac{\{2b_1 - \eta_1 \Delta a\}2\eta_2}{2\eta_1\{2b_2 + \eta_2 \Delta a\}} + \frac{2\eta_2 U_4}{B\{2b_2 + \eta_2 \Delta a\}} \tag{20}$$

which for deep-notch three-point bend, $\eta_1 = \eta_2 = 2$, and $U_4 = \Delta U$ gives

$$J_n = J_{n-1}\left(\frac{b_n}{b_{n-1}}\right) + \frac{2\Delta U_{n,n-1}}{Bb_{n-1}} \tag{21}$$

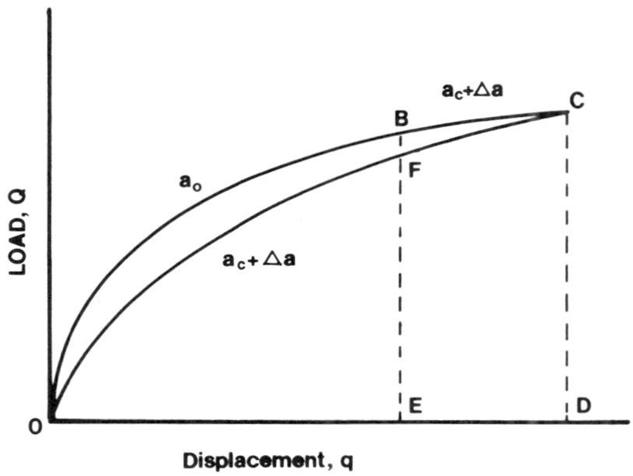

FIG. 12—*The "three parameter technique"; area OBE =* U_1, *area OFCD =* U_2, *area OBCF =* U_3, *and area BCDE =* U_4.

as reported first [9]. For any other case from Eq 20

$$\Delta J = J_2 - J_1 = J_1 \left\{ \left[\frac{\eta_2}{\eta_1} \right] \left[\frac{b_1 - \Delta a(\eta_1/2)}{b_2 + \Delta a \eta_2/2} - 1 \right] \right\} + \frac{\eta_2 U_4}{B\{b_2 + \Delta a \eta_2/2\}}$$

Assuming the incremental change in the value of a in the denominator is small gives

$$\Delta a \eta_2/2 = \text{negligible} \tag{22}$$

Therefore

$$\Delta J = J_1 \frac{1}{b_2} \left\{ \frac{b}{\eta} \bigg|_1 d\eta + (1 - \eta_2)\Delta a \right\} + \frac{\eta_2 U_4}{B b_2} \tag{23}$$

and hence

$$\Delta J_{n,(n-1)} = J_{(n-1)} \frac{\Delta a_{n,(n-1)}}{b_n} g(\eta)_n + \frac{\eta_n \Delta U_{n,(n-1)}}{B b_n} \tag{24}$$

giving Eq 25

$$J_{\mathrm{TPT}(n)} = J_{(n-1)} \left\{ 1 + \frac{\Delta a_{n,(n-1)} g(\eta)_n}{(W - a_n)} \right\} + \frac{\eta_n \Delta U_{n,(n-1)}}{B(W - a_n)} \tag{25}$$

APPENDIX II

Rising and Drooping R-Curves

The requirements for rising and drooping R-curves were derived [17], but they are refined and restated here since it is central to the present analysis. The starting point was based on the strict nonlinear elastic (NLE) definition of J for deeply cracked bend specimens (Eq 1)

$$J_{(w)} = \eta w/Bb \tag{26}$$

differentiated to give

$$dJ_{(w)}/da = (\eta/Bb) \frac{dw}{da} + (J_{(w)}/b) f(\eta) \tag{27}$$

where

$$f(\eta) = 1 + \frac{b}{\eta} \frac{d\eta}{da}$$

and using the physical relationship between work, U, and internal energy, w

$$dw = dU - BJ_{(w)} da \tag{28}$$

to state work rate as

$$dJ_{(w)}/da = (\eta/Bb) dU/da + (J_{(w)}/b) g(\eta) \tag{29}$$

where $g(\eta) = f(\eta) - \eta$. It is noted that $(dJ_{(w)}/da)$ can be zero if from Eqs 26 and 27

$$dw/da = -(w/b)f(\eta) \tag{30}$$

It follows that $(dJ_{(w)}/da)$ will remain positive (that is, a rising R-curve based on $J_{(w)}$, Eq 26) if

$$dw/da > -(w/b)f(\eta) \tag{31}$$

but will become negative (that is, a decreasing R-curve based on $J_{(w)}$ Eq 26) if

$$dw/da < -(w/b)f(\eta) \tag{32}$$

Note that for all cases of remote loading so far known, $f(\eta)$ is positive and there seems no physical reason why internal energy should not decrease with crack growth in suitable cases. It is also noted that $f(\eta)$ is a smaller quantity for compact tension specimens than for deep notch bending since $(d\eta/da)$ is slightly negative whereby $f(\eta) < 1$ for the former were as $d\eta/da = 0$ and $f(\eta) = 1$ for the latter.

If these arguments are expressed in terms of incremental work, Eq 28 giving Eq 29, then a rising R-curve, $R_{(u)}$, requires

$$dU/da \geq -(w/b)g(\eta) \tag{33}$$

Note that for all the known cases $g(\eta)$ is negative, that is, dU/da is essentially positive. This is consistent with $dU/da = 0$ for the instability condition of no external work input to cause cracking. A drooping R-curve is thus not possible when expressed in terms of $R_{(U)}$. If $R_{(u)}$ is based on the definition

$$J_{(u)} = \eta U/Bb$$

then

$$dJ_{(u)}/da = (\eta/Bb)dU/da + (J_{(u)}/b)f(\eta) \tag{34}$$

which by comparison with Eq 29 shows that

$$dJ_{(w)}/da = dJ_{(u)}/da - \eta(J_{(av)}/b)$$

where $J_{(av)} = (J_{(w)} + J_{(u)})/_2$. This is clearly indicative of why $dJ_{(w)}da$ may become negative while $dJ_{(u)}/da$ remains positive.

References

[1] Turner, C. E. in *Post-Yield Fracture Mechanics*, D. A. H. Latzko, Ed., Applied Science Publications, London, U.K., 1979, pp. 13–216.

[2] Turner, C. E. in *Post-Yield Fracture Mechanics*, 2nd ed., Elsevier Applied Science Publications, New York, 1984.

[3] Hutchinson, J. W. and Paris, P. C. in *Elastic-Plastic Fracture, ASTM STP 668*, American Society for Testing and Materials, Philadelphia, 1979, pp. 65–120.

[4] Rice, J. R., Drugan, W. J., and Sham, T. L. in *Fracture Mechanics (Twelfth Conference), ASTM STP 700*, American Society for Testing and Materials, Philadelphia, 1980, pp. 189–221.

[5] Krafft, J. M., Sullivan, A. M., and Boyle, R. W. in *Proceedings*, Crack Propagation Symposium, Cranfield, U. K., Vol. 1, 1961, pp. 8–38.

[6] Etemad, M. R. and Turner, C. E., "A Re-Interpretation of *R*-Curves Beyond J-Controlled Growth for Plane Stress and Fully Plastic Plane Strain," *International Journal of Pressure Vessels and Piping*, Vol. 20, 1985, pp. 1–5.

[7] Paris, P. C., Tada, H., Ernst, H., and Zahoor, A. in *Elastic-Plastic Fracture, ASTM STP 668*, American Society for Testing and Materials, Philadelphia, 1979, pp. 5–36 and pp. 251–265.

[8] Etemad, M. R. and Turner, C. E. in *Fracture Mechanics (Seventeenth Conference), ASTM STP 905*, J. H. Underwood, R. Chait, C. W. Smith, D. P. Wilhem, W. A. Andrews, and J. C. Newman, Eds., American Society for Testing and Materials, Philadelphia, 1986.

[9] Garwood, S. J., Robinson, J. N., and Turner, C. E., *International Journal of Fracture,* Vol. 11, 1985, pp. 528–530.
[10] Ernst, H. A., Paris, P. C., and Landes, J. D. in *Fracture Mechanics—13th Conference, ASTM STP 743,* American Society for Testing and Materials, Philadelphia, 1981, pp. 476–502.
[11] Etemad, M. R. and Turner, C. E., *Journal of Strain Analysis,* Vol. 20, No. 4, 1985, pp. 201–208.
[12] Shih, C. F., de Lorenzi, H. G., and Andrews, W. R. in *Elastic-Plastic Fracture, ASTM STP 668,* American Society for Testing and Materials, Philadelphia, 1979, pp. 65–120.
[13] John, S. J. and Turner, C. E., "Computer Interactive J Tests on Titanium Alloy (6Al-2Cb-1Ta-1Mo)," Internal Report, Imperial College, London, U.K., 1984.
[14] Wei, Y., Etemad, M. R., and Turner, C. E. in *Proceedings,* Sixth International Conference on Fracture (ICF6), Delphi, India, 1984, pp. 1103–1111.
[15] Hermann, L. and Rice, J. R. in *Proceedings,* Conference on Micro-Mechanisms of Crack Extension, Metal Society and Institute of Physics, Churchill College, Cambridge, U.K., 1980.
[16] Davis, D. A., Vassilaros, M. G., and Gudas, J. P. in *Elastic-Plastic Fracture: Second Symposium. Volume II: Fracture Curves and Engineering Applications, ASTM STP 803,* American Society for Testing and Materials, Philadelphia, 1983, pp. II-582–II-610.
[17] Etemad, M. R. and Turner, C. E., "Rising and Drooping R-Curves," *International Journal of Pressure Vessels and Piping,* Vol. 21, 1985, pp. 81–88.
[18] Garwood, S. J. and Turner, C. E. in *Proceedings,* Fourth International Conference on Fracture (ICF4), D. M. R. Taplin, Ed., 1977, pp. 279–284.
[19] McCabe, D. E. and Ernst, H. A. in *Fracture Mechanics: Fourteenth Symposium. Volume I: Theory and Analysis, ASTM STP 791,* American Society for Testing and Materials, Philadelphia, 1983, pp. I-561–I-584.
[20] Ernst, H. A. in *Elastic-Plastic Fracture: Second Symposium. Volume I: Inelastic Crack Analysis, ASTM STP 803,* American Society for Testing and Materials, Philadelphia, 1983, pp. I-191–I-213.
[21] Garwood, S. J., Ph.D. thesis, Imperial College, London, U.K., 1977.

Ludvik Hodulak[1] *and Johann G. Blauel*[1]

On the Accuracy of Ductile Failure Assessments

REFERENCE: Hodulak, L. and Blauel, J. G., "**On the Accuracy of Ductile Failure Assessments,**" *Fracture Mechanics: Eighteenth Symposium, ASTM STP 945,* D. T. Read and R. P. Reed, Eds., American Society for Testing and Materials, Philadelphia, 1988, pp. 1005–1022.

ABSTRACT: Ductile failure assessments of cracked specimens and components performed on the basis of the J-integral and the limit load concepts are discussed with particular interest in their ability and accuracy to predict actual failure loads. Possible sources of uncertainty in the modeling of material behavior and geometry are treated separately, comparing appropriate calculations with each other and with experimental data.

Especially approximate methods for evaluating J_{applied} are checked against finite-element calculations and are found accurate in all cases where plane models are relevant. The influence of the stress-strain law on the results of failure predictions is examined in detail, and a procedure is proposed for an appropriate evaluation of the Ramberg-Osgood constants in the Electric Power Research Institute estimation scheme. In addition, the influence of a variation of the material J-R curve and of the crack growth model in failure assessments is demonstrated for several examples.

KEY WORDS: failure assessment, ductile fracture, crack initiation, stable crack growth, tearing instability

The limit load concept and different concepts of elastic-plastic fracture mechanics can be used to predict the load carrying capacity of structural components with defects. An increasing number of results show that the results of fracture mechanics calculations agree better with experimental findings than limit load results do. The accuracy of a fracture-mechanics-based failure assessment depends, however, on many factors. It is the goal of this paper to analyze the influence of some of these factors on the final results.

The tearing instability concept based on the J-integral and material J-R curves [*1*] is considered in particular. It can describe the failure behavior of cracked specimens and components in the upper shelf regime of toughness. The use of this concept is no longer confined to research work since simplified conservative methods for safety assessment of pressurized components [*2*] have become available.

The capability of the tearing instability concept to predict failure was demonstrated for many cases by comparison of calculated and experimental data. One of such exercises was the international round robin on pretest estimates for the burst test of the HSST vessel ITV-8A performed at Oak Ridge National Laboratory in August 1983 [*3*]. The predictions turned out to be very good for both maximum pressure and amount of stable extension of the starter crack. Figure 1 shows the scheme of our pretest calculation as an example for applying the tearing instability concept. The main steps are as follows:

1. Material data such as initiation toughness (J_i, K_{Ji}, J_{Ic}), fracture resistance curve $J_R(\Delta a)$, and tensile data ($R_{p0.2}$, R_m) are determined in tests on small specimens.

[1]Senior scientist and head of Material and Component Assessment, respectively, Institut für Werkstoffmechanik, Wöhlerstrasse 11, D-7800 Freiburg, West Germany.

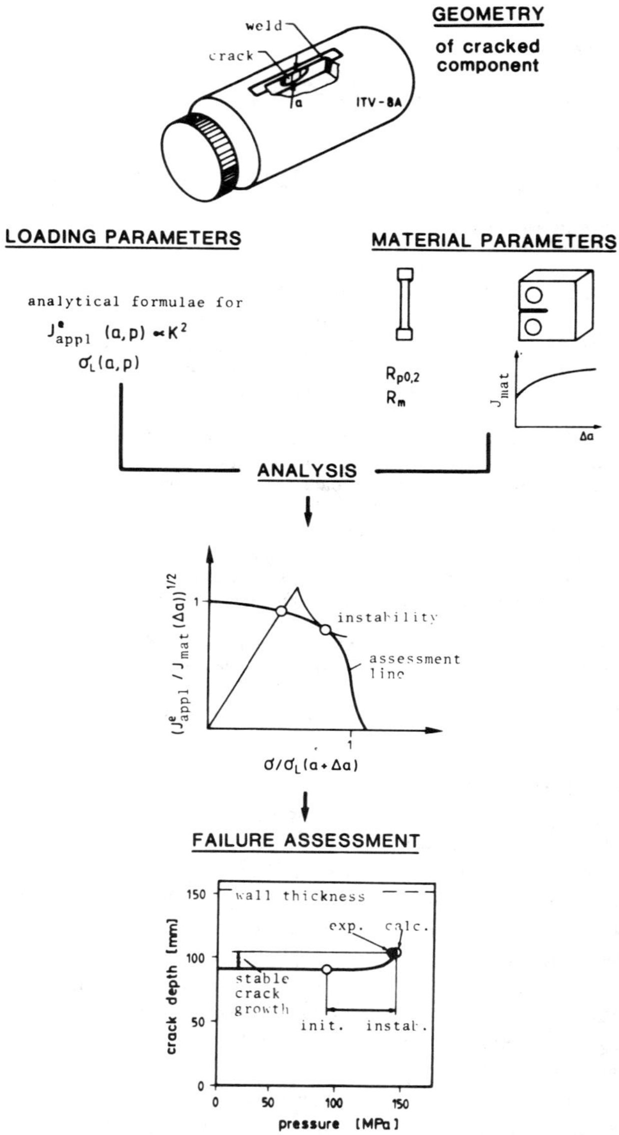

FIG. 1—*Principle of failure assessment using a J-R curve together with a failure assessment diagram* [5].

2. For the known geometry of the cracked component the loading parameter $J_{applied}$ is calculated as a function of load and crack dimensions.

3. For the analytical simulation of crack growth (including crack tip blunting) the crack is incrementally extended and in each step the load is adjusted in such a way that the condition

$$J_{applied} = J_{material}$$

is fulfilled.

4. Checking the instability criterion in each step

$$\frac{dJ_{\text{applied}}}{da} \geq \frac{dJ_{\text{material}}}{da}$$

critical values of load and crack length at instability are found.

Results of tearing instability assessments of several small- and large-scale specimens are shown in Fig. 2. The predictions have been made using the Electric Power Research Institute/General Electric (EPRI/GE) engineering approach (EA) [*4*] for calculation of J_{applied} with a J-R curve from a side-grooved compact tension (CT50) specimen. The ratio of the calculated instability load to the measured failure load for all specimens tested is somewhat higher than 1. It will be shown later that this systematic overestimation can be attributed to the plane-strain model used for the analysis of all specimens. Again the predictions made on the basis of limit load calculations are worse. The limit load method in particular does not account for the specimen size and stable crack growth in advance, whereas any tearing instability method does.

The two-dimensional modeling of the specimen geometry is not the only assumption causing uncertainty in tearing instability analyses; Table 1 gives others.

Despite careful measurements there will always be some scatter of the J_{material} (Δa) data as a consequence of experimental errors and of material inhomogeneity, which will influence the results of tearing instability analyses. Dependence of J_{material} on geometry (specimen type and size) is another source of inaccuracy.

The loading parameter J_{applied} will depend on the idealizations which necessarily have to be made in describing real cracked bodies, for example, the assumption of simple contours for the component, of semi-elliptical crack shapes, or of one of the limiting conditions of plane strain or plane stress.

Another influencing factor is the method of calculation, for example, complete finite element analysis (FE) or approximate analytical methods like the EA scheme; here the material stress-strain behavior, the accuracy of the data, and the way they are introduced in the calculations play an important role. Finally, the procedure of simulating crack growth, for example, taking into account one- or two-dimensional growth or using different step sizes in a step-by-step analysis, may influence the results. This paper investigates in detail some sources of uncertainty and inaccuracy.

For an approximate analysis the CEGB-R6 Routine failure assessment diagram (FAD) with an assessment line modified for strain hardening by Milne [*5*], taking into account stable crack growth, may also be used. Strictly speaking this method is not based on the J-integral concept and tearing instability but it can be interpreted in terms of them [*6–7*]. Although failure loads obtained by the FAD method will depend on the evaluation of the limit load, they agree better with measured failure loads than just collapse load predictions. This is demonstrated by results for pressure vessels with axial surface cracks given in Table 2.

Influence of Material Input Data on Accuracy of Failure Assessment

Stress-Strain Law

The value of the loading parameter J_{applied} depends strongly on the stress-strain relation of the material whenever extended plastic deformation occurs. In addition, J_{applied} cannot be always calculated on the basis of an accurate stress-strain relation, since either this relation is not known or the calculation method is too expensive or does not allow for a direct use. An example for the latter case is the evaluation of J_{applied} using the EA where only specific stress-strain functions, particularly the Ramberg-Osgood relation, can be chosen

$$\epsilon/\epsilon_0 = \sigma/\sigma_0 + \alpha(\sigma/\sigma_0)^n$$

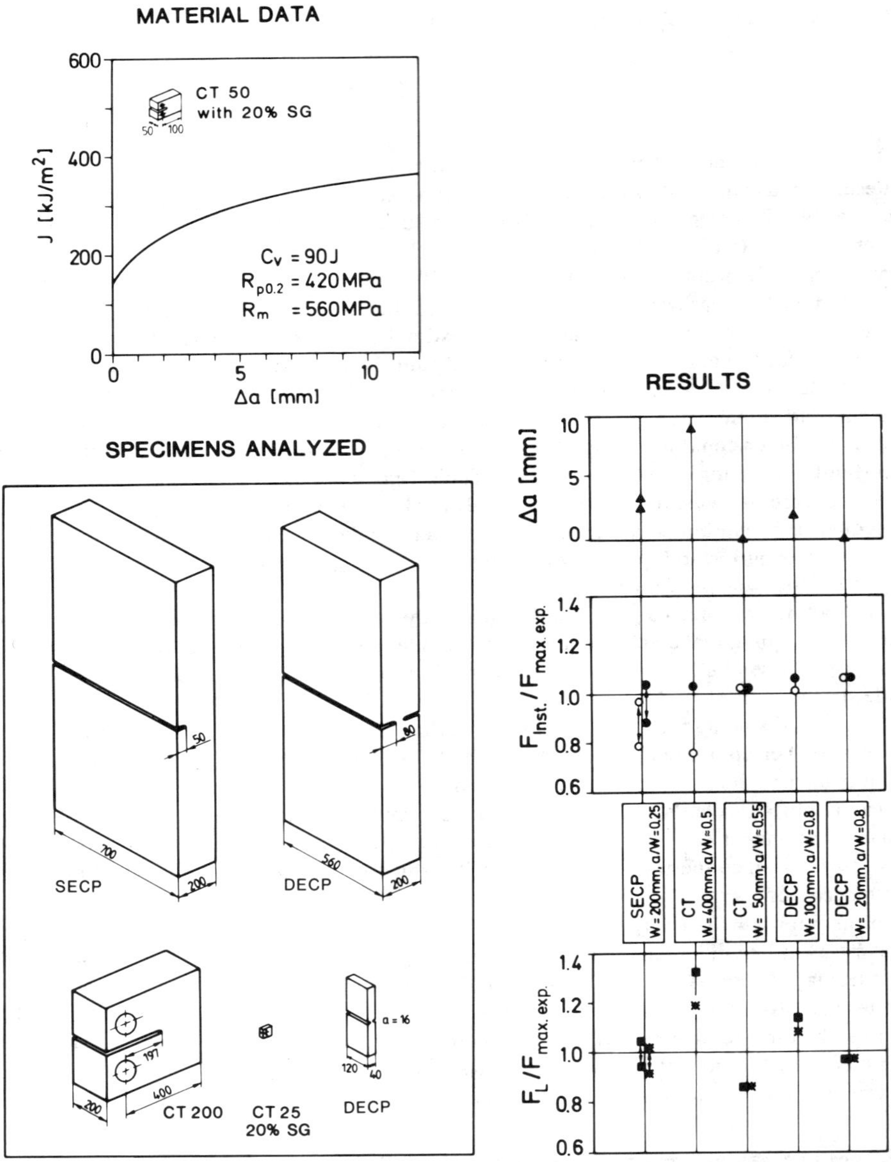

FIG. 2—*Comparison of calculated critical loads with measured maximum loads* F_{max}*;* F_{Inst} = *tearing instability load (EA);* $F_L(R_{p0.2})$ = *limit load;* Δa = *predicted stable crack extension.*

Early it was noticed [*11*] that no guides in the EA scheme demonstrate how stress-strain curves with pronounced Lüder's strains should be approximated by a Ramberg-Osgood relation, and it was shown [*12*] that this uncertainty in the approximation could severely influence failure predictions for large-scale specimens. In some investigations the agreement between calculated tearing instability and measured failure loads was taken as a criterion for a proper

TABLE 1—*Sources of inaccuracy in tearing instability analyses.*

$J_{material}$	material inhomogeneity	experimental scatter	geometry dependence of J-R curve
$J_{applied}$	structural modeling of cracked body	material modeling (stress-strain curve)	method of calculation
instability analysis	modeling of crack growth	choice of method	. . .

approximation of the material stress-strain behavior. Calculations are compared in this paper in order to eliminate possible sources of inaccuracy.

Two types of specimens in two sizes were chosen to achieve some generality of the results: CT specimens with and center-cracked tension panels (CCTP) without bending and showing different amounts of plastic deformation at failure depending on their size. Plane-strain calculations with the finite-element code ADINA extended for simulation of J-controlled crack growth (FE) [*13*] and with the EA method show (Fig. 3) that the load-displacement curves obtained by both methods are similar, provided the same stress-strain description [Ramberg-Osgood (R-O) law] is used. This is also true for the evaluation of the J-integral (see also Ref *14*).

An "original" true stress-strain curve with a pronounced Lüder's strain and two approximations of it as shown in Fig. 4 (different α and n-values for $\sigma_o = R_{p0.2}$) were used in further calculations: Curve 2 closely describes the elastic part of the original Curve 1 and underestimates strains in the hardening regime; Curve 3 closely describes the strain hardening part of the original Curve 1 and overestimates strains in the elastic part. Calculations using these different descriptions $\sigma(\epsilon)$ in the EA scheme are compared to complete FE results using the measured stress-strain relation No. 1 for two sizes of CCT panels in Fig. 5: For the large specimen, R-O-description 2 results in a better approximation in terms of F/V and J/F-diagrams than description 3; for the smaller specimen, however, the R-O description 3 delivers a better estimation of the load carrying capacity than description 2.

As a key to a better understanding of the influence of the stress-strain relation on deformation and fracture behavior of cracked specimens and components, normalized plots as in Fig. 6 may be used: through the normalization, unique load displacement curves for specimens of different size result in the FE and the EA analysis. But the points for initiation of stable crack growth ($J_{appl} = J_i$) and for instability ($F = F_{max}$) for specimens of different size lie on different parts of these curves. The shape of these "failure curves" and their mutual position look very much like those of the corresponding normalized stress strain curves.

Particularly, the R-O description 2 which underestimates stresses below and overestimates those above yield (see Fig. 4) delivers EA results underestimating the load carrying capacity of the specimens (given by the FE results) for loads below and near limit load and overestimating it in the regime above limit load.

The same kind of similarity can be seen between the normalized stress-strain and the normalized load J-integral curves (right part of Fig. 6) with a crossover again close to limit load.

Qualitatively the same behavior was also found for another specimen type. Therefore an optimum approximation of a stress-strain curve with Lüder's strain by the Ramberg-Osgood relation for fracture analysis by the EA method will depend on specimen geometry and material. Similarity considerations as in Fig. 6, however, suggest a simple stepwise procedure (see also Ref *15*):

Step 1: Use Ramberg-Osgood relation defined by points of the original true stress-strain curve

$$\epsilon_1 = R_{p0.2}/E + 0.002 \qquad \sigma_1 = R_{p0.2}$$

TABLE 2—*Results of failure evaluations for pressure vessels with external surface defects; crack growth only in the depth direction considered.*

	Vessel					
	HSST ITV-8A	EGF RR	MPa BVZ-22	MPa BVZ 30	MPa BVS 20	MPa BVS 30
Reference	[3]	[9]			[8]	
Vessel diameter/wall thickness, mm	686/152	1015/84			703/47.5	
Crack (notch) depth a/length $2c$, mm	91/279	66/200	38.3/728	36.2/1500	37.3/709	35/1100
Material	SA 533, weld	SA 533	20 MnMoNi 55		22 NiMoCr 37 mod.	
Charpy energy, J	59	199	200		51	
Test temperature, °C	150	40	305	300	320	305
Maximum pressure measured, P_{max}, MPa	140	80.5	21.9	19.5	14.8	13.1
$^{a}P_{Inst}/P_{max}$	1.04	0.99	1.15	1.14	1.12	1.12
$^{a}P_{coll}P/_{max}$	1.13	0.93	1.08	1.34	1.64	1.89
$^{b}P_{Inst}/P_{max}$	0.69	0.81	0.71	0.82	0.93	1.04
$^{b}P_{coll}/P_{max}$	1.04	0.77	0.67	0.82	0.96	1.21

$^{a}\sigma_{collapse} = R_{p0.2}\,(t/a - \pi/4)/[t/a - \pi/(4m)]$.

$^{b}\sigma_{collapse} = R_{p0.2}\,(t/a - 1)/[t/a - (1/m)]$ [*10*].

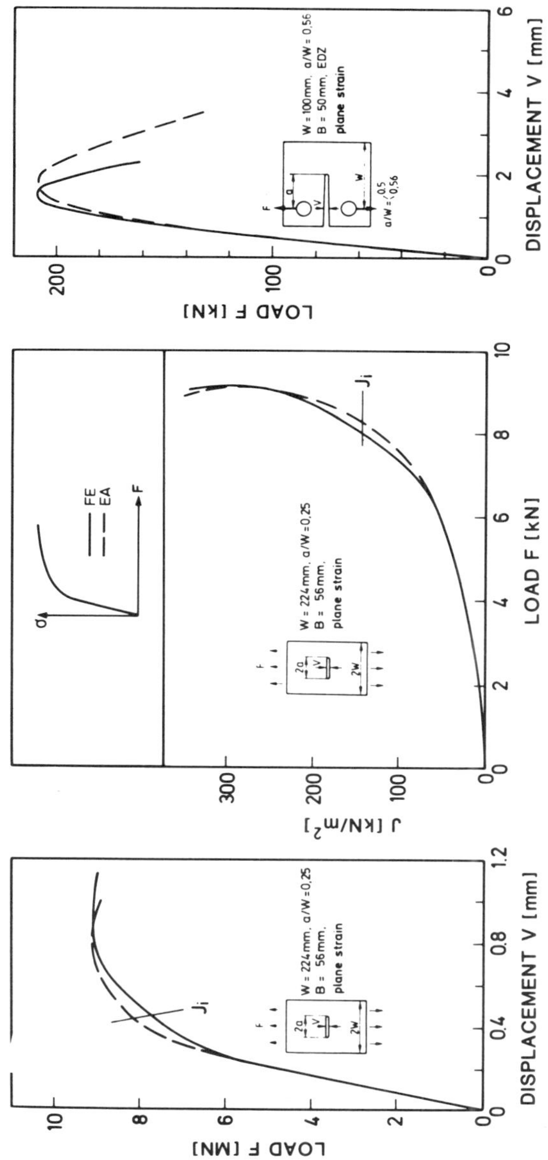

FIG. 3—*Examples of results calculated by FE and EA methods.*

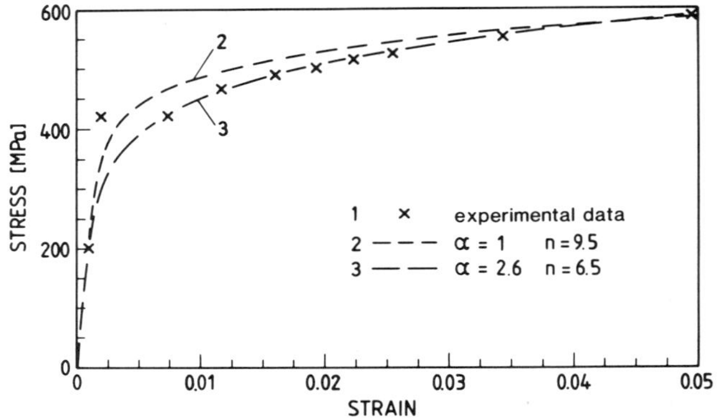

FIG. 4—*A true stress-strain curve and two Ramberg-Osgood approximations.*

$$\epsilon_2 = \epsilon(R_m) \qquad \sigma_2 = R_m$$

If calculated load $F_{\text{Inst}} \leq F_{\text{Limit}}$, the approximation is appropriate; otherwise proceed to Step 2.

Step 2: Use Ramberg-Osgood relation defined by points

$$\epsilon_1 = \epsilon_L \qquad \sigma_1 = R_{p0.2}$$

$$\epsilon_2 = \epsilon(R_m) \qquad \sigma_2 = R_m$$

where ϵ_L is the maximum strain at $R_{p0.2}$.

This two-step approximation in which either the elastic or the strain hardening part of the stress-strain curve is approximated is a conservative procedure as concerns load. If the specimen considered fails at a load equal or slightly higher than limit load a more realistic set of Ramberg-Osgood constants will be obtained by an average approximation using points

$$\epsilon_1 = (R_{p0.2}/E + 0.002 \times \epsilon_L)^{1/2} \qquad \sigma_1 = R_{p0.2}$$

$$\epsilon_2 = \epsilon(R_m) \qquad \sigma_2 = R_m$$

In the described procedure all input values can be obtained from standard tension tests with the exception of strain ϵ_2 at ultimate strength. The value of ϵ_2 is needed only for calculation of R-O constant n, since α follows from $[\epsilon_1, \sigma_1]$. An evaluation scheme for n also based only on standard tensile data was proposed in Ref *16*

$$R_{p0.2}/R_m = (n \cdot e \cdot \ln(1 + \epsilon_{1,pl})^{1/n} / (1 + \epsilon_{1,pl})$$

$$\epsilon_{1,pl} = 0.002$$

and may be used in the stepwise approximation procedure. A similar procedure was proposed in Ref *17*.

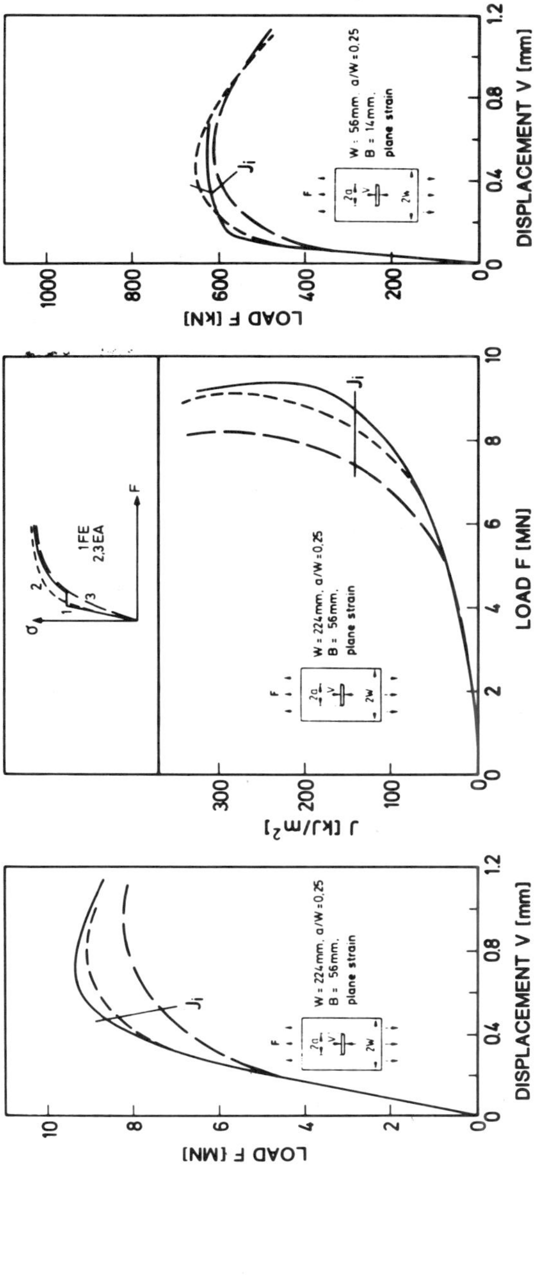

FIG. 5—*Influence of the stress-strain law on load-displacement and* $J_{applied}$*-load curves for center-cracked panels* ($\sigma_o = 450$ *MPa*).

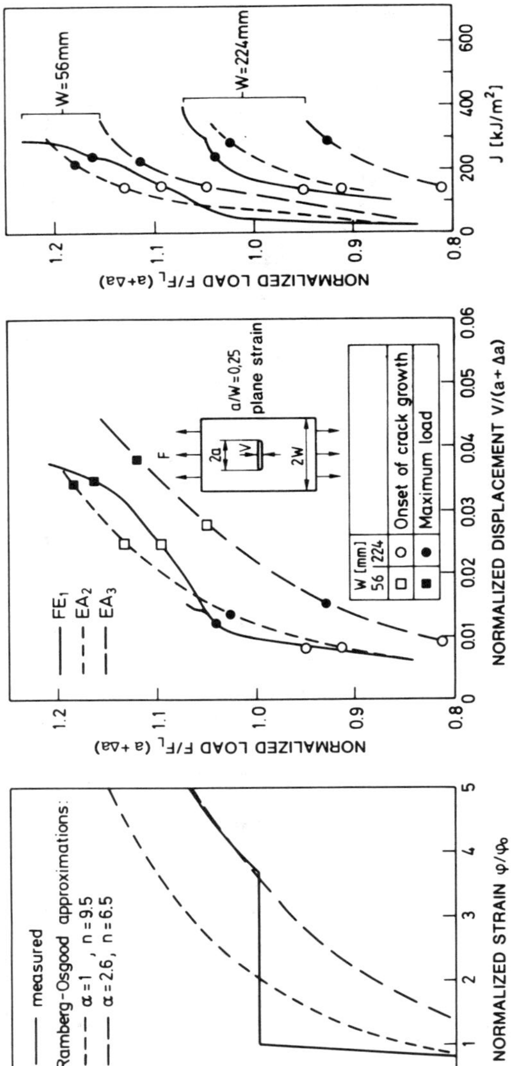

FIG. 6—*Normalized stress-strain, load-displacement, and load-J curves for center-cracked panels* [15].

Material J-R Curve

Besides the stress-strain data J-resistance curves [$J_{material} = J_R(\Delta a)$] are necessary input data for tearing instability analyses. Two examples are chosen to show the sensitivity of calculated critical loads to a variation of these material data.

In the first example different individual J-R curves determined for the same material 20 MnMoNi 55 at 80°C (Fig. 7*a*) are used to predict critical loads for a large scale CCTP and a CT specimen. The J-R curves were measured on smooth and side-grooved specimens with thicknesses ranging from 10 to 48 mm and having also different in plane dimensions [*18*]. A lower bound is formed by the 25-mm side-grooved and the smooth 48-mm CT specimens. Critical loads calculated by the EA method for $J_{applied}$ reaching values of J_i (true initiation) J^{Loss} and J^{ASTM} (technical initiation) and instability of the individual material curves are plotted in Figs. 7*b* and 7*c*. The loads F are normalized by the average load F_i for crack initiation ($J_{applied} = J_i$).

Critical loads obtained from both technical definitions of J_{Ic} are very close and appreciably higher than critical loads resulting from J_i. The scatter of critical loads due to different J-R curves used is only about $\pm 5\%$. The amount of stable crack growth shows more scatter.

In the second example the material data measured on three trial welds are used to calculate the growth of a part-through crack in the weld seam of the Heavy Section Steel Technology (HSST) intermediate vessel ITV-8A (see Fig. 1). As shown in Fig. 8, the difference in calculated maximum pressure is about 5% when using the J-R curves and the tensile data of welds V 852, V 8102, and the stable crack growth varies between 12 and 16 mm.

Influence of Structural Modeling on Accuracy of Failure Assessment

In calculating the tearing instability load for very thick large-scale specimens it seems natural to assume conditions of plane strain. Table 3 gives results of different authors using this assumption in FAD analyses of the double-edge notched specimen KS 01 AA, W = 200 mm, $a/W = 0.8$, and $B = 560$ mm (see insert in Fig. 2), tested at MPA Stuttgart [*19*]. There is some scatter because different J-R curves and different limit load formulae have been used. But all analyses overestimate the experimental findings. The evaluation of the notched specimens with J-R curves generated on precracked specimens is justified by negligible differences of J-R curves measured so far on cracked and notched compact specimens from this material. To check whether the overestimation is a consequence of the approximate method used, FE-calculations were performed for differently sized double-edge cracked (DEC) specimens [*20*].

In Fig. 9 measured and FE-calculated load-displacement curves are compared for a thin and a thick specimen from 20 MnMoNi 55 steel tested at 80°C. In the plane-stress calculation for the thin specimen, a moderate overestimation of the maximum load is found. This is presumably a result of a moment having existed in the experiment as indicated by the nonsymmetry of the COD values V_1 and V_2 which has not been taken into account in the analyses.

For the second DEN specimen with nearly 50 times larger thickness, the plane-stress solution is not appropriate at all and the plane-stress model clearly overestimates the measured F/V behavior (see Fig. 9, right). To prove the mixed state of stress in the specimen, an additional FE calculation was done modeling the crack front region in 3D and assuming plane stress for the rest of the specimen. Now the experimental results are well reproduced in the elastic regime, and they fall between the 3D/2D model and the plane-strain model for the regime of plastic deformations and stable crack growth. But it is suspected that in the 3D/2D analyses the part of the specimen modeled in 3D was not chosen large enough. Therefore, the loads may be underestimated, and the 3D/2D model seems to be correct for this specimen.

If conservative solutions from plane-stress models are not sufficient and three-dimensional solutions are too expensive to evaluate the load-carrying capacity, simple correlations between specimen geometry and relevant stress state may be helpful. This is why series of experiments on DEN specimens of different size and thickness and on compact specimens of different size are

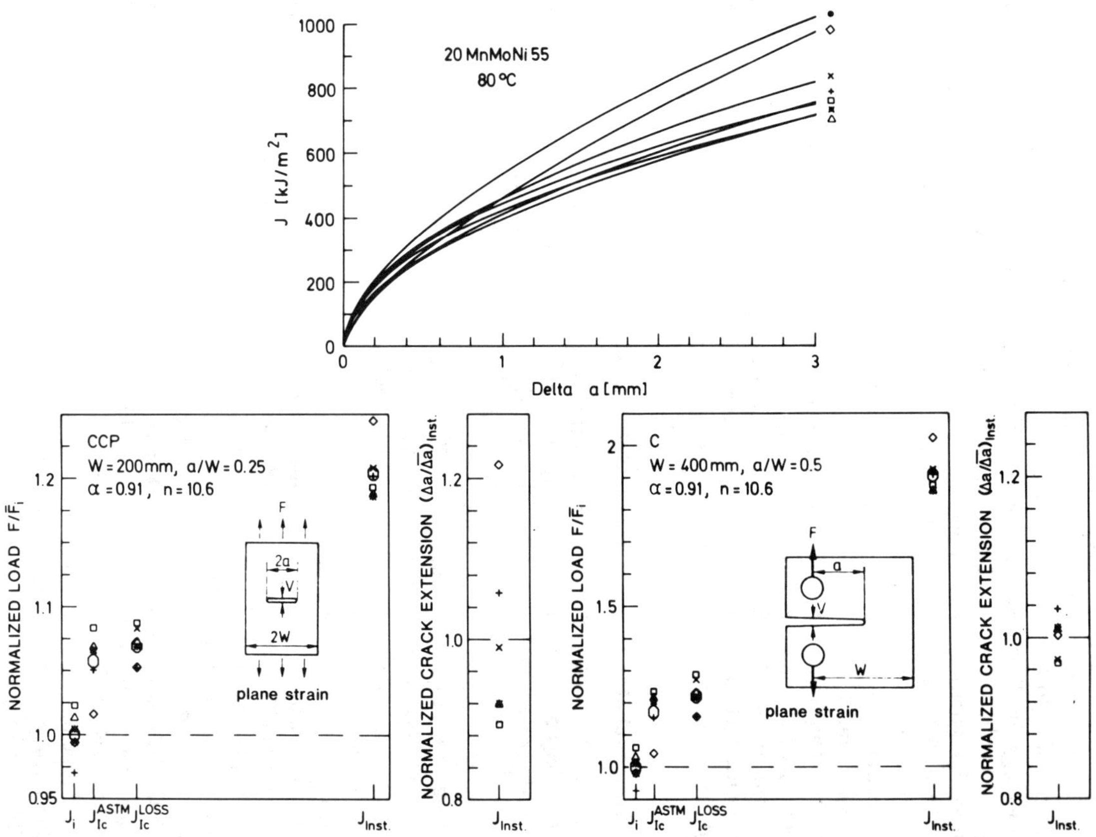

FIG. 7—*Influence of the variation of J-R curve* [18] *and the procedure for* J_{Ic} *determination on initiation and instability loads of two large-scale specimens (EA calculation).*

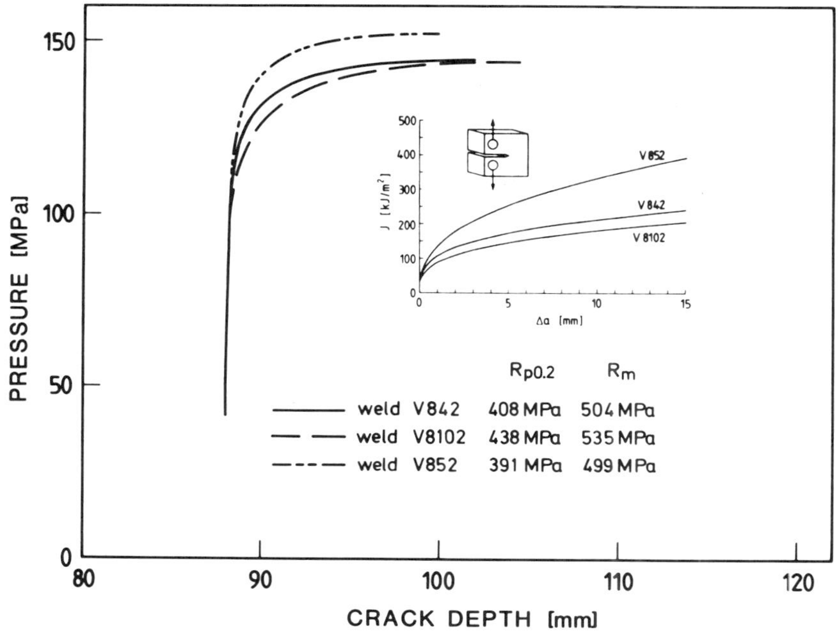

FIG. 8—*Influence of material properties on results of post-test estimates for HSST vessel ITV 8A. FAD calculation according to scheme in Fig. 1.*

TABLE 3—*Tearing instability load* F_{Inst} *normalized by the experimental load* F_{max} *for DEN-specimen KS 01 AA; analysis by FAD-method.*

	Krägeloh et al. [*19*]	Milne [*21*]	Hodulak and Blauel [*22*]
$F_{\text{Inst}}/F_{\text{max}}$	1.05	1.16 $[\sigma_F = (R_{p0.2} + R_m)/2]$ 1.00 $(\sigma_F = R_{p0.2})$	1.20 $[\sigma_F = (R_{p0.2} + R_m)/2]$

evaluated in the following way. In Fig. 10, tearing instability loads F_{Inst} calculated by the EA method are plotted versus the geometry parameter B/a for DEC specimens from Ref *23*. The normalization factor F_{max} is the experimentally determined maximum load. Despite of a size ratio of 9, the results for small and large specimens are almost identical, and a uniform decrease is found for increasing B/a in plane strain or plane stress. Similar correlations are obtained if the geometry parameters B/W or $B/(W - a)$ are used. The correlations seem to be independent of material toughness since two points for a brittle material with $C_v = 90$ J can be found on the direct prolongation of the curve generated for the "tough" material with $C_v = 160$ J.

In Fig. 11, corresponding results for smooth (without side grooves) compact specimens from various steels are shown. Since most of them are standard specimens with $B/W = 0.5$ and $a/W = 0.5$ or 0.6, a dependence on B/a is less well established. For standard specimens the plane-strain instability loads overestimate experimentally determined maximum loads by 3 to 20%.

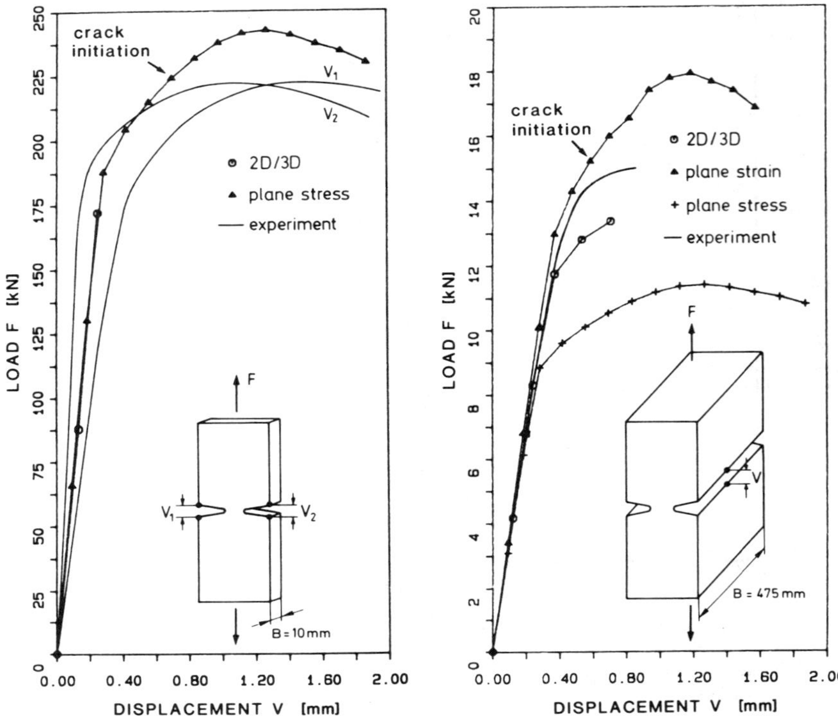

FIG. 9—*Load-displacement curves for a thin (*B = *10 mm) and a thick (*B = *475 mm) DECP of 20 MnMoNi 55 steel tested at 80°C. Finite-element calculations* [20] *with crack extension controlled by J-R curves generated on precracked compact specimens.*

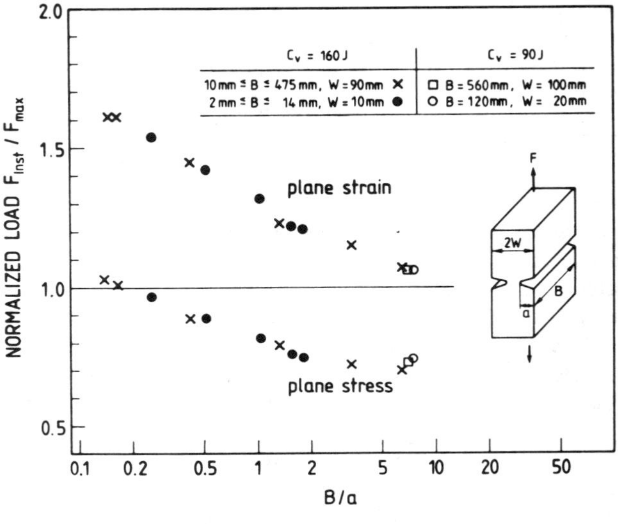

FIG. 10—*Comparison of calculated (EA method) instability loads* F_{Inst} *with experimental failure loads* F_{max} *for double-edge-cracked panels of different sizes.*

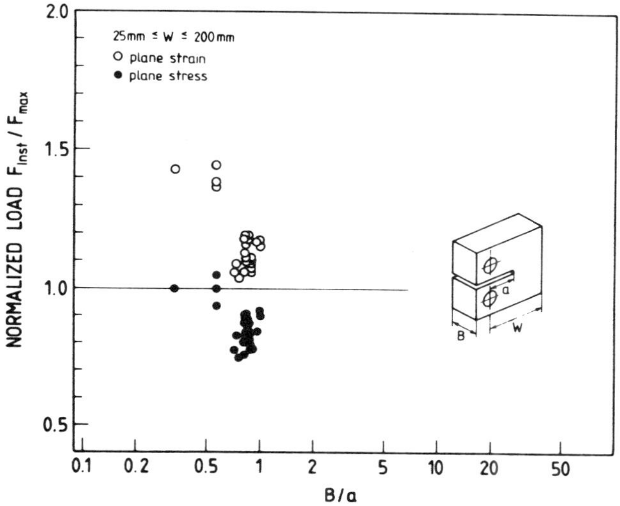

FIG. 11—*Comparison of calculated instability loads (EA method)* F_{Inst} *with experimental failure loads* F_{max} *for smooth-sided compact specimens from six different steels.*

In Fig. 12, experimentally determined maximum stresses ($\sigma_{max} = F_{max}/(B \times W)$) for compact specimens of a pressure vessel steel tested at 150°C are plotted as a function of normalized crack size a/W and compared with tearing instability stresses calculated for plane strain and plane stress. The dependence of σ_{inst} and σ_{max} on a/W has the same tendency. The values of σ_{max} for side-grooved specimens are shifted towards plane strain.

The results of this comparison of experimentally determined maximum loads with calculated plane-strain and plane-stress instability loads in Figs. 9 to 11 are encouraging although the variation of geometry parameters is still limited. They indicate that simple correlations between specimen geometry and relevant stress state can be found and may be used to improve calculational predictions. Neither plane stress nor plane strain works consistently—fracture is a 3D problem.

Influence of Crack Growth Model on Accuracy of Failure Assessment

In calculations of the stable growth of part-through cracks, the crack growth law has to be applied locally. To avoid finite-element calculations semi-elliptical crack shapes are assumed and averaged crack extensions are calculated for the length and depth direction only. Material resistance against crack growth, however, may vary along the crack front due to variations of the constraint conditions: Since presently no methods are available to measure J-R curve for different well-defined constraint conditions, the same J-R curve has to be used for both directions or assumptions have to be made for the lateral as compared to the through-thickness crack growth. The influence of such assumptions is shown again for the example of vessel ITV-8A (Fig. 13).

The solid line in Fig. 13*a* represents results of a calculation, in which lateral crack extension is calculated using the same J-R curve as for the depth direction; for the three dashed lines, different assumptions for the lateral crack growth have been made. Obviously the highest maximum load predicted to be carried by the vessel is obtained for constant crack length. This load is by 6% higher than that for the model attempting to match the final canoe-shaped crack

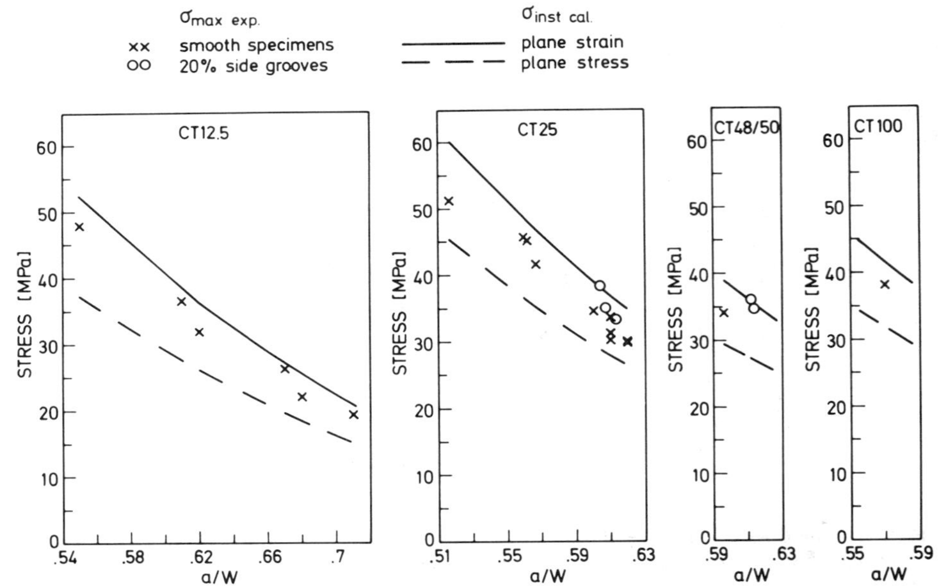

FIG. 12—*Comparison of calculated tearing instability stresses* σ_{Inst} *(EA method) with measured maximum stresses* σ_{max} *for compact specimens of different size (CT 25 = 25-mm-thick standard specimen) from a 20 MnMoNi 55 steel tested at 150°C.*

found on the fracture surface (Fig. 13*b*) by assuming $\Delta c = 2 \times \Delta a$ during stable crack growth. For comparison the instability point determined experimentally is also shown in Fig. 13*a*.

Conclusions

The investigations of the accuracy of ductile failure assessments show:

1. Fracture mechanics based tearing instability analyses give an improved description of component behavior compared with just limit-load analyses.
2. The calculated tearing instability load is sensitive to the stress-strain curve used as an input. Where the real curve cannot be used (for example, in EA method) an approximation of a specific part of it (dependent on the case investigated) is more appropriate than an average overall fitting. A proposal for a stepwise approximation is given which is based on the interpretation of normalized failure curves.
3. The influence of a possible variation of material J-R curve on tearing instability loads was found small for a homogeneous material. More inaccuracy has to be expected for the assessment of cracks in welds.
4. Most actual structures and laboratory specimens were not modeled accurately either by plane strain or plane stress. Therefore tearing instability load can be assessed with sufficient accuracy by approximate methods like EA and FAD only in cases where plane-strain or plane-stress models are relevant. With a plane-strain model some nonconservatism is possible even for extremely thick specimens. A possible uncertainty about the state of stress is the major factor of inaccuracy in tearing instability analyses and can be resolved only by three-dimensional FE calculations. Experimental correlations of size parameters and failure loads can be helpful in establishing rules for a possible use of two-dimensional models.

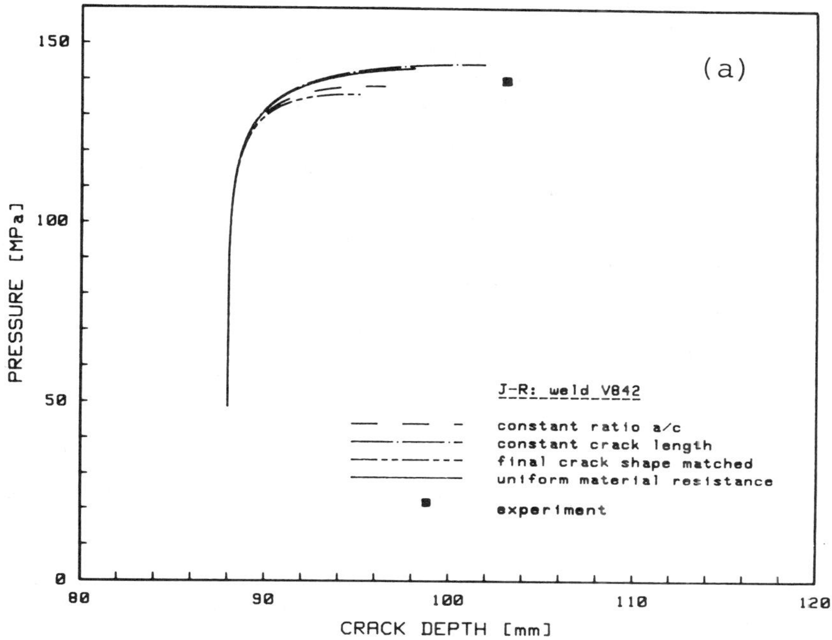

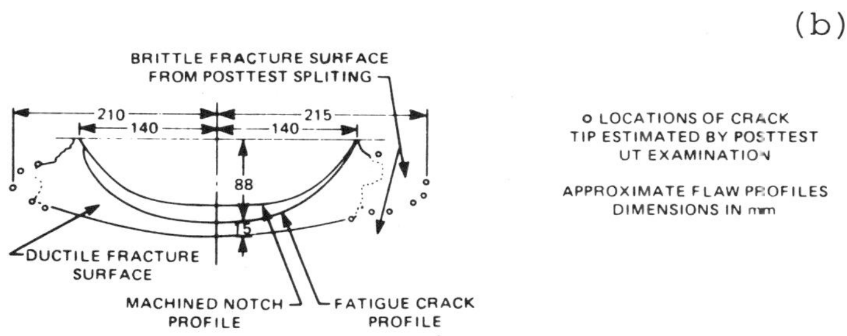

FIG. 13—(a) *Influence of crack growth model on results of post-test estimates for HSST vessel ITV-8A.* (b) *Real crack profiles in ITV-8A.*

5. A detailed modeling of the real crack growth in a component with surface crack is essential for a good prediction of its load-carrying capacity. A better understanding of predominant lateral crack growth as observed in ITV vessel 8A is required, taking into account the variation of constraint along the crack front.

References

[*1*] Paris, P. C. et al., "The Theory of Instability of the Tearing Mode of Elastic-Plastic Crack Growth" in *Elastic-Plastic Fracture, ASTM STP 668,* American Society for Testing and Materials, Philadelphia, 1979, pp. 5–36.

[*2*] Paris, P. C., "A Method of Application of Elastic-Plastic Fracture Mechanics to Nuclear Vessel Analysis," NUREG Report 0744, Nuclear Regulatory Commission, Washington, DC, Vol. 1, Revision 1, Appendix B, 1981.

[*3*] Merkle, J. G., "Preliminary Results of International Round Robin on ITV-8A," NUREG Report CP-0041, Nuclear Regulatory Commission, Washington, DC, Vol. 4, 1983.

[4] Kumar, V., German, M. D., and Shih, C. F., "An Engineering Approach for Elastic-Plastic Fracture Analysis," EPRI Report NP-1931, Electric Power Research Institute, Palo Alto, CA, July 1981.

[5] Milne, I., "Failure Assessment Diagrams and J Estimates: A Comparison for Ferritic and Austenitic Steels," *International Journal of Pressure Vessels and Piping,* Vol. 13, 1983, p. 107.

[6] Bloom, J. M., "Elastic-Plastic Failure Analysis" in *Proceedings,* 6th Materialprüfanstalt Seminar, Stuttgart, West Germany, Oct. 1980.

[7] Bloom, J. M., "Validation of a Deformation Plasticity Failure Assessment Diagram Approach to Flaw Evaluation" in *Elastic-Plastic Fracture: Second Symposium, Volume II: Fracture Curves and Engineering Applications, ASTM STP 803,* American Society for Testing and Materials, Philadelphia, 1983, pp. II-206–II-238.

[8] Sturm, D. et al., "Fracture Initiation and Fracture Opening under Light Water Reactor Operational Conditions" in *Proceedings,* 7th Materialprüfanstalt Seminar, Stuttgart, 8–9 Oct. 1981.

[9] Milne, I., "Preliminary Report on EGF Task Group 1 Exercise in Predicting Ductile Instability: Phase II, Experimental Cracked Pressure Vessel," European Group on Fracture Meeting, Paris, 27–28 Feb. 1984.

[10] Kiefner, J. F. et al., "Failure Stress Levels of Flaws in Pressurized Cylinders" in *Fracture Toughness Testing, ASTM STP 536,* American Society for Testing and Materials, Philadelphia, 1973, pp. 461–481.

[11] Issler, L., Birk, R., and Hund, R., "Application of a New Estimation Technique from the USA for the Prediction of Ductile Fracture" in *Proceedings,* 6th Materialprüfanstalt Seminar, Stuttgart, West Germany, Oct. 1980.

[12] Hodulak, L. and Blauel, J. G., "Application of Two Approximate Methods for Ductile Failure Assessment" in *Elastic-Plastic Fracture, Second Symposium, Volume II: Fracture Curves and Engineering Applications, ASTM STP 803,* American Society for Testing and Materials, Philadelphia, 1983, pp. II-103–II-114.

[13] Siegele, D., Kordisch, H., and Schmitt, W., "J-Integral-Berechnung für rißbehaftete Strukturen unter Temperatur- und Innendruckbelastung" in *Proceedings,* 16th Sitzung des Arbeitskreises Bruchvorgänge, Karlsruhe, West Germany, 1984.

[14] Hodulak, L. and Zhang, J., "Einfluss des Werkstoffgesetzes auf die Ergebnisse von Zähbruchanalysen," IWM Report W 1/83, Institut für Werkstoffmechanik, Freiburg, West Germany, Jan. 1983.

[15] Hodulak, L., Voss, B., and Zhang, J., "Zähbruchanalyse mit 'engineering approach' und FE-Berechnungen—Einfluss der Beschreibung der Spannungs-Dehnungs-Kurve" in *Proceedings,* 15th Sitzung des AK Bruchvorgänge, Darmstadt, West Germany, Feb. 1983.

[16] Voss, B. and Blauel, J. G., "Ersatzverfahren zur Ermittlung der Risszähigkeit K_{Ic}" in *Proceedings,* 12th Sitzung des AK Bruchvorgänge, Freiburg, West Germany, Oct. 1980.

[17] Bloom, J. M. and Malik, S. N., "A Procedure for the Assessment of the Integrity of Nuclear Pressure Vessels and Piping Containing Defects," EPRI Report NP-2431, Electric Power Research Institute, Palo Alto, CA, June 1982.

[18] Blauel, J. G. et al., "Material Characterization by J-R Curves for Component Safety Assessment—A Study on a 20 MnMoNi 55 Forging," *International Journal of Pressure Vessels and Piping,* Vol. 17, 1984, pp. 139–162.

[19] Krägeloh, E., Issler, L., and Zirn, R., "Large Scale Specimen Tests Analysed by Elastic-Plastic Fracture Concepts" in *Proceedings,* 6th Materialprüfanstalt Seminar, Stuttgart, West Germany, Oct. 1980.

[20] Hodulak, L., Zhang, J., and Schmitt, W., "Nachrechnung zum Versagensverhalten von DECP-Grossproben KS 14 G1 und KS 14 G7," IWM Report W 6/83, Institut für Werkstoffmechanik, Freiburg, West Germany, Nov. 1983.

[21] Milne, I., "Experimental Validation of Resistance Curve Analysis" in *Elastic-Plastic Fracture: Second Symposium, Volume II: Fracture Curves and Engineering Applications, ASTM STP 803,* American Society for Testing and Materials, Philadelphia, 1983, pp. II-657–II-685.

[22] Hodulak, L. and Blauel, J. G., "Ergänzende Untersuchungen zur Nachrechnung von KS 01-Proben," IWM Report V 27/80, Institute für Werkstoffmechanik, Freiburg, West Germany, Dec. 1980.

[23] Roos, E., "Erweiterte experimentelle und theoretische Untersuchungen zur Quantifizierung des Zähbruchverhaltens am Beispiel des Werkstoffs 20 MnMoNi 55," MPA Report 82-01, Materialprüfanstalt, Stuttgart, West Germany, 1982.

Fatigue II

Steven J. Gill,[1] Joseph A. Hauser, II,[1] Thomas W. Crooker,[2] Brian J. Kruse,[3] Ravi Menon,[4] and Carl D. Lundin[4]

Fatigue Behavior of 5Ni-Cr-Mo-V Steel Weldments Containing Fabrication Discontinuities

REFERENCE: Gill, S. J., Hauser, J. A., II, Crooker, T. W., Kruse, B. J., Menon, R., and Lundin, C. D., **"Fatigue Behavior of 5Ni-Cr-Mo-V Steel Weldments Containing Fabrication Discontinuities,"** *Fracture Mechanics: Eighteenth Symposium, ASTM STP 945,* D. T. Read and R. P. Reed, Eds., American Society for Testing and Materials, Philadelphia, 1988, pp. 1025–1049.

ABSTRACT: Fabrication discontinuities in weldments pose a serious engineering problem for structural applications involving cyclic loading. Often, the discovery of fabrication discontinuities through nondestructive inspection can pose a dilemma: if not repaired, failure can result from crack initiation and growth; however, if improperly repaired, post-repair discontinuities of even greater severity may result. Thus, rational decision-making criteria are needed to assess the mechanical severity of actual fabrication defects in weldments. This study was undertaken to explore the applicability of linear elastic fracture mechanics to characterize the fatigue behavior of high-strength steel weldments containing lack-of-penetration (LOP) and slag/lack-of-fusion (S/LOF) discontinuities. Full penetration, double-V butt welds with reinforcements removed were tested under zero-to-tension axial loading. Various filler metals and welding techniques were used. Both sound welds and welds containing discontinuities were cycled to failure. Whenever possible, cycles to crack initiation were estimated by strain gage measurements. The fracture mechanics approach was successful in correlating the fatigue lifetimes of specimens containing single LOP discontinuities of varying size. However, the fatigue behavior of specimens containing multiple S/LOF discontinuities proved to be much more complex and difficult to analyze.

KEY WORDS: fatigue, fracture mechanics, welding, discontinuities, lack-of-penetration, slag/lack-of-fusion, 5Ni-Cr-Mo-V steel, fatigue crack initiation, fatigue crack propagation

Nondestructive inspection (NDI) capabilities are steadily being improved. One result of this technological advancement is that fabrication discontinuities in weldments can now be more readily detected. In turn, this trend places greater emphasis on the need for rational accept/reject criteria for discontinuities found in weldments.

In many instances, existing fabrication standards are based upon arbitrary discontinuity acceptance criteria. Frequently, such standards are derived largely from workmanship considerations. For situations in which workmanship standards have a long history of satisfactory use, this is an acceptable approach. However, workmanship standards can also be costly or nonconservative or both when applied to the fabrication of failure-critical structures. Weld repairs to

[1]Metallurgist, Fatigue Criteria Section, Material Science and Technology Division, and Mechanical Engineer, Environmental Testing Section, Spacecraft Engineering Department, respectively, Naval Research Laboratory, Washington, DC 20375-5000.

[2]Program manager for Materials, National Aeronautics and Space Administration, Washington, DC 20546.

[3]Associate field engineer, Construction Department, Duke Power Company, Catawba Nuclear Project, Clover, SC 29710.

[4]Research associate and Professor, respectively, Chemical, Metallurgical and Polymer Engineering, The University of Tennessee, Knoxville, TN 37916.

remove fabrication discontinuities can be costly to perform and, if not properly carried out, can introduce discontinuities (and residual stresses) more severe than the ones being removed. Weld repairs are often conducted under adverse conditions and can easily result in a product which is inferior to the original.

Thus, rational criteria for discontinuity acceptance standards must be developed. This is often referred to as a fitness-for-purpose approach to discontinuity characterization. This approach has been applied in recent years to the development of several new codes and standards involving metal fatigue [*1–4*]. However, to date, this approach has yet to be developed for high-strength steels such as 5% nickel-chromium-molybdenum-vanadium (5Ni-Cr-Mo-V). Also, in some instances, the recent fitness-for-purpose criteria have been developed on a very conservative basis for application to situations in which failure would incur very grave consequences, and thus may be inappropriate for general engineering application.

Background

The present study was undertaken to explore the applicability of linear elastic fracture mechanics (LEFM) to characterize the fatigue behavior of 5Ni-Cr-Mo-V steel butt weldments containing internal lack-of-penetration (LOP) and slag/lack-of-fusion (S/LOF) discontinuities. For such welds, total fatigue life can be divided into a period of crack initiation in which the discontinuity does not enlarge perceptibly, and a subsequent period of crack propagation in which a crack originating at the flaw enlarges until fracture occurs. Crack initiation and crack propagation both play important roles in the overall fatigue failure process. The present study is aimed at understanding the relative importance of each phase of fatigue failure for high-strength steel weldments containing internal discontinuities, assessing the ability of fracture mechanics in analyzing both phases, and characterizing the effects of various discontinuity sizes and shapes on fatigue life.

Materials and Specimen Preparation

The test specimens used in this study were fabricated from 25-mm (1-in.)-thick 5Ni-Cr-Mo-V steel plate material joined by various weld metals, including 7018, 9018, 11018, and 14018. The chemistry and mechanical properties of the 5Ni-Cr-Mo-V plate material are given in Tables 1 and 2. Chemistries of the various weld metals are given in Table 3.

TABLE 1—*Chemical composition of base metal and weld metal.*

Element	Weight %	
	Base Metal	Weld Metal (E 14018)
Carbon	0.07	0.11
Manganese	0.72	0.92
Phosphorus	0.009	0.005
Sulfur	0.003	. . .
Silicon	0.29	0.20
Nickel	4.95	3.97
Chromium	0.48	0.53
Molybdenum	0.47	0.66
Vanadium	0.07	0.08
Titanium	0.005	. . .

TABLE 2—*Mechanical properties.*

Source	Region	Yield Strength		Ultimate Strength		Elongation, %	Reduction in Area, %	Charpy V Notch		
		MPa	(ksi)	MPa	(ksi)			J	(ft-lb)	°C (°F)
Manufacturer	base	962	(139.5)	990	(143.6)	19	70.8	210[a]	(155)[a]	21 (70)
		948	(137.5)	973	(141.1)	20	71.8	193[a]	(142)[a]	21 (70)
								198[a]	(146)[a]	21 (70)
								214[b]	(158)[b]	21 (70)
								201[b]	(148)[b]	21 (70)
								199[b]	(147)[b]	21 (70)
								156[a]	(115)[a]	−18 (0)
								172[a]	(127)[a]	−18 (0)
								164[a]	(121)[a]	−18 (0)
								187[b]	(138)[b]	−18 (0)
								187[b]	(138)[b]	−18 (0)
								201[b]	(148)[b]	−18 (0)
Fabricator	base							145[c]	(107)[c]	−7 (20)
	HAZ							118[c]	(87)[c]	−7 (20)
	weld							106[c]	(78)[c]	−7 (20)
	transverse			954	(138.3)					
	weld			929	(134.8)					

[a]Transverse.
[b]Longitudinal.
[c]Average of three tests.

TABLE 3—*Chemical composition of electrodes used.*

Manufacturer	AIRCO	AIRCO	AIRCO	Teledyne McKAY
Electrode designation	7018	9018B3	11018M	14018
Heat number	N/A	N/A	N/A	645K566
Electrode type	covered	covered	covered	covered
Electrode diameter, mm (in.)	3 (1/8)	3 (1/8)	3 (1/8)	3 (1/8) 4 (5/32)
Element	Chemical Composition, %			
Carbon	0.04	0.034	0.04	0.057
Manganese	0.65	0.53	1.60	0.96
Phosphorus	0.011	0.013	0.012	0.005
Sulfur	0.013	0.014	0.016	0.007
Silicon	0.55	0.66	0.40	0.21
Nickel	. . .	. . .	1.80	3.57
Chromium	. . .	2.40	0.25	0.50
Molybdenum	. . .	0.98	0.42	0.71
Vanadium	. . .	. . .	. . .	0.01

Full-penetration, double-V groove butt welds were used to fabricate the axially-loaded fatigue specimens (Fig. 1). Typical welding and post-weld heat treatment conditions for the fabrication of the test specimens, using the shielded metal-arc (SMA) process, are listed in Table 4. Several types of welds were prepared and studied. Discontinuity-free sound welds and welds containing LOP discontinuities were prepared using 14018 weld metal under contract by the Southwest Research Institute. Specimens containing S/LOF discontinuities using a variety of weld metals were prepared by the University of Tennessee under a joint cooperative research program. However, the intentional formation of specific discontinuities in welds is an unusual and highly selective procedure. LOP discontinuities were formed by butting large land areas which were left unjoined in subsequent weld passes; S/LOF discontinuities were formed by entrapping slag deposits. Both procedures were accomplished using multipass manual methods.

After welding, reinforcements were ground smooth to assure that internal discontinuity behavior would control the fatigue failure process. Also, the weld region was reduced slightly in width to further assure that fatigue failures would initiate in the desired location. Following preparation, all specimens were examined using radiographic inspection to determine the features of the internal discontinuities. The axial loading applied to the specimens assured that the cyclic stresses were uniform across the test section of the welds.

Test Procedure

The fatigue specimens were tested under constant-stress-amplitude zero-to-tension cyclic loading, with the load ratio $R = P_{min}/P_{max} = 0$. The specimens were loaded in a 2.7-MN (600 000-lb)-capacity closed-loop servohydraulic fatigue machine equipped with large hydraulic squeeze grips so that the plate specimens could be gripped directly without requiring any special machining or preparation.

All specimens were tested to complete fracture. However, the tests were monitored periodically to determine the onset of internal fatigue cracking. This was accomplished by placing strain gages on the specimen test section. Strain gage readings provided a sensitive indication of the point in the fatigue life at which significant enlargement of internal defects began to occur. This point is regarded as the fatigue crack initiation life. The sensitivity of this method depended

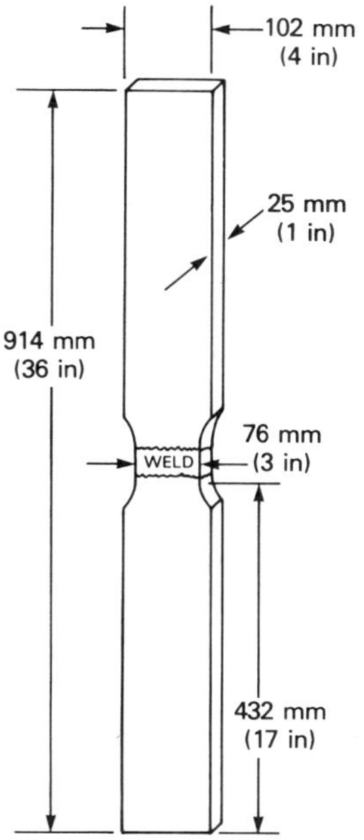

FIG. 1—*Welded fatigue test specimen.*

partly on minimizing the number of cycles between strain gage readings. It may have been possible to measure the crack growth by making crack front markings, but this was not attempted.

The post-failure fracture surfaces of each specimen were examined to various degrees, depending upon the particular specimen and its fatigue behavior. Initial discontinuity sizes were carefully measured and recorded for use in the fracture mechanics analysis of the fatigue life data. Crack initiation sites were noted, and microfracture modes were determined.

Results and Analysis

Fatigue Life Data

The raw fatigue life data are summarized in Table 5 and are plotted in Figs. 2 and 3, which are semilogarithmic plots of the range of stress (S) versus cycles-to-failure (N_f) and cycles-to-initiation (N_i), respectively. These plots include data from ten sound weld specimens, eleven specimens containing LOP discontinuities, and eighteen specimens containing S/LOF discontinuities from this investigation. In addition, selected data from a previous investigation by Lawrence and Radziminski [5] are also included for purposes of comparison.

The fatigue life data shown in Figs. 2 and 3 represent a broad range of stress levels and fatigue lifetimes. Ranges of stress vary from 276 to 827 MPa (40 to 120 ksi), approximately 30 to 90% of nominal yield strength, and the fatigue lifetimes vary from approximately 2×10^2 to 4×10^5 cycles. Most of these data fall within the fatigue life range of what is generally considered to be low-cycle fatigue, that is, less than 10^5 cycles to failure. However, the high-strength 5Ni-Cr-Mo-V steel studied in this investigation did not develop the cyclic plastic deformation normally associated with low-cycle fatigue testing as defined in the ASTM Standard Definitions of Terms Relating to Constant-Amplitude Low-Cycle Fatigue Testing (E 513-74).

TABLE 4—*Procedures for preparation of butt welds.*

Welding Parameter	Electrode			
	7018	9018B3	11018M	14018
Welding process	SMA (manual)	SMA (manual)	SMA (manual)	SMA (manual)
Electrode diameter, mm (in.)	3 (1/8)	3 (1/8)	3 (1/8)	3 (1/8) 4 (5/32)
Current, A	120–140	120–140	120–140	
Polarity	dc reversed polarity	dc reversed polarity	dc reversed polarity	dc reversed polarity
Voltage, V	21–25	21–25	21–25	
Preheat and interpass temperature, °C (°F)	107–149 (225–300)	107–149 (225–300)	107–149 (225–300)	107–149 (225–300)
Arc travel speed, m/min (in./min)	203–305 (8–12)	203–305 (8–12)	203–305 (8–12)	203–406 (8–16)
Maximum heat input, MJ/m (kJ/in.)	1.03 (26.25)	1.03 (26.25)	1.03 (26.25)	1.77 (45)
Shielding gas	...	...	...	...
Joint geometry	90 deg double V	90 deg double V	90 deg double V	90 deg double V
Discontinuity type	S/LOF	S/LOF	S/LOF	LOP, S/LOF
Post-weld heat treatment	621°C (1150°F) for 1 h, air cool	690°C (1275°F) for 1 h, air cool	566°C (1050°F) for 1 h, air cool	...

The data in Figs. 2 and 3 indicate a very broad spectrum of fatigue lifetimes as a function of the range of stress and discontinuity characteristics. The fatigue lifetimes, either N_f or N_i, can vary by as much as two orders of magnitude for a given range of stress depending upon discontinuity severity. Thus, it is apparent that discontinuity characteristics play a crucial role in the fatigue behavior of 5Ni-Cr-Mo-V steel weldments, and that fatigue life is a function of both the range of stress and discontinuity characterization parameters.

The relationship between N_f and N_i for the sound welds is shown in Fig. 4. This is consistent with previous observations that crack initiation constitutes a larger percentage of total fatigue life as fatigue lifetimes become longer [*6*].

Fracture Mechanics Methodology

LEFM provides a means of quantitatively assessing the behavior of cracks under stress [*6*]. Fracture mechanics procedures were used in this investigation to analyze two aspects of the fatigue behavior of the 5Ni-Cr-Mo-V steel weldments: first, to provide a comprehensive parameter which incorporates the range of stress, flaw size, and flaw shape in order to correlate initial as-fabricated discontinuities with fatigue life; and second, to provide a description of crack growth emanating from fabrication discontinuities. The fracture mechanics parameter used in both instances is the crack-tip stress-intensity factor, K_I [*6*]. K_I incorporates both stress and

TABLE 5—*Results of fatigue tests.*

ID	a, mm	b, mm	c, mm	t, mm	Range of Stress, MPa	N_i, cycles	N_f, cycles	Discontinuity Type
1	0	35.6	0	12.8	552	27 000	59 417	none
2	0	37.3	0	12.8	552	[a]	33 119	none
3	0	39.3	0	12.7	690	9 750	24 793	none
4	2.1	36.8	36.8	12.7	552	5 650	9 453	LOP
5	0	36.8	0	12.7	690	5 200	13 065	none
7	2.1	37.2	37.2	12.7	552	3 750	10 255	LOP
8	2	36.2	8.1	12.5	690	1 600	5 750	LOP
9	2.1	37.8	37.8	13	414	9 750	19 900	LOP
11	1.6	34.1	19.9	12.3	414	10 750	35 827	LOP
12	0	36.6	0	12.7	414	62 500	113 949	none
13	0	36.3	0	12.6	276	261 000	357 033	none
14	2.2	36.5	21.4	12.7	690	[a]	3 754	LOP
15	1.9	37.5	7.3	12.8	552	1 000	21 225	LOP
16	0	36.7	0	12.7	827	[a]	10 000	none
17	0	34.9	0	12.7	827	[a]	13 509	none
18	0	35.6	0	13.3	690	6 500	28 011	none
19	0	36.2	0	12.9	414	70 000	104 959	none
20	2.1	37.4	8.2	12.6	414	13 500	50 500	LOP
21	1.9	36.4	20.2	12.6	552	1 000	12 937	LOP
22	2	37.1	37.1	12.9	690	375	3 459	LOP
23	2.1	37.3	7.6	12.7	552	[a]	23 389	LOP
ND-20	0.7	50.8	3.8	12.7	345	5 000	36 000	LOP
ND-12	0.5	50.8	6.6	12.7	345	175 000	219 200	LOP
NA-1	0	31.7	0	12.7	552	[a]	64 200	none
NA-2	0	31.7	0	12.7	552	[a]	60 900	none
NC-13	0	31.7	0	12.7	552	[a]	59 800	none
NA-5	0	31.7	0	12.7	552	[a]	51 100	none
NC-14	0	31.7	0	12.7	552	[a]	50 500	none
7018N1	0.4	41.2	6.7	12.9	414	4 500	8 557	S/LOF
7018N2	1	42	4	12.9	448	[a]	6 300	S/LOF
7018N3	0	40.5	0	12.7	448	[a]	107 392	none
7018N4	0.6	42.8	22.3	12.7	379	15 000	73 664	S/LOF
9018N1	0.6	41.6	3.6	12.9	579	250	2 003	S/LOF
9018N2	0.2	41.4	6.8	12.9	345	20 000	56 800	S/LOF
9018N3	0	37.4	0	12.9	345	[a]	271 500	none
9018N4	0.5	38.4	10.4	12.9	483	500	16 804	S/LOF
11018N1	0.4	42.6	5.3	12.9	579	1 250	13 627	S/LOF
11018N2	0.1	39.6	9.2	12.9	379	6 250	21 500	S/LOF
11018N3	0.6	44.2	6.2	12.7	579	3 750	18 000	S/LOF
11018N4	0.6	42.6	14.8	13	345	55 000	82 400	S/LOF
14018N1	0.4	41.9	32.8	12.7	552	500	1 317	S/LOF
14018N2	1	39.5	21.2	13	552	[a]	1 849	S/LOF
14018N3	1	41.9	21.2	13	690	[a]	1 096	S/LOF
14018N5	1.3	41.3	8.6	12.8	414	[a]	5 896	S/LOF
14018N6	1.3	42.5	22.7	13	276	20 000	65 472	S/LOF
14018N7	3	44	20.3	12.9	414	[a]	7 531	S/LOF
ND-18	0.6	50.8	6.9	12.7	345	60 000	83 800	S/LOF
ND-17	1.2	50.8	12	12.7	345	27 500	71 300	S/LOF
ND-19	0.8	50.8	8.8	12.7	345	17 500	52 000	S/LOF
ND-16	0.2	31.7	3.8	12.7	552	[a]	6 200	S/LOF
ND-46	0.6	31.7	2.2	12.7	552	38 000	42 850	S/LOF
ND-47	0.6	31.7	0.7	12.7	552	5 450	10 400	S/LOF
ND-45	1	31.7	2.7	12.7	552	1 020	6 750	S/LOF
ND-38	0.5	31.7	0.7	12.7	552	500	5 700	S/LOF
NE-8	0.2	31.7	5.5	12.7	552	1 750	4 970	S/LOF
NE-7	1.1	31.7	1.5	12.7	552	250	2 870	S/LOF
NE-12	1	31.7	0.6	12.7	552	1 500	4 400	S/LOF

[a]Not measured.

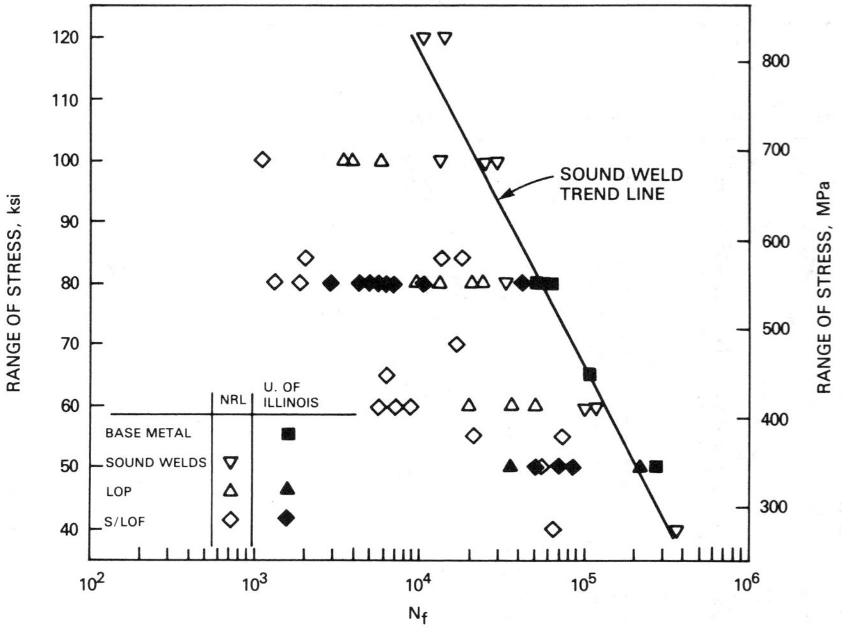

FIG. 2—*Total fatigue life of weldments.*

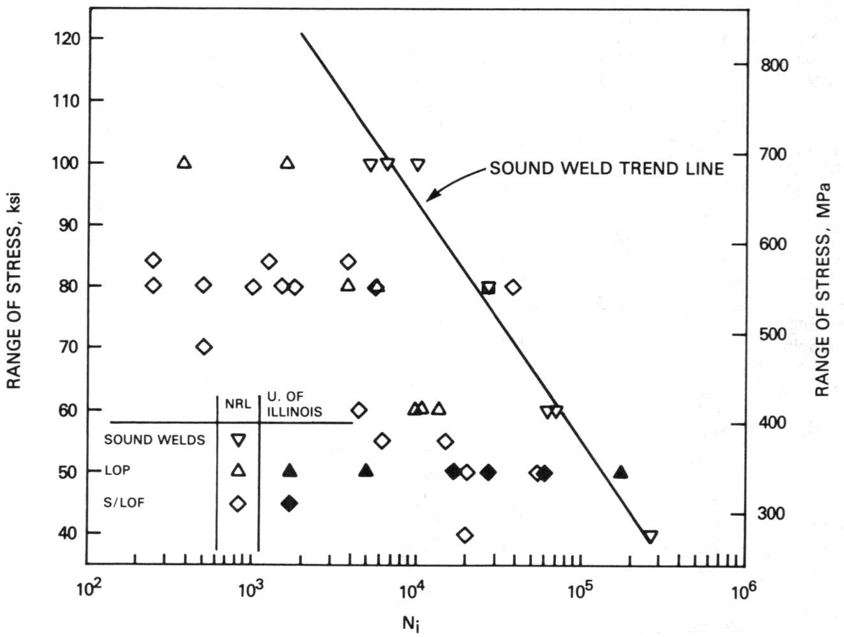

FIG. 3—*Fatigue initiation life of weldments.*

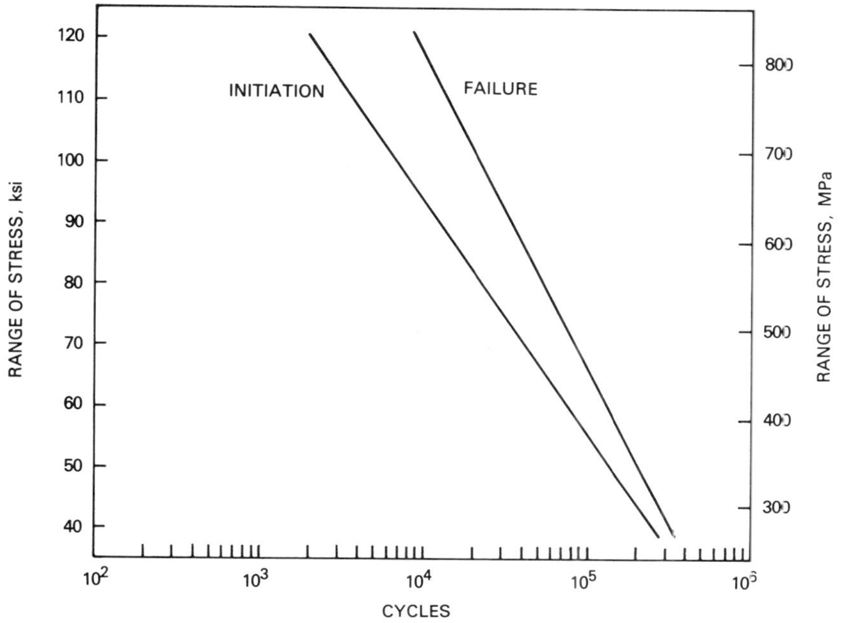

FIG. 4—*Trend lines for fatigue initiation and total fatigue life.*

flaw size terms and is proportional to the product of nominal stress and the square root of flaw size

$$K_{\mathrm{I}} = C\sigma\sqrt{\pi a} \tag{1}$$

where

C = a constant which is a function of specimen type and crack geometry,
σ = nominal stress,
a = flaw size, and
I = subscript denoting the opening mode of crack displacement.

For the purposes of this paper, it will be assumed that all references to K values will imply Mode I conditions.

The analysis of both aspects of fatigue behavior by fracture mechanics procedures was dependent upon the ability to assign an effective K value to weld discontinuities. Herein lies the focus of this investigation. Handbook stress-intensity factor solutions have been derived for infinitely sharp planar cracks of regular geometric proportions. In most instances, weld discontinuities fail to meet these criteria and the engineering application of fracture mechanics to defect characterization necessarily involves approximation and improvization. Details of these analytical procedures are described separately for LOP and S/LOF discontinuities in subsequent sections of this paper.

The general fracture mechanics methodology employed was to assign an initial K value to as-fabricated discontinuities. Initial K values were used in two ways: first, to correlate with corresponding N_i and N_f data in an attempt to provide a degree of quantitative discontinuity characterization lacking in the S-N plots of Figs. 2 and 3, and second to use in a crack growth rate analysis of the number of cycles spent in crack propagation ($N_f - N_i$).

The calculation of the number of cycles spent in crack propagation was performed by well-known procedures, as follows. The Paris law [6] was used to describe the average per cycle incremental crack growth advance by the following equation

$$\frac{da}{dN} = C(\Delta K)^n \tag{2}$$

where

da/dN = rate of crack advance per cycle,
C = material constant,
n = material constant, and
ΔK = range of stress-intensity factor = $K_{max} - K_{min}$.

Fatigue life spent in crack propagation may be estimated by integrating the Paris equation

$$N = \int_{a_0}^{a_f} \frac{1}{C(\Delta K)^n}\, da \tag{3}$$

where

N = cycles spent in advancing the crack from a_0 to a_f,
a_0 = initial half crack size (defect dimension in direction of subsequent crack growth), and
a_f = final half crack size.

In those cases for which the integration of Eq 3 is difficult, a finite-difference technique can be used and the integration performed numerically with the aid of a computer. Thus

$$N = \sum_{a_0}^{a_f} \frac{\Delta a}{C(\Delta K)^n} \tag{4}$$

where Δa is a small finite advance of the crack. It should be noted that the values obtained for N depended on the value chosen for Δa. It was found that the estimate for N increased linearly with Δa.

For each welded fatigue specimen studied in this investigation, measurements were made of the planar dimensions of the initial weld discontinuity which generated the largest final fatigue crack, that is, the dominant crack. These dimensions were then used to generate an effective initial discontinuity represented by a fully embedded elliptical discontinuity, for which fracture mechanics solutions are readily obtainable [7]. The critical dimensions of elliptical discontinuities are the minimum and maximum radii values. For the purposes of this investigation, post-failure measurements were made of the planar dimensions of initial discontinuities at 1 mm (0.040 in.) intervals to establish their size.

Typically, the discontinuities studied in this investigation were either roughly rectangular (LOP) or irregular and elongated (S/LOF). Examples are shown in Figs. 5 and 6. Numerous questions arise in equating such fabrication discontinuities to solutions for elliptical discontinuities. One such approach has been formalized in the American Society of Mechanical Engineers (ASME) Section XI Boiler and Pressure Vessel Code [1]. There, the procedure is an innately conservative one; that is, to define the effective fracture mechanics discontinuity as an ellipse which *surrounds or fully encloses* the fabrication discontinuities. However, for the purposes of the present analysis, more consistent results were obtained by choosing elliptical geometries which best fit *within* the major dimensions of the fabrication discontinuities. This latter procedure produced analytical results which gave much less scatter, as shown in Table 6.

FIG. 5—*Typical LOP discontinuity.*

It is recognized that discontinuity shape, as well as size, is an important variable in analyzing crack growth. For the purposes of this study, it was assumed that during crack growth both the major and minor radii of the fully embedded elliptical discontinuities varied as a function of the stress-intensity factor.

Fatigue life calculations using Eq 4 were made using the following Paris equation parameter values

$$C = 8.54 \times 10^{-7}(\text{m/cycle})/(\text{MPa}(\text{m})^{1/2})^n$$

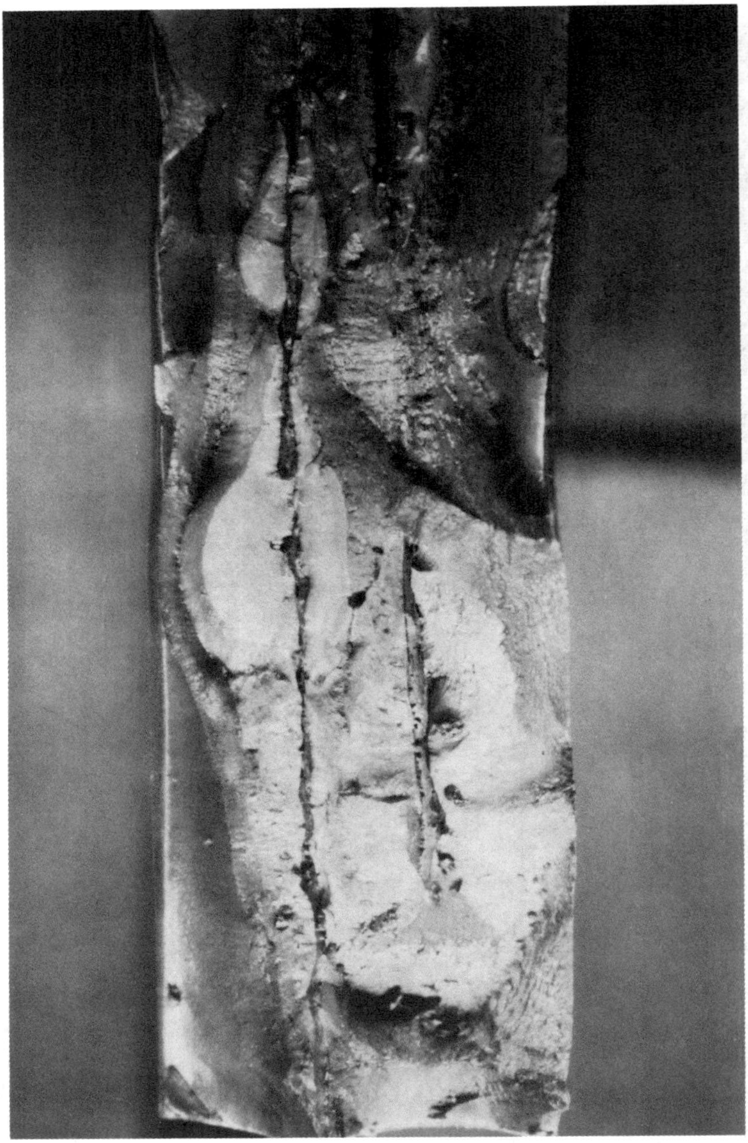

FIG. 6—*Typical S/LOF discontinuity.*

$$[= 4.13 \times 10^{-9}(\text{in./cycle})/(\text{ksi}(\text{in.})^{1/2})^n]$$

$$n = 2.176$$

These values were obtained from previous Naval Research Laboratory (NRL) studies on 5Ni-Cr-Mo-V steels [8].

In applying Eq 4, a_0 was taken to be half the maximum through-thickness dimension of the fabrication discontinuity and a_f was taken to be the half thickness, t, of the plate material.

TABLE 6—*Goodness of fit for* N *versus* $K_{initial}$ *data.*

		Location of Assumed Initial Ellipse Relative to Discontinuity					
		Inside			Outside		
Flaw	N	Correlation Coefficient	Coefficient of Determination	Standard Error of Estimate	Correlation Coefficient	Coefficient of Determination	Standard Error of Estimate
LOP	N_i	0.78177	0.61117	8.1508	0.61168	0.37415	13.092
	N_f	0.92507	0.85575	5.0519	0.60812	0.36981	12.833
	N_f-N_p	0.96361	0.92855	2.9873	0.50735	0.25741	13.754
S/LOF	N_i	0.40211	0.16169	4.3575	0.23000	0.05290	7.8116
	N_f	0.49873	0.24873	6.2275	0.13254	0.01756	20.099
	N_f-N_p	0.45410	0.20621	2.9732	0.28292	0.08004	29.119

However, there are two valid arguments for ending the summation even before the internal discontinuity has reached the surface, one related to net section stress and the other to fracture toughness. Accordingly, the computer program which performed the summation was written such that the summation would be ended if the net section stress exceeded 1034 MPa (150 ksi) or if the maximum stress intensity exceeded 180 MPa(m)$^{1/2}$ [150 ksi(in.)$^{1/2}$]. It is also recognized that, mathematically, a_0 exerts a much greater influence on cycles to failure than does a_f. Thus, efforts to characterize initial discontinuity severity have a greater potential to influence fatigue life estimates than do final failure criteria.

Finally, fracture mechanics methodology was used so that data from specimens which lacked intentional initial discontinuities could also be included in summary plots of initial K versus N_i and N_f. In conducting this analysis, it was assumed that the supposedly unflawed specimens behaved as if some finite initial discontinuity in fact did exist. The equivalent initial discontinuity size of the sound-weld specimens could then be defined as that which places the initial K-versus-N data on the same line as that which describes the flawed specimens.

This was done by first fitting an equation to the initial K-versus-N data for the LOP specimens as follows

$$\text{initial } K = m_1 \log_{10} N + b_1$$

Similarly, a line was fit to all the S-N data for sound weld specimens and was described by an equation of the following form

$$S = m_2 \log_{10} N + b_2$$

Letting initial $K = S(a_{\text{eq}})^{1/2}$, it was possible to estimate a_{eq} from

$$(a_{\text{eq}})^{1/2} = \frac{m_1 \log_{10} N + b_1}{m_2 \log_{10} N + b_2}$$

The resulting values of a_{eq} are tabulated in Table 7. Two results are apparent from these data. First, there is about 2:1 variation in a_{eq} depending upon whether the initial K values are calculated with the ellipse lying inside or outside fabrication discontinuities. Second, regardless of which assumption is made, a_{eq} values exhibit a good measure of consistency. If the preferred method of placing the ellipse inside the fabrication discontinuities is used, a_{eq} averages out to be on the order of 2.5 mm (0.10 in.) for the sound-weld specimens.

A fatigue notch factor analysis was made as well as an initial stress-intensity factor analysis. The fatigue notch factor was empirically estimated for all specimens containing discontinuities by using Peterson's equation [*9*]

$$K_f = 1 + \frac{K_t - 1}{1 + a/r} \tag{5}$$

where

K_t = theoretical (elastic) stress concentration factor,
r = notch root radius, and
a = an experimentally determined material parameter.

The parameter a is thought to be related to the ultimate tensile strength of the material S_u. The following relationship [*10*] was used to obtain rough estimates of a

TABLE 7—*Equivalent initial flaw sizes[a] for unflawed specimens.*

	Location of Assumed Initial Ellipse Relative to Discontinuity							
	Inside				Outside			
	$N = N_i$		$N = N_f$		$N = N_i$		$N = N_f$	
	Effective Initial Flaw Size							
Specimen	mm	(in.)	mm	(in.)	mm	(in.)	mm	(in.)
1	2.395	(0.09428)	2.289	(0.09013)	4.211	(0.1658)	4.831	(0.1902)
2			2.654	(0.1045)			4.615	(0.1817)
3	2.516	(0.09904)	2.807	(0.1105)	3.988	(0.1570)	4.531	(0.1784)
5			3.091	(0.1217)	3.886	(0.1530)	4.384	(0.1726)
12	2.251	(0.08861)	1.760	(0.06929)	4.496	(0.1770)	5.184	(0.2041)
13	1.802	(0.07095)	0.424	(0.01669)	5.509	(0.2169)	6.574	(0.2588)
16			3.193	(0.1257)			4.333	(0.1706)
17			3.079	(0.1212)			4.389	(0.1728)
18	2.553	(0.1005)	2.743	(0.1080)	3.919	(0.1543)	4.567	(0.1798)
19	2.227	(0.08766)	1.834	(0.0722)	4.544	(0.1789)	5.131	(0.2020)
NA-1			2.234	(0.08795)			4.864	(0.1915)
NA-2			2.272	(0.08944)			4.841	(0.1906)
NC-13			2.285	(0.08995)			4.834	(0.1903)
NA-5			2.392	(0.09418)			4.768	(0.1877)
NC-14			2.400	(0.09449)			4.763	(0.1875)

[a]Effective initial flaw size:

$$A_{\text{equivalent}} = \left[\frac{m_1 \log_{10} N + b_1}{m_2 \log_{10} N + b_2}\right]^2$$

where K_{initial} versus N data for LOP specimens is described by

$$K_{\text{initial}} = m_1 \log_{10} N + b_1$$

and S versus N data for unflawed specimens are described by

$$S = m \log_{10} N + b_2$$

$$a = 2.5 \times 10^{-5}(2068/S_u)^{1.8} \text{ (SI units)}$$
$$(a = 1.0 \times 10^{-3}(300/S_u)^{1.8}) \text{ (English units)} \tag{6}$$

in millimetres (or inches), where S_u is in MPa (or ksi). In general, K_t can be expressed in the form

$$K_t = A\left(\frac{C}{r}\right)^B + 1 \tag{7}$$

where A, B, and C are constants. Substituting Eq 7 into Peterson's equation (Eq 5), differentiating K_f with respect to r, and setting that expression equal to zero leads to the result that K_f is maximum when

$$r = r_{\text{critical}} = \frac{a}{B}(1 - B) \tag{8}$$

so that

$$(K_f)_{\max} = 1 + A\left\{\left[\frac{BC}{a(1 - B)}\right]^B \bigg/ \left[1 + \frac{B}{1 - B}\right]\right\} \tag{9}$$

If an elliptical flaw is assumed, then

$$K_t = 2(c/r)^{1/2} + 1 \tag{10}$$

Substitution of Eq 10 into Eq 9 yields

$$r_{\text{critical}} = a \tag{11}$$

and

$$(K_f)_{\max} = 1 + (c/a)^{1/2} \tag{12}$$

where c is the semimajor axis of the ellipse. The major dimension of the real discontinuity was used as the major dimension of the elliptical discontinuity assumed in order to simplify the analysis. For the $(K_f)_{\max}$ approach, essentially the same goodness-of-fit was obtained whether the elliptical discontinuity assumed surrounded or was surrounded by the real discontinuity.

It can be seen that, if the ultimate tensile strength S_u and the discontinuity length $2c$ are known, then $(K_f)_{\max}$ can be calculated. This was done for both the specimens containing LOP discontinuities and those containing S/LOF discontinuities.

LOP Discontinuity Analysis

The lack-of-penetration discontinuities were essentially rectangular and were either fully embedded or full-width internal discontinuities. They were approximated by the following two stress-intensity factor solutions. The first is for a fully embedded elliptical discontinuity and the second is for a rectangular through discontinuity.

A model for a discontinuity in a weld, an embedded elliptical discontinuity in a plate, is shown in Fig. 7. The stress intensity for a zero-to-tension stress condition (stress perpendicular to the plane of the crack) as defined in Ref *11* would be

$$K_{\text{I}} = S\left(\pi \frac{a}{Q}\right)^{1/2} F_e\left(\frac{a}{c}, \frac{a}{t}, \frac{c}{b}, \phi\right) \tag{13}$$

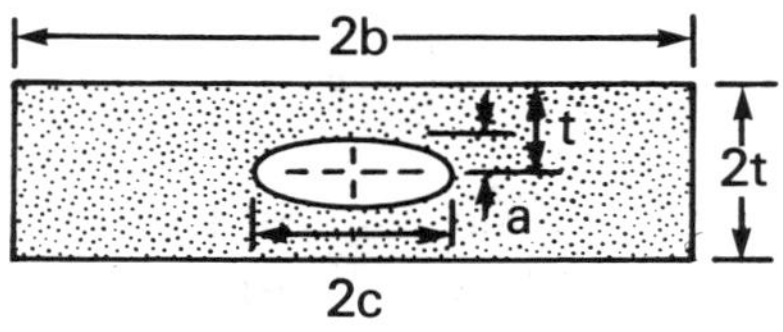

FIG. 7—*Coordinate system used to define dimensions of embedded elliptical discontinuities.*

for $0 \leq a/c \leq \infty$, $c/b < 0.5$, and $-\pi \leq \phi \leq \pi$ provided that a/t satisfies

$$\begin{aligned} \frac{a}{t} &< 1.25\left(\frac{a}{c} + 0.6\right) \quad \text{for} \quad 0 \leq \frac{a}{c} \leq 0.2 \\ \frac{a}{t} &\leq 1 \quad \text{for} \quad 0.2 \leq \frac{a}{t} \leq \infty \end{aligned} \tag{14}$$

where Q is the shape factor for an ellipse and is given by the square of the complete elliptic integral of the second kind. The width b and length h of the plate were taken to be large enough so that they would have a negligible effect on stress intensity. The boundary correction F accounts for the influence of various boundaries and is a function of crack depth, crack length, hole radius (where applicable), plate thickness, and the parametric angle of the ellipse. Figure 8 shows the coordinate system used to define the parametric angle.

Very useful empirical expressions for Q have been developed. For $a/c \leq 1.0$, which was the case for all of the specimens examined the expression is

$$Q = 1 + 1.464\left(\frac{a}{c}\right)^{1.65} \tag{15}$$

The maximum error in the stress-intensity factor caused by using these equations for Q was about 0.13% for all values of a/c.

The function F_e accounts for the influence of crack shape (a/c), crack size (a/t), finite width (c/b), and angular location (ϕ), and was chosen as

$$F_e = \left[M_1 + M_2\left(\frac{a}{t}\right)^2 + M_3\left(\frac{a}{t}\right)^4\right] g f_\phi f_w \tag{16}$$

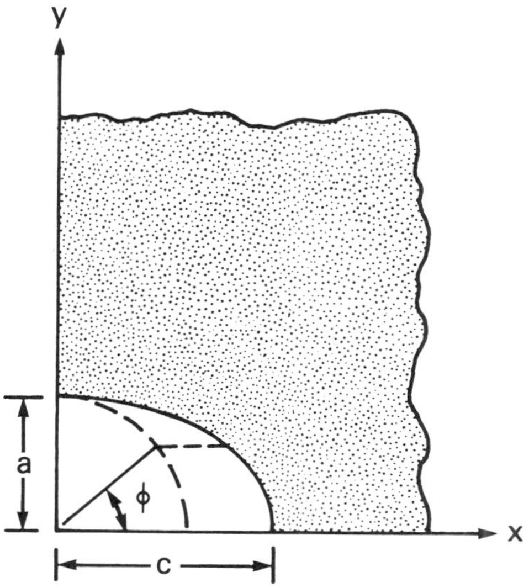

FIG. 8—*Coordinate system used to define the parametric angle for embedded elliptical discontinuities when* a/c ≤ *1.*

The term in brackets gives the boundary-correction factors at $\phi = \pi/2$ (where $g = f_\phi = 1$). The function f_ϕ was taken from the exact solution for an embedded elliptical discontinuity in an infinite solid, and f_w is a finite-width correction factor. The function g is a fine-tuning curve-fitting function. For $a/c \leq 1$

$$M_1 = 1 \tag{17}$$

$$M_2 = \frac{0.05}{0.11 + \left(\dfrac{a}{c}\right)^{3/2}} \tag{18}$$

$$M_3 = \frac{0.29}{0.23 + \left(\dfrac{a}{c}\right)^{3/2}} \tag{19}$$

$$g = 1 - \frac{\left(\dfrac{a}{t}\right)^4}{1 + 4\left(\dfrac{a}{c}\right)} |\cos \phi| \tag{20}$$

$$f_\phi = \left[\left(\frac{a}{c}\right)^2 \cos^2 \phi + \sin^2 \phi\right]^{1/4} \tag{21}$$

The finite-width correction f_w was

$$f_w = \left[\sec\left(\frac{\pi c}{2b}\left(\frac{a}{t}\right)^{1/2}\right)\right]^{1/2} \tag{22}$$

for $c/b < 0.5$. (Note that for the embedded discontinuity, t is defined as one half of the full plate thickness.)

A second model, that of a through discontinuity in a body which is finite in the direction of crack advance, is shown in Fig. 9. This model describes a situation which arises intentionally when partial-penetration welds are specified to obtain cost savings over full-penetration welds. The range of stress intensity for a zero-to-tension stress situation can be expressed according to Ref *12* as

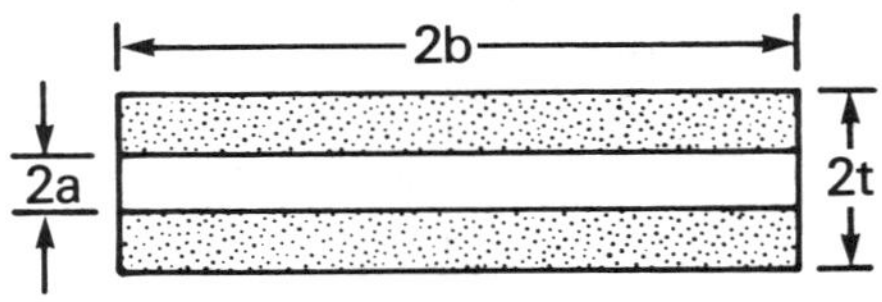

FIG. 9—*Coordinate system used to define dimensions of a through discontinuity in a finite body.*

$$\Delta K = \sigma(\pi a)^{1/2} \left(\sec \frac{\pi a}{2b}\right)^{1/2} \qquad 0 \leq a \leq 0.8b \tag{23}$$

where b is the plate half-thickness.

The initial K versus N_f and N_i data are plotted in Figs. 10 and 11, respectively. The data show a good agreement between the lifetimes of the welds containing LOP discontinuities and the models used. Compared with the data in Figs. 2 and 3, the use of a fracture mechanics approach has successfully normalized the data.

In addition to the initial K method of normalizing the data, a $(K_f)_{max}$ approach, as described earlier, was taken. The product of nominal stress and $(K_f)_{max}$ was plotted against cycles to failure. This plot is shown in Fig. 12. A comparison with Figs. 2 and 3 shows that this approach is also very successful in normalizing the data.

S/LOF Discontinuities Analysis

The slag/lack-of-fusion discontinuities were roughly elliptical and were fully embedded, so they were described by the solution for a fully embedded elliptical discontinuity given in the preceding section. The initial K versus N_f and N_i data are plotted in Figs. 13 and 14, respectively. The data show only a low correlation between lifetimes of welds containing multiple S/LOF discontinuities and the initial K values calculated by ignoring all S/LOF discontinuities except the one yielding the largest fatigue crack.

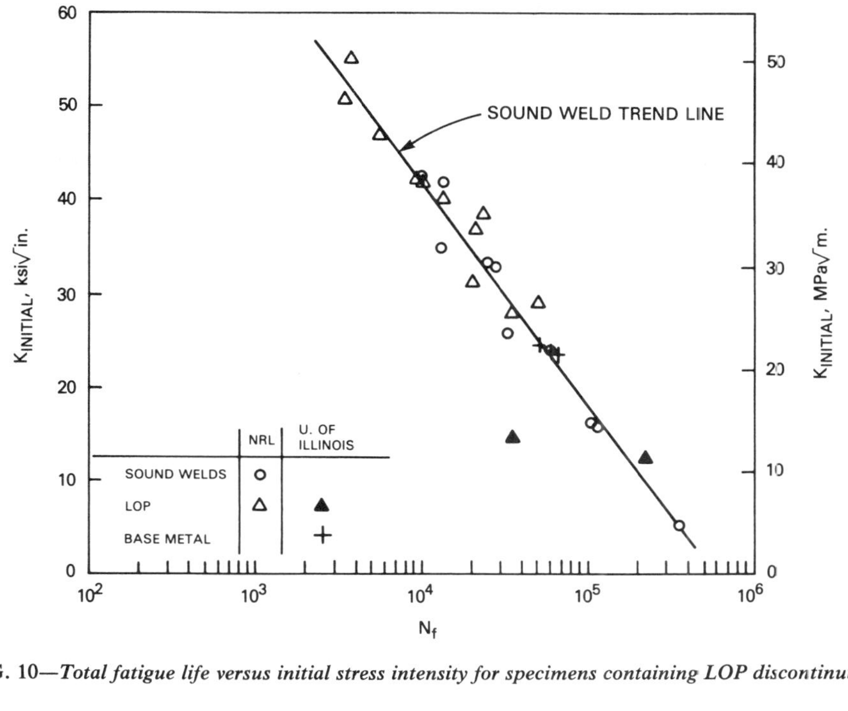

FIG. 10—*Total fatigue life versus initial stress intensity for specimens containing LOP discontinuities.*

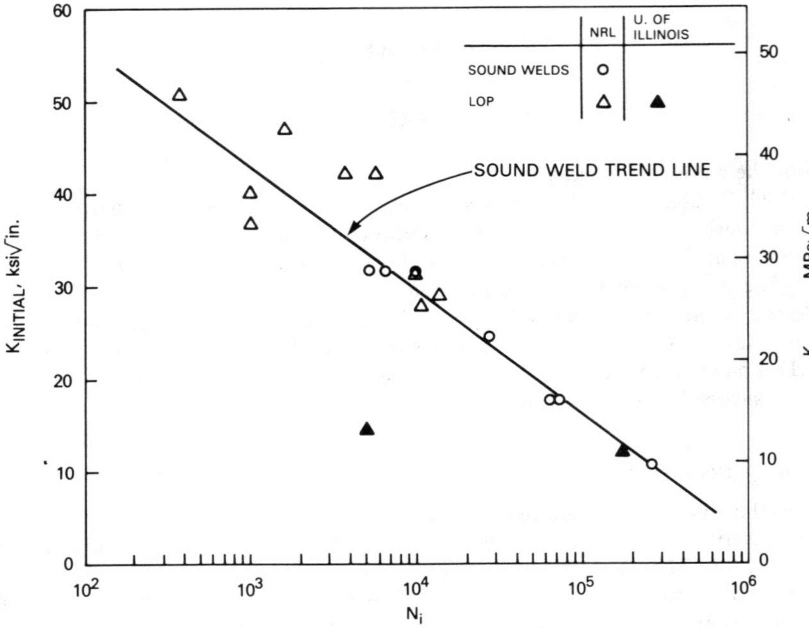

FIG. 11—*Strain gage estimates of fatigue initiation life versus initial stress intensity for specimens containing LOP discontinuities.*

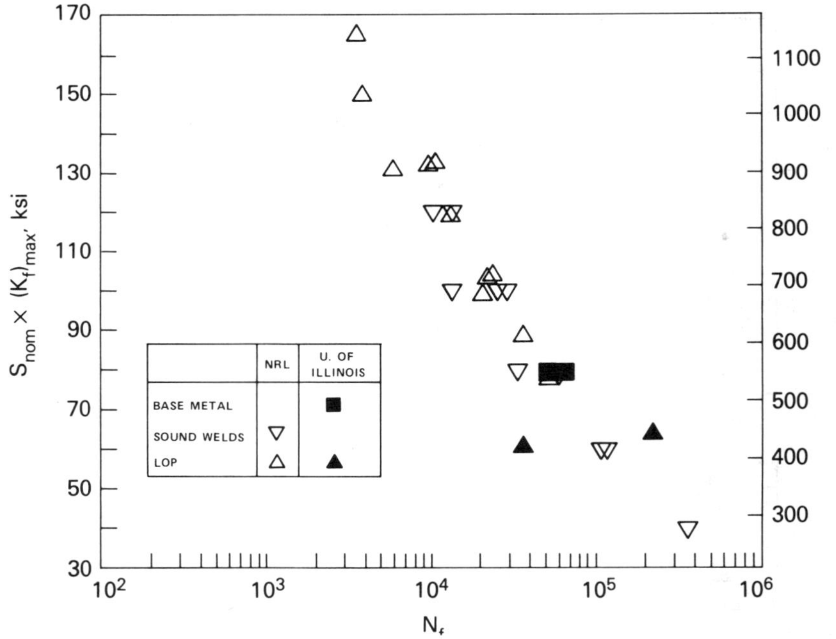

FIG. 12—*Total fatigue life versus nominal stress times* $(K_f)_{max}$ *for specimens containing LOP discontinuities.*

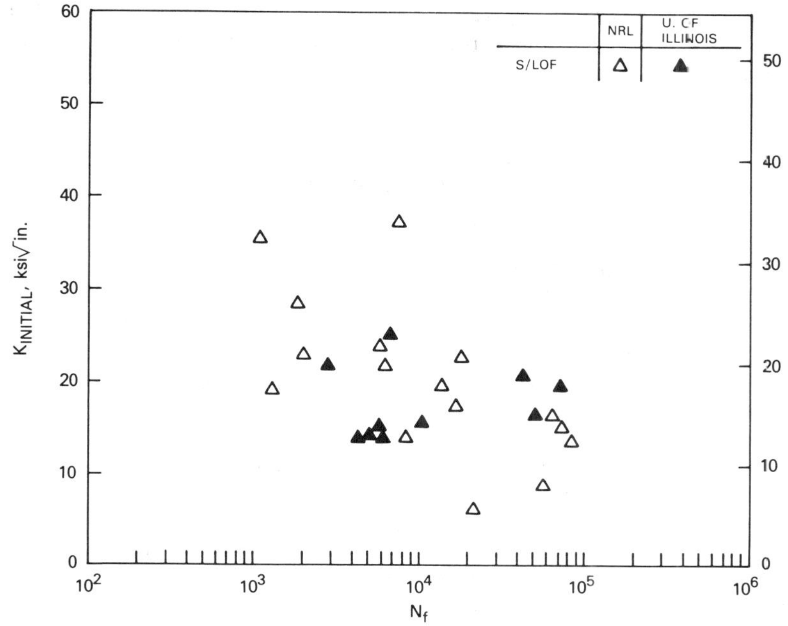

FIG. 13—*Total fatigue life versus initial stress intensity for specimens containing S/LOF discontinuities.*

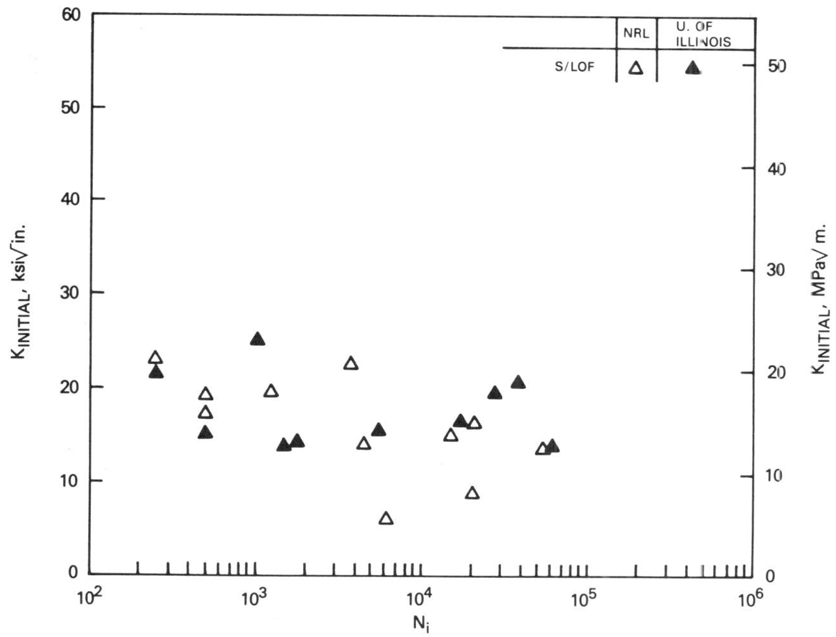

FIG. 14—*Strain-gage estimates of fatigue initiation life versus initial stress intensity for specimens containing S/LOF discontinuities.*

A plot of the product of nominal stress and $(K_f)_{max}$ versus cycles to failure is shown in Fig. 15. It can be seen that the $(K_f)_{max}$ approach is slightly more successful than the initial K approach in normalizing the data.

Discussion

The fracture mechanics analysis methods employed in this investigation were two-dimensional; that is, they dealt only with the planar dimensions of the discontinuities, length and width in the cross-sectional plane. Thus, discontinuity acuity or sharpness was not taken into consideration. For the fatigue behavior of the LOP discontinuities, this appears to have been a reasonably successful approach. The data in Figs. 10 and 11, initial K versus N_f and N_i, respectively, and in Fig. 12, nominal stress times $(K_f)_{max}$ versus N_f, for the LOP specimens, are nicely normalized with respect to stress level and discontinuity size. LOP discontinuities are categorized in the British Standards Institution document on discontinuity assessment for fatigue as being "planar" or crack-like [2], and thus should be largely amenable to a straightforward fracture mechanics approach.

However, it should be pointed up that some degree of conservatism is built into the assumptions underlying the approach used in the plots shown in Figs. 10, 11, and 12. Despite the planar nature of the LOP discontinuities, they are not fully crack-like, and some period of crack initiation life was observed in most instances. This is illustrated in Fig. 16, which is a plot of observed cycles-to-failure versus calculated cycles-to-failure using the crack growth rate methodology described earlier. Except for specimens which failed in very short fatigue lifetimes, the trend is for actual fatigue life to exceed calculated fatigue life, thus suggesting that even planar LOP discontinuities are not, in fact, truly crack-like with regard to their fatigue behavior.

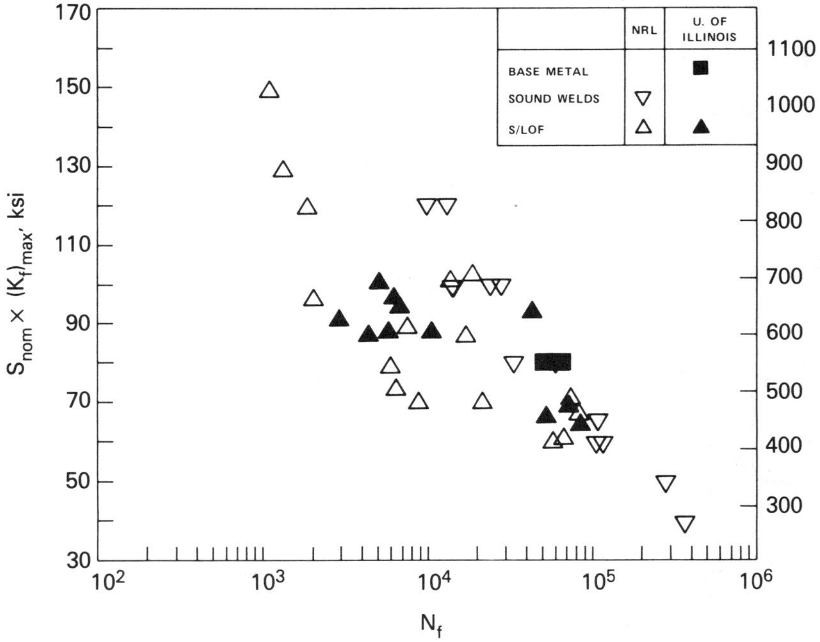

FIG. 15—*Total fatigue life versus nominal stress times* $(K_f)_{max}$ *for specimens containing S/LOF discontinuities.*

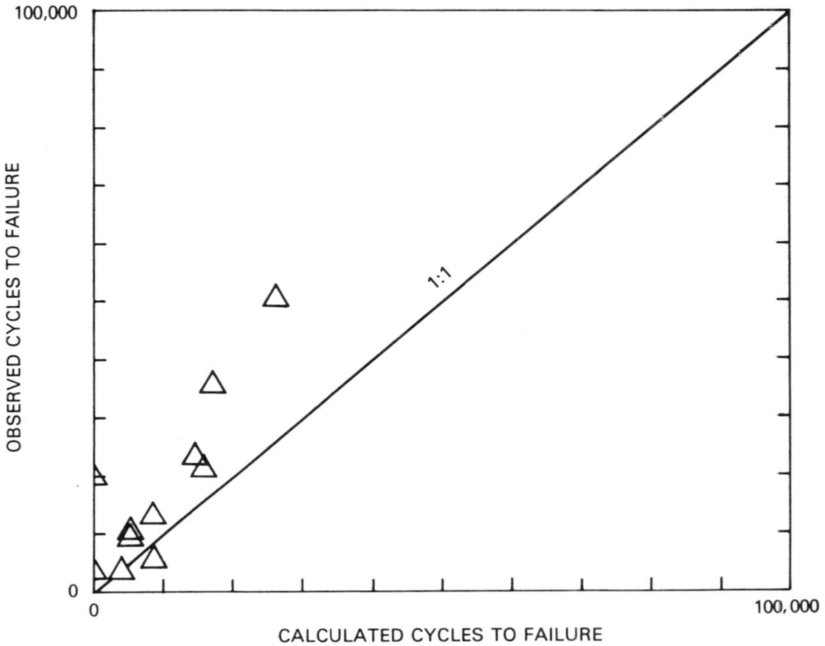

FIG. 16—*Observed versus calculated cycles to failure for specimens containing LOP discontinuities.*

The successful normalization of the LOP data is not matched for the S/LOF data (Figs. 13, 14, and 15). Here, unlike the LOP specimens, multiple initial discontinuities existed in each specimen. In undertaking a fracture mechanics analysis of the data, attention was focused on selecting the initial discontinuities which appeared to initiate crack growth first, and thus produced the largest fatigue crack at failure. Obviously, something other than this unrefined approach is needed. Only the broadest correlation trend between initial K and N_f and N_i is apparent in the data shown in Figs. 13 and 14. There is only a slightly better correlation in Fig. 15, in which nominal stress times $(K_f)_{max}$ is plotted against N_f.

Several complicating factors affect the S/LOF data. The S/LOF discontinuities were formed by intentionally leaving entrapped slag in the welds. The British Standards Institution document categorizes slag discontinuities as being "nonplanar" [*2*] and thus something less severe than crack-like. Post-failure examination of the S/LOF specimens revealed that, in many instances, the largest initial slag discontinuities did not lead most quickly to failure. Discontinuity acuity or the surrounding residual stress field or both played more important roles than discontinuity size per se. Obviously, a great deal more needs to be learned before the fatigue behavior of multiple S/LOF discontinuities can be predicted with confidence and accuracy.

Finally, the S/LOF specimens tested consisted of a mixture of weld metals (7018, 9018, 11018, and 14018) of varying yield strengths approximately 480 to 960 MPa (70 to 140 ksi). Among these were specimens tested in the as-welded (AW) and post-weld heat-treated (PWHT) conditions. Figure 17 shows range of stress versus N_f data for these specimens. These limited data do not suggest any significant effect of strength level *per se* on fatigue life. This is as expected for the fatigue life regime involved ($N_f < 10^5$ cycles to failure). However, the PWHT does appear to impart significant benefits to fatigue life, thus suggesting that residual tensile stress plays an important role in fatigue crack initiation at S/LOF discontinuities.

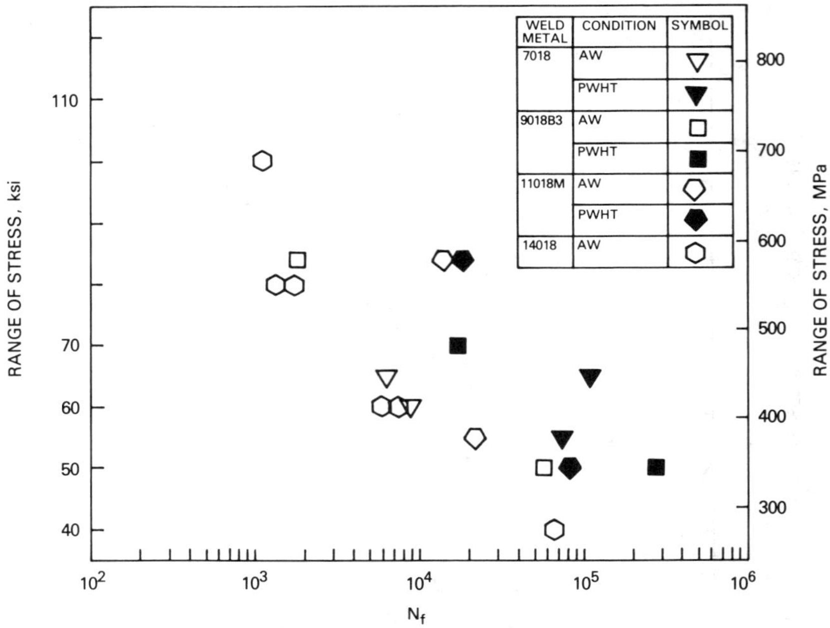

FIG. 17—*Cycles to failure versus stress for specimens containing S/LOF discontinuities.*

Summary and Conclusions

Butt-welded 5Ni-Cr-Mo-V steel specimens containing internal lack-of-penetration and slag/lack-of-fusion discontinuities were cycled to failure at nominal stress levels from 276 to 827 MPa (40 to 120 ksi) and resulting fatigue lifetimes between 10^2 and 10^6 cycles. Strain-gage instrumentation was employed to estimate the point of crack initiation during fatigue cycling. Attempts were made to analyze fatigue behavior on the basis of linear elastic fracture mechanics parameters. The following conclusions were reached as a result of this investigation.

1. For specimens containing single planar LOP discontinuities of varying sizes, a good correlation was obtained between the initial stress-intensity K and cycles-to-failure N_f and cycles-to-initiation N_i. A good correlation was also obtained between nominal stress times $(K_f)_{max}$ versus N_f. The rectangular LOP discontinuities were most successfully approximated for fracture mechanics analysis purposes by elliptical discontinuity solutions, for ellipses lying within the boundaries of the rectangular discontinuities. This approach embodies an inherent degree of conservatism because, even for the planar LOP discontinuities, a significant period of crack initiation occurred, except for very short fatigue lifetimes.
2. For specimens containing multiple nonplanar S/LOF discontinuities, a satisfactory correlation between initial K and N_f or N_i could not be readily obtained. The correlation obtained between nominal stress times $(K_f)_{max}$ versus N_f was only slightly less unsatisfactory. The effects of discontinuity acuity and surrounding residual stress fields appear to exert a greater influence than discontinuity size per se. Additional study is needed before reliable engineering criteria relating to S/LOF discontinuity behavior in fatigue can be developed.
3. For specimens containing S/LOF discontinuities prepared with a variety of weld metals of varying yield strengths, strength level *per se* did not exhibit any tangible influence on fatigue life. However, post-weld heat-treatment did have significant beneficial effects on fatigue life.

Acknowledgments

Financial support for this research was provided by the Office of Naval Research and the Welding Research Council. Experimental assistance was provided by M. L. Cigledy, G. W. Jackson, and J. L. Berman.

References

[*1*] "Rules for Inservice Inspection of Nuclear Reactor Coolant Systems," *ASME Boiler and Pressure Vessel Code,* Section XI, American Society of Mechanical Engineers, 1980.

[*2*] "Guidance on Some Methods for the Derivation of Acceptance Levels for Defects in Fusion Welded Joints," PD 6493, British Standards Institution, 1980.

[*3*] Reed, R. P., McHenry, H. I., and Kasen, M. B., "A Fracture Mechanics Evaluation of Flaws in Pipeline Girth Welds," *Welding Research Council Bulletin,* Vol. 245, 1979.

[*4*] Pellini, W. S., *Guidelines for Fracture-Safe and Fatigue-Reliable Design of Steel Structures,* The Welding Institute, Cambridge, U.K., 1983.

[*5*] Lawrence, F. V., Jr. and Radziminski, J. B., "Fatigue Crack Initiation and Propagation in High-Yield-Strength Steel Weld Metal," *Welding Journal, Research Supplement,* Vol. 49, No. 10, 1970, pp. 445-S–452-S.

[*6*] Rolfe, S. T. and Barsom, J. M., *Fracture and Fatigue Control in Structures,* Prentice-Hall, Englewood Cliffs, NJ, 1977.

[*7*] Pick, R. J., "Stress Intensity Factors: Their Prediction and Use in Design," *Canadian Metallurgical Quarterly,* Vol. 19, 1980, pp. 34–44.

[*8*] Sullivan, A. M. and Crooker, T. W., "Analysis of Fatigue-Crack Growth in a High-Strength Steel—I: Stress Level and Stress Ratio Effects at Constant Amplitude," *Transactions,* American Society of Mechanical Engineers, Series J, *Journal of Pressure Vessel Technology,* Vol. 98, 1976, pp. 179–184.

[*9*] Peterson, R. E. in *Metal Fatigue,* Sines and Waisman, Eds., McGraw-Hill, New York, Chapter 13, 1959.

[*10*] Peterson, R. E., "Analytical Approach to Stress Concentration Effect in Fatigue of Aircraft Structures," WADS Symposium, Wright Air Development Center, Wright-Patterson Air Force Base, Dayton, OH, Aug. 1959.

[*11*] Newman, J. C. and Raju, I. S., "Stress-Intensity Factor Equations for Cracks in Three-Dimensional Finite Bodies," *Fracture Mechanics: Fourteenth Symposium—Volume I: Theory and Analysis, ASTM STP 791,* American Society for Testing and Materials, Philadelphia, 1983, pp. I-238–I-265.

[*12*] Feddersen, C. E., Discussion to *Plane Strain Crack Toughness Testing of High Strength Metallic Materials, ASTM STP 410,* American Society for Testing and Materials, Philadelphia, 1966, pp. 77–79.

Avraham Berkovits[1] *and David Prinz*[2]

Growth of Parabolic Fatigue Cracks from Flaws at Fastener Holes

REFERENCE: Berkovits, A. and Prinz, D., "**Growth of Parabolic Fatigue Cracks from Flaws at Fastener Holes,**" *Fracture Mechanics: Eighteenth Symposium, ASTM STP 945,* D. T. Read and R. P. Reed, Eds., American Society for Testing and Materials, Philadelphia, 1988, pp. 1050–1069.

ABSTRACT: A crack of parabolic shape, emanating from corner and embedded flaws at a hole in a plate of finite width, was analyzed by the slice synthesis. In contrast with other crack shapes, equilibrium solutions for a parabolic crack are finite and continuous as the crack passes the bottom of the hole and starts to propagate along the rear surface. Fracture parameters derived were K_I for predicting fatigue crack growth along the plate surfaces and down the bore of the hole, and G_I for residual strength. The two cases studied were a flaw emanating from the corner of the hole and a flaw embedded in the side of the hole at the mid-depth, each case being either remotely loaded, or loaded at the hole. Fatigue crack curves were calculated with the aid of the derived fracture parameters, and fatigue crack curves were compared with experimental data and with calculated results obtained assuming an elliptically shaped crack. The parabolic crack analysis resulted in superior correspondence with measured data. Tables of constants for evaluating the fracture parameters are provided.

KEY WORDS: fracture mechanics, stress intensity factor, crack shape, crack propagation, open holes, lugs, aluminum alloys, fatigue crack growth, fatigue (materials)

Nomenclature

a	Crack length in width (x) direction
a_0	Crack length along front face
b	Crack length in thickness (z) direction
b_0	Crack length down drilled hole
d_0	Crack length along rear face
E	Elastic modulus
g	Universal weight function
G_I	Mode I energy release rate
K_I	Mode I stress intensity factor
l	Length of crack front
n	Cycles
p^0	Crackline loading
p^*	Spring stress
R	Hole radius
u, ξ	Length variables
x, y, z	Coordinate axes

[1] Associate professor, Department of Aeronautical Engineering, Technion-Israel Institute of Technology, Haifa 32000, Israel.
[2] Department of Defense, Israel.

Greeks

- μ Poisson's ratio
- σ Direct stress
- τ Shear stress
- ϕ Generalized weight function

Subscripts

- rc Radial crack
- s Spring

Regulations regarding repair of plate structure with fatigue cracks emanating from holes require accurate methods of estimating propagation rate, especially of small cracks. In this program, relationships between crack propagation rates at the surface, where cracks are generally measured, and within the structure, where maximum crack length often occurs, were derived by consideration of initial flaw position and of crack shape. A crack of parabolic shape, emanating from both corner and embedded flaws at a hole in a plate of finite thickness, was analyzed by the method of slices proposed by Fujimoto [*1*]. The solution was finite and continuous as the crack passed the bottom of the hole and started to propagate along the rear surface. Solutions based on an elliptical crack are divergent at this point.

Fracture parameters derived were K_I for predicting fatigue crack growth along the front and rear surfaces of the plate and down the bore of the hole, and G_I for residual strength. The two cases studied included a flaw emanating from the corner of the hole and a flaw embedded in the side of the hole at the mid-depth, each case being either remotely loaded, or loaded at the hole.

Consideration is first given to the physical aspects governing crack shape. The slicing model and method of calculating the fracture parameters on the plate surfaces and down the bore of the hole are detailed, with emphasis on the points where the development for the parabolic shaped crack differs from previous derivations for the elliptical crack. Next, the calculated fracture parameters are presented as functions of the geometry, in the form of correction factors to the stress-intensity factor and energy release rate of an infinite plate. Described are two series of tests, undertaken in order to develop fatigue crack propagation data to compare with calculated results. Specimens tested included open-hole plates and loaded-hole lugs of aluminum alloys. Both corner flaws and embedded flaws were tested. Finally, these and other experimental results, as well as calculated results obtained by assuming an elliptical crack, are compared with results calculated by the present parabolic crack approach.

Crack-Shape Considerations

Adequate prediction of fatigue crack propagation in a plate containing a hole was found to depend on consideration of the shape of the crack front.

Elliptical Cracks

A considerable volume of analytical work has been generated regarding an elliptically shaped fatigue crack emanating from a drilled hole. By methods such as that of Fujimoto [*1,2*] or that of Bowie [*3*] and with reference to Fig. 1, the length of an elliptical crack can be adequately predicted along the face of a plate a_0 and down the drilled hole b_0 until the crack length along the hole reaches the rear face of the plate. At this point, when the crack begins to propagate through a length d_0 along the rear face, the crack propagation rate dd_0/dn is indeterminate. A

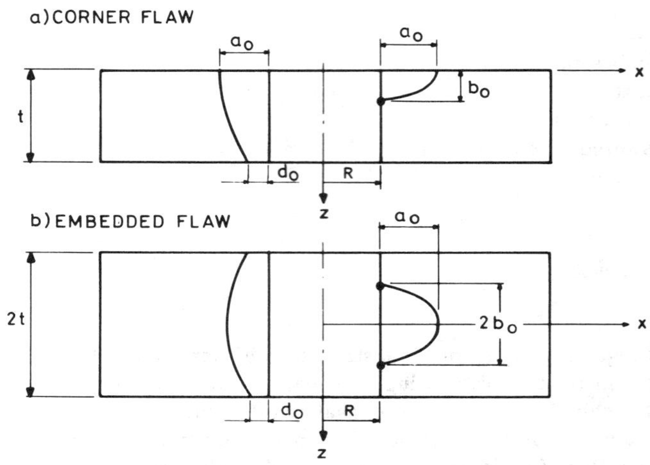

FIG. 1—*Definition of flaw geometries.*

common procedure is to assume dd_0/dn initially infinite, so that d_0 immediately becomes equal to a_0, the crack length of the front face, and thereafter progresses together with a_0.

For relatively large cracks and for failure prediction, this assumption is satisfactory. However, it may be inadequate for purposes of damage tolerance estimation and for determination of inspection and repair cycles, when accurate prediction of small cracks is decisive. A better estimate of dd_0/dn for small cracks can be made by consideration of the shape of actual fatigue cracks.

Parabolic Cracks

Representative fatigue crack data [*4,5*], shown in Figs. 2 and 3, clearly favor the parabolic shape, as do other data [*6*]. (In Fig. 2, parabolic and elliptical curves were fitted to the test data on the basis of the aspect ratio defined by the experimental endpoints.) However, the significance of the crack shape is more than geometric.

The important difference between the parabolic and elliptical crack front becomes clear when the mechanism of crack growth is considered. A typical fatigue-crack surface is replicated in Fig. 4. The shape of the crack front is generated at the end of local crack paths which originate at an initial flaw. The crack front is everywhere at right angles to the local crack paths. Crack paths extending along the edge of a hole tend away from the hole and into the material, and permit new crack paths to be generated adjacent to the hole. This mechanism of crack development results in a crack front which is parabolic in shape, rather than elliptical, as assumed in previous analytical work. The parabolic shape is maintained even after the crack becomes a through crack since a local cracking path running into a surface terminates there without affecting nearby cracking paths. The latter continue to advance without interference, as indicated at the right of Fig. 4. The mutual independence of the cracking paths permits a unified treatment of cracks emanating from corner flaws and from embedded flaws.

For the elliptical crack-front shape, on the other hand, the crack path which advances along the edge of the hole can never enter the interior of the material, since the crack front at right angles to it always remains perpendicular to the material edge. In order to create a continuous crack front, either crack paths within the material must diffuse or else secondary crack paths

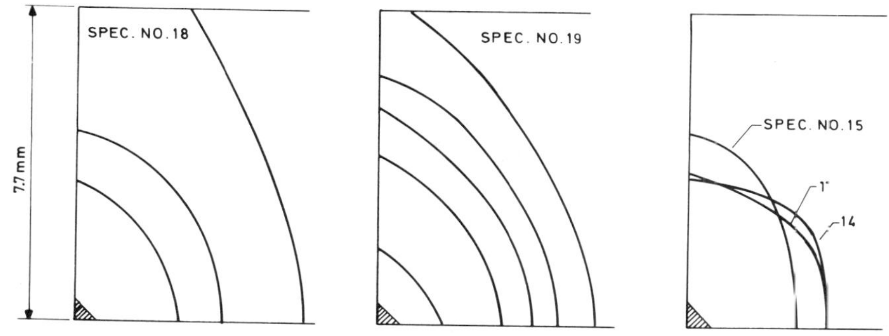

FIG. 2—*Marker bands in 2024-T351 aluminum-alloy open-hole specimens.*

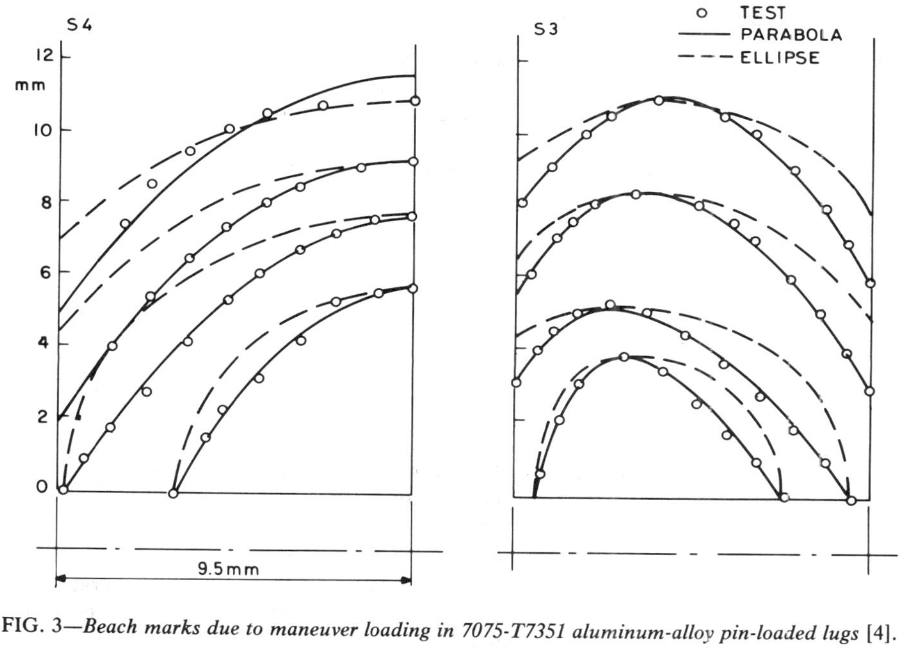

FIG. 3—*Beach marks due to maneuver loading in 7075-T7351 aluminum-alloy pin-loaded lugs* [4].

must start at sources within the material, since new crack paths cannot enter the material from the edge as the crack advances. Neither of these conditions is observed on fatigue fracture surfaces. Second, crack paths along and near the hole edge must meet the rear face of the plate at right angles to it, with no component along the rear face. The result is a crack propagation rate which is indeterminate along the rear face.

This difficulty does not arise with crack paths which form a parabolic crack front. Each crack path, including those nearest the edge of the hole, has a component perpendicular to the hole axis, parallel to the rear surface. As a result, each crack path displays a finite propagation rate component along the rear face when it reaches the surface, and continuity is preserved. For purposes of crack prediction, these phenomena must be expressed in mathematical terms.

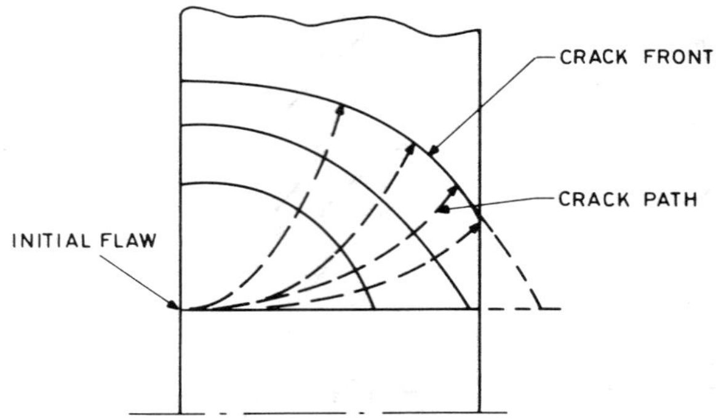

FIG. 4—*Parabolic crack showing crack paths and crack front.*

Mathematical Model—Flaw Idealization

Two types of flaws in a finite plate were considered; the corner flaw (Fig. 1*a*) and the embedded flaw (Fig. 1*b*), which was formed by reflecting the corner flaw across the *xy*-plane. A stress σ_0 was applied to the plate in the *y*-direction at a large distance from the hole, or at the hole itself. The position of any point along the parabolic crack front relative to the edge of the hole $a(z)$ and to the top surface of the plate $b(x)$ is described by the equations

$$a(z) = a_0\left[1 - \left(\frac{z}{b_0}\right)^2\right] \tag{1}$$

$$b(x) = b_0\sqrt{1 - \left(\frac{x - R}{a_0}\right)} \tag{2}$$

Slicing Procedure

The flawed plate was idealized as two series of infinitesimal slices, one oriented parallel to the *yz*-plane and the other parallel to the *xy*-plane. In this way each slice was loaded in plane stress, and coupling between slices arose only from the shear stress τ_{zy}. τ_{zy} was simulated in the idealized plate by slicing parallel to the *yz*-plane, as in the upper sketch of Fig. 5. For the corner-flaw plate, slices in the region of the crack appear as elements containing an edge crack of depth $b(x)$. (For the embedded flaw, center-crack elements were used instead of edge-crack slices.) The shear stress at the crack face then took the form of a compressive boundary stress $-p^*(x, z)$. This in turn was idealized as a system of line springs, shown in the *xy*-plane slice (Fig. 5), whose mechanical properties were identical to those of the *yz*-plane slices.

Since the tensile stress $p^*(x, z)$ transmitted across the springs was the negative of that acting on the faces of the idealization of Fig. 5, superposition of the stress fields associated with the figure resulted in a three-dimensional stress field which satisfied the required boundary condition of a force-free crack in a plate subjected to a uniform tension stress σ_0 far from the hole. Superposition of two-dimensional solutions in this manner enabled determination of the behavior of the three-dimensional plate.

Displacement Matching

The equation which matches the displacement of the radial crack to that of the springs over the crack face is of the form

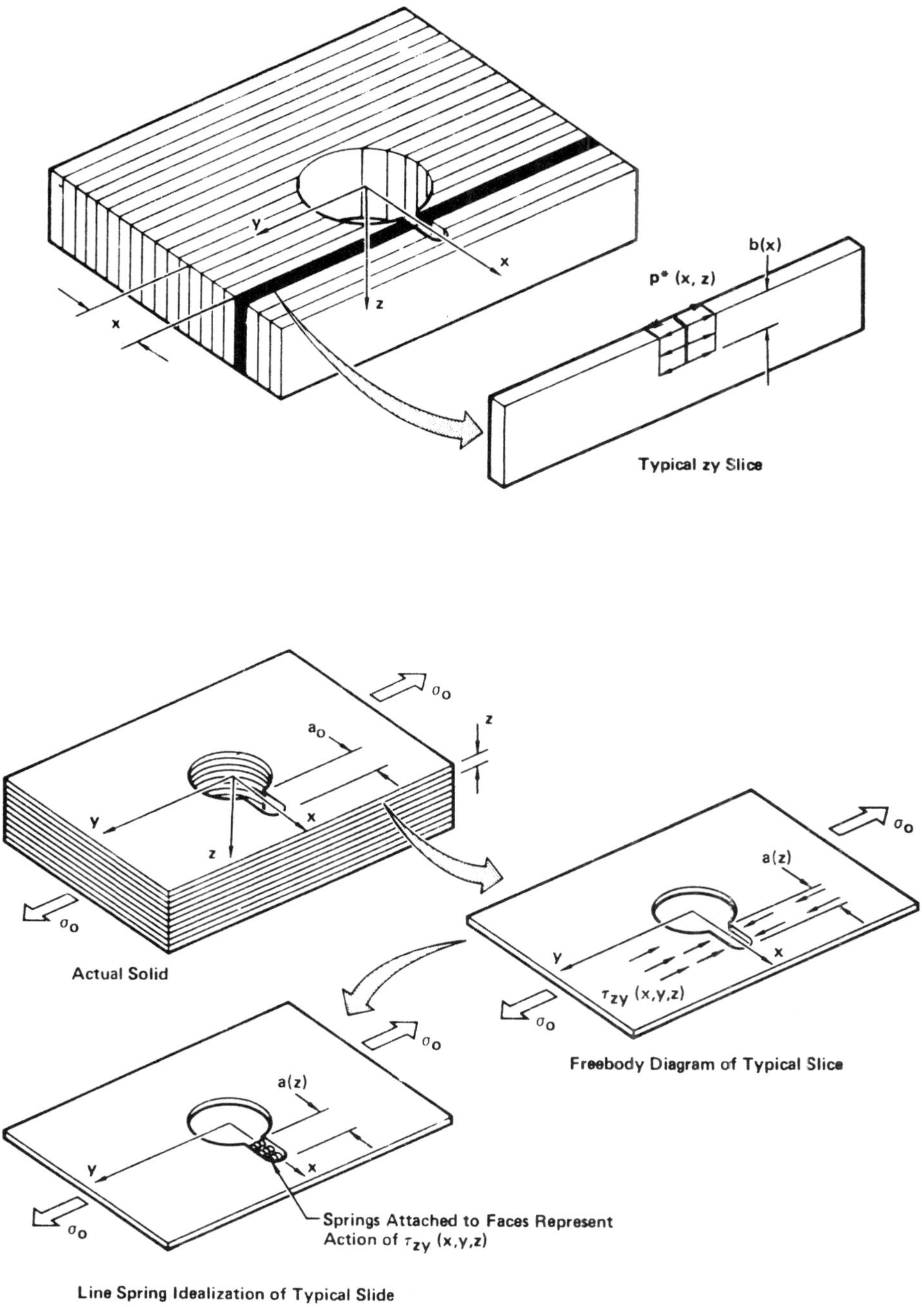

FIG. 5—*Plate idealization as vertical and horizontal slices* [1].

$$\frac{1}{E}\int_{x-R}^{a(z)}\int_{R}^{R+\xi}[p^0(u)-p^*(u,z)]g(u,\xi)_{\mathrm{rc}}du\,g(x,\xi)_{\mathrm{rc}}d\xi$$

$$=\frac{\Gamma}{E_s}\int_{z}^{b(x)}\int_{0}^{\xi}p^*(x,u)g(u,\xi)_s du\,g(z,\xi)_s d\xi \quad (3)$$

where

Γ = 1 for corner flaws and Γ = 2 for embedded flaws,
$p^0(x)$ = the crackline loading associated with the applied load system and is the negative of stresses existing along the xz-plane in the unflawed structure,
$p^*(x, z)$ = magnitude of the unknown stress transmitted across the spring,
E = elastic modulus of the flawed material,
E_s = elastic modulus of the spring elements,
$g(x, \xi)_{\mathrm{rc}}$ = universal weight function [7,8] for the through-the-thickness radial crack, and
$g(z, \xi)_s$ = universal weight function for the spring element.

Expressions used for the universal weight functions are given in Ref *1*.

Equation 3 contains two unknowns $p^*(x, z)$ and E_s. A second equation was obtained from the requirement that the springs attached to the faces of the radial crack elements describe only the additional resistance to crack opening due to transverse shear coupling between the elements. Constraints were placed upon the properties of the springs to assure that they would not counteract the surface crack displacement field which usually arises. The ratio of the effective spring modulus E_s and the elastic modulus of the plate E is a function of the shape and length ℓ of the crack front, and is given by

$$\frac{E_s}{E}=\left[\frac{\phi}{1-\mu^2}-1\right]\frac{b_0}{a_0} \quad (4)$$

where μ is Poisson's ratio and ϕ is the generalized weight function, which for the parabola ϕ becomes

$$\phi=\int\frac{d\ell}{b_0}=\frac{1}{2}\left\{\sqrt{\left(\frac{2a_0}{b_0}\right)^2+1}+\frac{b_0}{2a_0}\ln\left[\sqrt{\left(\frac{2a_0}{b_0}\right)^2+1}+\frac{2a_0}{b_0}\right]\right\} \quad (5)$$

The unknown stress in the springs $p^*(x, z)$ was defined for the embedded flaw as the polynomial

$$p^*(x,z)=\sigma_0(\alpha_{00}+\alpha_{10}x+\alpha_{20}x^2) \quad (6a)$$

and for the corner flaw as

$$p^*(x,z)=\sigma_0[\alpha_{00}+\alpha_{10}x+\alpha_{20}x^2+(\alpha_{01}+\alpha_{11}x+\alpha_{21}x^2)z] \quad (6b)$$

where σ_0 is nominal stress applied to the plate.

The crackline loading $p^0(x)$ is the negative of stresses existing along the plane formed by the xy-axes in the unflawed structure. The crackline loading was obtained from known solutions for open and loaded holes in a finite plate, with the assumption that crack growth occurred perpendicularly to the direction of loading. For the open (unloaded) hole, the classical solution [*9*] was used in the form

$$p^0(x) = \frac{\sigma_0}{2}\left[2 + \left(\frac{R}{x}\right)^2 + 3\left(\frac{R}{x}\right)^4\right] \tag{7}$$

Several recent studies [*10–12*] have shown that the load distribution at a loaded hole more nearly resembles a uniform pressure loading distributed over the loaded half of the hole. Therefore, the loaded hole case was taken to be a uniform pressure distributed over 46% of the circumference of the hole. The crackline loading was represented by the polynomial

$$p^0(x) = -\sigma_0\left[0.6663 - 8.5965\frac{R}{x} + 40.0271\left(\frac{R}{x}\right)^2 - 94.2467\left(\frac{R}{x}\right)^3 + 102.369\left(\frac{R}{x}\right)^4 - 43.0117\left(\frac{R}{x}\right)^5\right] \tag{8}$$

In Eqs 7 and 8, σ_0 varies by a factor of 2 depending on whether the corner flaw or embedded flaw is being treated. Equation 3 was solved numerically for the α-coefficients in Eqs 6 by linear regression. Values of α_{ij} were determined for a material of Poisson's ratio $\mu = 0.3$ for a wide range of crack geometry.

Fracture Parameters

With α_{ij} determined, stress-intensity factors $K_{\rm I}$ along both faces of the plate and down the bore of the hole, as well as the energy release rate $G_{\rm I}$, could be calculated. The value of $K_{\rm I}$ at any point along the crack front was obtained by asymptotic expansion of the displacement field. Accordingly, along the plate face

$$K_{\rm Ia} = \int_R^{R+a_0} [p^0(x) - p^*(x, 0)]g(x, a_0)_{\rm rc}dx \tag{9}$$

and down the bore of the hole

$$K_{\rm Ib} = \frac{E}{E_s}\int_{z_1}^{b_0} p^*(R, z)g(z, b_0)_s dz \tag{10}$$

The stress-intensity factor $K_{\rm Id}$ for the rear surface was calculated by integrating the x- and z-components of $K_{\rm I}$ along the crack path S to the rear surface. For embedded flaws, $K_{\rm Ia}$ and the x-component of $K_{\rm Id}$ must be multiplied by $1/1 - \mu^2$.

The energy release rate $G_{\rm I}$ was determined by differentiating the total strain energy U relative to crack path S, with the result

$$G_{\rm I} = \frac{dU}{dS} = \frac{3}{4\left(a_0 + b_0\dfrac{\partial a_0}{\partial b_0}\right)E_s}\int_R^{R+a_0}\frac{\partial b(x)}{\partial b_0}\left\{\int_0^{b(x)} p^*(x, z)g[z, b(x)]_s dz\right\}^2 dx$$
$$+ \frac{3}{4\left(a_0 + b_0\dfrac{\partial a_0}{\partial b_0}\right)E}\int_0^{b_0}\frac{\partial a(z)}{\partial b_0}\left\{\int_R^{R+a(z)} [p^0(x) - p^*(x, z)]g[x, a(z)]_{\rm rc}dx\right\}^2 dz \tag{11}$$

where

$$\frac{\partial b(x)}{\partial b_0} = \sqrt{1 - \left(\frac{x - R}{a_0}\right)} + \frac{b_0}{2a_0^2} \frac{(x - R)\dfrac{\partial a_0}{\partial b_0}}{\sqrt{1 - \left(\dfrac{x - R}{a_0}\right)}}$$

and

$$\frac{\partial a(z)}{\partial b_0} = \frac{\partial a_0}{\partial b_0}\left[1 - \left(\frac{z}{b_0}\right)^2\right] + \frac{2a_0 z^2}{b_0^3}$$

and where the shape change is described by the quantity

$$\frac{\partial a_0}{\partial b_0} = \left(\frac{K_{\mathrm{Ia}}}{K_{\mathrm{Ib}}}\right)^r \tag{12}$$

The exponent r in Eq 12 is related to that of the Paris crack propagation rate expression, and was taken to have a value of 4 as common for many alloys [*13*].

Calculated Results

Values of the fracture parameters were calculated for crack geometries in the range $0.5 \leq a_0/b_0 \leq 2.0$, $0.001 \leq a_0/R \leq 3.0$, and $0.01 \leq b_0/t \leq 3.0$. ($b_0/t \geq 1.0$ indicates a through-crack.) The various fracture parameters were expressed in polynomial form as

$$K_{\mathrm{Ia}}/\sigma_0\sqrt{\pi a_0} = \sum_{i=0}^{4}\sum_{j=0}^{2}\sum_{k=0}^{3} A_{ijk}\cdot\left(\frac{R}{R + a_0}\right)^i\cdot\left(\frac{a_0}{b_0}\right)^j\cdot\left(\frac{b_0}{t}\right)^k$$

$$K_{\mathrm{Ib}}/\sigma_0\sqrt{\pi b_0} = \sum_{i=0}^{4}\sum_{j=0}^{2}\sum_{k=0}^{3} B_{ijk}\cdot\left(\frac{R}{R + a_0}\right)^i\cdot\left(\frac{a_0}{b_0}\right)^j\cdot\left(\frac{b_0}{t}\right)^k$$

$$K_{\mathrm{Id}}/\sigma_0\sqrt{\pi d_0} = \sum_{i=0}^{4}\sum_{j=0}^{2}\sum_{k=0}^{3} D_{ijk}\cdot\left(\frac{R}{R + a_0}\right)^i\cdot\left(\frac{a_0}{b_0}\right)^j\cdot\left(\frac{b_0}{t}\right)^k$$

$$\frac{EG_{\mathrm{I}}}{(1 - \mu^2)\sigma_0^2\pi a_0} = \sum_{i=0}^{4}\sum_{j=0}^{2}\sum_{k=0}^{3} G_{ijk}\cdot\left(\frac{R}{R + a_0}\right)^i\cdot\left(\frac{a_0}{b_0}\right)^j\cdot\left(\frac{b_0}{t}\right)^k \tag{13}$$

The coefficients A_{ijk}, B_{ijk}, D_{ijk}, and G_{ijk} are tabulated in Tables 1 to 4, respectively.

The significance of the calculated results will be discussed in the light of elliptical crack-shape calculations as well as experimental results.

Experimental Program and Results

A number of fatigue tests were performed in order to produce crack propagation data with which to compare analytical results. Marker loads were applied to create beach marks on the crack surface. Two series of tests were conducted: one on open-hole plate specimens of 2024-T351 aluminum alloy, the other on loaded-hole lug specimens of 2024-T3, 2024-T351, and 7075-T7351 aluminum alloys. Both corner-flaw and embedded-flaw imperfections were tested. Details of specimen dimensions and loading program are available elsewhere [*4,5*].

TABLE 1—A_{ijk} *coefficients.*

$A_{(ijk)}$	Corner Flaw		Embedded Flaw	
	Open Hole	Loaded Hole	Open Hole	Loaded Hole
A000	−1.47212290	−0.36765143	−1.48176889	−0.31454637
A001	0.64470279	0.35271896	−0.04349280	−0.05406349
A002	−0.85266484	−0.43261770	0.02948717	0.03929750
A003	0.22047531	0.10946842	−0.00569561	−0.00772311
A010	1.01809798	0.16034064	0.32107518	0.07988223
A011	−2.93032174	−1.23357837	0.04163301	0.09091994
A012	2.92174769	1.44366424	−0.02263137	−0.06600463
A013	−0.68969976	−0.35217311	0.00396681	0.01296471
A020	−0.20890259	−0.02637021	−0.02317663	−0.00546641
A021	0.97531012	0.41291937	−0.00220736	−0.01823899
A022	−1.12273240	−0.52348984	−0.00027220	0.01328973
A023	0.27523258	0.13039923	0.00012535	−0.00261922
A100	16.97410074	2.30263282	17.31889654	1.75177602
A101	−5.26576808	−2.94546064	0.29284352	0.47564414
A102	6.56850994	3.49098445	−0.18923935	−0.34599335
A103	−1.67760847	−0.87458653	0.03620801	0.06802345
A110	−10.00070959	−1.41021180	−3.03432486	−0.66953368
A111	25.74926111	10.09058234	−0.10675713	−0.78063657
A112	−23.83168599	−11.54381957	0.00233974	0.56967473
A113	5.48040018	2.79523075	0.00365685	−0.11217941
A120	1.82558484	0.16279187	0.17652111	0.03672694
A121	−8.30199168	−3.28979356	−0.10041266	0.13706560
A122	9.19260694	4.15762108	0.09547733	−0.09963010
A123	−2.22206359	−1.03272759	−0.01959633	0.01962389
A200	−42.26209141	0.49329983	−42.63384712	2.60290110
A201	14.88035154	8.48742277	−0.85088919	−1.46275625
A202	−17.77129199	−9.87806538	0.54400415	1.06497040
A203	4.47823493	2.45227297	−0.10397140	−0.20947360
A210	25.94652330	1.41362937	8.24395769	1.34195690
A211	−67.15449788	−24.07148233	0.24622042	2.36593084
A212	61.76575510	29.55244591	0.04334128	−1.73616196
A213	−14.15762179	−7.27257810	−0.01955895	0.34285903
A220	−4.65087951	0.11013039	−0.55552312	−0.11577465
A221	21.37639247	7.69547649	0.49147703	−0.32886795
A222	−23.57646898	−10.33915166	−0.43040849	0.23862591
A223	5.68832636	2.60378071	0.08727200	−0.04700475
A300	46.79082981	−6.38537777	46.90584115	−9.22755978
A301	−16.96620262	−9.67652037	1.02122328	1.83322028
A302	19.45373834	11.15228510	−0.64673256	−1.33550339
A303	−4.84450395	−2.75452937	0.12343239	0.26277111
A310	−28.34692029	1.36588345	−9.43233356	−0.92009239
A311	73.08872419	22.83948977	−0.00445611	−2.75880680
A312	−66.34107471	−30.83284020	−0.27098500	2.03444134
A313	15.11587258	7.73830353	0.06602462	−0.40277668
A320	4.72897652	−0.88469231	0.63641634	0.11992790
A321	−22.81914094	−7.26450809	−0.87475341	0.26101270
A322	25.13716471	10.62869378	0.73093340	−0.18885647
A323	−6.05315997	−2.72085679	−0.14710789	0.03723528
A400	−16.65843812	7.09669804	−16.57902496	8.46840471
A401	6.76812186	3.83781015	−0.45957358	−0.79253823
A402	−7.60574943	−4.52494497	0.29429543	0.57795550
A403	1.87881616	1.11878991	−0.05633680	−0.11377838
A410	9.78039127	−3.02269065	3.47552222	−0.21672823
A411	−26.77090269	−5.77436657	−0.08413362	1.07776054
A412	24.77372073	10.71307269	0.17045881	−0.80230622
A413	−5.67363406	−2.83751410	−0.03825996	0.15959879
A420	−1.45345003	0.86242794	−0.23694282	−0.04320284
A421	8.21724474	1.93070104	0.44174721	−0.05575673
A422	−9.27869571	−3.59437504	−0.36059692	0.04061728
A423	2.24668837	0.95847995	0.07227225	−0.00808534

TABLE 2—B_{ijk} *coefficients.*

	Corner Flaw		Embedded Flaw	
$B_{(ijk)}$	Open Hole	Loaded Hole	Open Hole	Loaded Hole
B000	0.19615921	−0.10272850	0.04881679	0.03206847
B001	−0.75949076	−0.38951847	−0.24068457	−0.06784446
B002	5.14139088	3.06742444	0.88667425	0.22383864
B003	−4.44094660	−2.57608115	−0.67433830	−0.09981075
B010	0.41206957	0.33146992	−0.29875183	−0.02566494
B011	0.92650494	0.45067090	0.44210488	0.06912499
B012	−4.87969401	−2.92033213	−1.47855645	−0.19706538
B013	4.40118076	2.13970311	1.06586787	−0.02240903
B020	−0.14550537	−0.10038094	0.09947554	0.01366832
B021	−0.33430237	−0.14631855	−0.24830933	−0.04120030
B022	1.76546661	0.85922338	0.87591571	0.14134264
B023	−1.60747447	−0.58412346	−0.72794380	−0.08207584
B100	−1.17472428	1.06439420	−0.42374116	−0.29442868
B101	5.72067809	3.18367139	0.84697339	0.62000303
B102	−38.06242708	−24.92282911	−2.93436628	−2.04179430
B103	33.60144948	21.05129427	1.52951747	0.93347650
B110	−0.66491022	−1.55858595	5.56056541	1.14134395
B111	−9.95270181	−4.65274916	−1.88240764	−1.05352079
B112	64.78871328	32.80037704	4.48057549	3.13519445
B113	−56.79829427	−24.94348784	−2.54592013	−1.31437832
B120	0.86931706	0.72555252	−1.06039409	−0.16142979
B121	3.58592867	1.53536668	1.08375981	0.41097175
B122	−21.53650641	−9.35026627	−3.37769859	−1.40839870
B123	19.13161551	6.79708046	2.61861400	0.92264435
B200	2.71130702	−3.45310681	1.34648965	0.92347185
B201	−19.43392584	−10.98545896	−2.98414802	−1.98988265
B202	128.14419383	83.39330539	10.39294135	6.54161888
B203	−113.60415693	−70.81677473	−5.87343489	−3.06421270
B210	7.38594345	8.79293717	−13.50456771	−0.54145289
B211	31.16184563	13.54374443	6.99683086	2.98736506
B212	−191.47460054	−84.08993848	−20.14201283	−9.86271093
B213	169.13768966	64.21204903	13.23470271	3.96167934
B220	−3.50020020	−2.88758460	2.92369152	0.21513599
B221	−10.20602240	−3.90257456	−3.51491527	−1.11741962
B222	61.29518085	22.16620406	11.37577986	3.98832016
B223	−56.09238536	−16.60776105	−9.02412646	−2.59177513
B300	−1.70333671	5.21352610	−1.77532374	−1.22657847
B301	23.21947053	12.52222219	3.12320971	2.65987968
B302	−152.02171096	−95.11136856	−10.54023133	−8.72133304
B303	136.98190136	81.48860037	4.98756613	4.21946093
B310	−9.94080821	−10.74295110	17.61092758	0.52441534
B311	−33.23738351	−12.24897695	−7.81291551	−3.93078563
B312	207.82900701	74.75162929	21.63941114	13.16446836
B313	−186.74053493	−56.35063682	−13.95597630	−6.11402160
B320	4.29980209	3.47310346	−3.65220951	−0.10929042
B321	9.85469046	3.00270644	3.63087608	1.28912724
B322	−60.36007433	−14.89454202	−11.33743808	−4.70097988
B323	56.65376653	10.21787079	8.98357416	3.19950135
B400	0.06280794	−2.63453743	0.81712081	0.56340281
B401	−9.12587744	−4.68182870	−1.31458122	−1.24408141
B402	59.35516186	35.94069556	4.34145499	4.06090311
B403	−53.72016472	−30.23536222	−1.89908877	−2.03978598
B410	4.57316019	4.81640582	−6.91863296	1.21848235
B411	11.80632175	3.56031425	3.05556434	1.40308246
B412	−73.08155454	−17.55744402	−8.40593165	−5.04388810
B413	65.72840511	10.94554078	5.35393201	2.06270792
B420	−1.89434263	−1.55447446	1.36061773	−0.27828762
B421	−3.26475845	−0.82772949	−1.28839795	−0.34781653
B422	19.41098071	1.74847956	3.96776116	1.44726447
B423	−18.32252430	−0.03782336	−3.14575952	−0.91008261

TABLE 3—D_{ijk} *coefficients.*

	Corner Flaw		Embedded Flaw	
$D_{(ijk)}$	Open Hole	Loaded Hole	Open Hole	Loaded Hole
D000	−11.30209433	1.53918677	−11.37580430	1.31689204
D001	22.70090139	−1.58451821	−1.53135798	0.24286659
D002	−13.06716713	0.18668058	0.45194066	−0.01695835
D003	2.21031950	0.04020711	−0.05709691	−0.00283567
D010	−5.13183060	−5.55981904	−1.61852567	−2.77054536
D011	8.67693380	9.39254122	−0.12327185	−0.69219035
D012	−4.44570462	−4.80555757	0.03443353	0.21969632
D013	0.70765766	0.76392072	−0.00407024	−0.02812341
D020	1.51296783	1.79840315	0.16786405	0.37277415
D021	−2.65811352	−3.03764555	0.00578870	0.13418273
D022	1.31066360	1.55375624	0.00031777	−0.03944502
D023	−0.20862813	−0.24693353	−0.00009501	0.00495951
D100	150.09567866	30.81032857	153.14582900	23.43973850
D101	−263.09197265	−48.95884207	14.63043320	7.90601766
D102	145.24047077	27.84825284	−4.18441231	−2.76007104
D103	−24.19643430	−4.85902907	0.52226492	0.37779073
D110	48.03542315	49.17320442	14.57436120	23.34628610
D111	−81.21865742	−83.07763473	0.33674697	06.42684150
D112	41.61316701	42.51092505	−0.00408541	−2.11444887
D113	−6.62390868	−6.75862649	−0.00442004	0.27124347
D120	−14.50936302	−15.95839027	−1.40273800	−3.20643447
D121	24.53239256	26.95753368	0.29671142	−1.12318820
D122	−12.56936465	−13.79095084	−0.13054936	0.33047689
D123	2.00076967	2.19208131	0.01764447	−0.04164411
D200	−288.18538977	48.50006053	−290.71992300	255.91050100
D201	536.09888018	−73.66124325	−30.65596380	12.69506690
D202	−308.91006942	28.88298519	9.45625121	−3.11403068
D203	52.70314861	−3.36857356	−1.22360120	0.28773830
D210	−155.33336630	−150.67633701	−49.35394010	−143.04413700
D211	262.63904036	254.58513107	−0.96295817	−25.02328940
D212	−134.56592638	−130.28766237	−0.09442257	7.65421060
D213	21.42001912	20.71632335	0.02959366	−0.97663665
D220	47.85181383	49.00085902	5.71561269	−51.50939800
D221	−80.90807408	−82.78171482	−1.86025042	3.53771897
D222	41.45409905	42.35585462	0.75677968	−0.97756181
D223	−6.59862595	−6.73345458	−0.10123583	0.12154241
D300	195.86989277	−155.11519968	196.35056000	−224.15812600
D301	−393.10038049	255.98669533	23.66125480	−48.49692430
D302	242.52025091	−124.35044976	−8.06247447	14.89107380
D303	−42.84463491	18.74500518	1.09162858	−1.78819023
D310	199.96525643	186.94749541	66.53764140	−125.92040300
D311	−338.10442552	−315.88810838	0.02061319	38.15658540
D312	173.23252114	161.67590699	0.70764590	−10.66811420
D313	−27.57509806	−25.70956278	−0.12039971	1.33821043
D320	−62.12217586	−60.67691042	−8.36001364	8.22478590
D321	105.03734717	102.51429772	4.02651192	−3.68313266
D322	−53.81752344	−52.45800609	−1.56503479	0.93211943
D323	8.56670823	8.34031467	0.20818243	−0.11415242
D400	−43.28394057	77.26019275	−43.07769270	92.19335530
D401	97.38577560	−131.83745670	−6.61272618	27.22417700
D402	−65.79181218	67.45862849	2.54570504	−8.61631964
D403	12.13023537	−10.56161182	−0.36374505	1.07410582
D410	−87.56114109	−79.91215899	−31.11529250	−5.72978741
D411	148.05086203	135.03502181	0.46526174	−25.20354000
D412	−75.85675759	−69.11783165	−0.52194632	5.17626915
D413	12.07498897	10.99182743	0.07929965	−0.61826150
D420	27.26767677	25.83971412	4.44515383	−1.29435201
D421	−46.10522717	−43.65875927	−2.47854771	1.26083538
D422	23.62307678	22.34261365	0.91806344	−0.25247487
D423	−3.76039071	−3.55253611	−0.12095298	0.02996772

TABLE 4—G_{ijk} *coefficients.*

	Corner Flaw		Embedded Flaw	
$G_{(ijk)}$	Open Hole	Loaded Hole	Open Hole	Loaded Hole
G000	−2.28147030	2.90100253	−3.89053771	1.94477106
G001	−0.74986628	0.55341666	−3.93680042	0.16532207
G002	2.74057090	2.12289773	14.89677960	−2.65132687
G003	−1.74618695	−3.28997695	−13.12959645	1.74040264
G010	2.69233262	−2.18612181	1.10567088	−0.08476515
G011	1.10793786	−0.97040117	1.27048653	1.09630447
G012	−2.78401923	−3.30336874	−4.61404531	0.41167951
G013	1.03865879	7.15097694	5.00448891	0.22703359
G020	−0.74933457	0.46016748	0.00493429	−0.18130220
G021	−0.31183725	0.35254457	−0.35651497	−0.54824757
G022	0.58003348	1.11610169	1.19431291	0.41625607
G023	−0.07744028	−2.44823586	−1.63894010	−0.55686399
G100	21.05027387	−27.39370083	34.76386029	−19.74619607
G101	6.16762941	−4.95670278	20.95586637	−3.74609388
G102	−27.78719148	−18.57502434	−74.68975836	31.53641268
G103	19.90161384	29.31424364	65.38065867	−22.18894017
G110	−22.42184801	21.55847358	−9.43128578	3.48972853
G111	−8.95714025	8.82104902	0.46238379	−8.11912503
G112	38.05569802	28.28422903	−8.62613441	−11.99544686
G113	−25.28572967	−64.23314345	−0.13138639	6.03814719
G120	6.32540444	−4.72727136	0.53755626	0.75741841
G121	2.18380630	−3.21103413	−0.70731195	4.53659048
G122	−7.07961154	−9.44152544	4.68636711	−1.25759199
G123	4.05012201	21.89660467	1.15235969	2.44532548
G200	−49.89666389	93.44810067	−84.67635255	71.76530467
G201	−15.82069101	15.96481687	−37.20579344	17.76425585
G202	82.10195884	60.92706243	116.46960744	−117.99618260
G203	−61.94875506	−95.79248848	−105.20531946	84.24574185
G210	55.86478771	−73.95582941	23.73540156	−17.63429729
G211	23.24926851	−28.39098530	−11.31928267	21.48864854
G212	−119.43769087	−88.36281118	84.21338770	53.88534912
G213	91.71710049	206.53010370	−49.48787552	−33.38459921
G220	−16.05081131	16.60716636	−2.36474643	−0.20036887
G221	−4.95948825	10.23182008	4.20886393	−13.45611155
G222	22.15135200	30.29172492	−28.73002469	−1.13113793
G223	−17.33143112	−71.02928537	9.30347408	−2.31963757
G300	50.57779699	−133.99813200	87.88815012	−106.68238203
G301	16.22181477	−22.53437839	24.69476411	−29.18214663
G302	−109.87620327	−79.33381689	−50.86102767	174.10775172
G303	90.14601838	128.37326183	51.94052333	−125.00619247
G310	−55.81915362	109.15820644	−26.17480682	31.67043547
G311	−22.70680660	40.80540020	21.01643457	−25.47895146
G312	172.90265238	108.54457805	−155.90193695	−85.83852337
G313	−156.27580632	−275.85081038	103.88998641	57.14908957
G320	16.42814508	−25.24680673	4.73123926	−2.38683995
G321	3.38090431	−14.41476654	−6.63439340	17.48145486
G322	−30.11650505	−37.88836078	47.80846580	5.88720798
G323	31.26821713	95.01413962	−20.95782761	−1.90479996
G400	−11.35444627	71.99999130	−27.05128976	58.76795560
G401	−5.23080959	11.48224622	−2.12617127	17.25501251
G402	50.21483094	32.47735683	−12.68039150	−91.66384372
G403	−44.63026908	−56.97898055	5.05288745	65.37503353
G410	13.85366228	−59.59878353	9.52071427	−18.51988305
G411	6.50747150	−20.95613975	−11.90707891	10.08262329
G412	−75.01023696	−33.10708127	84.19307072	44.82655334
G413	78.93116591	117.62194282	−57.09594709	−29.89858480
G420	−4.43059200	14.22422760	−2.35800357	2.49493999
G421	−0.45231393	6.90196927	3.19576329	−8.06053122
G422	13.07059864	14.67682486	−24.31393695	−4.13708560
G423	−17.06295833	−42.64223702	12.09292302	3.00753399

Open-Hole Tests

Thirteen specimens contained EDM-machined corner flaws, and eight contained flaws similarly embedded in the side of the hole at the mid-thickness. The specimens were instrumented with crack-propagation gages mounted on the plate surface and in the hole, adjacent to the flaw. The measuring length of the gage along the plate surface was 15 mm, and that of the gage mounted in the bore of the hole was 2.5 mm. Crack growth beyond the range of the latter gage was determined by correlating beach marks on the fracture surface with gage readings obtained on the front surface. Details of the loading program are given in Table 5.

Beach marks occurred as a result of the intentional application of marker loads in most of the open-hole tests, in order to generate a record of crack-front propagation and shape. Several of the marker band patterns obtained from corner-flaw specimens are shown in Fig. 2, enlarged from fractographs. Three corner-flaw and two embedded-flaw specimens were tested under constant amplitude throughout as base tests for comparison with data obtained from marker-load tests. It was found that marker loads cycles at stress ratio of 0.7 and maximum loads equal to the mean fatigue load of the test did not alter overall test results.

Since space does not permit presentation of all crack growth data obtained, representative data from Ref 5 will be presented here. Crack growth from corner-flaw tests is plotted against total cycles in Fig. 6. In two cases (Specimens 18 and 19) it was possible to determine crack propagation down the complete length of the hole and along the rear surface of the plate. In both cases the initial crack propagation rate along the rear surface was approximately 2.5 times

TABLE 5—*Fatigue crack propagation test program for 2024-T351 aluminum-alloy plates containing a central unloaded hole.*

	Test Loading			Marker Loading				
Test No.	$\Delta\sigma$, kg/mm^2	Stress Ratio	Δn	$\Delta\sigma$, kg/mm^2	Stress Ratio	Δn	Total Life	Net Life
				CORNER FLAW				
12	10.9	0.01	...	...	...	...	24 410	24 410
3	7.2	0.01	...	...	...	...	227 940	227 940
7	13.4	−1.00	...	...	...	...	91 030	40 000[a]
13	18.8	0.01	200	9.5	0.5	200	4 600	2 300
14	16.8	0.01	2 500	8.5	0.5	2 500	7 190	3 690
15	12.9	0.01	5 000	6.5	0.5	5 000	22 170	12 170
11	10.9	0.01	5 000	5.5	0.5	5 000	30 290	15 290
23	10.9	0.01	2 500	3.4	0.7	2 500	41 500	21 500
19	10.9	0.01	2 500	1.7	0.7	2 500	51 580	26 680
10	7.9	0.01	25 000	4.0	0.5	25 000	50 410	25 410
22	7.0	0.12	25 000	1.4	0.7	25 000	54 300	34 300
18	5.4	0.10	7 500	1.1	0.7	7 500	80 560	40 560
9	3.6	0.10	25 000	0.7	0.7	25 000	>525 000	>275 000
				EMBEDDED FLAW				
4	18.0	0.10	...	...	...	...	5 990	5 990
6	7.2	0.10	...	...	...	...	62 260	62 260
1	22.5	0.10	100	4.2	0.7	100	3 086	1 586
8	18.0	0.10	500	3.3	0.7	500	5 380	2 880
2	11.7	0.10	1 500	2.2	0.7	1 500	20 500	10 500
16	11.7	0.10	1 500	2.2	0.7	1 500	22 335	11 885
5	7.2	0.10	5 000	1.3	0.7	5 000	83 900	43 900
17	5.4	0.10	12 500	1.0	0.7	12 500	261 680	136 680

[a] Goodman adjustment to stress ratio of 0.01.

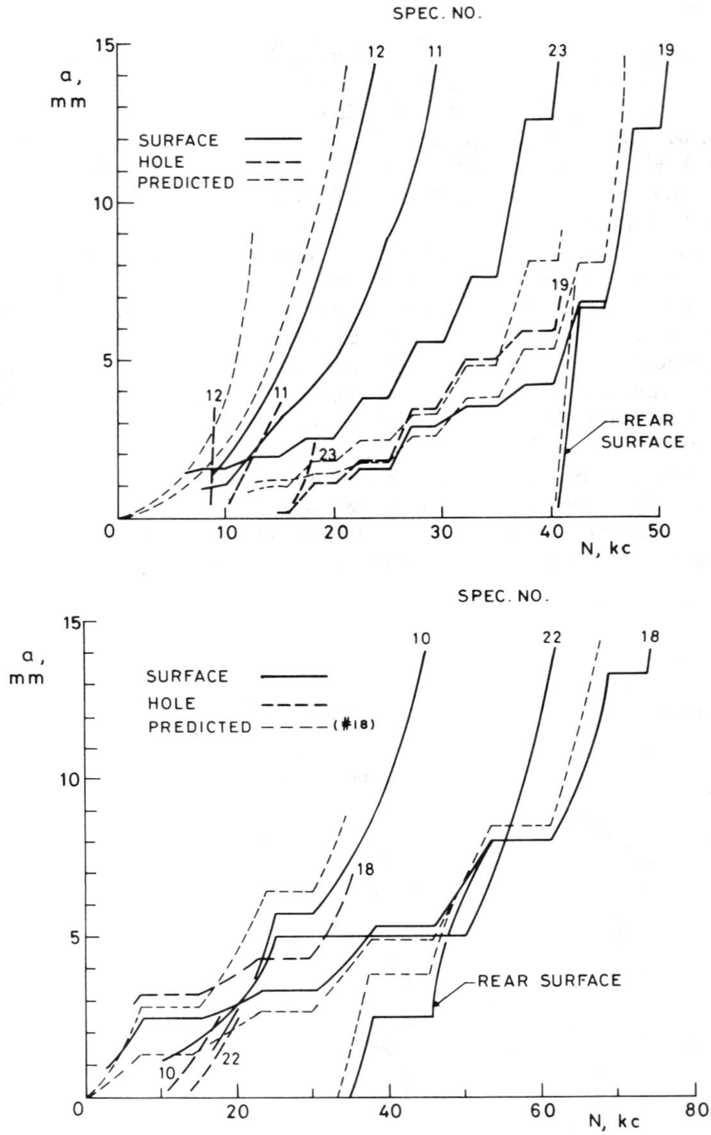

FIG. 6—*Crack growth from corner flaws in open-hole specimens of 2024-T351 aluminum-alloy.*

the corresponding propagation rate on the front surface. In general, for crack lengths greater than 2 mm, the crack edge down the hole led that along the front surface, although there appeared to be some impediment to the start of propagation down the hole. Embedded-flaw data were characterized by somewhat higher propagation rates.

The effect of marker load parameters are clearly seen in Fig. 6*a,* in which data from four tests under the same fatigue loads are shown. Specimen 12 was tested under constant amplitude,

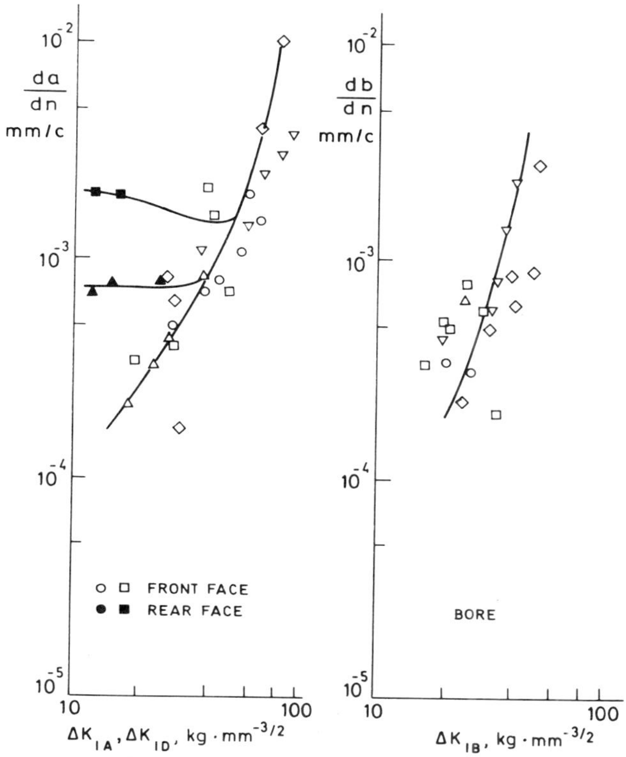

FIG. 7—*Crack propagation rate for open-hole, corner flaw specimens of 2024-T351 aluminum-alloy.*

while the other tests included marker loading. The figure is also an indication of the order of scatter obtained in the corner-flaw tests, and probably for embedded-flaw tests as well.

Crack propagation rates derived from the crack growth curves are shown in Fig. 7. At a given crack length and stress range, the propagation rate from an embedded crack was generally higher than that of a crack emanating from a corner flaw. Such data were used to compare with geometric correction factors for unloaded-hole plates obtained from calculated results. Similar data were also developed from loaded-hole lug specimens.

Lug Tests

Initial flaws were milled on one or both sides of the loaded hole of round headed lug specimens [*4*]. The lugs were instrumented with crack gages mounted on both lug faces adjacent to the loading holes. Fatigue loading consisted of either a constant amplitude of 6.3 kg/mm^2, or one of two quasi-randomized fighter maneuver spectra. The former were all single corner-flaw tests, and the latter embedded-flaw tests.

Crack growth obtained from maneuver-load tests is plotted against load cycles in Fig. 8. The scale of the abscissa of Fig. 8 is such that the curves appear smooth even though they represent maneuver-load data. Fatigue crack propagation rates obtained from the lug results showed that at a given stress-intensity factor range the crack propagation rate was somewhat higher for 2024 lugs than for 7075-T7351 lugs.

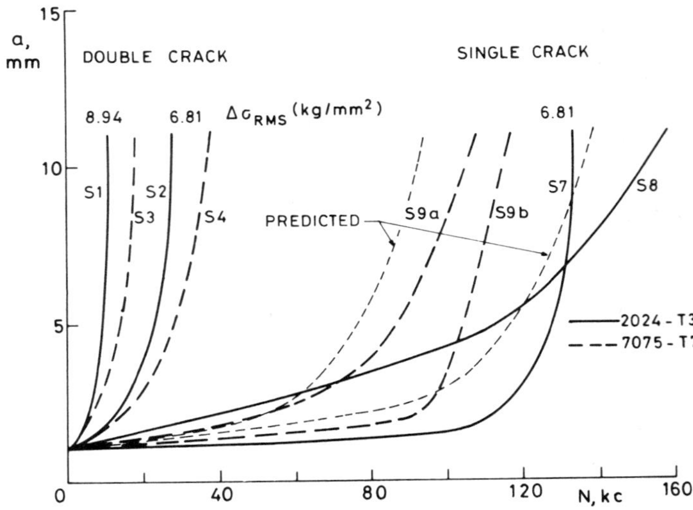

FIG. 8—*Crack growth from embedded flaws in aluminum-alloy lugs of 9.5 mm thickness under maneuver load spectra.*

Comparisons and Discussion

Predictions of fatigue crack growth based on stress-intensity factors estimated with the aid of geometric correction factors calculated by the parabolic-shaped crack front approach were superimposed as dotted lines on experimental data such as presented in Figs. 6 and 8. In calculating fatigue crack propagation, stress-intensity factors obtained from Eqs 13 were introduced into the Walker equation for crack propagation rate. Crack dimensions a_0 and b_0 were allowed to develop within the limits of Eq 12. Clearly, the predicted crack growth is reasonable in all cases, along the front and rear surfaces of the plate specimens as well as down the (unloaded) holes.

K_{Ia} calculated for the parabola and experimental values derived from marker bands initiating from corner flaws are compared for open-hole 2024-T351 plates and for pin-loaded lugs in Fig. 9. The lugs included aluminum alloy [*4,6*] and PSM-8 photoelastic specimens [*14*]. In general, the values calculated for $a_0/b_0 = 0.5$ fall in the vicinity of the experimental data.

In Fig. 10, geometric correction factors to K_{Ia} and K_{Ib} are plotted for the two approaches against experimental values obtained photoelastically [*15*] from open-hole specimens of PSM-8 material containing corner flaws. The experimental evidence strongly supports the parabolic approach. For K_{Ia}, points for the parabola fall within $\pm 10\%$ of the experimental results, whereas the ellipse predicts values which average 10% higher. Correlation between calculated and experimental results for K_{Ib} is more conservative than for K_{Ia}, but the results for the parabola are still reasonable. The parabola results are 15% higher than the test results, whereas the ellipse gives results which are 60% higher.

The physical arguments set forth earlier in support of the parabolic crack shape and the mechanism of crack growth at the wall of the hole seem to be justified by these results. The success of the parabolic approach in the straight hole cases presented in this report suggests that the method could be applicable to other types of holes. These might include the plastically cold-worked hole, the countersink hole and the blind hole. These cases will be the subject of future work.

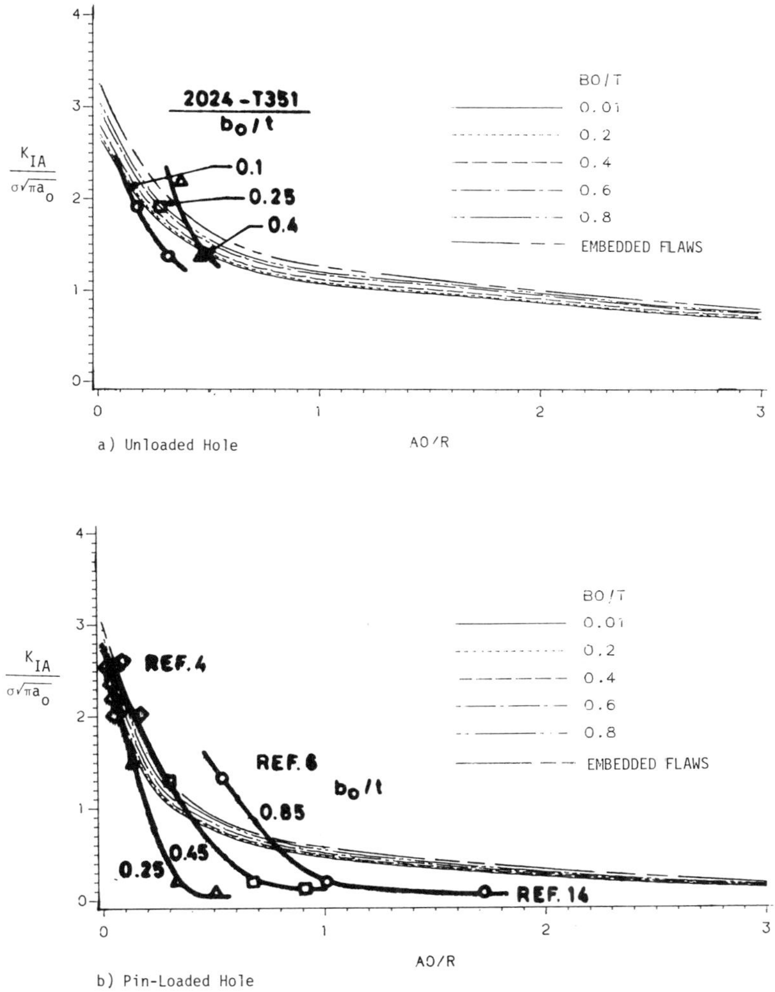

FIG. 9—*Calculated and experimental stress intensity factors along plate face for corner flaws.*

Conclusions

A crack of parabolic shape, emanating from corner and embedded flaws at a hole in a plate, was analyzed by the slice synthesis method. In contrast with other crack shapes, equilibrium solutions for a parabolic crack are finite and continuous as the crack passes the bottom of the hole and starts to propagate along the rear surface. Fracture parameters derived were K_I for predicting fatigue crack growth along the plate surfaces and down the bore of the hole, and G_I for residual strength. The cases studied included a flaw (*a*) emanating from the corner of the hole, or (*b*) embedded in the side of the hole at the mid-depth, each case being either (*c*) remotely loaded, or (*d*) loaded at the hole.

Fracture parameters derived on the basis of a parabolic crack shape were generally similar in value to those obtained using an elliptical crack. The parabolic crack permitted calculation of

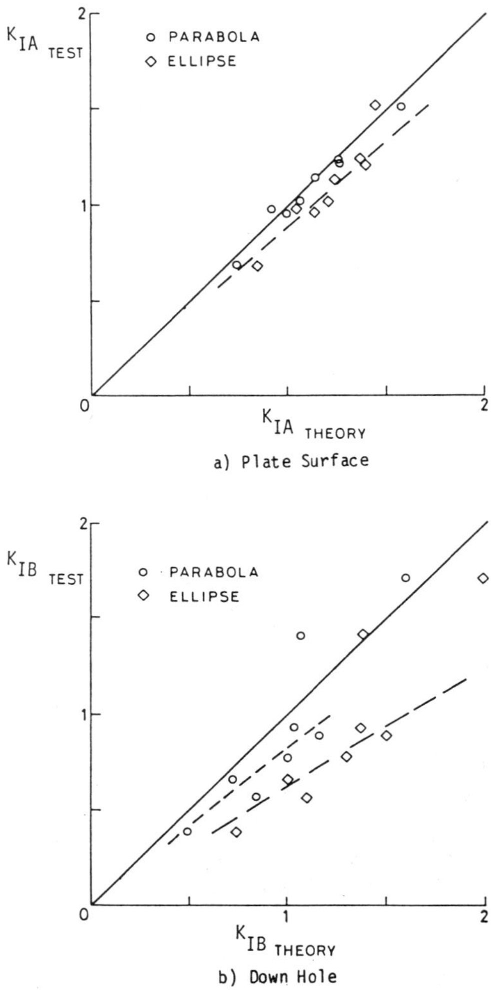

FIG. 10—*Calculated and experimental stress intensity factors for unloaded hole, corner flaw in PSM-8 material* [15].

stress intensity factors down to the end of the hole and then along the rear surface of the plate. Calculated values for the rear surface were found to be considerably higher than the corresponding front face values (for embedded flaws, the values at the plane of crack symmetry) when the crack breaks through, and approached the front face values as the crack advanced.

Correlation of calculated stress-intensity factors with values derived from marker bands in both unloaded-hole and loaded-hole cases indicated that the results of the parabola approach were in error within 15%, whereas results of the ellipse approach erred by 10 to 60%. Fatigue crack growth curves calculated by the parabolic-crack approach for a variety of geometries and loading cases were in good agreement with data obtained from tests.

Acknowledgment

This research was supported in part under USAF/EOARD Grant No. AFOSR-81-0242. The authors are indebted to S. Nahmani and A. Grunwald for their assistance in the laboratory and to A. Kapitanovsky for his perseverance at the computer terminal.

References

[*1*] Fujimoto, W. T. in *Proceedings,* 17th Structural Dynamics and Materials Conference, American Institute of Astronautics and Aeronautics/American Society of Mechanical Engineers/Society of Automotive Engineers, King of Prussia, PA, May 1976.

[*2*] Fujimoto, W. T. and Saff, C. R., *Experimental Mechanics,* Vol. 22, No. 4, April 1982, pp. 139–146.

[*3*] Bowie, O. L., *Journal of Mathematics and Physics,* Vol. 35, 1956, pp. 60–70.

[*4*] Buch, A. and Berkovits, A., "Fatigue Crack Propagation in Al-Alloy Lugs under Constant Amplitude and Maneuver Spectrum Loading," TAE Report No. 497, Technion—Israel Institute of Technology, Haifa, Israel, Feb. 1983.

[*5*] Berkovits, A. and Prinz, D., "Prediction of Fatigue Crack Growth from Flaws Emanating from Fastener Holes," TAE Report No. 554, Technion—Israel Institute of Technology, Haifa, Israel, Sept. 1984.

[*6*] Friedrich, S. and Schijve, J., "Fatigue Crack Growth of Corner Cracks in Lug Specimens," Report LR-375, Delft University of Technology, Delft, The Netherlands, Jan. 1983.

[*7*] Rice, J. R., *International Journal of Solids and Structures,* Vol. 8, 1972, pp. 751–758.

[*8*] Bueckner, H. F., *Zeitscrift fuer Angewandte Mathematik und Mechanik,* Vol. 50, 1970, pp. 529–546.

[*9*] Timoshenko, S. P. and Goodier, J. N., *Theory of Elasticity,* McGraw-Hill, New York, 1970.

[*10*] Kathiresan, K., Hsu, T. M., and Rudd, J. L., "Stress and Fracture Analyses of Tapered Attachment Lugs," Air Force Wright Aeronautical Laboratory, Wright-Patterson AFB, OH, 1982.

[*11*] Van der Velden, R. V. and Louwaard, E. P. in *Proceedings,* 12th International Committee of Aeronautical Fatigue Symposium, Toulouse, France, May 1983, pp. 1.5/1–1.5/52.

[*12*] Callinan, R. J., "Stress Analysis of a Lug Loaded by a Pin," ARL/Structural Note 439, Aeronautical Research Laboratories, Melbourne, Australia, July 1977.

[*13*] Paris, P. C. in *Proceedings,* 10th Sagamore Conference, Syracuse University Press, Syracuse, NY, 1964, p. 107.

[*14*] Smith, C. W., Jolles, M., and Peters, W. H. in *Flaw Growth and Fracture, ASTM STP 631,* American Society for Testing and Materials, Philadelphia, 1977, pp. 190–201.

[*15*] McGowan, J. J. and Smith, C. W. in *Mechanics of Crack Growth, ASTM STP 590,* American Society for Testing and Materials, Philadelphia, 1975, pp. 460–476.

Weldon W. Wilkening[1]

Computation of the Crack-Tip Energy Release Rate for Cyclic Crack Growth

REFERENCE: Wilkening, W. W., "**Computation of the Crack-Tip Energy Release Rate for Cyclic Crack Growth,**" *Fracture Mechanics: Eighteenth Symposium, ASTM STP 945,* D. T. Read and R. P. Reed, Eds., American Society for Testing and Materials, Philadelphia, 1988, pp. 1070–1082.

ABSTRACT: The elastic-plastic finite-element method, employing the mechanical sublayer theory of plasticity, was used to compute a local crack-tip parameter which shows significant promise for characterization of the cyclic crack growth process. The proposed cyclic crack growth rate characterizing parameter is the crack tip value of the energy release rate. This crack-tip energy release rate (G_0), has been shown to be equivalent to the global energy release rate (G), for the case of monotonic loading of a stationary crack. This work extends the proposed definition of the crack tip energy release rate to cyclic loading, including the important effects of crack growth.

Elastic-plastic incremental finite-element analyses of a model which represents the elastically-bounded crack-tip region for the case of small-scale-yielding (SSY) have been performed, for both stationary and growing cracks. The crack-tip energy release rate G_0, computed by the simple technique described here, was compared with the magnitude of the remote elastic K-field loading which was applied at the outer boundary of the SSY model, under plane-strain conditions.

The effects of plasticity-induced crack closure upon G_0 were investigated by allowing the crack tip to advance into the prior residual plastic zone in several steps, until a steady-state condition was achieved. The occurrence of crack face closure during the unloading portion of each loading cycle and of the subsequent reopening of the crack during the next loading cycle was accurately determined. From these computational results, a modified definition of G_0, embodying the effects of plasticity-induced crack closure, was developed. The results of these investigations show, generally, that the crack-tip energy release rate, when appropriately defined and computed in the context of cyclic loading, crack advance and crack face closure, provides a consistent and rational description of the local driving force acting upon the tip of a growing crack. As such, G_0 should prove to be a viable cyclic crack growth characterizing parameter.

KEY WORDS: crack tip energy release rate, cyclic loading, global release rate, small scale yielding, stress intensity factor

Prediction of the driving force for crack growth under cyclic loading conditions poses many problems not encountered under monotonic loading. For the elastic-plastic conditions pertinent to the cases of low-cycle fatigue of interest here, neither the stress-intensity factor, K, nor its cyclic range, ΔK, can be considered appropriate for use in characterizing the crack-tip stress field. Repeated unloading of the entire component at each load reversal and the nonproportional loading conditions resulting from the reversed yielding at the crack tip both lead to violation of the conditions for path-independence of the J-integral. Hence, the J-integral loses its significance as a measure of the crack-tip stress and strain singularities. Consequently, the J-integral is no longer viable as a crack-tip field characterizing parameter, either.

Therefore, neither K nor J can be expected to serve as a measure of the cyclic crack driving force. The present work has as its goal the development of an analytical technique for computing a candidate cyclic crack growth characterizing parameter, for general cyclic loading, which

[1]Mechanical engineer, General Electric Company, Schenectady, NY 12301.

is analogous to the monotonic J-integral. The proposed cyclic crack growth characterizing parameter is the crack-tip energy release rate, which will be denoted by G_0. The crack-tip value of the energy release rate, G_0, was selected because it is equivalent to the global energy release rate, G, for monotonic loading. Therefore, G_0 carries the alternate significance of the J-integral—the derivative of the pseudo-potential energy with respect to crack length—which is often used as the operational definition of J, in both analytical and experimental studies [*1,2*]. Also, since the elastic-plastic fracture methodology based upon J-controlled crack tip field parameters has been so successful in describing the crack driving force for monotonic loading [*3,4*], the present work seeks to develop the potential analogy of a "J-like" parameter for cyclic loading. The symbol G_0 is used here to represent the crack-tip energy release rate, in order to distinguish it from the J-integral and for the purpose of drawing the analogy to the global energy release rate, G.

Definition of the crack-tip energy release rate is made in terms of the virtual work required to close an infinitesimal portion of the crack, da, as described next. The analytical method proposed here for computing the crack-tip energy release rate is based upon two-dimensional elastic-plastic finite element techniques. The incremental cyclic plasticity analyses involved were accomplished with the use of the mechanical sublayer plasticity model [*5,6*], which was added to a special General Electric Company version of the nonlinear, incremental finite element code ADINA [*7*].

Background

The special finite-element tools and techniques required for handling the unique aspects of cyclic loading, except for crack advance, have been previously described by German, deLorenzi, and Wilkening [*8*]. They also first presented and discussed the basic definition of the crack-tip energy release rate, but only for the case of a stationary crack. The current work represents the extension of those concepts to the important general case of arbitrary cyclic loading, with associated crack advance. Crack advance into the plastic zone from prior loading cycles opens the possibility of plasticity-induced crack closure. Then only a portion of the loading cycle is expected to contribute to the driving force for continued crack growth.

Various definitions of the effective elastic stress-intensity factor range, or of the effective range of the J-integral, have been developed and used to explain a variety of cyclic crack growth effects. However, those definitions usually depend upon the calculation of ΔK or ΔJ from expressions which are based upon monotonic loading analyses. The typical approach involves the computation of ΔK or ΔJ from the portion of the remote load or displacement range for which the crack faces are not in contact [*9*].

Prediction or measurement of the occurrence of crack face closure and re-opening therefore have become the focus of many of these studies. The goal of this work is to study a candidate measure of the cyclic crack growth driving force which carries the physical significance of an energy release rate, and which is defined in terms of the local conditions at the crack tip. The crack-tip energy release rate fulfills both of these requirements.

The crack-tip energy release rate is simply the limit of the increment of the virtual crack closure work with respect to the crack closure increment, Δa, as Δa approaches zero

$$G_0 = \lim_{\Delta a \to 0} \Delta W/\Delta a = [\partial W/\partial a]_\delta \tag{1}$$

where δ is the remotely applied displacement.

A schematic description of the crack closure work calculation method, by analogy with the analytical computation of the virtual crack closure work, ΔW, is presented in Fig. 1. The left side of the figure illustrates the development of the expression for the elastic work required to

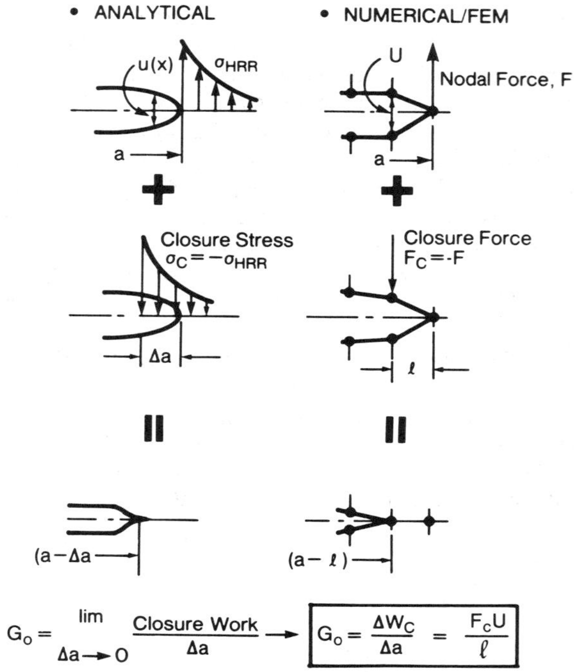

FIG. 1—*Calculation of the crack-tip energy release rate,* G_0.

close the crack over the incremental distance *da*. The right side of the figure illustrates the finite-element implementation of the virtual crack closure work calculation, for the simple but informative case of four-noded finite elements.

Provided the finite element mesh is sufficiently refined, the virtual work required to close the crack over the length of one crack tip element gives a sufficient approximation to that limit

$$G_0 \cong \Delta W/\ell = 2 \cdot \left[\int_0^U F du \right] / \ell \tag{2}$$

therefore

$$G_0 = FU/\ell \tag{3}$$

While a systematic study of the mesh refinement required to adequately satisfy the limit of Eq 1 has not been conducted, the development work described in Ref *8* has demonstrated that the degree of mesh refinement required for accurately computing G_0 is no greater than that required for the valid determination of J by the method of virtual crack extension [*8*].

For the case of 8-noded isoparametric finite elements, it is convenient to retain the crack-tip element size, ℓ, as the virtual crack advance increment, so that two virtual work terms are then involved in the closure work calculation, as illustrated in Fig. 2, and in the equations

$$\Delta W = 2 \cdot \left[\int_0^{U_1} F_1 du + \int_0^{U_2} F_2 du \right] \tag{4}$$

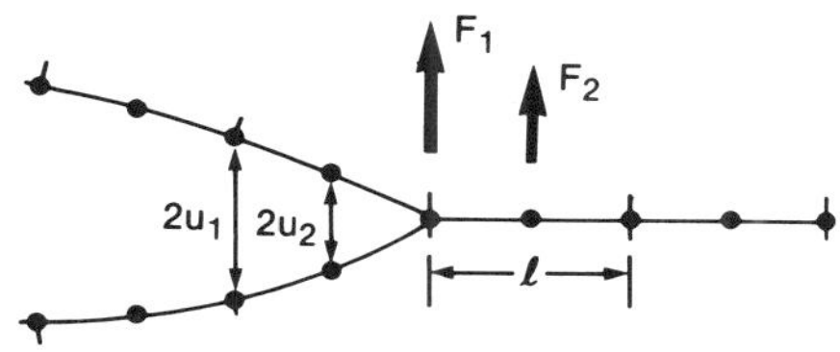

$$G_o = \frac{F_1u_1 + F_2u_2}{\ell}$$

FIG. 2—*Calculation of* G_0 *for the case of 8-noded finite elements.*

therefore

$$G_0 = [F_1U_1 + F_2U_2]/\ell \tag{5}$$

The monotonic results from Ref *8* demonstrated that the crack-tip energy release rate, computed from Eq 5, equals the global energy release rate, computed by the method of virtual crack extension, for monotonic loading. For the case of a stationary crack, the cyclic loading results from Ref *8* suggested that the simple operational definition of the monotonic crack-tip energy release rate could be extended to the case of cyclic loading in a logical fashion, by simply replacing each of the two reaction forces and the two crack opening displacements with their respective ranges

$$\Delta G_0 = [\Delta F_1 \Delta U_1 + \Delta F_2 \Delta U_2]/\ell \tag{6}$$

This approach was an attempt to define the range of the crack-tip energy release rate, relative to the starting point of the increasing-load portion of the loading cycle. The effects of prior yielding and the consequent residual stresses were essentially lumped together into a "zero shift" of the closure work. The logical extension of that concept to cases involving plasticity-induced crack closure would be to measure the range of each variable from the point at which reopening of the crack occurred. A different conclusion is to be presented here—one which leads to a much simpler and more physically appealing result.

Analytical Procedure

The constant-amplitude cyclic loading of a finite-element model representing the crack-tip region of a hypothetical large structural component subjected to cyclic loading within the small-scale-yielding (SSY) regime was chosen as the vehicle for demonstrating the proposed analytical technique for computing the crack tip energy release rate. The SSY finite element model used for the cyclic loading analyses to be described here is illustrated in Fig. 3*a*. By virtue of symmetry across the crack plane, only half of the circular region centered at the crack tip was modeled. The crack extended to the left in the model, with symmetry conditions applied along the uncracked ligament to the right of the crack tip.

The material model used in these analyses was representative of the aircraft turbine disk alloy DA718, at 398°C (750°F). The cyclic stress-strain curve was represented by the Ramberg-Osgood relation

$$\epsilon/\epsilon_0 = \sigma/\sigma_0 + \alpha \cdot [\sigma/\sigma_0]^n \tag{7}$$

where $\epsilon_0 = \sigma_0/E$. The reference stress, σ_0, was set equal to the yield stress, 344.7 MPa (50 ksi). The elastic modulus, E, was 181.3 GPa (26.3 $\times$ 10^6 psi), and Poisson's ratio, ν, was 0.3. The

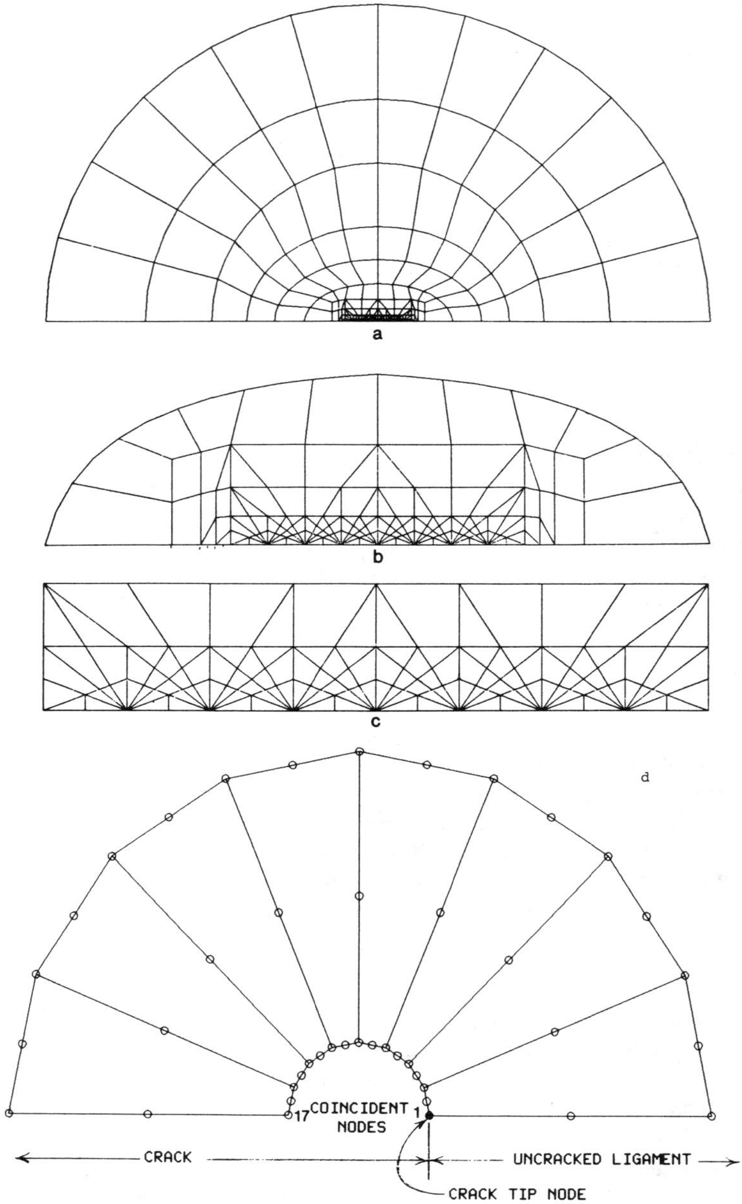

FIG. 3—*Finite element mesh for the small-scale-yielding (SSY) model:* (a) *complete model;* (b) *and* (c) *successive closeups of the refined region;* (d) *detail of the collapsed-node crack tip.*

power-law constants, α and n, were 0.01 and 8.2, respectively. The mechanical sublayer theory of plasticity [*5, 6*] was used to model the elastic-plastic behavior in terms of a multilinear representation of the constitutive behavior. More importantly, however, the mechanical sublayer plasticity model was used to provide the most faithful representation available for the loading history-dependence of strain hardening.

Remote K-controlled loading was produced by applying K-field displacements at the outer periphery of the model. This boundary represents the elastic boundary in a hypothetical structure which is many times larger than the finite element model, such that the remote loading is purely linear elastic. The analysis technique therefore provided an unambiguous connection between the remotely applied value of the stress-intensity factor (or the corresponding value of J) and the crack-tip energy release rate. Constant-amplitude cycling between $K = 0$ and $K = 54.94$ MPa $\sqrt{\text{m}}$ (50 ksi$\sqrt{\text{in.}}$), under plane-strain conditions, was applied to the model. Ten solution steps, with equal increments in K, were used for each of the increasing-load portions of the loading history.

The goal of this set of analyses was to advance the crack tip deeply into the plastic zone resulting from previous loading cycles, to establish a "steady-state" crack closure condition representative of a typical low-cycle fatigue experiment. Consequently, the finite-element mesh was constructed with a repetitive series of crack-tip regions. Each crack-tip region was separated from the next by a distance equal to 2.5% of the radius of the model. In the carefully controlled manner which is described in the following section, the crack tip was advanced from one of these locations to the next at each occurrence of the peak in the applied loading. While the total crack advance increment modeled in this example is large in absolute size, it is still quite small compared to the overall dimensions of the hypothetical structural component. The important point being investigated, however, is the effect of crack growth which is large compared to the initial (monotonic) plastic zone.

Closeup views of the crack-tip mesh structure are shown in Figs. 3*b* and 3*c*. Each of the crack-tip locations contained the collapsed-node structure, in which each of the pie-shaped crack tip elements contains three distinct nodes at its degenerate tip, as illustrated in Fig. 3*d*. Each of the extra crack tip nodes is free to move upon loading, which gives rise to the $1/r$ strain singularity, as first pointed out by Levy et al. [*10*] and discussed by deLorenzi and Shih [*11*].

The independent motion of the individual nodes at the crack tip also provides a somewhat realistic modeling of crack-tip blunting behavior. The collapsed-node crack tip structure, coupled with a reasonably refined finite element model, has been shown to be extremely useful in modeling crack tip behavior for general cases involving elastic-plastic behavior, including linear elastic behavior [*3,4,11*].

In the case of linear elastic fracture mechanics behavior (LEFM), the stress and strain singularities desired are both of $1/\sqrt{r}$ strength, while the general elastic-plastic case involves a stress singularity of strength $r^{1/(n+1)}$ and a strain singularity of strength $r^{n/(n+1)}$. These tend to r^0 and r^{-1}, respectively, in the limit of ideal plasticity (that is, $n \rightarrow \infty$). The quarter-point singularity elements of Barsoum [*12*] exhibit either a $1/\sqrt{r}$ strain singularity or combined $1/\sqrt{r}$ and $1/r$ strain singularities, depending upon whether or not the duplicate crack tip nodes are fixed together. The extensive crack tip modeling work summarized in Refs *3* and *4* has shown that use of the combined singularities results in an unrealistic oscillatory behavior of the crack face opening profile. Therefore the straight $1/\sqrt{r}$ singularity is preferred for LEFM, while the straight $1/r$ strain singularity is preferred for ideal plasticity. For intermediate elastic-plastic cases, and especially for cases involving the low-hardening materials of predominant interest in structural components, the nonhardening $1/r$ strain singularity seems to be more relevant than the $1/\sqrt{r}$ LEFM singularity.

With 8 elements distributed circumferentially about the crack tip in this model, 17 nodes were located at each crack-tip location. For each crack advance increment, 16 of the nodes (all but the crack-tip node), and the 4 nodes between adjacent crack tip locations, were released. Nodes currently on the uncracked ligament were held fixed in the vertical direction by means of a stiff truss element anchored to a fixed node below the model. Nodes along the crack face were each attached to a bilinear truss element. These elements were very stiff in compression, so that they acted to prevent the upper crack face from passing through the symmetry plane (that is, crack face interpenetration). The stiffness of these trusses in tension was zero, however, so that crack-opening displacements were not hampered. Each of these crack face closure trusses was assigned an element "birth time," and each of the reaction trusses along the uncracked ligament was assigned an element "death time." As the solution time for the crack advance process was reached, the next set of 20 nodes (which had been a part of the uncracked ligament) lost their connections to the reaction trusses and gained connections to the crack closure trusses. In addition, the reaction force which had been provided by the reaction truss was replaced by an equivalent concentrated nodal force, for each of the 20 nodes. The set of 20 nodal forces was then relaxed to zero, in four solution steps, while the remote load (actually, the K-controlled displacement field) was held fixed at its peak value, as shown in Fig. 4.

The fourth or central crack tip location was the "primary" crack-tip location. Ideally, steady-state conditions would be achieved by the time the crack advanced from the first station to the fourth. Three additional crack-tip locations were included in the model, however, in case further crack advance was needed to achieve steady state. It is important to note that the sequential analyses performed here were not intended to represent a realistic cyclic loading history, since the analyses began with a crack of considerable length, embedded in a fictitious material with no residual stress state. Consequently, the initial loading cycle can be viewed as directly analogous to a conventional monotonic fracture mechanics analysis, but the subsequent cycles are merely transitions toward the desired steady-state condition. It will be shown in the following that conditions which were essentially steady state were achieved after three crack advance increments, as desired. Therefore, the results to be discussed here are those pertaining to loading with the crack tip located at the central location.

Since the outer boundary of the semicircular model did not translate or distort with crack advance, the application of K-controlled remote loading was only approximately achieved for each of the crack-tip locations, except for the central one. Monotonic loading calibration analyses were conducted, with the crack tip initially located at each of the possible crack tip locations in the mesh. The results of these analyses showed that the error between the global energy release rate (which was computed by the method of virtual crack extension) and the J value corresponding to the remotely applied stress-intensity factor was less than 5% for the first and sev-

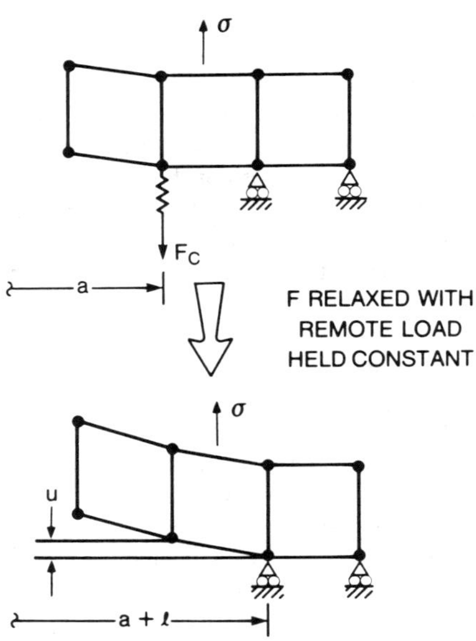

FIG. 4—*Schematic of the crack growth modeling technique.*

enth crack tip locations, for the range of loading used here. The error between the two measures of J was progressively less for the other locations, as the crack tip location approached the center of the model.

The plastic zones developed at each of the loading peaks are illustrated in Fig. 5. Each of the outlines represents the boundary containing all of the Gaussian integration points in the finite element model at which the current level of the von Mises equivalent stress is at or above the current (cyclicly hardened) yield stress. (Note that this is to be distinguished from the set of points at which the current equivalent stress is at or above the original yield stress, σ_0). The figure demonstrates the familiar plane-strain plastic zone for the initial loading cycle and shows dramatically the attainment of a steady-state cyclic plastic zone after the first few crack advance increments. Also, it is clear from the figure that the crack tip did advance completely through the initial plastic zone during the first three crack advance increments, as desired.

Results and Discussion

For each of the four increasing-load portions of the imposed loading history, the crack-tip energy release rate was computed, at each solution step, by means of the simple formula presented in Eq 5. The remote J value representing the applied loading was computed from the applied stress-intensity factor, K, as follows

$$J = (1 - \nu^2)\, K^2/E \tag{8}$$

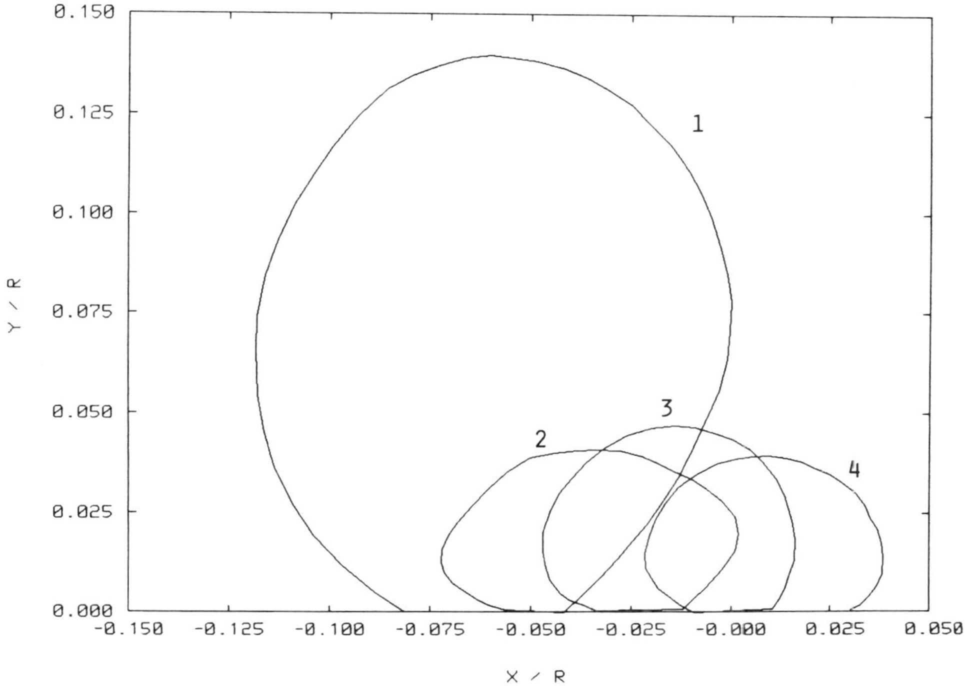

FIG. 5—*Plastic zone boundaries for each of the four loading cycles analyzed. The center of the model is at* X/R = *0,* Y/R = *0.*

The resulting crack-tip energy release rate values, G_0, are plotted against the corresponding remote J values in Fig. 6. The figure shows that the crack-tip energy release rate becomes positive at about 28% of the peak value of the applied K loading (or at about 8% of the peak value of the applied J loading), for the second through fourth loading cycles. Below that point, the residual compressive "crack closure" stresses result in negative values for the crack-tip reaction forces, F_1 and F_2, as the elastic material outside the cyclicly yielded region clamps the crack faces closed. Since u_1 and u_2 must be always positive (or zero), G_0 is negative while the crack faces are clamped shut. F_1 and F_2 become positive as the crack reopens (or, conversely, the crack reopens when F_1 and F_2 become positive!), which demonstrates the convenient and important result that this simple definition of G_0 leads to a natural zero value when the crack reopens. Therefore, the artificial "zero-shift" definition suggested in Ref *8* for the cyclic crack-tip energy release rate range, ΔG_0, is not needed—ΔG_0 is simply given by the maximum value of G_0, since G_0 is positive only when the crack is open.

Once the crack reopens, the relationship between G_0 and the remote value of J is essentially linear. Note that this does not mean that the crack tip behavior is linear, but only that the value of G_0 is approximately proportional to J. For larger amounts of crack advance than modeled here, it is likely that the relationship between G_0 and J would be also nonlinear. However, it is anticipated that a *unique* relationship between G_0 and J would still exist.

An apparent "shakedown" trend is also displayed by the data, but it must be remembered that there is also an error trend of a few percent due to the off-center nature of the applied K-controlled loading. Taking that into account, it can be shown that the desired steady-state condition was approximately reached by the fourth loading cycle, at which time the crack tip was at the central location.

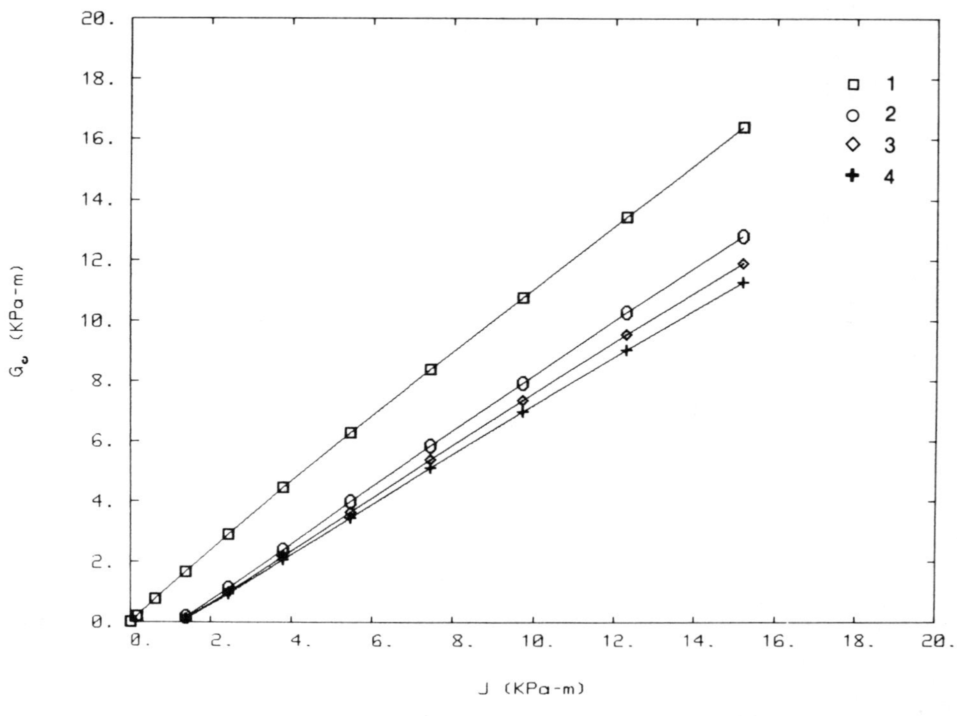

FIG. 6—*Variation of the crack-tip energy release rate,* G_0, *with remotely applied* J.

The corresponding variation of the crack opening displacement (COD) with the remotely applied J-value, for the four loading cycles, are presented in Fig. 7. The opening displacement of the seventeenth node of the collapsed crack-tip structure was used as a convenient and consistent measure of the crack-opening displacement (see Fig. 3*d*). Here the residual crack opening displacement produced by yielding at the crack tip is very evident, as is the significantly smaller COD amplitude for the second through fourth loading cycles. As with the crack tip energy release rate, the COD varies nearly linearly with the remotely applied J-value.

In Fig. 8, G_0 is plotted against COD for each of the four increasing-load portions of the loading history. Here again, as demonstrated for the case of monotonic loading under J-controlled conditions [3] and cyclic loading of a stationary crack [8], the relationship between the COD and the energy release rate is linear for each individual loading cycle

$$G_0/Y = \beta \cdot \mathrm{COD} \tag{9}$$

where Y is the yield strength and β is a material-dependent constant.

For monotonic loading, Y is simply the yield strength, σ_0. For cyclic loading, however, the effective value of Y is $2\sigma_0$. If the cyclic range of the crack tip energy release rate, ΔG_0, (which is simply equal to the value of G_0 here) is normalized by Y and plotted against the corresponding cyclic range of the COD, the constant β in Eq 9 is found to be the same for both the first (monotonic) and the subsequent (cyclic) loading cycles, as illustrated by the plots in Fig. 9.

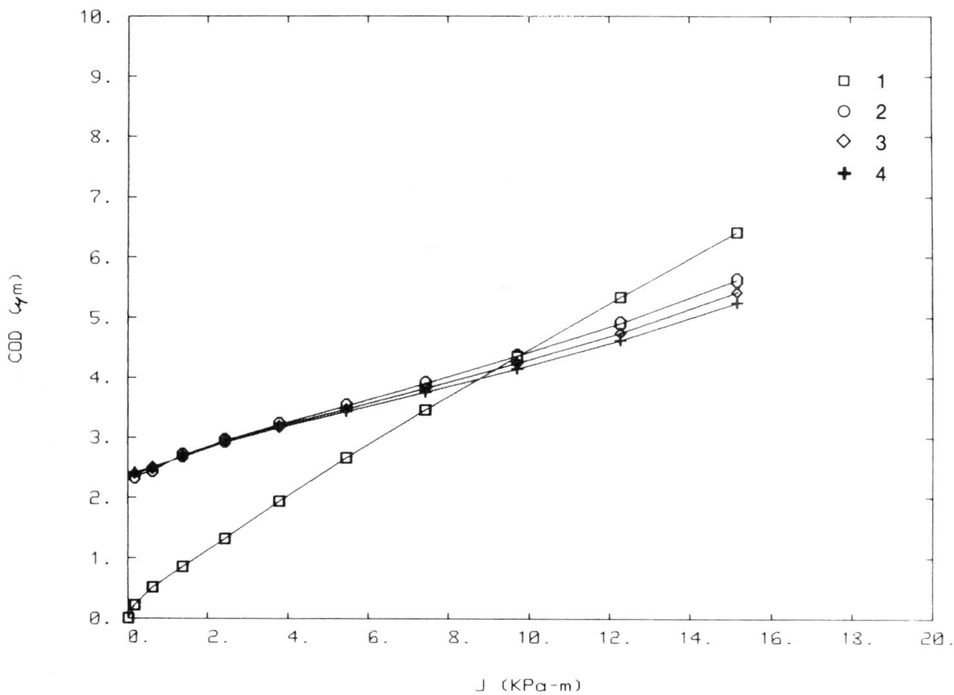

FIG. 7—*Variation of the crack opening displacement, COD, with remotely applied* J.

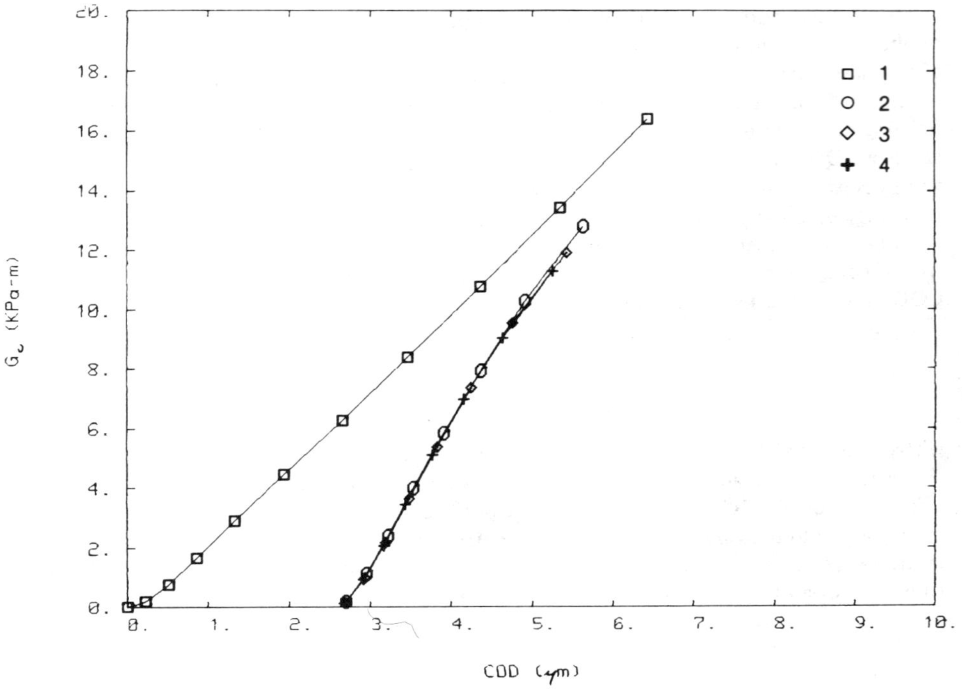

FIG. 8—*Variation of the crack-tip energy release rate,* G_0, *with the crack opening displacement, COD.*

Therefore, a unique relationship between the cyclic range of the COD and the range of the crack-tip energy release rate is demonstrated to hold, just as for the cases of monotonic *J*-controlled loading [*3*] and cyclic loading of a stationary crack [*8*]. The suggestion from this is that the simple definition of the crack-tip energy release rate presented here should serve as an equivalent to the COD as a cyclic crack growth characterizing parameter. Cyclic crack growth governed by a $\Delta G_0/Y$ versus Δa curve, in direct analogy with monotonic crack growth governed by a J versus Δa curve, then becomes a distinct possibility.

It is important to note that the difference between the value of the range of the crack-tip energy release rate, as defined here, and the corresponding range of the remote *J*-value is not slight. For the cyclic loading case illustrated by Figs. 5 to 9, the range of G_0/Y for the fourth ("steady-state") cycle is only about 40% of the range of the remotely applied J loading corresponding to the portion of the cycle for which the crack is open. Therefore, the difference between remote ΔJ-controlled growth and $\Delta G_0/Y$-controlled growth should be easily discriminated in a suitably designed low-cycle fatigue experiment.

Acknowledgments

The author is deeply indebted to Dr. Mark Snyder for providing all of the FORTRAN coding for the implementation of the mechanical sublayer plasticity theory into the General Electric Company's version of ADINA, as well as the special coding required for modeling many of the unique aspects of cyclic loading of cracked components. Also, the author would like to thank Professor C. F. Shih for suggesting this approach and for providing the technical guidance for

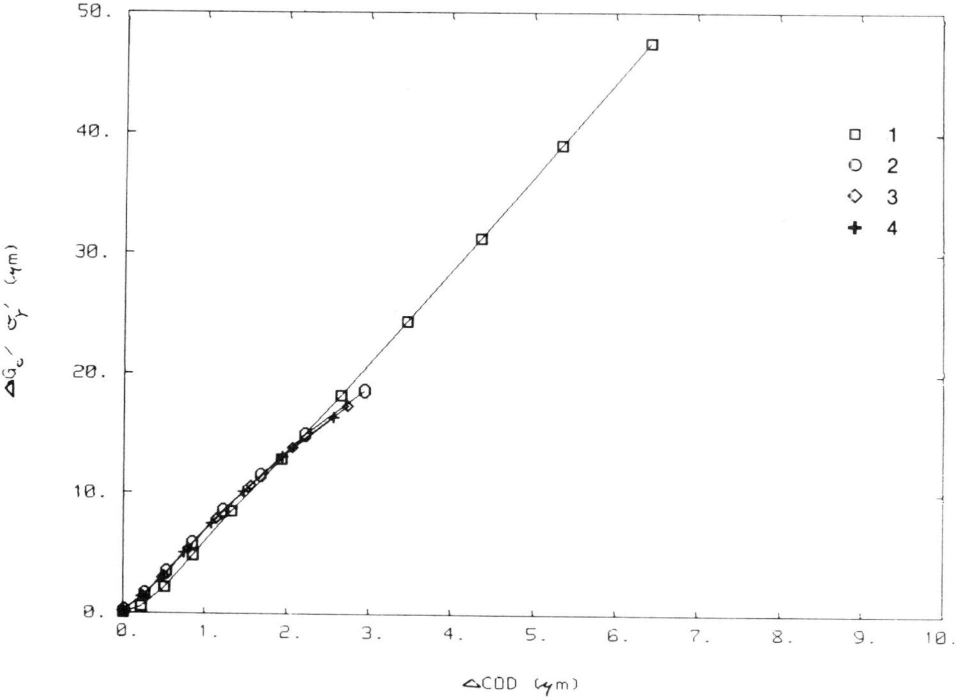

FIG. 9—*Variation of the normalized cyclic range of the crack-tip energy release rate,* $\Delta G_0/Y$, *with the cyclic range of the crack opening displacement,* ΔCOD.

this work in its initial stages. Dr. H. G. deLorenzi and Ms. M. D. German have also provided invaluable technical guidance, assistance, and moral support during my years of association with them at the Research and Development Center, for which I am sincerely grateful.

References

[1] Rice, J. R., *Journal of Applied Mechanics, Transactions,* American Society of Mechanical Engineers, Vol. 35, June 1968, pp. 379–386.

[2] Hutchinson, J. W., *Nonlinear Fracture Mechanics,* Department of Solid Mechanics, Technical University of Denmark, Copenhagen, Denmark, 1979.

[3] Shih, C. F., Andrews, W. R., deLorenzi, H. G., German, M. D., Van Stone, R. H., and Mowbray, D. F., "Methodology for Plastic Fracture," Final Report to Electric Power Research Institute, Contract No. RP601-2, General Electric Co., Schenectady, NY, 1981.

[4] Kumar, V., German, M. D., Wilkening, W. W., Andrews, W. R., deLorenzi, H. G., and Mowbray, D. F., "Further Developments in the Engineering Approach for Elastic-Plastic Fracture Analysis," Final Report to Electric Power Research Institute, Contract No. RP1237-1. General Electric Co., Schenectady, NY, 1983.

[5] Besseling, J. F., "A Theory of Plastic Flow for Anisotropic Hardening of an Initially Isotropic Material," Report S.410, National Aeronautical Research Institute, Amsterdam, The Netherlands, 1953.

[6] Besseling, J. F., "A Theory of Elastic, Plastic, and Creep Deformations of an Initially Isotropic Material Showing Anisotropic Strain Hardening, Creep Recovery, and Secondary Creep," *Journal of Applied Mechanics,* Vol. 25, 1958, pp. 529–536.

[7] *ADINA User's Manual,* Report No. AE81-1, ADINA Engineering, Watertown, MA, 1981.

[8] German, M. D., deLorenzi, H. G., and Wilkening, W. W., "Cyclic Elastic-Plastic Analyses of Cracks at Defects," *Engineering Fracture Mechanics,* Vol. 22, No. 5, 1985, pp. 871–883.

[9] Dowling, N. E. and Begley, J. A., "Fatigue Crack Growth During Gross Plasticity and the J-Integral," *Mechanics of Crack Growth, ASTM STP 590,* American Society for Testing and Materials, Philadelphia, 1976, pp. 82–103.

[10] Levy, N., Marcal, P. V., Ostergren, W. J., and Rice, J. R., "Small Scale Yielding Near a Crack in Plane Strain: A Finite Element Analysis," *International Journal of Fracture Mechanics,* Vol. 7, No. 2, June 1971, pp. 143–156.

[11] deLorenzi, H. G. and Shih, C. F., "On the Use of 2D Isoparametric Elements for Calculations in the Fully Plastic Range," *International Journal of Fracture,* Vol. 13, 1977, pp. 507–511.

[12] Barsoum, R. S., "Triangular Quarter-Point Elements as Elastic and Perfectly Plastic Crack Tip Elements," *International Journal of Numerical Methods in Engineering,* Vol. 11, 1977, pp. 85–98.

Robert J. Buzzard[1] *and Bernard Gross*[1]

Calibration of a Mode II Test Specimen

REFERENCE: Buzzard, R. J. and Gross, B., "**Calibration of a Mode II Test Specimen,**" *Fracture Mechanics: Eighteenth Symposium, ASTM STP 945,* D. T. Read and R. P. Reed, Eds., American Society for Testing and Materials, Philadelphia, 1988, pp. 1083–1088.

ABSTRACT: Mode II and Mode I calibrations of the NASA Lewis Research Center Mode II fatigue specimen were performed experimentally over crack length to specimen width ratios (a/W) of 0.5 to 0.9. Mode II displacements were measured both at the specimen notch mouth and at the intersection of the notch with the centerline of the loading pin holes. Mode I displacements were measured across the span of the specimen at the loading pins' centerline. Analytical stress intensity factor coefficients for both Mode II and Mode I are also presented.

KEY WORDS: Mode II, compliance, fatigue, edge-sliding mode

Nomenclature

- a Crack length
- B Specimen thickness
- d Loading pin hole diameter
- E Young's modulus of elasticity
- H Specimen arm height
- K_I Mode I stress intensity factor
- K_{II} Mode II stress intensity factor
- L Specimen length
- P Applied end load
- U Displacement along load line for Mode I loading
- V Displacement along load line for Mode II loading
- W Specimen width
- Y_I Mode I stress intensity factor coefficient, $K_I B\sqrt{W}/P$
- Y_{II} Mode II stress intensity factor coefficient, $K_{II} B\sqrt{W}/P$

A Mode II fatigue and fracture specimen has been developed at NASA Lewis Research Center for potential application to aircraft bearing race design [*1*]. Mode II refers to the edge sliding mode of crack displacement [2]. Primary to the development of this type of specimen is a definition of its compliance and of its stress intensity coefficients over a range of crack length to specimen width ratios. The objective of the present work was to obtain the Mode II and Mode I compliance calibrations of the specimen. While such calibration is an established procedure for determining stress intensity coefficients [*3–5*], the displacements herein were measured at the crack mouth and at the crack surface (since these are the most practical measurement locations for performing Mode II tests) and should not be used for such stress intensity factor coefficient determinations. These coefficients have been determined analytically [*6*], however, and are included herein.

[1]Materials engineer and aerospace engineer, respectively, National Aeronautics and Space Administration, Lewis Research Center, Cleveland, OH 44135.

Specimens

The specimen design used in this study was described in Ref *1* and is shown in Fig. 1*a*. Compliance calibration models of the specimen were made from 3.18-mm (1/8-in.)-thick 7075-T6 aluminum sheet with a reported modulus in tension of 71.66×10^3 MN/m^2 (10.4×10^6 psi). The notch representing a crack was made using a band saw. The resulting 0.51-mm (0.020 in.)-wide slot was finished at the tip to a vee-configuration using a hacksaw blade ground to that shape. Loading pins were ground to fit tightly into the 12.7-mm (0.500-in.) diameter loading holes, in order to reduce the likelihood of extraneous sideward movement of the specimen about the loading pins during load application. Dimensions given in Fig. 1 are for two specimen sizes. The larger size specimen is the original design. The smaller design requires less material, reduces the rotational moment on loading, is stiffer, and is therefore believed to be an improved design. As a result, more emphasis was placed upon calibration of the smaller specimen design than on the larger. Loading pin hole sizes are identical for both specimen sizes, but specimen proportions are slightly different.

Displacement Measurement

Two types of displacement measurements were performed on the Mode II specimen. The specimen was first loaded in Mode II in the fixture shown in Fig. 1*b*, and Mode II displacement

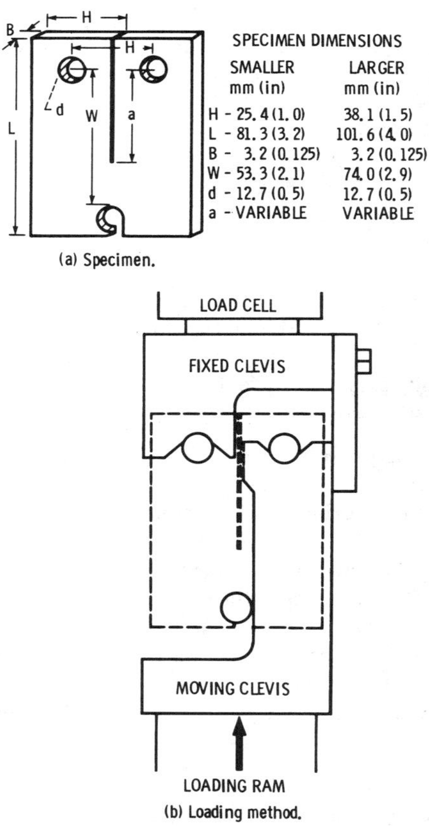

FIG. 1—*Mode II test specimen and loading method.*

values (V) were obtained. The specimen was subsequently transferred to another fixture (not shown) and loaded in the Mode I direction using conventional pin-type grips, in a manner similar to testing of standard compact tension specimens. Mode I displacement values (U) were thus obtained.

Mode II displacement measurements were performed by use of clip gages [*3*] attached to the specimen at the notch mouth, as shown in Fig. 2, and by photographic observation of a fine scribe line located where the loading pins' horizontal centerline intersects the notch (Fig. 2*b*). The scribe line was applied using the precision scribing device described in Ref *7* attached to a machinist's microscope. The photographic method was used only for the smaller specimen's calibration. Mode I displacements were measured by use of a cantilever beam type clip gage which spanned the outside of the specimen at the loading pins' horizontal centerline.

Test Procedure

The displacement gage was calibrated over the applicable range for each test run. Displacement calibrations were made using an extensometer calibrator reading to a least division of 0.00127 mm (5×10^{-5} in.). The 44.5-kN (10-kip) load cell used was initially calibrated using deadweight loading and subsequently checked by use of a calibrating resistor.

Load versus displacement slopes were recorded on an x-y recorder. Several slope determinations were made for each crack length. For Mode II runs, the specimen's notch width was checked using a feeler gage with the specimen at zero load and while under load to assure that no spreading or closure of the notch occurred as a result of side forces resulting from the application of the load. The validity of this method was verified by use of the Mode I clip gage. The maximum load for each series was such that the load versus displacement trace was well within the linear range for each crack length.

After the Mode II autographic test run series was completed for each crack length, the previously described scribe line zone was photographed with the specimen at zero load and at 890 N (200 lb) increments of loading. A final photograph was taken after unloading from the maximum load for each series. A microscope equipped with a camera was used. Photographs were taken at a magnification of ×26.4. Measurements on the photographs were performed with a machinist's microscope whose table was translated by means of a vernier calibrated in 0.0025-mm (0.0001-in.) increments. The magnification of each photograph was verified by measuring between fiducial marks scribed onto the specimen near the load line reference marks. The notch width at this zone was also verified photographically. Load line displacements for each load

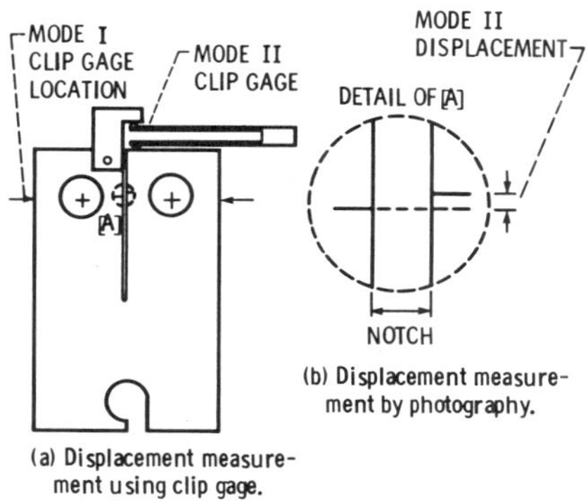

FIG. 2—*Displacement measurement methods used to perform Mode II compliance calibration.*

were measured as relative displacement of the scribed reference marks along the axis of loading as represented in Fig. 2*b*. The zero load displacement was checked using the pretest and post-test photographs, to ascertain that no discernible plastic deformation had occurred during loading.

Data Analysis

Load versus displacement slope determinations were obtained from the *x-y* recorder test records. For each crack length the slope measurements were averaged. In all cases, the maximum variation in slope measurements was always less than 1.1%.

Photographically obtained Mode II data points were plotted for each crack length, and a least-squares best fit was determined for each.

The resulting slopes were used to obtain *EVB/P* values, which are plotted in Fig. 3. A least-squares third-degree polynominal was used to correlate each set of *EVB/P*-versus-*a/W* data. The polynomials are described in Table 1.

Comparable finite-element results for *EVB/P* in Mode II loading have been obtained [*6*] and show very good agreement (within 3%) with the experimental results.

Experimentally determined dimensionless Mode I compliance values *EUB/P*, as a function of crack length to specimen width ratio *a/W*, were fit by a least-squares regression analysis resulting in the expression

$$\ln EUB/P = -8.236 + 55.434\,(a/W) - 86.827\,(a/W)^2 + 49.039\,(a/W)^3$$

Mode I compliance values *EUB/P* as a function of crack length to specimen width ratio *a/W*, were also determined by finite-element analysis. Both sets of results are shown in Fig. 4.

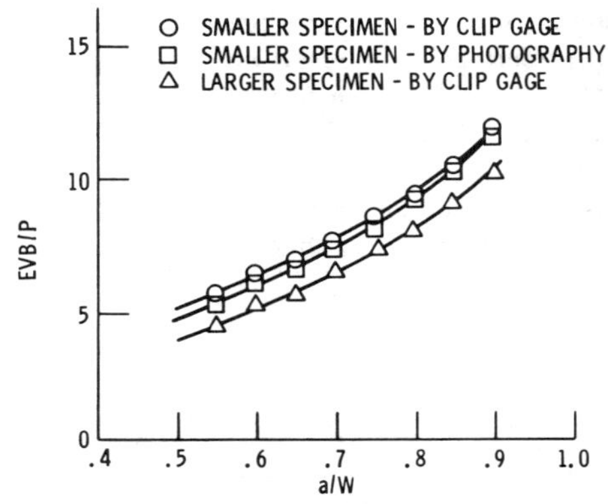

FIG. 3—*Experimentally determined Mode II compliance values for Mode II specimens.*

TABLE 1—EVB/P *versus* a/W *data correlated by least-squares third-degree polynomials.*

Results	*EVB/P* Polynomials
Clip gage (smaller specimen)	$-14.870 + 81.891\,(a/W) - 116.665\,(a/W)^2 + 65.202\,(a/W)^3$
Photographic (smaller specimen)	$-8.606 + 51.647\,(a/W) - 71.946\,(a/W)^2 + 43.802\,(a/W)^3$
Clip gage (larger specimen)	$-6.987 + 40.442\,(a/W) - 54.04\,(a/W)^2 + 33.774\,(a/W)^3$

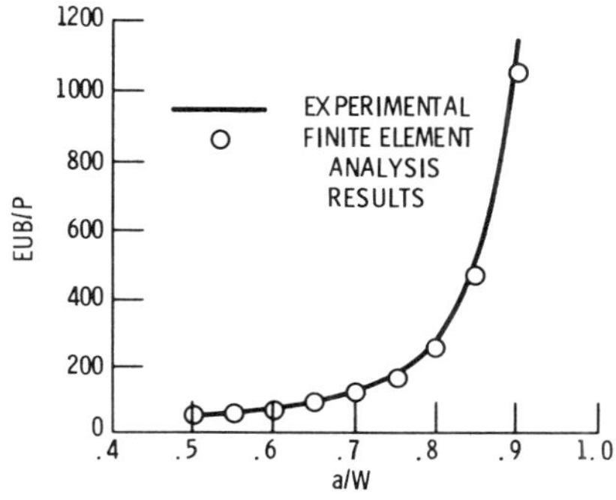

FIG. 4—*Mode I compliance values for Mode II specimens.*

In addition to the experimental and analytical compliance calibration, the finite-element approach was used to obtain stress intensity coefficients for both Mode II and Mode I. This analysis is described in detail in Ref *6*. The resulting Y values are presented herein as Table 2 and are plotted as Figs. 5 and 6. The Y-values are related to the stress intensity factor by the relationship

$$Y_j = K_j B \sqrt{W}/P$$

where j = I or II.

Summary

Both Mode II and Mode I compliance calibrations of the NASA Mode II test specimen were performed over crack length to specimen width ratios (a/W) of 0.5 to 0.9. Stress intensity factor coefficients over this range were obtained by finite-element analysis and are also presented. The information presented herein will allow the investigator to precrack the specimen in Mode I, if desired, and to perform Mode II tests for the determination of K_{II} values.

TABLE 2—*Stress-intensity coefficients for Mode II specimen obtained by finite-element analysis.*

Crack Length to Specimen Width Ratio, a/W	Stress-Intensity Coefficient, Y_{II}^{a} (Mode II loading)	Stress-Intensity Coefficient, Y_{I}^{b} (Mode I loading)
0.50	2.03	9.39
0.55	2.13	10.66
0.60	2.22	12.39
0.65	2.33	14.73
0.70	2.47	18.57
0.75	2.63	24.42
0.80	2.88	34.44
0.85	3.23	55.33
0.90	3.89	100.76

[a] $Y_{II} = 25.78 - 156.45\,(a/W) + 379.26\,(a/W)^2 - 403.13\,(a/W)^3 + 160.94\,(a/W)^4$ (from Ref *6*).

[b] $\text{Ln}\ Y_I = -3.14632 + 25.1261\,(a/W) - 41.7135\,(a/W)^2 + 25.9645\,(a/W)^3$.

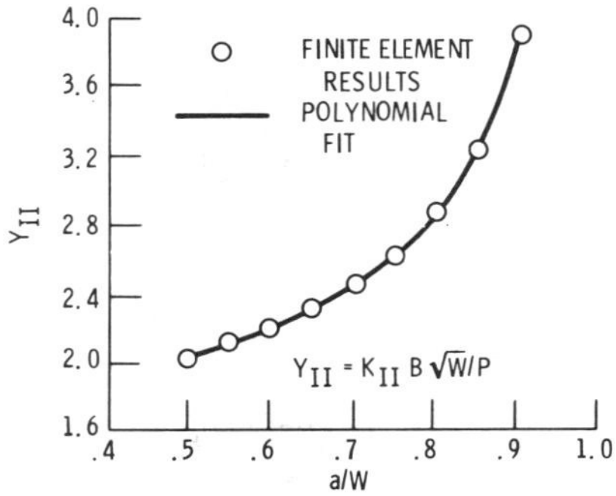

FIG. 5—*Analytical Mode II stress-intensity factor coefficients for Mode II test specimen.*

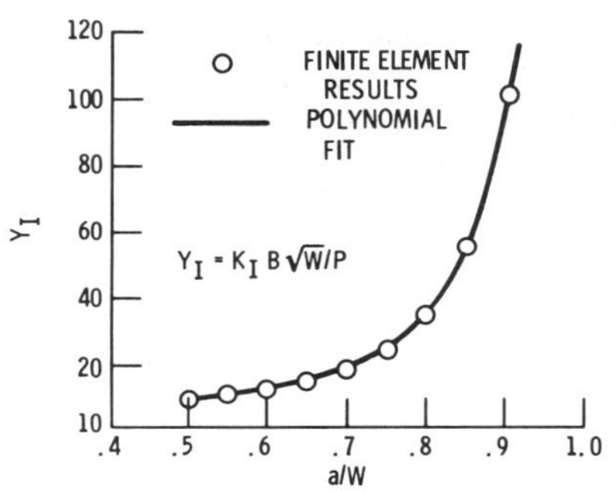

FIG. 6—*Analytical Mode I stress-intensity factor coefficients for Mode II specimen.*

References

[*1*] Buzzard, R. J., Gross, B., and Srawley, J. E., "Mode II Fatigue Crack Growth Specimen Development," NASA TM-83722, National Aeronautics and Space Administration, Washington, DC, Aug. 1984.

[*2*] Paris, P. C. and Sih, G. C., "Stress Analysis of Cracks," *Fracture Toughness Testing and Its Applications, ASTM STP 381,* American Society for Testing and Materials, Philadelphia, 1964, pp. 30–81.

[*3*] Bubsey, R. T., Fisher, D. M., Jones, M. H., and Srawley, J. E., "Compliance Measurements," *SESA Monograph No. 1: Experimental Techniques in Fracture Mechanics,* The Iowa State University Press, Iowa City, IA, 1973, pp. 76–95.

[*4*] Fisher, D. M. and Buzzard, R. J., "Comparison Tests and Experimental Compliance Calibration of the Proposed Standard Round Compact Plane Strain Fracture Toughness Specimen," NASA TM-81379, National Aeronautics and Space Administration, Washington, DC, Nov. 1979.

[*5*] Fisher, D. M. and Buzzard, R. J., "Experimental Compliance Calibration of the Compact Fracture Toughness Specimen," NASA TM-81665, National Aeronautics and Space Administration, Washington, DC, Dec. 1980.

[*6*] Gross, B., Buzzard, R. J., and Brown, W. F., "Elastic Analysis of a Mode II Fatigue Crack Test Specimen," *International Journal of Fracture,* Vol. 31, 1986, pp. 151–157.

[*7*] Buzzard, R. J., "Precision Scriber," NASA Technical Brief No. LEW 12976, National Aeronautics and Space Administration, Washington, DC, Vol. 4, No. 4, winter 1979, p. 563.

Summary

Summary

Summary

The various aspects of the interdisciplinary engineering science of fracture mechanics, including engineering mechanics, sophisticated laboratory testing, metallurgy, and practical engineering have all taken their turns in the limelight of research attention. All these areas are essential to fracture mechanics, and all are represented in this book.

Given this opportunity to emphasize a few key points of this symposium, we choose to highlight the papers on the ductile-to-brittle transition and the related areas of elastic-plastic fracture mechanics and J-integral test methods (noting that the standard J-integral methods available to date are not applicable to cleavage fracture). Fracture mechanics as an engineering science has been strongly influenced by the practical need to understand and overcome the dramatic ductile-to-brittle transition behavior of ferritic steels with decreasing temperatures. Other problems have been important, especially crack growth by fatigue and the need to use safely high-strength materials such as certain aluminum and titanium alloys and steels. But the ductile-to-brittle transition still poses a first order challenge to fracture mechanics, to penetrate the scatter in the test results in order to arrive at a correct and useful description of material fracture properties within the transition range. Heavy-section structures for exploiting petroleum resources in the arctic are a current application area in need of progress in the ductile to brittle transition. This volume samples the inventory of the physical understanding, elastic and elastic-plastic analysis methods, and elastic-plastic test methods accumulated by fracture mechanics.

In particular, the paper by Wellman, Sorem, Dodds, and Rolfe showed that the relative size of the highly constrained region ahead of a crack is only weakly dependent on the specimen thickness. This adds weight to the weak-link statistical approach to cleavage fracture. Tyson's paper on cleavage toughness variability makes a strong case for a specific weak-link theory relating toughness variability to the inclusion size distribution.

The ductile-to-brittle transition occurs under elastic-plastic conditions, so all the papers on elastic-plastic fracture mechanics contribute to its understanding. The J-integral and crack opening displacement test methods will continue to become more useful in this region.

Fatigue crack growth remains an important problem in high-performance components, and this symposium reflects the importance of this subject area. Progress continues in the understanding of the growth of very small fatigue cracks and frequency and hold-time effects on fatigue at high temperatures.

Analysis methods continue to improve in scope, and re-examination of existing fracture criteria continues to deepen our understanding. Other important areas emphasized in this proceedings are micromechanisms, crack arrest, and subcritical crack growth.

This Special Technical Publication extends to one ultimate goal of fracture mechanics, practical applications. Several distinct methods are sampled here, indicating the variety of possibilities available and leading toward more powerful and more universal methods.

On behalf of the organizing committee of this Symposium, the editors thank the authors of their papers for their contributions, the chairmen of the sixteen sessions for their services in facilitating orderly discussion, and all one hundred thirty-seven attendees for their discussions.

R. P. Reed

National Bureau of Standards, Boulder, CO 80303; symposium chairman and coeditor.

Author Index

Subject Index

A

G

H

K

L

M

N

O

P

T

U

V

W

Y

Z